CRC Series
Mechanics and Materials Science

Series Editor: Bharat Bhushan, Ph.D., D.Sc.

Handbook of
Micro/Nano
Tribology

SECOND EDITION

Edited by
Bharat Bhushan, Ph.D., D.Sc.

Computer Microtribology and Contamination Laboratory
Department of Mechanical Engineering
The Ohio State University
Columbus, Ohio

CRC Press
Boca Raton   London   New York   Washington, D.C.

Acquiring Editor:	Cindy Carelli
Project Editor:	Andrea Demby
Marketing Manager:	Jane Stark
Cover design:	Dawn Boyd
PrePress:	Carlos Esser
Manufacturing:	Lisa Spreckelson

Library of Congress Cataloging-in-Publication Data

Handbook of micro/nanotribology / edited by Bharat Bhushan. -- 2nd ed.
        p. cm.
    Includes bibliographical references and index.
    ISBN 9780849384028 (hbk)
    ISBN 9780367400170 (pbk)
    ISBN 9780367802523 (ebk)
    1. Tribology--Handbooks, manuals, etc.   I. Bhushan, Bharat, 1949-   .

TJ1075.H245   1999
621.8'9--dc21   98-24466
                                                                                                                        CIP

Foreword

The invention of the scanning tunneling microscope has led to an explosion of a family that is now called scanning probe microscopes (SPMs). The most popular instrument in this family is the atomic force microscope (AFM). According to some estimates, sales of SPMs in 1993 were about 100 million U.S. dollars (about 2,000 units installed to date) worldwide. The biggest portion of this results from AFM sales, although the first ideas and preliminary results were introduced to the scientific community only a few years ago (1986). The whole field of SPM is not very old (the first operation of STM was in 1981) and still is in a rapidly evolving state. The scientific industrial applications include quality control in the semiconductor industry and related research, molecular biology and chemistry, medical studies, materials science, and the field of information storage systems.

Soon after the invention of the AFM it was discovered that part of the information in the images resulted from friction and that the instrument could be used as a tool for tribology. In general, SPMs are now used intensively in this field. Researchers can image single lubricant molecules and their agglomeration and measure surface topography, adhesion, friction, wear, lubricant film thickness, and mechanical properties all on a micrometer to nanometer scale.

With the advent of more powerful computers, atomic-scale simulations have been conducted of tribological phenomena. Simulations have been able to predict the observed phenomena. Development of the field of micro/nanotribology has attracted numerous physicists and chemists. This is a field I personally know very little about. I am, however, very excited that SPMs have had such an immense impact on the field of tribology.

I congratulate Professor Bharat Bhushan in helping to develop this field of micro/nanotribology. The *Handbook of Micro/Nanotribology* is very timely and I expect that it will be well received by the international scientific community. With best wishes.

Prof. Dr. Gerd Binnig
IBM Research Division
Munich, Germany
Nobel Laureate Physics, 1986

Preface

Second Edition, 1999

The first edition of the *Handbook of Micro/Nanotribology* was published in the Spring of 1995. Soon after its publication, the first-of-a-kind monograph became a reference book for the novice, as well as experts, in the emerging field of micro/nanotribology. Since the field is evolving very rapidly, we felt that the monograph needed a second edition.

The second edition is totally revised. The scope of the first edition has been expanded. In the first part, Basic Studies, two new chapters on AFM Instrumentation and Tips and Surface Forces and Adhesion have been added. In the second part, Applications, four new chapters on Design and Construction of Magnetic Storage Devices, Microdynamic Devices and Systems, Mechanical Properties of Materials in Microstructure Technology, and Micro/Nanotribology and Micro/Nanomechanics of MEMS Devices have been added. The content of each original chapter has been revised and broadened. In many cases, new authors were invited to contribute chapters on topics covered in the first edition. Of the 16 chapters in the book — 11 in Basic Studies and five in Applications — 11 are written by new contributors, whereas five chapters have been thoroughly revised by the original authors.

The organization of the Handbook is straightforward. It is divided into two parts: Part I covers the basic studies and Part II encompasses design, construction, and applications to magnetic storage devices and MEMS. The introduction chapter starts out with a definition and the evolution of micro/nanotribology, description of various measurement techniques, and description of various industrial applications where the study of tribology, on a nanoscale, is critical. After this chapter, the subject matter is presented as follows: an overview of AFM instrumentation and tips; current understanding of surface physics and description of methods used to physically and chemically characterize solid surfaces; roughness characterization and static contact models using fractal analysis; surface forces and adhesion; introduction of sliding at the interface and study of friction on an atomic scale; study of scratching and wear as a result of sliding, applications to nanofabrication/nanomachining, as well as nano/picoindentation; study of lubricants used to minimize friction and wear; surface forces and microrheology of thin liquid films; measurement of nanomechanical properties of surfaces and thin films; and atomic-scale simulations of interfacial phenomena.

Part II includes material in the following order: design and construction of magnetic storage devices; microdynamic devices and systems; micro/nanotribology and micro/nano mechanics of magnetic storage devices; mechanical properties of materials in microstructure technology; and micro/nanotribology and micro/nanomechanics of MEMS devices.

The Handbook is intended for graduate students of tribology and researchers who are active or intend to become so in this field. This book should serve as an excellent text for a graduate course in micro/nanotribology. For a reduced scope of the course, Chapters 2, 3, and 10 (Part I) and Chapters 12 and 13 (Part II) can be eliminated. For a more concise course, Chapter 4 may also be eliminated.

I would like to thank the authors for their excellent contributions in a timely manner. My secretary, Kathleen Tucker, patiently typed six of the chapters contributed by me to this book. Last, but not least, I wish to thank my wife, Sudha, my son, Ankur, and my daughter, Noopur, who have been very forbearing during the preparation of this book.

Bharat Bhushan
Powell, Ohio

Preface

First Edition, 1995

Tribology is the science and technology of two interacting surfaces in relative motion and of related subjects and practices. The popular equivalent is friction, wear, and lubrication. The advent of new techniques to measure surface topography, adhesion, friction, wear, lubricant-film thickness, and mechanical properties all on a micro to nanometer scale, to image lubricant molecules and availability of supercomputers to conduct atomic-scale simulations has led to development of a new field referred to as Microtribology, Nanotribology, Molecular Tribology, or Atomic-Scale Tribology. This field is concerned with experimental and theoretical investigations of processes ranging from atomic and molecular scales to microscale, occurring during adhesion, friction, wear, and thin-film lubrication at sliding surfaces. These studies are needed to develop fundamental understanding of interfacial phenomena on a small scale and to study interfacial phenomena in micro- and nano structures used in magnetic storage systems, microelectromechanical systems (MEMS) and other industrial applications. The components used in micro- and nano structures are very light (on the order of few micrograms) and operate under very light loads (on the order of few micrograms to few milligrams). As a result, friction and wear (nanoscopic wear) of lightly loaded micro/nano components are highly dependent on the surface interactions (few atomic layers). These structures are generally lubricated with molecularly thin films. Micro- and nanotribological techniques are ideal to study friction and wear processes of micro/nano structures. These studies are also valuable in the fundamental understanding of interfacial phenomena in macrostructures to provide a bridge between science and engineering. Friction and wear on micro- and nanoscales have been found to be generally smaller compared to that at macroscales. Therefore, micro-nanotribological studies may identify regimes for ultra-low friction and zero wear.

The field of tribology is truly interdisciplinary. Until recently, it has been dominated by mechanical and chemical engineers who have conducted macro tests to predict friction and wear lives in machine components and devised new lubricants to minimize friction and wear. Development of the field of micro/nanotribology has attracted many more physicists and chemists who have significantly contributed to the fundamental understanding of friction and wear processes on an atomic scale. Thus, tribology is now studied by both engineers and scientists. The micro/nanotribology field is growing rapidly and it has become fashionable to call oneself a "tribologist". Since 1991, international conferences and courses have been organized on this new field of micro/nanotribology.

We felt a need to develop a **handbook of micro/nano tribology**. This first-of-a-kind monograph presents the state-of-the-art of the micro/nanotribology field by the leading international researchers. In each chapter we start with macroconcepts leading to microconcepts. We assume that the reader is not expert in the field of microtribology, but has some knowledge of macrotribology. It covers characterization of solid surfaces, various measurement techniques and their applications, and theoretical modeling of interfaces.

The organization of the handbook is straightforward. The book is divided into two parts. Part I covers the basic studies and Part II covers applications to magnetic storage devices and MEMS. Part I includes ten chapters. The introduction chapter starts out with a definition and evolution of micro/nanotribology, description of various measurement techniques, and description of various industrial applications where the study of tribology, on a nanoscale, is critical. Following the introduction chapter, subject matter is presented in the following order: description of methods used to physically and chemically characterize solid surfaces, roughness characterization and static contact models using fractal analysis, introduction of sliding at the interface and study of friction, study of wear as a result of sliding, study of lubricants used to minimize friction and wear, imaging of lubricant molecules, surface forces and microrheology of thin liquid films, measurement of micromechanical properties of surfaces and thin films, and atomic-scale simulations of interfacial phenomena. Part II includes three chapters. Subject matter is presented in the following order: micro/nanotribology and micro/nano mechanics of magnetic storage devices and MEMS, role of particulate contaminants in tribology and applications to magnetic storage devices and MEMS, and modeling of hydrodynamic lubrication with molecularly thin gas films and applications to MEMS.

The handbook is intended for graduate students of tribology and research workers who are active or intend to become active in this field. This book should serve as an excellent text for a graduate course in micro/nanotribology. For a reduced scope of the course, Chapters 2 and 9 from Part I and Chapters 11 to 13 from Part II can be eliminated. For a further reduction, Chapter 3 from Part I can also be eliminated.

I would like to thank the authors for their excellent contributions in a timely manner. And I wish to thank my wife, Sudha, my son, Ankur, and my daughter, Noopur, who have been very forbearing during the preparation of this book.

Bharat Bhushan
Powell, Ohio

The Editor

Bharat Bhushan, Ph.D., a pioneer in tribology and the mechanics of magnetic storage devices, is an internationally recognized expert in the general fields of conventional tribology and micro/nanotribology and one of the most prolific authors. He is presently an Ohio Eminent Scholar and The Howard D. Winbigler Professor in the Department of Mechanical Engineering and the Director of Computer Microtribology and Contamination Laboratory at The Ohio State University, Columbus.

He received a Masters in mechanical engineering from the Massachusetts Institute of Technology, a Masters in mechanics and a Ph.D. in mechanical engineering from the University of Colorado at Boulder, an MBA from Rensselaer Polytechnic Institute at Troy, NY, a Doctor Technicae from the University of Trondheim, Norway, and a Habilitate Doctor of Technical Sciences from the Warsaw University of Technology in Poland. He is a registered professional engineer (mechanical).

He has authored five technical books, 22 handbook chapters, more than 350 technical papers in refereed journals, more than 60 technical reports, edited more than 24 books, and holds seven U.S. patents. He is founding editor-in-chief of the *World Scientific Advances in Information Storage Systems Series, The CRC Press Mechanics and Materials Science Series*, and the *Journal of Information Storage and Processing Systems*. He has given more than 175 invited presentations on five continents, including several keynote addresses at major international conferences. He is an accomplished organizer. He organized the first symposium on Tribology and Mechanics of Magnetic Storage Systems in 1984 and the first international symposium on Advances in Information Storage Systems in 1990. He continues to organize and chair the AISS symposia annually. He founded an ASME Information Storage and Processing Systems Division and is the founding chair. His biography has been listed in over two dozen *Who's Who* books, including *Who's Who in the World*. He has received more than a dozen awards for his contributions to science and technology from professional societies, industry, and U.S. government agencies. He is a foreign member of the Byelorussian Academy of Engineering and Technology, the Russian Engineering Academy and the Ukrainian Academy of Transportation, a senior member of IEEE, and a member of STLE, NSPE, Sigma Xi, and Tau Beta Pi.

Dr. Bhushan has worked for the Department of Mechanical Engineering at the Massachusetts Institute of Technology, Cambridge, MA; Automotive Specialists, Denver, CO; the R & D Division of Mechanical Technology, Inc., Latham, NY; the Technology Services Division of SKF Industries, Inc., King of Prussia, PA; the General Products Division Laboratory of IBM Corporation, Tucson, AZ; the Almaden Research Center of IBM Corporation, San Jose, CA; and the Department of Mechanical Engineering at the University of California, Berkeley.

He is married and has two children. His hobbies include music, photography, hiking, and traveling.

Contributors

Phillip B. Abel, Ph.D.
NASA Lewis Research Center
Cleveland, Ohio

Alan Berman, Ph.D.
Tape Operations
Seagate Technology
Costa Mesa, California

Bharat Bhushan, Ph.D.
Department of Mechanical
    Engineering
Ohio State University
Columbus, Ohio

Donald Brenner, Ph.D.
Department of Materials Science
    and Engineering
North Carolina State University
Raleigh, North Carolina

Nancy A. Burnham, Ph.D.
Department of Physics
IGA – Ecole Polytechnique
Lausanne, Switzerland

Jaime Colchero, Ph.D.
Dep. Fisica de la Mat. Condensada
Universidad Autonoma de Madrid
Madrid, Spain

Fredric Ericson, Ph.D.
Department of Materials Science
Uppsala University
Uppsala, Sweden

John Ferrante, Ph.D.
Department of Physics
Cleveland State University
Cleveland, Ohio

Judith A. Harrison, Ph.D.
Chemistry Department
United States Naval Academy
Annapolis, Maryland

Jacob N. Israelachvili, Ph.D.
Department of Chemical
    Engineering
University of California
Santa Barbara, California

Hiroshi Kano, Ph.D.
Recording Media Division
Sony Corporation
Sakuragi Tagajo City, Japan

Hirofumi Kondo, Ph.D.
Recording Media Division
Sony Corporation
Sakuragi Tagajo City, Japan

Andrzej J. Kulik, Ph.D.
Department of Physics
IGA - Ecole Polytechnique
Lausanne, Switzerland

Othmar Marti, Ph.D.
Abteilung fuer Experimentelle
    Physik
Universitaet Ulm
Ulm, Germany

Ernst Meyer, Ph.D.
Institut fur Physik
Basel, Switzerland

Richard S. Muller, Ph.D.
Berkeley Sensor & Actuator Center
University of California
Berkeley, California

Hiroyuki Osaki, Ph.D.
Recording Media Division
Sony Corporation
Sakuragi Tagajo City, Japan

Norio Saito, Ph.D.
Recording Media Division
Sony Corporation
Sakuragi Tagajo City, Japan

Jan-Åke Schweitz, Ph.D.
Department of Materials Science
Uppsala University
Uppsala, Sweden

Steven J. Stuart, Ph.D.
Chemistry Department
United States Naval Academy
Annapolis, Maryland

Hiroshi Takino, Ph.D.
Recording Media Division
Sony Corporation
Sakuragi Tagajo City, Japan

Contents

Part I Basic Studies

Part II Applications

Part I
Basic Studies

1

Introduction — Measurement Techniques and Applications

Bharat Bhushan

In this chapter, we first present the history of macrotribology and micro/nanotribology and their industrial significance. Next, we describe various measurement techniques used in micro/nanotribological studies, then present the examples of magnetic storage devices and microelectromechanical systems (MEMS) where micro/nanotribological tools and techniques are essential for interfacial studies. Finally, we present examples of why micro/nanotribological studies are important in magnetic storage devices, MEMS, and other microcomponents.

1.1 History of Tribology and Its Significance to Industry

Tribology is the science and technology of two interacting surfaces in relative motion and of related subjects and practices. The popular equivalent is friction, wear, and lubrication. The word *tribology*, coined in 1966, is derived from the Greek word *tribos* meaning rubbing, thus the literal translation would be the science of rubbing (Jost, 1966). It is only the name tribology that is relatively new, because interest in the constituent parts of tribology is older than recorded history (Dowson, 1979). It is known that drills made during the Paleolithic period for drilling holes or producing fire were fitted with bearings made from antlers or bones, and potters' wheels or stones for grinding cereals, etc., clearly had a

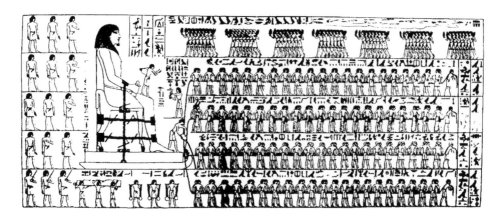

FIGURE 1.1   Egyptians using lubricant to aid movement of Colossus, El-Bersheh, circa 1800 BC.

requirement for some form of bearings (Davidson, 1957). A ball-thrust bearing dated about AD 40 was found in Lake Nimi near Rome.

Records show the use of wheels from 3500 BC, which illustrates our ancestors' concern with reducing friction in translationary motion. The transportation of large stone building blocks and monuments required the know-how of frictional devices and lubricants, such as water-lubricated sleds. Figure 1.1 illustrates the use of a sledge to transport a heavy statue by Egyptians circa 1880 BC (Layard, 1853). In this transportation, 172 slaves are being used to drag a large statue weighing about 600 kN along a wooden track. One man, standing on the sledge supporting the statue, is seen pouring a liquid into the path of motion; perhaps he was one of the earliest lubrication engineers. (Dowson, 1979, has estimated that each man exerted a pull of about 800 N. On this basis the total effort, which must at least equal the friction force, becomes 172 × 800 N. Thus, the coefficient of friction is about 0.23.) A tomb in Egypt that was dated several thousand years BC provides the evidence of use of lubricants. A chariot in this tomb still contained some of the original animal-fat lubricant in its wheel bearings.

During and after the glory of the Roman empire, military engineers rose to prominence by devising both war machinery and methods of fortification, using tribological principles. It was the renaissance engineer-artist Leonardo da Vinci (1452–1519), celebrated in his days for his genius in military construction as well as for his painting and sculpture, who first postulated a scientific approach to friction. Leonardo introduced, for the first time, the concept of coefficient of friction as the ratio of the friction force to normal load. In 1699, Amontons found that the friction force is directly proportional to the normal load and is independent of the apparent area of contact. These observations were verified by Coulomb in 1781, who made a clear distinction between static friction and kinetic friction.

Many other developments occurred during the 1500s, particularly in the use of improved bearing materials. In 1684, Robert Hooke suggested the combination of steel shafts and bell-metal bushes as preferable to wood shod with iron for wheel bearings. Further developments were associated with the growth of industrialization in the latter part of the 18th century. Early developments in the petroleum industry started in Scotland, Canada, and the U.S. in the 1850s (Parish, 1935; Dowson, 1979).

Although the essential laws of viscous flow had earlier been postulated by Newton, scientific understanding of lubricated bearing operations did not occur until the end of the nineteenth century. Indeed, the beginning of our understanding of the principle of hydrodynamic lubrication was made possible by the experimental studies of Tower (1884), the theoretical interpretations of Reynolds (1886), and related work by Petroff (1883). Since then, developments in hydrodynamic bearing theory and practice were extremely rapid in meeting the demand for reliable bearings in new machinery.

Wear is a much younger subject than friction and bearing development, and it was initiated on a largely empirical basis.

Since the beginning of the 20th century, from enormous industrial growth leading to demand for better tribology, our knowledge in all areas of tribology has expanded tremendously (Holm, 1946; Bowden and Tabor, 1950, 1964).

Tribology is crucial to modern machinery which uses sliding and rolling surfaces. Examples of productive wear are writing with a pencil, machining, and polishing. Examples of productive friction are brakes, clutches, driving wheels on trains and automobiles, bolts, and nuts. Examples of unproductive friction and wear are internal combustion and aircraft engines, gears, cams, bearings, and seals. According to some estimates, losses resulting from ignorance of tribology amount in the U.S. to about 6% of its gross national product or about $200 billion per year, and approximately one third of world energy resources in present use appear as friction in one form or another. In attempting to comprehend as enormous an amount as $200 billion, it is helpful to break it down into specific interfaces. It is believed that about $10 billion (5% of the total resources wasted at the interfaces) are wasted at the head–medium interfaces in magnetic recording. Thus, the importance of friction reduction and wear control cannot be overemphasized for economic reasons and long-term reliability. According to Jost (1966, 1976), the U.K. could save approximately £500 million per year, and the U.S. could save in excess of $16 billion per year by better tribological practices. The savings are both substantial and significant, and these savings can be obtained without the deployment of large capital investment.

The purpose of research in tribology is understandably the minimization and elimination of losses resulting from friction and wear at all levels of technology where the rubbing of surfaces are involved. Research in tribology leads to greater plant efficiency, better performance, fewer breakdowns, and significant savings.

1.2 Origins and Significance of Micro/Nanotribology

The advent of new techniques to measure surface topography, adhesion, friction, wear, lubricant film thickness, and mechanical properties, all on a micro- to nanometer scale, and to image lubricant molecules and the availability of supercomputers to conduct atomic-scale simulations has led to development of a new field referred to as microtribology, nanotribology, molecular tribology, or atomic-scale tribology (Bhushan et al., 1995a; Bhushan, 1997, 1998a). This field is concerned with experimental and theoretical investigations of processes ranging from atomic and molecular scales to microscales, occurring during adhesion, friction, wear, and thin-film lubrication at sliding surfaces. The differences between the conventional or macrotribology and micro/nanotribology are contrasted in Figure 1.2. In macrotribology, tests are conducted on components with relatively large mass under heavily loaded conditions. In these tests, wear is inevitable and the bulk properties of mating components dominate the tribological performance. In micro/nanotribology, measurements are made on components, at least one of the mating components, with relatively small mass under lightly loaded conditions. In this situation, negligible wear occurs and the surface properties dominate the tribological performance.

The micro/nanotribological studies are needed to develop fundamental understanding of interfacial phenomena on a small scale and to study interfacial phenomena in micro- and nanostructures used in

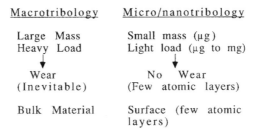

FIGURE 1.2 Comparisons between macrotribology and micro/nanotribology.

magnetic storage systems, MEMS, and other industrial applications. The components used in micro- and nanostructures are very light (on the order of a few micrograms) and operate under very light loads (on the order of a few micrograms to a few milligrams). As a result, friction and wear (on a nanoscale) of lightly loaded micro/nanocomponents are highly dependent on the surface interactions (few atomic layers). These structures are generally lubricated with molecularly thin films. Micro- and nanotribological techniques are ideal for studying the friction and wear processes of micro- and nanostructures. Although micro/nanotribological studies are critical to study micro- and nanostructures, these studies are also valuable in the fundamental understanding of interfacial phenomena in macrostructures to provide a bridge between science and engineering. At interfaces of technological innovations, contact occurs at multiple asperity contacts. A sharp tip of a tip-based microscope sliding on a surface simulates a single asperity contact, thus allowing high-resolution measurements of surface interactions at a single asperity contact. Friction and wear on micro- and nanoscales have been found to be generally small compared to that at macroscales. Therefore, micro/nanotribological studies may identify regimes for ultralow friction and near zero wear.

To give a historical perspective of the field, the scanning tunneling microscope (STM) developed by Dr. Gerd Binnig and his colleagues in 1981 at the IBM Zurich Research Laboratory, Forschungslabor, is the first instrument capable of directly obtaining three-dimensional images of solid surfaces with atomic resolution (Binnig et al., 1982). Binnig and Rohrer received a Nobel prize in physics in 1986 for their discovery. STMs can only be used to study surfaces which are electrically conductive to some degree. Based on their STM design in 1985, Binnig et al. developed an atomic force microscope (AFM) to measure ultrasmall forces (less than 1 μN) present between the AFM tip surface and the sample surface (Binnig et al., 1986a, 1987). AFMs can be used for measurement of *all engineering surfaces* which may be either electrically conducting or insulating. AFM has become a popular surface profiler for topographic measurements on micro- to nanoscale (Bhushan and Blackman, 1991; Oden et al., 1992; Ganti and Bhushan, 1995; Poon and Bhushan, 1995; Koinkar and Bhushan, 1997a; Bhushan et al., 1997c). Mate et al. (1987) were the first to modify an AFM in order to measure both normal and friction forces, and this instrument is generally called friction force microscope (FFM) or lateral force microscope (LFM). Since then, a number of researchers have used the FFM to measure friction on micro- and nanoscales (Erlandsson et al., 1988a,b; Kaneko, 1988; Blackman et al., 1990b; Cohen et al., 1990; Marti et al., 1990; Meyer and Amer, 1990b; Miyamoto et al., 1990; Kaneko et al., 1991; Meyer et al., 1992; Overney et al., 1992; Germann et al., 1993; Bhushan et al., 1994a–e, 1995a–g, 1997a–b; Frisbie et al., 1994; Ruan and Bhushan, 1994a–c; Koinkar and Bhushan, 1996a–c, 1997a,c; Bhushan and Sundararajan, 1998). By using a standard or a sharp diamond tip mounted on a stiff cantilever beam, AFMs can be used for scratching, wear, and measurements of elastic/plastic mechanical properties (such as indentation hardness and modulus of elasticity) (Burnham and Colton, 1989; Maivald et al., 1991; Hamada and Kaneko, 1992; Miyamoto et al., 1991, 1993; Bhushan, 1995; Bhushan et al., 1994b–e, 1995a–f, 1996, 1997a,b; Koinkar and Bhushan, 1996a,b, 1997b,c; Kulkarni and Bhushan, 1996a,b, 1997; DeVecchio and Bhushan, 1997).

AFMs and their modifications have also been used for studies of adhesion (Blackman et al., 1990a; Burnham et al., 1990; Ducker et al., 1992; Hoh et al., 1992; Salmeron et al., 1992, 1993; Weisenhorn et al., 1992; Burnham et al., 1993a,b; Hues et al., 1993; Frisbie et al., 1994; Bhushan and Sundararajan, 1998), electrostatic force measurements (Martin et al., 1988; Yee et al., 1993), ion conductance and electrochemistry (Hansma et al., 1989; Manne et al., 1991; Binggeli et al., 1993), material manipulation (Weisenhorn et al., 1990; Leung and Goh, 1992), detection of transfer of material (Ruan and Bhushan, 1993), thin-film boundary lubrication (Blackman et al., 1990a,b; Mate and Novotny, 1991; Mate, 1992; Meyer et al., 1992; O'Shea et al., 1992; Overney et al., 1992; Bhushan et al., 1995f,g; Koinkar and Bhushan, 1996b–c), to measure lubricant film thickness (Mate et al., 1989, 1990; Bhushan and Blackman, 1991; Koinkar and Bhushan, 1996c), to measure surface temperatures (Majumdar et al., 1993; Stopta et al., 1995), for magnetic force measurements including its application for magnetic recording (Martin et al., 1987b; Rugar et al., 1990; Schonenberger and Alvarado, 1990; Grutter et al., 1991, 1992; Ohkubo et al., 1991; Zuger and Rugar, 1993), and for imaging crystals, polymers, and biological samples in water (Drake et al., 1989; Gould et al., 1990; Prater et al., 1991; Haberle et al., 1992; Hoh and Hansma, 1992). STMs have been used in several different ways. They have been used to image liquids such as liquid crystals and lubricant molecules on graphite surfaces (Foster and Frommer, 1988; Smith et al., 1989, 1990; Andoh et al., 1992), to manipulate individual atoms of xenon (Eigler and Schweizer, 1990) and silicon (Lyo and Avouris, 1991), in formation of nanofeatures by localized heating or by inducing chemical reactions under the STM tip (Abraham et al., 1986; Silver et al., 1987; Albrecht et al., 1989; Mamin et al., 1990; Utsugi, 1990; Hosoki et al., 1992; Kobayashi et al., 1993), and nanomachining (Parkinson, 1990). AFMs have also been used for nanofabrication (Majumdar et al., 1992; Bhushan et al., 1994b–e, Bhushan, 1995, 1997; Boschung et al., 1994; Tsau et al., 1994) and nanomachining (Delawski and Parkinson, 1992).

Instruments that are able to measure tunneling current and forces simultaneously are being custom built (Specht et al., 1991; Anselmetti et al., 1992). Coupled AFM/STM measurements are made to distinguish between the topography of a sample and its electronic structure. Another aim is to determine the role of pressure in the tunnel junction in obtaining STM images.

Surface force apparatuses (SFA) are used to study both static and dynamic properties of the molecularly thin liquid films sandwiched between two molecularly smooth surfaces. Tabor and Winterton (1969) and later Israelachvili and Tabor (1972) developed apparatuses for measuring the van der Waals forces between two molecularly smooth mica surfaces as a function of separation in air or vacuum. These techniques were further developed for making measurements in liquids or controlled vapors (Israelachvili and Adams, 1978; Klein, 1980; Tonck et al., 1988; Georges et al., 1993). Israelachvili et al. (1988), Homola (1989), Gee et al. (1990), Homola et al. (1990, 1991), Klein et al. (1991), and Georges et al. (1994) measured the dynamic shear response of liquid films. Recently, new friction attachments were developed which allow for two surfaces to be sheared past each other at varying sliding speeds or oscillating frequencies, while simultaneously measuring both the friction forces and normal forces between them (Van Alsten and Granick, 1988, 1990a,b; Peachey et al., 1991; Hu et al., 1991). The distance between two surfaces can also be independently controlled to within ± 0.1 nm and the force sensitivity is about 10^{-8} N.

The SFAs are being used to study the rheology of molecularly thin liquid films; however, the liquid under study has to be confined between molecularly smooth, optically transparent surfaces with radii of curvature on the order of 1 mm (leading to poorer lateral resolution as compared with AFMs). SFAs developed by Tonck et al. (1988) and Georges et al. (1993, 1994) use an opaque and smooth ball with a large radius (~ 3 mm) against an opaque and smooth flat surface. Only AFMs/FFMs can be used to study *engineering surfaces* in the *dry and wet conditions* with *atomic resolution*.

The interest in the micro/nanotribology field grew from magnetic storage devices and its applicability to MEMS is clear. In this chapter, we first describe various measurement techniques, and then we present the examples of magnetic storage devices and MEMS where micro/nanotribological tools and techniques are essential for interfacial studies. We then present examples of why micro/nanotribological studies are important in magnetic storage devices, MEMS, and other microcomponents.

1.3   Measurement Techniques

The family of instruments based on STMs and AFMs are called scanning probe microscopes (SPMs). These include STM, AFM, FFM (or LFM), scanning magnetic microscopy (SMM) (or magnetic force microscopy, MFM), scanning electrostatic force microscopy (SEFM), scanning near-field optical microscopy (SNOM), scanning thermal microscopy (SThM), scanning chemical force microscopy (SCFM), scanning electrochemical microscopy (SEcM), scanning Kelvin probe microscopy (SKPM), scanning chemical potential microscopy (SCPM), scanning ion conductance microscopy (SICM), and scanning capacitance microscopy (SCM). The family of instruments which measures forces (e.g., AFM, FFM, SMM, and SEFM) are also referred to as scanning force microscopics (SFM). Although these instruments offer atomic resolution and are ideal for basic research, they are also used for cutting-edge industrial applications which do not require atomic resolution. Commercial production of SPMs started with STM in 1988 by Digital Instruments, Inc., and has grown to over $100 million in 1993 (about 2000 units installed to 1993) with an expected annual growth rate of 70%. For comparisons of SPMs with other microscopes, see Table 1.1 (Aden, 1994). The numbers of these instruments are equally divided among the U.S., Japan,

7

TABLE 1.1 Comparison of Various Conventional Microscopes with SPMs

	Optical	Confocal	SEM/TEM	SPM
Magnification	10^3	10^4	10^7	10^9
Instrument price, U.S. $	10k	30k	250k	100k
Technology age	200 yrs	10 yrs	30 yrs	9 yrs
Applications	Ubiquitous	New and unfolding	Science and technology	Cutting-edge
Market 1993	$800M	$80M	$400M	$100M
Growth rate	10%	30%	10%	70%

Data provided by Topometrix.

and Europe with the following industry/university and government laboratory splits: 50/50, 70/30, and 30/70, respectively. According to some estimates, over 3000 users of SPMs exist with $400 million in support. It is clear that research and industrial applications of SPMs are rapidly expanding.

STMs, AFMs, and their modifications can be used at extreme magnifications ranging from 10^3 to $10^9\times$ in x-, y-, and z-directions for imaging macro- to atomic dimensions with high-resolution information and for spectroscopy. These instruments can be used in any sample environment such as ambient air (Binnig et al., 1986a), various gases (Burnham et al., 1990), liquid (Marti et al., 1987; Drake et al., 1989; Binggeli et al., 1993), vacuum (Binnig et al., 1982; Meyer and Amer, 1988), low temperatures (Coombs and Pethica, 1986; Kirk et al., 1988; Giessibl et al., 1991; Albrecht et al., 1992; Hug et al., 1993), and high temperatures. Imaging in liquid allows the study of live biological samples, and it also eliminates water capillary forces present in ambient air present at the tip–sample interface. Low-temperature (liquid helium temperatures) imaging is useful for the study of biological and organic materials and the study of low-temperature phenomena such as superconductivity or charge density waves. Low-temperature operation is also advantageous for high-sensitivity force mapping due to the reduction in thermal vibration. These instruments are used for proximity measurements of magnetic, electrical, chemical, optical, thermal, spectroscopy, friction, and wear properties. Their industrial applications include micro-circuitry and semiconductor industry, information storage systems, molecular biology, molecular chemistry, medical devices, and materials science.

1.3.1 Scanning Tunneling Microscope

The principle of electron tunneling was proposed by Giaever (1960). He envisioned that if a potential difference is applied to two metals separated by a thin insulating film, a current will flow because of the ability of electrons to penetrate a potential barrier. To be able to measure a tunneling current, the two metals must be spaced no more than 10 nm apart. Binnig et al. (1982) introduced vacuum tunneling combined with lateral scanning. The vacuum provides the ideal barrier for tunneling. The lateral scanning allows one to image surfaces with exquisite resolution, lateral less than 1 nm and vertical less than 0.1 nm, sufficient to define the position of single atoms. The very high vertical resolution of STM is obtained because the tunnel current varies exponentially with the distance between the two electrodes, that is, the metal tip and the scanned surface. Typically, tunneling current decreases by a factor of 2 as the separation is increased by 0.2 nm. Very high lateral resolution depends upon the sharp tips. Binnig et al. (1982) overcame two key obstacles for damping external vibrations and for moving the tunneling probe in close proximity to the sample; their instrument is called the STM. Today's STMs can be used in the ambient environment for atomic-scale imaging of surfaces. Excellent reviews on this subject are presented by Pohl (1986), Hansma and Tersoff (1987), Sarid and Elings (1991), Durig et al. (1992), Frommer (1992), Guntherodt and Wiesendanger (1992), Wiesendanger and Guntherodt (1992), Bonnell (1993), Marti and Amrein (1993), Stroscio and Kaiser (1993), and Anselmetti et al. (1995) and the following dedicated issues of the *Journal of Vacuum Science Technology* (B9, 1991, pp. 401–1211) and *Ultramicroscopy* (Vol. 42–44, 1992).

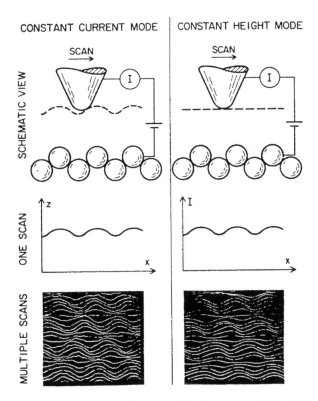

FIGURE 1.3 Scanning tunneling microscope can be operated in either the constant-current or the constant-height mode. The images are of graphite in air. (From Hansma, P. K. and Tersoff, J. (1987), *J. Appl. Phys.*, 61, R1–R23. With permission.)

The principle of STM is straightforward. A sharp metal tip (one electrode of the tunnel junction) is brought close enough (0.3 to 1 nm) to the surface to be investigated (second electrode) that, at a convenient operating voltage (10 mV to 1 V), the tunneling current varies from 0.2 to 10 nA, which is measurable. The tip is scanned over a surface at a distance of 0.3 to 1 nm, while the tunneling current between it and the surface is sensed. The STM can be operated in either the constant-current mode or the constant-height mode, Figure 1.3. The left-hand column of Figure 1.3 shows the basic constant current mode of operation. A feedback network changes the height of the tip z to keep the current constant. The displacement of the tip given by the voltage applied to the piezoelectric drives then yields a topographic picture of the surface. Alternatively, in the constant-height mode, a metal tip can be scanned across a surface at nearly constant height and constant voltage while the current is monitored, as shown in the right-hand column of Figure 1.3. In this case, the feedback network responds only rapidly enough to keep the average current constant (Hansma and Tersoff, 1987). A current mode is generally used for atomic-scale images. This mode is not practical for rough surfaces. A three-dimensional picture $[z(x, y)]$ of a surface consists of multiple scans $[z(x)]$ displayed laterally from each other in the y direction. It should be noted that if different atomic species are present in a sample, the different atomic species within a sample may produce different tunneling currents for a given bias voltage. Thus, the height data may not be a direct representation of the topography of the surface of the sample.[1]

---

[1]In fact, Marchon et al. (1989) STM imaged sputtered diamond-like carbon films in barrier-height mode by modulating the tip-to-surface distance, with lock-in detection of the tunneling current. The local barrier-height measurements give information on the local values of the work function, thus providing chemical information, in addition to the topographic map.

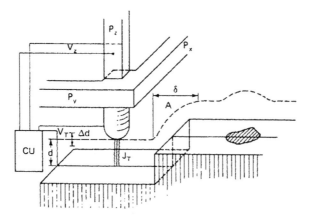

FIGURE 1.4 Principle of operation of the STM made by Binnig and Rohrer (1983).

1.3.1.1 Binnig et al.'s Design

Figure 1.4 shows a schematic of one of Binnig and Rohrer's designs for operation in an ultrahigh vacuum (Binnig et al., 1982; Binnig and Rohrer, 1983). The metal tip was fixed to rectangular piezodrives P_x, P_y, and P_z made out of commercial piezoceramic material for scanning. The sample is mounted on either a superconducting magnetic levitation or two-stage spring system to achieve the stability of a gap width of about 0.02 nm. The tunnel current J_T is a sensitive function of the gap width d; that is, $J_T \alpha V_T$ $\exp(-A\phi^{1/2}d)$, where V_T is the bias voltage, ϕ is the average barrier height (work function) and $A \sim 1$ if ϕ is measured in eV and d in Å. With a work function of a few eV, J_T changes by an order of magnitude for every angstrom change of h. If the current is kept constant to within, for example, 2%, then the gap h remains constant to within 1 pm. For operation in the constant-current mode, the control unit (CU) applies a voltage V_z to the piezo P_z such that J_T remains constant when scanning the tip with P_y and P_x over the surface. At the constant-work functions ϕ, $V_z(V_x, V_y)$ yields the roughness of the surface $z(x, y)$ directly, as illustrated at a surface step at A. Smearing the step, δ (lateral resolution) is on the order of $(R)^{1/2}$, where R is the radius of the curvature of the tip. Thus, a lateral resolution of about 2 nm requires tip radii on the order of 10 nm. A 1-mm-diameter solid rod ground at one end at roughly 90° yields overall tip radii of only a few hundred nanometers, but with closest protrusion of rather sharp microtips on the relatively dull end yielding a lateral resolution of about 2 nm. *In situ* sharpening of the tips by gently touching the surface brings the resolution down to the 1-nm range; by applying high fields (on the order of 10^8 V/cm) during, for example, half an hour, resolutions considerably below 1 nm could be reached. Most experiments were done with tungsten wires either ground or etched to a radius typically in the range of 0.1 to 10 μm. In some cases, *in situ* processing of the tips was done for further reduction of tip radii.

1.3.1.2 Commercial STMs

There are a number of commercial STMs available on the market. Digital Instruments, Inc., located in Santa Barbara, CA introduced the first commercial STM, the Nanoscope I, in 1987. In the Nanoscope III STM for operation in ambient air, the sample is held in position while a piezoelectric crystal in the form of a cylindrical tube scans the sharp metallic probe over the surface in a raster pattern while sensing and outputting the tunneling current to the control station, Figure 1.5 (Anonymous, 1992b). The digital signal processor (DSP) calculates the desired separation of the tip from the sample by sensing the tunneling current flowing between the sample and the tip. The bias voltage applied between the sample and the tip encourages the tunneling current to flow. The DSP completes the digital feedback loop by outputting the desired voltage to the piezoelectric tube. The STM operates in both the constant-height and constant-current modes depending on a parameter selection in the control panel. In the constant-current mode, the feedback gains are set high, the tunneling tip closely tracks the sample surface, and

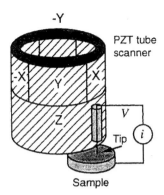

FIGURE 1.5 Principle of operation of a commercial STM, a sharp tip attached to a piezoelectric tube scanner is scanned on a sample.

the variation in the tip height required to maintain constant tunneling current is measured by the change in the voltage applied to the piezotube. In the constant-height mode, the feedback gains are set low, the tip remains at a nearly constant height as it sweeps over the sample surface, and the tunneling current is imaged. The following description of the instrument is almost exclusively based on Anonymous (1992b).

Physically, the Nanoscope STM consists of three main parts: the head which houses the piezoelectric tube scanner for three-dimensional motion of the tip and the preamplifier circuit (FET input amplifier) mounted on top of the head for the tunneling current, the base on which the sample is mounted, and the base support, which supports the base and head, Figure 1.6A. The assembly is connected to a control system that controls the operation of the microscope. The base accommodates samples up to 10×20 mm and 10 mm in thickness. The different scanning heads mount magnetically on the tripod formed by the front, coarse-adjust screws and the rear, find-adjust screws. Optional scan heads for the STM include 0.7 (for atomic resolution), 12, 75, and 125 µm square.

The scanning head controls the three-dimensional motion of tip. The removable head consists of a piezotube scanner, about 12.7 mm in diameter, mounted into an Invar shell used to minimize vertical thermal drifts because of good thermal match between the piezotube and the Invar. The piezotube has separate electrodes for X, Y, and Z which are driven by separate drive circuits. The electrode configuration (Figure 1.6B) provides X and Y motions, which are perpendicular to each other, minimizes horizontal and vertical coupling, and provides good sensitivity. The vertical motion of the tube is controlled by the Z-electrode which is driven by the feedback loop. The X and Y scanning motions are each controlled by two electrodes which are driven by voltages of the same magnitudes, but opposite signs. These electrodes are called −Y, −X, +Y, and +X. Applying complimentary voltages allows a short, stiff tube to provide a good scan range without large voltages. The motion of the tip due to external vibrations is proportional to the square of the ratio of vibrational frequency to the resonant frequency of the tube. Therefore, to minimize the tip vibrations, the resonant frequencies of the tube are high, about 60 kHz in the vertical direction and about 40 kHz in the horizontal direction. The tip holder is a stainless steel tube with a 300-µm inner diameter for 250-µm-diameter tips, mounted in ceramic in order to keep the mass on the end of the tube low. The tip is mounted either on the front edge of the tube (to keep mounting mass low and resonant frequency high) (Figure 1.5) or the center of the tube for large-range scanners, namely 75 and 125 µm (to preserve the symmetry of the scanning). This commercial STM will accept any tip with a 250-µm-diameter shaft. The piezotube requires X–Y calibration which is carried out by imaging an appropriate calibration standard. Cleaved graphite is used for the small-scan length head while two-dimensional grids (a gold-plated ruling) can be used for longer-range heads.

The Invar base holds the sample in position, supports the head, and provides X–Y motion for the sample, Figure 1.6C. A spring-steel sample clip with two thumbscrews holds the sample in place. An X–Y translation stage built into the base allows the sample to be repositioned under the tip. Three precision

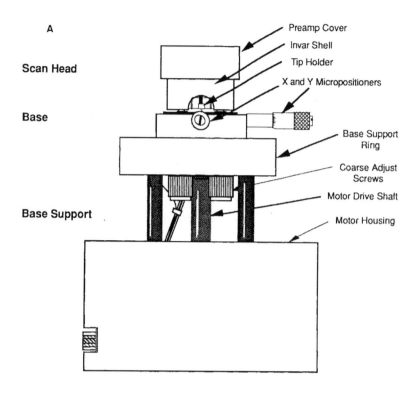

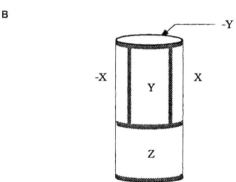

FIGURE 1.6 Schematics of a commercial STM made by Digital Instruments, Inc.: (A) front view, (B) general electrode configuration for piezoelectric tube scanner, and (C) front and top view of the STM base. (From Anonymous (1992), "Nanoscope III Scanning Tunneling Microscope, Instruction Manual," Courtesy of Digital Instruments, Inc., Santa Barbara, CA, 1992.)

screws arranged in a triangular pattern support the head and provide coarse and fine adjustment of the tip height. The base support consists of the base support ring and the motor housing. The base support ring cradles the base allowing access to the adjustment screws. The stepper motor enclosed in the motor housing allows the tip to be engaged and withdrawn from the surface automatically.

For measurements, the sample is placed under the sample-holding clip, with about half the sample extending forward of the wire using an appropriate scanner, and a tip is inserted in the tip-holding tube mounted on the piezotube. The tip is gripped with a tweezer near the sharp end and the blunt end of the tip is inserted into the tip holder. For the tip to be held in the tube, it is necessary to put a small bend in the tip before it is completely inserted. Next, the scanning head is placed on the three magnetic

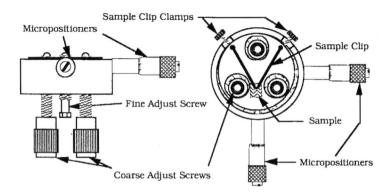

FIGURE 1.6(C)

balls mounted on the threaded screws in the base. The tip is lowered with the coarse-adjustment screws until there is only a slight gap, less than 0.25 mm (the tip will be damaged if brought into contact) between the end of the tip and its reflected image visible on the sample. Next the scan parameters are set, the motor is turned on, which engages the tip, and the scanning is initiated to form a desired image of the sample surface.

Samples to be imaged with STM must be conductive enough to allow a few nanoamperes of current to flow from the bias voltage source to the area to be scanned. In many cases, nonconductive samples can be coated with a thin layer of a conductive material to facilitate imaging. The bias voltage and the tunneling current depend on the sample. Usually they are set at a standard value for engagement and fine-tuned to enhance the quality of the image. The scan size depends on the sample and the features of interest. A maximum scan rate of 122 Hz can be used. The maximum scan rate is usually related to the scan size. Scan rate above 10 Hz is used for small scans (typically 60 Hz for atomic-scale imaging with a 0.7-µm scanner). The scan rate should be lowered for large scans, especially if the sample surfaces are rough or contain large steps. Moving the tip quickly along the sample surface at high scan rates with large scan sizes will usually lead to a tip crash. Essentially, the scan rate should be inversely proportional to the scan size (typically 2 to 4 Hz for 1 µm, 0.5 to 1 Hz for 12 µm, and 0.2 Hz for 125 µm scan sizes). Scan rate in length/time is equal to scan length divided by the scan rate in hertz. For example, for 10×10 µm scan size scanned at 0.5 Hz, the scan rate is 20 µm/s. Typically, 256×256 data formats are most commonly used. The lateral resolution at larger scans is approximately equal to scan length divided by 256.

Figure 1.7 shows an example of an STM image of freshly cleaved, highly oriented pyrolytic graphite (HOPG) surface taken at room temperature and ambient pressure (Binnig et al., 1986b; Park and Quate, 1986; Ruan and Bhushan, 1994b).

1.3.1.2.1 Electrochemical STM (ECSTM)

Electrochemical STM (ECSTM) allows the performance of the electrochemical reactions on the STM. It includes a microscope base with an integral potentiostat, a short head with a 0.7-µm scan range, and a differential preamp and the software required to operate the potentiostat and display the result of electrochemical reaction.

1.3.1.2.2 Stand-Alone STM

The stand-alone STMs are available to scan large samples which rest directly on the sample. From Digital Instruments, Inc., it is available in 12- and 75-µm scan ranges. It is similar to the standard STM except the sample base has been eliminated. Two coarse- and one fine-adjustment screws used to position the tip manually relative to the sample surface are mounted in the head shell.

1.3.1.3 Tip Construction

The STM cantilever should have a sharp metal tip with a low aspect ratio (tip length/tip shank) to minimize flexural vibrations. Ideally, the tip should be atomically sharp, but, in practice, most tip

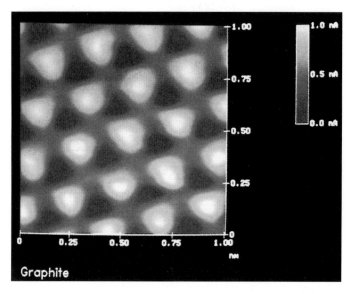

FIGURE 1.7 Typical STM image of freshly cleaved, HOP graphite taken using a mechanically sheared Pt–Ir (80–20) tip in constant-height mode (set point = 4 nA, bias = 16 mV, frequency = 20 Hz, 256 × 256 pixels, original scan size 3 × 3 nm). Bright spots correspond to the visible atoms.[*]

preparation methods produce a tip which is rather ragged and consists of several asperities with the one closest to the surface responsible for tunneling. STM cantilevers with sharp tips are typically fabricated from metal wires of tungsten (W), platinum–iridium (Pt–Ir), or gold (Au) and sharpened by grinding, cutting with a wire cutter or razor blade, field emission/evaporator, ion milling, fracture, or electrochemical polishing/etching (Ibe et al., 1990). The two most commonly used tips are made from either a Pt–Ir (80/20) alloy or tungsten wire. Iridium is used to provide stiffness. The Pt–Ir tips are generally mechanically formed and are readily available. The tungsten tips are etched from tungsten wire with an electrochemical process, for example, by using 1 mol KOH solution with a platinum electrode in an electrochemical cell at about 30 V. In general, Pt–Ir tips provide better atomic resolution than tungsten tips, probably due to the lower reactivity of Pt, but tungsten tips are more uniformly shaped and may perform better on samples with steeply sloped features. The wire diameter used for the cantilever is typically 250 μm with the radius of curvature ranging from 20 to 100 nm and a cone angle ranging from 10 to 60°, Figure 1.8. The wire can be bent in an L shape, if so required for use in the instrument. For calculations of normal spring constant and natural frequency of round cantilevers, see Sarid and Elings (1991).

Controlled geometry (CG) Pt-Ir probes are commercially available, Figure 1.9. These probes are electrochemically etched from Pt-Ir (80/20) wire and polished to a specific shape which is consistent from tip to tip. Probes have a full cone angle of approximately 15° and a tip radius of less than 50 nm. For imaging of deep trenches (>0.25 μm) and nanofeatures, focused ion beam (FIB) milled CG probes with an extremely sharp tip radius (<5 nm) are used. For electrochemistry, Pt-Ir probes are coated with a nonconducting film (not shown in the figure). These probes are available from Materials Analytical Services, Raleigh, NC.

Platinum alloy and tungsten tips are very sharp and have high resolution, but are fragile and sometimes break when contacting a surface. Diamond tips were used by Kaneko and Oguchi (1990), Figure 1.10. The diamond tip made conductive by boron ion implantation is found to be chip resistant. The diamond

---

[*] Color reproduction follows page 16.

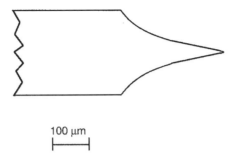

100 µm

FIGURE 1.8   Schematic of a typical tungsten cantilever with a sharp tip produced by electrochemical etching.

chip is brazed to a titanium shank having a tail diameter of 0.25 mm and total length of 10 mm. The diamond is ground to the shape of a three-sided pyramid whose point is sharpened to a radius of about 100 nm. The smallest apex angle to achieve a sharp point without chipping is 60°. Finally, boron ions are implanted in the diamond. Kaneko and Oguchi reported these tips as having a superior life.

1.3.2   Atomic Force Microscope

Like the STM, the AFM (a family of SFMs) relies on a scanning technique to produce very high resolution, three-dimensional images of sample surfaces. AFM measures ultrasmall forces (less than 1 nN) present between the AFM tip surface and a sample surface. These small forces are measured by measuring the motion of a very flexible cantilever beam having an ultrasmall mass. While the STM requires that the surface measured be electrically conductive, the AFM is capable of investigating surfaces of both conductors and insulators on an atomic scale if suitable techniques for measurement of cantilever motion are used. In the operation of high-resolution AFM, the sample is generally scanned instead of the tip as

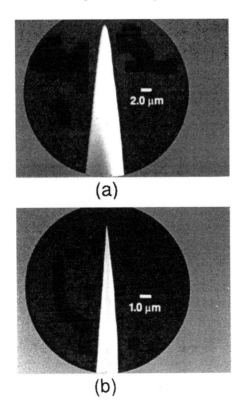

FIGURE 1.9   Schematics of (a) CG Pt–Ir probe and (b) CG Pt–Ir FIB-milled probe.

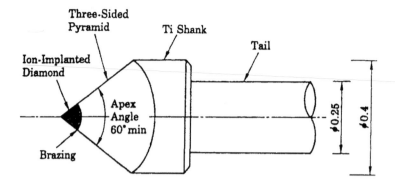

FIGURE 1.10 Schematic of a special diamond tip and shank with an overall length of 10 mm for use in STM. (From Kaneko, R. and Oguchi, S. (1990), *Jpn. J. Appl. Phys.*, 28, 1854–1855. With permission.)

in STM, because AFM measures the relative displacement between the cantilever surface and the reference surface, and any cantilever movement would add vibrations. However, AFMs are now available where the tip is scanned and the sample is stationary. As long as the AFM is operated in the so-called contact mode, little if any vibration is introduced.

The AFM combines the principles of the STM and the stylus profiler, Figure 1.11. In the AFM, the force between the sample and tip is detected rather than the tunneling current to sense the proximity of the tip to the sample. A sharp tip at the end of a cantilever is brought in contact with a sample surface by moving the sample with piezoelectric scanners. During initial contact, the atoms at the end of the tip experience a very weak repulsive force due to electronic orbital overlap with the atoms in the sample surface. The force acting on the tip causes a lever deflection which is measured by tunneling, capacitive, or optical detectors such as laser interferometry. The deflection can be measured to within ±0.02 nm, so for a typical lever force constant at 10 N/m a force as low as 0.2 nN (corresponding normal pressure ~200 MPa for an Si_3N_4 tip with a radius of about 50 nm against single-crystal silicon) could be detected. This operational mode is referred to as the "repulsive mode" or "contact mode" (Binnig et al., 1986a). An alternative is to use "attractive force imaging" or "noncontact imaging," in which the tip is brought in close proximity (within a few nanometers) to, and not in contact with, the sample (Martin et al., 1987a). A very weak van der Waals attractive force is present at the tip-sample interface. Although in this technique the normal pressure exerted at the interface is zero (desirable to avoid any surface deformation),

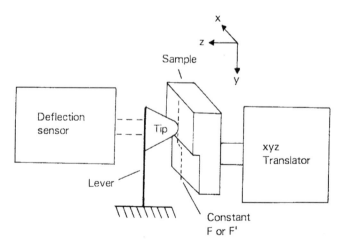

FIGURE 1.11 Principle of operation of the AFM. (From McClelland, G. E. et al. (1987), *Review of Progress in Quantitative Nondestructive Evaluation*, D. D. Thompson and D. E. Chimenti, eds., Vol. 6B, pp. 1307–1314, Plenum, New York. With permission.

it is slow and difficult to use and is rarely used outside research environments. In either mode, surface topography is generated by laterally scanning the sample under the tip while simultaneously measuring the separation-dependent force or force gradient (derivative) between the tip and the surface, Figure 1.11. The force gradient is obtained by vibrating the cantilever (Martin et al., 1987a; McClelland et al., 1987; Sarid and Elings, 1991) and measuring the shift of resonance frequency of the cantilever. To obtain topographic information, the interaction force is either recorded directly or used as a control parameter for a feedback circuit that maintains the force or force derivative at a constant value. The force derivative is normally tracked in noncontact imaging. With an AFM operated in the contact mode, topographic images with a vertical resolution of less than 0.1 nm (as low as 0.01 nm) and a lateral resolution of about 0.2 nm have been obtained (Albrecht and Quate, 1987; Binnig et al., 1987; Marti et al., 1987; Alexander et al., 1989; Meyer and Amer, 1990a; Weisenhorn et al., 1991; Bhushan et al., 1993; Ruan and Bhushan, 1994b). With a 0.01-nm displacement sensitivity, 10 nN to 1 pN forces are measurable. These forces are comparable to the forces associated with chemical bonding e.g., 0.1 μN for an ionic bond and 10 pN for a hydrogen bond (Binnig et al., 1986a). For further reading, see Rugar and Hansma (1990), Sarid (1991), Sarid and Elings (1991), Binnig (1992), Durig et al. (1992), Frommer (1992), Meyer (1992), Marti and Amrein (1993), and Guntherodt et al. (1995) and dedicated issues of *Journal of Vacuum Science Technology* (B9, 1991, pp. 401–1211) and *Ultramicroscopy* (Vols. 42–44, 1992).

Lateral forces being applied at the tip during scanning in the contact mode affect roughness measurements (den Boef, 1991). To minimize effects of friction and other lateral forces in the topography measurements in the contact-mode AFMs and to measure topography of soft surfaces, AFMs can be operated in the so-called force modulation mode or tapping mode (Maivald et al., 1991; Radmacher et al., 1992). In the force modulation mode, the tip is lifted and then lowered to contact the sample (oscillated at a constant amplitude) during scanning over the surface with a feedback loop keeping the average force constant. This technique eliminates frictional force entirely. The amplitude is kept large enough so that the tip does not get stuck to the sample because of adhesive attractions. The modulation mode can also be used to measure local variations in surface viscoelastic properties (Maivald et al., 1991; Salmeron et al., 1993).

STM is ideal for atomic-scale imaging. To obtain atomic resolution with AFM, the spring constant of the cantilever should be weaker than the equivalent spring between atoms. For example, the vibration frequencies ω of atoms bound in a molecule or in a crystalline solid are typically 10^{13} Hz or higher. Combining this with the mass of the atoms m, on the order of 10^{-25} kg, gives interatomic spring constants k, given by $\omega^2 m$, on the order of 10 N/m (Rugar and Hansma, 1990). (For comparison, the spring constant of a piece of household aluminum foil that is 4 mm long and 1 mm wide is about 1 N/m.) Therefore, a cantilever beam with a spring constant of about 1 N/m or lower is desirable. Tips have to be as sharp as possible. Tips with a radius ranging from 20 to 50 nm are commonly available.

Atomic resolution cannot be achieved with these tips at the normal force in the nanonewton range. Atomic structures obtained at these loads have been obtained from lattice imaging or by imaging of the crystal periodicity. Reported data show either perfectly ordered periodic atomic structures or defects on a larger lateral scale, but no well-defined, laterally resolved atomic-scale defects like those seen in images routinely obtained with STM. Interatomic forces with one or several atoms in contact are 20 to 40 or 50 to 100 pN, respectively. Thus, atomic resolution with AFM is only possible with a sharp tip on a flexible cantilever at a net repulsive force of 100 pN or lower (Ohnesorge and Binnig, 1993). Upon increasing the force from 10 pN, Ohnesorge and Binnig (1993) observed that monatomic steplines were slowly wiped away and a perfectly ordered structure was left. This observation explains why mostly defect-free atomic resolution has been observed with AFM. We note that for atomic-resolution measurements the cantilever should not be too soft to avoid jumps. We further note that measurements in the attractive-force imaging mode may be desirable for imaging with atomic resolution.

The key component in AFM is the sensor for measuring the force on the tip due to its interaction with the sample. A lever (with a sharp tip) with extremely low spring constants is required for high vertical and lateral resolutions at small forces (0.1 nN or lower), but at the same time a high-resonant frequency (about 10 to 100 kHz) in order to minimize the sensitivity to vibrational noise from the

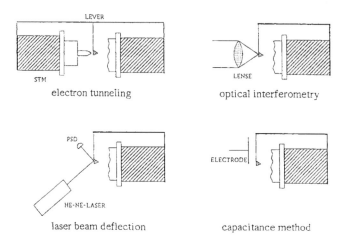

electron tunneling

optical interferometry

laser beam deflection

capacitance method

FIGURE 1.12 Geometries of the four commonly used detection systems for measurement of cantilever deflection. In each setup, the sample mounted on piezoelectric body is shown on the right, the cantilever in the middle, and the corresponding deflection sensor on the left. (From Meyer, E. (1992), *Surf. Sci.*, 41, 3–49. With permission.)

building near 100 Hz. This requires a spring with extremely low vertical spring constant (typically, 0.05 to 1 N/m) as well as low mass (on the order of 1 ng). Today, the most-advanced AFM cantilevers are microfabricated from silicon, silicon dioxide, or silicon nitride using photolithographic techniques. (For further details on cantilevers, see Section 1.3.2.6.) Typical lateral dimensions are on the order of 100 μm with the thicknesses on the order of 1 μm. The force on the tip due to its interaction with the sample is sensed by detecting the deflection of the compliant lever with a known spring constant. This lever deflection (displacement smaller than 0.1 nm) has been measured by detecting tunneling current similar to that used in STM in the pioneering work of Binnig et al. (1986a) and later used by Giessibl et al. (1991), by capacitance-detection (Neubauer et al., 1990; Goddenhenrich et al., 1990), and by four optical techniques, namely, (1) by optical interferometry (Mate et al., 1987; McClelland et al., 1987; Erlandsson et al., 1988a; Mate, 1992; Jarvis et al., 1993) and with the use of optical fibers (Ruger et al., 1989; Albrecht et al., 1992); (2) by optical polarization detection (Schonenberger and Alvarado, 1990); (3) by laser diode feedback (Sarid et al., 1988); and (4) by optical (laser) beam deflection (Meyer and Amer, 1988, 1990a,b; Marti et al., 1990). More recently, Smith (1994) used a piezoresistive cantilever beam which requires no external sensor. It makes the SPM design simpler and the STM and AFM functions can be combined readily. However, the piezoresistive beam needs power on the order of 10 mW and has less sensitivity. Geometries of the four more commonly used detection systems are shown in Figure 1.12. The tunneling method originally used by Binnig et al. (1986a) in the first version of AFM uses a second tip to monitor the deflection of the cantilever with its force-sensing tip. Tunneling is rather sensitive to contaminants and the interaction between the tunneling tip and the rear side of the cantilever can become comparable to the interaction between the tip and sample. Tunneling is rarely used and is mentioned earlier for historical purposes. Giessibl et al. (1991) recently used it for a low-temperature AFM/STM design. In contrast to tunneling, other deflection sensors are far away from the cantilever at distances of microns to tens of millimeters. The optical technique is believed to be a more sensitive, reliable, and easily implemented detection method than others (Sarid and Elings, 1991; Meyer, 1992). The optical beam deflection method has the largest working distance, is insensitive to distance changes, and is capable of measuring angular changes (friction forces); therefore, it is most commonly used in commercial SPMs.

Almost all AFMs use piezotranslators to scan the sample, or alternatively, to scan the tip. An electric field applied across a piezoelectric material causes a change in the crystal structure, with expansion in some directions and contraction in others. A net change in volume also occurs (Ashcroft and Mermin, 1976). The first STM used a piezotripod for scanning (Binnig et al., 1982). The piezotripod is one way to generate three-dimensional movement of a tip attached to its center. However, the tripod needs to be fairly large (~50 mm) to get a suitable range. Its size and asymmetric shape makes it susceptible to thermal drift. The tube scanners are widely used in AFMs (Binnig and Smith, 1986). These provide ample scanning range within a small size.

Control electronics systems for AFMs can use either analog or digital feedback. Digital feedback circuits might be better suited for ultralow noise operation.

Images from the AFMs need to be processed. An ideal AFM is a noise-free device that images a sample with perfect tips of known shape and has perfect linear scanning piezo. In reality, scanning devices are affected by distortions for which corrections must be made. The distortions can be linear and nonlinear. Linear distortions mainly result from imperfections in the machining of the piezotranslators causing cross talk between the Z-piezo to the X- and Y-piezos, and vice versa. Nonlinear distortions mainly result because of the presence of a hysteresis loop in piezoelectric ceramics. These may also result if the scan frequency approaches the upper frequency limit of the X- and Y-drive amplifiers or the upper frequency limit of the feedback loop (Z-component). In addition, electronic noise may be present in the system. The noise is removed by digital filtering in the real space (Park and Quate, 1987) or in the spatial frequency domain (Fourier space) (Cooley and Turkey, 1965).

Processed data consists of many tens of thousand of points per plane (or data set). The output of the first STM and AFM images were recorded on an X-Y chart recorder, with Z-value plotted against the tip position in the fast-scan direction. Chart recorders have slow response so storage oscilloscopes or computers are used for display of the data. The data are displayed as wire mesh display or gray scale display (with at least 64 shades of gray).

1.3.2.1 Binnig et al.'s Design

In the first AFM design developed by Binnig et al. (1986a), AFM images were obtained by measurement of the force on a sharp tip created by the proximity to the surface of the sample mounted on a three-dimensional piezoelectric scanner. The tunneling current between the STM tip and the backside of the cantilever beam with attached tip was measured to obtain the normal force. This force was kept at a constant level with a feedback mechanism. The STM tip was also mounted on a piezoelectric element to maintain the tunneling current at a constant level.

1.3.2.2 McClelland et al.'s Design

An AFM developed by Erlandsson et al. (1988a) for operation in ambient air is shown schematically in Figure 1.13. Following the STM design, the test sample was mounted on three orthogonal piezoelectric tubes (2 to 5 mm long), two of which (x, y) raster the sample in the surface plane while the third (z) moves the sample toward and away from the tip. The lever was made from a 70-µm-diameter, 3-mm-long tungsten microprobe with a 90° bend near one end that serves as the tip. (In most cases, the tip is electrochemically etched using a 12-V AC in 2 N (normal) NaOH solution to obtain a nominal tip radius between 150 to 300 nm.) The main resonant frequency of this lever was about 5 kHz and the force constant was about 30 N/m. The lever support was mounted on a piezoelectric transducer that makes it possible to oscillate the lever when needed. The lever motion was measured by optical interference. A light beam was focused on the backside of the lever by a microscope objective, and the interference pattern between the reflected beam and a reference beam reflected from an optical flat is projected on a photodiode that measures the instantaneous deflection of the lever as well as its vibration amplitude at high frequencies. The deflection can be detected within ±0.2 nm; so for a typical lever force constant of 10 N/m a force as low as 0.2 nN could be detected.

More recently, a high-sensitivity fiber-optic displacement sensor has been developed by the IBM group which is compact and does not require specular reflection and thus is compatible with both microfabricated thin-film cantilevers as well as fine wire cantilevers. All fiber construction results in smaller size and improved mechanical robustness (Rugar et al., 1989; Albrecht et al., 1992). Schematic design of the AFM with a fiber optic interferometer is shown in Figure 1.14. A multimode GaAlAs diode laser with a direct single-mode fiber output is used as a light source. The light is coupled into the input (labeled "1")

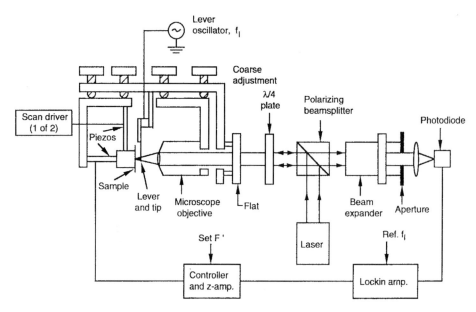

FIGURE 1.13 Schematic of an AFM which uses optical interference to detect the lever deflection. (Erlandsson, R. et al. (1988), *J. Vac. Sci. Technol.*, A6, 266–270. With permission.)

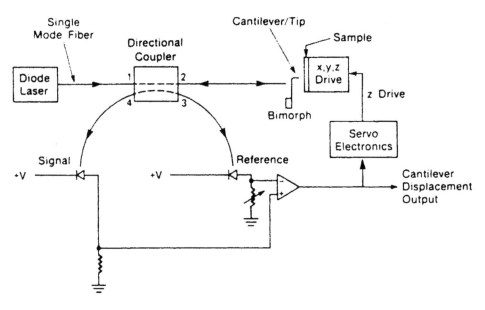

FIGURE 1.14 Schematic of an AFM with a fiber-optic interferometer. (From Rugar, D. et al. (1989), *Appl. Phys. Lett.*, 55, 2588–2590. With permission.)

of a 2×2 single-mode directional coupler. The coupler splits the incident optical power equally between leads 2 and 3, which carry the light to the AFM cantilever and the "reference" photodiode, respectively. Approximately 4% of the light in lead 2 is reflected from the glass–air interface at the cleaved end of the fiber. This reflected light comprises one of the two interfering beams. The other 96% of the light exits the fiber and impinges on the cantilever with a spot size of about 5 μm. Part of this light is scattered back into the fiber and interferes with the light reflected from the fiber end. The total optical power reflected back through the fiber depends on the phase difference between the fiber end reflection and

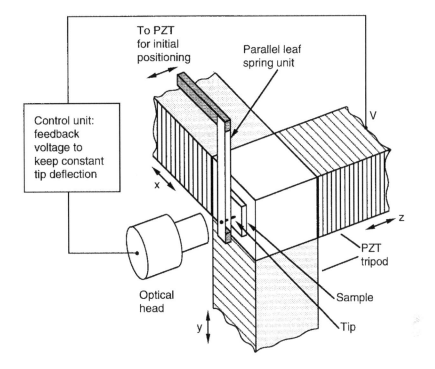

FIGURE 1.15 Schematic of an AFM in which the sample is mounted on a piezoelectric tripod and the tip is supported by a parallel-leaf spring unit. (From Kaneko, R. et al. (1988), *J. Vac. Sci. Technol.*, A6, 291–292. With permission.)

the cantilever reflection. The coupler directs half of the total reflected light to lead 4 and into the signal 2 photodiode where the intensity of the optical interference is measured. To reduce reflections from the ends of leads 3 and 4, the fibers were cleaved at a nonorthogonal angle and an index-matching liquid was placed between the photodiodes and the fiber ends. The output of the signal photodiode can be used directly as the AFM signal (Rugar et al., 1989).

AFMs can be used to obtain topographic images using repulsive contact forces as well as attractive electrostatic forces. Several methods have been used to detect the forces (Binnig et al., 1986a; Erlandsson et al., 1988a). In one force-detection method, the signal corresponding to the force can either be used as a control parameter for the feedback circuit to generate contours of equal force or be displayed directly without feedback while pressing the tip onto the sample with an average force larger than the recorded force variations. In another method, a small AC voltage is applied to the z-tube to induce an oscillation in the sample and, through the force coupling, to the lever. The resultant oscillation in the photodiode signal is converted by the lock-in amplifier to a voltage that is proportional to the derivative of the force, F'. The z-amplifier compares the voltage to some preset value and drives the z-tube to form a feedback loop to maintain F' constant, and a three-dimensional surface of F' can be obtained (Erlandsson et al., 1988a).

1.3.2.3 Kaneko et al.'s Design

In the AFM designs developed by Kaneko and co-workers for use in ambient air, the instrument consists of a piezoelectric tripod that holds the sample, a sharp diamond tip (to be presented later) supported by a parallel-leaf spring unit mounted on a laminated piezoelectric stack, and a focusing error detection type optical head, Figure 1.15 (Kaneko et al., 1988; Kaneko and Hamada, 1990; Miyamoto et al., 1990). This design was later modified by Kaneko et al. (1990, 1991). The major modifications were that a piezotube scanner was used to hold the sample and the tip was supported on a single-leaf spring, Figure 1.16. In their even newer design, they have incorporated a new tube scanner and an optical

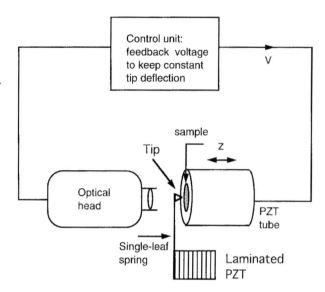

FIGURE 1.16 Schematic of an AFM in which the sample is mounted on a piezoelectric tube scanner and the tip is supported by a single-leaf spring. (From Kaneko, R. et al. (1990), *Tribology and Mechanics of Magnetic Storage Systems* (B. Bhushan, ed.) SP-29, pp. 31–34, STLE, Park Ridge, IL. With permission.)

multifunction sensor (Kaneko et al., 1992). The spring constants used in Figures 1.15 and 1.16 are 3 to 24 N/m and 0.3 to 3 N/m, respectively. A laminated piezoelectric stack was used for initial positioning of the tip and the piezoelectric tripod or piezotube scanner was used to place the tip in contact with the surface in the z-direction and to scan the surface in the x- or y-direction. For surface topography measurements, the sample was slowly moved in the z-direction until it contacted the tip then it was scanned in the x- or y-direction. Tip displacement during scanning was measured by a focusing error-detection-type optical head (with an accuracy of better than 1 nm), and the displacement signal was used as a control signal for the z-displacement of the piezoelectric tripold or tube scanner to keep the spring load constant. Variations in the vertical motion of the sample represented a roughness profile. Additional details of the construction of these designs can be found in a following FFM section.

1.3.2.4 Meyer and Amer's Design

In the AFM design developed by Meyer and Amer (1988) for operation in an ultrahigh vacuum (UHV), bending of a tungsten cantilever beam resulting from the normal force being applied at the tip was measured by detecting the deflection of a laser beam, which was reflected off its backside. The deflection was sensed with a segmented photodiode detector, typically a bicell, which consists of two photoactive (e.g., Si) segments (anodes) that are separated by about 10 μm and have a common cathode. This optical beam deflection technique is simple and sensitive and is used in a commercial AFM whose description follows.

1.3.2.5 Commercial AFMs

There are a number of commercial AFMs available on the market since 1989. Major manufacturers of AFMs for use in an ambient environment are Digital Instruments, Inc., 112 Robin Hill Road, Santa Barbara, CA; Park Scientific Instruments, 476 Ellis Street, Mountain View, CA; Topometrix, 5403 Betsy Ross Drive, Santa Clara, CA; Seiko Instruments, Japan; Olympus, Japan; and Centre Suisse D'Electronique et de Microtechnique (CSEM) S.A., Neuchâtel, Switzerland. In the CSEM design, both force sensors (using optical beam deflection method) and scanning unit are mounted on the microscope head; thus their AFM/FFM designs can be used as stand-alone (Hipp et al., 1992). UHV AFM/STMs are manufactured by Omicron Vakuumphysik GmbH, Idsteiner Strasse 78, D-6204, Taunusstein 4, Germany. Personal

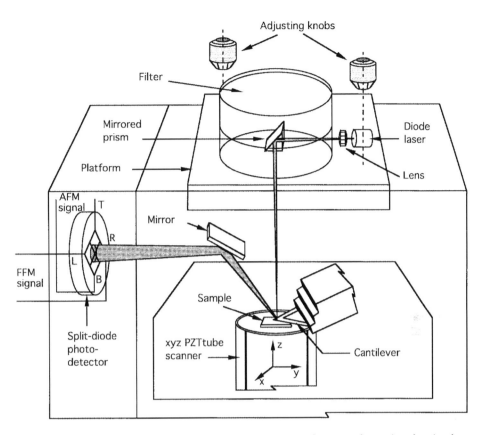

FIGURE 1.17 Principle of operation of a commercial AFM/FFM — sample mounted on a piezoelectric tube scanner is scanned against a sharp tip and the cantilever deflection is measured using a laser deflection technique.

STMs and AFMs for ambient environment and UHV/STMs are manufactured by Burleigh Instruments, Inc., Burleigh Park, Fishers, NY.

We describe here the commercial AFM for operation in ambient air, produced by Digital Instruments, Inc., with scanning lengths ranging from about 0.7 μm (for atomic resolution) to about 125 μm (Alexander et al., 1989; Anonymous, 1992b; Bhushan and Ruan, 1994a; Ruan and Bhushan, 1994a,b). This is the most commonly used design and the multimode AFM comes with many capabilities. The original design of this AFM version comes from Meyer and Amer (1988). Basically, the AFM scans the sample in a raster pattern while outputting the cantilever deflection error signal to the control station. The cantilever deflection (or the force) is measured using a laser deflection technique, Figure 1.17. The digital signal processor (DSP) in the workstation controls the Z-position of the piezo based on the cantilever deflection error signal. The AFM operates in both the constant-height and constant-force modes. The DSP always adjusts the height of the sample under the tip based on the cantilever deflection error signal, but if the feedback gains are low the piezo remains at a nearly constant height and the cantilever deflection data is collected. With the gains high, the piezo height changes to keep the cantilever deflection nearly constant (therefore, the force is constant), and the change in piezo height is collected by the system. The following description of the instrument is almost exclusively based on Anonymous (1992b).

To further describe the principle of operation of the commercial AFM, the sample is mounted on a piezoelectric tube scanner which consists of separate electrodes to scan the sample precisely in the X–Y plane in a raster pattern as shown in Figure 1.18 and to move the sample in the vertical (Z) direction. A sharp tip at the free end of a microfabricated flexible cantilever is brought in contact with the sample. Features on the sample surface cause the cantilever to deflect in the vertical direction as the sample moves

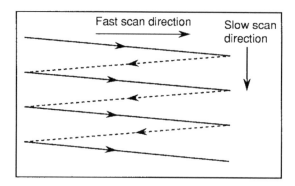

FIGURE 1.18 Schematic of triangular pattern trajectory of the AFM tip as the sample is scanned in two dimensions. During imaging, data are recorded only during scans along the solid scan lines whereas scratch and wear (to be described later) take place along both the solid and dotted lines.

under the tip. A laser beam generated from a diode laser (light-emitting diodes or LEDs with a 5-mW max peak output at 670 nm) is directed by a prism onto the back of the cantilever near its free end, tilted downward at about 10° with respect to a horizontal plane. The reflected beam from the vertex of the cantilever is directed through a mirror onto a quad photodetector (split photodiode detector with four quadrants commonly called position-sensitive detector or PSD), produced by Silicon Detector Corporation, 1240 Avenida Acasco, Camarillo, CA. The differential signal from the top and bottom photodiodes $[(T - B)/(T + B)]$ provides the AFM signal which is a sensitive measure of the cantilever vertical deflection. In the AFM operating mode, the "height mode," this AFM signal is used as the feedback signal to control the vertical position of the piezotube scanner and the sample, such that the cantilever vertical deflection (hence the normal force at the tip–sample interface) will remain (almost) constant as the sample is scanned. Thus, the vertical motion of the tube scanner relates directly to the topography of the sample surface. We note that normal force and vertical motion of the sample can be independently measured by the photodiode and the piezoelectric scanner, respectively. Topographic measurements are made using the height mode at any scanning angle. At a first instance, the scanning angle may not appear to be an important parameter. However, the friction force between the tip and the sample (which we will discuss later in the FFM section) will affect the topographic measurements in a parallel scan (scanning along the long axis of the cantilever), therefore, a perpendicular scan may be more desirable. Generally, one picks a scanning angle which gives the same topographic data in both directions; this angle may be slightly different than that for the perpendicular scan.

Physically, the AFM consists of three main parts: the optical head which senses the cantilever deflection, a piezoelectric tube scanner which controls the scanning motion of the sample mounted on its one end, and the base which supports the scanner and head and includes circuits for the deflection signal. The AFM connects directly to a control system. A front view of the AFM is shown in Figure 1.19A. Due to the weight of the optical head, the sensing system cannot be mounted on the piezo tube, therefore, the optical head and the cantilever are held stationary while the sample is scanned under them. The optical sensing system is packaged into the optical head of the microscope (Figure 1.19B). The head consists of a laser diode stage, a photodiode stage preamp board, a cantilever mount and its holding arm, and a deflection beam reflecting mirror. The laser diode stage is a tilt stage used to adjust the position of the laser beam relative to the cantilever. It consists of the laser diode, collimator, focusing lens, base plate, and the X and Y laser diode positioners. The positioners are used to place the laser spot on the end of the cantilever. The photodiode stage is an adjustable stage used to position the photodiode elements relative to the reflected laser beam. It consists of the split photodiode, the base plate, and the photodiode positioners. The top (AFM) positioner is used to adjust the AFM signal by moving the photodiode up and down. Similarly, the front (FFM) positioner adjusts the FFM signal by moving the photodiode elements in and out (used for FFM, to be described later). The preamp board contains preamplifier circuits for all four photodetecter signals and a laser diode power supply circuit. The cantilever mount

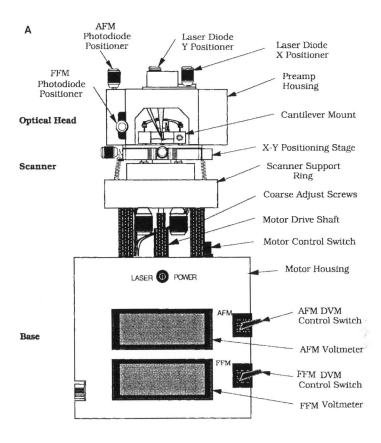

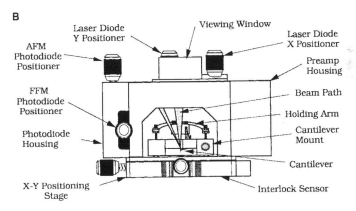

FIGURE 1.19 Schematics of a commercial AFM/FFM made by Digital Instruments, Inc.: (A) front view, (B) optical head, (C) base, and (D) cantilever substrate mounted on cantilever mount (not to scale). (From "Nanoscope III Atomic Force Microscope, Instruction Manual," Courtesy of Digital Instruments, Inc., Santa Barbara, CA., 1992)

is a metal (for operation in air) or glass (for water) block which holds the cantilever firmly at the proper angle, Figure 1.19D, and the deflection beam reflecting mirror is mounted on the upper left in the interior of the head which reflects the deflected beam toward the photodiode.

The scanner consists of an Invar cylinder holding a single tube made of piezoelectric crystal which provides the necessary three-dimensional motion to the sample, Figure 1.6B. Mounted on top of the tube is a magnetic cap on which the steel sample puck is placed. The tube is rigidly held at one end with the

C

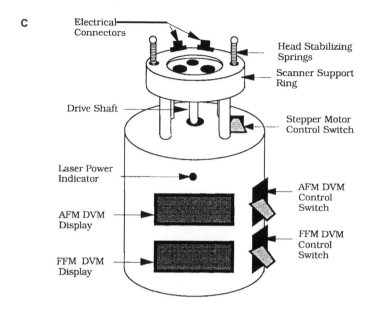

Electrical Connectors

Head Stabilizing Springs

Scanner Support Ring

Drive Shaft

Stepper Motor Control Switch

Laser Power Indicator

AFM DVM Control Switch

AFM DVM Display

FFM DVM Control Switch

FFM DVM Display

D

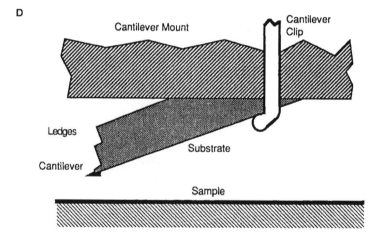

Cantilever Mount

Cantilever Clip

Ledges

Substrate

Cantilever

Sample

FIGURE 1.19 (continued)

sample mounted on the other end of the tube. Samples up to about 10×10 mm or about 15 mm in diameter and 10 mm in thickness can be used. The scanner also contains three fine-pitched screws which form the mount for the optical head. The optical head rests on the tips of the screws which are used to adjust the position of the head relative to the sample. The scanner fits into the scanner support ring mounted on the base of the microscope, Figure 1.19C. Two of the screws on the scanner are operated manually while the third is controlled with a stepper motor built into the base of the microscope. The stepper motor is controlled manually with the switch on the upper surface of the base and automatically by the computer during the tip-engage and tip-withdraw processes. The base also houses electronic circuits which are essential to the alignment and operation of the microscope. The two liquid-crystal (digital voltmeter or DVM) displays on the base show either the sum of the photodiode signals or the differential photodiode signals depending on the position of the respective control switches on the base. These voltages are required during the optical alignment of the system.

The scan sizes available for this instrument are 0.7, 12, and 125 μm. The scan rate must be decreased as the scan size is increased. A maximum scan rate of 122 Hz can be used. Scan rates of about 60 Hz should be used for small scan lengths (0.7 μm). Scan rates of 0.5 to 2.5 Hz should be used for large scans on samples with tall features. High scan rates help reduce drift, but they can only be used on flat samples with small scan sizes. Scan rate or scanning speed in length/time is equal to twice the scan length times the scan rate in Hz, and in the slow direction, it is equal to scan length times the scan rate in Hz divided by the number of data points in the transverse direction. For example, for 10×10 μm scan size scanned at 0.5 Hz, the scan rates in the fast and slow scan directions are 10 μm/s and 20 nm/s, respectively. Normally 256×256 data points are taken for each image. The lateral resolution at larger scans is approximately equal to scan length divided by 256. The piezotube requires X–Y calibration which is carried out by imaging an appropriate calibration standard. Cleaved graphite is used for small scan heads while two-dimensional grids (a gold-plating ruling) can be used for longer-range heads.

To prepare AFM for imaging, the following steps are required: installing a cantilever, loading a sample, aligning the optics, and doing the coarse approach of the tip to the sample. By loosening the cantilever holding-arm screw located on the back of the optical head, the cantilever mount is removed. The appropriate cantilever is mounted on the cantilever mount with a clip (Figure 1.19D), and the cantilever mount is replaced into the optical head. The AFM is provided with 12.7-mm-diameter steel pucks that can be attached to the magnetic cap on the end of the scanner tube. The sample is placed on the puck by using a sticky tab or a quick-drying glue and the puck is placed onto the magnetic cap on the top of the scanner tube. Next, the optical head is placed on the magnetic balls mounted on the ends of the three screws of a scanner on which the sample has already been loaded. When the head is in place, electrical connections are made. Next the laser, cantilever, and photodiode are aligned. While observing the substrate/cantilever through a magnifier, the laser spot is adjusted with the two positioning knobs on the top of the head so that it is positioned on the vertex of the cantilever. After the laser beam is properly aligned with the cantilever, photodiode positioners are adjusted to center the laser spot in the quad photodiode. As a first step, the laser spot is centered visually then centered more precisely to maximize the AFM sum signal (T + B), while setting the FFM difference signal (L − R) to zero (for friction measurements, to be discussed later). When the AFM sum signal is maximized, one should see a signal of 5 to 9 V. After optical alignment, the cantilever is lowered with the coarse-approach screw until the tip is about 0.1 mm above the sample, followed by the fine position of the tip by monitoring the reflection of the illuminated cantilever on the sample (tip must not touch the sample). A final step prior to engaging is the setting of the AFM control switch to difference signal (down position) and the adjustment of the photodiode position until the output of the preamp is set to a desirable value, between −1 and −4 V. Now the AFM is ready for scanning, which is initiated by engaging the microscope.

Examples of AFM images of freshly cleaved HOP graphite and mica surfaces are shown in Figure 1.20 (Albrecht and Quate, 1987; Marti et al., 1987; Ruan and Bhushan, 1994b).

Force calibration mode is used to study interaction between the cantilever and the sample surface. In the force calibration mode, the X- and Y-voltages applied to the piezotube are held at zero and a sawtooth voltage is applied to the Z-electrode of the piezotube, Figure 1.21A. The force measurement starts with the sample far away and the cantilever in its rest position. As a result of the applied voltage, the sample is moved up and down relative to the stationary cantilever tip. As the piezo moves the sample up and down, the cantilever deflection signal from the photodiode is monitored. The force curve, a plot of the cantilever deflection signal as a function of the voltage applied to the piezotube, is obtained. Figure 1.21B depicts a typical force–separation curve showing the various features of the curve. The arrow heads reveal the direction of piezo travel. At point 1, the tip is off the sample surface. From point 1 to point 2 there is no change in the deflection signal as the piezo extends, because the force is initially zero as the sample has not come into contact with the tip. At point 2 the tip is a fraction of a nanometer away from the sample, and the force between the tip and the sample suddenly becomes attractive. The cantilever bends toward the sample and the attractive force increases gradually until point 2′ of the sample and tip come into intimate contact and the force becomes repulsive. The maximum forward deflection of the cantilever

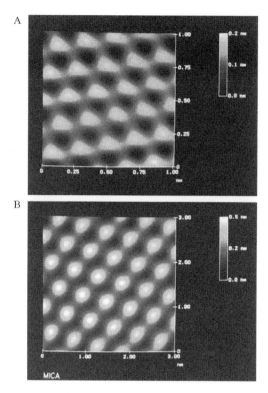

FIGURE 1.20 Typical AFM images of freshly cleaved (A) HOP graphite taken using a square pyramidal Si_3N_4 tip (frequency = 41 Hz, normal load = 20 nN, 256 × 256 pixels, original scan size = 10 × 10 nm) and (B) mica (frequency = 41 Hz, normal load = 12 nN, 256 × 256 pixels, original scan size 4 × 4 nm).*

multiplied by the spring constant of the cantilever is the pull-off force, point 2′. As we continue the forward position of the sample, it pushes the cantilever back through its original rest position (point of zero applied load) entering the repulsive region (or loading portion) of the force curve. The deflection signal reaches a maximum at point 3, the maximum piezo extension; then the piezo starts to retract (unloading portion).

The deflection signal decreases as the piezo and sample retract. Typically, the signal continues to decrease after the flat, zero deflection point of the force curve. At point 4 the cantilever is not deflected, but due to adhesion between the tip and the sample, the tip sticks to the sample and the cantilever is bent down as the piezo continues to retract. Eventually, the spring force of the bent cantilever overcomes the attractive forces and the cantilever quickly returns to its nondeflected, noncontact position. This is represented by points 5 and 6 on the example curve. At point 5, the spring force of the cantilever equals the attractive forces between the tip and the sample. At point 6 the cantilever has returned to its undeflected state. Then the cantilever deflection signal remains constant as the piezo continues to retract to point 7. In general, the pull-off force or adhesive force is always greater than the pull-on force. Because of creep of the piezotube material (lead zirconate titanate) during loading, the tip deflection is not the same during the extended and retracted mode, which is responsible for the horizontal shift between the loading and unloading curve. Upon unloading at point 6, the force curve may not return to the original

---

* Color reproduction follows page 16.

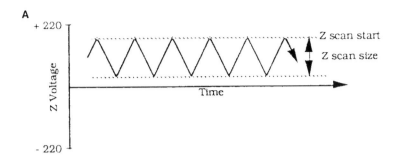

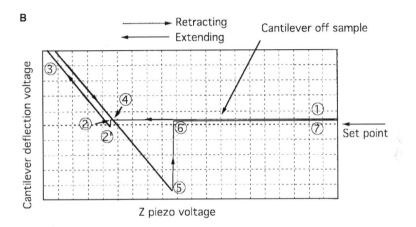

FIGURE 1.21 (A) Force calibration Z waveform and (B) a typical force–separation curve. The force between the cantilever tip and sample is shown as negative when attractive and positive when repulsive.

baseline because of thermal drift. By leaving DC power up for 30 min before starting the test, creep effects can be minimized.

The attractive forces experienced during loading include van der Waals forces (Burnham et al., 1991) and longer-range forces. A thin layer of liquid, such as condensation of water vapor from ambient air residing on the surface, will give rise to capillary forces that act to draw the tip and surface together at small separations (Mate et al., 1989; Blackman et al., 1990; O'Shea et al., 1992; Bhushan and Sundararajan, 1998). In general, any surface absorbate can potentially affect measurements, particularly if they alter the surface energy of the sample. To minimize liquid-mediated adhesive forces, scanning should be performed in dry nitrogen with partial pressure of water less than 0.1 Pa to minimize nanometer-scale capillary condensation (Burnham et al., 1990), or better still, under UHV conditions (Sugawara et al., 1993). Scanning in the presence of liquid (Marti et al., 1987; Drake et al., 1989; Giles et al., 1993) would also minimize liquid-mediated adhesion. An imaging technique in water has been developed to study biological subjects in real environments. For a detailed discussion of forces, see Burnham et al. (1991), Hues et al. (1993), Burnham and Colton (1993a), and Burnham et al. (1993b).

1.3.2.5.1 Multimode Capabilities

In the multimode, AFM can be used for topography measurements in the "tapping mode," and for measurements of lateral (friction) force (to be described later), electric force gradients, and magnetic force gradients.

In the tapping mode, during scanning over the surface, an oscillating tip slightly taps the surface at about 300 kHz with a 20- to 100-nm amplitude introduced in the vertical direction with a feedback loop keeping the average force constant. Oscillation to the cantilever beam is provided by oscillating a bio-morph mounted on the beam. The oscillating amplitude is kept large enough so that the tip does not get stuck to the sample because of adhesive attractions. The tapping mode is used in topography

29

measurements to minimize effects of friction and other lateral forces and to measure topography of soft surfaces. The tapping mode is also referred to as dynamic force microscopy.

The multimode AFM, used with a grounded conducting tip, can measure electric field gradients by oscillating the tip near its resonant frequency. When the lever encounters a force gradient from the electric field, the effective spring constant of the cantilever is altered, changing its resonant frequency. Depending on which side of the resonance curve is chosen, the oscillation amplitude of the cantilever increases or decreases due to the shift in the resonant frequency. By recording the amplitude of the cantilever, an image revealing the strength of the electric field gradient is obtained.

In its simplest form, MFM used with a magnetically coated tip detects static cantilever deflection that occurs when a magnetic field exerts a force on the tip, and the MFM images of magnetic materials can be produced. Multimode AFM enhances MFM sensitivity by oscillating the cantilever near its resonant frequency. When the tip encounters a magnetic force gradient, the effective spring constant, and hence the resonant frequency, is shifted. By driving the cantilever above or below the resonant frequency, the oscillation amplitude varies as the resonance shifts. An image of magnetic field gradients is obtained by recording the oscillation amplitude as the tip is scanned over the sample.

Topographic information is separated from the electric field gradients and magnetic field images by using a so-called lift mode. Measurements in lift mode are taken in two passes over each scan line. On the first pass, topographical information is recorded in the standard tapping mode where the oscillating cantilever lightly taps the surface. On the second pass, the tip is lifted to a user-selected separation (typically 20 to 200 nm) between the tip and local surface topography. By using the stored topographical data instead of the standard feedback, the separation remains constant without sensing the surface. At this height, cantilever amplitudes are sensitive to electric field force gradients or relatively weak but long-range magnetic forces without being influenced by topographic features. Two-pass measurements are taken for every scan line, producing separate topographic and magnetic force images.

1.3.2.5.2   Electrochemical AFM (ECAFM)

This option allows us to perform electrochemical reactions on the AFM. It includes a potentiostat, a fluid cell with a transparent cantilever holder and electrodes, and the software required to operate the potentiostat and display the results of the electrochemical reaction.

1.3.2.5.3   Stand-Alone AFM

Digital Instruments, Inc., also manufactures a stand-alone AFM which measures the topography of a sample with subnanometer resolution regardless of the size of the sample (Anonymous, 1991). The stand-alone AFM can be placed directly on large samples (larger than about 10×10 mm) which cannot be fitted into the AFM assembly, Figure 1.22. Either the sample must be larger in diameter than the three support posts or the sample must be rigidly mounted to a larger substrate. Scan lengths of this instrument are 75 and 125 μm. In these units, the sample is stationary. The cantilever beam and the compact assembly of laser source and detector are attached to the free end of a piezoelectric transducer, which drives the tip over the stationary sample, Figure 1.22A and B. Because the cantilever beam and detector assembly are scanned instead of the sample, some vibration is introduced and lateral resolution of this instrument is reduced. In the stand-alone AFMs, a single photodetector instead of split photodiode detector is used. As a result, friction force measurement (to be discussed later) cannot be made.

A cylindrical piezoelectric tube scans a very sharp tip which is mounted on a flexible cantilever over the sample surface. A compact interferometric detection system mounted on the end of the piezotube senses the deflection of the cantilever as features in the sample are encountered. In the most common operating mode, the control system varies the Z-voltage applied to the piezo to keep the cantilever deflection nearly constant as the tip is scanned over the sample surface in a raster pattern. The variation in the Z-voltage applied to the piezo translates directly into the variation in height across the sample.

The interference system used to detect cantilever deflection can be made quite compact and therefore is mounted directly on the piezotube. Figure 1.22C shows the cantilever deflection detection system. The laser diode emits light from both the top, beam 2, and the bottom, beam 1. The light emitted from the bottom of the laser is reflected off the cantilever and back into the laser. The reflective cantilever forms

A

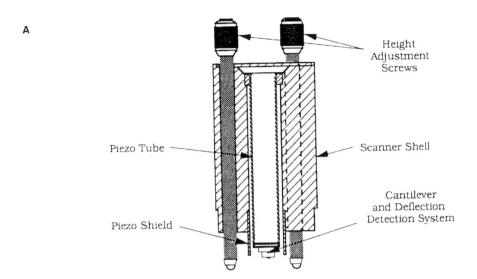

Height Adjustment Screws

Piezo Tube

Scanner Shell

Cantilever and Deflection Detection System

Piezo Shield

B

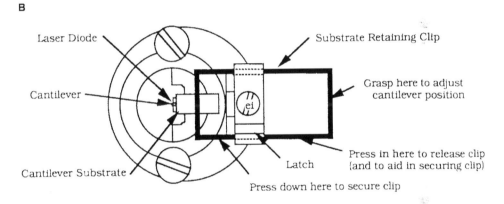

Laser Diode

Substrate Retaining Clip

Grasp here to adjust cantilever position

Cantilever

Cantilever Substrate

Latch

Press in here to release clip (and to aid in securing clip)

Press down here to secure clip

C

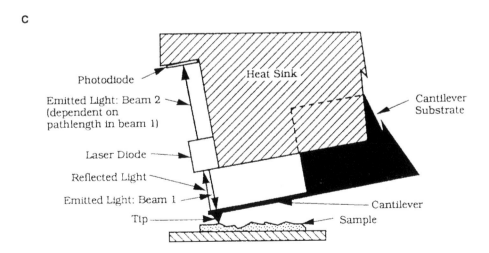

Photodiode

Heat Sink

Emitted Light: Beam 2 (dependent on pathlength in beam 1)

Cantilever Substrate

Laser Diode

Reflected Light

Emitted Light: Beam 1

Tip

Cantilever

Sample

FIGURE 1.22 Schematics of a stand-alone AFM: (A) cross-sectional view, (B) top view, and (C) cantilever deflection detection system. (From "Stand Alone Atomic Force Microscope, User's Manual," Courtesy of Digital Instruments, Inc., Santa Barbara, CA, 1991.)

an external resonant cavity with the laser. The efficiency of the laser, and hence the intensity of the beam emitted from the top of the laser, varies according to the phase difference in the light returned from the external resonant cavity. The phase difference depends on the path length between the cantilever and the laser diode. Therefore, the light detected by the photodiode provides a measure of the variation in the path length of the reflected beam. As the cantilever deflects, the path length changes, causing a change in the signal from the photodiode. Due to the interference between the internal beam and the reflected beam, the photodiode signal varies sinusoidally with cantilever deflection. The signal from the photodiode is used to sense the cantilever deflection.

Before the feedback loop can control the cantilever deflection, the tip must be brought into contact with the sample surface. The three height-adjustment screws control the tip-to-sample spacing. To facilitate the tip engagement process, the force calibration mode displays the photodiode signal vs. the Z-position as the piezo is modulated in Z. As the piezo moves the tip up and down, the three height-adjustment screws are used to bring the tip into contact with the sample. The signal from the photodiode changes as the tip contacts the surface.

Large-sample AFMs are available which can scan samples as large as 200×200 μm without cutting the sample or touching its surface. In these instruments, the sample is mounted on a motorized X-Y stage.

1.3.2.6 Tip Construction

Now we discuss the various cantilevers and tips used for AFM and FFM (to be described later) studies. The cantilever stylus used in the AFM/FFM should meet the following criteria: (1) low spring constant (stiffness); (2) a high resonant frequency; (3) a high mechanical Q; (4) high lateral spring constant (stiffness); (5) short lever length; (6) incorporation of components (such as mirror) for deflection sensing; and (7) a sharp protruding tip (Albrecht et al., 1990; Marti and Colchero, 1995). In order to register a measurable deflection with small forces, the cantilever must flex with a relatively low force (on the order of few nanonewtons) requiring vertical spring constants of 10^{-2} to 10^2 N/m for atomic resolution in the contact-profiling mode. The data rate or imaging rate in the AFM is limited by the mechanical resonant frequency of the cantilever. To achieve a large imaging bandwidth, AFM cantilevers should have a resonant frequency greater than about 10 kHz (preferable is 30 to 100 kHz) in order to make the cantilever the least sensitive part of the system. Fast imaging rates are not just a matter of convenience, since the effects of thermal drifts are more pronounced with slow scanning speeds. The combined requirements of a low spring constant and a high resonant frequency is met by reducing the mass of the cantilever. The mechanical Q (relative amplitude at the resonant frequency) of the cantilever should have a high value for some applications. For example, resonance curve detection is a sensitive modulation technique for measuring small force gradients in noncontact imaging. Increasing the Q increases the sensitivity of the measurements. Mechanical Q values of 100 to 1000 are typical. In contacting modes, the Q is of less importance. A high lateral spring constant in the cantilever is desirable to reduce the effect of lateral forces in the AFM as frictional forces can cause appreciable lateral bending of the cantilever. Lateral bending results in error in the topography measurements. For friction measurements, cantilevers with less lateral rigidity are preferred. A sharp protruding tip must be formed at the end of the cantilever to provide a well-defined interaction with sample over a small area. The tip radius should be much smaller than the radii of corrugations in the sample in order for these to be measured accurately. The lateral spring constant depends critically on the tip length. Additionally, the tip should be centered at the free end.

In the past, cantilevers have been cut by hand from thin metal foils or formed from fine wires. Tips for these cantilevers were prepared by attaching diamond fragments to the ends of the levers by hand, or in the case of wire cantilevers, electrochemically etching the wire to a sharp point. Several cantilever geometries for wire cantilevers have been used. The simplest geometry is the L-shaped cantilever, usually made by bending a wire at a 90° angle. Other geometries include single- and double-V geometries with a sharp tip attached at the apex of V, and a double-X configuration with a sharp tip attached at the intersection (Marti et al., 1988; Burnham and Colton, 1989). These cantilevers can be constructed with high vertical spring constants. For example, a double-cross cantilever with an effective spring constant of 250 N/m was used by Burnham and Colton (1989). The small size and low mass needed in the AFM make hand fabrication of the cantilever a difficult process with poor reproducibility. Conventional

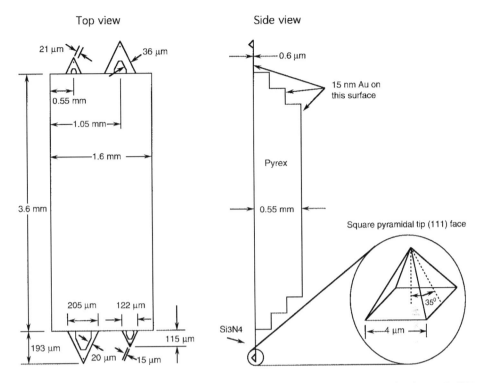

FIGURE 1.23 Schematic of triangular cantilever beam with square pyramidal tips made of PECVD Si_3N_4.

TABLE 1.2 Measured Vertical Spring Constants and Natural Frequencies of Triangular (V-Shaped) Cantilevers Made of PECVD Si_3N_4

Cantilever Dimension	Spring Constant (k_z), N/m	Natural Frequency (ω_0), kHz
115 μm long, narrow leg	0.38	40
115 μm long, wide leg	0.58	40
193 μm long, narrow leg	0.06	13–22
193 μm long, wide leg	0.12	13–22

Data provided by Digital Instruments, Inc.

microfabrication techniques are ideal for constructing planar thin-film structures which have submicron lateral dimensions. The triangular (V-shaped) cantilevers have an improved (higher) lateral spring constant in comparison to rectangular cantilevers. The triangular cantilevers are approximately equivalent to two rectangular cantilevers in parallel (Albrecht et al., 1990). Although the macroscopic radius of a photolithographically patterned corner is seldom much less than about 50 nm, microscopic asperities on the etched surface provide tips with near atomic dimensions.

The cantilevers used most commonly for topography measurements are microfabicated silicon nitride triangular beams with integrated square pyramidal tips made of plasma-enhanced chemical vapor deposition (PECVD) using photolithographic techniques (Albrecht et al., 1990).[2] These are marketed by Digital Instruments, Inc., Santa Barbara, CA and Park Scientific Instruments, Mountain View, CA. Four cantilevers with different sizes and spring constants on each cantilever substrate made of boron silicate

---

[2]Some of the best force sensors for magnetic and electrostatic imaging have been made from fine, electrochemically etched wires. The etched wires have a tapered geometry that varies from about 10 μm in diameter at the point of attachment to less than 50 nm at the end. The end of the wire may be bent with a knife to form a tip.

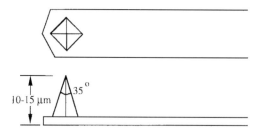

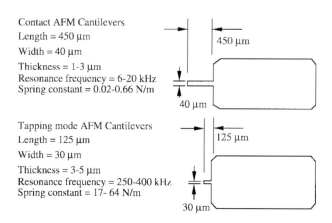

Contact AFM Cantilevers
Length = 450 μm
Width = 40 μm
Thickness = 1-3 μm
Resonance frequency = 6-20 kHz
Spring constant = 0.02-0.66 N/m

450 μm

40 μm

Tapping mode AFM Cantilevers
Length = 125 μm
Width = 30 μm
Thickness = 3-5 μm
Resonance frequency = 250-400 kHz
Spring constant = 17- 64 N/m

125 μm

30 μm

Material: Etched single-crystal n-type silicon;
resistivity = 0.01-0.02 ohm/cm
Tip shape: 10 nm radius of curvature, 35° interior angle

FIGURE 1.24 Schematic of rectangular cantilever beams with square pyramidal tips made of single-crystal silicon.

glass (Pryex®) are shown in Figure 1.23. Two pairs of the cantilevers on each substrate measure about 115 and 193 μm from the substrate to the apex of the triangular cantilever with base widths of 122 and 205 μm, respectively. Both cantilever legs with the same thicknesses (0.6 μm) of all the cantilevers are available with wide and narrow legs. Only one cantilever is selected and used from each substrate. Calculated spring constants and measured natural frequencies for each of the configurations are listed in Table 1.2. The most commonly used cantilever beam is the 115-μm-long, wide-legged cantilever (vertical spring constant = 0.58 N/m). Cantilevers with smaller spring constants should be used on softer samples. The pyramidal tips are highly symmetric with their ends having a radius of about 20 to 50 nm. The tip side walls have a slope of 35° and the length of the edges of the tip at the cantilever base is about 4 μm. Ducker et al. (1992) glued a 3.5-μm-radius glass sphere to the free end of the triangular Si_3N_4 cantilever (with tip removed) for their measurement of colloidal forces. Digital Instruments, Inc., also markets etched single-crystal, n-type silicon rectangular cantilevers with square pyramidal tips with a radius of about 10 nm for contact and tapping mode AFMs, Figure 1.24. Spring constants and resonant frequencies are also presented in Figure 1.24. Park Scientific Instruments markets PECVD Si_3N_4 rectangular cantilevers with square pyramidal tips with a radius of about 40 nm. Table 1.3A lists the spring constants and natural frequencies of the beams with their full length used.

Commercial cantilevers have a typical width–thickness ratio of 10:30 which results in 100 to 1000 times stiffer spring constants in the lateral direction compared to the normal direction. Therefore, these cantilevers are well suited for torsion. For friction measurements, the torsional spring constant should be minimized in order to be sensitive to the lateral forces. Rather long cantilevers with small thickness and large tip length are most suitable. Rectangular beams have lower torsional spring constants in comparison to the triangular (V-shaped) cantilevers. Meyer and Amer (1990b) used a rectangular beam

TABLE 1.3(A) Vertical Spring Constants and Natural Frequencies of Rectangular Beams Made of PECVD Si_3N_4

Cantilever Dimensions (μm)			Vertical Spring Constant (k_z) (N/m)	Natural Frequency (ω_0) (kHz)
L	W	T		
100	10	0.6	0.08	66
100	20	0.6	0.17	66
100	10	0.3	0.010	33
100	20	0.3	0.021	33

Note: $k_z = EWT^3/4L^3$, and $\omega_0 = [k_z/(m_c + 0.24WTL\rho)]^{1/2}$ where E is the Young's modulus, m_c is the concentrated mass of the tip, and ρ is the mass density of the cantilever (Sarid and Elings, 1991). For Si_3N_4, $E = 150$ GPa and $\rho = 3100$ kg/m^3.

Data provided by Park Scientific Instruments.

TABLE 1.3(B) Vertical (k_z), Lateral (k_y), and Torsional (k_{yT}) Spring Constants of Rectangular Cantilevers Made of Si (IBM) and PECVD Si_3N_4

Dimensions/Stiffness	Si Cantilever	Si_3N_4 Cantilever
Length (L), μm	100	100
Width (W), μm	10	20
Thickness (T), μm	1	0.6
Tip length (l), μm	5	3
k_z, N/m	0.4	0.15
k_y, N/m	40	175
k_{yT}, N/m	120	116

Note: $k_z = EWT^3/4L^3$, $k_y = EW^3T/4l^2$, and $k_{yT} = GWT^3/3Ll^2$, where E is Young's modulus and G is the modulus of rigidity [$= E/2(1 + \nu)$, where ν is Poisson's ratio]. For Si, $E = 130$ GPa and $G = 50$ GPa.

From Park Scientific Instruments and Meyer, G. and Amer. N.M. (1990), *Appl. Phys. Lett.*, 57, 2089–2091. With permission.

made of Si or Si_3N_4 for topography and friction studies. Table 1.3B lists the spring constants (with full length of the beam used) in three directions of the typical beams used by them. We note that lateral and torsional spring constants are about two orders of magnitude larger than the normal spring constants. Thicker silicon cantilevers made by Digital Instruments, Inc. (Figure 1.24) and Nanosensors GmbH (Dr. Olaf Wolter), Aidlingen, Germany, are also used for topography and friction measurements. Etched silicon beams have finer tips compared to those of Si_3N_4 beams. A cantilever beam required for the tapping mode is quite stiff and may not be sensitive enough for friction measurements. Meyer et al. (1992) used a specially designed rectangular silicon cantilever with length = 200 μm, width = 21 μm, thickness = 0.4 μm, tip length = 12.5 μm, and shear modulus = 50 GPa, giving a normal spring constant of 0.007 N/m and torsional spring constant of 0.72 N/m, which gives a lateral force sensitivity of 10 pN and an angle of resolution of 10^{-7} rad. With this particular geometry, sensitivity to lateral forces could be improved by about a factor of 100 compared with commercial V-shaped Si_3N_4 or rectangular Si or Si_3N_4 cantilevers used by Meyer and Amer (1990b) with a torsional spring constant of ~100 N/m. Ruan and Bhushan (1994a) and Bhushan and Ruan (1994a) used 115-μm-long, wide-legged V-shaped cantilevers made of Si_3N_4 for friction measurements.

For scratching, wear, and indentation studies, Hamada and Kaneko (1992), Miyamoto et al. (1991, 1993), Bhushan (1994), and Bhushan et al. (1994b–e; 1995a–e, 1997a) have used single-crystal natural diamond tips ground to the shape of a three-sided pyramid with an apex angle of either 60° or 80° whose point is sharpened to a radius of about 100 nm (Figure 1.25). The tips are bonded with conductive epoxy to a gold-plated 304 stainless steel spring sheet (length = 20 mm, width = 0.2 mm, thickness = 20 to

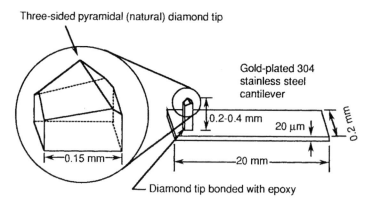

Three-sided pyramidal (natural) diamond tip

Gold-plated 304 stainless steel cantilever

0.2-0.4 mm

20 μm

0.2 mm

0.15 mm

20 mm

Diamond tip bonded with epoxy

FIGURE 1.25 Schematic of rectangular cantilever stainless steel beam with three-sided pyramidal natural diamond tip.

60 μm) which acts as a cantilever. Free length of the spring is varied to change the beam stiffness. The normal spring constant of the beam ranges from about 5 to 600 N/m for a 20-μm-thick beam. The tips are produced by Advanced Film Technology, Inc., Tokyo. Bhushan (1995) used a spring constant of about 25 N/m for studies of magnetic media.

For imaging within trenches by AFM, high-aspect-ratio tips (HART) are used. Examples of the two probes are shown in Figure 1.26. The HART probes are produced by starting with a conventional Si_3N_4 pyramidal probe. Through a combination of FIB and high-resolution scanning electron microscopy (SEM) techniques, a thin filament is grown at the apex of the pyramid. The probe filament is approximately 1 μm long and 0.1 μm in diameter. It tapers to an extremely sharp point (radius better than the resolution of most SEMs). The long, thin shape and sharp radius make it ideal for imaging within "vias" of microstructures and trenches (>0.25 μm). Because of flexing of the probe, it is unsuitable for imaging structures at the atomic level since the flexing of the probe can create image artifacts. For atomic-scale imaging, an FIB-milled probe is used that is relatively stiff yet allows for closely spaced topography. These probes start out as conventional Si_3N_4 pyramidal probes, but the pyramid is FIB milled until a small cone shape is formed which has a high aspect ratio that is 0.2 to 0.3 μm in length. The milled probes allow nanostructure resolution without sacrificing rigidity. These probes are manufactured by Materials Analytical Services, Raleigh, NC.

Ruger et al. (1992) and Sidles and Rugar (1993) developed a cantilever beam which allowed the measurements of forces on the order of 10^{-16} to 10^{-18} N. They used an Si_3N_4 beam 10 μm in length and about 20 nm thick with a spring constant of about 10^{-5} N/m.

Binh and Garcia (1991, 1992) made nanocantilevers with high resonant frequencies on the order of tens of kilohertz and a spring constant of around 1 N/m from W and Cu. They produced these with an extremely thin round cantilever with a spherical tip at its end for measurement of very small van der Waals forces (Garcia and Binh, 1992). For a cantilever beam radius, cantilever length, and the ball tip radius of 10, 200, and 150 nm, respectively, made of W, they achieved a resonant frequency of 60 kHz and a spring constant of 0.3 N/m.

1.3.3   Friction Force Microscope (FFM)

1.3.3.1   Mate et al.'s Design

The first FFM was developed by Mate et al. (1987) at IBM Almaden Research Center, San Jose, CA. In their setup, the sample was mounted on three orthogonal piezoelectric tubes (25 mm long), two of which (x-, y-axes) raster the sample in the surface plane with the third (z) moving the sample toward and away from the tip, Figure 1.27A. A tungsten wire (12 mm long, 0.25 or 0.5 mm in diameter) was used as the cantilever, one end (the free end) of which was bent at a right angle and was electrochemically etched in NaOH solution to obtain a sharp point (150 to 300 nm in radius) to serve as the tip. The spring

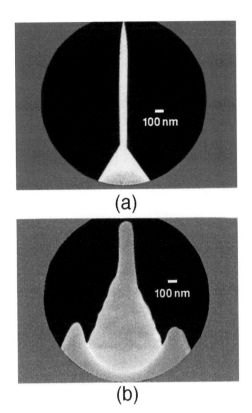

(a)

(b)

FIGURE 1.26 Schematics of (a) HART Si_3N_4 probe and (b) FIB-milled Si_3N_4 probe.

constants were 150 and 2500 N/m for the 0.25- and 0.5-mm-diameter wires, respectively. A laser beam was used to monitor cantilever deflection in the lateral direction. A light beam was focused on the edge of the lever by a microscope objective. The interference pattern between the reflected and the reference beams reflected from the tungsten wire and an optical flat was projected on a photodiode to measure the instantaneous deflection of the lever. Normal force was approximated by using a calibrated piezo-electric tube extension. The force was determined by multiplying the cantilever deflection by the spring constant of the cantilever in the lateral direction. Later, Erlandsson et al. (1988b) used two independent laser beams to monitor cantilever deflections in the normal and lateral directions to measure normal and friction forces, Figure 1.27B. They included another laser beam in the lateral direction (Figure 1.27B) to their original AFM design shown in Figure 1.13. Mate et al. (1987) measured the friction of a tungsten tip sliding against a freshly cleaved HOP graphite by pushing the tip against the sample in the z-direction at desired loads (ranging from 7.5 to 56 μN) by moving the sample back and forth parallel to the surface plane at a velocity of 40 nm/s, and repeating the scanning by stepping the sample (for three-dimensional profiling). Erlandsson et al. (1988b) measured the atomic-scale friction of muscovite mica. Germann et al. (1993) measured the atomic-scale friction of diamond surfaces.

1.3.3.2 Kaneko et al.'s Design

The second type of FFM was developed by Kaneko and his co-workers (Kaneko, 1988; Kaneko et al., 1991). Their earlier design is shown in Figure 1.28 (Kaneko, 1988). A diamond tip was held by a parallel-leaf spring unit (length = 10 mm, width = 1 mm, thickness = 20 μm, spring constant = 3N/m). The sample was mounted on another Parallel-leaf spring unit (length = 10 mm, width = 3 mm, thickness = 20 to 30 μm, spring constant 9 to 30 N/m). These parallel-leaf springs have greater torsional rigidity than single-leaf springs with the same spring constants. Thus deflection errors caused by tip movement are reduced by using parallel-leaf springs (Miyamoto et al., 1990). A piezoelectric tripod (16 μm stroke at 100 V) was used for loading the sample against the tip as well as moving it in the two directions in the sample plane. A focusing error-detection-type optical head (resolution <1 nm) was used to measure the

37

A

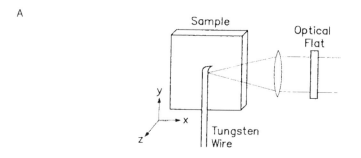

B

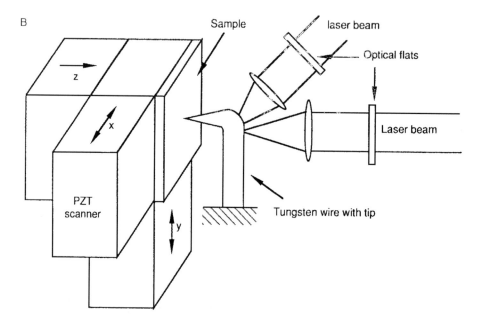

FIGURE 1.27 Schematics of the FFM designs (A) in which friction force is measured by measuring the cantilever deflection in the lateral direction by optical interference (Mate et al., 1987) and (B) in which friction and normal forces are measured by measuring the cantilever deflections in both lateral and normal directions.

tip displacement in the lateral (x) direction. A U-shaped electromagnet was set to pull the tip assembly to overcome the frictional force. (An amorphous iron alloy plate was attached to the spring to obtain an effective pulling force.) Thus the friction force was measured by measuring the current that was required to hold the tip stationary.

This design was modified later by Kaneko et al. (1991). The major modifications in the design are that (1) the sample is no longer supported by a parallel-leaf spring unit and is directly mounted to a piezotube scanner (which may give more stability while scanning the sample), and the tube scanner is used instead of piezoelectric tripod to move the sample, (2) the friction force (tip motion) is sensed by the voltage difference applied between two parallel electrodes rather than the current passing a coil around a magnet, and (3) the tip is mounted on a single-leaf spring. Figure 1.29 shows the new FFM design presented by Kaneko et al. (1991). The piezoelectric tripods have larger z-travel (on the order of 10 μm) than piezotube scanners (couple of microns); however, tube scanners were used in their newer designs so that commercial controllers designed for tube scanners could be used. This tube scanner had an outer diameter of 10 mm, an inner diameter of 8 mm, and an effective length of 40 mm. For friction measurements, a diamond tip was used which was ground to the shape of a three-sided pyramid, whose point was sharpened to a radius of 0.1 mm with an apex of 90° (Figure 1.25). The tip was mounted on one end of a single-leaf

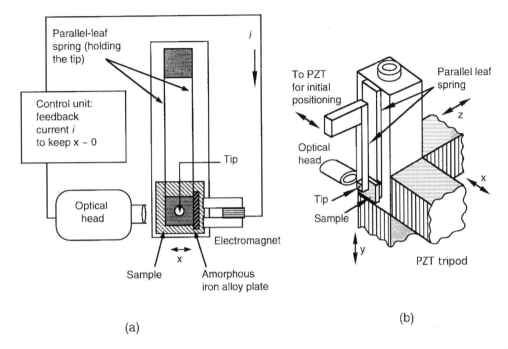

FIGURE 1.28 Schematics of an FFM. (a) The overall setup showing the tip assembly (sample support assembly not shown) and the associated instrumentation and (b) parallel spring unit for supporting and loading the sample. Normal force is measured by the deflection of the parallel spring unit in (b). Friction force is measured by measuring the current that is required to hold the tip stationary; the tip displacement is sensed by the optical head. (From Kaneko, R. (1988), *J. Micros.*, 152(2), 363–369. With permission.)

spring (length = 3 to 6.5 mm, width = 0.2 mm, thickness = 20 μm, and spring constant = 0.3 to 3 N/m). The single-leaf spring was mounted perpendicular to a parallel-leaf spring unit (length = 5 to 10 mm, width = 1 mm, thickness = 20 μm, and spring constant = 3 to 24 N/m). The tip-to-sample contact was established by observing the parallel-leaf spring vibration resulting from the vibrating tip; when in contact, there is an absence of parallel-leaf spring vibration. Applied normal force was obtained by the tube scanner displacement and the stiffness of the single-leaf spring. The tip assembly shown in Figure 1.29 consists of two flat electrodes (2 mm square) attached to the ends of the parallel-leaf spring unit and an elastic member. The gap between the electrodes (typically 0.1 mm) is adjusted by a screw. The attractive force between the electrodes is controlled by the control unit to move the parallel-leaf spring to the zero-friction position. For friction measurement, the sample is scanned against the tip. The friction force being applied at the tip deflects the parallel-leaf spring which is sensed by an optical head. A control unit generates a voltage signal applied to the electrodes in order to move the associated leaf-spring back to zero displacement by overcoming the friction force. Thus, friction force is measured by measuring the required voltage difference between the electrodes.

1.3.3.3 Meyer and Amer's and Fujisawa et al.'s Designs

Meyer and Amer (1990b) modified their AFM design to measure both surface topography and friction forces simultaneously. In the case of surface topography, the bending of the cantilever was detected with a segmented photodiode detector, typically a bicell. Additionally, lateral forces induce a torsion of the cantilever which, in turn, causes the reflected laser beam to undergo a change in a direction perpendicular to that due to surface topography. Thus, with a simple combination of two orthogonal bicells i.e., a quadrant photodiode detector, one is able to measure, simultaneously yet independently, lateral forces while imaging. Thus, the optical beam deflection method allows measurements of both orientation and displacement of the cantilever beam. Marti et al. (1990) independently developed the same measurement technique.

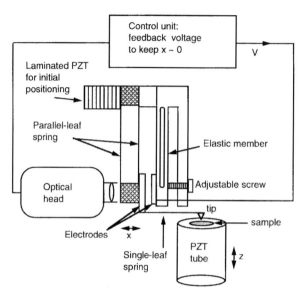

FIGURE 1.29 Schematic of an FFM in which the sample is mounted directly on the piezoscanner as opposed to on a parallel spring unit and the tip is mounted on a single-leaf spring instead of on a parallel-leaf spring unit as shown in Figure 1.28. (From Kaneko, R. et al. (1991), *Adv. Inf. Storage Syst.*, 1, 267–277. With permission.)

Fujisawa et al. (1994) used a combination of optical beam deflection and optical interferometry methods to measure force components being applied at the tip in the three directions, Figure 1.30b. In general, friction force has two components F_X and F_Y with tip sliding on a rough surface, Figure 1.30a. One normal and one of the lateral components of the forces (F_X and F_Y) were measured using the AFM signal. The lateral force (F_X) was measured using the LFM signal. They used the optical interference method to independently measure normal force component F_Z. In this method, the distance between the cleaved end of the optical fiber and the rear of the cantilever is measured, which uses the interference between the light reflected at the end of the optical fiber and at the rear of the cantilever (Rugar et al., 1989). With the measurements of AFM signal and optical interference signal, one can then obtain F_Y and F_Z independently.

1.3.3.4 Marti et al.'s Design

Hipp et al. (1992) mounted both forces' sensors and scanning unit on the microscope head, Figure 1.31a. In this design, the sample is separated from the scanning piezo to accommodate any kind and size of samples. This design is referred to as a stand-alone AFM/FFM. The optical beam deflection was used in a collinear arrangement in order to detect normal and friction forces acting on the cantilever beam, Figure 1.31b. The adjustment of the optics and the calibration of forces were performed with accessible micrometer screws. The microscope includes an automatic coarse and fine approach facility. This machine is commercially available from CSEM, Neuchâtel, Switzerland.

The three force measurement techniques described thus far have their advantages and disadvantages. For example, the use of two laser beams in Mate et al.'s design adds an additional complexity in the design of the apparatus; however, the friction and normal forces are measured independently. Kaneko et al.'s design does not have the capability of measuring the surface topography and friction force simultaneously. The topography and friction force of their samples were measured separately (Kaneko et al., 1991). This is a significant drawback as any correlation between local variations in friction force and the surface topography cannot be easily observed. In the Meyer and Amer and Marti et al. designs, both friction and normal forces can be measured simultaneously by using a single laser beam. Their technique is compact and cheaper to fabricate and is commercially used. Friction force can be measured and calibrated relatively easily for all of these instruments.

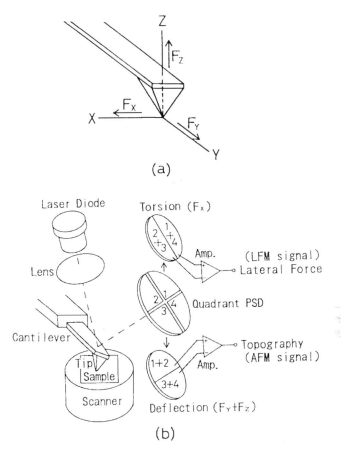

FIGURE 1.30 (a) Schematic representation of X, Y, and Z components of force (F_X, F_y, and F_Z) acting at the end of the cantilever tip and (b) schematic of an FFM used to measure force components using optical beam deflection and optical interferometry methods. (From Fujisawa, S. et al. (1994), *Rev. Sci. Instrum.,* 65, 644–647. With permission.)*

1.3.3.5 Commercial FFMs

Most commercial AFMs listed under the AFM section come with FFM capability. We now describe a commercial FFM from Digital Instruments, Inc. (NanoScope III; Anonymous, 1992c; Bhushan and Ruan, 1994a; Ruan and Bhushan, 1994a–c). This instrument can provide simultaneous measurements of friction force and surface topography. The schematic of this instrument is shown in Figure 1.17 and has been described in detail previously. The original design of this AFM version comes from Meyer and Amer (1988). The hardware modification of this instrument in order to measure friction force is also based on the same authors' suggestion (Meyer and Amer, 1990b). Bhushan (1995), Bhushan et al. (1994a–e, 1995a,b), and Ruan and Bhushan, (1994b,c) have used a 115-µm wide-legged triangular cantilever with square pyramidal PECVD Si_3N_4 tips (see Figure 1.23 and Table 1.2) for friction measurements. Square pyramidal Si tips with a rectangular beam with 450-µm length, 40-µm width, and 2-µm thickness have also been used for friction measurements.

Friction force can also be measured in the height mode defined earlier in the AFM section (referred to as method 1 in Figure1.32). In addition to T and B photodiodes present with AFM design (Figure 1.17),

* Color reproduction follows page 16.

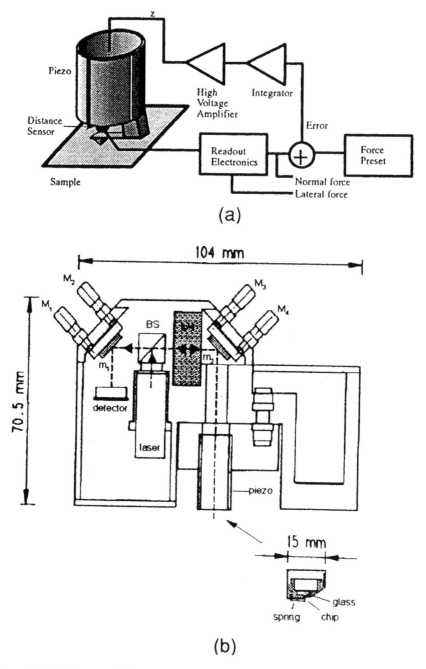

(a)

(b)

FIGURE 1.31 (a) Block schematic of the stand-alone AFM/FFM and (b) schematic of the collinear detection system. (From Hipp, M. et al. (1992), *Ultramicroscopy,* 42–44, 1498–1503. With permission.)

the other two photodiodes ("L" and "R") are arranged horizontally. Their purpose is to measure the "lateral" or "frictional" force exerted on the cantilever by the sample in the so-called aux mode (referred to as method 2 in Figure 1.32). If the sample is scanned back and forth in the x-direction as shown in Figure 1.32, friction force between the sample and the tip will produce a twisting of the cantilever in the x-direction, schematically shown in Figure 1.32b (right). The laser beam will be reflected out of the plane *defined by the incident beam and the vertically reflected beam from an untwisted cantilever.* This produces an intensity difference of the laser beam received between the left and right (L and R) photodiodes. The

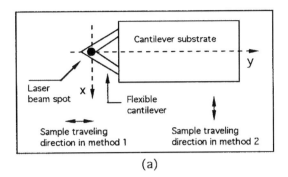

(a)

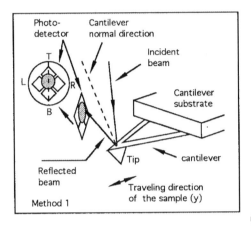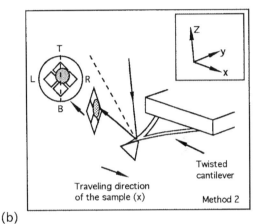

(b)

FIGURE 1.32 (a) Schematic defining the x- and y-directions relative to the cantilever and showing the sample traveling direction in two different measurement methods discussed in the text. (b) Schematic of deformation of the tip and cantilever shown as a result of sliding in the x- and y-directions. A twist is introduced to the cantilever if the scanning is in the x-direction (right). (From Ruan, J. and Bhushan, B. (1994), *ASME J. Tribol.*, 116, 378–388. With permission.)

differential signal from the left and right photodiodes $[(L-R)/(L+R)]$ provides the "FFM" signal which reflects being applied at the tip–sample interface.

1.3.3.5.1 Friction Measurement Methods

Based on the work by Ruan and Bhushan (1994a), we now describe in more detail the two methods for friction measurements. (Also see Meyer and Amer, 1990b.) We define a scanning angle to be the angle relative to the y-axis in Figure 1.32a. This is also the long axis of the cantilever. A 0° scanning angle corresponds to the sample scanning in the y-direction, and a 90° scanning angle corresponds to the sample scanning perpendicular to this axis in the xy-plane (in x-axis). If the scanning angle is in both y– and $-y$-directions, we call this "parallel scan." Similarly, a "perpendicular scan" means the scanning direction is in the x- and $-x$-directions. The sample traveling direction for each of these two methods is illustrated in Figure 1.32b.

In method 1 (using the height mode with parallel scans) in addition to topographic imaging, it is also possible to measure friction force when the scanning direction of the sample is parallel to the y-direction (parallel scan). If there was no friction force between the tip and the moving sample, the topographic feature would be the only factor which could cause the cantilever to be deflected vertically. However, friction force does exist on all contact surfaces where one object is moving relative to another. The friction force between the sample and the tip will also cause a cantilever deflection. We assume that the normal force between the sample and the tip is W_0 when the sample is stationary (W_0 is typically in the range of 10 to 200 nN), and the friction force between the sample and the tip is W_f as the sample scans against

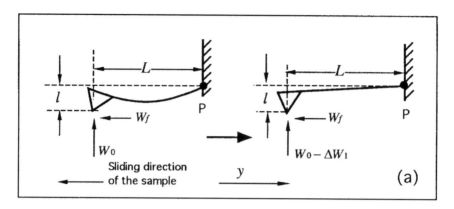

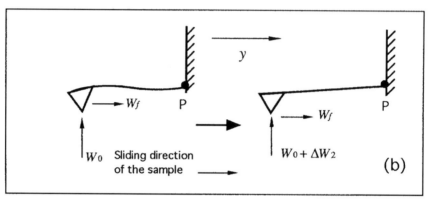

FIGURE 1.33 Schematic showing an additional bending of the cantilever due to friction force when the sample is scanned in the y- or $-y$-direction (left). This effect will be canceled by adjusting the piezo height by a feedback circuit (right). (From Ruan, J. and Bhushan, B. (1994), *ASME J. Tribol.*, 116, 378–388. With permission.)

the tip. The direction of friction force (W_f) is reversed as the scanning direction of the sample is reversed from positive (y) to negative ($-y$) directions ($\vec{W}_{f(y)} = -\vec{W}_{f(-y)}$).

When the vertical cantilever deflection is set at a constant level, it is the total force (normal force and friction force) applied to the cantilever that keeps the cantilever deflection at this level. Because the friction force is in opposite directions as the traveling direction of the sample is reversed, the normal force will have to be adjusted accordingly when the sample reverses its traveling direction, so that the total deflection of the cantilever will remain the same. We can calculate the difference of the normal force between the two traveling directions for a given friction force W_f. First, by means of a constant deflection, the total moment applied to the cantilever is constant. If we take the reference point to be the point where the cantilever joins the cantilever holder (substrate), point P in Figure 1.33, we have the following relationship:

$$\left(W_0 - \Delta W_1\right) L + W_f l = \left(W_0 + \Delta W_2\right) L - W_f l, \tag{1.1}$$

or

$$\left(\Delta W_1 + \Delta W_2\right) L = 2\, W_f\, l. \tag{1.2}$$

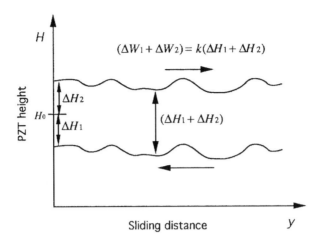

FIGURE 1.34 Schematic illustration of the height difference of the piezotube scanner as the sample is scanned in the y- and $-y$-directions. (From Ruan, J. and Bhushan, B. (1994), *ASME J. Tribol.,* 116, 378–388. With permission.)

Thus,

$$W_f = \left(\Delta W_1 + \Delta W_2\right) L \big/ \left(2l\right), \tag{1.3}$$

where ΔW_1 and ΔW_2 are the absolute value of the changes of normal force when the sample is traveling in $-y$ and y-directions, respectively, as shown in Figure 1.33; L is the length of the cantilever; and l is vertical distance between the end of the tip and point P. The coefficient of friction (μ) between the tip and the sample is then given as

$$\mu = W_f \big/ W_0 = \left[\left(\Delta W_1 + \Delta W_2\right) \big/ W_0\right]\left(L \big/ 2l\right). \tag{1.4}$$

In all circumstances, there are adhesive and interatomic attractive forces between the cantilever tip and the sample. The adhesive force can be due to water from the capillary condensation and other contaminants present at the surface which form meniscus bridges (Mate et al., 1989; Blackman et al., 1990; O'Shea et al., 1992) and the interatomic attractive force includes van der Waals attraction (Burnham et al., 1991). If these forces (and indentation effect as well, which is usually small for rigid samples) can be neglected, the normal force W_0 is then equal to the initial cantilever deflection H_0 multiplied by the spring constant of the cantilever. ($\Delta W_1 + \Delta W_2$) can be measured by multiplying the same spring constant by the height difference of the piezotube between the two traveling directions (y- and $-y$-directions) of the sample. This height difference is denoted as ($\Delta H_1 + \Delta H_2$), shown schematically in Figure 1.34. Thus, Equation 1.4 can be rewritten as

$$\mu = W_f \big/ W_0 = \left[\left(\Delta H_1 + \Delta H_2\right) \big/ H_0\right]\left(L \big/ 2l\right). \tag{1.5}$$

Because the piezotube vertical position is affected by the surface topographic profile of the sample in addition to the friction force being applied at the tip, this difference has to be taken point by point at the same location on the sample surface, as shown in Figure 1.34. Subtraction of point-by-point measurements may introduce errors, particularly for rough samples. We will come back to this point later. In addition, precise measurement of L and l (which should include the cantilever angle) are also required.

If the adhesive forces between the tip and the sample are large enough that they cannot be neglected, one should include them in the calculation. However, there could be a large uncertainty in determining this force, and thus an uncertainty in using Equation 1.5. An alternative approach is to make the measurements at different normal loads and to use $\Delta(H_0)$ and $\Delta(\Delta H_1 + \Delta H_2)$ from the measurements in Equation 1.5. Another comment on Equation 1.5 is that, since only the ratio between $(\Delta H_1 + \Delta H_2)$ and H_0 comes into this equation, the piezotube vertical position H_0 and its position difference $(\Delta H_1 + \Delta H_2)$ can be in volts as long as the vertical traveling distance of the piezotube and the voltage applied to it have a linear relationship. However, if there is a large nonlinearity between the piezotube traveling distance and the applied voltage, this nonlinearity must be included in the calculation.

It should also be pointed out that Equations 1.4 and 1.5 are derived under the assumption that the friction force W_f is the same for the two scanning directions of the sample. This is an approximation since the normal force is slightly different for the two scans and there may also be a directionality effect in friction. However, this difference is much smaller than W_0 itself. We can ignore the second order correction.

Method 2 (aux mode with perpendicular scan) to measure friction was suggested by Meyer and Amer (1990b). The sample is scanned perpendicular to the long axis of the cantilever beam (i.e., to scan along the x- or $-x$-direction in Figure 1.32a) and the output of the horizontal two quadrants of the photodiode detector is measured. In this arrangement, as the sample moves under the tip, the friction force will cause the cantilever to twist. Therefore the light intensity between the left and right (L and R in Figure 1.32b, right) detectors will be different. The differential signal between the left and right detectors is denoted as FFM signal $[(L - R)/(L + R)]$. This signal can be related to the degree of twisting, hence to the magnitude of friction force. Again, because of a possible error in determining normal force due to the presence of an adhesive force at the tip–sample interface, the slope of the friction data (FFM signal vs. normal load) needs to be taken for an accurate value of coefficient of friction.

While friction force contributes to the FFM signal, friction force may not be the only contributing factor in commercial FFM instruments (for example, Nanoscope III). One can notice this fact by simply engaging the cantilever tip with the sample. Before engaging, the left and right detectors can be balanced by adjusting the position of the detectors so that the intensity difference between these two detectors is zero (FFM signal is zero). Once the tip is engaged with the sample, this signal is no longer zero even if the sample is not moving in the xy-plane with no friction force applied. This would be a detrimental effect. It has to be understood and eliminated from the data acquisition before any quantitative measurement of friction force becomes possible.

One of the fundamental reasons for this observation is the following. The detectors may not have been properly aligned with respect to the laser beam. To be precise, the vertical axis of the detector assembly (the line joining T–B in Figure 1.35) is not in the plane defined by the incident laser beam and the beam reflected from an untwisted cantilever (we call this plane the "beam plane"). When the cantilever vertical deflection changes due to a change of applied normal force (without having the sample scanned in the xy-plane), the laser beam will be reflected up and down and form a projected trajectory on the detector. (Note that this trajectory is in the defined beam plane.) If this trajectory is not coincident with the vertical axis of the detector, the laser beam will not evenly bisect the left and right quadrants of the detectors, even under the condition of no torsional motion of the cantilever, see Figure 1.35. Thus, when the laser beam is reflected up and down due to a change of the normal force, the intensity difference between the left and right detectors will also change. In other words, the FFM signal will change as the normal force applied to the tip is changed, even if the tip is not experiencing any friction force. This (FFM) signal is unrelated to friction force or to the actual twisting of the cantilever. We will call this part of the FFM signal FFM$_F$, and the part which is truly related to friction force, FFM$_T$.

The FFM$_F$ signal can be eliminated. One way of doing this is as follows. First, the sample is scanned in both the x- and $-x$-directions and the FFM signal for scans in each direction is recorded. Because friction force reverses its directions when the scanning direction is reversed from x- to $-x$-direction, the FFM$_T$ signal will have opposite signs as the scanning direction of the sample is reversed (FFM$_T(x) = -$FFM$_T(-x)$). Hence, the FFM$_T$ signal will be canceled out if we take the sum of the FFM signals for the two scans. The average value of the two scans will be related to FFM$_F$ due to the misalignment,

46

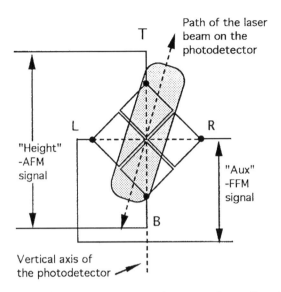

T

Path of the laser
beam on the
photodetector

L   R

"Height"
-AFM
signal

"Aux"
-FFM
signal

B

Vertical axis of
the photodetector

FIGURE 1.35 The trajectory of the laser beam on the photodetectors in the cantilever is vertically deflected (with no torsional motion) for a misaligned photodetector with respect to the laser beam. For a change of normal force (vertical deflection of the cantilever), the laser beam is projected at a different position on the detector. Due to a misalignment, the projected trajectory of the laser beam on the detector is not parallel with the detector vertical axis (the line joint T–B). (From Ruan, J. and Bhushan, B. (1994), *ASME J. Tribol.*, 116, 378–388. With permission.)

$$\mathrm{FFM}\left(x\right)+\mathrm{FFM}\left(-x\right)=2\mathrm{FFM}_F \qquad (1.6)$$

This value can therefore be subtracted from the original FFM signals of each of these two scans to obtain the true FFM signal (FFM_T), or, alternatively, by taking the difference of the two FFM signals, one directly gets the FFM_T value

$$\mathrm{FFM}\left(x\right)-\mathrm{FFM}\left(-x\right)=\mathrm{FFM}_T\left(x\right)-\mathrm{FFM}_T\left(-x\right)$$
$$=2\mathrm{FFM}_T\left(x\right) \qquad (1.7)$$

Ruan and Bhushan (1994a) have been shown that error signal (FFM_F) can be very large compared to friction signal FFM_T; thus correction is required.

Now we compare the two methods. The method of using height mode and parallel scan (method 1) is very simple to use. Technically, this method can provide three-dimensional friction profiles and the corresponding topographic profiles; however, there are some problems with this method. Under most circumstances, the piezoscanner displays a hysteresis when the traveling direction of the sample is reversed. Therefore, the measured surface topographic profiles will be shifted relative to each other along the y-axis for the two opposite (y and $-y$) scans. This would make it difficult to measure the local height difference of the piezotube for the two scans. However, the average height difference between the two scans and hence the average friction can still be measured. The measurement of average friction can serve as an internal means of friction force calibration. Method 2 is a more desirable approach. The subtraction of FFM_F signal from FFM for the two scans does not introduce error to local friction force data. An ideal approach in using this method would be to add the average value of the two profiles in order to get the error component (FFM_F) and then subtract this component from either profiles to get true friction profiles in either direction. By making measurements at various loads, we can get the average value of

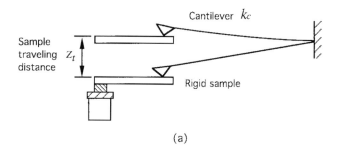

(a)

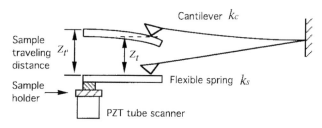

FIGURE 1.36 Illustration showing the deflection of cantilever as it is pushed by (a) a rigid sample or by (b) a flexible spring sheet. (From Ruan, J. and Bhushan, B. (1994), *ASME J. Tribol.*, 116, 378–388. With permission.)

the coefficient of friction which then can be used to convert the friction profile to the coefficient of friction profile. Thus any directionality and local variations in friction can be easily measured. In this method, since topography data are not affected by friction, accurate topography data can be measured simultaneously with friction data and a better localized relationship between the two can be established.

1.3.3.5.2 Normal Force and Friction Force Calibrations

Based on Ruan and Bhushan (1994a), we now discuss normal force and friction force calibrations. In order to calculate the absolute value of normal and friction forces in newtons using the measured AFM and FFM_T voltage signals, it is necessary to first have an accurate value of the spring constant of the cantilever (k_c). The spring constant can be calculated using the geometry and the physical properties of the cantilever material (Albrecht et al., 1990; Meyer and Amer, 1990b; Sarid and Elings, 1991). However, the properties of the PECVD Si_3N_4 (used in fabricating cantilevers) could be different from those of bulk material. For example, by using an ultrasonic measurement, we found the Young's modulus of the cantilever beam to be about 238 ± 18 GPa, which is less than that of bulk Si_3N_4 (310 GPa). Furthermore, the thickness of the beam is nonuniform and difficult to measure precisely. Because the stiffness of a beam goes as the cube of thickness, minor errors in precise measurements of thickness can introduce substantial stiffness errors. Thus one should experimentally measure the spring constant of the cantilever. Cleveland et al. (1993) measured the normal spring constant by measuring resonant frequencies of the beams.

For normal spring constant measurement, we used a stainless steel spring sheet of known stiffness (width = 1.35 mm, thickness = 15 μm, free hanging length = 5.2 mm) (Ruan and Bhushan, 1994a). One end of the spring was attached to the sample holder and the other end was made to contact with the cantilever tip during the measurement, see Figure 1.36. We measured the piezo traveling distance for a given cantilever deflection. For a rigid sample (such as diamond), the piezo traveling distance Z_t (measured from the point where the tip touches the sample) should equal the cantilever deflection. To keep the cantilever deflection at the same level using a flexible spring sheet, the new piezo traveling distance $Z_{t'}$. would be different from Z_t. The difference between $Z_{t'}$. and Z_t corresponds to the deflection of the spring sheet. If the spring constant of the spring sheet is k_s, the spring constant of the cantilever k_c can be calculated by

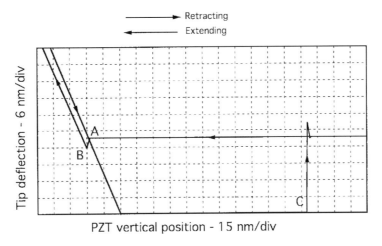

FIGURE 1.37 Displacement curve of the cantilever tip as it is pushed toward (extending) and pulled away from (retracting) a silicon sample in measurements made in the ambient environment. The large separation between point B where the tip is touching the sample and point C where the tip is pulled off the sample is due to a large pull-off (adhesive) force between the tip and the sample. (From Ruan, J. and Bhushan, B. (1994), *ASME J. Tribol.*, 116, 378–388. With permission.)

$$\left(Z_{t'} - Z_t\right)k_s = Z_t k_c$$

or

$$k_c = k_s \left(Z_{t'} - Z_t\right)/Z_t \qquad (1.8)$$

The spring constant of the spring sheet (k_s) used in this study is calculated to be 1.54 N/m. For a wide-legged cantilever used in our study (length = 115 μm, base width = 122 μm, leg width = 21 μm, and thickness = 0.6 μm), k_c was measured to be 0.40 N/m instead of 0.58 N/m reported by its manufacturer, Digital Instruments Inc. To relate photodiode detector output to the cantilever deflection in nanometers, we used the same rigid sample to push against the AFM tip. Because in a rigid sample the cantilever vertical deflection equals the sample traveling distance measured from the point where the tip touches the sample, the photodiode output as the tip is pushed by the sample can be converted directly to cantilever deflection. For our measurements, we found the conversion to be 20 nm/V.

The normal force applied to the tip can be calculated by multiplying the cantilever vertical deflection by the lever spring constant for samples which have very small adhesive force with the tip. If the adhesive force between the sample and the tip is large, it should be included in the normal force calculation. This is particularly important in atomic-scale force measurement because in this region, the typical normal force that we measure is in the range of a few hundred nanonewtons to a few millinewtons. The adhesive force could be comparable to the applied force. The magnitude of the adhesive force is determined by multiplying the maximum cantilever deflection in the downward direction before the tip is pulled off the sample surface by the spring constant of the cantilever. Figure 1.37 shows an example of cantilever deflection as a function of sample position (height). "Extending" means the sample is pushed toward the tip, and "retracting" means the sample is pulled away from the tip. As the sample surface approaches the tip within a few nanometers (point A), an attractive force exists between the atoms of the tip surface and the atoms of the sample surface. The tip is pulled toward the sample and the contact occurs at point B. As the sample is pushed further against the tip, the force at the interface increases and the cantilever is deflected upward. This deflection equals the sample traveling distance measured from point B for a rigid sample. As the sample is retracted, the force is reduced. At point C in the retracting curve, the sample is disengaged from the tip. Before the disengagement, the tip is pulled downward due to the attractive force. The force that is required to pull the tip away from the sample is the force that equals (but in the opposite direction) the adhesive force. This force is calculated by the maximum cantilever deflection in the downward direction times the spring constant of the cantilever. The maximum cantilever deflection in downward direction is just the horizontal distance between points B and C in this curve. We measured this distance to be about 200 nm in this curve, which corresponds to an adhesive force of 80 nN for a spring constant of 0.4 N/m. Friction force at a zero cantilever deflection is associated with this force between the sample and the tip. Because the calculation of both the externally applied and adhesive forces involves the same spring constant of the cantilever, the total normal force (once the sample and the tip are in contact) is equal to the spring constant times the cantilever "deflection" measured right before the pull-off point in the retracting curve. This "deflection" is also the piezo traveling distance measured from point C toward the tip for a rigid sample. Although the calculation of adhesive force is important in the calculation of normal force, it is not important in the calculation of the coefficient of friction if we take the slope of friction force vs. normal force curve.

The conversion of the friction signal (from FFM_T to friction force) is not as straightforward. For example, one can calculate the degree of twisting for a given friction force using the geometry and the physical properties of the cantilever (Meyer and Amer, 1988; O'Shea et al., 1992). One would need the information on the detectors such as the quantum efficiency of the detector, the laser power, the instrument gain, etc. in order to be able to convert the signal into the degree of twisting. Generally speaking, this procedure cannot be accomplished without having some detailed information about the instrument. This information is not usually provided by the manufacturers. Even if this information is readily available, errors may still occur in using this approach because there will always be variations as a result of the instrumental setup. For example, it has been noticed that the measured FFM_T signal could be different for the same sample when different AFM microscopes of the same kind are used. The essence is that one cannot calibrate the instrument experimentally using this calculation. O'Shea et al. (1992) did perform a calibration procedure in which the torsional signal was measured as the sample is displaced a known distance laterally while ensuring that the tip does not slide over the surface. However, it is difficult to verify that the tip sliding does not occur.

Apparently, a new method of calibration is required. There is a more direct and simpler way of doing this. The first method described (method 1) to measure friction can directly provide an absolute value of the coefficient of friction. It can therefore be used just as an internal means of calibration for the data obtained using method 2 or for a polished sample which introduces least error in friction measurement using method 1. Method 1 can be used to obtain calibration for friction force for method 2. Then this calibration can be used for measurement on all samples using method 2. In method 1, the length of the cantilever required can be measured using an optical microscope; the length of the tip can be measured using an SEM. The relative angle between the cantilever and the horizontal sample surface can be measured directly. Thus, the coefficient of friction can be measured with few unknown parameters. The friction force can then be calculated by multiplying the coefficient of friction by the normal load. The FFM_T signal obtained using method 2 can then be converted into friction force. For our instrument, we found the conversion to be 8.6 nN/V.

1.3.3.5.3 Typical Friction Data

Ruan and Bhushan (1994a) measured the friction of Pt (calibration grid with 10×10 μm grid dimension from Digital Instruments, Inc., with rms roughness of 0.22 nm over 1-mm^2 area), single-crystal silicon (rms roughness of 0.14 nm over 1-μm^2 area), polished natural diamond (IIa); (rms roughness of 2.3 nm over 1-μm^2 area), and HOP graphite against an Si_3N_4 tip using both methods 1 and 2. An engineering material Al was also measured for reference. Samples were ultrasonically cleaned in alcohol and dried for a few hours before measurement. The result on Pt (measured using both methods 1 and 2) is used to calibrate the friction data of other samples obtained using method 2. These data were compared with

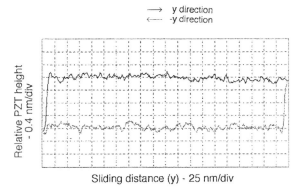

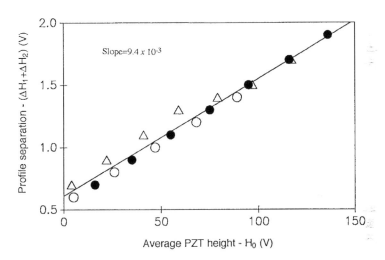

FIGURE 1.38 A typical profile of the piezo height as a Pt sample is scanned back and forth in the y- and $-y$-directions. The normal load is about 50 nN. The cantilever stiffness is 0.4 N/m. (From Ruan, J. and Bhushan, B. (1994), *ASME J. Tribol.*, 116, 378–388. With permission.)

FIGURE 1.39 The vertical height difference (separation of the surface profile, $\Delta H_1 + \Delta H_2$) as a function of the piezo center position (piezo height H_0) between the two traveling directions (y and $-y$) of a Pt sample. The three symbols represent three sets of repeated measurements. The slope of the linear fit is proportional to the coefficient of friction between the Si_3N_4 tip and Pt. (From Ruan, J. and Bhushan, B. (1994), *ASME J. Tribol.*, 116, 378–388. With permission.)

those obtained using method 1 for each sample (except graphite and aluminum) to examine the consistency between these two methods.

We first show in Figure 1.38 an example of a typical trajectory of the piezo vertical height as the AFM tip is scanned in both the y- and $-y$-directions across a Pt surface. For a smooth sample like this, the separation between these two trajectories can easily be measured. We have measured the height difference of the piezo ($\Delta H_1 + \Delta H_2$) at different normal loads (H_0), and used $\Delta (\Delta H_1 + \Delta H_2)$ and $\Delta(H_0)$ in Equation 1.5 to calculate μ. Figure 1.39 shows the data from three sets of measurements at various loads on Pt. The horizontal axis is the center position of the piezo scanner and the vertical axis is the average (over a 1-mm scan length) height difference of the piezo between the y and $-y$ scans. During the measurement, the piezotube center position is changed by setting the $[(T - B)/(T + B)]$ signal at different values for the feedback circuit. The resulting height difference of the piezotube between the two scans

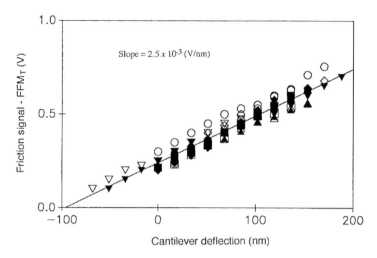

FIGURE 1.40 Friction signal as a function of cantilever vertical deflection for Pt. Different symbols represent 11 sets of repeated measurements. The slope of the linear fit is proportional to the coefficient of friction between the Si_3N_4 tip and Pt. Vertical cantilever vertical spring constant is 0.4 N/m with conversion of AFM signal to be 20 nm/V and the FFM (friction) signal is 8.6 nN/V. (From Ruan, J. and Bhushan, B. (1994), *ASME J. Tribol.*, 116, 378–388. With permission.)

(y and $-y$) was measured and averaged over the scan length. We see in Figure 1.30 that all data fall closely on a straight line. The slope (best linear fit) of this curve is 9.4×10^{-3}. Using the geometry of the cantilever and the tip and attitude of the cantilever with respect to the sample ($L = 115$ mm, $l = 10$ mm in Equation 1.4), a value of the coefficient of friction of 0.054 is obtained.

Figure 1.40 is obtained for Pt using method 2. Here, the cantilever vertical deflection vs. the friction signal (half of the difference of the FFM signals between the x- and $-x$-scans, or $FFM_T(x)$ signal) is plotted. The sample was moved in the x- and $-x$-direction at a velocity of 4 µm/s and slowly stepped (by 4 nm/step) in the y-direction (16 nm/s) to scan the whole area (1×1 µm). The data (averaged over the scan area) for each measurement can be fitted into a straight line. To show the statistical variation between different measurements, 11 sets of measurements at various loads were made and were plotted in this figure. Again, a good linear fit of the data has been observed with a slope of 2.5 mV/nm. Using the value of the coefficient of friction calculated previously, we calculate that a slope of 1 mV/nm in this curve corresponds to a coefficient of friction 0.022. Using the spring constant of the cantilever (0.4 N/m), we then calculated that 1 V in this curve (FFM_T signal) corresponds to a friction force of 8.6 nN.

Ruan and Bhushan (1994a) made similar measurements on silicon and diamond. They used the values obtained above for Pt, and the slopes of the curves of these two samples (silicon and diamond) obtained using method 2 to calculate their coefficient of friction. The data were compared with those obtained directly using method 1 for the corresponding samples. For simplicity, we summarize the result in Table 1.4. A reasonable agreement between the measurements using the two methods was obtained. Friction data for graphite and aluminum are also included in Table 1.4. (For graphite data, also see Ruan and Bhushan, 1994b–c.) Method 1 was not used for these samples since a good agreement has already been obtained between methods 1 and 2 using other samples. In addition, the aluminum sample was not polished and had a rough surface. Method 1 is difficult to apply to this sample. The friction force for all tips against graphite was so small that it could not be directly measured using method 1. Even with method 2, we found that the signal variation at different applied normal forces was also within the range of experimental uncertainty. The slope that best fits to the data with an uncoated tip is 3×10^{-4} (V/nm), which corresponds to a coefficient of friction 0.006. This value is listed in Table 1.4. (The coefficient of friction of graphite against a tungsten tip was reported to be 0.012, according to Mate et al., 1987). Friction force vs. normal force plots for diamond, HF-cleaned silicon, and graphite are shown in Figure 1.41.

TABLE 1.4   Roughness and Microscale Friction Data of Various Samples

Sample	Roughness[a] (nm)	Coefficient of Friction FFM Measurements	
		Method 1	Method 2
Platinum	0.22	0.05	0.05
Single-crystal silicon (HF cleaned)	0.14	0.07	0.08
Single-crystal silicon (before HF cleaning)	0.14	0.06	0.06
Polished natural diamond (IIa)	2.3	0.04	0.05
HOP graphite	0.09		~0.006
Aluminum	40		0.08

[a] Root mean square value measured over a 1×1 μm area using AFM.

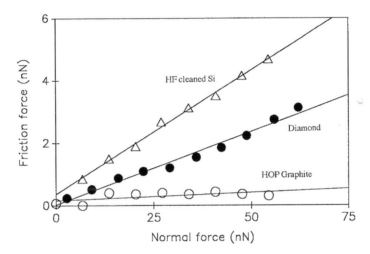

FIGURE 1.41   Friction force as a function of normal force for an Si_3N_4 tip sliding against silicon, diamond, and graphite. (From Ruan, J. and Bhushan, B. (1994), *ASME J. Tribol.*, 116, 378–388. With permission.)

Figure 1.42 shows atomic-scale friction profiles of freshly cleaved HOP graphite. It is clearly seen that the friction profile exhibits the periodicity of the graphite surface structure (Figure 1.20A). However, the high points in the topography and the high friction points are shifted relative to each other (Ruan and Bhushan, 1994b).

1.3.4   Surface Force Apparatus

SFAs are used to study both static and dynamic properties of the molecularly thin liquid films sandwiched between two molecularly smooth surfaces. The SFAs were originally developed by Tabor and Winterton (1969) and later by Israelachvili and Tabor (1972) to measure van der Waals forces between two mica surfaces as a function of separation in air or vacuum. Israelachvili and Adams (1978) developed a more-advanced apparatus to measure normal forces between two surfaces immersed in a liquid so thin that their thickness approaches the dimensions of the liquid molecules themselves. A similar apparatus was also developed by Klein (1980). The SFAs, originally used in studies of adhesive and static interfacial forces, were first modified by Chan and Horn (1985) and later by Israelachvili et al. (1988) and Klein et al. (1991) to measure the dynamic shear (sliding) response of liquids confined between molecularly smooth optically transparent mica surfaces. Optically transparent surfaces are required because the surface separation is measured using an optical interference technique. Van Alsten and Granick (1988) and Peachey et al. (1991) developed a new friction attachment which allows for the two surfaces to be

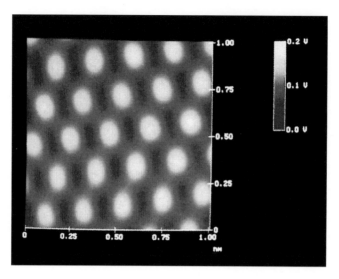

FIGURE 1.42 Gray scale plot of friction profile of a 1×1 nm area of a freshly cleaved HOP graphite showing the atomic-scale variation of topography and friction. Data were taken using a square pyramidal Si_3N_4 tip. Higher points are shown by lighter color. (From Ruan, J. and Bhushan, B. (1994), *J. Appl. Phys.*, 76, 5022–5035. With permission.) For atomic-scale variations in topography, see Figure 1.20a.

sheared past each other at varying sliding speeds or oscillating frequencies while simultaneously measuring both the friction force and normal force between them. Israelachvili (1989) and Luengo et al. (1997) have also presented modified SFA designs for dynamic measurements at oscillating frequencies. Because the mica surfaces are molecularly smooth, the actual area of contact is well defined and measurable, and asperity deformation does not complicate the analysis. During sliding experiments, the area of parallel surfaces is very large compared to the thickness of the sheared film and this provides an ideal condition for studying shear behavior, because it permits one to study molecularly thin liquid films whose thickness is well defined to the resolution of an angstrom. Molecularly thin liquid films cease to behave as a structural continuum with properties different from that of the bulk material (Van Alsten and Granick, 1988, 1990b; Homola et al., 1989; Gee et al., 1990; Granick, 1991).

Tonck et al. (1988) and Georges et al. (1993) developed an SFA used to measure the static and dynamic forces (in the normal direction) between a smooth fused borosilicate glass against a smooth and flat silicon wafer. They used the capacitance technique to measure surface separation; therefore, use of optically transparent surfaces was not required. Among others, metallic surfaces can be used at the interface. Georges et al. (1994) modified the original SFA so that a sphere can be moved toward and away from a plane and can be sheared at constant separation from the plane, for interfacial friction studies.

For a detailed review of various types of SFAs, see Israelachvili (1989, 1991), Horn (1990), and Homola (1993). SFAs based on their design are commercially available from Anutech Pte Ltd., GPO Box 4, Canberra, Australia 2601.

1.3.4.1 Israelachvili's and Granick's Design

The following review is primarily based on the papers by Israelachvili (1989) and Homola (1993). Israelachvili et al.'s design, later followed by Granick et al., for oscillating shear studies is most commonly used by researchers around the world.

1.3.4.1.1 Classical SFA

The classical apparatus developed for measuring equilibrium or static intersurface forces in liquids and vapors by Israelachvili and Adams (1978) consists of a small, airtight stainless steel chamber in which two molecularly smooth curved mica surfaces can be translated toward or away from each other, see Figure 1.43. The distance between the two surfaces can also be independently controlled to within

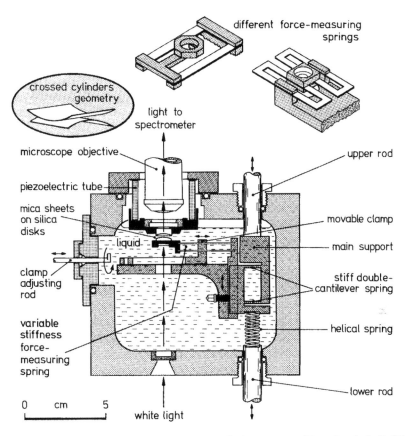

different force-measuring springs

crossed cylinders geometry

light to spectrometer

microscope objective

piezoelectric tube

mica sheets on silica disks

liquid

clamp adjusting rod

variable stiffness force-measuring spring

upper rod

movable clamp

main support

stiff double-cantilever spring

helical spring

lower rod

0   cm   5

white light

FIGURE 1.43 Schematic of the SFA that employs the cross-cylinder geometry. (From Israelachvili, J.N. and Adams, G.E. (1978), *Chem. Soc. J. Faraday Trans. I*, 74, 975–1001; and Israelachvili, J.N. (1989), *Chemtracts Anal. Phys. Chem.*, 1, 1–12. With permission.)

±0.1 nm and the force sensitivity is about 10 nN. The technique utilizes two molecularly smooth mica sheets, each about 2 μm thick, coated with a semireflecting 50- to 60-nm layer of pure silver, glued to rigid cylindrical silica disks of radius about 10 mm (silvered side down) mounted facing each other with their axes mutually at right angles (crossed-cylinder position), which is geometrically equivalent to a sphere contacting a flat surface. The adhesive glue which is used to affix the mica to the support is sufficiently compliant, so the mica will flatten under the action of adhesive forces or applied load to produce a contact zone in which the surfaces are locally parallel and planar. Outside of this contact zone the separation between surfaces increases and the liquid, which is effectively in a bulk state, makes a negligible contribution to the overall response. The lower surface is supported on a cantilever spring which is used to push the two surfaces together with a known load. When the surfaces are forced into contact, they flatten elastically so that the contact zone is circular for the duration of the static or sliding interactions. The surface separation is measured using optical interference fringes of equal chromatic order (FECO) which enables the area of molecular contact and the surface separation to be measured to within 0.1 nm. For measurements, white light is passed vertically up through the two mica surfaces and the emerging beam is then focused onto the slit of a grating spectrometer. From the positions and shapes of the colored FECO fringes in the spectrogram, the distance between the two surfaces and the exact shape of the two surfaces can be measured (as illustrated in Figure 1.44), as can the refractive index of the liquid (or material) between them. In particular, this allows for reasonably accurate determinations of the quantity of material deposited or adsorbed on the surfaces and the area of contact between two molecularly smooth surfaces. Any changes may be readily observed in both static and sliding conditions

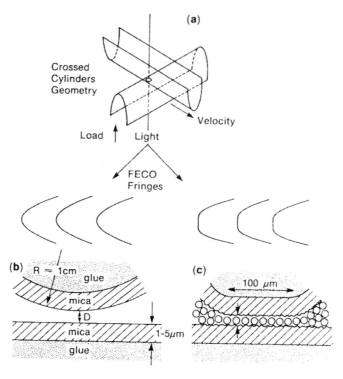

FIGURE 1.44 (a) Cross-cylinder configuration of mica sheet, showing formation of contact area. Schematic of the fringes of equal chromatic order (FECO) observed when two mica surfaces are (b) separated by distance D and (c) are flattened with a monolayer of liquid between them. (From Homola, A. M. et al. (1990), *Wear*, 136, 65–83. With permission.)

in real time (applicable to the design shown in Figure 1.44) by monitoring the changing shapes of these fringes.

The distance between the two surfaces is controlled by use of a three-stage mechanism of increasing sensitivity: coarse control (upper rod) allows positioning of within about 1 μm, the medium control (lower rod, which depresses the helical spring and which in turn bends the much stiffer double-cantilever spring by 1/1000 of this amount) allows positioning to about 1 nm, and the piezoelectric crystal tube — which expands or controls vertically by about 0.6 nm/V applied axially across the cylindrical wall — is used for final positioning to 0.1 nm.

The normal force is measured by expanding or contracting the piezoelectric crystal by a known amount and then measuring optically how much the two surfaces have actually moved; any difference in the two values when multiplied by the stiffness of the force-measuring spring gives the force difference between the initial and final positions. In this way both repulsive and attractive forces can be measured with a sensitivity of about 10 nN. The force-measuring springs can be either single-cantilever or double-cantilever fixed-stiffness springs (as shown in Figure 1.43), or the spring stiffness can be varied during an experiment (by up to a factor of 1000) by shifting the position of the dovetailed clamp using the adjusting rod. Other spring attachments, two of which are shown at the top of the figure, can replace the variable stiffness spring attachment (top right: nontilting, nonshearing spring of fixed stiffness). Each of these springs are interchangeable and can be attached to the main support, allowing for greater versatility in measuring strong or weak and attractive or repulsive forces. Once the force F as a function of distance D is known for the two surfaces of radius R, the force between any other curved surfaces simply scales by R. Furthermore, the adhesion energy (or surface or interfacial free energy) E per unit area between two flat surfaces is simply related to F by the so-called Derjaguin approximation (Israelachvili, 1991) $E = F/2\pi R$. We note that SFA is one of the few techniques available for directly measuring equilibrium force laws (i.e., force vs. distance at constant chemical potential of the surrounding solvent medium) (Israelachvili, 1989). The SFA allows for both weak or strong and attractive or repulsive forces.

Mostly the molecularly smooth surface of mica is used in these measurements (Pashley, 1981); however, silica (Horn et al., 1989) and sapphire (Horn and Israelachvili, 1988) have also been used. It is also possible to deposit or coat each mica surface with metal films (Christenson, 1988; Smith et al., 1988), carbon and metal oxides (Hirz et al., 1992), adsorbed polymer layers (Patel and Tirrell, 1989), and surfactant monolayers and bilayers (Christenson, 1988; Israelachvili, 1987, 1991; Israelachvili and McGuiggan, 1988). The range of liquids and vapors that can be used is almost endless.

1.3.4.1.2 Sliding Attachments for Tribological Studies

Thus far we have described a measurement technique which allows measurements of the normal forces between surfaces, that is, those occurring when two surfaces approach or separate from each other. However, in tribological situations, it is the transverse or shear forces that are of primary interest when two surfaces slide past each other. There are essentially two approaches used in studying the shear response of confined liquid films. In the first approach (constant-velocity friction or steady-shear attachment), the friction is measured when one of the surfaces is traversed at a constant speed over a distance of several hundreds of microns (Israelachvili et al., 1988; Homola, 1989; Gee et al., 1990; Homola et al., 1990, 1991; Klein et al., 1991; Hirz et al., 1992). The second approach (oscillatory-shear attachment) relies on the measurement of viscous dissipation and elasticity of confined liquids by using periodic sinusoidal oscillations over a range of amplitudes and frequencies (Van Alsten and Granick, 1988, 1990a,b; Peachey et al., 1991; Hu et al., 1991; Luengo et al., 1997).

For the constant-velocity friction (steady-shear) experiments, the SFA was outfitted with a lateral sliding mechanism (Israelachvili et al., 1988; Israelachvili, 1989; Homola, 1989; Gee et al., 1990; Homola et al., 1990, 1991) allowing measurements of both normal and shearing forces (Figure 1.45). The piezoelectric

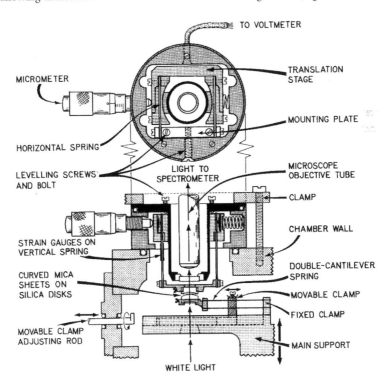

FIGURE 1.45 Schematic of shear force apparatus. Lateral motion is initiated by a variable-speed motor-driven micrometer screw that presses against the translation stage which is connected through two horizontal double-cantilever strip springs to the rigid mounting plate. (From Israelachvili, J.N. et al. (1988), *Science*, 240, 189–190; and Israelachvili, J.N. (1989), *Chemtracts Anal. Phys. Chem.*, 1, 1–12. With permission.)

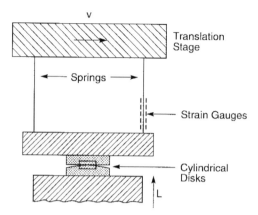

FIGURE 1.46 Schematic of the sliding attachment. The translation stage also supports two vertical double-cantilever springs, which at their lower end are connected to a steel plate supporting the upper silica disk. (From Gee, M.L. et al. (1990), *J. Chem. Phys.*, 93, 1895–1906. With permission.)

crystal tube mount supporting the upper silica disk of the basic apparatus shown in Figure 1.43 is replaced. Lateral motion is initiated by a variable-speed motor-driven micrometer screw that presses against the translation stage, which is connected via two horizontal double-cantilever strip springs to the rigid mounting plate. The translation stage also supports two vertical double-cantilever springs (Figure 1.46) that at their lower end are connected to a steel plate supporting the upper silica disk. One of the vertical springs acts as a frictional force detector by having four resistance strain gauges attached to it, forming the four arms of a Wheatstone bridge and electrically connected to a chart recorder. Thus, by rotating the micrometer, the translation stage deflects, causing the upper surface to move horizontally and linearly at a steady rate. If the upper mica surface experiences a transverse frictional or viscous shearing force, this will cause the vertical springs to deflect, and this deflection can be measured by the strain gauges. The main support, force-measuring double-cantilever spring, movable clamp, white light, etc. are all parts of the original basic apparatus (Figure 1.43) whose functions are to control the surface separation, vary the externally applied normal load, and measure the separation and normal force between the two surfaces, as already described. Note that during sliding, the distance between the surfaces, their true molecular contact area, their elastic deformation, and their lateral motion can all be simultaneously monitored by recording the moving FECO fringe pattern using a video camera and recording it on a tape (Gee et al., 1990).

The two surfaces can be sheared past each other at sliding speeds which can be varied continuously from 0.1 to 20 μm/s while simultaneously measuring both the transverse (frictional) force and the normal (compressive or tensile) force between them. The lateral distances traversed are on the order of several hundreds of micrometers which correspond to several diameters of the contact zone.

With an oscillatory shear attachment, developed by Granick et al., viscous dissipation and elasticity and dynamic viscosity of confined liquids by applying periodic sinusoidal oscillations of one surface with respect to the other can be studied (Van Alsten and Granick, 1988, 1990a,b; Peachey et al., 1991; Hu et al., 1991). This attachment allows for the two surfaces to be sheared past each other at varying sliding speeds or oscillating frequencies while simultaneously measuring both the transverse (friction or shear) force and the normal load between them. The externally applied load can be varied continuously, and both positive and negative loads can be applied. Finally, the distance between the surfaces, their true molecular contact area, their elastic (or viscoelastic) deformation, and their lateral motion can all be simultaneous by recording the moving interference fringe pattern using a video camera–recorder system.

To produce shear while maintaining constant film thickness or constant separation of the surfaces, the top mica surface is suspended from the upper portion of the apparatus by two piezoelectric bimorphs. A schematic description of the SFA with the installed shearing device is shown in Figure 1.47 (Van Alsten and Granick, 1988, 1990a,b; Peachey et al., 1991; Hu et al., 1991). Israelachvili (1989) and Luengo et al.

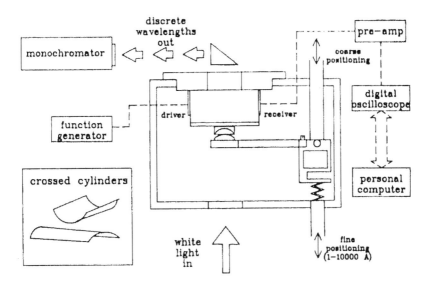

FIGURE 1.47 Schematic of the oscillatory shearing apparatus. (From Van Alsten, J. and Granick, S. (1988), *Phys. Rev. Lett.*, 61, 2570–2573. With permission.)

(1997) have also presented similar designs. The lower mica surface, as in the steady-shear sliding attachment, is stationary and sits at the tip of a double-cantilever spring attached at the other end to a stiff support. The externally applied load can be varied continuously by displacing the lower surface vertically. An AC voltage difference applied by a signal generator (driver) across one of the bimorphs tends to bend it in an oscillatory fashion while the frictional force resists that motion. Any resistance to sliding induces an output voltage across the other bimorphs (receiver), which can be easily measured by a digital oscilloscope. The sensitivity in measuring force is on the order of a few micronewtons and the amplitudes of measured lateral displacement can range from a few nanometers to 10 μm. The design is flexible and allows the inducement of time-varying stresses with different characteristic wave shapes simply by changing the waveform of the input electrical signal. For example, when measuring the apparent viscosity, a sine wave input is convenient to apply. Figure 1.48a shows an example of the raw data, obtained with a hexadecane film at a moderate pressure, when a sine wave was applied to one of the bimorphs (Van Alsten and Granick, 1990a). By comparing the calibration curve with the response curve, which was attenuated in amplitude and lagged in phase, an apparent dynamic viscosity can be inferred. On the other hand, a triangular waveform is more suitable when studying the yield stress behavior of solidlike films as in Figure 1.48b. The triangular waveform, showing a linear increase and decrease of the applied force with time, is proportional to the driving force acting on the upper surface. The response waveform, which represents a resistance of the interface to shear, remains very small indicating that the surfaces are in stationary contact with respect to each other until the applied stress reaches a yield point. At the yield point the slope of the response curve increases dramatically, indicating the onset of sliding.

Homola (1994) compared the two approaches — steady-shear attachment and oscillating-shear attachment. In experiments conducted by Israelachvili and his co-workers, the steady-shear attachment was employed to focus on the dynamic frictional behavior of the film after a sufficiently high shear stress was applied to exceed the yield stress and to produce sliding at a constant velocity. In these measurements, the film was subjected to a constant shearing force for a time sufficiently long enough to allow them to reach a dynamic equilibrium; i.e., the molecules, within the film, had enough time to order and align with respect to the surface, both normally and tangentially. Under these conditions, dynamic friction was observed to be "quantized" according to the number of liquid layers between the solid surfaces and independent of the shear rate (Israelachvili et al., 1988). Clearly, in this approach, the molecular ordering is optimized by a steady shear which imposes a preferred orientation on the molecules in the direction of shear.

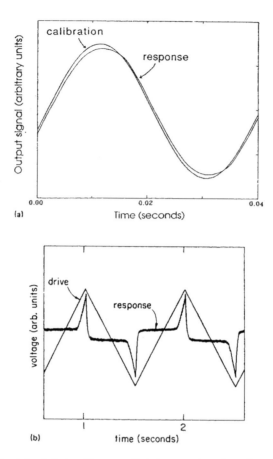

FIGURE 1.48 (a) Two output signals induced by an applied sine wave (not shown) are displaced. The "calibration" waveform is obtained with the mica sheets completely separated. The response waveform is obtained with a thin liquid film between the sheets, which causes it to lag the calibration waveform. (b) The oscilloscope trace of the drive and response voltages used to determine critical shear stress. The drive waveform shows voltage proportional to induced stress on the sheared film and the response waveform shows voltage proportional to resulting velocity. Spikes in the response curve correspond to the stick-slip event. (From Van Alsten, J. and Granick, S. (1990), *Tribol. Trans.*, 33, 436–446. With permission.)

The above mode of sliding is particularly important when the sheared film is made of long-chain lubricant molecules requiring a significantly long sliding time to order and align and even a longer time to relax (disorder) when sliding stops. This suggests that a steady-state friction is realized only when the duration of sliding exceeds the time required for an ensemble of the molecules to order fully in a specific. shear field. It also suggests that static friction should depend critically on the sliding time and the extent of the shear-induced ordering (Homola, 1993).

In contrast, the oscillatory-shear method, which utilizes periodic sinusoidal oscillations over a range of amplitudes and frequencies, addresses a response of the system to rapidly varying strain rates and directions of sliding. Under these conditions, the molecules, especially those exhibiting a solid-like behavior, cannot respond sufficiently fast enough to stress and are unable to order fully during duration of a single pass i.e., their dynamic and static behavior reflects an oscillatory shear-induced ordering which might or might not represent an equilibrium dynamic state. Thus, the response of the sheared film will depend critically on the conditions of shearing i.e., the strain, the pressure, and the sliding conditions (amplitude and frequency of oscillations) which in turn will determine a degree of molecular ordering. This may explain the fact that the layer structure and "quantization" of the dynamic and static friction were not observed in these experiments in contrast to results obtained when velocity was kept constant.

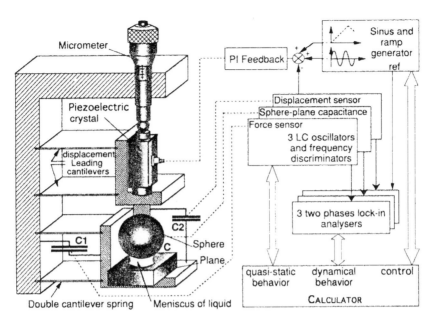

FIGURE 1.49 Schematic of the SFA that employs a sphere–plane arrangement. (From Georges, J.M. et al. (1993), *J. Chem. Phys.*, 98, 7345–7360. With permission.)

Intuitively, this behavior is expected considering that the shear-ordering tendency of the system is frequently disturbed by a shearing force of varying magnitude and direction. Nonetheless, the technique is capable of providing an invaluable insight into the shear behavior of molecularly thin films subjected to nonlinear stresses as it is frequently encountered in practical applications. This is especially true under conditions of boundary lubrication, where interacting surface asperities will be subjected to periodic stresses of varying magnitudes and frequencies (Homola, 1993).

1.3.4.2 Georges et al.'s Design

The SFA developed by Tonck et al. (1988) and Georges et al. (1993) to measure static and dynamic forces in the normal direction, between surfaces in close proximity, is shown in Figure 1.49. In their apparatus, a drop of liquid is introduced between a macroscopic spherical body and a plane. The sphere is moved toward and away from a plane using the expansion and the vibration of a piezoelectric crystal. Piezoelectric crystal is vibrated at low amplitude around an average separation for dynamic measurements to provide dynamic function of the interface. The plane specimen is supported by a double-cantilever spring. Capacitance sensor C_1 measures the elastic deformation of the cantilever and thus the force transmitted through the liquid to the plane. Second capacitance sensor C_2 is designed to measure the relative displacement between the supports of the two solids. The reference displacement signal is the sum of two signals: first, a ramp provides a constant normal speed from 50 to 0.01 nm/s, and, second, the piezoelectric crystal is designed to provide a small sinusoidal motion in order to determine the dynamic behavior of sphere–plane interactions. A third capacitance sensor C measures the electrical capacitance between the sphere and the plane. In all cases, the capacitance is determined by incorporating the signal of an oscillator in the inductive capacitance (L-C) resonant input stage of an oscillator to give a signal-dependent frequency in the range of 5 to 12 MHz. The resulting fluctuations in oscillation frequency are detected using a low-noise-frequency discriminator. Simultaneous measurements of sphere–plane displacement, surface force, and the damping of the interface allow an analysis of all regimes of the interface (Georges et al., 1993). Loubet et al. (1993) used SFA in the crossed-cylinder geometry using two freshly cleaved mica sheets similar to the manner used by Israelachvili and co-workers.

Georges et al. (1994) modified their original SFA to measure friction forces. In this apparatus, in addition to having the sphere move normal to the plane, the sphere can be sheared at constant separation

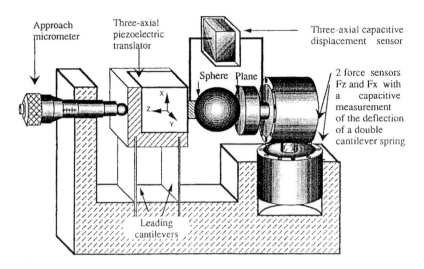

FIGURE 1.50 Schematic of shear force apparatus. (From Georges, J.M. et al. (1994), *Wear*, 175, 59–62. With permission.)

from the plane. The shear force apparatus is shown in Figure 1.50. Three piezoelectric elements controlled by three capacitance sensors permit accurate motion control and force measurement along three orthogonal axes with displacement sensitivity of 10^{-3} nm and force sensitivity of 10^{-8} N. Adhesion and normal deformation experiments are conducted in the normal approach (Z-axis). Friction experiments are conducted by introducing displacement in the X-direction at a constant normal force. In one of the experiments, Georges et al. (1994) used a 2.95-mm-diameter sphere made of cobalt-coated fused borosilicate glass and a silicon wafer for the plane.

1.3.5   Vibration Isolation

STM, AFM, and SFA should be isolated from sources of vibration in the acoustic and subacoustic frequencies, especially for atomic-scale measurements. Vibration isolation is generally provided by placing the instrument on a vibration isolation air table. For further isolation, the instrument should be placed on a pad of soft silicone rubber. A cheaper alternative consists of a large mass of 100 or more newtons suspended from elastic "bungee" cords. The mass should stretch the cords at least 0.3 m, but not so much that the cords reach their elastic limit. The instrument should be placed on the large mass. The system, including the microscope, should have a natural frequency of about 1 Hz or less both vertically and horizontally. Test this by gently pushing on the mass and measure the rate at which it swings or bounces (Anonymous, 1992a).

1.4   Magnetic Storage and MEMS Components

1.4.1   Magnetic Storage Devices

Magnetic storage devices used for storage and retrieval are tape and flexible and rigid disk drives. These devices are used for audio, video, and data storage applications. The magnetic storage industry is a $90 billion a year industry with $30 billion for audio and video recording (almost all tape drives/media) and $60 billion for data storage. In the data storage industry, magnetic rigid disk drives/media, tape drives/media, flexible disk drives/media, and optical disk drive/media account for about $42, $9, $5, and $4 billion, respectively. Magnetic recording and playback involves the relative motion between a magnetic medium against a read–write magnetic head. Heads are designed so that they develop a (load-carrying) hydrodynamic air film under steady operating conditions to minimize head–medium contact. However, physical contact between the medium and head occurs during starts and stops (Bhushan, 1992, 1996a). In the modern magnetic storage devices, the flying heights (head-to-medium separation) are on the order of 50 nm and roughnesses of head and medium surfaces are on the order of 2 nm rms. The need for ever-increasing recording densities requires that surfaces be as smooth as possible and the flying heights be as low as possible. In fact, several contact or near-contact recording devices are at various stages of development. High static friction (stiction) at the head–medium interface (HMI) after storage and increase in kinetic friction after use as a result of medium wear and head wear from usage are the major impediments to the commercialization of the contact recording.

Magnetic media fall into two categories: (1) particulate media, where magnetic particles (γ-Fe_2O_3, Co-γFe_2O_3, CrO_2, Fe or metal, barium ferrite) are dispersed in a polymeric matrix and coated onto a polymeric substrate for flexible media (tape and floppy disks) or onto a rigid substrate (typically aluminum and more recently glass or glass ceramic) and (2) thin-film media, where continuous films of magnetic materials are deposited onto the substrate by vacuum techniques. The most commonly used thin magnetic films for tapes are evaporated Co–Ni (82–18 at %). Typical magnetic films for rigid disks are metal films of cobalt-based alloys (such as sputtered Co–Pt–Ni, Co–Ni, Co–Pt–Cr, Co–Cr, and Co–NiCr). For high recording densities, trends have been to use thin-film media. Magnetic heads used to date are either conventional inductive or thin-film inductive and magnetoresistive (MR) heads. The air-bearing surfaces (ABS) of tape heads are cylindrical in shape. For dual-sided floppy-disk heads, two heads are either spherically contoured and slightly offset (to reduce normal pressure) or are flat and loaded against each other. The ABS of rigid-disk heads are a two- or three-rail taper flat design supported by a leaf-spring (flexure) suspension. The ABS of magnetic heads are almost exclusively made of Mn–Zn ferrite, Ni–Zn ferrite, Al_2O_3–TiC, and calcium titanate. The ABS of some conventional heads are made of plasma-sprayed coatings of hard materials such as Al_2O_3–TiO_2 and ZrO_2 (Bhushan, 1992, 1996a).

Figure 1.51 shows the schematic of a data-processing linear tape drive (IBM 3490) which uses a rectangular tape cartridge ($100 \times 125 \times 25$ mm). Figure 1.52A shows the sectional views of particulate and thin-film magnetic tapes. Almost exclusively, the base film is made of semicrystalline, biaxially oriented poly (ethylene terephthalate) or PET. The particulate coating formulation consists of binder (typically polyester polyurethane), submicron magnetic particles, submicron head-cleaning agents (typically alumina), and lubricant (typically fatty acid ester). The thin-film tape is topically lubricated typically with a perfluoropolyether lubricant. Figure 1.52B shows the schematic of a 36-track, thin-film read–write head (with a radius of cylindrical contour of about 20 mm) whose ABS is made of Ni–Zn ferrite with multilayered thin-film head structures. A tape tension of about 2 N over a 12.7-mm-wide tape (normal pressure ~14 kPa) is used during use. The rms roughnesses of ABS of the heads and tape surfaces typically are 1 to 1.5 nm and 5 to 10 nm, respectively.

Figure 1.53 shows the schematic of a data-processing rigid-disk drive with 48-, 65-, 95-, or 130-mm form factor. A nonremovable stack of multiple disks mounted on a ball-bearing spindle is rotated by an electric motor at constant angular speed ranging from about 3000 to 10,000 rpm, dependent upon the disk size. Head slider–suspension assembly (allowing one slider for each disk surface) is actuated by a stepper motor or a voice coil motor using a linear or rotary actuator. Generally a rotary actuator is used to save space, as shown in Figure 1.53. Figure 1.54A shows the sectional views of particulate and thin-film rigid disks. The substrate for rigid disks is generally the nonheat-treatable aluminum–magnesium alloy 5086. For metal thin-film disks, the substrate is coated with a hard coating to improve its surface hardness to Knoop 600 to 800 kg/mm^2 and smoothness. The particulate coating formulation consists of binder (typically hard epoxy), submicron magnetic particles, and submicron head-cleaning agent (typically alumina). The protective overcoat commonly used for thin-film disks is generally either sputtered diamondlike carbon (DLC) or silicon dioxide. Both types of disks are topically lubricated with perfluoropolyether types of lubricants. Lubricants with polar-end groups are generally used for thin-film disks in order to provide chemical bonding to the overcoat surface. Figure 1.54B shows the schematic of a thin-film head nanoslider with a three-rail, taper-flat design (MR-read, inductive-write). These sliders generally use Al_2O_3–TiC (70–30 wt%) as the substrate material with multilayered thin-film head structure. In the MR head design, the ABS and the MR structure are generally coated with about 10-nm-thick DLC

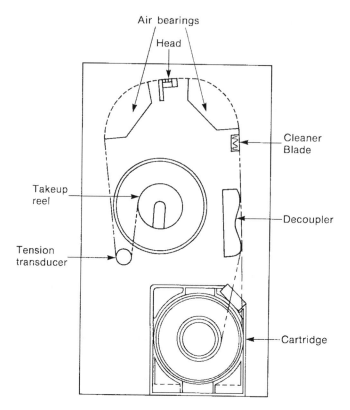

FIGURE 1.51   Tape path in an IBM 3490 data-processing tape drive.

coating. A nanoslider commonly used today has the dimensions of $2 \times 1.6 \times 0.425$ mm with a mass of about 7.7 mg, Table 1.5. A normal load of about 3.5 g is applied during use. Future trends are to use smaller drives with ultrasmall sliders and an ultrasmooth (1 to 2 nm rms) and flat disk substrate (made of glass or glass ceramic) with about 5- to 10-nm-thick DLC overcoat and 0.5- to 2-nm-thick bonded perfluoropolyether lubricant.

Horizontal thin-film heads with a single-crystal silicon substrate referred to as silicon planar head (SPH) sliders are under development which can be mass produced inexpensively and miniaturized using silicon integrated-circuit technology, Figure 1.55 (Lazzari and Deroux-Dauphin, 1989; Bhushan et al., 1992). Several integrated suspension/head devices have been proposed. For example, Fujitsu's integrated suspension with a so-called GUPPY nanoslider attached to an MR head, Figure 1.56 (Ohwe et al., 1993). Suspension design is a miniaturized monolithic flexible body with a load beam and a gimbal. The integrated suspension is 25-μm thick, 10-mm long, 4-mm wide, and the beam width of the gimbal is about 200 μm.

Various contact recording devices for ultimate recording densities are at various stages of development. Censtor Corporation in San Jose, CA is developing an integrated head/flexure/conductor structure called a microflex head which is substantially smaller than the conventional head (Hamilton et al., 1991). The microflex head is supposed to be in continuous sliding contact with the disk surface during use. A single-pole perpendicular-probe-type transducer is used so as to tolerate several microns of pole-tip wear without significant signal degradation. In order to achieve low stiction and low head wear, the slider suspension mass, the footprint of the contact pad, and the normal load are reduced by two to three orders of magnitude from current designs. In the microflex head design, a read–write transducer is fabricated at one end of a rectangular (30-μm-thick) sputtered amorphous alumina flexure beam containing two parallel copper leads connected to the read–write element at one end and bonding pads at

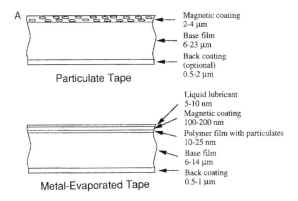

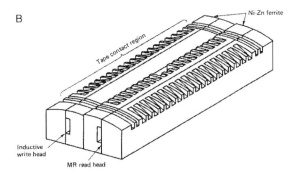

FIGURE 1.52 (A) Sectional views of particulate and thin-film magnetic tapes and (B) schematic of a 36-track magnetic thin-film read–write head (with a radius of cylindrical contour of about 20 mm) for an IBM 3490 tape drive.

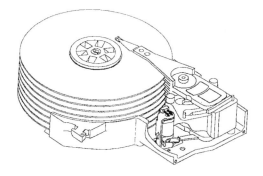

FIGURE 1.53 Schematic of a data-processing rigid-disk drive.

the other end, Figure 1.57. The length, width, and thickness of the initial configurations of the flexure beam are 10 mm, 500 μm, and 30 μm, respectively, the spring constant is about 0.6 N/m, and the mass is about 600 μg. A perpendicular-probe-type head with a layered pancake coil design and flux return pole is fabricated in the end of the beam. The main pole and yoke reside on the vertical end of the beam. The pole tip (about 7 μm in height) is encapsulated in the wear-resistant amorphous hydrogenated carbon material, which forms a contact pad whose nominal length, width, and depth (height) are 40, 20, and 7 μm, respectively. This configuration provides a large tolerance for head wear and negligible air-bearing lift of the flexure beam. The pole material is an ion-beam-sputtered Co–Nb–Zr alloy of about 275-nm thickness, surrounded on both sides by about 200 nm of rf-sputtered alumina. There is also an rf-sputtered SiC adhesion layer of about 70-nm thickness between the rf-PECVD-deposited amorphous hydrogenated carbon wear pad and the alumina. On the side nearest the free end of the beam, there is an rf-sputtered Si adhesion layer of about 100 nm between the alumina and the carbon. The progression

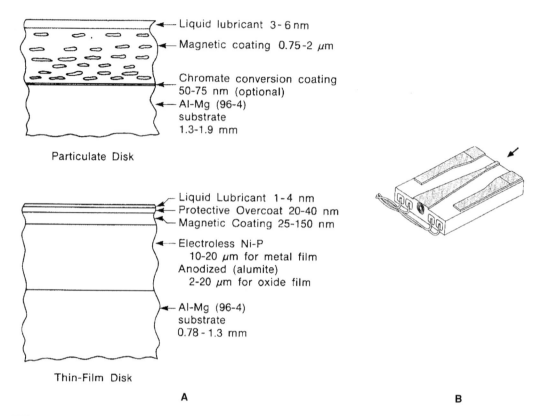

FIGURE 1.54 (A) Sectional views of particulate and thin-film rigid disks and (B) schematic of a thin-film slider with a three-rail, taper-flat design (arrow indicates the disk rotation).

TABLE 1.5 Physical Dimensions of Rigid-Disk Head Sliders

| | Typical Dimensions | | | | |
Slider Type	$L \times W \times H$ (mm)	Rail Width (mm)	Volume (mm^3)	Mass (mg)	Load (g)
Regular slider	$4.00 \times 3.20 \times 0.850$	420	10.88	50	9.5
Microslider (70%)	$2.80 \times 2.24 \times 0.595$	330	3.73	17	6.0
Nanoslider (50%)	$2.00 \times 1.60 \times 0.425$ or	250	1.36	6	3.5
(small slider)	$2.50 \times 1.70 \times 0.425$		1.81	8	
Under development					
Picoslider (35%)	$L = 1.25$ mm (flying slider)				1
	$L = 1.00$ mm (contact slider)				
Femtoslider (30%)					
Attoslider (10%)					

Note: L = length, W = width, and H = height.

of layers is carbon, SiC, alumina, Co–Nb–Zr, alumina, Si, and carbon, going from the fixed to free end of the beam. Typical normal load applied is about 40 mg. We note that the microflex head consists of nongimbaled stiff suspension, which requires appropriate mounting.

These small structures are fabricated by photolithography and etching on a host substrate. In a departure from convention, thin-film fabrication is accomplished on two orthogonal surfaces. The magnetic core, coil, conductors, and flux return pole are constructed within the dielectric flexure on the host wafer. The wafer is then sliced into bars and the pole/yoke structures are fabricated on the cut surface of the bar. The finished structures are chemically released from the host substrate, the entire process

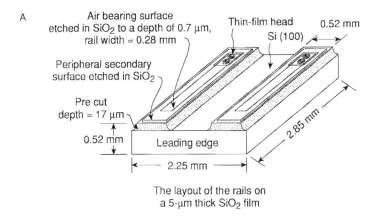

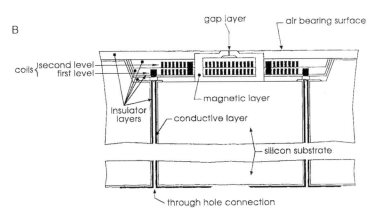

FIGURE 1.55 (a) Isometric view (not drawn to scale) and (b) cross section of the silicon planar head slider. (From Bhushan, B. et al. (1992), *IEEE Trans. Magn.*, 28, 2874–2876. With permission.)

holding the potential to all but eliminate the machining process. The integrated flexure/head almost looks like an AFM cantilever, except the dimensions are much larger.

1.4.2   MEMS

In recent years, the emerging field of MEMS has received increasing attention. MEMS are fabricated using the technology developed for integrated circuits. The total size of the component can be as small as 100 μm or even smaller. Microsensors, fine-polishing microactuators, microgrippers, micromotors, micropumps, gear trains, cranks, manipulators, nozzles, and valves a fraction of a millimeter in size are being fabricated in laboratories around the world (Howe, 1988; Tai and Muller, 1989; Fan et al., 1989; Tang et al., 1990; Bhushan and Venkatesan, 1993; Miu et al., 1993; Bhushan, 1996b, 1998b; Bhushan et al.,. 1997a,b; Sundararajan and Bhushan, 1998). Figure 1.58 shows an example of a surface micromachined structure — a micromotor-driven electrostatically (by electrostatic attraction between positive and negatively charged surfaces) which has been fabricated at MIT and UC Berkeley. The UC Berkeley micromotor consists of 12 stators and a four-pole rotor of 120-μm diameter with an air gap between the rotor and stator of 2 μm. Structural material for the hub, stator, and rotor is polysilicon film deposited by LPCVD. The vertical walls of the stator and rotor contain 340-nm-thick LPCVD silicon nitride film for low friction and wear at high operating speeds, Figure 1.44. The motor built at MIT delivered a torque of 12 pNm and synchronous motor speeds of up to 2500 rpm. The wear resistance of rotor/stator interface and reduction of friction at the hub/rotor interface is critical for high-speed interfaces.

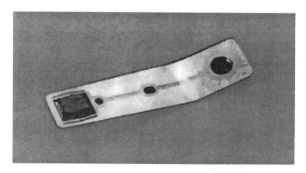

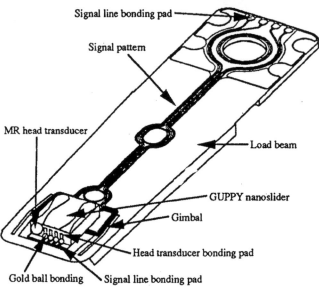

FIGURE 1.56 Schematic and a photograph of an integrated head/suspension device. (From Ohwe, T. et al. (1993), *IEEE Trans. Magn.*, 29, 3924–3926. With permission.)

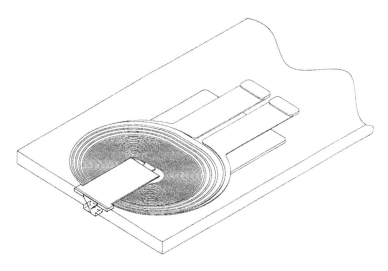

FIGURE 1.57 Schematic of probe-type perpendicular head on end of flexure beam (Microflex head) with layered pancake coil, conductors, and bonding pads with the contact pad of about 20×40 μm. (Censtor Corporation, San Jose, CA.)

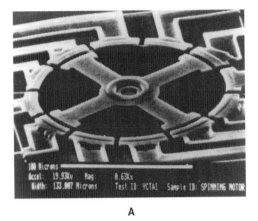
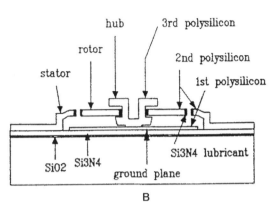

A   B

FIGURE 1.58 (A) Schematic of an electrostatic (variable-capacitance) motor fabricated of polysilicon film and (B) schematic cross section of the micromotor. (Tai, Y.C. and Mueller, R.S. (1989), *Sensors Actuators*, 20, 49–55. With permission.)

Several international journals are devoted to micromachines and sensor technology: *Journal of Micromechanics and Microengineering Structures, Devices, and Systems; International Journal of Sensor Technology: Sensors and Materials;* and *Journal of Microelectromechanical Systems.*

1.5 Role of Micro/Nanotribology in Magnetic Storage Devices, MEMS, and Other Microcomponents

The magnetic storage devices and MEMS are two examples where micro/nanotribological tools and techniques are essential for studies of micro/nanoscale phenomena. Magnetic storage components continue to shrink in physical dimensions; for example, rigid-disk slider dimensions are shrinking as shown in Table 1.5. Thicknesses of hard solid coating and liquid lubricant overlays on the magnetic disk surface (Figure 1.54A) continue to decrease; these are expected to approach 5 and 1 nm, respectively. A number of contact recording devices are at various stages of development. Surface roughness of the storage components continues to decrease and is expected to approach about 0.5 nm rms. Interface studies of ultrasmall components with ultrathin coatings can be ideally performed using micro/nanotribological tools and techniques

In the case of MEMS, the friction and wear problems of ultrasmall moving components generally made of polysilicon films need to be addressed for high performance, long life, and reliability. For example, in a micromotor shown earlier (Figure 1.58), the rotor is separated from the stator, but in practice there may be some physical contact. In addition, reduction of friction at the hub/rotor interface is crucial for high-speed performance. Molecularly thin films of solid and/or liquids may be required for low friction and wear. Again, interfacial phenomena in MEMS can be ideally studied using micro/nanotribological tools and techniques.

Magnetic storage devices and MEMS are the two examples of components where micro/nanotribological tools and techniques are ideal for interfacial studies. Studies of other microcomponents would benefit from these tools and techniques. In addition, micro/nanotribological studies on macrocomponents provide a fundamental understanding of tribological phenomena.

References

Abraham, D.W., Mamin, H.J., Ganz, E., and Clark, J. (1986), "Surface Modification with the Scanning Tunneling Microscope," *IBM J. Res. Dev.* 30, 492–499.

Aden, G.D. (1994), Topometrix, 5403 Betsy Ross Drive, Santa Clara, CA 95054, personal communications.

Albrecht, T.R. and Quate, C.F. (1987), "Atomic Resolution Imaging of a Nonconductor by Atomic Force Microscopy," *J. Appl. Phys.* 62, 2599–2602.

Albrecht, T.R., Dovek, M.M., Kirk, M.D., Lang, C.A., Quate, C.F., and Smith, D.P.E. (1989), "Nanometer-Scale Hole Formation on Graphite Using a Scanning Tunneling Microscope," *Appl. Phys. Lett.* 55, 1727–1729.

Albrecht, T.R., Akamine, S., Carver, T.E., and Quate, C.F. (1990), "Microfabrication of Cantilever Styli for the Atomic Force Microscope, *J. Vac. Sci. Technol.* A8, 3386–3396.

Albrecht, T.R., Grutter, P., Rugar, D., and Smith, D.P.E. (1992), "Low Temperature Force Microscope with All-Fiber Interferometer," *Ultramicroscopy* 42–44, 1638–1646.

Alexander, S., Hellemans, L., Marti, O., Schneir, J., Elings, V., and Hansma, P.K. (1989), "An Atomic-Resolution Atomic-Force Microscope Implemented Using an Optical Lever," *J. Appl. Phys.* 65, 164–167.

Andoh, Y., Oguchi, S., Kaneko, R., and Miyamoto, T. (1992), "Evaluation of Very Thin Lubricant Films," *J. Phys. D: Appl. Phys.* 25, A71–A75.

Anonymous (1991), "Stand Alone Atomic Force Microscope, User's Manual," Digital Instruments, Inc., Santa Barbara, CA.

Anonymous (1992a), "Nanoscope III Scanning Tunneling Microscope, Instruction Manual," Digital Instruments, Inc., Santa Barbara, CA.

Anonymous (1992b), "Nanoscope III Atomic Force Microscope, Instruction Manual," Digital Instruments, Inc., Santa Barbara, CA.

Anonymous (1992c), "Nanoscope III Lateral Force Microscope, Instruction Manual," Digital Instruments, Inc., Santa Barbara, CA.

Anselmetti, D., Gerber, Ch., Michel, B., Guntherodt, H., and Rohrer, H. (1992), "Compact, Combined Scanning Tunneling/Force Microscope," *Rev. Sci. Instrum.* 63, 3003–3006.

Ashcroft, N.W. and Mermin, N.D. (1976), *Solid State Physics*, Holt Reinhart and Winston, New York..

Bhushan, B. (1992), *Mechanics and Reliability of Flexible Magnetic Media*, Springer-Verlag, New York.

Bhushan, B. (1995), "Micro/Nanotribology and its Applications to Magnetic Storage Devices and MEMS," *Tribol. Int.* 28, 85–96.

Bhushan, B. (1996a), *Tribology and Mechanics of Magnetic Storage Devices*, 2nd ed., Springer-Verlag, New York.

Bhushan, B. (1996b), "Nanotribology and Nanomechanics of MEMS Devices," in *Proc. MEMS 1996*, IEEE, New York.

Bhushan, B. (1997), *Micro/Nanotribology and Its Applications*, E330, Kluwer Academic, Dordrecht, The Netherlands.

Bhushan, B. (1998a), "Micro/Nanotribology Using Atomic Force/Friction Force Microscopy. State of the Art," *Proc. Inst. Mech. Eng. Part J: J. Eng. Tribol. 212*, 1-18.

Bhushan, B. (1998b), *Tribology Issues and Opportunities in MEMS*, Kluwer Academic Publishers, Dordrecht, The Netherlands.

Bhushan, B. and Blackman, G.S. (1991), "Atomic Force Microscopy of Magnetic Rigid Disks and Sliders and Its Applications to Tribology," *ASME J. Tribol. Trans. ASME* 113, 452–458.

Bhushan, B., Dominiak, M., and Lazzari, J.P. (1992), "Contact-Start-Stop Studies with Silicon Planar Head Sliders against Thin-Film Disks," *IEEE Trans. Magn.* 28, 2874–2876.

Bhushan, B., Ruan, J., and Gupta, B.K. (1993), "A Scanning Tunneling Microscopy Study of Fullerene Films," *J. Phys. D: Appl. Phys.* 26, 1319–1322.

Bhushan, B. and Venkatesan, S. (1993), "Mechanical and Tribological Properties of Silicon for Micromechanical Applications — A Review," *Adv. Info. Storage Syst.* 5, 211–239.

Bhushan, B. and Ruan, J. (1994a), "Atomic-Scale Friction Measurements Using Friction Force Microscopy: Part II — Application to Magnetic Media," *ASME J. Tribol.* 116, 389–396.

Bhushan, B., Koinkar, V.N., and Ruan, J. (1994b), "Microtribology of Magnetic Media," *Proc. Inst. Mech. Eng. Part J: J. Eng. Tribol.* 208, 17–29.

Bhushan, B. and Koinkar, V.N. (1994c), "Tribological Studies of Silicon for Magnetic Recording Applications," *J. Appl. Phys.* 75, 5741–5746.

Bhushan, B. and Koinkar, V.N. (1994d), "Nanoindentation Hardness Measurements Using Atomic Force Microscopy," *Appl. Phys. Lett.* 64, 1653–1655.

Bhushan, B. and Koinkar, V.N. (1994e), "Microtribological Studies by Using Atomic Force and Friction Force Microscopy and Its Applications," *Determining Nanoscale Physical Properties of Materials by Microscopy and Spectroscopy* (M. Sarikaya, H.K. Wickramasinghe, and M. Isaacson, Eds.), Vol. 332, pp. 93–98, Materials Research Society, Pittsburgh.

Bhushan, B., Israelachvili, J.N., and Landman, U. (1995a), "Nanotribology: Friction, Wear and Lubrication at the Atomic Scale," *Nature* 374, 607–616.

Bhushan, B. and Koinkar, V.N. (1995b), "Microtribology of PET Polymeric Films," *Tribol. Trans.* 38, 119–127.

Bhushan, B. and Koinkar, V.N. (1995c), "Microscale Mechanical and Tribological Characterization of Hard Amorphous Carbon Coatings as Thin as 5 nm for Magnetic Disks," *Surf. Coat. Technol.* 76–77, 655–669.

Bhushan, B. and Koinkar, V.N. (1995d), "Microtribology of Metal Particle, Barium Ferrite and Metal Evaporated Tapes," *Wear* 181–183, 360–370.

Bhushan, B. and Koinkar, V.N. (1995e), "Macro and Microtribological Studies of CrO_2 Video Tapes," *Wear* 180, 9–16.

Bhushan, B., Kulkarni, A.V., Koinkar, V.N., Boehm, M., Odoni, L., Martelet, C., and Belin, M. (1995f), "Microtribological Characterization of Self-Assembled and Langmuir–Blodgett Monolayers by Atomic and Friction Force Microscopy," *Langmuir* 11, 3189–3198.

Bhushan, B., Miyamoto, T., and Koinkar, V.N. (1995g), "Microscopic Friction between a Sharp Tip and Thin-Film Magnetic Rigid Disks by Friction Force Microscopy," *Adv. Info. Storage Syst.* 6, 151–161.

Bhushan, B., Kulkarni, A.V., Bonin, W., and Wyrobek, J.T. (1996), "Nano-Picoindentation Measurements Using Capacitive Transducer System in Atomic Force Microscopy," *Philos. Mag. A* 74, 1117–1128.

Bhushan, B. and Koinkar, V.N. (1997a), "Microtribological Studies of Doped Single-Crystal Silicon and Polysilicon Films for MEMS Devices," *Sensors Actuators A* 57, 91–102.

Bhushan, B. and Li, X. (1997b), "Micromechanical and Tribological Characterization of Doped Silicon and Polysilicon Films for MEMS Devices," *J. Mater. Res.* 12, 54–63.

Bhushan, B., Sundararajan, S., Scott, W.W., and Chilamakuri, S. (1997c), "Stiction Analysis of Magnetic Tapes," *IEEE Trans. Magn.* 33, 3211–3213.

Bhushan, B. and Sundararajan, S. (1998), "Micro/Nanoscale Friction and Wear Mechanisms of Thin Films Using Atomic Force and Friction Force Microscopy," *Acta Mater.* 46, 3793-3804.

Binggeli, M., Christoph, R., Hintermann, H.E., Colchero, J., and Marti, O. (1993), "Friction Force Measurements on Potential Controlled Graphite in an Electrolytic Environment," *Nanotechnology* 4, 59–63.

Binh, Vu T. and Garcia, N. (1991), *J. Phys. I (Paris)* 1, 605.

Binh, Vu T. and Garcia, N. (1992), "On the Electron and Metallic Ion Emission from Nanotips Fabricated by Field-Surface-Melting Technique: Experiments on W and Au Tips," *Ultramicroscopy* 42–44, 80.

Binnig, G. (1992), "Force Microscopy," *Ultramicroscopy* 42–44, 7–15.

Binnig, G., Rohrer, H., Gerber, Ch., and Weibel, E. (1982), "Surface Studies by Scanning Tunneling Microscopy," *Phys. Rev. Lett.* 49, 57–61.

Binnig, G. and Rohrer, H. (1983), "Scanning Tunneling Microscopy," *Surf. Sci.* 126, 236–244.

Binnig, G. and Smith, D.P.E. (1986), "Single-Tube Three-Dimensional Scanner for Scanning Tunneling Microscopy," *Rev. Sci. Instrum.* 57, 1688.

Binnig, G., Quate, C.F., and Gerber, Ch. (1986a), "Atomic Force Microscope," *Phys. Rev. Lett.* 56, 930–933.

Binnig, G., Fuchs, H., Gerber, Ch., Rohrer, H., Stoll, E., and Tosatti, E. (1986b), "Energy-Dependent State-Density Corrugation of a Graphite Surface as Seen by Scanning Tunneling Microscopy," *Europhys. Lett.* 1, 31–36.

Binnig, G., Gerber, Ch., Stoll, E., Albrecht, T.R., and Quate, C.F. (1987), "Atomic Resolution with Atomic Force Microscope," *Europhys. Lett.* 3, 1281–1286.

Blackman, G.S., Mate, C.M., and Philpott, M.R. (1990a), "Interaction Forces of a Sharp Tungsten tip with Molecular Films on Silicon Surface," *Phys. Rev. Lett.* 65, 2270–2273.

Blackman, G.S., Mate, C.M., and Philpott, M.R. (1990b), "Atomic Force Microscope Studies of Lubricant Films on Solid Surfaces," *Vacuum* 41, 1283–1286.

Bonnell, D.A. (ed.) (1993), *Scanning Tunneling Microscopy and Spectroscopy — Theory, Techniques, and Applications,* VCH Publishers, New York.

Boschung, E., Heuberger, M., and Dietler, G. (1994), "Energy Dissipation during Nanoscale Indentation of Polymers with an Atomic Force Microscope," *Appl. Phys. Lett.* 64, 1794–1796.

Bowden, F.P. and Tabor, D. (1950 and 1964), *The Friction and Lubrication of Solids* Vol. 1(1950) and Vol. 2(1964), Clarendon Press, Oxford.

Burnham, N.A. and Colton, R.J. (1989), "Measuring the Nanomechanical Properties and Surface Forces of Materials Using an Atomic Force Microscope," *J. Vac. Sci. Technol.* A7, 2906–2913.

Burnham, N.A., Domiguez, D.D., Mowery, R.L., and Colton, R.J. (1990), "Probing the Surface Forces of Monolayer Films with an Atomic Force Microscope," *Phys. Rev. Lett.* 64, 1931–1934.

Burnham, N.A., Colton, R.J., and Pollock, H.M. (1991), "Interpretation Issues in Force Microscopy," *J. Vac. Sci. Technol.* A9, 2548–2556.

Burnham, N.A. and Colton, R.J. (1993a), "Force Microscopy," in *Scanning Tunneling Microscopy and Spectroscopy — Theory, Techniques, and Applications* (D.A. Bunnell, ed.), pp. 191–249, VCH Publishers, New York.

Burnham, N.A., Colton, R.J., and Pollock, H.M. (1993b), "Interpretation of Force Curves in Force Microscopy," *Nanotechnology* 4, 64–80.

Chan, D.Y.C. and Horn, R.G. (1985), "The Drainage of Thin Liquid Films between Solid Surfaces," *J. Chem. Phys.* 83, 5311–5324.

Christenson, H.K. (1988), "Adhesion between Surfaces in Unsaturated Vapors — A Reexamination of the Influence of Meniscus Curvature and Surface Forces," *J. Colloid Interface Sci.* 121, 170–178.

Cleveland, J.P., Manne, S., Bocek, D. and Hansma, P.K. (1993), "A Nondestructive Method for Determining the Spring Constant of Cantilevers for Scanning Force Microscopy," *Rev. Sci. Instrum.* 64, 403–405.

Cohen, S.R., Neubauer, G., and McClelland, G.M. (1990), "Nanomechanics of a Au-Ir Contact Using a Bidirectional Atomic Force Microscope," *J. Vac. Sci. Technol.* A8, 3449–3454.

Cooley, J.W. and Tukey, J.W. (1965), "An Algorithm for Machine Calculation of Complex Fourier Series," *Math. Computation* 19, 297.

Coombs, J.H. and Pethica, J.B. (1986), "Properties of Vacuum Tunneling Currents: Anomalous Barrier Heights," *IBM J. Res. Dev.* 30, 455–459.

Davidson, C.S.C. (1957), "Bearings since the Stone Age," *Engineering* 183, 2–5.

Delawski, E. and Parkinson, B.A. (1992), "Layer-by-Layer Etching of Two-Dimensional Metal Chalcogenides with the Atomic Force Microscope," *J. Am. Chem. Soc.* 114, 1661–1667.

den Boef, A.J. (1991), "The Influence of Lateral Forces in Scanning Force Microscopy," *Rev. Sci. Instrum.* 62, 88–92.

DeVecchio, D. and Bhushan, B. (1997), "Localized Surface Elasticity Measurements Using an Atomic Force Microscope," *Rev. Sci. Instrum.* 68, 4498–4505.

Dowson, D. (1979), *History of Tribology,* Longman, London.

Drake, B., Prater, C.B., Weisenhorn, A.L., Gould, S.A.C., Albrecht, T.R., Quate, C.F., Cannell, D.S., Hansma, H.G., and Hansma, P.K. (1989), "Imaging Crystals, Polymers and Processes in Water with the Atomic Force Microscope," *Science* 243, 1586–1589.

Ducker, W.A., Senden, T.J., and Pashley, R.M. (1992), "Measurement of Forces in Liquids Using a Force Microscope," *Langmuir* 8, 1831–1836.

Durig, U., Zuger, O., and Stalder, A. (1992), "Interaction Force Detection in Scanning Probe Microscopy: Methods and Applications," *J. Appl. Phys.* 72, 1778–1797.

Eigler, D.M. and Schweizer, E.K. (1990), "Positioning Single Atoms with a Scanning Tunneling Microscope," *Nature* 344, 524–528.

Erlandsson, R., McClelland, G.M., Mate, C.M., and Chiang, S. (1988a), "Atomic Force Microscopy Using Optical Interferometry," *J. Vac. Sci. Technol.* A 6, 266–270.

Erlandsson, R., Hadzioannou, G., Mate, C.M., McClelland, G.M., and Chiang, S. (1988b), "Atomic-Scale Friction between the Muscovite Mica Cleavage Plane and a Tungsten Tip," *J. Chem. Phys.* 69, 5190–5193.

Fan, L.S., Tai, Y.C., and Muller, R.S. (1989), "IC-Processed Electrostatic Micromotors," *Sensors Actuators* 20, 41–47.

Foster, J. and Frommer, J. (1988), "Imaging of Liquid Crystal Using a Tunneling Microscope," *Nature* 333, 542–547.

Frisbie, C.D., Rozsnyai, L.F., Noy, A., Wrighton, M.S., and Lieber, C.M. (1994), "Functional Group Imaging by Chemical Force Microscopy," *Science* 265, 2071–2074.

Frommer, J. (1992), "Scanning Tunneling Microscopy and Atomic Force Microscopy in Organic Chemistry," *Angew. Chem. Int. Ed. Engl.* 31, 1298–1328.

Fujisawa, S., Ohta, M., Konishi, T., Sugawara, Y., and Morita, S. (1994), "Difference between the Forces Measured by an Optical Lever Deflection and by an Optical Interferometer in an Atomic Force Microscope," *Rev. Sci. Instrum.* 65, 644–647.

Ganti, S. and Bhushan, B. (1995), "Generalized Fractal Analysis and Its Applications to Engineering Surfaces," *Wear* 180, 17–34.

Garcia, N. and Binh, Vu T. (1992), "van der Waals Forces in Atomic Force Microscopy Operating in Liquids: A Spherical-Tip Model," *Phys. Rev. B* 46, 7946–7948.

Gee, M.L., McGuiggan, P.M., Israelachvili, J.N., and Homola, A.M. (1990), "Liquid to Solid-like Transitions of Molecularly Thin Films under Shear," *J. Chem. Phys.* 93, 1895–1906.

Georges, J.M., Millot, S., Loubet, J.L., and Tonck, A. (1993), "Drainage of Thin Liquid Films between Relatively Smooth Surfaces," *J. Chem. Phys.* 98, 7345–7360.

Georges, J.M., Tonck, A., and Mazuyer, D. (1994), "Interfacial Friction of Wetted Monolayers," *Wear* 175, 59–62.

Germann, G.J., Cohen, S.R., Neubauer, G., McClelland, G.M., and Seki, H. (1993), "Atomic-Scale Friction of a Diamond Tip on Diamond (100) and (111) Surfaces," *J. Appl. Phys.* 73, 163–167.

Giaever, I. (1960), "Energy Gap in Superconductors Measured by Electron Tunneling," *Phys. Rev. Lett.* 5, 147–148.

Giessibl, F.J., Gerber, Ch., and Binnig, G. (1991), "A Low-Temperature Atomic Force/Scanning Tunneling Microscope for Ultrahigh Vacuum," *J. Vac. Sci. Technol.* B9, 984–988.

Giles, R., Cleveland, J.P., Manne, S., Hansma, P.K., Drake, B., Maivald, P., Boles, C., Gurley, J., and Elings, V. (1993), "Noncontact Force Microscopy in Liquids," *Appl. Phys. Lett.* 63, 617–618.

Goddenhenrich, T., Lemke, H., Hartmann, U., and Heiden, C. (1990), "Force Microscope with Capacitive Displacement Detection," *J. Vac. Sci. Technol.* A8, 383–387.

Gould, S.A.C., Drake, B., Prater, C.B., Weisenhorn, A.L., Manne, S., Hansma, H.G., Hansma, P.K. et al. (1990), "From Atoms to Integrated Circuit Chips, Blood Cells, and Bacteria with the Atomic Force Microscope," *J. Vac. Sci. Technol.* A8, 369–373.

Granick, S. (1991), "Motions and Relaxations of Confined Liquids," *Science* 253, 1374–1379.

Grutter, P., Ruger, D., Mamin, H.J., Castillo, G., Lin, C.-J., McFadyen, I.R., Valletta, R.M., Wolter, O., Bayer, T., and Greschner, J. (1991), "Magnetic Force Microscopy with Batch-Fabricated Force Sensors," *J. App. Phys.* 69, 5883–5885.

Grutter, P., Mamin, H.J., and Rugar, D. (1992), "Magnetic Force Microscopy (MFM)," in *Scanning Tunneling Microscopy II* (R. Wiesendanger and H.J. Guntherodt, eds.), pp. 151–207, Springer-Verlag, Berlin.

Guntherodt, H.J. and Wiesendanger, R. (eds.) (1992), *Scanning Tunneling Microscopy I: General Principles and Applications to Clean and Adsorbate-Covered Surfaces,* Springer-Verlag, Berlin.

Guntherodt, H.J., Anselmetti, D., and Meyer, E., (Eds.) (1995), *Forces in Scanning Probe Methods,* Vol.E 286, Kluwer Academic, Dordrecht, The Netherlands.

Haberle, W., Horber, J.K.H., Ohnesorge, F., Smith, D.P.E., and Binnig, G. (1992) "*In Situ* Investigations of Single Living Cells Infected by Viruses," *Ultramicroscopy* 42-44, 1161–1167.

Hamada, E. and Kaneko, R. (1992), "Microdistortion of Polymer Surfaces by Friction," *J. Phys. D: Appl. Phys.* 25, A53–A56.

Hamilton, H., Anderson, R., and Goodson, K. (1991), "Contact Perpendicular Recording on Rigid Media," *IEEE Trans. Magn.* 27, 4921–4926.

Hansma, P.K. and Tersoff, J. (1987), "Scanning Tunneling Microscopy," *J. Appl. Phys.* 61, R1–R23.

Hansma, P.K., Drake, B., Marti, O., Gould, S.A.C., and Prater, C.B. (1989), "The Scanning Ion-Conductance Microscope," *Science* 243, 641–643.

Hipp, M., Bielefeldt, H., Colchero, J., Marti, O., and Mlynek, J. (1992), "A Stand Alone Scanning Force and Friction Microscope," *Ultramicroscopy* 42–44, 1498–1503.

Hirano, M. and Shinjo, K. (1990), "Atomistic Locking and Friction," *Phys. Rev. B* 41, 11837–11851.

Hirz, S.J., Homola, A.M., Hadzioannou, G., and Frank, C.W. (1992), "Effect of Substrate on Shearing Properties of Ultrathin Polymer Films," *Langmuir* 8, 328–333.

Hoh, J.H., Cleveland, J.P., Prater, C.B., Revel, J.P., and Hansma, P.K. (1992), "Quantized Adhesion Detected with the Atomic Force Microscope," *J. Am. Chem. Soc.* 114, 4917–4918.

Hoh, J.H. and Hansma, P.K. (1992), "Atomic Force Microscopy for High Resolution Imaging in Cell Biology," *Trends Cell Biol.* 2, 208–213.

Holm, R. (1946), *Electrical Contacts,* Springer-Verlag, New York.

Homola, A.M. (1989), "Measurement of and Relation between the Adhesion and Friction of Two Surfaces Separated by Thin Liquid and Polymer Films," *ASME J. Tribol.* 111, 675–682.

Homola, A.M. (1993), "Interfacial Friction of Molecularly Thin Liquid Films," in *Surface Diagonistics in Tribology* (K. Miyoshi and Y.W. Chung, eds.), pp. 271–298, World Scientific Publishing, River Edge, NJ.

Homola, A.M., Israelachvili, J.N., Gee, M.L., and McGuiggan, P.M. (1989), "Measurement of and Relation between the Adhesion and Friction of Two Surfaces Separated by Thin Liquid and Polymer Films," *ASME J. Tribol.* 111, 675–682.

Homola, A.M., Israelachvili, J.N., McGuiggan, P.M., and Gee, M.L. (1990), "Fundamental Experimental Studies in Tribology: The Transition from Interfacial Friction of Undamaged Molecularly Smooth Surfaces," *Wear* 136, 65–83.

Homola, A.M., Nguyen, H.V., and Hadzioannou, G. (1991), "Influence of Monomer Architecture on the Shear Properties of Molecularly Thin Polymer Melts," *J. Chem. Phys.* 94, 2346–2351.

Horn, R.G. (1990), "Surface Forces and Their Action in Ceramic Materials," *Am. Ceram. Soc.* 73, 1117–1135.

Horn, R.G. and Israelachvili, J.N. (1988), "Molecular Organization and Viscosity of a Thin Film of Molten Polymer between Two Surfaces as Probed by Force Measurements," *Macromolecules* 21, 2836–2841.

Horn, R.G., Smith, D.T., and Haller, W. (1989), "Surface Forces and Viscosity of Water Measured between Silica Sheets," *Chem. Phys. Lett.* 162, 404–408.

Hosoki, S., Hosaka, S., and Hasegawa, T. (1992), "Surface Modification of MoS_2 Using an STM," *Appl. Surf. Sci.* 60/61, 643–647.

Howe, R.T. (1988), "Surface Micromachining for Microsensors and Microactuators," *J. Vac. Sci. Technol.* B6, 1809–1813.

Hu, H.W., Carson, G.A., and Granick, S. (1991), "Relaxation Time of Confined Liquids under Shear," *Phys. Rev. Lett.* 66, 2758–2761.

Hues, S.M., Colton, R.J., Meyer, E., and Guntherodt, H.-J. (1993), "Scanning Probe Microscopy of Thin Film," *MRS Bull.* January, 41–49.

Hug, H.J., Moser, A., Jung, Th., Fritz, O., Wadas, A., Parashikor, I., and Guntherodt, H.J. (1993), "Low Temperature Magnetic Force Microscopy," *Rev. Sci. Instrum.* 64, 2920–2925.

Ibe, J.P., Bey, P.P., Brandon, S.L., Brizzolara, R.A., Burnham, N.A., DiLella, D.P., Lee, K.P., Marrian, C.R.K., and Colton, R.J. (1990), "On the Electrochemical Etching of Tips for Scanning Tunneling Microscopy," *J.Vac. Sci. Technol.* A8, 3570–3575.

Israelachvili, J.N. (1987), "Solvation Forces and Liquid Structure — As Probed by Direct Force Measurements," *Acc. Chem. Res.* 20, 415–421.

Israelachvili, J.N. (1989), "Techniques for Direct Measurements of Forces between Surfaces in Liquids at the Atomic Scale," *Chemtracts Anal. Phys. Chem.* 1, 1–12.

Israelachvili, J.N. (1991), *Intermolecular and Surface Forces*, 2nd ed., Academic Press, London.

Israelachvili, J.N. and Adams, G.E. (1978), "Measurement of Friction between Two Mica Surfaces in Aqueous Electrolyte Solutions in the Range 0-100 nm," *Chem. Soc. J. Faraday Trans. I* 74, 975–1001.

Israelachvili, J.N. and McGuiggan, P.M. (1988), "Forces between Surface in Liquids," *Science* 241, 795–800.

Israelachvili, J.N., McGuiggan, P.M., and Homola, A.M. (1988), "Dynamic Properties of Molecularly Thin Liquid Films," *Science* 240, 189–190.

Israelachvili, J.N. and Tabor, D. (1972), "The Measurement of van der Waals Dispersion Forces in the Range of 1.5 to 130 nm," *Proc. R. Soc. London* A331, 19–38.

Jarvis, S.P., Oral, A., Weihs, T.P., and Pethica, J.B. (1993), "A Novel Force Microscope and Point Contact Probe," *Rev. Sci. Instrum.* 64, 3515–3520.

Jost, P. (1966), Lubrication (tribology), *A Report on the Present Position and Industry's Needs*, Department of Education and Science, H.M. Stationery Office, London.

Jost, P. (1976), "Economic Impact of Tribology, *Proc. Mechanical Failures Prevention Group*, NBS Spec. Pub. 423, Gaithersburg, Maryland.

Kaneko, R. (1988), "A Friction Force Microscope Controlled with an Electromagnet," *J. Micros.* 152(2), 363–369.

Kaneko, R., Nonaka, K., and Yasuda, K. (1988), "Scanning Tunneling Microscopy and Atomic Force Microscopy for Microtribology," *J. Vac. Sci. Technol.* A6, 291–292.

Kaneko, R. and Hamada, E. (1990), "Local Modification of Organic Dye Materials by Dielectric Breakdown," *J. Vac. Sci. Technol.* A8, 577–580.

Kaneko, R. and Oguchi, S. (1990), "Ion-Implanted Diamond Tip for a Scanning Tunneling Microscope," *Jpn. J. Appl. Phys.* 28, 1854–1855.

Kaneko, R., Oguchi, S., Miyamoto, T., Andoh, Y., and Miyake, S. (1990), "Micro-Tribology of Magnetic Recording," in *Tribology and Mechanics of Magnetic Storage Systems* (B. Bhushan, ed.), SP-29, pp. 31-34, STLE, Park Ridge, IL.

Kaneko, R., Miyamoto, T., and Hamada, E. (1991), "Development of a Controlled Friction Force Microscope and Imaging of Recording Disk Surfaces," *Adv. Inf. Storage Syst.* 1, 267–277.

Kaneko, R., Oguchi, S., Hara, S., Matsuda, R., Okada, T., Ogawa, H., and Nakamura, Y. (1992), "Atomic Force Microscope Coupled with an Optical Microscope," *Ultramicroscopy* 42-44, 1542–1548.

Kirk, M.D., Albrecht, T., and Quate, C.F. (1988), "Low-Temperature Atomic Force Microscopy," *Rev. Sci. Instrum.* 59, 833–835.

Klein, J. (1980), "Forces between Mica Surfaces Bearing Layers of Adsorbed Polystyrene in Cyclohexane," *Nature* 288, 248–250.

Klein, J., Perahia, D., and Warburg, S. (1991), "Forces between Polymer-Bearing Surfaces Undergoing Shear," *Nature* 352, 143–145.

Kobayashi, A., Grey, F., Williams, R.S., and Ano, M. (1993), "Formation of Nanometer-Scale Grooves in Silicon with a Scanning Tunneling Microscope," *Science* 259, 1724–1726.

Koinkar, V.N. and Bhushan, B. (1996a), "Microtribological Studies of Al_2O_3, Al_2O_3–TiC, Polycrystalline and Single-Crystal Mn–Zn Ferrite and SiC Head Slider Materials, *Wear* 202, 110–122.

Koinkar, V.N. and Bhushan, B. (1996b), "Microtribological Studies of Unlubricated and Lubricated Surfaces Using Atomic Force/Friction Force Microscopy," *J. Vac. Sci. Technol.* A 14, 2378–2391.

Koinkar, V.N. and Bhushan, B. (1996c), "Micro/Nanoscale Studies of Boundary Layers of Liquid Lubricants for Magnetic Disks," *J. Appl. Phys.* 79, 8071–8075.

Koinkar, V.N. and Bhushan, B. (1997a), "Effect of Scan Size and Surface Roughness on Microscale Friction Measurements," *J. Appl. Phys.* 81, 2472–2479.

Koinkar, V.N. and Bhushan, B. (1997b), "Scanning and Transmission Electron Microscopies of Single-Crystal Silicon Microworn/Machined Using Atomic Force Microscopy," *J. Mater. Res.* 12, 3219–3224.

Koinkar, V.N. and Bhushan, B. (1997c), "Microtribological Properties of Hard Amorphous Carbon Protective Coatings for Thin-Film Magnetic Disks and Heads," *Proc. Inst. Mech. Eng. Part J: J. Eng. Tribol.* 211, 365-372.

Kulkarni, A.V. and Bhushan, B. (1996a), "Nanoscale Mechanical Property Measurements Using Modified Atomic Force Microscopy," *Thin Solid Films* 290–291, 206–210.

Kulkarni, A.V. and Bhushan, B. (1996b), "Nano/Picoindentation Measurement on Single-Crystal Aluminum Using Modified Atomic Force Microscopy," *Mater. Lett.* 29, 221–227.

Kulkarni, A.V. and Bhushan, B. (1997), "Nanoindentation Measurements of Amorphous Carbon Coatings," *J. Mater. Res.* 12, 2707–2714.

Layard, A.G. (1853), *Discoveries in the Ruins of Nineveh and Babylon,* I and II, John Murray, Albemarle Street, London.

Lazzari, J.P. and Deroux-Dauphin, P. (1989), "A New Thin Film Head Generation IC Head," *IEEE Trans. Magn.* 25, 3190–3193.

Leung, O.M. and Goh, M.C. (1992), "Orientational Ordering of Polymers by Atomic Force Microscope Tip-Surface Interactions," *Science* 225, 64–66.

Loubet, J.L., Bauer, M., Tonck, A., Bec, S., and Gauthier-Manuel, B. (1993), "Nanoindentation with a Surface Force Apparatus," in *Mechanical Properties and Deformation Behavior of Materials Having Ultra-Fine Microstructures* (M. Nastasi et al., eds.), pp. 429-447, Kluwer Academic Publishers, Dordrecht.

Luengo, G., Schmitt, F.J., Hill, R., and Israelachvili, J.N. (1997), "Thin Film Bulk Rheology and Tribology of Confined Polymer Melts: Contrasts with Bulk Properties," *Macromolecules* 30, 2482–2494.

Lyo, I.W. and Avouris, Ph. (1991), "Field-Induced Nanometer-to-Atomic-Scale Manipulation of Silicon Surfaces with the STM," *Science* 253, 173–176.

Maivald, P., Butt, H.J., Gould, S.A.C., Prater, C.B., Drake, B., Gurley, J.A., Elings, V.B., and Hansma, P.K. (1991), "Using Force Modulation to Image Surface Elasticities with the Atomic Force Microscope," *Nanotechnology* 2, 103–106.

Majumdar, A., Oden, P.I., Carrejo, J.P., Nagahara, L.A., Graham, J.J., and Alexander, J. (1992), "Nanometer-Scale Lithography Using the Atomic Force Microscope," *Appl. Phys. Lett.* 61, 2293–2295.

Majumdar, A., Carrejo, J.P., and Lai, J. (1993), "Thermal Imaging Using the Atomic Force Microscope," *Appl. Phys. Lett.* 62, 2501–2503.

Mamin, H.J., Guethner, P.H., and Rugar, D. (1990), "Atomic Emission from a Gold Scanning-Tunneling-Microscope Tip," *Phys. Rev. Lett.* 65, 2418–2421.

Manne, S., Hansma, P.K., Massie, J., Elings, V.B., and Gewirth, A.A. (1991), "Atomic-Resolution Electrochemistry with the Atomic Force Microscope: Copper Deposition on Gold," *Science* 251, 183–186.

Marchon, B., Salmeron, M., and Siekhaus, W. (1989), "Observation of Graphite and Amorphous Structures on the Surface of Hard Carbon Films by Scanning Tunneling Microscopy," *Phys. Rev. B* 39, 12907–12910.

Marti, O., Drake, B., and Hansma, P.K. (1987), "Atomic Force Microscopy of Liquid-Covered Surfaces: Atomic Resolution Images," *Appl. Phys. Lett.* 51, 484–486.

Marti, O., Drake, B., Gould, S., and Hansma, P.K. (1988), "Atomic Resolution Atomic Force Microscopy of Graphite and the Native Oxide on Silicon," *J. Vac. Sci. Technol.* A6, 287–290.

Marti, O., Colchero, J., and Mlynek, J. (1990), "Combined Scanning Force and Friction Microscopy of Mica," *Nanotechnology* 1, 141.

Marti, O. and Amrein, M. (eds.) (1993), *STM and SFM in Biology,* Academic Press, San Diego.

Marti, O. and Colchero, J. (1995), "Scanning Probe Microscopy Instrumentation," in *Forces in Scanning Probe Methods* (H.J. Guntherodt et al., eds.), Vol. E286, pp. 15-34, Kluwer Academic, Dordrecht, The Netherlands.

Martin, Y., Williams, C.C., and Wickramasinghe, H.K. (1987a), "Atomic Force Microscope-Force Mapping and Profiling on a Sub 100-A Scale," *J. Appl. Phys.* 61, 4723–4729.

Martin, Y. and Wickramasinghe, H.K. (1987b), "Magnetic Imaging by Force Microscopy with 1000 A Resolution," *Appl. Phys. Lett.* 50, 1455–1457.

Martin, Y., Abraham, D.W., and Wickramasinghe, H.K. (1988), "High-Resolution Capacitance Measurement and Potentiometry by Force Microscopy," *Appl. Phys. Lett.* 52, 1103–1105.

Mate, C.M. (1992), "Atomic-Force-Microscope Study of Polymer Lubricants on Silicon Surfaces," *Phys. Rev. Lett.* 68, 3323–3326.

Mate, C.M., McClelland, G.M., Erlandsson, R., and Chiang, S. (1987), "Atomic-Scale Friction of a Tungsten Tip on a Graphite Surface," *Phys. Rev. Lett.* 59, 1942–1945.

Mate, C.M., Lorenz, M.R., and Novotny, V.J. (1989), "Atomic Force Microscopy of Polymeric Liquid Films," *J. Chem. Phys.* 90, 7550–7555.

Mate, C.M., Lorenz, M.R., and Novotny, V.J. (1990), "Determination of Lubricant Film Thickness on a Particulate Disk Surface by Atomic Force Microscopy," *IEEE Trans. Magn.* 26, 1225–1228.

Mate, C.M. and Novotny, V.J. (1991), "Molecular Conformation and Disjoining Pressures of Polymeric Liquid Films," *J. Chem. Phys.* 94, 8420–8427.

McClelland, G.M., Erlandsson, R., and Chiang, S. (1987), "Atomic Force Microscopy: General Principles and a New Implementation, in *Review of Progress in Quantitative Nondestructive Evaluation* (D.O. Thompson and D.E. Chimenti, eds.), Vol. 6B, pp. 1307-1314, Plenum, New York.

Melmed, A.J. (1991), "The Art and Science and Other Aspects of Making Sharp Tips," *J. Vac. Sci. Technol. B.* 9, 601–608.

Meyer, E. (1992), "Atomic Force Microscopy," *Surf. Sci.*, 41, 3–49.

Meyer, G. and Amer, N.M. (1988), "Novel Optical Approach to Atomic Force Microscopy," *Appl. Phys. Lett.* 53, 1045–1047.

Meyer, G. and Amer, N.M. (1990a), "Optical-Beam-Deflection Atomic Force Microscopy: The NaCl (001) Surface," *Appl. Phys. Lett.* 56, 2100–2101.

Meyer, G. and Amer, N.M. (1990b), "Simultaneous Measurement of Lateral and Normal Forces with an Optical-Beam-Deflection Atomic Force Microscope," *Appl. Phys. Lett.* 57, 2089–2091.

Meyer, E., Overney, R., Luthi, R., Brodbeck, D., Howald, L., Frommer, J., Guntherodt, H.-J., Wolter, O., Fujihira, M., Takano, T., and Gotoh, Y. (1992), "Friction Force Microscopy of Mixed Langmuir-Blodgett Films," *Thin Solid Films* 220, 132–137.

Miu, D.K., Wu, S., Temesvary, V., and Tai, Y.C. (1993), "Silicon Microgimbals for Super-Compact Magnetic Recording Rigid Disk Drives," *Adv. Info. Storage Syst.* 5, 139–152.

Miyamoto, T., Kaneko, R., and Andoh, Y. (1990), "Interaction Force between Thin Film Disk Media and Elastic Solids Investigated by Atomic Force Microscope," *ASME J. Tribol.* 112, 567–572.

Miyamoto, T., Kaneko, R., and Miyake, S. (1991), "Tribological Characteristics of Amorphous Carbon Films Investigated by Point Contact Microscopy," *J. Vac. Sci. Technol.* B9 1336–1339.

Miyamoto, T., Miyake, S., and Kaneko, R. (1993), "Wear Resistance of C^+-Implanted Silicon Investigated by Scanning Probe Microscopy," *Wear* 162–164, 733–738.

Neubauer, G., Cohen, S.R., McClelland, G.M., Horne, D., and Mate, C.M. (1990), "Force Microscopy with a Bidirectional Capacitance Sensor," *Rev. Sci. Instrum.* 61, 2296–2308.

Oden, P.I., Majumdar, A., Bhushan, B., Padmanabhan, A., and Graham, J.J. (1992), "AFM Imaging, Roughness Analysis and Contact Mechanics of Magnetic Tape and Head Surfaces," *ASME J. Tribol.* 114, 666–674.

Ohkubo, T., Kishigami, J., Yanagisawa, K., and Kaneko, R. (1991), "Submicron Magnetizing and Its Detection Based on the Point Magnetic Recording Concept," *IEEE Trans. Magn.* 27, 5286–5288.

Ohnesorge, F. and Binnig, G. (1993), "True Atomic Resolution by Atomic Force Microscopy through Repulsive and Attractive Forces," *Science* 260, 1451–1456.

Ohwe, T., Mizoshita, Y., and Yoneoka, S. (1993), "Development of Integrated Suspension System for a Nanoslider with an MR Head Transducer," *IEEE Trans. Magn.* 29, 3924–3926.

O'Shea, S.J., Welland, M.E., and Rayment, T. (1992), "Atomic Force Microscope Study of Boundary Layer Lubrication," *Appl. Phys. Lett.* 61, 2240-2242.

Overney, R.M., Meyer, E., Frommer, J., Brodbeck, D., Luthi, R., Howard, L., Guntherodt, H.-J., Fujihira, M., Takano, H., and Gotoh, Y. (1992), "Friction Measurements on Phase-Separated Thin Films with a Modified Atomic Force Microscope," *Nature* 359, 133-135.

Parish, W.F. (1935), "Three Thousand Years of Progress in the Development of Machinery and Lubricants for the Hand Crafts," *Mill Factory* Vols. 16 and 17.

Park, S. and Quate, C.F. (1986), "Tunneling Microscopy of Graphite in Air," *Appl. Phys. Lett.* 48, 112-114.

Park, S.I. and Quate, C.F. (1987), "Digital Filtering of STM Images," *J. Appl. Phys.* 62, p. 312.

Parkinson, B. (1990), "Layer-by-Layer Nanometer Scale Etching of Two-Dimensional Substrates Using the Scanning Tunneling Microscopy," *J. Am. Chem. Soc.* 112, 7498-7502.

Pashley, R.M. (1981), "Hydration Forces between Solid Surfaces in Aqueous Electrolyte Solutions," *J. Colloid Interface Sci.* 80, 153-162.

Patel, S.S. and Tirrell, M. (1989), "Measurement of Forces between Surfaces in Polymer Fluids," *Annu. Rev. Phys. Chem.* 40, 597-635.

Peachey, J., Van Alsten, J., and Granick, S. (1991), "Design of an Apparatus to Measure the Shear Response of Ultrathin Liquid Films," *Rev. Sci. Instrum.* 62, 463-473.

Petroff, N.P. (1883), "Friction in Machines and the Effects of the Lubricant," *Eng. J.* (in Russian), St. Petersburg, pp. 71-140, 228-279, 377-436, 535-564.

Pohl, D.W. (1986), "Some Design Criteria in STM," *IBM J. Res. Dev.* 30, 417.

Poon, C.Y. and Bhushan, B. (1995), "Comparison of Surface Roughness Measurements by Stylus Profiler, AFM and Non-contact Optical Profiler," *Wear* 190, 76–88.

Prater, C.B., Wilson, M.R., Garnaes, J., Massie, J., Elings, V.B., and Hansma, P.K. (1991), "Atomic Force Microscopy of Biological Samples at Low Temperature," *J. Vac. Sci. Technol.* B9, 989–991.

Radmacher, M., Tillman, R.W., Fritz, M., and Gaub, H.E. (1992), "From Molecules to Cells: Imaging Soft Samples with the Atomic Force Microscope," *Science* 257, 1900–1905.

Reynolds, O.O. (1886), "On the Theory of Lubrication and Its Application to Mr. Beauchamp Tower's Experiments," *Philos. Trans. R. Soc. (London)*, 177, 157–234.

Ruan, J. and Bhushan, B. (1993), "Nanoindentation Studies of Fullerene Films Using Atomic Force Microscopy," *J. Mater. Res.* 8, 3019-3022.

Ruan, J. and Bhushan, B. (1994a), "Atomic-Scale Friction Measurements Using Friction Force Microscopy: Part I — General Principles and New Measurement Techniques," *ASME J. Tribol.* 116, 378–388.

Ruan, J. and Bhushan, B. (1994b), "Atomic-Scale and Microscale Friction of Graphite and Diamond Using Friction Force Microscopy," *J. Appl. Phys.* 76, 5022–5035.

Ruan, J. and Bhushan, B. (1994c), "Frictional Behavior of Highly Oriented Pyrolytic Graphite," *J. Appl. Phys.* 76, 8117–8120.

Rugar, D., Mamin, H.J., and Guethner, P. (1989), "Improved Fiber-Optic Interferometer for Atomic Force Microscopy," *Appl. Phys. Lett.* 55, 2588–2590.

Rugar, D. and Hansma, P.K. (1990), "Atomic Force Microscopy," *Phys. Today* 43, 23–30.

Rugar, D., Mamin, H.J., Guethner, P., Lambert, S.E., Stern, J.E., McFadyen, I., and Yogi, T. (1990), "Magnetic Force Microscopy: General Principles and Application to Longitudinal Recording Media," *J. Appl. Phys.* 63, 1169–1183.

Rugar, D., Yannoni, C.S., and Sidles, J.A. (1992), "Mechanical Detection of Magnetic Resonance," *Nature* 360, 563–566.

Salmeron, M., Folch, A., Neubauer, G., Tomitori, M., Ogletree, D.F., and Kolbe, W. (1992), "Nanometer Scale Mechanical Properties of Au(111) Thin Films," *Langmuir* 8, 2832–2842.

Salmeron, M., Neubauer, G., Folch, A., Tomitori, M., Ogletree, D.F., and Sautet, P. (1993), "Viscoelastic and Electrical Properties of Self-Assembled Monolayers on Au(111) Films," *Langmuir* 9, 3600–3611.

Sarid, D. (1991), *Scanning Force Microscopy*, Oxford University Press, New York.

Sarid, D. and Elings, V. (1991), "Review of Scanning Force Microscopy," *J. Vac. Sci. Technol.* B 9, 431–437.

Sarid D., Iams, D., Weissenberger, V., and Bell, L.S. (1988), "Compact Scanning-Force Microscope Using Laser Diode," *Optics Lett.* 13, 1057–1059.

Schonenberger, C. and Alvarado, S.F. (1990), "Understanding Magnetic Force Microscopy, *Z. Phys. B.* (Germany) 80, 373–383.

Sidles, J.A. and Rugar, D. (1993), "Signal-to-Noise Ratios in Inductive and Mechanical Detection of Magnetic Resonance," *Phys. Rev. Lett.* 70, 3506–3509.

Silver, R.M., Ehrichs, E.E., and de Lozanne, A.L. (1987), "Direct Writing of Submicron Metallic Features with a Scanning Tunnelling Microscope," *Appl. Phys. Lett.* 51, 247–249.

Smith, C.P., Maeda, M., Atanasoska, L., and White, H.S. (1988), "Ultrathin Platinum Films on Mica and Measurement of Forces at the Platinum/Water Interface," *J. Phys. Chem.* 95, 199–205.

Smith D.P.E. (1994), "AFM at Low Temperature Using Piezoresistive Sensors," presented at a NATO ASI on *Forces in Scanning Probe Methods* Schluchsee, Germany, March 7-18, 1994.

Smith, D., Horber, H., Gerber, C., and Binnig, G. (1989), "Smectic Liquid Crystal Monolayers on Graphite Observed by Scanning Tunneling Microscopy," *Science* 245, 43–45.

Smith, D., Horber, J., Binnig, G., and Nejoh, H. (1990), "Structure, Registry and Imaging Mechanism of Alkylcyanobiphenyl Molecules by Tunneling Microscopy," *Nature* 344, 641–644.

Specht, M., Ohnesorge, F., and Heckl, W. (1991), *Surf. Sci. Lett.* 65, p. 2418.

Stopka, M., Hadjiiski, L., Oesterschulze, E., and Kassing, R. (1995), "Surface Investigations by Scanning Thermal Microscopy," *J. Vac. Sci. Technol. B* 13, 2153–2156.

Stroscio, J.A. and Kaiser, W.J. (eds.) (1993), *Scanning Tunneling Microscopy*, Academic Press, Boston.

Sugawara, Y., Ohta, M., Konishi, T., Morita, S., Suzuki, M., and Enomoto, Y. (1993), "Effects of Humidity and Tip Radius on the Adhesive Force Measured with Atomic Force Microsopy," *Wear* 168, 13–16.

Sundararajan, S. and Bhushan, B. (1998), "Micro/Nanotribological Studies of Polysilicon and SiC Films for MEMS Applications," *Wear* (in press).

Tabor, D. and Winterton, R.H.S. (1969), "The Direct Measurement of Normal and Retarded van der Waals Forces," *Proc. R. Soc. London* A312, 435–450.

Tai, Y.C., and Mueller, R.S. (1989), "IC-Processed Electrostatic Synchronous Micromotors," *Sensors and Actuators* 20, 49–55.

Tang, W.C., Lim, T.G., and Howe, R.T. (1990), "Electrostatically Balanced Comb Drive for Controlled Levitation," *Technical Digests of IEEE Solid-State Sensors and Actuators Workshop*, held at Hilton Head Island, SC, June.

Tonck, A., Georges, J.M., and Loubet, J.L. (1988), "Measurements of Intermolecular Forces and the Rheology of Dodecane between Alumina Surfaces," *J. Colloid Interface Sci.* 126, 1540–1563.

Tower, B. (1884), "Report on Friction Experiments," *Proc. Inst. Mech. Eng.* p. 632.

Tsau, L., Wang, D., and Wang, K.L. (1994), "Nanometer Scale Patterning of Silicon (100) Surface by an Atomic Force Microscope Operating in Air," *Appl. Phys. Lett.* 64, 2133–2135.

Utsugi, Y. (1990), "Nanometer-Scale Chemical Modification Using a Scanning Tunneling Microscope," *Nature* 347, 747–749.

Van Alsten, J. and Granick, S. (1988), "Molecular Tribology of Ultrathin Liquid Films," *Phys. Rev. Lett.* 61, 2570–2573.

Van Alsten, J. and Granick, S. (1990a), "Tribology Studied Using Atomically Smooth Surfaces," *Tribol. Trans.* 33, 436–446.

Van Alsten, J. and Granick, S., (1990b), "Shear Rheology in a Confined Geometry — Polysiloxane Melts," *Macromolecules* 23, 4856–4862.

Weisenhorn, A.L., MacDougall, J.E., Gould, S.A.C., Cox, S.D., Wise, W.S., Massie, J., Maivald, P., Elings, V.B., Stucky, G.D., and Hansma, P.K. (1990), "Imaging and Manipulating Molecules on a Zeolite Surface with an Atomic Force Microscope," *Science* 247, 1330–1333.

Weisenhorn, A.L., Egger, M., Ohnesorge, F., Gould, S.A.C., Heyn, S.P., Hansma, H.G., Sinsheimer, R.L., Gaub, H.E., and Hansma, P.K. (1991), "Molecular Resolution Images of Langmuir-Blodgett Films and DNA by Atomic Force Microscopy," *Langmuir* 7, 8–12.

Weisenhorn, A.L., Maivald, P., Butt, H.J., and Hansma, P.K. (1992), "Measuring Adhesion, Attraction, and Repulsion between Surfaces in Liquid with an Atomic-Force Microscope," *Phys. Rev. B*, 45, 11226–11232.

Wiesendanger, R. and Guntherodt, H.J. (eds.) (1992), *Scanning Tunneling Microscopy, II: Further Applications and Related Scanning Techniques,* Springer-Verlag, Berlin.

Yee, S., Kanazawa, K.K., and Sonnenfeld, R.G. (1993), "Simultaneous Distance and Potential Servo for the Scanning Probe Microscope," *Proc. 183rd Meeting of The Electrochemical Society.*

Zuger, O. and Rugar, D. (1993), "First Images from a Magnetic Resonance Force Microscope," *Appl. Phys. Lett.* 63, 2496–2498.

2

AFM Instrumentation and Tips

Othmar Marti

Introduction

The performance of AFMs and the quality of AFM images greatly depend on the instruments available and the sensors (tips) in use. To utilize a microscope to its fullest, it is necessary to know how it works and where its strong points and its weaknesses are. This chapter describes the instrumentation of force detection, of cantilevers, and of the instruments themselves.

2.1 Force Detection

Atomic force microscopy (AFM)(Binnig et al., 1986) was an early offspring of scanning tunneling microscopy (STM). The force between a tip and the sample was used to image the surface topography. The force between the tip and the sample, also called the tracking force, was lowered by several orders of magnitude compared with the profilometer (Jones, 1970). The contact area between the tip and the sample was reduced considerably. The force resolution was similar to that achieved in the surface force apparatus (Israelachvili, 1985). Soon thereafter atomic resolution in air was demonstrated (Binnig et al., 1987), followed by atomic resolution of liquid covered surfaces (Marti et al., 1987) and low-temperature (4.2 K) operation (Kirk et al., 1988). The AFM measures either the contours of constant force, force gradients or the variation of forces or force gradients, with position, when the height of the sample is not adjusted by a feedback loop. These measurement modes are similar to the ones of the STM, where contours of constant tunneling current or the variation of the tunneling current with position at fixed sample height are recorded.

The invention of the AFM demonstrated that forces could play an important role in other scanned probe techniques. It was discovered that forces might play an important role in STM. (Anders and Heiden, 1988; Blackman et al., 1990). The type of force interaction between the tip and the sample surface can be used to characterize AFMs. The highest resolution is achieved when the tip is at about zero external force, i.e., in light contact or near contact resonant operation. The forces in these modes basically stem from the Pauli exclusion principle that prevents the spatial overlap of electrons. As in the STM, the force applied to the sample can be constant, the so-called constant-force mode. If the sample z-position is not adjusted to the varying force, we speak of the constant z-mode. However, for weak cantilevers (0.01 N/m spring constant) and a static applied load of 10^{-8} N we get a static deflection of 10^{-6} m, which means that even structures of several nanometers height will be subject to an almost constant force, whether it is controlled or not. Hence, for the contact mode with soft cantilevers the distinction between constant-force mode and constant z-mode is rather arbitrary. Additional information on the sample surface can be gained by measuring lateral forces (friction mode) or modulating the force to get dF/dz, which is nothing else than the stiffness of the surfaces. When using attractive forces, one normally measures also dF/dz with a modulation technique. In the attractive mode the lateral resolution is at least one order of magnitude worse than for the contact mode. The attractive mode is also referred to as the noncontact mode.

We will first try to estimate the forces between atoms to get a feeling for the tolerable range of interaction forces and, derived from them, the compliance of the cantilever.

For a real AFM tips the assumption of a single interacting atom is not justified. Attractive forces like van der Waals forces reach out for several nanometers. The attractive forces are compensated by the repulsion of the electrons when one atom tries to penetrate another. The decay length of the interaction and its magnitude depend critically on the type of atoms and the crystal lattice they are bound in. The shorter the decay length, the smaller is the number of atoms which contribute a sizable amount to the total force. The decay length of the potential, on the other hand, is directly related to the type of force. Repulsive forces between atoms at small distances are governed by an exponential law (like the tunneling current in the STM), by an inverse power law with large exponents, or by even more complicated forms. Hence, the highest resolution images are obtained using the repulsive forces between atoms in contact or near contact. The high inverse power exponent or even exponential decay of this distance dependence guarantees that the other atoms beside the apex atom do not significantly interact with the sample surface. Attractive van der Waals interactions on the other hand, are reaching far out into space. Hence, a larger number of tip atoms take part in this interaction so that the resolution cannot be as good. The same is true for magnetic potentials and for the electrostatic interaction between charged bodies.

A crude estimation of the forces between atoms can be obtained in the following way: assume that two atoms with mass m are bound in molecule. The potential at the equilibrium distance can be approximated by a harmonic potential or, equivalently, by a spring constant. The frequency of the vibration f of the atom around its equilibrium point is then a measure for the spring constant k:

$$k = \omega^2 \frac{m}{2} \qquad (2.1)$$

where we have to use the reduced atomic mass. The vibration frequency can be obtained from optical vibration spectra or from the vibration quanta $\hbar\omega$

$$k = \left(\frac{\hbar\omega}{\hbar}\right)^2 \frac{m}{2} \qquad (2.2)$$

As a model system, we take the hydrogen molecule H_2. The mass of the hydrogen atom is $m = 1.673 \times 10^{-27}$ kg and its vibration quantum is $\hbar\omega = 8.75 \times 10^{-20}$ J. Hence, the equivalent spring constant is $k = 560$ N/m. Typical forces for small deflections (1% of the bond length) from the equilibrium position are $\propto 5 \times 10^{-10}$ N. The force calculated this way is an order of magnitude estimation of the forces between two atoms. An atom in a crystal lattice on the surface is more rigidly attached since it is bound to more than one other atom. Hence, the effective spring constant for small deflections is larger. The limiting force is reached when the bond length changes by 10% or more, which indicates that the forces used to image surfaces must be of the order of 10^{-8} N or less. The sustainable force before damage is dependent on the type of surfaces. Layered materials like mica or graphite are more resistant to damage than soft materials like biological samples. Experiments have shown that on selected inorganic surfaces such as mica one can apply up to 10^{-7} N. On the other hand, forces of the order of 10 to 9 N destroy some biological samples.

2.2 The Mechanics of Cantilevers

2.2.1. Compliance and Resonances of Lumped Mass Systems

Any one of the building blocks of an AFM, be it the body of the microscope itself or the force measuring cantilevers, is a mechanical resonator. These resonances can be excited either by the surroundings or by the rapid movement of the tip or the sample. To avoid problems due to building or air-induced oscillations, it is of paramount importance to optimize the design of the scanning probe microscopes for high resonance frequencies; which usually means decreasing the size of the microscope (Pohl, 1986). By using cubelike or spherelike structures for the microscope, one can considerably increase the lowest eigenfrequency. The eigenfrequency of any spring is given by

$$f = \frac{1}{2\pi}\sqrt{\frac{k}{m_{\text{eff}}}} \qquad (2.3)$$

where k is the spring constant and m_{eff} is the effective mass. The spring constant k of a cantilevered beam with uniform cross section is given by (Thomson, 1988)

$$k = \frac{3EI}{\ell^3}, \qquad (2.4)$$

where E is the Young's modulus of the material, ℓ the length of the beam, and I the moment of inertia. For a rectangular cross section with a width b (perpendicular to the deflection) and a height h, one obtains for I

$$I = \frac{bh^3}{12} \qquad (2.5)$$

83

Combining Equations 2.3 through 2.5, and we get the final result for f:

$$f = \frac{1}{2\pi}\sqrt{\frac{3EI}{\ell^3 m_{eff}}} = \frac{1}{2\pi}\sqrt{\frac{Ebh^3}{4\ell^3 m_{eff}}} \tag{2.6}$$

The effective mass can be calculated using Rayleigh's method. The general formula using Rayleigh's method for the kinetic energy T of a bar is

$$T = \frac{1}{2}\int_0^\ell \frac{m}{\ell}\left(\frac{\partial z(x)}{\partial t}\right)^2 dx \tag{2.7}$$

For the case of a uniform beam with a constant cross section and length L, one obtains for the deflection $z(x) = z_{max}(1 - (3x)/(2\ell) + (x^3)/(2\ell^3))$. Inserting z_{max} into Equation 2.7 and solving the integral gives

$$T = \int_0^\ell \frac{m}{\ell}\left[\frac{\partial z_{max}(x)}{\partial t}\left(1 - \frac{3x}{2\ell}\right) + \left(\frac{x^3}{\ell^3}\right)\right] dx$$

$$= \frac{1}{2}m_{eff}\left(z_{max}t\right)^2 \tag{2.8}$$

and

$$m_{eff} = \frac{9}{20}m$$

for the effective mass.

Combining Equations 2.4 and 2.8 and noting that $m = \rho\ell bh$, where ρ is the density of mass, one obtains for the eigenfrequency

$$f = \left(\frac{1}{2\pi}\frac{\sqrt{5}}{3}\sqrt{\frac{E}{\rho}}\right)\frac{h}{\ell^2} \tag{2.9}$$

Further reading on how to derive this equation can be found in the literature (Thomson, 1988). It is evident from Equation 2.9, that one way to increase the eigenfrequency is to choose a material with as high a ratio E/ρ. Another way to increase the lowest eigenfrequency is also evident in Equation 2.9. By optimizing the ratio h/ℓ^2 one can increase the resonance frequency. However, it does not help to make the length of the structure smaller than the width or height. Their roles will just be interchanged. Hence, the optimum structure is a cube. This leads to the design rule, that long, thin structures like sheet metal should be avoided. For a given resonance frequency the quality factor should be as low as possible. This means that an inelastic medium such as rubber should be in contact with the structure to convert kinetic energy into heat.

2.2.2 Cantilevers

Cantilevers are mechanical devices specially shaped to measure tiny forces. The analysis given in the previous chapter is applicable. However, to understand better the intricacies of force detection systems we will discuss the example of a straight cantilevered beam (Figure 2.1).

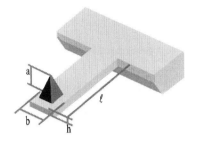

FIGURE 2.1 A typical force microscope cantilever with a length ℓ, a width b, and a height h. The height of the tip is a. The material is characterized by Young's modulus E, the shear modulus $G = E/(2(1 + \sigma))$, where σ is the Poisson number, and a density ρ.

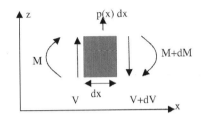

FIGURE 2.2 Moments and forces acting on an element of the beam.

The bending of beams with a cross section $A(x)$ is governed by the Euler equation (Thomson, 1988):

$$\frac{d^2}{dx^2}\left(EI(x)\frac{d^2}{dx^2}z\right) = p(x) \tag{2.10}$$

where E is Young's modulus, $I(x)$ the flexure moment of inertia defined by

$$I(x) = \int_{A(x)} z^2 dydz \tag{2.11}$$

Equations 2.10 and 2.11 can be derived by evaluating torsion moments about an element of infinitesimal length at position x.

Figure 2.2 shows the forces and moments acting on an element of the beam. V is the shear moment, M the bending moment, and $p(x)$ the position-dependent load per unit length. Summing forces in the z-direction, one obtains

$$dV - p(x)dx = 0 \tag{2.12}$$

Summing moments on the right face of the element gives

$$dM - Vdx - \tfrac{1}{2}p(x)(dx)^2 = 0 \tag{2.13}$$

Finally, one obtains for the shear and bending moments

$$\frac{dV}{dx} = p(x)$$

$$\frac{dM}{dx} = V \tag{2.14}$$

Combining both parts of Equation 2.14, one obtains the following result

$$\frac{d^2 M}{dx^2} = \frac{dV}{dx} = p(x) \tag{2.15}$$

Using the flexure equation to express the bending moment, one obtains

$$M = EI \frac{d^2 z}{dx^2} \tag{2.16}$$

Combining Equations 2.15 and 2.16, and one obtains the Euler Equation 2.10. Beams with a nonuniform cross section are difficult to calculate. Let us, therefore, concentrate on straight beams. These cantilever beams are widely used for friction mode as well as for noncontact experiments.

A force acting on the cantilever at a position x_0 can be handled by the Dirac function $\delta(x - x_0)$, for which one has

$$\int_{-\infty}^{\infty} f(x)\delta(x - x_0)dx = f(x_0) \tag{2.17}$$

Hence, one sets

$$p(x) = F\delta(\ell) \tag{2.18}$$

where ℓ is the length of the cantilever. Integrating M twice from the beginning to the end of the cantilever, one obtains

$$M(x) = (\ell - x)F \tag{2.19}$$

since the moment must vanish at the end point of the cantilever. Integrating twice more and observing that EI is a constant for beams with an uniform cross section, one gets

$$\frac{d^2 z}{dx^2} = \frac{M(x)}{EI}$$
$$\Rightarrow z(x) = \frac{\ell^3}{6EI} \left(\frac{x}{\ell}\right)^2 \left(\frac{x}{\ell} - 3\right)F \tag{2.20}$$

The slope of the beam is

$$z'(x) = \frac{dz}{dx} = \frac{\ell^2}{2EI} \frac{x}{\ell} \left(\frac{x}{\ell} - 2\right)F = \frac{\ell x}{2EI} \left(\frac{x}{\ell} - 2\right)F \tag{2.21}$$

Evaluating this and Equation 2.20 at the end of the cantilever, i.e., for $x = \ell$, one gets

$$z(\ell) = -\frac{\ell^3}{3EI}F$$
$$z'(\ell) = -\frac{\ell^2}{2EI}F = \frac{3}{2} \frac{z(\ell)}{\ell} \tag{2.22}$$

$z'(\ell)$ is also the tangent of the deflection angle. Using the definition of the moment of inertia for a beam with a rectangular cross section,

$$I = \tfrac{1}{12} bh^3 \qquad (2.23)$$

where b is the width and h the thickness of the lever, one gets for the deformation z at the end of the cantilever is related to the applied normal force F by

$$z = \frac{4}{Eb}\left(\frac{\ell}{h}\right)^3 F \qquad (2.24)$$

Hence, the compliance k_N is

$$k_N = \frac{F}{z} = \frac{Eb}{4}\left(\frac{h}{\ell}\right)^3 \qquad (2.25)$$

and a change in angular orientation of the end of

$$\Delta\alpha = \frac{6}{Ebh}\left(\frac{\ell}{h}\right)^2 F_N = \frac{3}{2}\frac{\Delta z}{\ell} \qquad (2.26)$$

We can ask ourselves what will, to first order, happen if we apply a lateral force F_L to the end of the cantilever. The cantilever will bend sideways and it will twist. The sideways bending can be calculated with Equation 2.24 by exchanging b and h

$$k_{L,b} = \frac{F_L}{\Delta z} = \frac{Eh}{4}\left(\frac{b}{\ell}\right)^3 \qquad (2.27)$$

Therefore, the compliance for bending in lateral direction is larger than the compliance for bending in the normal direction by $(b/h)^2$. The twisting or torsion on the other side is more complicated to handle. For wide, thin cantilevers ($b \gg h$), we obtain

$$k_{L,tor} = \frac{Gbh^3}{3\ell a^2} \qquad (2.28)$$

The ratio of the torsion compliance to the bending compliance is (Colchero, 1993)

$$\frac{k_{L,tor}}{k_{L,b}} = \frac{1}{2}\left(\frac{ab}{h\ell}\right)^2 \qquad (2.29)$$

where we assumed a Poisson ratio $s = 0.333$. We see that thin, wide cantilevers with long tips favor torsion while cantilevers with square cross sections and short tips favor bending. Finally, we calculate the ratio between the torsion compliance and the normal mode-bending compliance.

$$\frac{k_{L,tor}}{k_N} = 2\left(\frac{\ell}{a}\right)^2 \qquad (2.30)$$

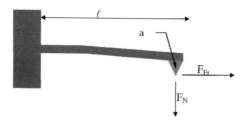

FIGURE 2.3 The effect of normal F_N and frontal forces F_{Fr} on a cantilever.

Equations 2.28 to 2.30 hold in the case where the cantilever tip is exactly in the middle axis of the cantilever. Triangular cantilevers and cantilevers with tips not on the middle axis can be dealt with by finite-element methods.

The third possible deflection mode is the one from the forces along the cantilever axis. Their effect on the cantilever is a torque. The boundary condition for the free end of the cantilever is $M_0 = a^*F_{Fr}$ (see Figure 2.3). This leads to the following modification of Equation 2.19:

$$M(x) = (\ell - x)F_N + F_{Fr}a \tag{2.31}$$

Integration of Equation 2.31 now leads to

$$z'(x) = \frac{dz}{dx} = \frac{1}{EI}\left[\frac{\ell x}{2}\left(\frac{x}{\ell} - 2\right)F_N + axF_{Fr}\right] \tag{2.32}$$

A second integration gives the deflection

$$z(x) = \frac{1}{2EI}\left[\ell x^2\left(\frac{x}{3\ell} - 1\right)F_N + ax^2 F_{Fr}\right] \tag{2.33}$$

Evaluating Equations 2.32 and 2.33 at the end of the cantilever, we get the deflection and the tilt due to the normal force F_N and the force from the front F_{Fr}

$$
\begin{aligned}
z(\ell) &= -\frac{\ell^3}{3EI}F_N + \frac{a\ell^2}{2EI}F_{Fr} = \frac{\ell^2}{EI}\left(\frac{a}{2}F_{Fr} - \frac{\ell}{3}F_N\right) \\
z'(\ell) &= -\frac{\ell^2}{2EI}F_N + \frac{a\ell}{EI}F_{Fr} = \frac{\ell}{EI}\left(aF_{Fr} - \frac{\ell}{2}F_N\right)
\end{aligned}
\tag{2.34}
$$

These equations can be inverted. One obtains the two:

$$
F_N = -\frac{12EI}{\ell^3}\left(z(\ell) - \frac{\ell z'(\ell)}{2}\right)
\tag{2.35}
$$

$$
F_{Fr} = -\frac{2EI}{a\ell^2}\left(3z(\ell) - 2\ell z'(\ell)\right)
$$

A second class of interesting properties of cantilevers is their resonance behavior. For cantilevered beams one can calculate that the resonance frequencies are (Colchero, 1993)

$$\omega_n^{\text{free}} = \frac{\lambda_n^2}{2\sqrt{3}} \frac{h}{\ell^2} \sqrt{\frac{E}{\rho}}$$

(2.36)

with $\lambda_0 = (0.596864\ldots)\pi$, $\lambda_1 = (1.494175\ldots)\pi$, $\lambda_n \to (n + \frac{1}{2})\pi$.

A similar Equation 2.36 as holds for cantilevers in rigid contact with the surface. Since there is an additional restriction on the movement of the cantilever, namely, the location of its end point, the resonance frequency increases. Only the λ_n's terms change to (Colchero, 1993)

with $\qquad \lambda'_0 = (1.2498763\ldots)\pi$, $\quad \lambda'_1 = (2.2499997\ldots)\pi$, $\quad \lambda'_n \to (n + \frac{1}{4})\pi \qquad$ (2.37)

The ratio of the fundamental resonance frequency in contact to the fundamental resonance frequency not in contact is 4.3851. For the torsion mode, we can calculate the resonance frequency to

$$\omega_0^{\text{tors}} = 2\pi \frac{h}{\ell b} \sqrt{\frac{G}{\rho}}$$

(2.38)

for thin, wide cantilevers. In contact, we obtain

$$\omega_0^{\text{tors,contact}} = \frac{\omega_0^{\text{tors}}}{\sqrt{1 + 3(2a/b)^2}}$$

(2.39)

The amplitude of the thermally induced vibration can be calculated from the resonance frequency using

$$\Delta z_{\text{therm}} = \sqrt{\frac{k_B T}{k}}$$

(2.40)

where k_b is Boltzmann's factor and k the compliance of the cantilever. Since force microscope cantilevers are resonant structures, sometimes with rather high qualities Q, the thermal noise is not evenly distributed as Equation 2.40 suggests. The spectral noise density below the peak of the response curve is

$$z_0 = \sqrt{\frac{4k_B T}{k\omega_0 Q}} \quad \text{in m } \sqrt{\text{Hz}}$$

(2.41)

2.2.3 Tips and Cantilevers

The key to the successful operation of an AFM is the measurement of the interaction forces between the tip and the sample surface. The tip would ideally consist of only one atom, which is brought in the vicinity of the sample surface. The interaction forces between the AFM tip and the sample surface must be smaller than about 10^{-7} N for bulk materials and preferably well below 10^{-9} N for organic macromolecules. To obtain a measurable deflection larger than the inevitable thermal drifts and noise the cantilever deflection for static measurements should be at least 10 nm. Hence, the spring constants should be less than 10 N/m for bulk materials and less than 1 N/m for organic macromolecules. Experience shows that cantilevers with spring constants of about 0.01 N/m work best in liquid environments, whereas stiffer cantilevers excel in resonant detection methods.

TABLE 2.1 Material Properties of Cantilevers

	Diamond	Si_3N_4	Si	W	Ir	Steel	Au	Al	PMMA
E in GPa	1,000	300/180	110	410	530	200	80	70	2.5-3
Mohs Hardness	10	9	7	6	6–6.5	5–8.5	2.5–3	2–3	<1
c_{long} in m/s	17,500	10,000	5,970	5,400	4,860	6,000	3,240	6,420	1,600

Building vibrations usually have frequencies in the range from 10 to 100 Hz. These vibrations are coupled to the cantilever. To get an estimate, we use Equation 2.3. Inserting 100 Hz for the resonance frequency and a spring constant of 0.1 N/m, we obtain an upper limit of the lumped effective mass m_{eff} of 0.25 mg. The quality factor of this resonance in air is typically between 10 and 100. To get a reasonable suppression of the excitation of cantilever oscillations, the resonance frequency of the cantilever has to be at least a factor of 10 higher than the highest of the building vibration frequencies. This means, that m_{eff} has to be under any circumstances no larger than 0.25 mg/100 = 2.5 µg. It would be preferable to limit the mass to 0.1 µg. This lumped mass m_{eff}, however, is smaller than the real mass m, by a factor that depends on the geometry of the cantilever.

A good rule of thumb says that the effective mass m_{eff} is 9/20 of the real mass. Today, micro-machined cantilevers are commercially available and are used almost exclusively.

2.2.4 Materials and Geometry

Cantilevers have been made from a whole range of materials (Pitsch et al., 1989; Akamine et al., 1990; Grütter et al., 1990; Wolter et al., 1991; Colchero, 1993). Most common are cantilevers made of Si and of Si_3N_4. As has been shown in Equation 2.25, Young's modulus E and the density ρ are the material parameters determining the resonance frequency, besides the geometry. The realizable thickness depends on the fabrication process and the material properties. Grown materials such as Si_3N_4 can be made thinner than those fabricated out of the bulk.

The first row of Table 2.1 shows the different materials. The second row gives Young's modulus. The third row is the hardness, a quantity that is important to judge the durability of the cantilevers. The last row, finally, shows the speed of sound, indicative of the resonance frequency for a given shape.

Cantilevers come basically in two shapes (Figure 2.4). Straight types are preferentially used for lateral force measurements and noncontact modes. Their properties are rather easy to calculate. Triangular-shaped cantilevers are easier to align. They are mostly made of silicon nitride. Their response to lateral forces is more complicated.

Whereas type b must be calculated using finite-element methods, one can get a good estimate of the normal force compliance of type c in Figure 2.4 using analytical methods. Using Equation 2.25 and observing that the length of the two joined cantilever beams are $\ell_{eff}^2 = \ell^2 + (w/2)^2$, where w is the width of the base of the cantilever, one gets for the compliance:

$$k_N = \frac{Ebh^3}{2}\left(\ell^2 + \frac{w^2}{4}\right)^{3/2}$$

(2.42)

a)   b)   c)

FIGURE 2.4 Shapes of cantilevers: (a) is preferentially used for lateral force measurements and for noncontact measurements; (b) and (c) are two types, mostly fabricated from silicon nitride.

2.2.5 Outline of Fabrication

Most force sensors in use today or commercially available are manufactured either from silicon or from silicon nitride. These two material systems are compatible with standard integrated circuit processing techniques. The shape and the thickness are easily controlled with sub-100 nm precision. This is necessary because the largest extension of the cantilevers is typically smaller than 300 µm. Microfabrication techniques and batch processing are important prerequisites for any successful large-scale production of force sensors.

The first published production recipe for cantilevers (Akamine et al., 1990) was for a sensor made of silicon nitride. All silicon nitride levers available today are made more or less along the guidelines outlined there. A silicon (100) wafer is thinned. Next, the tips are defined by masking the topside of the wafer with oxide, leaving square openings with about 4-µm-long sides. They have to be oriented parallel to the (110) directions. The silicon in these openings is attacked by the anisotropic etchant KOH. The etch process is fastest parallel to the (111) surfaces. Therefore, a pyramidal-shaped depression is etched away. Since the anisotropy of the etch rate is of the order of 100, the process slows down considerably once all the sides of the pyramid meet. The etch process is then terminated.

In the next step the silicon nitride is grown on top of the silicon, on the side with the etch pits. The thickness of the layer, together with the shape of the cantilever, determines the resonance frequency and the compliance. Since the silicon nitride is grown, one has a very good control on the layer thickness. Typically, cantilevers are 300 nm thick, or more. Calculated and experimentally verified spring constants are of the order of 0.01 to 1 N/m. In a next step, Pyrex glass with openings for the cantilevers is bonded from the topside onto the wafer. The remaining silicon is dissolved, leaving the cantilevers free. In the last manufacturing step the cantilevers are coated with a thin reflective film, since most microscopes use light reflected off the back of the cantilever to detect its deflection. Gold is usually used as the coating material, together with a 1-nm layer of chromium as an adhesion layer.

The radius of curvature of silicon nitride cantilevers is limited to about 30 to 50 nm, because of the manufacturing process. The imperfections of the etch pits and the filled-in silicon nitride limit the sharpness. Silicon nitride tips can be sharpened during the production by thermal oxidation (Akamine and Quate, 1992). Instead of directly depositing silicon nitride on the wafers with the pyramidal etch pits, an oxide layer is deposited first. Then, the silicon nitride is added. When the oxide was removed with buffered oxide etch, a sharpening effect was observed. Details of the process are described by the inventors (Akamine and Quate, 1992). A second method is to grow in an electron microscope a so-called supertip on top of the silicon nitride. It is well known that in scanning electron microscopes with a base pressure of more than 10^{-10} mbar hydrocarbon residues are present. These residues are cracked at the surface of the sample by the electron beam, leaving carbon in a presumed amorphous state on the surface. It is known that prolonged imaging in such an instrument degrades the surface. If the electron beam is not scanned, but stays at the same place, one can build up tips with a diameter comparable with the electron beam diameter and with a height determined by the dwell time. These tips are extremely sharp; they can reach radii of curvature of a few nanometers. They allow therefore an imaging with a very high resolution. In addition, they enable the microscope to image the bottoms of small crevasses and ditches on samples. Unprocessed silicon nitride tips are not able to do this, since their sides enclose an angle of 90°, due to the crystal structure of the silicon.

Silicon nitride cantilevers are less expensive than those made of other materials. They are very rugged and well suited to imaging in almost all environments. They are especially compatible to organic and biological materials.

Alternatives to silicon nitride cantilevers are those made of silicon. The basic manufacturing idea is the same as for silicon nitride. Masks determine the shape of the cantilevers. Processes from the microelectronics fabrication are used. Since the thickness of the cantilevers is determined by etching and not by growth, wafers have to be more precise as for the manufacturing of the silicon nitride cantilevers.

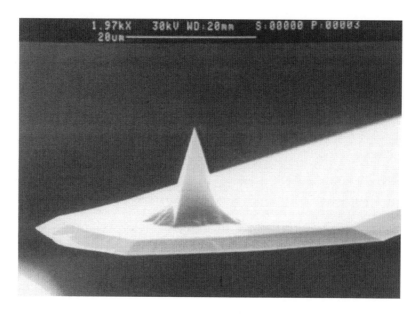

FIGURE 2.5 A commercial cantilever from Nanosensors. (Courtesy of Nanosensors.)

The first step in the process is a wet chemical etch to thin the wafer to a thin membrane (Wolter et al., 1991; Kassing and Oesterschulze, 1997). The membrane thickness is adjusted such that it corresponds to the lever thickness and the tip height (10 to 30 μm). The resulting membrane must be free of stresses. The next step is to define the cantilever layout by reactive ion etching and by chemical etching, which already creates freestanding cantilevers, as used later on in the microscopes. The third step is to define the tip. One way starts with a small oxide cap at the place where the tip should be. The silicon is then attacked by KOH. Its anisotropical etching characteristics then attack the silicon such that the protective oxide cap is underetched. The art of cantilever manufacturing consists in timing this process such that the silicon under the tip is just about to be etched away. The caps then fall down, and the rupture site produces cantilevers with well-defined radii of curvature of 2 to 5 nm. An example of such a cantilever is shown in Figure 2.5. The end section of the cantilever is shown. The lever is rounded to minimize unwanted contacts of the lever edge with the sample. Improved fabrication processes have made it possible to produce regularly tips with a radius of curvature of 2 nm. Tips, such as the one shown in Figure 2.6, permit imaging at the highest resolution in all known imaging modes.

Since the thickness of the cantilever is determined by etching, it cannot be made as thin as in the silicon nitride case. The lower limit is typically 1 μm. Therefore, the stiffness of the silicon cantilevers is higher, ranging from 1 to 100 N/m. Since the material is a single crystal, unlike the silicon nitride, it has a very high quality of the resonance. Values exceeding 100,000 have been observed in vacuum. Therefore, silicon cantilevers are often used for noncontact or tapping mode experiments. The cantilevers have two drawbacks when working in the contact mode. First, they have a very high affinity to organic materials. They often destroy such samples. Second, their index of refraction matches the one of water rather closely. Silicon cantilevers have a very poor reflectivity in aqueous environments.

There are efforts under way to make cantilevers of GaAs (Kassing and Oesterschulze, 1997). This material is more difficult to process, but it would offer new advantages. GaAs is a direct band gap material. Optoelectronic functions could be easily integrated into such cantilevers. The investigation of magnetic properties could be improved by the use of spin-polarized tunneling.

Occasionally, cantilevers are made with tungsten wire or thin metal foils, with tips of diamond or other materials glued to it.

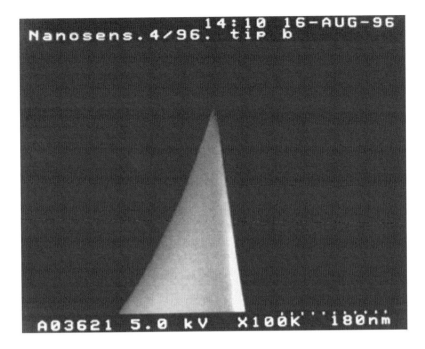

FIGURE 2.6　A SuperSharpSilicon™ tip from Nanosensors. The distance between two points on the scale in the image is 18 nm. (Courtesy of Nanosensors. Used with permission.)

2.3　Optical Detection Systems

2.3.1　Interferometer

Soon after the first papers on the AFM (Binnig et al., 1986), which used a tunneling sensor, an instrument based on an interferometer was published (McClelland et al., 1987). The sensitivity of the interferometer depends on the wavelength of the light employed in the apparatus. Figure 2.7 shows the principle of such an interferometric design. The light incident from the left is focused by a lens on the cantilever. The

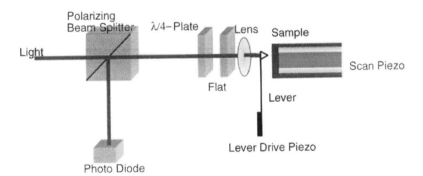

FIGURE 2.7　Principle of an interferometric AFM. The light of the laser light source is polarized by the polarizing beam splitter and focused on the back of the cantilever. The light passes twice through a quarter wave plate and is hence orthogonally polarized to the incident light. The second arm of the interferometer is formed by the flat. The interference pattern is modulated by the oscillating cantilever.

reflected light is collimated by the same lens and interferes with the light reflected at the flat. To separate the reflected light from the incident light, a $\lambda/4$-plate converts the linear polarized incident light to circular polarization. The reflected light is made linear polarized again by the $\lambda/4$-plate, but with a polarization orthogonal to that of the incident light. The polarizing beam splitter then deflects the reflected light to the photodiode.

2.3.1.1 Homodyne Interferometer

To improve the signal-to-noise ratio of the interferometer, the lever is driven by a piezoactuator near its resonance frequency. The amplitude Δz of the lever is

$$\Delta z\left(\Omega\right) = \Delta z_0 \frac{\Omega_0^2}{\sqrt{\left(\Omega^2 - \Omega_0^2\right)^2 + \frac{\Omega^2 \Omega_0^2}{Q^2}}} \tag{2.43}$$

where Δz_0 is the constant drive amplitude, Ω_0 the resonance frequency of the lever, Q the quality of the resonance, and Ω the drive frequency. The resonance frequency of the lever is given by the effective potential

$$\Omega_0 = \sqrt{\left(k + \frac{\partial^2 U}{\partial z^2}\right) \frac{1}{m_{\text{eff}}}} \tag{2.44}$$

where k is the spring constant of the free lever, U the interaction potential between the tip and the sample, and m_{eff} the effective mass of the cantilever. Equation 2.44 shows that an attractive potential decreases the resonance frequency Ω_0. The change in the resonance frequency Ω_0 in turn results in a change of the lever amplitude Δz (see Equation 2.43).

The movement of the cantilever changes the path difference in the interferometer. The light reflected from the lever with the amplitude $A_{l,0}$ and the reference light with the amplitude $A_{r,0}$ interfere on the detector. The detected intensity $I(t) = \{A_l(t) + A_r(t)\}^2$ consists of two constant terms and a fluctuating term:

$$2A_l\left(t\right)A_r\left(t\right) = A_{l,0}A_{r,0}\sin\left[\omega t + \frac{4\pi\delta}{\lambda} + \frac{4\pi\Delta z}{\lambda}\sin\left(\Omega t\right)\right]\sin\left(\omega t\right) \tag{2.45}$$

Here ω is the frequency of the light, δ the path difference in the interferometer, and Δz is the instantaneous amplitude of the lever, given according to Equations 2.43 and 2.44 as a function of the driving frequency Ω, the spring constant k, and the interaction potential U. The time average of Equation 2.45 then becomes

$$\left\langle 2A_l\left(t\right)A_r\left(t\right)\right\rangle_T \propto \cos\left[\frac{4\pi\delta}{\lambda} + \frac{4\pi\Delta z}{\lambda}\sin\left(\Omega t\right)\right]$$

$$\approx \cos\left(\frac{4\pi\delta}{\lambda}\right) - \sin\left[\frac{4\pi\Delta z}{\lambda}\sin\left(\Omega t\right)\right] \tag{2.46}$$

$$\approx \cos\left(\frac{4\pi\delta}{\lambda}\right) - \frac{4\pi\Delta z}{\lambda}\sin\left(\Omega t\right)$$

Here all small quantities have been omitted and functions with small arguments have been linearized. The amplitude of the lever oscillation Δz can be recovered with a lock-in technique. However, Equation 2.46 shows that the measured amplitude is also a function of the path difference δ in the

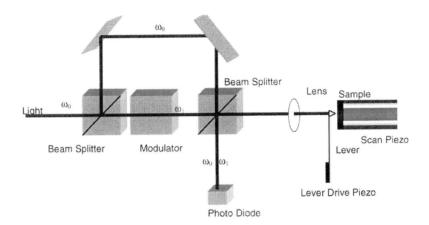

FIGURE 2.8 Heterodyne interferometer AFM. Light with the frequency ω_0 is split into a reference path (upper path) and a measurement path. The light in the measurement path is frequency shifted to ω_1 by an acousto-optical modulator (or an electro-optical modulator). The light reflected from the oscillating cantilever interferes with the reference beam on the detector.

interferometer. Hence, this path difference δ must be very stable. The best sensitivity is obtained when $\sin(4\delta/\lambda) \approx 0$.

2.3.1.2 Heterodyne Interferometer

This influence is not present in the heterodyne detection scheme shown in Figure 2.8. Light incident from the left with a frequency ω is split in a reference path (upper path in Figure 2.8) and a measurement path. Light in the measurement path is shifted in frequency to $\omega_1 = \omega + \Delta\omega$ and focused on the cantilever. The cantilever oscillates at the frequency Ω, as in the homodyne detection scheme. The reflected light $A_l(t)$ is collimated by the same lens and interferes on the photodiode with the reference light $A_r(t)$. The fluctuating term of the intensity is given by

$$2A_l\!\left(t\right)A_r\!\left(t\right) = A_{l,0}A_{r,0}\sin\!\left[\left(\omega+\Delta\omega\right)t+\frac{4\pi\delta}{\lambda}+\frac{4\pi\Delta z}{\lambda}\sin\!\left(\Omega t\right)\right]\sin\!\left(\omega t\right) \qquad (2.47)$$

where the variables are defined as in Equation 2.45. Setting the path difference $\sin(4\pi\delta/\lambda) \approx 0$ and taking the time average, omitting small quantities and linearizing functions with small arguments, we get

$$\left\langle 2A_l\!\left(t\right)A_r\!\left(t\right)\right\rangle_T \propto \cos\!\left[\Delta\omega t+\frac{4\pi\delta}{\lambda}+\frac{4\pi\Delta z}{\lambda}\sin\!\left(\Omega t\right)\right]$$

$$= \cos\!\left(\Delta\omega t+\frac{4\pi\delta}{\lambda}\right)\cos\!\left[\frac{4\pi\Delta z}{\lambda}\sin\!\left(\Omega t\right)\right]$$

$$-\sin\!\left(\Delta\omega t+\frac{4\pi\delta}{\lambda}\right)\sin\!\left[\frac{4\pi\Delta z}{\lambda}\sin\!\left(\Omega t\right)\right] \qquad (2.48)$$

$$\approx \cos\!\left(\frac{4\pi\delta}{\lambda}\right)-\sin\!\left[\frac{4\pi\Delta z}{\lambda}\sin\!\left(\Omega t\right)\right]$$

$$\approx \cos\!\left(\Delta\omega t+\frac{4\pi\delta}{\lambda}\right)\!\left[1-\frac{8\pi^2\Delta z^2}{\lambda^2}\sin\!\left(\Omega t\right)\right]$$

$$-\frac{4\pi\Delta z}{\lambda}\sin\left(\Delta\omega t\ \frac{4\pi\delta}{\lambda}\right)\sin\left(\Omega t\right)$$

$$=\cos\left(\Delta\omega t+\frac{4\pi\delta}{\lambda}\right)-\frac{8\pi^2\Delta z^2}{\lambda^2}\cos\left(\Delta\omega t+\frac{4\pi\delta}{\lambda}\right)\sin\left(\Omega t\right)$$

$$-\frac{4\pi\Delta z}{\lambda}\sin\left(\Delta\omega t+\frac{4\pi\delta}{\lambda}\right)\sin\left(\Omega t\right)$$

$$=\cos\left(\Delta\omega t+\frac{4\pi\delta}{\lambda}\right)-\frac{4\pi^2\Delta z^2}{\lambda^2}\cos\left(\Delta\omega t+\frac{4\pi\delta}{\lambda}\right)$$

$$-\frac{4\pi^2\Delta z^2}{\lambda^2}\cos\left(\Delta\omega t+\frac{4\pi\delta}{\lambda}\right)\cos\left(2\Omega t\right)$$

$$-\frac{4\pi\Delta z}{\lambda}\sin\left(\Delta\omega t+\frac{4\pi\delta}{\lambda}\right)\sin\left(\Omega t\right)$$

$$=\cos\left(\Delta\omega t+\frac{4\pi\delta}{\lambda}\right)\left(1-\frac{4\pi^2\Delta z^2}{\lambda^2}\right)$$

$$+\frac{2\pi^2\Delta z^2}{\lambda^2}\left\{\cos\left[\left(\Delta\omega+2\Omega\right)t+\frac{4\pi\delta}{\lambda}\right]+\cos\left[\left(\Delta\omega-2\Omega\right)t+\frac{4\pi\delta}{\lambda}\right]\right\}$$

$$+\frac{2\pi\Delta z}{\lambda}\left\{\cos\left[\left(\Delta\omega+\Omega\right)t+\frac{4\pi\delta}{\lambda}\right]+\cos\left[\left(\Delta\omega-\Omega\right)t+\frac{4\pi\delta}{\lambda}\right]\right\}$$

Multiplying electronically the components oscillating at $\Delta\omega$ and $\Delta\omega+\Omega$ and rejecting any product except the one oscillating at Ω, we obtain

$$A=\frac{2\Delta z}{\lambda}\left(1-\frac{4\pi^2\Delta z^2}{\lambda^2}\right)\cos\left[\left(\Delta\omega+2\Omega\right)t+\frac{4\pi\delta}{\lambda}\right]\cos\left(\Delta\omega t+\frac{4\pi\delta}{\lambda}\right)$$

$$=\frac{\Delta z}{\lambda}\left(1-\frac{4\pi^2\Delta z^2}{\lambda^2}\right)\left\{\cos\left[\left(2\Delta\omega+\Omega\right)t+\frac{8\pi\delta}{\lambda}\right]+\cos\left(\Omega t\right)\right\} \qquad (2.49)$$

$$\approx\frac{\pi\Delta z}{\lambda}\cos\left(\Omega t\right)$$

Unlike in the homodyne detection scheme, the recovered signal is independent from the path difference δ of the interferometer. Furthermore, a lock-in amplifier with the reference set $\sin(\Delta\omega t)$ can measure the path difference δ independent of the cantilever oscillation. If necessary, a feedback circuit can keep $\delta = 0$.

2.3.1.3 Fiber-Optic Interferometer

The fiber-optic interferometer (Rugar et al., 1989) is one of the simplest interferometers to build and use. Its principle is sketched in Figure 2.9. The light of a laser is fed into an optical fiber. Laser diodes with integrated fiber pigtails are convenient light sources. The light is split in a fiber-optic beam splitter into two fibers. One fiber is terminated by an index-matching oil to avoid any reflections back into the fiber. The end of the other fiber is brought close to the cantilever in the AFM. The emerging light is partially reflected back into the fiber by the cantilever. Most of the light, however, is lost. This is not a

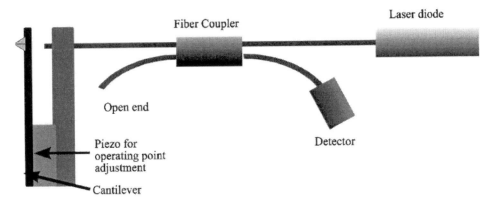

FIGURE 2.9 A typical setup for a fiber-optic interferometer readout.

big problem since only 4% of the light is reflected at the end of the fiber, at the glass–air interface. The two reflected light waves interfere with each other. The product is guided back into the fiber coupler and again split into two parts. One half is analyzed by the photodiode. The other half is fed back into the laser. Communications-grade laser diodes are sufficiently resistant against feedback to be operated in this environment. They have, however, a bad coherence length, which in this case does not matter, since the optical path difference is in any case no larger than 5 μm. Again, the end of the fiber has to be positioned on a piezo drive to set the distance between the fiber and the cantilever to λ ($n + \frac{1}{4}$).

2.3.1.4   Nomarski Interferometer

Another solution to minimize the optical path difference uses the Nomarski (Schönenberger and Alvarado, 1989). Figure 2.10 depicts a sketch of the microscope. The light of a laser is focused on the cantilever by a lens. A birefringent crystal (for instance, calcite) between the cantilever and the lens with its optical axis 45° off the polarization direction of the light splits the light beam into two paths, offset by a distance given by the length of the birefringent crystal. Birefringent crystals have varying indexes of refraction. In calcite, one crystal axis has a lower index than the other two. This means that certain light rays will propagate at a different speed through the crystal than others. By choosing a correct polarization, one can select the ordinary ray, the extraordinary ray, or one can get any distribution of the intensity among those two rays. A detailed description of birefringence can be found in textbooks (Shen, 1984). A calcite crystal deflects the extraordinary ray at an angle of 6° within the crystal. By choosing a suitable length of the calcite crystal, any separation can be set.

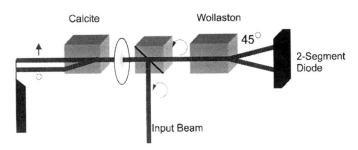

FIGURE 2.10 Principle of the Nomarski AFM (Schönenberger and Alvarado, 1989, 1990). The circular polarized input beam is deflected to the left by a nonpolarizing beam splitter. The light is focused onto a cantilever. The calcite crystal between the lens and the cantilever splits the circular polarized light into two spatially separated beams with orthogonal polarizations. The two light beams reflected from the lever are superimposed by the calcite crystal and collected by the lens. The resulting beam is again circular polarized. A Wollaston prism produces two interfering beams with a π/2 phase shift between them. The minimal path difference accounts for the excellent stability of this microscope.

TABLE 2.2   Noise in Interferometers

	Homodyne Interferometer, Fiber-Optic Interferometer	Heterodyne Interferometer	Nomarski Interferometer
Laser noise $\langle \delta i^2 \rangle_L$	$\frac{1}{4} \eta^2 F^2 P_i^2 \, \mathrm{RIN}$	$\eta^2 \left(P_R^2 + P_S^2 \right) \mathrm{RIN}$	$\frac{1}{16} \eta^2 P^2 \delta\Theta$
Thermal noise $\langle \delta i^2 \rangle_t$	$\frac{16\pi^2}{\lambda^2} \eta^2 F^2 P_i^2 \frac{4k_B TBQ}{\omega_0 k}$	$\frac{4\pi^2}{\lambda^2} \eta^2 P_d^2 \frac{4k_B TBQ}{\omega_0 k}$	$\frac{\pi^2}{\lambda^2} \eta^2 P^2 \frac{4k_B TBQ}{\omega_0 k}$
Shot noise $\langle \delta i^2 \rangle_S$	$4 e\eta P_d B$	$2 e\eta (P_R + P_S) B$	$\frac{1}{2} e\eta PB$

F is the finesse of the cavity in the homodyne interferometer, P_i is the incident power, P_d is the power on the detector, η is the sensitivity of the photodetector, and RIN is the relative intensity noise of the laser. P_R and P_S are the power in the reference and sample beam in the heterodyne interferometer. P is the power in the Nomarsky interferometer, and $\delta\Theta$ is the phase difference between the reference and the probe beam in the Nomarsky interferometer. B is the bandwidth and e the electron charge. λ is the wavelength of the laser and k the stiffness of the cantilever, T is the temperature.

The focus of one light ray is positioned near the free end of the cantilever while the other is placed close to the clamped end. Both arms of the interferometer pass through the same space, except for the distance between the calcite crystal and the lever. The closer the calcite crystal is placed to the lever, the less influence disturbances like air currents have.

2.3.2   Sensitivity

Sarid (1991) has given values for the sensitivity of the different interferometeric detection systems. Table 2.2 shows a summary of his results.

2.4   Optical Lever

The most common cantilever deflection detection system is the optical lever (Meyer and Amer, 1988; Alexander et al., 1989). This method, depicted in Figure 2.11, employs the same technique as light beam deflection galvanometers. A fairly well collimated light beam is reflected off a mirror and projected to a receiving target. Any change in the angular position of the mirror will change the position where the light ray hits the target. Galvanometers use optical path lengths of several meters and scales projected to the target wall as a readout help.

For the AFM using the optical lever method, a photodiode segmented into two (or four) closely spaced devices detects the orientation of the end of the cantilever (see Figure 2.11). Initially, the light ray is set to hit the photodiodes in the middle of the two subdiodes. Any deflection of the cantilever will cause an imbalance of the number of photons reaching the two halves. Hence, the electrical currents in the

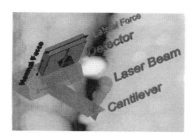

FIGURE 2.11   Optical lever setup.

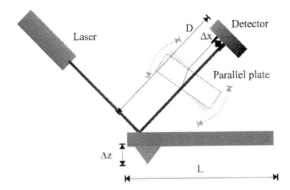

FIGURE 2.12   The setup of optical lever detection microscope.

photodiodes will be unbalanced, too. The difference signal is further amplified and is the input signal to the feedback loop. Unlike the interferometric AFMs, where often a modulation technique is necessary to get a sufficient signal-to-noise ratio, most AFMs employing the optical lever method are operated in a static mode. AFMs based on the optical lever method are universally used. It is the simplest method to construct an optical readout and it can be confined in volumes smaller than 5 cm on the side.

The optical lever detection system is a simple yet elegant way to detect normal and lateral force signals simultaneously (Meyer and Amer, 1988, 1990; Alexander et al., 1989; Marti, Colchero et al., 1990). It has the additional advantage that it is a remote detection system.

2.4.1   Implementations

Light from a laser diode or from a superluminescent diode is focused on the end of the cantilever. The reflected light is directed onto a quadrant diode that measures the direction of the light beam. A Gaussian light beam far from its waist is characterized by an opening angle β. The deflection of the light beam by the cantilever surface tilted by an angle α is 2α. The intensity on the detector then shifts to the side by the product of 2α and the separation between the detector and the cantilever. The readout electronics calculates the difference of the photocurrents. The photocurrents, in turn, are proportional to the intensity incident on the diode.

The output signal is hence proportional to the change in intensity on the segments:

$$I_{sig} \propto 4\frac{\alpha}{\beta}I_{tot} \tag{2.50}$$

Figure 2.12 shows a schematic drawing of the optical lever setup. For the sake of simplicity, we assume that the light beam is of uniform intensity with its cross section increasing proportionally with the distance between the cantilever and the quadrant detector. The movement of the center of the light beam is then given by

$$\Delta x_{Det} = \Delta z \frac{d}{\ell} \tag{2.51}$$

The photocurrent generated in a photodiode is proportional to the number of incoming photons hitting it. If the light beam contains a total number of N_0 photons, then the change in difference current becomes

$$\Delta\left(I_R - I_L\right) = \Delta I = \text{const } \Delta z d N_0 \tag{2.52}$$

99

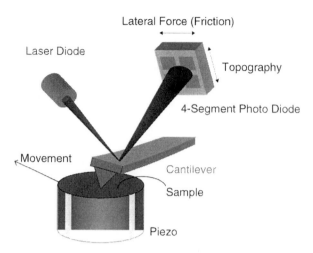

Lateral Force (Friction)

Laser Diode

Topography

4-Segment Photo Diode

Movement

Cantilever

Sample

Piezo

FIGURE 2.13 Scanning force and friction microscope (SFFM). The lateral forces exerted on the tip by the moving sample cause a torsion of the lever. The light reflected from the lever is deflected orthogonally to the deflection caused by normal forces.

Combining Equations 2.51 and 2.52, one obtains that the difference current ΔI is independent of the separation of the quadrant detector and the cantilever. This relation is true if the light spot is smaller than the quadrant detector. If it is greater, the difference current ΔI becomes smaller with increasing distance. In reality, the light beam has a Gaussian intensity profile. For small movements Δx (compared with the diameter of the light spot at the quadrant detector), Equation 2.52 still holds. Larger movements Δx, however, will introduce a nonlinear response. If the AFM is operated in a constant-force mode, only small movements Δx of the light spot will occur. The feedback loop will cancel out all other movements.

The scanning of a sample with an AFM can twist the microfabricated cantilevers because of lateral forces (Mate et al., 1987; Marti et al., 1990; Meyer and Amer, 1990) and affect the images (den Boef, 1991). When the tip is subjected to lateral forces, it will twist the lever, and the light beam reflected from the end of the lever will be deflected perpendicular to the ordinary deflection direction. For many investigations, this influence of lateral forces is unwanted. The design of the triangular cantilevers stems from the desire to minimize the torsion effects. However, lateral forces open up a new dimension in force measurements. They allow, for instance, a distinction of two materials because of the different friction coefficient, or the determination of adhesion energies. To measure lateral forces the original optical lever AFM has to be modified; Figure 2.13 shows a sketch of the instrument. The only modification compared with Figure 2.12 is the use of a quadrant detector photodiode instead of a two-segment photodiode and the necessary readout electronics. The electronics calculates the following signals:

$$U_{\text{Normal Force}} = \alpha\left[\left(I_{\text{Upper Left}} + I_{\text{Upper Right}}\right) - \left(I_{\text{Lower Left}} + I_{\text{Lower Right}}\right)\right]$$

$$U_{\text{Lateral Force}} = \beta\left[\left(I_{\text{Upper Left}} + I_{\text{Upper Left}}\right) - \left(I_{\text{Lower Right}} + I_{\text{Lower Right}}\right)\right]$$

(2.53)

The calculation of the lateral force as a function of the deflection angle does not have a simple solution for cross sections other than circles. An approximate formula for the angle of twist for rectangular beams is (Baumeister and Marks, 1967)

$$\Theta = \frac{M_t \ell}{\beta G b^3 h}$$

(2.54)

where $M_t = Fa$ is the external twisting moment due to friction, ℓ is the length of the beam, b and h the sides of the cross section, G the shear modulus, and β a constant determined by the value of h/b. For the equation to hold, h has to be larger than b.

Inserting the values for a typical microfabricated lever with integrated tips

$$
\begin{aligned}
b &= 6 \times 10^{-7} \, \text{m} \\
h &= 10^{-5} \, \text{m} \\
\ell &= 10^{-4} \, \text{m} \\
a &= 3.3 \times 10^{-6} \, \text{m} \\
G &= 5 \times 10^{10} \, \text{Pa} \\
\beta &= 0.333
\end{aligned}
\tag{2.55}
$$

into Equation 2.54, we obtain the relation

$$
F_{\text{Lateral Force}} = 1.1 \times 10^{-4} \, N \times \Theta
\tag{2.56}
$$

Typical lateral forces are of order 10^{-10} N.

2.4.2 Sensitivity

The sensitivity of this setup has been calculated in various papers (Colchero et al., 1991; Sarid, 1991; Colchero, 1993), to name just three examples. Assuming a Gaussian beam, the resulting output signal as a function of the deflection angle is dispersion like. Equation 2.50 shows that the sensitivity can be increased by increasing the intensity of the light beam I_{tot} or by decreasing the divergence of the laser beam. The upper bound of the intensity of the light I_{tot} is given by saturation effects on the photodiode. If we decrease the divergence of a laser beam, we automatically increase the beam waist. If the beam waist becomes larger than the width of the cantilever, we start to get diffraction. Diffraction sets a lower bound on the divergence angle. Hence, one can calculate the optimal beam waist w_{opt} and the optimal divergence angle β (Colchero et al., 1991; Colchero, 1993)

$$
\begin{aligned}
w_{\text{opt}} &\approx 0.36 b \\
\theta_{\text{opt}} &\approx 0.89 \frac{\lambda}{b}
\end{aligned}
\tag{2.57}
$$

where b is the width of the cantilever and λ is the wavelength of the light. The optimal sensitivity of the optical lever then becomes

$$
\varepsilon[\text{mW}/\text{rad}] = 1.8 \frac{b}{\lambda} I_{\text{tot}}[\text{mW}]
\tag{2.58}
$$

The angular sensitivity optical lever can be measured by introducing a parallel plate into the beam. A tilt of the parallel plate results in a displacement of the beam, mimicking an angular deflection.

Additional noise source can be considered. Of little importance is the quantum mechanical uncertainty of the position (Colchero et al., 1991; Colchero, 1993), which is for typical cantilevers at room temperature

$$\Delta z = \sqrt{\frac{\hbar}{2m\omega_0}} = 0.05 \text{ fm} \tag{2.59}$$

At very low temperatures and for high-frequency cantilevers, this could become the dominant noise source. A second noise source is the shot noise of the light. The shot noise is related to the particle number. We can calculate the number of photons incident on the detector

$$n = \frac{I\tau}{\hbar\omega} = \frac{I\lambda}{2\pi B\hbar c} = 1.8 \times 10^9 \frac{I[\text{W}]}{B[\text{Hz}]} \tag{2.60}$$

where I is the intensity of the light, τ the measurement time, $B = 1/\tau$ the bandwidth, c the speed of light, and λ the wavelength of the light. The shot noise is proportional to the square root of the number of particles. Equating the shot noise signal with the signal resulting for the deflection of the cantilever, one obtains

$$\Delta z_{\text{shot}} = 68 \frac{\ell}{w} \sqrt{\frac{B[\text{kHz}]}{I[\text{mW}]}} [\text{fm}] \tag{2.61}$$

where w is the diameter of the focal spot. Typical AFM setups have a shot noise of 2 pm. The thermal noise can be calculated from the equipartition principle. The amplitude at the resonance frequency is

$$\Delta z_{\text{therm}} = 129 \sqrt{\frac{B}{k[\text{N/m}]\omega_0 Q}} [\text{pm}] \tag{2.62}$$

where Q is the quality of the cantilever resonance, ω_0 the resonance frequency, and k is the stiffness of the cantilever spring. A typical value is 16 pm. Upon touching the surface, the cantilever increases its resonance frequency by a factor of 4.39. This results in a new thermal noise amplitude of 3.2 pm for the cantilever in contact with the sample.

2.5 Piezoresistive Detection

2.5.1 Implementations

An alternative detection system which is not as widespread as the optical detection schemes are piezoresistive cantilevers (Ashcroft and Mermin, 1976; Stahl et al., 1994; Kassing and Oesterschulze, 1997). These levers are based on the fact that the resistivity of certain materials, in particular of Si, changes with the applied stress. Figure 2.14 shows a typical implementation of a piezoresistive cantilever. Four resistances are integrated on the chip, forming a Wheatstone bridge. Two of the resistors are in unstrained parts of the cantilever; the other two are measuring the bending at the point of the maximal deflection. For instance, when an AC voltage is applied between terminals A and C one can measure the detuning of the bridge between terminals B and D. With such a connection, the output signal varies only due to bending, but not due to changing of the ambient temperature and thus the coefficient of the piezoresistance.

2.5.2 Sensitivity

The resistance change is (Kassing and Oesterschulze, 1997)

$$\frac{\Delta R}{R_0} = \Pi\delta \tag{2.63}$$

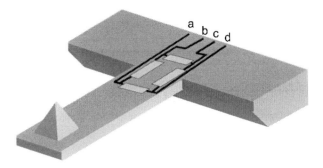

FIGURE 2.14 A typical setup for a piezoresistive readout.

where Π is the tensor element of the piezoresistive coefficients, δ the mechanical stress tensor element, and R_0 the equilibrium resistance. For a single resistor, they separate the mechanical stress and the tensor element in longitudinal and transversal components.

$$\frac{\Delta R}{R_0} = \Pi_t \delta_t + \Pi_l \delta_l \qquad (2.64)$$

The maximum value of the stress components are $\Pi_t = -64.0 \times 10^{-11}$ m²/N and $\Pi_l = 71.4 \times 10^{-11}$ m²/N for a resistor oriented along the (110) direction in silicon (Kassing and Oesterschulze, 1997). In the resistor arrangement of Figure 2.14 two of the resistors are subject to the longitudinal piezoresistive effect and two of them are subject to the transversal piezoresistive effect. The sensitivity of that setup is about four times that of a single resistor, with the advantage that temperature effects cancel to first order. It is then calculated that

$$\frac{\Delta R}{R_0} = \Pi \frac{3Eh}{2\ell^2} \Delta z = \Pi \frac{6\ell}{bh^2} F \qquad (2.65)$$

where the geometric constants are defined in Figure 2.1, F is the normal force applied to the end of the cantilever, Δz is the deflection resulting from this force, and $\Pi = 67.7 \times 10^{-11}$ m²/N is the averaged piezoresistive coefficient. Plugging in typical values for the dimensions ($\ell = 100$ μm, $b = 10$ μm, $h = 1$ μm), one obtains that

$$\frac{\Delta R}{R_0} = \frac{4 \times 10^{-5}}{nN} F \qquad (2.66)$$

The sensitivity can be tailored by optimizing the dimensions of the cantilever.

2.6 Capacitive Detection

The capacitance of an arrangement of conductors depends on the geometry. Generally speaking, the capacitance increases for decreasing separations. Two parallel plates form a simple capacitor (see Figure 2.15, upper left), with the capacitance

$$C = \frac{\varepsilon \varepsilon_0 A}{x} \qquad (2.67)$$

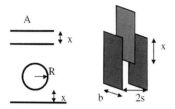

FIGURE 2.15 Three possible arrangements of a capacitive readout. The upper left shows the cross section through a parallel plate capacitor. The lower left shows the geometry sphere vs. the plane. The right side shows the more-complicated, but linear capacitive readout.

where A is the area of the plates, assumed equal, and x the separation. Alternatively, one can consider a sphere vs. an infinite plane (see Figure 2.15, lower left). Here, the capacitance is (Sarid, 1991)

$$C = 4\pi\varepsilon_0 R \sum_{n=2}^{\infty} \frac{\sinh(\alpha)}{\sinh(n\alpha)} \tag{2.68}$$

where R is the radius of the sphere and α is defined by

$$\alpha = \ln\left[1 + \frac{z}{R} + \sqrt{\frac{z^2}{R^2} + 2\frac{z}{R}} \right] \tag{2.69}$$

One has to keep in mind that capacitance of a parallel plate capacitor is a nonlinear function of the separation. Using a voltage divider, one can circumvent this problem.

Figure 2.16a shows a low-pass filter. The output voltage is given by

$$U_{out} = U_\approx \frac{\frac{1}{j\omega C}}{R + \frac{1}{j\omega C}} = U_\approx \frac{1}{j\omega CR + 1} \cong \frac{U_\approx}{j\omega CR} \tag{2.70}$$

Here, C is given by Equation 2.67, ω is the excitation frequency, and j is the imaginary unit. The approximate relation in the end is true when $\omega CR \gg 1$. This is equivalent to the statement that C is fed by a current source, since R must be large in this setup. Plugging this equation into Equation 2.70 and neglecting the phase information, one obtains

$$U_{out} = \frac{U_\approx x}{\omega R\varepsilon\varepsilon_0 A} \tag{2.71}$$

which is linear in the displacement x.

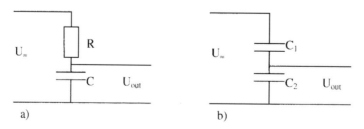

FIGURE 2.16 Measuring the capacitance. The left side (a) shows a low-pass filter; the right side (b) shows a capacitive divider. C (left) or C_2 are the capacitances under test.

Figure 2.16b shows a capacitive divider. Again, the output voltage U_{out} is given by

$$U_{out} = U_{\approx} \frac{C_1}{C_2 + C_1} = U_{\approx} \frac{C_1}{\dfrac{\varepsilon\varepsilon_0 A}{x} + C_1} \tag{2.72}$$

If there is a stray capacitance C_s, then it modifies Equation 2.72 to

$$U_{out} = U_{\approx} \frac{C_1}{\dfrac{\varepsilon\varepsilon_0 A}{x} + C_s + C_1} \tag{2.73}$$

Provided $C_s + C_1 \ll C_2$, one has a system that is linear in x. The driving voltage $U_{\approx}$ has to be large (more than 100 V) to have the output voltage in the range of 1 V. The linearity of the readout depends on the capacitance C_1 (Figure 2.17).

Another idea is to keep the distance constant and to change the relative overlap of the plates (see Figure 2.15, right side). The capacitance of the moving center plate vs. the stationary outer plates becomes

$$C = C_{stray} + 2\frac{\varepsilon\varepsilon_0 bx}{s} \tag{2.74}$$

where the variables are defined in Figure 2.15. The stray capacitance C_{stray} comprises all effects, including the capacitance of the fringe fields. When length x is comparable to the width b of the plates, one can safely assume that the stray capacitance C_{stray} is constant, independent of x. The main disadvantage of this setup is that it is not as easily incorporated in a microfabricated device as the others.

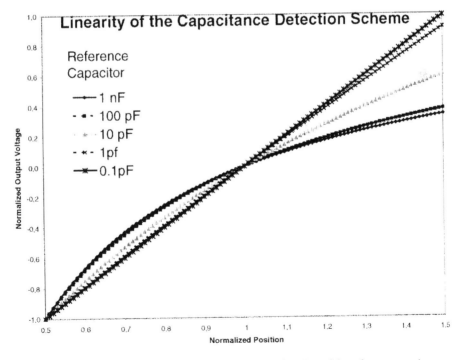

FIGURE 2.17 Linearity of the capacitance readout as a function of the reference capacitor.

2.6.1 Sensitivity

The capacitance itself is not a measure of the sensitivity, but its derivative is indicative of the signals one can expect. Using the situation described in Figure 2.15, upper left, and in Equation 2.67, one obtains for the parallel plate capacitor

$$\frac{dC}{dx} = -\frac{\varepsilon\varepsilon_0 A}{x^2} \tag{2.75}$$

Assuming a plate area A of 20 μm by 40 μm and a separation of 1 μm, one obtains a capacitance of 31 fF (neglecting stray capacitance and the capacitance of the connection leads) and a dC/dx of 3.1×10^{-8} F/m = 31 fF/μm. Hence, it is of paramount importance to maximize the area between the two contacts and to minimize the distance x. The latter, however, is far from being trivial. One has to go to the limits of microfabrication to achieve a decent sensitivity.

If the capacitance is measured by the circuit shown in Figure 2.16, one obtains for the sensitivity

$$\frac{dU_{out}}{U_\approx} = \frac{dx}{\omega R \varepsilon \varepsilon_0 A} \tag{2.76}$$

Using the same value for A as above, setting the reference frequency to 100 kHz, and selecting $R = 1$ GΩ, we get the relative change of the output voltage U_{out} to

$$\frac{dU_{out}}{U_\approx} = \frac{22.5 \times 10^{-6}}{\text{Å}} \times dx \tag{2.77}$$

A driving voltage of 45 V then translates to a sensitivity of 1 mV/Å. A problem in this setup are the stray capacitances. They are in parallel to the original capacitance and decrease the sensitivity considerably.

Alternatively, one could build an oscillator with this capacitance and measure the frequency. RC-oscillators typically have an oscillation frequency of

$$f_{res} \propto \frac{1}{RC} = \frac{x}{R \varepsilon \varepsilon_0 A} \tag{2.78}$$

Again, the resistance R must be of the order of 1 GΩ, when stray capacitances C_s are neglected. However, C_s is of the order of 1 pF. Therefore, one gets $R = 10$ MΩ. By using these values, the sensitivity becomes

$$df_{res} = \frac{C dx}{R(C + C_s)^2 x} \approx \frac{0.1 \text{ Hz}}{\text{Å}} dx \tag{2.79}$$

The problem is that the stray capacitances have made the signal nonlinear again. The linearized setup in Figure 2.15 has a sensitivity of

$$\frac{dC}{dx} = 2\frac{\varepsilon\varepsilon_0 b}{s} \tag{2.80}$$

Plugging in typical values, $b = 10$ μm, $s = 1$ μm one gets $dC/dx = 1.8 \times 10^{-10}$ F/m. It is noteworthy that the sensitivity remains constant for scaled devices.

2.6.2 Implementations

The readout of the capacitance can be done in different ways. All include an alternating current or voltage with frequencies in the 100 kHz to the 100 MHz range. One possibility is to build a tuned circuit with the capacitance of the cantilever determining the frequency. The resonance frequency of a high-quality Q tuned circuit is

$$\omega_0 = \left(LC\right)^{-\frac{1}{2}} \tag{2.81}$$

where L is the inductance of the circuit. The capacitance C includes not only the sensor capacitance but also the capacitance of the leads. The precision of a frequency measurement is mainly determined by the ratio of L and C

$$Q = \left(\frac{L}{C}\right)^{\frac{1}{2}} \frac{1}{R} \tag{2.82}$$

Here R symbolizes the losses in the circuit. The higher the quality, the more precise the frequency measurement. For instance, a frequency of 100 MHz and a capacitance of 1 pF gives an inductance of 250 μH. The quality becomes then 2.5×10^8. This value is an upper limit, since losses are usually too high.

Using a value of $dC/dx = 31$ fF/μm, one gets $\Delta C/\text{Å} = 3.1$ aF/Å. With a capacitance of 1 pF, one finally gets

$$\frac{\Delta\omega}{\omega} = \frac{\Delta\upsilon}{\upsilon} = \frac{1}{2}\frac{\Delta C}{C}$$

$$\Delta\upsilon = 100 \text{ MHz} \times \frac{1}{2}\frac{3.1 \text{ aF}}{1 \text{ pF}} = 155 \text{ Hz} \tag{2.83}$$

This is the frequency shift for 1 Å deflection. The calculation shows that this is a measurable quantity. The quality also indicates that there is no physical reason why this scheme should not work.

2.7 Combinations for Three-Dimensional Force Measurements

Three-dimensional force measurements are essential if one wants to know all the details of the interaction between the tip and the cantilever. The straightforward attempt to measure three forces is complicated, since force sensors such as interferometers or capacitive sensors need a minimal detection volume, which often is too large. The second problem is that the force-sensing tip has to be held by some means. This implies that one of the three Cartesian axes is stiffer than the others.

However, by the combination of different sensors one can achieve this goal. Straight cantilevers are employed for these measurements, because they can be handled analytically. The key observation is that the optical lever method does not determine the position of the end of the cantilever. It measures the orientation. In the previous sections use has always been made of the fact that for a force along one of the orthogonal symmetry directions at the end of the cantilever (normal force, lateral force, force coming from the front) there is a one-to-one correspondence of the tilt angle and the deflection. The problem is that the force coming from the front and the normal force create a deflection in the same direction. Hence, what is called the normal force component is actually a mixture of two forces. The deflection of the cantilever is the third quantity, which is not considered in most AFMs. A fiber-optic interferometer in parallel to the optical lever measures the deflection. Three measured quantities then allow the separation

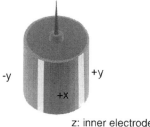

-y   +y

+x

z: inner electrode

FIGURE 2.18 Schematic drawing of a piezotube. The piezoceramic is molded into a tube form. The outer electrode is separated into four segments and connected to the scanning voltages. The z-voltage is applied to the inner electrode.

of the three orthonormal force directions, as is evident from Equations 2.28 and 2.35 (Fujisawa et al., 1994a,b; Fujisawa, Grafström et al., 1994; Overney et al., 1994; Warmack et al., 1994).

Alternatively, one can put the fast scanning direction along the axis of the cantilever. Forward and backward scans then exert opposite forces F_{Fr}. Provided that the piezo movement is linearized, this allows the determination of both components in AFMs based on the optical lever detection. In this cast the normal force is simply the average of the forces in the forward and backward direction. The force form the front, F_{Fr}, is the difference of the forces measured in forward and backward directions.

2.8 Scanning and Control Systems

Almost all scanning probe microscopes (SPMs) use piezotranslators to scan the tip or the sample. Even the first STM (Binnig and Rohrer, 1982; Binnig et al., 1982) and some of the predecessor instruments (Young et al., 1971, 1972) used them. Other materials or setups for nanopositioning have been proposed, but were not successful (Gerber and Marti, 1985; Garcìa Cantù and Huerta Garnica, 1990).

2.8.1 Piezotubes

A popular solution is use of tube scanners (Figure 2.18) They are now widely used in SPMs because of their simplicity and their small size (Binnig and Smith, 1986; Chen, 1992a,b). The outer electrode is segmented in four equal sectors of 90°. Opposite sectors are driven by signals of the same magnitude, but opposite sign. This gives, through bending, a two-dimensional movement on, approximately, a sphere. The inner electrode is normally driven by the z-signal. It is possible, however, to use only the outer electrodes for scanning and for the z-movement. The main drawback of applying the z-signal to the outer electrodes is that the applied voltage is the sum of both the x- or y-movement and the z-movement. Hence, a larger scan size effectively reduces the available range for the z-control.

2.8.2 Piezoeffect

An electric field applied across a piezoelectric material causes a change in the crystal structure, with expansion in some directions and contraction in others. Also, a net volume change occurs (Ashcroft and Mermin, 1976). Many SPMs use the transverse piezoelectric effect, where the applied electric field $\vec{E}$ is perpendicular to the expansion/contraction direction.

$$\Delta \ell = \ell \left(\vec{E} \cdot \vec{n} \right) d_{31} = \ell \frac{V}{t} d_{31} \tag{2.84}$$

where d_{31} is the transverse piezoelectric constant, V the applied voltage, and t the thickness of the piezoslab or the distance between the electrodes where the voltage is applied. $\vec{n}$ is the direction of polarization. Piezotranslators based on the transverse piezoelectric effect have a wide range of sensitivities, limited mainly by mechanical stability and breakdown voltage.

2.8.3   Scan Range

The calculation of the scanning range of a piezotube is difficult (Carr, 1988; Chen, 1992a,b). The bending of the tube depends on the electric fields and the nonuniform strain induced. A finite-element calculation where the piezotube was divided into 218 identical elements was used (Carr, 1988) to calculate the deflection. On each node the mechanical stress, stiffness, strain, and piezoelectric stress were calculated when a voltage was applied on one electrode. The results were found to be linear on the first iteration, and higher-order corrections were very small even for large electrode voltages. It was found that to first order the x- and z-movement of the tube could be reasonably well approximated by assuming that the piezotube is a segment of a torus. Using this model, one obtains

$$dx = \left(V_+ - V_-\right)|d_{31}|\frac{\ell^2}{2td} \qquad (2.85)$$

$$dz = \left(V_+ + V_- - 2V_z\right)|d_{31}|\frac{\ell^2}{2t} \qquad (2.86)$$

where $|d_{31}|$ is the coefficient of the transversal piezoelectric effect, ℓ is the tube free length, t is the tube wall thickness, d is the tube diameter, V_+ is the voltage on positive outer electrode while V_- is the voltage of the opposite quadrant negative electrode, and V_z is voltage of inner electrode.

The cantilever or sample mounted on the piezotube has an additional lateral movement because the point of measurement is not in the end plane of the piezotube. The additional lateral displacement of the end of the tip is $\ell_S \sin \varphi \approx \ell_S \varphi$, where ℓ_S is the tip length and φ is the deflection angle of the end surface. Assuming that the sample or cantilever is always perpendicular to the end of the walls of the tube and calculating with the torus model, one gets for the angle

$$\varphi = \frac{\ell}{R} = \ell\frac{2dx}{\ell^2} = \frac{2dx}{\ell} \qquad (2.87)$$

where R is the radius of curvature the piezotube. Using the result of Equation 2.87, one obtains for the additional x-movement

$$dx_{add} = \ell_S \varphi = \frac{2dx\ell_S}{\ell} = \left(V_+ - V_-\right)|d_{31}|\frac{\ell\ell_S}{td} \qquad (2.88)$$

and for the additional z-movement due to the x-movement

$$dz_{add} = \ell_S - \ell_S \cos \varphi = \frac{\ell_S \varphi^2}{2} = \frac{2\ell_S\left(dx\right)^2}{\ell^2} = \left(V_+ - V_-\right)^2|d_{31}|^2\frac{\ell_S\ell^2}{2t^2d^2} \qquad (2.89)$$

Carr assumed for his finite-element calculations that the top of the tube was completely free to move and, as a consequence, the top surface was distorted, leading to a deflection angle about half that of the geometric model. Depending on the attachment of the sample or the cantilever, this distortion may be smaller, leading to a deflection angle in between that of the geometric model and the one of the finite-element calculation.

2.8.4   Nonlinearities, Creep

Piezomaterials with a high conversion ratio, i.e., a large d_{31} or small electrode separations, with large scanning ranges are hampered by substantial hysteresis resulting in a deviation from linearity by more than 10%. The sensitivity of the piezoceramic material (mechanical displacement divided by driving voltage) decreases with reduced scanning range, whereas the hysteresis is reduced. A careful selection of the material for the piezoscanners, the design of the scanners, and of the operating conditions is necessary to get optimum performance.

2.8.5 Linearization Strategies

2.8.5.1 Passive Linearization: Calculation

The analysis of images affected by piezo nonlinearities (Libioulle et al., 1991; Stoll, 1992; Durselen et al., 1995; Fu, 1995) shows that the dominant term is

$$x = AV + BV^2 \qquad (2.90)$$

where z is the excursion of the piezo, V the applied voltage, and A and B two coefficients describing the sensitivity of the material. Equation 2.90 holds for scanning from $V = 0$ to large V. For the reverse direction, the equation becomes

$$x = \tilde{A}V = \tilde{B}\left(V - V_{\text{max}}\right)^2 \qquad (2.91)$$

where $\tilde{A}$ and $\tilde{B}$ are the coefficients for the back scan and V_{max} is the applied voltage at the turning point. Both equations demonstrate that the true x-travel is small at the beginning of the scan and becomes larger toward the end. Therefore, images are stretched at the beginning and compressed at the end.

Similar equations hold for the slow scan direction. The coefficients, however, are different. The combined action causes a greatly distorted image. This distortion can be calculated. The data acquisition systems record the signal as a function of V. However, the data are measured as functions of x. Therefore, we have to distribute the x-values evenly across the image, which can be done by inverting an approximation of Equation 2.90. First, we write

$$x = AV\left(1 - \frac{B}{A}V\right) \qquad (2.92)$$

For $B \ll A$, we can approximate

$$V = \frac{x}{A} \qquad (2.93)$$

We now substitute Equation 2.93 into the nonlinear term of Equation 2.92. This gives

$$x = AV\left(1 + \frac{Bx}{A^2}\right)$$

$$V = \frac{x}{A}\frac{1}{1 + Bx/A^2} \approx \frac{x}{A}\left(1 - \frac{Bx}{A^2}\right) \qquad (2.94)$$

Hence, an equation of the type

$$x_{\text{true}} = x\left(\alpha - x/x_{\text{max}}\right) \quad \text{with } \beta = \quad - \qquad \alpha \qquad (2.95)\beta$$

takes out the distortion of an image. α and β are dependent on the scan range, the scan speed, and on the scan history and have to be determined with exactly the same settings as for the measurement. x_{max} is the maximal scanning range. The condition for α and β guarantees that the image is transformed onto itself.

Similar equations as the empirical one shown above Equation 2.95 can be derived by analyzing the movements of domain walls in piezoceramics.

2.8.5.2 Passive Linearization: Measuring the Position

An alternative strategy is to measure the position of the piezotranslators. Several possibilities exist.

- The interferometers described above can be used to measure the elongation of the piezoelongation. The fiber-optic interferometer is especially easy to implement. The coherence length of the laser only limits the measurement range. However, the signal is of periodic nature. Hence, a direct use of the signal in a feedback circuit for the position is not possible. However, as a measurement tool and, especially, as a calibration tool the interferometer is without competition. The wavelength of the light, for instance, in an HeNe laser is so well defined that the precision of the other components determines the error of the calibration or measurement.

- The movement of the light spot on the quadrant detector can be used to measure the position of a piezo (Barrett and Quate, 1991). The output current changes by 0.5 A/cm $\times$ P [W]/R [cm].

- Typical values ($P = 1$ mW, $R = 0.001$ cm) give 0.5 A/cm = 5×10^{-8} A/nm.

- Again, this means that the laser beam above would have a 0.1-nm noise limitation for a bandwidth of 21 Hz. The advantage of this method is that, in principle, one can linearize two axes with only one detector.

- A knife-edge blocking part of a light beam incident on a photodiode can be used to measure the position of the piezo. This technique, commonly used in optical shear force detection (Betzig et al., 1992; Toledo-Crow et al., 1992), has a sensitivity of better than 0.1 nm.

- The capacitive detection (Griffith et al., 1990; Holman et al., 1996) of the cantilever deflection can be applied to the measurement of the piezoelongation. Equations 2.67 to 2.82 apply to the problem. This technique is commonly used in commercial instruments. The difficulties lie in the avoidance of fringe effects at the borders of the two plates. While conceptually simple, one needs the latest technology in surface preparation to get a decent linearity. The electronic circuits used for the readout are often proprietary.

- Linear variable differential transformers (LVDT) are a convenient means to measure positions down to 1 nm. They can be used together with a solid-state joint setup, as often used for large scan range stages. Unlike the capacitive detection, there are few difficulties to implementation. The sensors and the detection circuits for LVDTs are available commercially.

- A popular measurement technique is use of strain gauges. They are especially sensitive when mounted on a solid-state joint where the curvature is maximal. The resolution depends mainly on the induced curvature. A precision of 1 nm is attainable. The signals are low — a Wheatstone bridge is needed for the readout.

2.8.5.3 Active Linearization

Active linearization is done with feedback systems. Sensors need to be monotonic. Hence, all the systems described above, with the exception of the interferometers, are suitable. The most common solutions include the strain gauge approach, the capacitance measurement, or the LVDT, which are all electronic solutions. Optical detection systems have the disadvantage that the intensity enters into the calibration.

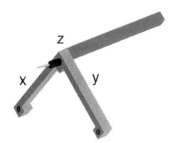

FIGURE 2.19   An alternative type of piezoscanners: the tripod.

2.8.6   Alternative Scanning Systems

The first STMs were based on piezotripods (Binnig and Rohrer, 1982). The piezotripod (Figure 2.19) is an intuitive way to generate the three-dimensional movement of a tip attached to its center. However, to get a suitable stability and scanning range, the tripod needs to be fairly large (about 5 cm). Some instruments use piezostacks instead of monolithic piezoactuators. They are arranged in the tripod arrangement. Piezostacks are thin layers of piezoactive materials glued together to form a device with up to 200 μm of actuation range. Preloading with a suitable metal casing reduces the nonlinearity.

If construction of a home-built scanning system is attempted, using linearized scanning tables is recommended. They are built around solid-state joints and actuated by piezostacks. The joints guarantee that the movement is parallel with little deviation from the predefined scanning plane. Because of the construction, it is easy to add measurement devices such as capacitive sensors, LVDT, or strain gauges, which are essential for a closed-loop linearization. Two-dimensional tables can be purchased from several manufacturers. They have a linearity of better than 0.1% and a noise level of 10^{-4} to 10^{-5} of the maximal scanning range.

2.8.7   Control Systems

2.8.7.1   Basics

The electronics and software play an important role in the optimal performance of an SPM. Control electronics and software are supplied nowadays with commercial AFMs. Control electronic systems can use either analog or digital feedback. While digital feedback offers greater flexibility and ease of configuration, analog feedback circuits might be better suited for ultralow noise operation. We will describe here the basic setups for atomic force microscopy.

Figure 2.20 shows a block schematic of a typical AFM feedback loop. The signal from the force transducer is fed into the feedback loop consisting mainly of a subtraction stage to get an error signal and an integrator. The gain of the integrator (high gain corresponds to short integration times) is set as high as possible without generating more than 1% overshoot. High gain minimizes the error margin of the current and forces the tip to follow the contours of constant density of states as well as possible. This operating mode is known as the "constant-force mode." A high-voltage amplifier amplifies the outputs of the integrator. As AFMs using piezotubes usually require ±150 V at the output. The output of the integrator needs to be amplified by a high-voltage amplifier.

In order to scan the sample, additional voltages at high tension are required to drive the piezo. For example, with a tube scanner, four scanning voltages are required, namely, $+V_x$, $-V_x$, $+V_y$, and $-V_y$. The x- and y-scanning voltages are generated in a scan generator (analog or computer controlled). Both voltages are input to the two respective power amplifiers. Two inverting amplifiers generate the input voltages for the other two power amplifiers. The topography of the sample surface is determined by recording the input- voltage to the high-voltage amplifier for the z-channel as a function of x and y (constant-force mode).

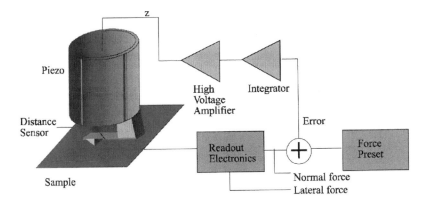

FIGURE 2.20 Block schematics of the feedback control loop of AFMs.

Another operating mode is the "variable-force mode." The gain in the feedback loop is lowered and the scanning speed increased such that the force on the cantilever is no longer constant. Here the force is recorded as a function of x and y.

2.8.7.2 Force Spectroscopy

Four modes of spectroscopic imaging are in common use with force microscopes: measuring lateral forces, $\partial F/\partial z$, $\partial F/\partial x$ spatially resolved, and measuring force vs. distance curves. Lateral forces can be measured by detecting the deflection of a cantilever in a direction orthogonal to the normal direction. The optical lever deflection method does this most easily. Lateral force measurements give indications of adhesion forces between the tip and the sample.

$\partial F/\partial z$ measurements probe the local elasticity of the sample surface. In many cases the measured quantity originates from a volume of a few cubic nanometers. The $\partial F/\partial z$, or local stiffness signal, is proportional to Young's modulus, as far as one can define this quantity. Local stiffness is measured by vibrating the cantilever by a small amount in the z-direction. The expected signal for very stiff samples is zero; for very soft samples one also gets, independent of the stiffness, a constant signal. This signal is again zero for the optical lever deflection and equal to the driving amplitude for interferometric measurements. The best sensitivity is obtained when the compliance of the cantilever matches the stiffness of the sample.

A third spectroscopic quantity is the lateral stiffness. It is measured by applying a small modulation in the x-direction on the cantilever. The signal is again optimal when the lateral compliance of the cantilever matches the lateral stiffness of the sample. The lateral stiffness is, in turn, related to the shear modulus of the sample.

Detailed information on the interaction of the tip and the sample can be gained by measuring force vs. distance curves. It is necessary to have cantilevers with high enough compliance to avoid instabilities due to the attractive forces on the sample.

2.8.7.3 Using the Control Electronics as a Two-Dimensional Measurement Tool

Usually, the control electronics of an AFM is used to control the x- and y-piezosignals while several data acquisition channels record the position-dependent signals. The control electronics can be used in another way: it can be viewed as a two-dimensional function generator. What is normally the x- and y-signal can be used to control two independent variables of an experiment. The control logic of the AFM then ensures that the available parameter space is systematically probed at equally spaced points.

2.8.7.3.1 *Friction Force Curves on a Line*

An example is a friction force curve measured along a line across a step on graphite. Figure 2.21 shows the connections. The z-piezo is connected as usual, like the x-piezo. However, the y-output is used to command the electrochemical potential.

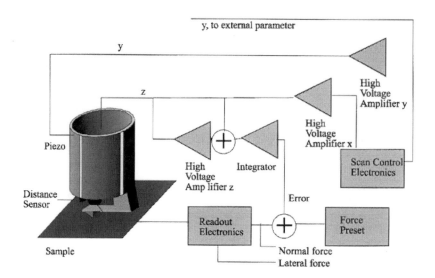

FIGURE 2.21 Wiring of an AFM to measure friction force curves along a line.

The offset of the y-channel determines the position of the tip on the sample surface, together with the x-channel. Figure 2.22 shows a typical result. The image shows the lateral force on one scan line across a step on highly pyrolytic graphite (HOPG), as a function of the applied electrochemical potential. A discussion of this result can be found in the literature (Binggeli et al., 1993; Weilandt et al., 1995a,b).

2.9   AFMs

2.9.1   Special Design Considerations

The size of an AFM is typically of the order of several centimeters. To illustrate the problem of thermal drift, we calculate the requirements on the temperature stability for a microscope working with repulsive forces and which does not employ heterodyne detection. If we assume a cantilever spring with a 1 N/m

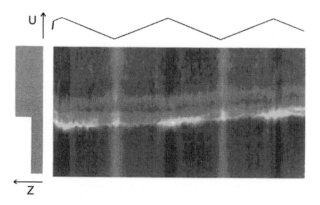

FIGURE 2.22 Friction measurement curve along a sample line. Bright colors encode high lateral forces. Friction was modified by changing the applied potential.

spring constant and if we set the force to 10^{-8} N, then the static deflection of the cantilever is 10 nm. The typical size of a tunneling force detector is 1-cm length from the tunnel junction to the common attachment plane. A design with well-compensated thermal expansion coefficients will have a remaining thermal expansion coefficient of 10^{-6} 1/K. This means that keeping the force within 10% requires a thermal stability of the microscope of 0.1 K. In less well compensated design, the allowable temperature fluctuations might be as low as 0.01 K.

If the temperature stability of the setup is not sufficient, one can either use larger static deflections, which means larger forces, or a softer cantilever spring, which means degraded frequency response. For measurements with the smallest possible forces, a careful design of the force sensor with respect to thermal drift is a prerequisite.

The classical Michelson or Mach–Zehnder interferometers have the worst temperature drift, since their relevant distances for differential thermal expansions may be more than 10 cm long. The fiber-optic interferometer is comparable to the tunneling detector in its thermal performance, since the distances needed to position the end of the fiber are of order 1 cm. Much better is the Nomarski detector for the cantilever deflection. This detector is only sensitive to a thermally induced rotation of the cantilever spring. A crude estimate gives relevant distances for the thermal expansion of a few 10 mm. This increases the allowable temperature variations to more than 1 K.

Equally well suited is the optical lever method. This method is, to first order, only sensitive to the tilt of the reflecting mirror. For small angles between the incident and the reflected light beam, the change in distance between the plane defined by the quadrant detector and the light source is negligible. Any distance change between the light source and the quadrant detector directly affects the output signal. However, the deflection of the cantilever is amplified by a factor of up to 1000 due to the geometric amplification. Hence, the optical lever method is, to first order, insensitive to thermal drift.

2.9.2 Classical Setup

Scanning tunneling microscopes are almost exclusively built according to one principle: the tunneling tip is mounted on some translation stage and scanned past the sample. The reason for this is twofold. First, most STMs were built for ultrahigh vacuum operation, where it is of paramount importance that the sample can be exchanged. Since piezos are rather fragile, it was the natural choice to scan the tip. The second reason is that the tip is lighter than the sample. As one can deduce from the equations describing cantilevers, reduced masses mean increased resonance frequency and hence increased scanning speed. By following these rules, AFMs were built with scanned samples. Most AFMs operate at ambient conditions. The force detection mechanism is bulky; especially after the implementation of the optical lever (Meyer and Amer, 1988; Alexander et al., 1989), where the detection unit is of the size of several centimeters, microscope designers preferred to scan the sample instead of the sensor..

Figure 2.23 shows an example of such a microscope. It consists of two parts, the base and force sensor head. The base houses the scanning piezo, the motor driving the approach screw, and the sample. All connectors from the control unit to the microscope end here. The base acts as an additional vibration damper, shielding the sensitive parts from the environment. The force sensor head includes the cantilever, a laser diode with suitable optics, and a position-sensitive detector. The preamplifier for this detector is often located on the head. This minimizes the stray capacitance and hence the noise and maximizes the bandwidth. The microscope shown in Figure 2.23 relies on external vibration damping. Common means are an air-cushioned table or a concrete platform suspended by bungee cords.

If one is not satisfied with the scan range or the linearity of a microscope, then there is the possibility to replace the base by a commercial linearized scanning table. The geometry is the same; all that is needed is to adapt the support and the approach system of the force-sensing head.

AFMs operating in air usually have their vibration damping system built-in. Figure 2.25 shows the multimode ultrahigh vacuum SPM of Omicron (Omicron; Howald et al., 1993). The instrument is based on the optical lever principle (see Figure 2.24). The light beam coming from the laser diode is directed by a mirror mounted on a piezomotor with two tilt directions to the cantilever. The reflected light is

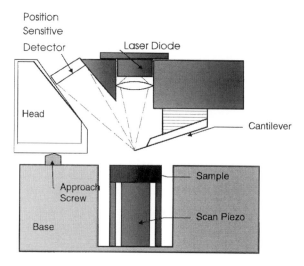

FIGURE 2.23   An example of the basic AFM setup.

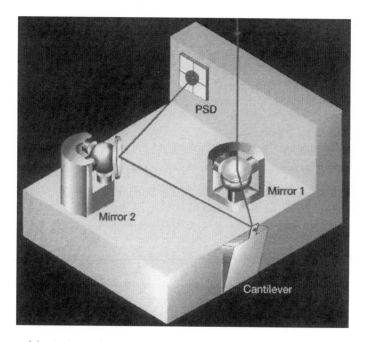

FIGURE 2.24   Setup of the Omicron ultrahigh vacuum AFM. (Courtesy of Omicron. Used with permission.)

steered onto the quadrant detector by another mirror mounted on a piezomotor. The piezomotors are essential for the design because it is not possible to have an adjustment from the exterior side in the ultrahigh vacuum.

The optics shown in Figure 2.24 is mounted on a vibration isolation platform, as can be seen in Figure 2.25. The three pillars house a spring each. The microscope platform is supported by these springs. Damping is provided by the magnetic eddy current brakes, shown at the rim of the instrument.

The microscope in Figure 2.25 is capable of imaging delicate samples such as the silicon (111) surface at atomic resolution in noncontact mode. Figure 2.26 is an example of such a measurement. The resolution is proved by the fact that defects are visible.

FIGURE 2.25 View of a commercial AFM for ultrahigh vacuum operation. (Courtesy of Omicron. Used with permission.)

2.9.3 Stand-Alone Setup

The disadvantage of the classical AFM setup is that the sample size is limited, both by the acceleration power of the scanning system and by the allowable size. Many samples are of extended size and often it is not desirable to cut them into small pieces. Therefore, a design where the force sensor is scanned past the sample is desirable. A first implementation (Bryant et al., 1988) used tunneling from the force sensor to the tip to detect the deflection of the cantilever. In environments such as the ambient where adsorbate layers cover both the cantilever and the tunneling tip, there might be problems adjusting the force (Marti et al., 1988; Meyer and Amer, 1988; Alexander et al., 1989) The adsorbate layer between the tip and the cantilever can stiffen the cantilever and induce unwanted large forces.

By using the interaction of light backscattered by a cantilever into a laser diode (Sarid et al., 1988; Sarid et al., 1989, 1990), it was possible to build a compact AFM, where the cantilever was scanned. This system, however, was not able to measure lateral forces.

Realizing that the divergence angle of a light beam reflected from a cantilever is given either by the focal diameter or by the width of the cantilever (whichever is smaller) (Colchero, 1993), it was found that by using a rather broad laser beam with a focal diameter of 50 µm one could safely displace the cantilever for more than 10 µm laterally without any unwanted cross talk, provided the light was incident perpendicularly to the cantilever (Hipp et al., 1992) (see Figure 2.27). This can be achieved with the setup shown in Figure 2.28. A laser diode serves as a light source, and the linearly polarized light passes through a polarizing beam splitter and a quarter wave plate. The laser is focused onto the cantilever. Since the light is incident perpendicularly on the cantilever, its polarization state will not change. For all other angles of incidence, the *s*- and *p*-polarized light will have different reflectivities. The cantilever now deflects the light according to the force acting on it. The deflection angles are of the order of microradians and, hence, negligible. The returning circularly polarized light is converted to linear polarized light by

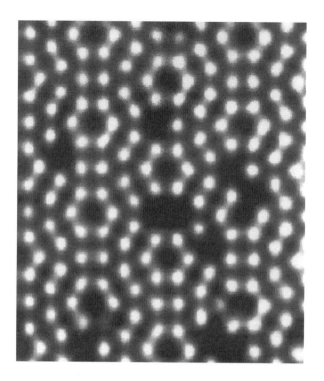

FIGURE 2.26 Silicon (111) surface measured by noncontact atomic force microscopy. (Courtesy of Omicron. Used with permission.) The color scale corresponds to the frequency shift in the oscillation of the cantilever. It is a measure of the interaction strength.

FIGURE 2.27 An implementation of the stand-alone principle discussed above. (From Hipp, M. et al., 1992, *Ultramicroscopy*, 42–44, 1498–1503. With permission.)

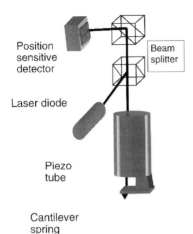

Position
sensitive
detector

Beam
splitter

Laser diode

Piezo
tube

Cantilever
spring

FIGURE 2.28 Optical path in a stand-alone AFM. The incident light impinges perpendicularly onto the cantilever. A quarter wave plate serves as a separator for the incoming and the outgoing light.

the quarter wave plate. The linear polarization state is orthogonal to the incident one. Therefore, the polarizing beam splitter cube now keeps the returning light away from the laser and directs it to the quadrant diode. The main limitation of this setup is the limited scanning range, due to the focal diameter of the laser beam.

As was discussed in the section on scanning systems, there are powerful long-range scanners available nowadays using piezostacks and solid-state joints. These scanners can also serve as a mount for the optical beam deflection system. Since the entire optics are scanned, the problem of the cantilever moving out of the focus of the laser beam is absent. The scanning range is only limited by the available stroke of the translation table. Some commercial instruments are based on this principle.

An exciting new development is the integration of an AFM into the lens holder of a classical microscope, as shown in Figure 2.29. A detailed view of this principle is shown in Figure 2.30. This microscope has a scan range of larger than 30 μm, combined with a resolution of better than 1 nm. Only a clever arrangement of components and extreme miniaturization made it possible to design it. The objective contains everything, from scanners to the approach unit.

2.9.4 Data Acquisition

The interaction between the tip and the sample is usually of very short range. The range of the interaction is inversely proportional to the potentially achievable resolution. The tunneling current in the STM, in particular, changes approximately by an order of magnitude for a distance change of the separation of

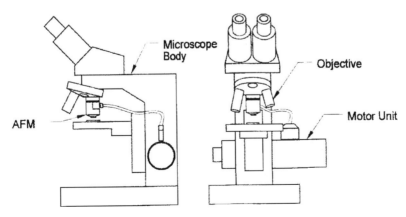

Microscope
Body

Objective

Motor Unit

AFM

FIGURE 2.29 A sketch of the mounting principle of the SIS stand-alone instrument. (Courtesy of SIS. With permission.)

FIGURE 2.30 The stand-alone microscope Ultra-Objective™ from SIS. (Courtesy of SIS. With permission.)

100 pm. However, many surfaces have surface features with height differences that are much larger than the range of the control interaction. Therefore, the probe height above the sample has to be adjusted to avoid uncontrolled contact with the sample or the loss of information.

The height of the probe above the sample is controlled by a feedback loop (Park and Quate, 1987; Marti et al., 1988; DiLella et al., 1989; Piner and Reifenberger, 1989; Troyanovskii, 1989; Grafström et al., 1990; Jeon and Willis, 1991; Schummers et al., 1991). A schematic sketch of such a feedback loop was shown in Figure 2.20. The probe tip or, alternatively, the sample is connected to a converter which outputs a voltage proportional to the interaction between the tip and the sample. The output of the converter is low-pass filtered and fed into a linearizing network, whose output is compared with a preset voltage. The difference voltage or error voltage is fed into a feedback control amplifier. The simplest of those amplifiers consists of an integrator. To speed up the response of the instrument, proportional and differential gain can be added (DiStefano et al., 1976). The resulting signal controls the high-voltage amplifiers for the z-piezotranslator. The feedback systems has stringent requirements: in an STM it is not tolerable if the overshoot at a step is larger than 100 pm. Set into relation with a typical z-range of 1 μm the precision must be 0.01%. Likewise, the noise level must be of even better quality. High gain minimizes the error margin of the control interaction and forces the tip to follow the contours of constant-interaction signal magnitude as well as possible. This mode of operation is known as the "constant-interaction" mode. The information on the surface topography can be found in the control signal for the z-piezo. The "constant-height" mode is used for flat samples. The probe tip is scanned across the sample without feedback control. The variations of the interaction signal reflect the topography. Linear interactions such as that in contact-mode AFM are easy to translate into true heights of the sample surface.

Signal generators produce the drive voltages for the x- and y-scan piezos. Usually the fast scan direction is labeled by x. Its scanning frequency might be as high as 1000 Hz. The y-scan frequency is lower by the number of lines in the image.

The feedback control and scan generation system can be implemented with analog electronics or with digital signal processors. While digital feedback control offers greater flexibility in the choice of the frequency response, allows adaptive controls, and is easy to reconfigure (Aguilar et al., 1986, 1987; Becker, 1987; Brown and Cline, 1990; Baselt et al., 1993), analog feedback circuits might be better suited for ultralow noise operation.

The basic feedback loops can be expanded to allow a wide variety of imaging modes. Often the feedback loop can be interrupted and the position of the probe tip or the control voltages or signal are varied. The response of the probe signal is sampled. This mode of imaging is called "spectroscopic imaging"

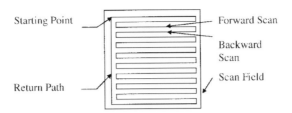

FIGURE 2.31 Scan pattern in an AFM.

(Baratoff et al., 1986; Hamers et al., 1986; Stroscio et al., 1986; Bell et al., 1990; van Kempen, 1990; Schummers et al., 1991).

2.9.4.1 Sampling Theorem Applied to AFM

AFMs acquire the data in a raster scan fashion. This means that the surface is probed in lines, one adjacent to the other. In each line there are only a limited number of points. Figure 2.31 shows a typical scan pattern. There are variations to this pattern. The backward scan can be on the same line as the forward scan. The sample can be scanned backward from the bottom to return to the starting point.

The scanning pattern is characterized by the number of points along the line N_p and the number of lines N_L. If the size of the scanning pattern is x_p by y_L, then one can calculate the distance between two points or two lines. We obtain

$$\Delta x = \frac{x_P}{N_P}$$

$$\Delta y = \frac{y_L}{N_L}$$

(2.96)

Assume that the sample has features with a characteristic size of x_S and y_S. These characteristic sizes can only be resolved if the following inequalities hold:

$$x_S > 2\Delta x_P$$

$$y_S > 2\Delta y_L$$

(2.97)

Equation 2.97 is a fundamental limitation of sampling data (Yaroslavski, 1985). It is intuitively clear that, if one wants to measure a sine, one has to sample at least two points per period. If this effect is not observed, one can get data at the wrong positions. Figure 2.32 shows a simulation of this effect. It is obvious that, for instance, the sine with period 5 is not sampled correctly at one quarter of the sampling points. The dangerous thing is that due to aliasing a wrong period is appearing. A similar effect can be seen for the square waves with a period 4. Even at one third the sampling rate, one misses the true periodicity, since the sampling theorem is violated. The single point spikes above 450 are imaged or not, depending on the sampling rate.

The conclusion from this experiment is that if one wants to image something with a characteristic distance of, for example, 1 nm, then one needs at least 200, better 600, data points if the field of view is 100 nm on the side. Otherwise, one is bound to have aliasing effects and, consequently, moiré patterns and a loss of information.

Since a sampling spacing in real space is translated to a frequency by the scanning speed, one can also apply the sampling theorem to frequency.

$$f_x = \frac{v_x}{\Delta x_P}$$

(2.98)

Sampling Theorem

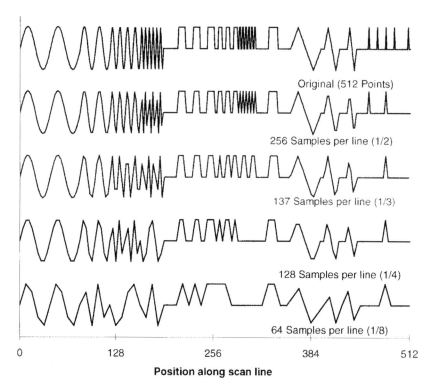

Original (512 Points)

256 Samples per line (1/2)

137 Samples per line (1/3)

128 Samples per line (1/4)

64 Samples per line (1/8)

0   128   256   384   512

Position along scan line

FIGURE 2.32   Effect of the sampling theorem. The successive lines from the top to the bottom are the original and sampled data. The original contains the following objects (from the left side): sine with period 40 (from 0 to 80), sine with period 20 (from 80 to 120), sine with period 10 (from 120 to 160), sine with period 5 (from 160 to 190), a square wave with period 18 (190 to 256), a square wave with period 10 (256 to 286), a square wave with period 4 (286 to 310), a single square pulse of width 12 between 310 and 356, a triangle of period 40 between 356 and 396, a triangle of period 20 between 400 and 420, a triangle of period 12 between 430 and 442, and, finally, delta pulses (1 wide) at 458, 469, 480, 491, and 509.

Equation 2.98 shows this effect for the fast scanning direction. In frequency the sampling theorem means that the digitizing rate needs to be at least twice as high as the highest frequency in the signal. An example demonstrates this. If one wants to measure something with a 1-nm periodicity at a scanning speed of 1 µm/s at least 2000 samples per second are needed, better yet 12,000 samples per second. In the time/frequency domain it is easy to prevent aliasing. Just use a low-pass filter with the maximum allowable frequency (the Nyquist frequency) in front of the analog-to-digital converter. It prevents aliasing. If the sampling theorem is violated, then nothing will be seen.

2.9.5   Typical Setups

The analysis of the sampling theorem suggests that a typical setup for the analog-to-digital conversion should look like Figure 2.33 or Figure 2.34. The difference between the two figures lies in the fact that for an analog feedback circuit one has to digitize the topographical signal, which is at the output of the feedback controller, whereas in a digital feedback circuit the computer calculates the z-signal from the error signal. In both cases, one must match the cutoff frequency of the low-pass filters to the conversion rate of the digital-to-analog (D/A) or analog-to-digital (A/D) converters.

   The precision of the converters limits the resolution; 16 bits are equal to one part in 65,536. By observing the sampling theorem, this means that the smallest object must be 32,768 times smaller than

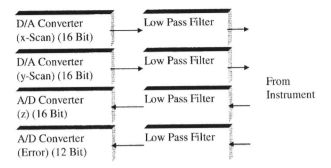

FIGURE 2.33 Typical setup of a data acquisition system with an analog feedback circuit.

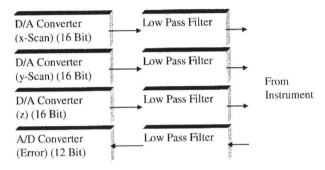

FIGURE 2.34 Typical setup of a data acquisition system with a digital feedback system.

TABLE 2.3 Sampling Resolution (in nanometers) as a
Function of the Scan Range and the Converter Resolution

	8 Bit	12 Bit	16 Bit	18 Bit	20 Bit
1 nm	0.008	4.9E-04	3.1E-05	7.6E-06	1.9E-06
10 nm	0.078	4.9E-03	3.1E-04	7.6E-05	1.9E-05
100 nm	0.78	0.05	0.003	7.63E-04	1.91E-04
1 μm	7.81	0.49	0.031	0.008	0.002
10 μm	78.13	4.88	0.31	0.08	0.02
100 μm	781.25	48.83	3.05	0.76	0.19
250 μm	1953.13	122.07	7.63	1.91	0.48

the scanning range. Table 2.3 shows a compilation of the resolution as a function of the most common maximal scanning ranges and some D/A (or A/D) converter resolutions. Since the 16-bit converters are very common (because of the CD–audio market), most instruments use this resolution. A 100-μm scanner is capable of resolving a little bit more than 1 nm.

2.9.6 Data Representation

The visualization and interpretation of images from SPMs is intimately connected to the processing of these images.

An ideal scanning probe microscope is a noise-free device that images a sample with perfect tips of known shape and has perfect linear scanning piezos. In reality, AFMs are not that ideal. The scanning device in AFMs is affected by distortions. To do quantitative measurements, such as determining the unit cell size, these distortions have to be measured on test substances and have to be corrected for. The distortions are both linear and nonlinear. Linear distortions mainly result from imperfections in the machining of the piezotranslators causing cross talk from the z-piezo to the x- and y-piezos, and vice versa. Among the linear distortions, there are two kinds which are very important: first, scanning piezos invariably have different sensitivities along the different scan axes due to the variation of the piezo material and uneven sizes of the electrode areas. Second, the same reasons might cause the scanning axis to not be orthogonal. Furthermore, the plane in which the piezoscanner moves for constant z is hardly ever coincident with the sample plane. Hence, a linear ramp is added to the sample data. This ramp is especially bothersome when the height z is displayed as an intensity map, also called a top-view display.

The nonlinear distortions are harder to deal with. They can affect AFMs for a variety of reasons. First, piezoelectric ceramics do have a hysteresis loop, much like ferromagnetic materials. The deviations of piezoceramic materials from linearity increase with increasing amplitude of the driving voltage. The mechanical position for one voltage depends on the voltages applied to the piezo before. Hence, to get the best position accuracy, one should always approach a point on the sample from the same direction. Another type of nonlinear distortion of the images occurs when the scan frequency approaches the upper frequency limit of the x- and y-drive amplifiers or the upper frequency limit of the feedback loop (z-component). This distortion, due to the feedback loop, can only be minimized by reducing the scan frequency. On the other hand, there is a simple way to reduce distortions due to the x- and y-piezo drive amplifiers. To keep the system as simple as possible, one normally uses a triangular waveform for driving the scanning piezos. However, triangular waves contain frequency components at multiples of the scan frequency. If the cutoff frequency of the x- and y-drive electronics or of the feedback loop is too close to the scanning frequency (two to three times the scanning frequency), the triangular drive voltage is rounded off at the turning points. This rounding error causes, first, a distortion of the scan linearity and, second, through phase lags, the projection of part of the backward scan onto the forward scan. This type of distortion can be minimized by carefully selecting the scanning frequency and by using driving voltages for the x- and y-piezos with waveforms like trapezoidal waves, which are closer to a sine wave. The values measured for x, y, or z are affected by noise. The origin of this noise can be either electronic, some disturbances, or a property of the sample surface due to adsorbates. In addition to this incoherent noise, interference with mains and other equipment nearby might be present. Depending on the type of noise, one can filter it in the real space or in Fourier space. The most important part of image processing is to visualize the measured data. Typical AFM data sets can consist of many thousands to over a million points per plane. There may be more than one image plane present. The AFM data represents a topography in various data spaces.

We use data from a combined measurement of topography, stiffness, and adhesion to outline the different points of data processing. Figure 2.35a shows the topography, Figure 2.35b the local stiffness image, and Figure 2.35c the adhesion image acquired simultaneously, as described in (Marti et al., 1997). The same data can be rendered as a series of consecutive cross sections. Figure 2.36 is an example. Line renderings can be excellent to judge the general form of the topography, independent of color settings. The original data set contains 256 lines with 256 points each. Since it is not possible to draw 256 distinct lines, the data in Figure 2.36 show every sixth line. Along the lines, all the points are drawn.

Similarly, the data can also be rendered in a wire mesh fashion (Figure 2.37). Again, a line is drawn at every sixth point both in the horizontal and vertical direction. It is especially suitable for monochrome display systems with only two colors. The number of scan lines that can be displayed is usually well below 100 and the display resolution along the fast scanning axis x is much better than along y.

If the computer display is capable of displaying at least 64 shades of gray, then top-view images can be created (Figure 2.38). In these images, the position on the screen corresponds to the position on the sample and the height is coded as a shade of gray. Usually, the convention is that the brighter a point, the higher it is. The number of points that can be displayed is only limited by the number of pixels available. This view of the data is excellent for measuring distances between surface features. Periodic structures show up particularly well on such a top view. The human eye is not capable of distinguishing more than 64 shades of gray. If the average z-height of the tip varies from one side of the image to the other, then the interesting features usually have too little contrast. Hence, contrast equalization is needed.

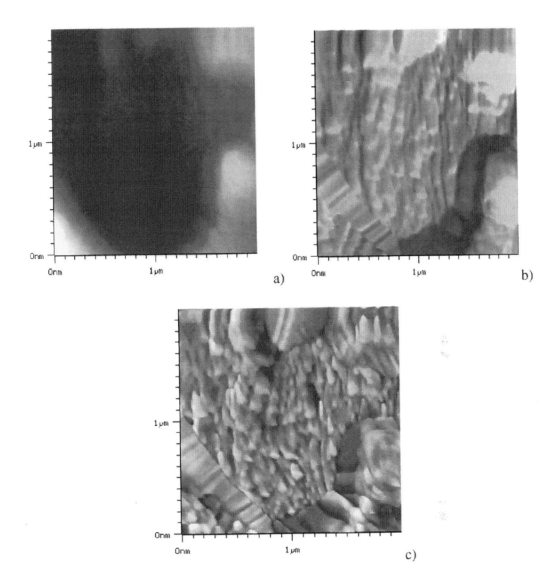

FIGURE 2.35 A rubber sample: (a) shows the topography, (b) the local stiffness, and (c) the adhesion.

For data being affected by a large background slope, it is often still possible to detect some features in the line scan view. Some researchers prefer a simultaneous display of both line scan images and top-view images to get the most information in the shortest time. Top views use much less calculation time than line scan images. Hence, computerized fast data acquisition systems usually display the data as a top view first.

Figure 2.35 shows in a detailed way what the topography of the sample is. However, the fine details in the low-lying parts are obscured. There are simply not enough shades which the eye can distinguish. The display can be enhanced by adding some illumination to it. In illuminated data sets one depicts the cosine of the angle between the surface normal and some predefined direction, the illumination direction. This technique is a powerful tool to enhance the appearance of a data set, but it can be abused! Changing the direction of the light source, as shown in Figure 2.39, can obscure some undesired features.

The effect of the illumination is similar to displaying the magnitude of the gradient of the sample surface along the direction to the light source. Features perpendicular to the illumination cannot be seen.

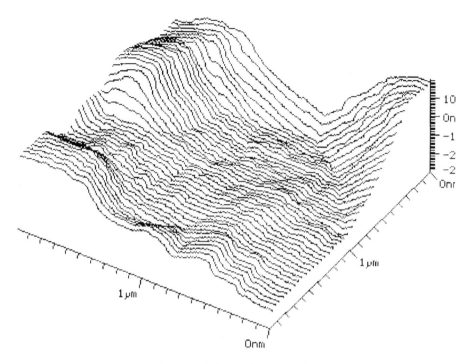

FIGURE 2.36   A line rendering of the topography (Figures 2.35).

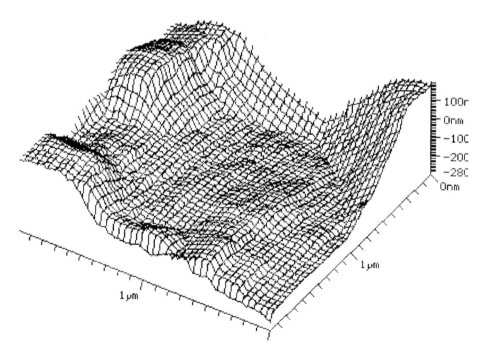

FIGURE 2.37   A wire mesh rendering of the topography of a rubber sample (Figure 2.35).

Multiple light sources or extended light sources diminish this effect, but the illumination is much more complicated to calculate. If not shown in conjunction with some other display method, one is not able to judge the validity of such an image.

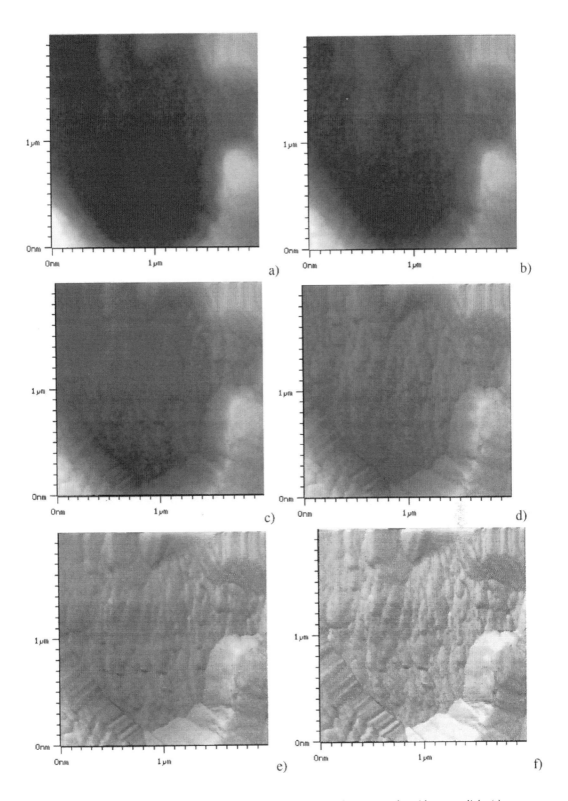

FIGURE 2.38 Variation of the light added to a topography. (a) is the topography without any light (the same as Figure 2.35). (c) is the same data, but rendered with a light source from the right at 50° elevation. (b) has 20% light added, (c) 40%, (d) 60%, (e) 80%, and (f) 100%.

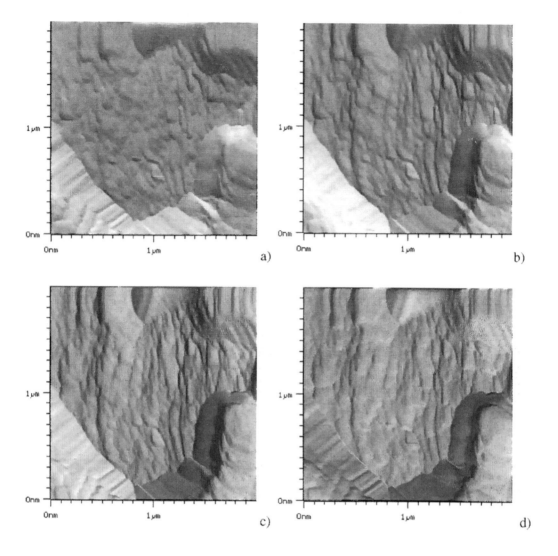

FIGURE 2.39(a–d)

Figure 2.40 shows the influence of the elevation angle on the displayed image. It is clear that an illumination from about 30° to 60° is optimal. One can combine top views or illuminated top views and wire mesh scan displays to form solid surface models of the sample surface. Such images are usually only generated in the final processing stage before publication, because they need quite a lot of computing time.

Continuous shading techniques sometimes make it difficult to analyze the topography. By using discontinuous color tables, one can create something similar to contour maps. An example is shown in Figure 2.41.

Figure 2.42 shows a combination of the top view display and the wire mesh display, a three-dimensional model where the height is coded as a shade of gray. As with the top view (Figure 2.35), one can vary the amount of illumination added to the data. Figures 2.42 through 2.47 show the influence of the amount of illumination in the display.

Data consisting of more than one channel can be packed into one three-dimensional image. One channel, usually the topography, is used to model the height, whereas the second channel is used to produce the shades. Figure 2.48 shows the topography shaded with the stiffness. Figure 2.49 shows the same data, but colored with the topography.

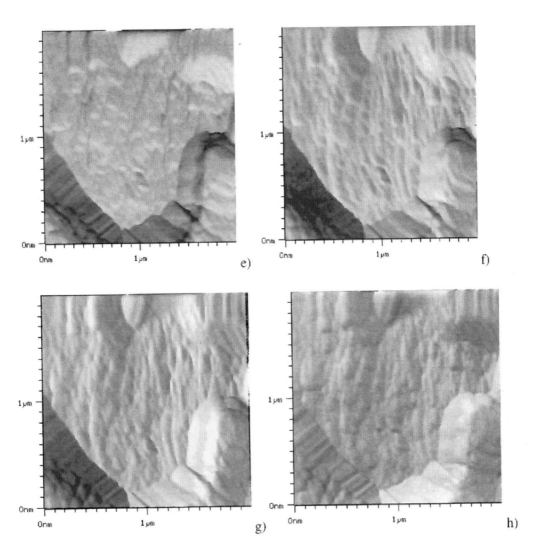

FIGURE 2.39 Illuminated surface: (a) shows an illumination from the top (0°) with an elevation of 50°. (b) is from 45° (upper right) — (b) and all the other illuminations have an elevation of 50°. (c) is at 90°, (d) at 135°, (e) at 180°, (f) at 225°, (g) at 270°, and (h) at 315°.

Additional information can be packed into an image by using color. Assume that an image has two planes of data. We can display the first plane with shades of green and the second one with shades of red on top of each other. Where the magnitude of both planes is high, one gets an orange color; where both are low, one gets black. But if the magnitude of one plane is larger than that of the other plane on one pixel, one gets red or green colors. This way, one can display the registry of two different quantities in the same image.

2.9.7 The Two-Dimensional Histogram Method

To analyze quantitatively friction microscopy data, one would like to calculate local friction coefficients. Normally, one operates a friction microscope in the constant-force mode, recording the topography $z(x,y)$ and the lateral force $F_L(x,y)$. If the normal force F_N is kept constant, one can get a local friction coefficient $\gamma(x,y) = F_L(x,y)/F_N$. $\gamma(x,y)$ is well defined, if the normal force F_N is truly constant, if the calibrations of both forces are known, and, most crucially, if the zero points of the force scales are known and stable.

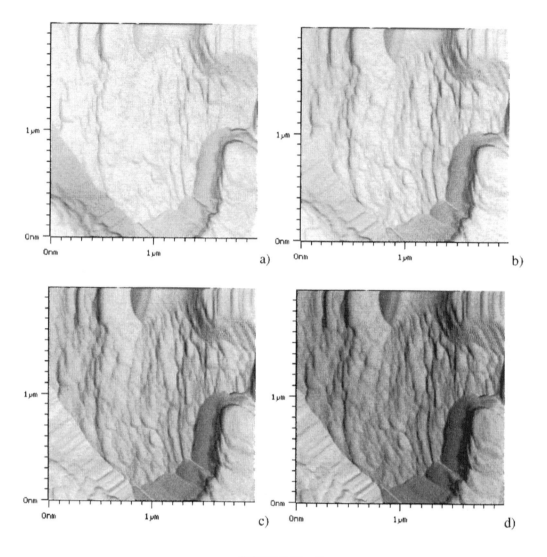

FIGURE 2.40(a–d)

In many force microscopes the scales can be calibrated rather accurately, for instance, using the procedure outlined above, but the zero points are quite unreliable. The zeros usually do not drift very much, but it is hard to determine their exact position on the force scales.

Figure 2.50 shows the two-dimensional histograms calculated with data from Figure 2.35. These histograms, first used as a data reduction means in communications with satellites (Yaroslavski, 1985), show correlations between different data channels. In the case of friction data, the lateral force $F_L(x,y)$ and the normal force $F_N(x,y)$ are used as an address to bins arranged in a two-dimensional array (Marti, 1993a,b; Marti et al., 1993). For every data point (x,y) the bin addressed by $(F_N(x,y), F_L(x,y))$ is incremented by one. The two-dimensional histogram then shows the statistical weight of every combination. For a constant friction coefficient and a varying normal force, one obtains $(F_N(x,y), F_L(x,y))$ pairs located on a straight line, with the slope equal to the friction coefficient. Since the creation of a two-dimensional histogram is independent of the zero point values, it is not necessary to know the offsets. If different materials are present on the sample surface and if they have friction coefficients of sufficiently different magnitudes, then the two materials will fall into different groups of bins in the two-dimensional histogram. Displaying only points falling into one group of bins will select all the locations where one material is present (Marti 1993a,b; Lüthi et al., 1995; Meyer et al., 1996).

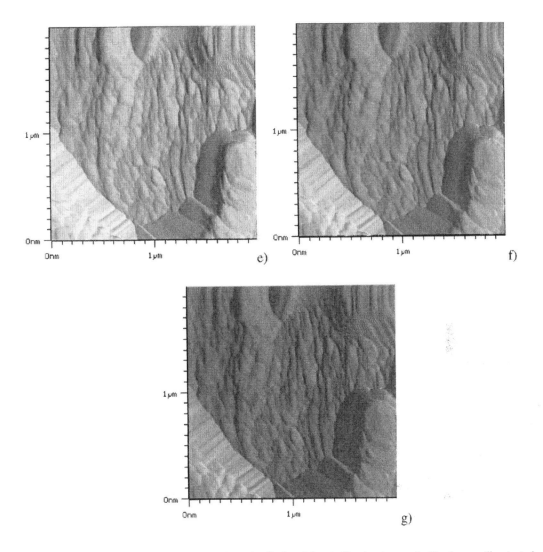

FIGURE 2.40 Influence of the elevation angle on the displayed data in illumination mode. The data are illuminated from the right. (a) has an elevation angle of 0° (from top). (b) is at 15°, (c) at 30°, (d) at 45°, (e) at 60°, (f) at 75°, and (g) at 90° (parallel to the surface).

As was shown above, the two-dimensional histogram analysis is not limited to an analysis of friction force data. It could be applied to other problems in scanning probe microscopy, for instance, relating spectroscopic data to the tunneling current or to determine the locations of excessive forces or tunneling currents, to name a few.

2.9.8   Some Common Image-Processing Methods

Since many commercial data acquisition systems use implicitly some kind of data processing, we will describe some of the effects of these procedures. Since the original data is commonly subject to slopes on the surface, most programs use some kind of slope correction. Figure 2.51 shows the effect of such algorithms. In Figure 2.51a there is the original data, without any correction. The least disturbing way is to subtract a plane $z(x,y) = a_x x + a_y y + a_0$ from the data. The coefficients are determined by fitting $z(x,y)$ to the data. As can be seen in Figure 2.51b, the main effect is to lower the right side and to lift up the left side. A more severe option is to subtract a second order function such as $\bar{z}(x,y) = a_{xx} x^2 + a_{yy} y^2 + a_{xy} xy + a_x x + a_y y + a_0$. Again, the parameters are determined with a fit. This function is appropriate for

131

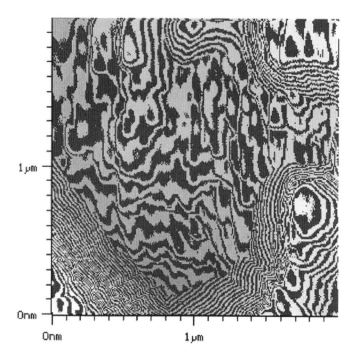

FIGURE 2.41 A discontinuous color table can generate a data display similar to contour maps. Here we have used a color table with 32 steps.

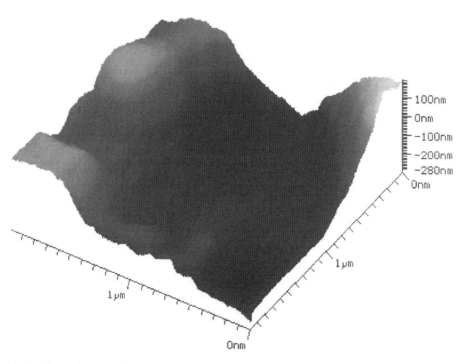

FIGURE 2.42 Three-dimensional rendering of the data of Figure 2.35. The shading of the surface is purely topographical.

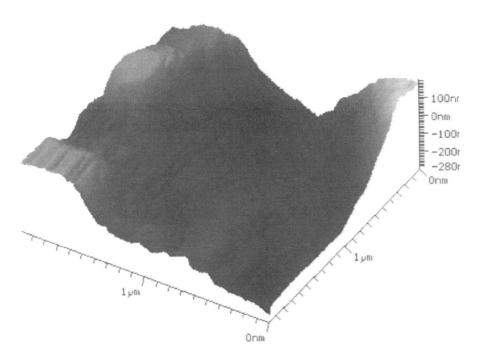

FIGURE 2.43 Three-dimensional rendering of the data of Figure 2.35. The shading of the surface is 80% topographical and 20% illumination.

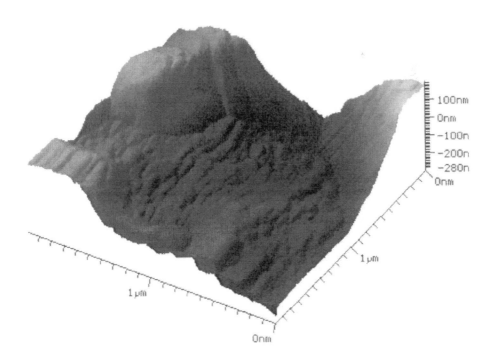

FIGURE 2.44 Three-dimensional rendering of the data of Figure 2.35. The shading of the surface is 60% topographical and 40% illumination.

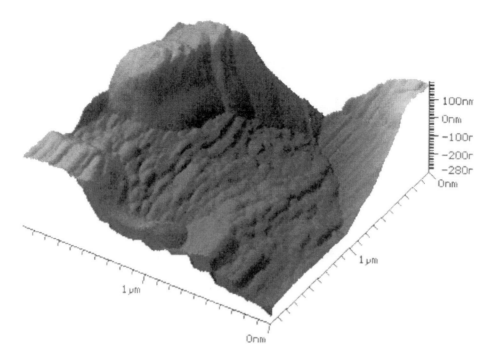

FIGURE 2.45 Three-dimensional rendering of the data of Figure 2.35. The shading of the surface is 40% topographical and 60% illumination.

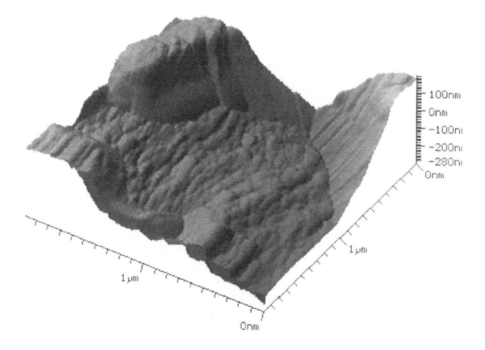

FIGURE 2.46 Three-dimensional rendering of the data of Figure 2.35. The shading of the surface is 20% topographical and 80% illumination.

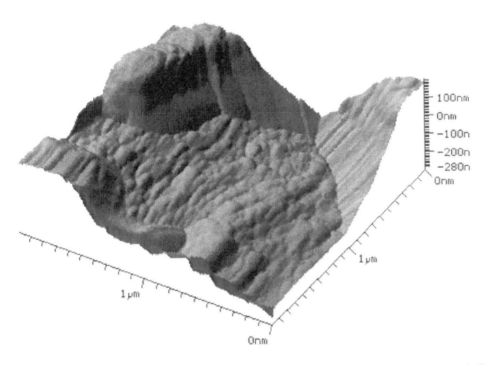

FIGURE 2.47   Three-dimensional rendering of the data of Figure 2.35. The shading of the surface is purely illumination.

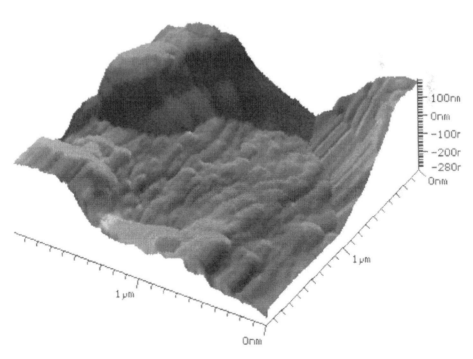

FIGURE 2.48   Three-dimensional rendering of the data of Figure 2.35. The topography channel determines the height, whereas the stiffness channel gives the color.

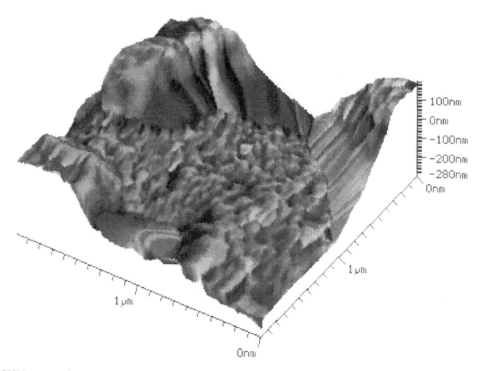

FIGURE 2.49 Three-dimensional rendering of the data of Figure 2.35. The topography channel determines the height, whereas the adhesion channel gives the color.

almost plane data, where the nonlinearity of the piezos caused such a distortion. It is obvious that the data are now grossly distorted. Both algorithms have two disadvantages. They can only be used for data that are already known and they can be calculation intensive.

During data acquisition, only part of the data is known. Therefore, other algorithms which only work on single rows or columns are preferred. Figure 2.51d shows the effect of subtracting the average of a row from every point in the row. Data sets that have fine structure and no long-range variations of the slope can safely be subject to this algorithm. These data, however, are affected and change its character. Figure 2.51e shows the same procedure, but now applied to the columns. Figure 2.51f is an extension to subtracting the data. Here not only the average value is subtracted, but also the slope of a line fitted to one row. Hence, all the tilt is also taken out of the data. This is a dangerous function, since it can completely change the data appearance. However, most data acquisition programs use exactly this function. Figure 2.51g, finally, is the same procedure applied to rows.

Figure 2.51h shows the effect of an unsharp mask. This procedure first calculates a low-pass filtered image by averaging all the points in the neighborhood and then subtracting this value. As can be seen, the effect is a high-pass filter where all the slow variations are gone. While this technique is ideally suited to finding out if there are short-range variations in the data.

With data sets such as that in Figure 2.35 where low-lying data are almost not visible, one might be tempted to apply a technique widely used in image processing: histogram equalization (Figure 2.52). This technique calculates a cumulative histogram of the z-values. The number as a function of the height class defines a function. The inverse of this function is then applied to the data. After this operation the histogram has an equal number of points in the height classes. Since it is a nonlinear function, we do not recommend that it be used on AFM data.

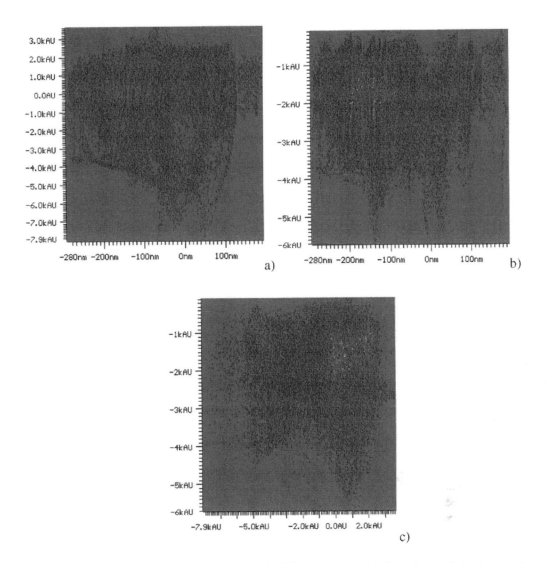

FIGURE 2.50 Two-dimensional histograms between the different channels. (a) shows the correlation between the topography (horizontal) and the local stiffness. (b) shows the correlation between the topography (horizontal) and the adhesion. (c), finally, is the correlation between the local stiffness (horizontal) and the adhesion.

Acknowledgments

The author thanks his colleagues and co-workers for many enlightening discussions. Special thanks go to Jaime Colchero, Michael Hipp, Eva Weilandt, Sabine Hild, Armin Rosa, Bernd Zink, Georg Krausch, Bernd Heise, Martin Pietralla, Gerd-Ingo Asbach, Roland Winkler, and Joachim Spatz. Many of the insights have been gained while preparing lectures for very interesting students.

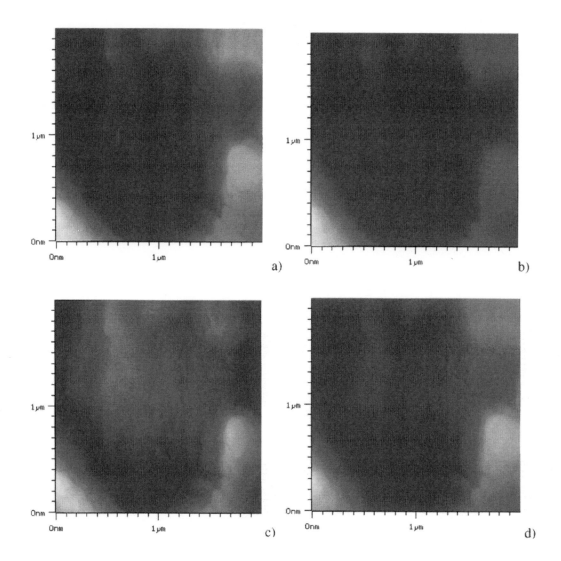

FIGURE 2.51(a-d)

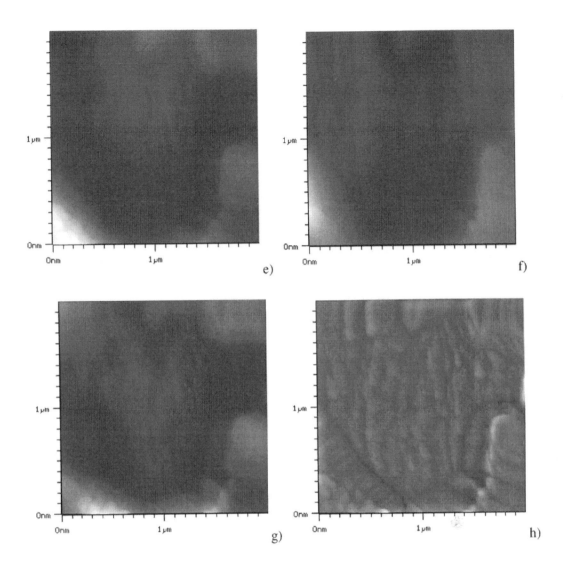

e)

f)

g)

h)

FIGURE 2.51 The effect of different slope correction algorithms. (a) shows the original data. In (b) a plane was fitted to the data and subtracted. (c) show the effect of subtracting a second-order surface from the data. (c) shows the effect of setting the average of each row to 0. In (e) the same has been done with the columns. (f) shows the effect of setting the average and the average slope of each row to 0. (g), finally, shows the same, but for the columns. (h) shows the effect of unsharp masking with an averaging area of about one fifth of the image size.

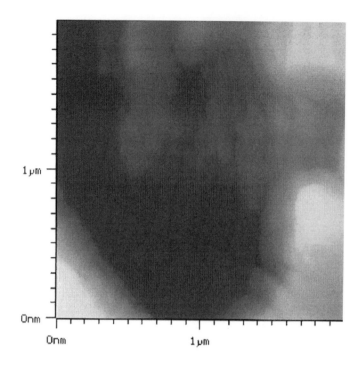

FIGURE 2.52 Application of histogram equalization.

References

Aguilar, M., Pascual, P. J., and Santisteban, A. (1986). "Scanning Tunneling Microscope Automation." *IBM J. Res. Dev.* 30(5), 525–532.

Aguilar, M., Garcìa, A., Pascual, P. J., Presa, J., and Santisteban, A., (1987). "Computer System for Scanning Tunneling Microscope Automation," *Surf. Sci.* 181, 191.

Akamine, S. and Quate, C. F. (1992). "Low Temperature Thermal Oxidation Sharpening of Microcast Tips." *J. Vac. Sci. Technol.* B10(6), 2307–2311.

Akamine, S., Albrecht, T. R., Zdeblick, M. J., and Quate, C. F. (1990a). "A Planar Process for Microfabrication of a Scanning Tunneling Microscope," *Sensors and Actuators*, A21-A23, 964–970.

Akamine, S., Barrett, R. C., and Quate, C. F. (1990b). "Improved Atomic Force Microscope Images Using Microcantilevers with Sharp Tips." *Appl. Phys. Lett.* 57, 316.

Alexander, S., Hellemans, L., Marti, O., Schneir, J., Elings, V., Hansma, P. K., Longmire, M., and Gurley, J. (1989). "An Atomic-Resolution Atomic-Force Microscope Implemented Using an Optical Lever." *J. Appl. Phys.* 65, 164.

Anders, M. and Heiden, C. (1988). "Imaging of Tip-Sample Compliance in STM." *J. Microsc.* 152, 643.

Ashcroft, N. W. and Mermin, N. D. (1976). *Solid State Physics,* Holt, Rinehart, and Winston, New York.

Baratoff, A., Binnig, G., Fuchs, H., Salvan, F., and Stoll, E. (1986). "Tunneling Microscopy and Spectroscopy of Semiconductor Surfaces and Interfaces." *Surf. Sci.* 168, 734.

Barrett, R. C. and Quate, C. F. (1991). "Optical Scan-Correction System Applied to Atomic Force Microscopy." *Rev. Sci. Instrum.* 62(6), 1393.

Baselt, D. R., Clark, S. M., Youngquist, M. G., Spence, C. F., and Baldeschwieler, J. D. (1993). "Digital Signal Processor Control of Scanned Probe Microscope." *Rev. Sci. Instrum.* 64(7), 1874–1883.

Baumeister, T. and Marks, S. L. (1967). *Standard Handbook for Mechanical Engineers,* McGraw-Hill, New York.

Becker, J. (1987). "Scanning Tunneling Microscope Automation." *Surf. Sci.* 181, 200.

Bell, L. D., Hecht, M. H., Kaiser, W. J., and Davis, L. C. (1990). "Direct Spectroscopy of Electron and Hole Scattering." *Phys. Rev. Lett.* 64, 2679.

Betzig, E., Finn, P. L., and Weiner, J. S. (1992). "Combined Shear Force and Near-Field Scanning Optical Microscopy." *Appl. Phys. Lett.* 60(20), 2484.

Binggeli, M., Christoph, R., Hintermann, H.-E., Colchero, J., and Marti, O. (1993). "Friction Force Measurements at Potential Controlled Graphite in Electrolytic Environment." *Nanotechnology* 4, 59-63.

Binnig, G. and Rohrer, H. (1982). "Scanning Tunneling Microscopy." *Helv. Phys. Acta* 55, 726.

Binnig, G. and Smith, D. P. E. (1986). "Single-Tube Three-Dimensional Scanner for Scanning Tunneling Microscopy." *Rev. Sci. Instrum.* 57, 1688.

Binnig, G., Rohrer, H., Gerber, C., and Weibel, E. (1982). "Vacuum Tunneling." *Physica* 109&110B, 2075.

Binnig, G., Quate, C. F. and Gerber, C. (1986). "Atomic Force Microscope." *Phys. Rev. Lett.* 56(9), 930–933.

Binnig, G., Gerber, C., Stoll, E., Albrecht, T. R., and Quate, C. F. (1987). "Atomic Resolution with Atomic Force Microscope." *Europhys. Lett.* 3(12), 1281–1286.

Blackman, G. S., Mate, C. M., and Philpott, M. R. (1990). "Interaction Forces of a Sharp Tungsten Tip with Molecular Films on Silicon Surface." *Phys. Rev. Lett.* 65, 2270.

Brown, A. and Cline, R. W. (1990). "A Low Cost, High Performance Imaging System for Scanning Tunneling Microscopy." *Rev. Sci. Instrum.* 61, 1484.

Bryant, P. J., Miller, R. G., Deeken, R., Yang, R., and Zheng, Y. C. (1988). "Scanning Tunneling and AFM Performed with the Same Probe in One Unit." *J. Microsc.* 152, 871.

Carr, R. G. (1988). *J. Microsc.* 152, 379.

Chen, C. J. (1992a). "Electromechanical Deflections of Piezoelectric Tubes with Quartered Electrodes." *Appl. Phys. Lett.* 60(1), 132.

Chen, C. J. (1992b). "In Situ Testing and Calibration of Tube Piezo Piezoelectric Scanners." *Ultramicroscopy* 42–44, 1653–1658.

Colchero, J. (1993). Reibungskraftmikroskopie. *Physics Faculty.* Konstanz, University of Konstanz: 198.

Colchero, J., Marti, O., Bielefeldt, H., and Mlynek, J. (1991). "Scanning Force and Friction Microscopy." *Phys. Status Solid.* (a) 131, 73–75.

den Boef, A. J. (1991). "The Influence of Lateral Forces in Scanning Force Microscopy." *Rev. Sci. Instrum.* 62, 88.

DiLella, D. P., Wandass, J. H., Colton, R. J., and Marrian, C. R. K. (1989). "Control Systems for Scanning Tunneling Microscopes with Tube Scanners." *Rev. Sci. Instrum.* 60(6), 997–1002.

DiStefano, J. J., III, Stubberud, A. R. and Williams, I. J. (1976). *Theory and Problems of Feedback and Control Systems,* McGraw-Hill, New York.

Durselen, R., Grunewald U., and Preuss W. (1995). "Calibration and Applications of a High Precision Piezo Scanner for Nanometrology." *Scanning.* 17, 91–96.

Fu, J. (1995). "In Situ Testing and Calibrating of *z*-Piezo of an Atomic Force Microscope." *Rev. Sci. Instrum.* 66, 3785–3788.

Fujisawa, S., Kishi, E., Sugawara, Y., and Morita, S. (1994). "Fluctuation in 2-Dimensional Stick-Slip Phenomenon Observed with 2-Dimensional Frictional Force Microscope." *Jpn. J. Appl. Phys. Pt.1.* 33, 3752–3755.

Fujisawa, S., Ohta, M., Konishi, T., Sugawara, Y., and Morita, S. (1994b). "Difference between the Forces Measured by an Optical Lever Deflection and by Optical Interferometer in an Atomic Force Microscope." *Rev. Sci. Instrum.* 65, 644–647.

Garcìa Cantù, R. and Huerta Garnica, M. A. (1990). "Long-Scan Imaging by STM." *J. Vac. Sci. Technol.* A8, 354.

Gerber, C. and Marti, O. (1985). "Magnetostrictive Positionner." *IBM Tech. Disclosure Bull.* 27, 6373.

Grafström, S., Kowalski, J., and Neumann, R. (1990). "Design and Detailed Analysis of a Scanning Tunnelling Microscope." *Meas. Sci. Technol.* 1, 139–146.

Grafstrom, S., Ackermann, J., Hagen, T., Neumann, R., and Probst, O. (1994). "Analysis of Lateral Force Effects on the Topography in Scanning Force Microscopy." *J. Vac. Sci. Technol. B.* 12, 1559–1564.

Griffith, J. E., Miller, G. L., and Green, C. A. (1990). "A Scanning Tunneling Microscope with a Capacitance-Based Position Monitor." *J. Vac. Sci. Technol. B* 8(6), 2023–2027.

Grütter, P., Rugar, D. et al. (1990). "Batch Fabricated Sensors for Magnetic Force Microscopy." *Appl. Phys. Lett.* 57, 1820.

Hamers, R. J., Tromp, R. M., and Demuth, J. E. (1986). "Surface Electronic Structure of Si(111)-7×7 Resolved in Real Space." *Phys. Rev. Lett.* 56, 1972.

Hipp, M., Bielefeldt, H., Colchero, J., Marti, O., and Mlynek, J. (1992). "A Stand-Alone Scanning Force and Friction Microscope." *Ultramicroscopy* 42–44, 1498–1503.

Holman, A. E., Laman, C. D., Scholte, P. M. L. O., Heerens, W. C., and Tuinstra, F. (1996). "A Calibrated Scanning Tunneling Microscope Equipped with Capacitive Sensors." *Rev. Sci. Instrum.* 67(6), 2274–2280.

Howald, L., Haefke, H., Lüthi, R., Meyer, E., Gerth, G., Rudin, H., and Güntherodt, H.-J. (1993). "Ultrahigh Vacuum Scanning Force Microscopy: Atomic Resolution at Monoatomic Cleavage Steps." *Phys. Rev. B* 49, 5651–5656.

Israelachvili, J. N. (1985). *Intermolecular and Surface Forces.* Academic Press, London.

Jeon, D. and Willis, R. F. (1991). "Feedback System Response in a Scanning Tunneling Microscope." *Rev. Sci. Instrum.* 62, 1650.

Jones, R. V. (1970). *Proc. IEEE* 17, 1185.

Kassing, R. and Oesterschulze, E. (1997). "Sensors for Scanning Probe Microscopy," in *Micro/Nanotribology and Its Applications.* (B. Bhushan, ed.), 330, 35–54, Kluwer Academic Publishers, Dordrecht.

Kirk, M. D., Albrecht, T. R., and Quate, C. F. (1988). "Low Temperature Atomic Force Microscopy." *Rev. Sci. Instrum.* 59(6), 833–835.

Libioulle, N., Ronda, A., Taborelli, M., and Gilles, J. M. (1991). "Deformations and Nonlinearity in Scanning Tunneling Microscope Images." *J. Vac. Sci. Technol. B* 9(2), 655–658.

Lüthi, R., Meyer, E., Haefke, H., Howald, L., Gutmannsbauer, W., Guggisberg, M., Bammerlin, M., and Güntherodt, H.-J. (1995). "Nanotribology: An UHV-SFM Study on Thin Films of C_{60} and AgBr." *Surf. Sci.* 338, 247–260.

LVDT Part E115.21 from Physik Instrumente GmbH&Co, Polytec Platz 5-7, D-76377 Waldbrnn; AD598 from Analog Devices, One Technology Way, P.O. Box 9106, Norwood, MA 02062-9106, U.S.A.

Marti, O. (1993a). "Friction and Measurement of Friction on a Nanometer Scale." *Surf. Coatings Technol.* 62, 510–516.

Marti, O. (1993b). "Nanotribology: Friction on a Nanometer Scale." *Phys. Scr.* T49, 599–604.

Marti, O., Drake, B., and Hansma, P. K. (1987). "Atomic Force Microscopy of Liquid-Covered Surfaces: Atomic Resolution Images." *Appl. Phys. Lett.* 51(7), 484–486.

Marti, O., Gould, S., and Hansma, P. K. (1988). "Control Electronics for Atomic Force Microscopy." *Rev. Sci. Instrum.* 59(6), 836–839.

Marti, O., Colchero, J., and Mlynek, J. (1990). "Combined Scanning Force and Friction Microscopy of Mica." *Nanotechnology* 1, 141–144.

Marti, O., Colchero, J., Bielefeldt, H., Hipp, M., and Linder, A. (1993). "Scanning Probe Microscopy — Applications in Biology and Physics." *Microsc. Microanal. Microstruct.* 4, 429–440.

Marti, O., Hild, S., Staud, J., Rosa, A., and Zink, B. (1997). "Nanomechanical Interactions of Scanning Force Microscope Tips with Polymer Surfaces," in *Micro/Nanotribology and Its Applications.* (B. Bhushan, ed.), E:330, 455–456, Kluwer Scientific Publishers, Dordrecht.

Mate, C. M., McClelland, G. M., Erlandsson, R., and Chiang, S. (1987). "Atomic-Scale Friction of a Tungsten Tip on a Graphite Surface." *Phys. Rev. Lett.* 59(17), 1942–1945.

McClelland, G. M., Erlandsson, R., and Chiang, S. (1987). "Atomic Force Microscopy: General Principles and a New Implementation." *Rev. Progr. Quant. Non-Destr. Eval.* 6, 1307.

Meyer, E., Lüthi, R., Howald, L., Bammerlin, M., Guggisberg, M., and Güntherodt, H.-J. (1996). "Friction Force Spectroscopy," in *Physics of Sliding Friction* (B. N. J. Persson and E. Tosatti, eds.), E311, 349, Kluwer Academic Publishers, Dordrecht.

Meyer, G. and Amer, N. M. (1988). "Novel Optical Approach to Atomic Force Microscopy." *Appl. Phys. Lett.* 53(12), 1045–1047.

Meyer, G. and Amer, N. M. (1990). "Simultaneous Measurement of Lateral and Normal Forces with an Optical-Beam-Deflection AFM." *Appl. Phys. Lett.* 57(20), 2089–2091.

Omicron Omicron Vakuumphysik GmbH, Idsteiner Strasse 78, D-65232 Taunusstein.

Overney, R. M., Takano, H., Fujihira, M., Paulus, W., and Ringsdorf, H. (1994). "Anisotropy in Friction and Molecular Stick-Slip Motion." *Phys. Rev. Lett.* 72, 3546–3549.

Park, S.-I. and Quate, C. F. (1987). "Theories of the Feedback and Vibration Isolation Systems for the STM." *Rev. Sci. Instrum.* 58(11), 2004–2009.

Piner, R. and Reifenberger, R. (1989). "Computer Control of the Tunnel Barrier Width for the STM." *Rev. Sci. Instrum.* 60(10), 3123–3127.

Pitsch, M., Metz, O., Kohler, H.-H., Heckmann, K., and Strnad, J. (1989). "Atomic Resolution with a New Atomic Force Tip." *Thin Solid Films* 175, 81.

Pohl, D. W. (1986). "Some Design Criteria in STM." *IBM J. Res. Dev.* 30, 417.

Rugar, D., Mamin, H. J., and Güthner, P. (1989). "Improved Fiber-Optic Interferometer for Atomic Force Microscopy." *Appl. Phys. Lett.* 55(25), 2588–2590.

Sarid, D. (1991). *Scanning Force Microscopy with Applications to Electric, Magnetic and Atomic Forces,* Oxford University Press, Oxford.

Sarid, D., Iams, D., Weissenberger, V., and Bell, L. S. (1988). "Compact Scanning-Force Microscope Using a Laser Diode." *Opt. Lett.* 13(12), 1057–1059.

Sarid, D., Weissenberger, V., Iams, D. A., and Ingle, J. T. (1989). "Theory of the Laser Diode Interaction in Scanning Force Microscopy." *IEEE J. Quantum Electr.* 25, 1968.

Sarid, D., Iams, D. A., Ingle, J. T., Weissenberger, V., and Ploetz, J. (1990). "Performance of a Scanning Force Microscope Using a Laser Diode." *J. Vac. Sci. Technol.* 8, 378.

Schönenberger, C. and Alvarado, S. F. (1990). "Understanding Magnetic Force Microscopy," *Zeitschrift für Physik* B80, 373–383.

Schönenberger, C. and Alvarado, S. F. (1989). "A Differential Interferometer for Force Microscopy." *Rev. Sci. Instrum.* 60(10), 3131–3135.

Schummers, A., Halling, H., Besocke, K. H., and Cox, G. (1991). "Controls and Software for Tunneling Spectroscopy." *J. Vac. Sci. Technol.* B9, 615.

Shen, Y.R. (1984). *The Principles of Nonlinear Optics,* John Wiley & Sons, New York.

Stahl, U., Yuan, C. W., Delozanne, A. L., and Tortonese, M. (1994). "Atomic Force Microscope Using Piezoresistive Cantilevers and Combined with a Scanning Electron Microscope." *Appl. Phys. Lett.* 65, 2878–2880.

Stoll, E. P. (1992). "Restoration of STM Images Distorted by Time-Dependent Piezo Driver Aftereffects." *Ultramicroscopy* 42–44, 1585–1589.

Stroscio, J. A., Feenstra, R. M., and Fein, A. P. (1986). "Electronic Structure of the Si(111)-2×1 Surface by Scanning Tunneling Microscopy." *Phys. Rev. Lett.* 57, 2579.

Thomson, W. T. (1988). *Theory of Vibration with Applications,* Unwin Hyman, London.

Toledo-Crow, R., Yang, P. C., Chen, Y., and Vaez-Iravani, M. (1992). "Near-Field Differential Scanning Optical Microscope with Atomic Force Regulation." *Appl. Phys. Lett.* 60(24), 2957–2959.

Troyanovskii, A. M. (1989). "Feedback Control System for STM." *Instrum. Exp. Tech.* (USSR) 32, 188.

van Kempen, H. (1990). "Spectroscopy Using Conduction Electrons" in *Scanning Tunneling Microscopy and Related Methods* (R. J. Behm, N. Garcìa, and H. Rohrer, eds.), E 184, 242–267, Kluwer Academic Publishers, Dordrecht.

Warmack, R. J., Zheng, X. Y., Thundat, T., and Allison, D. P. (1994). "Friction Effects in the Deflection of Atomic Force Microscope Cantilevers." *Rev. Sci. Instrum.* 65, 394–399.

Weilandt, E., Menck, A., Binggeli, M., and Marti, O. (1995a). "Friction Force Measurements on Graphite Steps under Potential Control." in *Electrochemistry* (A. A. Gewirth and H. Siegenthaler, eds.) E:288, 307–315, Kluwer, Doordrecht.

Weilandt, E., Menck, A., and Marti, O. (1995b). "Friction Studies at Steps with Friction Force Microscopy." *Surf. Interface Anal.* 23, 428–430.

Wolter, O., Bayer, T., and Gerschner, J. (1991). "Micromachined Silicon Sensors for Scanning Force Microscopy." *J. Vac. Sci. Technol.* B9, 1353.

Yaroslavski, L. P. (1985). *Digital Picture Processing*, Springer Verlag, Berlin.

Young, R., Ward, J., and Scire, F. (1971). "Observation of Metal-Vacuum-Metal Tunneling, Field Emission, and the Transition Region." *Phys. Rev. Lett.* 27, 922.

Young, R., Ward, J., and Scire, F. (1972). "The Topographiner: An Instrument for Measuring Surface Microtopography." *Rev. Sci. Instrum.* 43, 999.

3

Surface Physics in Tribology

John Ferrante and Phillip B. Abel

3.1 Introduction

Tribology, the study of the interaction between surfaces in contact, spans many disciplines from physics and chemistry to mechanical engineering and material science. Besides the many opportunities for interesting research, it is of extreme technological importance. The key word in this chapter is surface. The chapter will be rather ambitious in scope in that we will attempt to cover the range from microscopic considerations to the macroscopic experiments used to examine the surface interactions. We will approach this problem in steps, first considering the fundamental idea of a surface and next recognizing its atomic character and the expectations of a ball model of the atomic structures present, viewed as a terminated bulk. We will then consider a more realistic description of a relaxed surface and then consider how the class of surface, i.e., metal, semiconductor, or insulator affects these considerations. Finally, we will present what is expected when a pure material is alloyed, as well as the effects of adsorbates.

Following these more fundamental descriptions, we will give brief descriptions of some of the experimental techniques used to determine surface properties and their limitations. The primary objective here will be to provide a source for more thorough examination by the interested reader.

145

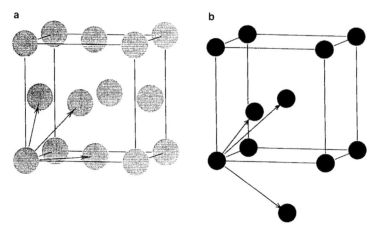

FIGURE 3.1   (a) Unit cube of fcc crystal structure with primative cell basis vectors indicated. (b) Unit cube of bcc crystal structure, with primative cell basis vectors indicated.

Finally, we will examine the relationship of tribological experiments to these more fundamental atomistic considerations. The primary goals of this section will be to again provide sources for further study of tribological experiments and to raise critical issues concerning the relationship between basic surface properties with regard to tribology and the ability of certain classes of experiments to reveal the underlying interactions. We will attempt to avoid overlapping the material that we present with that presented by other authors in this publication. This chapter cannot be a complete treatment of the physics of surfaces due to space limitations. We recommend an excellent text by Zangwill (1988) for a more thorough treatment. Instead, we concentrate on techniques and issues of importance to tribology on the nanoscale.

3.2   Geometry of Surfaces

We will now discuss simply from a geometric standpoint what occurs when you create two surfaces by dividing a solid along a given plane. We limit the discussion to single crystals, since the same arguments apply to polycrystalline samples except for the existence of many grains, each of which could be described by a corresponding argument. This discussion will start by introducing the standard notation for describing crystals given in many solid-state texts (Ashcroft and Mermin, 1976; Kittel, 1986). It is meant to be didactic in nature and because of length limitations will not attempt to be comprehensive. To establish notation and concepts we will limit our discussion to two of the possible Bravais lattices, face-centered cubic (fcc) and body-centered cubic (bcc), which are the structures often found in metals. The unit cells, i.e., the structures which most easily display the symmetries of the crystals, are shown in Figure 3.1. The other descriptions that are frequently used are the primitive cells, which show the simplest structures that can be repeated to create a given structure. In Figure 3.1 we also show the primitive cell basis vectors, which can be used to generate the entire structure by the relation

$$\vec{R} = n_1\vec{a} + n_2\vec{a}_2 + n_3\vec{a}_3 \tag{3.1}$$

where n_1, n_2, and n_3 are integers, and $\vec{a}_1$, $\vec{a}_2$, and $\vec{a}_3$ are the unit basis vectors.

Since we are interested in describing surface properties, we want to present the standard nomenclature for specifying a surface. The algebraic description of a surface is usually given in terms of a vector normal to the surface. This is conveniently accomplished in terms of vectors that arise naturally in solids, namely, the reciprocal lattice vectors of the Bravais lattice (Ashcroft and Mermin, 1976; Kittel, 1986). This is

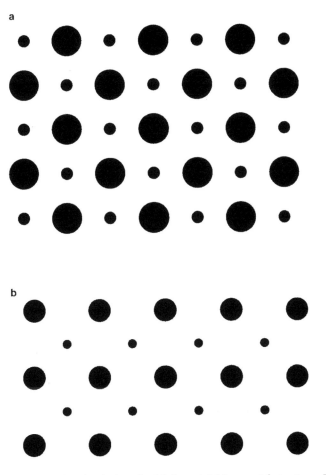

FIGURE 3.2 Projection of cubic face (100) plane for (a) fcc and (b) bcc crystal structures. In both cases, smaller dots represent atomic positions in the next layer below the surface.

convenient since these vectors are used to describe the band structure and diffraction effects in the solid. They are usually given in the form

$$\vec{K} = h\vec{b}_1 + k\vec{b}_2 + l\vec{b}_3 \tag{3.2}$$

where h, k, and l are integers. The reciprocal lattice vectors are related to the basis vectors of the direct lattice by

$$\vec{b}_i = 2\pi \frac{\vec{a}_j \times \vec{a}_k}{\vec{a}_1 \left(\vec{a}_2 \times \vec{a}_3 \right)} \tag{3.3}$$

where a cyclic permutation of i, j, k are used in the definition. Typically, parentheses are used in the definition of the plane, e.g., (h,k,l). The (100) planes for fcc and bcc lattices are shown in Figure 3.2 where dots are used to show the location of the atoms in the next plane down.

This provides the simplest description of the surface in terms of terminating the bulk. There is a rather nice NASA publication by Bacigalupi (1964) which gives diagrams of many surfaces and subsurface structures for fcc, bcc, and diamond lattices, in addition to a great deal of other useful information such

FIGURE 3.3 Representation of fcc (110) face with an additional "2 × 2" layer, in which the species above the surface atoms have twice the spacing of the surface. Atomic positions in the next layer below the surface are presented by smaller dots.

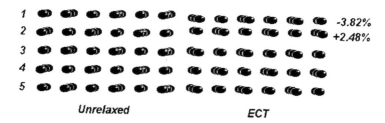

FIGURE 3.4 Side view of nickel (100) surface. On the left, the atoms are positioned as if still within a bulk fcc lattice ("unrelaxed"). On the right, the surface planes have been moved to minimize system energy. The percent change in lattice spacing is indicated, with the spacing in the image exaggerated to illustrate the effect. (From Bozzolo, G. et al. (1994), *Surf. Sci.* 315, 204–214. With permission.)

as surface density and interplanar spacings. A modern reprinting of this NASA publication is called for. In many cases, this simple description is not adequate since the surface can reconstruct. The two most prominent cases of surface reconstruction are the Au(110) surface (Good and Banerjea, 1992) for metals and the Si(111) surface (Zangwill, 1988) for semiconductors. In addition, adsorbates often form structures with symmetries different from the substrate, with the classic example the adsorption of oxygen on W(110) (Zangwill, 1988). Wood (1963) in a classic publication gives the nomenclature for describing such structures. In Figure 3.3 we show an example of 2 × 2 structure, where the terminology describes a surface that has a layer with twice the spacings of the substrate. There are many other possibilities, such as structures rotated with respect to the substrate and centered differently from the substrate. These are also defined by Wood (1963).

The next consideration is that the interplanar spacing can vary, and slight shifts in atomic positions can occur several planes from the free surface. A recent paper by Bozzolo et al. (1994) presents the results for a large number of metallic systems and serves as a good review of available publications. Figure 3.4 shows some typical results for Ni(100). The percent change given represents the deviation from the equilibrium interplanar spacing. The drawing in Figure 3.4 exaggerates these typically small differences to elucidate the behavior. Typically, this pattern of alternating contraction and expansion diminishing as the bulk is approached is found in most metals. It can be understood in a simple manner (Bozzolo et al., 1994). The energy for the bulk metal is a minimum at the bulk metallic density. The formation of the surface represents a loss of electron density because of the missing neighbors for the surface atoms. Therefore, this loss of electron density can be partially offset by a contraction of the interplanar spacing between the first two layers. This construction causes an electron density increase between layers 2 and 3, and thus the energy is lowered by a slight increase in the interplanar spacing. There are some exceptions

148

a   b

FIGURE 3.5 Side view of gold (110) surface: (a) unreconstructed; (b) 1 × 2 missing row surface reconstruction. (From Good, B. S. and Banerjea, A. (1992), *Mater. Res. Soc. Symp. Proc.*, 278, 211–216. With permission.)

to this behavior where the interplanar spacing increases between the first two layers due to bonding effects (Needs, 1987; Feibelman, 1992). However, the pattern shown in Figure 3.4 is the usual behavior for most metallic surfaces. There can be similar changes in position within the planes; however, these are usually small effects (Rodriguez et al., 1993; Foiles, 1987). In Figure 3.5, we show a side view of a gold (110) surface (Good and Banerjea, 1992). Figure 3.5a shows the unreconstructed surface and Figure 3.5b shows a side view of the (2 × 1) missing row reconstruction. Such behavior indicates the complexity that can arise even for metal surfaces and the danger of using ideas which are too simplistic, since more details of the bonding interactions are needed in this case and those of Needs (1987) and Feibelman (1992).

Crystal surfaces encountered typically are not perfectly oriented nor atomically flat. Even "on-axis" (i.e., within a fraction of a degree) single-crystal low-index faces exhibit some density of crystallographic steps. For a gold (111) face tilted one half degree toward the (011) direction, evenly spaced single atomic height steps would be only 27 nm apart. Other surface-breaking crystal defects such as screw and edge dislocations may also be present, in addition to whatever surface scratches, grooves, and other polishing damage which remain in a typical single-crystal surface. Surface steps and step kinks would be expected to show greater reactivity than low-index surface planes. During either deposition or erosion of metal surfaces, one expects incorporation into or loss from the crystal lattice preferentially at step edges. More generally on simple metal surfaces, lone atoms on a low-index crystal face are expected to be most mobile (i.e., have the lowest activation energy to move). Atoms at steps would be somewhat more tightly bound, and atoms making up a low-index face would be least likely to move. High-index crystal faces can often be thought of as an ordered collection of steps on a low-index face. When surface species and even interfaces become mobile, consolidation of steps may be observed. Alternating strips of two low-index crystal faces can then develop from one high-index crystal plane, with lower total surface energy but with a rougher, faceted topography. Much theoretical and experimental work has been done over the last decade on nonequilibrium as well as equilibrium surface morphology (e.g., Redfield and Zangwill, 1992; Vlachos et al., 1993; Conrad and Engel, 1994; Bartelt et al., 1994; Williams, 1994; Kaxiras, 1996).

Semiconductors and insulators generally behave differently. Unlike most metals for which the electron gas to some degree can be considered to behave like a fluid, semiconductors have strong directional bonding. Consequently, the loss of neighbors leaves dangling bonds which are satisfied in ultrahigh vacuum by reconstruction of the surface. The classic example of this is the silicon (111) 7 × 7 structure, where rebonding and the creation of surface states gives a complex structure. Until STM provided real-space images of this reconstruction (Binnig et al., 1983) much speculation surrounded this surface. Zangwill (1988) shows both the terminated bulk structure of Si(111) and the relaxed 7 × 7 structure. It is clear that viewing a surface as a simple terminated bulk can lead to severely erroneous conclusions. The relevance to tribology is clear since the nature of chemical reactions between surfaces, lubricants, and additives can be greatly affected by such radical surface alterations.

There are other surface chemical state phenomena, even in ultrahigh vacuum, just as important as the structural and bonding states of the clean surface. Surface segregation often occurs to metal surfaces and interfaces (Faulkner, 1996, and other reviews cited therein). For example, trace quantities of sulfur often segregate to iron and steel surfaces or to grain boundaries in polycrystalline samples (Jennings et al., 1988). This can greatly affect results since sulfur, known to be a strong poisoning contaminant in catalysis, can affect interfacial bond strength. Sulfur is often a component in many lubricants. For alloys similar geometric surface reconstructions occur (Kobistek et al., 1994). Again, alloy surface composition can vary dramatically from the bulk, with segregation causing one of the elements to be the only component on a surface. In Figure 3.6 we show the surface composition for a CuNi alloy as a function of bulk composition with both a large number of experimental results and some theoretical predictions for the composition

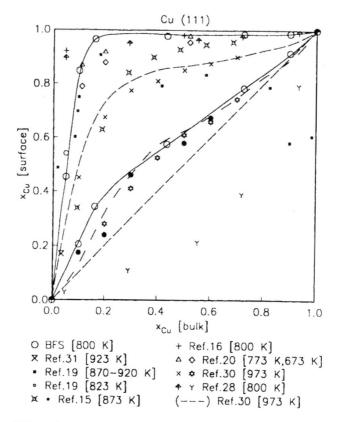

O   BFS [800 K]   + Ref.16 [800 K]
X   Ref.31 [923 K]   △ ◇ Ref.20 [773 K,673 K]
■   Ref.19 [870–920 K]   × ✿ Ref.30 [973 K]
∘   Ref.19 [823 K]   ✦ Y Ref.28 [800 K]
⋈ • Ref.15 [873 K]   (–––) Ref.30 [973 K]

FIGURE 3.6 Copper (111) surface composition vs. copper-nickel alloy bulk composition: comparison between the experimental and theoretical results for the first and second planes. (See Good et al., 1993, and references therein.)

(Good et al., 1993). In addition, nascent surfaces typically react with the ambient, giving monolayer films and oxidation even in ultrahigh vacuum, producing even more pronounced surface composition effects. In conclusion, we see that even in the most simple circumstances, i.e., single-crystal surfaces, the situation can be very complicated.

3.3 Theoretical Considerations

3.3.1 Surface Theory

We have shown how the formation of a surface can affect geometry. We now present some aspects of the energetics of surfaces from first-principles considerations. For a long time, calculations of the electronic structure and energetics of the surface had proven to be a difficult task. The nature of theoretical approximations and the need for high-speed computers limited the problem to some fairly simple approaches (Ashcroft and Mermin, 1976). The advent of better approximations for the many body effects, namely, for exchange and correlation, and the improvements in computers have changed this situation in the not too distant past. One aspect of the improvements was density functional theory and the use of the local density approximation (LDA) (Kohn and Sham, 1965; Lundqvist and March, 1983). Difficulties arise because in the creation of the surface, periodicity in the direction perpendicular to the surface is lost. Periodicity simplifies many problems in solid-state theory by limiting the calculation to a single unit cell with periodic boundary conditions. With a surface present the wave vector perpendicular to the surface, $\vec{k}_\perp$, is not periodic, although the wave vector parallel to the surface, $\vec{k}_\parallel$, still is.

The process usually proceeds by solving the one-electron Kohn–Sham equations (Kohn and Sham, 1965; Lundqvist and March, 1983), where a given electron is treated as though it is in the mean field of all of the other electrons. The LDA represents the mean field in terms of the local electron density at a given location. The Kohn–Sham equations are written in the form (using atomic units where the constants appearing in the Schroedinger equation along with the electron charge and the speed of light, $\hbar = m_e = e = c = 1$).

$$\left[-1/2\nabla^2 + V\left(\vec{r}\right)\right]\Psi_i\left(\vec{k}_\parallel, \vec{r}\right) = \epsilon_i\left(\vec{k}_\parallel\right)\Psi_i\left(\vec{k}_\parallel, \vec{r}\right) \tag{3.4}$$

where Ψ_i and ϵ_i are the one-electron wave function and energy, respectively, and

$$V\left(\vec{r}\right) = \Phi\left(\vec{r}\right) + V_{xc}\left[\rho\left(\vec{r}\right)\right] \tag{3.5}$$

where $V_{xc}[\rho(\vec{r})]$ is the exchange and correlation potential, $\rho(\vec{r})$ is the electron density (the brackets indicate that it is a functional of the density), and $\Phi(\vec{r})$ is the electrostatic potential given by

$$\Phi\left(\vec{r}\right) = \int d\vec{r}' \frac{\rho\left(\vec{r}\right)}{\left|\vec{r} - \vec{r}'\right|} - \Sigma_j \frac{Z_j}{\left|\vec{r} - \vec{R}_j\right|} \tag{3.6}$$

in which the first term is the electron–electron interaction and the second term is the electron–ion interaction, Z_j is the ion charge, and the electron density is given by

$$\rho\left(\vec{r}\right) = \Sigma_{occ} \left|\Psi_i\left(\vec{k}_\parallel, \vec{r}\right)\right|^2 \tag{3.7}$$

where occ refers to occupied states. The calculation proceeds by using some representation for the wave functions such as the linear muffin tin orbital approximation (LMTO), and iterating self-consistently. Self-consistency is obtained when either the output density or potential agree to within some specified criterion with the input. These calculations are not generally performed for the semi-infinite solid. Instead, they are performed for slabs of increasing thickness to the point where the interior atoms have essentially bulk properties. Usually, five planes are sufficient to give the surface properties. The values of $\epsilon_i(\vec{k}_\parallel)$ give the surface band structure and surface states, localized electronic states created because of the presence of the surface.

The second piece of information given is the total energy in terms of the electron density, as obtained from density functional theory. This is represented schematically by the expression

$$E\left[\rho\right] = E_{ke}\left[\rho\right] + E_{es}\left[\rho\right] + E_{xc}\left[\rho\right] \tag{3.8}$$

where E_{ke} is the kinetic energy contribution to the energy, E_{es} is the electrostatic contribution, E_{xc} is the exchange correlation contribution, and the brackets indicate that the energy is a functional of the density. Thus, the energy is an extremum of the correct density. Determining the surface energy accurately from such calculations can be quite difficult since the surface energy, or indeed any of the energies of various structures of interest, are obtained as the difference of big numbers. For example, for the surface the energy would be given by

$$E_{surface} = \frac{E\left(a\right) - E\left(\infty\right)}{2A} \tag{3.9}$$

where a is the distance between the surfaces (a = 0 to get the surface energy) and A is the cross-sectional area.

The initial and classic solutions of the Kohn–Sham equations for surfaces and interfaces were accomplished by Lang and Kohn (1970) for the free surface and Ferrante and Smith for interfaces (Ferrante and Smith, 1985; Smith and Ferrante, 1986). The calculations were simplified by using the jellium model to represent the ionic charge. In the jellium model the ionic charge is smeared into a uniform distribution. Both sets of authors introduced the effects of discreteness on the ionic contribution through perturbation theory for the electron–ion interaction and through lattice sums for the ion–ion interaction. The jellium model is only expected to give reasonable results for the densest packed planes of simple metals.

In Figures 3.7 and 3.8 we show the electron distribution at a jellium surface for Na and for an Al(111)–Mg(0001) interface (Ferrante and Smith, 1985) that is separated a small distance. In Figure 3.7 we can see the characteristic decay of the electron density away from the surface. In Figure 3.8 we see the change in electron density in going from one material to another. This characteristic tailing is an indication of the reactivity of the metal surface.

In Figure 3.9 we show the electron distribution for a nickel (100) surface for the fully three-dimensional calculations performed by Arlinghouse et al. (1980) and that for a silver layer adsorbed on a palladium (100) interface (Smith and Ferrante, 1985) using self-consistent localized orbitals (SCLO) for approximations to the wave functions. First, we note that for the Ni surface we see there is a smoothing of the surface density characteristic of metals. For the adsorption we can see that there are localized charge transfers and bonding effects indicating that it is necessary to perform three-dimensional calculations in order to determine bonding effects. Hong et al. (1995) have also examined metal–ceramic interfaces and the effects of impurities at the interface on the interfacial strength.

In Figure 3.10 we schematically show the results of determining the interfacial energies as a function of separation between the surfaces with the energy in Figure 3.10a and the derivative curves giving the interfacial strength. In Figure 3.11 we show Ferrante and Smith's results for a number of interfaces of jellium metals (Ferrante and Smith, 1985; Smith and Ferrante, 1986; Banerjea et al., 1991). Rose et al. (1981, 1983) found that these curves would scale onto one universal curve and indeed that this result applied to many other bonding situations including results of fully three-dimensional calculations. We show the scaled curves from Figure 3.11 in Figure 3.12. Somewhat surprisingly because of large charge transfer, Hong et al. (1995) found that this same behavior is applicable to metal–ceramic interfaces. Finnis (1996) gives a review of metal–ceramic interface theory.

The complexities that we described earlier with regard to surface relaxations and complex structures can also be treated now by modern theoretical techniques. Often in these cases it is necessary to use "supercells" (Lambrecht and Segall, 1989). Since these structures are extended, it would require many atoms to represent a defect. Instead, in order to model a defect and take advantage of the simplicities of periodicities, a cell is created selected at a size which will mimic the main energetics of the defects. In conclusion, we can see that theoretical techniques have advanced substantially and are continuing to do so. They have and will shed light on many problems of interest experimentally.

3.3.2 Friction Fundamentals

Friction, as commonly used, refers to a force resisting sliding. It is of obvious importance since it is the energy loss mechanism in sliding processes. In spite of its importance, after many centuries friction surprisingly has still avoided a complete physical explanation. An excellent history of the subject is given in a text by Dowson (1979). In this section we will outline some of the basic observations and give some recent relevant references treating the subject at the atomic level, in keeping with the theme of this chapter, and since the topic is much too complicated to treat in such a small space.

There are two basic issues, the nature of the friction force and the energy dissipation mechanism. There are several commonplace observations, often considered general rules, regarding the friction force as outlined in the classic discussions of the subject by Bowden and Tabor (1964):

1. The friction force does not depend on the apparent area of contact.
2. The friction force is proportional to the normal load.
3. The kinetic friction force does not depend on the velocity and is less than the static friction force.

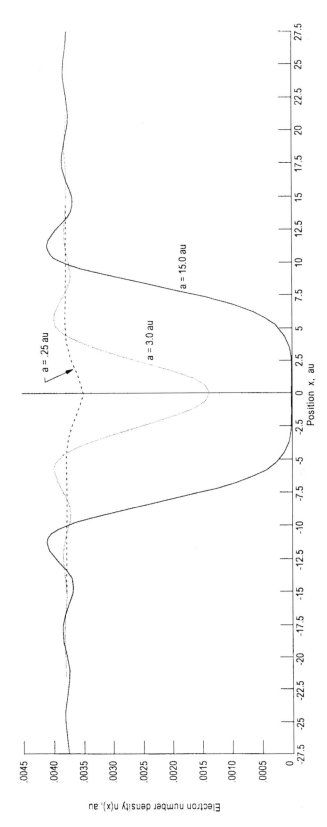

FIGURE 3.7 Electron density at a jellium surface vs. position for a Na(011)–Na(011) contact for separations of 0.25, 3.0, and 15.0 au. (From Ferrante, J. and Smith, J. R. (1985). *Phys. Rev. B* 31, 3427–3434. With permission.)

153

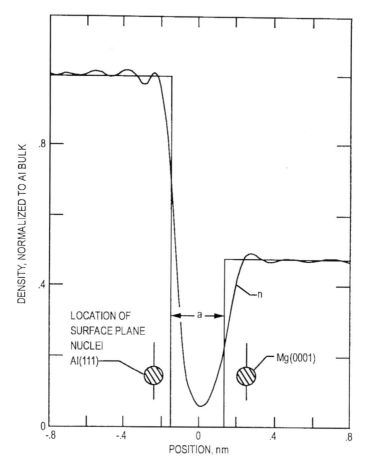

FIGURE 3.8 Electron number density n and jellium ion charge density for an aluminum (111)–magnesium (0001) interface. (From Ferrante, J. and Smith, J. R. (1985), *Phys. Rev. B* 31, 3427–3434. With permission.)

Historically, Coulomb (Bowden and Tabor, 1964; Dowson, 1979), realizing that surfaces were not ideally flat and were formed by asperities (a hill-and-valley structure), proposed that interlocking asperities could be a source of the friction force. This model has many limitations. For example, if we picture a perfectly sinusoidal interface there is no energy dissipation mechanism, since once the top of the first asperity is attained the system will slide down the other side, thus needing no additional force once set in motion. Bowden and Tabor (1964), recognizing the existence of interfacial forces, proposed another mechanism based on adhesion at interfaces. Again, recognizing the existence of asperities, they propose that adhesion occurs at asperity surfaces and that shearing occurs on translational motion. This model explains a number of effects such as the disparity between true area of contact and apparent area of contact and the tracking of friction force with load, since the asperities and thus the true area of contact change with asperity deformation (load). The actual arguments are more complex than indicated here and require reading of the primary text for completeness. These considerations also emphasize the basic topic of this chapter, i.e., the important effect of the state of the surface and interface on the friction process. Clearly, adsorbates, the differences of materials in contact, and lubricants greatly affect the interaction.

We now proceed to briefly outline some models of both the friction force and frictional energy dissipation. As addressed elsewhere in this book, there have recently been a number of attempts to model theoretically the friction interaction at the atomic level. The general approaches have involved assuming a two-body interaction potential at an interface, which in some cases may only be one dimensional, and

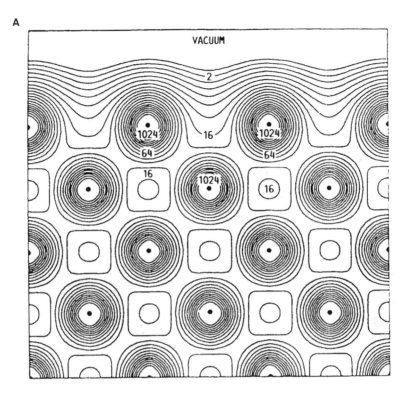

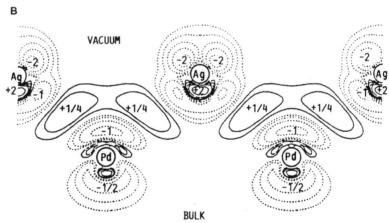

FIGURE 3.9 (a) Electronic charge density contours at a nickel (100) surface. (From Arlinghaus, F. J. et al. (1980), *Phys. Rev. B* 21, 2055–2059. With permission.) (b) Charge transfer of the palladium [100] slab upon silver adsorption. (From Smith, J. R. and Ferrante, J. (1985), *Mater. Sci. Forum*, 4, 21–38. With permission.)

allowing the particles to interact across an interface, allowing motion of internal degrees of freedom in either one or both surfaces. Hirano and Shinjo (1990) examine a quasi-static model where one solid is constrained to be rigid and the second is allowed to adapt to the structure of the first, interacting through a two-body potential as translation occurs. No energy dissipation mechanism is included. They conclude that two processes occur, atomic locking where the readjusting atoms change their positions during sliding, and dynamic locking where the configuration of the surface changes abruptly due to the dynamic process if the interatomic potential is stronger than a threshold value. The latter process they conclude is unlikely to happen in real systems. They also conclude that the adhesive force is not related to the

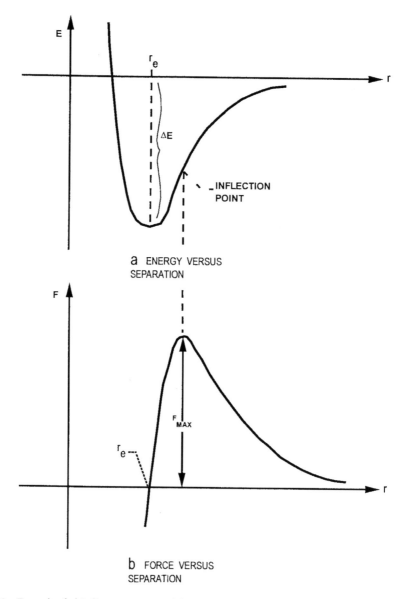

FIGURE 3.10 Example of a binding energy curve: (a) energy vs. separation; (b) force vs. separation. (From Banerjea, A. et al. (1991), in *Fundamentals of Adhesion* (Liang-Huang Lee, ed.), Plenum Press, New York. With permission.)

friction phenomena, and discuss the possibility of a frictionless "superlubric" state (Shinjo and Hirano, 1993; Hirano et al., 1997). Matsukawa and Fukuyama (1994) carry the process further in that they allow both surfaces to adjust and examine the effects of velocity with attention to the three rules of friction stated above. They argue, not based on their calculations, that the Bowden and Tabor argument is not consistent with flat interfaces having no asperities. Since an adhesive force exists, there is a normal force on the interfaces with no external normal load. Consequently, rules of friction 1 and 2 break down. With respect to rule 3, they find it restricted to certain circumstances. They found that the dynamic friction force, in general, is sliding velocity dependent, but with a decreasing velocity dependence with increasing maximum static friction force. Hence, for systems with large static friction forces, the kinetic friction force shows behavior similar to classical rule 3, above. Finally, Zhong and Tomanek (1990) performed a first-principles calculation of the force to slide a monolayer of Pd in registry with the graphite surface.

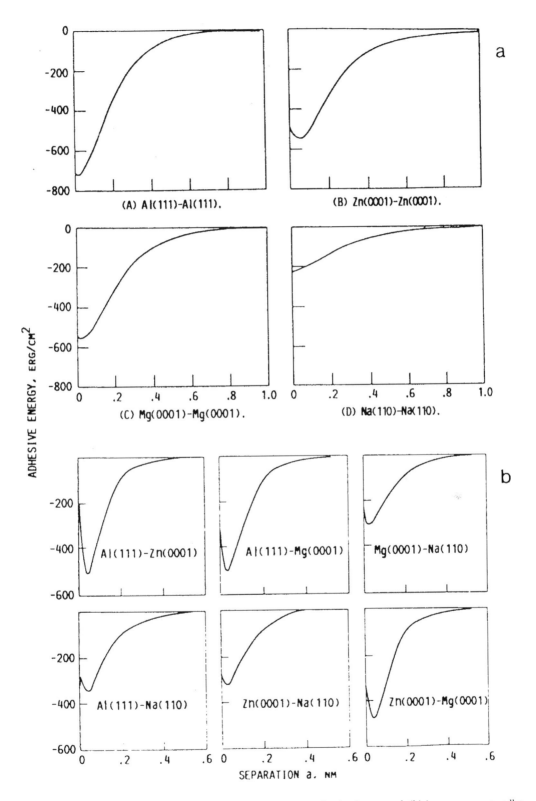

FIGURE 3.11 Adhesive energy vs. separation: (a) commensurate adhesion is assumed; (b) incommensurate adhesion is assumed. (From Rose, J.H. et al. (1983), *Phys. Rev. B* 28, 1835–1845. With permission.)

157

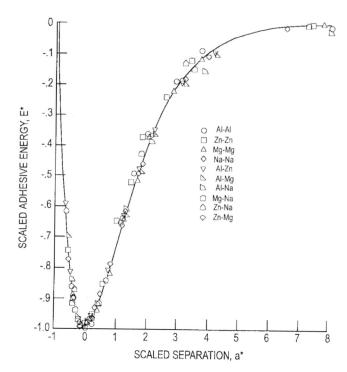

FIGURE 3.12 Scaled adhesive binding energy as a function of scaled separation for systems in Figure 3.11. (From Rose, J.H. et al. (1983), *Phys. Rev. B* 28, 1835–1845. With permission.)

Assuming some energy dissipation mechanism to be present, they calculated tangential force as a function of load and sliding position.

Sokoloff (1990, 1992, and references therein) addresses both the friction force and frictional energy dissipation. He represents the atoms in the solids as connected by springs, thus enabling an energy dissipation mechanism by way of lattice vibrations. He also looks at such issues as the energy to create and move defects in the sliding process and examines the velocity dependence of kinetic friction based on the possible processes present, including electronic excitations (Sokoloff, 1995). Persson (1991) also proposes a model for energy dissipation due to electronic excitations induced within a metallic surface. Persson (1993, 1994, 1995) addresses in addition the effect of a boundary lubricant between macroscopic bodies, modeling fluid pinning to give the experimentally observed logarithmic time dependence of various relaxation processes. Finally, as more fully covered in other chapters of this book, much recent effort has gone into modeling specifically the lateral force component of the probe tip interaction with a sample surface in scanning probe microscopy (e.g., Hölscher et al., 1997; Diestler et al., 1997, and references therein; Lantz et al., 1997).

In conclusion, while these types of simulations may not reflect the fully complexity of real materials, they are necessary and useful. Although limited in scope, it is necessary to break down such complex problems into isolated phenomena which it is hoped can result in the eventual unification to the larger picture. It simply is difficult to isolate the various components contributing to friction experimentally.

3.4 Experimental Determinations of Surface Structure

In this section we will discuss three techniques for determining the structure of a crystal surface, low-energy electron diffraction (LEED), high-resolution electron microscopy (HREM), and field ion microscopy (FIM). The first, LEED, is a diffraction method for determining structure and the latter two are methods to view the lattices directly. There are other methods for determining structure such as ion

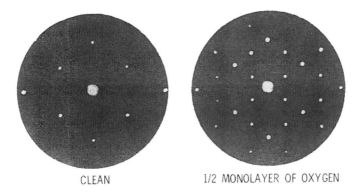

CLEAN   1/2 MONOLAYER OF OXYGEN

FIGURE 3.13 LEED pattern for (a) clean and (b) oxidized tungsten (110) with one half monolayer of oxygen. The incident electron beam energy for both patterns is 119 eV. (From Ferrante, J. et al. (1973), in *Microanalysis Tools and Techniques* (McCall, J. L. and Mueller, W. M., eds.), Plenum Press, New York. With permission.)

scattering (Niehus et al., 1993), low-energy backscattered electrons (De Crescenzi, 1995), and even secondary electron holography (Chambers, 1992), which we will not discuss. Other contributors to this book address scanning probe microscopy and tribology, which are also nicely covered in an extensive review article by Carpick and Salmeron (1997).

3.4.1   Low-Energy Electron Diffraction

Since LEED is a diffraction technique, when viewing a LEED pattern, you are viewing the reciprocal lattice structure and not the atomic locations on the surface. A LEED pattern typically is obtained by scattering a low-energy electron beam (0 to 300 eV) from a single-crystal surface in ultrahigh vacuum. In Figure 3.13 we show the LEED pattern for the W(110) surface with a half monolayer of oxygen adsorbed on it (Ferrante et al., 1973). We can first notice in Figure 3.13a that the pattern looks like the direct lattice W(110) surface, but this only means that the diffraction pattern reflects the symmetry of the lattice. Notice that in Figure 3.13b extra spots appear at ½ order positions upon adsorption of oxygen. Since this is the reciprocal lattice, this means that the spacings of the rows of the chemisorbed oxygen actually are at double the spacing of the underlying substrate. In fact, the interpretation of this pattern is more complicated since the structure shown would not imply a ½ monolayer coverage, but is interpreted as an overlapping of domains at 90° from one another. In this simple case the coverage is estimated by adsorption experiments, where saturation is interpreted as a monolayer coverage. The interpretation of patterns is further complicated, since with complex structures such as the silicon 7×7 pattern, the direct lattice producing this reciprocal lattice is not unique. Therefore, it is necessary to have a method to select between possible structures (Rous and Pendry, 1989).

We now digress for a moment in order to discuss the diffraction process. The most familiar reference work is X-ray diffraction (Kittel, 1986). We know that for X rays the diffraction pattern of the bulk would produce what is known as a Laue pattern where the spots represent reflections from different planes. The standard diffraction condition for constructive interference of a wave reflected from successive planes is given by the Bragg equation

$$2d \sin \theta = n\lambda \tag{3.10}$$

where d is an interplanar spacing, θ is the diffraction angle, λ is the wavelength of the incident radiation, and n is an integer indicating the order of diffraction. Only certain values of θ are allowed where diffractions from different sets of parallel planes add up constructively. There is another simple method for picturing the diffraction process known as the Ewald sphere construction (Kittel, 1986), where it can be easily shown that the Bragg condition is equivalent to the relationship

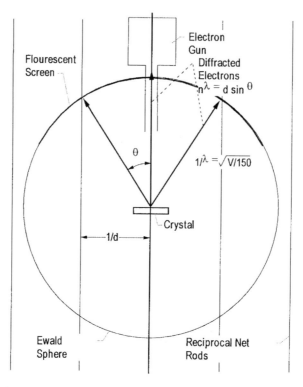

FIGURE 3.14 Ewald sphere construction for LEED. (From Ferrante, J. et al. (1973), in *Microanalysis Tools and Techniques* (McCall, J. L., and Mueller, W. M., eds.), Plenum Press, New York. With permission.)

$$\vec{k} - \vec{k}' = \vec{G} \tag{3.11}$$

where $\vec{k}$ is the wave vector ($2\pi/\lambda$) of the incident beam, $\vec{k}'$ is the wave vector of the diffracted beam, and $\vec{G}$ is a reciprocal lattice vector. The magnitude of the wave vectors $k = k'$ are equal since momentum is conserved; i.e., we are only considering elastic scattering. Therefore, a sphere of radius k can be constructed, which when intersecting a reciprocal lattice point indicates a diffracted beam. This is equivalent to the wave vector difference being equal to a reciprocal lattice vector, with that reciprocal lattice vector normal to the set of planes of interest, and θ the angle between the wave vectors. In complex patterns, spot intensities are used to distinguish between possible structures. The equivalent Ewald construction for LEED is shown in Figure 3.14. We note that the reciprocal lattice for a true two-dimensional surface would be a set of rods instead of a set of points. Consequently, the Ewald sphere will always intersect the rods and give diffraction spots resulting from interferences due to scattering between rows of surface atoms, with the number of spots changing with electron wavelength and incident angle. However, for LEED complexity results from spot intensity modulation by the three-dimensional lattice structure, and determining that direct lattice from the spot intensities. In X-ray diffraction the scattering is described as kinematic, which means that only single scattering events are considered. With LEED, multiple scattering occurs because of the low energy of the incident electrons; thus structure determination involves solving a difficult quantum mechanics problem. Generally, various possible structures are constructed and the multiple scattering problem is solved for each proposed structure. The structure that minimizes the difference between the experimental intensity curves and the theoretical calculations is the probable structure. There are a number of parameters involved with atomic positions and electronic properties, and the best fit parameter is denoted as the "R-factor." In spite of the seeming complexity, considerable progress has been made and computer programs for performing the analysis are available (Van Hove et al., 1993). The LEED structures give valuable information about adsorbate binding which can be used in the energy calculations described previously.

3.4.2 High-Resolution Electron Microscopy

Fundamentally, materials derive their properties from their makeup and structure, even down to the level of the atomic ordering in alloys. To understand fully the behavior of materials as a function of their composition, processing history, and structural characteristics, the highest resolution examination tools are needed. In this section we will limit the discussion to electron microscopy techniques using commonly available equipment and capable or achieving atomic-scale resolution. Traditional scanning electron microscopy (SEM), therefore, will not be discussed, although in tribology SEM has been and should continue to prove very useful, particularly when combined with X-ray spectroscopy. Many modern Auger electron spectrometers (discussed in the next section on surface chemical analysis) also have high-resolution scanning capabilities, and thus can perform imaging functions similar to a traditional SEM. Another technique not discussed here is photoelectron emission microscopy (PEEM). While PEEM can routinely image photoelectron yield (related to the work function) differences due to single atomic layers, lateral resolution typically suffers in comparison to SEM. PEEM has been applied to tribological materials, however, with interesting results (Montei and Kordesch, 1996).

Both transmission electron microscopy (TEM) and scanning transmission electron microscopy (STEM) make use of an electron beam accelerated through a potential of, typically, up to a few hundred thousand volts. Generically, the parts of a S/TEM consist of an electron source such as a hot filament or field emission tip, a vacuum column down which the accelerated and collimated electrons are focused by usually magnetic lenses, and an image collection section, often comprising a fluorescent screen for immediate viewing combined with a film transport and exposure mechanism for recording images. The sample is inserted directly into the beam column and must be electron transparent, both of which severely limit sample size. There are numerous good texts available about just TEM and STEM (e.g., Hirsch et al., 1977; Thomas and Goringe, 1979).

An advantage to probing a sample with high-energy electrons lies in the De Broglie formula relating the motion of a particle to its wavelength

$$\lambda = \frac{h}{\left(2mE_k\right)^{1/2}} \qquad (3.12)$$

where λ is the electron wavelength, h is the Planck constant, m is the particle mass, and E_k is the kinetic energy of the particle. An electron accelerated through a 100-kV potential then has a wavelength of 0.04 Å, well below any diffraction limitation on atomic resolution imaging. This is in contrast with LEED, for which electron wavelengths are typically of the same order as interatomic spacings. As the electron beam energy increases in S/TEM, greater sample thickness can be penetrated with a usable signal reaching the detector. Mitchell (1973) discusses the advantages of using very high accelerating voltages, which at the time included TEM voltages up to 3 MV.

As the electron beam traverses a sample, any crystalline regions illuminated will diffract the beam, forming patterns characteristic of the crystal type. Apertures in the microscope column allow the diffraction patterns of selected sample areas to be observed. Electron diffraction patterns combined with an ability to tilt the sample make determination of crystal type and orientation relatively easy, as discussed in Section 4.1 above for X-ray Ewald sphere construction. Electrons traversing the sample can also undergo an inelastic collision (losing energy), followed by coherent rescattering. This gives rise to cones of radiation which reveal the symmetry of the reflecting crystal planes, showing up in diffraction images as "Kikuchi lines," named after the discoverer of the phenomenon. The geometry of the Kikuchi lines provides a convenient way of determining crystal orientation with fairly high accuracy. Another technique for illuminating sample orientation uses an aperture to select one of the diffracted beams to form the image, which nicely highlights sample area from which that diffracted beam originates ("darkfield" imaging technique).

One source of TEM image contrast is the electron beam interacting with crystal defects such as various dislocations, stacking faults, or even strain around a small inclusion. How that contrast changes with microscope settings can reveal information about the defect. For example, screw dislocations may "disappear" (lose contrast) for specific relative orientations of crystal and electron beam. An additional tool in examining the three-dimensional structures within a sample is stereomicroscopy, where two images of the same area are captured tilted from one another, typically by around 10°. The two views are then simultaneously shown each to one eye to reveal image feature depth.

For sample elemental composition, both an X-ray spectrometer and/or an electron energy-loss spectrometer can be added to the S/TEM. Particularly for STEM, due to minimal beam spreading during passage through the sample the analyzed volume for either spectrometer can be as small as tens of nanometers in diameter. X-ray and electron energy-loss spectrometers are somewhat complementary in their ranges of easily detected elements. Characteristic X rays are more probable when exciting the heavier elements, while electron energy losses due to light element K-shell excitations are easily resolvable.

Both TEM and STEM rely on transmission of an electron beam through the sample, placing an upper limit on specimen thickness which depends on the accelerating voltage available and on specimen composition. Samples are often thinned to less than a micrometer in thickness, with lateral dimensions limited to a few millimeters. An inherent difficulty in S/TEM sample preparation thus is locating a given region of interest within the region of visibility in the microscope, without altering sample characteristics during any thinning process needed. For resolution at an atomic scale, columns of lighter element atoms are needed for image contrast, so individual atoms are not "seen." Samples also need to be somewhat vacuum compatible, or at least stable enough in vacuum to allow examination. The electron beam itself may alter the specimen by heating, by breaking down compounds within the sample, or by depositing carbon on the sample surface if there are residual hydrocarbons in the microscope vacuum. In short, S/TEM specimens should be robust under high-energy electron bombardment in vacuum.

3.4.3 Field Ion Microscopy

For many decades, FIM has provided direct lattice images from sharp metal tips. Some early efforts to examine contact adhesion used the FIM tip as a model asperity, which was brought into contact with various surfaces (Mueller and Nishikawa, 1968; Nishikawa and Mueller, 1968; Brainard and Buckley, 1971, 1973; Ferrante et al., 1973). As well, FIM has been applied to the study of friction (Tsukizoe et al., 1985), the effect of adsorbed oxygen on adhesion (Ohmae et al., 1987), and even direct examination of solid lubricants (Ohmae et al., 1990).

In FIM a sharp metal tip is biased to a high negative potential relative to a phosphor-coated screen in an evacuated chamber backfilled to about a millitorr with helium or other noble gas. A helium atom impinging on the tip experiences a high electric field due to the small tip radius. This field polarizes the atom and creates a reasonable probability that an electron will tunnel from the atom to the metal tip leaving behind a helium ion. Ionization is most probable directly over atoms in the tip where the local radius of curvature is highest. Often, only 10 to 15% of the atoms on the tip located at the zone edges and at kink sites are visible. The helium ions are then accelerated to a phosphorescent screen at some distance from the tip, giving a large geometric magnification. Uncertainty in surface atom positions is often reduced by cooling the tip to liquid helium temperature. Figure 3.15 is an FIM pattern for a clean tungsten tip oriented in the (110) direction. The small rings are various crystallographic planes that appear on a hemispherical single-crystal surface. A classic discussion of FIM pattern interpretation can be found in Mueller (1969), a recent review has been published by Kellogg (1994), and a more extensive discussion of FIM in tribology can be found in Ohmae (1993).

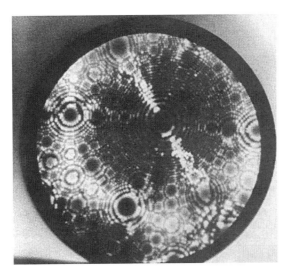

FIGURE 3.15 Field ion microscope pattern of a clean tungsten tip oriented in the (110) direction. (From Ferrante, J. et al. (1973), in *Microanalysis Tools and Techniques* (McCall, J. L. and Mueller, W. M., eds.), Plenum, Press New York. With permission.)

3.5   Chemical Analysis of Surfaces

In this section we will discuss four of the many surface chemical analytic tools which we feel have had the widest application in tribology, Auger electron spectroscopy (AES), X-ray photo-electron spectroscopy (XPS), secondary ion mass spectroscopy (SIMS), and infrared spectroscopy (IRS). AES gives elemental analysis of surfaces, but in some cases will give chemical compound information. XPS can give compound information as well as elemental. SIMS can exhibit extreme elemental sensitivity as well as "fingerprint" lubricant molecules. IR can identify hydrocarbons on surfaces, which is relevant because most lubricants are hydrocarbon based. Hantsche (1989) gives a basic comparison of some surface analytic techniques. Before launching into this discussion we wish to present a general discussion of surface analyses. We use a process diagram to describe them given as

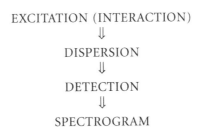

The first step, excitation in interaction, represents production of the particles or radiation to be analyzed. In light or photon emission spectroscopy a spark causes the excitation of atoms to higher energy states, thus emitting characteristic photons. The dispersion stage could be thought of as a filtering process where the selected information is allowed to pass and other information is rejected. In light spectroscopy this would correspond to the use of a grating or prism, for an ion or electron it might be an electrostatic analyzer. Next is detection of the particle which could be a photographic plate for light or an electron multiplier for ions or electrons. And, finally, the spectrogram tells what materials are present and, it is hoped, how much is there.

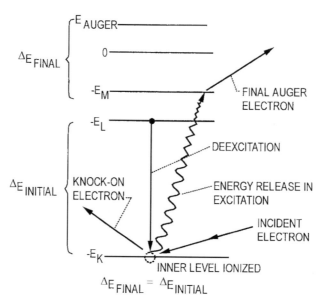

FIGURE 3.16 Auger transition diagram for an atom. (From Ferrante, J. et al. (1973), in *Microanalysis Tools and Techniques* (McCall J. L. and Mueller W. M., eds.), Plenum Press, New York. With permission.)

3.5.1 Auger Electron Spectroscopy

The physics of the Auger emission process is shown in Figure 3.16. An electron is accelerated to an energy sufficient to ionize an inner level of an atom. In the relaxation process an electron drops into the ionized energy level. The energy that is released from this de-excitation is absorbed by an electron in a higher energy level, and if the energy is sufficient it will escape from the solid. The process shown is called a KLM transition, i.e., a level in the K-shell is ionized, an electron decays from an L-shell, and the final electron is emitted from an M-shell. Similarly, a process involving different levels will have corresponding nomenclature. The energy of the emitted electron has a simple relationship to the energies of the levels involved, depending only on differences between these levels. The relationships for the process shown are

$$\Delta E_{\text{final}} = \Delta E_{\text{initial}} \tag{3.13}$$

giving

$$E_{\text{Auger}} = E_K - E_L - E_M \tag{3.14}$$

Consequently, since the energy levels of the atoms are generally known, the element can be identified. There are surprisingly few overlaps for materials of interest. When peaks do overlap, other peaks peculiar to the given element along with data manipulation can be used to deconvolute peaks close in energy. AES will not detect hydrogen, helium, or atomic lithium because there are not enough electrons for the process to occur. AES is surface sensitive because the energy of the escaping electrons is low enough they cannot originate from very deep within the solid without detectable inelastic energy losses. The equipment is shown schematically in Figure 3.17. The dispersion of the emitted electrons is usually accomplished by any of a number of electrostatic analyzers, e.g., cylindrical mirror or hemispherical analyzers. Although the operational details of the analyzers differ somewhat, the net result is the same.

An example spectrum is shown in Figure 3.18 for a wear scar on a pure iron pin worn with dibutyl adipate with 1 wt. % zinc-dialkyl-dithiophosphate (ZDDP). This spectrum corresponds to the first derivative of the actual spectral lines (peaks) in the spectrum (Brainard and Ferrante, 1979).

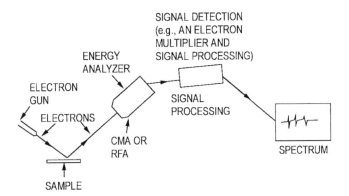

FIGURE 3.17 Schematic diagram of AES apparatus. (From Ferrante, J. (1982), *J. Am. Soc. Lubr. Eng.* 38, 223–236. With permission.)

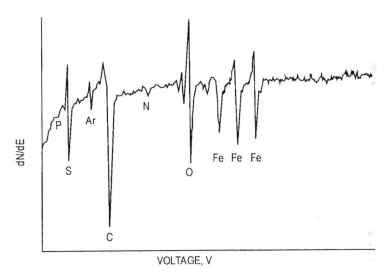

FIGURE 3.18 Auger spectrum of wear scar on pure iron pin run against M2 tool steel disk in dibutyl adipate containing 1 wt% ZDPP. Sliding speed, 2.5 cm/s; load, 4.9 N; atmosphere, dry air. (From Brainard, W. A. and Ferrante, J. (1979), NASA TP-1544, Washington, D.C.)

Historically, first-derivative spectra were taken because the actual peaks were very small compared with the slowly varying background, posing signal-to-noise problems when amplification was sufficient to bring out the peak. The derivative emphasized the more rapidly changing peak, but made quantification more difficult, since the AES peaks are not a simple shape such as Gaussian, where a quantitative relationship exists between the derivative peak-to-peak height and the area under the original peak. The advent of dedicated microprocessors and the ability to digitize the results enable more-sophisticated treatment of the data. The signal-to-background problem can now be handled by modeling the background and subtracting it, leaving an enhanced AES peak. Thus, the number of particles present can be obtained by finding the area under the peak, enhancing the quantitative capability of AES. AES can be chemically sensitive in that energy levels may shift when chemical reactions occur. Large shifts can be detected in the AES spectrum, or alternatively peak shapes may change with chemical reaction. Some examples of these effects will be given later in the chapter.

There are two other techniques that are used in conjunction with AES that should be mentioned, scanning Auger microscopy (SAM) and depth profiling. SAM is simply "tuning" to a particular AES peak and rastering the electron beam in order to obtain an elemental map of a surface. This can be particularly

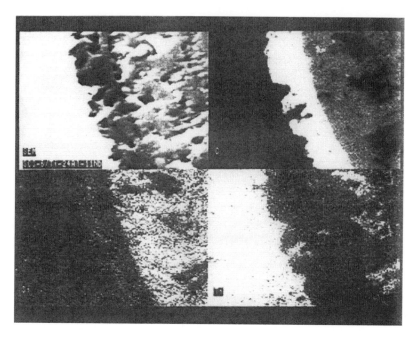

FIGURE 3.19 Example scanning Auger microscopy results. Sample is silicon carbide fiber-reinforced titanium aluminide matrix composite. Single element images as labeled, with higher concentrations represented as brighter regions. (Courtesy of Darwin Boyd).

useful in tribology since you are often dealing with rough, inhomogeneous surfaces. We show a sample SAM map in Figure 3.19.

Depth profiling is the process of sputter-eroding a sample by bombarding the surface with ions while simultaneously obtaining AES or other spectra. This enables one to obtain the composition of reaction-formed or deposited films on a surface as a function of sputter time or depth. Consequently, AES has many applications for studying tribological and other surfaces. Some examples will be given in subsequent sections.

3.5.2 X-Ray Photoelectron Spectroscopy

The physical processes involved in XPS are diagrammed in Figure 3.20. XPS is a simpler process than AES. An X-ray photon ionizes the inner level of an atom and in this case the emitted electron from the ionization is itself detected, as opposed to AES where several levels are involved in the final electron production. The dispersion and detection methods are similar to AES.

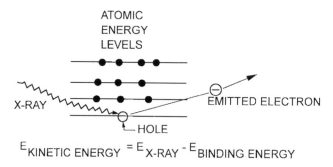

FIGURE 3.20 XPS transition diagram for an atom. (From Ferrante, J. (1993), in *Surface Diagnostics in Tribology* (K. Miyoshi and Y. W. Chung, eds.), World Scientific, Singapore. With permission.)

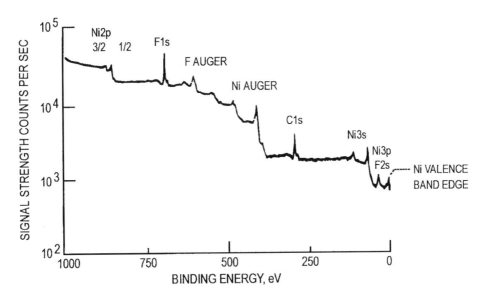

FIGURE 3.21 Example XPS spectrum. (From Ferrante, J. (1993), in *Surface Diagnostics in Tribology* (K. Miyoshi and Y. W. Chung, eds.), World Scientific, Singapore. With permission.)

Monochromatic, incoming X-ray photons are generated from an elemental target such as magnesium or aluminum. Measurement of the energy distribution of emitted electrons from the sample permits the identification of the ionized levels by the simple relation

$$E_{\text{final}} = E_{\text{xray}} - E_{\text{binding energy}}$$ (3.15)

Since the final energy is measured and the X-ray energy is known, one can determine the binding energy and consequently the material. AES peaks are also present in the XPS spectrum. AES peaks can be distinguished from the fact that the energies of the Auger electrons are fixed because they depend on a difference in energy levels, whereas the XPS electron energies depend on the energy of the incident X ray. A sample XPS spectrum is shown in Figure 3.21 and a schematic diagram of the apparatus is shown in Figure 3.22.

XPS can perform chemical as well as elemental analysis. As stated earlier, when an element is in a compound, there is a shift in energy levels relative to the unreacted element. Unlike AES, where energy level differences are detected, a chemical reaction results in an energy shift of the element XPS peak. For example, the iron peak from Fe_2O_3 is shifted by nearly 4 eV from an elemental iron peak. Not only are the shifts simpler to interpret, but it is easier to detect peaks directly (as opposed to AES derivative mode measurements) since the signal-to-background in XPS is greater than in AES. In addition, the mode of operation in the dispersion step typically enables higher resolution. The surface sensitivity of XPS is similar to AES because the energies of the emitted electrons are similar. Figure 3.23 shows some examples of oxygen and sulfur peak shifts resulting from reactions with iron and chromium for wear scars on a steel pin run with dibenzyl-disulfide as the lubricant additive (Wheeler, 1978).

3.5.3 Secondary Ion Mass Spectroscopy

The physical process involved in SIMS differs from both AES and XPS in that both the excitation source and detected quantity are ions. Rather than illuminate the sample surface with either electrons (AES) or photons (XPS), ions are used to bombard the sample surface and knock off (sputter) surface particles. The dispersion phase analyzes the emitted particle masses, instead of energy analyzing the emitted electrons as in AES or XPS. Although using sputtering implies an erosion of the sample surface, a

167

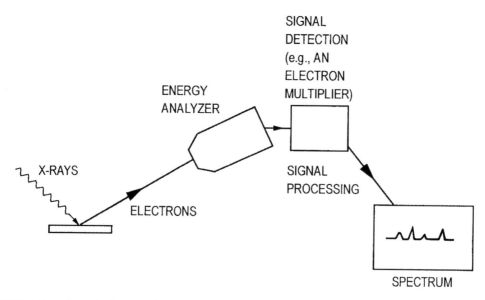

FIGURE 3.22 Schematic diagram of XPS apparatus. (From Ferrante, J. (1993), in *Surface Diagnostics in Tribology* (K. Miyoshi and Y. W. Chung, eds.), World Scientific, Singapore. With permission.)

compensating advantage for SIMS is extreme sensitivity. Under advantageous conditions, as few as 10^{12} atoms per cm^3 (ppb) have been detected (Gnaser, 1997), with more typical sensitivities for most elements in the ppm range (Wilson et al., 1989). A comprehensive discussion of the SIMS technique has been published by Benninghoven et al. (1987).

The SIMS technique typically used in surface studies gives partial monolayer sensitivity using small incident ion currents ("static" SIMS). Higher ion beam currents, often rastered, give species information as a function of sputter depth ("dynamic" SIMS or SIMS depth profiling). SIMS instrumentation can be roughly categorized by the type of ion detector used, e.g., quadrupole, magnetic sector, or time-of-flight, with their inherent differences in sensitivity and lateral and mass resolution. As well, the incident angle, energy, and type (e.g., noble gas, cesium, or oxygen) of the primary ion sputtering beam employed can greatly affect the magnitude and character of the secondary ion yield.

SIMS has several complexities. SIMS only detects secondary ions, rather than all of the sputtered species, which can lead to difficulty in quantification. Large molecules on the surface such as hydrocarbon lubricants or typical additives can exhibit complex patterns of possible fragments. A knowledge of the adsorbate and cracking patterns is often needed for interpretation. As well, multiply ionized fragments or simply different species may overlap in the spectra, having nearly identical charge-to-mass ratios. As a simple example, carbon monoxide (CO) and diatomic nitrogen (N_2) overlap, requiring examination of other mass fragments to distinguish between the two. As with depth profiling for either AES or XPS, depth resolution "smearing" can occur either due to ion beam mixing of near-surface species or due to the development of surface topography after long times under the ion beam. Despite these potential limitations, SIMS should remain the technique of choice for many low detection limit, high surface sensitivity studies (Zalm, 1995).

3.5.4 Infrared Spectroscopy

IRS is particularly useful in detecting lubricant films on surfaces. It can provide binding and chemical information for adsorbed large molecules. It has an additional advantage in that it is nondestructive. Incident electrons in AES can cause desorption and decomposition even for aluminum oxide, and can be very destructive for polymers. Similarly, the emitted electrons can cause destruction of some films for both AES and XPS. In IRS, the specimen is illuminated with infrared light of well-defined energy. If the

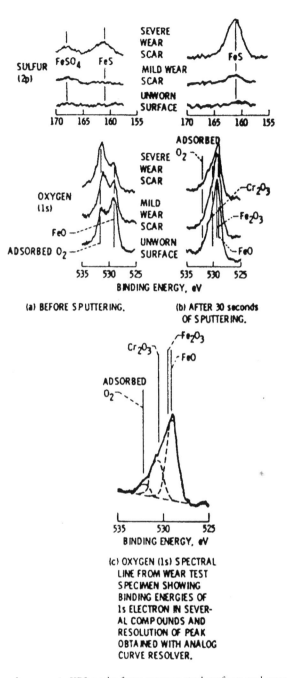

FIGURE 3.23 Sulfur $2p$ and oxygen $1s$ XPS peaks from unworn steel surfaces and wear scars run in mineral oil with 1% dibenzyl disulfide. (From Wheeler, D. R. (1978), *Wear* 47, 243–254. With permission.)

energy of the incident light corresponds to a transition between vibrational energy levels in the specimen, the light can be absorbed. When compared to the reference light beam that has not passed through a sample, the infrared light interacting with the sample will appear at reduced intensity at these vibrational excitation energies. The dispersion step is similar to dispersion in photon spectroscopy in that a grating or prism is used to isolate the wavelengths of interest. A variation of IRS which has advantages in sensitivity and resolution is called Fourier transform infrared spectroscopy (FTIR). In FTIR, the incident beams are passed through a Michelson interferometer in which one of the paths is modulated by moving a

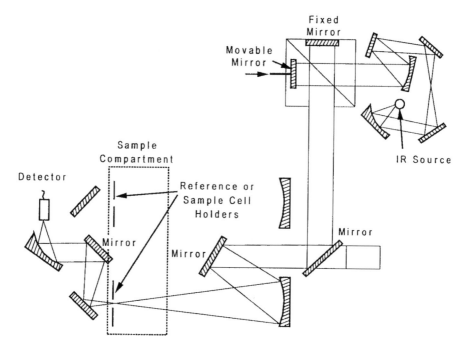

FIGURE 3.24 Optical arrangement for a FTIR spectrometer. (From Ferrante, J. (1993), in *Surface Diagnostics in Tribology* (K. Miyoshi and Y. W. Chung, eds.), World Scientific, Singapore. With permission.)

mirror. As before, one of the modulated beams is passed through the sample and impinges directly on the detector, which is a heat-sensitive device. When nonmonochromatic radiation is used, the Fourier transform of the spatially modulated beam contains all of the information in one signal, as opposed to the dispersion method where each beam must be analyzed separately. A schematic diagram of the equipment is shown in Figure 3.24.

IRS is surface sensitive and has been used in a diverse range of analytical science applications (McKelvy et al., 1996), which we will not address. A primary interest in tribology is the detection of hydrocarbon or additive films on surfaces. AES and XPS, for contrast, would primarily be useful in detecting elemental species and are limited to use in ultrahigh vacuum. IRS can be used in air as well as vacuum. The surface sensitivity of IRS can be enhanced by multiple reflections in the surface films and by using grazing incidence angles. Orientation of adsorbed molecules can be obtained by examining the polarization dependence of the spectrum. There are selection rules for what materials can be detected depending on whether the molecule has a dipole moment and the orientation of the molecule on a metal surface. Sample spectra for micron thick and less than 200 Å thick Krytox films on a metal substrate are shown in Figure 3.25 (Herrera-Fierro, 1993).

3.5.5 Thermal Desorption

We mention briefly at this point another useful tool for examining the behavior of adsorbates on surfaces, thermal desorption spectroscopy (TDS). We give only a brief description and for a more complete treatment refer the readers to Zangwill (1988). The methods described so far give little information concerning binding energies of adsorbates to surfaces, a topic of importance when choosing stable lubricants or additives. Many species adsorb strongly but do not react chemically, e.g., by forming an oxide. When the surface is heated they can be removed intact or in some decomposed form. A simple view would be that the binding energy would resemble the curves presented in Figure 3.12, and the adsorbates could be removed by giving them sufficient energy to overcome the energy well depth. This would be accomplished by heating a sample following adsorption and then either observing the surface

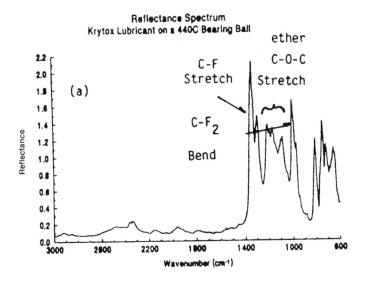

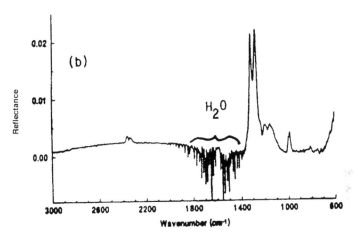

FIGURE 3.25 IR spectrum of a krytox film on a 440C bearing showing characteristic absorption corresponding to certain functional groups: (a) micron-thick film; (b) less than 200-Å-thick film, note the increased sensitivity and the bend from absorbed water from 1400–2000 cm^{-1}. (Courtesy of Herrera-Fierro, 1993).

coverage via AES or monitoring pressure increases in the vacuum system. There can be a variety of bonding states depending on the structure of the surface; e.g., at a step edge one would expect a different binding energy from a surface site. Although the real situation may be quite complex, we describe the simplest case, a single bonding state, and no decomposition. The rate of desorption can be described by (Meyers et al., 1996)

$$ r = d\theta/dt = \theta\nu \exp\left(-E_d/kT\right) \tag{3.16} $$

where θ is the coverage, ν is the pre-exponential frequency factor, E_d is the desorption energy, T is the temperature, and k is the Boltzmann constant. By heating to various temperatures and performing an Arrhenius plot, one can extract E_d and determine the strength of bonding to a surface. This technique will be seen to be useful in studying monolayer and submonolayer adsorbed film effects on friction between otherwise clean metal surfaces.

3.6 Surface Effects in Tribology

In this section we will deal with a number of issues, analyzing evidence first for effects at the atomic monolayer or submonolayer level in tribology. We will examine monolayer effects both from a fundamental standpoint as well as from a more practical viewpoint. This will not be a comprehensive review of the literature, but in keeping with the objectives of this chapter, again will be didactic in nature. The references selected, however, will have references to the relevant literature. In this chapter we are not concerned with the effects of lubricants other than their effects of changing shear strength or adhesion at the interface. Consequently, we are interested in issues involving boundary lubrication and interfacial properties (Ferrante and Pepper, 1989; Gellman 1992; Carpick and Salmeron, 1997).

There are a number of issues to address which emphasize the difficulties involved in answering fundamental questions concerning bonding (Ferrante and Pepper, 1989). One clear difficulty is the fact that one cannot observe the interface during the interaction. It is necessary to infer what happened at the interface by examining the states of the surfaces before and after interaction to the extent analytical techniques can determine. There are often situations where the locus of failure is not the interface, e.g., shear or adhesive failure can occur in the bulk of one of the materials rather than at the interface, or both effects can occur depending on the region in contact. There are uncertainties regarding the measurement of forces, although some recent efforts nicely relate lateral force sensitivity to normal force sensitivity for a commercial scanning probe microscope (SPM) cantilever beam, i.e., for a single asperity contact (Ogletree et al., 1996). Clearly, there are elastic effects in any measuring apparatus and in the materials involved that make measurement of the force distribution at the interface difficult. Materials can change mechanical properties as a result of the forces applied. For example, such properties as hardness, ductility, defect formation, plasticity, strain hardening, and creep must be considered. Surface properties may be altered just by contact with the counterface material (Carpick et al., 1996). Generally, even the true area of contact is not known in macroscopic studies due to the fact that asperities determine the contact area on most practical materials.

There is a great deal yet to be learned concerning the basic interfacial properties in tribology. This is in part due to the complexities involved. As an example of such complexities, in Table 3.1 we show some results of Buckley and Pepper (1971) for metallic transfer of dissimilar metals in sliding contact performed using a pin-on-disk apparatus in an ultrahigh vacuum system with AES analysis in the wear track.

Both pin and disk specimens were ion sputter cleaned. As we can see all metals transferred to tungsten, and cobalt transferred in all cases. However, iron and nickel did not transfer to tantalum, molybdenum, or niobium. The surprising result is that the softer metals did not transfer to the harder in all cases. Pepper explained these results in terms of the mechanical properties of the materials. Tungsten which is the hardest of the materials fits the expected pattern. However, since nickel and iron strain harden, transfer

TABLE 3.1 Metallic Transfer for Dissimilar
Metals in Sliding Contact

Disk	Rider	Transfer of Metal from Rider to Disk
Tungsten	Iron	Yes
	Nickel	Yes
	Cobalt	Yes
Tantalum	Iron	No
	Nickel	No
	Cobalt	Yes
Molybdenum	Iron	No
	Nickel	No
	Cobalt	Yes
Niobium	Iron	No
	Nickel	No
	Cobalt	Yes

and deformation are minimized. Cobalt, which has a hexagonal close-packed structure, has easy slip planes and thus transferred in all cases. Thus, simple explanations based on cohesive and interfacial energies can be misleading if mechanical properties are not taken into account.

We give a second example by Pepper (1974) demonstrating the care necessary in performing studies of polymer films transferred from polymer pins sliding on a S-Monel disk in ultrahigh vacuum. Figure 3.26 shows the AES spectra for polytetrafluoroethylene (PTFE), polyvinyl chloride (PVC) and polychlorotrifluoroethylene (PCTFE) pins sliding on S-Monel. The PTFE spectrum shows large fluorine and carbon peaks and large attenuation of the metal peaks (care had to be taken due to electron bombardment desorption of the fluorine). The friction coefficient was low and smooth suggesting slip. The combined results indicated that PTFE strands were transferring to the metal surface consistent with the models of Pooley and Tabor (1972). For PVC the AES spectrum shows a large chlorine peak and small attenuation of the metal peaks suggesting decomposition and chlorine adsorption rather than polymer transfer. The friction coefficient, although reduced, remained large and exhibited some stick slip. For PCTFE the spectrum shows chlorine, carbon, and intermediate attenuation of the metal peaks, but with stability under electron bombardment suggesting the possibility of both decomposition and some polymer transfer. The friction coefficient was high with stick slip. Consequently, it is difficult to anticipate what is happening, again demonstrating the need for more extensive surface characterization.

3.6.1 Monolayer Effects in Adhesion and Friction

The bulk of the discussion in this section will be based on the recent high-quality experiments of Gellman and collaborators (Gellman, 1992; McFadden and Gellman, 1995a,b), while we acknowledge the pioneering contributions of Buckley and the members of his group (Buckley, 1981). Gellman has performed a number of adhesion and friction experiments on single crystals in contact, both for clean and adsorbate covered interfaces, namely, Cu(111)–Cu(111) and Ni(100)–Ni(100). The apparatus for the friction and adhesion experiments is shown in Figure 3.27 (Gellman, 1992).

The contacting crystals had a slight curvature in order to prevent contact at the edges of the samples where large concentrations of steps were expected, and to ensure point contact. The vacuum system was also vibration isolated which is very important for reasons to be discussed below. Normally, the experiments were repeated for ten different points. The normals of the crystals were aligned by lasers. The apparatus provided the ability to sputter clean both contacting surfaces, and LEED and AES could be performed on the surfaces in order to guarantee the crystallinity and to measure contamination and determine concentration of adsorbates. The AES measurements were performed with LEED optics in the old retarding field analyzer mode. Following sputter cleaning, the samples were annealed and analyzed with LEED in order to verify the crystal structure and with AES to verify cleanliness.

On Cu(111) McFadden and Gellman (1995a) examined the effects of sulfur adsorption, since sulfur is a common antiwear additive. It is often used along with phosphorous in order to prevent metal to metal contact when the lubricant breaks down. Both surfaces were dosed with sulfur using hydrogen sulfide, which decomposes upon heating, desorbing the hydrogen and leaving sulfur behind on the surface. At saturation, sulfur forms an ordered superlattice which has a $\sqrt{7} \times \sqrt{7}$ R19 degree LEED pattern. This corresponds to a close-packed monolayer of sulfide ions (S^{2-}) on top of the Cu(111) lattice with an absolute coverage of 0.43 monolayers relative to the Cu(111) substrate. The results of the adhesion experiments are shown in Figure 3.28.

The adhesion coefficient, $\mu_{ad} = F_{ad}/F_N$, which is the ratio of pull-off force to normal force was found to be 0.69 ± 0.21. As little as 0.05 monolayers (11% of saturation) gave a substantial percentage reduction. The saturation value at one monolayer was 0.26 ± 0.07. They also found that there was no dependence of the adhesion coefficient on contact time, separation time, temperature, or normal force. There would be no expected dependence on normal force since higher loads would simply increase the contact area, and the adhesion coefficient, therefore, would be expected to track with the normal force. Buckley (1981) reported adhesion coefficients much greater than 1.0 for clean single crystals in contact. As mentioned earlier, it is important to control vibrations in the experiments. Bowden and Tabor (1964) showed that

173

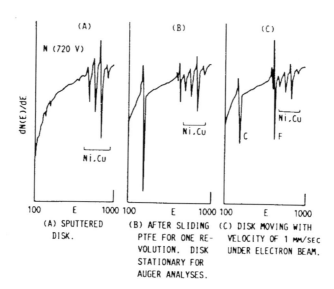

N (720 V)

$\partial N(E)/\partial E$

Ni,Cu

Ni,Cu

Ni,Cu

C

F

100   E   1000   100   E   1000   100   E   1000

(A) SPUTTERED
DISK.

(B) AFTER SLIDING
PTFE FOR ONE RE-
VOLUTION. DISK
STATIONARY FOR
AUGER ANALYSES.

(C) DISK MOVING WITH
VELOCITY OF 1 ᴍᴍ/sᴇᴄ
UNDER ELECTRON BEAM.

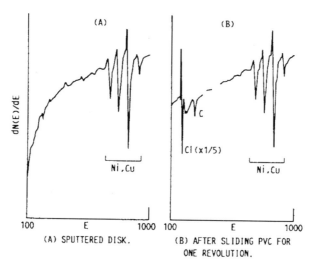

$\partial N(E)/\partial E$

Ni,Cu

Ni,Cu

C

Cl (x1/5)

100   E   1000   100   E   1000

(A) SPUTTERED DISK.

(B) AFTER SLIDING PVC FOR
ONE REVOLUTION.

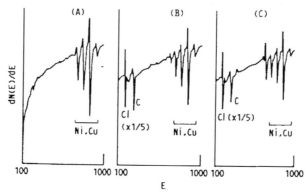

$\partial N(E)/\partial E$

Ni,Cu

Ni,Cu

Ni,Cu

C

C

Cl
(x1/5)

Cl (x1/5)

100   1000   100   1000   100   1000

E

(A) SPUTTERED DISK.

(B) AFTER SLIDING
PCTFE FOR ONE
REVOLUTION.
DISK STATIONARY
FOR AUGER
ANALYSES.

(C) DISK MOVING WITH
VELOCITY OF 1 ᴍᴍ/sᴇᴄ
UNDER ELECTRON BEAM.

when clean metal surfaces were plastically loaded and then translated, the junction grew, i.e., the true contact area increased until shear occurred. In all probability, the large adhesion coefficients observed by Buckley were caused by increased contact area from junction growth caused by vibrations. In any event, there is strong bonding at clean metal interfaces and, as Gellman and others have shown, the adhesion coefficient can be decreased by submonolayer films.

McFadden and Gellman (1995a) performed a somewhat different experiment on the Cu(111) surface with regard to static friction measurements. The surfaces were exposed to laboratory atmosphere for several days. The resulting AES spectrum is shown in Figure 3.29. As can be seen, the primary contaminants were sulfur and carbon. Although tenuous, they estimated the coverage to be between 10 and 15 Å thick. They then executed a number of sputtering cycles followed by annealing with static friction measurements after each cycle. The results are shown in Figure 3.30. The friction coefficient, μ_s, increased with removal of contaminant to their clean surface value of 4.4 ± 1.3. Again these results imply submonolayer effects by these contaminants.

Wheeler (1976) earlier performed static friction experiments with both chlorine and oxygen adsorbed on polycrystalline surfaces of copper, iron, and steel. The measurements were performed with a pin-on-disk apparatus, with AES used to monitor coverage. The results of these studies are shown in Figure 3.31. Wheeler found that there were no differences between the effects of chlorine and oxygen on any of the surfaces if adsorbate coverage was taken into account. Adsorption at partial monolayer coverages reduced the coefficient of friction in all cases. Wheeler found that he could get a good correlation with a junction growth model where the surfaces were partially covered during translation. The values of the coefficient were of comparable magnitude to those of Gellman.

3.6.2 Atomic Effects Due to Adsorption of Hydrocarbons

Our primary objective in this and following sections is to explore the evidence for atomic effects on friction. Again, we start by referring to recent publications by Gellman et al. (Gellman, 1992; McFadden and Gellman, 1995a,b; Meyers et al., 1996) describing a number of friction experiments performed with hydrocarbon adsorbates. First, we present the effects of ethanol adsorption on a sulfur-covered Ni(100) surface. The sulfur was adsorbed as previously described and produced a c 2 × 2 structure on the nickel surface. The sulfur overlayer was used since stable results could not be obtained with ethanol alone, which Gellman found desorbed at very low temperatures. There were a number of different friction states depending on ethanol coverage, shown in Figure 3.32 (Gellman, 1992).

These are the classic behaviors observed in friction (Buckley, 1981). The first shown is "slip," where once a certain transverse force is applied there is simple sliding. The second is "stick-slip," where rebonding occurs and the simple slip process is repeated between rebonding events, with poor definition of a friction coefficient. This behavior probably results from junction growth with the adsorbate only effective in certain regions, as observed by Wheeler (1976). Finally, "stick" is where the surfaces weld and it is similar to shearing a bulk solid. The results of these studies are shown in Figure 3.33 where the key indicates differences in behavior.

Unlike the results for Cu(111) the sulfur coating did not reduce μ_s and gave stick behavior. Partial coverages of ethanol gave erratic behavior typically exhibiting stick-slip. Finally at one monolayer the ethanol effectively lubricated the surface giving the desired slip as well as low adhesion. Consequently, a monolayer film can effectively lubricate. Similar behavior was observed for 2,2,2-trifluoroethanol adsorbed on clean Cu(111) surfaces (McFadden and Gellman, 1995b). Again stick-slip was observed for coverages less than a monolayer and slip for coverages greater than or equal to one monolayer. Similarly, high adhesion was observed for low coverages and low adhesion for higher coverages.

FIGURE 3.26 AES spectra from an S-Monel disk with transfer films from different polymers. (From Pepper, S. V. (1974), *J. Appl. Phys.* 45, 2947–2956. With permission.)

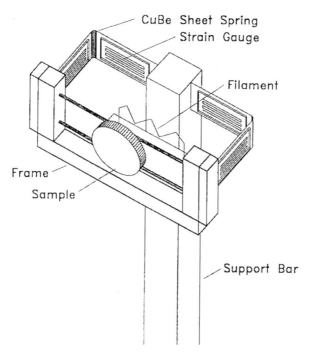

FIGURE 3.27 Diagram of UHV tribometer by Gellman (1992). The sample mounted to the tribometer is ≈1 cm in diameter. The support bar is mounted onto a liquid nitrogen reservoir. (From Gellman, A.J. (1992), *J. Vac. Sci. Techol. A* 10, 180–187. With permission.)

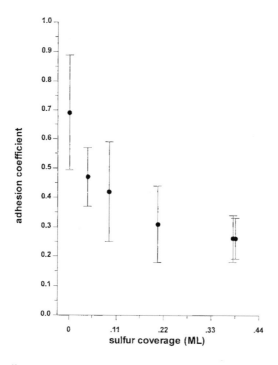

FIGURE 3.28 Adhesion coefficient vs. sulfur coverage by McFadden and Gellman (1995a). Experimental conditions: temperature = 300 K; normal force = 40–60 mN; contact time before separation = 15–20 s; separation speed = 50 μm/s. At each coverage the standard deviation of the adhesion coefficient calculated from ten measurements is plotted. (From McFadden, C.F. and Gellman, A.J. (1995), *Tribol. Lett.* 1, 201–210. With permission.)

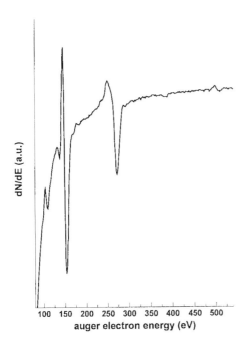

FIGURE 3.29 AES spectrum of air-exposed copper (111) surface reintroduced to vacuum chamber and annealed to 1000 K to remove weakly adsorbed species showing the primary contaminants sulfur and carbon. (From McFadden, C. F. and Gellman, A. J. (1995), *Tribol. Lett.* 1, 201–210. With permission.)

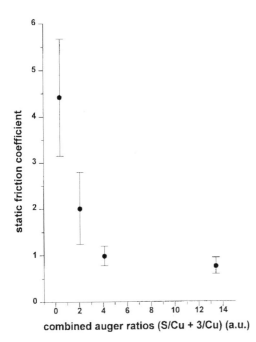

FIGURE 3.30 Static friction coefficient vs. Auger peak ratios as the contaminated copper (111) surfaces are gradually cleaned by ion bombardment. For each contamination level the average and standard deviation of the friction coefficient calculated from ten measurements are presented. (From McFadden, C. F. and Gellman, A. J. (1995), *Tribol. Lett.* 1, 201–210. With permission.)

We conclude this discussion with some TDS and IRS studies of fluorinated hydrocarbons on Cu(111) surfaces by Meyers et al. (1996). Fluorinated hydrocarbons are of interest because of their thermal stability and, therefore, are useful for higher-temperature lubrication. All of the ethers were found to adsorb molecularly and reversibly on the Cu(111) surfaces, exhibiting first-order desorption kinetics. These studies found that fluorination of perfluoropolyalkyl ethers (PFPAE) reduced the desorption energy. The general reason for this behavior was felt to be that bonding occurred with the oxygens on the PFPAEs. Fluorine is thought to weaken this interaction. In addition, fluorination changed the adsorbate–adsorbate

177

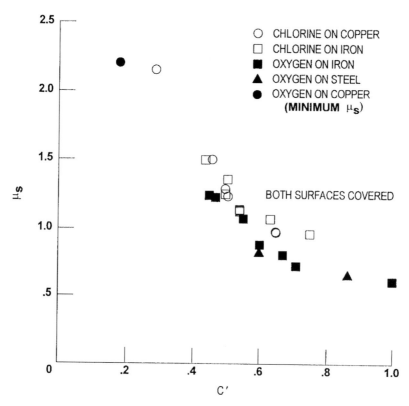

FIGURE 3.31 The effects of oxygen and chlorine adsorption on static friction for a metal–metal contact as a function of coverage. (From Wheeler, D. R. (1976), *J. Appl. Phys.* 47, 1123–1130. With permission.)

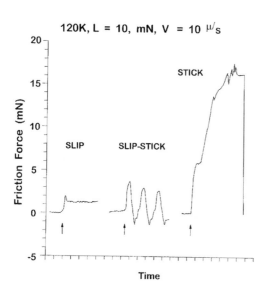

FIGURE 3.32 Friction force vs. time for three types of sliding behavior observed during shearing of interfaces between ethanol-charged nickel (100) surfaces, by Gellman (1992). The loads on the interface are all 10 mN at the beginning of each trace and shearing begins at the points marked with the arrows. Temperature = 120 K, load = 10 mN, velocity = 10 μm/s. (From Gellman, A.J. (1992), *J. Vac. Sci. Techol. A* 10, 180–187. With permission.)

interaction from repulsive to attractive. FTIR studies shows that the diethyl ethers are oriented with their molecular axes parallel to the Cu(111) surface.

We conclude this section with some elegant experiments performed by Krim et al. (Sokoloff et al., 1993; Mak et al., 1994; Krim et al., 1995; Daly and Krim, 1996) for which Widom and Krim (1986; also Krim and Widom, 1988) originally adapted an apparatus for sliding monolayer and thicker physisorbed

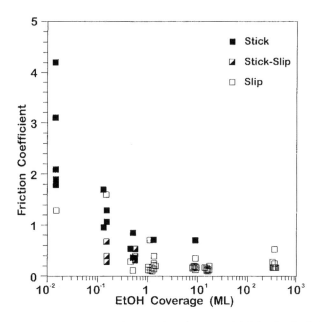

FIGURE 3.33 Coefficients of friction measured at interfaces between nickel (100)-c(2×2)–S surface modified by adsorption of ethanol at coverages from 0 to 300 ML. (From Gellman, A. J. (1992), *J. Vac. Sci. Technol. A* 10, 180–187. With permission.)

films on silver substrates. The true contact area is known in these studies and there is no wear of the surfaces, which are silver deposited with a (111) face on a quartz crystal microbalance. These experiments were designed to investigate the nature of the friction forces. Slipping of physisorbed films, e.g., ethane and ethylene, on the silver surfaces causes shifts in the frequency and amplitude of the quartz crystal vibrations, and film characteristic slip times are obtained from shifts in the quality factor, Q, of the oscillator circuit. Friction results from energy losses due to the interactions at an interface. Typically, it is assumed that these losses occur through phonons, vibrations of the lattice which dissipate the energy (Sokoloff et al., 1993), although recently Persson (1993), followed by Sokoloff (1995), demonstrated that electronic excitations could also be an energy dissipation mechanism. In these experiments the average friction force per unit area is related to the slip time by

$$F_f = \rho v / \tau \qquad (3.17)$$

where F_f is the friction force, ρ and v are the film density and velocity, respectively, and τ is the slip time. Therefore, a longer slip time implies a smaller friction force. Based on electronic effects alone (Persson, 1994) ethane on silver is expected to have a longer slip time than ethylene and thus a higher friction. Krim's results are consistent with this prediction, although the ability of the theory to predict hard numbers is limited. Monolayer oxygen adsorption on the surface increases the slip time and thus lowers the friction coefficient substantially, consistent with the electronic excitation model. Although these experiments are not completely definitive at this time, they open new areas of investigation for probing the little-understood energy loss mechanisms in friction.

3.6.3 Atomic Effects in Metal–Insulator Contacts

We now discuss a number of results where monolayer effects have been observed in metal–insulator friction experiments performed by Pepper (1976, 1979, 1982). First, we discuss static friction experiments performed with a copper ball on the basal plane of a diamond flat. The diamond surface is known to be terminated by hydrogen atoms. Pepper (1982) found that the hydrogen could be removed by either

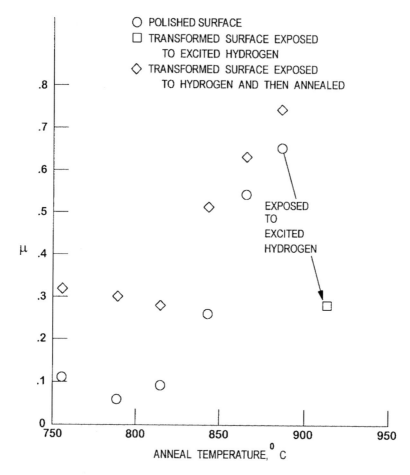

FIGURE 3.34 Copper–diamond static friction coefficient as a function of diamond-annealing temperature. (From Pepper, S. V. (1982), *J. Vac. Sci. Technol.* 20, 643–646. With permission.)

electron bombardment or heating in ultrahigh vacuum. Following removal of the hydrogen, the LEED pattern for the surface changed from a 1×1 to 1×2 pattern. The interesting feature is shown in the static friction results for the two situations given in Figure 3.34. We can see that removal of the hydrogen caused an increase in μ_s. Pepper found that exposing the surface to excited hydrogen could cause readsorption. Following hydrogen readsorption the static friction was again reduced, although not to its original value. Pepper also performed core-level ionization energy loss spectroscopy on the diamond surface and found changes in the electronic structure of the valence band. Therefore, these experiments showed results sensitive to the electronic structure of the surface.

In Figure 3.35 we show similar studies performed by Pepper (1979) for copper and nickel balls on the basil plane of sapphire with chlorine and oxygen as adsorbates. Again the balls were cleaned by sputtering and the sapphire was cleaned by heating as verified by AES. In both cases, we see that oxygen increased the interfacial friction and chlorine decreased it compared to the clean metal. For the increased shear strength case, one cannot identify the locus of failure, whether at the interface or in the bulk of one of the materials. Unfortunately, the wear scar area was too small to analyze. In any event, again monolayer adsorption was detectable in friction experiments. Perhaps even more surprising were effects shown in Figure 3.36. Here dynamic friction experiments were performed by Pepper (1976) with a polycrystalline sapphire pin on a polycrystalline iron disk in ultrahigh vacuum. We see the same behavior reproduced with dynamic friction. After removing the oxygen or chlorine atmospheres, the friction coefficients returned to their original values.

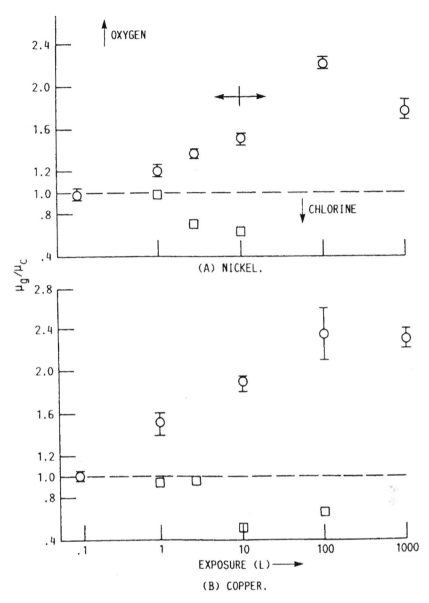

FIGURE 3.35 Static friction coefficient after exposure to gas (μ_g) ratio to static friction coefficient of clean contact (μ_c), plotted vs. exposure to oxygen or chlorine for (A) nickel and (B) copper. (From Pepper, S. V. (1979), *J. Appl. Phys.* 50, 8062–8065. With permission.)

3.7 Concluding Remarks

We have attempted to present a description of issues and techniques in the physics of surfaces of interest in nanotribology. This field of research is exceedingly difficult with the effects of a wide variety of materials and phenomena to understand. It is further complicated by the inability to observe the interactions directly, resulting in conclusions from inference. It is clear that there is a great deal of research necessary to obtain a comprehensive understanding of tribology at the nanoscale. It is hoped that the demonstrations in this chapter and in this volume indicate that there is the possibility of gaining rational understanding leading to the design of better materials in this technologically important field.

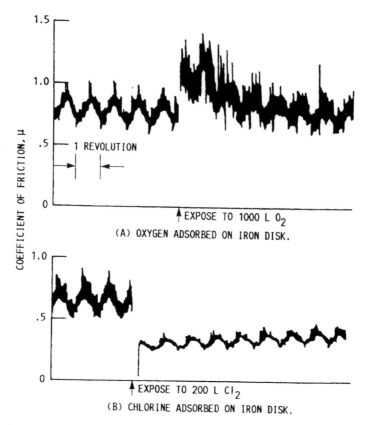

FIGURE 3.36 Effect of oxygen or chlorine on the dynamic friction of an iron disk sliding on sapphire. (From Pepper, S. V. (1976), *J. Appl. Phys.* 47, 2579–2583. With permission.)

References

Arlinghaus, F. J., Gay, J. G., and Smith, J. R. (1980). "Self-Consistent Local-Orbital Calculation of the Surface Electronic Structure of Ni(100)," *Phys. Rev. B*, 21, 2055–2059.

Ashcroft, N. W. and Mermin, N. D. (1976), *Solid State Physics*, Holt, Rinehart and Winston, New York.

Bacigalupi, R. J. (1964), "Surface Topography of Single Crystals of Face-Centered-Cubic, Body-Centered-Cubic, Sodium Chloride, Diamond, and Zinc-Blende Structures," NASA TN D-2275, National Aeronautics and Space Administration, Washington, D.C.

Banerjea, A., Ferrante, J., Smith, J. R. (1991), "Adhesion at Metal Interfaces," in *Fundamentals of Adhesion* (Liang-Huang Lee, ed.), Plenum, New York, 325–348.

Bartelt, N. C., Einstein, T. L., and Williams, E. D. (1994), "Measuring Surface Mass Diffusion Coefficients by Observing Step Fluctuations," *Surf. Sci.*, 312, 411–421.

Benninghoven, A., Rüdenauer, F. G., and Werner, H. W. (1987). *Secondary Ion Mass Spectroscopy*, Wiley, New York.

Binnig, G., Rohrer, H., Gerber, Ch., and Weibel, E. (1983), "7 × 7 Reconstruction on Si(111) Resolved in Real Space," *Phys. Rev. Lett.*, 50, 120–123.

Bowden, F. P. and Tabor, D. (1964), *The Friction and Lubrication of Solids*, Oxford University Press, London.

Bozzolo, G., Rodriguez, A. M., and Ferrante, J. (1994), "Multilayer Relaxation and Surface Energies of Metallic Surfaces," *Surf. Sci.*, 315, 204–214.

Brainard, W. A. and Buckley, D. H. (1971), "Preliminary Studies by Field Ion Microscopy of Adhesion of Platinum and Gold to Tungsten and Iridium," NASA TN D-6492, National Aeronautics and Space Administration, Washington, D.C.

Brainard, W. A. and Buckley, D. H. (1973). "Adhesion and Friction of PTFE in Contact with Metals as Studied by Auger Spectroscopy, Field Ion, and Scanning Electron Microscopy," *Wear*, 26, 75–93.

Brainard, W. A. and Ferrante, J. (1979), "Evaluation and Auger Analysis of a Zinc-Dialkyl-Dithioiphosphate Antiwear Additive in Several Diester Lubricants," NASA TP-1544, National Aeronautics and Space Administration, Washington, D.C.

Buckley, D. H. (1981). *Surface Effects in Adhesion, Friction, Wear, and Lubrication,* Elsevier Scientific Publishing, Amsterdam, The Netherlands.

Buckley, D. H. and Pepper, S. V. (1971), "Elemental Analysis of a Friction and Wear Surface during Sliding using Auger Spectroscopy," NASA TN D-6497, National Aeronautics and Space Administration, Washington, D.C.

Carpick, R. W. and Salmeron, M. (1977), "Scratching the Surface: Fundamental Investigations of Tribology with Atomic Force Microscopy," *Chem. Rev.* 97, 1163–1194.

Carpick, R. W., Agrait, N., Ogletree, D. F., and Salmeron, M. (1996), "Variation of the Interfacial Shear Strength and Adhesion of a Nanometer-Sized Contact," *Langmuir,* 12, 3334–3340.

Chambers, S. A. (1992), "Elastic Scattering and Interference of Backscattered Primary, Auger and X-Ray Photoelectrons at High Kinetic Energy: Principles and Applications," *Surf. Sci. Rep.* 16, 261–331.

Conrad, E. H. and Engel, T. (1994), "The Equilibrium Crystal Shape and the Roughening Transition on Metal Surfaces," *Surf. Sci.* 299/300, 391–404.

Daly, C. and Krim, J. (1996), "Sliding Friction of Solid Xenon Monolayers and Bilayers on Ag(111)," *Phys. Rev. Lett.* 76, 803–806.

De Crescenzi, M. (1995), "Structural Surface Investigations with Low-Energy Backscattered Electrons," *Surf. Sci. Rep.* 21, 89–175.

Diestler, D. J., Rajasekaran, E., and Zeng, X. C. (1997), "Static Frictional Forces at Crystalline Interfaces," *J. Phys. Chem. B.* 101, 4992–4997.

Dowson, D. (1979), *History of Tribology,* Longman, New York.

Faulkner, R. G. (1996), "Segregation to Boundaries and Interfaces in Solids," *Int. Mater. Rev.* 41, 198–208.

Feibelman, P. J. (1992), "First-Principles Calculation of the Geometric and Electronic Structure of the Be(0001) Surface," *Phys. Rev. B.* 46, 2532–2539.

Ferrante, J. (1982), "Practical Applications of Surface Analysis Tools in Tribology," *J. Am. Soc. Lubr. Eng.* 38, 223–236.

Ferrante, J. (1989), "Applications of Surface Analysis and Surface Theory in Tribology," *Surf. Int. Anal.* 14, 809–822.

Ferrante, J. (1993), "Surface Analysis in Applied Tribology," in *Surface Diagnostics in Tribology* (K. Miyoshi and Y. W. Chung, eds.), pp. 19–32, World Scientific, Singapore.

Ferrante, J. and Pepper, S. V. (1989), "Fundamentals of Tribology at the Atomic Level," *Mater. Res. Soc. Symp. Proc.* 140, 37–50.

Ferrante, J. and Smith, J. R. (1985), "Theory of the Bimetallic Interface," *Phys. Rev. B.* 31, 3427–3434.

Ferrante, J., Buckley, D. H., Pepper, S. V., and Brainard, W. A. (1973), "Use of LEED, Auger Emission Spectroscopy and Field Ion Microscopy in Microstructural Studies," in *Microstructural Analysis Tools and Techniques* (J. L. McCall and W. M. Mueller, eds.), pp. 241–279, Plenum Press, New York.

Ferrante, J., Bozzolo, G. H., Finley, C. W., and Banerjea, A. (1988), "Interfacial Adhesion: Theory and Experiment," *Mater. Res. Soc. Symp. Proc.* 118, 3–16.

Finnis, M. W. (1996), "The Theory of Metal-Ceramic Interfaces," *J. Phys. Condens. Matter* 8, 5811–5836.

Foiles, S. M. (1987), "Reconstruction of fcc (110) Surfaces," *Surf. Sci.* 191, L779–L786.

Gellman, A. J. (1992), "Lubrication by Molecular Monolayers at Ni–Ni Interfaces," *J. Vac. Sci. Technol. A* 10, 180–187.

Gnaser, H. (1997), "SIMS Detection in the 10^{12} Atoms cm^{-3} Range," *Surf. Int. Anal.* 25, 737–740.

Good, B. S. and Banerjea, A. (1992), "Monte Carlo Study of Reconstruction of the Au(110) Surface Using Equivalent Crystal Theory," *Mater. Res. Soc. Symp. Proc.* 278, 211–216.

Good, B., Bozzolo, G., and Ferrante, J. (1993), "Surface Segregation in Cu–Ni Alloys," *Phys. Rev. B.* 48, 18284–18287.

Hantsche, H. (1989), "Comparison of Basic Principles of the Surface-Specific Analytical Methods: AES/SAM, ESCA(XPS), SIMS, and ISS with X-ray Microanalysis, and Some Applications in Research and Industry," *Scanning*, 11, 257–280.

Herrera-Fierro, P. C., private communication, first published in Ferrante (1993).

Hirano, M. and Shinjo, K. (1990), "Atomistic Locking and Friction," *Phys. Rev. B* 41, 11837–11851.

Hirano, M., Shinjo, K., Kaneko, R., and Murata, Y. (1997), "Observation of Superlubricity by Scanning Tunneling Microscopy," *Phys. Rev. Lett.* 78, 1448–1451.

Hirsch, P., Howie, A., Nicholson, R. B., Pashley, D. W., and Whelan, M. J. (1977), *Electron Microscopy of Thin Crystals,* Robert E. Krieger, Malabar, FL.

Hölscher, H., Schwarz, U. D., and Wiesendanger, R. (1997), "Modeling of the Scan Process in Lateral Force Microscopy," *Surf. Sci.* 375, 395–402.

Hong, T., Smith, J. R., and Srolovitz, D. J. (1995), "Theory of Metal-Ceramic Adhesion," *Acta Met. Mater.* 43, 2721–2730.

Jennings, W. D., Chottiner, G. S., and Michal, G. M. (1988), "Sulpher Segregation to the Metal Oxide Interface during the Early Stages in the Oxidation of Iron," *Surf. Int. Anal.* 11, 377–382.

Kaxiras, E. (1996), "Review of Atomistic Simulations of Surface Diffusion and Growth on Semiconductors," *Computational Mater. Sci.* 6, 158–172.

Kellogg, G. L. (1994), "Field Ion Microscope Studies of Single-Atom Surface Diffusion and Cluster Nucleation on Metal Surfaces," *Surf. Sci. Rep.* 21, 1–88.

Kittel, C. (1986), *Introduction to Solid State Physics,* 6th ed., John Wiley & Sons, New York.

Kobistek, R. J., Bozzolo, G., Ferrante, J., and Schlosser, H. (1994), "Multilayer Relaxation and Surface Structure of Ordered Alloys," *Surf. Sci.* 307–309, 390–395.

Kohn, W. and Sham, L. J. (1965), "Self-Consistent Equations Including Exchange and Correlation Effects," *Phys. Rev.* 140, A1133–A1138.

Krim, J. and Widom, A. (1988), "Damping of a Crystal Oscillator by an Absorbed Monolayer and its Relation to Interfacial Viscosity, "*Phys. Rev. B* 38, 12184–12189.

Krim, J., Daly, C., and Dayo, A. (1995), "Electronic Contributions to Sliding Friction," *Tribol. Lett.* 1, 211–218.

Lambrecht, W. R. L. and Segall, B. (1989), "Efficient Direct Calculation Method for Dielectric Response in Semiconductors," *Phys. Rev. B* 40, 7793–7801.

Lang, N. D. and Kohn, W. (1970), "Theory of Metal Surfaces: Charge Density and Surface Energy," *Phys. Rev. B* 1, 4555–4568.

Lantz, M. A., O'Shea, S. J., Hoole, A. C. F., and Welland, M. E. (1997), "Lateral Stiffness of the Tip and Tip-Sample Contact in Frictional Force Microscopy," *Appl. Phys. Lett.* 70, 970–972.

Lundqvist, S. and March, N. H., Eds. (1983), *Theory of Inhomogeneous Electron Gas,* Plenum Press, New York.

Mak, C., Daly, C., and Krim, J. (1994), "Atomic-Scale Measurements on Silver and Chemisorbed Oxygen Surfaces," *Thin Solid Films* 253, 190–193.

Matsukawa, H. and Fukuyama, H. (1994), "Theoretical Study of Friction: One-Dimensional Clean Surfaces," *Phys. Rev. B* 49, 17286–17292.

McFadden, C. F. and Gellman, A. J. (1995a), "Effect of Surface Contamination on the UHV Tribological Behavior of the Cu(111)/Cu(111) Interface," *Tribol. Lett.* 1, 201–210.

McFadden, C. F. and Gellman, A. J. (1995b), "Ultrahigh Vacuum Boundary Lubrication of the Cu–Cu Interface by 2,2,2-Trifluoroethanol," *Langmuir* 11, 273–280.

McKelvy, M. L., Britt, T. R., David, B. L., Gillie, J. K., Lentz, L. A., Leugers, A., Nyquist, R. A., and Putzig, C. L. (1996), "Infrared Spectroscopy," *Anal. Chem.* 68, 93R–160R.

Meyers, J. M., Street, S. C., Thompson, S., and Gellman, A. J. (1996), "Effect of Fluorine on the Bonding and Orientation of Perfluoroalkyl Ethers on the Cu(111) Surface," *Langmuir* 12, 1511–1519.

Mitchell, T.E. (1973), "High Voltage Electron Microscopy of Microstructural Analysis," in *Microstructural Analysis Tools and Techniques* (J.L. McCall and W.M. Mueller, eds.), pp. 125–152, Plenum Press, New York.

Montei, E. L. and Kordesch, M. E. (1996), "Detection of Tribochemical Reactions Using Photoelectron Emission Microscopy," *J. Vac. Sci. Technol. A* 14, 1352–1356.

Mueller, E. W. (1969), *Field Ion Microscopy,* American Elsevier, New York.

Mueller, E. W. and Nishikawa, O. (1968), "Atomic Surface Structure of the Common Transition Metals and the Effect of Adhesion as Seen by Field Ion Microscopy," *Adhesion or Cold Welding of Materials in Space Environment,* ASTM Special Tech. Publ. No. 431, pp. 67–87, American Society for Testing and Materials, Philadelphia.

Needs, R. J. (1987), "Calculations of the Surface Stress Tensor at Aluminum (111) and (110) Surfaces," *Phys. Rev. Lett.* 58, 53–56.

Niehus, H., Heiland, W., and Taglauer, E. (1993), "Low-Energy Ion Scattering at Surfaces," *Surf. Sci. Rep.* 17, 213–303.

Nishikawa, O. and Mueller, E. W. (1968), "Field Ion Microscopy of Contacts," *Proc. of the Holm. Seminar on Electric Contact Phenomena,* Illinois Institute of Technology, Chicago, pp. 193–206.

Ogletree, D. F., Carpick, R. W., and Salmeron, M. (1996), "Calibration of Frictional Forces in Atomic Force Microscopy," *Rev. Sci. Instrum.* 67, 3298–3306.

Ohmae, N. (1993), "Field Ion Microscopy in Tribology Studies," in *Surface Diagnostics in Tribology* (K. Miyoshi and Y. W. Chung, eds.), pp. 47–74, World Scientific, Singapore.

Ohmae, N., Umeno, M., and Tsubouchi, K. (1987), "Effect of Oxygen Adsorption on Adhesion of W to Au Studied by Field Ion Microscopy," *ASLE Trans.* 30, 409–418.

Ohmae, N., Tagawa, M., Umeno, M., and Koike, S. (1990), "High Voltage Field Ion Microscopy of Solid Lubricants," *Proc. Jpn. Int. Tribol. Conf.* Nagoya, pp. 1827–1832.

Pepper, S. V. (1974), "Auger Analysis of Films on Metals in Sliding Contact with Halogenated Polymers," *J. Appl. Phys.* 45, 2947–2956.

Pepper, S. V. (1976), "Effect of Adsorbed Films on Friction of Al_2O_3-Metal Systems," *J. Appl. Phys.* 47, 2579–2583.

Pepper, S. V. (1979), "Effect of Interfacial Species on Shear Strength of Metal–Sapphire Contacts," *J. Appl. Phys.* 50, 8062–8065.

Pepper, S. V. (1982), "Effect of Electronic Structure of the Diamond Surface on the Strength of the Diamond-Metal Interface," *J. Vac. Sci. Technol.* 20, 643–646.

Persson, B. N. J. (1991), "Surface Resistivity and Vibrational Damping in Adsorbed Layers," *Phys. Rev. B* 44, 3277–3296.

Persson, B. N. J. (1993), "Theory of Friction and Boundary Lubrication," *Phys. Rev. B* 48, 18140–18158.

Persson, B. N. J. (1994), "Theory of Friction: The Role of Elasticity in Boundary Lubrication," *Phys. Rev. B* 50, 4771–4786.

Persson, B. N. J. (1995), "Theory of Friction: Stress Domains, Relaxation, and Creep," *Phys. Rev. B* 51, 13568–13585.

Pooley, C. M. and Tabor, D. (1972), "Friction and Molecular Structure: the Behaviour of Some Thermoplastics," *Proc. R. Soc.* A329, 251–274.

Redfield, A. C. and Zangwill, A. (1992), "Attractive Interactions between Steps," *Phys. Rev. B* 46, 4289–4291.

Richter, R., Gay, J. G., and Smith, J. R. (1985), "Spin Separations in a Metal Overlayer," *Phys. Rev. Lett.* 54, 2704–2707.

Rodriguez, A. M., Bozzolo, G., and Ferrante, J. (1993), "Multilayer Relaxation and Surface Energies of fcc and bcc Metals Using Equivalent Crystal Theory," *Surf. Sci.* 289, 100–126.

Rose, J. H., Ferrante, J., and Smith, J. R. (1981), "Universal Binding Energy Curves for Metals and Bimetallic Interfaces," *Phys. Rev. Lett.* 47, 675–678.

Rose, J. H., Smith, J. R., and Ferrante, J. (1983), "Universal Features of Bonding in Metals," *Phys. Rev. B* 28, 1835–1845.

Rous, P. J. and Pendry, J. B. (1989), "Applications of Tensor LEED," *Surf. Sci.* 219, 373–394.

Shinjo, K. and Hirano, M. (1993), "Dynamics of Friction: Superlubric State," *Surf. Sci.* 283, 473–478.

Smith, J. R. and Ferrante, J. (1985), "Materials in Intimate Contact," *Mater. Sci. Forum* 4, 21–38.

Smith, J. R. and Ferrante, J. (1986), "Grain-Boundary Energies in Metals from Local-Electron-Density Distributions," *Phys. Rev. B* 34, 2238–2245.

Sokoloff, J. B. (1990), "Theory of Energy Dissipation in Sliding Crystal Surfaces," *Phys. Rev. B.* 42, 760–765.

Sokoloff, J. B. (1992), "Theory of Atomic Level Sliding Friction between Ideal Crystal Interfaces," *J. Appl. Phys.* 72, 1262–1269.

Sokoloff, J. B. (1995), "Theory of the Contribution to Sliding Friction from Electronic Excitations in the Microbalance Experiment," *Phys. Rev. B* 52, 5318–5322.

Sokoloff, J. B., Krim, J., and Widom, A. (1993), "Determination of an Atomic-Scale Frictional Force Law through Quartz-Crystal Microbalance Measurements," *Phys. Rev. B* 48, 9134–9137.

Thomas, G. and Goringe, M. J. (1979), *Transmission Electron Microscopy of Materials,* John Wiley & Sons, New York.

Tsukizoe, T., Tanaka, S., Nishizaki, K., and Ohmae, N. (1985), "Field Ion Microscopy of Metallic Adhesion and Friction," *Proc. JSLE Int. Tribol. Conf.,* Tokyo, pp. 121–126.

Van Hove, M. A., Moritz, W., Over, H., Rous, P. J., Wander, A., Barbieri, A., Materer, N., Starke, U., and Somorjai, G. A. (1993), "Automated Determination of Complex Surface Structures by LEED," *Surf. Sci Rep.* 19, 191–229.

Vlachos, D. G., Schmidt, L. D., Aris, R. (1993), "Kinetics of Faceting of Crystals in Growth, Etching, and Equilibrium," *Phys. Rev. B* 47, 4896–4909.

Wheeler, D. R. (1976), "Effect of Adsorbed Chlorine and Oxygen on the Shear Strength of Iron and Copper Junctions," *J. Appl. Phys.* 47, 1123–1130.

Wheeler, D. R. (1978), "X-Ray Photoelectron Spectroscopic Study of Surface Chemistry of Dibenzyl Disulfide on Steel under Mild and Severe Wear Conditions," *Wear* 47, 243–254.

Widom, A. and Krim, J. (1986), "Q Factors of Quartz Oscillator Modes as a Probe of Submonolayer-film Dynamics, "*Phys. Rev B* 34, 1403–1404.

Williams, E. D. (1994), "Surface Steps and Surface Morphology: Understanding Macroscopic Phenomena from Atomic Observations," *Surf. Sci.* 299/300, 502–524.

Wilson, R. G., Stevie, F. A., and Magee, C. W. (1989), *Secondary Ion Mass Spectroscopy,* Wiley, New York.

Wood, E. A. (1963), "Vocabulary of Surface Crystallography," *J. Appl. Phys.* 35, 1306–1312.

Zalm, P. C. (1995), "Ultra Shallow Doping Profiling with SIMS," *Rep. Prog. Phys.* 58, 1321–1374.

Zangwill, A. (1988), *Physics at Surfaces,* Cambridge University Press, Cambridge, U.K.

Zhong, W. and Tomanek, D. (1990), "First-Principles Theory of Atomic-Scale Friction," *Phys. Rev. Lett.* 64, 3054–3057.

Zhu, X. Y., Hermanson, J., Arlinghaus, F. J., Gay, J. G., Richter, R., and Smith, J. R. (1984), "Electronic Structure and Magnetism of Ni(100) Films: Self-Consistent Local-Orbital Calculations," *Phys. Rev. B* 29, 4426–4438.

4

Characterization and Modeling of Surface Roughness and Contact Mechanics

Arun Majumdar and Bharat Bhushan

Abstract

Almost all surfaces found in nature are observed to be rough at the microscopic scale. Contact between two rough surfaces occurs at discrete contact spots. During sliding of two such surfaces, interfacial forces that are responsible for friction and wear are generated at these contact spots. Comprehensive theories of friction and wear can be developed if the size and the spatial distributions of the contact spots are known. The size of contact spots ranges from nanometers to micrometers, making tribology a multiscale phenomena. This chapter develops the framework to include interfacial effects over a whole range of length scales, thus forming a link between nanometer-scale phenomena and macro-scopically observable friction and wear. The key is in the size and spatial distributions, which depend not only on the roughness but also on the contact mechanics of surfaces. This chapter reexamines the intrinsic nature of surface roughness as well as reviews and develops techniques to characterize rough-ness in a way that is suitable to model contact mechanics. Some general relations for the size distri-butions of contact spots are developed that can form the foundations for theories of friction and wear.

4.1   Introduction

Friction and wear between two solid surfaces sliding against each other are encountered in several day-to-day activities. Sometimes they are used to our advantage such as the brakes in our cars or the sole of our shoes where higher friction is helpful. In other instances such as the sliding of the piston against the cylinder in our car engine, lower friction and wear are desirable. In such cases, lubricants are often used. Friction is usually quantified by a coefficient μ, which is defined as

$$\mu = \frac{F_f}{F_n} \tag{4.1.1}$$

where F_f is the frictional force and F_n is the normal compressive force between the two sliding bodies. The basic problem in all studies on friction is to determine the coefficient μ. With regard to wear, it is necessary to determine the volume rate of wear, $\dot{V}$, and to establish the conditions when catastrophic failure may occur due to wear.

Despite the common experiences of friction and wear and the knowledge of its existence for thousands of years, their origins and behavior are still not well understood. Although the effects of friction and wear can sometimes be explained post-mortem, it is normally very difficult to predict the value of μ and the wear characteristics of the two surfaces. One could therefore characterize tribology as a field that is perhaps in its early stages of scientific development, where phenomena can be observed but can rarely be predicted with reasonable accuracy. The reason for this lies in the extreme complexity of the surface phenomena involved in tribology. Three types of surface characteristics contribute to this complexity: (1) surface geometric structure; (2) the nature of surface forces; and (3) material properties of the surface itself.

The lack of predictability in tribology lies in the convolution of effects of surface chemistry, mechanical deformations, material properties, and complex geometric structure. It is very difficult to say which of these is more dominant than the others. Physicists and chemists normally focus on the surface physics and chemistry aspects of the problem, whereas engineers study the mechanics and structural aspects. In a real situation of two macroscopic surfaces sliding in ambient conditions, it is very difficult to separate or isolate the different effects and then study their importance. They can indeed be isolated under controlled conditions such as in ultrahigh vacuum, but how those results relate to real situations is not clearly understood. It is, therefore, of no surprise that the tribology literature is replete with different theories of friction and wear that are applicable at different length scales — macroscopic to atomic scales. It is also of no surprise that a single unifying theory of tribology has not yet been developed.

In this chapter we will examine only one aspect of tribology and that is the effect of surface geometric structure. It will be shown that friction and wear depend on surface phenomena that occurs over several length scales, starting from atomic scales and extending to macroscopic "human" scales, where objects, motion, and forces can be studied by the human senses. Atomic-scale studies focus on the nature of surface forces and the displacements that atoms undergo during contact and sliding of two surfaces. This book has several chapters devoted to this important new field of *nanotribology*. One must, however, remember that friction is still a macroscopically observable phenomena. Hence, there must be a link or a bridge between the atomic-scale phenomena and the macroscopically observable motion and measur-able forces. This chapter takes a close look at surface geometric structure, or surface roughness, and attempts to formulate a methodology to form this link between all the length scales.

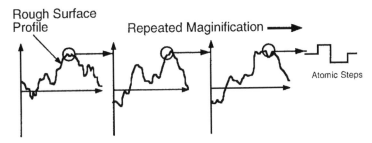

FIGURE 4.1 Appearance of surface roughness under repeated magnification up to the atomic scales, where atomic steps are observed.

First, the influence of surface roughness in tribology is established. Next, the complexities of the surface microstructure are discussed, and then techniques to quantify the complex structures are developed. The final discussion will demonstrate how to combine the knowledge of surface microstructure with that of surface forces and properties to develop comprehensive models of tribology. The reader will find that the theories and models are not all fully developed and much research remains to be done to understand the effects of surface roughness, in particular, and tribology, in general. This, of course, means that there is tremendous opportunity for contributions to understand and predict tribological phenomena.

4.2   Why Is Surface Roughness Important?

Solid surfaces can be formed by any of the following methods: (1) fracture of solids; (2) machining such as grinding or polishing; (3) thin-film deposition; and (4) solidification of liquids. It is found that most solid surfaces formed by these methods are not smooth. Perfectly flat surfaces that are smooth even on the atomic scale can be obtained only under very carefully controlled conditions and are very rare in nature. Therefore, it is of most practical importance to study the characteristics of naturally occurring or processed surfaces which are inevitably rough. However, the first question that the reader may ask is how rough is rough and when does one call something smooth?

4.2.1   How Rough Is Rough?

Smoothness and roughness are very qualitative and subjective terminologies. A polished metal surface may appear very smooth to the touch of a finger, but an optical microscope can reveal hills and valleys and appear rough. The finger is essentially a sensor that measures surface roughness at a lateral length scale of about 1 cm (typical diameter of a finger) and a vertical scale of about 100 μm (typical resolution of the finger). A good optical microscope is also a roughness sensor that can observe lateral length scales of the order of about 1 μm and can distinguish vertical length scales of about 0.1 μm. If the polished metal surface has a vertical span of roughness (hills and valleys) of about 1 μm, then the person who uses the finger would call it a smooth surface and the person using the microscope would call it rough. This leads to what one may call a "roughness dilemma." When someone asks whether a surface is rough or smooth, the answer is — it depends!! It basically depends on the length scale of the roughness measurement.

This problem of scale-dependent roughness is very intrinsic to solid surfaces. If one uses a sequence of high-resolution microscopes to zoom in continuously on a region of a solid surface, the results are quite dramatic. For most solid surfaces it is observed that under repeated magnification, more and more roughness keeps appearing until the atomic scales are reached where roughness occurs in the form of atomic steps (Williams and Bartlet, 1991). This basic nature of solid surfaces is shown graphically for a surface profile in Figure 4.1. Therefore, although a surface may appear very smooth to the touch of a finger, it is rough over all lateral scales starting from, say, around 10^{-4} m (0.1 mm) to about 10^{-9} m (1 nm). In addition, the roughness often appears random and disordered, and does not seem to follow

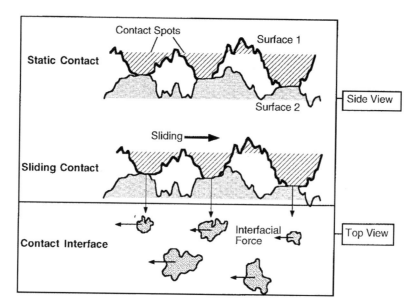

FIGURE 4.2 (a) Schematic diagram of two surfaces in static contact against each other. Note that the contact takes place at only a few discrete contact spots. (b) When the surfaces start to slide against each other, interfacial forces act on the contact spots.

any particular structural pattern (Thomas, 1982). The *randomness* and the *multiple roughness scales* both contribute to the *complexity* of the surface geometric structure. It is this complexity that is partly responsible for some of the problems in studying friction and wear.

The multiscale structure of surface roughness arises due to the fundamentals of physics and thermo-dynamics of surface formation, which will not be discussed in this chapter. What will be discussed is the following. Given the complex multiscale roughness structure of a surface, (1) how does it influence tribology; (2) how does one quantify or characterize the structure; and (3) how does one use these characteristics to understand or study tribology?

4.2.2 How Does Surface Roughness Influence Tribology?

Consider two multiscale rough surfaces (belonging to two solid bodies), as shown in Figure 4.2a, in contact with each other without sliding and under a static compressive force of F_n. Since the surfaces are not smooth, contact will occur only at discrete points which sustain the total compressive force. Figure 4.3 shows a typical contact interface which is formed of contact spots of different sizes that are spatially distributed randomly over the interface. The spatial randomness comes from the random nature of surface roughness, whereas the different sizes of spots occur due to the multiple scales of roughness. For a given load, the size of spots depends on the surface roughness and the mechanical properties of the contacting bodies. If this load is increased, the following would happen. The existing spots will increase in size, new spots will appear, and two or more spots may coalesce to form a larger spot. This is depicted by computer simulations of real surfaces in Figure 4.3. The surface in this case has isotropic statistical properties; that is, it does not have any texture or bias in any particular direction. It is evident that even for an isotropic surface the shapes of the contact spots are not isotropic and can be quite irregular and complex. In addition, when the load is increased, there are no set rules that the contact spots follow. Thus, the static problem itself is quite difficult to analyze. But one must, nevertheless, attempt to do so since these contact spots play a critical role in friction as explained below.

Consider the two surfaces to slide against each other. To do so, one must overcome a resistive tangential or frictional force F_f. It is clear that this frictional force must arise from the force interactions between the two surfaces that act only at the contact spots, as shown in Figure 4.2b. Since the normal load-bearing

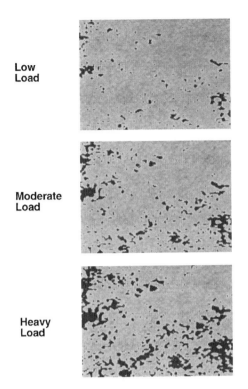

**Low
Load**

**Moderate
Load**

**Heavy
Load**

FIGURE 4.3 Qualitative illustration of behavior of contact spots (dark patches) on a contact interface under different loads. (a) At very light loads only few spots support the load; (b) at moderate loads the contact spots increase in size and number; (c) at high loads the contact spots merge to form larger spots and the number further increases.

capacity depends on the contact spot size, it is reasonable to assume that the tangential force is also size dependent. Therefore, to predict the total frictional force, it is very important to determine the *size distribution*, $n(a)$, of the contact spots such that the number of spots between area a and $a + da$ is equal to $n(a)da$. In addition to the interactions at each spot, there could be tangential force interactions between two or more contact spots. This is because the contact spots cannot operate independently of each other since they are connected by the solid bodies that can sustain some elastic or plastic deformation. So one can imagine the contact spots to be connected by springs whose spring constant depends on the elasticity/plasticity of the contacting materials. Because the number of contact spots is very large, the mesh of contact spots and springs thus forms a very complicated dynamic system. The deformations of the springs are usually localized around the contact spot and so the proximity of two spots influences their dynamic interactions. Therefore, it is also important to determine the *spatial distribution*, $\Delta(a_i, a_j)$, of contact spots where Δ is equal to the average closest distance between a contact spot of area a_i and a spot of area a_j. In other words, the frictional force, F_f, is a cumulative effect that arises due to force interactions at each spot and also dynamic force interactions between two or more spots. This can be written in a mathematical form as

$$F_f = \int_0^{a_L} \tau(a)\, n(a)\, a\, da + \{\text{dynamic interaction terms}\} \tag{4.2.1}$$

where $\tau(a)$ is the shear stress on a contact spot of area a, a_L is the area of the largest contact spot, and $n(a)$ is the size distribution of contact spots. Similarly, the total volume rate $\dot{V}$ of wear that is removed from the surface can be written as

$$\dot{V} = \int_0^{a_L} \dot{v}(a)\, n(a)\, da \tag{4.2.2}$$

where $\dot{v}(a)$ is the volume rate of wear at the microcontact of area a. It can be seen that as long as the size distribution $n(a)$ is known, tribological phenomena can be studied at the scale of the contact spots.

Let us concentrate on the first term on the right-hand side of Equation 4.2.1. This term adds up the tangential force on each contact spot starting from areas that tend to zero to the upper limit a_L, which is the area of the largest contact spot. Recent studies have shown that the shear stress $\tau(a)$ is not a constant and can be size dependent. In other words, the frictional phenomena at the nanometer scale can be quite different from that at macroscales (Mate et al., 1987; Israelachvili et al., 1988; McGuiggan et al., 1989; Landman et al., 1990). In addition, the shear stress is strongly influenced by the different types of surface forces (Israelachvili, 1992). Some of the chapters in this book have concentrated on studying the nature of $\tau(a)$ when a is at the atomic or nanometer scales.

The second term on the right hand side of Equation 4.2.1 represents the dynamic spring–mass interactions between the contact spots. Although this depends on the spatial and size distributions, it is unclear what the functional form would be. However, it is not insignificant since collective phenomena such as onset of sliding and stick-slip depend upon these types of interactions. Recently, there has been some interest in studying this as a percolation or a self-organized critical phenomena (Bak et al., 1988). The onset of sliding friction can be pictured as follows. When an attempt is made to slide one surface against another, the force on a contact spot can be released and distributed among neighboring spots. The forces in at least one of these spots may exceed a critical level creating a cascade or an avalanche. The avalanche may turn out to be limited to a small region or become large enough so that the whole surface starts sliding. During this process, the interface evolves into a self-organized critical system insensitive to the details of the distribution of initial disorder. This type of analysis has been used to provide a physical interpretation of the Guttenberg–Richter relation between earthquake magnitude and its frequency (Sornette and Sornette, 1989; Knopoff, 1990; Carlson et al., 1991).

In summary, the basic problem of tribology can be divided as follows: (1) to determine the size, $n(a)$, and spatial distribution of contact spots which depends on the surface roughness, normal load, and mechanical properties; (2) to find the tangential surface forces at each spot; (3) to determine the dynamic interactions between the spots; and, finally, (4) to find the cumulative effect in terms of the frictional force, F_f.

4.3   Surface Roughness Characterization

A rough surface can be written as a mathematical function: $z = f(x,y)$, where z is the vertical height and x and y are the coordinates of a point on the two-dimensional plane, as shown in Figure 4.4a. This is typically what can be obtained by a roughness-measuring instrument. The surface is made up of hills and valleys often called surface asperities of different lateral and vertical sizes, and are distributed randomly on the surface as shown in the surface profiles in Figure 4.4b. The randomness suggests that one must adopt statistical methods of roughness characterization. It is also important to note that because of the involvement of so many length scales on a rough surface, the characterization techniques must be independent of any length scale. Otherwise, the characterization technique will be a victim of the "rough or smooth" dilemma as discussed in Section 4.2.1.

4.3.1   Probability Height Distribution

One of the characteristics of a rough surface is the probability distribution (Papoulis, 1965) $g(z)$, of the surface heights such that the probability of encountering the surface between height z and $z + dz$ is equal to $g(z)dz$. Therefore, if a rough surface contacts a hard perfectly flat surface* and it is assumed that the

---

*Although hard flat surfaces are rarely found in nature, we make the assumption because contact between two rough surfaces can be reduced to the contact between an equivalent surface and a hard flat surface (see Section 4.4).

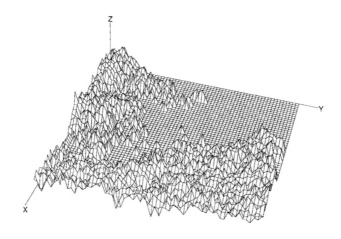

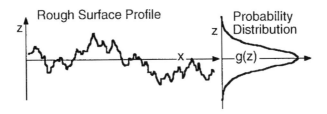

FIGURE 4.4 (a) Schematic diagram of a rough surface whose surface height is $z(x, y)$ at a coordinate point (x, y). (b) A vertical cut of the surface at a constant y gives surface profile $z(x)$ with a certain probability height distribution.

distribution $g(z)$ remains unchanged during the contact process, then the ratio of real area of contact, A_r, to the apparent area, A_a, can be written as

$$\frac{A_r}{A_a} = \int_d^\infty g(z)\,dz = \sigma \int_{d/\sigma}^\infty g(\bar{z})\,d\bar{z} \qquad (4.3.1)$$

where d is the separation between the flat surface, σ is the standard deviation of the surface heights, and $\bar{z} = z/\sigma$ is the nondimensional surface height. The real area of contact, A_r, is usually about 0.1 to 10% of the apparent area and is the sum of the areas of all the contact spots. Therefore, the probability distribution, $g(z)$, can be used to determine the sum of the contact spot areas but does not provide the crucial information on the size distribution, $n(a)$. In addition, it contains no information concerning the shape of the surface asperities.

It is often found that the normal or Gaussian distribution fits the experimentally obtained probability distribution quite well (Thomas, 1982; Bhushan, 1990). In addition, it is simple to use for mathematical calculation. The bell-shaped normal distribution (Papoulis, 1965) which has a variance of unity is given as

$$g(\bar{z}) = \frac{1}{\sqrt{2\pi}} \exp\left[-\frac{(\bar{z} - \bar{z}_m)^2}{2} \right] \quad -\infty < \bar{z} < \infty \qquad (4.3.2)$$

where $\bar{z}_m$ is the nondimensional mean height. The mean height and the standard deviation can be found from a roughness measurement $z(x,y)$ as

193

$$z_m = \frac{1}{L_x L_y} \int_0^{L_x} \int_0^{L_y} z(x, y)\, dx\, dy = \frac{1}{N_x N_y} \sum_{i=1}^{N_x} \sum_{j=1}^{N_y} z(x_i, y_j) \qquad (4.3.3)$$

$$\sigma = \sqrt{\frac{1}{L_x L_y} \int_0^{L_x} \int_0^{L_y} \left[z(x, y) - z_m \right]^2 dx\, dy} = \sqrt{\frac{1}{N_x N_y} \sum_{i=1}^{N_x} \sum_{j=1}^{N_y} \left[z(x_i, y_j) - z_m \right]^2} \qquad (4.3.4)$$

Here, L_x and L_y are the lengths of surface sample, whereas N_x and N_y are the number of points in the x and y lateral directions, respectively. The integral formulation is for theoretical calculations, whereas the summation is used for calculating the values from finite experimental data.

Although used extensively, the normal distribution has limitations in its applicability. For example, it has a finite nonzero probability for surface heights that go to infinity, whereas a real surface ends at a finite height, z_{max}, and has zero probability beyond that. Therefore, the normal distribution near the tail is not an accurate representation of real surfaces. This is an important point since it is usually the tail of the distribution that is significant for calculating the real area of contact. Other distributions, such as the inverted chi-squared (ICS) distribution, fit the experimental data much better near the tail of the distribution (Brown and Scholz, 1985). This is given for zero mean and in terms of nondimensional height, $\bar{z}$, as

$$g(\bar{z}) = \frac{(v/2)^{v/4}}{\Gamma(v/2)} \left(\bar{z}_{max} - \bar{z} \right)^{(v/2)-1} e\left(\bar{z} - \bar{z}_{max} \right) \sqrt{v/2} \quad -\infty < \bar{z} < \bar{z}_{max} \qquad (4.3.5)$$

which has a variance of $2v$ and a maximum height $\bar{z}_{max} = \sqrt{v/2}$. The advantage of the ICS distribution is it has a finite maximum height, as does a real surface, and has a controlling parameter v, which gives a better fit to the topography data. The Gaussian and the ICS distributions are shown in Figure 4.5. Note that as v increases, the ICS distribution tends toward the normal distribution. Brown and Scholz (1985)

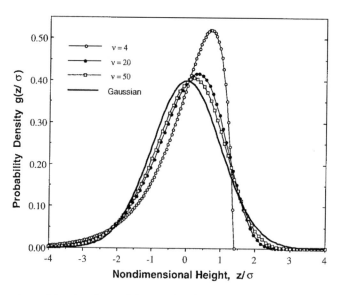

FIGURE 4.5 Comparison of the Gaussian and the ICS distributions for zero mean height and nondimensional surface height $\bar{z}$.

found that the surface heights of a ground-glass surface were not symmetric like the normal distribution but were best fitted by an ICS distribution with $\nu = 21$.

4.3.2 RMS* Values and Scale Dependence

A rough surface is often assumed to be a statistically *stationary* random process (Papoulis, 1965). This means that the measured roughness sample is a true statistical representation of the entire rough surface. Therefore, the probability distribution and the standard deviation of the measured roughness should remain unchanged, except for fluctuations, if the sample size or the location on the surface is altered. The properties derived from the distribution and the standard deviation are therefore unique to the surface, thus justifying the use of such roughness characterization techniques.

Because of simplicity in calculation and its physical meaning as a reference height scale for a rough surface, the rms height of the surface is used extensively in tribology. However, it was shown by Sayles and Thomas (1978) that the variance of the height distribution is a function of the sample length and in fact suggested that the variance varied as

$$\sigma^2 \approx L \tag{4.3.6}$$

where L is the length of the sample. This behavior implies that any length of the surface cannot fully represent the surface in a statistical sense. This proposition was based on the fact that beyond a certain length, L, the surface heights of the same surface were uncorrelated such that the sum of the variances of two regions of lengths L_1 and L_2 can be added up as

$$\sigma^2\left(L_1 + L_2\right) = \sigma^2\left(L_1\right) + \sigma^2\left(L_2\right) \tag{4.3.7}$$

They gathered roughness measurements of a wide range of surfaces to show that the surfaces follow the nonstationary behavior of Equation 4.3.6. However, Berry and Hannay (1978) suggested that the variance can be represented in a more general way as follows:

$$\sigma^2 \approx L^n \tag{4.3.8}$$

where n varies between 0 and 2.

If the exponent n in Equation 4.3.8 is not equal to zero of a particular surface, then the standard deviation or the rms height, σ, is scale dependent, thus making a rough surface a *nonstationary* random process. This basically arises from the multiscale structure of surface roughness where the probability distribution of a small region of the surface may be different from that of the larger surface region as depicted in Figure 4.4b. If the larger segment follows the normal distribution, then the magnified region may or may not follow the same distribution. Even if it does follow the normal distribution, the rms σ can still be different.

Other statistical parameters that are also used in tribology (Nayak, 1971, 1973) are the rms slope, σ'_x, and rms curvature, σ''_x, defined as

$$\sigma'_x = \sqrt{\frac{1}{L_x} \int_0^{L_x} \left(\frac{\partial z(x, y)}{\partial x}\right)^2 dx} = \sqrt{\frac{1}{N-1} \sum_{i=i}^{N-1} \left[\frac{z(x_{i+1}, y_j) - z(x_i, y_j)}{\Delta x}\right]^2} \tag{4.3.9}$$

---

*The rms values (of height, slope, or curvature) are related to the corresponding standard deviation, σ, of a surface in the following way: $\text{rms}^2 = \sigma^2 + z_m^2$, where z_m is the mean value. In this chapter it will be assumed that $z_m = 0$; that is, the mean is taken as the reference, such that $\text{rms} = \sigma$.

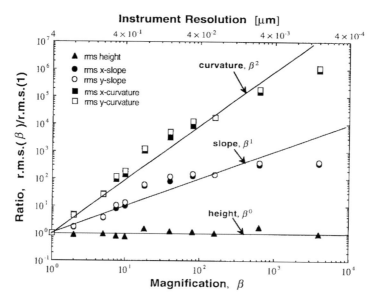

FIGURE 4.6 Variation of rms height, slope, and curvature of a magnetic tape surface as a function of magnification, β, or instrument resolution. The vertical axis is the ratio of an rms quantity at a magnification β to the rms quantity at magnification of unit, which corresponds to an instrument resolution of 4 μm. Roughness measurements of β < 10 were obtained by optical interferometry (Bhushan et al., 1988), whereas that for β > 10 were obtained by atomic force microscopy (Oden et al., 1992).

$$\sigma''_x = \sqrt{\frac{1}{L_x}\int_0^{L_x}\left(\frac{\partial^2 z(x,y)}{\partial x^2}\right)^2 dx} = \sqrt{\frac{1}{N-2}\sum_{i=i}^{N-2}\left[\frac{z(x_{i+2},y_j)+z(x_i,y_j)-2z(x_{i+1},y_j)}{\Delta x}\right]^2} \qquad (4.3.10)$$

Here, although the rms slope and the rms curvature are expressed only for the x-direction, these values can similarly be obtained for the y-direction. These parameters are extensively used in contact mechanics (McCool, 1986) of rough surfaces.

The question that now remains to be answered is whether the rms parameters σ, σ', and σ" vary with the statistical sample size or the instrument resolution. Figure 4.6 shows the rms data for a magnetic tape surface (Bhushan et al., 1988; Majumdar et al., 1991). Along the ordinate is plotted the ratio of the rms value at a magnification, β, to the rms value at magnification of unity. The magnification β = 1 corresponds to an instrument resolution of 4 μm and scan size of 1024 × 1024 μm containing 256 × 256 roughness data points. The roughness data in the range 1 < β < 10 were obtained by optical interferometry (Bhushan et al., 1988), whereas for β > 10, the data were obtained by atomic force microscopy (Majumdar et al., 1991; Oden et al., 1992). An increase in β corresponds to an increase in instrument resolution with the highest being equal to 1 nm. The data clearly show that the rms height does not change over five decades of length scales and can therefore be considered *scale independent* over this range of length scales. However, the rms slope increases with magnification as β^1 and the rms curvature increases as β^2. Figure 4.7 shows similar variations for a polished aluminum nitride surface where the roughness data was obtained by atomic force microscopy. In this case, the rms height σ reduces with decreasing sample size but does not follow the trend $\sigma \approx \sqrt{L}$ as suggested by Sayles and Thomas (1978). Nevertheless, the variation does make the surface a nonstationary random process. The rms slope and the rms curvature, on the other hand, increase with the instrument resolution, as observed in Figure 4.7.

Although Figures 4.6 and 4.7 show statistics for specific surfaces, the trends are typical for most rough surfaces that have been examined. The following can be concluded from these trends. The rms height,

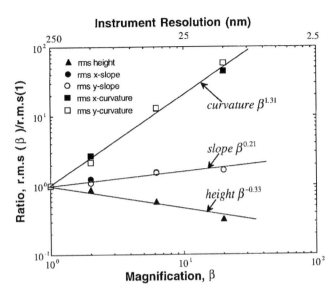

FIGURE 4.7 Variation of rms height, slope, and curvature of a polished aluminum nitride surface as a function of magnification, β, or instrument resolution.

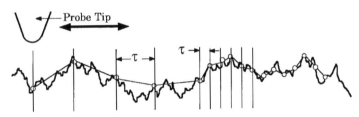

FIGURE 4.8 Illustration of roughness measurements at different instrument resolution τ. As τ is reduced, the surface that is measured is quite different, as qualitatively shown. The average slope and the average curvature of the profile is higher for smaller τ.

σ, is a parameter which could be scale independent for some surfaces but is not necessarily so for other surfaces. The rms slope, σ', and the rms curvature, σ'', on the other hand, always tend to be scale dependent. Therefore, the rms height can be used to characterize a rough surface uniquely *if* it is scale independent, as is the case of the magnetic tape surface in Figure 4.6. However, it is not clear under what conditions the rms height is scale dependent or independent. These conditions will be explored in the Section 4.3.3. However, the reasons can be qualitatively shown by the self-repeating nature of the surface roughness depicted in Figure 4.8.

Given a rough surface, an instrument with resolution τ will measure the surface height of points that are separated by a distance τ. If τ is reduced, new locations on the surface are accessed. Due to the multiple scales of roughness present, a reduction in τ makes the measured profile look different for the same surface. When τ is reduced, it is found that the straight line joining two neighboring points becomes steeper on an average, as qualitatively observed in Figure 4.8. This increases the average slope and the curvature of the surface. Therefore, the slope and the curvature fall victim to the "rough or smooth" dilemma that is qualitatively discussed in Section 4.2. Figures 4.6 and 4.7 quantitatively exhibit scale dependence of the rms slope and curvature. One can conclude these parameters cannot be used to characterize a rough surface uniquely since they are scale dependent; that is, the use of these parameters in any statistical theory of tribology can lead to erroneous results. It is thus necessary to obtain some scale-independent techniques for roughness characterization.

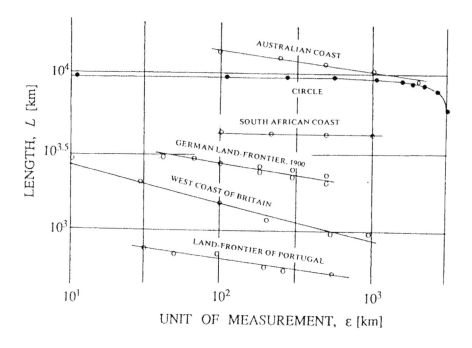

$$L \sim \varepsilon^{1-D}$$

FIGURE 4.9 Dependence of the length of different coastlines and curves on the unit ε of measurement. Note the power law dependence of the length on ε.

4.3.3 Fractal Techniques

4.3.3.1 A Primer for Fractals

The self-repeating nature of surface roughness has not only been found in surfaces but also in several objects found in nature. In his classic paper, Mandelbrot (1967) showed that the coastline of Britain has self-similar features such that the more the coastline is magnified, the more features and wiggliness are observed. In fact, the answer to the question — "How long is the coastline of Britain?" — is it depends on the unit of measurement and is not unique. This is shown in Figure 4.9 for several coastlines and also for a circle. The fundamental problem of this scale dependence is that "length" as measured by a ruler or a straightedge is a measure of only one-dimensional objects. No matter how small a unit you take for the measurement, the length would still come out the same. In other words, if you take a straight line, then the length would be the same whether you take 1 mm or 1 μm as the unit of measurement. The reason for the scale independence at a very minute scale is that the line or the curve is made up of *smooth* and straight line segments. However, if an object is never smooth no matter what length scale you choose, then repeated magnifications will reveal different levels of wiggliness as shown in Figure 4.10. Large units of measurement fail to measure the small wiggliness of the curve, whereas the small units of measurement will measure them. In other words, different units of measurement will measure only some levels of the wiggliness but not all levels. Thus, one would get a different number for the length of the object as the unit of measurement is changed.

Since objects of the dimension unity are defined to have their lengths independent of the unit of measurement, an object with scale-dependent length is *not* one-dimensional. Similarly, if the area of a surface depends on the unit of measurement, then it is not a two-dimensional object.

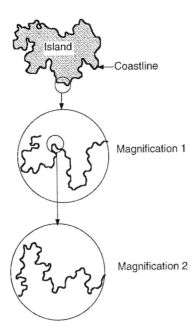

Island
Coastline

Magnification 1

Magnification 2

FIGURE 4.10 Repeated magnification of a coastline produces an increased amount of wiggliness without any appearance of smoothness at any scale. Note that the magnification is equal in all directions.

One of the properties of naturally occurring wiggly objects is that if a small part of the object is enlarged sufficiently, then statistically it appears very similar to the whole object. For example, if you look at the photograph of hills and valleys (with appropriate color), then unless the scale is given it will be very difficult to say whether it is a photograph of the Rocky Mountains or a micrograph of a surface obtained by a scanning electron microscope. This feature is called "self-similarity." To characterize such wiggly and complex objects which display self-similarity, Mandelbrot (1967) generalized the definition of *dimension* to take fractional values such that a wiggly curve like the coastline will have a dimension D between 1 and 2. Under such a generalized definition, the specific integer values of 0, 1, 2, and 3 correspond to smooth objects such as a point, line, surface, and sphere (or any three-dimensional object), whereas the generalized noninteger values correspond to wiggly and complex objects which show self-similar behavior. Self-similar objects that contain nonsmooth self-similar features over all length scales are called *fractals* and the noninteger dimension characterizing it is called the *fractal dimension*. Detailed discussions on fractal geometry can be found in several books (Mandelbrot, 1982; Peitgen and Saupe, 1988; Barnsley, 1988; Feder, 1988; Vicsek, 1989; Avnir, 1989).

A rough surface, as shown in Figure 4.1, has fractallike features — it has wiggly features appearing over a large range of length scales and, as will be shown later, they sometimes do follow the self-similar hierarchy. Whereas mathematical fractals follow self-repetition over all length scales, rough surfaces have a higher and lower length scale limit between which the fractal behavior is observed. Analogous to the nonuniqueness of the length of a coastline, we have already seen the nonuniqueness of the rms height, the rms slope, and the rms curvature. The question that a reader can ask is, can the fractallike behavior of a rough surface be utilized to develop a characterization technique that will be independent of length scales? Recent work (Kardar et al., 1986; Gagnepain, 1986; Jordan et al., 1986; Meakin, 1987; Voss, 1988; Majumdar and Tien, 1990; Majumdar and Bhushan, 1990) has shown that this is sometimes possible and is discussed below.

Figure 4.9 shows that if the length, L, of a coastline is plotted against the unit of measurement, ε, then the length follows a power law of the form (Mandelbrot, 1967)

$$L \approx \varepsilon^{1-D} \qquad (4.3.11)$$

199

where D is called the fractal dimension of the coastline. If $D = 1$, then the length is independent of ε and it can be called a one-dimensional object. It is observed that this power law behavior remains unchanged over several decades of length scales such that the value of D, which in some sense measures the wiggliness of the curve, remains constant and independent of ε. Therefore, D is one parameter that can be used to characterize a coastline. Another way of looking at this behavior is the following — although the coastline seems a rather convoluted and complex geometric structure, the power law behavior represents a pattern or order in this chaotic structure.

4.3.3.2 Fractal Characterization of Surface Roughness

The same concept can be used to characterize a rough surface. However, there is a difference between a coastline and a rough surface. To show the self-similarity of a coastline, one needs to take a small part and enlarge it *equally* in all directions to resemble the full coastline statistically, as qualitatively shown in Figure 4.9. However, for a small region of a rough surface to statistically resemble* a larger region, the enlargement should be done *unequally* in the vertical (z) and lateral (x and y) directions. Such objects, which *scale* differently in different directions, are called *self-affine* (Mandelbrot, 1982, 1985; Voss, 1988). To characterize a self-affine object one cannot use the length of the surface profile or the area of the surface as a measure (Mandelbrot, 1985). There are two other ways to characterize it — the power spectrum $P(\omega)$ and the structure function, $S(\tau)$.

4.3.3.2.1 Power Spectrum

Consider a surface profile, $z(x)$ in the x-direction. The power spectrum of the profile can be found by the relation (Blackman and Tuckey, 1958; Papoulis, 1965):

$$P(\omega) = \frac{1}{L} \left| \int_0^L z(x) \exp(i\omega x) \, dx \right|^2 \tag{4.3.12}$$

where the coordinate x ranges from 0 to L. The power spectrum can be obtained from a measured roughness profile by a simple fast Fourier transform routine (Press et al., 1992). The square of the amplitude of $z(x)$ or the power at a frequency ω is equal to $P(\omega)d\omega$. The rms height, the rms slope, and the rms curvature can be obtained from the power spectrum (McCool, 1987; Majumdar and Bhushan, 1990):

$$\sigma = \sqrt{\int_{\omega_l}^{\omega_h} P(\omega) \, d\omega} \tag{4.3.13}$$

$$\sigma' = \sqrt{\int_{\omega_l}^{\omega_h} \omega^2 P(\omega) \, d\omega} \tag{4.3.14}$$

$$\sigma'' = \sqrt{\int_{\omega_l}^{\omega_h} \omega^4 P(\omega) \, d\omega} \tag{4.3.15}$$

where ω_l and ω_h are the low-frequency and the high-frequency cutoffs, respectively. For a roughness measurement, the low-frequency cutoff is equal to the reciprocal of the sample length, $\omega_l = 1/L$ and the high-frequency cutoff is equal to the Nyquist frequency or equal to $\omega_h = \frac{1}{2}\tau$, where τ is the distance between two adjacent points of the data sample. It is evident that the power spectrum is a more

---

*Statistical resemblance is for the power spectrum or structure function of the rough surface as shown later.

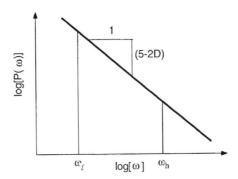

FIGURE 4.11 Qualitative description of a fractal power spectrum plotted on a log–log plot. Note that the spectrum is a straight line whose slope depends on the fractal dimension. A roughness measurement contains a lower, ℓ, and upper limit, L, of length scales which correspond to the frequency window between $\omega_l = 1/L$ and $\omega_h = 1/(2\ell)$.

fundamental quantity than the rms values since the rms values can be obtained from the spectrum, and not vice versa.

For a fractal surface profile, the power spectrum follows a power law of the form (Mandelbrot, 1982; Voss, 1988; Majumdar and Tien, 1990; Majumdar and Bhushan, 1990):

$$P(\omega) = \frac{C}{\omega^{(5-2D)}} \tag{4.3.16}$$

where $1 < D < 2$ is the fractal dimension of the profile and C is a scaling constant, which depends on the amplitude of the rough surface. If the power spectrum of a measured surface profile is found and plotted against the frequency in a log–log plot, then the surface profile can be called fractal if the spectrum follows a straight line, as qualitatively shown in Figure 4.11. The dimension D can be obtained from the slope and the constant C from the power. Since the profile is a vertical cut through a surface, the dimension of the surface, D_s is equal $D_s = D + 1$ only for an isotropic surface. For anisotropic surfaces one needs to determine the fractal dimensions of surface profiles in different directions. For a fractal profile, the independence of D and C from the length scale ω make them unique to a surface and can therefore be used for roughness characterization. When the rms quantities are obtained from the fractal spectrum by using Equations 4.3.13 through 4.3.15), they exhibit the following behavior: $\sigma = \sqrt{C}\omega_l^{-(2-D)} = \sqrt{C}L^{(2-D)}$; $\sigma' = \sqrt{C}\,\omega_h^{(D-1)}$; $\sigma'' = \sqrt{C}\omega_h^{D}$. It is evident that the rms values depend either on the low-frequency or high-frequency cutoff and are therefore scale dependent. Figures 4.6 and 4.7 confirm this experimentally and, in fact, show the decrease in exponent by 1 as we go from the rms curvature to the rms slope and finally to the rms height. The only difference that one finds in the rms quantities is that the rms slope and the curvature depend on the high-frequency cutoff, whereas the rms height depends on the low-frequency cutoff. The relation $\sigma \approx L^{(2-D)}$ is exactly the same as suggested in Equation 4.3.8 with $n = 2(2 - D)$. In fact, the relation suggested by Sayles and Thomas (1978) in Equation 4.3.6 is a special case when $D = 1.5$.

The variance of the height distribution, σ^2, is equal to the area under the power spectral curve as mathematically shown in Equation 4.3.13. When the variance (or the rms height) is independent of the sample size or any length scale, as demonstrated in Figure 4.6, the area under the power spectrum must be constant and independent of ω_l and ω_h. Therefore, the fractal power law variation of the spectrum in Equation 4.3.16 is clearly not valid for such a case since it always leads to ω_l-dependence of the rms height. One must note, then, that the fractal behavior is not followed all the time.

One of the practical difficulties of using the power spectrum to obtain the values of D and C is that for a single measured roughness profile, the calculated spectrum turns out to be very noisy. This is because the roughness profile is not bandwidth limited and is in fact a broad-band spectrum. However, the power spectrum of any measured roughness will be limited to the Nyquist frequency ω_n, on the high-frequency side. This gives rise to the problem of aliasing (Press et al., 1992) which falsely translates the power of frequencies in the range $\omega > \omega_n$ into the range $\omega < \omega_n$. The problem comes about due to the discreteness of the roughness measurement. To overcome this problem, we have found that the structure function can yield more accurate estimation of D and C.

4.3.3.2.2 Structure Function

The structure function (Mandelbrot, 1982; Voss, 1988) is defined as

$$S(\tau) = \frac{1}{L}\int_0^L \left[z(x+\tau) - z(x)\right]^2 dx = \frac{1}{(N-\tau)/\Delta x} \sum_{i=1}^{(N-\tau)/\Delta x} \left[z(x_i+\tau) - z(x_i)\right]^2 \qquad (4.3.17)$$

The summation on the right-hand side can be used for calculation of a measured surface profile containing N points. As one can see, the structure function is easy to calculate since it does not involve any transformation but simple height differences and averages. It is sometimes used in experimental and theoretical analysis of velocity and scalar fluctuations in turbulent fluid dynamics (Kolmogoroff, 1941). In turbulence, the fluctuating quantity varies with time and space, whereas for rough surface, the same varies with space. The problems are quite similar since in turbulence, too, the power spectrum of the fluctuations is broadband and follows the power law behavior of Equation 4.3.16.

It is interesting to note that in some ways the structure function and the variance, σ^2, of height in Equation 4.3.4 are similar since both involve finding the average of the square of height differences. However, the structure function uses height differences with points separated by a distance τ, whereas for the variance, the height differences are with the mean height z_m. The structure function yields much more information than the rms height since by varying τ, one can study the roughness structure at different length scales. This is, of course, not possible for the variance, σ^2, which finds the average height difference from the mean over the whole surface. In addition, the variance of the profile slope, $S'(\tau)$, can be found as

$$S'(\tau) = \frac{1}{L}\int_0^L \left(\frac{z(x+\tau) - z(x)}{\tau}\right)^2 dx = \frac{S(\tau)}{\tau^2} \qquad (4.3.18)$$

A surface profile is said to be fractal if the structure function follows a power law behavior as (Mandelbrot, 1985; Voss, 1988; Majumdar et al., 1991)

$$S(\tau) = G^{2(D-1)}\tau^{2(2-D)} \qquad (4.3.19)$$

This can also be derived from the power spectrum by the relation (Berry, 1978)

$$S(\tau) = \int_{-\infty}^{\infty} P(\omega)\left[\exp(i\omega\tau) - 1\right]d\omega \qquad (4.3.20)$$

where D is the fractal dimension and G is a scaling constant that has units of length. When $S(\tau)$ is plotted against τ on a log–log plot, the curve will be a straight line for a fractal profile. The dimension can be obtained from the slope and G from the intercept at a certain value of τ. The two characterization parameters, D and G, are unique for a fractal profile and are independent of any length scale τ. Thus, they form the fundamental set of parameters for a rough surface profile. By using the fractal power law spectrum of Equation 4.3.16 in Equation 4.3.20, the structure function becomes (Berry, 1978)

$$S(\tau) = \frac{C}{(2-D)} \sin\left(\frac{\pi(2D-3)}{2}\right) \Gamma(2D-3) \, \tau^{2(2-D)} \tag{4.3.21}$$

such that the factor C of the power spectrum is related to the scaling constant G of the structure function as

$$C = \frac{(2-D) G^{2(D-1)}}{\sin\left(\dfrac{\pi(2D-3)}{2}\right) \Gamma(2D-3)} \tag{4.3.22}$$

Berry and Blackwell (1981) follow a slightly different definition of a fractal surface — a surface profile is said to be a self-affine fractal when

$$S(\tau) = G^{2(D-1)} \, \tau^{2(2-D)} \quad \text{for} \quad \tau \to 0 \tag{4.3.23}$$

where the parameter G is called "topothesy" following the term coined by Sayles and Thomas (1978). This definition is valid in the limit $\tau \to 0$ and the fractal dimension D so obtained is called the Hausdorff–Besicovitch dimension (Mandelbrot, 1982). For larger-scale roughness, Berry and Blackwell (1981) suggest a simple model for $S(\tau)$ as

$$S(\tau) = 2\sigma^2 \left[1 - \exp\left(-\frac{G^{2(D-1)} \, \tau^{2(2-D)}}{2\sigma^2} \right) \right] \tag{4.3.24}$$

As $\tau \to 0$, Equation 4.3.24 is recovered and when $\tau \gg G(2\sigma/G)^{1/(2-D)}$, $S(\tau) = 2\sigma^2$. In this case it is assumed that the rms height, σ, is independent of the sample size and can be obtained from roughness data for a sample size larger than the correlation length, $\tau_c = G(2\sigma/G)^{1/(2-D)}$. Experimental data will show that the behavior of Equation 4.3.24 is followed by several surfaces and can be used as a good model for surfaces. However, if the rms height is scale dependent, as observed by Sayles and Thomas (1978), then the model breaks down.

In the rest of the chapter the structure function will be used to study the statistical properties of rough surfaces. This is due to its simplicity of use and the roughness information it reveals at different length scales.

4.3.3.3 Roughness Measurements

Typically, roughness between 1 cm to about 10 μm is measured by stylus profilometers, between 500 and 1 μm by optical interferometry and between 100 μm and 1 Å by scanning tunneling or atomic force microscopy. The overlaps in the length scales between these instruments are used to corroborate the roughness measured by different techniques.

4.3.3.3.1 Stylus Profilometry

The roughness of machined (lapped, ground, and shape turned) stainless steel surfaces was measured by a contact stylus profiler (Majumdar and Tien, 1990). The instrument used a diamond stylus of radius 2.5 μm and had a vertical resolution of 0.5 nm. The scan lengths ranged from 50 to 30 mm with each scan having 800 to 1000 evenly spaced points. Figure 4.12 shows the roughness profile of a lapped stainless steel surface.

The structure functions, $S(\tau)$, of these surface profiles are plotted on a log–log plot in Figure 4.13. Also shown is the straight line, $S(\tau) \approx \tau^1$, which corresponds to a fractal dimension of $D = 1.5$. It is

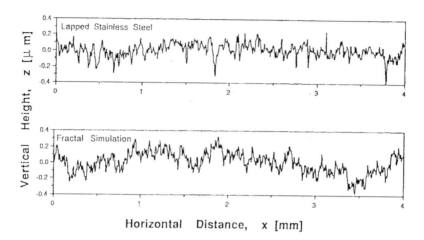

FIGURE 4.12 Profile of lapped stainless steel surface measured by a stylus profilometer. Fractal simulation of the profile was conducted by the Weierstrass–Mandelbrot function. (From Majumdar, A. and Tien, C. L. (1990), *Wear* 136; 313–327. With permission.)

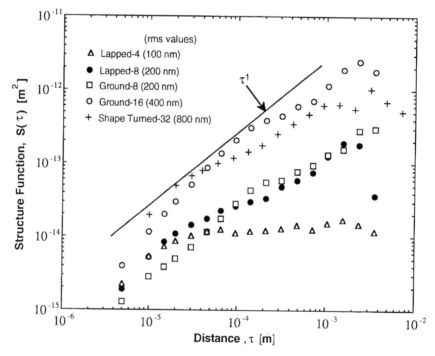

FIGURE 4.13 Structure function of machined stainless steel surfaces whose roughness was measured by stylus profilometry.

evident that the experimental structure functions do follow a power law at small length scales. In fact, although they do not coincide, they all tend to follow the same slope, that of $D = 1.5$. The higher value of $S(\tau)$ for the rougher surfaces leads to a higher value of G. The structure function for the lapped-4*

---

*The number associated with each process is the rms roughness produced in microinches. One microinch is equal to 25.4 nm.

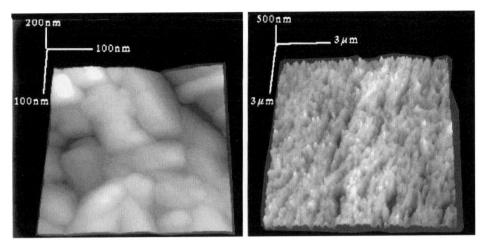

FIGURE 4.14 Images of a magnetic tape A (Bhushan et al., 1988) surface obtained by atomic force microscopy. Note the magnetic particles, which are oblong in shape with aspect ratio 10 and a diameter of about 100 nm.

profile departs from the $D = 1.5$ behavior at about 30 µm and those of ground-8 and lapped-8 profile depart at about 100 µm. This behavior is probably due to the following reason. For any machining process, there exists a critical length scale below which the surface remains unaffected during machining. For grinding, this length scale is the grain size of the abrasive material, whereas for turning it is the tool radius. Below this scale, the surface is formed by a natural process such as fracture. This natural process seems to lead to the same type of surface fractal behavior with $D = 1.5$. At length scales larger than the critical one, the machining processes flattens the surface and thus reduces the height differences between two points on the surface. Thus, the structure function decreases at larger scales. As shown earlier, the rms height depends on the total length, L, of the roughness sample as $\sigma \approx \omega_l^{D-2} = L^{2-D}$. Although the structure function of the different surface profiles at small length scales are nearly the same, their rms heights are quite different. This is because at larger length scales, which control the value of σ, the structure functions are different with smoother surfaces having smaller values of $S(\tau)$.

4.3.3.3.2 Atomic Force Microscopy

Oden et al. (1992) measured the surface roughness of magnetic tape A (Bhushan et al., 1988) at four different resolutions by atomic force microscopy. Figure 4.14 shows the image of the tape obtained from a 0.4 × 0.4 µm scan and 2.5 × 2.5 µm scan. The accicular magnetic particles, typically 0.1 µm in diameter with an aspect ratio of about 10, are clearly visible. Figure 4.15 shows the structure function of all the four scans, including the two in Figure 4.14 and for 10 × 10 µm and 40 × 40 µm scans. The overlap between the two structure functions indicates that scan rates did not influence the roughness measurement. The slight anisotropy in the x- and the y-directions correspond to the marginal bias in the orientation of the magnetic particles along the length of the tape. It is interesting to note that the structure function has two regions with a knee at around 0.1 µm. This suggests that the behavior $S(\tau) \sim \tau^{1.23}$ for scales smaller than 0.1 µm correspond to the roughness within a single particle. The $S(\tau) \sim \tau^0$ behavior for larger scales probably arise due to the fact that these particles lie adjacent to each other, much like a single layer of pencils on a flat surface. Since the diameters of the particles are nearly the same, the height difference $(z(x + \tau) - z(x))$ remains independent of τ for $\tau > 0.1$ µm. The $S(\tau) \sim \tau^0$ behavior corresponds to a dimension of $D = 2$ for surface profiles. This, therefore, explains the variation of rms curvature as $\sigma'' = \sqrt{C}\,\omega_h^D \propto \beta^2$, rms slope as $\sigma' = \sqrt{C}\,\omega_h^{(D-1)} \propto \beta^1$; and rms height as $\sigma = \sqrt{C}\,L^{(2-D)} \propto \beta^0$ in Figure 4.4.

The scale independence of the rms height, and the general behavior of the structure function data of the magnetic tape, suggests that this surface is a perfect example of the model proposed by Berry and Blackwell (1981), given in Equation 4.3.24 — power law behavior of $D = 1.39$ as $\tau \to 0$ and a saturation behavior as $\tau \to \infty$.

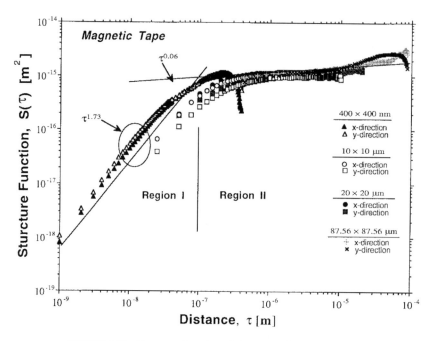

FIGURE 4.15 Structure function of the magnetic tape A surface.

TABLE 4.1 Construction of the Magnetic Rigid Disks

Disk Designation	Substrate (Ni–P on Al–Mg)	Construction of Magnetic Layer	Overcoat
A	Polished	γ-Fe$_2$O$_3$ particles in epoxy binder	Perfluoropolyether (PFPE) lubricant (liquid)
B	Textured	Sputtered metal film	Sputtered+ PFPE
C	Polished	Sputtered metal film	Sputtered+ PFPE
D	Textured	Plated metal film	Sputtered+ PFPE
E	Polished	Plated metal film	Sputtered+PFPE

From Bhushan, B. and Doerner, M. F. (1989), *J. Tribol.* 111, 452–458. With permission.

The surface topography of several magnetic thin-film rigid disks was also studied by atomic force microscopy (Bhushan and Blackman, 1991). The manufacturing process for these disks are discussed by Bhushan and Doerner (1989) and are summarized in Table 4.1. Figure 4.16 is an example of an AFM image of magnetic disk C (Bhushan and Doerner, 1989) for which the surface is composed of columnar grains of about 0.1 to 0.2 μm width which form during sputter-deposition. Since the substrate was untextured, the roughness of the films appeared quite isotropic. This is a 2.5×2.5 μm image which has a resolution of 12.5 nm. To check whether or not surface roughness appears at even smaller scales, a 0.4×0.4 μm scan, having a resolution of 2 nm, was obtained for the same surface. The structure function of the surface profiles for both scans revealed that roughness does appear fractal at nanometer scales as shown in Figure 4.17. The power law behavior of $S(\tau) \sim \tau^{1.49}$ suggests a fractal dimension of $D = 1.26$ for the surface profiles. The structure function deviates from this power law behavior at about 0.2 μm. It is interesting to note that this corresponds to the size of the columnar grains that are visible in the atomic force microscopy image. Therefore, this power law behavior corresponds to intergranular surface roughness. It is difficult to obtain any meaningful information for larger length scales when τ is comparable to the sample size, L. This is because the number of data points available is not good enough for statistical averaging required to obtain the structure function.

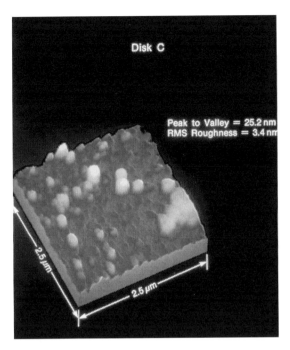

FIGURE 4.16 Surface image of magnetic rigid disk C (Bhushan and Doerner, 1989) obtained by atomic force microscopy (Bhushan and Blackman, 1991).

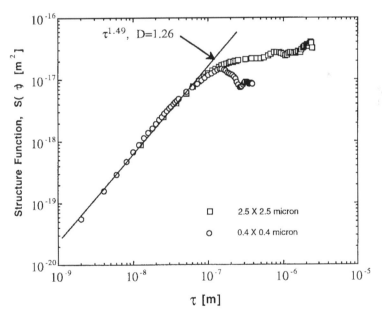

FIGURE 4.17 Structure function of the magnetic rigid disk C surface.

Figure 4.18 shows the structure function for a particulate disk. Note the vertical scale is higher than that of Figure 4.17, suggesting that the particulate magnetic disk A is much rougher than the sputter-deposited one. In this case again, the power law behavior at smaller scales suggests a fractal behavior. Deviations at larger length scale could be due to nonfractal characteristics or lack of statistical average.

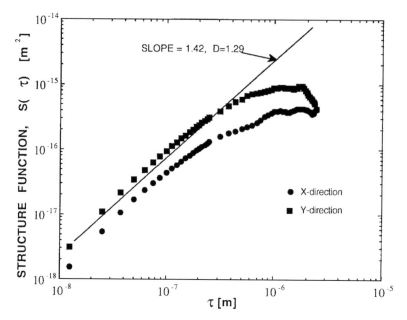

FIGURE 4.18 Structure function of a particulate magnetic rigid disk A (Bhushan and Doerner, 1989) whose surface roughness was measured by atomic force microscopy.

Figure 4.19 shows the structure function for magnetic rigid disk B which is textured in the circumferential direction. The magnetic thin films were sputter-deposited on the textured substrate. Note the differences in the structure function in the circumferential and radial directions. The profile in the radial direction goes across all the quasi-periodic texture marks, which leads to oscillations in the structure function. Such oscillations cannot be modeled by fractals and must be handled by a more general technique, as discussed in Section 4.3.4. Figure 4.20 shows the structure function of the textured magnetic rigid disk D, whereas Figure 4.21 shows that of the untextured rigid disk E. In both these cases the magnetic thin films were electroless plated on to the substrate. Note that the data levels off for $\tau > 50$ nm, which is probably a characteristic length scale for the plating process.

It is clear from the structure function data that there normally exists a transition length scale, ℓ_{12}, which demarcates two regimes of power law behavior. At scales smaller than ℓ_{12}, the fractal power law behavior is generally followed for all of the surfaces. At larger length scales, the structure function of the polished (or untextured) disks either saturates such that the Berry–Blackwell model can be easily applied or, in some cases, it follows a different power law behavior that can be characterized by another fractal dimension. If the surface is textured, however, the structure function at larger length scales oscillates and does not follow a scaling power law behavior. Such nonfractal behavior cannot be characterized by the fractal techniques and a more general method is needed. This is discussed in detail in Section 4.3.4. The transition length scale, ℓ_{12}, usually corresponds to a surface machining or growth process. For polycrystalline surfaces this may be the grain size, whereas for machining it is the characteristic tool size.

Recent experiments by Ganti and Bhushan (1995) showed that when a surface is imaged with atomic force microscopy and an optical profiler, values of D of a wide variety of surfaces fall in a close range but the values of G can vary a lot for the same surface. This is in contrast with the data presented above. However, the check for reliable data is to see whether or not the structure functions of the roughness measured at different resolutions and by different instruments overlap over common length scales. Inspection of their data showed that although the structure functions of the atomic force microscopy measurements of different scan sizes for the same surface seem to overlap over the common length scales, there was large discrepancy between the structure functions obtained from atomic force microscopy and optical profiler data. Therefore, it is inconclusive whether the discrepancy is due to the measurement technique or due to the characterization method.

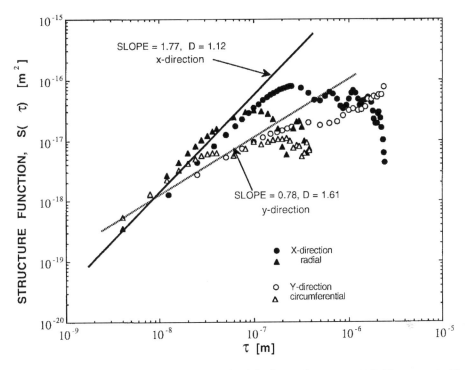

FIGURE 4.19 Structure function of magnetic rigid disk B (Bhushan and Doerner, 1989). The magnetic thin films were sputter-deposited on a textured substrate. The triangles are for a 0.8 × 0.8 μm atomic force microscopy scan containing 200 × 200 points. The circles are for a 2 × 2 μm atomic force microscopy scan containing 200 × 200 points.

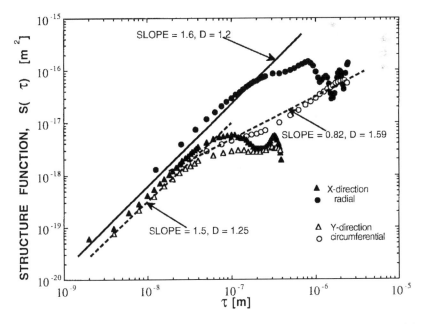

FIGURE 4.20 Structure function of textured magnetic rigid disk D (Bhushan and Doerner, 1989) in which the magnetic thin films were electroless plated on to the substrate. The triangles are for a 0.4 × 0.4 μm atomic force microscopy scan containing 200 × 200 points. The circles are for a 2.5 × 2.5 μm atomic force microscopy scan containing 200 × 200 points.

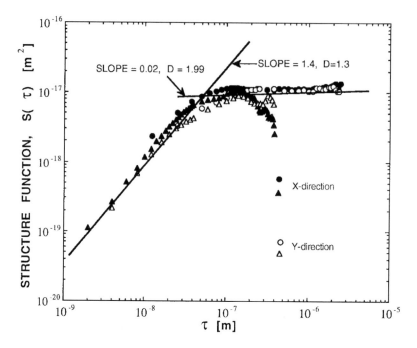

FIGURE 4.21 Structure function of untextured magnetic rigid disk E (Bhushan and Doerner, 1989) in which the magnetic thin films were electroless plated on a polished substrate. The triangles are for a 0.4×0.4 µm atomic force microscopy scan containing 200×200 points. The circles are for a 2.5×2.5 µm atomic force microscopy scan containing 200×200 points.

It is evident that fractal characterization is valid in certain regimes of surface length scales. In these regimes, the fractal techniques prove to be superior to conventional characterization techniques that use rms values. It is therefore instructive to understand what the fractal parameters D and G really mean.

4.3.3.4 What Do *D* and *G* Really Mean?

Since a rough surface is self-affine, thereby scaling differently in the two orthogonal directions, it needs two parameters for characterization. These are D and G. At this point, the reader may ask what do a surface profiles look like for different values of D and G. Figure 4.22* shows that when D is close to unity, the profile is smooth having more amplitude for long wavelength undulations and low amplitude for short wavelength undulations. As D is increased, the profile gets more wiggly and jagged. When D reaches close to 2, the profile becomes nearly space filling and therefore more like a surface. Therefore, a decrease in D effectively *stretches* the profile along the lateral direction and therefore changes the spatial frequency. So the value of D controls the *relative* amplitude of roughness at different length scales. In contrast, an increase in G stretches the curve in the vertical direction as shown in Figure 4.22. So the value of G controls the absolute amplitude of the roughness over *all* length scales.

The concept of roughness and smoothness, as discussed in Section 4.2.1, becomes quite ambiguous under these conditions. Should a surface with a higher G, and thus more amplitude, be called rougher or should a surface with higher D, and therefore more jagged, be called rougher? The problem is that the concepts of rough and smooth are too crude to distinguish between amplitude variations and frequency variation (or jaggedness) and so it is difficult to say which can be called rougher or smoother. It could be a combination of both, but at present it is unknown what this combination is.

---

*These are fractal simulations of rough surfaces obtained by using the Weierstrass–Mandelbrot function. Details of the simulation procedure is discussed in detail elsewhere (Voss, 1988; Majumdar and Bhushan, 1990; Majumdar and Tien, 1990).

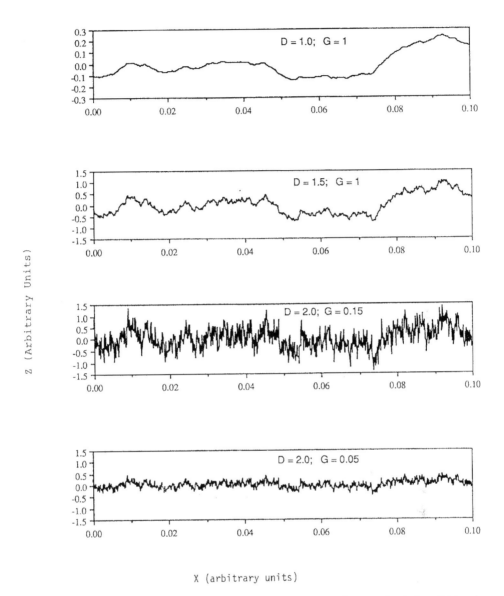

FIGURE 4.22 Effect of varying D and G on the profiles of rough surfaces. These are simulations (Majumdar and Tien, 1990) of rough surfaces obtained from the Weierstrass–Mandelbrot function. The effect of increasing D is to make the profile more jagged, or in other words, a lateral compression. An increase in G increases the amplitude of roughness over all length scales.

4.3.3.5 rms Values and D and G

The rms parameters, σ, σ', and σ'' have been used extensively in the tribology literature and therefore researchers are more familiar with them. Although they show scale-dependent characteristics, it is instructive to know how they relate to fractal characterization parameters D and G. The rms parameters are related to the power spectrum as shown in Equations 4.3.13 through 4.3.15.

Consider the rms height, σ, first. If σ is scale independent, then the model of Berry and Blackwell (1981) for the structure function as given in Equation 4.3.24 is valid. In this case, G and D correspond to the limit of $\tau \rightarrow 0$, whereas σ corresponds to the scales much larger than the correlation length τ_c. In other words, the rms height is unrelated to the fractal parameters G and D since the structure function does not follow the same behavior in the two limits.

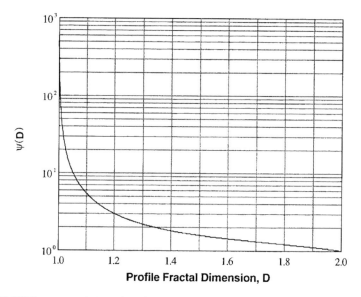

FIGURE 4.23 Variation of the factor $\psi(D)$ with the profile fractal dimension D.

The scale dependence of the rms height, σ, comes from the fractal power law variation of the power spectrum as given in Equation 4.3.16. The variance, σ^2, can be found as

$$\sigma^2 = \int_{1/L}^{\infty} \frac{C}{\omega^{(5-2D)}}\, d\omega = \frac{C}{2(2-D)}\, L^{2(2-D)} = \frac{G^{2(D-1)} L^{2(2-D)}}{2 \sin\left(\dfrac{\pi(2D-3)}{2}\right)\Gamma(2D-3)} \qquad (4.3.25)$$

The variation of the factor

$$\Psi(D) = \sin\left(\frac{\pi(2D-3)}{2}\right)\Gamma(2D-3)$$

with the fractal dimension D is shown in Figure 4.23. As $D \to 1$, ψ tends to ∞, whereas when $D = 2$, the ψ is equal to unity. It is clear that σ, G, and D are related in a complicated manner. But what is important is that σ does not depend on the instrument resolution but on the sample length, L. If σ is obtained for a wide range of varying sample length, L, then the fractal dimension D can be found from the slope of the log–log plot of σ vs. L. Once the D is found, the factor ψ can be found from Figure 4.23. With D and ψ known, the scaling constant G can be obtained from the relation in Equation 4.3.25. If the structure function or the power spectrum follows different power laws in different length scales, Equation 4.3.25 must be modified. This is discussed in Appendix 4.1.

The relation between rms slope, σ', and the fractal parameters G and D can also be obtained from the power spectrum as follows:

$$\sigma'^2 = \int_{1/L}^{1/\ell} \frac{\omega^2 C}{\omega^{(5-2D)}}\, dw = \frac{C}{2(D-1)}\left[\left(\frac{1}{\ell}\right)^{2(D-1)} - \left(\frac{1}{L}\right)^{2(D-1)}\right] \qquad (4.3.26)$$

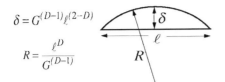

$$\delta = G^{(D-1)}\ell^{(2-D)}$$

$$R = \frac{\ell^D}{G^{(D-1)}}$$

FIGURE 4.24 Geometry of an asperity from a fractal surface of profile dimension D. The asperities are modeled as hemispheres with radius R and base diameter, ℓ, such that the base area is equal to $a = \ell^2$.

where ℓ is the smallest length scale that is measured by the instrument. For $D > 1$ and $L \gg \ell$, Equation 4.3.26 can be simplified to

$$\sigma'^2 = \frac{(2-D)}{2(D-1)\Psi(D)}\left(\frac{G}{\ell}\right)^{2(D-1)} \tag{4.3.27}$$

It is clear that the rms slope does not depend on the sample length, L, but on the instrument resolution, ℓ as $\sigma' \approx \ell^{-(D-1)}$. For the limiting case, $D \to 1$, we find that $\psi(D) \to 1/2(D-1)$ such that the denominator in Equation 4.3.27 is equal to unity. For a surface with multifractal regimes, see Appendix 4.1 for the rms slope.

The rms curvature, σ'', can be found similarly as

$$\sigma''^2 = \int_{1/L}^{1/\ell} \frac{\omega^4 C}{\omega^{(5-2D)}}\, d\omega = \frac{C}{2D}\left[\left(\frac{1}{\ell}\right)^{2D} - \left(\frac{1}{L}\right)^{2D}\right] \approx \frac{(2-D)}{2D\Psi(D)G^2}\left(\frac{G}{\ell}\right)^{2D} \tag{4.3.28}$$

where it is assumed that $L \gg \ell$. It is evident that $\sigma'' \approx \ell^{-D}$ such that it depends only on the instrument resolution and not on the sample length, L.

The autocorrelation function is also often used to characterize rough surface profiles. The relation between the autocorrelation function and the fractal parameters D and G are given in Appendix 4.2.

4.3.3.6 Asperity Geometry from Fractal Characteristics

One of the advantages of using the fractal roughness characterization and, in particular, the structure function technique is that the geometric shape of asperities can be described at all length scales in the fractal regime. Thus, for an asperity that has a base diameter of ℓ, the height of the asperity, δ, follows from the structure function as

$$\delta = G^{(D-1)}\ell^{(2-D)} = G^{(D-1)}a^{(2-D)/2} \tag{4.3.29}$$

Here the diameter ℓ is used as a characteristic length scale such that the area, a, of the asperity base can be written as, $a = \ell^2$. The geometry is schematically shown in Figure 4.24 for different length scales ℓ. If the shape is assumed spherical, the radius of curvature, R, for the asperities can be found to follow the relation

$$R = \frac{\ell^D}{G^{(D-1)}} = \frac{a^{D/2}}{G^{(D-1)}} \tag{4.3.30}$$

Thus the surface can be imagined to be a collection of asperities where small asperities are mounted on larger asperities that are in turn mounted on larger asperities in a hierarchical manner. Once the geometric structure is determined, the mechanics of contact and the surface force interactions can be modeled.

4.3.4 Generalized Technique for Fractal and Nonfractal Surfaces

The fractal characterization techniques, discussed in detail in Section 4.3.3, overcome some of the short-comings of conventional methods that use σ, σ', and σ''. One of the requirements of the fractal technique is that the structure function or the power spectrum must follow power law scaling behavior, that is, $S(\tau) \approx \tau^{2(2-D)}$ or $P(\omega) \approx \omega^{-(5-2D)}$. If this is satisfied, then the asperity height, δ, and the base size, ℓ, follow the scaling relation $\delta \approx \ell^{(2-D)}$. This is particularly useful in tribology since only two parameters, G and D, need to be known to study tribological phenomena at all length scales in the fractal regime. However, the experimental data in Figures 4.13 and 4.15 through 4.21 show that although the scaling behavior is followed in some cases, it is *not* universal. In addition, the power law can change at a transition length scale and is not universal over all length scales. Yet, the rms slope and curvature cannot be used to characterize them since the surface can have multiple scales, which, although they do not follow the scaling behavior $\delta \approx \ell^{(2-D)}$, can lead to scale-dependent rms values. A technique must therefore be developed that will work for both fractal and nonfractal surfaces and yet be scale independent. This section introduces a *new* method with these issues in mind.

It is necessary to identify first how the surface characteristics will be used. As discussed in Section 4.2, knowledge of the surface structure is important for predicting the size and spatial distributions of contact spots as well as for the mechanics of asperity sliding. Since these spots are formed by asperities, what is important is the asperity geometry at relevant length scales and its size and spatial distributions on the surface. The conventional techniques, which use rms height, slope and curvature, find an average asperity shape, whereas the fractal techniques determine the shape at all length scales by the scaling law. Both techniques can be combined to form a general method for roughness characterization as follows.

The roughness is characterized by two parameters — $V(\ell)$ and $K(\ell)$ — which are found in the x- and y-directions* by the following relations.

$$V_x\left(\ell\right) = \sqrt{\left\langle \left[z\left(x+\ell,\, y\right) - z\left(x,\, y\right)\right]^2 \right\rangle}$$
$$V_y\left(\ell\right) = \sqrt{\left\langle \left[z\left(x,\, y+\ell\right) - z\left(x,\, y\right)\right]^2 \right\rangle}$$

(4.3.31)

$$K_x\left(\ell\right) = \sqrt{\left\langle \frac{\left[z\left(x+\ell,\, y\right) + z\left(x-\ell,\, y\right)\right]^2}{\ell^2} \right\rangle}$$
$$K_y\left(\ell\right) = \sqrt{\left\langle \frac{\left[z\left(x,\, y+\ell\right) + z\left(x,\, y-\ell\right) - 2z\left(x,\, y\right)\right]^2}{\ell^2} \right\rangle}$$

(4.3.32)

Here the $\langle\ \rangle$ symbol implies averaging over the measured data points. It is evident that the function $V(\ell)$ is the square root of the structure function $S(\ell)$. For a fractal surface $V(\ell) = G^{(D-1)}\ell^{(2-D)}$. The function $K(\ell)$ is the rms curvature of asperities of lateral scale ℓ which in the fractal model is assumed to vary as $K(\ell) = G^{(D-1)}/\ell^D$. In the generalized model, such power laws will not be assumed and instead the raw data

---

*The x- and y-directions are chosen to be the principal directions of an anisotropic surface. The principal directions can be found by first obtaining V in all directions to get a V vs. l surface. If one takes a horizontal cut of the surface for V = constant, then one can connect the loci of the intersecting l values into a curve which in general can be approximated by an ellipse. Then the x- and the y-directions correspond to the major and the minor axes of the ellipse. The assumption made in this model is that the major and minor axes remain the same at all length scales.

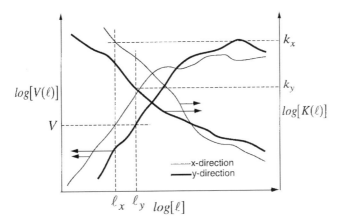

FIGURE 4.25 Qualitative demonstration of a typical V vs. ℓ and K vs. ℓ plot for the generalized roughness characterization technique. Note that, in general, the surface is anisotropic such that the curves are different in the x- and y-directions.

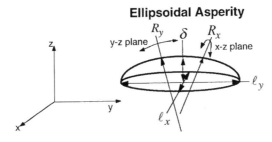

FIGURE 4.26 (a) Schematic diagram of an ellipsoidal asperity of an anistropic surface. (b) Equivalent hemispherical asperity that can be used to study the mechanics of contact with a flat hard plane.

of $V(\ell)$ and $K(\ell)$ will be used. There is no need to prove whether the surface is fractal or nonfractal. If the surface is fractal, both these parameters will show scaling behavior. If the surface is nonfractal and yet contains multiple length scales, this technique will allow one to incorporate the scale-dependent information contained in $V(\ell)$ and $K(\ell)$. When the surface is perfectly periodic with wavelength, λ, then $K(\ell)$ will show a peak when $\ell = \lambda/2$. In general, $V(\ell)$ and $K(\ell)$ neither follow scaling behavior nor do they show sudden jumps in the data but are a combination of both. It will be shown in Section 4.5 that these characteristics can be used to develop theories of contact mechanics and other tribological phenomena. Before that, it is necessary to understand how this technique can be used to characterize anisotropic surfaces.

Consider a typical plot of V_x, V_y, K_x, and K_y as a function of ℓ for a general anisotropic surface in Figure 4.25. An asperity on this surface will have an elliptic base with major and minor axes ℓ_x and ℓ_y and curvatures k_x and k_y as shown in Figure 4.26. The values of ℓ_x and ℓ_y can be found from Figure 4.25 by taking the intersection of a horizontal line with the V_x and V_y curves. Intersections of the vertical lines through ℓ_x and ℓ_y with the K_x and K_y curves give the respective curvature values of k_x and k_y. Note that it is possible to have more than one intersection of a horizontal line with the curves V_x and V_y producing

215

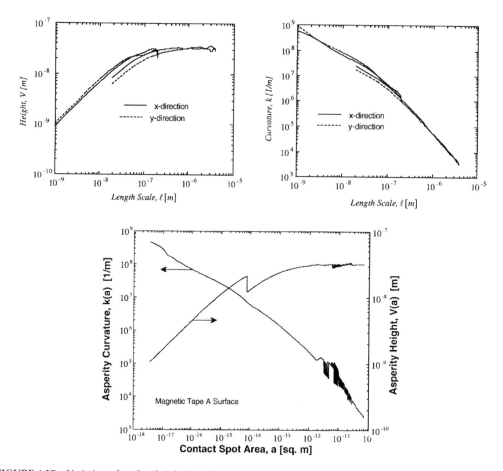

FIGURE 4.27 Variation of surface height V and curvature K for the magnetic tape A surface. (a) V and K vs. ℓ for two orthogonal directions; (b) V and K vs. the asperity base area. This collapses the curves in (a) by the procedure discussed in the text.

more than one value of ℓ_x and ℓ_y, as shown in Figure 4.25. The area of an asperity base is $a = \pi \ell_x \ell_y$. For more than one value of ℓ_x and ℓ_y, the product $\ell_x \ell_y$ can be obtained for different combinations, and therefore for a particular value of V = constant (horizontal line) it is possible to have different asperity base areas. For each combination of $\ell_x \ell_y$, the effective curvature of an ellipsoidal asperity of base area a is $k(a) = 0.5(k_x + k_y)$. This will be used in the contact mechanics of ellipsoidal asperities (Timoshenko and Goodier, 1970) with a flat plate. The procedure followed above collapses the anisotropic roughness data in the x- and y-directions in Figure 4.25 into a plot of V and k vs. the asperity base area a.

Roughness data of a magnetic tape surface will be taken as an example to demonstrate the application of the generalized contact mechanics model. Figure 4.27a shows the plots of V_x, V_y, K_x, and K_y as a function of ℓ for a magnetic tape, whereas Figure 4.27b shows the collapsed data of $V(a)$ and $K(a)$ as a function of asperity base area, a. The roughness was measured by an atomic force microscope and the details of the measurement as well as the surface images are described elsewhere (Oden et al., 1992).

It is evident that this generalized roughness characterization technique does not demand any requirements for either a scaling fractal surface structure or a surface with single scale of asperities. In that respect, it is more powerful than the other techniques, as will be shown in Section 4.5. However, the simplicity of the rms characterization or the fractal techniques that use only G and D is more than that of the generalized technique. As will be shown later, numerical integration must be performed to get any meaningful results for contact mechanics and size distribution.

4.4   Size Distribution of Contact Spots

As discussed in Section 4.2, our goal in characterizing rough surfaces is to determine the size and spatial distribution of contact spots due to asperity–asperity contact. Conventional methods in tribology (Bhushan, 1990) use an asperity density, η, as a measure of number of spots per unit area and a mean diameter, $\bar{d}$, as a measure of the average spot size. Although η counts the number, it does not involve the size distribution which may be useful for scale-dependent tribological phenomena. In addition, the asperity density under elastic contact theory is found by its statistical relationship to the rms height and the curvature as (Bhushan, 1984a,b)

$$\eta \approx \frac{\sigma''\left(F/EA_a\right)}{\sigma\sqrt{\sigma''\sigma}} \tag{4.4.1}$$

$$\bar{d} \approx \sqrt{\frac{\sigma}{\sigma''}} \tag{4.4.2}$$

where F is the normal load, E is the elastic modulus, and A_a is the apparent area. As has been proved earlier, the rms parameters are not unique for a surface. Therefore, any estimate of the asperity density using these parameters is also not unique.

4.4.1   Observations of Size Distribution for Fractal Surfaces

The multiscale nature of surface roughness suggests that there will be spots of different sizes. As can be observed in Figure 4.3, which is a simulation of a real surface contact, the spots come in all sizes and shapes. As a general trend it is evident that there is a greater number of smaller spots than larger ones. Consider first a fractal surface of profile dimension D and surface dimension $D + 1$. In the study of geomorphology of Earth, Mandelbrot (1975) found that the cumulative size distribution of islands on Earth's surface follows the power law $N(a) \approx a^{-D/2}$, where N is the total number of islands with area larger than a and D is the fractal dimension of the coastline of the islands. The coastline is a self-similar curve for which the dimension D is related to the surface dimension D_s by the relation $D = D_s - 1$. If the surface of, for example, a machined metal or a thin-film magnetic disk is enlarged enough, the hills and valleys would appear similar to that found on Earth's surface. The contact spots formed by a horizontal cut of such a surface by a contact plane would be analogous to the islands formed by cutting the Earth's surface by that of the ocean. Therefore, it can be expected that the contact spots would also follow a power law relation of the form:

$$N(a) = \left(\frac{a_L}{a}\right)^{D/2} \tag{4.4.3}$$

where the distribution is normalized by the area of the largest contact spot a_L. The size distribution of contact, $n(a)$, is defined such that the number of contact spots between a and $a + da$ is equal to $n(a)da$. For a fractal surface, the distribution $n(a)$ can be obtained from $N(a)$ of Equation 4.4.3 as

$$n(a) = -\frac{dN}{da} = \frac{D}{2a}\left(\frac{a_L}{a}\right)^{D/2} \tag{4.4.4}$$

217

If any instrument is used to count the number of spots and measure their diameters, the values would obviously be finite and nonzero, respectively. Therefore, conventional parameters such as the contact spot density η and the mean diameter $\bar{d}$ can be used to quantify the size distribution of contact spots. However, note that in the distribution of Equation 4.4.3 the number of the largest spot is unity, whereas the number of spots of area $a \to 0$ would tend to infinity. There is obviously a discrepancy between the two methods. One must note that since any measuring instrument is limited by its resolution, not all spots on the contact interface can be detected. In other words, both η and the mean spot diameter are instrument-dependent parameters as has been analytically shown (Majumdar and Bhushan, 1990). The fractal model does not involve these parameters and allows the smallest spot area to tend to zero with their number tending to infinity.

4.4.2 Derivation of Size Distribution for Any Surface

The power law behavior of Equation 4.4.4 is not only an experimental observation, but can be proved rigorously as follows. Consider a rough surface to contact a hard flat plane such that the surface mean and the plane is separated by a distance, d. Then the real area of contact, A_r, can be written in two ways:

$$A_r = \int_0^{a_L} n(a)\, a\, da = A_a \int_d^{\infty} g(z)\, dz \qquad (4.4.5)$$

The first integral is the summation over all the contact spots where a_L is the area of the largest contact spot. The second integral is over the probability distribution $g(z)$ of the surface heights as described in Section 4.3.1, where A_a is the apparent surface area. The size distribution $n(a)$ can, hence, be found as

$$n(a) = \frac{1}{a} \frac{dA_r}{da} = \frac{1}{a} \frac{dA_r}{dz} \frac{dz}{da} \qquad (4.4.6)$$

where the chain rule is used for the right-hand side.

Now dA_r/dz is related to $g(z)$ as

$$\left. \frac{dA_r}{dz} \right|_{z=d} = -A_a g(d) \qquad (4.4.7)$$

The factor dz/da must be derived from the deformation of a single asperity. Consider an asperity of curvature $k(a)$ in contact with a flat hard plane such that the contact area is equal to a and the deformation is equal to δ. Since δ increases in the negative z-direction, $dz/da = -d\delta/da$.

4.4.2.1 Size Distribution under Elastic Deformation

If the contact is under elastic deformation, then according to Hertzian theory (Timoshenko and Goodier, 1970; Johnson, 1985) the asperity deformation δ is related to the area a of contact and curvature k as

$$\delta = \frac{a k(a)}{\pi} \qquad (4.4.8)$$

such that dz/da is equal to

$$\frac{dz}{da} = -\frac{k(a)}{\pi} \left[1 + \frac{a}{k(a)} \frac{dk}{da} \right] \qquad (4.4.9)$$

Combining this with dA_r/dz, the size distribution $n_e(a)$ can be written as

$$n_e(a) = \frac{A_a\, k(a)\, g(d)}{\pi a}\left[1 + \frac{a}{k(a)}\frac{dk}{da}\right] \tag{4.4.10}$$

This is an important result, because the only assumption that has been made in deriving this result is that the asperity has an ellipsoidal shape of an effective curvature $k(a)$. It is a general relation for *any* surface — fractal or nonfractal. If the surface deformations are predominantly elastic, then the largest contact spot area, a_L, can be found by using Equation 4.4.10 in the equality of Equation 4.4.5 to get

$$\int_0^{a_L} k(a)\left[1 + \frac{a}{k(a)}\frac{dk}{da}\right]da = \frac{\pi\Omega(d)}{g(d)} \tag{4.4.11}$$

where $\Omega(d)$ is the cumulative probability distribution defined as

$$\Omega(d) = \int_d^{\infty} g(z)\,dz \tag{4.4.12}$$

To use the size distribution, $n_e(a)$, of Equation 4.4.10 in any tribological analysis, it is necessary to find the probability distribution, $g(z)$, and the curvature, $k(a)$, as a function of spot area a, and the separation d between the two surfaces. The probability distribution, $g(z)$, can be assumed to be a Gaussian or perhaps some nonsymmetric function. The curvature, $k(a)$, can be obtained by the generalized roughness characterization technique discussed in Section 4.3.4. The separation, d, can be found by studying the contact mechanics which depends on the applied normal load on the surface, the size distribution, $n_e(a)$, the surface roughness, and the material properties such as hardness and elastic modulus. This will be dealt with in Section 4.5.

4.4.2.1.1   Gaussian Height Distribution
If the surface heights have a Gaussian distribution, then the size distribution reduces to

$$n_e(a) = \frac{A_a k(a)}{\pi a}\left[1 + \frac{a}{k(a)}\frac{dk}{da}\right]\frac{e^{-d^2/2\sigma^2}}{\sigma\sqrt{2\pi}} \tag{4.4.13}$$

where σ is the standard deviation of the surface heights. The largest contact spot area, a_L, can be found from Equation 4.4.12 to be

$$\int_0^{a_L} k(a)\left[1 + \frac{a}{k(a)}\frac{dk}{da}\right]da = \pi\sqrt{\frac{\pi}{2}}\,\frac{\sigma\cdot\text{erfc}\left(d/\sigma\sqrt{2}\right)}{e^{-d^2/2\sigma^2}} \tag{4.4.14}$$

where erfc( ) is the complementary error function. Thus one would get a relation between a_L and the nondimensional height d/σ.

4.4.2.1.2   Fractal Surface
If the surface is a fractal with a profile dimension of D, then we saw earlier that the radius of curvature, R, behaved as $R = a^{D/2}/G^{D-1}$ such that $k(a) = G^{D-1}/a^{D/2}$. Therefore the size distribution, $n_e(a)$, varies with spot size as

$$n_e(a) = \frac{A_a\, G^{(D-1)}\, g(d)}{\pi a^{(D/2)+1}} \left(\frac{2-D}{2} \right) \qquad (4.4.15)$$

It is evident that the variation of $n_e(a) \approx a^{-(1+D/2)}$ is exactly what was observed by Mandelbrot (1975) for the distribution of islands on Earth's surface and given in Equation 4.4.4. This is a mathematical derivation of the experimentally observed distribution.

The largest island size, a_L, can be found for a Gaussian fractal surface by using the relation $k(a) = G^{D-1}/a^{D/2}$ in Equation 4.4.14 to get the relation

$$a_L^{(2-D)/2} = \pi \sqrt{\frac{\pi}{2}} \frac{\sigma}{G^{(D-1)}} \frac{\mathrm{erfc}\left(d\,\sigma\sqrt{2} \right)}{e^{-d^2/2\sigma^2}} \qquad (4.4.16)$$

Using the fact that $\sigma = G^{(D-1)}\, L^{(2-D)} \Big/ \sqrt{2\Psi(D)}$, we find that

$$a_L = L^2 \left[\frac{\pi\sqrt{\pi}}{2\sqrt{\Psi(D)}} \frac{\mathrm{erfc}\left(d\,\sigma\sqrt{2} \right)}{e^{-d^2/2\sigma^2}} \right]^{2/(2-D)} \qquad (4.4.17)$$

4.4.2.2 Size Distribution under Plastic Deformation

When the contact is under plastic deformation, then using volume conservation the area of contact is found to follow (Chang et al., 1987):

$$\delta = \frac{\delta_c(a)}{2} + \frac{ak(a)}{2\pi} \qquad (4.4.18)$$

where $\delta_c(a)$ is the critical deformation such that when $\delta > \delta_c$, the deformation is plastic and when $\delta < \delta_c$ the deformation is elastic. It is clear that when $\delta = \delta_c$, the relation for elastic deformation in Equation 4.4.8 is retrieved. In this case, dz/da is equal to

$$\frac{dz}{da} = -\frac{1}{2}\left[\frac{d\delta_c}{da} + \frac{k(a)}{\pi}\left(1 + \frac{a}{k(a)} \frac{dk}{da} \right) \right] \qquad (4.4.19)$$

More will be discussed on the critical deformation δ_c in the section on contact mechanics whereas now it will be stated that it depends on the contact area a and curvature as $\delta_c = \phi/k(a)$, where ϕ depends on the material properties. By using this, the size distribution under plastic deformation is

$$n_p(a) = \frac{A_a\, k(a)\, g(d)}{\pi a}\left[-\frac{\pi\phi}{2k^3} \frac{dk}{da} + \frac{1}{2}\left(1 + \frac{a}{k}\frac{dk}{da} \right) \right] \qquad (4.4.20)$$

For a Gaussian height distribution, one can follow the analysis for elastic deformation to obtain the relation for the largest island. For a fractal surface with profile dimension, D, such that $k(a) = G^{D-1}/a^{D/2}$, the size distribution can be shown to follow the relation:

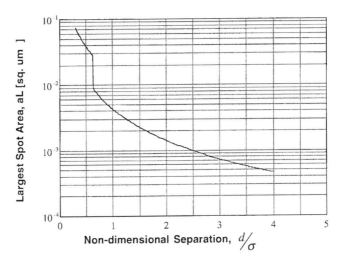

$$n_p(a) = \frac{A_a \, G^{(D-1)} \, g(d)}{\pi a^{1+(D/2)}} \left[\frac{\pi D \phi \, a^{(D-1)}}{4G^{2(D-1)}} + \frac{2-D}{4} \right] \qquad (4.4.21)$$

4.4.2.3 Size Distribution under Elastic-Plastic Deformations

In the next section it will be shown that when surface contact involves a combination of elastic and plastic deformations there exists a critical contact spot area, a_c, such that all spots smaller than the critical area, $a < a_c$, deform plastically and all spots larger than a_c, $a > a_c$, deform elastically. Under these circumstances, the largest contact spot area, a_L, must be found by the equation:

$$\int_0^{a_c} n_p(a) a \, da + \int_{a_c}^{a_L} n_e(a) a \, da = A_a \Omega(d) \qquad (4.4.22)$$

where $n_e(a)$ and $n_p(a)$ are given by Equations 4.4.10 and 4.4.20, respectively. If the surface is assumed to have a Gaussian surface height distribution, Equation 4.4.22 reduces to

$$\int_0^{a_c} k(a) \left[-\frac{\pi \phi}{2k^3} \frac{dk}{da} + \frac{1}{2}\left(1 + \frac{a}{k}\frac{dk}{da}\right) \right] da + \int_0^{a_c} k(a)\left(1 + \frac{a}{k}\frac{dk}{da}\right) da = \pi \sqrt{\frac{\pi}{2}} \; \frac{\sigma \, \mathrm{erfc}\left(d/\sqrt{2}\sigma\right)}{e^{-d^2/2\sigma^2}} \qquad (4.4.23)$$

The relation between the largest contact spot, a_L, and the nondimensional separation d/σ for a magnetic tape surface in Figure 4.28. It is assumed here that $\phi = 0.072$ and $a_c = 0.35 \ \mu m^2$ as will be shown in the next section. The relation between curvature $k(a)$ as a function of a is obtained from the generalized roughness characterization technique discussed in Section 4.3.4.

4.5 Contact Mechanics of Rough Surfaces

In Section 4.4, the size distributions of contact spots were obtained as a function of the separation d between two surface mean planes as well as the height distribution, $g(z)$, and standard deviation, σ, of

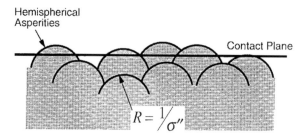

FIGURE 4.29 Model of surface roughness used in the GW (1966) model of contact mechanics.

the equivalent surface. However, a relation between the load F and the separation d occurs is still not yet obtained. To do so, it is necessary to understand the elastic and the plastic deformations of the surface asperities in contact. Then one can finally determine the size distributions of contact spots for a given load and material properties. That will become the starting point to study friction, wear, and other tribological phenomena.

Static contact between two surfaces can be modeled as the contact between an equivalent surface and a flat hard plane as discussed in Appendix 4.3. In the past, several theories of elastic–plastic contact between a rough surface and a flat hard plane have been proposed. The most popular of these is the Greenwood–Williamson (GW) (1966) model. A brief description of this model is presented here and more descriptions are available in several review papers and textbooks. This subsection will present a critical review of when the GW model can and cannot be applied. Note that the GW model is taken as representative of the several "constant-curvature" models available in the literature (McCool, 1986).

4.5.1 Greenwood–Williamson Model

4.5.1.1 The Model

The GW model assumes the surface to be composed of hemispherical asperities all having the same radius of curvature R. The summit heights or asperity peaks are distributed randomly about a mean summit plane and follow a Gaussian distribution with a standard deviation, σ_s, as shown in Figure 4.29. Note that the distribution of the summits is not the same as that of the surface. The summits are presumed to be uniformly distributed over the rough surface with a known density D_{sum} of summits per unit area.

The probability, Ω_0, that a randomly selected summit has a height in excess of some value d, can be found from the cumulative probability distribution as

$$\Omega_0\left(\frac{d}{\sigma_s}\right) = \int_{d/\sigma_s}^{\infty} g(\bar{z}_s)\, d\bar{z}_s, \qquad (4.5.1)$$

where $\bar{z}_s$ is the summit height normalized by the summit rms height, σ_s. This is the probability that a randomly selected summit will contact a smooth hard plane when the distance between the plane and the surface mean is equal to d. Since the surface density of summits is D_{sum}, the average expected number of contacts, n, per unit area is

$$n = D_{\text{sum}}\Omega_0\left(\frac{d}{\sigma_s}\right) \qquad (4.5.2)$$

For a summit of height, z_s, which exceeds the separation, d, the deformation, w, is equal to $w = z_s - d$. Then according to Hertz's theory of elastic contact (Timoshenko and Goodier, 1970; Johnson, 1985) between a sphere and flat plane, the area of contact for this summit is

$$a = \pi R w = \pi R \left(z_s - d \right) \qquad (4.5.3)$$

The corresponding asperity load is

$$f = \tfrac{4}{3} E R^{1/2} \, w^{3/2} = \tfrac{4}{3} E R^{1/2} \left(z_s - d \right)^{3/2} \qquad (4.5.4)$$

where E is the elastic modulus of the equivalent surface and is given as

$$\frac{1}{E} = \frac{1 - v_1^2}{E_1} + \frac{1 - v_2^2}{E_2} \qquad (4.5.5)$$

where v is the Poisson's ratio and the subscripts 1 and 2 correspond to the two surfaces.

Using these expressions for microcontact of a single asperity, the expected values of the total area of contact, A_r, and load, F_n, can be found as

$$\frac{A_r}{A_a} = \pi R \sigma_s D_{\text{sum}} \Omega_1 \left(\frac{d}{\sigma_s} \right)$$

$$\frac{F_n}{A_a} = \frac{4}{3} E R^{1/2} \sigma_s^{3/2} D_{\text{sum}} \Omega_{3/2} \left(\frac{d}{\sigma_s} \right) \qquad (4.5.6)$$

where $\Omega_n(t)$ is equal to

$$\Omega_n\left(t \right) = \int_t^{\infty} \left(x - t \right)^n g\left(x \right) dx \qquad (4.5.7)$$

The GW model contains three parameters that need input from the roughness characteristics — R, σ_s, and D_{sum}. Nayak (1971, 1973) used statistical analysis to show that their values depend on the variances of the surface height, σ^2, the surface slope, σ'^2, and the surface curvature, σ''^2. The final expression for the real area of contact and load reduces to

$$\frac{A_r}{A_a} = 0.064 \left(\alpha - 0.897 \right)^{1/2} \Omega_1 \left(\frac{d}{\sigma_s} \right)$$

$$\frac{F_n}{A_a} = 0.033 \, E \sigma' \left(\alpha - 0.897 \right)^{3/4} \Omega_{3/2} \left(\frac{d}{\sigma_s} \right) \qquad (4.5.8)$$

where α is a nondimensional parameter defined as

$$\alpha = \left(\frac{\sigma \sigma''}{\sigma'^2} \right)^2 \qquad (4.5.9)$$

It is interesting to note that a plot of A_r vs. F_n is essentially a plot of Ω_1 vs. $\Omega_{3/2}$. If a Gaussian distribution is used for the surface height distribution, it can be shown that this follows a power law $A_r \approx F_n^{0.95}$. More details of the GW model, in particular for plastic deformation, and other theories of contact can be found in McCool (1986).

4.5.1.2 A Critique

The GW model is based on the assumption that all the asperities have the same radius of curvature R. Therefore, the requirement of the GW model is that the surface must be made up of asperities of a single length scale. The question is whether or not real surfaces follow this behavior. Polycrystalline materials, which have a very narrow distribution of grain sizes, fit this requirement since they can have surfaces whose asperities are made up of a single grain. Also, machining processes such as turning, shaping, or milling can produce a textured surface that contain grooves made by moving the tool at a certain feed rate. Although the surface will be highly anisotropic, with texture marks only in one direction, the profile in one direction will be sinusoidal-like and can fit the description of the GW model. In both these cases, there exists a dominant surface length scale — the grain size for a polycrystalline surface or the texture grooves for a machined surface. The assumption of a constant-radius asperities of the GW model suggests that the model is applicable when a surface contains such a dominant length scale.

However, when a surface contains multiple length scales with no dominant scale, the GW model cannot be applicable. This is because the rms slope and curvature depend on the instrument resolution for surfaces with multiple length scales. In addition, the rms height can sometimes be instrument dependent as we saw earlier. Note that multiscale surfaces can, but need not be, fractal in structure. In these cases, the assumption of constant-radius asperities of the GW model is not valid. In addition, the parameter α in Equation 4.5.9 is also scale dependent, making the predictions of the GW model questionable for multiscale surfaces.

In Section 4.2, it was emphasized that an important parameter necessary for tribological analysis is the size distribution, $n(a)$. However, the GW model does not directly provide this information but gives the total real area of contact A_r. This can only be useful in studying friction and wear if $\tau(a)$ and $\dot{v}(a)$ in Equations 4.2.1 and 4.2.2 are independent of the contact spot area a. In general, $\tau(a)$ and $\dot{v}(a)$ do depend on the area, and therefore the size distribution is necessary. We can derive the distribution using Equation 4.4.10 to be

$$n\!\left(a\right) = \frac{A_a g\!\left(d\right)}{\pi R a}$$

(4.5.10)

where R is the radius of asperity curvature and is assumed constant. By using this, the area of the largest island can be found as

$$a_L = \pi \sqrt{\frac{\pi}{2}}\, \sigma R \frac{\operatorname{erfc}\!\left(d\,\sqrt{2}\sigma\right)}{e^{-d^2/2\sigma^2}}$$

(4.5.11)

if the height distribution is assumed to be Gaussian. Note that for multiscale surfaces, the radius of curvature R is not a constant, thereby making Equations 4.5.10 and 4.5.11 scale dependent.

To overcome the difficulties of the constant-radius assumption of the GW model, Majumdar and Bhushan (1991) proposed a model of elastic and plastic contact between rough fractal surfaces. The Majumdar–Bhushan (MB) model allows for asperity curvature to vary with size. The results obtained by this model are quite interesting and at times drastically different from the GW model. The following analysis gives the development of this model.

4.5.2 Majumdar–Bhushan Model

Consider an asperity of radius of curvature R being deformed by a hard plane as shown in Figure 4.26b. Initially, the asperity will deform elastically but beyond a critical deformation, δ_c, the material will deform inelastically. For ductile materials this would lead to plastic deformations, whereas for brittle materials fracture would occur. The critical deformation can be written as (Greenwood and Williamson, 1966)

$$\delta_c = \left(\frac{\pi H}{2E}\right)^2 R = \phi R = \frac{\phi}{k(a)} \qquad (4.5.12)$$

where $k(a) = 1/R$ is the curvature as a function of spot area, H is the surface hardness of the softer of the two surfaces, and E is the elastic modulus of the equivalent surface (given in Equation 4.5.5) such that $\phi = (\pi H/2E)^2$ is a nondimensional material property. It is assumed that ϕ is independent of the spot area a. This may not be a good assumption when the material hardness varies with the depth from the surface as found in surface-hardened materials (Suh, 1986). In such cases, the properties of the smaller spots are determined by the material close to the surface, whereas those of larger spots are dictated by the hardness deeper in the material. For simplicity, however, it will be assumed that the hardness H is a constant although its dependence on depth or spot area can be included.

The criterion for the inception of plastic deformation is that $\delta > \delta_c$. When the deformation is further increased, the fraction of the load carried by the elastic deformation diminishes whereas that by plastic deformation increases until the asperity is completely under plastic deformation (Johnson, 1985). Therefore, the transition between purely elastic to purely plastic is not drastic and occurs over a range of load and deformation. In this study, however, we will assume the transition to be drastic to keep the analysis simple. Since the deformation δ depends on the asperity size, a statistical estimate of d from the fractal surface analysis is $\delta = G^{(D-1)} a^{(2-D)/2}$. In addition, the curvature varies as $k(a) = G^{D-1}/a^{D/2}$. By using this in Equation 4.5.12 and comparing it with δ, it can be derived that there exists a critical contact spot area a_c such that all spots smaller than a_c deform plastically whereas those that are larger than a_c deform elastically. The relation for a_c is given as follows:

$$a < a_c \ \text{ plastic; } \ a > a_c \ \text{ elastic; } \ a_c = \frac{G^2}{\phi^{1/(D-1)}} \qquad (4.5.13)$$

The derivation above is true for fractal surfaces As discussed in Section 4.3.4, not all surfaces are fractal and therefore a new method of characterizing rough surfaces was introduced. Using this technique a good statistical estimate of the asperity deformation δ is $V(a)$. By using Equation 4.5.12 for the critical deformation δ_c, the criterion for plastic deformation can be written as

$$V(a) > \frac{\phi}{k(a)} \ \text{ or } \ \Psi > \phi \ \text{ where } \Psi = V(a)k(a) \qquad (4.5.14)$$

Here $\Psi(a) = V(a)k(a)$ is a nondimensional number which we call the "plasticity index." Therefore, plastic deformation occurs when the plasticity index is larger than a critical number ϕ which depends only on material properties as

$$\phi = \left(\frac{\pi H}{2E}\right)^2 \qquad (4.5.15)$$

In this analysis, what is needed is dependence of $V(a)$ and $k(a)$ on the asperity base area a. The generalized roughness analysis of Section 4.3.4 provides this data from the measured roughness. Figure 4.27 shows the V vs. a and the k vs. a curves for the magnetic tape A surface. There exists a critical contact spot area, a_c, which demarcates the elastic and the plastic regime.

Figure 4.30 shows a plot of Ψ as a function of spot area, a, for the magnetic tape A whose roughness statistics are shown in Figure 4.27. By using the values of $H = 0.25$ GPa and $E = 1.75$ GPa for the magnetic tape contacting a flat hard plane, the critical plasticity index is found to be $\phi = 0.4$ which is a horizontal

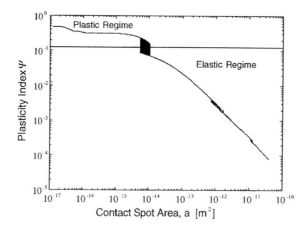

FIGURE 4.30 Variation of the plasticity index, Ψ, on the contact spot area, a, for the magnetic tape A surface whose roughness was measured by atomic force microscopy.

line in Figure 4.27. It is also found that the critical contact spot area is $a_c = 7 \times 10^{-15}$ m^2 or 0.7 μm^2 such that all contact spots $a < a_c$ deform plastically whereas for those satisfying $a > a_c$ deform elastically. The magnetic particles present on the magnetic tape are observed to be oblong with a length-to-width ratio of about 10. Thus, contact would produce an approximately elliptical spot with the major axis L about ten times the minor axis d. The area of the spot is therefore $a = \pi L d = 10\pi d^2$. By equating this to 0.35 μm^2, the minor axis is found to be equal to $d = 0.15$ μm. This is approximately the diameter of the magnetic particles observed by atomic force microscopy as shown in Figure 4.14. Thus, it can be concluded that the individual magnetic particles deform plastically, whereas, when deformed together with other particles, they deform elastically. Such an analysis to determine the nature of deformation is very useful to predict the type of surface wear and the failure mechanisms.

It is important to note that both the MB model and the generalized analysis show that smaller spots tend to deform plastically whereas larger spots deform elastically. One may argue that this may be due to the assumption of the fractal scaling behavior which surfaces do not always follow. However, the generalized analysis makes no such assumptions and is obtained directly from *raw data* of roughness measurements. Figure 4.30 clearly shows that the plasticity index for smaller contact spots are larger than that of bigger spots. This prediction is radically different from the classical theories such as the GW model which predicts just the opposite. The question is why there is so much of discrepancy.

The problem lies in the assumption of a constant radius of curvature R or a curvature k in the GW model. Equation 4.5.12 will still hold since no assumption is made in its derivation. Plasticity will occur when the asperity deformation $V(a)$ is larger than δ_c, $V(a) > \delta_c$. Now by Hertzian analysis, $V(a) = a/\pi R$ for an elastic contact (see Equation 4.5.6) with R assumed to be constant. Therefore, for plastic deformation, the following condition must be met

$$\frac{a}{\pi R} > \phi R \qquad (4.5.16)$$

It follows that all spots larger than a critical spot $a_c = \pi R^2 \phi$ deforms plastically and all spots smaller than this deform elastically, summarized as follows

$$\text{GW predictions: } a > \pi R^2 \phi \quad \text{plastic}$$

$$a < \pi R^2 \phi \quad \text{elastic} \qquad (4.5.17)$$

This is exactly opposite of what the MB model and the generalized analysis predict. Since roughness measurements at different resolutions will produce different values of R, as shown clearly in Figure 4.27, the predictions of the elastic–plastic regime by Equation 4.5.17 will be misleading.

With regard to the predictions of the MB model and the generalized analysis, there is one question that must be resolved. One may ask how is it that, under increasing load, smaller asperities under plastic deformation may merge to form a larger spot which can be under elastic deformation. A qualitative answer can be provided now since there is no experimental evidence yet. When the loads are low, only small asperities are deformed. Since the radius of curvature of these asperities is small, these points have high stress concentrations making plastic deformations likely. When the load is increased, they can merge to form a larger spot with a higher radius of curvature. This can lead to stress relief and therefore a change from plastic to elastic deformation. One must note that although the stress has reduced, the load-bearing capacity can be increased since the area is increased by merging smaller spots into one large one. Thus, increase in load does not necessarily mean higher *local* stresses.

When an asperity deforms elastically under a compressive load from a flat hard plane, the elastic load, f_e, and the asperity deformation, δ, can be obtained from Hertz's analysis (Timoshenko and Goodier, 1970; Johnson, 1985) as

$$f_e = \frac{4}{3} E R^{1/2} \delta^{3/2} = \frac{4 E \delta^{3/2}}{3 k^{1/2}(a)} \tag{4.5.18}$$

By using Equation 4.4.8 for the microcontact area, a, the elastic load can be written as a function of contact spot area a as

$$f_e(a) = \frac{4 E k(a) a^{3/2}}{3 \pi^{3/2}} \tag{4.5.19}$$

When an asperity is under plastic deformation, we will assume a simple linear relation between the load and the area as

$$f_p(a) = Ha \tag{4.5.20}$$

where H is the surface microhardness. It is assumed here that H is a constant and does not vary with depth. Nevertheless, a functional dependence of H with deformation δ and in turn the contact spot area, a, can easily be incorporated. The total normal load on the surface can be calculated to be

$$F_n = \int_0^{a_c} f_p(a) n_p(a) da + \int_{a_c}^{a_L} f_e(a) n_e(a) da \tag{4.5.21}$$

Here a_c is the critical spot area demarcating the elastic and the plastic regimes and a_L is the largest spot area. The first integral represents the total plastic load and the second term represents the total elastic load. In Equation 4.5.21 it is assumed that $a_L > a_c$. If $a_L < a_c$, then the surface deformation will be predominantly plastic and only the first term will remain in Equation 4.5.21.

The model for contact between fractal surfaces was first developed by Majumdar and Bhushan (1991). This subsection gives a synopsis of the model. A fractal surface is characterized by scaling behavior such that the curvature follows the power law $k(a) = G^{D-1}/a^{D/2}$, where D is the profile dimension. Using this in Equation 4.5.19, the elastic load on a single contact spot of area a becomes

$$f_e(a) = \frac{4 E G^{(D-1)} a^{(3-D)/2}}{3 \pi^{3/2}} \tag{4.5.22}$$

227

It is interesting to note that for a fractal surface, the load and the area are related by the power law $f_e \approx a^{(3-D)/2}$, whereas under the GW model the relation is $f_e \approx a^{3/2}$. The plastic load remains the same as that given in Equation 4.5.20. The size distribution of contact spots for a fractal surface of profile dimension D is given by Equation 4.4.4. By using this in conjunction with the elastic and microcontact load in Equation 4.5.21 the total load can be found as

$$F_n = \frac{DHa_L^{D/2}}{2}\int_0^{a_c}\frac{da}{a^{D/2}} + \frac{2DEG^{(D-1)}a_L^{D/2}}{3\pi^{3/2}}\int_{a_c}^{a_L}\frac{da}{a^{D-\frac{1}{2}}}$$

$$= \frac{DHa_L^{D/2}a_c^{(2-D)/2}}{(2-D)} + \frac{4DEG^{(D-1)}a_L^{(3-D)/2}}{3\pi^{3/2}(3-2D)}\left[1-\left(\frac{a_c}{a_L}\right)^{(3-2D)/2}\right] \quad \text{for } D \neq 1.5 \quad (4.5.23)$$

$$= 3Ha_L^{3/4}a_c^{1/4} + \frac{EG^{1/2}a_L^{3/4}}{\pi^{3/2}}\ln\left(\frac{a_L}{a_c}\right) \quad \text{for } D = 1.5$$

The real area of contact, A_r, can be found using the fractal size distribution as

$$A_r = \int_0^{a_L} n(a)a\,da = \frac{D}{2-D}a_L \quad (4.5.24)$$

The linear relationship between A_r and a_L suggests that when $a_L > a_c$, the elastic load depends on the real area of contact as $F_e \approx A_r^{(3-D)/2}$, whereas the plastic load varies as $F_p \approx A_r^{D/2}$. When $a_L < a_c$, then the deformation is predominantly plastic and the load varies linearly with area as $F_p \approx A_r$.

The main results of the fractal model are listed in Table 4.2. The predictions for the real area of contact and fraction of elastic contact area are graphically shown in Figures 4.31a and 4.32a for variation of one parameter keeping the others constant. Figures 4.31b and 4.32b correspondingly show the fraction of the real area of contact in elastic deformation. For $D = 1.5$ and $\phi = 0.01$ in Figure 4.31a, it is evident that as $G^* = G/\sqrt{A_a}$ is increased, the load required to produce a particular real area of contact increases. An increase in G^* implies, an increase in asperity deformation δ, which therefore requires a higher load. Figure 4.31b shows that as G^* is increased, the percentage of elastic contact decreases. This is because increase in G^* increases the critical spot area, as given by Equation 4.5.13, thus increasing the plastic regime and reducing the elastic regime. Also, when the real area of contact is increased, the fraction of elastic contact area increases. This results from merging of small contact spots in plastic deformation to form larger spots in elastic deformation.

The effects of fractal dimension on the real area of contact and elastic contact area is interesting indeed and are shown in Figure 4.32 for $G^* = 10^{-10}$ and $\phi = 0.01$. It is observed that as the value of dimension is increased, the real area of contact first increases until $D = 1.5$ and then decreases for $D = 1.9$ for a given load. A similar reversal is observed for the fraction of elastic contact area in Figure 4.32b. The reason for this behavior is that as the dimension is increased from 1.1 to 1.9, the value of $a_c^* = a_c/A_a$ decreases from 10^{16} to 10^{-16}, respectively, for the above values of G and ϕ. Therefore, for $D = 1.1$, all the contact spots are smaller than the critical contact area and so in plastic contact as evident in Figure 4.32b. As D is increased, the value of a_c^* decreases which results in a larger number of elastic contact spots. Therefore, the fraction of elastic contact area for $D = 1.5$ is nearly unity. As the value of D is further increased, the critical area decreases further and one may expect that the fraction of elastic contact should increase. However, as D is increased, the number and the contribution of smaller contact spots to the total area become significant as suggested by the size distribution. The increase in the value of fractal dimension therefore leads to two competing effects — the critical contact area decreases and the number of contact spots below the critical increases. This results in the reversal behavior.

TABLE 4.2 Profile Structure Function, Critical Contact Spot, Size-Distribution of Spots, Real Area of Contact, Fraction of Elastic Contact Area and Total Load — Predictions of Fractal Contact Model

Surface profile structure function	$$F_f = \int_0^L \tau(a)\, n(a$$
Critical contact for spot area	$$a_c = \frac{G^2}{\left[\left(\pi H / 2E \right)^{1/(D-1)} \right]}$$
Size-distribution of contact spots, $n(a)$	$$n(a) = \frac{D}{2} \frac{a_L^{D/2}}{a^{D/2+1}}$$
Real area of contact	$$A_r = \frac{D}{2-D}\, a_L$$
Fraction of elastic contact area	$$\frac{A_{re}}{A_r} = 1 - \left[\frac{Da_c}{(2-D)A_r} \right]^{(2-D)/2}$$
Total elastic-plastic load $(a_L > a_c)$	$$F_n = \frac{DHa_L^{D/2}\, a_c^{(2-D)/2}}{(2-D)} + \frac{4DEG^{(D-1)}\, a_L^{(3-D)/2}}{3\pi^{3/2}(3-2D)} \left[1 - \left(\frac{a_c}{a_L} \right)^{(3-2D)/2} \right] \quad \text{for } D \neq 1.5$$
	$$= 3Ha_L^{3/4}\, a_c^{1/4} + \frac{EG^{1/2}\, a_L^{3/4}}{\pi^{3/2}} \ln\left(\frac{a_L}{a_c} \right) \quad \text{for } D = 1.5$$
Total plastic load $(a_L < a)$	$$F_n = HA_r$$

Figure 4.33 shows the experimental results for two Pyrex glass surfaces compressed against each other (Yamada et al., 1978). It can be observed that the area–load relation is $A_r \approx F^{1.19}$. The exponent corresponds to a fractal dimension of the surface profile to be 1.4 and of the surface to be 2.4. Note that for $F_n/EA_a > 2 \times 10^{-3}$ the experimental results start to deviate from the predictions. It is speculated that this may be due to interactions between asperities in contact which has been neglected in the fractal model. Figure 4.34 shows the comparison between the predictions of the GW model, the fractal model, and experiments on contact between a magnetic thin-film rigid disk and a planoconvex lens (Bhushan and Dugger, 1990). The predictions of the fractal model are evaluated for experimentally observed values of $D = 1.38$ (Majumdar and Bhushan, 1990) and $\phi = 0.05$ (Bhushan and Doerner, 1989) for disk type C, and an assumed value for $G = 10^{-10}$ for curve fitting. It can be seen that, first, the trends of the fractal model are in better agreement than that of the GW model. In addition, the GW model predicts a linear load–area relation which is not observed experimentally. However, one should note that having four data points in Figure 4.34 may not provide enough evidence for the validity of the fractal model. In that respect, Figure 4.33 provides a more conclusive agreement between theory and experiments.

4.5.3   Generalized Model for Fractal and Nonfractal Surfaces

The structure function data of several types of machined metals, thin-film disks, and magnetic tape in Section 4.3.3.3 show such surfaces have a multiscale structure and yet do not necessarily follow a fractal scaling behavior over all length scales. For such surfaces neither the GW model nor the fractal model can be used. A more generalized analysis is therefore necessary and is developed here for the first time.

In Section 4.4.2 we saw that the size distributions for elastically and plastically deformed asperities are not the same as was assumed in the fractal analysis. By using the size distributions for elastic

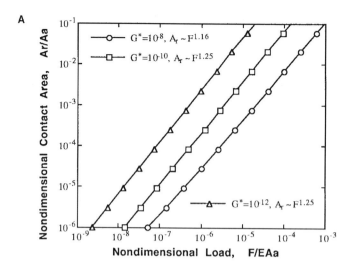

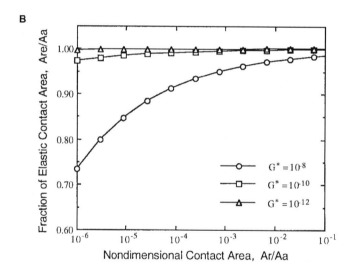

FIGURE 4.31 Effect of varying roughness parameter G on (a) real area of contact (b) fraction of real area of contact in elastic deformation.

(Equation 4.4.10) and plastic (Equation 4.4.20) contact spots, the total load can be found from Equation 4.5.23 to be

$$
F_n = g(d) A_a \left\{ \frac{H}{\pi} \int_0^{a_c} k(a) \left[-\frac{\pi \phi}{2k^3(a)} \frac{dk}{da} + \frac{1}{2} \left(1 + \frac{a}{k(a)} \frac{dk}{da} \right) \right] da \right.
$$

$$
\left. + \frac{4E}{3\pi^{5/2}} \int_{a_c}^{a_L} k^2(a) a^{1/2} \left(1 + \frac{a}{k(a)} \frac{dk}{da} \right) da \right\}
$$

(4.5.25)

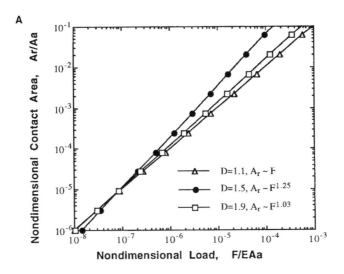

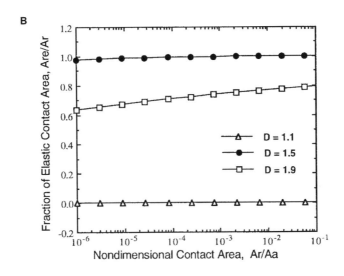

FIGURE 4.32 Effect of varying roughness parameter D on (a) real area of contact (b) fraction of real area of contact in elastic deformation.

We saw earlier in Section 4.4.2 that the largest contact spot area, a_L, is related to the surface mean separation d/σ by Equation 4.4.22. Therefore, Equation 4.5.25 establishes a relation between the total compressive load, F_n, and the separation d/σ. If a Gaussian height distribution is assumed for the magnetic tape, the real area of contact is related to the separation as

$$A_r = \frac{A_a}{2}\operatorname{erfc}\left(\frac{d}{\sqrt{2}\sigma}\right) \qquad (4.5.26)$$

To solve the integrals in Equation 4.5.25 we need the relation between the curvature $k(a)$ and the contact spot area, a. In the fractal model a scaling relation of the form $k(a) = G^{D-1}/a^{D/2}$ is assumed. In contrast, the generalized model does not make any assumption for the curvature $k(a)$ but *directly* uses the experimental roughness data that is processed by the generalized characterization techniques

231

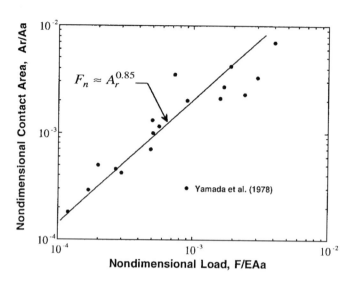

FIGURE 4.33 Comparison between predictions of fractal model and experiments of Yamada et al. (1978).

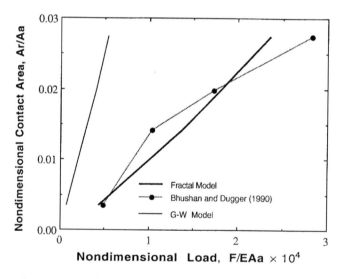

FIGURE 4.34 Comparison among predictions of the fractal model, the GW model, and experiments of Bhushan and Dugger (1990) for contact of a magnetic rigid disk.

described in Section 4.3.4. We demonstrate the application of the generalized analysis by studying the contact between magnetic thin-film rigid disk C and a hard Pyrex glass surface. The hardness of the magnetic film is $H = 6.2$ GPa and the elastic modulus of the equivalent surface is $E = 52$ GPa. Figure 4.16 shows the surface image of disk C, whereas Figure 4.17 shows its structure function. Figure 4.35 plots the surface height, $V(a)$ and the curvature $k(a)$ as a function of spot area, a, for this rough surface. This data can be used in Equation 4.5.25 along with numerical integration of the integrals to obtain the normal load. Figure 4.36 shows the elastic–plastic regime map for the this surface. Using the values of hardness and elastic modulus, the critical plasticity index is equal to $\phi = 0.035$. The plasticity index, $\Psi = V(a)k(a)$ is found to be always less than ϕ, thereby leading to the conclusion that all the asperities deform elastically.

Figure 4.37 compares the results of the generalized theory and the experiments (Bhushan and Dugger, 1990) in terms of the apparent pressure, F_n/A_a and the nondimensional real area of contact, A_r/A_a. Not only are the trends quite similar, the values for the predicted A_r/A_a are reasonably close to those measured.

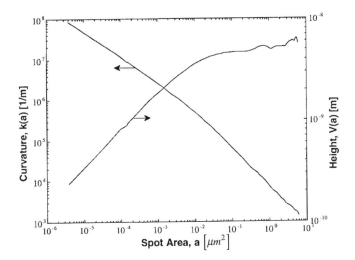

FIGURE 4.35 Plot of surface height, $V(a)$ and curvature $k(a)$ for the magnetic thin-film rigid disk C surface used in the computations of the generalized model.

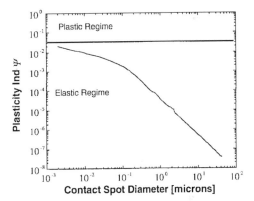

FIGURE 4.36 Elastic–plastic regime of the magnetic thin-film rigid disk surface. The critical plasticity index $\phi = 0.035$.

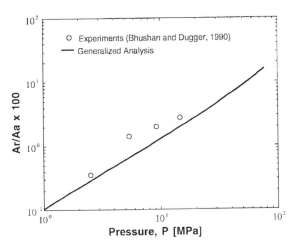

FIGURE 4.37 Comparison of the experimental data (Bhushan and Dugger, 1990) and the predictions of the generalized theory for the magnetic thin-film rigid disk C in terms of compressive apparent pressure, F_n/A_a, and the nondimensional real area of contact, A_r/A_a.

233

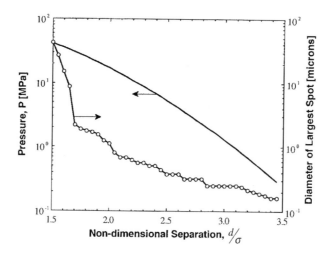

FIGURE 4.38 Plot of the apparent pressure, F_n/A_a, and the diameter of the largest spots size as a function of the nondimensional surface separation, d/σ.

Keeping in mind the variability of surface mechanical properties, particularly for thin films, the agreement is quite good and any better agreement would only be by chance. Figure 4.38 shows the predicted compressive pressure and the diameter of the largest contact spot as a function of the nondimensional separation, d/σ. It is evident that the largest spot size decreases drastically as the separation is increased slightly.

Figure 4.39a shows histograms of the experimentally (Bhushan and Dugger, 1990) obtained size distribution $n(a)\Delta a/A_a$ per 1 mm^2 of the apparent area, A_a, of rigid disk C. The observations were made by an optical microscope. The contact spots are placed into bins of different spot areas and then the numbers are counted for each bin. The two sets of data are for A_r/A_a = 1.64 and 2.51%. The general trend is that as the size of the contact spot decreases, the number of contact spots increases. To obtain the contribution of contact spots of a particular size to the real area of contact, A_r, Figure 4.39b plots the product of $an(a)\Delta a/A_a$ as a percentage. Therefore, this plot shows that when A_r/A_a = 2.51%, contact spots of area 100 μm^2 contribute 1.46%, those of area 67.83 μm^2 contribute 0.35%, and those of 30.91 μm^2 contribute 0.91%, and so on. If the percentages of each bin are added up, it should equal to the real area of contact A_r/A_a = 2.51%. However, an addition of the percentages of individual bins equals 3.33%. Similarly when the real area of contact is claimed to be A_r/A_a = 1.64% the addition shows the fraction to be 4.32%. Therefore, there could be some margin of error in these measurements. Nevertheless, we compared the cumulative size distribution obtained from the experimental data with that of the predictions of the generalized theory in Figure 4.40. The agreement between the theory and experiments is quite poor. It is difficult to say which one is more accurate. The experiments may have flaws as discussed in Figure 4.39b. In addition, contact spots smaller than 1 μm are near the limit of optical diffraction. Therefore, the accuracy in observing and counting spots may be poor in this range. This could perhaps explain the discrepancy for small length scales. With regard to the predictions, they depend on accurate statistics of the surface roughness. The statistics seem to be accurate for small length scales since the data set needed for making roughness averages described in Section 4.3.4 is quite large. In contrast, the roughness statistics may not be accurate for large length scales since the averages needed for accurate roughness characterization are made over very few data points and therefore may not be good representative numbers. However, the optical observations of the contact spots are quite accurate in this size range since this is well above the optical diffraction limit. In summary, the discrepancy for small and large spots could be attributed to measurements and observations. It is encouraging to see that at intermediate length scales, the agreement between theory and experiments is not that terrible.

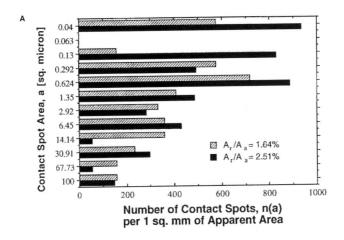

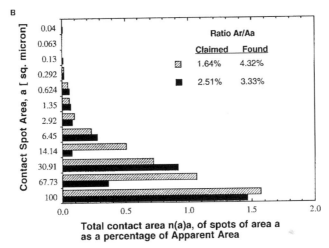

FIGURE 4.39 (a) Histograms of the number of contact spots, $n(a)\Delta a$ found of a particular range of spot areas for two different real contact areas. The numbers $n(a)\Delta a$ area for 1 mm² of apparent area of rigid disk C. (b) Histogram of the total contact area of spots in a certain size range as a percentage of the apparent area, $an(a)\Delta a/A_n$. Note that the percentages of the real area of contact found by adding up the contributions of each bin are not equal to those claimed. (From Bhushan, B. and Dugger, M. T. (1990), *Wear* 137, 41–50. With permission.)

4.5.4 Cantor Set Contact Models

A new approach to modeling contact mechanics between two rough surfaces has recently been explored (Warren et al., 1996; Warren and Krajcinovic, 1996a). This approach models the surface as a *self-affine* Cantor set (see Figure 4.41), such that laterally and vertically the surface is divided further and further by the recursive algorithm

$$\text{Lateral:} \quad L_{n+1} = \frac{L_n}{f_x}$$

$$\text{Vertical:} \quad h_{n+1} = \frac{h_n}{f_z}$$

(4.5.27)

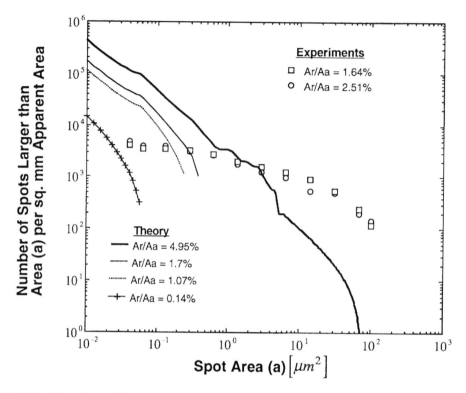

FIGURE 4.40 Comparison of the cumulative size distributions experimentally obtained and theoretically predicted.

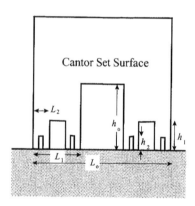

FIGURE 4.41 Illustration of a Cantor set surface constructed by dividing the bottom surface length, L_0, into three segments and removing the middle segment of length $L_0/3$ height, h_0. The two other segments are again broken into three more segments and the middle segment removed. This recursive algorithm can be continued to obtain a fractal surface. (From Warren, T. L. et al. (1996), *J. Appl. Mech.* 63, 47–53. With permission.)

The fractal dimension of the surface obtained through this algorithm was derived to be

$$D = 1 + \frac{\ln(s)}{\ln(sf_x)} - \frac{\ln(f_z)}{\ln(sf_x)} \qquad (4.5.28)$$

where s is the number of remaining asperities after the recursive algorithm breaks up a big asperity into s-smaller ones. For example, the surface in Figure 4.41 uses $s = 2$. The second term in Equation 4.5.28 is the dimension of the Cantor set, $D_c = \ln(s)/\ln(sf_x)$, which breaks up a line into a set of points.

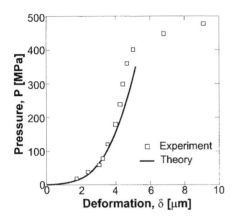

FIGURE 4.42 Comparison between the predictions of the Cantor set model (Warren and Krajcinovic, 1996b) and the experiments (Handzel-Powierza et al., 1992).

Warren et al. (1996) first developed a model of contact between rigid–perfectly plastic Cantor surface with a hard flat surface. They found that the load and the real area of contact depended on the deformation, δ, as follows

$$F \propto \delta^{\alpha}; \quad A_r \propto \delta^{\alpha} \tag{4.5.29}$$

where

$$\alpha = \frac{1 - D_c}{1 + D_c - D} \tag{4.5.30}$$

In a following paper, Warren and Krajcinovic (1996a) considered the contact between *randomized* Cantor surfaces which behaved as elastic–perfectly plastic. They found similar correlations as above. A comparison of their predictions with experimental data of Handzel-Powierza et al. (1992) is shown in Figure 4.42.

The advantage of the generalized Cantor surface models over previous fractal models is that the size and spatial distributions of contact spots is a natural outcome of the recursive algorithm for producing the surface. This can be very useful for studying the interaction of contact spots during sliding for phenomena (Warren and Krajcinovic, 1996b) such as stick-slip as well as static and kinetic friction. However, the models still need to be developed further so as to study realistic surfaces.

4.6 Summary and Future Directions

In this chapter, it was established that surface roughness plays a vital role in microtribology. Rough surfaces are ubiquitous in nature, and in fact it is rare that a surface is perfectly flat. Therefore, any realistic study of tribology must include the effects of surface roughness. Contact between two rough surfaces occurs in the form of discrete spots. It is at these contact spots that all the forces responsible for friction and wear are generated. These spots are not of the same size and are not uniformly distributed over the contact interface. Due to the occurrence of surface roughness over several decades of length scales, ranging from nanometer to millimeter scales, the contact spots have a wide distribution of sizes. Since interfacial forces have been found to be size dependent, it is very important to determine the size distribution of contact spots. In addition, dynamic and force interactions between neighboring spots are also important thus requiring knowledge of the spatial distribution of contact spots.

In order to study the influence of surface roughness on microtribology, it is first necessary to characterize surface roughness in a way that can be easily used to develop models and theories of friction and wear. What is of vital importance is the size and spatial distribution of contact spots. Such distributions are influenced both by roughness and the contact mechanics of surfaces. In this chapter it was shown that conventional methods of roughness characterization using rms values of surface height, slope, and curvature can depend on the resolution of the roughness-measuring instrument and are not unique to a surface. Therefore, theories based on such parameters can produce misleading results. Roughness measurements show that most surfaces are composed of small asperities that sit on larger asperities, which sit on even larger asperities in a hierarchical manner. To characterize this intrinsic multiscale structure of surface roughness, one must develop techniques that are independent of any length scale. The hierarchical structure allows fractal geometry to be used to characterize a surface by two scale-independent parameters — fractal dimension D and amplitude scaling constant G. The latter has units of length. The relation between the fractal parameters (D and G) and the conventional rms quantities were also developed. Roughness measurements also showed that there exist some surfaces that do not follow the scaling hierarchical behavior that fractal characterization demands. A generalized technique was therefore developed which can characterize both fractal and nonfractal surfaces. This technique can be used directly in theories of contact mechanics that finally yield the size distribution of contact spots needed to develop theories of friction and wear.

A new relation for the size distribution of contact spots was developed in terms of surface height probability distribution and asperity curvature for different asperity sizes. This relation is general and can be applied to both fractal and nonfractal surfaces. It was encouraging to find that application of the fractal scaling law for asperity curvature in this general size distribution produced a well-known empirical power law behavior.

Since the size distribution of contact spots depends not only on the roughness but on the contact mechanics, the last section dealt with some contact mechanics theories. As representative of classical theories, the Greenwood–Williamson model was discussed and critiqued. It was shown that the GW model is applicable only when the surface contains a single dominant roughness scale. Machined surfaces with periodic grooves or polycrystalline surfaces with a narrow grain size distribution are typical examples. However, roughness measurements show that many surfaces do not follow this behavior and are best described by fractal scaling laws. A fractal contact theory showed that for elastic contact between surfaces, the real area of contact A_r and the compressive load F_n follow a power law of the form $F_n \approx A_r^\alpha$, where the exponent α varies between 0.75 and 1 depending on the fractal dimension D. Since several surfaces were found to be fractal in certain length regimes and nonfractal in others, there was a need to develop a generalized theory of contact mechanics. This was developed and was based on the generalized roughness characterization technique. To demonstrate the use of this theory, contact of a magnetic thin-film rigid disk was studied. The prediction for real area of contact as a function of compressive load was in close agreement with experimental results. The theory gave the load, real area of contact, and the size of the largest contact spot as a function of the surface separation. In addition, the size distribution of contact spots was also determined as a function of the real area of contact, but was found to be in poor agreement with experimental observations. The reasons for this disagreement were explored.

Despite recent progress in surface roughness characterization and contact mechanics, there still exist several unanswered questions and unaddressed issues. We have assumed that contact spots of isotropic surfaces are circular and those of anisotropic surfaces are elliptical. This is not necessarily true since even for isotropic surfaces, the spots can be elongated in one direction. However, the direction of elongation is random for isotropic surfaces and nonrandom in anisotropic surfaces. The noncircular contact spots for isotropic surfaces is clearly shown in the simulation of Figure 4.3. Therefore, the assumption of circular contact spots is not correct and must be changed. In this chapter we have considered asperity deformation due to a flat hard plane. The problem of asperity indentation into a flat but softer surface has not been addressed.

Although this chapter establishes the importance of size and spatial distributions of contact spots in the development of tribological theories, the issue of spatial distribution has not been addressed at all. The spatial distribution is needed to study stress interactions and for dynamic interaction between contact spots. This is very important for high surface deformation cases and for the transition between static and kinetic friction. The problem lies not in developing a relation for the spatial distribution but developing a method that can be used in a theory or model of neighboring asperity interaction. That itself has not been developed or properly understood. Therefore, the question of neighboring asperity interaction and the spatial distribution is an open problem that needs future attention.

References

Avnir, D. (1989), *The Fractal Approach to Heterogeneous Chemistry*, J. Wiley, New York.

Bak, P., Tang, C., and Wiesenfeld, K. (1988), "Self-Organized Criticality," *Phys. Rev. A*, 38, 364–374.

Barnsley, M. (1988), *Fractals Everywhere*, Academic Press, San Diego.

Beckmann, P. and Spizzichino, A. (1963), *The Scattering of Electromagnetic Waves from Rough Surfaces*, Macmillan, New York.

Berry, M. V. and Hannay, J. H. (1978), "Topography of Random Surfaces," *Nature*, 271, 573

Berry, M. V. (1978), "Diffractals," *J. Phys. A*, 12, 781–797.

Berry, M. V. and Blackwell, T. M. (1981), "Diffractal Echoes," *J. Phys. A Math. Gen.*, 14, 3101–3110.

Bhushan, B. (1984a), "Prediction of Surface Parameters in Magnetic Media," *Wear*, 95, 19–27.

Bhushan, B. (1984b), "Analysis of the Real Area of Contact between a Polymeric Magnetic Medium and a Rigid Surface," *J. Tribol.*, 106, 26–34.

Bhushan, B. (1990), *Tribology and Mechanics of Magnetic Storage Devices*, Springer-Verlag, New York.

Bhushan, B. and Blackman, G. (1991), "Atomic Force Microscopy of Magnetic Rigid Disks and Sliders and Its Applications to Tribology," *Trans. ASME, J. Tribol.*, 113, 452–457.

Bhushan, B. and Doerner, M. F. (1989), "Role of Mechanical Properties and Surface Texture in the Real Area of Contact of Magnetic Rigid Disks," *J. Tribol.*, 111, 452–458.

Bhushan, B. and Dugger, M. T. (1990), "Real Contact Area Measurements on Magnetic Rigid Disks," *Wear*, 137, 41–50.

Bhushan, B., Wyant, J. C., and Meiling, J. (1988), "A New Three-Dimensional Non-Contact Digital Optical Profiler," *Wear*, 122, 301–312.

Blackman, R. B. and Tuckey, J. W. (1958), *The Measurement of Power Spectra*, Dover, New York.

Brown, S. R. and Scholz, C. H. (1985), "Closure of Random Elastic Surfaces in Contact," *J. Geophys. Res.*, 90(B7), 5531–5545.

Carlson, J. M., Langer, J. S., Shaw, B. E., and Tang, C. (1991), "Intrinsic Properties of a Burridge–Knopoff Model of an Earthquake Fault," *Phys. Rev. A*, 44, 884–897.

Chang, W. R., Etsion, I., and Bogy, D. B. (1987), "An Elastic-Plastic Model for the Contact of Rough Surfaces," *J. Tribol.*, 109, 257–263.

Church, E. L. (1988), "Fractal Surface Finish," *Appl. Opt.*, 27, 1518–1526.

Feder, J. (1988), *Fractals*, Plenum Press, New York.

Gagnepain, J. J. (1986), "Fractal Approach to Two-Dimensional and Three-Dimensional Surface Roughness," *Wear*, 109, 119–126.

Ganti, S. and Bhushan, B. (1995), "Generalized Fractal Analysis and Its Applications to Engineering Surfaces," *Wear*, 180, 17–34.

Greenwood, J. A. and Williamson, J. B. P. (1966), "Contact of Nominally Flat Surfaces," *Proc. R. Soc. London*, A 295, 300–319.

Handzel-Powierza, Z., Klimczak, T., and Polijaniuk, A. (1992), "On the Experimental Verification of the Greenwood-Williamson Model for the Contact of Rough Surfaces," *Wear*, 154, 115–124.

Israelachvili, J. N. (1992), *Intermolecular and Surface Forces*, 2nd ed., Academic Press, San Diego.

Israelachvili, J. N., McGuiggan, P. M., and Homola, A. M. (1988), "Dynamic Properties of Molecularly Thin Liquid Films," *Science*, 240, 189–191.

Johnson, K. L. (1985), *Contact Mechanics*, Cambridge University Press, Cambridge.

Jordan, D. L., Hollins, R. C., and Jakeman, E. (1986), "Measurement and Characterization of Multiscale Surfaces," *Wear*, 109, 127–134.

Kardar, M., Parisi, G., and Zhang, Y. C. (1986), "Dynamic Scaling of Growing Interfaces," *Phys. Rev. Lett.*, 56, 889–892.

Knopoff, L. (1990), "The Modeling of Earthquake Occurrence," in *Disorder and Fracture*, (J. C. Charmet, S. Rou, and E. Guyon, eds.), pp. 284–300 Plenum Press, New York.

Kolmogoroff, A. N. (1941), "Dissipation of Energy in the Locally Isotropic Turbulence," *CR* (Doklady) *Acad. Sci. URSS*, 32, 16–18. Also in *Turbulence — Classic Papers in Statistical Theory* (S. K. Friedlander and L. Topper, eds.), pp. 159–161, Interscience Publishers, New York, 1961.

Landman, U., Luedtke, W. D., Burnham, N. A., and Colton, R. J. (1990), "Atomistic Mechanisms and Dynamics of Adhesion, Nanoindentation and Fracture," *Science*, 248, 454–461.

Majumdar, A. and Bhushan, B. (1990), "Role of Fractal Geometry in Roughness Characterization and Contact Mechanics of Surfaces," *ASME J. Tribol.*, 112, 205–216.

Majumdar, A. and Bhushan, B. (1991), "Fractal Model of Elastic-Plastic Contact between Rough Surfaces," *ASME J. Tribol.*, 113, 1–11.

Majumdar, A. and Tien, C. L. (1990), "Fractal Characterization and Simulation of Rough Surfaces," *Wear*, 136, 313–327.

Majumdar, A., Bhushan, B., and Tien, C. L. (1991), "Role of Fractal Geometry in Tribology," *Adv. Inf. Storage Syst.*, 1., 231–266.

Mandelbrot, B. B. (1967), "How Long is the Coast of Britain? Statistical Self-Similarity and Fractional Dimension," *Science*, 155, 636–638.

Mandelbrot, B. B. (1975), "Stochastic Models for the Earth's Relief, the Shape and the Fractal Dimension of the Coastlines, and the Number-Area Rule for Islands," *Proc. Natl. Acad. Sci. U.S.A.*, 72, 3825–3828.

Mandelbrot, B. B. (1982), *The Fractal Geometry of Nature*, W. H. Freeman, New York.

Mandelbrot, B. B. (1985), "Self-Affine Fractals and Fractal Dimension," *Phys. Scr.*, 32, 257–260.

Mate, C. M., McClelland, G. M., Erlandsson, R., and Chiang, S. (1987), "Atomic-Scale Friction of a Tungsten Tip on a Graphite Surface," *Phys. Rev. Lett.*, 59, 1942–1945.

McCool, J. I. (1986), "Comparison of Models for the Contact of Rough Surfaces," *Wear*, 107, 37–60.

McCool, J. I. (1987), "Relating Profile Instrument Measurements to the Functional Performance of Rough Surfaces," *J. Tribol.*, 109, 264–270.

McGuiggan, P., M., Israelachvili, J. N., Gee, M. L., and Homola, A. M. (1989), "Measurements of Static and Dynamic Interactions of Molecularly Thin Liquid Films between Solid Surfaces," in *New Materials Approaches to Tribology: Theory and Applications*, L. E. Pope, L. L. Fehrenbacher, and W. O. Winer, eds), *Materials Research Society Symposium*, 140, 79–88.

Meakin, P. (1987), "Fractal Scaling in Thin Film Condensation and Material Surfaces," *CRC Crit. Rev. Solid State Mater. Sci.*, 13, 143–189.

Nayak, P. R. (1971), "Random Process Model of Rough Surfaces," *ASME J. Lubr. Technol.*, 93, 398–407.

Nayak, P. R. (1973), "Random Process Model of Rough Surfaces in Plastic Contact," *Wear*, 26, 305–333.

Oden, P. I., Majumdar, A., Bhushan, B., Padmanabhan, A., and Graham, J. J. (1992), "AFM Imaging, Roughness Analysis and Contact Mechanics of Magnetic Tape and Head Surfaces," *ASME J. Tribol.*, 114, 666–674.

Papoulis, A. (1965), *Probability, Random Variables and Stochastic Processes*, McGraw Hill, New York.

Peitgen, H. O. and Saupe, D. (1988), *The Science of Fractal Images*, Springer-Verlag, New York.

Press, W. H., Teukolsky, S. A., Vetterling, W. T., and Flannery, B. P. (1992), *Numerical Recipes*, 2nd ed., Cambridge University Press, New York.

Sayles, R. S. and Thomas, T. R. (1978), "Surface Topography as a Nonstationary Random Process," *Nature*, 271, 431–434.

Sornette, A. and Sornette, D. (1989), "Self-Organized Criticality and Earthquakes," *Europhys. Lett.*, 9, 197–202.

Suh, N. P. (1986), *Tribophysics*, pp. 35–41, Prentice-Hall, NJ.

Thomas, T. R. (1982), *Rough Surfaces*, Longman, New York.

Timoshenko, S. and Goodier, J. N. (1970), *Theory of Elasticity*, 3rd ed., McGraw-Hill, New York.

Vicsek, T, (1989), *Fractal Growth Phenomena*, World Scientific, New Jersey.

Voss, R. F. (1988), "Fractals in Nature: From Characterization to Simulation," in *The Science of Fractal Images*, (H. O. Peitgen and D. Saupe, eds.), pp. 21–70, Springer-Verlag, New York.

Warren, T. L. and Krajcinovic, D. (1996a), "Random Cantor Set Models for the Elastic Perfectly Plastic Contact of Rough Surfaces," *Wear*, 196, 1–15.

Warren, T. L. and Krajcinovic, D. (1996b), "A Fractal Model of Static Coefficient of Friction at the Fiber-Matrix Interface," *Composites: Part B*, 27, 421–430.

Warren, T. L., Majumdar, A., and Krajcinovic, D. (1996), "A Fractal Model for the Rigid-Perfectly Plastic Contact of Rough Surfaces," *J. Appl. Mech.*, 63, 47–54.

Williams, E. D. and Bartlet, N. C. (1991), "Thermodynamics of Surface Morphology," *Science*, 251, 393–400.

Yamada, K., Takeda, N., Kagami, J., and Naoi, T. (1978), "Mechanisms of Elastic Contact and Friction between Rough Surfaces," *Wear*, 48, 15–34.

Appendix 4.1 — RMS Values for Multifractal Surfaces

Roughness measurements have shown that structure function does not always follow a single power law over a wide range of length scales, but may follow two or more power laws at different length regimes. This can be mathematically written as

$$P(\omega) = \frac{C_1}{\omega^{(5-2D_1)}} \quad \text{for } \omega > 1/\ell_{12} \quad \text{regime 1}$$

$$= \frac{C_2}{\omega^{(5-2D_2)}} \quad \text{for } \omega < 1/\ell_{12} \quad \text{regime 2}$$

(A4.1.1)

where ℓ_{12} is a transition length scale where the fractal behavior changes. The question is how the rms height is related to the fractal parameters — D_1, D_2, G_1, and G_2. If the length, L, of the measured roughness sample is smaller than ℓ_{12}, that is, $L < \ell_{12}$, then regime 2 is never accessed and the problem can be analyzed as a single fractal behavior, as shown with the help of Equation 4.3.25. If $L > \ell_{12}$, then rms height can be found as

$$\sigma^2 = \int_{1/L}^{1/\ell_{12}} \frac{C_2}{\omega^{(5-2D_2)}} \, d\omega + \int_{1/\ell_{12}}^{\infty} \frac{C_1}{\omega^{(5-2D_1)}} \, d\omega = \frac{C_2 L^{2(2-D_2)}}{2(2-D_2)} + \frac{C_1 \ell_{12}^{2(2-D_1)}}{2(2-D_1)} - \frac{C_2 \ell_{12}^{2(2-D_2)}}{2(2-D_2)} \quad \text{(A4.1.2)}$$

When the length, L, is much larger than the critical length, ℓ_{12}, that is, $L \gg \ell_{12}$, then the first term in Equation A4.1.2 is dominant. In such a case, the rms height, σ, is dominated by large-scale roughness of regime 2 and one can again use the relation of Equation 4.3.25 for a single fractal behavior. Note that in such a case, σ is only related to G_1 and D_2 of regime 2 and not to D_1 and G_1, of regime 1.

If the surface profile follows two fractal regimes as shown in Equation A4.1.1 then the rms slope is equal to

$$\sigma'^2 = \int_{1/L}^{1/\ell_{12}} \frac{\omega^2 C_2}{\omega^{(5-2D_2)}} \, d\omega + \int_{1/L}^{1/\ell_{12}} \frac{\omega^2 C_1}{\omega^{(5-2D_1)}} \, d\omega$$

$$= \frac{C_2}{2(D_2 - 1)} \left[\left(\frac{1}{\ell_{12}}\right)^{2(D_2-1)} - \left(\frac{1}{L}\right)^{2(D_2-1)} \right] + \frac{C_1}{2(D_1 - 1)} \left[\left(\frac{1}{\ell}\right)^{2(D_1-1)} - \left(\frac{1}{\ell_{12}}\right)^{2(D_1-1)} \right]$$

(A4.1.3)

where it is assumed that $L > \ell_{12} > \ell$ such that the sample length, L, and the instrument resolution, ℓ, fall on the different fractal regimes demarcated by the critical length scale, ℓ_{12}. If they fall in the same regime, then the rms slope is obtained by Equation 4.3.27. Often it is found that $L \gg \ell_{12} \gg \ell$, in which case the rms slope varies as

$$\sigma'^2 = \frac{C_1}{2(D_1 - 1)} \left(\frac{1}{\ell}\right)^{2(D_1-1)}$$

(A4.1.4)

Here it is found that the rms slope depends on the fractal parameters G_1 and D_1 of only fractal regime 1. This is in contrast with the rms height which depends on the parameters of fractal regime 2.

Appendix 4.2 — Autocorrelation Function and Fractal Parameters

The structure function, $S(\tau)$, can be broken up and written as

$$S(\tau) = \frac{1}{L} \int_0^L \left[\left(z(x+\tau) - z_m \right) - \left(z(x) - z_m \right) \right]^2 dx$$

$$= \frac{1}{L} \int_0^L \left[\left(z(x+\tau) - z_m \right)^2 + \left(z(x) - z_m \right)^2 - 2 \left(z(x+\tau) - z_m \right) \left(z(x) - z_m \right) \right] dx \quad (A4.2.1)$$

$$= 2 \left[\sigma^2(L) - R(\tau, L) \right]$$

where z_m is the surface mean height and $R(\tau, L)$ is the *autocovariance* function defined as

$$R(\tau, L) = \frac{1}{L} \int_0^L \left(z(x+\tau) - z_m \right) \left(z(x) - z_m \right) dx = \frac{1}{N - (\tau/\Delta x)} \sum_{i=1}^{N-(\tau/\Delta x)} \left(z(x_i + \tau) - z_m \right) \left(z(x_i) - z_m \right) \quad (A4.2.2)$$

It is evident that $R(0,L) = \sigma^2(L)$. Therefore, the autocovariance function is often written as $R(\tau,L) = \sigma^2(L) f(\tau,L)$, where $f(\tau,L)$ is called the *autocorrelation* function and is equal to unity for $\tau = 0$. The autocorrelation function is related to the structure function and the rms height as

$$f(\tau, L) = 1 - \frac{S(\tau)}{2\sigma^2(L)} \quad (A4.2.3)$$

When the rms height, σ, is independent of the sample size, L, then it can be characterized by the structure function model proposed by Berry and Blackwell (1981) given in Equation 4.3.24. By using Equation 4.3.24 in Equation A4.2.3, the autocorrelation function can be obtained to be

$$f(\tau) = \exp\left(-\frac{G^{2(D-1)} \tau^{2(2-D)}}{2\sigma^2} \right) = \exp\left[-\left(\frac{\tau}{\tau_c} \right)^{2(2-D)} \right] \quad (A4.2.4)$$

where the correlation length is $\tau_c = G(2\sigma/G)^{1/(2-D)}$.

It is interesting to note that an approximate autocorrelation function that is often used to model rough surface profiles, for optical scattering by rough surfaces in particular (Beckmann and Spizzichino, 1963; Church, 1988) is the exponential decay function of the form:

$$f(\tau) = \exp\left(-\frac{\tau}{\tau_c} \right) \quad (A4.2.5)$$

where τ_c is called the correlation length. Comparison with Equation A4.2.4 shows that this approximate function corresponds to a specific case of $D = 1.5$ — a Brownian fractal. This can also be verified another way. The autocovariance function, $\sigma^2 f(\tau)$, and the power spectrum are Fourier transforms of each other such the power spectrum can be written as

$$P(\omega) = \int_{-\infty}^{\infty} \sigma^2 \, f(\tau) \exp(-i\omega\tau) \, d\tau \qquad (A4.2.6)$$

By using this, the power spectrum corresponding to the autocorrelation function of Equation A4.2.5 is found to be

$$P(\omega) = \frac{4\sigma^2 \tau_c}{1 + (2\pi\tau_c\omega)^2} \qquad (A4.2.7)$$

Note that for high spatial frequency, $\omega \gg \frac{1}{2}\pi\tau_c$, this spectrum follows the power law behavior of $P(\omega) \approx \omega^{-2}$. When compared to the fractal power spectrum, $P(\omega) \approx \omega^{-(5-2D)}$, this corresponds to the Brownian fractal, $D = 1.5$.

Appendix 4.3

Consider a reference plane such that the height of surface 1 is $z_1(x)$ and that of surface 2 is $z_2(x)$ as shown in Figure A4.3.1. The surface means are separated by a distance d. One can construct an equivalent surface $z(x) = z_1(x) - z_2(x)$. The mean height of the equivalent surface is located at a vertical position $\langle z \rangle = \langle z_1 \rangle - \langle z_2 \rangle = -d$ from a flat hard plane at $z = 0$. This hard plane is the contact plane since contact is made between the two surfaces at any position x when $z(x) = z_1(x) - z_2(x) \geq 0$.

Normalized by the surface means, the equivalent surface is related to surfaces 1 and 2 as follows:

$$z(x) + d = \left(z_1(x) - \langle z_1 \rangle \right) - \left(z_2(x) - \langle z_2 \rangle \right) \qquad (A4.3.1)$$

The variance of the equivalent surface can be found as

$$\begin{aligned}
\sigma^2 &= \frac{1}{L} \int_0^L \left(z(x) + d \right)^2 dx \\
&= \frac{1}{L} \int_0^L \left[\left(z_1(x) - \langle z_1 \rangle \right)^2 + \left(z_2(x) - \langle z_2 \rangle \right)^2 \right] - \frac{2}{L} \int_0^L \left(z_1(x) - \langle z_1 \rangle \right)\left(z_2(x) - \langle z_2 \rangle \right) dx
\end{aligned} \qquad (A4.3.2)$$

If the surfaces are statistically uncorrelated, then the last integral in Equation 4.5.2 is equal to zero. Then the variances of the equivalent surface is equal to the sum of the variances of the two surfaces as $\sigma^2 = \sigma_1^2 + \sigma_2^2$. Similarly, for uncorrelated surfaces, the structure function and the power spectrum of the equivalent surface can be written as

$$\begin{aligned}
S(\tau) &= S_1(\tau) + S_2(\tau) \\
P(\omega) &= P_1(\omega) + P_2(\omega)
\end{aligned} \qquad (A4.3.3)$$

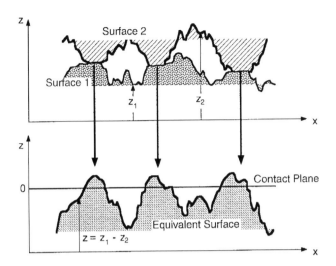

FIGURE A4.3.1   Static contact between two surfaces can be modeled as contact between an equivalent surface and a flat hard plane.

5

Surface Forces
and Adhesion

Nancy A. Burnham and Andrzej J. Kulik

5.1 Introduction

5.1.1 Goals and Motivations

Small bits of eraser cling to your homework assignments, yet the eraser itself easily slides off a piece of paper. Why? What is different about the interaction of small things with a surface as opposed to big things? **The goal of this chapter is to familiarize you, on a conceptual basis, with the forces acting between asperities, or between an asperity and a flat surface.**

Asperity behavior is thought to determine the most famous relationship in macrotribology — Amonton's law $F_f = \mu N$. As the normal load N is increased, the frictional force F_f also increases, with a constant proportionality factor μ, the coefficient of friction. Remembering that even "atomically flat" surfaces have finite corrugation, and that most surfaces exhibit roughnesses well in excess of atomic dimensions, increasing the normal load causes more asperities to touch, which as a consequence augments the real area of contact between the two bodies.

In nanotribology, where one considers the interaction of a single asperity contact, the researcher has the luxury of studying an ideologically far simpler system. Not only is this an area of intensive research in materials science and physics, but also its major applications area — microelectromechanical systems, where moving parts touch only at one or few point contacts — has a commercially lucrative future. One of the fascinations with nanotribology is accurately expressed by the example of the eraser above. **The behavior of small things is different from the behavior of big ones.** Let us perform a dimensional analysis for the case of an eraser and its residue.

A particle of eraser residue may be roughly spherical, with a radius R of the order of 100 μm, whereas the eraser itself may have dimensions in the range of 1 cm. The surface-to-volume ratio for spheres equals $4\pi R^2/{}^4/_3\pi R^3 = 3/R$, which means that the residue will have a surface-to-volume ratio 100 times greater than for the eraser. The properties of the surface and near-surface region are important for small particles, as will be emphasized in Sections 5.3 and 5.4. The weight of a sphere of density ρ is ${}^4/_3\rho\pi R^3$ and its attraction to a flat piece of paper is $2\pi R\varpi$, where ϖ is the work of adhesion (Sections 5.3.1 and 5.4.2). Therefore, the ratio between the surface forces and the weight for a spherical particle near a flat surface is $3\varpi/2\rho R^2$. The value of this ratio for our residual particle is 10,000 times larger than that for the eraser, and we might predict that the residue will cling to the paper if the value is greater than one. As long as ϖ is nonzero (the usual case), there is always an R at which surface forces are stronger than gravity. In summary, **surface forces predominate at small enough scales.**

5.1.2   Surface Forces vs. Adhesion

Throughout this chapter, we shall distinguish between the forces that are present when two bodies are brought together (*surface forces*) and those that work to hold two bodies in contact (*adhesive forces* or *adhesion*). Other authors have differentiated them by using the nomenclature *advancing/receding* or *loading/unloading*. Surface forces are in general attractive, but under some conditions can be repulsive. Adhesive forces, as the name implies, tend to hold two bodies together. If a process between two bodies is perfectly elastic, that is, if no energy dissipates during their interaction, the adhesive and surface forces are equal in magnitude. Normally, however, the adhesion is greater than any initial attraction, giving rise to *adhesion hysteresis*. Why this is so is one of the subjects of Section 5.4.

5.1.3   Previous Knowledge Assumed

In this chapter, the assumption is that the reader is already familiar with first-year college physics, chemistry, and calculus, and Chapter 2 of this book. We draw broadly from a variety of existing texts and conference proceedings listed at the end of this chapter, wherein many detailed references are given. We concentrate on the surface forces and adhesion that act between an asperity and a flat surface, because this is a configuration likely to occur in microelectromechanical systems, and is the most common situation in scanning probe microscopy studies which are used to probe materials properties with nanometer-scale lateral resolution.

5.1.4   Carte Routière

To aid the reader, important concepts are emphasized by **boldface** type, and significant terminology by *italics*. This chapter is intended to be complementary to Chapter 9, "Surface forces and microrheology of molecularly thin liquid films." Here, we first cover some aspects of instrumentation that may not be discussed in other parts of this textbook, then subsequent sections elaborate surface forces, adhesion, and the interpretation of experimental data, before a final summary.

5.2   Pertinent Instrumental Background

5.2.1   The Instrument Family

The correct usage of scanning probe microscopes (SPMs) to study surface forces and adhesion shall be the focus of this section. Chapter 2 details the instrumentation of atomic force microscopes (AFMs), one

TABLE 5.1 Tabular Comparison of SFA, Indentor, SAM, and SPM

	Surface Forces and Adhesion	Mechanical Properties	Imaging	Lateral Resolution
SFA	√	—	—	—
Indentor	—	√	—	~1 μm
SAM	—	√	√	~1 μm
SPM	√	√	√	~1 nm

Checks mean that the instrument was designed to measure surface forces and adhesion, or mechanical properties, or is optimized for imaging.

of the many varieties of SPMs. The researcher should bear in mind that SPMs have many features in common with other instruments, notably the surface force apparatus (SFA), the indentor, and the scanning acoustic microscope (SAM). The overlap extends from the materials properties desired, to how force and displacement are controlled and measured, to calibration procedures, to the ease with which imaging is performed. It can be seen from Table 5.1 that SPMs are capable of measuring surface forces and adhesion, of determining mechanical properties such as elasticity and hardness, and are optimized for imaging surfaces. **Overly enthusiastic readers must be chided into remembering that an instrument optimized for imaging is not necessarily the best for surface forces, adhesion, or mechanical properties measurements.** Issues concerning SPM usage for all materials properties measurements are found on pp. 421–454 in Bhushan (1997) and references therein.

Scanning probe microscopy will excel in applications where changes in materials properties vary over scales less than a micron, for example, in new composite materials, or across a cell membrane. So although imaging to capture the lateral variations in properties is ultimately desired, we restrict our discussion to the SPM mode of operation most closely related to SFA and the indentor — that of force curve acquisition using an AFM.

5.2.2 What Are You Measuring?

Care must be taken to avoid artifacts and to calibrate the instrument properly. Then the researcher still must avoid the trap of measuring a property of the instrument, rather than of the sample or its interaction with the tip. One can model the AFM–sample system as two springs and two dashpots in series. The springs symbolize energy storage or stiffness, and the dashpots symbolize energy dissipation. One spring and dashpot combination represents the AFM cantilever and the other the tip–sample interaction, as represented in Figure 5.1. Normally, a cantilever with an effective spring constant not too different from that of a Slinky™— approximately 1 N/m — is used. The spring stiffness representing the interaction between the tip and the sample is usually significantly larger. **Thus, during force curve acquisition, as the sample and cantilever are first brought together and then separated, the weaker spring (the cantilever) suffers most of the deflection, and the properties of the stiffer spring (the interaction of the sample with the tip) are not observable.** This is an important concept, one that deserves further development.

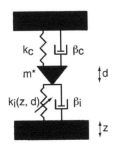

FIGURE 5.1 Schematic diagram of the cantilever and its interaction with the sample. The cantilever moves a distance of d in response to the movement of the sample z. The dashpots β_c and β_i symbolize energy dissipation. Energy storage is represented by the springs with stiffnesses k_c and $k_i(z, d)$ (cantilever and interaction, respectively). Note that $k_i(z, d)$ may vary. It depends on the position of the tip relative to the sample. For example, when the tip is far from the sample, the stiffness is zero, but when the tip is indented into a sample, the stiffness is high.

249

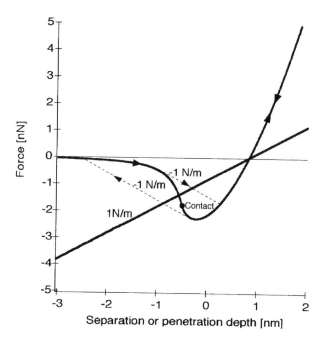

FIGURE 5.2 A typical force–distance curve (curved line) and the force applied by the cantilever (straight line). The force is plotted as a function of tip–sample separation or penetration depth. At the point labeled "Contact" (the inflection of the curve), the tip touches the sample. The dashed lines represent the path that a 1 N/m cantilever would follow if it were used for data acquisition, i.e., it jumps over sections of the curve.

5.2.2.1 Cantilever Instabilities and Mechanical Hysteresis

Figure 5.2 displays a typical theoretical force curve (curved line) of an elastic tip–sample interaction, with force plotted as a function of the separation between tip and sample, or as the penetration of the tip into the sample. The curve has an attractive region near contact, a repulsive region when the tip is firmly indented into the sample, and no interaction when the tip is far removed from the sample. The origin has been placed such that the area below (above) the x-axis represents a net attractive (repulsive) tip–sample force, and the negative (positive) values of separation imply that the tip is separated from (indented into) the sample. Hence, the lower-left quadrant corresponds roughly to the forces and separations investigated with a surface forces apparatus, and the upper-right quadrant to penetration, or indentation, experiments. Note that in this example, contact, indicated by the appropriately labeled point in the lower-left quadrant, commences at negative values of separation. As in human relationships, attraction causes two bodies to reach out and touch each other.

The straight line in Figure 5.2 represents the spring constant k_c of the cantilever. In this example, k_c = 1 N/m. The force is determined using Hooke's law $F = -k_c d$, where d is the deflection of the cantilever. The tip of the cantilever will find an equilibrium position such that the cantilever restoring force balances that of the tip–sample interaction. Summed, the two curves in Figure 5.2 become the plot of Figure 5.3, in which a region that is triple valued in force exists. Should the tip be approaching the sample (left-to-right on the graph), it is accelerated over the triple-valued region, following the upper dashed line to the right, until it finds the new equilibrium position at the same force magnitude at the right-hand termination of the upper dashed line. Similarly, if the scanner is withdrawn such that the tip moves right-to-left on the graph, the tip follows the thick line until the triple-valued region, where once again the cantilever restoring forces do not balance those of the interaction, and the tip finds its new steady-state position at the left-hand end of the lower dashed line.

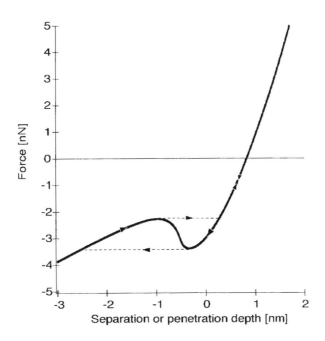

FIGURE 5.3 The two curves in Figure 5.2 are summed to obtain the total force for the tip–sample system. As the separation or penetration depth is changed, the force increases or decreases in a nonmonotonic fashion, such that there exist three points with the same force value in the region between the dashed lines. The tip does not always follow the thick line, but rather it follows the dashed lines over the triple-valued region — the upper one upon loading and the lower one upon unloading.

These are examples of *cantilever instabilities,* giving rise to *mechanical hysteresis* in the force curve. **The weaker the cantilever, the larger is the region of triple-valued force and the greater is the mechanical hysteresis.** Indeed, for a hypothetical cantilever of zero stiffness, the triple-valued region corresponds to everything below the x-axis of Figure 5.2. For a sufficiently stiff (high k_c) cantilever, the shape of the curve in Figure 5.3 would become single valued everywhere, and no instabilities due to the compliance of the cantilever could occur. Cantilever instabilities are also referred to as jump-to-contact or snap-in events, and they are often equivalently graphically (but perhaps more confusingly) explained using the dashed lines labeled –1 N/m in Figure 5.2.

Mathematically, the reason for the instabilities can be seen from the following equations. The first is the simple harmonic oscillator expression for the cantilever, set equal to the forces acting upon it by the tip–sample interaction.

$$m^\star \ddot{d} + 2m^\star \beta_c \dot{d} + k_c d = k_i\left(z,d\right)\left[z-d\right] + 2m^\star \beta_i\left[\dot{z}-\dot{d}\right]. \qquad (5.1)$$

The cantilever is assumed to act as a spring, with an effective mass of $m^\star$. Its position is represented by d, its damping by β_c, and its spring constant by k_c. The distance $[z-d]$ represents the separation between tip and sample, or the indentation of the tip into the sample. The tip–sample interaction has damping β_i, and its interaction stiffness can be written as $k_i(z, d)$. It should be emphasized that the stiffness, graphically the slope at a given point on a properly presented force–distance curve, is a function of the separation $[z-d]$, and may be positive, zero, or negative. If we for the moment suppose that damping is negligible, the above can be rewritten as

$$m^\star \ddot{d} + \left[k_c + k_i\left(z, d\right)\right] d = k_i\left(z, d\right) z. \qquad (5.2)$$

251

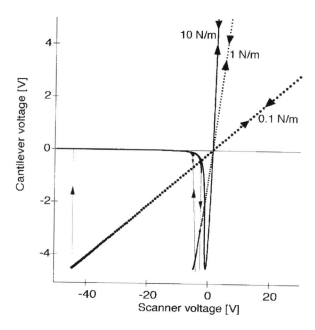

FIGURE 5.4 How the force curve of Figure 5.2 would appear when measured by cantilevers of 0.1, 1.0, and 10 N/m stiffnesses. The axes are labeled in terms of the data before conversion into distance and force, that is, the voltage driving the scanner, and the voltage corresponding to the cantilever position. The thin lines indicate the cantilever instabilities, and the arrows the motion of the tip.

During quasi-static (slow enough such that equilibrium conditions apply) data acquisition, the acceleration term $m^*\ddot{d}$ will be equal to zero most of the time because the cantilever can find some position d that satisfies the requirement $d = k_i(z, d) \, z / \, [k_c + k_i(z, d)]$. But when $k_i(z, d)$ is the negative of the cantilever spring constant k_c, the $[k_c + k_i(z, d)]d$ term in Equation 5.2 equals zero, and the cantilever is accelerated by the force $k_i(z, d) \, z$.

5.2.2.2 Measured and Processed Force Curve Data

The raw data as recorded by an AFM with different cantilever stiffnesses appear as in Figure 5.4. The voltage corresponding to the cantilever deflection is plotted as a function of the scanner position voltage. We use the terminology *force curve* to embrace both *force-separation* or *penetration depth curves* (the theoretical or processed data), as in Figure 5.2, and *force-scanner position curves* (the measured data), as in Figure 5.4. The weaker the cantilever, the greater the mechanical hysteresis, and the more linear the cantilever response upon contact with the sample. The raw data reflect neither the actual tip–sample separation, nor the penetration of the tip into the sample. For this one must subtract the cantilever position from the x-axis of the raw data, in order to obtain curves such as those shown in Figure 5.5.

There are two striking features of Figure 5.5. One is that much of the 0.1 N/m curve has an infinite slope and appears linear. The other is that many data points do not exist for the 0.1 N/m curve, fewer for the 1.0 N/m curve, and none for the 10 N/m curve. **The missing data correspond to those points omitted because of cantilever instabilities. The infinite slope of the linear curve indicates that the microscope was operated outside of its detection limits.**

Two factors limit detectability for the case of the 0.1 N/m cantilever. The first is the noise of the system. Figure 5.4 has ±10 pm noise added to both x- and y-coordinates — an amount hardly discernible in that figure. The compliant cantilever deflects almost as much as the scanner moves. For the processed data of Figure 5.5, one of these two very close values has been subtracted from the other, and the noise becomes significant, obscuring the nonlinearity of the data. The second limiting factor is that weak cantilevers are often used to calibrate the detection system response of the microscope. It is assumed that a compliant cantilever moves as much as the scanner does. If the scanner is calibrated, then by placing a weak cantilever

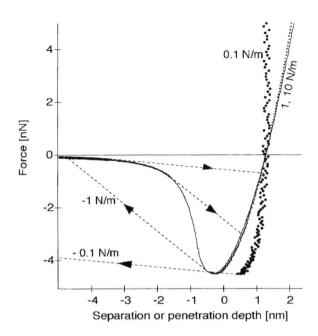

FIGURE 5.5 How the data of Figure 5.4 would appear after processing. The cantilever voltage is converted to force, and the cantilever position is subtracted from the scanner position in order to arrive at the separation or penetration depth. The solid line represents the data as taken with a 10 N/m cantilever, the small squares with a 1 N/m cantilever, and the dots a 0.1 N/m cantilever. The dashed lines show the paths taken by weaker cantilevers. Comparing with the original force interaction of Figure 5.2, one can see that the stiffer cantilevers can reproduce the original data well. The shape of the force curve is lost with the compliant cantilever. In general, if the slope of the measured force curve is linear, the cantilever is too weak for all but a pull-off force measurement.

in contact with a rigid sample, the detection system response voltage is taken to be equal to the scanner movement. In fact, for this example, the 0.1 N/m cantilever did not constitute 100% of the compression, but rather 98.3%. This small error of 1.7% leads to the infinite slope in Figure 5.5.

5.2.2.3 Where's the Beef?

As mentioned above, AFMs have been designed and optimized for imaging sample surfaces, and often employ a compliant cantilever to enhance force resolution and to avoid compressing the sample surface, which distorts topographic features and may permanently disfigure them. Therefore, there has been until recently (pp. 421–454 in Bhushan, 1997) a dilemma between using stiff cantilevers for materials properties measurements and compliant cantilevers for imaging surface topography.

Figures 5.2 through 5.5 emphasize the intractable nature of the force curve measurement process—the measurement sensor, i.e., the cantilever—is influencing your results. The instabilities and mechanical hysteresis caused by the weakness of the cantilever in comparison with the tip–sample interaction lead to loss of important data and insensitivity to exactly what you would like to observe. Under no circumstances should the mechanical hysteresis of the entire cantilever be confused with the possible and inherently more interesting adhesion hysteresis of the tip–sample contact (Section 5.4). Nor should the presence of instabilities be automatically associated with the existence of water layers on surfaces under ambient conditions. Any attractive interaction gives rise to instabilities and hysteresis if the cantilever is sufficiently compliant.

The oft-found linearity in the raw force-scanner position curves usually implies that the only thing you are recording is the stiffness of the cantilever itself. **Indeed, the only tip–sample interaction characteristic that can be readily measured with a weak cantilever using force curve acquisition is the pull-off force** — the maximum adhesive force during retraction of the scanner. (Observe that the most negative values of the curves in Figure 5.5 are almost the same.) With a well-calibrated instrument and a sufficiently stiff cantilever, it is possible to determine the operative surface forces and adhesion, as well as the elastic and plastic response of the sample. With a weak cantilever, there's hardly any beef.

5.2.3 Probe Geometry

Cantilever tips come in a variety of shapes and sizes. (See Chapter 2.) **Because the magnitude and functional dependence of surface forces often depend on the shape of the tip, it is important to calibrate the tip.** Although surface forces can be calculated numerically for any given tip–sample geometry, analytical modeling calls for tips and samples of easily defined form: spherical, hyperbolic, parabolic, cone-shaped, or a flat-ended punch against a flat sample. The assumption of a spherically shaped tip end and a flat sample will be used in this chapter. Let the reader beware that if the range of force interaction extends beyond the spherical part of the tip, or if the sample is very rough, this assumption will no longer be valid. Another important assumption that may or may not hold in a given experiment is that the distance over which the forces act is much less than the tip radius. Nevertheless, the sphere–flat geometry is illustrative.

5.3 Surface Forces

After stating the Derjaguin approximation, four broad classes of long-ranged surface forces will be presented: electrostatic, electrodynamic, electromagnetic, and liquid forces. Because of the breadth and depth of this subject, this section is necessarily written in summary form; for full details, consult Israelachvili (1992). Users new to SPM should become familiar with the functional dependencies of the force interactions (Figures 5.6 through 5.8), and their typical relative strengths, as well as be exposed to the wide variety of possible sources of the forces.

Short-ranged forces, that is, those due to chemical or metallic bonding, will not be discussed in detail here, although they can greatly change the overall measured attractive and adhesive interactions. One layer of a nonmetallic film can completely destroy the welding, or junction formation, that would normally occur between clean metal tips and samples.

5.3.1 The Derjaguin Approximation

The *Derjaguin approximation* is a useful method by which to arrive at a force law for the sphere–flat geometry. It states that **if the interaction energy per unit area, ϖ, as a function of separation for two semi-infinite parallel planes is known, then the force law for a sphere near a flat surface becomes**

$$F\big(\delta\big)_{\text{sphere–flat}} = 2\pi R \varpi \big(\delta\big)_{\text{planes}}. \tag{5.3}$$

The assumptions used to obtain this expression are that both the range of the forces and the separation δ are much shorter than the tip radius. In some scanning probe microscope geometries, these assumptions may not hold.

5.3.2 Electrostatic Forces

Electrostatic forces include those due to charges, image charges, and dipoles. Electric fields polarize molecules and atoms, so that there exist forces that act between the electric field and the polarized object. Electric fields can purposefully be applied between tip and sample, or may exist due to differences in the work function between them. Moreover, electric fields may surround the tip and/or sample due to variations in the work function over their surfaces.

5.3.2.1 Charges and Image Charges

The expression for the force between two charges should be well known to you. The force F is equal to the charges multiplied together, divided by a proportionality constant and the square of the distance between them, r. The proportionality factor is $1/4 \pi \varepsilon \varepsilon_0 = 8.99 \times 10^9$ Nm^2/C, where ε_0 is the permittivity constant factor which has the value 8.85×10^{-12} C^2/Nm^2, and ε is the relative permittivity for the medium across which the force acts.

$$F = \frac{q_1 q_2}{4 \pi \varepsilon \varepsilon_0 r^2}. \tag{5.4}$$

If, as an illustration, we set $q_1 = q_2 = 1.6 \times 10^{-19}$ C, the charge of one electron, and solve for the force when two electrons are an atomic distance 0.2 nm apart, we find that the resulting force is about 6 nN. This is a magnitude that can be detected with most AFMs. One must also remember that free charges induce surface charge on nearby surfaces that acts as an image charge buried within the material. Image charges always carry the opposite sign of the original charge. In this case, the force relationship becomes

$$F = \frac{-q_1 q_2}{4 \pi \varepsilon_m \varepsilon_0 r^2} \left(\frac{\varepsilon_s - \varepsilon_m}{\varepsilon_s + \varepsilon_m} \right), \tag{5.5}$$

with ε_m and ε_s representing the permittivities of the medium and sample, respectively. For metals, where the permittivities are infinite, the term in parentheses approaches one. Because of the high permittivity of liquids, if the system is immersed in a liquid, the force is drastically reduced and can even be repulsive (i.e., positive), depending on the relative values of ε_m and ε_s.

5.3.2.2 Dipoles

Molecules can have positive and negative ends to them, due to one atom or atomic group having a stronger electronegativity than the others. Dipoles have associated electric fields, and these electric fields interact with other charges and dipoles. The magnitude of a dipole moment of a molecule or an atomic bond is $p = q\ell$, where charges $\pm q$ lie a distance ℓ apart. Ubiquitous water's dipole moment is 6.18×10^{-30} Cm, which is a modestly high value, exceeded in general only by strongly ionic pairs such as NaCl.

It is interesting that the interaction potential of a dipole with a charge, another dipole, or a polarizable atom or molecule is related to whether the dipole is fixed or free to rotate. The functional dependence upon distance changes. Although the force between two dipoles can be quite weak, collective effects may be large enough to be measurable using SPM techniques. Typically, one integrates the force or potential over the volumes where the charges, molecules, or dipoles are located. Once again, this changes the functional dependence of the force law. For example, the interaction potential between a fixed dipole and a polar molecule is proportional to $1/r^6$. If the fixed dipole finds itself in front of a semi-infinite half-space of polar molecules (a flat sample), the interaction potential is then a function of $1/r^3$. The derivations may be found in Israelachvili (1992).

5.3.2.3 Polarizability

All atoms and molecules are polarizable. The effect originates from the charged nature of atoms. In an electric field, the positively charged nucleus moves slightly in the direction of the field, and the electrons against it, until the force exerted by the electric field is balanced by the internal restoring forces of the atom or molecule. This is similar to the dipole moment $p = q\ell$, but it is an induced, rather than permanent, dipole moment. The relation among the induced dipole moment μ, the electric field E, and the polarizability α is simply $\mu = \alpha E$. Polarizabilities are of the order of 10^{-40} C^{2}m^2/J. Because electric field strengths and functional dependencies on distance depend on whether the source of the field is a dipole or charge, the interaction potentials between two individual atoms or molecules exhibit either $1/r^4$ or $1/r^6$ proportionalities.

5.3.2.4 Applied Electrostatic Fields

An easy way in which the experimentalist can actively control an SPM measurement is to apply a voltage between the tip and sample, forming a capacitor between them. The energy stored in such an electric field is equal to $W = -\frac{1}{2}CV^2$, where C is the tip–sample capacitance and V the applied voltage. The capacitance of two parallel plates separated by a distance δ is $\varepsilon_0 A/\delta$, where A is the area of the plates. Therefore, the work per unit area for two planes is $\varpi(\delta)_{\text{planes}} = -\frac{1}{2}\varepsilon_0 V/\delta$, and by using the Derjaquin approximation above (Equation 5.3), we arrive at

$$F\big(\delta\big)_{\text{sphere-flat}} = -\frac{\varepsilon_0 \pi R V^2}{\delta}.$$ (5.6)

The presence of a dielectric material with dielectric constant ε of thickness $b/2$ upon each electrode (tip and sample) will modify the equation to

$$F\big(\delta\big)_{\text{sphere-flat}} = -\frac{\varepsilon\varepsilon_0 \pi R V^2}{\big[\varepsilon\delta + b\big(1-\varepsilon\big)\big]}.$$ (5.7)

Try taking the limits $b \to 0$ and $b \to \delta$ in order to check this latter result with the former one. A typical value for the force at contact ($\delta \approx 0.2$ nm) with an applied voltage of 10 V and tip radius of 10 nm is -14 nN. The functional dependence on δ for a capacitive interaction is plotted in Figure 5.6.

5.3.2.5 Innate Electrostatic Fields

If there is no voltage applied between two conductors that form a capacitor, there may still be an electrostatic field between them. If the work functions Φ_i (nominally, the potential difference between the Fermi level and the vacuum level) of the two materials are not equal, when electrically connected the Fermi levels equilibrate, and the voltage difference is $\Phi_1 - \Phi_2$. The value can be of the order of a volt, giving rise to a nonnegligible force. An offset voltage may be applied to compensate this *contact potential difference*.

Even when the contact potential difference has been removed, the system may continue to be affected by innate electrostatic fields due to *work function anisotrophies*. The work function is very sensitive to perturbations at a surface. Surface preparation, uneven distribution of adsorbates, crystallographic orientation, and the presence of surface steps, hillocks, pits, or defects can all influence the work function and make its value change with position on the surface.

Let us now perform a short gedanken experiment. We assume a metallic material, the surface of which incorporates a patch of adsorbates having work function Φ_A and the rest of the surface, bare, having work function Φ_B. We know that energy must be conserved, and that everywhere inside the metal, the electron has a mean energy equal to the Fermi level. An electron taken out of the material via the patch and put back into it through the bare surface will not conserve energy ($\Phi_A - \Phi_B \neq 0$) unless there exist electric fields external to the material. The electric fields emanate from the adsorbate patches, which can be modeled as dipole sheets. The fields are strong enough to affect electron trajectories in field emission microscopy, and they induce surface charge in the sample. The resulting tip–sample forces are known as *patch charge forces*.

The calculation of patch charge forces is complex, first, because of the nontrivial nature of the electric field associated with the dipole sheet, and second because the distribution of image charges induced in nearby bodies is heavily dependent on their geometry. Nonetheless, the resulting force is notable not only in that it can initially be repulsive, as in Figure 5.6, or attractive, but also because induced image charges are always of the opposite sign, and therefore the force always turns attractive as the distance between tip and sample is reduced (Section 5.3.2.1). The normal component of the electric field E_z along the central axis of a dipole disk with dipole moment per unit area μ and diameter ρ is

$$E_z = \pm \frac{\mu}{2\varepsilon_0}\left[\frac{1}{\sqrt{\rho^2 + z^2}} - \frac{z^2}{\big(\rho^2 + z^2\big)^{3/2}}\right].$$ (5.8)

256

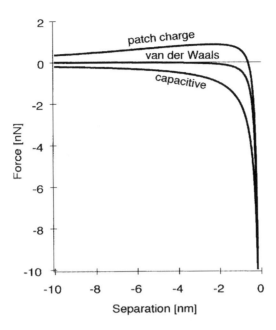

FIGURE 5.6 General behavior of electrostatic and electrodynamic forces as a function of tip–sample separation δ. Capacitive forces approximately follow a $1/\delta$ force, with van der Waals forces following $1/\delta^2$. Patch charge forces can first be repulsive before becoming attractive.

In the limits of either an infinitely large disk or at an infinite distance from the patch, $E_z \rightarrow 0$. The magnitude of the electric field is maximum at about $\rho \approx z$. If we assume that there is one extra electron on the tip, then the force felt by that electron 10 nm from a patch on the sample of diameter $\rho = 10$ nm and dipole moment per unit area $\mu = 1.6 \times 10^{-8}$ C/m will be $F_z = qE_z = \pm 5$ nN, and its image force will be -2 pN. As the tip is lowered onto the sample, the image force grows and the charge-patch field force will drop. There will also be image charges built up in the tip from the patch field, which is attractive to the patch.

5.3.3   Electrodynamic Forces

5.3.3.1   The Dispersion Force

Now we need to consider oscillating electrons. The position of an electron about the nucleus of an atom is not fixed with time; it oscillates, generating a fluctuating dipole field. The field interacts with nearby atoms, inducing the appropriate instantaneous dipole moments in them that are always attractive. This is known as the *dispersion force* because the frequencies correspond to those of visible and ultraviolet light, which the fluctuations disperse. The London equation gives the dispersion potential $W(r)$ between two nonpolar molecules a distance r apart:

$$W\left(r\right) = -\frac{3h\nu\alpha^{2}}{4\left(4\pi\varepsilon_{0}\right)^{2}r^{6}} . \tag{5.9}$$

The electrons orbit the nuclei at a frequency ν, h is the Planck constant, and α is the polarizability of the atom. **The dispersion force acts between all materials, because all atoms have fluctuating electrons.** For two atoms with $\alpha/4\pi\varepsilon_0 = 10^{-30}$ m³, $h\nu = 10^{-18}$ J, and interatomic distance of 2×10^{-10} m, $W = -1.2 \times 10^{-20}$ J. This is three times more energy than the thermal energy kT at 300 K, so the dispersion force may not be ignored at room temperature. The London constant $3h\nu\alpha^2/4 \ (4\pi\varepsilon_0)^2$ for Ar–Ar is approximately 50×10^{-79} Jm^6, whereas for CCl_4–CCl_4 it is about 1500×10^{-79} Jm^6, the difference being largely due to the sixfold change in polarizability of the molecular pairs.

The electronic motion is correlated at visible and ultraviolet frequencies, which are of the order of 10^{15} Hz. During one period of the fluctuation frequency, the electromagnetic field travels 300 nm. Correlated electron motion does not occur if upon leaving one atom, the field travels to another atom and returns to the original one to find that the motion is out of phase. Therefore, at distances greater than approximately 100 nm, the dispersion potential for two atoms drops off at a rate of $1/r^7$, instead of $1/r^6$. This is known as the *retardation effect*. Integrated over the volume of a flat surface and an approaching spherical tip, the distance dependence for nonretarded dispersion forces become $1/\delta^2$; that for retarded dispersion forces is $1/\delta^3$.

5.3.3.2   van der Waals Forces

There is not just one van der Waals force, but rather there are van der Waals forces. The terminology *van der Waals forces* encompasses three forces of different origin. The dominant contribution is the dispersion, or *London force*, due to the nonzero instantaneous dipole moments of all atoms and molecules, as described in the previous section. The second contribution is the *Keesom force*, which originates from the attraction between rotating permanent dipoles. The interaction between rotating permanent dipoles and the polarizability of all atoms and molecules generates the third contribution, the *Debye force*. The interaction potential between atoms or molecules of each force is a function of $1/r^6$. The dispersion force is the most important component of van der Waals forces because all materials are polarizable, whereas for Keesom and Debye forces, there must be permanent dipoles present.

The *Hamaker constant, A,* reflects the strength of the van der Waals interaction for two bodies 1 and 2 in medium 3, with permittivities ε_i and indexes of refraction n_i. The first term includes Keesom and Debye interactions, the second the London interaction.

$$A = \frac{3}{4} kT \left(\frac{\varepsilon_1 - \varepsilon_3}{\varepsilon_1 + \varepsilon_3} \right) \left(\frac{\varepsilon_2 - \varepsilon_3}{\varepsilon_2 + \varepsilon_3} \right) + \frac{3hv}{8\sqrt{2}} \frac{\left(n_1^2 - n_3^2 \right) \left(n_2^2 - n_3^2 \right)}{\left(n_1^2 + n_3^2 \right)^{1/2} \left(n_2^2 + n_3^2 \right)^{1/2} \left[\left(n_1^2 + n_3^2 \right)^{1/2} + \left(n_2^2 + n_3^2 \right)^{1/2} \right]}. \quad (5.10)$$

Inspection of this equation reveals that for identical materials across any medium ($\varepsilon_1 = \varepsilon_2$ and $n_1 = n_2$), A is positive (attractive), whereas if ε_3 and n_3 are intermediate to ε_1, n_1 and ε_2, n_2, A is negative (repulsive). If ε_3 and n_3 equal the permittivity and index of refraction of either of the two bodies, A vanishes. For $\varepsilon_3 = n_3 = 1$ (air or vacuum), A is always positive. In other words, **van der Waals forces can be attractive, repulsive, or zero. The judicious choice of the medium in which an SPM experiment is carried out helps control the van der Waals forces between tip and sample.**

The form of the van der Waals interaction potential for two flat surfaces is $W(\delta)_{\text{planes}} = -A/12\pi\delta^2$. Using Derjaguin's equation from Section 5.3.1, we find that the force between a sphere and a flat is $F(\delta)_{\text{sphere–flat}} = -AR/6\delta^2$, as seen in Figure 5.6. A typical value for A is 10^{-19} J; in air or vacuum a force at contact of the order of -4 nN would be expected for a 10 nm tip radius. In water, A is drastically reduced because of the high permittivity of water (≈ 80).

5.3.4   Electromagnetic Forces

There are three classes of magnetism — *diamagnetism, paramagnetism,* and *ferromagnetism.* In paramagnetic and diamagnetic materials, electron spins are randomly oriented due to thermal fluctuations, yielding no net permanent magnetic moment for the material. However, in ferromagnetic materials, a strong quantum effect called exchange coupling causes the spins to align, giving the ferromagnet a permanent magnetic moment.

Spins will respond to an external magnetic field. In all materials, the change in electron orbital moment is opposite to the external field, giving a repulsive force. (This is an atomic analog of Lenz's law, which

258

says that it takes work to move a magnet toward a current loop.) Materials in which this is the only effect are diamagnets. The weak diamagnetic interaction can easily be overridden. In paramagnetism, the electron spins in atoms when magnetic moments line up in the same direction as the external field, yielding a net attraction. Paramagnetism is stronger than diamagnetism, but weaker than ferromagnetism by several orders of magnitude.

The force on a magnetic dipole with magnetic moment $\vec{m}$ is $F = \nabla \vec{m} \cdot \vec{B}$, where $\vec{B}$ is the magnetic flux density. For constant magnetic moment $\vec{m}$, the force depends on the gradient of the flux density. In magnetic force microscopy, if the tip is ferromagnetic, and the sample is paramagnetic or diamagnetic, then the diverging field from the tip interacts with the induced magnetic moment of the sample. If the sample is ferromagnetic, then a para- or diamagnetic tip would detect magnetic force only when the tip is over part of the sample where the field has a gradient, for example, at the edge of a magnetic domain. If both sample and tip are ferromagnets, then a magnetic force of some magnitude and direction will be detected over all of the sample.

The interpretation of magnetic force microscope images is not simple, and requires considerable theoretical effort to understand the data. They depend on the magnetic structure of both tip and sample. The orientation of the tip with respect to the sample determines its sensitivity to the in-plane or perpendicular components of the gradient of the field. Topography may contribute to the overall "magnetic" image. The functional dependence on tip–sample distance will vary with induction effects. (Paramagnetism and diamagnetism are induced magnetic effects.) The field from a ferromagnetic tip may alter the domain structure of the sample. Yet another complication is that of magnetic surface charge which gives rise to an apparent double image of thin domains.

5.3.5 Forces in and Due to Liquids

The high permittivities of liquids have a strong affect on the strength of the force interaction between tip and sample. Equations 5.5, 5.7, and 5.10 all explicitly state the influence of the permittivity on the force magnitude. Also, there are other effects due to the presence of liquids. They are the double-layer, charge regulation, hydration, structural, and capillary forces.

5.3.5.1 Double-Layer, Charge Regulation, Hydration, and Structural Forces

When immersed in a polar liquid such as water, surface charge on the tip and sample may be induced by the fluid. This may occur by either ionization or dissociation of the surface species, or by adsorption of ions from solution. To maintain electrical neutrality, ions of the opposite charge gather near the surface of the tip and sample to form a diffuse electrical *double layer*. If the tip and sample are pushed together, a strong, long-ranged repulsive force is observed because of the overlap of the electrical double layers (Figure 5.7). This does not necessarily remain repulsive all the way to contact, however. *Charge regulation* — the readsorption of counterions onto the surface, reducing the surface charge density — diminishes the repulsive force in concert with attractive van der Waals forces. Acting to keep the force repulsive is the *hydration force*, which comes from the repellent interaction between hydrated ions bound to the tip and sample surfaces. Whether attractive or repulsive forces dominate near contact depends on the specific materials and medium involved.

In nonpolar liquids, much more subtle effects due to molecular ordering or *structure* at the liquid–solid interface can be detected because of the low surface tensions and thereby reduced tip–sample force interactions when nonpolar fluids are present. Specifically, when tip and sample are within ten or fewer molecular diameters of each other, it is possible to observe oscillations between attractive and repulsive forces that display a period equal to the molecular diameter. Almost a standard experiment using a surface forces apparatus, it is less reliable in SPM setups, probably because of the small curvature radius at the end of the cantilever tip.

5.3.5.2 The Capillary Force

If a thin wetting film (that is, no droplets, but rather a uniform layer) of water or another liquid covers the tip and sample, one would expect a force–distance relationship as presented in Figure 5.8. The process

259

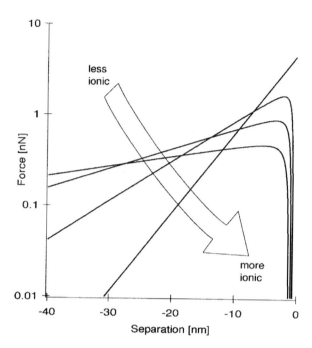

FIGURE 5.7 Typical shapes of force curves taken in liquids. The logarithmic dependence on distance is characteristic of the double-layer force. The higher the concentration of ions in the solution, the steeper the curve. Charge regulation and van der Waals forces can overcome the repulsive interactions at small separations, whereas the most ionic solution remains repulsive all the way to contact because of the hydration force.

by which the peculiar shape is formed in the following. The tip approaches the sample, and when the liquids touch, the system is driven to minimize the fluid surface area. The liquids suddenly draw the tip toward the sample (marked "A" in the figure). Although the work of adhesion between the tip and sample is generally modified by the presence of the fluid, in the displayed case there remains enough tip–sample attraction such that the force becomes more negative until tip–sample contact is achieved. Upon retraction, the force curve shows extreme hysteresis. This corresponds to the elongation of a liquid bridge, or capillary, between the tip and sample. If the microscope were stable enough, the liquid bridge would thin until the waist of the capillary were of atomic dimensions, but usually mechanical or thermal vibrations cause it to break, generating the second instability labeled "C" in the figure.

The overall magnitude of the *capillary force* can be large enough to obscure van der Waals forces. The liquid films surely make a large contribution to any damping measurements that are performed. The permittivities of the films must be considered if any attempt is made to quantify capacitive or other electrical forces. (See Equations 5.7 and 5.10.) Any SPM imaging mode — contact, noncontact, or intermittent contact — is subject to the effect of the capillary force via the change in work of adhesion, the modified contact stiffness due to the presence of the liquid films, or the viscous properties of the films. **Experiments where quantitative results are desired should be performed in dry nitrogen, in liquids, or in vacuum to temper the strong capillary influence on the data.**

5.3.6 Overview

Figures 5.6 through 5.8 compare the distance dependencies of the forces discussed in this section. One can see that the shapes of the curves can be quite different. **The shape of the curve, plus the magnitude of the force, is an aid to the interpretation of the data.** But remember that the shape will be accurately revealed only if a stiff cantilever is used. Of course, more than one surface force may be operative, in which case the data present more of a challenge.

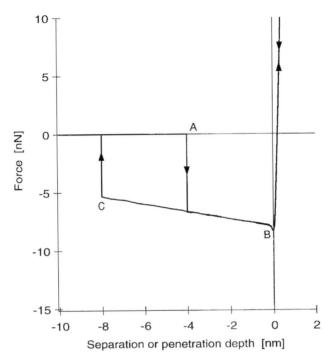

FIGURE 5.8   Capillary force as a function of separation or penetration depth. As the tip approaches the sample, the fluid layers touch, and the tip is suddenly accelerated into the fluid by surface tension (A). The attractive force increases until tip–sample contact is established at B. Upon withdrawal, there is a lot of hysteresis in the curve due to the extension of a liquid bridge between the tip and sample. Typically, a mechanical vibration causes the film to break at a finite bridge diameter (C).

We remind the reader that the discussion above has been limited to the situation where the tip and sample may be modeled as a sphere and a flat. This may or may not apply well to a given experiment.

In Section 5.1, we stated that surface forces predominate at small enough scales. The order of magnitude of the forces from the examples given in this section is approximately 10^{-8} N. We know that this is enough force to attract an AFM tip to a sample. What radius spherical particles would stick to a surface if they had the density of water? Of plutonium?

5.4   Adhesive Forces

One of the first questions that comes to mind upon viewing a hysteretic force curve is "What is the origin of the hysteresis?" Or, in other words, "Why is adhesion greater than attraction?" We shall see in this section that the chemical and mechanical properties of the surface and near-surface region of the tip and sample determine the adhesive behavior. A compliant tip–sample system of large curvature radius and large mutual attraction is more likely to be hysteretic than a stiff system with small curvature radius and negligible mutual attraction. Also, materials are not elastic for pressures approaching infinity; at high enough loads all materials permanently deform, and hysteresis results.

We first cover *anelasticity*, where energy is dissipated as a function of relative tip–sample velocity. There is no permanent damage. We then treat *elastic continuum contact mechanics*, where no permanent deformation occurs and the tip and sample are assumed to be continuous media. The origins of adhesion hysteresis will be examined in this limit. Subsequently, the role of surface forces in generating permanent deformation will be introduced. This becomes important at the nanoscale. At the macroscale, permanent deformation occurs only when force is purposefully applied. Then the central ideas of *molecular dynamics simulations*, where the atomistic nature of material is incorporated into the calculations, will be presented.

261

5.4.1 Anelasticity

Taking the resonant behavior of a cantilever as an example, we can ask ourselves why its resonant peak is not infinitely sharp. It is because of the presence of the dashpot in the model for the system, visible in Figure 5.1, and represented in Equation 5.1 as β_c. Energy is absorbed by the dashpot. The dashpot is not easily related to a specific physical process, but a result is the generation of heat. As seen in Equation 5.1, the dissipative force is a function of the effective mass m^*, β_c, and $\dot{d}$, the velocity. Anelasticity is usually studied by means of modulation techniques, where the phase lag between excitation and response is monitored as a function of frequency or temperature. For velocities (frequencies) approaching zero, this term may be ignored. Because force curve acquisition should be performed under equilibrium (quasi-static) conditions, we shall not focus on anelastic behavior.

5.4.2 Adhesion Hysteresis in Elastic Continuum Contact Mechanics

The work of adhesion must be distinguished from adhesion which must be distinguished from adhesion hysteresis. The work of adhesion is the energy per unit area required to separate two flat surfaces in vacuum from contact to infinity. Adhesion, as we use the word here, is the maximum force needed to separate two bodies. If one of the bodies is curved and indented into the second, then extra force is required to separate them. In force curve analysis, the maximum negative force upon separation of the tip and sample is often referred to as the *pull-off force*. Unfortunately, measuring the pull-off force does not directly give the work of adhesion. The pull-off force is a function of the local curvature radius, the equilibrium interatomic distance, the reduced elastic modulus of the tip and sample, as well as the work of adhesion. Details are forthcoming in Section 5.4.2.6.

If the maximum observed force upon tip–sample approach (the *pull-on force*) in the absence of cantilever instabilities is the same as the pull-off force, there is no adhesion hysteresis. Adhesion hysteresis is the energy difference between loading and unloading, and is proportional to the difference between pull-on and pull-off forces. The energy lost is presumed to be transformed into heat via the generation of phonons. Recent publications indicate that friction and adhesion hysteresis are strongly linked, much more so than friction and adhesion.

Although there are many different sources of adhesion hysteresis including plasticity, chemical bonding, and molecular entanglement, its fundamental behavior can be understood from classical contact mechanical theories. In classical continuum mechanics, no atomistic structure to the materials is evident. Five theories are summarized below. The first three behave nonhysteretically; that is, the loading curve is the same as the unloading curve, but the second two may. Before proceeding, we define some notation common to all five. The five approaches are discussed in increasing order of sophistication.

5.4.2.1 Notation

The Poisson ratio ν and the Young's modulus E are needed to write about the *reduced elastic modulus K* for the tip–sample system

$$\frac{1}{K} = \frac{3}{4}\left[\left(\frac{1-\nu_{\text{tip}}^2}{E_{\text{tip}}}\right) + \left(\frac{1-\nu_{\text{sample}}^2}{E_{\text{sample}}}\right)\right].$$

Other variables used in this section are ξ_0 the interatomic equilibrium distance at the tip–sample interface, a the radius of the contact area between tip and sample, ϖ the work of adhesion evaluated at contact, $1/R = 1/R_{\text{tip}} + 1/R_{\text{sample}}$ the *reduced tip–sample curvature*, P the load, and δ the penetration depth of the tip into the sample. The load P for various contact mechanical theories is plotted as a function of δ in Figures 5.9 and 5.10.

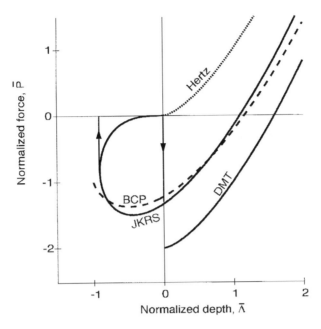

FIGURE 5.9 Normalized force plotted as a function of normalized penetration depth for Hertz, DMT, BCP, and JKRS. The normalization factors are given in Equations 5.18. Hertz, DMT, BCP are not hysteretic, whereas JKRS is. Even a hypothetical infinitely stiff cantilever, indicated by the vertical lines with arrows, would detect adhesion hysteresis for the JKRS case.

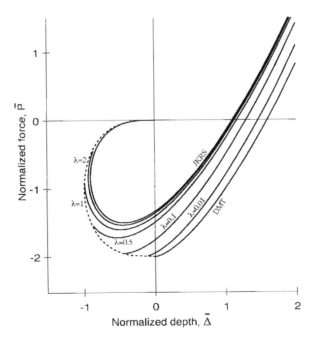

FIGURE 5.10 Maugis contact mechanics, where normalized force is displayed as a function of normalized penetration depth, using the notation of Equations 5.18. Low λ values approach DMT mechanics, and high λ values JKRS. Adhesion hysteresis appears for $\lambda > 0.94$.

5.4.2.2 Hertzian Mechanics

The Hertzian theory goes back over a hundred years to 1888. Neither surface forces nor adhesion hysteresis are assumed ($\varpi \to 0$). The relationship between the load P and the system parameters is

$$P = \frac{Ka^3}{R}, \tag{5.11}$$

and between the contact radius a and the indentation depth δ,

$$\delta = \frac{a^2}{R}. \tag{5.12}$$

One can see from the above two equations that P is a function of $\delta^{3/2}$. In the limit of high loads or low surface forces, Hertzian mechanics can be applied to SPM experiments. But since an SPM experiment is typically run under conditions where the cantilever is near its free equilibrium position in the presence of nonnegligible attractive forces, one of the four following approaches should be more applicable.

5.4.2.3 DMT Mechanics

DMT (Derjaguin, Muller, Toporov) equations date back to 1975. They apply to rigid systems, low adhesion, and small radii of curvature, but may underestimate the true contact area. They account for long-ranged attraction around the periphery of the contact area, but constrain the tip–sample geometry to remain Hertzian. The equations take the following form:

$$P = \frac{Ka^3}{R} - 2\pi R\varpi,$$

$$\delta = \frac{a^2}{R}. \tag{5.13}$$

In other words, DMT is Hertz with an offset due to surface forces. There is no hysteresis between loading and unloading. At $\delta = a = 0$, $P = -2\pi R\varpi$. When one tries to match attractive forces that exist up to the point of contact at $\delta = 0$, the slope at zero is not continuous. The next theory attempts to rectify this unphysical behavior.

5.4.2.4 BCP Mechanics

The BCP (Burnham, Colton, Pollock) semiempirical approach was formulated in 1991, after several years of force curve study revealed that existing possibilities (Hertz, DMT, and JKRS) did not match the experimental SPM data. No surface forces are incorporated into Hertzian mechanics, DMT mechanics predict a sharp discontinuity in slope at contact, and JKRS mechanics predict no detectable attractive forces before contact. Experimentally, with stiff cantilevers, one observes long-ranged attractive forces well before contact, and then a gradual transition between negative and positive slopes in the curves. BCP behavior is representative of many combinations of tip–sample materials.

$$P = \frac{Ka^3}{R} - \sqrt{\frac{3\pi\varpi Ka^3}{2}} - \pi R\varpi,$$

$$\delta = \frac{a^2}{R} - \left(\frac{\pi^2 R\varpi^2}{K^2}\right)^{1/3}. \tag{5.14}$$

The second term in the equation for P accounts for the attractive forces which act to deform the tip–sample geometry, causing the nonrigid materials to bulge out toward each other during the tip–sample approach. The third term has a prefactor of 1 instead of 2, as is the case for DMT, because of the reduction in attractive force due to the bulged geometry. That is, the bulge prevents the full attractive force from being attained. In the equation for δ, the second term represents the distance that the surfaces deform toward each other. However, the Hertzian functional relationship between penetration depth and contact radius, $\delta \sim a^2$, is retained, and there is no adhesion hysteresis.* **The imagery associated with loading and unloading for Hertz, DMT, and BCP is that they are smooth processes, whereas in the JKRS model, both loading and unloading are abrupt.**

A good exercise is to plot P and δ as a function of a. (P is shown as a function of δ in Figure 5.9.) How do R, ϖ, and K influence a?

5.4.2.5 JKRS Mechanics

In the JKRS analysis (Johnson, Kendall, Roberts, Sperling — 1964 and 1971) there are no forces between the surfaces when they are not in contact, upon contact short-ranged attractive forces suddenly operate within the contact area, and the tip–sample geometry is not constrained to remain Hertzian. This set of assumptions applies well to highly adhesive systems that have large radii of curvature and low stiffness. During unloading, a connective neck is formed between the tip and sample, and contact is ruptured at negative loads. This causes adhesion hysteresis.

$$P = \frac{Ka^3}{R} - \sqrt{6\pi\varpi Ka^3},$$

$$\delta = \frac{a^2}{R} - \frac{2}{3}\sqrt{\frac{6\pi\varpi a}{K}}. \tag{5.15}$$

What are the values of a when $P = 0$ and $\delta = 0$? How do they compare to BCP, DMT, and Hertz?

In the example in Section 5.1, we considered the residue of an eraser binding to paper. This very compliant material sticking to a nominally flat surface is a good example of a typical JKRS system, where the material deforms locally in the contact region in response to surface forces, thereby increasing the residue–paper adhesion over what would be expected for Hertz, DMT, or BCP.

5.4.2.6 Maugis Mechanics

A more complex, yet more accurate, description of sphere–flat mechanics was formulated by Maugis in 1992. It is more accurate in that one does not have to assume a particular limit for the materials properties; Maugis mechanics applies to all systems, from large compliant spheres with strong adhesion, to small diamond/diamond systems with low adhesion. The parameter λ,

$$\lambda = \frac{2.06}{\xi_0}\left(\frac{R\varpi^2}{\pi K^2}\right)^{1/3}, \tag{5.16}$$

is used to characterize this spectrum of materials possibilities. As can be seen, large λ is for the more compliant, adhesive combinations, and vice versa, for small λ. In the limit of infinite λ, the Maugis equation approaches those of JKRS theory, and the pull-off force goes to $-1.5\pi R\varpi$. For the limit of λ equals zero, the system responds like the DMT mechanics, and the pull-off force approaches $-2\pi R\varpi$. **Measuring the pull-off force is therefore not an accurate method to determine ϖ.**

---

*The values for P and δ at which contact is made and broken, i.e., $P = -\pi R\varpi$ and $\delta = -(\pi^2 R\varpi^2/K^2)$, correspond to the case for the Maugis $\lambda = 0.8785$, defined in Section 5.4.2.6. This choice results in a large outward bulge so that no sharp discontinuities exist in the force curve, yet the mathematics remain simple, in that there is no hysteresis. A good fit to data is obtained at low loads, less so at higher loads.

TABLE 5.2 Comparison of the Contact Mechanical Theories Discussed in this Chapter

Theory	Assumptions	Limitations
Hertz	No surface forces	Not appropriate for low loads if surface forces present
DMT	Long-ranged surface forces act only *outside* contact area	May underestimate contact area due to restricted geometry
	Geometry constrained to be Hertzian	Applies to *low* λ systems only
BCP	Long-ranged surface forces act only *outside* contact area	May underestimate pull-off force due to Hertzian function for geometry
	Hertzian functional dependence for geometry	Applies to *moderate* λ systems
JKRS	Short-ranged surface forces act only *inside* contact area	May underestimate loading due to surface forces
	Contact geometry allowed to deform	Applies to *high* λ systems only
Maugis	Periphery of tip–sample interface modeled as a crack that fails at its theoretical strength	Solution analytical, but parametric equations
		Applies to *all* values of λ

In the Maugis model, by analogy with the plastic zone ahead of a crack, the adhesion is represented by a constant additive traction acting over an annular region around the contact area. The ratio of the width of the annular region to the radius of the contact area is denoted by m. The set of equations relating λ, m, a, δ, and P is

$$1 = \frac{\lambda a^2}{2} \left(\frac{K}{\pi R^2 \varpi} \right)^{2/3} \left[\sqrt{m^2 - 1} + (m^2 - 2) \arctan \sqrt{m^2 - 1} \right]$$

$$+ \frac{4\lambda^2 a}{3} \left(\frac{K}{\pi R^2 \varpi} \right)^{1/3} \left[1 - m + \sqrt{m^2 - 1} \arctan \sqrt{m^2 - 1} \right],$$

$$\delta = \frac{a^2}{R} - \frac{4\lambda a}{3} \left(\frac{\pi \varpi}{RK} \right)^{1/3} \sqrt{m^2 - 1} ,$$

$$P = \frac{Ka^3}{R} - \lambda a^2 \left(\frac{\pi \varpi K^2}{R} \right)^{1/3} \left[\sqrt{m^2 - 1} + m^2 \arctan \sqrt{m^2 - 1} \right].$$

(5.17)

Additionally, the theory predicts that an instability always occurs when the tip–sample force is controlled (with the exception of λ = 0), and, when the tip–sample displacement is controlled, the instability occurs for λ larger than about 0.94. The instability, the jump from a finite contact radius to none, generates adhesion hysteresis.

5.4.2.7 Comparison of the Five Theories

In each previous subsection one theory was presented. **There may be cases when the assumptions made for a given approach do not exactly describe the materials combinations or the tip–sample geometry.** The Maugis mechanics clearly has an advantage here, as all systems, from rubber/rubber to diamond/diamond are adequately described. Table 5.2 presents the major assumptions and limitations inherent to each theory, while Table 5.3 summarizes the pertinent normalized equations. You should be able to progress from Equations 5.18, below, and Equations 5.11 through 5.17 to Table 5.3 by substitution. The definitions of the normalized contact radius $\bar{A}$, load $\bar{P}$, and the penetration depth $\bar{\Delta}$ are

$$\text{radius } \bar{A} = \frac{a}{\left(\dfrac{\pi R^2 \varpi}{K} \right)^{1/3}} ,$$

(5.18)

TABLE 5.3 Summary Table of the Contact Mechanical Theories

Theory	Normalized equations
Hertz	$\overline{P} = \overline{A}^3$
	$\overline{\Delta} = \overline{A}^2$
DMT	$\overline{P} = \overline{A}^3 - 2$
	$\overline{\Delta} = \overline{A}^2$
BCP	$\overline{P} = \overline{A}^3 - \overline{A}\sqrt{\dfrac{3\overline{A}}{2} - 1}$
	$\overline{\Delta} = \overline{A}^2 - 1$
JKRS	$\overline{P} = \overline{A}^3 - \overline{A}\sqrt{6\overline{A}}$
	$\overline{\Delta} = \overline{A}^2 - \dfrac{2\sqrt{6\overline{A}}}{3}$
Maugis	$1 = \dfrac{\lambda\overline{A}^2}{2}\left[\sqrt{m^2-1} + \left(m^2-2\right)\arctan\sqrt{m^2-1}\right] + \dfrac{4\lambda^2\overline{A}}{3}\left[1 - m + \sqrt{m^2-1}\,\arctan\sqrt{m^2-1}\right],$
	$\overline{P} = \overline{A}^3 - \lambda\overline{A}^2\left[\sqrt{m^2-1} + m^2\arctan\sqrt{m^2-1}\right]$
	$\overline{\Delta} = \overline{A}^2 - \dfrac{4\lambda\overline{A}}{3}\left[\sqrt{m^2-1}\right]$

$$\text{load }\ \overline{P} = \frac{P}{\pi R\varpi},$$

$$\text{depth }\ \overline{\Delta} = \frac{\delta}{\left(\dfrac{\pi^2 R\varpi^2}{K^2}\right)^{1/3}},$$

Figures 5.9 and 5.10, in which the normalized load is plotted as a function of normalized indentation depth, compare the five theories that we have summarized. Much is missing from what is presented here: the effects of plastic deformation, viscoelasticity, surface roughness, and chemical bonding, to name just a few. On the other hand, the equations in this section can be applied to a wide variety of materials systems, and the plots of Figures 5.9 and 5.10 show how even completely elastic systems give rise to adhesion hysteresis and therefore energy dissipation. This may be a useful way of understanding frictional behavior.

5.4.3 Adhesion in Nanometer-Sized Contacts

In the previous subsection on contact mechanics and adhesion hysteresis, two important assumptions to the discussions were (1) that even though adhesion hysteresis was possible, there was no permanent damage to either the tip or sample; and (2) that the materials involved were not composed of discrete particles, but rather were continuous media. These can be dangerous assumptions for the case of nanometer-sized contacts. In this section, we shall show how the tip–sample system can widely stray from continuum elastic theory.

267

5.4.3.1 Surface Forces Alone Can Induce Plasticity

If surface forces are high, and the contact radius small, a great deal of pressure can build up underneath the tip even in the absence of an externally applied load. In some cases, **surface forces alone can induce plasticity.** *Plastic deformation,* that is, permanent deformation, can generate hysteresis in both the positive and negative load sections of force–indentation depth curves. Whereas energy loss due to elastic deformation occurs in the last few nanometers before contact is broken, plasticity-related energy dissipation can happen at both high and low indentation depths.

We define Π as an effective peripheral attractive force, as in Equation 5.14 above. U_s represents the surface energy, equal to $-\pi a^2 \varpi$, and δ is the indentation depth changed by the load P.

$$\Pi = -\left(\frac{\partial U_s}{\partial \delta}\right)_P = \sqrt{\frac{3\pi\varpi K a^3}{2}}.$$

It can be shown that the contact radius a increases in such a way that it is as if a load of $P + 2\Pi$ were acting upon it, as opposed to just $P + \Pi$.

$$P + 2\Pi = \frac{K a^3}{R}.$$

For a wide range of materials parameters, the following approximation predicts a load for the onset of plasticity.

$$P + 1.5\Pi = 1.1\pi a^2 Y,$$

where Y is the elastic limit (*yield stress*). The smaller the curvature radius, the higher the pressure beneath the tip. From the last three equations, a radius can be calculated at which and below which plasticity can be expected. It is $R \approx 0.7\varpi K^2 / Y^3$, an expression that depends heavily on the modulus and yield stress. As an example, consider clean metals in vacuum: their moduli are of the order of 100 GPa, their yield stresses about 10 GPa, their works of adhesion approximately 3 J/m². We therefore expect that for tip radii less than or equal to around 20 nm, the metal will plastically deform. Covering the surface with even one layer of a low-surface-energy material can reduce the work of adhesion by one or even two orders of magnitude, so that radii of a few tens of nanometers no longer permanently deform the materials. But the main point here is that the assumption of perfectly elastic behavior for SPM experiments and many tribological systems is not necessarily valid.

5.4.3.2 Molecular Dynamics Simulations

In Section 5.4.2, we introduced continuum contact mechanical theories, where the tip and sample were modeled as continuous media. At the scale of SPM experiments, this view is not always appropriate. Matter is composed of atoms, and its fundamental structure should be reflected in theoretical models of tip–sample interactions. One way of accounting for the atomistic nature of matter is with molecular dynamics simulations (Rapaport, 1995).

In a molecular dynamics simulation, a set of atoms is modeled. They are arranged into a desired pattern, e.g., a tip held over a sample, as in Figure 5.11. Typically, a few layers of atoms are held fixed to a specified position, but allowed to vibrate about their respective origins so that their fluctuations reflect the system temperature. The remaining atoms are free to respond to whatever forces may act upon them. They may change their position or energy. The fixed layers of atoms may be moved toward and then away from each other so as to model indentation experiments; or, if they are slid sideways with respect to each other, they model friction. The velocity and position of each atom are calculated as a function of time; these data are stored and later analyzed to yield snapshots, or even movies, of the system evolution. Simulated force curves can also be obtained.

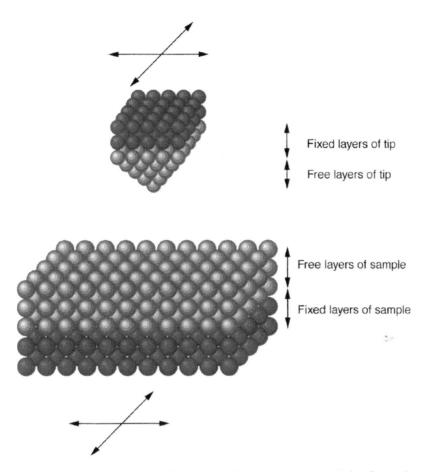

FIGURE 5.11 Typical setup for a molecular dynamics simulation. Atoms are arranged to form a tip and sample, where several layers of atoms far from the contact point act as thermal sinks and are free only to vibrate around their fixed central positions. The free atoms move in response to the forces acting upon them. Two-dimensional boundary conditions are indicated by the crossed arrows.

Keeping track of the position and velocity of several thousand atoms requires an enormous amount of computer time. A major consideration in molecular dynamics simulations is how many atoms can one afford to study, and what should the simulated time be? "Afford," in this sense, is how long the researcher is willing to wait for the results? The more atoms, the more realistic the simulations because the typical real-world tip of 50 nm radius may comprise a million atoms in its lower hemisphere. Similarly, the typical physical experiment takes place over orders of magnitude longer times than the usual computer experiment.

Another important decision in molecular dynamics simulations is the choice of interatomic potentials. If a simple interatomic potential, such as Lennard–Jones ($\alpha/r^{12} - \beta/r^6$), is used, the results apply very well to noble gas solids, where the atoms only marginally interact with each other. The results do not reflect the directionality of covalent chemical bonds, or the sea of electrons surrounding the atomic nuclei in metals. Lennard–Jones potentials are employed because they are computationally "inexpensive," in other words, fast. Computationally expensive potentials are implemented when behavior specific to bonding or the plasticity of metals is of interest.

Comparing predictions from continuum mechanics and molecular dynamics is very interesting. Both approaches yield about the same results for the expected attractive force and neck height for the tip–sample contact. Some further examples of what is predicted from molecular dynamics are the transfer of atoms from the lower surface energy material to the higher, the ductile extension of the more plastic material upon tip retraction, and the elongation of the resulting connective neck between tip and sample by atomic units. **By simulating nanoscale tip–sample interactions via molecular dynamics, more physical phenomena can be revealed than by using continuum mechanics theories, but the effort required for the "computer experiments" is much greater than for traditional modeling.**

5.4.3.3 Experimental Observations

The values for the parameters in Equation 5.16 usually give a Maugis λ of the order of 1 for AFM experiments in air or dry nitrogen. This inconveniently lies between the JKRS and DMT extremes. Accordingly, **it is not easy to predict if there will be significant elastic adhesion hysteresis during an experiment.**

Gross plastic behavior can usually be observed in SPM images. If permanent damage occurs, images evolve as a function of time. Often people zoom out after scanning a small area to compare the original scan region with neighboring ones. Small changes, such as plastic deformation below the sample surface, are more difficult to spot by this imaging method. **If good force curve acquisition is possible with your instrument, it is best to check the reversibility of the force curve up to the load at which the images are collected.**

It has been observed in ultrahigh vacuum experiments that just one layer of adsorbate on a clean surface is enough to change adhesional behavior drastically from plastic to elastic. In general, one can apply elastic theory to adsorbate-covered surfaces, and reserve the effects of plasticity for very clean, high-surface-energy systems such as metals in vacuum. One known exception to this is gold, which has a low yield stress and is very ductile, and forms long extruded "atomic wires" between the tip and samples upon unloading.

5.4.4 Overview

In this section, we have tried to distinguish among (1) anelastic energy dissipation, where the near-surface region absorbs energy as does a dashpot; (2) elastic energy dissipation, where the near-surface region of a complaint sample is temporarily deformed into a connective neck; and (3) plastic energy dissipation, where the near-surface region is permanently changed. **All three processes create adhesion hysteresis in force curves, with the result that adhesive forces are greater in magnitude than attractive forces.** If the force curve data are acquired quasi-statically, then anelasticity is not a factor. The first surface layer has an important role in determining the work of adhesion for the asperity and flat. The yield stress of the near surface, relative to the work of adhesion and the local curvature radius, fixes the dissipation process, i.e., elastic or plastic. The elastic modulus of the near surface, relative to the work of adhesion and the local curvature radius, determines the amount of energy dissipated for the elastic case.

5.5 Closing Words

5.5.1 Interpreting Your Data

It is essential that your instrument is well calibrated, that you understand the sources of error, and that you avoid, or at least understand, artifacts. As you interpret your data, pay attention not only to the magnitude of the surface and adhesive forces, but also to the force–distance relationship (Figures 5.6 through 5.8). This can help you decide which forces or processes are active. Naturally, more than one process may be operative at the same time. When considering the results of molecular dynamics simulations, remember that computer experiments are a logical result of the assumptions that were made when the program was written. For example, many simulations neglect the fact that electrons have spin, so that normally magnetic materials may not have magnetic moments in a simulation. Magnetic forces can overpower other effects, and this may or may not lead to difficulties when comparing computer with real-world experiments.

The concepts discussed in this chapter will be most helpful to you if you are interpreting your nanotribological results in terms of adhesion hysteresis. **One should keep in mind that there are other** approaches to understanding friction; more than one conceptual framework may be useful for a given situation. Coulomb, 200 years ago, proposed the "cobblestone" theory of friction, where surface asperities act as cobblestones in a street; the bigger the asperities, the rougher the ride, and the more energy is used for travel. Modern experiments show that even on atomically flat surfaces the molecular packing and tilt of long-chained hydrocarbon monolayers affect the frictional response. Bowden and Tabor (1950, 1964) understood the friction and wear of metal–metal sliding in terms of the plastic deformation of asperity contacts.

5.5.2   Outlook

Part of the excitement in nanotribology is due to the great intellectual simplification associated with the routine way in which we can now make single point contacts between the tip of an SPM and a sample surface. In this chapter, we have tried to introduce to readers those ideas, equations, and concepts that should help speed them on the way to a good grasp of the surface forces and adhesion interactions relevant to this growing field. Of course, the gap between laboratory experiments that study one single contact and the real world where thousands or millions of asperities are involved remains enormous, and it is not clear if the lubrication engineer will ever make practical use of nanotribology. But it is hoped that at least we have broadened the reader's appreciation of asperity interactions.

We close with a list of questions that may stimulate future work.

- What are the effects of water vapor and surface adsorbates in friction force measurements?
- How does friction affect the imaging mechanisms in scanning probe microscopy?
- Do existing theories quantitatively account for all of the features of force curves?
- To what extent do traditional continuum mechanics explain nanometer-scale tribological processes, and how well do they agree with atomistic simulations?
- Can we develop fundamental understanding of friction, lubrication, and wear?

The interpretation of surface forces and adhesion measurements is challenging. Even if you understand your instrumental artifacts, and are confident about the interpretation of your data, you must still put your work in the context of the research and current thinking of other groups. The future prospects for continued intriguing research in nanotribology are excellent.

Acknowledgments

The timely work of F. Oulevey on some of the graphics is very much appreciated. F. Hutson was kind to review the manuscript.

References

Bhushan, B. (1997), *Micro/Nanotribology and Its Applications*, Vol. E330, Kluwer Academic, Dordrecht.

Bowden, F. P., and Tabor D. (1950, 1964), *The Friction and Lubrication of Solids*, Parts I & II. Clarendon Press, Oxford.

Briggs, G. A. D. (1995), *Advances in Acoustic Microscopy*, Plenum, New York.

Burnham N. A. and Colton, R. J. (1993), "Force Microscopy," in *Scanning Tunneling Microscopy and Spectroscopy: Theory, Techniques and Applications* (D. A. Bonnell, ed.), pp. 191–249, VCH Publishers, New York.

Israelachvili, J. N. (1992), *Intermolecular and Surface Forces*, Academic Press, New York

Johnson, K. L. (1985), *Contact Mechanics*, Cambridge University Press, New York.

Rapaport, D. C. (1995), *The Art of Molecular Dynamics Simulation*, Cambridge University Press, New York.

Singer, I. and Pollock, H. M. (1992), *Fundamentals of Friction: Macroscopic and Microscopic Processes*, Vol. E220, Kluwer Academic, Dordrecht.

6

Friction on an Atomic Scale

Jaime Colchero, Ernst Meyer,
and Othmar Marti

6.1   Introduction

The science of friction, i.e., tribology, is possibly together with astronomy one of the oldest sciences. Human interest in astronomy has many reasons, the awe experienced when observing the dark and endless sky, the fear associated with phenomena such as eclipses, meteorites, or comets, and perhaps also practical issues such as the prediction of seasons, tides, or possible floods. By contrast, the interest in tribology is purlye practical: to move mechanical pieces past each other as easily as possible. This goal has not changed essentially since tribology was born. Ultimately, the person who a few thousand years ago had the brilliant idea to pour water between two mechanical pieces was working on the same problem as the expert tribologist today, the only difference being their level of knowledge. A better understanding of friction and wear could save an enormous amount of energy and money, which would be positive for economy and ecology. On the other hand, friction is not only negative, since it is fundamental for basic technological applications: brakes as well as screws are based on friction.

   The first approach to tribology is due to Leonardo da Vinci at the beginning of the 15th century. In a certain sense he introduced the idea of a friction coefficient. For smooth surfaces he found that "friction corresponds to one fourth its weight"; in other words, he assumed a friction coefficient of 0.25. To appreciate these tribological studies one should bear in mind that the modern concept of force was not introduced until about 200 years later. The next tribologist was Amontons around the year 1700. Surprisingly, the model he proposed to explain the origin of friction is still quite modern. According to Amontons, surfaces are tilted on a microscopic scale. Therefore, when two surfaces are pressed against each other and moved, a certain lateral force is needed to lift the surfaces against the loading force. Assuming that no friction occurs between the tilted surfaces, one immediately finds from purely geometric arguments

$$F_{lat} = \tan(\alpha) \cdot F_{load}$$

where α is the tilting angle on a microscopic scale. This model relates the friction to the microscopic structure of the surface. Today we know that this model is too simple to explain the friction on a macroscopic scale, i.e., everyday friction. In fact, it is well known that surfaces touch each other at many microasperities and that the shearing of these microasperities is responsible for friction (Bowden and Tabor, 1950). Within this model the friction coefficient is related to such parameters as shear strength and hardness of the surfaces. On an atomic scale, however, the mechanism responsible for friction is different. As will be discussed in more detail in this chapter, the model for explaining energy dissipation in a scanning force microscope (SFM) is that the tip has to overcome the potential well between adjacent atoms of the surface. For certain experimental conditions, which are in practice almost always realized, the tip jumps from one stable equilibrium position on the surface to another. This process is not reversible, leads to energy dissipation, and, therefore, on average to a friction force. The similarity between Amontons' model of friction and these modern models for friction on an atomic scale is evident. In both cases asperities have to be passed, the only difference is the length scale of these asperities, in the first case assumed to be microscopic, in the second case atomic.

Although tribology is an old science, and in spite of the efforts and progress made by scientists and engineers, tribology is still far from being a well-understood subject, in fact (Maugis, 1982),

"It is incredible that, all properties being known (surface energy, elastic properties, loss properties), a friction coefficient cannot be found by an *a priori* calculation."

This is in contrast to other fields in physics, such as statistical physics, quantum mechanics, relativity, or gauge field theories, which in spite of being much younger are already well established and serve as fundamental theories for more complex problems such as solid state physics, astronomy and cosmology, or particle physics. A fundamental theory of friction does not exist. Moreover, and although recently considerable progress has been made, the determination of relevant tribological phenomena from first principles is right now a very complicated task, indeed (Anonymous, 1995):

"What is needed … would be to calculate the results of moving a probe of known Miller surface of a perfect crystal and calculate how energy is generated in the various phonon modes of the crystal as a function of time."

From another point of view, the difficulties encountered in tribology are not so surprising taking into account the diversity of phenomena which in principle can contribute to the process of friction. In fact, for a detailed understanding of friction the precise nature of the surfaces and their mutual interaction have to be known. Adsorbed films which can serve as lubricants, surface roughness, oxide layers, and maybe even defects and surface reconstructions determine the tribological properties of surfaces. The essential complexity of friction has been described very accurately by Dowson (1979):

"… If an understanding of the nature of surfaces calls for such sophisticated physical, chemical, mathematical, materials and engineering studies in both macro and molecular terms, how much more challenging is the subject of … interacting surfaces in relative motion."

An additional problem in tribology is that until recently it has not been possible to find a simple experimental system which would serve as a model system. This contrasts with other fields in physics. There, complex physical situations can usually be reduced to much simpler and basic ones where theories can be developed and tested under well-defined experimental conditions. Note that it is not enough if such a system can be thought of theoretically. For testing the theory this system has to be constructed experimentally. The lack of such a system had slowed progress in tribology considerably. Recently, however, with the development of such techniques as the surface force apparatus, the quartz microbalance, and most recently the SFM, we consider that such simple systems can be prepared, which in turn has also triggered theoretical interest and progress. In recent years this has led to a new field, termed nanotribology, which is one of the subjects of the present book.

Within this new field, the SFM and the scanning force and friction microscope (SFFM), which is essentially an SFM with the additional ability to measure lateral forces, have probably drawn the most attention, even though in some respects, namely, reproducibility and precision, the surface force apparatus as well as the quartz microbalance might at the moment be superior. Presumably the interest which has accompanied the SFFM is due to its great potential in tribology. The most dramatic manifestation of this potential is its ability to resolve the atomic periodicity of the topography and of the friction force as the tip moves over a flat sample surface.

An important feature of modern tribological instruments is that wear can be excluded down to an atomic scale. Under appropriate experimental conditions this is true for the SFFM as well as for the surface force apparatus and the quartz microbalance. In general, wear can lead to friction, but it is known that wear is usually not the main process that leads to energy dissipation. Otherwise, the lifetime of mechanical devices — a car, for example — would be only a fraction of what it is in reality. In most technical applications — excluding, of course, grinding and polishing — the lifetime of devices is fundamental; therefore, surfaces are needed where friction is not due to wear, even though in some cases wear can actually reduce friction. Research in wearless friction of a simple contact is thus of technical as well as of fundamental interest. From a fundamental point of view, wearless friction of a single contact is possibly the conceptually simple and controlled system needed for the well-established interplay between experiment and theory: development of models and theories which are then tested under well-defined experimental conditions.

Four features makes the SFFM a unique instrument as compared with other tribological instruments:

1. The SFFM is capable of measuring simultaneously the three most relevant quantities in tribological processes, namely, topography, normal force, and lateral force.
2. The SFFM has a resolution which is orders of magnitude higher than that of classical tribological instruments. Topography can be determined with nanometer resolution, and forces can be measured in the nanonewton or even piconewton regime.
3. Experiments with the SFFM can be performed with and without wear. However, due to its imaging capability, wear on the sample is easily controlled. Therefore, operation in the wearless regime, where tip and sample are only elastically but not plastically deformed, is possible.
4. In general, an SFFM setup can be considered a single asperity contact (see, however, Section 6.3.3).

While some instruments used in tribology share some of these features with the SFFM, we believe that the combination of all these properties makes the SFFM a unique tool for tribology. Of these four features, the last might be the most important one. Of course, it is always valuable to be able to measure as many quantities with the highest possible resolution. The fact that an SFFM setup is a simple contact — which can also be achieved with the surface force apparatus — is a qualitative improvement as compared with other tribological systems, where it is well known that contact between the sliding surfaces occurs at many, usually ill-defined asperities.

Classic models of friction propose that the friction is proportional to the real contact area. We will see that this seems to be also the case for single asperity contacts with nanometer dimension. It is evident that roughness is a fundamental parameter in tribological processes (see Chapter 4 by Majumdar and Bhushan). On the other hand, a simple gedanken experiment shows that the relation between roughness and friction cannot be trivial: very rough surfaces should show high friction due to locking of the asperities. As roughness decreases, friction should decrease as well. Absolutely smooth surfaces, however, will again show a very high friction, since the two surfaces can approach each other so that the very strong surface forces act between all the atoms of the surfaces. In fact, two ideally flat surfaces of the same material brought together in vacuum will join perfectly. To move these surfaces past each other, the material would have to be torn apart. This has been observed on a nanoscale and will be discussed in Section 6.3.1.3.

In conclusion, it seems reasonable that for a better understanding of friction in macroscopic systems one should first investigate friction of a single asperity contact, a field where the surface force apparatus and, more recently, the SFFM have led to important progress. Macroscopic friction could then possibly be explained by taking all possible contacts into account, that is, by adding the interaction of the individual contacts which form due to the roughness of the surfaces.

As discussed above, three instruments can be considered to be "simple" tribological systems: the surface force apparatus, the SFFM, and the quartz microbalance. All three represent single-contact instruments, the last being in a sense an "infinite single contact." Since experiments with the surface force apparatus are discussed in detail in Chapter 9 by Berman and Israelachvili, we will limit our discussion to the last two and mainly to the SFFM. Accordingly, in the next section we will describe the main features of an SFFM, then present experiments which we feel are especially relevant to friction on an atomic scale, and finally try to explain these experiments in a more theoretical section.

6.2 Instrumentation

An SFM (Binnig et al., 1986) and an SFFM (Mate et al., 1987) consist essentially of four main components: a tip which interacts with the sample, a force-sensing element which detects the force acting on the tip, a piezoelectric element which can move the tip and the sample relative to each other in all three directions of space, and control electronics including the data acquisition system as well as the feedback system which nowadays is usually realized with the help of a computer. A detailed description of the instrument can be found in this book in Chapter 2 by Marti. Therefore, we will limit the discussion of the instrument only to the first two components, the tip and the force-sensing element, which we consider especially relevant to friction on an atomic scale. For many applications a thorough understanding of how the SFFM works is essential to the understanding and correct interpretation of data. Moreover, in spite of the impressive performance of this instrument, the SFFM is unfortunately still far from being ideal and the experimentalist should be aware of its limitations and of possible artifacts.

6.2.1 The Force-Sensing System

The force-sensing system is the central part of an SFFM. Usually, it is made up of two distinct elements: a small cantilever which converts the force acting on the tip into a displacement and a detection system which measures this often very small displacement. The force is then given by

$$F = c \cdot \Delta$$

where c is the force constant of the cantilever and Δ the displacement which is measured. The fact that the force is not measured directly but through a displacement has important consequences. The first one is evident: for an exact determination of the force, the force constant has to be known precisely and this is quite often a problem in SFM. Another implication is that an SFM setup is not stiff. If a force acts on the tip, the cantilever bends and the tip moves to a new equilibrium position. Therefore, especially in a strongly varying force field, the tip position cannot be controlled directly. Moreover, a spring in a mechanical system subject to friction forces can modify its behavior substantially (see the Chapter 9 by Berman and Israelachvili). This is specially important in SFM: since the resolution is limited by the minimum displacement that can be measured, a force measurement gives high resolution if the force constant is low. With a low force constant, however, the tip–sample distance is less easily controlled. Finally, for a low force constant the properties of the system are increasingly determined by the force constant of the macroscopic cantilever and not by the intrinsic properties of the tip–sample contact, which is the system to be studied. Therefore, a reasonable trade-off between resolution and control of the tip–sample distance has to be found for each experiment. Although some schemes, such as feedback

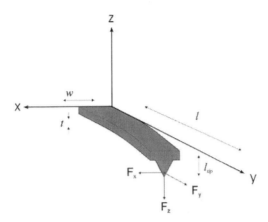

FIGURE 6.1 Geometry and coordinate system for a typical cantilever. Its length is l, its width w, its thickness t, and the tip length l_{tip}. The y-axis is oriented in the direction corresponding to the long axis of the cantilever. Forces act at the tip apex and not directly at the free end of the cantilever. This induces bending and twisting moments as discussed in the main text.

control of the cantilever force constant (Mertz et al., 1993) and displacement controlled SFMs (Joyce and Houston, 1991; Houston and Michalske, 1992; Jarvis et al. 1993; Kato et al. 1997), have been proposed to avoid this problem, up to now these schemes have not been commonly used.

6.2.1.1   The Cantilever — The Force Transducer

The cantilever serves as a force transducer. In SFFM not only the force normal to the surface, but also forces parallel to it have to be considered; therefore, the response of the cantilever to all three force components has to be analyzed. In principle, the cantilever can be approximated by three springs characterized by the corresponding force constants. Within this model, the tip is attached to the rest of the rigid microscope through these three springs, one in each direction of space. The force acting on the tip causes a deflection of these springs. To determine the force and the exact behavior of the microscope, their spring constants have to be known.

A cantilever is a complex mechanical system; therefore, calculation of these force constants can be a difficult problem (Neumeister and Drucker, 1994; Sader, 1995), in some cases requiring numerical computation. Most SFFM experiments are done with rectangular cantilevers of uniform cross section, since they have a higher sensitivity for lateral forces than triangular ones, which are commonly used in SFM. Moreover, for rectangular cantilevers the relevant force constants can be calculated analytically. We will limit the following discussion to these cantilevers. The equation describing the deflection of a cantilever is (see, for example, Feynmann, 1964)

$$z''(y) = M(y)/(E \cdot I),\qquad(6.1)$$

where E is the Young's modulus of the material, $I = \int z^2 dA$ the moment of inertia of the cantilever and $M(y)$ the bending moment acting on the surface which cuts the cantilever at the position $z(y)$ in the direction perpendicular to the long axis of the cantilever (see Figure 6.1). For a cantilever of rectangular cross section of width w and thickness t the moment of inertia is $I = w \cdot t^3/12$. Solving Equation 6.1 with the correct boundary conditions one finds the bending line

$$z(y) = \left(\frac{y}{l}\right)^2 \left(\frac{y}{l} - 3\right) \frac{l^3 \cdot F_z}{6 \cdot E \cdot I},\qquad(6.2)$$

where l is the length of the cantilever and F_z the force acting at its end. From this bending line, the force constant is read off as

$$c = \frac{F_z}{z(l)} = \frac{3 \cdot E \cdot I}{l^3} = \frac{1}{4} E \cdot w \cdot \left(\frac{t}{l}\right)^3.$$

(6.3)

This is the "normal" force constant in a double sense: it is the force constant associated with a deflection in a direction normal to the surface, and also the force constant generally used to characterize a cantilever. However, other force constants are also relevant in an SFM and an SFFM setup. Exchanging t and w in the above equation gives the force constant corresponding to the bending due to lateral force F_x (see Figure 6.1):

$$c_x^{bend} = \frac{1}{4} E \cdot t \cdot \left(\frac{w}{l}\right)^3 = \left(\frac{w}{t}\right)^2 \cdot c,$$

(6.4)

where c is the normal force constant (Equation 6.3). Since the lateral force acts at the end of the tip and not at the end of the cantilever directly, this force exerts a moment $M = F_x \cdot l_{tip}$ which twists the cantilever. This twisting angle ϑ causes an additional lateral displacement $\Delta_x = \vartheta \cdot l_{tip}$ of the tip. The corresponding force constant is (Saada, 1974)

$$c_x^{tors} = \frac{K}{3} G \cdot w \cdot \frac{t^3}{l \cdot l_{tip}^2},$$

where G is the shear modulus and $K \simeq 1$ for cantilevers that are much wider than thick ($w \gg t$), which is the usual case in SFFM. It is useful to relate this force constant to the normal force constant c. With the relation $G = E/2(1 + v)$ and assuming a Poisson factor $v = \frac{1}{3}$, one obtains

$$c_x^{tors} = \frac{2}{3} \frac{K}{(1+v)} \left(\frac{l}{l_{tip}}\right)^2 \cdot c \simeq \frac{1}{2} \left(\frac{l}{l_{tip}}\right)^2 \cdot c.$$

(6.5)

Both lateral bending and torsion of the cantilever contribute to the total lateral force constant which is calculated from the relation of two springs in series (see Section 4.3.1, Equation 6.22):

$$\frac{1}{c_x^{tot}} = \frac{1}{c_x^{bend}} + \frac{1}{c_x^{tors}} = \frac{1}{c} \cdot \left[\left(\frac{t}{w}\right)^2 + 2\left(\frac{l_{tip}}{l}\right)^2\right].$$

(6.6)

The last case is that of a force F_y acting in the direction of the long axis of the cantilever (y-direction). This force induces a moment $M = F_y \cdot l_{tip}$ on the cantilever which causes it to bend in a way similar but not equal to the bending induced by a normal force. Solving Equation 6.1 one finds the new bending line:

$$\tilde{z}(y) = \frac{1}{2} \frac{l_{tip} \cdot F_y}{E \cdot I} \cdot y^2.$$

(6.7)

This bending has two effects. First, the tip is displaced an amount, $\delta_z = \tilde{z}(l) = (3/2) \cdot (l_{tip}/l) \cdot (F_y/c)$ in the z direction. Second, the tip is displaced an amount $\delta_y = \alpha \cdot l_{tip}$ in the y direction, where α is the bending angle $\alpha = \tilde{z}'(l)$, which follows from Equation 6.7. The corresponding force constant for bending due to the force F_y is then

$$c_y = \frac{F_y}{\delta y} = \frac{1}{3} \cdot \left(\frac{l}{l_{\text{tip}}} \right)^2 \cdot c \, . \tag{6.8}$$

We note that the displacement of the tip in the z-direction due to a force F_y implies that the model describing the movement of the tip by three independent springs is not completely correct. The correct description of an SFM setup is in terms of a symmetric tensor $\hat{\mathbf{C}}$ which relates the two vectors force Δ and displacement $\mathbf{F}$:

$$\Delta = \hat{\mathbf{C}}^{-1} \circ \mathbf{F}$$

$$\hat{\mathbf{C}}^{-1} = \begin{pmatrix} c_{xx} & c_{xy} & c_{xz} \\ c_{yx} & c_{yy} & c_{yz} \\ c_{zx} & c_{zy} & c_{zz} \end{pmatrix} = \frac{1}{c} \cdot \begin{pmatrix} 2l_{\text{tip}}^2 / l^2 + t^2 / w^2 & 0 & 0 \\ 0 & 3l_{\text{tip}}^2 / l^2 & 3l_{\text{tip}} / 2l \\ 0 & 3l_{\text{tip}} / 2l & 1 \end{pmatrix} \, .$$

The terms c_{yz} corresponds to the displacement $\delta_y = \vartheta \cdot l_{\text{tip}}$ of the tip in the y-direction due to bending induced by a normal force F_z (Equation 6.3). If the off-diagonal terms are neglected, the relation between forces and displacements is determined by the diagonal terms, the three force constants, which can then be related to three independent springs.

We finally note that usually the cantilever is tilted with respect to the sample. This directly affects the relation between the different components of the forces, and has to be taken into account if the tilting angle is significant (Grafström et al., 1993, 1994; Aimé et al., 1995).

6.2.1.2 Measuring Forces

Force is a vector and therefore in our three-dimensional world it has three components. A classical SFM measures the component normal to the surface, while an SFFM measures at least one of the components parallel to the surface. Since normal force and lateral force are usually intimately related, the simultaneous measurement of both is fundamental in tribological studies. In fact, nowadays practically all commercial SFMs offer this possibility. The optimum solution is, of course, the determination of the complete force vector, that is, of all three force components, and in fact such a system has been proposed (Fujisawa et al., 1994) but is not widely used. As described in Chapter 2 by Marti, the simultaneous detection of normal force and the x-component of the lateral force is easy with the optical beam deflection technique (Meyer and Amer, 1990b; Marti et al., 1990), see Figure 6.2. Since this detection technique is most commonly used in SFFM, we will briefly recall some of its properties. A very particular feature of the optical beam deflection technique is that it is inherently two dimensional: the motion of the reflected beam in response to a variation in orientation of the reflecting surface is described by a two-dimensional vector. In the case of SFFM, if the cantilever and the optical components are aligned correctly, and if the sample is scanned perpendicular to the long axis of the cantilever (x-axis), then normal and lateral forces cause motions of the reflected beam which are perpendicular to each other (see Figure 6.3). This motion can then easily be measured with a four-segment photodiode or a two-dimensional position sensitive device (PSD).

Another important feature of the optical beam deflection method is that unlike other detection techniques, angles and not displacements are measured. Moreover, due to the reflection properties, the angles that are detected on the photodiode are twice the bending or twisting angles of the cantilever. This has to be taken into account when signals are converted into forces.

One consequence of measuring angles instead of displacements is that, in the case of a lateral force acting on the tip, only the displacement corresponding to the torsion of the cantilever is detected. However, the tip is also displaced due to lateral bending which does not result in a variation of the

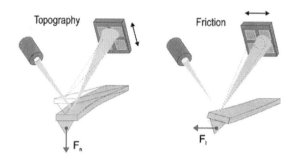

FIGURE 6.2 Schematic setup of the optical beam deflection method. With a four-segment photodiode, the two-dimensional motion of the reflected beam is measured. Therefore, normal and lateral forces can be detected simultaneously. Bending of the cantilever due to a normal force causes a vertical motion of the reflected beam. Torsion of the cantilever due to a lateral force causes a horizontal motion.

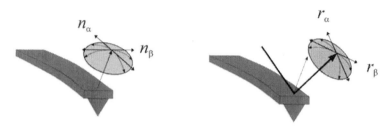

FIGURE 6.3 If the cantilever and the optical setup are aligned correctly, the motions n_α and n_β induced by normal and lateral forces cause perpendicular movements r_α and r_β of the reflected laser spot on the photodiode. This is not the case for arbitrary alignment of the optical axes.

measured angle. Therefore, this motion is not detected. Depending on the calibration procedure used, this might lead to errors in the estimation of the lateral force when the cantilever is displaced more due to bending than due to torsion. From Equations 6.4 and 6.5 we see that this is the case for cantilevers with $t/w \gg l_{\mathrm{tip}}/l$.

The technique for measuring friction forces with the optical beam deflection method just described assumes scanning in a direction perpendicular to the long axis of the cantilever (x-axis). However, friction forces can also be measured in the other direction parallel to the surface (Radmacher et al., 1992; Ruan and Bhushan, 1994a). In this different mode for measuring friction, the sample is scanned back and forth in a direction parallel to the long axis of the cantilever (y-axis). As discussed previously, the friction force acting at the end of the cantilever then bends it in a similar way as when induced by a normal force. From Equations 6.2 and 6.7 the bending line corresponding to the back-and-forth scan can be calculated. One obtains

$$z_{\mathrm{tot}}(y) = \frac{1}{2c}\left(\frac{y}{l}\right)^2\left(\left(\frac{y}{l}-3\right)\cdot F_z + 3\frac{l_{\mathrm{tip}}}{l}\cdot F_y\right).$$

Note that the sign of F_y depends on the scan direction. A technique is needed to discriminate between bending due to a normal force and bending due to a lateral force. The friction force changes sign when the scanning direction is reversed, while the normal force remains unchanged; therefore the difference signal corresponds to the effect caused by friction and the mean signal is due to the normal force. It should be noted, however, that usually the microscope is operated in the so-called constant-force mode. In the present case, this mode is better called the constant-deflection mode, since the deflection (more precisely, the bending angle) and not the (normal) force is kept constant. To maintain a constant

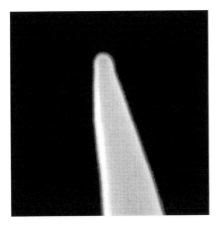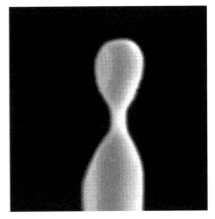

FIGURE 6.4 Well-defined spherical tip ends of tungsten cantilevers produced by heating the cantilever as described in the text. The formation of these tips is controlled by the balance between surface diffusion and surface energy. By carefully tuning the experimental conditions, tip ends of different shapes can be obtained. (Courtesy of Augustina Asenjo Barahona, Universidad Autónoma de Madrid.)

deflection, the feedback adjusts the height of the sample to correct for the difference in bending due to friction while scanning back and forth; that is, the feedback adjusts the height so that

$$z'_{\text{tot}+}(l) - z'_{\text{tot}-}(l) = 2\,\tilde{z}'(l, F_y),$$

where $z'_{\text{tot}+}(l)$ is the angle of the free end of the cantilever during the forward scan, $z'_{\text{tot}-}(l)$ the angle of the cantilever during the backward scan and $\tilde{z}(l, F_y)$ the angle at the free end of the cantilever induced by a force F_y according to Equation 6.7. The friction is related to the difference in height of the topographic images corresponding to the back-and-forth scan. Solving the above equation for the friction force $F_{\text{fric}} = F_y$ as a function of the height difference Δ_z between back-and-forth scan, one finally finds

$$F_{\text{fric}} = \frac{c}{4} \cdot \frac{1}{l_{\text{tip}}} \cdot \Delta z \,,$$

with $\Delta_z = z_{\text{tot}+}(l) - z_{\text{tot}-}(l)$.

6.2.2 The Tip

One of the great merits of the SFFM, the nanometric size of the contact, is on the other hand a serious experimental problem, since it is almost impossible to characterize the tip and thus the contact down to an atomic scale. Different schemes have been proposed to solve this problem. One possibility is to use electron microscopy not only to image, but also to grow a well-defined tip (Schwarz et al., 1997). If the electron beam is focused on the tip, molecules from the residual gas are ionized and accelerated towards its end, where they spread out due to their charge. The result is a well-defined spherical tip end. A similar procedure is to heat a very sharp metallic tip in high vacuum (Binh and Vzan, 1987; Binh and García, 1992). Surface diffusion will induce migration of atoms from regions of high curvature to regions of lower curvature. Again, to control the process, an electron microscope is needed. As in the previous case, this process will form a well-defined and smooth tip (see Figure 6.4). These preparation methods are very effective but also have disadvantages, the first one being the immense effort needed to fabricate just one single tip. Moreover, modification of the tip during transfer, and, even more critical, during the SFFM experiment due to wear cannot be excluded. To control possible wear, the tips should be imaged before and after the measurement.

Not only the geometry of the tip, but also its chemical composition is important for tribological studies, since the tip represents half of the sliding interface. Commercial microfabricated cantilevers are usually made of silicon or silicon nitride (Si_3N_4), which are oxidized on the surface under ambient conditions. Therefore, most SFFM experiments are performed with this material. To vary the chemical composition of the tip, a metal film can be evaporated onto the cantilever (Carpick et al., 1996a) or alternatively thin films such as Teflon (Howald et al., 1995) or self-assembling monolayers (Ito et al., 1997) can be adhered to the tip. Finally, the tip may even be biologically functionalized by attaching antibodies.

In most SFFM experiments, the tip is not prepared. Instead, it is used as delivered on the commercial microfabricated cantilever and some method is used to characterize the geometry of the tip end. One possibility is to image a very sharp object in the normal SFM mode. If the radius of curvature of this object is smaller than that of the tip, then the tip is imaged by this even sharper object (Sheiko et al., 1993).

Another possibility to characterize the tip is to assume a spherical tip end and estimate its radius through the interaction with the sample. In most cases, this interaction is proportional to the tip radius. Typically, to determine the tip radius the variation of the interaction is measured as the tip sample distance is varied. If all other parameters are known, the tip radius can be extracted from a fit to the experimental data points. For adhesion in air due to a liquid meniscus one has for example

$$F_{meniscus} = 4\pi\gamma R \cos(\varphi),$$

where R is the tip radius, γ the surface energy (for water $4\pi\gamma \approx 0.88$ N/m) and ϑ the contact angle, and for van der Waals interaction

$$F_{vdW} = \frac{6 \cdot A \cdot R}{z},$$

where A is the Hamaker constant and z the tip–sample distance. Other interactions which can be used include electrostatic forces and friction force. The last option, however, is not very useful if the friction force itself is to be investigated.

It should be noted that the determination of the tip using some interaction law can only be considered an estimation, since on the one hand a spherical tip end is assumed *a priori* and, on the other hand, macroscopic values are used for physical properties such as the surface energy, the contact angle, or the Hamaker constant. On an atomic scale, the values of these properties might change or not even by defined. Finally, in air many interactions are affected by the films adsorbed between tip and sample.

6.3 Experiments

This section will, of course, present the perhaps most dramatic progress in nanotribology, namely, imaging the atomic periodicity of the lateral force as a sharp tip scans over a sample surface. However, the scope of this section will be extended also to other experiments that shed light on the imaging mechanisms and in general on tribological processes on an atomic scale, which are as yet very poorly understood. Even the fact that the resolution of the atomic periodicity of some surfaces is quite easy — with modern commercial instruments this should be standard provided the vibration isolation of the instrument is good enough — is quite intriguing. In fact, from simple continuum theories for elastic bodies one finds that the contact area between tip and sample is much larger than the atomic periodicity that is measured. The Hertz theory (see Section 6.4.2) is commonly used to estimate the contact radius between tip and sample. According to this model, the contact radius r_c is given by

$$r^3 = \frac{3}{4} \frac{R \cdot F_n}{E^*} \simeq \frac{2}{3} \cdot \left(\frac{1}{E_{tip}} + \frac{1}{E_{sample}} \right) \cdot R \cdot F_n,$$

where F_n is the loading force, R the tip radius, and E^* an effective modulus of elasticity (see Equation 6.16). The approximation is valid assuming a Poisson ratio of ⅓. For an Si_3N_4 tip on mica ($E^* \simeq 150$ GPa), with a sharp tip ($R \simeq 25$ nm) and a loading force of 20 nN, one finds $r \simeq 2.0$ nm, a contact radius corresponding to an area of about 75 unit cells. Even for the hardest possible contact, namely, a diamond–diamond contact, this radius is of the order of 1 nm. Therefore, the contact radius is usually much bigger than the smallest features that are resolved, and the possible mechanisms leading to this apparent high resolution should be explained.

To investigate the contrast mechanism, as well as the fundamental tribological properties of small contacts, experiments have been made to measure the friction as the loading force is varied. Unlike for macroscopic friction, the friction of nanoscale contacts does not increase linearly with load, but seems to increase linearly with the contact area.

Another important feature of nanoscale contacts is that, although the forces are usually very small, the interaction and thus the pressures are very high due to the small contact area. For the Si_3N_4–mica contact above, one finds pressures of about 1.5 GPa. Generally, the pressures in nanoscale contacts can be much higher than the bulk yield pressure and of the order of magnitude of the theoretical yield pressure of defect-free materials (Agraït et al., 1996). Therefore, although the SFFM can be operated in the wearless regime, care has to be taken not to exceed this regime if the aim is to investigate wearless friction. Moreover, as the contact radius is decreased to increase resolution, this problem becomes more critical.

The strong interaction of tip and sample is usually considered a disadvantage of scanning probe microscopy. In the case of the SFFM, however, strong interaction is evidently inherent to friction. Moreover, in technologically relevant situations the interaction of the surfaces in contact is also very strong. Sometimes, however, it is interesting to study friction when the interaction is weak. This is achieved with quartz microbalance experiments where a thin film is adsorbed on a moving substrate. As will be discussed in more detail, the essential physics of the SFFM and the quartz microbalance is similar, although the pressures and timescales are vastly different. In both cases, energy is dissipated as the atomically corrugated surfaces are moved relative to each other. In the case of the quartz microbalance interaction is very weak, while in the case of the SFFM this interaction is usually much stronger due to long-range forces between tip and sample. In a certain way, a quartz microbalance experiment can be interpreted as an SFFM experiment but with only a few last-tip atoms.

6.3.1 Atomic-Scale Imaging of the Friction Force

6.3.1.1 First Experiments

SFM was introduced in 1986 to measure the topography of nonconducting surfaces (Binnig et al., 1986). Only 1 year later, the potential of the SFM to measure forces was applied successfully to image the atomic-scale variation of the friction force as a sharp tip scans over a surface (Mate et al., 1987). Essentially, Mate et al. had the simple but clever idea to turn their SFM around by 90° in order to measure the lateral force instead of the normal force and so SFFM was born. They used an interferometric detection scheme to measure the displacement of a tungsten wire. As shown in Figure 6.5, the free end of this wire was bent and electrochemically etched to serve as a probing tip. With a typical length of 12 mm and diameters of 0.25 and 0.5 mm, they obtained force constants of 150 and 2500 N/m, respectively, which is considerably high for SFM and SFFM standards. As a consequence, the loading force of the tip on the sample was in the millinewton regime, which is also very high. In spite of this high load, atomic resolution of the friction force on a HOPG sample was observed. Figure 6.6 illustrates how the lateral force varied as the sample is scanned back and forth in a direction perpendicular to the long axis of the cantilever. Three of these so-called friction loops are shown, each measured at a different loading force. Since this kind of curve is quite general in SFFM, we will discuss them in some detail. At the beginning of each scan, which can be considered to start either left or right, the tip first sticks to the sample. Its position with respect to the sample is therefore fixed. Since the sample is moved, the cantilever is bent. As long as the lateral force is lower than the force needed to shear the tip–sample junction, the signal corresponding to the lateral

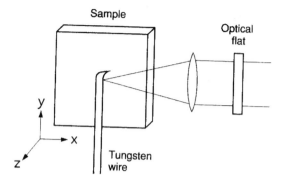

FIGURE 6.5 SFM setup used to measure atomic-scale friction. An interferometer detects the small lateral deflection of the cantilever due to the friction force between the tip and the sample. (From Mate, C. M. et al. (1987), *Phys. Rev. Lett.* 59, 1942–1945. With permission.)

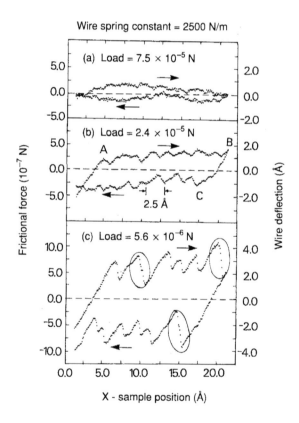

FIGURE 6.6 Variation of the lateral force between a tungsten tip and a graphite surface as the tip is scanned laterally over the surface. Three of these so-called lateral force curves are shown for different loading forces. The lower curve shows the typical stick-slip behavior most clearly. (From Mate, C. M. et al. (1987), *Phys. Rev. Lett.* 59, 1942–1945. With permission.)

force increases linearly with the scanned distance. However, at a certain critical force the junction is sheared, the tip then "slips" into a new equilibrium position, and the lateral force decreases. In its new position, the tip first sticks until the critical force is reached again; then another "slip" will occur. This process repeats itself as long as the scanning direction is not reversed. The number of discontinuous slips therefore depends on the total scan range. If the scanning direction is reversed, the lever first unbends;

FIGURE 6.7 Two-dimensional map of the lateral force recorded as the tip is moved 2 nm from left to right. The spatial variation of the lateral force has the periodicity of the HPOG surface. (From Mate, C. M. et al. *Phys. Rev. Lett.* 59, 1942–1945. With permission.)

then bends again in the new scan direction — the signal polarity therefore changes sign — until the critical force is reached. Then the whole stock-slip process starts again but with opposite polarity. We note that the area enclosed by the lateral force curve has the dimension of energy and, in fact, this area represents the energy dissipated during each scanning cycle. A more precise description of stick-slip behavior is given in Section 6.4.3.3.

If the sample is scanned slowly in the direction perpendicular to the fast scan which corresponds to the acquisition of the lateral force curves, then two-dimensional maps of the lateral force are obtained, as shown in Figure 6.7. These two figures illustrate the two usual ways of representing data in SFFM. From the lateral force curve the friction is directly read off: the friction corresponds to half the height of the lateral force curve. Two-dimensional images, on the other hand, show the variation of the friction on different spots on the sample. Most conveniently, two-dimensional images with the data corresponding to the back-and-forth scan are acquired simultaneously; then all data is available and one can choose between the most convenient representation.

The amazing and puzzling feature about the two figures shown is the fact that they show a variation of the lateral force that corresponds to the atomic periodicity. For a tungsten tip with a typical radius of 50 nm on graphite ($E_{HOPG} \simeq 5\,GPa$) and loading forces of up to 100 μN, Equation 6.16 leads to a contact radius of almost 100 nm, corresponding to a contact of more than 100,000 unit cells. One possible explanation for this evident misfit between contact area and apparent resolution is that imaging is due to a flake of surface material which adheres to the tip and is dragged over the surface. Since the periodicity of the "tip"-flake and the sample is equal, this would lead to a coherent interaction and thus to the observed atomic periodicity. This explanation is very plausible for the present experiment and generally for experiments involving layered surface materials and high loads. Similar experiments performed by the same group on mica, which is also a layered material, showed again atomic resolution of the friction force (Erlandson et al., 1988).

These first experiments had two major difficulties. First, the normal force could not be controlled directly but had to be estimated. Second, the cantilevers used had a high force constant. The rapid development of SFM led, on the one hand, to microfabricated cantilevers with integrated tips (Albrecht, 1989; Albrecht et al., 1990; Akamine et al., 1990; Wolter et al., 1991) and, on the other hand, to new detection schemes, in particular to the optical beam deflection method (Meyer and Amer, 1988, 1990b; Alexander et al., 1989; Marti et al., 1990). Historically, it is interesting to note that both developments — and not only the second as is commonly assumed — were equally important for the success of SFFM. In fact, a year before the successful application of the optical beam deflection method for measuring lateral forces two of the authors had already tried this technique with the first microfabricated cantilevers. However, since these first cantilevers lacked the tip which induces the bending moment that causes the cantilever to twist, no reasonable signal corresponding to lateral forces was detected.

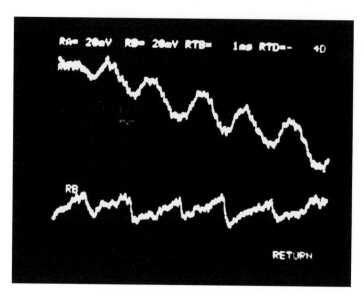

FIGURE 6.8 Oscilloscope traces corresponding to the topography (upper trace) and to the lateral force (lower trace) taken as the tip scans over a mica surface. Both traces were acquired simultaneously. The corrugation is about 0.2 nm for the topography and 1 nN for friction. (From Marti, O. et al. (1990), *Nanotechnology* 1, 141–144. With permission.)

The optical beam deflection method in combination with microfabricated cantilevers and integrated tips not only allowed the simultaneous measurement of normal and lateral forces, it also increased the lateral force resolution by more than one order of magnitude. Figure 6.8 shows an image corresponding to the topographic signal measured simultaneously with the lateral force taken at an estimated loading force of about 20 nN. As in the previously described experiment, the lateral force signal shows the typical stick-slip behavior. Surprisingly, the slip motion occurs near a minimum in the topographic signal and not near its maximum, which is what is predicted by usual models. One possible explanation of this behavior is that a delay in the topographic signal is introduced due to the finite response time of the feedback loop.

A detailed study of the relative phase between the topographic and the lateral force signal is due to Ruan and Bushan (1994b). In this work topographic and lateral force images of a HPOG surface were acquired simultaneously. The surface was imaged with commercial microfabricated Si_3N_4 cantilevers under ambient conditions. The corresponding raw data as well as the Fourier filtered images are seen in Figure 6.9. From the filtered images, the relative displacement of the two images is easily determined: the lattices corresponding to topographic and lateral force signals are shifted by about one third unit cell. To explain this phenomenon, the authors argue that the lateral force signal is not necessarily always due to stick-slip motion and to dissipative phenomena. In fact, the lateral force can be decomposed in a conservative and a nonconservative component. The latter component is due to energy dissipation and is proportional to the area enclosed by the lateral force curve (see Figure 6.6 or 6.32). Only this nonconservative component can be considered a friction force. The conservative component, on the other hand, is not related to energy dissipation. For example, if lateral forces act on the tip in noncontact SFM in such a way that no stick-slip is observed and correspondingly no energy is dissipated, then this lateral force would be truly conservative.

If stick-slip occurs, then the slip distance is smaller than the interatomic distance, but of this order. From the data shown in Figure 6.9 the lateral displacement of the tip during slip is calculated to be 0.01 nm and thus much less than the 0.1 nm between maxima and minima in the images shown, which can be considered a typical slipping distance. From this the authors conclude that the lateral force signal measured is not simply due to the stick-slip motion of the tip. The authors propose that the signal observed is due to a conservative interatomic interaction which results in an atomic-scale variation of

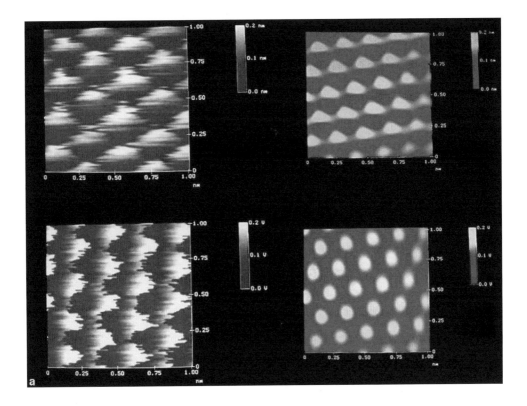

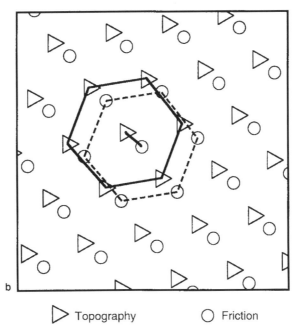

▷ Topography   ○ Friction

FIGURE 6.9   Set of atomically resolved images taken on HOPG with an Si_3N_4 cantilever and a microfabricated tip. The left images are raw data and the right images have been Fourier filtered to show the different positions of the maxima in the topographic and the friction images. The bottom image shows the relative positions of these maxima, which are shifted with respect to each other. (From Ruan, J. and Bhushan, B. (1994), *J. Appl. Phys.* 76, 5022–5035. With permission.)

the lateral force. By using Fourier expansion of the interaction potential, the normal and lateral forces between tip and sample are calculated (Ruan and Bhushan, 1994b). For one lateral dimension (x-axis) the argument elaborated on in this work can be summarized as follows: for a surface potential of type,

$$V_{surf}(x,z) = V_0 \cos(k_x x) \cdot e^{-k_z z},$$

which describes, on the one hand, the atomic corrugation of the surface and, on the other, the decrease of the normal force for increasing tip–sample distance, normal and lateral forces are calculated as

$$F_z = -\frac{\partial V_{surf}}{\partial z} = k_z \cdot V_0 \cos(k_x x) \cdot e^{-k_z z},$$

$$F_x = -\frac{\partial V_{surf}}{\partial x} = k_x \cdot V_0 \sin(k_x x) \cdot e^{-k_z z}.$$

Therefore, minima and maxima of the normal and lateral forces are shifted by one quarter lattice spacing with respect to each other. In the constant-force mode, topographic image is obtained by adjusting the position of the sample to maintain a constant force. Hence, minima and maxima of the normal force and of the topography coincide and are both shifted with respect to the lateral force. This explains qualitatively the relative phase between the topographic image and the lateral force in the case of purely conservative forces. For a detailed understanding of the lateral force and its relation to the normal force and the topography, the exact shape of the interaction potential has to be known and other phenomena such as the elasticity and deformation of the tip–sample contact (see Section 6.4.3.5) have to be taken into account.

Two-dimensional images of the mica surface are shown in Figure 6.10. The images correspond to the topography, the lateral force, and the normal force (from left to right). Essentially all images have the sixfold symmetry of the hexagonal mica lattice, even though in the topographic image clearly some directions are more pronounced. The lateral force image shows the typical stick-slip behavior and the increase in lateral force at the beginning of each line discussed above. It is interesting that an effect due to the friction force can be observed also in the topographic image. In fact, at the beginning of each line some atoms seem to be "stretched." This stretching is about one lattice constant long (0.52 nm). In principle, two effects can be responsible for this stretching: bending and torsion of the macroscopic cantilever at the beginning of the lateral force curve — this was discussed above — or the deformation of the microscopic tip–sample contact (Colchero, 1993). In the case of the images shown, the displacement of the cantilever was estimated to be only 0.1 nm and is thus too slight to explain the observed effect. Therefore, the second option seems more probable. We would like to stress that this stretching is not unique to the images shown but, on the contrary, quite common and is even seen in scanning tunneling microscopy, where it is also explained by a sticking effect of the tip–sample contact (Albrecht, 1989).

Another interesting feature is seen in the image corresponding to the normal force, which shows rather sudden peaks with the lattice periodicity. To understand this, we first note that the topographic image and the normal force image are complementary: if the feedback system does not appropriately correct the height of the piezo (topographic signal), then the cantilever will be deflected (normal force signal). Images taken in the constant-height mode show no contrast in the topographic image, in this case all information is in the normal force image. Images taken in the constant-deflection mode, on the other hand, should show no contrast in the normal force image, all information being in the topographic image. Since feedback systems are never ideal, in this second case usually a small amount of structure is visible also in the normal force image. However, in the case of the normal force image shown in Figure 6.10 the magnitude of the variation as well as its shape cannot be explained only by assuming as low feedback. This is seen as follows: if a tip scans over a corrugated surface assumed to be approximately harmonic, then the tip will "see" a harmonic variation of the surface height. Due to the finite bandwidth of the

FIGURE 6.10 Two-dimensional images of the mica surface taken in air with an Si_3N_4 cantilever and a microfabricated tip. Atomic resolution of the lattice periodicity on mica is seen in all three images. The upper images represent raw data (in the case of the topographic image a plane has been substracted) and the lower images have been Fourier filtered to enhance the atomic periodicity. The left images correspond to the topography, the center ones to the lateral force, and the right ones to the normal force. The scan range is about 5 nm.

feedback loop, the topographic and the normal force signal correspond to low-pass and high-pass filteredimages of the real surface. Therefore, both images should again be harmonic. Their amplitudes and their respective phase will depend on the time constant of the feedback system. The topographic and the normal force images shown, however, have a different structure. While the topographic image is indeed rather smooth — which can be explained by filtering due to the feedback system — the normal force image shows sudden peaks. These peaks are explained by an effect of the stick-slip motion. In fact, if we assume that the tip sticks to some point on the surface until the lateral force exceeds some critical value, whereupon the system becomes unstable and jumps to a new position, then it seems reasonable that the normal force varies as the tip jumps into the new equilibrium position. This has important consequences for the correct interpretation of images: if stick-slip occurs, the tip jumps over part of the unit cell which accordingly is not imaged. Moreover, since the lattice spacing of the unit cell is reproduced in the images, this further implies that the part of the unit cell which is imaged is stretched. A more elaborate explanation for these sudden jumps has been proposed by Fujisawa et al. (1993) and will be discussed in the next section. However, this different explanation does not modify the main message: when stick-slip is observed, which is equivalent to a nonzero friction force and thus to energy dissipation, then only a fraction of the unit cell is imaged, since the tip rapidly jumps over the other part of the unit cell (Colchero, 1993). A more detailed description of this process will be presented in Section 6.4.3.3.

At this point again the question can be raised whether or not in the present case atomic resolution is possible taking into account the finite contact radius. Taking again Equation 6.16 and assuming an Si_3N_4 tip of about 30 nm radius, we estimate a contact radius of 2.5 nm, which is roughly the size of the image shown and therefore again much larger than the periodicity resolved. Therefore, in this context, the high resolution still has to be explained. Moreover, the above considerations regarding imaging within the unit cell, although in principle important, are rather academic at the present point.

6.3.1.2 Two-Dimensional Stick-Slip

In the preceding discussion the frictional force was assumed to act only in the direction of the (fast) scan. This is analogous to macroscopic friction, where the friction force is parallel to the relative velocity of the sliding bodies. According to the simple model discussed above, the tip sticks to potential minima on the surface until the lateral force built up due to the scanning motion of the tip exceeds the force needed to shear the tip–sample junction. The potential minima were assumed to lie along the scanning direction. A surface is, however, a two-dimensional structure and accordingly the potential minima do not have to lie necessarily on the line defined by the scan, that is, the line which the tip would follow if no friction forces act on the tip. We will call this line the scan line. Depending on the symmetry of the surface, the minima of the surface potential can be arranged in a very complex way. The tip, on the other hand, can be deflected in principle in any direction, as was discussed in detail in Section 6.2.1.1. Therefore, if the tip is scanned along an arbitrary line over a surface, the tip will not only stick to points exactly on the scan line — in fact, for most scan lines there might not be any sticking points exactly on the scan line — but will "look" for the most favorable sticking points off the scan line. Since the tip is then deflected from the scan line, this induces lateral forces which are perpendicular to the direction of motion of the tip and thus to the usual friction force. Therefore, for a real two-dimensional surface and a real SFFM setup the tip motion is expected to be much more complex than the one-dimensional stick-slip motion described usually. This two-dimensional stick-slip motion has been studied by Fujisawa et al. (1993) in detail and published in a long series of papers (for a review, see Morita et al., 1996).

The first question that arises in this context is how to detect this two-dimensional motion. With the optical beam deflection method it is possible to detect simultaneously bending and torsion of a cantilever. A lateral force causes a torsion of the cantilever if this force acts along the x-axis (see Figure 6.1 for the convention used) and a bending if this force is along the y-axis (see Section 6.2.1.1). In the latter case, the friction force can be separated from the normal force by taking the difference of the back and the forward scan. Finally, we recall that in the case of typical rectangular cantilevers the force constants for displacements along the x-axis and the y-axis are of similar magnitude, namely,

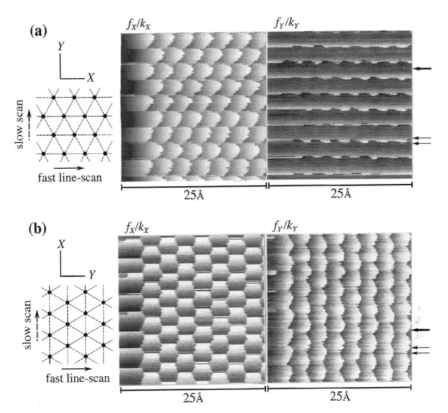

FIGURE 6.11 Images illustrating the two-dimensional stick-slip behavior taken with an Si_4N_3 tip on an MoS_2 surface. The images labeled f_x/k_x correspond to a twisting of the cantilever and the images labeled f_y/k_y to a bending. For the upper images (a), the fast scan is perpendicular to the cantilever (this is the usual imaging mode), and for the lower ones (b) the fast scan is parallel to the cantilever (y-axis). The cantilever is aligned along the y-direction (see Figure 6.1 for the convention used). The thick arrows mark the positions of the sections corresponding to the lateral force curve shown in Figure 6.14, and the thin arrows to lines from which the sections shown in Figure 6.12 were obtained. (From Morita, S. et al. (1996), *Surf. Sci. Rep.* 23, 1–41. With permission.)

$$c_x^{\text{tors}} \simeq \frac{1}{2} \cdot \left(\frac{l}{l_{\text{tip}}} \right)^2 \cdot c \quad \text{and} \quad c_y = \frac{1}{3} \cdot \left(\frac{l}{l_{\text{tip}}} \right)^2 \cdot c \,,$$

where c is the force constant for bending due to a normal force (see Equations 6.3, 6.5, and 6.8) and the convention is used that the long axis of the cantilever is along the y-direction. Therefore, with the optical beam deflection method and with appropriate cantilevers it is possible to detect simultaneously lateral forces in both directions parallel to the sample (see Section 6.2.1.1).

Figure 6.11 illustrates the two-dimensional stick-slip behavior for an Si_3N_4 tip on an MoS_2 surface, which is a layered material with a lattice periodicity of 0.274 nm. The upper images (Figure 6.11a) correspond to the usual setup in SFFM where the fast scan is along the x-axis. The left image shows the torsion of the cantilever — this is the image which is usually acquired as the friction image — and the right image the bending of the cantilever. The lower images (Figure 6.11b) were taken with the fast scan along the y-direction. Again, the left image shows the torsion and the right image the bending of the cantilever. The fast scan in all images is from left to right. The first and the last images are easily understood: they reflect the typical one-dimensional stick-slip behavior. While the first image is measured in the usual torsion mode, the last image is measured by scanning along the cantilever. The other two images cannot be explained within the one-dimensional stick-slip behavior. In particular, note the square-wave shape of the third image.

To understand the two-dimensional stick-slip motion in detail, one should remember that the surface can be described by a two-dimensional potential with a symmetrical arrangement of minima and maxima. If no external forces act on an ideally sharp tip, then the last tip atom will move to the energetically most favorable position which is a minimum of the tip–sample potential and the whole tip will correspondingly be caught in this minimum. To move the tip from this minimum, a shearing force is needed. We consider first two special cases in which the tip is scanned parallel to the symmetry axes of the crystal which contains the potential minima (see Figure 6.13). In the first case, the tip is moved on a line going through these minima (for example, the lines ζ or η in Figure 6.13), and thus in an energy "valley." The tip will then stick in the nearest minimum until the shearing force built up during the scanning motion is high enough so that the tip jumps into the next minimum along the scan line. Since the tip is caught in an energy valley, only forces parallel to the scan line are measured and the stick-slip behavior is that of one-dimensional stick-slip. Such a scan line is marked with a thick arrow in Figure 6.11a. The measured deflection of the cantilever corresponding to such a line scan is shown in Figure 6.12a and j, as well as

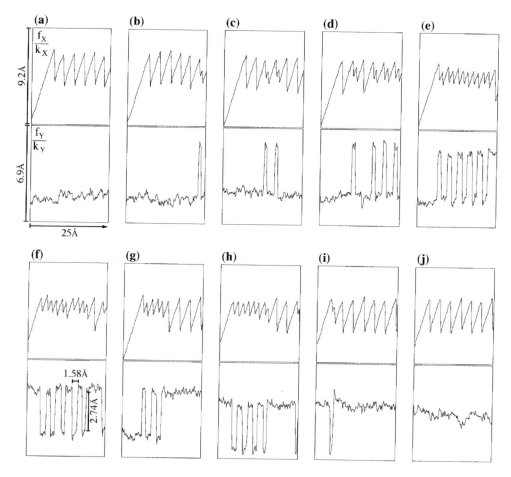

FIGURE 6.12 Signals corresponding to the line scans between the two thin arrows in Figure 6.11a. The signals labeled f_x/k_x correspond to the twisting of the cantilever and the signals labeled f_y/k_y to its bending. Curves (a) and (j) can be considered to show the "classical" one-dimensional stick-slip behavior: the cantilever is only twisted as the tip is scanned perpendicular to the cantilever (x-axis) in an energy valley. (From Morita, S. et al. (1996), *Surf. Sci. Rep.* 23, 1–41. With permission.)

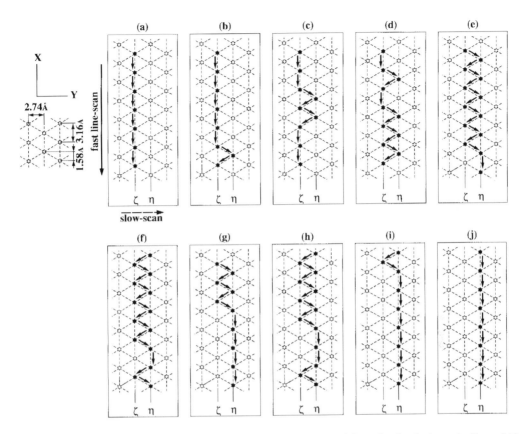

FIGURE 6.13 Two-dimensional stick-slip motion of the tip reconstructed from the signals shown in Figure 6.12. (From Morita, S. et al. (1996), *Surf. Sci. Rep.* 23, 1–41. With permission.)

in Figure 6.14a. Now let us consider a second case, where the tip is assumed to move between the lines containing the potential minima (for example, a scan line halfway between the positions ζ and η in Figure 6.13). The tip can then be considered to move over the maxima of the surface potential and the minima lie on both sides of the scan line. If the cantilever is soft enough in the direction perpendicular to the scan line, the tip will move to the nearest minimum off the scan line which induces a force perpendicular to the scan direction. As the tip is scanned but still sticks to this point, a lateral force is built up oriented along the scan direction (*x*-axis). This force increases linearly with the scanned distance, while the lateral force perpendicular to scan line remains constant. When the (total) lateral force built up during the scanning motion is high enough, the tip snaps into the nearest minimum, which now is on the other side of the scan line. This behavior leads to a lateral force signal which is triangular-shaped in the case of the component along the scan line and rectangular-shaped in the case of the component perpendicular to this line (see Figure 6.12e and f). The intermediate cases where the tip is scanned neither through the potential minima nor through the potential maxima are a combination of the two cases discussed. Figure 6.12a to j shows the signals corresponding to scan lines between the thin arrows in Figure 6.11a, and Figure 6.13a to j the motion of the tip reconstructed from these signals. Interestingly, a region exists around the lines containing the potential minima where the tip is caught in the valleys containing the potential minima, so that no jumps perpendicular to the scan line are observed. This is seen in Figure 6.11, for example, around the position marked with the thick arrow (and also the corresponding signals in Figure 6.14).

The cases described above assumed scanning parallel to the symmetry axis of the surface containing the stick points. If the scan is perpendicular to this direction, essentially the same behavior is observed. However, then the trivial case — scanning through the stick points — does not occur; therefore a

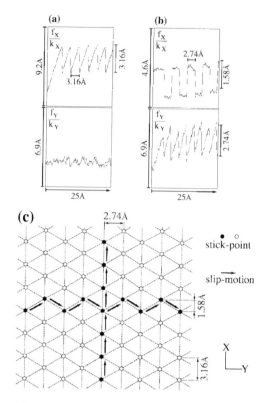

FIGURE 6.14 (a) and (b) Friction signals measured in the two acquisition channels corresponding to twisting of the cantilever (f_x/k_x) and to bending of the cantilever (f_y/k_y) shown for two different orthogonal scan lines. (c) Motion of the tip reconstructed from the measured deflection of the cantilever. The signals shown correspond to the scan lines at the positions marked by the thick arrows in Figure 6.11a and b. (From Morita, S. et al. (1996), *Surf. Sci. Rep.* 23, 1–41. With permission.)

rectangular-shaped signal is always observed in the direction perpendicular to the scan line. This is seen in Figure 6.14, which shows the signals for two different orthogonal scan lines and the corresponding paths reconstructed from the measured motion of the tip.

In the images shown, the axes of the crystal, of the motion of the tip, and of the cantilever were aligned with respect to each other. More specifically, the x-axis of the cantilever was parallel to the symmetry axis of the surface containing the stick points and this axis in turn defined one of the two perpendicular fast-scan directions. If the scan is not aligned with respect to the crystal axis, the two-dimensional stick-slip is even more complex than described above (Figure 6.15). In this case, the square wave signal corresponding to the jumping of the tip between points on both sides of the central scan line is not flat as in the images before, but has a slope. This is seen as follows: let us assume that the tip sticks to some point and that the sticking points are a long the x-axis. If the tip is now moved along a scan line which is not parallel to the x-axis of the cantilever, then the tip will be moved away from the sticking point not only in the x-direction, but also in the y-direction. This will induce lateral force components F_x and F_y which increase linearly with the scanned distance until the total lateral force built up during the scanning motion is high enough to induce a slip into the next sticking point. Moreover, the values of the lateral force corresponding to the stick-slip points are not all equal as in the previous case. However, they can be calculated from the exact geometry of the experimental setup (Gyalog et al., 1995; Morita et al., 1996).

In conclusion, although the general two-dimensional stick-slip behavior can be very complex, it can be understood within a simple model of the surface assuming that the potential minima of the two-dimensional surface potential correspond to sticking points to which the tip adheres. It is evident that

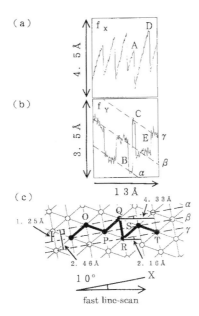

(a)

(b)

(c)

fast line-scan

FIGURE 6.15 Friction signals for scan lines where the crystal axes are not aligned with respect to the axes of the cantilever. In this case, the tilting angle was 10°. The corresponding motion is more complex than in the previous cases. (From Morita, S. et al. (1996), *Surf. Sci. Rep.* 23, 1–41. With permission.)

the model described here for a hexagonal lattice can be generalized to a surface of arbitrary symmetry. A more elaborate model will be presented in Section 6.4.3.4.

6.3.1.3 SFFM in Ultrahigh Vacuum

In air, adhesive forces are mainly due to the liquid meniscus which condenses around the tip–sample contact. Because adsorbed water is always present under ambient conditions, the tip–sample contact is not a well-defined system. Therefore, although experiments in air are generally much more relevant to technical problems, experiments performed in ultrahigh vacuum (UHV) are easier to understand and interpret from a fundamental point of view. Also, most theoretical models on atomic-scale friction assume a tip–sample contact under UHV conditions, again because this is conceptually the simplest system. This explains the great efforts being made to set up SFFM in UHV.

The first observation of atomic-scale friction in UHV is due to German et al. (1993), who chose hydrogen-terminated diamond (100) and (111) surfaces as a sample and a diamond tip grown by chemical vapor deposition (CVD) on the end of a tungsten cantilever. The tip–sample contact was therefore the hardest possible contact that can be made with known materials. Moreover, the passivated diamond surface was extremely inert, and, finally, the surface was a nonlayered material in contrast to many other samples commonly used in SFFM, such as mica, HOPG, or MoS_2. The system chosen by the authors therefore seems ideal to study wearless friction on an atomic scale.

Figure 6.16 shows two images of the friction force corresponding to the (100) and the (111) surface taken at an estimated loading force of 15 nN. Both images show stick-slip-like variations of the lateral force with a periodicity of atomic dimensions. While the variation in the first image was reported to be consistent with a known 2 × 1 reconstruction of the diamond surface (Figure 6.16b), the second image bears no clear relation to the corresponding lattice (Figure 6.16c). The normal force which was acquired simultaneously did not show any structure within the resolution limit (0.07 nm peak to peak noise). In addition to the friction force, the interaction of tip and sample was studied as a function of distance. The corresponding force vs. distance showed very little hysteresis, an adhesive force of about 8 nN and a distance dependence in accordance with a pure van der Waals interaction of tip and sample assuming a tip radius of 30 nm. This was also the radius of curvature which the authors estimated from scanning

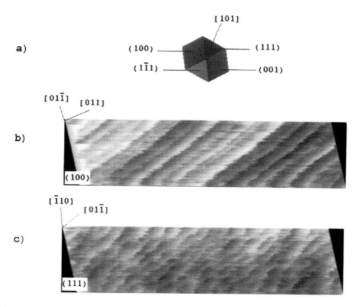

FIGURE 6.16 (a) Schematic view of the diamond tip, seen as if looking through the tip along the surface normal. This view is aligned correctly with respect to the two lateral force images shown below. (b) Lateral force image of a hydrogen-terminated diamond (100) surface. (c) Lateral force image of a hydrogen-terminated diamond (111) surface. For both images, the scan size is 5.8 × 1.25 nm. The gray scale of the lateral force images corresponds to a total variation of 11 nN, and the loading force is about 15 nN. (From Germann, G. J. et al. (1993), *J. Appl. Phys.* 73, 163–167. With permission).

electron microscopy images. In this experiment, the tip–sample contact was much better defined than in the previously discussed experiments. Assuming that the Hertz theory is still valid at this small scale, a contact radius of about 1 nm can be estimated, indicating that most probably the contact was not through a single atom, but was at least of near atomic dimensions.

Another important observation in these experiments is that the friction did not increase linearly with load. Instead, it first increased rather sharply and then remained approximately constant so that the "differential friction coefficient" was essentially zero. For the range of forces measured, and given the low precision of the experimental data, we believe that this result is consistent with recent experiments which show that friction is essentially proportional to the contact area (see Section 6.3.3).

A series of experiments performed explicitly to study the resolution of SFM and SFFM have been made by Howald et al. (1994a,b, 1995). In their first work, the (001) surface of NaF was imaged at room temperature under UHV conditions. The crystal was cleaved in UHV and examined by low-energy electron diffraction to ensure its correct orientation and a structurally good cleavage face. The measured diffraction patterns were found to agree with the cubic cell mesh of the unreconstructed (001) surface. Figures 6.17a and b show large-scale images of the surface as seen by SFFM. Different cleavage steps are visible, most clearly in the topographic image. Their height as measured by SFFM were 0.25 and 0.5 nm, which was in good agreement with the expected value of 0.23 nm for a monatomic step. The lateral force image showed an increase when the tip moved up a step. Similar behavior has also been found by other groups, and under different conditions, as, for example, in an electrochemical cell (Weilandt et al., 1997), and is still not understood in detail. Processes at steps are complex, since a variety of parameters such as contact area and the normal force may vary. Moreover, atoms at steps have a different coordination number than on terraces, which in turn might lead to different chemical and physical behavior. Step edges such as those seen in Figure 6.17 are, however, an ideal structure to test resolution on an atomic scale. On the one hand, a step edge is an extended object which is easily resolved by SFFM, in contrast to a single point defect. On the other hand, due to the discrete structure of the lattice, it represents a well-defined structure of known atomic dimensions. Usually, it is assumed that the tip images the surface.

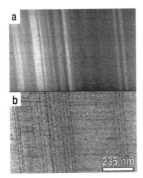

FIGURE 6.17 Large-scale images of the NaF(001) surface taken under UHV conditions with an SFFM in contact mode. Image (a) shows the topography of the surface and image (b) the corresponding lateral force. (From Howald, L. et al. (1994), *Phys. Rev. B* 49, 5651–5656. With permission.)

FIGURE 6.18 High-resolution lateral force images of the NaF(001) surface. The left image was taken on a flat terrace; the total scan size is about 3 nm. The right image was taken at a monatomic step with an approximate loading force of 5 nN and the total scan size is about 5.6 nm. The white and black lines in the right image serve as guidelines to show that the lattices in the left and right part of this image are shifted by half a lattice constant. This shift, as well as the vertical stripes in this image is caused by a monatomic step which is not atomically resolved due to the extension of the contact area. The lattice periodicity is not resolved in the middle of the image. The corresponding diffuse region of about 1 nm width is a measure of the contact radius. The periodicity of both images is measured to be 0.45 nm. (From Howald, L. et al. (1994), *Phys. Rev. B* 49, 5651–5656. With permission.)

This, however, is not always correct. Quite generally, in SFM and SFFM the sharper object essentially images the blunter one. The tip only images the surface if the radius of curvature of the tip is much smaller than the radius of curvature of the sample features. At an atomic step on the sample, however, the step is usually sharper than the tip; therefore, the surface is imaging the tip. From a simple geometric analysis, the relation

$$R = \frac{S^2 + H^2}{2H}$$

between the radius R, measured step width S, and known step height H is obtained.

Figure 6.18 shows high-resolution lateral force images. The left image was taken on a flat terrace and the right image at a monoatomic step. The loading force was about 5 nN. The periodicity of both images was measured to be 0.45 nm, which corresponds well with the length of the unit cell. This implies that only one species of the ions within the unit cell were imaged. Similar results have been also observed with SFM in UHV for ionic crystals, such as NaCl (Meyer and Amer, 1990a), LiF (Meyer, 1991b), and AgBr (Haefke et al., 1992), while in the case of KBr both kinds of ions have been resolved (Giessibl and Binnig, 1992). As in the experiments with diamond on diamond described above, simultaneously with the lateral force an image of the normal force was acquired, but showed no resolution within the noise level of the instrument. The reason for this is not well understood.

In the right image in Figure 6.18, at first sight no clear evidence for a step is seen. The most evident features are vertical stripes, which look like noise. However, these stripes are aligned parallel to the step edges and were not seen on flat terraces (see the left image in Figure 6.18). Analyzing this image in more detail it can be observed that the center of the image shows a diffuse region without clear periodicity. Moreover, the periodicity of the lattice corresponding to the upper and lower terrace is shifted by half a lattice constant. This is exactly what one expects when, as described above, the tip images only one species of ions. In a cubic lattice of the NaCl type, two atomic layers corresponds to one lattice spacing. Therefore, the positions of ions which are alike are shifted by half a lattice constant in the layer below and are again in the same position two layers below. On a monatomic step, the layer below is seen as the lower terrace; therefore, if only one sort of ions is resolved, a shift of the atomic periodicity has to be observed, as in fact is the case. From the image discussed the important consequence is drawn that in this case, which we think is representative for all SFFM images taken up to the present date, the resolution is lower than the periodicity seen. Taking the width of the diffuse region corresponding to the step as an estimate for resolution, one finds a value of about 1 nm, which is consistent with the contact radius that one would expect for a tip of 10 nm radius (see below) on a ionic crystal at the reported loading force of 5 nN.

From the topographic image at this monatomic step (not shown) a width of about 2 nm is measured, and the corresponding radius of curvature is calculated to be 8 nm. However, higher steps give values for the tip radius which are considerably larger: on a biatomic step the width was 6 nm and $R = 40$ nm. This implies that on a larger scale, the tip is considerably larger, which in turn seems very reasonable.

Summarizing, this study proves that the resolution of an SFFM tip in contact with the sample is lower than the periodicity seen and of the order of 1 nm or more, even under UHV conditions. Again, it is interesting to observe that while an atomic periodicity is seen in the lateral force image, the periodicity in the normal force is not resolved. This is rather surprising, since a simple hard sphere model predicts signals of similar structure as well as magnitude for the lateral force and the normal force.

Another important study dealing with friction on an atomic scale is again due to Howald et al. (1994b, 1995). In this study the Si(111) 7×7 surface was investigated by SFFM. This surface is well known from scanning tunneling microscopy (STM) experiments and has a very characteristic structure. Moreover, height differences are much more pronounced than on other surfaces, due to the complex surface reconstruction. In the noncontact mode (see discussion below) the authors achieved large-scale images resolving individual steps; atomic resolution, however, was not obtained (Figure 6.19a). In the contact regime, on the other hand, the surface was severely damaged. This is reasonable taking into account the high reactivity of the surface due to the dangling bonds. Under UHV conditions, where contamination

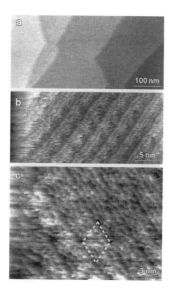

FIGURE 6.19 (a) Large-scale topographic image taken in noncontact mode of the Si(111) 7×7 surface. Images (b) and (c): high-resolution lateral force images taken in contact mode which show the periodicity and some internal structure of the Si(111) 7×7 lattice. (From Howald, L. et al. (1994), *Phys. Rev. B* 49, 5651–5656. With permission.)

layers which passivate the surfaces are absent, one might expect a silicon tip on a silicon surface to simply weld together to form a nanocontact. Tip and surface atoms will merge and lose memory of where they came from. This simple view is in fact supported by molecular dynamics simulation of a silicon tip on the Si(111) 7 × 7 surface (Landman et al., 1989a,b). These simulations also predict nanoneck formation for other tip–sample systems, in particular for metal–metal contacts (Landmann et al., 1990). If nanoneck formation occurs, atomic-scale friction is not only limited to the tip–sample interface, which is not well defined any more, but to atomic rearrangement which can happen within the whole nanoneck. These processes are discussed in detail in Chapter 11 by Harrison et al.

To avoid welding of tip and sample Howald et al. (1995) covered the tip with PTFE (Teflon). This covering was obtained by imaging a PTFE surface prior to the experiments on silicon. It is known that this procedure results in the transfer of PTFE onto the tip, to which it adheres as a thin film (Wittmann and Smith, 1991). Other evaporated coverings such as Pt, Au, Ag, Cr, and Pt/C were reported to offer no improvement as compared with untreated tips. With the PTFE-covered tips, adhesion as well as friction were reduced significantly. The maximum adhesive forces were of the order of 10 nN. Under best conditions, the atomic periodicity of the Si(111) 7 × 7 surface could be resolved and the typical stick-slip behavior of the lateral force was observed. The images shown in Figure 6.19b and c were taken at an approximate loading force of 10 nN, and the total lateral force is 50 nN, whereas its variation due to the stick-slip motion is about 10 nN.

This study shows the importance of the chemical nature of the tip and of the tip–sample contact and that in reactive systems a thin passivating film is needed to avoid welding of tip and sample.

6.3.1.4 Atomic Resolution in SFM and SFFM

It is important to note that while "true" atomic resolution is quite difficult in SFM and SFFM, it is standard in the case of STM in vacuum. In the case of SFM, "true" atomic resolution is much more difficult, but has been achieved in UHV by several groups using STM detection (Giessibl and Binnig, 1992), high-amplitude modulation techniques (Sugawara et al., 1995; Giessibl, 1995), as well as with low-amplitude modulation and careful tuning of the tip–sample interaction. True atomic resolution has also been observed in liquids, again with careful adjustment of the tip–sample interaction (Ohnesorge and Binnig, 1993). The main reason for this difficulty in SFM as compared to STM is twofold. On the one hand, the tunneling current decreases much faster than typical surface forces. Since the lateral resolution in any scanning probe microscope depends not only on the tip radius, but also on the decay length of the interaction used for imaging, this implies a higher resolution for an STM as compared with SFM or SFFM. On the other hand, an STM tip is usually a very stiff system and can therefore be positioned at (almost) any tip–sample distance. An SFM, however, needs a soft cantilever to convert forces into displacements which can then be measured. In consequence, instabilities occur and the tip can usually not be positioned easily very close to the sample, which is the region needed for high-resolution imaging. These two factors essentially explain the difficulty in obtaining true atomic resolution in SFM and SFFM. One approach to achieve true atomic resolution in SFM and SFFM, therefore, seems to be to use stiff cantilevers and to start from STM techniques. In a certain respect, this approach is contrary to that described in this chapter up to this point. The experiments described above were performed in the contact regime, and the goal was to decrease the normal force and thus interaction as much as possible. The approach to be described now starts from the tunneling regime with essentially no interaction and tries to increase interaction to measure normal forces and possibly lateral forces. As we will see, imaging of normal force with true atomic resolution has been achieved, but not the corresponding imaging of lateral forces.

Mostly, in the "almost contact regime" the force is not measured directly, but through the force gradient. The force gradient is measured by detecting a small shift in resonance frequency of the tip–sample system. The measurement of the force gradient is much more convenient for a variety of reasons. First, resonant techniques can be applied which result in an increased resolution. Second, since the snapping of the cantilever has to be avoided, stiff cantilevers have to be used. Therefore, the resolution in direct force measurements is low (since $\Delta F_{noise} = C \cdot \Delta z_{noise}/c$, where c is the force constant of the cantilever), while the resolution measuring the force gradient is less strongly affected. And finally, the force gradient has a stronger variation near the surface than the force itself: if $F(z) \propto 1/z^n$, then $F(z) \propto 1/z^{n+1}$. Moreover, long-range force components such as van der Waals forces or electrostatic forces are in a sense low-pass filtered if the force gradient is measured, so that essentially only the short-range interaction is detected. The first measurement of forces in scanning probe microscopy is due to Dürig et al. (1986), who measured how the spectrum of thermal noise changes as an STM tip approaches a conducting surface. Dürig et al. (1988, 1990) also measured how the tunneling current I and the force gradient F_n vary with the tip–sample distance. A simple relation between tunneling current and force was proposed by Chen (1991):

$$F_n\left(z\right) \propto \frac{dI}{dz}\left(z\right).$$

(6.9)

This relation essentially holds because, within perturbation theory, the tunneling current is proportional to the matrix element between the electronic wave functions $|\mathcal{S}_T\rangle$ and $|\mathcal{S}s/\rangle$ of the tip and sample, respectively: $I(z) \propto /< \cdot \mathcal{S}_T(z)| H_{int}|\mathcal{S}s(z)\rangle$, where H_{int} is the interaction Hamiltonian (Tersoff, 1990). Since this matrix element can be interpreted in terms of an interaction energy between tip and sample, one finds Equation 6.9 by differentiation. This equation establishes a relation between STM and SFM experiments and shows that essentially the same physical quantity is measured with these different instruments. Note, however, that Equation 6.9 is only correct within perturbation theory when tip and sample are far away (for STM standards) and therefore only in weak interaction. For strong interaction, that is, in the "near contact" regime, substantial electronic rearrangement and a lowering of the work function occurs and correspondingly perturbation theory fails. This is the interesting regime for combined STM and SFM experiments: since the tunneling current and the forces are not easily related, they can be assumed to be complementary magnitudes.

A detailed analysis of the physics of combined tunneling and force gradient measurements is due to Dürig et al. (1992). Essentially the proposed technique is to monitor the tunneling current while a very small oscillation is applied to the tip. It is important to keep this oscillation smaller than the typical length scale on which the surface forces vary; otherwise, the interaction is averaged over the distance which the tip moves during oscillation. This oscillation appears in the tunneling current and can be analyzed by appropriate frequency modulation (FM) techniques (Albrecht et al., 1991, Dürig et al., 1992). With this measurement technique, two quantities are measured simultaneously, the tunneling current which can be compared to other STM experiments and the force gradient as an additional source of information.

This method has been applied successfully to obtain true atomic resolution of the variation of the normal force on the Si(111) 7×7 surface (Lüthi, 1996). This is seen in Figure 6.20, and it is evident from the resolved step and the defect that the resolution is "true" in the sense discussed above. The image shown corresponds to a variation of the resonance frequency of +2 Hz over the adatom sites and of −6 Hz over

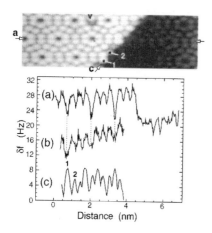

FIGURE 6.20 True atomic resolution image of the Si(111) 7×7 surface taken with dynamic force microscopy. The upper image corresponds to the frequency shift of the tip–sample system as the average tunneling current was kept constant at 25 pA. Note the defect in the upper part of the image at the position marked with the V which demonstrates that the atomic resolution is indeed true. The lower graph shows line profiles of the measured frequency shift. The upper profile (a) in this graph corresponds to a horizontal line at the position **a** in the image and was recorded (presumably) with a one-atom tip, while profile (b) results from a multiatom tip. Profile (c) corresponds to the frequency along the step (marked with the lower arrow and the positions **c**, 1, 2 in the image). (From Lüthi, R. et al. (1996), *Z. Phys. B* 100, 165–167. With permission.)

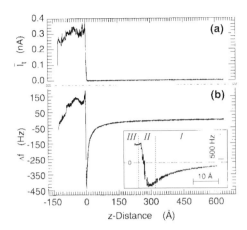

FIGURE 6.21 Simultaneously recorded traces of the time-averaged tunneling current I_t (a) and of the frequency shift δv (b) taken as the tip approaches the Si(111) 7×7 surface. The onset of the tunneling current is shifted slightly to the left of the maximum negative frequency shift δv. Inset: the frequency shift on an expanded scale corresponding to the last few nanometers. (From Lüthi, R. et al. (1996), *Z. Phys. B* 100, 165–167. With permission.)

the corner holes. The mean tunneling current was 25 pA which caused a total frequency shift of approximately –70 Hz. This kind of image is only possible after careful tuning of the interaction parameters. This is illustrated in Figure 6.21 which shows the simultaneous recording of the average tip current as well as the frequency shift vs. the tip–sample distance. The inset corresponds to an expansion of the critical region which shows approximately the last three nanometers before contact. Three regions can be distinguished as the tip approaches the sample. First, a long-range attractive region due to van der Waals and electrostatic interaction. This regime is suitable for noncontact measurements, but atomic resolution is not obtained since the tip is too far to detect the atomic corrugation. The second regime starts with the appearance of tunneling current. This is the regime of typical STM operation. Within this regime, the slope of the frequency shift changes, which has to be taken into account if feedback is performed on this signal. In this region, stable atomic resolution force (gradient) microscopy is possible. As the tip approaches further, the frequency shift becomes zero and even positive, due to repulsive interaction. This is the third region. The point of zero frequency shift corresponds to the maximum adhesive force, and therefore to a point of strong short-range interaction. As the tip approaches even further, a point of vanishing force is reached with positive force gradient. At this point the tip–sample system can be interpreted as being chemically bound at its equilibrium position. Finally, when not only the force gradient, but also the force itself is positive, a true mechanical contact has been made between tip and sample. In this region, only lattice periodicity is resolved, but no true atomic resolution is possible because of the extended radius of the tip–sample contact. As discussed above, in the case of reactive tip–sample configurations, tip and sample will weld together and very strong mechanical interaction will occur.

Up to now, lateral forces have only been measured in this true contact regime. For a detailed understanding of the atomic processes governing friction it would be of great importance to operate the SFFM if possible also in the second regime corresponding to strong tip–sample interaction but not to real mechanical contact. Hopefully, the experimental problems can be solved and this will be achieved in the near future.

6.3.2 Thin Films and Boundary Lubrication

The previous sections have dealt mainly with atomic-scale resolution in the direction parallel to the surface. However, SFM and SFFM also have an extreme resolution in the direction perpendicular to the surface. SFFM can therefore in principle also image what is "on top" of the surface and investigate how the properties of the surface are modified by adsorbents and thin films. The term *thin films* applies in the present section usually to coverings merely one or two monolayers thick and therefore to "really thin films." In the context of tribology, the question of how these very thin films can modify the tribological properties of surfaces leads to the important and vast topic of boundary lubrication, a field of great interest due to its evident technological applications. Accordingly, a lot of research has been made in this field, recently also with SFFM techniques. A detailed analysis of this field is beyond the scope of this section. For details see, for example, reviews by Bhushan et al. (1995), Fujihira (1997), and Carpick and Salmeron (1997). We will only attempt to show how SFFM can contribute to this field.

6.3.2.1 Adsorbed Liquid Films

The most widely used lubricant is probably water in the form of a thin liquid film which adsorbs on most surfaces. This film chemically passivates surfaces and lowers long-range adhesion forces since the van der Waals interaction is weaker through water than through air or vacuum (Israelachvili, 1992). The superb imaging performance of SFM and SFFM in the contact mode under ambient conditions is probably mainly due to the lubricating properties of this adsorbed film. It prevents tip and sample from welding together into a strong contact. As discussed above, imaging in UHV is generally much more difficult. On the other hand, the meniscus force seems to increase the adhesion force and, therefore, the total force exerted on the surface by the tip. Consequently, the contact area and the friction of the tip–sample system also increase. Similar effects are known from micromechanics (Matthewson and Marmin, 1988; Legtenberg et al., 1994; Tas et al., 1997) and magnetic tape disk sliders (Mate, 1992; Mate and Homola, 1997), where stiction due to adhesion caused by meniscus forces is often an important problem. In ambient conditions the effect of relative humidity in an SFM and SFFM setup is twofold. On the one hand, a film of thickness (see, for example, Israelachvili, 1992)

$$d = \left(\frac{A}{6\pi nkT} \frac{1}{\ln(x)} \right)^{1/3} \tag{6.10}$$

with A the Hamaker constant of the tip–sample system, n the number density of the liquid (molecules/m^3), kT the thermal energy, and x the relative humidity, condenses on the tip as well as on the sample. On the other hand, a liquid meniscus of radius (Israelachvili, 1992)

$$\kappa = \frac{\gamma}{nkT} \frac{1}{\ln(x)}, \tag{6.11}$$

the so-called Kelvin radius, where γ is the surface energy of the liquid ($\gamma_{H_2O} \simeq 72$ mJ/m^2), can form around the tip–sample contact. For water in air at room temperature those equations give $d \simeq 0.1$ nm/log $(x)^{1/3}$ and $\kappa \simeq 0.5$ nm/ln (x). The water meniscus results in an adhesive force whose magnitude is approximately (Israelachvili, 1992)

$$F_{\text{men}} = 4\pi\gamma R \cos(\vartheta) \tag{6.12}$$

where R is the tip radius and ϑ the contact angle of the meniscus with tip and sample. According to this equation, the adhesion force should not depend on the relative humidity. However, in typical SFM experiments the adhesion force does seem to depend on the relative humidity (see below). A more-detailed analysis of tip–sample capillary interaction can be found in Mamur (1993).

In spite of its importance, we think that rather little research has been done on the effect of this liquid film in SFM and SFFM. Binggeli and Mate (1994) used an SFFM system similar to the one built by German et al. (1993) to measure adhesion and friction at different humidities. With two fibers normal and lateral forces could be measured simultaneously. Cantilevers were made by bending tungsten wires and etching their free end to form a sharp tip of approximately 100-nm radius. Three representative surfaces were used as samples: SiO_2 (high hydrophilicity, contact angle of approximately 2°), an amorphous carbon film (medium hydrophilicity, contact angle of approximately 45°), and a perfluoropolyether lubricant (Fomblin Z-Dol, low hydrophilicity, contact angle of approximately 90°), which is typically used as a magnetic disk lubricant. The largest effect of humidity on friction and adhesion was found on SiO_2, as might be expected. Essentially, the adhesion force is reduced by approximately a factor of two on SiO_2 as well as on the carbon surface if the humidity is increased from 75 to 90%. This implies that Equation 6.12 has to be used with caution in SFM experiments. The friction, on the other hand, is reduced by the same amount only on the SiO_2 surface, but stays constant on the amorphous carbon film.

Similar results have been obtained for an Si_3N_4 tip on mica (Hu et al., 1995). In this system, the friction coefficient decreased by about a factor of two in the range between 7 and 95% relative humidity. The friction force itself decreased by almost one order of magnitude in this range. Therefore, the lubricant effect of water in the Si_3N_4–mica system seems to be twofold: on the one hand, the absolute friction force is reduced and, on the other hand, the friction coefficient is lowered.

Experiments to measure the role of humidity on friction and adhesion were also performed by Fuhihira et al. (1996). In the range between 10 and 60% they found that neither the friction nor the adhesion depended significantly on the humidity on a hydrophobic surface. On a hydrophilic SiO_2 surface, however, both friction and adhesion increased by about a factor of two. We note that this result is not in contradiction with the experiment of Binggeli and Mate discussed above (they find a decrease of adhesion and friction at higher humidity), since they cover different humidity ranges. The increase of friction at higher humidity is explained as a result of a higher adhesion force due to a larger liquid meniscus around tip and sample. A higher adhesion force implies a large contact area and ultimately a higher friction force (see Sections 6.3.3 and 6.4.2). Interestingly, the authors propose to use this dependence of friction on humidity in the case of hydrophilic surfaces to implement a "scanning hydrophilicity microscope" (Fujihira et al., 1996).

6.3.2.2   Boundary Lubrication

Although adsorbed water on surfaces acts as a lubricant, it is not what is usually understood as a boundary lubricant. This term is generally used for organic molecules such as Langmuir–Blodgett (LB) films or self-assembling monolayers which are deliberately deposited on surfaces to improve their tribological properties, that is, to provide better sliding of the surfaces and reduce wear. Boundary lubricants are essentially organic molecules with carbon chains and some head group. Typically, boundary lubricants are amphiphilic molecules. The hydrophobic part consists of a hydrocarbon chain with approximately 10 to 30 methylene groups. Alternatively, fluorinated carbon chains are used. The hydrophilic part consists of a polar end group, such as a carboxylic acid head group. The end group can form strong ionic bonds with metal ions. Dissolved in water, in the form of fatty acids or as metallic soaps, the molecules form aggregates with the hydrophilic part exposed to water. With the LB technique, the molecules can be assembled as two-dimensional aggregates and molecular layers are transferred to a substrate. Alternatively, self-assembly techniques can be used where the film is directly grown at the liquid–solid interface.

Due to this structure, boundary lubricants can be synthesized in an almost infinite variety: the chain length and its bond character can be varied as well as the head group. Each compound may have distinct physical and chemical properties. Accordingly, one of the aims of boundary lubrication research is to characterize the physical and chemical properties that improve lubrication and wear resistance in order to design and optimize technically relevant boundary lubricants. Typical self-assembling monolayers are thiols (–SH end group), which bind chemically to gold, and silanes (–SiR_3 end groups). Very common LB films are Cd–arachidate films, which are a well-established standard for boundary lubrication.

The first SFFM experiment on boundary lubrication was performed on these films (Meyer et al., 1992a). Figure 6.22 shows a Cd–arachidate film of both 1- and 2-bilayer height on an Si wafer substrate, which is visible on the lowest level. With an applied normal force of 4 nN, the frictional force on the silicon substrate was found to be 3 nN and about 0.2 nN on areas that were covered with the organic

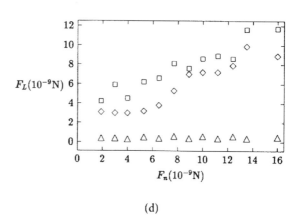

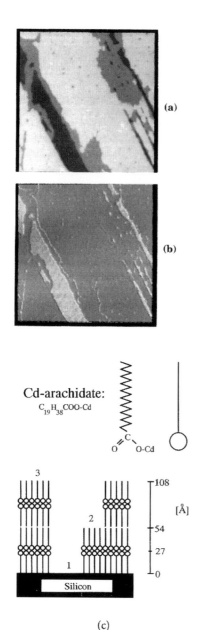

Cd-arachidate:

$C_{19}H_{38}COO\text{-}Cd$

(a)

(b)

$F_L(10^{-9}N)$

$F_n(10^{-9}N)$

(d)

FIGURE 6.22 SFFM study of a double bilayer of Cd–arachidate. Scan size is 2×2 μm². Image (a) shows the topography of these films and image (b) the measured friction force. The total vertical scale in the topographic image is 24 nm and in the friction image 14 nN. In the topographic image the lowest level corresponds to the silicon substrate, the next level to the first bilayer, and the highest level to the second bilayer of the LB film. The friction force image shows a higher friction on the substrate and the same friction on the single- and double-bilayer surfaces. Figure (c) is a schematic diagram of the Cd–arachidate molecules on the substrate. The graph (d) shows how the lateral force varies with normal load on the silicon substrate ($\diamond$) on the first and second bilayer of the LB film ($\triangle$) as well as on the step edges ($\square$). (From Meyer, E. et al. (1992), *Phys. Rev. Lett.* 69(12), 1777–1780.

film. No difference in friction was measured between the first and second bilayer within an accuracy of 10%. The friction was thus reduced by about a factor of ten on film-covered surfaces, which is in agreement with the magnitude of reduction observed on the macroscopic scale (Bailey and Courtney-Pratt, 1954). This observation shows that the LB film acts as a lubricant on a nanoscopic scale.

At loads between 1 and 10 nN, only small changes in the frictional force on the LB films were measured and are taken in a first approximation to the constant (see Figure 6.22d). Wear processes started at forces above 10 nN. Above this critical value small islands were moved in their entirety, which were then absent in the topography image. Figure 6.23a and b shows an area on a bilayer film, imaged with an nondestructive load of 4 nN, while Figure 6.23c and d was taken on the same area imaged with an increased force of 16 nN. The island indicated by an arrow was sheared to the upper margin of the scanned area. In a subsequent wide-area topographic scan, the sheared particle was found again. It still had bilayer

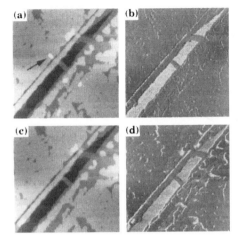

FIGURE 6.23 Topographic , (a) and (c), and friction (b) and (d), images of a double bilayer of Cd–arachidate. Scan size is $2 \times 2 \ \mu m^2$, the loading force is 4 nN for the upper images and 16 nN for the lower ones. The total vertical scale in the topographic image is 24 nm and in the friction image 14 nN. No damage is observed in the upper images at low loading forces. At higher loading, however, small islands and flakes are sheared away. For example, the island indicated by the arrow in image (a) is not observed in (c). (From Meyer, E. et al. (1992), *Phys. Rev. Lett.* 69(12), 1777–1780. With permission.)

height, which demonstrates that the movement of these particles does not change the normal orientation of the aliphatic chains. From this observation the shear strength at the interface between the substrate and the LB film can be determined. The contact area is given by the geometric dimensions of the sheared island. In the case of Figure 6.23 the island area is 4900 nm². The lateral force to shear the island is 5 nN. A shear strength of $\tau = 1 \pm 0.2$ MPa is deduced.

Another class of boundary lubricants are fluorocarbons. Direct comparison of fluorocarbons and hydrocarbons by SFFM on a nanometer scale provides a very effective way to study the lubricating properties of these compounds. In one particular study, a 1:1 molar mixture of arachidic acid ($C_{19}H_{39}COOH$), AA, and partially fluorinated ether carboxylic acid ($C_9F_{19}C_2H_4$–O–C_2H_4COOH), PFCEA, complexed to a polymeric counterion (polyvinylpyridine), was deposited by LB technique. Monolayers are formed on oxidized silicon (Meyer et al., 1992b) while bilayers are formed on hydrophobized silicon (100) (Overney et al., 1992). Figure 6.24 shows a ($2 \times 2 \ \mu m^2$) area of a mixed bilayer of AA and PFECA. Circular domains are surrounded by a sealike, flat region. The step height is 1.6 nm, which corresponds to approximately twice the difference between the longer hydrocarbons and the fluorocarbons (≈ 1 nm). The deviation from the expected value is possibly related to a certain tilt of the fluorocarbons which increases the step height. Therefore, the higher, circular domains are assigned to the hydrocarbons, while the surrounding flat regions are formed of fluorinated bilayers. Within the circular domains, holes and grooves of 5 nm depth are observed, which is consistent with the height of a bilayer of AA. The holes reach all the way to the silica substrate, confirmed by the large frictional forces recorded on these regions. By contrast, the fluorinated regions are free of holes. With higher forces, we observe that particles of hydrocarbons are pulled out and sheared out of the scanning area while the fluorocarbons remain undeformed. Control films of pure AA and pure PFECA show the same tendency. The hydrocarbon films have many defects and are easily deformed while the fluorinated films are rather uniform and cannot be broken by the probing tip. The deformation conditions can be controlled by adjusting normal force and scan speed. Therefore, nanometer lithography can be performed and "smileys" can be drawn, as shown in Figure 6.25

The lateral force measurements indicate high friction on the substrate, intermediate friction on the fluorocarbon and small friction on the hydrocarbon domains. The ratio is approximately 10:4:1. At first sight the larger friction on the fluorinated regions is surprising. From compounds such as Teflon (PTFE), reduced friction is expected. However, it is known from surface force apparatus measurements that the shear strength on fluorinated LB films is increased compared to the hydrogenated films (Briscoe and Evans, 1982). Nevertheless, fluorinated films are good boundary lubricants. They combine intermediate friction with high resistance to rupture.

Finally, another interesting study on boundary lubrication is due to Salmeron et al. (1995). This group prepared a alkythiol monolayer on an Au(111) surface. From electron-, helium-, and X-ray diffraction

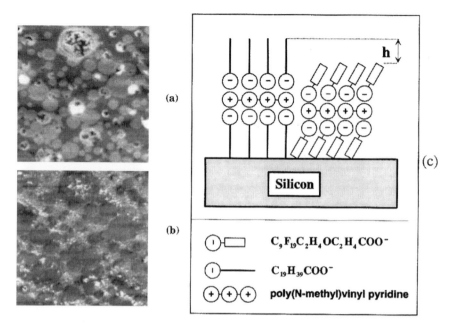

(a)

(b)

(c)

$C_9F_{19}C_2H_4OC_2H_4COO^-$

$C_{19}H_{39}COO^-$

poly(N-methyl)vinyl pyridine

FIGURE 6.24 SFFM images of a bilayer of a 1:1 mixture of PFECA and AA; the scan size is 4×4 μm². The circular islands in the topographic image (a) correspond to the hydrocarbons (AA) and the surrounding "sea" is composed of fluorocarbons (PFCEA). The friction image (b) shows low friction on the hydrocarbon phase, medium friction on the fluorocarbon phase, and high friction on the substrate. The ratio of the friction force on the different materials is 1:4:10 and friction forces are in the nN range. (c) shows a schematic diagram of the arrangement of AA and PFCEA on the hydrophobized silicon substrate. (From Overney, R. M. et al. (1992), *Nature 359, 133–135*. With permission.)

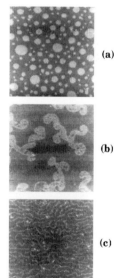

(a)

(b)

(c)

FIGURE 6.25 SFFM images of a monolayer of a 1:1 mixture of PFCEA and AA; the scan size is 2.4×2.4 μm². The topographic image (a) shows the circular islands of AA which are also seen in Figure 6.24. If the forces applied with the SFFM tip are high enough, "smileys" can be written. Image (b) shows the corresponding friction force image. (From Meyer, E. et al. (1992), *Thin Solid Films*, 220, 132–137. With permission.)

studies it is known that these thiols arrange in a $\sqrt{3} \times \sqrt{3}$ r30° structure with respect to the gold substrate. This structure was also found by SFFM, which is noteworthy, since it shows that the structure of the thiols remains stable in spite of the high tip–sample interaction which is typical for SFM and SFFM. As the loading force was increased over a certain threshold value, which was of the order of a few hundred

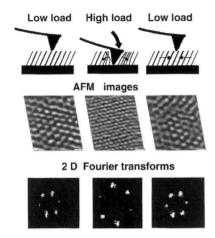

Low load   High load   Low load

AFM images

2 D Fourier transforms

FIGURE 6.26 SFM study of a C_{18} alkanethiol monolayer at low load (about 30 nN, left images), high load (about 300 nN, middle images), and again low load (right images). The top images show a schematic setup of the tip–sample system and the images in the middle the acquired topographic signals. The lower images are two-dimensional Fourier transforms of the corresponding topographic images showing the relation between the alkanethiol monolayer and the Au(111) lattices. (From Salmeron, M. et al. (1995), in *Forces in Scanning Probe Methods*, pp. 593–598, NATO ANSI Series Kluwer, Dordrecht. With permission.)

nanonewtons, a transition takes place and the Au(111) lattice of the underlying surface is imaged (see Figure 6.26). This transition is reversible: as the force was lowered about 100 nN below the critical force of the first transition, the image corresponding to the $\sqrt{3} \times \sqrt{3}\, r30°$ structure was recovered. The exact values for the critical forces corresponding to the transitions varied from experiment to experiment and depended mainly on the tip radius. Moreover, for tip radii larger than about 100 nm, these transitions were not observed. The pressure at which the transitions occur was approximately constant, with a value of about 1 GPa for the transition thiols–Au(111). This is a huge pressure, much higher than that at which bulk gold yields and only slightly lower than the observed yield pressure of nanoscale gold contacts.

These observations are explained qualitatively as follows. At the critical pressure, the thiol chains are pushed aside by the tip, but the head groups still remain chemically bound to the gold–substrate (see Figure 6.26). Therefore, around the tip the thiol molecules are more densely packed, which results in a radial inward pressure. This pressure can lift the tip again and restore the original configuration if the pressure is reduced. A more-detailed analysis is found in Salmeron et al. (1995).

Summarizing, we have discussed some applications of SFFM that are relevant for a better understanding of lubrication. Adsorbed water can act as a lubricant also on a nanoscale. Films of one molecular height are found to be effective lubricants. The observation that one bilayer of lubricant causes about the same reduction of friction as two or more layers is of interest for the fundamental understanding of boundary lubrication. Thiol monolayers on gold show stability up to very high pressures.

6.3.3 Nanocontacts

The experiments discussed previously made use of the imaging capabilities of the SFFM. SFFM and SFM, however, are not limited to the acquisition of images. Essentially, an SFM and quite generally a scanning probe microscope can be operated in two modes, the well-known imaging mode where the sample is scanned, and a nonimaging mode where the decay of the interaction is studied by varying the tip–sample distance over a fixed location. The latter mode is sometimes referred to as the spectroscopic mode. In this section, we will discuss the application of this mode to study the dependence of friction on the normal force. For tribological studies, this mode is experimentally more complicated than in other cases, since the tip evidently has to move in order to measure a friction force. Different experiment techniques have been developed to measure friction at a point while moving the tip. On the one hand, several friction images of the same region may be taken at different loads. By taking the friction value corresponding to the same point in the different images, a friction vs. load curve is obtained. With this technique, friction vs. load curves can be constructed at every point of the image (Meyer et al., 1992a). This method is especially convenient if the sample is inhomogeneous and the variation of friction on different regions of the sample is to be studied. Another method is to assume that the sample is homogeneous over a certain area and take one friction image in which the loading force is increased or decreases for each new scan line (Hu et al., 1995). In this case from each line a friction value is obtained as the average friction force along the line. The whole "image" therefore gives a friction vs. load curve. This method gives a higher signal-to-noise ratio, since all the points along a line are averaged to obtain one data point on the friction vs. load curve. Finally, modulation techniques (Yamanaka and Tomita, 1995; Colchero et al., 1996a) as well as histogram techniques (Marti, 1993; Lüthi et al., 1995) have also been proposed.

In macroscopic friction, the relation between friction and load is usually linear and is expressed through the well-known friction coefficient: $F_{fric} = \mu \cdot F_n$. On the other hand, it is known experimentally that the true contact area is also proportional to the loading force (Bowden and Tabor, 1964). Moreover, for certain statistical distribution of roughness, this relation can be shown to be theoretically correct for elastic contacts (Greenwood, 1992a,b). Therefore, on a macroscopic scale the linear relations

$$F_n \propto F_{fric}$$

$$F_n \propto A$$

$$F_{fric} \propto A$$

between loading force F_n, friction force F_{fric}, and contact area A hold. For low loads, we know from different continuum models that for single asperity contacts, the relation between contact area and loading force is not linear (see Section 6.4.2). The question arises, whether for nanoscale contacts the last proportionality — $F_{fric} \propto A$ — is correct. This seems to be the case and correspondingly the friction vs. load curve increases in a nonlinear way. Therefore, the notion of friction coefficient is no longer well defined. Instead, it appears that shear strength is the fundamental parameter which describes friction in nanoscale contacts (see discussion below).

The importance of studying frictional behavior with varying load was recognized already in the first SFFM experiments. Mate et al. (1987) found a linear relation between friction and load for a tungsten tip on graphite, in contrast to what has just been discussed. Essentially, the same behavior was found on other surfaces in a series of early SFFM experiments. Putman et al. (1995) studied the friction vs. load curve in various environments. In ambient conditions they found a nonlinear relation, whereas in an N_2 or an Ar atmosphere, where the liquid film covering tip and sample is reduced, a linear behavior is found. Putman et al. propose that this dependence of friction on environmental conditions is due to the surface roughness of the tip and to contaminants: in ambient conditions with about 55% relative humidity, the liquid film and possibly also other contaminants may smooth the tip surface, so that the tip forms a single contact with the sample. If the liquid film is absent, the contact may be due to microasperities which lead to the same behavior as in macroscopic friction. The linear behavior of friction vs. load curves is therefore considered to be an artifact. In a way, the tip–sample contact in this case is still "macroscopic," that is, at many ill-defined contacts and not at one single and well-defined contact. This study shows the importance of having a well-defined tip.

To separate the dependence of the friction on the area from intrinsic properties of the tip–sample contact, it is useful to rewrite the friction as (Schwarz et al., 1996)

$$F_{fric} = S_0 \cdot A_c = S_0 \cdot \pi \cdot \left(\frac{R \cdot F_n}{E^*} \right)^{2/3}, \qquad (6.13)$$

where S_0 is a frictional force per area, A_c is the contact area, R is the tip radius, and E^* is an effective modulus of elasticity which is defined in Equation 6.16. The last relation follows assuming a Hertzian contact and a normal force F_n which is corrected for the adhesion between tip and sample: $F_n = F_{load} + F_{ad}$, with F_{load} external loading force and F_{ad} the adhesion force between tip and sample. As discussed in Section 6.4.2, other relations describing the load-area dependence can also be assumed. The frictional force per unit area can be interpreted as the shear strength of the tip–sample junction, according to the macroscopic model of friction by Bowden and Tabor (1964). If this shear strength is assumed constant, then the friction is only determined by the contact area. Another simple relation is

$$S(p) = S_0 + \alpha \cdot p,$$

where p is the pressure on the contact, which can be shown to follow from a thermally activated model of the Erying type (Eyring, 1935; Briscoe and Evans, 1982). With this dependence, Equation 6.13 can be written as

$$F_{\text{fric}} = a \cdot F_n^{2/3} + b \cdot F_n. \tag{6.14}$$

The first term is proportional to the contact area, while the second is proportional to the normal force, as in the case of macroscopic friction. Note that this conceptually simple linear dependence on normal force is due to a nontrivial dependence of the friction on the area and on the shear strength.

In their experiments Schwarz et al. (1996) investigated the friction of C_{60} on a GeS substrate as a function of load. Interestingly, they found a different behavior on these two materials. While the friction vs. load curve on the C_{60} followed excellently an $F^{2/3}$ dependence, the friction vs. load curve on the GeS substrate was approximately linear. For a good fit to the experimental data on the GeS substrate, however, the friction had to be described by both factors of Equation 6.14, which implies a complex relation.

In a similar experiment, Carpick et al. (1996a,b) found that the friction vs. load curve of a Pt tip on mica was described almost perfectly assuming a constant shear strength and a load–area relation as predicted by the JKR theory (see Section 6.4.2, Equation 6.20) of elastic contacts. To avoid possible contamination, the experiments were performed in UHV, including the cleaving of the mica to expose a fresh surface. To ensure a well-defined tip shape, the tip was imaged using a faceted $SrTiO_3$ (305) surface, as proposed by Sheiko et al. (1993), and found to be paraboloidal to a very good approximation. To acquire one friction vs. load curve, images were taken by varying the load for every line, as discussed above. Typical friction vs. load curves are shown in Figure 6.27. The nonlinear behavior is evident. The experimental data points have been fitted to a contact area vs. load curve as predicted by the JKR theory. Since the agreement between the fit and the data is very good, the shear strength of the contact can be assumed to be constant, so that the dependence on friction is determined only by the contact area.

Another result of these experiments was that adhesion and friction decreased by more than an order of magnitude during the experiments, that is, as more and more curves were acquired (Carpick et al., 1996b). The authors checked that this decrease was not due to wear of the tip, which in any case would increase the radius of curvature and thus also adhesion as well as friction. Every friction vs. load curve could be fitted by a JKR curve as described above (see Figure 6.28). From this, the authors deduce that the decrease in friction is due to changes in the chemistry of the interface, which appears in the JKR equation through the surface energy (see Equation 6.20). Since the experiments were performed in UHV, not many mechanisms can explain this behavior. The authors propose either progressive graphitization of hydrocarbon residues on the tip, or adsorption of K from the mica on the tip as possible explanations. Nevertheless, it is quite surprising that even under UHV conditions, where contaminant layers are absent, the friction of the tip–sample junction can be affected in such a dramatic way.

To investigate the effect of tip shape on the friction vs. load curve, Carpick et al. deliberately blunted the tip and showed that unlike in the previous experiment the corresponding friction vs. load curve did not follow the JKR area–load relation for a spherical tip in contact with a flat sample (Carpick et al., 1996a). As shown in Figure 6.27, an appropriate extension of the JKR theory for a tip with a general shape $y(r) = c \cdot r^n$, however, could be fitted almost perfectly to the experimental data. The best fit corresponded to the exponent n which was deduced from the measured tip shape by imaging a faceted $SrTiO_3$ (305) surface. Therefore, although tip shape determines the friction vs. load curve in a nontrivial way, if the tip shape is known its contribution can be taken into account.

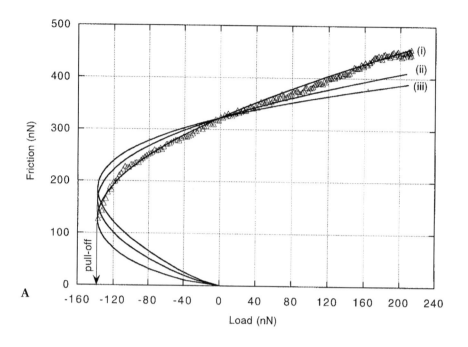

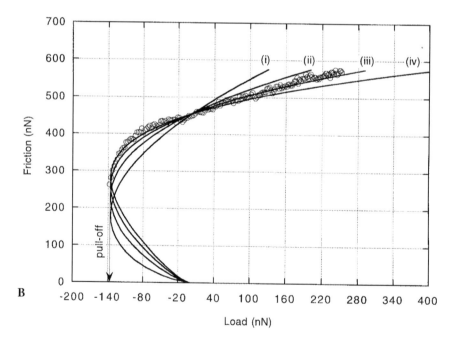

FIGURE 6.27

Similar results have been reported recenty by Schwarz et al. (1997). They show that even though the absolute friction scales with the tip radius, its fundamental behavior does not vary for tip radii between 10 and 125 nm. In these experiments they also find a shear strength which is independent of pressure.

The experiments described above were limited to the study of the frictional behavior of the tip–sample contact. Experiments have also been performed to investigate other properties of tip–sample contacts with scanning probe methods. More specifically, the formation of nanonecks has been investigated by

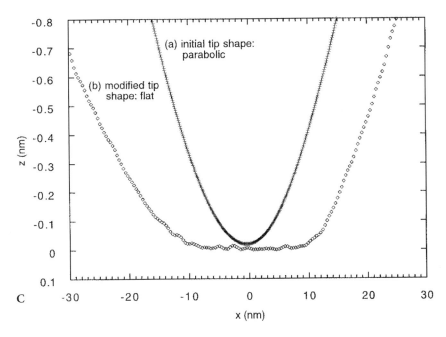

FIGURE 6.27 A: Friction vs. load curve of a Pt tip on a mica surface taken under UHV conditions. The curve marked (i) corresponds to an area vs. load curve as predicted from the JKR theory for a spherical tip. B: Friction vs. load curve of a Pt tip on mica after blunting of the tip. The curves shown corresponds to area vs. load curves according to JKR theory for increasingly flatter tip profiles: curve (i) is for a spherical or paraboloidal tip ($z(r) \sim r^2$), curve (ii) for $z(r) \sim r^4$, curve (iii) for $z(r) \sim r^6$, and curve (iv) for $z(r) \sim r^8$. Each of these curves has been least-square fitted to the experimental data, the best fit being for the shape measured experimentally and shown below. C: Experimentally determined tip profiles (see text) showing the initial and final tip shape which agree very well with the corresponding friction vs. load curves. (From reference Carpick, R. W. et al. (1996), *J. Vac. Sci. Technol. B* 14(2), 1289. With permission.)

STM by measuring the current as a sharp tip is indented into a sample surface. The conductance of these necks has been shown to vary in integer steps of the quantum resistance (Agraït et al., 1993; Pascual et al., 1993). This can be explained in terms of the number of conductance channels which fit in the minimum cross section of the tip–sample contact. Other experiments have been performed to study the physics of tip–sample contact not only with STM but also in combination with SFM (Rubio et al., 1996; Stalder, 1996). These studies suggest that during neck formation very strong pressures, compressive as well as tensile, act in the neck and that substantial atomic rearrangement occurs, as predicted by molecular dynamics simulations. Moreover, on a nanoscale materials seem to be much harder than on a macroscopic scale (Agraït, 1996), which is in accordance with other studies (Bhushan and Koinkar, 1994) and can be explained by the absence of defects in small nanometric volumes. These studies on the formation of nanonecks have had up to now no direct consequences on atomic-scale tribology; we believe, however, that both topics are very strongly related and that in the near future similar experiments will be done which are indeed directly connected to tribology on a nanoscale.

6.3.4 Quartz Microbalance Experiments in Tribology

In SFM and SFFM the interaction between tip and sample is usually very strong and correspondingly the pressures are very high. To study tribological processes at very low interaction and pressures, the SFFM is at the moment not the ideal instrument. In this regime, the quartz microbalance (QCM) has been applied successfully to study energy dissipation on an atomic scale. A QCM is essentially a crystalline piece of quartz glass. Since this material is piezoelectric, if it is cut and segmented appropriately, it can

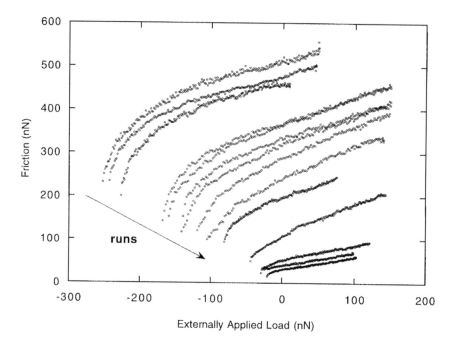

FIGURE 6.28 Series of friction vs. load curves of a Pt tip in contact with mica taken under UHV conditions. As more and more curves are acquired, the pull-off force as well as the friction force decreased. Each of these curves can be fitted to a JKR curve. The reduction of adhesion and friction observed in this experiment is not due to a change in tip shape as in the previous figure, but due to a reduction in surface energy and shear stress which is caused probably by variations of the structure or chemistry of the interface. (From reference Carpick, R. W. et al. (1996), *Langmuir* 12(3), 3334–3340. With permission.)

be oscillated mechanically by applying a harmonic signal of correct frequency. Commonly a QCM is used in vacuum chambers to measure evaporation rates: when material is deposited on the QCM, due to the increment in mass its resonance shifts and from this shift the mass can be calculated (Lu and Czanderna, 1984).

To study tribological processes, Krim and co-workers (Krim et al., 1991; Daly and Krim, 1996) use a QCM which is driven in transverse mode, the motion of the QCM being then parallel to its surface. Since a QCM is essentially a harmonic oscillator, three parameters determine its behavior: the resonance frequency, a driving amplitude, and the quality factor Q of the resonance, which is a measure of how much energy is lost during one oscillation. Typical values for the resonance frequency f_0, driving amplitude a_0, and the Q-factor in the experiments to be described are $f_0 \approx 8$ MHz, $a_0 \approx 20$ nm, and $Q \approx 10^5$, respectively.

A metal film — usually Au or Ag — is evaporated prior to the experiment as the substrate for the friction measurements. The surface roughness of these films was measured by STM, X-ray reflectivity, and liquid nitrogen adsorbtion measurements. Since the estimated sliding distance of the films on the substrate is of the order of 2 nm and therefore considerably smaller than the typical grain size, the effect of surface roughness is assumed to be small. Experiments are performed in a vacuum chamber in a liquid nitrogen bath. A noble gas — Kr or Xe — is brought into the chamber and adsorbs under equilibrium conditions onto the metal surface on the QCM. At 77.4 K these films are known to adsorb as one- or two-monolayer-thick films. The first monolayer adsorbs first as a liquid, which then solidifies as the partial pressure of the gas is increased and the monolayer is compressed.

During an experiment, a film is deposited by slowly increasing the partial pressure of the gas, while the shift in resonance frequency and the variation of the Q-factor is measured. The shift in resonance frequency δf is proportional to the (areal) mass density ρ (unit: Kg/m²) deposited on the QCM, and the variation of the *Q*-factor is a measure of the energy dissipated by the presence of the film. From these measured quantities, the so-called slip time,

$$\tau = \frac{1}{4\pi} \frac{\delta\left(1/Q\right)}{\delta f},$$

of the film is calculated (Krim and Widom, 1988). This slip time can be interpreted as a characteristic time until friction stops the relative motion of the film, i.e., a damping time. Since the total mass is proportional to the frequency shift, the slip time can be considered to be normalized with respect to the number of atoms which have been adsorbed. Assuming a friction force proportional to the sliding velocity of the film (Stokes law), it can be shown that $1/\tau$ is proportional to the mean energy dissipated per atom (Krim et al., 1991). The friction itself can be written in terms of the shear stress *S* needed to slide the film over the surface:

$$S = \frac{\rho}{\tau} \cdot v$$

where ρ is the areal mass density and *v* the velocity of the film. This relation implies a friction which is proportional to velocity and therefore of Stokes-type as is typical for gas or liquid flow. In fact, the coefficient ρ/τ can be interpreted as interfacial viscosity. The velocity dependence of the friction has been confirmed experimentally (Krim et al., 1991).

Figure 6.29 shows the result of an experiment where xenon is adsorbed on a silver (111) surface (Daly and Krim, 1996, 1997). The slip time τ decreases until the first monolayer is completed. As this monolayer is compressed and solidifies, the slip time increases; therefore, the friction of the film decreases (not shown). The same behavior is observed for Kr on Ag and Au (Krim et al., 1991). This is interesting and counterintuitive because it means that, unlike in macroscopic friction, a solid noble gas film on a metal substrate slides easier than a liquid one. The films are thus "slippier when dry" (Krim, 1996). Similar results have been obtained theoretically by Cieplak et al. (1994), who attribute the friction to phonon processes within the film–substrate system (see also Solokoff, 1990). Other possible contributions to the friction are due to the electronic resistance felt by the electrons in the metals as they are influenced by the moving film (Levitov, 1989; Persson, 1991, 1993). The experiments by Krim and co-workers, including some recent results (Daly and Krim, 1996, 1997), seem to indicate that phonon processes are mainly responsible for the friction observed. This is in accordance with friction in SFFM, which can also be interpreted in terms of phonon creation (see Section 6.4.3.3). The underlying mechanism would then be similar even though the range of shear stress is almost ten orders of magnitudes different.

6.4 Modeling of an SFFM

In the preceding sections, many experiments have been described which we believe are relevant to the processes governing friction on an atomic scale. We have tried to explain most results with simple arguments along the way. No deeper knowledge of theoretical work in nanotribology has been assumed. In this section, the attempt will be made to explain some of the addressed experiments in more detail and to describe basic concepts involved in SFM and SFFM. Since SFM, SFFM, and nanotribology in general are part of a still very young field, many phenomena are still far from being fully understood, both from an experimental as well as from a theoretical point of view. The fundamental question in tribology is, of course, how energy is dissipated. Essentially, two different mechanisms have been proposed, one involves phonic and the other one electronic excitation. Very much has also been learned from molecular dynamics simulations as well as from first principles calculations. To study tribological phenomena, these latter approaches are very difficult and complex, due to the large number of atoms

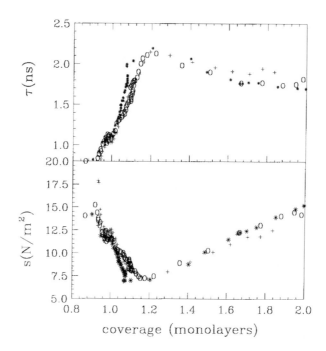

FIGURE 6.29 Top: Slip time vs. coverage for a xenon film deposited on a silver surface for three different surfaces. One monolayer has been defined as 5.970 atoms/nm², the spacing of the compressed monolayer, and the slip time is normalized with respect to this point. Bottom: Corresponding shear stress for the xenon film on the silver surfaces as a function of coverage. This shear stress is the force per unit area required to slide the film at a constant speed of 1 cm/s (the order at which the films are moving). (From Daly, C. and Krim, J. (1996), *Phys. Rev. Lett.* 76(5), 803–806. With permission.)

involved in tribological processes: while the main interaction might involve only a few atoms on each of the sliding surfaces, the elastic deformation reaches much farther into the two interacting bodies and finally, the volume into which elastic waves — that is, phonons — can be scattered is even larger. Therefore, very different length scales are typically involved in tribological processes even if interaction is through a small, well-defined nanometer contact. Similar arguments lead to the conclusion that also very different timescales can be involved. Recently, mixed atomistic and continuum models have been proposed (Tadmor et al., 1996), but the situations treated up to now with this approach are very simple. Molecular dynamics and surface physics are treated in detail in this book in Chapters 11 and 3 by Harrison et al. and by Ferrante and Abel, respectively. Essentially, the scope of this section is to describe the processes encountered in SFFM. In particular, we will try to address the following main questions in SFFM:

What is the reason for the high apparent resolution?
What is the mechanism for the stick-slip motion? and
How and where is energy dissipated?

We will limit the present section to simple models which, although much less precise, still explain some fundamental aspects of friction on an atomic scale (Zhong and Tománek, 1990; Tománek et al., 1991; Colchero and Marti, 1995; Colchero et al., 1996b; Gyalog et al., 1995). These models can be considered an extension of a molecular model of friction proposed by Tomlison (1929).

6.4.1   Resolution in SFFM

Mostly, the total interaction of tip and sample in SFM can be calculated by adding all the interactions between infinitesimal volumes of tip and sample. Here, a similar approach will be presented, which however will assume that the interaction takes place between the surfaces of tip and sample (for a similar discussion, see Pethica and Sutton, 1995). The decay of this interaction is described by a function $i(z)$, where z is the distance between the surfaces. The surfaces of tip and sample are parameterized by $T(x)$ and $S(x)$, respectively. To simplify the model further, we will restrict the analysis to one dimension. The total interaction between tip and sample is then

$$I_{tot}(\Delta) = \int_{-\infty}^{+\infty} i\left(T(x) - S(x - \Delta)\right) dx. \tag{6.15}$$

The term Δ takes account of the scanning, that is, by varying the distance Δ, the relative position of tip and sample is changed. Assuming some geometry for the tip and for the sample, this integral can be solved. A general geometry for an SPM setup on an atomic scale is a slowly varying corrugated surface

$$S(x - \Delta) = a_S \cos\left(k_S(x - \Delta)\right) + h(x - \Delta),$$

where a_s is the corrugation of the surface, k_s the corresponding reciprocal lattice vector, and h a function which describes the slow variation of the surface height (that is, $h'(x) \ll k_s$). The tip is assumed parabolic with an atomic corrugation,

$$T(x) = a_T \cos(k_T x) + x^2/2R + a_{TS} \cos(k_S x),$$

where a_T is the corrugation of the tip, R the tip radius, and $a_{TS} \cos(k_s x)$ a term which is introduced phenomenologically. Such a term could arise from interaction with the surface and would describe, for example, a flake of surface material picked up by the tip or some sort of reconstruction on the tip induced by the surface. This term has the same periodicity as the surface and it seems reasonable that only one of the terms a_T or a_{TS} is important for a given tip–sample system. In the case of contact mode SFM and SFFM the tip and the surface are deformed and a mechanical contact of radius r_c is formed (see Section 6.4.2). Then, the main interaction is through the contact area; therefore, the term describing the parabolic tip can be omitted and the integral can be evaluated only cover the contact radius. A Taylor expansion of the interaction around a mean height $h_0(x - \Delta)$ up to second order leads to

$$I_{tot}(\Delta) \simeq \int_{-r_c}^{+r_c} \left(i\left(h_0(x - \Delta)\right) + \frac{\partial i}{\partial z} \delta h(x - \Delta) + \frac{1}{2} \frac{\partial^2 i}{\partial z^2} \delta h^2(x - \Delta) \right) dx,$$

where $h_0(x - \Delta)$ is a mean height and $\delta h(x - \Delta) = a_s \cos(k_s(x - \Delta)) - a_T \cos(k_T(x)) - a_{TS} \cos(k_s(x))$. This integral can be interpreted in terms of the mean value of the integrand averaged over the contact radius r_c:

$$I_{tot}(\Delta) \simeq 2r_c \left(\langle i(h_0(\Delta)) \rangle_{r_c} + \frac{\partial i}{\partial z} \langle \delta h(\Delta) \rangle_{r_c} + \frac{1}{2} \frac{\partial^2 i}{\partial z^2} \langle \delta h^2(\Delta) \rangle_{r_c} \right).$$

If the contact radius is large compared to the periodicity of the lattice, then the term $\langle \delta h(\Delta) \rangle_{r_c}$ vanishes and the only nonvanishing term in $\langle \delta h^2(\Delta) \rangle_{r_c}$ is due to $\langle a_s \cos(k_s(x - \Delta)) \cdot a_{TS} \cos(k_s(x)) \rangle_{r_c}$. Therefore, the total interaction is

$$I_{tot}(\Delta) \simeq 2r_c \left(\langle i(h_0(\Delta)) \rangle_{r_c} + \frac{1}{2} \frac{\partial^2 i}{\partial z^2} a_s a_{TS} \cos(k_s \Delta) \right),$$

from which we see that, under the assumptions made, the atomic periodicity is "resolved" as the tip scans the sample — that is, as Δ is varied — even though the "true" resolution is much less, namely, of the order of the contact radius. To resolve the atomic periodicity with a large contact radius, the term a_{TS} which was introduced above has to be nonzero; the tip therefore has to have the same periodicity as the surface. To achieve "true" atomic resolution within this simple model, the contact radius would have to be reduced to atomic dimensions and the tip would have to be sharp enough, which is what is expected intuitively. If this is the case, however, the term $\langle \delta h\,(\Delta)\rangle_{r_c}$ does not necessarily vanish and thus the integral 15 should be computed without approximations.

6.4.2 Deformation of Tip and Sample

The deformation of tip and sample is fundamental to the adhesion and friction of any contact, including the special case of an SFM or SFFM setup. To describe these deformations, continuum theories are generally used. A detailed description of contact mechanics can be found in Johnson (1985). The first approach to contact mechanics is due to Hertz (1881). He showed that two elastic solids of radii R_1 and R_2 pressed against each other with a loading force F_n deform elastically and touch on a circle of radius

$$r_c = \left(\frac{3}{4} \frac{R^\star}{E^\star} F_n \right)^{1/3},\qquad (6.16)$$

where $R^\star$ is an effective radius such that $1/R^\star = 1/R_1 + 1/R_2$ and $E^\star$ is an effective modulus of elasticity such that $1/E^\star = (1-\nu_1^2)/E_1 + (1-\nu_2^2)/E_2$, with E the modulus of elasticity and ν Poisson's ratio. In the case of SFM and SFFM, the sample is assumed flat and the tip of radius R_{tip}, and correspondingly $R^\star = R_{tip}$. Due to the elastic deformation, distant parts of the two bodies approach each other by a distance $\delta = r_c^2/R^\star$. With Equations 6.16 this relation can be restated as

$$F_n\left(\delta\right) = \frac{4}{3}\sqrt{R^\star} \cdot E^\star \cdot \delta^{3/2},$$

from which the amount of elastic energy $V_{el}\,(\delta)$ stored in the system is computed by integration:

$$V_{el}\left(\delta\right) = \frac{8}{15}\sqrt{R^\star} \cdot E^\star \cdot \delta^{5/2},$$

and the stiffness c_{con} of the contact by differentiation:

$$c_{con}\left(\delta\right) = 2\sqrt{R^\star} \cdot E^\star \cdot \delta^{1/2} = 2\,E^\star \cdot r_c.\qquad (6.17)$$

The elastic energy $V_{el}\,(\delta)$ can be interpreted as a repulsive surface potential which acts only after tip and sample have touched. It should be noted, however, that the distance δ is not directly the indentation distance of the tip. Another important magnitude is the pressure distribution $p(r)$ in the contact region:

$$p(r) = \frac{3}{2} \frac{F_n}{\pi \cdot r_c^2} \sqrt{1 - \left(\frac{r}{r_c}\right)^2}.$$

If in addition to the normal loading force a lateral force F_{lat} is applied to the tip and if the friction is high enough to prevent slip, then the tip–sample contact will be deformed laterally. The corresponding lateral displacement of the contact is (Johnson, 1985):

$$\delta_{\text{lat}} = \frac{F_{\text{lat}}}{8 \, r_c \cdot G^\star} \,,$$

where $G^\star$ is an effective modulus of elasticity such that $1/G^\star = (2 - \nu_T)/G_T + (2 - \nu_S)/G_S$, G_T and G_S are the shear moduli of tip and sample, respectively, and ν_T and ν_S the corresponding Poisson ratios. The lateral stiffness of the contact is then

$$c_{\text{lat}} = 8 \, r_c \cdot G^\star. \tag{6.18}$$

And, finally, the shear stress in the contact region is

$$S(r) = \frac{F_{\text{lat}}}{2 \, \pi \cdot r_c^2} \left(1 - \left(\frac{r}{r_c} \right)^2 \right)^{-1/2} .$$

The Hertz theory does not account for surface forces and adhesion. Due to these surface forces, bodies adhere to each other and will be deformed slightly by adhesion forces. The adhesion force of a sphere or a parabolic tip on a flat surface is (Israelachvili, 1992)

$$F_{\text{ad}} = 4 \, \pi \cdot \gamma \cdot R \,, \tag{6.19}$$

where R is the tip radius and γ is the surface energy. This attractive adhesion force acts in addition to any external load. Therefore, the simplest way to take into account adhesion in the deformation of the tip–sample contact is to define an effective normal force as $F_n = F_{\text{load}} + F_{\text{ad}}$, and to apply Equation 6.16 with this effective normal force. At zero load the contact radius is then $(3\pi\gamma R^2/E^\star)^{1/3}$. This approach is appropriate if the deformation of the tip–sample contact induced by surface forces is small compared with the deformation due to the contact forces, which is the case if the two bodies are sufficiently rigid. If, on the contrary, the surface forces are not negligible, then the exact shape of the contact will differ from the one predicted by the Hertz theory. One way to account for the surface forces is to allow the tip–sample contact to deform in order to increase the contact area and thus the surface energy. The shape of the tip–sample contact is then determined by the balance between the elastic energy needed to deform the contact and the surface energy gained as the contact forms. This is the basic idea behind the JKR theory (Johnson, 1971), which predicts a contact radius

$$r_c = \left(\frac{3}{4} \frac{R \cdot F_{\text{ad}}}{E^\star} \left(2 + \frac{F}{F_{\text{ad}}} + 2 \sqrt{1 + \frac{F}{F_{\text{ad}}}} \right) \right)^{1/3} , \tag{6.20}$$

with $F_{\text{ad}} = 3\pi\gamma R$. For vanishing surface energy, the Hertz relation is recovered. For a nonvanishing surface energy, the JKR theory predicts an adhesion force $-F_{\text{ad}}$, which is lower than the adhesion force predicted by the Hertz theory. According to Equation 6.20, the minimum contact radius $r_{\text{min}} = (3R \cdot F_{\text{ad}}/4E^\star)^{1/3}$ is obtained with a tensile force $F = -F_{\text{ad}}$. Due to surface forces, a neck forms between tip and sample which becomes unstable if the minimum contact radius is reached. This is especially relevant in SFM and SFFM because it means that, unlike in the case of the Hertz theory, the contact radius cannot be reduced to atomic dimensions by compensating the adhesion force adequately.

The JKR theory describes adhesion by means of the surface energy. For a complete description of the contact problem, the finite range of the surface forces has to be taken into account. Forces act not only in the contact region, but also further out, where the two bodies are not in direct mechanical contact but separated by distances so small that interaction forces are still present. The exact shape of the whole contact is then determined by the balance of forces due to the surface potential and due to the elastic deformation of tip and sample. Furthermore, most approaches neglect lateral forces, which are only treated in very recent studies (Johnson, 1997a,b). A more-detailed description of these theories can be found in Chapter 5 by Burnham.

6.4.3 Modeling of SFM and SFFM: Energy Dissipation on an Atomic Scale

For a detailed understanding of an instrument, a model is usually developed that retains its fundamental properties but is still as simple as possible. In the present section, such a model will be discussed for SFM and SFFM (Zhong and Tománek, 1990; Tománek et al., 1991; McClelland and Gosli, 1992; Colchero and Marti, 1995; Colchero et al., 1996b; as well as Gyalog et al., 1995). To keep ideas simple, the model will be restricted first to one dimension; therefore the behavior normal to the surface and the behavior lateral to the surface will be analyzed separately. From an experimental point of view, in an SFM only the behavior normal to the surface is important. On the other hand, from the theoretical point of view, any SFM is also an SFFM: for nonvanishing friction, lateral forces always act on the tip independently of whether these forces are measured or not.

6.4.3.1 Modeling Energy Dissipation

Energy dissipation in SFM and SFFM can be understood in terms of a simple spring model: two springs held together, for example, by magnets at one end and attached to rigid supports at the other. As the supports are separated, potential energy is stored in the springs. Since the force on each spring is the same, one finds $F = \Delta_1/c_1 = \Delta_2/c_2$. The energy stored E_i in each spring can therefore be written as

$$E_1 = \frac{c_1}{2} \cdot \Delta_1^2 = \frac{1}{2} \frac{F^2}{c_1} \quad \text{and, correspondingly,} \quad E_2 = \frac{1}{2} \frac{F^2}{c_2}, \quad (6.21)$$

and the total energy is $E_{tot} = E_1 + E_2 = (1/c_1 + 1/c_2)F^2/2$, from which the effective spring constant of the total system is determined as

$$1/c_{tot} = 1/c_1 + 1/c_2 . \quad (6.22)$$

The energy stored in each spring (see Equation 6.21) scales as

$$\frac{E_1}{E_2} = \frac{c_2}{c_1} . \quad (6.23)$$

Correspondingly, most energy is stored in the weakest spring. If the springs are separated enough and the springs break apart, the elastic energy stored in each spring will be converted into kinetic energy. Since the total system is now separated into two springs that cannot interact with each other any longer, the energy is finally dissipated within each spring. The energy is therefore dissipated according to Equation 6.23.

This simple model can be applied directly to an SFM setup; each component that can store elastic energy is represented by a spring with the corresponding force constant (see also Equation 6.30). Usually, in SFM the cantilever is the softest spring, so that the most energy is dissipated in the cantilever. However, the other extreme — the cantilever being the stiffest spring — is also possible as, for example, in STM. In STM the tip is ideally stiff, but snapping still occurs due to the intrinsic elasticity of the tip–sample contact (Smith et al., 1989). In this case all the energy dissipated during snapping into and out of contact is dissipated in the microscopic degrees of freedom of the tip–sample system.

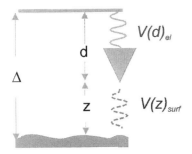

FIGURE 6.30 Simple model for a typical SFM setup. The surface potential $V_{surf}(z)$ and the potential $V_{el}(d)$, which represents the elastic energy stored in the system, are represented by springs. (From Colchero, J. et al. (1996), *Tribol. Lett.* 2, 327–343. With permission.)

6.4.3.2 SFM and Normal Forces

A simple model for an SFM is shown in Figure 6.30. The main components are the tip, a spring, its rigid support, and the sample to be studied. It is fundamental to note that the rigid support and not the tip is moved with respect to the sample surface. Three different distances are relevant in this model: the tip–sample distance z, the deflection d of the spring, and the separation Δ between the rigid support and the sample. However, only two of these distances are independent, since $d = z - \Delta$. The separation Δ is the distance that is controlled experimentally, z the distance usually used to describe the tip–sample interaction as a function of distance, and d the deflection measured. In the present discussion, z and Δ will be chosen as independent parameters which determine the state of the SFM setup. To determine the behavior of the system, a standard analysis in terms of classical mechanics for zero temperature will be presented (for $T \neq 0$, see Dürig et al., 1992). The total energy of the tip–sample contact is determined by the elastic energy $V_{el}(d)$ stored in the system as well as by the potential energy $V_{surf}(z)$ of the tip in the surface potential. The elastic energy will first be assumed to be due to the elastic energy of the spring representing the cantilever, $V_{el}(d) = c \cdot d^2/2$. In this case, the total energy is

$$V_{tot}\left(z, d\right) = V_{surf}\left(z\right) + \tfrac{1}{2}c \cdot d^2 = V_{surf}\left(z\right) + \tfrac{1}{2}c \cdot \left(z - \Delta\right)^2 = V_{tot}\left(z, \Delta\right). \qquad (6.24)$$

In the right part of the equation, the total energy has been written in terms of the two independent parameters z and Δ. In a typical SFM setup, the separation Δ is controlled experimentally, but the tip is free to move to an equilibrium position by varying its tip–sample distance, z; therefore, the force equilibrium condition is

$$F_{tot}\left(z, \Delta\right) = -\frac{dV_{tot}\left(z, \Delta\right)}{dz} = -\frac{dV_{surf}\left(z\right)}{dz} - c \cdot \left(z - \Delta\right) = 0. \qquad (6.25)$$

This force equilibrium condition determines the tip–sample distance z_{eq} for each separation Δ. If this separation is varied, in general the equilibrium distance also varies. By solving Equation 6.25 for different separations Δ, the curve $z_{eq}(\Delta)$ is evaluated, which is fundamental to the behavior of the tip–sample contact. To have stable equilibrium, Equation 6.25 is not sufficient; in addition,

$$c_{tot}\left(z_{eq}\left(\Delta\right)\right) = \frac{d^2 V_{tot}\left(z_{eq}\left(\Delta\right), \Delta\right)}{dz^2} = \frac{d^2 V_{surf}\left(z_{eq}\left(\Delta\right)\right)}{dz^2} + c > 0 \qquad (6.26)$$

has to hold. For repulsive potentials, that is, if the force increases as the distance is decreased, this condition is always fulfilled. However, for sufficiently attractive potentials, this relation will fail. Then, the tip jumps onto the surface.

The tip–sample behavior can be illustrated as follows. Consider the two-dimensional energy surface $V_{tot}(z, \Delta)$ shown in Figure 6.31. Any possible configuration of the tip–sample contact is represented by

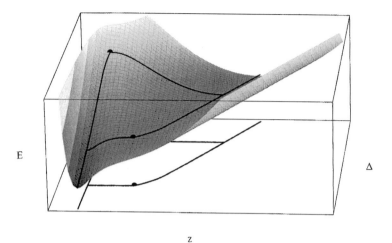

E

Δ

z

FIGURE 6.31 Energy surface that describes the state of an SFM tip–sample system. The coordinate Δ describes the experimentally controlled distance between the sample and the base of the cantilever. The coordinate z represents the tip–sample distance. As the distance Δ is varied, the system evolves along the solid curve in one of the two valleys. The system becomes unstable at the high end of a valley and moves to the lower valley. The projected curve $z(\Delta)$ in the (Δ, z)-plane visualizes the hysteresis of the system. (From Colchero, J. et al. (1996), *Tribol. Lett.* 2, 327–343. With permission.)

a point on this energy surface corresponding to the coordinate (z, Δ) and to an energy $V_{tot} (z, \Delta)$. For each fixed separation Δ_0, the curve $V_{tot} (z, \Delta_0)$ is an energy curve on which the tip can move to minimize its energy. The equilibrium condition, together with the stability condition (Equation 6.26) and the additional assumption that the tip movement is "quasi-static" (the tip does not oscillate), restricts the possible configurations to the local minima of these curves, that is, to the valleys seen in Figure 6.31. As the separation Δ is varied, the tip–sample contact will evolve to a new equilibrium position in the same valley. If the system, however, is moved past the end of a valley, it will not be stable and move to the other valley. Since this other valley is lower than the end point of the first valley, the system will oscillate around the new minimum until its kinetic energy is damped. We will assume that some damping mechanism exists and will discuss its nature below. If the separation Δ is varied in a sufficiently large cycle, the system will evolve along the curve $V_{tot} (z_{eq}(\Delta), \Delta)$ indicated by the solid line in Figure 6.31. The projection of this curve onto the (z, Δ)-plane defines the curve $z_{eq}(\Delta)$, which is the tip–sample separation as a function of the separation Δ.

For a given surface potential the tip–sample distance $z_{eq}(\Delta)$ can be constructed uniquely from the equilibrium condition and from the stability condition. In a typical SFM experiment, the force — more precisely, the deflection of the cantilever — is measured as the separation Δ is varied. This so-called force vs. distance curve (Weisenhorn et al., 1989) does not correspond to the force curve $F(z)$, but to

$$F\left(\Delta\right) = c \cdot \left(z_{eq}\left(\Delta\right) - \Delta\right) = -\frac{dV_{surf}\left(z_{eq}\left(\Delta\right)\right)}{dz}.$$

Another important question is the energy that is dissipated during the acquisition of one force vs. distance curve. This energy is the area enclosed by the force vs. distance curve and can be written as

$$E = \oint F\left(\Delta\right) d\Delta.$$

In the discussion above, the deformation of the tip–sample contact was not taken into account. The elastic energy was described only by the harmonic potential of the spring. The model can be generalized to account for the elastic energy stored in the tip–sample contact by introducing the corresponding potential (Colchero et al., 1996b). Within this generalized model, the total elastic energy stored in the SFM setup can be approximated by assuming that this energy is stored in an effective spring with an effective force constant c_{eff} defined by

$$\frac{1}{c_{eff}} = \frac{1}{c_{tip}} + \frac{1}{c_{con}},$$

where $c_{con} = 2E' \cdot r_c$ is the stiffness of the tip–sample contact (see Equation 6.17).

To illustrate the formal description of the SFM setup presented above and to analyze the dissipation of energy, a typical force vs. distance curve will be discussed in physical terms (see also Tabor, 1992). First, the tip is far from the sample, feels no surface potential, and is not deflected. As the (base of the) cantilever approaches the sample, surface forces begin to act. Due to the attractive surface potential, the tip approaches the surface more than the increase of the separation: $\delta z = \delta \Delta + \delta d$, where δz is the increase in tip–sample distance, δd the increase in deflection, and $\delta \Delta$ the increase in separation. Eventually, the spring is not "strong" enough to hold the tip and the tip jumps onto the surface. During this process, the tip is first accelerated, then hits the surface, and is suddenly stopped. This will induce vibrations of the cantilever, but also a sudden elastic deformation of the microscopic tip–sample contact, which in turn will result in elastic waves propagating out of the tip–sample contact. After some time, the kinetic energy gained by the tip during the snapping process will be effectively removed from the tip–sample system either into the macroscopic cantilever or into the two solids tip and sample. A similar process happens when the tip is separated from the sample. Due to adhesion, the tip sticks to the sample. Elastic energy is built up in the deformation of the cantilever, but also in the microscopic deformation of the tip–sample contact. When this contact breaks, the cantilever will vibrate and the elastic deformation of the tip–sample contact will relax. Again, the elastic energy stored in the system will be converted into kinetic energy of the cantilever, or into waves in the tip and the sample, and will in the end be dissipated due to some kind of friction (for example, internal friction in the solids or air damping in the case of the cantilever).

6.4.3.3 SFFM and Lateral Forces

In the preceding section a general model for an SFM setup was described. However, this model was applied only to study the behavior of the tip–sample system as the tip approaches the sample. In the present section, the model will be applied to study the behavior of the tip–sample contact as the separation between the support and the sample is varied in a direction parallel to the surface (Zhong and Tománek, 1990; Tománek et al., 1991). This is the situation that corresponds to the usual scanning motion of the tip. Essentially, the description is as before: the tip is attached to a rigid support by means of a spring and the sample is moved relative to this support by varying the separation Δ_x. Again, the tip position is not controlled directly, but follows from the equilibrium condition. The only difference compared to the description above is that in order to describe the atomic corrugation of the surface, a surface potential with the periodicity of the sample is assumed. Ideally, this is the potential that a perfect tip with only one atom at its apex would "see," and can be computed from first principles. However, also if a nonideal tip is in contact with the sample on an area with radius r_c the effective interaction between tip and sample will show a modulation with lattice spacing (see Section 6.4.1). The total potential seen by the tip is therefore (Tománek et al., 1991).

$$V_{tot}\left(x, \Delta_x\right) = E_0 \cos\left(kx\right) + \frac{c_x}{2} \cdot \left(x - \Delta_x\right)^2, \tag{6.27}$$

where E_0 is the amplitude of the surface potential, k the reciprocal lattice vector of the surface, and c_x the force constant of the system. The stability condition for this potential is $c_x > E_0 \cdot k^2$; therefore, if the amplitude of the surface potential is sufficiently large, the range of postions with $|x| < \cos^{-1}(c/E_0 \cdot k^2)/k$

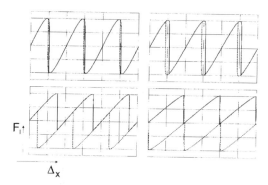

F_l↑

$\vec{\Delta}_x$

FIGURE 6.32   Calculated lateral force curve $F(\Delta)$ — also called friction loop — for different values of the ratio $\gamma = c/E_0 k^2$ (see text). For each curve, the lateral force is plotted vs. the (lateral) displacement Δ of the tip. Top: $\gamma = 1.25$ (left) and $\gamma = \pi/2$ (right); bottom: $\gamma = \pi$ (left) and $\gamma = 4.5$ (right). Note the stick-slip-like behavior for high values of γ. (From reference Colchero, J. et al. (1996), *Tribol. Lett.* 2, 327–343. With permission.)

will be unstable. From the equilibrium condition and the stability condition, the curve $x(\Delta_x)$ as well as the lateral force curve $F(x(\Delta_x))$ can be calculated as described in the previous section. The lateral force curve is equivalent to the force vs. distance curve described above and is the experimental curve which is measured as the tip scans over the surface. Four of these curves are shown in Figure 6.32 for different values of $c/E_0 \cdot k^2$. The solid lines represent the forward scan, the dotted curve the backward scan. As in the case of a force vs. distance curve, the area enclosed by these lateral force curves corresponds to the energy dissipated during one scan cycle. As the ratio $c/E_0 \cdot k^2$ decreases, the dissipated energy increases and the lateral force curves look increasingly sawtooth-like with the typical stick-slip behavior observed experimentally. For low values of $c/E_0 \cdot k^2$ the lateral force curve can be described by a curve which is determined by the lattice spacing l, a maximum force F_0, and a force jump ΔF. For this type of curve, the energy dissipated can be calculated using simple geometric arguments as

$$E = n \cdot l \cdot \left(2F_0 - \Delta F\right), \tag{6.28}$$

where n is the number of stick-slip processes, and the corresponding mean friction force is

$$F_{\text{fric}} = \frac{1}{2}\left\langle \frac{\partial E}{\partial x} \right\rangle = F_0 - \Delta F / 2 \,.$$

The factor ½ takes account of the fact that the energy given in Equation 6.28 is due to a whole scan cycle, while the friction force is only half the mean total height of the friction loop. Equation 6.28 can be shown to follow also from the relation describing the energy dissipation of a general lateral force curve (Colchero et al., 1996b).

6.4.3.4   Two-Dimensional Stick-Slip

The modeling of an SFFM setup above has been restricted to one dimension. However, experimentally a two-dimensional stick-slip motion is observed (see Section 6.3.1.2). Therefore, in this section the model described above will be generalized to a real two-dimensional surface (Gyalog et al., 1995). A schematic view of the corresponding SFFM setup is shown in Figure 6.33, together with a typical lateral force curve. In analogy with the one-dimensional model, the tip–sample position is described by a vector $\mathbf{X} = (x, y)$ and the separation between the rigid support of the cantilever and the surface by a vector $\Delta = (\Delta_x, \Delta_y)$. The surface potential is a periodic function $V_{\text{surf}}(\mathbf{X})$ in the x- and y-direction and describes the atomic

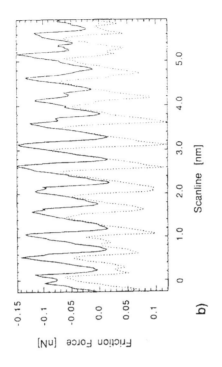

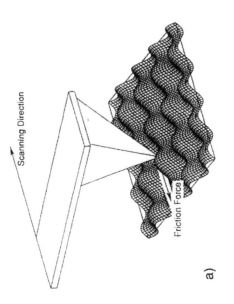

FIGURE 6.33 Schematic setup of an SFFM scanning a two-dimensional surface and a typical lateral force curve. (From Gyalog, T. et al. (1995), *Europhys. Lett.* 31(5–6), 269–274. With permission.)

corrugation of the surface. The elastic energy stored in the system is determined by an elasticity matrix $\hat{\mathbf{C}}$ (see Section 6.2.1.1) and is

$$V_{el}\left(\mathbf{X},\,\boldsymbol{\Delta}\right)=\tfrac{1}{2}\left(\mathbf{X}-\boldsymbol{\Delta}\right)\circ\hat{\mathbf{C}}\circ\left(\mathbf{X}-\boldsymbol{\Delta}\right),$$

where $\mathbf{X}-\boldsymbol{\Delta}$ is the deflection of the tip from its zero-force position. This relation is the generalization of Equation 6.27 for a one-dimensional harmonic potential. The total energy of the tip–sample system is $V_{tot}\,(\mathbf{X},\,\Delta)=V_{surf}\,(\mathbf{X})+V_{el}\,(\mathbf{X},\,\Delta)$. For a fixed separation Δ, the tip can minimize its energy by varying the tip–sample position $\mathbf{X}$. The corresponding two-dimensional equilibrium and stability conditions are

$$\nabla_{\mathbf{X}}V_{tot}\left(\mathbf{X},\,\boldsymbol{\Delta}\right)=0\Leftrightarrow\nabla_{\mathbf{X}}V_{surf}\left(\mathbf{X}\right)=\hat{\mathbf{C}}\circ\left(\mathbf{X}-\boldsymbol{\Delta}\right)$$

$$\lambda_{1,2}\geq0,$$

where the numbers $\lambda_{1,2}$ are the eigenvalues of the Hessian matrix $H_{ij}=\partial^2\,V_{tot}/\partial x_i\partial x_j$. From the equilibrium condition, the separation

$$\boldsymbol{\Delta}\left(\mathbf{X}\right)=\mathbf{X}-\hat{\mathbf{C}}^{-1}\circ\nabla_{\mathbf{X}}V_{surf}\left(\mathbf{X}\right)\tag{6.29}$$

as a function of the tip–sample distance $\mathbf{X}$ is determined. However, since the separation Δ and not the position $\mathbf{X}$ is controlled experimentally, the function $\mathbf{X}(\Delta)$ is needed to specify the behavior of the tip–sample contact. If the surface potential is sufficiently corrugated, the function $\Delta(\mathbf{X})$ has the same value for different $\mathbf{X}$ values; therefore, the inverse function $\mathbf{X}(\Delta)$ is not well defined. Physically, this means that the tip can be in more than one position for a given separation Δ. As in the one-dimensional case, the exact configuration of the tip–sample contact is determined by the history of the system, which leads to hysteresis of the lateral force curve and to energy dissipation. To characterize the two-dimensional stick-slip motion of the tip, the areas (x, y) of the surface, as well as the separations (Δ_x, Δ_y) which correspond to instable positions, have to be determined. These positions (x, y) on the surface are given by all the points where the Hessian matrix H_{ij} has a nonpositive eigenvalue. The borders of these areas form closed curves in the (x, y)-plane. To determine the separations where the tip jumps, these curves are imaged into the (Δ_x, Δ_y)-plane by applying Equation 6.29. The corresponding curves in the (Δ_x, Δ_y)-plane are again closed curves, so-called critical curves. A scanning path is described by a line in the (Δ_x, Δ_y)-plane. If this line crosses such a critical curve, the system becomes unstable and jumps into a new (tip–sample) position. These critical curves are shown in Figure 6.34 for a hard and soft spring (top) in the case of isotropic support ($\hat{\mathbf{C}}=c\cdot\hat{\mathbf{I}}$), as well as for a hard and soft spring (bottom) in the case of asymmetric support. For soft springs, the areas enclosed by the critical curves overlap; then all paths show friction, and slip motion occurs when the critical curves are crossed. For hard springs, the areas enclosed by the critical curves do not overlap. Then paths exists with and without friction.

The lateral force curve is obtained from the computed equilibrium position $\mathbf{X}(\Delta)$ as in the one-dimensional case:

$$\mathbf{F}\left(\mathbf{X}\left(\boldsymbol{\Delta}\right)\right)=\hat{\mathbf{C}}\circ\left(\mathbf{X}\left(\boldsymbol{\Delta}\right)-\boldsymbol{\Delta}\right).$$

At points Δ_0 on the critical curve, the tip–sample contact is not stable and moves into the next stable minimum which is found by following the curve of steepest descent on the potential surface $V_{tot}\,(\mathbf{X},\Delta_0)$. After the excess energy has been damped, the tip–sample contact is in this nearest minimum and continues to evolve along the curve until the next critical curve is crossed.

In conclusion, this two-dimensional stick-slip model generalizes the one-dimensional model discussed earlier and explains all features of the observed two-dimensional lateral force curves.

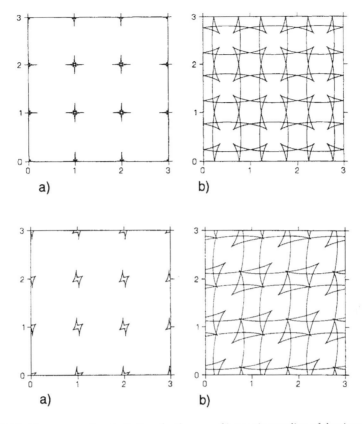

FIGURE 6.34 Critical force curves (see text). Top: for the case of isotropic coupling of the tip to its support (that is, the force constants of the tip in the *x*- and *y*-direction are equal). Left: hard lever, right: soft lever. Bottom: for the case of anisotropic coupling of the tip to its support. Left: hard lever, right: soft lever. When the separation Δ is moved so that the scan line crosses a critical curve, the tip–sample system becomes unstable and the tip snaps into a new position on the sample. (From Gyalog, T. et al. (1995), *Europhys. Lett.* 31(5–6), 269–274. With permission.)

6.4.3.5 Energy Dissipation and Friction

The model presented explains the observed lateral force curve and the fact that energy is dissipated during the scanning motion of the tip. However, the central physical question still remains: exactly how and where is this energy dissipated?

If the force constant c_x in Equation 6.27 is assumed to be simply the lateral force constant of the cantilever, then this energy is dissipated only in the macroscopic cantilever. From a fundamental point of view, this would be rather disappointing. If friction is to be modeled on an atomic scale, it would be more appealing if the energy is lost within the microscopic tip–sample contact. This should also be the case if friction in an SFFM setup is to be considered as a model system for the friction of a single asperity contact. To address the question of where and how the energy is dissipated, we recall the spring model presented above: in a complex system energy is stored in each different subsystem. Each of these subsystems can be described by a spring with a (differential) force constant c_i and the ratio of energy stored in this system is proportional to $1/c_i$. If an instability occurs, the energy stored in each subsystem is converted into kinetic energy which is finally dissipated. Most energy is dissipated in the subsystem with the lowest force constant. In an SFFM setup elastic energy is built up during the scanning motion not only in the cantilever, but also in the tip–sample contact. Each of these subsystems can in turn be divided into further subsystems. The cantilever stores energy not only in the torsion and bending mode, but also in the bending of the tip itself (Fujihira, 1997; Lantz et al., 1997), and the energy stored in the tip–sample contact is due to deformation of the tip as well as of the sample surface. Therefore, the force constant is in fact an effective force constant c_{eff} which can be decomposed into different force constants according to

$$\frac{1}{c_{eff}} = \frac{1}{c_{lever}} + \frac{1}{c_{cont}} = \frac{1}{c_x^{bend}} + \frac{1}{c_x^{tors}} + \frac{1}{c_x^{tip}} + \frac{1}{c_{cont}^{tip}} + \frac{1}{c_{cont}^{sample}}, \qquad (6.30)$$

where c_x^{bend} is the force constant describing the lateral bending of the cantilever, c_x^{tors} the force constant describing its lateral twisting, c_x^{tip} the force constant describing the (macroscopic) bending of the tip, c_{cont}^{tip} the force constant describing the microscopic deformation of the tip, and correspondingly, c_{cont}^{sample} the force constant describing the microscopic deformation of the sample.

Depending on the value of these force constants energy can therefore be dissipated mainly in the macroscopic degrees of freedom of the cantilever or in the microscopic degrees of freedom of the tip–sample contact. Since the stiffness of the tip–sample contact depends on the normal force (see Equations 6.17 and 6.18), for sufficiently low loads the lateral force constant will be determined mainly by this local stiffness and correspondingly energy is dissipated mostly in the microscopic degrees of freedom of the tip–sample sample (Colchero and Marti, 1995; Colchero et al., 1996b). This dependence of the local stiffness on the normal force has been verified experimentally for the case of an Si_3N_4 tip on mica (Carpick et al., 1997). Energy dissipation due to the intrinsic local stiffness of the tip–sample contact is the more fundamental process, since it makes it possible to model friction of a single asperity contact.

Continuum theories have been applied to study the friction of such a single asperity contact (Johnson, 1996, 1997a,b). According to those models, once a critical stress is reached, complete sliding of the tip–sample contact will occur and therefore all energy stored in its elastic deformation will be released. By contrast, due to the atomic corrugation within the interface, the model discussed above predicts that a new stable position will be found after the tip has slipped about one lattice constant. In this case only part of the elastic energy stored in the elastic deformation is released. It would be interesting to try to combine atomistic and continuum models.

By computing the positions of all the atoms of tip and sample before and after the slip, the exact shape of the elastic wave generated in the contact can in principle be calculated. This wave propagates out of the contact area into the tip and the sample and is finally thermalized. In a phonon picture this process can be interpreted as creating (a superposition of) phonons in the tip–sample contact during the slip event, which are then radiated into the bulk solids of the tip and sample, where they are finally scattered and thermalized. In the end, the elastic energy built up during the scanning motion of the tip — or within any of the multiple asperities in a macroscopic tribological contact — is converted into heat, which is what is usually associated with friction.

The ideas discussed in the present section are a qualitative description of what we consider to be the most fundamental processes associated with energy dissipation in an SFFM setup. However, for a detailed and quantitative analysis of friction, much more elaborate models and exact theories involving molecular dynamics simulations and first principles calculations have to be developed.

6.5   Summary

The most striking and impressive performance of SFFM is imaging the lateral force with the atomic periodicity of the surface. The first interpretation is that these images correspond to the friction force as an atomically sharp tip passes over the surface atoms. However, this simple view is incorrect, since both experimental and theoretical arguments show that in a typical SFFM setup the radius of the tip–sample contact is considerably larger than the atomic periodicity observed. Many experiments are described in this chapter to illustrate this point. "True" atomic resolution of the lateral force, as demonstrated, for example, by imaging defects or steps, has not yet been reported. "Atomically resolved" friction images should therefore be interpreted with great care. The question of how this apparent atomic resolution can be achieved has still not been settled. Different mechanisms have been proposed, but none seems to apply generally and others are simply a mathematical restatement of the problem. On the other hand, if an atomically corrugated potential between tip and sample is assumed *a priori,* a simple mechanical model of the SFFM setup can explain the features observed in SFFM, i.e., the shape of the friction loop as well as the stick-slip motion with the atomic periodicity of the surface.

From a fundamental point of view, the question could be posed as whether atomic resolution of the friction force is reasonable at all. Evidently, mechanical friction of a single isolated atom is not defined at all. "True" atomic resolution of the friction force in an SFFM would require an interface with a single atom between the sliding surfaces, but it is not clear that such a one-atom contact is a stable configuration that can be realized experimentally. And even in this hypothetical case, friction would occur between the interfaces consisting of many atoms. Correspondingly, a large number of atoms are always involved in friction.

The question of whether true atomic resolution is possible or not is indeed intriguing. However, from a pragmatic point of view it is rather irrelevant. Compared with other tribological instruments, SFFM achieves a resolution which is order of magnitudes higher; lateral and normal forces can be measured with nano- and even piconewton precision and with nanometer spatial resolution, and the topography of the surface can be determined with (sub)atomic resolution. Moreover, an SFFM can measure simultaneously these three tribologically most relevant quantities. The amount of new tribological phenomena that can be addressed is therefore immense. We have described recent progress in two important fields, boundary lubrication and the friction of a single asperity contact.

In air, as well as in many tribological applications, surfaces are covered by a thin liquid film which modifies their tribological properties. Liquid menisci may form increasing the adhesive force between the sliding surfaces. The physical and chemical interactions between the surfaces are affected by the presence of water which can act as a lubricant. The tribology of the contact can also be changed deliberately by covering the surface with a few monolayers of organic materials, that is, by boundary lubrication. SFFM has resolved the structure of these films and already shown that one monolayer of LB films or self-assembling monolayers can significantly reduce friction and wear on a nanoscale.

The other important matter addressed in this chapter is the tribology of a single-asperity contact. The behavior of such a contact is rather well understood by now. Essentially, the shear stress within a single-asperity contact seems to be constant. Therefore, the friction is proportional to the contact area, as in the case of macroscopic friction. By contrast, the corresponding friction–load relation is nonlinear and determined by the tip geometry and the area–load relation as predicted by continuum theories. Further experiments have to show up to what limit the shear stress can really be assumed constant.

Typical for SFFM are high pressures in the contact easily exceeding 1 GPa and thus the yield pressure of many materials. In the context of tribology this is not a disadvantage, since high pressures are usually inherent to friction; therefore, those occurring in an SFFM contact are necessary in order to study tribology under "normal" conditions. On the other hand, to investigate some fundamental aspects of friction a system with much lower interaction is easier to understand. In tribology, the friction of an atomically thin film sliding on the moving surface of a quartz microbalance represents such a low-interaction system. A lot has been learned, theoretically as well as experimentally, from this system. It seems that the dominant factor for energy dissipation is phononic excitation even though electronic excitation is important in the case of conducting substrates. This is in agreement with the corresponding processes in SFFM, where energy dissipation is assumed to occur as elastic waves — phonons — are scattered into the sample and the tip during the sudden stick-slip transition.

Summarizing, nanotribology is a young and emerging field that is maturing fast, as the experiments described show. Due to the never-ending trend to miniaturization, understanding friction on a nanoscale will become of increasing importance, since as the length scale is reduced, friction forces become stronger relative to other forces. On the other hand, the development of simple systems such as the (single-asperity contact) SFFM, the quartz microbalance, but also the surface force apparatus, is important for the progress of basic research in tribology, since these well-defined and conceptually simple systems are needed for the classical interplay between theory and experiment: developing theories that can be tested with simple model systems. Possibly, if a bridge between nano- and macroscopic friction is found, nanotribology might improve our understanding of classical macroscopic friction and thus help to improve the efficiency and lifetime of everyday devices.

We hope to have convinced the reader that friction on an atomic scale is not only a intriguing field, but also an increasingly important one.

Acknowledgments

This work was supported by the Ministerio de Educación y Cultura (proyecto CICYT, Modalidad C Referencia PB 95-0169), by the Swiss National Science Foundation and the Kommission zur Förderung der Wissenschaftlichen Forschung. We thank all the colleagues who have generously supplied images from their work. We are indebted to many people for stimulating discussions, valuable suggestions, and proofreading of the manuscript. E. Paetz, R. M. Overney, J. Frommer, A. M. Baró, N. Agaït, M. Salmeron, M. Luna, E. Barrena, T. Bonner, W. Gutmannsbauer, L. Howald, R. Lüthi, H. Haefke, M. Rüetschi, H.-J. Güntherodt, and H. Rudin, among others.

References

Agraït, N., Rodrigo, J. G., and Vieira, S. (1993). "Conductance Steps and Quantization in Atomic Scale Contacts, *Phys. Rev. B,* 47, 12345.

Agraït, N., Rubio, G., and Vieira, S. (1996). "Plastic Deformation in Nanometer Scale Contacts," *Langmuir* 12, 4505–4509.

Aimé, J. P., Elkaakour, Z., Gauthier, S., Michel, D., Bouhacina, T., and Curély, J. (1995). "Role of Force of Friction on Curved Surfaces in Scanning Force Microscopy," *Surf. Sci.* 329, 149–156.

Akamine, S., Barett, R. C., and Quate, C. F. (1990), "Improved Atomic Force Microscope Images Using Microcantilevers with Sharp Tips," *Appl. Phys. Lett.* 57(3), 316–318.

Albrecht, T. R. (1989), "Advances in Atomic Force Microscopy and Scanning Tunneling Microscopy," Ph.D. Thesis, G.L. No. 4529, Stanford University, Palo Alto, CA.

Albrecht, T. R., Akamine, S., Carver, T. E., and Quate, C. F. (1990), "Microfabrication of Cantilever Styli for the Atomic Force Microscopy," *J. Vac. Sci. Technol. A* 6, 271–281.

Albrecht, T. R., Grütter, P., Horne, D., and Rugar, D. (1991), "Frequency Modulation Detection Using High-Q Cantilevers for Enhanced Force Microscopy Sensitivity," *J. Appl. Phys.* 69(2), 668–673.

Alexander, S., Hellemans, L., Marti, O., Schneir, J., Elings, V., Hansma, P. K., Longmire, M., and Gurley, J. (1989), "An Atomic Resolution AFM Implemented Using an Optical Lever," *J. Appl. Phys.* 65, 164–167.

Anonymous (1995), a referee's reply.

Bailey, A. I. and Courtney-Pratt, J. S. (1954), "The Area of Real Contact and the Shear Strength of Monomolecular Layers of a Boundary Lubricant," *Proc. R. Soc. A* 227, 501–515.

Bhushan, B. and Koinkar, V. N. (1994), "Nanoindentation Hardness Measurements Using Atomic Force Microscopy," *Appl. Phys. Lett.* 64(13), 1653–1655.

Bhushan, B., Israelachvili, J., and Landman, U. (1995), "Nanotribology: Friction, Wear and Lubrication at the Atomic Scale," *Nature* 374, 607–616.

Binggeli, M. and Mate, C. M. (1994), "Influence of Capillary Condensation of Water on Nanotribology Studied by Force Microscopy," *Appl. Phys. Lett.* 65(4), 415–417.

Binh, V. T. and García, N. (1992), "On the Electron and Metallic Ion Emission from Nanotips Fabricated by Field-Surface-Melting Technique: Experiments on W and Au Tips," *Ultramicroscopy,* 42–44, 80–90.

Binh, V. T. and Uzan, R. (1987), "Tip-Shape Evolution: Capillary Induced Matter Transport by Surface Diffusion," *Surf. Sci.* 179, 540–560.

Binnig, G., Quate, C. F., and Gerber, Ch. (1986), "Atomic Force Microscope," *Phys. Rev. Lett.* 56, 930–933.

Bowden, F. P. and Tabor, D. (1950), *The Friction and Lubrication of Solids — Part 1,* Clarendon Press, Oxford, U.K.

Bowden, F. P. and Tabor, D. (1964), *The Friction and Lubrication of Solids — Part 2,* Clarendon Press, Oxford, U.K.

Briscoe, B. J. and Evans, D. C. B. (1982), "The Shear Properties of Langmuir-Blodgett Layers," *Proc. R. Soc. Lond. A* 380, 389–407.

Carpick, R. W. and Salmeron, M. (1997), "Scratching the Surface: Fundamental Investigations of Tribology with Atomic Force Microscopy," *Chem. Rev.* 97, 1163–1194.

Carpick, R. W., Agraït, N., and Salmeron, M. (1996a). "Measurement of Interfacial Shear (Friction) with an Ultrahigh Vacuum Atomic Force Microscopy," *J. Vac. Sci. Technol. B* 14(2), 1289.

Carpick, R. W., Agraït, N., Ogletree, D. F., and Salmeron, M. (1996b), "Variations of the Interfacial Shear Strength and Adhesion of a Nanometer Sized Contact," *Langmuir,* 12(3), 3334–3340.

Carpick, R. W., Ogletree, D. F., and Salmeron, M. (1997), "Lateral Stiffness: A New Nanomechanical Measurement for the Determination of Shear Strengths with Friction Force Microscopy," *Appl. Phys. Lett.* 70(12), 1548–1550.

Chen, C. J. (1991), "Attractive Interatomic Force as a Tunneling Phenomenon," *J. Phys. Condens. Matter,* 3, 1227.

Cieplak, M., Smith, E. D., and Robbins, M. (1994), "Molecular Origins of Friction: The Force on Adsorbed Layers," *Science* 265, 1209–1211.

Colchero, J. (1993), "Reibungskraftmikroskopie," Konstanzer Dissertationen Band 404, Hartung Gore Verlag, Konstanz, Germany.

Colchero, J. and Marti, O. (1995), "Friction on an Atomic Scale," in *Forces in Scanning Probe Methods,* pp. 345–352, Kluwer Academic Publishers, Dordrecht.

Colchero, J., Luna, M., and Baró, A. M., and Marti, O. (1996a), "Lock-In Technique for Measuring Friction on a Nanometer Scale," *Appl. Phys. Lett.* 68(20), 2896–2898.

Colchero, J., Baró, A. M., and Marti, O. (1996b), "Energy Dissipation in Scanning Force Microscopy — Friction on an Atomic Scale," *Tribol. Lett.* 2, 327–343.

Daly, C. and Krim, J. (1996), "Sliding Friction of Solid Xenon Monolayers and Bilayers on Ag(111)," *Phys. Rev. Lett.* 76(5), 803–806.

Daly, C. and Krim, J. (1997), "Sliding Friction of Compressing Xenon Monolayers," in *Micro/Nanotribology and Its Applications,* (B. Bhushan, ed.), pp. 311–316. NATO ASI Series, Kluwer Academic Publishers, Dordrecht, The Netherlands.

Dowson, D. (1979), *History of Tribology,* p. 3, Longman, London.

Dürig, U., Gimzewski, J. K., and Pohl, D. W. (1986), "Experimental Observation of Forces Acting during Scanning Tunneling Microscopy," *Phys. Rev. Lett.* 57, 2403–2406.

Dürig, U., Züger, O., and Pohl, D. (1988), "Force Sensing in STM: Observation of Adhesion Forces on Clean Metal Surfaces," *J. Microsc.* 152, 259.

Dürig, U., Züger, O., and Pohl, D. (1990), "Observation of Metallic Adhesion Using the Scanning Tunneling Microscopy," *Phys. Rev. Lett.* 65(3), 349–352.

Dürig, U., Züger, O., and Stalder, A. (1992), "Interaction Force Detection in Scanning Probe Microscopy: Methods and Applications," *J. Appl. Phys.* 72, 1778–1798.

Erlandson, R., Hadziioannou, G., Mate, C. M., McClelland, G. M., and Chiang, S. (1988), "Atomic-Scale Friction between the Muscovite Mica Cleavage Plane and a Tungsten Tip," *J. Chem. Phys.,* 89, 5190–5193.

Eyring, H. (1935), "The Activated Complex in Chemical Reactions," *J. Chem. Phys.* 3, 107–115.

Feynman, R. (1964), *Lectures on Physics,* Vol. II, Chap. 38, Addison-Wesley, Reading, MA.

Fujihira, M. (1997), "Friction Force Microscopy of Organic Thin Films and Crystals," in *Micro/Nanotribology and Its Applications,* (B. Bhushan, ed.), pp. 239–260, NATO ASI Series, Kluwer Academic Publishers, Dordrecht.

Fujihira, M., Aoki, D., Okabe, Y., Takano, H., Hokari, H., Frommer, J., Nagatani, Y., and Sakai, F. (1996), "Effect of Capillary Force on Friction Force Microscopy: A Scanning Hydrophilicity Microscope," *Chem. Lett.* 499–500.

Fujisawa, S., Sugawara, Y., Ito, S., Mishima, S., Okada, T., and Morita, S. (1993), "The Two-Dimensional Stick-Slip Phenomenon with Atomic Resolution," *Nanotechnology,* 4(3), 138–142.

Fujisawa, S., Ohta, M., Konishi, T., Sugawara, Y., and Morita, S. (1994), "Difference between the Forces Measured by an Optical Lever Deflection and by an Optical Interferometer in an Atomic Force Microscope," *Rev. Sci. Instrum.* 65(3), 644–647.

Germann, G. J., Cohen, S. R., Neubauer, G., McClelland, G. M., and Seki, H. (1993), "Atomic-Scale Friction of a Diamond Tip on Diamond (100) and (111) Surfaces," *J. Appl. Phys.* 73, 163–167.

Giessibl, F. J. (1995), "Atomic Resolution of the Silicon (111)-7×7 Surface by Atomic Force Microscopy," *Science,* 267, 1451–1454.

Giessibl, F. J. and Binnig, G. (1992), "Investigation of the (001) Cleavage Plane of Potassium Bromide with an Atomic Force Microscope at 4.2K in Ultrahigh Vacuum," *Ultramicroscopy,* 42–44, 281–289.

Grafström, S., Neitzert, M., Hagen, T., Ackermann, J., Neumann, R., Probst, O., and Wörtge, M. (1993), "The Role of Topography and Friction for the Image Contrast in Lateral Force Microscopy," *Nanotechnology,* 4, 143–151.

Grafström, S., Ackermann, J., Hagen, T., Neumann, R., and Probst, O. (1994), "Analysis of Lateral Force Effects on the Topography," *J. Vac. Sci. Technol. B* 12(3), 1559–1564.

Greenwood, J. A. (1992a), "Contact of Rough Surfaces," in *Fundamentals of Friction: Macroscopic and Microscopic Processes,* NATO ANSI Series, pp. 37–56, Kluwer Academic Publishers, Dordrecht.

Greenwood, J. A. (1992b), "Problems with Surface Roughness," in *Fundamentals of Friction: Macroscopic and Microscopic Processes,* NATO ANSI Series, pp. 57–76, Kluwer Academic Publishers, Dordrecht.

Gyalog, T., Bammerlin, M., Lüthi, R., Meyer, E., and Thomas, H. (1995), "Mechanism of Atomic Friction," *Europhys. Lett.* 31(5–6), 269–274.

Haefke, H., Meyer, E., Schwarz, U., Gerth, G., and Krohn, M. (1992), "Atomic Surface and Lattice Structures of AgBr Thin Films," *Ultramicroscopy* 42–44, 290–297.

Hertz, J. (1881), "Über die Behührung fester elastischer Körper," *Reine Angew. Math.* 92, 156–171.

Houston, J. E. and Michalske, T. A. (1992), "The Interfacial-Force Microscope," *Nature* 356, 266–268.

Howald, L., Haefke, H., Lüthi, R., Meyer, E., Gerth, G., Rudin, H., and Güntherodt, H.-J. (1994a), "Ultrahigh-Vacuum Scanning Force Microscopy: Atomic-Scale Resolution at Monoatomic Cleavage Steps," *Phys. Rev. B* 49(8), 5651–5656.

Howald, L., Lüthi, R., Meyer, E., Güthner, P., and Güntherodt, H.-J. (1994b), "Scanning Force Microscopy on the Si(111) 7×7 Surface Reconstruction," *Z. Phys. B* 93, 267–268.

Howald, L., Lüthi, R., Meyer, E., and Güntherodt, H.-J. (1995), "Atomic Force Microscopy on the Si(111) 7×7 Surface," *Phys. Rev. B* 51(8), 5484–5487.

Hu, J., Xiao, X. D., Ogletree, D. F., and Salmeron, M. (1995), "Atomic Scale Friction and Wear of Mica," *Surf. Sci.* 327, 358–370.

Israelachvili, J. N. (1992), *Intermolecular and Surface,* Academic Press, San Diego.

Ito, T., Namba, M., Bühlmann, Ph., and Umezawa, Y. (1997), "Modification of Silicon Nitride Tips with Trichorosilane Self-Assembled Monolayers (SAMs) for Chemical Force Microscopy," *Langmuir* 13, 4323–4332.

Jarvis, S. P., Oral, A., Weihs, T. P., and Pethica, J. B. (1993), "A Novel Force Microscope and Point Contact Probe," *Rev. Sci. Instrum.* 64(12), 3515–3520.

Johnson, K. L. (1985), *Contact Mechanics,* Cambridge University Press, Cambridge, U.K.

Johnson, K. L. (1996), "Continuum Mechanics Modeling of Adhesion and Friction," *Langmuir,* 12(19), 4537–4542.

Johnson, K. L. (1997a), "A Continuum Mechanics Model of Adhesion and Friction in a Single Asperity Contact," in *Micro/Nanotribology and Its Applications,* (B. Bhushan, ed.), pp. 151–168, NATO ASI Series, Kluwer Academic Publishers, Dordrecht.

Johnson, K. L. (1997b), "Adhesion and Friction between a Smooth Elastic Spherical Asperity and a Plane Surface," *Proc. R. Soc. Lond. A* 453, 163–179.

Johnson, K. L., Kendall, K., and Roberts, A. D. (1971), "Surface Energy and the Contact of Elastic Solids," *Proc. R. Soc. Lond. A* 324, 301–313.

Joyce, S. A. and Houston, J. E. (1991), "A New Force Sensor Incorporating Force Feedback Control for Interfacial Force Microscopy," *Rev. Sci. Instrum.* 62(3), 710–715.

Kato, N., Suzuki, I., Kikuta, H., and Iwata (1997), "Force Balancing Microforce Sensor with an Optical Interferometer," *Rev. Sci. Instrum.* 68(6), 2475–2478.

Krim, J. (1996), "Friction of the Atomic Scale," *Sci. Am.* Oct., 48–56.

Krim, J. and Widom, A. (1988), "Damping of a Crystal Oscillator by an Adsorbed Monolayer and Its Relation to Interfacial Viscosity," *Phys. Rev. B* 38(17), 12184–12189.

Krim, J., Solina, D. H., and Chiarello, R. (1991), "Nanotribology of a Kr Monolayer: A Quartz-Crystal Microbalance Study of Atomic-Scale Friction," *Phys. Rev. Lett.* 66, 181–184.

Landman, U., Luedtke, W. D., and Nitzan, A. (1989a), "Dynamics of Tip-Substrate Interaction in Atomic Force Microscopy," *Surf. Sci. Lett.* 210, L177–184.

Landman, U., Luedtke, W. D., and Ribarsky, M. W. (1989b), "Structural and Dynamical Consequences of Interactions in Interfacial Systems," *J. Vac. Sci. Technol. A* 7, 2829–2839.

Landman, U., Luedtke, W. D., Burnham, N. A., and Colton, R. J. (1990), "Atomic Mechanisms and Dynamics of Adhesion, Nanoindentation and Fracture," *Science,* 248, 454–461.

Lantz, M. A., O'Shea, S. J., Hoole, A. C., and Welland, M. E. (1997), "Lateral Stiffness of the Tip and the Tip-Sample Contact in Frictional Force Microscopy," *Appl. Phys. Lett.* 70(8), 970–972.

Legtenberg, R., Tilmans, H. A. C., Elders, J., and Elwenspoek, M. (1994), "Stiction of Surface Micromachined Structures after Rinsing and Drying: Model and Investigation of Adhesion Mechanisms," *Sensors Actuators A* 43, 230–238.

Levitov, L. S. (1989), "Van der Waals' Friction," *Europhys. Lett.* 8, 499–502.

Lu, L. and Czanderna, A. (1994), *Applications of Piezoelectric Quartz Microbalance,* Elsevier, Amsterdam.

Lüthi, R., Meyer, E., Haefke, H., Howald, L., Gutmansbauer, W., Guggisberg, M., Bammerlin, M., and Güntherodt, H.-J. (1995), "Nanotribology: An UHV-SFM Study on Thin Films of C60 and AgBr," *Surf. Sci.* 338, 247–260.

Lüthi, R., Meyer, E., Bammerlin, M., Baratoff, A., Lehmann, T., Howald, L., Gerber, Ch., and Güntherodt, H.-J. (1996), "Atomic Resolution in Dynamic Force Microscopy across Steps on Si(111) 7×7," *Z. Phys. B* 100, 165–167.

Mamur, A. (1993), "Tip-Surface Capillary Interactions," *Langmuir,* 9, 1922–1926.

Marti, O. (1993), "Nanotribology: Friction on a Nanometer Scale," *Phys. Scr.* T49, 599–604.

Marti, O., Colchero, J., and Mlynek, J. (1990), "Combined Scanning Force and Friction Microscopy of Mica," *Nanotechnology* 1, 141–144.

Marti, O., Colchero, J., and Mlynek, J. (1993), "Friction and Forces on an Atomic Scale," in *Nanostructures and Manipulations of Atoms under High Fields and Temperatures: Applications,* pp. 253–269, Kluwer Academic Publishers, Dordrecht.

Mate, C. M. (1992), "Application of Disjoining and Capillary Pressure to Liquid Lubricant Film in Magnetic Recording," *J. Appl. Phys.* 72, 3084–3098.

Mate, C. M. and Homola, A. M. (1997), "Molecular Tribology of Disk Drives," in *Micro/Nanotribology and Its Applications,* (B. Bhushan, ed.), pp. 647–661, NATO ASI Series, Kluwer Academic Publishers, Dordrecht.

Mate, C. M., McClelland, G. M., Erlandsson, R., and Chiang, S. (1987), "Atomic-Scale Friction of a Tungsten Tip on a Graphite Surface," *Phys. Rev. Lett.* 59, 1942–1945.

Mattewson, M. J. and Mamin, H. J. (1988), "Liquid Mediated Adhesion of Ultra-Flat Solid Surfaces," *Proc. Mater. Res. Soc. Symp.* 119, 87–92.

Maugis, D. (1982), in *Microscopic Aspects of Adhesion and Lubrication* (Georges, J.M. ed.), Elsevier, Amsterdam.

McClelland, G. M. and Gosli, J. N. (1992), "Friction at the Atomic Scale," in *Fundamentals of Friction: Macroscopic and Microscopic Processes,* pp. 405–425, NATO ANSI Series, Kluwer Academic Publishers, Dordrecht.

Mertz, J., Marti, O., and Mlynek, J. (1993), "Regulation of a Microcantilever Response by Force Feedback," *Appl. Phys. Lett.* 62(19), 2344–2346.

Meyer, E., Heinzelmann, H., Rudin, H., and Güntherodt, H.-J. (1991a), "Atomic Resolution on LiF (100) by Atomic Force Microscopy," *Z. Phys. B* 79, 3.

Meyer, E., Heinzelmann, H., and Brodbeck, D. (1991b), "Atomic Resolution on LiF (100) by Atomic Force Microscopy," *J. Vac. Sci. Technol. B* 9(2), 1329.

Meyer, E., Overney, R., Brodbeck, D., Howald, L., Lüthi, R., Frommer, J., and Güntherodt, H.-J. (1992a), "Friction and Wear of Langmuir Blodgett Films Observed by Friction Force Microscopy," *Phys. Rev. Lett.* 69(12), 1777–1780.

Meyer, E., Overney, R., Brodbeck, D., Howald, L., Lüthi, R., Frommer, J., Güntherodt, H.-J., Wolter, O., Fujihira, M., Takano, H., and Gotoh, Y. (1992b), "Friction Force Microscopy of Mixed Langmuir-Blodgett Films," *Thin Solids Films* 220, 132–137.

Meyer, E., Lüthi, R., Howald, L., Bammerlin, M., Guggisberg, M., and Güntherodt, H.-J. (1997), "Instrumental Aspects and Contrast Mechanisms of Friction Force Microscopy," in *Micro/Nanotribology and Its Applications,* (B. Bhushan, ed.), pp. 193–214, NATO ASI Series, Kluwer Academic Publishers, Dordrecht.

Meyer, G. and Amer, N. M. (1988), "Novel Optical Approach to Atomic Force Microscopy," *Appl. Phys. Lett.* 53, 1045–1047.

Meyer, G. and Amer, N. M. (1990a), "Optical Beam Deflection Microscopy: The NaCl(001) Surface," *Appl. Phys. Lett.* 56(21), 2100–2101.

Meyer, G. and Amer, N. M. (1990b), "Simultaneous Measurement of Lateral and Normal Forces with an Optical Beam Deflection Atomic Force Microscope," *Appl. Phys. Lett.* 57(20), 2089–2091.

Morita, S., Fujisawa, S., and Yasuhiro, S. (1996), "Spatially Quantized Friction with a Lattice Periodicity," *Surf. Sci. Rep.* 23(1), 1–41.

Neumeister, J. M. and Ducker, W. A. (1994), "Lateral, Normal and Longitudinal Spring Constants of Atomic Force Microscopy Cantilevers," *Rev. Sci. Instrum.* 65(8), 2527–2531.

Ohnesorge, F. and Binnig, G. (1993), "True Atomic Resolution by Atomic Force Microscopy through Repulsive and Attractive Forces," *Science* 260, 1451.

Overney, R. M., Meyer, E., Frommer, J., Brodbeck, D., Lüthi, R., Howald, L., Güntherodt, H.-J., Fujihira, M., Takano, H., and Gotoh, Y. (1992), "Friction Measurements of Phase-Separated Thin Films with a Modified Atomic Force Microscope," *Nature* 359, 133–135.

Pascual, J. I., Méndez, J., Gómez-Herrero, J., Baró, A. M., García, N., and Vu Thien Binh, (1993), "Quantum Contact in Gold Nanostructures by Scanning Tunneling Microscopy," *Phys. Rev. Lett.* 71, 1852–1855.

Persson, B. N. J. (1991), "Surface Resistivity and Vibrational Damping in Adsorbed Layers," *PRB* 44(7), 3277–3296.

Persson, B. N. J. (1993), "Theory and Simulation of Sliding Friction," *Phys. Rev. Lett.* 71(8), 1212–1215.

Pethica, J. B. and Liver, W. C. (1987), "Tip Surface Interactions on STM an AFM," *Phys. Scr.* T19, 61–66.

Pethica, J. B. and Sutton, A. P. (1995), "Nanomechanics: Atomic Resolution and Frictional Energy Dissipation in Atomic Force Microscopy," in *Forces in Scanning Probe Methods,* pp. 353–366, NATO ANSI Series, Kluwer Academic Publishers, Dordrecht, The Netherlands.

Putman, C. A. J., Igarashi, M., and Kaneko, R. (1995), "Single-Asperity Friction in Friction Force Microscopy: The Composite-Tip Model," *Appl. Phys. Lett.* 66(23), 3221–3223.

Radmacher, M., Tillmann, R. W., Fritz, M., and Gaub, H. E. (1992), "From Molecules to Cells: Imaging Soft Samples with the Atomic Force Microscope," *Science* 257, 1900–1905.

Ruan, J. and Bhushan, B. (1994a), "Atomic-Scale Friction Measurements Using Friction Force Microscopy: Part I — General Principles and New Measurement Techniques," *ASME J. Tribol.* 116, 378–388.

Ruan, J. and Bhushan, B. (1994b), "Atomic-Scale and Microscale Friction on Graphite and Diamond Using Friction Force Microscopy," *J. Appl. Phys.* 76, 5022–5035.

Ruan, J. and Bhushan, B. (1994c), "Frictional Behaviour of Highly Oriented Pyrolytic Graphite," *J. Appl. Phys.* 76, 8117–8120.

Rubio, G., Agraït, N., and Vieira, S. (1996), "Atomic-Sized Metallic Contacts: Mechanical Properties and Electronic Transport," *Phys. Rev. Lett.* 76, 2602–2605.

Saada, A. S. (1974), "Elasticity: Theory and Applications," Pergamon Press, Oxford.

Sader, E. J. (1995), "Parallel Beam Approximation for V-Shaped Atomic Force Cantilevers," *Rev. Sci. Instrum.* 66(9), 4583–4587.

Salmeron, M., Liu, Y., and Ogletree, D. F. (1995), "Molecular Arrangement and Mechanical Stability of Self-Assembled Monolayers on Au(111) under Applied Load," in *Forces in Scanning Probe Methods,* pp. 593–598, NATO ANSI Series, Kluwer Academic Publishers, Dordrecht, The Netherlands.

Schwarz, U. D., Allers, W., Gersterblum, G., and Wiesendanger, R. (1996), "Low-Load Friction Behavior of Epitaxial C60 Monolayers under Herztian Contact," *Phys. Rev. B* 52, 14976–14986.

Schwarz, U. D., Zwörner, O., Köster, P., and Wiesendanger, R. (1997), "Friction Force Spectroscopy in the Low-Load Regime with Well-Defined Tips," in *Micro/Nanotribology and Its Applications,* (B. Bhushan, ed.), pp. 150–160, NATO ANSI Series, Kluwer Academic Publishers, Dordrecht.

Sheiko, S. S., Möller, M., Reuvecamp, E. M., and Zandbergen, H. W. (1993), "Calibration and Evaluation of Scanning-Force-Microscopy Probes," *Phys. Rev. B* 48, 5675–5680.

Smith, J. R., Bozzolo, G., Banerja, A., and Ferrante, J. (1989), "Avalanche in Adhesion," *Phys. Rev. Lett.* 63, 1269–1272.

Solokoff, J. B. (1990), "Theory of Energy Dissipation in Sliding Crystal Surfaces," *Phys. Rev. B* 42, 760–765.

Stalder, U. (1996), "Study of Yielding Mechanics in Nanometer-Sized Au Contacts," *Appl. Phys. Lett.* 68(5), 637–639.

Sugawara, Y., Ohta, M., Ueyama, H., and Morita, S. (1995), "Defect Motion on an InP (100) Surface Observed with Noncontact Atomic Force Microscopy," *Science* 270, 1646.

Tabor, D. (1992), "Problems with Surface Roughness," in *Fundamentals of Friction: Macroscopic and Microscopic Processes,* pp. 3–24 and 580, NATO ANSI Series, Kluwer Academic Publishers, Dordrecht.

Tadmor, E. B., Phillips, R., and Ortiz, M. (1996), "Mixed Atomistic and Continuum Models of Deformation in Solids," *Langmuir* 12, 4529–4534.

Tas, N., Vogelzang, B., Elwenspoek, M., and Legtenberg, R. (1997), "Adhesion and Friction in MEMS," in *Micro/Nanotribology and Its Applications* (B. Bhushan, ed.), pp. 621–628, NATO ANSI Series, Kluwer Academic Publishers, Dordrecht.

Tersoff, J. (1990), *Scanning Tunneling Microscopy and Related Methods,* Kluwer Academic Publishers, Dordrecht.

Tomanek, D., Zhong, W., and Thomas, H. (1991). "Calculations of an Atomically Modulated Friction Force in Atomic-Force Microscopy," *Europhys. Lett.* 15, 887.

Tomlinson, G. A. (1929), "A Molecular Theory of Friction," *Philos. Mag.* 7, 905–939.

Weilandt, E., Zinl, B., Stifter, Th., and Marti, O. (1997), "Nanotribology in Electrolytic Environments," in *Micro/Nanotribology and Its Applications,* (B. Bhushan, ed.), pp. 283–297, NATO ANSI Series, Kluwer Academic Publishers, Dordrecht.

Weisenhorn, A. L., Hansma, P. K., and Albrecht, T. R. (1989), "Forces in Atomic Force Microscopy in Air and Water," *Appl. Phys. Lett.* 54(26), 2651–2653.

Wittman, J. C. and Smith, P. (1991), "Highly Oriented Thin Films of Polytetrafluoroethylene as a Substrate for Oriented Growth of Materials," *Nature* 352, 414–417.

Wolter, O., Bayer, Th., and Gerschner, J. (1991), "Micromachined Silicon Sensors for Scanning Force Microscopy," *J. Vac. Sci. Technol.* 9, 1353.

Yamanaka, K. and Tomita, E. (1995), "Lateral Force Modulation Atomic Force Microscope for Selective Imaging of Friction Forces," *Jpn. J. Appl. Phys.* 34, 2279–2882.

Zhong, W. and Tománek, D. (1990), "First-Principles Theory of Atomic-Scale Friction," *Phys. Rev. Lett.* 64(25), 3054–3057.

7

Microscratching/ Microwear, Nanofabrication/ Nanomachining, and Nano/Picoindentation Using Atomic Force Microscopy

Bharat Bhushan

7.1 Introduction

Wear of sliding surfaces can occur by one or more wear mechanisms, including adhesive, abrasive, fatigue, impact, corrosive, and fretting. The wear rate, a measure of wear, generally needs to be minimized (Bhushan, 1996). As the dimensions of components and loads used continue to decrease (such as in microelectromechanical systems or MEMS), scratching/wear and mechanical properties at the micro- to nanoscales become very important. With the advent of the newly developed scanning probe microscopes (SPMs), particularly the atomic force microscope (AFM), it is possible to study the interfacial phenomena at a small scale and light load. The AFM/FFM tip simulates a sharp single asperity traveling over a surface. Scratching and wear processes at different normal loads are studied in AFMs by using a sharp diamond tip. This tip can also be used for nanofabrication/nanomachining. AFMs, in conjunction with special sensors, are used for measurement of mechanical properties on nano- to picoscales.

335

This chapter presents an overview of microscratching/microwear, nanofabrication/nanomachining, and nano/picoindentation using AFM and related instrumentation.

7.2 Experimental Techniques

7.2.1 AFM for Microscratching/Microwear and Nanoindentation

Commercial AFMs are commonly used to conduct microscratching/microwear and nanoindentation. Special sensors may be used in conjunction with AFMs for nano/picoindentation studies.

For microscale scratching, microscale wear, nanoscale indentation hardness measurements, and nano-fabrication/nanomachining, a three-sided pyramidal single-crystal natural diamond tip with an apex angle of 80° and a radius of about 100 nm mounted on a stainless steel cantilever beam with high normal stiffness of about 25 N/m is used at relatively high loads (1 to 150 μN); see Chapter 1 for further details (Bhushan et al., 1994a). For scratching and wear studies, the sample is generally scanned in a direction orthogonal to the long axis of the cantilever beam (typically at a rate of 0.5 Hz) so that friction force can be measured during scratching and wear. The tip is mounted on the beam such that one of its edges is orthogonal to the long axis of the beam; therefore, wear during scanning along the beam axis is higher (about 2 to 3 times) than that during scanning orthogonal to the beam axis. For wear studies, typically an area of 2 × 2 μm is scanned at various normal loads (ranging from 1 to 100 μN) for a selected number of cycles. For nanofabrication/nanomachining, the nanoscratching operation is extended.

For nanoindentation hardness measurements, the scan size is set to zero and normal load is applied to make the indents (Bhushan et al., 1994a,d). During this procedure, the diamond tip is continuously pressed against the sample surface for about 2 s at various indentation loads. Nanohardness is calculated by dividing the indentation load by the projected residual area of the indents.

Sample surface is scanned before and after the scratching, wear, or indentation to obtain the initial and the final surface topography, at a low normal load of about 0.3 μN using the same diamond tip. An area larger than the scratching, wear, and indentation region is scanned to observe the marks.

7.2.2 Nano/Picoindenter

As stated earlier, conventional AFMs have been used for indentation studies on nanometer-scale depths. In these studies, the hardness value is based on the projected residual area after imaging the indent. Direct imaging of the indent allows one to quantify piling up of ductile material around the indenter. However, it becomes difficult to identify the boundary of the indentation mark with great accuracy. This makes the direct measurement of the contact area somewhat inaccurate (Bhushan et al., 1994a,d). A technique with the dual capability of depth sensing as well as *in situ* imaging is most appropriate in nanomechanical property studies (Bhushan et al., 1996). This indentation system is used to make load-displacement measurement and subsequently carry out *in situ* imaging of the indent. A schematic of the nano/picoindenter system used is shown in Figure 7.1. The indentation system consists of a three-plate transducer with electrostatic actuation hardware used for direct application of normal load and a capacitive sensor used for measurement of vertical displacement. The AFM head is replaced with this transducer assembly while the specimen is mounted on the piezoelectric scanner which remains stationary during indentation experiments. The transducer consists of a three (Be–Cu) plate capacitive structure which provides high sensitivity, large dynamic range, and a linear output signal with respect to load or displacement. The tip is mounted on the center plate. The upper and lower plates serve as drive electrodes. Load is applied by applying appropriate voltage to the drive electrodes, thereby generating an electrostatic force between the center plate and the drive electrodes. Vertical displacement of the tip (indentation depth) is measured by measuring the displacement of the center plate relative to the two outer electrodes using capacitance technique. The load resolution is 100 nN or better, and the displacement resolution is 0.1 nm. At present, a load range of 1 μN to 10 mN can be employed. Loading rates can be varied by changing the load/unload period. The AFM functions as the platform providing an *in situ* image of the indent with a lateral resolution of 1 nm and a vertical resolution of 0.2 nm. The load–displacement data

336

can be acquired and displayed on the display monitor. Hardness value can be obtained from the load–displacement data, as well as from direct measurement of the projected residual area of the indent after imaging. Young's modulus of elasticity is obtained from the slope of the unloading curve.

A three-sided Berkovich indenter with tip radius of about 100 nm is generally used for the measurements, see Chapter 10 on Nanomechanical Properties in this book. Sharper diamond tips with included angle of 60 to 90° and tip radii of 30 to 60 nm are sometimes employed for shallower indentation (on the order of 1 nm). To obtain an accurate relation between the indentation depth and the projected contact area, tip shape calibration needs to be done. Also for surfaces with rms roughness on the order of indentation depth, the original (unindented) profile is subtracted from the indented profile (Bhushan et al., 1994a,d).

In a typical indentation experiment, the tip is lowered close to the sample (ideally <100 μm). Scan size and scan rate are selected. The tip is engaged to the sample surface by a stepper motor with a set point of 1 nA (about 1 μN). A desired image area is captured prior to indentation. The feedback is set to zero to disable the scanner; the scan size is set to zero so that the indenter will be positioned at the center of the image. An appropriate set point for the preload condition is selected. The indentation rate can be varied by changing the load/unload period.

7.3   Microscratching/Microwear Studies

By using a standard or sharp diamond tip mounted on a stiff cantilever beam, AFMs can be used to investigate how surface materials can be moved or removed on micro- to nanoscales, for example, in scratching and wear (where these things are undesirable) and nanofabrication/nanomachining (where they are desirable) (Hamada and Kaneko, 1992; Miyamoto et al., 1991; Bhushan and Koinkar, 1994b, 1995a–c,e, 1997; Bhushan et al., 1994 a,c; 1995d; Bhushan, 1995a,b, 1997, 1998a,b; Koinkar and Bhushan, 1997a,b). A variety of polymers and ceramics and hard coatings have been studied. Many examples of scratching/wear of magnetic recording materials have been presented in Chapter 14 on magnetic storage devices in this book. This chapter focuses on the studies with silicon material.

Figure 7.2a shows microscratches made on Si(111) at various loads (Bhushan and Koinkar, 1994b). As expected, the scratch depth increases linearly with load. Such microscratching measurements can be used to study failure mechanisms on the microscale and to evaluate the mechanical integrity (scratch resistance) of ultrathin films at low loads. To study the effect of scanning velocity, unidirectional scratches, 5 μm in length, were generated at scanning velocities ranging from 1 to 100 μm/s at normal loads ranging from 40 to 140 μN. There is no effect of scanning velocity obtained at a given normal load. For representative scratches profiles at 80 μN, see Figure 7.2b (Koinkar, 1997). Insensitivity to scanning velocity may be because of a small effect of frictional heating with the change in scanning velocity used here. Furthermore, for a small change in interface temperature, there is a large underlying volume to dissipate the heat generated during scratching (Bhushan, 1998a).

By scanning the sample in two dimensions with the AFM, wear scars are generated on the surface. Figure 7.3 shows the effect of normal load on the wear rate. We note that wear rate is very small below 20 μN of normal load. A normal load of 20 μN corresponds to contact stresses comparable to the hardness of the silicon. Primarily, elastic deformation at loads below 20 μN is responsible for low wear (Bhushan et al., 1995d; Bhushan and Kulkarni, 1995f; Koinkar and Bhushan, 1997b).

Typical wear mark generated at a normal load of 40 μN for one scan cycle and imaged using AFM at 300 nN load is shown in Figure 7.4a (Koinkar and Bhushan, 1997b). The inverted map of a wear mark shown in Figure 7.4b indicates the uniform material removal at the bottom of the wear mark. Next we examine the mechanism of material removal on microscale at low loads, in AFM wear experiments. Figure 7.5 shows a secondary electron image of a wear mark and associated wear particles. The specimen used for the SEM was not scanned after initial wear, to retain wear debris in the wear region. Wear debris is clearly observed. An AFM image of the wear mark shows small debris at the edges, swiped during AFM scanning. Thus, the debris is "loose" (not sticky) and can be removed during the AFM scanning. SEM micrographs show both cutting-type and ribbonlike debris. TEM studies were performed to understand

337

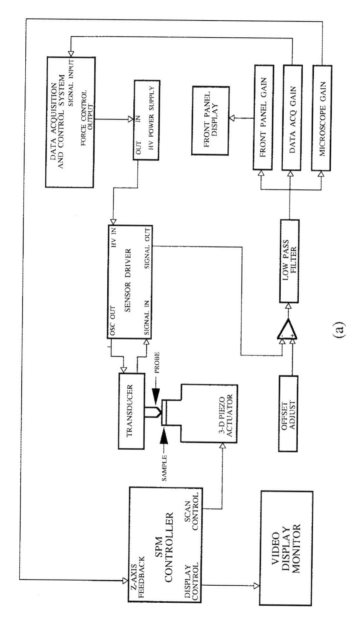

FIGURE 7.1 Schematics of (a) indentation system, (b) three-plate transducer with electrostatic actuation hardware and capacitance sensor, and (c) tip-holder mount assembly. (From Bhushan, B. et al. (1996). *Philos. Mag.*, 74, 117–1128. With permission.)

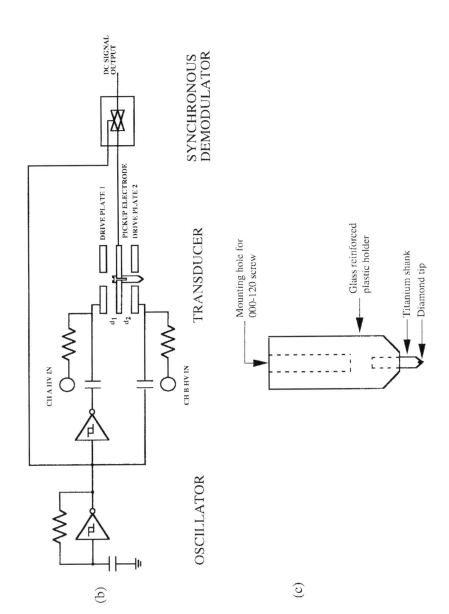

SYNCHRONOUS
DEMODULATOR

TRANSDUCER

OSCILLATOR

(b)

(c)

FIGURE 7.1

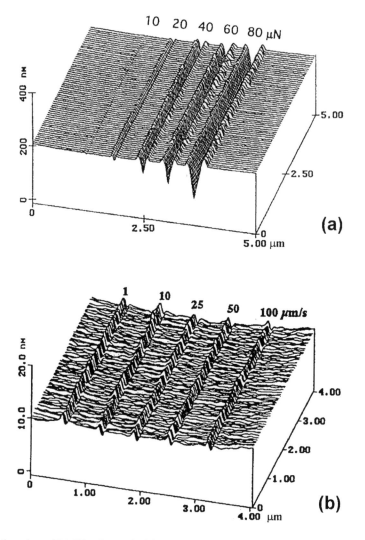

FIGURE 7.2 Surface plots of (a) Si(111) scratched for ten cycles at various loads and a scanning velocity of 2 μm/s (note that the *x*- and *y*-axes are in μm and the *z*-axis is in nm), and (b) Si(100) scratched in one unidirectional scan cycle at a normal load of 80 μN and different scanning velocities.

the material removal process. The TEM micrograph of the worn region in Figure 7.6 shows evidence of bend contours passing through the wear mark. The bend contours around and inside the wear mark suggest that there are some residual stresses around and inside the wear mark region. There is no dislocation activity or cracks observed inside the wear track. The dislocation activity and/or cracking probably occurs at the subsurface. Based on SEM and TEM studies, it is believed that the material in the experiment described here is removed in a brittle manner without much plastic deformation (dislocation activity).

Finally, we study evolution of wear of a diamondlike carbon (DLC) coated disk substrate, Figure 7.7 (Bhushan et al., 1994a). The data illustrate how the microwear profile for a load of 20 μN develops as a function of the number of scanning cycles. Wear is not uniform, but is initiated at the nanoscratches indicating that the nanoscratches (with high surface energy) and nonuniform coverage of DLC at nanoscratches act as initiation sites. Thus, scratch-free surfaces will be relatively resistant to wear.

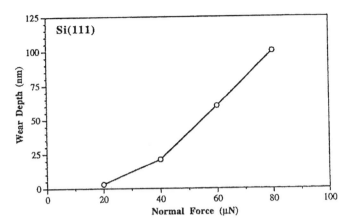

FIGURE 7.3   Wear depth as a function of normal load for Si(111) after one cycle. (From Koinkar, V. N. and Bhushan, B. (1997), *J. Mater. Res.*, 12, 3219–3224. With permission.)

7.4   Nanofabrication/Nanomachining Studies

Scanning tunneling microscopes (STMs) have been used to form nanofeatures by localized heating or by inducing chemical reactions under the STM tip (Abraham et al., 1986; Silver et al., 1987; Albrecht et al., 1989; Utsugi, 1990; Kobayashi et al., 1993), and nanomachining (Parkinson, 1990). AFMs have also been used for nanofabrication (Majumdar et al., 1992; Bhushan et al., 1994a,c; Bhushan, 1995a,b, 1998a,b; Tsau et al., 1994) and nanomachining (Delawski and Parkinson, 1992).

Figure 7.8 shows an example of nanofabrication. The word "OHIO" was written on a (100) single-crystal silicon wafer by scratching the sample surface with a diamond tip at specified locations and scratching angles (Bhushan, 1995a). The normal load used for scratching (writing) was 50 μN and the writing speed was 0.2 μm/s. Each line is scribed manually and debris at the ends of each line is visible. A few lines are not connected to each other because of the PZT drift and hysteresis. Sufficient time should be given for the thermal stabilization of the PZT scanner so that the hysteresis effect is small during the nanofabrication. Next, more complex patterns were generated at a normal load of 15 μN and a writing speed of 0.5 μm/s, Figure 7.9 (Koinkar, 1997). Such a type of patterns is useful for resistor trimming (to increase the path resistor) on a small scale. The separation between lines is about 50 nm. In Figure 7.9a, the variation in line width is due to the tip asymmetry. A spiral pattern generated as shown in Figure 7.9b. Nanofabrication parameters — normal load, scanning speed, and tip geometry — can be controlled precisely to control depth and length of the devices.

Nanofabrication using mechanical scratching has several advantages over other techniques (Koinkar, 1997). Better control over the applied normal load, scan size, and scanning speed can be used for nanofabrication of devices. Using this technique, nanofabrication can be performed on any engineering surface. Use of chemical etching or reactions is not required, and this dry nanofabrication process can be used where use of chemicals and electric field is prohibited. One disadvantage of this technique is the formation of debris during scratching. At light loads, debris formation is not a problem compared with high-load scratching. However, debris can be removed easily out of the scan area at light loads during scanning.

7.5   Nano/Picoindentation

Nanohardness measurements using conventional AFMs is covered in Chapter 14 on magnetic storage devices. In this chapter, we will limit the discussion to the application of the three-plate transducer with

Si(111)

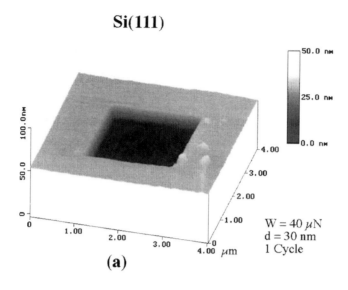

(a)

W = 40 μN
d = 30 nm
1 Cycle

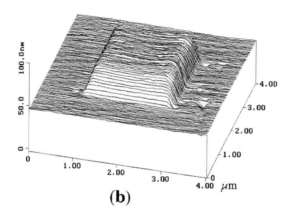

(b)

FIGURE 7.4 (a) Typical gray scale and (b) inverted AFM images of a wear mark created using a diamond tip at a normal load of 40 μN and one scan cycle on Si(111) surface. (From Koinkar, V. N. and Bhushan, B. (1997), *J. Mater. Res.*, 12, 3219–3224. With permission.)

electrostatic actuation hardware used in conjunction with conventional AFMs (Bhushan et al., 1996; Kulkarni et al., 1996a,b, 1997; Bhushan, 1997, 1998a,b; Bhushan and Koinkar, 1997; and Koinkar and Bhushan, 1997a).

Figure 7.10a shows the load–displacement curves at different peak loads for Si(100). Load–displacement data at residual depth as low as about 1 nm can be obtained. Loading/unloading curves are not smooth, but exhibit sharp discontinuities particularly at high loads (shown by arrows in the figure). Any discontinuities in the loading part of the curve probably result from slip of the tip. The sharp discontinuities in the unloading part of the curves are believed to be due to formation of lateral cracks that form at the base of median crack, which results in the surface of the specimen being thrust upward. From the load–displacement curves in Figure 7.10 the indentation hardness of surface films with an indentation depth of as small as about 1 nm has been measured. Triangular indentations are observed for shallow penetration depths, Figure 7.10b.

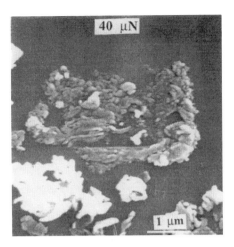

FIGURE 7.5 Secondary electron image of wear mark and debris for Si(111) produced at a normal load of 40 μN and one scan cycle. (From Koinkar, V. N. and Bhushan, B. (1997), *J. Mater. Res.*, 12, 3219–3224. With permission.)

Tip Sliding Direction

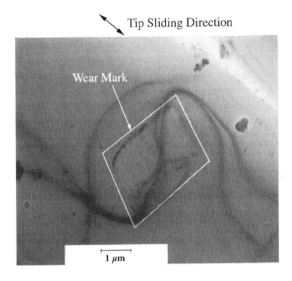

FIGURE 7.6 Bright-field TEM micrograph showing wear mark and bend contour around and inside the wear mark in Si(111) produced at a normal load of 40 μN and one scan cycle. (From Koinkar, V. N. and Bhushan, B. (1997), *J. Mater. Res.*, 12, 3219–3224. With permission.)

Figure 7.11 shows the load–displacement curves during three loading and unloading cycles for single-crystal silicon. The unloading and reloading curves reveal a large hysteresis, which shows no sign of degeneration through three cycles of deformation and the peak load displacement shift to higher values in successive loading–unloading cycles. Pharr et al. (1989, 1990), Page et al. (1992), and Pharr (1992) have also observed hysteresis behavior in silicon at similar loads using a nanoindenter. The fact that the curves are highly hysteretic implies that deformation is not entirely elastic. Pharr (1992) concluded that large hysteresis is due to a pressure-induced phase transformation from its normal diamond cubic form to a β-tin metal phase.

Table 7.1 summarizes the hardness and Young's modulus of eleasticity data at various depths for single-crystal silicon (Bhushan et al., 1996). Comparison of nanohardness values with that of bulk hardness

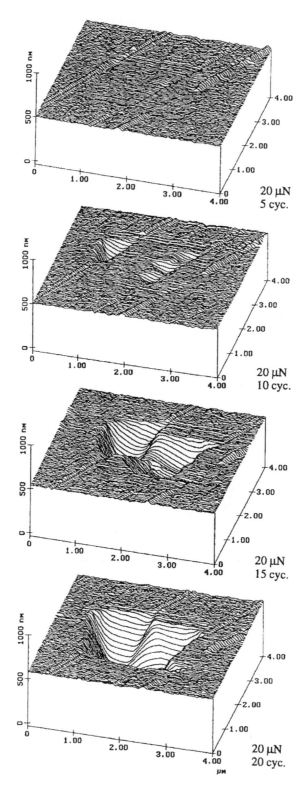

FIGURE 7.7 Surface plots of DLC-coated thin-film disk showing the worn region; the normal load and number of test cycles are indicated. (Bhushan, B. et al. (1994), *Proc. Inst. Mech. Eng. Part J: J. Eng. Tribol.* 208, 17–29. With permission.)

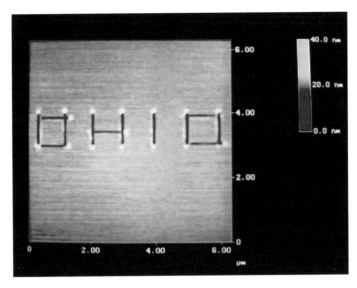

FIGURE 7.8   Example of nanofabrication. The letters "OHIO" (which stands for the state of Ohio) were generated by scratching a Si(111) surface using a diamond tip at a normal load of 50 μN and writing speed of 0.2 μm/s. (Bhushan, 1995b).

shows that nanohardness of silicon is higher that its bulk hardness. Figure 7.12 shows the plot of hardness and Young's modulus of elasticity as a function of residual depth. Note that hardness increases with a decrease in the residual depth. Similar results on single-crystal silicon have been reported by Pethica et al. (1983) and Page et al. (1992) based on nanoindentation data obtained using commercial nanoindentors, and by Bhushan and Koinkar (1994d) based on AFM data. Gane and Cox (1970) reported similar results on gold based on microindenter data. This increase in hardness with a decrease in indentation depth can be rationalized on the basis that, as the volume of deformed material increases, there is a higher probability of encountering material defects (Bhushan and Koinkar, 1995d). The increase in hardness also could be due to surface-localized cold work resulting form polishing.

The nano/picoindentation system has been used to study the creep and strain rate effects of ceramics. Most materials, including ceramics and even diamond, are known to creep at temperatures well below half their melting points, even at room temperature. Indentation creep and strain rate sensitivity experiments were conducted on single-crystal silicon. Figure 7.13 shows the load–displacement curves for various peak loads held at 180 s. To demonstrate the creep effects, the load–displacement curves for a 500-μN peak load held at 0 and 30 s are also shown as an inset. Note that significant creep occurs at room temperature. Nanoindenter experiments conducted by Li et al. (1991) exhibited significant creep only at high temperatures (greater than or equal to 0.25 times the melting point of silicon). The mechanism of dislocation glide plasticity is believed to dominate the indentation creep process. To study strain rate sensitivity of silicon, experiments were conducted at two different (constant) rates of loading (Figure 7.14). Note that a change in the loading rate of a factor of about five results in a change in the load–displacement data. Strain rate sensitivity to single-crystal aluminum (111) has been reported by LaFontaine et al. (1990).

Hardness of 5-nm- to 100-nm-thick DLC (a–c:H) films have been measured by Kulkarni and Bhushan (1997) and Koinkar and Bhushan (1997a). Figure 7.15a shows the load–displacement plots of silicon (100) and various 20-nm coatings at 100-μN peak load. Peak indentation depth for silicon (100) at 100-μN load is 8.7 nm and those for sputtered, ion beam and cathodic arc carbon coatings are 7.8, 8.2, and 6.4 nm, respectively. The residual depth is less than 1 nm for cathodic arc carbon coating at this load. From the comparison of these coatings just on the basis of load–displacement data, cathodic arc carbon exhibits the least displacement at the peak load. Representative load–displacement plots for

345

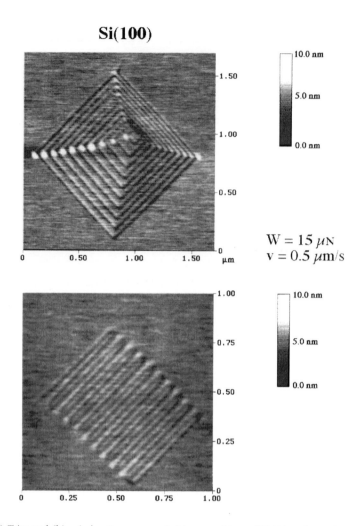

Si(100)

$W = 15\ \mu\text{N}$
$v = 0.5\ \mu\text{m/s}$

FIGURE 7.9 (a) Trim and (b) spiral patterns generated by scratching a Si(100) surface using a diamond tip at a normal load of 15 μN and writing speed of 0.5 μm/s.

100-nm-thick carbon coatings at peak indentation load of 500 μN are shown in Figure 7.15b. Note again that the cathodic arc carbon specimen exhibits the lowest indentation depth, about 20 nm, at the peak load as compared with other coatings.

Figure 7.16a shows the plot of hardness as a function of residual depth for silicon(100) and various carbon coating. The indentation loads used in each case ranged from 100 to 2000 μN. The cathodic arc carbon coating exhibits the highest hardness of about 24.9 GPa, whereas the sputtered and ion beam carbon coatings exhibit hardness values of 17.2 and 15.2 GPa, respectively. The hardness of silicon(100) is 13.2 GPa. The elastic modulus as a function of residual depth is plotted in Figure 7.16b. The cathodic arc carbon coating exhibits a decrease in the elastic modulus with increasing residual depth, while the elastic moduli for the other carbon coatings remain almost constant.

In general, hardness and the elastic modulus of coatings are strongly influenced by their crystalline structure, stoichiometry, and growth characteristics, which usually depend on the deposition parameters. Mechanical properties of a–C:H coatings have been known to change over a wide range with sp^3 to sp^2 bonding ratio and amount of hydrogen. Hydrogen is believed to play an important role in the bonding configuration of carbon atoms by helping to stabilize tetrahedral coordination of carbon atoms.

7.6 Closure

AFM with suitable diamond tips has been used successfully to measure microscratching/microwear and nano/picoindentation behavior of solid surfaces and thin films. AFM has also been used for nanofabrication/nanomachining purposes. Scratch and wear properties of single-crystal silicon have been measured. Scratch and wear depths increase with an increase of normal load. Wear rate for single-crystal silicon is found to be negligible below 20 μN, is much higher at higher loads, and increases almost linearly with load. Elastic deformation at low loads is responsible for negligible wear. The mechanism of material removal on the microscale is studied. At the loads used in the study, material is removed by the plowing mode in a brittle manner without much plastic deformation. Most of the wear debris is loose. Evolution of the wear of thin DLC coatings has also been studied using AFM. Wear is found to be initiated at nanoscratches. AFM has also been demonstrated to be useful for nanofabrication. AFM has been modified to obtain load–displacement curves and for measurement of nanoindentation hardness and Young's modulus of elasticity, with depth of indentation as low as 1 nm. Hardness of ceramics on nanoscales is found to be higher than that on the microscale. Ceramics exhibit significant plasticity and creep on the nanoscale.

Scratching and indentation on nanoscales are the powerful ways to screen for adhesion and resistance to deformation of ultrathin films. These studies are also directly applicable to interfacial phenomena of MEMS devices.

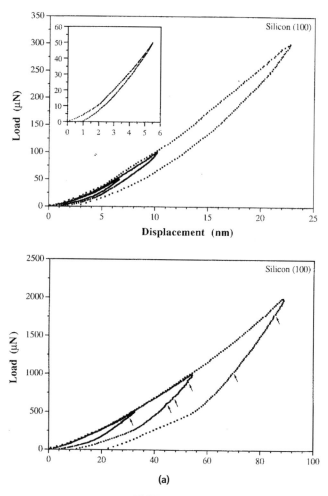

FIGURE 7.10

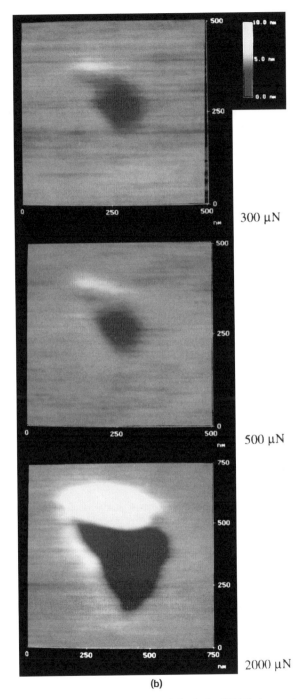

300 µN

500 µN

2000 µN

(b)

FIGURE 7.10 (a) Load–displacement curves at various peak loads for Si(100). Arrows indicate the discontinuities in the displacement; (b) AFM gray scale images of the indentations made at these peak loads. (Kulkarni, A. V. and Bhushan, B. (1997), *J. Mater. Res.*, 12, 2707–2714. With permission.)

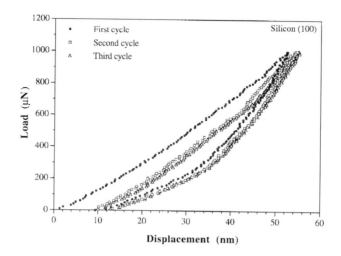

FIGURE 7.11 Load–displacement curves during repeated loading–unloading cycles for Si(100) (Bhushan, B. et al. (1996), *Philos. Mag.*, 74, 1117–1128. With permission.)

TABLE 7.1 Comparison of Nanohardness and Young's Modulus of Elasticity of Single-Crystal Silicon (100) as a Function of Normal Load

Normal Load (μN)	Residual Depth (nm)	Contact Depth (nm)	Hardness[a] (GPa)	Young's Modulus of Elasticity[a]
500	2.8	17.4	13.0	160
700	7.0	25.4	12.7	141
1000	8.5	32.2	13.2	143
1200	9.5	37.0	13.4	141
2000	15.0	55.3	13.1	141

[a]Bulk values of hardness and Young's modulus of elasticity are 9 to 10 and 130 GPa, respectively.

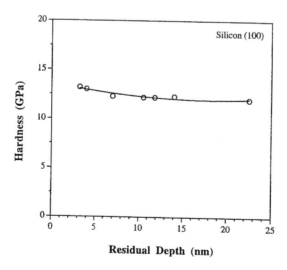

FIGURE 7.12 Indentation hardness as a function of residual indentation depth for Si(100). (Bhushan, B. et al. (1996), *Philos. Mag.*, 74, 1117–1128. With permission.)

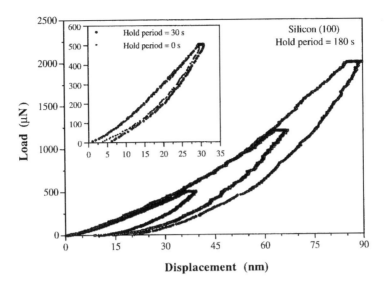

FIGURE 7.13 Creep behavior of single-crystal silicon. (Bhushan, B. et al. (1996), *Philos. Mag.,* 74, 1117–1128. With permission.)

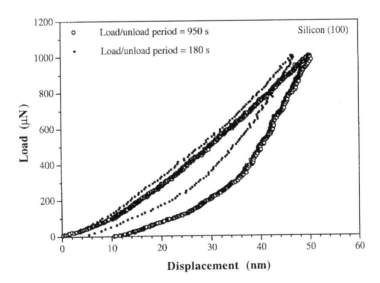

FIGURE 7.14 Strain rate sensitivity of single-crystal silicon. (Bhushan, B. et al. (1996), *Philos. Mag.,* 74, 1117–1128. With permission.)

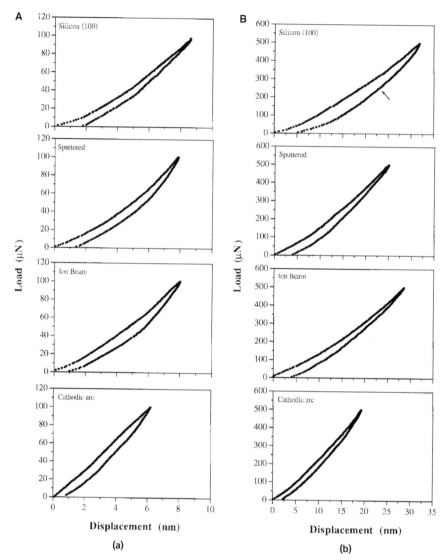

FIGURE 7.15 Load–displacement curves for (a) Si(100) and various 20-nm-thick DLC coatings at peak indentation load of 100 μN and (b) for Si(100) and 100-nm-thick DLC coatings at peak indentation load of 500 μN. Arrow indicates the discontinuity in the displacement. (Kulkarni, A. V. and Bhushan, B. (1997), *J. Mater. Res.* 12, 2707–2714. With permission.)

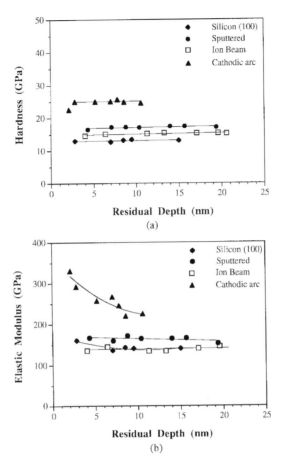

FIGURE 7.16 (a) Hardness and (b) Young's modulus of elasticity as a function of residual depth for silicon (100) and various 100-nm-thick DLC coatings. (Kulkarni, A. V. and Bhushan, B. (1997), *J. Mater. Res.* 12, 2707–2714. With permission.)

References

Abraham, D.W., Mamin, H.J., Ganz, E., and Clark, J. (1986), "Surface Modification with the Scanning Tunneling Microscope," *IBM J. Res. Dev.* 30, 492–499.

Albrecht, T.R., Dovek, M.M., Kirk, M.D., Lang, C.A., Quate, C.F., and Smith, D.P.E. (1989), "Nanometer-Scale Hole Formation on Graphite Using a Scanning Tunneling Microscope," *Appl. Phys. Lett.* 55, 1727–1729.

Bhushan, B. (1995a), "Nanotribology and Its Applications to Magnetic Storage Devices and MEMS," *Forces in Scanning Probe Methods* (H.J. Guntherodt, D. Anselmetti, and E. Meyer, eds.), Vol. E 286, pp. 367–395, Kluwer Academic, Dordrecht, Netherlands.

Bhushan, B. (1995b), "Micro/Nanotribology and Its Applications to Magnetic Storage Devices and MEMS," *Tribol. Int.* 28, 85–95.

Bhushan, B. (1996), *Tribology and Mechanics of Magnetic Storage Devices*, 2nd ed., Springer-Verlag, New York.

Bhushan, B. (1997), *Micro/Nanotribology and Its Application*, Vol. E330, Kluwer Academic, Dordrecht, Netherlands.

Bhushan, B. (1998a), "Micro/Nanotribology Using Atomic Force/ Friction Force Microscopy: State of the Art," *Proc. Inst. Mech Eng., Part J: J. Eng. Tribol.* 212, 1-18.

Bhushan, B. (1998b), *Tribology Issues and Opportunities in MEMS,* Kluwer Academic, Dordrecht, The Netherlands.

Bhushan, B., Koinkar, V.N., and Ruan, J. (1994a), "Microtribology of Magnetic Media," *Proc. Inst. Mech Eng., Part J: J. Eng. Tribol.* 208, 17–29.

Bhushan, B. and Koinkar, V.N. (1994b), "Tribological Studies of Silicon for Magnetic Recording Applications," *J. Appl. Phys.* 75, 5741–5746.

Bhushan, B., Koinkar, V.N., and Ruan, J. (1994c), "Microtribological Studies by Using Atomic Force and Friction Force Microscopy and Its Applications," in *Determining Nanoscale Physical Properties of Materials by Microscopy and Spectroscopy* (M. Sarikaya, H.K. Wickramasinghe and M. Isaacson, eds.), Vol. 332, pp. 93–98, Materials Research Society, Pittsburgh.

Bhushan, B. and Koinkar, V.N. (1994d), "Nanoindentation Hardness Measurements Using Atomic Force Microscopy," *Appl. Phys. Lett.* 64, 1653–1655.

Bhushan, B. and Koinkar, V.N. (1995a), "Microtribology of PET Polymeric Films," *Trib. Trans.* 38, 119–127.

Bhushan, B. and Koinkar, V.N. (1995b), "Macro and Microtribological Studies of CrO_2 Video Tapes," *Wear* 180, 9–16.

Bhushan, B. and Koinkar, V.N. (1995c), "Microtribology of Metal Particle, Barium Ferrite and Metal Evaporated Magnetic Tapes," *Wear* 183, 360–370.

Bhushan, B., Israelachvili, J.N., and Landman, U. (1995d), "Nanotribology: Friction, Wear and Lubrication at the Atomic Scale," *Nature* 374, 607–616.

Bhushan, B. and Koinkar, V.N. (1995e), "Microscale Mechanical and Tribological Characterization of Hard Amorphous Carbon Coatings as Thin as 5 nm for Magnetic Disks," *Surf. Coatings Technol.* 76–77, 655–669.

Bhushan, B., and Kulkarni, A.V. (1995f), "Effect of Normal Load on Microscale Friction Measurements," *Thin Solid Films* 278, 49–56.

Bhushan, B., Kulkarni, A.V., Bonin, W., and Wyrobek, J.T. (1996), "Nano/Picoindentation Measurement Using a Capacitive Transducer System in Atomic Force Microscopy," *Philos. Mag.* 74, 1117–1128.

Bhushan, B., and Koinkar, V.N. (1997), "Microtribological Studies of Doped Single-Crystal Silicon and Polysilicon Films for MEMS Devices," *Sensors Actuators A* 57, 91–102.

Delawski, E. and Parkinson, B.A. (1992), "Layer-by-Layer Etching of Two-Dimensional Metal Chalcogenides with the Atomic Force Microscope," *J. Am. Chem. Soc.* 114, 1661–1667.

Gane, N. and Cox, J.M. (1970), "The Micro-Hardness of Metals at Very Light Loads," *Philos. Mag.* 22, 881–891.

Hamada, E. and Kaneko, R. (1992), "Microdistortion of Polymer Surfaces by Friction," *J. Phys. D: Appl. Phys.* 25, A53–A56.

Kobayashi, A., Grey, F., Williams, R.S., and Ano, M. (1993), "Formation of Nanometer-Scale Grooves in Silicon with a Scanning Tunneling Microscope," *Science* 249, 1724–1726.

Koinkar, V.N. (1997), "Micro/Nanotribology and Its Applications to Magnetic Media, Heads and MEMS," Ph.D. Dissertation, The Ohio State University, Columbus.

Koinkar, V.N., and Bhushan, B. (1997a), "Microtribological Properties of Hard Amorphous Carbon Protective Coatings for Thin-Film Magnetic Disks and Heads," *Proc. Inst. Mech. Eng. Part J: J. Eng. Tribol.* 211, 365–372.

Koinkar, V.N. and Bhushan, B. (1997b), "Scanning and Transmission Electron Microscopies of Single-Crystal Silicon Microworn/Machined Using Atomic Force Microscopy," *J. Mater. Res.* 12, 3219–3224.

Kulkarni, A.V. and Bhushan, B. (1996a), "Nanoscale Mechanical Property Measurements Using Modified Atomic Force Microscopy," *Thin Solid Films* 290–291, 206–210.

Kulkarni, A.V. and Bhushan, B. (1996b), "Nano/Picoindentation Measurements on Single-Crystal Aluminum Using Modified Atomic Force Microscopy," *Mater. Lett.* 29, 221–227.

Kulkarni, A.V. and Bhushan, B. (1997), "Nanoindentation Measurements of Amorphous-Carbon Coatings," *J. Mater. Res.,* 12, 2707–2714.

LaFontaine, W.R., Yost, B., Black, R.D., and Li, C.Y. (1990), "Indentation Load Relaxation Experiments with Indentation Depth in the Submicron Range," *J. Mater. Res.* 5, 2100–2106.

Li, W.B., Henshall, J.L., Hooper, R.M., and Easterling, K.E. (1991), "The Mechanism of Indentation Creep," *Acta Metall. Mater.* 39, 3099–3110.

Majumder, A., Oden, P.I., Carrejo, J.P., Nagahara, L.A., Graham, J.J., and Alexander, J. (1992), "Nanometer-Scale Lithography Using the Atomic Force Microscope", *Appl. Phys. Lett.* 61, 2293–2295.

Miyamoto, T., Kaneko, R., and Miyake, S. (1991), "Tribological Characteristics of Amorphous Carbon Films Investigated by Point Contact Microscopy," *J. Vac. Sci. Technol. B* 9, 1338–1339.

Page, T.F., Oliver, W.C., and McHargue, C.J. (1992), "The Deformation Behavior of Ceramic Crystals Subjected to Very Low Load (Nano) Indentations," *J. Mater. Res.* 7, 450–473.

Parkinson, B. (1990), "Layer-by-Layer Nanometer Scale Etching of Two-Dimensional Substrates Using the Scanning Tunneling Microscopy," *J. Am. Chem. Soc.* 112, 7498–7502.

Pethica, J.N., Hutchings, R., and Oliver, W.C. (1983), "Hardness Measurements at Penetration Depths as Small as 20 nm," *Philos. Mag. A* 48, 598–606.

Pharr, G.M. (1992), "The Anomalous Behavior of Silicon During Nanoindentation," in *Thin Films: Stresses and Mechanical Properties,* III (W.D. Nix, J.C. Braveman, E. Arzt, and L.B. Freund, eds.), Vol. 239, pp. 301–312, MRS, Pittsburgh.

Pharr, G.M., Oliver, W.C., and Clarke, D.R. (1989), "Hysteresis and Discontinuity in the Indentation Load-Displacement Behavior of Silicon," *Scrip. Metall.* 23, 1949–1952.

Pharr, G.M., Oliver, W.C., and Clarke, D.R. (1990), "The Mechanical Behavior of Silicon during Small-Scale Indentation," *J. Electr. Mater.* 19, 881–887.

Silver, R.M., Ehrichs, E.E., and deLozanne, A.L. (1987), "Direct Writing of Submicron Metallic Features with a Resonance," *Phys. Rev. Lett.* 70, 3506–3509.

Tsau, L., Wang, D., and Wang, K. L. (1994), "Nanometer Scale Patterning of Silicon(100) Surface by an Atomic Force Microscope Operative in Air," *Appl. Phys. Lett.* 64, 2133–2135.

Utsugi, Y. (1990), "Nanometer-Scale Chemical Modification Using a Scanning Tunneling Microscope," *Nature* 347, 747–749.

8

Boundary Lubrication Studies Using Atomic Force/Friction Force Microscopy

Bharat Bhushan

8.1 Introduction

Boundary films are formed by physical adsorption, chemical adsorption, and chemical reaction. The physisorbed film can be either monomolecular or polymolecular thick. The chemisorbed films are monomolecular, but stoichiometric films formed by chemical reaction can have a large film thickness. In general, the stability and durability of surface films decrease in the following order: chemical reaction films, chemisorbed films, and physisorbed films. A good boundary lubricant should have a high degree of interaction between its molecules and the sliding surface. As a general rule, liquids are good lubricants when they are polar and thus able to grip solid surfaces (or be adsorbed). Polar lubricants contain reactive functional groups with low ionization potential or groups having high polarizability (Bhushan, 1993). Boundary lubrication properties of lubricants are also dependent upon the molecular conformation and lubricant spreading (Novotny et al., 1989; Novotny, 1990; Mate and Novotny, 1991; Mate, 1992a).

This chapter presents an overview of lubrication studies of polar and nonpolar lubricants and Langmuir–Blodgett and chemically grafted films, using atomic force/friction force microscopy.

8.2 Nanodeformation, Adhesive Forces, and Molecular Conformation

Nanodeformation behavior of the bonded lubricant was studied using atomic force microscopy (AFM) by Blackman et al. (1990a). They used Si(100) substrate with about 1.5 nm of native oxide. Just prior to

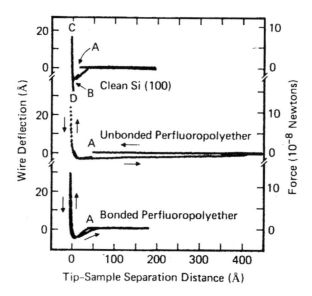

FIGURE 8.1   Wire deflection (normal load) as a function of tip–sample separation distance curves comparing the behavior of clean Si(100) surface to a surface lubricated with free and unbonded PFPE lubricant, and a surface where the PFPE lubricant film was thermally bonded to the surface. (From Blackman, G. S. et al. (1990), *Phys. Rev. Lett.* 65, 3189–3198. With permission.)

the application of lubricant, the surface was cleaned with methylene chloride, spun dried, and followed by exposure to ultraviolet-created ozone for several minutes to remove the remaining adsorbates. Liquid films of the perfluoropolyether Z-Dol of about 4-nm thickness were deposited by a dip-coating method. The lubricant molecules were bonded to the substrate via the reactive end groups by heating at 150°C for 1 h, followed by rinsing with a freon solvent to remove any unbonded molecules, leaving behind about 2-nm-thick film. Before bringing a tungsten tip into contact with a molecular overlayer, it was first brought into contact with a bare clean-silicon surface, Figure 8.1. As the sample approaches the tip, the force initially is zero, but at point A the force suddenly becomes attractive (top curve) which increases until at point B where the sample and tip come into intimate contact and the force becomes repulsive. As the sample is retracted, a pull-off force of 5×10^{-8} N (point D) is required to overcome adhesion between the tungsten tip and the silicon surface. The deformation is reversible (elastic) since the retracting (outgoing) portion of the curve (C to D) follows the extending (ingoing) portion. When an AFM tip is brought into contact with a molecularly thin film of a nonreactive lubricant, a sudden jump into adhesive contact is observed. The adhesion is initiated by the formation of a lubricant meniscus surrounding the tip pulling the surfaces together by Laplace pressure. However, when the tip was brought into contact with a lubricant film which was firmly bonded to the surface, the liquidlike behavior disappears. The initial attractive force (point A) is no longer sudden as with the liquid film, but, rather, gradually increases as the tip penetrates the film. Meniscus formation is suppressed because the polymer molecules are no longer free to move about on the surface as at least one end is attached.

According to Blackman et al. (1990a), if the substrate and tip were infinitely hard with no compliance in the tip and sample supports, the line for B to C would be vertical with an infinite slope. The tangent to the force–distance curve at a given point is referred to as the stiffness at that point and was determined by fitting a least-squares line through the nearby data points. For bonded lubricant film, at the point where slope of the force changes gradually from attractive to repulsive, the stiffness changes gradually, indicating compression of the molecular film. As the load is increased, the slope of the repulsive force eventually approaches that of the bare surface. The bonded film was found to respond elastically up to the highest loads of 5 μN which could be applied. Thus, bonded lubricant behaves as a soft polymer solid.

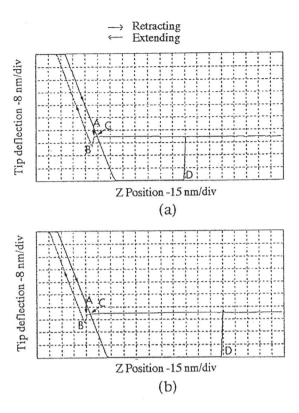

FIGURE 8.2 Tip deflection (normal load) as a function of Z (separation distance) curve for (a) unlubricated and (b) lubricated thin-film magnetic rigid disks. The pull-off force is 42 nN for the unlubricated disk and 64 nN for the lubricated disk calculated from the horizontal distance between points C and D and the cantilever spring constant of 0.4 N/m. (From Bhushan, B. and Ruan, J. (1994), *ASME J. Tribol.* 116, 389–396. With permission.)

The attractive adhesive forces at different parts of the surface can be estimated by bringing the sample into contact with the tip and then measuring the maximum force needed to pull the tip and sample apart, Mate (1993) and Bhushan and Ruan (1994). Figure 8.2 shows typical normal load as a function of separation distance (Z) curves for unlubricated and lubricated (with about 2-nm thickness of a perfluoropolyether lubricant with an alcohol group, Z-Dol) magnetic disk samples. In these experiments, the disks are first brought into contact and then withdrawn from the tip. The presence of the water vapor for the unlubricated disk and along with the lubricant film for the lubricated disk causes a sudden attractive force to occur at point A due to a meniscus of liquid forming around the top of the liquid film and a long break-free distance out to point B where the liquid meniscus breaks as the sample is withdrawn. The major difference between the two curves is that the pull-off force is lower for the unlubricated surface compared with lubricated surface. Pull-off force is determined by multiplying the cantilever spring constant (0.4 N/m) by the horizontal distance between points C and D, which corresponds to the maximum cantilever deflection toward the disks before the tip is disengaged. The horizontal distance/pull-off force is 105 nm/42 nN for the unlubricated disk and 160 nm/64 nN for the lubricated disk. The higher value of the pull-off force in the case of lubricated disk arises from the larger meniscus contribution from the liquid films (Bhushan and Ruan, 1994).

Figure 8.3 illustrates two extremes for the conformation on a surface of a linear liquid polymer without any reactive end groups and at submonolayer coverages (Novotny et al., 1989; Mate and Novotny, 1991). At one extreme, the molecules lie flat on the surface, reaching no more than their chain diameter δ above the surface. This would be the case if a strong attractive interaction exists between the molecules and the

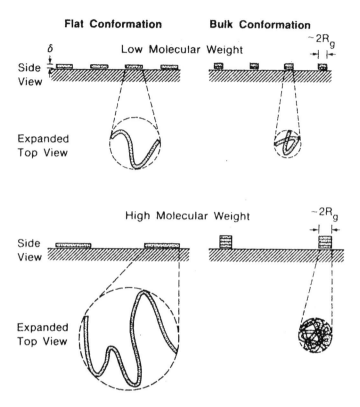

FIGURE 8.3 Schematic representation of two extreme liquid conformations at the surface of the solid for low and high molecular weights at low surface coverage. δ is the cross-sectional diameter of the liquid chain and R_g is the radius of gyration of the molecules in the bulk. (From Mate, C. M. and Novotny, V. J. (1991), *J. Chem. Phys.* 92, 3189–3196. With permission.)

solid. On the other extreme, when a weak attraction exists between polymer segments and the solid, the molecules adopt conformation close to that of the molecules in the bulk, with the microscopic thickness equal to about the radius of gyration R_g. Mate and Novotny (1991) used AFM to study conformation of 0.5- to 13-nm-thick perfluoropolyether molecules on clean Si(100) surfaces. They found that the thickness measured by AFM is thicker than that measured by ellipsometry with the offset ranging from 3 to 5 nm. They found that the offset was the same for very thin submonolayer coverages. If the coverage is submonolayer and inadequate to make a liquid film, the relevant thickness is then the height (h_e) molecules extended above the solid surface. The offset should then equal $2h_e$, assuming that the molecules extend the same height above both the tip and silicon surfaces. They thus concluded that the molecules do not extend more than 1.5 to 2.5 nm above a solid or liquid surface, much smaller than the radius of gyration of the lubricants ranging between 3.2 and 7.3 nm, and to the approximate cross-sectional diameter of 0.6 to 0.7 nm for the linear polymer chain. Consequently, the height that the molecules extend above the surface is considerably less than the diameter of gyration of the molecules and only a few molecular diameters in height, implying that the physisorbed molecules on a solid surface have an extended, flat conformation. They also determined the disjoining pressure of these liquid films from AFM measurements of the distance needed to break the liquid meniscus that forms between the solid surface and the AFM tip. (Also see Mate, 1992a). For a monolayer thickness of about 0.7 nm, the disjoining pressure is about 5 MPa, indicating strong attractive interaction between the liquid molecules and the solid surface. The disjoining pressure decreases with increasing film thickness in a manner consistent with a strong attractive van der Waals interaction between the liquid molecules and the solid surface.

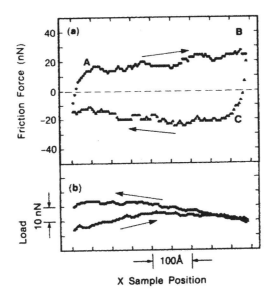

FIGURE 8.4   (a) Friction force and (b) normal load over an oscillation of the X-sample position during sliding of the tungsten tip on an Si(100) surface coated with perfluoropolyether lubricant with alcohol end group. (From Mate, C. M. (1992), *Phys. Rev. Lett.* 68, 3323–3326. With permission.)

Attempts to measure mechanical properties of self-assembled monolayer films on Au(111) films have been made by Salmeron et al. (1993). They have used AFM in the tapping mode. This technique has the potential of measuring local viscoelastic properties of lubricant films.

8.3   Boundary Lubrication Studies

8.3.1   Liquid Lubricants

Mate (1992b), O'Shea et al. (1992), Bhushan et al. (1995a–c), and Koinkar and Bhushan (1996a,b) used AFM to provide insight into how lubricants function at the molecular level. Mate (1992b) conducted friction experiments on Si(100) surface lubricated with a lubricant with alcohol end group (Z-Dol). In these experiments, the sample was moved rapidly back and forth in the X-direction at a velocity of 1 μm/s, while the normal load on the tip was slowly increased to some maximum value, then decreased back to zero by moving the sample in the Z-direction. Figure 8.4 shows an example of the friction force on the tip during one complete X oscillation of the sample. Initially, the tip moves with the sample, until, at point A, the cantilever wire exerts enough force to overcome the static frictional force and the tip starts to slide across the surface. When the X sample direction is reversed at point B, the tip again moves with the sample until it starts to slide at point C. The upward shift in the normal load over the cycle comes with the slight increase in load as the sample is slowly pushed up against the tip. The slight variation in load during the cycle correspond to a surface roughness of about 0.1 nm.

Figure 8.5 shows the average normal load and friction forces during sliding as a function of the Z-sample position for the 3-nm-thick films of different types of lubricants. Each data point in the figure represents the average over the sliding portion of a cycle in the X-direction like the one shown in Figure 8.4. For the liquid film in Figure 8.5a and c, Mate (1992b) noted that, just before the hard wall contact is made, the normal load during sliding becomes more attractive for nonalcohol end group (Z03) than that for alcohol end group (Z-Dol) in the same manner as when no sliding occurs. When the sample is withdrawn, the friction force returns to zero when the hard wall contact is broken. This regime is called full-film lubrication, where shearing of a liquid film takes place resulting in a negligible friction

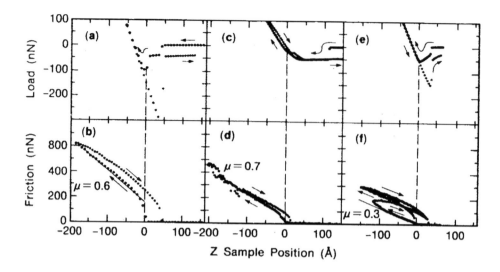

FIGURE 8.5 The friction force and normal load as a function of the Z sample position for 3-nm-thick films on Si(100) of (a), (b) unbonded lubricant with unreactive end groups, (c), (d) unbonded lubricant with alcohol end groups and (e), (f) bonded lubricant. The open circles in (f) show the friction force when the experiment is repeated in the same spot on the bonded lubricant. (From Mate, C. M. (1992), *Phys. Rev. Lett.* 68, 3323–3326. With permission.)

(Bowdon and Tabor, 1950). After hard wall contact, one is in boundary lubrication regime, where solid–solid shearing takes place. The transition between the two regimes is very sharp, requiring a change in separation distance less than a chain diameter. In the boundary lubrication regime, the friction force for the liquid films initially rises quickly, but soon increases linearly with load (coefficient of friction in the range of 0.5 to 0.8). Similar values of the coefficient of friction are observed for unlubricated silicon surfaces, indicating that most of the unbonded liquid lubricant is squeezed out from the rubbing surfaces. Some liquid molecules may still be trapped among the microasperities of the tip and contribute to the solid-solid shearing.

Figure 8.5e and f show the results from sliding on the bonded lubricant film prepared by conveniently bonding of the alcohol-ended lubricant (by heating Z-Dol film at 150°C for 1 h in order to react the end groups with hydroxyl groups on the silicon surface with native oxide and washing off nonreactive portion with freon). As was the case for the unbonded lubricant, no significant friction is observed until the hard wall contact was made. So, even though the ends of the polymer are rigidly attached to the substrate, the backbone of the polymer apparently has enough flexibility to offer little resistance to the sliding tip except when rigidly compressed between the two surfaces. For the bonded lubricant, the initial coefficient of friction is about 0.3, which is about half that for the unbonded liquid films. The lower coefficient of friction indicates that significantly more molecules are trapped between the rubbing surfaces than for the unbonded lubricant. With repeated traversals of the sliding tip, these attached molecules eventually wear away and the coefficient of friction increases with increasing load.

Mate (1992b) concluded that the liquid films have negligible shear stress to applied shear strains until the molecules are completely compressed or squeezed out from between the sliding surfaces, showing that hydrodynamic lubrication can occur for surfaces separated by only a few cross-sectional diameters of the polymer backbone. (Also see O'Shea et al., 1992, 1994). The addition of alcohol end groups greatly improves the resistance of the molecules to being squeezed out from between the sliding surfaces.

Koinkar and Bhushan (1996a,b) studied the friction and wear performance of Si(100) sample lubricated with about 2-nm-thick Z-15 and Z-Dol perfluoropolyether (PFPE) lubricants. Z-Dol film was thermally bonded at 150°C for 30 min and washed off with a solvent to provide a chemically bonded layer of the lubricant film. Data showing the effect of environment on the unlubricated and lubricated samples are summarized in Figure 8.6. Note that lubricated silicon samples show a lower value of

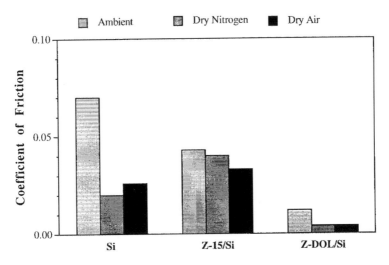

FIGURE 8.6 Coefficient of friction for unlubricated and lubricated samples in ambient (~50% RH), dry nitrogen (~5% RH), and dry air (~5% RH). (From Koinkar, V. N. and Bhushan, B. (1996), *J. Vac. Sci. Technol. A* 14, 2378–2391. With permission.)

coefficient of friction than that of unlubricated sample. Furthermore, sample lubricated with Z-Dol

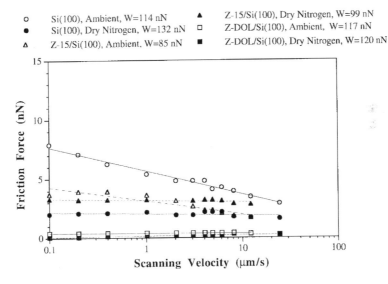

FIGURE 8.7 Friction force as a function of scanning velocity for unlubricated and lubricated samples in ambient and dry nitrogen environments. Normal loads used are given in the figure. (From Koinkar, V. N. and Bhushan, B. (1996), *J. Vac. Sci. Technol. A* 14, 2378–2391. With permission.)

exhibits a lower value of the coefficient of friction than that of the Z-15 lubricated sample. For the unlubricated and lubricated samples, the coefficient of friction in a dry environment is lower than at ambient of about 50% relative himidity. At high humidity, the condensed water film from the environment results in dewetting of the lubricant film (or water film for the unlubricated sample) resulting in poorer lubrication performance. Figure 8.7 shows the effect of scanning velocity on the coefficient of

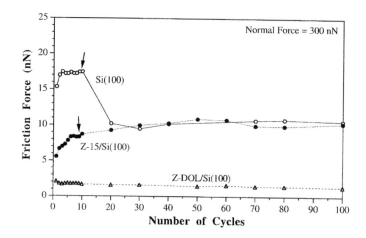

FIGURE 8.8 Friction force as a function of number of cycles using an Si$_3$N$_4$ tip at a normal load of 300 nN for unlubricated and lubricated samples in ambient environment. Arrows in the figure indicate significant changes in the friction force because of removal of surface or lubricant film. (From Koinkar, V. N. and Bhushan, B. (1996), *J. Vac. Sci. Technol. A* 14, 2378–2391. With permission.)

friction in ambient and dry nitrogen environments. The coefficient of friction of unlubricated silicon sample and lubricated with Z-15 decreases with an increase of the scanning velocity in the ambient environment, whereas the sample lubricated with Z-15 in the dry nitrogen and the sample lubricated with Z-Dol in both ambient and dry nitrogen environments do not show any velocity dependence. We believe that alignment of free liquid molecules at higher scanning velocities results in lower values of the coefficient of friction.

To study lubricant depletion during microscale measurements, Koinkar and Bhushan (1996a,b) conducted nanowear studies using Si$_3$N$_4$ tips. They measured friction as a function of the number of cycles for virgin silicon and silicon surfaces lubricated with Z-15 and Z-Dol lubricants, Figure 8.8. An area of 1×1 μm was scanned at a normal force of 300 nN. Note that the friction force in a virgin silicon surface decreases in a few cycles after the natural oxide film present on silicon surface gets removed. In the case of Z-15-coated silicon sample, the friction force starts out to be low and then approaches that of an unlubricated silicon surface after a few cycles. The increase in friction of the lubricated sample suggests that the lubricant film gets worn and the silicon underneath is exposed. In the case of the Z-Dol-coated silicon sample, the friction force starts out to be low and remains low during the 100-cycles test. It suggests that Z-Dol does not get displaced/depleted as readily as Z-15. (Also see Bhushan et al., 1995a.) Microwear studies were also conducted using the diamond tip at various loads. Figure 8.9 shows the plots of wear depth as a function of normal force and Figure 8.10 shows the wear profiles of worn samples at 40 μN normal force. The Z-Dol-lubricated sample exhibits better wear resistance than the unlubricated and Z-15-lubricated silicon samples. Wear resistance of the Z-15-lubricated sample is little better than that of the unlubricated sample. The Z-15-lubricated sample shows debris inside the wear track. Since Z-15 is a liquid lubricant, debris generated is held by the lubricant and it becomes sticky, which moves inside the wear track and does damage, Figure 8.10.

8.3.2 LB and Self-Assembled Monolayers

Organized and dense molecular-scale layers of, preferably, long-chain molecules have shown to be superior lubricants on both macro- and microscales as compared with freely supported multimolecular layers (Bhushan et al., 1995b,c). Common methods to produce monolayers and thin films (Ulman, 1991) are the Langmuir–Blodgett (LB) deposition (Roberts, 1990) and self-assembled films by chemical grafting of molecules (Jaffrezic-Renault and Martelet, 1992). The LB films are bonded to the substrate by weak

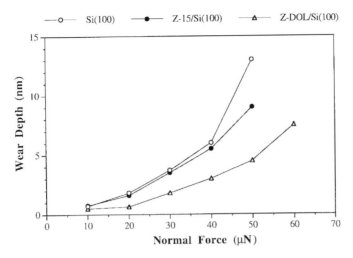

FIGURE 8.9 Wear depth as a function of normal force using a diamond tip for unlubricated and lubricated samples after one cycle. (From Koinkar, V. N. and Bhushan, B. (1996), *J. Vac. Sci. Technol. A* 14, 2378–2391. With permission.)

van der Waals attraction; whereas the grafting process forms directly covalently bonded dense films of long-chain molecules.

Bhushan et al. (1995b,c) conducted micro- and macroscale friction and wear studies on single-grafted and double-grafted octodecyl (C_{18}) and LB films and their subsurfaces. C_{18} films were produced by grafting of long-chain organic molecules onto an Si(100) wafer covered with a thermally grown SiO_2 layer, Figure 8.11. The structure of the LB film consisted of an octadecylthiol (ODT) coated gold sample (gold films thermally evaporated onto single-crystal silicon) on top of which a single, upper inverted bilayer of zinc arachidate (C_{20}, ZnA) was deposited by the LB technique, Figure 8.12. Macro- and microscale friction and wear data are summerized in Table 8.1 and Figure 8.13. We note that C_{18} double-grafted film exhibits the lowest coefficient of friction of 0.018 as compared with other samples measured in this study (the coefficient of friction of ZnA films is ~0.03). The coefficient of friction for LB film is comparable to that of the ODT layer and lower than the Au film. The wear resistance of C_{18} double-grafted film is much better than C_{18} single-grafted, LB, and Au films and is comparable with that of SiO_2 film. The C_{18} double-grafted films can withstand much higher normal force of 40 μN as compared with 200 nN for the case of LB films. Surface profiles showing the worn regions after one scan cycle for C_{18} double-grafted and ZnA films are shown in Figure 8.14. Nanoindentation studies conducted by Bhushan et al. (1995c) showed that the C_{18} double-grafted films are more rigid than the LB films which may be responsible for the high wear resistance of C_{18} films. Flexibility of choosing the alkyl chain length, functional terminal group, and cross-linking enables the adaptability of the grafting process for lubrication of microcomponents.

8.4 Closure

In nanodeformation experiments, bonded lubricants behave as soft polymer solids. AFM/friction force microscopy friction experiments show that lubricants with polar (reactive) end groups dramatically increase the load or contact pressure that a liquid film can support before solid–solid contact and these exhibit long durability. Chemically grafted films may be suitable for lubrication of microcomponents.

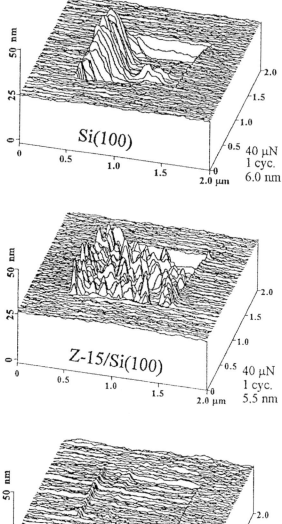

FIGURE 8.10 Wear profiles for unlubricated and lubricated samples after microwear using diamond tip. Normal force used and wear depths are listed in the figure. (From Koinkar, V. N. and Bhushan, B. (1996), *J. Vac. Sci. Technol. A* 14, 2378–2391. With permission.)

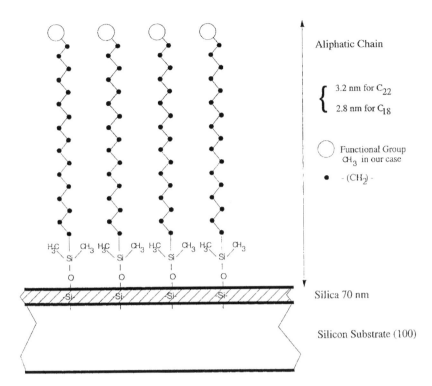

FIGURE 8.11 Schematic of the ODT (C$_{18}$) and docosyl (C$_{22}$) grafted film on silica.

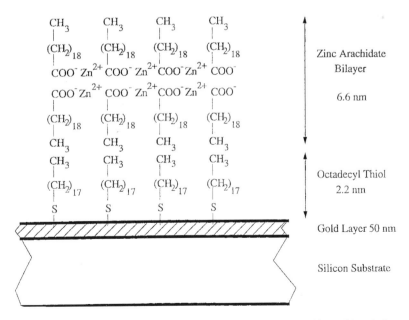

FIGURE 8.12 Schematic of the zinc arachidate LB film on ODT with a gold underlayer.

TABLE 8.1   Typical rms Roughness and Coefficients of Microscale and Macroscale Friction Values of Various Samples

Sample	rms Roughness (nm)[a]	Coefficient of microscale friction[b]	Coefficient of macroscale friction[c]
Si(100)	0.12	0.03	0.18
SiO$_2$/Si	0.21	0.03	0.19
C$_{18}$ single-grafted/SiO$_2$/Si	0.16	0.04	0.23
C$_{18}$ double-grafted/SiO$_2$/Si	0.16	0.018	0.07
Au/Si	1.16	0.04	0.13
ODT/Au/Si	0.92	0.03	0.14
ZnA/ODT/Au/Si	0.55	0.03	0.16

[a]Measured on 1×1 μm scan area using AFM.

[b]Si$_3$N$_4$ tip with a radius of about 30–50 nm at normal load in the range of 10–200 nN and scanning speed of 4 μm/s on a 1×1 μm scan area.

[c]Alumina ball with 3-mm radius at normal loads of 0.1 N and average sliding speed of 0.8 mm/s.

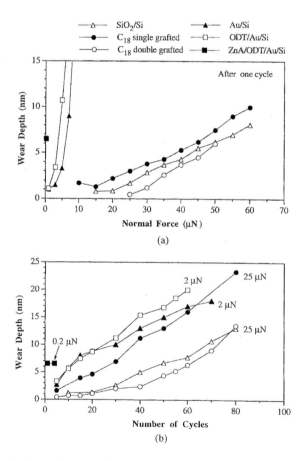

FIGURE 8.13   (a) Wear depth as a function of normal force after one scan cycle (b) wear depth as a function of number of scan cycles at the normal force indicated for SiO$_2$/Si, C$_{18}$ single-grafted, C$_{18}$ double-grafted, Au/Si, ODT/Au/Si, and ZnA/ODT/Au/Si. (From Bhushan, B. et al. (1995), *Langmuir* 11, 3189–3198. With permission.)

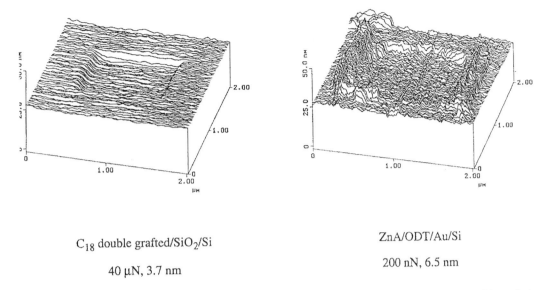

C_{18} double grafted/SiO_2/Si

40 μN, 3.7 nm

ZnA/ODT/Au/Si

200 nN, 6.5 nm

FIGURE 8.14 Surface profile showing the worn region (center 1×1 μm) after one scan cycle for C_{18} double-grafted and ZnA films. Normal force and the wear depths are indicated in the figure.

References

Bhushan, B. (1993), "Magnetic Recording Surfaces," in *Characterization of Tribological Materials* (W.A. Glaeser, ed.), pp. 116–133, Butterworth-Heinemann, Boston.

Bhushan, B. and Ruan, J. (1994), "Atomic-Scale Friction Measurements Using Friction Force Microscopy: Part II — Application to Magnetic Media," *ASME J. Tribol.* 116, 389–396.

Bhushan, B., Miyamoto, T., and Koinkar, V. N. (1995a), "Microscopic Friction between a Sharp Diamond Tip and Thin-Film Magnetic Rigid Disks by Friction Force Microscopy," *Adv. Info. Storage Syst.* 6, 151–161.

Bhushan, B. Israelachvili, J. N., and Landman, U. (1995b), "Nanotribology: Friction, Wear and Lubrication at the Atomic Scale," *Nature* 374, 607–616.

Bhushan, B., Kulkarni, A. V., Koinkar, V. N., Boehm, M., Odoni, L., Martelet, C., and Belin, M. (1995c), "Microtribological Characterization of Self-Assembled and Langmuir-Blodgett Monolayers by Atomic and Friction Force Microscopy," *Langmuir* 11, 3189–3198.

Blackman, G. S., Mate, C. M., and Philpott, M. R. (1990a), "Interaction Forces of a Sharp Tungsten Tip with Molecular Films on Silicon Surface," *Phys. Rev. Lett.* 65, 2270–2273.

Blackman, G. S., Mate, C. M., and Philpott, M. R. (1990b), "Atomic Force Microscope Studies of Lubricant Films on Solid Surfaces," *Vacuum* 41, 1283–1286.

Bowden, F. P. and Tabor, D. (1950), *The Friction and Lubrication of Solids*, Part 1, Clarendon Press, Oxford.

Jaffrezic-Renault, N. and Martelet, C. (1992), "Preparation of Well-Engineered Thin Molecular Layer on Semiconductor–Based Transducers," *Sensors Actuators* A 32, 307–312.

Koinkar, V. N. and Bhushan, B. (1996a), "Microtribological Studies of Unlubricated and Lubricated Surfaces Using Atomic Force/Friction Force Microscopy," *J. Vac. Sci. Technol.* A 14, 2378–2391.

Koinkar, V. N. and Bhushan, B. (1996b), "Micro/Nanoscale Studies of Boundary Layers of Liquid Lubricants for Magnetic Disks," *J. Appl. Phys.* 79, 8071–8075.

Mate, C. M. (1992a), "Application of Disjoining and Capillary Pressure to Liquid Lubricant Films in Magnetic Recording," *J. Appl. Phys.* 72, 3084–3090.

Mate, C. M. (1992b), "Atomic-Force-Microscope Study of Polymer Lubricants on Silicon Surface," *Phys. Rev. Lett.* 68, 3323–3326.

Mate, C. M. (1993), "Nanotribology of Lubricated and Unlubricated Carbon Overcoats on Magnetic Disks Studied by Friction Force Microscopy," *Surf. Coat. Technol.* 62, 373–379.

Mate, C. M. and Novotny, V. J. (1991), "Molecular Conformation and Disjoining Pressures of Polymeric Liquid Films," *J. Chem. Phys.* 94, 8420–8427.

Novotny, V. J. (1990), "Migration of Liquid Polymers on Solid Surfaces," *J. Chem. Phys.* 92, 3189–3196.

Novotny, V. J., Hussla, I., Turlet, J. M., and Philpott, M. R. (1989), "Liquid Polymer Conformation on Solid Surfaces," *J. Chem. Phys.* 90, 5861–5868.

O'Shea, S. J., Welland, M. E., and Rayment, T. (1992), "Atomic Force Microscope Study of Boundary Layer Lubrication," *Appl. Phys. Lett.* 61, 2240–2242.

O'Shea, S. J., Welland, M. E., and Pethica, J. B. (1994), "Atomic Force Microscopy of Local Compliance at Solid-Liquid Interface," *Chem. Phys. Lett.* 223, 336–340.

Roberts, G. G. (1990), *Langmuir-Blodgett Films,* Plenum, New York.

Salmeron, M., Neubauer, G., Folch, A. Tomitori, M., Ogletree, D. F., and Sautet, P. (1993), "Viscoelastic and Electrical Properties of Self-Assembled Monolayers on Au(111) Films," *Langmuir* 9, 3600–3611.

Ulman, A. (1991), *An Introduction to Ultrathin Organic Films,* Academic Press, Boston.

9

Surface Forces and Microrheology of Molecularly Thin Liquid Films

Jacob N. Israelachvili and
Alan D. Berman

9.1 Introduction

In this chapter the most important types of surface forces are described and the relevant equations for the force laws given. A number of attractive and repulsive forces operate between surfaces and particles. Some of these occur only in vacuum, for example, attractive van der Waals and repulsive hard-core interactions. Others can arise only when the interacting surfaces are separated by another condensed phase, which is usually a liquid medium. The most common types of surface forces and their main characteristics are list in Table 9.1.

In *vacuum*, the two main long-ranged interactions are the attractive van der Waals and electrostatic (coulombic) forces, while at smaller surface separations — corresponding to molecular contacts at surface separations of $D \approx 0.2$ nm — additional attractive forces can come into play, such as covalent or metallic bonding forces. These attractive forces are stabilized by the hard-core repulsion, and together they determine the surface and interfacial energies of planar surfaces as well as the strengths of materials and adhesive junctions. Adhesion forces are often strong enough to deform the shapes of two bodies or particles elastically or plastically when they come into contact for the first time.

When exposed to *vapors* (e.g., atmospheric air containing water and organic molecules), two solid surfaces in or close to contact will generally have a surface layer of chemisorbed or physisorbed molecules, or a capillary condensed liquid bridge between them. These effects can drastically modify their adhesion. The adhesion usually falls, but in the case of capillary condensation the additional Laplace pressure or attractive "capillary" force between the surfaces may make the adhesion stronger than in inert gas or vacuum.

When totally immersed in a *liquid*, the force between two surfaces is once again completely modified from that in vacuum or air (vapor). The van der Waals attraction is generally reduced, but other forces can now arise which can qualitatively change both the range and even the sign of the interaction. The overall attraction can be either stronger or weaker than in the absence of the intervening liquid medium, for example, stronger in the case of two hydrophobic surfaces in water, but weaker for two hydrophilic surfaces. Since a number of different forces may be operating simultaneously in solution, the overall force law is not generally monotonically attractive, even at long range: it can be repulsive, oscillatory, or the force can change sign at some finite surface separation. In such cases, the potential energy minimum,

TABLE 9.1   Types of Surface Forces

Type of Force	Subclasses and Alternative Names	Main Features
		Attractive
van der Waals	Dispersion force (v & s) Induced dipole force (v & s) Casimir force (v & s)	Ubiquitous force, occurs both in vacuum and in liquids
Electrostatic	Coulombic force (v & s) Ionic bond (v) Hydrogen bond (v) Charge–transfer interaction (v & s) "Harpooning" interaction (v)	Strong, long-ranged force arising in polar solvents; requires surface charging or charge-separation mechanism
Quantum mechanical	Covalent bond (v) Metallic bond (v) Exchange interaction (v)	Strong short-ranged forces responsible for contact binding of crystalline surfaces
Hydrophobic	Attractive hydration force (s)	Strong, apparently long-ranged force; origin not yet understood
Ion–correlation	van der Waals force of polarizable ions (s)	Requires mobile charges on surfaces in a polar solvent
Solvation	Oscillatory force (s) Depletion force (s)	The oscillatory force generally alternates between attraction and repulsion; mainly entropic in origin
Specific binding	"Lock and key" binding (v & s) Receptor–ligand interaction (s) Antibody–antigen interaction (s)	Subtle combination of different noncovalent forces giving rise to highly specific binding; main "recognition" mechanism of biological systems.
		Repulsive
Quantum mechanical	Hard-core (v) Steric repulsion (v) Born repulsion (v)	Forces stabilizing attractive covalent and ionic binding forces, effectively determine molecular size and shape
van der Waals	van der Waals disjoining pressure (s)	Arises only between dissimilar bodies interacting in a medium
Electrostatic		Arises only for certain constrained surface charge distributions.
Solvation	Oscillatory solvation force (s) Structural force (s) Hydration force (s)	Monotonically repulsive forces, believed to arise when solvent molecules bind strongly to surfaces.
Entropic	Osmotic repulsion (s) Double-layer force (s) Thermal fluctuation force (s) Steric polymer repulsion (s) Undulation force (s) Protrusion force (s)	Forces due to confinement of molecular or ionic species between two approaching surfaces. Requires a mechanism which keeps trapped species between the surfaces.
		Dynamic Interactions
Nonequilibrium	Hydrodynamic forces (s) Viscous forces (s) Friction forces (v & s) Lubrication forces (s)	Energy-dissipating forces occurring during relative motion of surfaces or bodies.

Note: v, Applies only to interactions in *vacuum*; s, applies only to interactions in *solution*, or to surfaces separated by a liquid; v & s, applies to interactions occurring both in vacuum and in solution.

which determines the adhesion force or energy, occurs not at true molecular contact but at some small distance farther out.

The forces between two surfaces in a liquid medium can be particularly complex at *short range*, i.e., at surface separations below a few nanometers or 5 to 10 molecular diameters. This is partly because, with increasing confinement, a liquid ceases to behave as a structureless continuum with properties

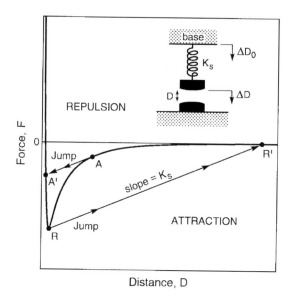

FIGURE 9.1 Schematic attractive force law between two macroscopic objects, such as two magnets, or between two microscopic objects such as the van der Waals force between a metal tip and a surface. On lowering the base supporting the spring, the latter will expand or contract such that at any equilibrium separation D the attractive force balances the elastic spring restoring force. However, once the gradient of the attractive force between the surfaces dF/dD exceeds the gradient of the spring restoring force, defined by the spring constant K_S, the upper surface will jump from A into contact at A' (A for advancing). On separating the surfaces by raising the base, the two surfaces will jump apart from R to R' (R for receding). The distance $R-R'$ multiplied by K_S gives the adhesion force, i.e., the value of F at R.

determined solely by its bulk properties; the size and shape of its molecules begin to play an important role in determining the overall interaction. In addition, the surfaces themselves can no longer be treated as inert and structureless walls (i.e., mathematically flat) — their physical and chemical properties at the atomic scale must also now be taken into account. Thus, the force laws will now depend on whether the surface lattices are crystallographically matched or not, whether the surfaces are amorphous or crystalline, rough or smooth, rigid or soft (fluidlike), hydrophobic or hydrophilic.

In practice, it is also important to distinguish between *static* (i.e., equilibrium) forces and *dynamic* (i.e., nonequilibrium) forces such as viscous and friction forces. For example, certain liquid films confined between two contacting surfaces may take a surprisingly long time to equilibrate, as may the surfaces themselves, so that the short-range and adhesion forces appear to be time dependent, resulting in "aging" effects.

9.2 Methods for Measuring Static and Dynamic Surface Forces

9.2.1 Adhesion Forces

The simplest and most direct way to measure the adhesion of two solid surfaces, such as two spheres or a sphere on a flat surface, is to suspend one on a spring and measure — from the deflection of that spring — the adhesion or "pull-off" force needed to separate the two bodies. Figure 9.1 illustrates the principle of this method when applied to the interaction of two magnets. However, the method is applicable even at the microscopic or molecular level, and it forms the basis of all direct force-measuring apparatuses such as the surface forces apparatus (SFA) (Israelachvili, 1989, 1991) or the atomic force microscope (AFM) (Ducker et al., 1991).

If K_S is the stiffness of the force-measuring spring and ΔD the distance the two surfaces jump apart when they separate, then the adhesion force F_S is given by

$$F_S = F_{max} = K_S \cdot \Delta D, \tag{9.1}$$

where we note that in liquids the maximum or minimum in the force may occur at some nonzero surface separation (see Figures 9.3 and 9.4 below).

From F_S one may also calculate the surface or interfacial energy γ. However, this depends on the geometry of the two bodies. For a sphere of radius R on a flat surface, or for two crossed cylinders of radius R, we have (Israelachvili, 1991)

$$\gamma = F_s / 3\pi R, \tag{9.2}$$

while for two spheres of radii R_1 and R_2

$$\gamma = \frac{F_s}{3\pi} \left(\frac{1}{R_1} + \frac{1}{R_2} \right), \tag{9.3}$$

where γ is in units of J/m^2.

9.2.2 Force Law

The full force law $F(D)$ between two surfaces, that is, the force F as a function of surface separation D, can be measured in a number of ways. The simplest is to move the base of the spring (see Figure 9.1) by a known amount, say, ΔD_0. If there is a detectable force between the two surfaces, this will cause the force-measuring spring to deflect by, say, ΔD_S, while the surface separation changes by ΔD. These three displacements are related by

$$\Delta D_S = \Delta D_0 = \Delta D. \tag{9.4}$$

The force difference ΔF between the initial and final separations is given by

$$\Delta F = K_S \Delta D_S. \tag{9.5}$$

The above equations provide the basis for measuring the force difference between any two surface separations. For example, if a particular force-measuring apparatus can measure ΔD_0, ΔD_S, and K_S, then by starting at some large initial separation where the force is zero ($F = 0$) and measuring the force difference ΔF between this initial or reference separation D and $(D - \Delta D)$, then working one's way in increasing increments of $\Delta D = (\Delta D_0 - \Delta D_S)$, the full force law $F(D)$ can be constructed over any desired distance regime.

Whenever an equilibrium force law is required, it is essential to establish that the two surfaces have stopped moving before the "equilibrium" displacements are measured. When displacements are measured while two surfaces are still in relative motion, one also measures a viscous or frictional contribution to the total force. Such dynamic force measurements have enabled the viscosities of liquids near surfaces and in thin films to be accurately measured (Israelachvili, 1989).

In practice, it is difficult to measure the forces between two perfectly flat surfaces because of the stringent requirement of perfect alignment for making reliable measurements at the angstrom level. It is far easier to measure the forces between curved surfaces, for example, two spheres, a sphere and a flat, or two crossed cylinders. As an added convenience, the force $F(D)$ measured between two curved surfaces can be directly related to the energy per unit area $E(D)$ between two flat surfaces at the same separation, D. This is given by the so-called "Derjaguin" approximation:

$$E\left(D\right) = \frac{F\left(D\right)}{2\pi R},\tag{9.6}$$

where R is the radius of the sphere (for a sphere and a flat) or the radii of the cylinders (for two crossed cylinders).

9.2.3 The Surface Force Apparatus and the Atomic Force Microscope

In a typical force-measuring experiment, two or more of the above displacement parameters: ΔD_0, ΔD_S, ΔD, and K_S, are directly or indirectly measured, from which the third displacement and resulting force law $F(D)$ are deduced using Equations 9.4 and 9.5. For example, in SFA experiments, ΔD_0 is changed by expanding a piezoelectric crystal by a known amount and the resulting change in surface separation ΔD is measured optically, from which the spring deflection ΔD_S is obtained. In contrast, in AFM experiments, ΔD_0 and ΔD_S are measured using a combination of piezoelectric, optical, capacitance, or magnetic techniques, from which the surface separation ΔD is deduced. Once a force law is established, the geometry of the two surfaces must also be known (e.g., the radii R of the surfaces) before one can use Equation 9.6 or some other equation that enables the results to be compared with theory or with other experiments.

Israelachvili (1989, 1991), Horn (1990), and Ducker et al. (1991) have described various types of SFAs suitable for making adhesion and force law measurements between two curved molecularly smooth surfaces immersed in liquids or controlled vapors. The optical technique used in these measurements employs multiple beam interference fringes which allows for surface separations D to be measured to ±1 Å. From the shapes of the interference fringes, one also obtains the radii of the surfaces, R, and any surface deformation that arises during an interaction (Israelachvili and Adams, 1978; Chen et al., 1992). The distance between the two surfaces can also be independently controlled to within 1 Å, and the force sensitivity is about 10^{-8} N (10^{-6} g). For the typical surface radii of R ≈ 1 cm used in these experiments, γ values can be measured to an accuracy of about $\pm 10^{-3}$ mJ/m^2 ($\pm 10^{-3}$ erg/m^2).

Various surface materials have been successfully used in SFA force measurements including mica (Pashley, 1981, 1982, 1985), silica (Horn et al., 1989b), and sapphire (Horn et al., 1988). It is also possible to measure the forces between adsorbed polymer layers (Klein, 1983, 1986; Patel and Tirrell, 1989; Ploehn and Russel, 1990), surfactant monolayers and bilayers (Israelachvili, 1987, 1991; Christenson, 1988a; Israelachvili and McGuiggan, 1988), and metal and metal oxide layers deposited on mica (Coakley and Tabor, 1978; Parker and Christenson, 1988; Smith et al., 1988; Homola et al., 1993; Steinberg et al., 1993). The range of liquids and vapors that can be used is almost endless, and so far these have included aqueous solutions, organic liquids and solvents, polymer melts, various petroleum oils and lubricant liquids, and liquid crystals.

Recently, new friction attachments were developed suitable for use with the SFA (Homola et al., 1989; Van Alsten and Granick, 1988, 1990b; Klein et al., 1994; Luengo et al., 1997). These attachments allow for the two surfaces to be sheared past each other at varying sliding speeds or oscillating frequencies while simultaneously measuring both the transverse (frictional or shear) force and the normal force or load between them. The externally applied load, L, can be varied continuously, and both positive and negative loads can be applied. Finally, the distance between the surfaces D, their true molecular contact area A, their elastic (or viscoelastic or elastohydrodynamic) deformation, and their lateral motion can all be monitored simultaneously by recording the moving interference fringe pattern using a video camera–recorder system.

TABLE 9.2 van der Waals Interaction Energy and Force Between Macroscopic Bodies of Different Geometries

Geometry of Bodies With Surfaces D Apart ($D \ll R$)	van der Waals Interaction	
	Energy	Force
Two flat surfaces (per unit area)	$E = A/12pD^2$	$F = A/6pD^3$
Sphere of radius R near flat surface	$E = AR/6D$	$F = AR/6D^2$
Two identical spheres of radius R	$E = AR/12D$	$F = AR/12D^2$
Cylinder of radius R near flat surface (per unit length)	$E = \dfrac{A\sqrt{R}}{12\sqrt{2}\ D^{3/2}}$	$F = \dfrac{A\sqrt{R}}{8\sqrt{2}\ D^{5/2}}$
Two identical parallel cylinders of radius R (per unit length)	$E = \dfrac{A\sqrt{R}}{24\ D^{3/2}}$	$F = \dfrac{A\sqrt{R}}{16\ D^{5/2}}$
Two identical cylinders of radius R crossed at 90°	$E = AR/6D$	$F = AR/6D^2$

9.3 van der Waals and Electrostatic Forces between Surfaces in Liquids

9.3.1 van der Waals Forces

Table 9.2 lists the van der Waals force laws for some common geometries. The van der Waals interaction between macroscopic bodies is usually given in terms of the Hamaker constant, A, which can either be measured or calculated in terms of the dielectric properties of the materials (Israelachvili, 1991). The Lifshitz theory of van der Waals forces provides an accurate and simple approximate expression for the Hamaker constant for two bodies 1 interacting across a medium 2:

$$A = \frac{3}{4}\,kT\left(\frac{\varepsilon_1 - \varepsilon_2}{\varepsilon_1 + \varepsilon_2}\right)^2 + \frac{I}{16\sqrt{2}}\,\frac{\left(n_1^2 - n_2^2\right)^2}{\left(n_1^2 + n_2^2\right)^{3/2}}, \tag{9.7}$$

where ε_1, ε_2, and n_1, n_2 are the static dielectric constants and refractive indexes of the two phases and where I is their ionization potential which is close to 10 eV or 2×10^{-18} J for most materials. For nonconducting liquids and solids interacting in vacuum or air ($\varepsilon_2 = n_2 = 1$), their Hamaker constants are typically in the range (5 to 10) $\times 10^{-20}$ J, rising to about 4×10^{-19} J for metals, while for interactions in a liquid medium, the Hamaker constants are usually about an order of magnitude smaller.

For inert nonpolar surfaces, e.g., of hydrocarbons or van der Waals solids and liquids, the Lifshitz theory has been found to apply even at molecular contact, where it can predict the surface energies (or tensions) of solids and liquids. Thus, for hydrocarbon surfaces the Hamaker constant is typically $A = 5 \times 10^{-20}$ J. Inserting this value into the appropriate equation for two flat surfaces (Table 9.2) and using a "cut-off" distance of $D = D_0 \approx 0.15$ nm when the two surfaces are in contact, we obtain for the surface energy γ (which is conventionally defined as half the interaction energy):

$$\gamma = \tfrac{1}{2}E = \frac{A}{24\pi D_0^2} \approx 30 \ \mathrm{mJ/m}^2, \tag{9.8}$$

a value that is typical for hydrocarbon solids and liquids (for liquids, γ is sometimes referred to as the surface tension and is expressed in units of mN/m).

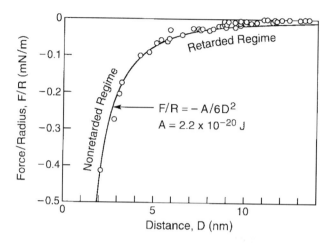

FIGURE 9.2 Attractive van der Waals force F between two curved mica surfaces of radius $R \approx 1$ cm measured in water and various aqueous electrolyte solutions. The measured nonretarded Hamaker constant is $A = 2.2 \times 10^{-20}$ J. Retardation effects are apparent at distances above 5 nm, as expected theoretically. Agreement with the continuum Lifshitz theory of van der Waals forces is generally good at all separations down to five to ten solvent molecular diameters (e.g., $D \approx 2$ nm in water) or down to molecular contact ($D = D_0$) in the absence of a solvent (in vacuum).

If the adhesion force is measured between a spherical surface of radius $R = 1$ cm and a flat surface using an SFA, we expect the following value for the adhesion force (see Table 9.2):

$$F = \frac{AR}{6D_0^2} = 4\pi R\gamma$$

$$\approx 3.7 \times 10^{-3} \text{ N} \left(\text{about 0.4 grams}\right).$$

(9.9)

Using the SFA with a spring constant of $K_S = 100$ N/m, such an adhesive force will cause the two surfaces to jump apart by $\Delta D = F/K_S = 3.7 \times 10^{-5}$ m = 37 µm, which can be accurately measured (actually, for elastic bodies that deform on coming into adhesive contact, their radius R changes during the interaction and the measured adhesion force is 25% lower — see Equation 9.21). The above example shows how the surface energies of solids can be directly measured with the SFA and, in principle, with the AFM (if the geometry of the tip and surface at the contact zone can be quantified). The measured values are generally in good agreement with calculated values based on the known surface energies γ of the materials and, for nonpolar low-energy solids, are well accounted for by the Lifshitz theory (Israelachvili, 1991).

For adhesion measurements in vacuum or inert atmosphere to be meaningful, the surfaces must be both atomically smooth and clean. This is not always easy to achieve, and for this reason only inert, low-energy surfaces, such as hydrocarbon and certain polymeric surfaces, have had their true adhesion forces and surface energies directly measured so far. Other smooth surfaces have also been studied, such as bare mica, metal, metal oxide, and silica surfaces but these are high-energy surfaces, so that it is difficult to prevent them from physisorbing a monolayer of organic matter or water from the atmosphere or from getting an oxide monolayer chemisorbed on them, all of which affects their adhesion.

Many contaminants that physisorb onto solid surfaces from the ambient atmosphere usually dissolve away once the surfaces are immersed in a liquid, so that the short-range forces between such surfaces can usually be measured with great reliability. Figure 9.2 shows results of measurements of the van der Waals forces between two crossed cylindrical mica surfaces in water and various salt solutions, showing the good agreement obtained between experiment and theory (compare the solid curve, corresponding to $F = AR/6D^2$, where $A = 2.2 \times 10^{-20}$ J is the fitted value, which is within about 15% of the theoretical nonretarded Hamaker constant for the mica–water–mica system). Note how at larger surface separations, above about 5 nm, the measured forces fall off faster than given by the inverse-square law. This, too, is predicted by Lifshitz theory and is known as the "retardation effect."

From Figure 9.2 we may conclude that at separations above about 2 nm, or 8 molecular diameters of water, the *continuum* Lifshitz theory is valid. This can be expected to mean that water films as thin as 2 nm may be expected to have bulklike properties, at least as far as their interaction forces are concerned. Similar results have been obtained with other liquids, where in general for films thicker than 5 to 10 molecular diameters their continuum properties, both as regards their interactions and other properties such as viscosity, are already manifest.

9.3.2 Electrostatic Forces

Most surfaces in contact with a highly polar liquid such as water acquire a surface charge, either by the dissociation of ions from the surfaces into the solution or the preferential adsorption of certain ions from the solution. The surface charge is balanced by an equal but opposite layer of oppositely charged ions (counterions) in the solution at some small distance away from the surface. This distance is known as the Debye length which is purely a property of the electrolyte solution. The Debye length falls with increasing ionic strength and valency of the ions in the solution, and for aqueous electrolyte (salt) solutions at 25°C the Debye length is

$$
\begin{aligned}
\kappa^{-1} &= 0.304 \big/ \sqrt{M_{1:1}} \quad \text{for 1:1 electrolytes such as NaCl} \\
&= 0.174 \big/ \sqrt{M_{1:2}} \quad \text{for 1:2 or 2:1 electrolytes such as CaCl}_2 \\
&= 0.152 \big/ \sqrt{M_{2:2}} \quad \text{for 2:2 electrolytes such as MgSO}_4,
\end{aligned}
\tag{9.10}
$$

where the salt concentration M is in moles. The Debye length also relates the surface charge density σ of a surface to the electrostatic surface potentials ψ_0 via the Grahame equation:

$$
\sigma = 0.117 \sinh\left(\psi_0 \big/ 51.4\right) \left[M_{1:1} + M_{2:2} \left(2 + e^{-\psi_0/25.7}\right) \right]^{1/2},
\tag{9.11}
$$

where the concentrations $[M_{1:1}]$ and $[M_{2:2}]$ are again in M, ψ_0 in mV, and σ in C m^{-2} (1 C m^{-2} corresponds to one electronic charge per 0.16 nm^2 or 16 Å^2). For example, for NaCl solutions, $1/\kappa \approx 10$ nm at 1 mM, and 0.3 nm at 1 M. In totally pure water at pH 7, where $[M_{1:1}] = 10^{-7}$ M, the Debye length is 960 nm, or about 1 μm.

The Debye length, being a measure of the thickness of the diffuse atmosphere of counterions near a charged surface, also determines the range of the electrostatic "double-layer" interaction between two charged surfaces. The repulsive energy E per unit area between two similarly charged planar surfaces is given by the following approximate expressions, known as the "weak overlap approximations":

$$
\begin{aligned}
E &= 0.0482 \left[M_{1:1} \right]^{1/2} \tanh^2 \left[\psi_0 (\text{mV}) \big/ 103 \right] e^{-\kappa D} \ \text{J m}^{-2} \quad \text{for monovalent salts} \\
&= 0.0211 \left[M_{2:2} \right]^{1/2} \tanh^2 \left[2\psi_0 (\text{mV}) \big/ 103 \right] e^{-\kappa D} \ \text{J m}^{-2} \quad \text{for divalent salts,}
\end{aligned}
\tag{9.12}
$$

where the concentration $[M_{1:1}]$ and $[M_{2:2}]$ are again in moles.

Using the Derjaguin approximation, Equation 9.6, we may immediately write the expression for the force F between two spheres of radius R as $F = \pi RE$, from which the interaction free energy is obtained by a further integration as

$$W = 4.61 \times 10^{-11} R \tanh^2 \left[\psi_0 (\text{mV}) / 103 \right] e^{-\kappa D} \text{ J} \quad \text{for 1:1 electrolytes.} \tag{9.13}$$

The above approximate expressions are accurate only for surface separations beyond about one Debye length. At smaller separations one must resort to numerical solutions of the Poisson–Boltzmann equation to obtain the exact interaction potential for which there are no simple expressions (Hunter, 1987). In the limit of small D, it can be shown that the interaction energy depends on whether the surfaces remain at constant potential ψ_0 (as assumed in the above equations) or at constant charge σ (when the repulsion exceeds that predicted by the above equations), or somewhere in between these two limits. In the "constant charge limit," since the total number of counterions between the two surfaces does not change as D falls, the number density of ions is given by $2\sigma/eD$, so that the limiting pressure P (or force per unit area, F) in this case is the osmotic pressure of the confined ions, given by

$$F = kT \times \text{ion number density} = 2\sigma kT / zeD \quad \text{for } D \ll \kappa^{-1}, \tag{9.14}$$

that is, as $D \to 0$ the double-layer pressure becomes infinitely repulsive and independent of the salt concentration. However, the van der Waals attraction, which goes as $1/D^2$ between two spheres or as $1/D^3$ between two planar surfaces (see Table 9.2) actually wins out over the double-layer repulsion as $D \to 0$. At least this is the theoretical prediction, which forms the basis of the so-called Derjaguin–Landau–Verwey–Overbeek (DLVO) theory, illustrated in Figure 9.3. In practice, other forces (described below) often come in at small separations, so that the full force law between two surfaces or colloidal particles in solution can be more complex than might be expected from the DLVO theory.

9.4 Solvation and Structural Forces: Forces Due to Liquid and Surface Structure

When a liquid is confined within a restricted space, for example, a very thin film between two surfaces, it ceases to behave as a structureless continuum. Likewise, the forces between two surfaces close together in liquids can no longer be described by simple continuum theories. Thus, at small surface separations — below about 10 molecular diameters — the van der Waals force between two surfaces or even two solute molecules in a liquid (solvent) is no longer a smoothly varying attraction. Instead, there now arises an additional "solvation" force that generally oscillates with distance, varying between attraction and repulsion, with a periodicity equal to some mean dimension σ of the liquid molecules (Horn and Israelachvili, 1981). Figure 9.4 shows the force law between two smooth mica surfaces across the hydrocarbon liquid tetradecane whose inert chainlike molecules have a width of $\sigma \approx 0.4$ nm.

The short-range oscillatory force law, varying between attraction and repulsion with a molecular-scale periodicity, is related to the "density distribution function" and "potential of mean force" characteristic of intermolecular interactions in liquids. These forces arise from the confining effect that two surfaces have on the liquid molecules between them, forcing them to order into quasi-discrete layers which are energetically or entropically favored (and correspond to the free energy minima) while fractional layers are disfavored (energy maxima). The effect is quite general and arises with all simple liquids when they are confined between two smooth surfaces, both flat and curved.

Oscillatory forces do not require that there be any attractive liquid–liquid or liquid–wall interaction. All one needs is two hard walls confining molecules whose shapes are not too irregular and that are free to exchange with molecules in the bulk liquid reservoir. In the absence of any attractive forces between the molecules, the bulk liquid density may be maintained by an external hydrostatic pressure. In real liquids, attractive van der Waals forces play the role of the external pressure, but the oscillatory forces are much the same.

Oscillatory forces are now well understood theoretically, at least for simple liquids, and a number of theoretical studies and computer simulations of various confined liquids, including water, which interact

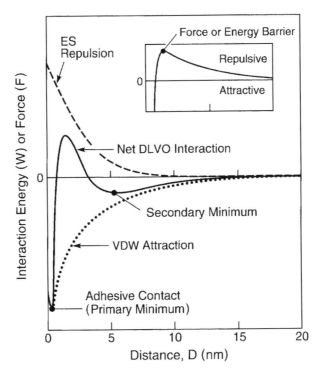

FIGURE 9.3 Classical DLVO interaction potential energy as a function of surface separation between two flat surfaces interacting in an aqueous electrolyte (salt) solution via an attractive van der Waals (VDW) force and a repulsive screened electrostatic (ES) double-layer force. The double-layer potential (or force) is repulsive and roughly exponential in distance dependence. The attractive van der Waals potential has an inverse power law distance dependence (see Table 9.2) and it therefore "wins out" at small separations, resulting in strong adhesion in a "primary minimum". The inset shows a typical interaction potential between surfaces of high surface charge density in dilute electrolyte solution. All curves are schematic. Note that the force F between two *curved* surfaces of radius R is directly proportional to the interaction energy E or W between two *flat* surfaces according to the Derjaguin approximation, Equation 9.6.

via some form of the Lennard–Jones potentials have invariably led to an oscillatory solvation force at surface separations below a few molecular diameters (Snook and van Megan, 1979, 1980, 1981; van Megan and Snook, 1979, 1981; Kjellander and Marcelja, 1985a,b; Tarazona and Vincente, 1985; Henderson and Lozada-Cassou, 1986; Evans and Parry, 1990).

In a first approximation the oscillatory force laws may be described by an exponentially decaying cosine function of the form

$$E \approx E_0 \cos\left(2\pi D / \sigma\right) e^{-D/\sigma}, \qquad (9.15)$$

where both theory and experiments show that the oscillatory period and the characteristic decay length of the envelope are close to σ (Tarazona and Vincent, 1985).

It is important to note that once the solvation zones of two surfaces overlap, the mean liquid density in the gap is no longer the same as that of the bulk liquid. And since the van der Waals interaction depends on the optical properties of the liquid, which in turn depend on the density, one can see why the van der Waals and oscillatory solvation forces are not strictly additive. Indeed, it is more correct to think of the solvation force as *the* van der Waals force at small separations with the molecular properties and density variations of the medium taken into account.

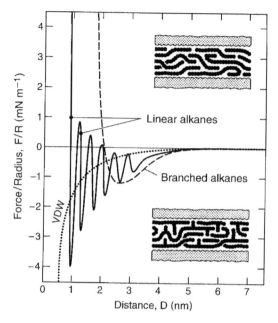

FIGURE 9.4 *Solid curve:* Forces between two mica surfaces across saturated linear-chain alkanes such as *n*-tetradecane (Christenson et al., 1987; Horn and Israelachvili, 1988; Israelachvili and Kott, 1988; Horn et al., 1989a). The 0.4-nm periodicity of the oscillations indicates that the molecules align with their long axis preferentially parallel to the surfaces, as shown schematically in the upper insert. The theoretical continuum van der Waals force is shown by the dotted line. *Dashed line:* Smooth, nonoscillatory force law exhibited by irregularly shaped alkanes, such as branched isoparaffins, that cannot order into well-defined layers (lower insert) (Christenson et al., 1987). Similar nonoscillatory forces are also observed between rough surfaces, even when these interact across a saturated linear chain liquid. This is because the irregularly shaped surfaces (rather than the liquid) now prevent the liquid molecules from ordering in the gap.

It is also important to appreciate that solvation forces do not arise simply because liquid molecules tend to structure into semiordered layers at surfaces. They arise because of the disruption or *change* of this ordering during the approach of a second surface. If there were no change, there would be no solvation force. The two effects are, of course, related: the greater the tendency toward structuring at an isolated surface, the greater the solvation force between two such surfaces, but there is a real distinction between the two phenomena that should always be borne in mind.

Concerning the adhesion energy or force of two smooth surfaces in simple liquids, a glance at Figure 9.4 and Equation 9.15 shows that oscillatory forces lead to multivalued, or "quantized," adhesion values, depending on which energy minimum two surfaces are being separated from. For an interaction energy that varies as described by Equation 9.15, the quantized adhesion energies will be E_0 at $D = 0$ (primary minimum), E_0/e at $D = \sigma$ (second minimum), E_0/e^2 at $D = 2\sigma$, etc. Such multivalued adhesion forces have been observed in a number of systems, including the interactions of fibers. Most interesting, the depth of the potential energy well at contact ($-E_0$ at $D = 0$) is generally deeper but of similar magnitude to the value expected from the *continuum* Lifshitz theory of van der Waals forces (at a cutoff separation of $D_0 \approx 0.15 - 0.20$ nm), even though the continuum theory fails to describe the shape of the force law at intermediate separations.

There is a rapidly growing literature on experimental measurements and other phenomena associated with short-range oscillatory solvation forces. The simplest systems so far investigated have involved measurements of these forces between molecularly smooth surfaces in organic liquids. Subsequent measurements of oscillatory forces between different surfaces across both aqueous and nonaqueous liquids have revealed their subtle nature and richness of properties (Christenson, 1985, 1988a; Christenson and Horn, 1985; Israelachvili, 1987; Israelachvili and McGuiggan, 1988), for example, their great sensitivity

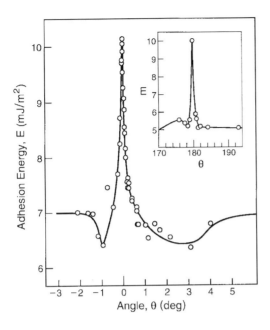

Adhesion energy for two mica surfaces in a primary minimum contact in water as a function of the mismatch angle θ about θ = 0° between the two contacting surface lattices (McGuiggan and Israelach-vili, 1990). Similar peaks are obtained at the other coincidence angles: θ = ±60°, ±120°, and 180° (inset).

to the shape and rigidity of the solvent molecules, to the presence of other components, and to the structure of the confining surfaces. In particular, the oscillations can be smeared out if the molecules are irregularly shaped (e.g., branched) and therefore unable to pack into ordered layers, or when the inter-acting surfaces are rough or fluidlike (e.g., surfactant micelles or lipid bilayers in water) even at the angstrom level (Gee and Israelachvili, 1990).

9.4.1 Effects of Surface Structure

It has recently been appreciated that the structure of the confining surfaces is just as important as the nature of the liquid for determining the solvation forces (Rhykerd et al., 1987; Schoen et al., 1987, 1989; Landman et al., 1990; Thompson and Robbins, 1990; Han et al., 1993). Between two surfaces that are completely smooth or "unstructured," the liquid molecules will be induced to order into layers, but there will be no lateral ordering within the layers. In other words, there will be positional ordering normal but not parallel to the surfaces. However, if the surfaces have a crystalline (periodic) lattice, this will induce ordering parallel to the surfaces as well, and the oscillatory force then also depends on the structure of the surface lattices. Further, if the two lattices have different dimensions ("mismatched" or "incommen-surate" lattices), or if the lattices are similar but are not in register but are at some "twist angle" relative to each other, the oscillatory force law is further modified.

McGuiggan and Israelachvili (1990) measured the adhesion forces and interaction potentials between two mica surfaces as a function of the orientation (twist angle) of their surface lattices. The forces were measured in air, in water, and in an aqueous salt solution where oscillatory structural forces were present. In air, the adhesion was found to be relatively independent of the twist angle θ due to the adsorption of a 0.4-nm-thick amorphous layer of organics and water at the interface. The adhesion in water is shown in Figure 9.5. Apart from a relatively angle-independent "baseline" adhesion, sharp adhesion peaks (energy minima) occurred at θ = 0°, ±60°, ±120°, and 180°, corresponding to the "coincidence" angles of the surface lattices. As little as ±1° away from these peak, the energy decreases by 50%. In aqueous salt (KCl) solution, due to potassium ion adsorption the water between the surfaces becomes ordered, resulting in an oscillatory force profile where the adhesive minima occur at discrete separations of about 0.25 nm, corresponding to integral numbers of water layers. The whole interaction potential was now found to depend on orientation of the surface lattices, and the effect extended at least four molecular layers.

Although oscillatory forces are predicted from Monte Carlo and molecular dynamic simulations, no theory has yet taken into account the effect of surface structure, or atomic "corrugations," on these forces, nor any lattice mismatching effects. As shown by the experiments, within the last 1 or 2 nm, these effects can alter the adhesive minima at a given separation by a factor of two. The force barriers, or maxima, may also depend on orientation. This could be even more important than the effects on the minima. A high barrier could prevent two surfaces from coming closer together into a much deeper adhesive well. Thus, the maxima can effectively contribute to determining not only the final separation of two surfaces, but also their final adhesion. Such considerations should be particularly important for determining the thickness and strength of inter-granular spaces in ceramics, the adhesion forces between colloidal particles in concentrated electrolyte solutions, and the forces between two surfaces in a crack containing capillary condensed water.

The intervening medium profoundly influences how one surface interacts with the other. As experimental results show (McGuiggan and Israelachvili, 1990), when two surfaces are separated by as little as 0.4 nm of an amorphous material, such as adsorbed organics from air, then the surface granularity can be completely masked and there is no mismatch effect on the adhesion. However, with another medium, such as pure water which is presumably well ordered when confined between two mica lattices, the atomic granularity is apparent and alters the adhesion forces and whole interaction potential out to $D > 1$ nm. Thus, it is not only the surface structure but also the liquid structure, or that of the intervening film material, which together determine the short-range interaction and adhesion.

On the other hand, for surfaces that are *randomly* rough, the oscillatory force becomes smoothed out and disappears altogether, to be replaced by a purely monotonic solvation force. This occurs even if the liquid molecules themselves are perfectly capable of ordering into layers. The situation of *symmetric* liquid molecules confined between *rough* surfaces is therefore not unlike that of *asymmetric* molecules between *smooth* surfaces (see Figure 9.4).

To summarize some of the above points, for there to be an oscillatory solvation force, the liquid molecules must be able to be correlated over a reasonably long range. This requires that both the liquid molecules and the surfaces have a high degree of order or symmetry. If either is missing, so will the oscillations. A roughness of only a few angstroms is often sufficient to eliminate any oscillatory component of a force law.

9.4.2 Effect of Surface Curvature and Geometry

It is easy to understand how oscillatory forces arise between two *flat*, plane parallel surfaces (Figure 9.5). Between two curved surfaces e.g., two spheres, one might imagine the molecular ordering and oscillatory forces to be smeared out in the same way that they are smeared out between two randomly rough surfaces. However, this is not the case. Ordering can occur so long as the curvature or roughness is itself regular or uniform, i.e., not random. This interesting matter is due to the Derjaguin approximation, Equation 9.6, which relates the force between two curved surfaces to the energy between two flat surfaces. If the latter is given by a decaying oscillatory function, as in Equation 9.15, then the energy between two curved surfaces will simply be the integral of that function, and since the integral of a cosine function is another cosine function, with some appropriate phase shift, we see why periodic oscillations will not be smeared out simply by changing the surface curvature. Likewise, two surfaces with regularly curved regions will also retain their oscillatory force profile, albeit modified, *so long as the corrugations are truly regular*, i.e., periodic. On the other hand, surface roughness, even on the nanometer scale, can smear out any oscillations if the roughness is random and the liquid molecules are smaller than the size of the surface asperities.

9.5 Thermal Fluctuation Forces: Forces between Soft, Fluidlike Surfaces

If a surface or interface is not rigid but very soft or even fluidlike, this can act to smear out any oscillatory solvation force. This is because the thermal fluctuations of such interfaces make them dynamically "rough" at any instant, even though they may be perfectly smooth on a time average. The types of surfaces that fall into this category are fluidlike amphiphilic surfaces of micelles, bilayers, emulsions, soap films, etc.,

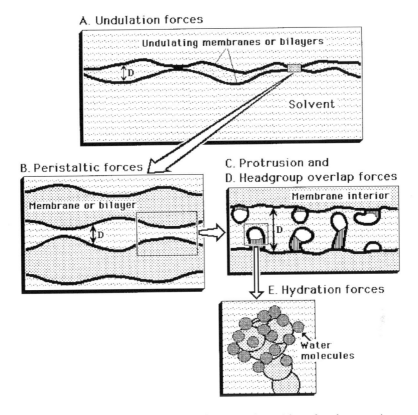

A. Undulation forces

Undulating membranes or bilayers

D

Solvent

B. Peristaltic forces

Membrane or bilayer

D

C. Protrusion and
D. Headgroup overlap forces

Membrane interior

D

E. Hydration forces

Water
molecules

FIGURE 9.6   The four most common types of *thermal fluctuation* forces (also referred to as *steric* or *entropic* forces) between fluid-like, usually amphiphilic, surfaces and membranes in liquids.

but also solid colloidal particle surfaces that are coated with surfactant monolayers, as occur in lubricating oils, paints, toners, etc.

Thermal fluctuation forces are usually of short range and repulsive, and are very effective at stabilizing the attractive van der Waals forces at some small but finite separation which can reduce the adhesion energy or force by up to three orders of magnitude. It is mainly for this reason that fluidlike micelles and bilayers, biological membranes, emulsion droplets (in salad dressings), or gas bubbles (in beer) adhere to each other only very weakly (Figure 9.6).

Because of their short range, it was, and still is, commonly believed that these forces arise from water ordering or "structuring" effects at surfaces, and that they reflect some unique or characteristic property of water (see Section 9.6). However, it is now known that these repulsive forces also exist in other liquids. Moreover, they appear to become stronger with increasing temperature, which is unlikely for a force that originates from molecular ordering effects at surfaces. Recent experiments, theory, and computer simulations (Israelachvili and Wennerström, 1990, 1996; Granfeldt and Miklavic, 1991) have shown that these repulsive forces have an entropic origin — arising from the osmotic repulsion between exposed thermally mobile surface groups once these overlap in a liquid.

9.6   Hydration Forces: Special Forces in Water and Aqueous Solutions

9.6.1   Repulsive Hydration Forces

The forces occurring in water and electrolyte solutions are more complex than those occurring in nonpolar liquids. According to continuum theories, the attractive van der Waals force is always expected to win over the repulsive electrostatic double-layer force at small surface separations (Figure 9.3). However, certain surfaces (usually oxide or hydroxide surfaces such as clays and silica) swell spontaneously or repel each other in aqueous solutions even in very high salt. Yet in all these systems one would expect the surfaces or particles to remain in strong adhesive contact or coagulate in a primary minimum if the only forces operating were DLVO forces.

There are many other aqueous systems where DLVO theory fails and where there is an additional short-range force that is not oscillatory but smoothly varying, i.e., monotonic. Between hydrophilic surfaces this force is exponentially repulsive and is commonly referred to as the *hydration* or *structural force*. The origin and nature of this force has long been controversial especially in the colloidal and biological literature. Repulsive hydration forces are believed to arise from strongly H-bonding surface groups, such as hydrated ions or hydroxyl (–OH) groups, which modify the H-bonding network of liquid water adjacent to them. Since this network is quite extensive in range (Stanley and Teixeira, 1980), the resulting interaction force is also of relatively long range.

Repulsive hydration forces were first extensively studied between clay surfaces (van Olphen, 1977). More recently they have been measured in detail between mica and silica surfaces (Pashley, 1981, 1982, 1985; Horn et al., 1989b) where they have been found to decay exponentially with decay lengths of about 1 nm. Their effective range is about 3 to 5 nm, which is about twice the range of the oscillatory solvation force in water. Empirically, the hydration repulsion between two hydrophilic surfaces appears to follow the simple equation (over a limited range)

$$E = E_0 e^{-D/\lambda_0} \tag{9.16}$$

where $\lambda_0 \approx 0.6 - 1.1$ nm for 1:1 electrolytes, and where $E_0 = 3$ to 30 mJ m^{-2} depending on the hydration (hydrophilicity) of the surfaces, higher E_0 values generally being associated with lower λ_0 values.

In a series of experiments to identify the factors that regulate hydration forces, Pashley (1981, 1982, 1985) found that the interaction between molecularly smooth mica surfaces in dilute electrolyte solutions obeys the DLVO theory. However, at higher salt concentrations, specific to each electrolyte, hydrated cations bind to the negatively charged surfaces and give rise to a repulsive hydration force (Figure 9.7). This is believed to be due to the energy needed to dehydrate the bound cations, which presumably retain some of their water of hydration on binding. This conclusion was arrived at after noting that the strength and range of the hydration forces increase with the known hydration numbers of the electrolyte cations in the order: $Mg^{2+} > Ca^{2+} > Li^+ \sim Na^+ > K^+ > Cs^+$. Similar trends are observed with other negatively charged colloidal surfaces.

While the hydration force between two mica surfaces is overall repulsive below about 4 nm, it is not always monotonic below about 1.5 nm but exhibits oscillations of mean periodicity 0.25 ± 0.03 nm, roughly equal to the diameter of the water molecule. This is shown in Figure 9.7, where we may note that the first three minima at $D \approx 0, 0.28$, and 0.56 nm occur at negative energies, a result that rationalizes observations on certain colloidal systems. For example, clay platelets such as motmorillonite often repel each other increasingly strongly as they come closer together, but they are also known to stack into stable aggregates with water interlayers of typical thickness 0.25 and 0.55 nm between them (Del Pennino et al., 1981; Viani et al., 1984), suggestive of a turnaround in the force law from a monotonic repulsion to discretized attraction. In chemistry we would refer to such structures as stable hydrates of fixed stochiometry, while in physics we may think of them as experiencing an oscillatory force.

Both surface force and clay-swelling experiments have shown that hydration forces can be modified or "regulated" by exchanging ions of different hydrations on surfaces, an effect that has important practical applications in controlling the stability of colloidal dispersions. It has long been known that colloidal particles can be precipitated (coagulated or flocculated) by increasing the electrolyte concentration — an effect that was traditionally attributed to the reduced screening of the electrostatic double-layer repulsion between the particles due to the reduced Debye length. However, there are many examples where colloids become stable — not at lower salt concentrations — but at high concentrations. This effect is now recognized as being due to the increased hydration repulsion experienced by certain surfaces

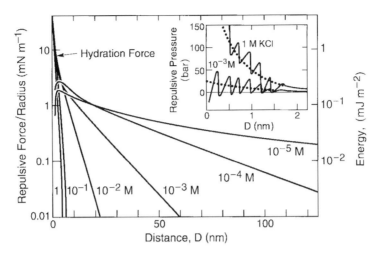

FIGURE 9.7 Measured forces between charged mica surfaces in various dilute and concentrated KCl solutions. In dilute solutions (10^{-5} and 10^{-4} M) the repulsion reaches a maximum and the surfaces jump into molecular contact from the tops of the force barriers (see also Figure 9.3). In dilute solutions the measured forces are excellently described by the DLVO theory, based on exact numerical solutions to the nonlinear Poisson–Boltzmann equation for the electrostatic forces and the Lifshitz theory for the van der Waals forces (using a Hamaker constant of $A = 2.2 \times 10^{-20}$ J). At higher electrolyte concentrations, as more hydrated K^+ cations adsorb onto the negatively charged surfaces, an additional hydration force appears superimposed on the DLVO interaction. This has both an oscillatory and a monotonic component. Inset: Short-range hydration forces between mica surfaces plotted as pressure against distance. Lower curve: force measured in dilute 1 mM KCl solution where there is one K^+ ion adsorbed per 1.0 nm² (surfaces 40% saturated with K^+ ions). Upper curve: force measured in 1 M KCl where there is one K^+ ion adsorbed per 0.5 nm² (surfaces 95% saturated with adsorbed cations). At larger separations the forces are in good agreement with the DLVO theory. The right-hand ordinate gives the corresponding interaction energy according to Equation 9.6.

when they bind highly hydrated ions at higher salt concentrations. "Hydration regulation" of adhesion and interparticle forces is an important practical method for controlling various processes such as clay swelling (Quirk, 1968; Del Pennino et al., 1981; Viani et al., 1983), ceramic processing and rheology (Horn, 1990; Velamakanni et al., 1990), material fracture (Horn, 1990), and colloidal particle and bubble coalescence (Lessard and Zieminski, 1971; Elimelech, 1990).

9.6.2 Attractive Hydrophobic Forces

Water appears to be unique in having a solvation (hydration) force that exhibits both a monotonic and an oscillatory component. Between hydrophilic surfaces the monotonic component is repulsive (Figure 9.7), but between hydrophobic surfaces it is attractive and the final adhesion in water is much greater than expected from the Lifshitz theory.

A hydrophobic surface is one that is inert to water in the sense that it cannot bind to water molecules via ionic or hydrogen bonds. Hydrocarbons and fluorocarbons are hydrophobic, as is air, and the strongly attractive hydrophobic force has many important manifestations and consequences, some of which are illustrated in Figure 9.8.

In recent years there has been a steady accumulation of experimental data on the force laws between various hydrophobic surfaces in aqueous solutions. These surfaces include mica surfaces coated with surfactant monolayers exposing hydrocarbon or fluorocarbon groups, or silica and mica surfaces that had been rendered hydrophobic by chemical methylation or plasma etching (Israelachvili and Pashley, 1982; Pashley et al., 1985; Claesson et al., 1986; Claesson and Christenson, 1988; Rabinowich and Derjaguin, 1988; Parker et al., 1989; Christenson et al., 1990; Kurihara et al., 1990). These studies have found that the hydrophobic force law between two macroscopic surfaces is of surprisingly long range, decaying exponentially with a characteristic decay length of 1 to 2 nm in the range 0 to 10 nm, and then more

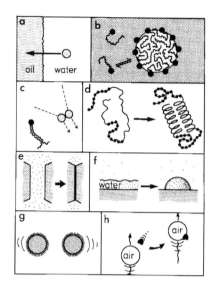

FIGURE 9.8 Examples of attractive hydrophobic interactions in aqueous solutions. (a) Low solubility/immiscibility of water and oil molecules; (b) micellization; (c) dimerization and association of hydrocarbon chains in water; (d) protein folding; (e) strong adhesion of hydrophobic surfaces; (f) nonwetting of water on hydrophobic surfaces; (g) rapid coagulation of hydrophobic or surfactant-coated surfaces; (h) hydrophobic particle attachment to rising air bubbles (basic mechanism of "froth flotation" process used to separate hydrophobic and hydrophilic minerals).

gradually farther out. The hydrophobic force can be far stronger than the van der Waals attraction, especially between hydrocarbon surfaces for which the Hamaker constant is quite small.

As might be expected, the magnitude of the hydrophobic attraction falls with the decreasing hydrophobicity (increasing hydrophilicity) of surfaces. Helm et al. (1989) measured the forces between uncharged but hydrated lecithin bilayers in water as a function of increasing hydrophobicity of the bilayer surfaces. This was achieved by progressively increasing the head group area per amphiphilic molecule exposed to the aqueous phase, i.e., by progressively exposing more of the hydrocarbon chains. The results showed that with increasing hydrophobic area the forces became progressively more attractive at longer range, that the adhesion increased, and that the stabilizing repulsive short-range hydration forces decreased. This shows how the overall force curve changes as an initially hydrophilic surface becomes progressively more hydrophobic.

For two surfaces in water their purely hydrophobic interaction energy (i.e., ignoring DLVO and oscillatory forces) in the range 0 to 10 nm is given by

$$E = -2\gamma_i e^{-D/\lambda_0}, \qquad (9.17)$$

where, typically, $\lambda_0 = 1$ to 2 nm, and $\gamma_i = 10$ to 50 mJ m^{-2}, where the higher value corresponds to the interfacial energy of a pure hydrocarbon–water interface.

At a separation below 10 nm the hydrophobic force appears to be insensitive or only weakly sensitive to changes in the type and concentration of electrolyte ions in the solution. The absence of a "screening" effect by ions attests to the nonelectrostatic origin of this interaction. In contrast, some experiments have shown that at separations greater than 10 nm the attraction does depend on the intervening electrolyte, and that in dilute solutions, or solutions containing divalent ions, it can continue to exceed the van der Waals attraction out to separations of 80 nm (Christenson et al., 1989, 1990).

The long-range nature of the hydrophobic interaction has a number of important consequences. It accounts for the rapid coagulation of hydrophobic particles in water, and may also account for the rapid folding of proteins. It also explains the ease with which water films rupture on hydrophobic surfaces. In this, the van der Waals force across the water film is repulsive and therefore favors wetting, but this is more than offset by the attractive hydrophobic interaction acting between the two hydrophobic phases across water. Finally, hydrophobic forces are increasingly being implicated in the adhesion and fusion of biological membranes and cells. It is known that both osmotic and electric field stresses enhance membrane fusion, an effect that may be due to the increase in the hydrophobic area exposed between two adjacent surfaces.

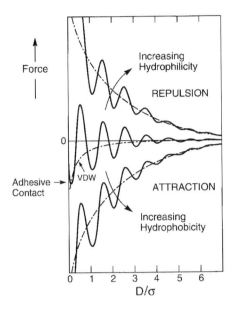

FIGURE 9.9 Typical short-range solvation (hydration) forces in water as a function of distance, D, normalized by the diameter of the water molecule, σ (about 0.25 nm). The hydration forces in water differ from those in other liquids in that there is a monotonic component in addition to the normal purely oscillatory component. For hydrophilic surfaces the monotonic component is repulsive (upper dashed curve), whereas for hydrophobic surfaces it is attractive (lower dashed curve). For simpler liquids there are no such monotonic components, and both theory and experiments show that the oscillations decay with distance with the maxima and minima, respectively, above and below the baseline of the van der Waals force (middle dashed curve) or superimposed on the continuum DLVO interaction.

9.6.3 Origin of Hydration Forces

From the previous discussions we can infer that hydration forces are not of a simple nature, and it may be fair to say that this interaction is probably the most important yet the least understood of all the forces in liquids. Clearly, the very unusual properties of water are implicated, but the nature of the surfaces is equally important. Some particle surfaces can have their hydration forces regulated, for example, by ion exchange. Other surfaces appear to be intrinsically hydrophilic (e.g., silica) and cannot be coagulated by changing the ionic conditions. However, such surfaces can often be rendered hydrophobic by chemically modifying their surface groups. For example, on heating silica to above 600°C, two surface silanol –OH groups release a water molecule and combine to form a hydrophobic siloxane –O– group, whence the repulsive hydration force changes into an attractive hydrophobic force.

How do these exponentially decaying repulsive or attractive forces arise? Theoretical work and computer simulations (Christou et al., 1981; Jönsson, 1981; Kjellander and Marcelja, 1985a,b; Henderson and Lozada-Cassou, 1986) suggest that the solvation forces in water should be purely oscillatory, while other theoretical studies (Marcelja and Radic, 1976; Marcelja et al., 1977; Gruen and Marcelja, 1983; Jönsson and Wennerström, 1983; Schiby and Ruckenstein, 1983; Luzar et al., 1987; Attard and Batchelor, 1988; Marcelja, 1997) suggest a monotonic exponential repulsion or attraction, possibly superimposed on an oscillatory profile. The latter is consistent with experimental findings, as shown in the inset to Figure 9.7, where it appears that the oscillatory force is simply additive with the monotonic hydration and DLVO forces, suggesting that these arise from essentially different mechanisms.

It is probable that the intrinsic hydration force between all smooth, rigid, or crystalline surfaces (e.g., mineral surfaces such as mica) has an oscillatory component. This may or may not be superimposed on a monotonic repulsion (Figure 9.9) due to image interactions (Jönsson and Wennerström, 1983), structural or H-bonding interactions (Marcelja and Radic, 1976; Marcelja et al., 1977; Gruen and Marcelja, 1983) or — as now seems more likely — steric and entropic forces (Israelachvili and Wennerström, 1996; Marcelja, 1997).

Like the repulsive hydration force, the origin of the hydrophobic force is still unknown. Luzar et al. (1987) carried out a Monte Carlo simulation of the interaction between two hydrophobic surfaces across water at separations below 1.5 nm. They obtained a decaying oscillatory force superimposed on a monotonically attractive curve, i.e., similar to Figure 9.9.

It is questionable whether the hydration or hydrophobic force should be viewed as an ordinary type of solvation or structural force — simply reflecting the packing of the water molecules. It is important to note that for any given positional arrangement of water molecules, whether in the liquid or solid state, there is an almost infinite variety of ways the H-bonds can be interconnected over three-dimensional space while satisfying the Bernal–Fowler rules requiring two donors and two acceptors per water molecule. In other words, the H-bonding structure is actually quite distinct from the molecular structure. The energy (or entropy) associated with the H-bonding network, which extends over a much larger region of space than the molecular correlations, is probably at the root of the long-range solvation interactions of water. But whatever the answer, it is clear that the situation in water is governed by much more than the simple molecular-packing effects that seem to dominate the interactions in simpler liquids.

9.7 Adhesion and Capillary Forces

When considering the adhesion of two solid surfaces or particles in air or in a liquid, it is easy to overlook or underestimate the important role of capillary forces, i.e., forces arising from the Laplace pressure of curved menisci which have formed as a consequence of the condensation of a liquid between and around two adhering surfaces (Figure 9.10).

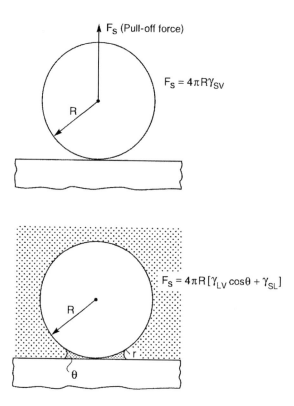

FIGURE 9.10 Sphere on flat in an inert atmosphere (top), and in an atmosphere containing vapor that can "capillary condense" around the contact zone (bottom). At equilibrium the concave radius, r, of the liquid meniscus is given by the Kelvin equation. The radius r increases with the relative vapor pressure, but for condensation to occur the contact angle θ must be less than 90° or else a concave meniscus cannot form. The presence of capillary condensed liquid changes the adhesion force, as given by Equations 9.18 and 9.19. Note that this change is independent of r so long as the surfaces are perfectly smooth. Experimentally, it is found that for simple inert liquids such as cyclohexane, these equations are valid already at Kelvin radii as small as 1 nm — about the size of the molecules themselves. Capillary condensation also occurs in binary liquid systems e.g., when small amounts of water dissolved in hydrocarbon liquids condense around two contacting hydrophilic surfaces, or when a vapor cavity forms in water around two hydrophobic surfaces.

The adhesion force between a spherical particle of radius R and a flat surface in an inert atmosphere is

$$F_s = 4\pi R\gamma_{sv}, \tag{9.18}$$

but in an atmosphere containing a condensable vapor, the above becomes replaced by

$$F_s = 4\pi R\left(\gamma_{LV}\cos\theta + \gamma_{SL}\right), \tag{9.19}$$

where the first term is due to the Laplace pressure of the meniscus and the second is due to the direct adhesion of the two contacting solids within the liquid. Note that the above equation does not contain the radius of curvature, r, of the liquid meniscus (Figure 9.10). This is because for smaller r the Laplace pressure γ_{LV}/r increases, but the area over which it acts decreases by the same amount, so the two effects cancel out. A natural question arises as to the smallest value of r for which Equation 9.19 will apply. Experiments with inert liquids, such as hydrocarbons, condensing between two mica surfaces indicate that Equation 9.19 is valid for values of r as small as 1 to 2 nm, corresponding to vapor pressures as low as 40% of saturation (Fisher and Israelachvili, 1981; Christenson, 1988b). With water condensing from vapor or from oil it appears that the bulk value of γ_{LV} is also applicable for meniscus radii as small as 2 nm.

The capillary condensation of liquids, especially water, from vapor can have additional effects on the whole physical state of the contact zone. For example, if the surfaces contain ions, these will diffuse and build up within the liquid bridge, thereby changing the chemical composition of the contact zone as well as influencing the adhesion. More dramatic effects can occur with amphiphilic surfaces, i.e., those containing surfactant or polymer molecules. In dry air, such surfaces are usually nonpolar — exposing hydrophobic groups such as hydrocarbon chains. On exposure to humid air, the molecules can overturn so that the surface nonpolar groups become replaced by polar groups, which renders the surfaces hydrophilic. When two such surfaces come into contact, water will condense around the contact zone and the adhesion force will also be affected — generally increasing well above the value expected for inert hydrophobic surfaces.

It is clear that the adhesion of two surfaces in vapor or a solvent can often be largely determined by capillary forces arising from the condensation of liquid that may be present only in very small quantities e.g., 10 to 20% of saturation in the vapor, or 20 ppm in the solvent.

9.7.1 Adhesion Mechanics

Modern theories of the adhesion mechanics of two contacting solid surfaces are based on the Johnson–Kendall–Roberts (JKR) theory (Johnson et al., 1971, Pollock et al., 1978; Barquins and Maugis, 1982). In the JKR theory two spheres of radii R_1 and R_2, bulk elastic moduli K, and surface energy γ per unit area will flatten when in contact. The contact area will increase under an external load or force F, such that at mechanical equilibrium the contact radius r is given by

$$r^3 = \frac{R}{K}\left[F + 6\pi R\gamma + \sqrt{12\pi R\gamma F + \left(6\pi R\gamma\right)^2}\right], \tag{9.20}$$

where $R = R_1 R_2/(R_1 + R_2)$. Another important result of the JKR theory gives the adhesion force or pull off force:

$$F_S = -3\pi R\gamma_S, \tag{9.21}$$

where, by definition, the surface energy γ_S, is related to the reversible work of adhesion W, by $W = 2\gamma_S$. Note that according to the JKR theory a finite elastic modulus, K, while having an effect on the load–area curve, has no effect on the adhesion force — an interesting and unexpected result that has nevertheless been verified experimentally (Johnson et al., 1971; Israelachvili, 1991).

Equations 9.20 and 9.21 are the basic equations of the JKR theory and provide the framework for analyzing the results of adhesion measurements of contacting solids, known as contact mechanics (Pollock et al., 1978; Barquins and Maugis, 1982), and for studying the effects of surface conditions and time on adhesion energy hysteresis (see next section).

9.8   Nonequilibrium Interactions: Adhesion Hysteresis

Under ideal conditions the adhesion energy is a well-defined thermodynamic quantity. It is normally denoted by E or W (the work of adhesion) or γ (the surface tension, where $W = 2\gamma$), and it gives the reversible work done on bringing two surfaces together or the work needed to separate two surfaces from contact. Under ideal, equilibrium conditions these two quantities are the same, but under most realistic conditions they are not: the work needed to separate two surfaces is always greater than that originally gained on bringing them together. An understanding of the molecular mechanisms underlying this phenomenon is essential for understanding many adhesion phenomena, energy dissipation during loading–unloading cycles, contact angle hysteresis, and the molecular mechanisms associated with many frictional processes. It is wrong to think that hysteresis arises because of some imperfection in the system, such as rough or chemically heterogeneous surfaces, or because the supporting material is viscoelastic; adhesion hysteresis can arise even between perfectly smooth and chemically homogeneous surfaces supported by perfectly elastic materials, and can be responsible for such phenomena as "rolling" friction and elastoplastic adhesive contacts (Bowden and Tabor, 1967; Greenwood and Johnson, 1981; Maugis, 1985; Michel and Shanahan, 1990) during loading–unloading and adhesion–decohesion cycles.

Adhesion hysteresis may be thought of as being due to mechanical or chemical effects, as illustrated in Figure 9.11. In general, if the energy change, or work done, on separating two surfaces from adhesive contact is not fully recoverable on bringing the two surfaces back into contact again, the adhesion hysteresis may be expressed as

$$W_R \; > \; W_A$$
$$\text{Receding} \quad \text{Advancing}$$

or

$$\Delta W = \left(W_R - W_A \right) > 0, \tag{9.22}$$

where W_R and W_A are the adhesion or surface energies for receding (separating) and advancing (approaching) two solid surfaces, respectively. Figure 9.12 shows the results of a typical experiment that measures the adhesion hysteresis between two surfaces (Chaudhury and Whitesides, 1991; Chen et al., 1991). In this case, two identical surfactant-coated mica surfaces were used in an SFA. By measuring the contact radius as a function of applied load both for increasing and decreasing loads two different curves are obtained. These can be fitted to the JKR equation, Equation 9.20, to obtain the advancing (loading) and receding (unloading) surface energies.

Hysteresis effects are also commonly observed in wetting/dewetting phenomena (Miller and Neogi, 1985). For example, when a liquid spreads and then retracts from a surface the advancing contact angle θ_A is generally larger than the receding angle θ_R. Since the contact angle, θ, is related to the liquid–vapor surface tension, γ, and the solid–liquid adhesion energy, W, by the Dupré equation:

$$\left(1 + \cos \theta \right) \gamma_L = W, \tag{9.23}$$

we see that *wetting hysteresis* or *contact angle hysteresis* ($\theta_A > \theta_R$) actually implies adhesion hysteresis, $W_R > W_A$, as given by Equation 9.22.

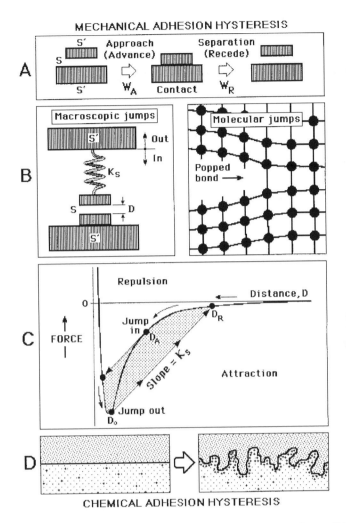

MECHANICAL ADHESION HYSTERESIS

CHEMICAL ADHESION HYSTERESIS

FIGURE 9.11 Origin of adhesion hysteresis during the approach and separation of two solid surfaces. (A) In all realistic situations the force between two solid surfaces is never measured at the surfaces themselves, S, but at some other point, say S', to which the force is elastically transmitted via the backing material supporting the surfaces. (B, left) "Magnet" analogy of how *mechanical adhesion hysteresis* arises for two approaching or separating surfaces, where the lower is fixed and where the other is supported at the end of a spring of stiffness K_S. (B, right) On the molecular or atomic level, the separation of two surfaces is accompanied by the spontaneous breaking of bonds, which is analogous to the jump apart of two macroscopic surfaces or magnets. (C) Force–distance curve for two surfaces interacting via an attractive van der Waals–type force law, showing the path taken by the upper surface on approach and separation. On approach, an instability occurs at $D = D_A$, where the surfaces spontaneously jump into contact at $D \approx D_0$. On separation, another instability occurs where the surfaces jump apart from ~D_0 to D_R. (D) *Chemical adhesion hysteresis* produced by interdiffusion, interdigitation, molecular reorientations and exchange processes occurring at an interface after contact. This induces roughness and chemical heterogeneity even though initially (and after separation and reequilibration) both surfaces are perfectly smooth and chemically homogeneous.

Energy dissipating processes such as adhesion and contact angle hysteresis arise because of practical constraints of the *finite time* of measurements and the *finite elasticity* of materials which prevent many loading–unloading or approach–separation cycles to be thermodynamically irreversible, even though if these were carried out infinitely slowly they would be. By thermodynamic irreversibly one simply means that one cannot go through the approach–separation cycle via a continuous series of equilibrium states

393

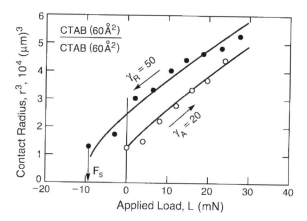

FIGURE 9.12   Measured advancing and receding radius vs. load curves for two surfactant-coated mica surfaces of initial, undeformed radii $R \approx 1$ cm. Each surface had a monolayer of CTAB (cetyl-trimethyl-ammonium-bromide) on it of mean area 60 Å^2 per molecule. The solid lines are based on fitting the advancing and receding branches to the JKR equation, Equation 9.20), from which the indicated values of γ_A and γ_R were determined, in units of mJ/m^2 or erg/cm^2. The advancing/receding rates were about 1 μm/s. At the end of each unloading cycle the pull-off force, F_s, can be measured, from which another value for γ_R can be obtained using Equation 9.21).

because some of these are connected via spontaneous — and therefore thermodynamically irreversible — instabilities or transitions (Figure 9.11C) where energy is liberated and therefore "lost" via heat or phonon release (Israelachvili and Berman, 1995). This is an area of much current interest and activity, especially regarding the fundamental molecular origins of adhesion and friction, and the relationships between them.

9.9   Rheology of Molecularly Thin Films: Nanorheology

9.9.1   Different Modes of Friction: Limits of Continuum Models

Most frictional processes occur with the sliding surfaces becoming damaged in one form or another (Bowden and Tabor, 1967). This may be referred to as "normal" friction. In the case of brittle materials, the damaged surfaces slide past each other while separated by relatively large, micron-sized wear particles. With more ductile surfaces, the damage remains localized to nanometer-sized, plastically deformed asperities.

There are also situations where sliding can occur between two perfectly smooth, undamaged surfaces. This may be referred to as "interfacial" sliding or "boundary" friction, which is the focus of the following sections. The term *boundary lubrication* is more commonly used to denote the friction of surfaces that contain a thin protective lubricating layer, such as a surfactant monolayer, but here we shall use this term more broadly to include any molecularly thin solid, liquid, surfactant, or polymer film.

Experiments have shown that as a liquid film becomes progressively thinner, its physical properties change, at first quantitatively then qualitatively (Van Alsten and Granick, 1990a,b, 1991; Granick, 1991; Hu et al., 1991; Hu and Granick, 1992; Luengo et al., 1997). The quantitative changes are manifested by an increased viscosity, non-Newtonian flow behavior, and the replacement of normal melting by a glass transition, but the film remains recognizable as a liquid. In tribology, this regime is commonly known as the "mixed lubrication" regime, where the rheological properties of a film are intermediate between the bulk and boundary properties. One may also refer to it as the "intermediate" regime (Table 9.3).

For even thinner films, the changes in behavior are more dramatic, resulting in a qualitative change in properties. Thus, first-order phase transitions can now occur to solid or liquid-crystalline phases (Gee et al., 1990; Israelachvili et al., 1990a,b; Thompson and Robbins, 1990; Yoshizawa et al., 1993; Klein and Kumacheva, 1995), whose properties can no longer be characterized — even qualitatively — in terms of bulk or *continuum* liquid properties such as viscosity. These films now exhibit yield points (characteristic

TABLE 9.3 The Three Main Tribological Regimes Characterizing the Changing Properties of Liquids Subjected to an Increasing Confinement between Two Solid Surfaces[a]

Regime	Conditions for Getting into this Regime	Static/Equilibrium Properties[b]	Dynamic Properties[c]
Bulk	• Thick films ($>10\sigma$, $\gg R_g$) • Low or zero loads • High shear rates	Bulk, continuum properties: • Bulk liquid density • No long-range order	Bulk, continuum properties: • Newtonian viscosity • Fast relaxation times • No glass temperature • No yield point • EHD lubrication
Mixed or intermediate	• Intermediately thick films (4–10 molecular diameters $\sim R_g$ for polymers) • Low loads	Modified fluid properties include: • Modified positional and orientational order[a] • Medium to long-range molecular correlations • Highly entangled states	Modified rheological properties include: • Non-Newtonian flow • Glassy states • Long relaxation times • Mixed lubrication
Boundary	• Molecularly thin films (<4 molecular diameters) • High loads • Low shear rates • Smooth surfaces or asperities	Onset of non-fluidlike properties: • Liquidlike to solidlike phase transitions • Appearance of new liquid-crystalline states • Epitaxially induced long-range ordering	Onset of tribological properties: • No flow until yield point or critical shear stress reached • Solidlike film behavior characterized by defect diffusion, dislocation motion, shear melting • Boundary lubrication

Based on work by Granick (1991), Hu and Granick (1992), and others (Gee et al., 1990; Hirz et al., 1992; Yoshizawa et al., 1993) on the dynamic properties of short-chain molecules such as alkanes and polymer melts confined between surfaces.

[a] Confinement can lead to an increased or decreased order in a film, depending both on the surface lattice structure and the geometry of the confining cavity.

[b] In each regime both the static and dynamic properties change. The static properties include the film density, the density distribution function, the potential of mean force, and various positional and orientational order parameters.

[c] Dynamic properties include viscosity, viscoelastic constants, and tribological yield points such as the friction coefficient and critical shear stress.

of fracture in solids) and their molecular diffusion and relaxation times can be ten orders of magnitude longer than in the bulk liquid or even in films that are just slightly thicker. The three friction regimes are summarized in Table 9.3.

9.9.2 Viscous Forces and Friction of Thick Films: Continuum Regime

Experimentally, it is usually difficult to unambiguously establish which type of sliding mode is occurring, but an empirical criterion, based on the Stribeck curve (Figure 9.13) is often used as an indicator. This curve shows how the friction force or the coefficient of friction is expected to vary with sliding speed depending on which type of friction regime is operating. For thick liquid lubricant films whose behavior can be described by bulk continuum properties, the friction forces are essentially the hydrodynamic or viscous drag forces. For example, for two plane parallel surfaces of area A separated by a distance D and moving laterally relative to each other with velocity v, if the intervening liquid is *Newtonian*, i.e., if its viscosity η is independent of the shear rate, the frictional force experienced by the surfaces is given by

$$F = \frac{\eta A v}{D},\qquad(9.24)$$

where the shear rate $\dot{\gamma}$ is defined by

$$\dot{\gamma} = \frac{V}{D}.\qquad(9.25)$$

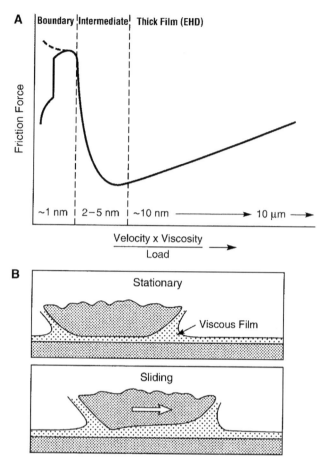

FIGURE 9.13 (A) Stribeck curve: empirical curve giving the trend generally observed in the friction forces or friction coefficients as a function of sliding velocity, the bulk viscosity of the lubricating fluid, and the applied load. The three friction/lubrication regimes are known as the thick-film or EHD lubrication regime (see B below), the intermediate or mixed lubrication regime (see Figure 9.14), and the boundary lubrication regime (see Figure 9.24). The film thicknesses, believed to correspond to each of these regimes, are also shown. For thick films the friction force is purely viscous e.g., Couette flow at low shear rates, but may become complicated at higher shear rates where EHD deformations of surfaces can occur during sliding, as shown in (B).

At higher shear rates, two additional effects often come into play. First, certain properties of liquids may change at high $\dot{\gamma}$ values. In particular, the "effective" viscosity may become non-Newtonian, one form given by

$$\eta_{eff} = \dot{\gamma}^n \tag{9.26}$$

where $n = 0$ for Newtonian fluids, $n > 0$ for shear thickening (dilatant) fluids, and $n < 0$ for shear thinning (pseudoplastic) fluids (the latter become less viscous, i.e., flow more easily, with increasing shear rate). An additional effect on η can arise from the higher local stresses (pressures) experienced by the liquid film as $\dot{\gamma}$ increases. Since the viscosity is generally also sensitive to the pressure (usually increasing with P), this effect also acts to increase η_{eff} and thus the friction force.

A second effect that occurs at high shear rates is surface deformation, arising from the large hydrodynamic forces acting on the sliding surfaces. For example, Figure 9.13B shows how two surfaces elastically deform when the sliding speed increases to a high value. These deformations alter the hydrodynamic

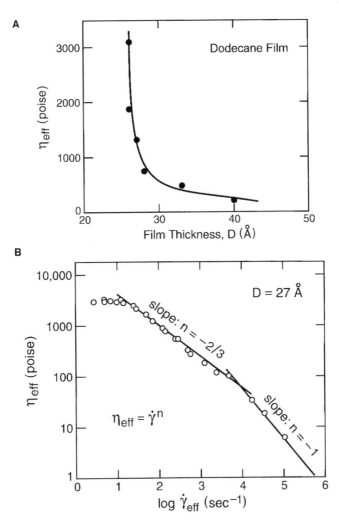

FIGURE 9.14 Typical rheological behavior of liquid film in the mixed lubrication regime. (A) Increase in effective viscosity of dodecane film between two mica surfaces with decreasing film thickness (Granick, 1991). Beyond 40 to 50 Å, the effective viscosity η_{eff} approaches the bulk value η_{bulk}. (B) Non-Newtonian variation of η_{eff} with shear rate of a 27-Å-thick dodecane film (from Luengo et al., 1996). The effective viscosity decays as a power law, as in Equation 9.26. In this example, $n = 0$ at the lowest $\dot{\gamma}$, then transitions to $n = -1$ at higher $\dot{\gamma}$. For bulk thick films, dodecane is a low-viscosity Newtonian fluid ($n = 0$).

friction forces, and this type of friction is often referred to as *elastohydrodynamic lubrication* (EHD or EHL) as mentioned in Table 9.3.

One natural question is: How thin can a liquid film be before its dynamic e.g., viscous flow, behavior ceases to be described by bulk properties and continuum models? Concerning the *static* properties, we have already seen that films composed of simple liquids display continuum behavior down to thicknesses of 5 to 10 molecular diameters. Similar effects have been found to apply to the dynamic properties, such as the viscosity, of simple liquids in thin films. Concerning viscosity measurements, a number of dynamic techniques were recently developed (Chan and Horn, 1985; Israelachvili, 1986a; Van Alsten and Granick, 1988; Israelachvili and Kott, 1989) for directly measuring the viscosity as a function of film thickness and shear rate across very thin liquid films between two surfaces. By comparing the results with theoretical predictions of fluid flow in thin films, one can determine the effective positions of the shear planes and the onset of non-Newtonian behavior in very thin films.

The results show that for simple liquids including linear-chain molecules such as alkanes, their viscosity in thin films is the same, within 10%, as the bulk even for films as thin as ten molecular diameters (or segment widths) (Chan and Horn, 1985; Israelachvili, 1986a; Israelachvili and Kott, 1989). This implies that the shear plane is effectively located within one molecular diameter of the solid liquid interface, and these conclusions were found to remain valid even at the highest shear rates studied (of $\sim 2 \times 10^5$ s^{-1}). With water between two mica or silica surfaces (Chan and Horn, 1985; Israelachvili, 1986a; Horn et al., 1989b; Israelachvili and Kott, 1989), this has been found to be the case (to within $\pm 10\%$) down to surface separations as small as 2 nm, implying that the shear planes must also be within a few angstrom of the solid–liquid interfaces. These results appear to be independent of the existence of electrostatic double-layer or hydration forces. For the case of the simple liquid toluene confined between surfaces with adsorbed layers of C_{60} molecules, this type of viscosity measurement has shown that the traditional no-slip assumption for flow at a solid interface does not always hold (Campbell et al., 1996). For this system, the C_{60} layer at the mica–toluene interface results in a "full-slip" boundary, which dramatically lowers the viscous drag or effective viscosity for regular Couette or Poiseuille flow.

With polymeric liquids (polymer melts) such as polydimethylsiloxanes (PDMS) and polybutadienes (PBD), or polystyrene (PS) adsorbed onto surfaces from solution, the far-field viscosity is again equal to the bulk value, but with the no-slip plane (hydrodynamic layer thickness) being located at $D = 1$ to $2 R_g$ away from each surface (Israelachvili, 1986b; Luengo et al., 1997), or at $D = L$ for polymer brush layers of thickness L per surface (Klein et al., 1993). In contrast, the same technique was used to show that for nonadsorbing polymers in solution, there is actually a depletion layer of nearly pure solvent that exists at the surfaces that affects the confined solution flow properties (Kuhl et al., 1998). These effects are observed from near contact to surface separations in excess of 200 nm.

Further experiments with surfaces closer than a few molecular diameters ($D < 20$ to 40 Å for simple liquids, or $D < 2$ to $4 R_g$ for polymer fluids) indicate that large deviations occur for thinner films, described below. One important conclusion from these studies is therefore that the dynamic properties of simple liquids, including water, near an *isolated* surface are similar to those of the bulk liquid *already within the first layer of molecules adjacent to the surface*, only changing when another surface approaches the first. In other words, the viscosity and position of the shear plane near a surface are not simply a property of that surface, but of how far that surface is from another surface. The reason for this is because when two surfaces are close together, the constraining effects on the liquid molecules between them are much more severe than when there is only one surface. Another obvious consequence of the above is that one should not make measurements on a single, isolated solid–liquid interface and then draw conclusions about the state of the liquid or its interactions in a thin film *between* two surfaces.

9.9.3 Friction of Intermediate Thickness Films

For liquid films in the thickness range between 6 and 10 molecular diameters, their properties can be significantly different from those of bulk films. But the fluids remain recognizable as fluids; in other words, they do not undergo a phase transition into a solid or liquid-crystalline phase. This regime has recently been studied by Granick and co-workers (Van Alsten and Granick, 1990a,b, 1991; Granick, 1991; Hu et al., 1991; Hu and Granick, 1992; Klein and Kumacheva, 1995) who used a different type of friction attachment (Van Alsten and Granick, 1988, 1990b) to the SFA where the two surfaces are made to vibrate laterally past each other at small amplitudes. This method provides information on the real and imaginery parts (elastic and dissipative components, respectively) of the shear modulus of thin films at different shear rates and film thickness. Granick (1991) and Hu et al. (1991) found that films of simple liquids become non-Newtonian in the 25 to 50 Å regime (about ten molecular diameters), whereas polymer melts become non-Newtonian at much thicker films, depending on their molecular weight (Luengo et al., 1997).

A generalized friction map (Figure 9.16) has been proposed by Luengo et al. (1996) that illustrates the changes in η_{eff} from bulk Newtonian behavior ($n = 0$, $\eta_{eff} = \eta_{bulk}$) through the transition regime where

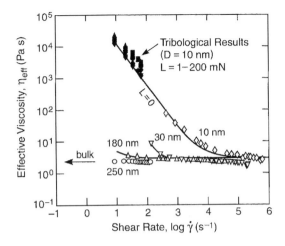

FIGURE 9.15 Effective viscosity plotted against effective shear rate on log–log scales for polybutadiene (MW = 7000) at four different separations, D (adapted from Luengo et al., 1997). Open data points were obtained from sinusoidally applied shear at zero load ($L = 0$) at the indicated separations. Solid points were obtained from friction experiments at constant-sliding velocities. These tribological results extrapolate, at high shear rate, to the bulk viscosity.

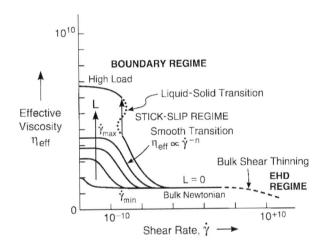

FIGURE 9.16 Proposed generalized friction map of effective viscosity plotted against effective shear rate on a log–log scale. (From Luengo, G. et al., 1996, *Wear* 200, 328–335. With permission.) Three main classes of behavior emerge: (1) Thick films; EHD sliding. At zero load ($L = 0$), approximating bulk conditions, η_{eff} is independent of shear rate except when shear thinning may occur at sufficiently large $\dot{\gamma}$. (2) Boundary layer films, intermediate regime. A Newtonian regime is again observed (η_{eff} = constant, $n = 0$ in Equation 9.26) at low loads and low shear rates, but η_{eff} is much higher than the bulk value. As the shear rate $\dot{\gamma}$ increases beyond $\dot{\gamma}_{min}$, the effective viscosity starts to drop with a power law dependence on the shear rate (see Figure 9.14B), with n in the range of $-\frac{1}{2}$ to -1 most commonly observed. As the shear rate $\dot{\gamma}$ increases still more, beyond $\dot{\gamma}_{max}$, a second Newtonian plateau is again encountered. (3) Boundary layer films, high load. The η_{eff} continues to grow with load and to be Newtonian provided that the shear rate is sufficiently low. Transition to sliding at high velocity is discontinuous ($n < -1$) and usually of the stick-slip variety.

n reaches a minimum of -1 with decreasing shear rate, to the solidlike creep regime at very low $\dot{\gamma}$, where n returns to 0. The data in Figure 9.15 show the transition for thicker polymer films from bulk behavior to the tribological regime where n reaches -1 (Luengo et al., 1997). With further decreasing shear rates the exponent n increases from -1 to 0, as illustrated in Figure 9.14B for a dodecane system. A number of results from experimental, theoretical, and computer simulation work have shown values of n from $-\frac{1}{2}$ to -1 for this transition regime for a variety of systems and assumptions (Hu and Granick, 1992; Granick, 1991; Thompson et al., 1992, 1995; Urbakh et al., 1995; Rabin and Hersht, 1993).

The intermediate regime appears to extend over a narrow range of film thickness, from about four to ten molecular diameters or polymer radii of gyration. Thinner films begin to adopt "boundary" or "interfacial" friction properties (described below, see also Table 9.3). Note that the intermediate regime is actually a very narrow one when defined in terms of film thickness, for example, varying from about $D = 20$ to 40 Å for hexadecane films (Granick, 1991).

The effective viscosity η_{eff} of a fluid in the intermediate regime is usually higher than in the bulk, but η_{eff} usually *decreases* with increasing sliding velocity, $\mathbf{v}$ (known as *shear thinning*). When two surfaces slide in the intermediate regime, the motion tends to thicken the film (dilatency). This sends the system into the bulk EHL regime where, as indicated by Equation 9.24, the friction force now *increases* with velocity. This initial decrease, followed by an increase, in the frictional forces of many lubricant systems is the basis for the empirical Stribeck curve of Figure 9.13A. In the transition from bulk to boundary behavior there is first a quantitative change in the material properties (viscosity and elasticity) which can be continuous, to discontinuous qualitative changes which result in yield stresses and non-liquidlike behavior.

The rest of this chapter is devoted to friction in the interfacial and boundary regimes. The former (interfacial friction) may be thought of as applying to the sliding of two dry, unlubricated surfaces in true molecular contact. The latter (boundary friction) may be thought of as applying to the case where a lubricant film is present, but where this film is of molecular dimensions — a few molecular layers or less.

9.10 Interfacial and Boundary Friction: Molecular Tribology

9.10.1 General Interfacial Friction

When a lateral force, or shear stress, is applied to two surfaces in adhesive contact, the surfaces initially remain "pinned" to each other until some critical shear force is reached. At this point, the surfaces begin to slide past each other either smoothly or in jerks. The frictional force needed to initiate sliding from rest is known as the *static* friction force, denoted by F_s, while the force needed to maintain smooth sliding is referred to as the *kinetic* or *dynamic* friction force, denoted by F_k. In general, $F_s > F_k$. Two sliding surfaces may also move in regular jerks, known as "stick-slip" sliding, which is discussed in more detail in Section 9.13. Note that such friction forces cannot be described by equations, such as Equation 9.26, used for thick films that are viscous and therefore shear as soon as the smallest shear force is applied.

Experimentally, it has been found that during both smooth and stick-slip sliding the local geometry of the contact zone remains largely unchanged from the static geometry, and that the contact area vs. load is still well described by the JKR equation, Equation 9.20.

The friction force of two molecularly smooth surfaces sliding while in adhesive contact with each other is not simply proportional to the applied load, L, as might be expected from Amontons' law. There is an additional adhesion contribution that is proportional to the area of contact, A, which is described later. Thus, in general, the interfacial friction force of dry, unlubricated surfaces sliding smoothly past each other is given by

$$F = F_k = S_c A + \mu L. \tag{9.27}$$

where S_c is the critical shear stress (assumed to be constant), $A = \pi r^2$ is the contact area of radius r given by Equation 9.20, and μ is the coefficient of friction. For low loads we have

$$F = S_c A = S_c \pi r^2 = S_c \pi \left[\frac{R}{K} \left(L + 6\pi R\gamma + \sqrt{12\pi R\gamma L + \left(6\pi R\gamma\right)^2} \right) \right]^{2/3}, \tag{9.28}$$

while for high loads, Equation 9.27 reduces to Amontons' law:

$$F = \mu L. \tag{9.29}$$

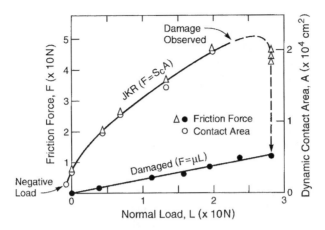

FIGURE 9.17 Friction force F and contact area A vs. load L for two mica surfaces sliding in adhesive contact in dry air. The contact area is well described by the JKR theory, Equation 9.20), even during sliding, and the frictional force is found to be directly proportional to this area, Equation 9.28. The vertical dashed line and arrow show the transition from interfacial to normal friction with the onset of wear (lower curve).

Depending on whether the friction force, F, in Equation 9.27 is dominated by the first or second terms, one may refer to the friction as adhesion controlled or load controlled, respectively.

Figure 9.17 shows a plot of contact area, A, and friction force, F, both plotted against the applied load, L, in an experiment where two molecularly smooth surfaces of mica in adhesive contact were slid past each other in an atmosphere of dry nitrogen gas. This is an example of the low-load, adhesion-controlled limit, which is excellently described by Equation 9.28. In a number of different experiments, S_c was measured to be 2.5×10^7 N/m², and to be independent of the sliding velocity. Note that there is a friction force even at negative loads, where the surfaces are still sliding in adhesive contact.

9.10.2 Boundary Friction of Surfactant Monolayer–Coated Surfaces

The high friction force of unlubricated sliding can often be reduced by treating the solid surface with a boundary layer of some other solid material that exhibits lower friction, such as a surfactant monolayer, or by ensuring that during sliding, a thin liquid film remains between the surfaces. The effectiveness of a solid boundary lubricant layer on reducing the forces of friction is illustrated in Figure 9.18. Comparing this with the friction of the unlubricated/untreated surfaces (Figure 9.17) shows that the critical shear stress has been reduced by a factor of about 10: from 2.5×10^7 to 3.5×10^6 N/m². At much higher applied loads or pressures, the friction force is proportional to the load rather than the area of contact (Briscoe et al., 1977), as expected from Equation 9.27.

9.10.3 Boundary Lubrication of Molecularly Thin Liquid Films

A liquid lubricant film is usually much more effective at lowering the friction of two surfaces than a solid boundary lubricant layer. However, to use a liquid lubricant successfully, it must "wet" the surfaces; that is, it should have a high affinity for the surfaces so that the liquid molecules do not become squeezed out when the surfaces come close together, even under a large compressive load. Another important requirement is that the liquid film remains a liquid under tribological conditions, i.e., that it does not epitaxially solidify between the surfaces.

Effective lubrication usually requires that the lubricant be injected between the surfaces, but in some cases the liquid can be made to condense from the vapor. Both of these effects are illustrated in Figure 9.19 for two untreated mica surfaces sliding with a thin layer of water between them. We may note that a monomolecular film of water (of thickness 2.5 Å per surface) has reduced S_c by a factor of more than 30,

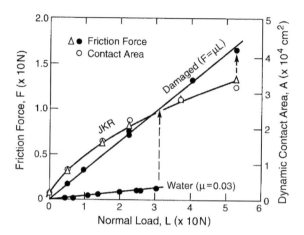

FIGURE 9.18 Sliding of mica surfaces each coated with a 25-Å-thick monolayer of calcium stearate surfactant in the absence of damage (obeying JKR-type boundary friction) and in the presence of damage (obeying Amontons-type normal friction). (Homola et al., 1989, 1990). At much higher applied loads the undamaged surfaces also follow Amontons-type sliding, but for different reasons (see next section). Lower line: interfacial sliding with a monolayer of water between the mica surfaces, shown for comparison. Note that in both cases, after damage occurs, the friction force obeys Amontons' law with the same coefficient of friction of $\mu \approx 0.3$.

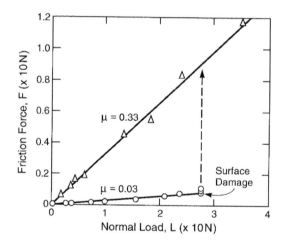

FIGURE 9.19 Two mica surfaces sliding past each other while immersed in a 0.01 M KCl salt solution. In both cases the water film is molecularly thin: 2.5 to 5.0 Å thick, and the interfacial friction force is very low: $S_c \approx 5 \times 10^5$ N/m², $\mu \approx 0.015$ (before damage occurs).

which may be compared with the factor of 10 attained with the boundary lubricant layer (of thickness 25 Å per surface).

The effectiveness of a water film only 2.5 Å thick to lower the friction force by more than an order of magnitude is attributed to the "hydrophilicity" of the mica surface (mica is "wetted" by water) and to the existence of a strongly repulsive short-range hydration force between such surfaces in aqueous solutions (see Section 9.6) which effectively removes the adhesion-controlled contribution to the friction force (Berman et al., 1998a). It is also interesting that a 2.5-Å-thick water film between two mica surfaces is sufficient to bring the coefficient of friction down to 0.01 to 0.02, a value that corresponds to the unusually low friction of ice. Clearly, a single monolayer of water can be a very good lubricant — much better than most other monomolecular liquid films, for reasons that will be discussed below.

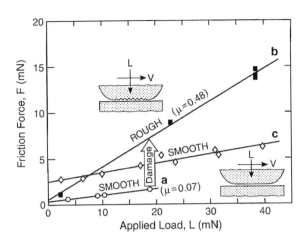

FIGURE 9.20 Friction forces as a function of load for smooth (undamaged) and rough (damaged) surfaces (Berman et al., 1998b). Untreated alumina surfaces (curve **a**) exhibit the lowest friction due to a thin physisorbed layer of lubricating contaminants, but these are easily damaged upon sliding, resulting in rough surface (curve **b**) with a higher friction. Monolayer coated surfaces (curve **c**) slide with higher friction at lower loads than untreated alumina, but remain undamaged even after prolonged sliding, keeping the friction and wear substantially lower than rough surfaces at high loads. Smooth sliding was "adhesion controlled," i.e., the contact area A is well described by the JKR equation (Equation 9.20), and $F/A =$ constant (Equation 9.28). In addition, F is finite at $L = 0$. Rough, damaged sliding was "load controlled"; i.e., the real contact area is undefined, and $F \propto L$ (Equation 9.29) with $F \approx 0$ at $L = 0$. Experimental conditions: sliding velocity $V = 0.05$ to 0.5 μm/s; undeformed radius of curved surfaces, $R \approx 1$ cm; temperature $T = 25°C$; contact pressure range, $P = 0$ to 10 MPa; relative humidity, RH $= 0\%$ (curves **a** and **c**), RH $= 0\%$ and 100% (curve **b**).

9.10.4 Transition from Interfacial to Normal Friction (with Wear)

Frictional damage can have many causes such as adhesive tearing at high loads or overheating at high sliding speeds. Once damage occurs, there is a transition from "interfacial" to "normal" or load-controlled friction as the surfaces become forced apart by the torn-out asperities (wear particles). For low loads, the friction changes from obeying $F = S_c A$ to obeying Amontons' law: $F = \mu L$, as shown in Figures 9.17 through 9.20, and sliding now proceeds smoothly with the surfaces separated by a 100 to 1,000 Å forest of wear debris (mica or alumina flakes in this case). The wear particles keep the surfaces apart over an area that is much greater than their size, so that even one submicroscopic particle or asperity can cause a significant reduction in the area of contact and therefore in the friction (Homola et al., 1990). For this type of frictional sliding, one can no longer talk of the molecular contact area of the two surfaces, although the macroscopic or "apparent" area is still a useful parameter. A further discussion on the impact of normal load and contact area is found in Section 9.11.4.

One remarkable feature of the transition from interfacial to normal friction of brittle surfaces is that while the strength of interfacial friction, as reflected in the values of S_c, is very dependent on the type of surface and on the liquid film between the surfaces, this is not the case once the transition to normal friction has occurred. At the onset of damage, it is the material properties of the underlying substrates that control the friction. In Figures 9.17 through 9.19 the friction for the damaged surfaces is that of any damaged mica–mica system, while in Figure 9.20 the damaged surfaces friction is that of general alumina–alumina sliding (with a friction coefficient that agrees with literature values for the bulk materials), *independent of the initial surface coatings or liquid films between the surfaces.*

In order to modify practically the frictional behavior of such brittle materials, it is important to use coatings that will both alter the interfacial tribological character and remain intact and protect the surfaces from damage during sliding (Berman et al., 1998b). An example of the friction behavior of a strongly bound octadecyl phosphonic acid monolayer on alumina surfaces is shown in Figure 9.20. In this case, the friction is higher than untreated undamaged α-alumina surfaces, but the bare surfaces easily damage upon sliding, resulting in an ultimately higher friction system with greater wear rates than the more robust monolayer-coated surfaces.

Clearly, the mechanism and factors that determine *normal* friction are quite different from those that govern *interfacial* friction. These mechanisms are described in the theoretical section below, but one should point out that this effect is not general and may only apply to brittle materials. For example, the friction of ductile surfaces is totally different and involves the continuous plastic deformation of contacting surface asperities during sliding rather than the rolling of two surfaces on hard wear particles (Bowden and Tabor, 1967). Furthermore, in the case of ductile surfaces, water and other surface-active components do have an effect on the friction coefficients under normal sliding conditions.

9.11   Theories of Interfacial Friction

9.11.1   Theoretical Modeling of Interfacial Friction: Molecular Tribology

The following friction model, first proposed by Tabor (1982) and developed further by Sutcliffe et al. (1978), McClelland (1989), and Homola et al. (1989), has been quite successful at explaining the interfacial and boundary friction of two solid crystalline surfaces sliding past each other in the absence of wear. The surfaces may be unlubricated, or they may be separated by a monolayer or more of some boundary lubricant or liquid molecules. In this model, the values of the critical shear stress S_c, and the coefficient of friction μ, of Equation 9.27 are calculated in terms of the energy needed to overcome the attractive intermolecular forces and compressive externally applied load as one surface is raised and then slid across the molecular-sized asperities of the other.

This model (variously referred to as the *interlocking asperity model, coulomb friction,* or the *cobblestone model*) is akin to pushing a cart over a road of cobblestones where the cartwheels (which represent the molecules of the upper surface or film) must be made to roll over the cobblestones (representing the molecules of the lower surface) before the cart can move. In the case of the cart, the downward force of gravity replaces the attractive intermolecular forces between two material surfaces. When at rest the cartwheels find grooves between the cobblestones where they sit in potential energy minima and so the cart is at some stable mechanical equilibrium. A certain lateral force (the "push") is required to raise the cartwheels against the force of gravity in order to initiate motion. Motion will continue as long as the cart is pushed, and rapidly stops once it is no longer pushed. Energy is dissipated by the liberation of heat (phonons, acoustic waves, etc.) every time a wheel hits the next cobblestone. The cobblestone model is not unlike the old coulomb and interlocking asperity models of friction (Dowson, 1979) except that it is being applied at the molecular level and where the external load is augmented by attractive intermolecular forces.

There are thus two contributions to the force pulling two surfaces together: the externally applied load or pressure, and the (internal) attractive intermolecular forces which determine the adhesion between the two surfaces. Each of these contributions affects the friction force in a different way, and we start by considering the role of the internal adhesion forces.

9.11.2   Adhesion Force Contribution to Interfacial Friction

Consider the case of two surfaces sliding past each other as shown in Figure 9.21. When the two surfaces are initially in adhesive contact, the surface molecules will adjust themselves to fit snugly together, in an analogous manner to the self-positioning of the cartwheels on the cobblestone road. A small tangential force applied to one surface will therefore not result in the sliding of that surface relative to the other. The attractive van der Waals forces between the surfaces must first be overcome by having the surfaces separate by a small amount. To initiate motion, let the separation between the two surfaces increase by a small amount ΔD, while the lateral distance moved is Δd. These two values will be related via the

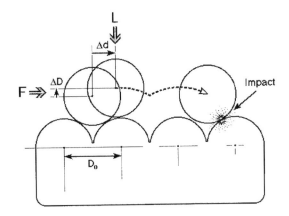

FIGURE 9.21 Schematic illustration of how one molecularly smooth surface moves over another when a lateral force F is applied. As the upper surface moves laterally by some fraction of the lattice dimension Δd, it must also move up by some fraction of an atomic or molecular dimension ΔD, before it can slide across the lower surface. On impact, some fraction ε of the kinetic energy is "transmitted" to the lower surface, the rest being "reflected" back to the colliding molecule (upper surface).

geometry of the two surface lattices. The energy put into the system by the force F acting over a lateral distance Δd is

$$\text{Input energy: } F \times \Delta d. \tag{9.30}$$

This energy may be equated with the change in interfacial or surface energy associated with separating the surfaces by ΔD, i.e., from the equilibrium separation $D = D_0$ to $D = (D_0 + \Delta D)$. Since $\gamma \propto D^{-2}$, the surface energy cost may be approximated by

$$\text{Surface energy change: } 2\gamma A \left(1 - D_0^2 \middle/ \left(D_0 + \Delta D \right)^2 \right) \approx 4\gamma A \left(\Delta D / D_0 \right), \tag{9.31}$$

where γ is the surface energy, A the contact area, and where D_0 is the surface separation at equilibrium. During steady-state sliding (kinetic friction), not all of this energy will be lost or absorbed by the lattice every time the surface molecules move by one lattice spacing: some fraction will be reflected during each impact of the cartwheel molecules (McClelland, 1989). Assuming that a fraction ε of the above surface energy is lost every time the surfaces move across the characteristic length Δd (Figure 9.21), we obtain after equating the above two equations

$$S_c = \frac{F}{A} = \frac{4\gamma\varepsilon \cdot \Delta D}{D_0 \cdot \Delta d} . \tag{9.32}$$

For a typical hydrocarbon or a van der Waals surface, $\gamma \approx 25 \times 10^{-3}$ J/m^2. Other typical values would be $\Delta D \approx 0.5$ Å, $D_0 \approx 2$ Å, $\Delta d \approx 1$ Å, and $\varepsilon \approx 0.1$ to 0.5. By using the above parameters, Equation 9.32 predicts $S_c \approx (2.5 \text{ to } 12.5) \times 10^7$ N/m^2 for van der Waals surfaces. This range of values compares very well with typical experimental values of 2×10^7 N/m^2 for hydrocarbon or mica surfaces sliding in air or separated by one molecular layer of cyclohexane (Homola et al., 1989).

The above model may be extended, at least semiquantitatively, to lubricated sliding, where a thin liquid film is present between the surfaces. With an increase in the number of liquid layers between the surfaces, D_0 increases while ΔD decreases, hence the lower the friction force. This is precisely what is observed. But with more than one liquid layer between two surfaces, the situation becomes too complex to analyze

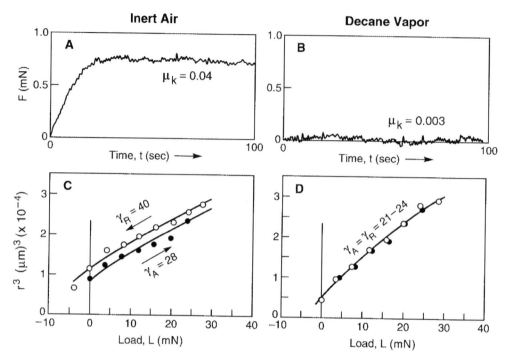

FIGURE 9.22 Top: Friction traces for two fluid-like monolayer-coated surfaces at 25° C showing that the friction force is much higher between dry monolayers (A) than between monolayers whose fluidity has been enhanced by hydrocarbon penetration from vapor (B). (From Chen et al., 1991.) Bottom: Contact radius vs. load (r^3–L) curves measured for the same two surfaces as above and fitted to the JKR equation (Equation 9.20 — shown by the solid lines. For dry monolayers (C) the adhesion energy on unloading ($\gamma_R = 40$ mJ/m^2) is greater than that on loading ($\gamma_A = 28$ mJ/m^2), indicative of an adhesion energy hysteresis of $\Delta\gamma = \gamma_R - \gamma_A = 12$ mJ/m^2. For monolayers exposed to saturated decane vapor (D) their adhesion hysteresis is zero ($\gamma_A = \gamma_R$), and both the loading and unloading curves are well fitted by the thermodynamic value of the surface energy of fluid hydrocarbon chains, $\gamma \approx 24$ mJ/m^2.

analytically (actually, even with one or no interfacial layers, the calculation of the fraction of energy dissipated per molecular collision ε is not a simple matter). Sophisticated modeling based on computer simulations is now required, as described in the following section.

9.11.3 Relation between Boundary Friction and Adhesion Energy Hysteresis

While the above equations suggest that there is a direct correlation between friction and adhesion, this is not the case. The correlation is really between friction and adhesion hysteresis, described in Section 9.8. In the case of friction, this subtle point is hidden in the factor ε, which is a measure of the amount of energy absorbed (dissipated, transferred, or lost) by the lower surfaces when it is impacted by a molecule from the upper surface. If $\varepsilon = 0$, all the energy is reflected and there will be no kinetic friction force, nor any adhesion hysteresis, but the absolute magnitude of the adhesion force or energy would remain finite and unchanged. This is illustrated in Figure 9.22.

The following simple model shows how adhesion hysteresis and friction may be quantitatively related. Let $\Delta\gamma = (\gamma_R \text{ to } \gamma_A)$ be the adhesion energy hysteresis per unit area, as measured during a typical loading–unloading cycle (see Figure 9.22C and D). Now consider the same two surfaces sliding past each other and assume that frictional energy dissipation occurs through the same mechanism as adhesion energy dissipation, and that both occur over the same characteristic molecular length scale σ. Thus, when the two surfaces (of contact area $A = \pi r^2$) move a distance σ, equating the frictional energy ($F \times \sigma$) to the dissipated adhesion energy ($A \times \Delta\gamma$), we obtain

$$\text{Friction force:} \quad F = \frac{A \times \Delta\gamma}{\sigma} = \frac{\pi r^2}{\sigma}\left(\gamma_R - \gamma_A\right) \tag{9.33}$$

or

$$\text{Friction stress:} \quad S_c = F/A = \Delta\gamma/\sigma \tag{9.34}$$

which is the desired expression and which has been found to give order of magnitude agreement between measured friction forces and adhesion energy hysteresis (Chen et al., 1989). If we equate Equation 9.34 with Equation 9.32, since $4\Delta D/D_0 \Delta d \approx \sigma$, we obtain the intuitive relation:

$$\varepsilon \approx \frac{\Delta\gamma}{\gamma}. \tag{9.35}$$

Figure 9.22 illustrates the relationship between adhesion hysteresis and friction for surfactant-coated surfaces under different conditions. This effect, however, is much more general, and has been shown to hold for other surfaces as well (Vigil et al., 1994; Israelachvili et al., 1994, 1995). Direct comparisons between absolute adhesion energies and friction forces show little correlation. In some cases higher adhesion energies for the same system under different conditions correspond with lower friction forces. For example, for hydrophilic silica surfaces it was found that with increasing relative humidity the adhesion energy *increases*, but the adhesion energy hysteresis measured in a loading–unloading cycle *decreases*, as does the friction force (Vigil et al., 1994). For hydrophobic silica surfaces under dry conditions, the friction at load $L = 5.5$ mN was $F = 75$ mN. For the same sample, the adhesion energy hysteresis was $\Delta\gamma = 10$ mJ/m^2, with a contact area of $A \approx 10^{-8}$ m^2 at the same load. Assuming a value for the characteristic distance σ on the order of one lattice spacing, $\sigma \approx 1$ nm. By inserting these values into Equation 9.33, the friction force is predicted to be $F \approx 100$ mN for the kinetic friction force, which is close to the measured value of 75 mN. Alternatively, we may conclude that the dissipation factor is $\varepsilon = 0.75$, i.e., that almost all the energy is dissipated as heat at each molecular collision.

9.11.4 External Load Contribution to Interfacial Friction

When there is no interfacial adhesion, S_c is zero. Thus, in the absence of any adhesive forces between two surfaces, the only "attractive" force that needs to be overcome for sliding to occur is the externally applied load or pressure.

For a preliminary discussion of this question, it is instructive to compare the magnitudes of the *externally* applied pressure to the *internal* van der Waals pressure between two smooth surfaces. The internal van der Waals pressure is given by $P = \mathbf{A}/6\pi D_0^3 \approx 10^4$ atm (using a typical Hamaker constant of $\mathbf{A} = 10^{-12}$ erg, and assuming $D_0 \approx 2$ Å for the equilibrium interatomic spacing). This implies that we should not expect the externally applied load to affect the interfacial friction force F, as defined by Equation 9.27, until the externally applied pressure L/A begins to exceed ~1000 atm. This is in agreement with experimental data (Briscoe et al., 1977), where the effect of load became dominant at pressures in excess of 1000 atm.

For a more general semiquantitative analysis, again consider the cobblestone model as used to derive Equation 9.32 but now include an additional contribution to the surface energy change of Equation 9.31 due to the work done against the external load or pressure, $L\Delta D = P_{ext}A \cdot \Delta D$ (this is equivalent to the work done against gravity in the case of a cart being pushed over cobblestones). Thus,

$$S_c = \frac{F}{A} = \frac{4\gamma\varepsilon \cdot \Delta D}{D_0 \cdot \Delta d} + \frac{P_{ext}\varepsilon \cdot \Delta D}{\Delta d}, \tag{9.36}$$

which gives the more general relation:

$$S_c = F/A = C_1 + C_2 P_{ext}, \tag{9.37}$$

where $P_{ext} = L/A$ and where C_1 and C_2 are constants characteristic of the surfaces and sliding conditions. The constant $C_1 = (4\gamma\varepsilon \cdot \Delta D/D_0 \cdot \Delta d)$ depends on the mutual adhesion of the two surfaces, while both C_1 and $C_2 = \varepsilon \cdot \Delta D/\Delta d$ depend on the topography or atomic bumpiness of the surface groups (Figure 9.21) — the smoother the surface groups the smaller the ratio $\Delta D/\Delta d$ and hence the lower the value of C_2. In addition, both C_1 and C_2 depend on ε — the fraction of energy dissipated per collision which depends on the relative masses of the shearing molecules, the sliding velocity, the temperature, and the characteristic molecular relaxation processes of the surfaces. This is by far the most difficult parameter to compute, and yet it is the most important since it represents the energy transfer mechanism in any friction process. Further, since ε can vary between 0 and 1, it determines whether a particular friction force will be large or close to zero. Molecular simulations offer the best way to understand and predict the magnitude of ε, but the complex multibody nature of the problem makes simple conclusions difficult to draw. Some of the basic physics of the energy transfer and dissipation of the molecular collisions can be drawn from simplified models (Urbakh et al., 1995; Rozman et al., 1996, 1997) such as a one-dimensional three-body system (Israelachvili and Berman, 1995). This system, described in more detail in Section 9.11.5, offers insight into the mechanisms of energy transfer and relates them back to the familiar parameter De, the Deborah number.

Finally, the above equation may also be expressed in terms of the friction force F:

$$F = S_c A = C_1 A + C_2 L. \tag{9.38}$$

Equations similar to Equations 9.37 and 9.38 were previously derived by Derjaguin (1988, 1934) and by Briscoe and Evans (1982), where the constant C_1 and C_2 were interpreted somewhat differently than in this model.

In the absence of any attractive interfacial force, we have $C_1 \approx 0$, and the second term in Equations 9.37 and 9.38 should dominate. Such situations typically arise when surfaces repel each other across the lubricating liquid film, for example, when two mica surfaces slide across a thin film of water (Figure 9.18). In such cases the total frictional force should be low and it should increase *linearly* with the external load according to

$$F = C_2 L. \tag{9.39}$$

An example of such lubricated sliding occurs when two mica surfaces slide in water or in salt solution, where the short-range hydration forces between the surfaces are repulsive. Thus, for sliding in 0.5 M KCl it was found that $C_2 = 0.015$ (Berman et al., 1998a). Another case where repulsive surfaces eliminate the adhesive contribution to friction is for tethered polymer chains attached to surfaces at one end and swollen by a good solvent (Klein et al., 1994). For this class of systems $C_2 < 0.001$ for a finite range of polymer layer compressions (normal loads, L). The low friction between the surfaces in this regime is attributed to the entropic repulsion between the opposing brush layers with a minimum of entanglement between the two layers. However, with higher normal loads, the brush layers become compressed and begin to entangle, resulting in higher friction.

It is important to note that Equation 9.39 has exactly the same form as Amontons' law:

$$F = \mu L, \tag{9.40}$$

where μ is the coefficient of friction. When damage occurs, there is a rapid transition to normal sliding in the presence of wear debris. As previously described, the mechanisms of interfacial friction and normal friction are vastly different on the submicroscopic and molecular levels. However, under certain circumstances both may appear to follow a similar equation (see Equations 9.39 and 9.40) even though the friction coefficients C_2 and μ are determined by quite different material properties in each case.

At the molecular level a thermodynamic analog of the coulomb or cobblestone models (see Section 9.11.1) based on the contact value theorem (Israelachvili, 1991; Berman et al., 1998a; Berman and Israelachvili, 1997) can explain why $F \propto L$ also holds at the microscopic or molecular level. In this analysis we consider the surface molecular groups as being momentarily compressed and decompressed as the surfaces move along. Under irreversible conditions, which always occur when a cycle is completed in a finite amount of time, the energy lost in the compression/decompression cycle is dissipated as heat. For two nonadhering surfaces, the stabilizing pressure P_i acting locally between any two elemental contact points i of the surfaces may be expressed by the contact value theorem (Israelachvili, 1992):

$$P_i = \rho_i k_B T = k_B T / V_i, \tag{9.41}$$

where $\rho_i = 1/V_i$ is the local number density (per unit volume) or activity of the interacting entities, be they molecules, atoms, ions, or the electron clouds of atoms. This equation is essentially the osmotic or entropic pressure of a gas of confined molecules. As one surface moves across the other, as local regions become compressed and decompressed by a volume ΔV_i, the work done per cycle can be written as $\varepsilon \, P_i \, \Delta V_i$, where ε ($\varepsilon \leq 1$) is the fraction of energy per cycle lost as heat, as defined earlier. The energy balance shows that for each compression/decompression cycle, the dissipated energy is related to the friction force by

$$F_i x_i = \varepsilon \, P_i \Delta V_i, \tag{9.42}$$

where x_i is the lateral distance moved per cycle, which can be the distance between asperities or the distance between surface lattice sites. The pressure at each contact junction can be expressed in terms of the local normal load L_i and local area of contact A_i as $P_i = L_i/A_i$. The volume change over a cycle can thus be expressed as $\Delta V_i = A_i z_i$, where z_i is the vertical distance of confinement. Plugging these back into Equation 9.42, we get

$$F_i = \varepsilon \, L_i \left(z_i / x_i \right), \tag{9.43}$$

which is independent of the local contact area A_i. The total friction force is thus

$$F = \Sigma \, F_i = \Sigma \, \varepsilon \, L_i \left(z_i / x_i \right) = \varepsilon \left\langle z_i / x_i \right\rangle \Sigma \, L_i = \mu \, L, \tag{9.44}$$

where it is assumed that on average, the local values of L_i and P_i are independent of the local "slope" z_i/x_i. Therefore, the friction coefficient μ is a function only of the average surface topography and the sliding velocity, but is independent of the local (real) or macroscopic (apparent) contact areas.

While this analysis explains nonadhering surfaces, there is still an additional explicit contact area contribution for the case of adhering surfaces, as in Equation 9.38. The distinction between the two cases arises because the initial assumption of the contact value theorem, Equation 9.41, is incomplete for adhering systems. A more appropriate starting equation would reflect the full intermolecular interaction potential, including the attractive interactions in addition to the purely repulsive contributions of Equation 9.41, much as the van der Waals equation of state modifies the ideal gas law.

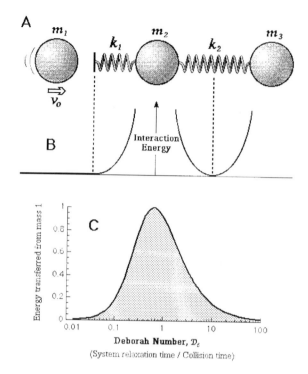

FIGURE 9.23 (A) Schematic diagram of a basic three-body collision. Mass **1** approaches the other masses with an initial velocity, v_0. The kinetic energy of mass **1** after the collision is compared to its initial kinetic energy. The characteristic times of the m_1–m_2 and m_2–m_3 interactions are functions of the respective masses and the strengths of the connecting springs k_1 and k_2 (B). The Deborah number for this system is the ratio of the collision time between masses **1** and **2** to the harmonic oscillating frequency of masses **2** and **3**. (C) The energy transferred from mass **1** to the rest of the system (masses **2** and **3**) is plotted as a function of the Deborah number. In this calculation $k_1 = 1$, $k_2 = 10$, $m_1 = 1$, and masses $m_2 = m_3$ were varied.

9.11.5 Simple Molecular Model of Energy Dissipation ε

The simple three-body system of balls and springs illustrated in Figure 9.23A provides molecular insight into dissipative processes such as sliding friction and adhesion hysteresis. Mass m_1 may be considered to constitute one body or surface that approaches another at velocity v_0. Masses m_2 and m_3 represent the other surface. As shown in Figure 9.23B, the masses of the second surface are bound together via a parabolic (Hookian spring) potential, while masses m_1 and m_2 interact via a nonadhesive repulsion modeled as a half parabola. Although this system is simple in appearance, it is rich in its physics. Different relative values of the masses and spring constants lead to very different collision outcomes where the initial translational or kinetic energy of m_1 is distributed among the final kinetic, translational and vibrational energies of m_1 and the m_2–m_3 couple.

The collision is a molecular analogy, and perhaps a good microscopic representation of an adhesive loading–unloading cycle (see Figure 9.11) or a frictional sliding process. The finer molecular model of Figure 9.23A allows us to determine exactly *how* the energy is being dissipated in these cycles. In a loading–unloading process, m_1 represents the surface molecules of one body approaching the surface molecules of a second body (m_2 and m_3). Solutions to the equations of motion show that in the collision the first mass can be reflected, stopped, or even continue forward at a different velocity. In the case of lateral sliding, the molecules of the two surfaces can still be considered to interact in this way; the surface groups from the slider continually collide with those of the opposing surface, with the amount of the energy transferred during the collisions defining the friction. Analysis of this problem shows that the essential determinant of the amount of energy transferred from m_1 is based on the ratio of the collision time to the characteristic relaxation time of the system, in other words, to the Deborah number:

$$De = \text{Relaxation time}/\text{Measuring time}. \qquad (9.45)$$

Thus, it is found that m_1 loses most of its energy to the m_2–m_3 couple when the collision time is close to the characteristic vibration time of the m_2–m_3 harmonic system, which corresponds to De = 1. When the collision time is much larger or smaller than the system characteristic time, mass m_1 is found to retain most of its original kinetic energy (Figure 9.23C). In this simple example, the interaction times are functions only of the three masses, the intermolecular potential between m_1 and m_2, and the potential or spring constant of the m_2–m_3 couple. In more complex systems with more realistic interaction potentials (i.e., attractive interactions between m_1 and m_2), the velocity v_0 becomes an important factor as well, and also affects the Deborah number.

It is important to note that in this simple one-dimensional analysis, additional energy modes and degrees of freedom of the molecules have not been considered. These modes, when present, will also be involved in the interaction, affecting the energy transferred from m_1 and sharing in the final distribution of the energy transferred. In addition, different types of energy modes (e.g., rotational modes) will generally have different relaxation times, so that their energy peaks will occur at different measuring times. Such real systems may be considered to have more than one Deborah number.

The three-body system sheds some light on the molecular mechanisms of energy dissipation and the impact of the Deborah number on the dissipation parameter ε, but because of its simplicity, does not offer predictive capabilities for real systems. More-sophisticated models have been presented by Urbakh et al. (1995) and Rozman et al. (1996, 1997).

9.12   Friction and Lubrication of Thin Liquid Films

When a liquid is confined between two surfaces or within any narrow space whose dimensions are less than five to ten molecular diameters, both the static (equilibrium) and dynamic properties of the liquid, such as its compressibility and viscosity, can no longer be described even qualitatively in terms of the bulk properties. The molecules confined within such molecularly thin films become ordered into layers ("out-of-plane" ordering), and within each layer they can also have lateral order ("in-plane" ordering). Such films may be thought of as behaving more like a liquid crystal or a solid than a liquid.

As described in Section 9.4, the measured normal forces between two solid surfaces across molecularly thin films exhibit exponentially decaying oscillations, varying between attraction and repulsion with a periodicity equal to some molecular dimension of the solvent molecules. Thus, most liquid films can sustain a finite normal stress, and the adhesion force between two surfaces across such films is quantized, depending on the thickness (or number of liquid layers) between the surfaces. The structuring of molecules in thin films and the oscillatory forces it gives rise to are now reasonably well understood, both experimentally and theoretically, at least for simple liquids.

Work has also recently been done on the dynamic e.g., viscous or shear, forces associated with molecularly thin films. Both experiments (Israelachvili et al., 1988; Gee et al., 1990; Hirz et al., 1992; Homola et al., 1993), and theory (Schoen et al., 1989; Thompson and Robbins, 1990; Thompson et al., 1992) indicate that even when two surfaces are in steady-state sliding they still prefer to remain in one of their stable potential energy minima, i.e., a sheared film of liquid can retain its basic layered structure. Thus, even during motion the film does not become totally liquidlike. Indeed, if there is some in-plane ordering within a film, it will exhibit a yield point before it begins to flow. Such films can therefore sustain a finite shear stress, in addition to a finite normal stress. The value of the yield stress depends on the number of layers comprising the film and represents another quantized property of molecularly thin films.

The dynamic properties of a liquid film undergoing shear are very complex. Depending on whether the film is more liquidlike or solidlike, the motion will be smooth or of the stick-slip type. During sliding,

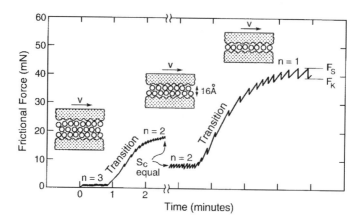

FIGURE 9.24 Measured change in friction during interlayer transitions of the silicone liquid octamethylcyclotet-rasiloxane (OMCTS, an inert liquid whose quasi-spherical molecules have a diameter of 8 Å) (Gee et al., 1990). In this system, the shear stress $S_c = F/A$ was found to be constant so long as the number of layers n remained constant. Qualitatively similar results have been obtained with other quasi-spherical molecules such as cyclohexane (Israelachvili et al., 1988). The shear stresses are only weakly dependent on the sliding velocity v. However, for sliding velocities above some critical value v_c, the stick-slip disappears and sliding proceeds smoothly in the purely kinetic value.

transitions can occur between n layers and $(n − 1)$ or $(n + 1)$ layers, and the details of the motion depend critically on the externally applied load, the temperature, the sliding velocity, the twist angle between the two surface lattices and the sliding direction relative to the lattices.

9.12.1 Smooth and Stick-Slip Sliding

Recent advances in friction-measuring techniques have enabled the interfacial friction of molecularly thin films to be measured with great accuracy. Some of these advances have involved the SFA technique (Israelachvili et al., 1988; Gee et al., 1990; Hirz et al., 1992; Homola *et al.*, 1989, 1990, 1993), while others have involved the AFM (McClelland, 1989; McClelland and Cohen, 1990). In addition, molecular dynamics computer simulations (Schoen et al., 1989; Landman et al., 1990; Thompson and Robbins, 1990; Robbins and Thompson, 1991) have become sufficiently sophisticated to enable fairly complex tribological systems to be studied for the first time. All these advances are necessary if one is to probe such subtle effects as smooth or stick-slip friction, transient and memory effects, and ultralow friction mechanisms at the molecular level.

Figure 9.24 shows typical results for the friction traces measured as a function of time (after commencement of sliding) between two molecularly smooth mica surfaces separated by three molecular layers of the liquid OMCTS, and how the friction increases to higher values in a quantized way when the number of layers falls from $n = 3$ to $n = 2$ and then to $n = 1$.

With the many added insights provided by recent computer simulations of such systems, a number of distinct molecular processes have been identified during smooth and stick-slip sliding. These are shown schematically in Figure 9.25 for the case of spherical liquid molecules between two solid crystalline surfaces. The following regimes may be identified:

Surfaces at rest — Figure 9.25a: Even with no externally applied load, solvent–surface epitaxial interactions can induce the liquid molecules in the film to solidify. Thus at rest the surfaces are stuck to each other through the film.

Sticking regime (frozen, solidlike film) — Figure 9.25b: A progressively increasing lateral shear stress is applied. The film, being solid, responds elastically with a small lateral displacement and a small increase or "dilatency" in film thickness (less than a lattice spacing or molecular dimension, σ). In this regime the film retains its frozen, solidlike state — all the strains are elastic and reversible, and the surfaces remain effectively stuck to each other. However, slow creep may occur over long time periods.

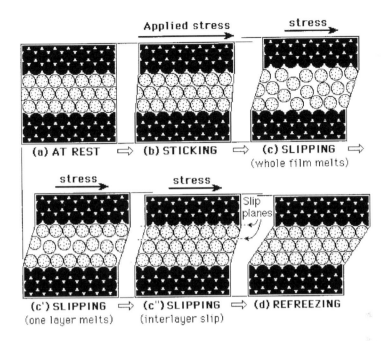

FIGURE 9.25 Idealized schematic illustration of molecular rearrangements occurring in a molecularly thin film of spherical or simple chain molecules between two solid surfaces during shear. Note that, depending on the system, a number of different molecular configurations within the film are possible during slipping and sliding, shown here as stages (c) — total disorder as whole film melts, (c′) — partial disorder, and (c″) — order persists even during sliding with slip occurring at a single slip plane either within the film or at the walls.

Slipping and sliding regimes (molten, liquidlike film) — Figure 9.25c, c′, c″: When the applied shear stress or force has reached a certain critical value F_s, the *static* friction force, the film suddenly melts (known as shear melting) or rearranges to allow for wall slip or film-slip to occur, at which point the two surfaces begin to slip rapidly past each other. If the applied stress is kept at a high value, the upper surface will continue to slide indefinitely.

Refreezing regime (resolidification of film) — Figure 9.25d: In many practical cases, the rapid slip of the upper surface relieves some of the applied force, which eventually falls below another critical value F_k, the kinetic friction force, at which point the film resolidifies and the whole stick-slip cycle is repeated. On the other hand, if the slip rate is smaller than the rate at which the external stress is applied, the surfaces will continue to slide smoothly in the kinetic state and there will be no more stick-slip. (Figure 9.26). The critical velocity at which stick-slip disappears is discussed in more detail in Section 9.13.

Experiments with linear-chain (alkane) molecules show that the film thickness remains quantized during sliding, so that the structure of such films is probably more like that of a nematic liquid crystal where the liquid molecules have become shear aligned in some direction enabling shear motion to occur while retaining some order within the film. Computer simulations for simple spherical molecules (Thompson and Robbins, 1990) further indicate that during the slip, the film thickness is roughly 15% higher than at rest (i.e., the film density falls), and that the order parameter within the film drops from 0.85 to about 0.25. Both of these are consistent with a disorganized liquidlike state for the whole film during the slip, as illustrated schematically in Figure 9.25c. At this stage, we can only speculate on other possible configurations of molecules in the slipping and sliding regimes. This probably depends on the shapes of the molecules (e.g., whether spherical or linear or branched), on the atomic structure of the surfaces, on the sliding velocity, etc. Figure 9.25c, c′, and c″ show three possible sliding modes wherein the shearing film either totally melts or where the molecules retain their layered structure and where slip occurs between two or more layers. Other sliding modes, for example, involving the movement of dislocations or disclinations are also possible, and it is unlikely that one single mechanism applies in all cases.

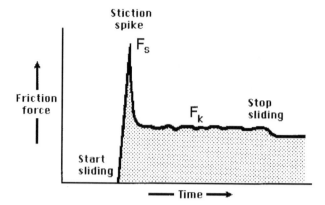

FIGURE 9.26 Stiction is the high starting frictional force F_s experienced by two moving surfaces which causes them to jerk forward rather than accelerate smoothly from rest. It is a major cause of surface damage and wear. The figure shows a typical "stiction spike" or "starting spike," followed by smooth sliding in the kinetic state.

TABLE 9.4   Effect of Molecular Shape and Short-Range Forces on Tribological Properties[a]

Liquid (Dry)	Short-Range Force	Type of Friction	Friction Coefficient	Bulk Liquid Viscosity (cP)
		Organic		
Cyclohexane	Oscillatory	Quantized stick-slip	»1	0.6
Octane	Oscillatory	Quantized stick-slip	1.5	0.5
Tetradecane	Oscil↔smooth	Stick-slip↔smooth	1.0	2.3
Octadecane (branched)	Oscil↔smooth	Stick-slip↔smooth	0.3	5.5
PDMS (M = 3700, melt)	Oscil↔smooth	Smooth	0.4	50
PBD (M = 3500, branched)	Smooth	Smooth	0.03	800.0
		Water		
Water (KCl solution)	Smooth	Smooth	0.01–0.03	1.0

[a]For molecularly thin liquid films between two shearing mica surfaces at 20°C.
Note: PDMS: Polydimethylsiloxane, PBD: Polybutadiene.

9.12.2   Role of Molecular Shape and Liquid Structure

The above scenario is already quite complicated, and yet this is the situation for the simplest type of experimental system. The factors that appear to determine the critical velocity v_c depend on the type of liquid between the surfaces (as well as on the surface lattice structure). Small spherical molecules such as cyclohexane and OMCTS have been found to have very high v_c, which indicates that these molecules can rearrange relatively quickly in thin films. Chain molecules and especially branched chain molecules have been found to have much lower v_c, which is to be expected, and such liquids tend to slide smoothly rather than in a stick-slip fashion (see Table 9.4). With highly asymmetric molecules, such as multiply branched isoparaffins and polymer melts, no regular spikes or stick-slip behavior occurs at any speed since these molecules can never order themselves sufficiently to "solidify." Examples of such liquids are perfluoropolyethers and polydimethylsiloxanes (PDMS).

Table 9.4 shows the trends observed with some organic and polymeric liquid between smooth mica surfaces. Also listed are the bulk viscosities of the liquids. From the data of Table 9.4 it appears that there is a direct correlation between the shapes of molecules and their coefficient of friction or effectiveness as lubricants (at least at low shear rates). Small spherical or chain molecules have high friction with stick-slip because they can pack into ordered solidlike layers. In contrast, longer-chained and irregularly shaped molecules remain in an entangled, disordered, fluidlike state even in very thin films and these give low friction and smoother sliding. It is probably for this reason that irregularly shaped branched chain molecules are usually better lubricants. It is interesting to note that the friction coefficient generally decreases as the bulk viscosity of the liquids *increases*. This unexpected trends occurs because the factors that are conducive to low friction are generally conducive to high viscosity. Thus, molecules with side groups such as branched alkanes and polymer melts usually have higher bulk viscosities than their linear homologues for obvious reasons. However, in thin films the linear molecules have higher shear stresses because of their ability to become ordered. The only exception to the above correlations is water, which has been found to exhibit both low viscosity *and* low friction (see Figure 9.19). In addition, the presence of water can drastically lower the friction and eliminate the stick-slip of hydrocarbon liquids when the sliding surfaces are hydrophilic.

If an effective viscosity η_{eff} were to be calculated for the liquids of Table 9.1, the values would be many orders of magnitude higher than those of the bulk liquids. This can be demonstrated by the following simple calculation based on the usual equation for Couette flow (see Equation 9.24):

$$\eta_{eff} = F_k D / A v, \qquad (9.46)$$

where F_k is the kinetic friction force, D is the film thickness, A the contact area, and v the sliding velocity. By using typical values for experiments with hexadecane (Yoshizawa and Israelachvili, 1993) — $F_k = 5$ mN, $D = 1$ nm, $A = 3 \times 10^{-9}$ m^2 and $v = 1$ μm/s — yields $\eta_{eff} \approx 2000$ N m^{-2} s, or 20,000 P, which is $\sim 10^6$ times higher than the bulk viscosity η_{bulk} of the liquid. It is instructive to consider that this very high effective viscosity nevertheless still produces a low friction force or friction coefficient μ of about 0.25. It is interesting to speculate that if a 1-nm film were to exhibit bulk viscous behavior, the friction coefficient under the same sliding conditions would be as low as 0.000001. While such a low value has never been reported for any tribological system, one may consider it as a theoretical lower limit that could, conceivably, be attained under certain experimental conditions.

Various studies (Van Alsten and Granick, 1990a,b, 1991; Granick, 1991; Hu and Granick, 1992) have shown that confinement and load generally increase the effective viscosity and/or relaxation times of molecules, suggestive of an increased glassiness or solidlike behavior. This is in marked contrast to studies of liquids in small confining capillaries where the opposite effects have been observed (Warnock et al., 1986; Awschalom and Warnock, 1987). The reason for this is probably because the two modes of confinement are different. In the former case (confinement of molecules between two structured solid surfaces) there is generally little opposition to any lateral or vertical displacement of the two surface lattices relative to each other. This means that the two lattices can shift in the *x-y-z* plane (Figure 9.27A) to accommodate the trapped molecules in the most crystallographically commensurate or "epitaxial" way, which would favor an ordered, solidlike state. In contrast, the walls of capillaries are rigid and cannot easily move or adjust to accommodate the confined molecules (Figure 9.27B), which will therefore be forced into a more disordered, liquidlike state (unless the capillary wall geometry and lattice is *exactly* commensurate with the liquid molecules, as occurs in certain zeolites).

Experiments have demonstrated the effects of surface lattice mismatch on the friction between surfaces (Hirano et al., 1991; Berman, 1996). Similar to the effects of lattice mismatch on adhesion (Figure 9.5, Section 9.4.1), the static friction of a confined liquid film is maximum when the lattices of the confining surfaces are aligned. For OMCTS confined between mica surfaces (Berman, 1996) the static friction was found to vary by more than a factor of 4 (Figure 9.28), while for bare mica surfaces the variation was by a factor of 3.5 (Hirano et al., 1991). In contrast to the sharp variations in adhesion energy over small twist angles, the variations in friction as a function of twist angle were much more broad — both in magnitude and angular spread. Similar variations in friction as a function of twist or misfit angles have also been observed in computer simulations (Gyalog and Thomas, 1997).

With rough surfaces, i.e., those that have *random* protrusions rather than being periodically structured, we expect a smearing out of the correlated intermolecular interactions that are involved in film freezing and melting (and in phase transitions in general). This should effectively eliminate the highly regular

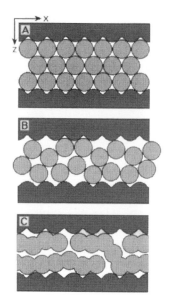

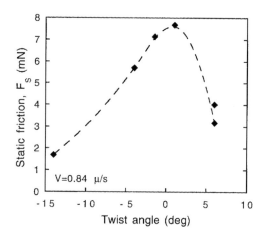

FIGURE 9.27 Schematic view of interfacial film composed of spherical molecules under a compressive pressure between two solid crystalline surfaces. If the two surface lattices are free to move in the *XYZ* directions so as to attain the lowest energy state, they could equilibrate at values of *X*, *Y*, and *Z* which induce the trapped molecules to become "epitaxially" ordered into a solidlike film. (B) Similar view of trapped molecules between two solid surfaces that are not free to adjust their positions, for example, as occurs in capillary pores or in brittle cracks. (C) Similar to (A) but with chain molecules replacing the spherical molecules in the gap. These may not be able to order as easily as do spherical molecules even if *X*, *Y*, and *Z* can adjust, resulting in a situation that is more akin to (B).

FIGURE 9.28 Static friction of a 2-nm-thick OMCTS film as a function of the lattice twist angle between the two confining crystalline mica surfaces. The variation in friction is comparable to variations in an unlubricated mica–mica system (Hirano et al., 1991).

stick-slip and may also affect the location of the slipping planes. The stick-slip friction of "real" surfaces, which are generally rough, may therefore be quite different from those of perfectly smooth surfaces composed of the same material (see next section). We should note, however, that even between rough surfaces, most of the contacts occur between the tips of microscopic asperities, which may be smooth over their microscopic contact area.

9.13 Stick-Slip Friction

An understanding of stick-slip is of great practical importance in tribology (Rabinowicz, 1965) since these spikes are the major cause of damage and wear of moving parts. But stick-slip motion is a much more common phenomenon and is also the cause of sound generation (the sound of a violin string, a squeaking door, or the chatter of machinery), sensory perception (taste, texture, and feel), earthquakes, granular flow, nonuniform fluid flow, such as the "spurting" flow of polymeric liquids, etc. In the previous section the stick-slip motion arising from freezing–melting transitions in thin interfacial films was described. But there are other mechanisms that can give rise to stick-slip friction, which will now be considered. However, before proceeding with this, it is important to clarify exactly what one is measuring during a friction experiment.

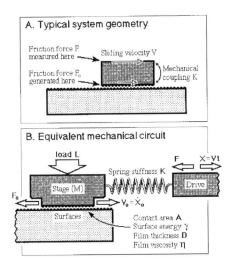

A. Typical system geometry

Friction force F
measured here — Sliding velocity V

Friction force F_0
generated here

Mechanical
coupling K

B. Equivalent mechanical circuit

load L

Spring stiffness K

F   X=Vt

Stage (M)

Drive

F_0

$V_0 = \dot{X}_0$

Surfaces

Contact area A
Surface energy γ
Film thickness **D**
Film viscosity η

FIGURE 9.29 (A) Schematic geometry of two shearing surfaces illustrating how the friction force, F_0, which is generated at the surfaces, is generally measured as F at some other place. The mechanical coupling between the two may be described in terms of a simple elastic stiffness or compliance, K, or in terms of more complex nonelastic coefficients, depending on the system. Here, the mechanical coupling is simply via the backing material supporting one of the surfaces. (B) Equivalent mechanical circuit for the above setup applicable to most tribological systems. Note that F_0 is the force generated at the surfaces, but that the measured or detected force is $F = (X - X_0)K$. The differences between the forces, the velocities and the displacements at the surfaces and at the drive (or detector) are illustrated graphically in Figures 9.31 through 9.34.

Figure 9.29 shows the basic mechanical coupling and equivalent mechanical circuit characteristic of most tribological systems and experiments. The distinction between F and F_0 is important because in almost all practical cases, the applied, measured, or detected force, F, is *not* the same as the "real" or "intrinsic" friction force, F_0, generated at the surfaces. F and F_0 are coupled in a way that depends on the mechanical construction of the system, for example, the axle of a car wheel which connects it to the engine. In Figure 9.29A, this coupling is shown to act via the material supporting the upper surface, which can be modeled (Figure 9.29B) as an elastic spring of stiffness K and mass M. This is the simplest type of mechanical coupling and is also the same as in SFA and AFM experiments, illustrated in Figure 9.30. More-complicated real systems can be reduced to a system of springs and dashpots as described by Luengo et al. (1997).

We now consider four different models of stick-slip friction. These are illustrated in Figures 9.31 through 9.34, where the mechanical couplings are assumed to be of the simple elastic spring type as

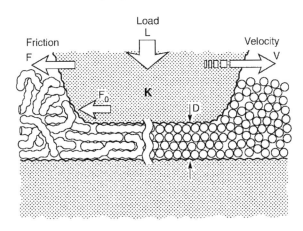

Load
L

Friction   Velocity

F   V

F_0

K

D

FIGURE 9.30 (A) Schematic geometry of two contacting asperities separated by a thin liquid film. This is also the geometry adopted in most pin-on-disk, SFA, and AFM experiments. In the SFA experiments described here, typical experimental values were: undeformed radius of surfaces, $R \approx 1$ cm; radius of contact area, $r = 10$ to 40 μm; film thickness, $D \approx 10$Å; externally applied load, $L = -10$ to $+100$ mN; measured friction forces, $F = 0.001$ to 100 mN; sliding or driving velocity, $V = 0.001$ to 100 μm/s; surface or interfacial energy, $\gamma = 0$ to 30 mJ/m² (erg/cm²); effective elastic constant of supporting material, $K = 10^8$ N/m; elastic constant of friction-measuring spring, $K = 500$ N/m; temperature, $T = 15$ to 40°C.

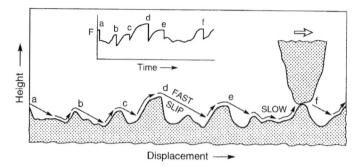

FIGURE 9.31 Figures 9.31 through 9.34 show three different models of friction and the stick-slip friction force vs. time traces they give rise to. Figure 9.31 shows the Rabinowicz model (Rabinowicz, 1965) for rough surfaces which produces irregular stick-slip (inset) when the elastic stiffness of the system (reflected by the slopes of the SLIP lines) is not too high.

shown in Figures 9.29 and 9.30. The first two mechanisms (Figures 9.31 and 9.33) may be considered as the "traditional" or "classical" mechanisms or models (Rabinowicz, 1965), the third (Figure 9.34) is essentially the same as the freezing–melting phase-transition model described in Section 9.12.

9.13.1 Rough Surfaces Model

Rapid slips can occur whenever an asperity on one surface goes over the top of an asperity on the other surface. As shown in Figure 9.31, the extent of the "slip" will depend on asperity heights and slopes, on the speed of sliding, and on the elastic compliance of the surfaces and the moving stage. We may note that, as in all cases of stick-slip motion, the driving velocity (V) may be constant but the resulting motion at the surfaces (V_0) will display large slips as shown in the inset. This type of stick-slip has been described by Rabinowicz (1965). It will not be of much concern here since it is essentially a noise-type fluctuation, resulting from surface imperfections rather than from the intrinsic interaction between two surfaces. Actually, at the atomic level, the regular atomic-scale corrugations of surfaces can lead to periodic stick-slip motion of the type shown here. This is what is sometimes measured by AFM tips (McClelland, 1989; McClelland and Cohen, 1990).

9.13.2 Distance-Dependent Model

Another theory of stick-slip, observed in solid-on-solid sliding, is one that involves a characteristic *distance* (but also a characteristic time, τ_s, this being the characteristic time required for two asperities to increase their adhesion strength after coming into contact). Originally proposed by Rabinowicz (1958, 1965), this model suggests that two rough macroscopic surfaces adhere through their microscopic asperities of characteristic length D_c. During shearing, each surface must first creep a distance D_c — the size of the contacting junctions — after which the surfaces continue to slide, but with a lower (kinetic) friction force than the original (static) value. The reason for the decrease in the friction force is that even though, on average, new asperity junctions should form as rapidly as the old ones break, the time-dependent adhesion and friction of the new ones will be lower than the old ones. This is illustrated in Figure 9.32.

The friction force therefore remains high during the creep stage of the slip, but once the surfaces have moved the characteristic distance D_c, the friction rapidly drops to the kinetic value. For any system where the kinetic friction is less than the static force (or one that has a negative slope over some part of its F_0–V_0 curve) will exhibit regular stick-slip sliding motion for certain values of K, m, and driving velocity, V.

This type of friction has been observed in a variety of dry (unlubricated) systems such as paper-on-paper (Baumberger et al., 1994; Heslot et al., 1994) and steel-on-steel (Sampson et al., 1943; Heymann et al., 1954; Rabinowicz, 1958). This model is also used extensively in geologic systems to analyze rock-on-rock sliding (Dieterich, 1978, 1979).

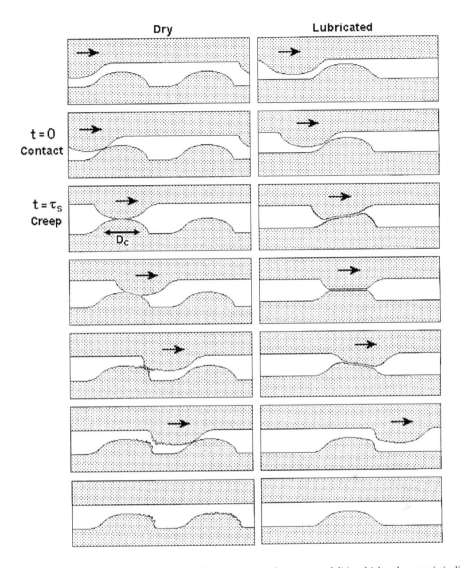

Dry   **Lubricated**

t = 0
Contact

t = τ_s
Creep

D_c

FIGURE 9.32 Distance-dependent friction model (also known as the creep model) in which a characteristic distance D_c has to be moved to break adhesive junctions. The model also has a characteristic, τ_s, this being the time needed for the adhesion and friction forces per junction to equilibrate after each contact is made.

While originally described for adhering macroscopic asperity junctions, the distance-dependent model may also apply to molecularly smooth surfaces. For example, for polymer lubricant films, the characteristic length D_c would now be the chain–chain entanglement length, which could be much larger in a confined geometry than in the bulk.

9.13.3 Velocity-Dependent Friction Model

This is the most-studied mechanism of stick-slip and, until recently, was considered to be the only cause of intrinsic stick-slip. If a friction force decreases with increasing sliding velocity, as occurs with boundary films exhibiting shear thinning, the force (F_s) needed to initiate motion will be higher than the force (F_k) needed to maintain motion. Such a situation is depicted in Figure 9.33 (Case c), where a decreasing intrinsic friction force F_0 with sliding velocity V_0 results in the sliding surface or stage moving in a periodic fashion where during each cycle rapid acceleration is followed by rapid deceleration (see curves for X_0

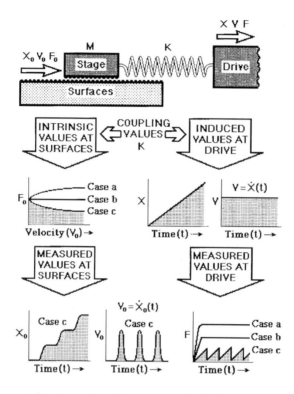

FIGURE 9.33 The lower part of the figure shows the stage or surface displacement X_0, surface velocity $V_0 (= dX_0/dt$ or $\dot{X}_0$), and measured friction force F, as functions of time t for surfaces whose intrinsic friction force is F_0. In general, F_0 is a function of X_0, V_0, and t. Three specific examples are shown here corresponding to F_0 either increasing monotonically (Case a), remaining constant (Case b), or decreasing monotonically (Case c) with V_0. The latter case corresponds to a film that exhibits "shear thinning." Only when $F_0(V_0)$ has a negative slope is the resulting motion of the stick-slip type, characterized by very different motions and friction forces being detected at the surfaces (the Stage) and the detector (the Drive).

and V_0 in Figure 9.33). So long as the drive continues to move at a fixed velocity V, the surfaces will continue to move in a periodic fashion punctuated by abrupt stops and starts whose frequency and amplitude depend not only on the function $F_0(V_0)$ but also on the stiffness K and mass M of the moving stage, and on the starting conditions at $t = 0$.

More precisely, referring to Figures 9.29 and 9.33, the motion of the sliding surface or stage can be determined by solving the following differential equation:

$$M\ddot{X}_0 = \left(F_0 - F\right) = F_0 - \left(X_0 - X\right)K \tag{9.47}$$

or

$$M\ddot{X}_0 + \left(X_0 - X\right)K - F_0 = 0, \tag{9.48}$$

where $F_0 = F_0(X_0, V_0, t)$ is the intrinsic or real friction force at the shearing surfaces which is generally a function of X_0, $V_0 = \dot{X}_0$ and t, F is the force on the spring (the externally applied or measured force), and $F_s = (F_0 - F)$ is the force on the stage. To solve fully Equation 9.47, one must also know the initial (starting) conditions at $t = 0$, and the driving or steady-state conditions at finite t. For example, in the present experiments, the driving condition is

$$X = 0 \qquad \text{for } t < 0,$$
$$X = Vt \qquad \text{for } t > 0, \text{ where } V = \text{constant.} \qquad (9.49)$$

In other systems, the appropriate driving condition may be $F = $ constant.

Various forms for $F_0 = F_0(X_0, V_0, t)$ have been proposed, mainly phenomenological, to explain various kinds of stick-slip phenomena. These models generally assume a particular functional form for the friction as a function of velocity only, $F_0 = F_0(V_0)$, and they may also contain a number of mechanically coupled elements comprising the stage (Tomlinson, 1929; Carlson and Langer, 1989). One version is a two-state model characterized by two friction forces, F_s and F_k, which is a simplified version of the phase-transitions model (next section). More complicated versions can have a rich F–v spectrum, as proposed by Persson (1994). Unless the experimental data is very detailed and extensive, these models cannot generally distinguish between different types of mechanisms. Neither do they address the basic question of the *origin* of the friction force, since this is assumed to begin with.

Experimental data have been used to calculate the friction force as a function of velocity *within* an individual stick-slip cycle (Nasuno et al., 1997). For a macroscopic granular material confined between solid surfaces, the data show a velocity-weakening friction force during the first half of the slip. However, the data also show a hysteresis loop in the friction–velocity plot, with a different behavior in the deceleration half of the slip phase. Similar results were observed for a 1 to 2 nm liquid lubricant film between mica surfaces (Berman, Carlson and Ducker, unpublished results). These results indicate that a purely velocity-dependent friction law is insufficient to describe such systems, and an additional element such as the *state* of the confined material must be considered (next section).

9.13.3 Phase Transitions Model

Recent molecular dynamics computer simulations have found that thin interfacial films undergo first-order phase transitions between solidlike and liquidlike states during sliding (Thompson and Robbins, 1990; Robbins and Thompson, 1991) and have suggested this is responsible for the observed stick-slip behavior of simple isotropic liquids between two solid crystalline surfaces. With this interpretation, stick-slip is seen to arise because of the abrupt change in the flow properties of a film at a transition (Israelachvili et al., 1990; Thompson et al., 1992) rather than the gradual or continuous change as occurs in the previous example. Such simulations have accounted for many of the observed properties of shearing liquids in ultrathin films between molecularly smooth surfaces, and have so far offered the most likely explanation for experimental data on stick-slip friction, such as shown in Figure 9.36.

A novel interpretation of the well-known phenomenon of decreasing coefficient of friction with increasing sliding velocity has been proposed by Thompson and Robbins (1990) based on their computer simulation. This postulates that it is not the friction that changes with sliding speed v, but rather the time various parts of the system spend in the sticking and sliding modes. In other words, at any instant during sliding, the friction at any local region is always F_s or F_k, corresponding to the static or kinetic values. The measured frictional force, however, is the sum of all these discrete values averaged over the whole contact area. Since as v increases, each local region spends more time in the sliding regime (F_k) and less in the sticking regime (F_s) the overall friction coefficient falls. One may note that this interpretation reverses the traditional way that stick-slip has been explained, for rather than invoking a decreasing friction with velocity to explain stick-slip, it is now the more fundamental stick-slip phenomenon that is producing the apparent decrease in the friction force with increasing sliding velocity. This approach has been studied analytically by Carlson and Batista (1996), with a comprehensive rate- and state-dependent friction force law. This model includes an analytic description of the freezing–melting transitions of a film, resulting in a friction force that is a function of sliding velocity in a natural way. This model predicts a full range of stick-slip behavior observed experimentally.

An example of the rate- and state-dependent model is observed when shearing thin films of OMCTS between mica surfaces (Berman et al., 1996a,b). In this case the static friction between the surfaces is dependent on the time that the surfaces are at rest with respect to each other, while the intrinsic kinetic friction F_{k0} is relatively constant over the range of velocities (Figure 9.35). At slow driving velocities, the

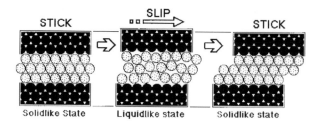

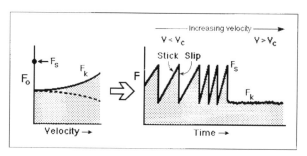

FIGURE 9.34 Phase-transition model of stick-slip, for example, where a thin liquid film alternately freezes and melts as it shears, shown here for 22 spherical molecules confined between two solid crystalline surfaces. This model differs from that of Figure 9.33 in that the intrinsic friction force F_0 is here assumed to change abruptly (at the transitions) rather than smoothly or continuously. The resulting stick-slip is also different, for example, the peaks are sharper and the stick-slip disappears above some critical velocity, V_c. Note that while the slip displacement is here shown to be only two lattice spacings; in most practical situations it is much larger.

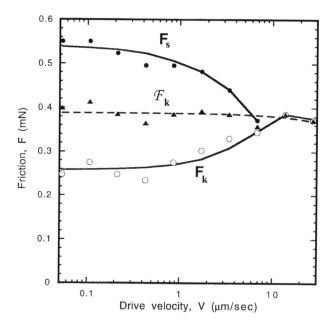

FIGURE 9.35 Measured friction F_s and F_k for increasing drive velocity V with OMCTS confined between two mica surfaces. F_s (solid circles) decreases with increasing velocity because at higher drive speeds the sticking time is shorter, resulting in less complete freezing of the lubricant layer. The observed F_k (open circles) increases as F_s decreases, resulting in a nearly constant intrinsic friction force F_{k0} (triangles) $\approx (F_s + F_k)/2$ for these underdamped conditions. The critical velocity is reached when F_s decreases to the intrinsic friction F_{k0}.

system responds with stick-slip sliding with the surfaces reaching maximum static friction before each slip event, and the amplitude of the stick-slip, $F_s - F_k$, is relatively constant. As the driving velocity increases, the static friction decreases as the time at relative rest becomes shorter with respect to the characteristic time of the lubricant film. As the static friction decreases with increasing drive velocity, it eventually equals the intrinsic kinetic friction F_{k0}, which defines the critical velocity V_c, above which the surfaces slide smoothly, without the jerky stick-slip motion.

The above classifications of stick-slip are not exclusive, and molecular mechanisms of real systems may exhibit aspects of different models simultaneously. They do, however, provide a convenient classification of existing models and indicate which experimental parameters should be varied to test the different models.

9.13.4   Critical Velocity for Stick-Slip

For any given set of conditions, stick-slip disappears above some critical sliding velocity V_c, above which the motion continues smoothly in the liquidlike or kinetic state. The critical velocity is found to be well described by two simple equations. Both are based on the phase-transition model, and both include some parameter associated with the inertia of the measuring instrument. The first equation is based on both experiments and simple theoretical modeling (Yoshizawa and Israelachvili, 1993):

$$
V_c \approx \frac{\left(F_s - F_k\right)}{5K\tau_0},
\tag{9.50}
$$

where τ_0 is the *characteristic nucleation time* or freezing time of the film. For example, inserting the following typically measured values for $a \sim 10$ Å thick hexadecane film between mica: $(F_s - F_k) \approx 5$ mN, spring constant $K \approx 500$ N/m, and nucleation time (Yoshizawa and Israelachvili, 1993) $\tau_0 \approx 5$ s, we obtain $V_c \approx 0.4$ μm/s, which is close to typically measured values (Figure 9.36). The second equation is based on computer simulations (Robbins and Thompson, 1991):

$$
V_c \approx 0.1 \sqrt{\frac{\Delta F \cdot \sigma}{M}},
\tag{9.51}
$$

where σ is a molecular dimension and M is the mass of the stage. Again, inserting typical experimental values into this equation, specifically, $M \approx 20$ gm, $\sigma \approx 0.5$ nm, and $(F_s - F_k) \approx 5$ mN as before, we obtain $V_c \approx 0.3 \propto$μm/s, which is also close to measured values.

Stick-slip also disappears above some critical temperature T_c, which is not the same as the melting temperature of the bulk fluid. Certain correlations have been found between V_c and T_c, and between various other tribological parameters, that appear to be consistent with the principle of "time-temperature superposition," similar to that occurring in viscoelastic polymer fluids (Ferry, 1980). We end by considering these correlations.

9.13.5   Dynamic Phase Diagram Representation of Tribological Parameters

Both friction and adhesion hysteresis vary nonlinearly with temperature, often peaking at some particular temperature, T_0. The temperature dependence of these forces can therefore be represented on a dynamic phase diagram such as shown in Figure 9.37. Experiments have shown that T_0, and the whole bell-shaped curve, are shifted along the temperature axis (as well as in the vertical direction) in a systematic way when the load, sliding velocity, etc. are varied. These shifts also appear to be highly correlated with one another, for example, an increase in temperature producing effects that are similar to *decreasing* the sliding speed or load.

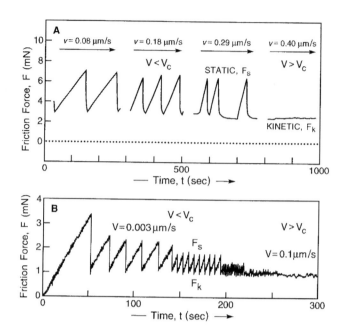

FIGURE 9.36 Exact reproductions of chart-recorded traces of friction forces at increasing sliding velocities V (in µm/s) plotted as a function of time for a hydrocarbon liquid (A) and a boundary monolayer (B). In general, with increasing sliding speed, the stick-slip spikes increase in frequency and decrease in magnitude. As the critical sliding velocity V_c is approached, the spikes become erratic, eventually disappearing altogether at $V = V_c$. At higher sliding velocities the sliding continues smoothly in the kinetic state. Such friction traces are fairly typical for simple liquid lubricants and dry boundary lubricant systems, and may be referred to as the "conventional" type of static-kinetic friction. (A) Liquid hexadecane ($C_{16}H_{34}$) film between two untreated mica surfaces. Experimental conditions (see Figure 9.30): contact area, $\pi r^2 = 4 \times 10^{-9}$ m^2; load, $L \approx 1$ g; film thickness, $D = 0.4$ to 0.8 nm; $V = 0.08$ to 0.4 µm/s; $V_c \approx 0.3$ µm/s; atmosphere: dry N_2 gas; $T = 18°C$. (B) Close-packed surfactant monolayers on mica (dry boundary lubrication) showing qualitatively similar behavior to that obtained above with a liquid hexadecane film. In this case, $V_c \approx 0.1$ µm/s.

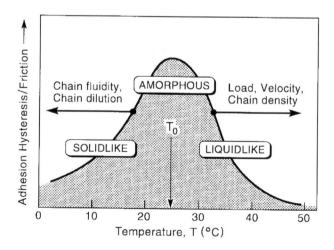

FIGURE 9.37 Schematic "friction phase diagram" representing the trends observed in the boundary friction of a variety of different surfactant monolayers (Yoshizawa et al., 1993). The characteristic bell-shaped curve also correlates with the monolayer adhesion energy hysteresis. The arrows indicate the direction in which the whole curve is dragged when the load, velocity, etc., are increased.

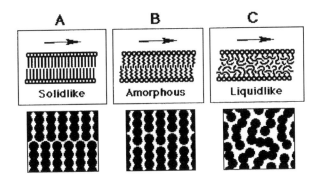

FIGURE 9.38 Different dynamic-phase states of boundary monolayers during adhesive contact and/or frictional sliding. Low adhesion hysteresis and friction is exhibited by solidlike and liquidlike layers (A and C). High adhesion hysteresis and friction is exhibited by amorphous layers (B). Increasing the temperature generally shifts a system from the left to the right. Changing the load, sliding velocity and other experimental conditions can also change the dynamic-phase state of surface layers as shown in Figure 9.37.

Such effects are also commonly observed in other energy-dissipating phenomena such as polymer viscoelasticity (Ferry, 1980), and it is likely that a similar physical mechanism is at the heart of all such phenomena. A possible molecular process underlying the energy dissipation of chain molecules during boundary layer sliding is illustrated in Figure 9.38, which shows the three main dynamic phase states of boundary monolayers.

Acknowledgment

This work was supported by ONR grant N00014-931D269.

References

Andrade, J. D., Smith, L. M., and Gregonis, D. E. (1985), "The Contact Angle and Interface Energetics," in *Surface and Interfacial Aspects of Biomedical Polymers*, Vol. 1 (J.D. Andrade, ed.), pp. 249–292, Plenum, New York.

Attard, P. and Batchelor, M. T. (1988), "A Mechanism for the Hydration Force Demonstrated in a Model System," *Chem. Phys. Lett.*, 149, 206–211.

Awschalom, D. D. and Warnock, J. (1987), "Supercooled Liquids and Solids in Porous Glass," *Phys. Rev. B*, 35, 6792–6799.

Barquins, M. and Maugis, D. (1982), "Adhesive Contact of Axisymmetric Punches on an Elastic Half-Space: The Modified Hertz-Herber's Stress Tensor for Contacting Spheres," *J. Mec. Theor. Appl.* 1, 331–357.

Baumberger, T., Heslot, F., and Perrin., B. (1994), "Crossover from Creep to Inertial Motion in Friction Dynamics," *Nature*, 367, 544–546.

Berman, A. (1996), "Dynamics of Molecules at Surface," Ph.D. dissertation, University of California, Santa Barbara.

Berman, A. D. and Israelachvili, J. N. (1997), "Control and Minimization of Friction via Surface Modification," in *Micro/Nanotribology and Its Applications*, B. Bhushan, ed., NATO Advanced Institute Series, Kluwer Academic Publishers, Dordrecht, pp. 317–329.

Berman, A., Ducker, W., and Israelachvili, J. (1996a), "Origin and Characterization of Different Stick-Slip Friction Mechanisms," *Langmuir* 12, 4559–4563.

Berman, A., Ducker, W., and Israelachvili, J. (1996b), "Experimental and Theoretical Investigations of Stick-Slip Friction," in *Physics of Sliding Friction* (B. Persson and E. Tosati, eds.), NATO Advanced Science Institute Series, Kluwer Academic Publishers, Dordrecht, pp. 51–67.

Berman, A., Drummond, C., and Israelachvili, J. (1998a), "Amontons' Law at the Molecular Level," *Tribol. Lett.*, 4, 95–101.

Berman, A., Steinberg, S., Campbell, S., Ulman, A., and Israelachvili, J. (1998b), "Controlled Microtribology of a Metal Oxide Surface," *Tribol. Lett.*, 4, 43–48.

Bowden, F. P. and Tabor, D. (1967), *Friction and Lubrication*, 2nd ed., Methuen, London.

Briscoe, B. J. and Evans, D. C. (1982), "The Shear Properties of Langmuir Blodgett Layers," *Proc. R. Soc. London*, A380, 389–407.

Briscoe, B. J., Evans, D. C., and Tabor, D. (1977), "The Influence of Contact Pressure and Saponification on the Sliding Behavior of Steric Acid Monolayers," *J. Colloid Interface Sci.*, 61, 9–13.

Campbell, S. E., Luengo, G., Srdanov, V. I., Wudl, F., and Israelachvili, J. N. (1996), "Very Low Viscosity at the Solid-Liquid Interface Induced by Adsorbed C_{60} Monolayers," *Nature* 382, 520–522.

Carlson, J. M. and Batista, A. A. (1996), "A Constituative Relation for the Friction between Lubricated Surfaces," *Phys. Rev. E*, 53, 4153–4165.

Carlson, J. M. and Langer, J. S. (1989), "Mechanical Model of an Earthquake Fault," *Phys. Rev. A*, 40, 6470–6484.

Chan, D. Y. C. and Horn, R. G. (1985), "The Drainage of Thin Liquid Films between Solid Surfaces," *J. Chem. Phys.*, 83, 5311–5324.

Chaudhury, M. K. and Whitesides, G. M. (1991), "Direct Measurement of Interfacial Interactions between Semispherical Lenses and Flat Sheets of Poly(Dimethylsiloxane) and Their Chemical Derivatives," *Langmuir*, 7, 1013–1025.

Chen, Y. L., Gee, M. L., Helm, C. A., Israelachvili, J. N., and McGuiggan, P. M. (1989), "Effects of Humidity on the Structure and Adhesion of Amphiphilic Monolayers on Mica," *J. Phys. Chem.*, 93, 7057–7059.

Chen, Y. L., Helm, C., and Israelachvili, J. N. (1991), "Molecular Mechanisms Associated with Adhesion and Contact Angle Hysteresis of Monolayer Surfaces," *J. Phys. Chem.*, 95, 10736–10747.

Chen, Y. L., Kuhl, T., and Israelachvili, J. N. (1992), "Mechanism of Cavitation Damage in Thin Liquid Films: Collapse Damage vs. Inception Damage," *Wear*, 153, 31–51.

Christenson, H. K. (1985), "Forces between Solid Surfaces in a Binary Mixture of Non-Polar Liquids," *Chem. Phys. Lett.*, 118, 455–458.

Christenson, H. K. (1988a), "Non-DLVO forces between surfaces — solvation, hydration and capillary effects," *J. Disp. Sci. Technol.*, 9, 171–206.

Christenson, H. K. (1988b), "Adhesion between Surfaces in Unsaturated Vapors — A Reexamination of the Influence of Meniscus Curvature and Surface Forces," *J. Colloid Interface Sci.*, 121, 170–178.

Christenson, H. K. and Horn, R. G. (1985), "Solvation Forces Measured in Non-Aqueous Liquids," *Chem. Scr.*, 25, 37–41.

Christenson, H. K., Gruen, D. W. R., Horn, R. G., and Israelachvili, J. N. (1987), "Structuring in Liquid Alkanes Between Solid Surfaces: Force Measurements and Mean-Field Theory," *J. Phys. Chem.* 87, 1834–1841.

Christenson, H. K., Claesson, P. M., Berg, J., and Herder, P. C. (1989), "Forces between Fluorocarbon Surfactant Monolayers: Salt Effects on the Hydrophobic Interaction," *J. Phys. Chem.* 93, 1472–1478.

Christenson, H. K., Fang, J., Ninham, B. W., and Parker, J. L. (1990), "Effect of Divalent Electrolyte on the Hydrophobic Attraction," *J. Phys Chem.* 94, 8004–8006.

Christou, N. I., Whitehouse, J. S., Nicholson, D., and Parsonage, N. G. (1981), "A Monte Carlo Study of Fluid Water in Contact with Structureless Walls," *Symp. Faraday Soc.* 16, 139–149.

Claesson, P. M. and Christenson, H. K. (1988), "Very Long-Range Attractive Forces between Uncharged Hydrocarbon and Fluorocarbon Surfaces in Water," *J. Phys. Chem.* 92, 1650–1655.

Claesson, P. M., Blom, C. E., Herder, P. C., and Ninham, B. W. (1986), "Interactions between Water-Stable Hydrophobic Langmuir-Blodgett Monolayers on Mica," *J. Colloid Interf. Sci.* 114, 234–242.

Coakley, C. J. and Tabor, D. (1978), "Direct Measurement of van der Waals Forces between Solids in Air," *J. Phys. D* 11, L77–L82.

Del Pennino, U., Mezzega, E., Valeri, S., Alietti, A., Brigatti, M. F., and Poppi, L. (1981), "Interlayer Water and Swelling Properties of Monoionic Montmorillonites," *J. Colloid Interf. Sci.* 84, 301–309.

Derjaguin, B. V. (1966), *Research in Surface Forces,* Vol. 2, p. 312, Consultants Bureau, New York.

Derjaguin, B. V. (1988), "Mechanical Properties of the Boundary Lubrication Layer," *Wear* 128, 19–27.

Dieterich, J. H. (1978), "Time-Dependent Friction and the Mechanics of Stick-Slip," *Pure Appl. Geophys.* 116, 790–806.

Dieterich, J. H. (1979), "Modeling of Rock Friction 1. Experimental Results and Constitutive Equations," *J. Geophys. Res.* 84, 2162–2168.

Dowson, D. (1979), *History of Tribology,* Longman, London.

Drake, B., Prater, C. B., Weisenhorn, A. L., Gould, S. A. C., Albrecht, T. R., Quate, C. F., Cannell, D. S., Hansma, H. G., and Hansma, P. K. (1989), "Imaging Crystals, Polymers, and Processes in Water with the Atomic Force Microscope," *Science* 243, 1586–1589.

Ducker, W. A., Senden, T. J., and Pashley, R. M. (1991), "Direct Measurement of Colloidal Forces Using an Atomic Force Microscope," *Nature* 353, 239–241.

Elimelech, M. (1990), "Indirect Evidence for Hydration Forces in the Deposition of Polystyrene Latex Colloids on Glass Surfaces," *J. Chem. Soc. Faraday Trans.* 86, 1623–1624.

Ellul, M. D. and Gent, A. N. (1984), "The Role of Molecular Diffusion in the Adhesion of Elastomers," *J. Polym. Sci. Polym. Phys.* 22, 1953–1968: (1985), "The Role of Molecular Diffusion in the Adhesion of EPDM and EPR Elastomers," *J. Polym. Sci. Polym. Phys.* 23, 1823–1850.

Evans, R. and Parry, A. O. (1990), "Liquids at Interfaces: What can a Theorist Contribute?" *J. Phys. (Condens. Matter)* 2, SA15–SA32.

Ferry, J. D. (1980), *Viscoelastic Properties of Polymers,* 3rd ed., John Wiley, New York.

Fisher, L. R. and Israelachvili, J. N. (1981), "Direct Measurements of the Effect of Meniscus Forces on Adhesion: A Study of the Applicability of Macroscopic Thermodynamics to Microscopic Liquid Interfaces," *Colloids Surf.* 3, 303–319.

Gee, M. L. and Israelachvili, J. N. (1990), "Interactions of Surfactant Monolayers across Hydrocarbon Liquids," *J. Chem. Soc. Faraday Trans.* 86, 4049–4058.

Gee, M., McGuiggan, P., Israelachvili, J., and Homola, A. (1990), "Liquid to Solid-Like Transitions of Molecularly Thin Films under Shear," *J. Phys. Chem.* 93, 1895–1906.

Granfeldt, M. K. and Miklavic, S. J. (1991), "A Simulation Study of Flexible Zwitterionic Monolayers. Interlayer Interaction and Headgroup Conformation," *J. Phys. Chem.* 95, 6351–6360.

Granick, S. (1991), "Motions and Relaxations of Confined Liquids," *Science* 253, 1374–1379.

Green, N. M. (1975), "Avidin," *Adv. Protein Chem.* 29, 85–133.

Greenwood, J. A. and Johnson, K. L. (1981), "The Mechanics of Adhesion of Viscoelastic Solids," *Philos. Mag. A* 43, 697–711.

Gruen, D. W. R. and Marcelja, S. (1983), "Spatially Varying Polarization in Water: A Model for the Electric Double Layer and the Hydration Force," *J. Chem. Soc. Faraday Trans II* 79, 225–242.

Gyalog, T. and Thomas, H. (1997), "Friction between Atomically Flat Surfaces," *Europhys. Lett.* 37, 195–200.

Han, K. K., Cushman, J. H., and Diestler, D. J. (1993), "Grand Canonical Monte Carlo Simulations of a Stockmayer Fluid in a Slit Micropore," *Mol. Phys.* 79, 537–545.

Helm, C. A., Israelachvili, J. N., and McGuiggan, P. M. (1989), "Molecular Mechanisms and Forces Involved in the Adhesion and Fusion of Amphiphilic Bilayers," *Science* 246, 919–922.

Helm, C., Knoll, W., and Israelachvili, J. N. (1991), "Measurement of Ligand-Receptor Interactions," *Proc. Natl. Acad. Sci. U.S.A.* 88, 8169–8173.

Henderson, D. and Lozada-Cassou, M. J. (1986), "A Simple Theory for the Force between Spheres Immersed in a Fluid," *J. Colloid Interf. Sci.* 114, 180–183.

Heslot, F., Baumberger, T., Perrin, B., Caroli, B., and Caroli, C. (1994), "Creep, Stick-Slip, and Dry-Friction Dynamics — Experiments and a Heuristic Model," *Phys. Rev. E* 49, 4973–4988.

Heymann, F., Rabinowicz, E., and Rightmire, B. (1954), "Friction Apparatus for Very Low-Speed Sliding Studies," *Rev. Sci. Instrum.* 26, 56–58.

Hirz, S. J., Homola, A. M., Hadziioannou, G., and Frank, C. W. (1992), "Effect of Substrate on Shearing Properties of Ultrathin Polymer Films," *Langmuir* 8, 328–333.

Holly, F. J. and Refojo, M. J. (1975), "Wettability of Hydrogels. I. Poly (2-Hydroxyethyl Methylcrylate)," *J. Biomed. Mater. Res.* 9, 315–326.

Homola, A. M., Israelachvili, J. N., Gee, M. L., and McGuiggan, P. M. (1989), "Measurements of and Relation between the Adhesion and Friction of Two Surfaces Separated by Thin Liquid and Polymer Films," *J. Tribol.* 111, 675–682.

Homola, A. M., Israelachvili, J. N., McGuiggan, P. M., and Gee, M. L. (1990), "Fundamental Experimental Studies in Tribology: The Transition from 'Interfacial' Friction of Undamaged Molecularly Smooth Surfaces to 'Normal' Friction with Wear," *Wear* 136, 65–83.

Homola, A. M., Nguyen, H. V., and Hadziioannou, G. (1993), *J. Phys. Chem.* 94, 2346.

Horn, R. G. (1990), "Surface Forces and Their Action in Ceramic Materials," *Am. Ceram. Soc.* 73, 1117–1135.

Horn, R. G. and Israelachvili, J. N. (1981), "Direct Measurement of Structural Forces between Two Surfaces in a Nonpolar Liquid," *J. Phys. Chem.* 75, 1400–1411.

Horn, R. G. and Israelachvili, J. N. (1988), "Molecular Organization and Viscosity of a Thin Film of Molten Polymer between Two Surfaces as Probed by Force Measurements," *Macromolecules* 21, 2836–2841.

Horn, R. G., Clarke, D. R., and Clarkson, M. T. (1988), "Direct Measurements of Surface Force between Sapphire Crystals in Aqueous Solutions," *J. Mater. Res.* 3, 413–416.

Horn, R. G., Hirz, S. J., Hadziioannou, G. H., Frank, C. W., and Catala, J. M. (1989a), "A Reevaluation of Forces Measured across Thin Polymer Films: Nonequilibrium and Pinning Effects," *J. Phys. Chem.* 90, 6767–6774.

Horn, R. G., Smith, D. T., and Haller, W. (1989b), "Surface Forces and Viscosity of Water Measured between Silica Sheets," *Chem. Phys. Lett.* 162, 404–408.

Hu, H.-W. and Granick, S. (1992), "Viscoelastic Dynamics of Confined Polymer Melts," *Science* 258, 1339–1342.

Hu, H. W., Carson, G., and Granick, S. (1991), "Relaxation Time of Confined Liquids under Shear," *Phys. Rev. Lett.* 66, 2758–2761.

Hunter, R. J. (1987), *Foundations of Colloid Science*, Vol. I, Ch. 7, Clarendon Press, Oxford, U.K.

Israelachvili, J. N. (1986a), "Measurement of the Viscosity of Liquids in Very Thin Films," *J. Colloid Interf. Sci.* 110, 263–271.

Israelachvili, J. N. (1986b), "Measurements of the Viscosity of Thin Films between Two Surfaces with and without Absorbed Polymers," *Colloid Polym. Sci.* 264, 1060–1065.

Israelachvili, J. N. (1987), "Solvation Forces and Liquid Structure — As Probed by Direct Force Measurements," *Acc. Chem. Res.* 20, 415–421.

Israelachvili, J. N. (1988), "Measurements and Relation between the Dynamic and Static Interactions between Surfaces Separated by Thin Liquid and Polymer Films," *Pure Appl. Chem.* 60, 1473–1478.

Israelachvili, J. N. (1989), "Techniques for Direct Measurements of Forces between Surfaces in Liquids at the Atomic Scale," *Chemtracts Anal. Phys. Chem.* 1, 1–12.

Israelachvili, J. N. (1991), *Intermolecular and Surface Forces*, 2nd ed., Academic Press, London.

Israelachvili, J. N. and Adams, G. E. (1978), "Measurement of Forces between Two Mica Surfaces in Electrolyte Solutions in the Range 0–100 nm," *J. Chem. Soc. Faraday Trans. I* 74, 975–1001.

Israelachvili, J. and Berman, A. (1995), "Irreversibility, Energy Dissipation, and Time Effects in Intermolecular and Surface Interactions," *Isr. J. Chem.* 35, 85–91.

Israelachvili, J. N. and Kott, S. J. (1988), "Liquid Structuring at Solid Interfaces as Probed by Direct Force Measurements: The Transition from Simple to Complex Liquids and Polymer Fluids," *J. Phys. Chem.* 88, 7162–7166.

Israelachvili, J. N. and Kott, S. (1989), "Shear Properties and Structure of Simple Liquids in Molecularly Thin Films: The Transition from Bulk (Continuum) to Molecular Behavior with Decreasing Film Thickness," *J. Colloid Interf. Sci.* 129, 461–467.

Israelachvili, J. N. and McGuiggan, P. M. (1988), "Forces between Surfaces in Liquids," *Science* 241, 795–800.

Israelachvili, J. N. and McGuiggan, P. M. (1990), "Adhesion and Short-Range Forces between Surfaces. I — New Apparatus for Surface Force Measurements," *J. Mater. Res.* 5, 2223–2231.

Israelachvili, J. N. and Pashley, R. M. (1982), "The Hydrophobic Interaction Is Long Range, Decaying, Exponentially with Distance," *Nature* 300, 341–342.

Israelachvili, J. N. and Wennerström, H. K. (1990), "Hydration or Steric Forces between Amphiphilic Surfaces?" *Langmuir* 6, 873–876.

Israelachvili, J. N. and Wennerström, H. K. (1992), "Entropic Forces between Amphiphilic Surfaces in Liquids," *J. Phys. Chem.* 96, 520–531.

Israelachvili, J. and Wennerström, H. (1996), "Role of Hydration and Water Structure in Biological and Colloidal Interactions," *Nature* 379, 219–225.

Israelachvili, J. N., Homola, A. M., and McGuiggan, P. M. (1988), "Dynamic Properties of Molecularly Thin Liquid Films," *Science* 240, 189–191.

Israelachvili, J. N., Gee, M. L., McGuiggan, P., Thompson, P., and Robbins, M. (1990a), "Melting-Freezing Transitions in Molecularly Thin Liquid Films during Shear," in *Dynamics in Small Confining Systems, Proc. 1990 Fall Meeting of the Materials Research Society* (J. M. Drake, J. Klafter, and R. Kopelman, eds.), pp 3–6, MRS Publications, Pittsburgh, PA.

Israelachvili, J. N., McGuiggan, P., Gee, M., Homola, A., Robbins, M., and Thompson, P. (1990b), "Liquid Dynamics in Molecularly Thin Films," *J. Phys. Condens. Matter* 2, SA89–SA98.

Israelachvili, J., Chen, Y.-L., and Yoshizawa, H. (1994), "Relationship between Adhesion and Friction Forces," *J. Adhesion Sci. Technol.* 8, 1–18.

Israelachvili, J., Chen, Y.-L., and Yoshizawa, H. (1995), "Relationship between Adhesion and Friction Forces," in *Fundamentals of Adhesion and Interfaces* (D. S. Rimai, L. P. DeMejo, and K. L. Mittal, eds.), 261–279, VSP, Utrecht, The Netherlands.

Johnson, K. L., Kendall, K., and Roberts, A. D. (1971), "Surface Energy and the Contact of Elastic Solids," *Proc. R. Soc. London A* 324, 301–313.

Jönsson, B. (1981), "Monte Carlo Simulations of Liquid Water between Two Rigid Walls," *Chem. Phys. Lett.* 82, 520–525.

Jönsson, B. and Wennerström, H. (1983), "Image-Charge Forces in Phospholipid Bilayer Systems," *J. Chem. Soc. Faraday Trans II* 79, 19–35.

Kjellander, R. and Marcelja, S. (1985a), "Perturbation of Hydrogen Bonding in Water Near Polar Surfaces," *Chem. Phys. Lett.* 120, 393–396.

Kjellander, R. and Marcelja, S. (1985b), "Polarization of Water between Molecular Surfaces: A Molecular Dynamics Study," *Chem. Scr.* 25, 73–80.

Klein, J. (1983), "Forces between Mica Surfaces Bearing Adsorbed Macromolecules in Liquid Media," *J. Chem. Soc. Faraday Trans. I* 79, 99–118.

Klein, J. (1986), "Surface Forces with Adsorbed Polymers: Direct Measurements and Model Calculations," *Macromol. Chem. Macromol. Symp.* 1, 125–137.

Klein, J. and Kumacheva, E. (1995), "Confinement-Induced Phase Transitions in Simple Liquids," *Science* 269, 816–819.

Klein, J., Kamiyama, Y., Yoshizawa, H., Israelachvili, J. N., Fredrickson, G. H., Pincus, P., and Fetters, L. J. (1993), "Lubrication Forces between Surfaces Bearing Polymer Brushes," *Macromolecules* (in press).

Klein, J., Kumacheva, E., Mahalu, D., Perahia, D., and Fetters, L. J. (1994), "Reduction of Frictional Forces between Solid Surfaces Bearing Polymer Brushes," *Nature* 370, 634–636.

Kuhl, T. L., Berman, A. D., Israelachvili, J. N., and Hui, S. W. (1998), "Part I: Direct Measurement of Depletion Attraction and Thin Film Viscosity between Lipid Bilayers in Aqueous Polyethylene Oxide Solutions," *Macromolecules*, (in press).

Kurihara, K., Kato, S., and Kunitake, T. (1990), "Very Strong Long Range Attractive Forces between Stable Hydrophobic Monolayers of a Polymerized Ammonium Surfactant," *Chem. Lett.* (*Chem. Soc. Jpn.*) 1990, 1555–1558.

Landman, U., Luedtke, W. D., Burnham, N. A., and Colton, R. J. (1990), "Atomistic Mechanisms and Dynamics of Adhesion, Nanoindentation, and Fracture," *Science* 248, 454–461.

Lawn, B. R. and Wilshaw, T. R. (1975), *Fracture of Brittle Solids*, Cambridge University Press, London.

Leckband, D. E., Schmitt, F.-J., Knoll, W., and Israelachvili, J. N. (1992), "Long-Range Attraction and Molecular Rearrangements in Receptor-Ligand Interactions," *Science* 255, 1419–1421.

Lessard, R. R. and Zieminski, S. A. (1971), "Bubble Coalescence and Gas Transfer in Aqueous Electrolytic Solutions," *Ind. Eng. Chem. Fundam.* 10, 260–269.

Luengo, G., Israelachvili, J., and Granick, S. (1996), "Generalized effects in confined fluids: new friction map for boundary lubrication," *Wear* 200, 328–335.

Luengo, G., Schmitt, F.-J., Hill, R., and Israelachvili, J. (1997), "Thin Film Rheology and Tribology of Confined Polymer Melts: Contrasts with Bulk Properties," *Macromolecules* 30, 2482–2494.

Luzar, A., Bratko, D., and Blum, L. J. (1987), "Monte Carlo Simulation of Hydrophobic Interaction," *J. Phys. Chem.* 86, 2955–2959.

Marcelja, S. (1997), "Hydration in Electrical Double Layers," *Nature* 385, 688–690.

Marcelja, S. and Radic, N. (1976), "Repulsion of Interfaces Due to Boundary Water," *Chem. Phys. Lett.* 42, 129–130.

Marcelja, S., Mitchell, D. J., Ninham B. W., and Sculley, M. J. (1977), "Role of Solvent Structure in Solution Theory," *J. Chem. Soc. Faraday Trans. II* 73, 630–648.

Maugis, D. (1985), "Subcritical Crack Growth, Surface Energy, Fracture Toughness, Stick-Slip, and Embrittlement," *J. Materials Sci.* 20, 3041–3073.

McClelland, G. M. (1989), in *Adhesion and Friction. Springer Series in Surface Sciences* (M. Grunze and H. J. Kreuzer, eds.), Vol. 17, 1–16, Springer, Berlin.

McClelland, G. M. and Cohen, S. R. (1990), *Chemistry and Physics of Solid Surfaces VIII*, pp. 419–445, Springer, Berlin.

McGuiggan, P. and Israelachvili, J. N. (1990), "Adhesion and Short-Range Forces between Surfaces. Part II: Effects of Surface Lattice Mismatch," *J. Mater. Res.* 5, 2232–2243.

Michel, F. and Shanahan, M. E. R. (1990), "Kinetics of the JKR Experiment," *C. R. Acad. Sci. Paris* 310II, 17–20.

Miller, C. A. and Neogi, P. (1985), *Interfacial Phenomena: Equilibrium and Dynamic Effects*, Marcel Dekker, New York.

Nasuno, S., Kudrolli, A., and Gollub, J. (1997), "Friction in Granular Layers: Hysteresis and Precursors," *Phys. Rev. Lett.* 79, 949–952.

Parker, J. L. and Christenson, H. K. (1988), "Measurements of the Forces between a Metal Surface and Mica across Liquids," *J. Chem Phys.* 88, 8013–8014.

Parker, J. L., Cho, D. L., and Claesson, P. M. (1989), "Plasma Modification of Mica: Forces between Fluorocarbon Surfaces in Water and a Nonpolar Liquid," *J. Phys. Chem.* 93, 6121–6125.

Pashley, R. M. (1981), "DLVO and Hydration Forces between Mica Surfaces in Li$^+$, Na$^+$, K$^+$, and Cs$^+$ Electrolyte Solutions: A Correlation of Double-Layer and Hydration Forces with Surface Cation Exchange Properties," *J. Colloid Interf. Sci.* 83, 531–546. Also "Hydration Forces between Solid Surfaces in Aqueous Electrolyte Solutions," *J. Colloid Interf. Sci.* 80, 153–162.

Pashley, R. M. (1982), "Hydration Forces between Mica Surfaces in Electrolyte Solutions," *Adv. Colloid Interf. Sci.* 16, 57–62.

Pashley, R. M. (1985), "The Effects of Hydrated Cation Adsorption on Surface Forces between Mica Crystals and Its Relevance to Colloidal Systems," *Chem. Scr.* 25, 22–27.

Pashley, R. M., McGuiggan, P. M., Ninham, B. W., and Evans, D. F. (1985), "Attractive Forces between Uncharged Hydrophobic Surfaces: Direct Measurements in Aqueous Solutions," *Science* 229, 1088–1089.

Patel, S. S. and Tirrell, M. (1989), "Measurement of Forces between Surfaces in Polymer Fluids," *Annu. Rev. Phys. Chem.* 40, 597–635.

Persson, B. N. J. (1994), "Theory of Friction — The Role of Elasticity in Boundary Lubrication," *Phys. Rev. B* 50, 4771–4786.

Ploehn, H. J. and Russel, W. B. (1990), "Interactions between Colloidal Particles and Soluble Polymers," *Adv. Chem. Eng.* 15, 137–227.

Pollock, H. M., Barquins, M., and Maugis, D. (1978), "The Force of Adhesion between Solid Surfaces in Contact," *Appl. Phys. Lett.* 33, 798–799.

Quirk, J. P. (1968), "Particle Interaction and Soil Swelling," *Isr. J. Chem.* 6, 213–234.

Rabin, Y. and Hersht, I. (1993), "Thin Liquid Layers in Shear — Non-Newtonian Effects," *Physica A* 200, 708–712.

Rabinovich, Ya. I. and Derjaguin, B. V. (1988), "Interaction of Hydrophobized Filaments in Aqueous Electrolyte Solutions," *Colloids Surf.* 30, 243–251.

Rabinowicz, E. (1958), "The Intrinsic Variables Affecting the Stick-Slip Process," *Proc. Phys. Soc.* 71, 668–675.

Rabinowicz, E. (1965), *Friction and Wear of Materials*, John Wiley, New York, chap. 4.

Rhykerd, C., Schoen, M., Diestler, D., and Cushman, J. (1987), "Epitaxy in Simple Classical Fluids in Micropores and Near-Solid Surfaces," *Nature* 330, 461–463.

Robbins, M. O. and Thompson, P. A. (1991), "Critical Velocity of Stick-Slip Motion," *Science* 253, 916.

Rozman, M. G., Urbakh, M., Klafter, J. (1996), "Origin of Stick-Slip Motion in a Driven Two-Wave Potential," *Phys. Rev. E* 54, 6485–6494.

Rozman, M. G., Urbakh, M., and Klafter, J. (1997), "Stick-Slip Dynamics as a Probe of Frictional Forces," *Europhys. Lett.* 39, 183–188.

Sampson, J., Morgan, F., Reed, D., and Muskat, M. (1943), "Friction Behavior during the Slip Portion of the Stick-Slip Process," *J. Appl. Phys.* 14, 689–700.

Schiby, D. and Ruckenstein, E. (1983), "The Role of the Polarization Layers in Hydration Forces," *Chem. Phys. Lett.* 95, 435–438.

Schoen, M., Dietsler, D. J., and Cushman, J. H. (1987), "Fluids in Micropores. I. Structure of a Simple Classical Fluid in a Slit-Pore," *J. Phys. Chem.* 87, 5464–5476.

Schoen, M., Rhykerd, C., Diestler, D., and Cushman, J. (1989), "Shear Forces in Molecularly Thin Films," *Science* 245, 1223–1225.

Shanahan, M. E. R., Schreck, P., and Schultz, J. (1988), "Effets de la Réticulation Sur L'Adhésion d'un Élastomére," *C.R. Acad. Sci. Paris* 306II, 1325–1330.

Smith, C. P., Maeda, M., Atanasoska, L., White, H. S., and McClure, D. J. (1988), "Ultrathin Platinum Films on Mica and the Measurement of Forces at the Platinum/Water Interface," *J. Phys. Chem.* 92, 199–205.

Snook, I. K. and van Megen, W. (1979), "Structure of Dense Liquids at Solid Interfaces," *J. Phys. Chem.* 70, 3099–3105.

Snook, I. K. and van Megen, W. (1980), "Solvation Forces in Simple Dense Fluids. I," *J. Phys. Chem.* 72, 2907–2913.

Snook, I. K. and van Megen, W. (1981), "Calculation of Solvation Forces between Solid Particles Immersed in Simple Liquid," *J. Chem. Soc. Faraday Trans. II* 77, 181–190.

Stanley, H. E. and Teixeira, J. (1980), "Interpretation of the Unusual Behavior of H_2O and D_2O at Low Temperatures: Tests of a Perculation Model," *J. Phys. Chem.* 73, 3404–3422.

Steinberg, S., Ducker, W., Vigil, G., Hyukjin, C., Frank, C., Tseng, W., Clarke, D. R., and Israelachvili, J. N. (1993), "Van der Waals Epitaxial Growth of α — Alumina Nanocrystals on Mica," *Science* 260, 656–659.

Sutcliffe, M. J., Taylor, S. R., and Cameron, A. (1978), "Molecular Asperity Theory of Boundary Friction," *Wear* 51, 181–192.

Tabor, D. (1982), "The Role of Surface and Intermolecular Forces in Thin Film Lubrication," in *Microscopic Aspects of Adhesion and Lubrication*, pp. 651–679, Societe de Chimie Physique, Paris.

Tarazona, P. and Vicente, L. (1985), "A Model for the Density Oscillations in Liquids between Solid Walls," *Mol. Phys.* 56 557–572.

Thompson, P. and Robbins, M. (1990), "Origin of Stick-Slip Motion in Boundary Lubrication," *Science* 250, 792–794.

Thompson, P. A., Grest, G. S., and Robbins, M. O. (1992), "Phase Transitions and Universal Dynamics in Confined Films," *Phys. Rev. Lett.* 68, 3448–3451. Also (1993), in *Thin Films in Tribology, Proceedings of the 19th Leeds-Lyon Symposium on Tribology* (D. Dawson, C. M. Taylor, and M. Godet, eds.), pp. 1–14, Elsevier, Amsterdam.

Thompson, P. A., Robbins, M. O., and Grest, G. S. (1995), "Structure and Shear Response in Nanometer-Thick Films," *Isr. J. Chem.* 35, 93–106.

Tomlinson, G. A. (1929), "A Molecular Theory of Friction," *Philos. Mag.* 7, 905–939.

Urbakh, M., Daikhin, L., and Klafter, J. (1995), "Dynamics of Confined Liquids under Shear," *Phys. Rev. E* 51, 2137–2141.

Van Alsten, J. and Granick, S. (1988), "Molecular Tribometry of Ultrathin Liquid Films," *Phys. Rev. Lett.* 61, 2570–2573.

Van Alsten, J. and Granick, S. (1990a), "The Origin of Static Friction in Ultrathin Liquid Films," *Langmuir* 6, 876–880.

Van Alsten, J. and Granick, S. (1990b), "Shear Rheology in a Confined Geometry — Polysiloxane Melts," *Macromolecules* 23, 4856–4862.

van Megen, W. and Snook, I. K. (1979), "Calculation of Solvation Forces between Solid Particles Immersed in a Simple Liquid," *J. Chem. Soc. Faraday Trans. II* 75, 1095–1102.

van Megen, W. and Snook, I. K. (1981), "Solvation Forces in Simple Dense Fluids. II. Effect of Chemical Potential," *J. Phys. Chem.* 74, 1409–1411.

van Olphen, H. (1977), *An Introduction to Clay Colloid Chemistry,* 2nd ed., Wiley, New York, chap. 10.

Velamakanni, B. V., Chang, J. C., Lange, F. F., and Pearson, D. S. (1990), "New Method for Efficient Colloidal Particle Packing via Modulation of Repulsive Lubricationg Hydration Forces," *Langmuir* 6, 1323–1325.

Viani, B. E., Low, P. F., and Roth, C. B. (1983), "Direct Measurement of the Relation between Interlayer Force and Interlayer Distance in the Swelling of Montmorillonites," *J. Colloid Interf. Sci.* 96, 229–244.

Vigil, G., Xu, Z., Steinberg, S., and Israelachvili, J. (1994), "Interactions of Silica Surfaces," *J. Colloid Interf. Sci.* 165, 367–385.

Warnock, J., Awschalom, D. D., and Shafer, M. W. (1986), "Orientational Behavior of Molecular Liquids in Restricted Geometries," *Phys. Rev. B* 34, 475–478.

Weisenhorn, A. L., Hansma, P. K., Albrecht, T. R., and Quate, C. F. (1989), "Forces in Atomic Force Microscopy in Air and Water," *Appl. Phys. Lett.* 54, 2651–2653.

Wilchek, M. and Bayer, E. B. (1990), *Methods in Enzymology* Vol. 184: *Avidin-Biotin Technology,* Academic Press, San Diego, 5–45.

Yoshizawa, H. and Israelachvili, J. N. (1993), "Fundamental Mechanisms of Interfacial Friction II: Stick-Slip Friction of Spherical and Chain Molecules," *J. Phys. Chem.* 97, 11300–11313.

Yoshizawa, H., Chen, Y. L., and Israelachvili, J. (1993), "Fundamental Mechanisms of Interfacial Friction I: Relation between Adhesion and Friction," *J. Phys. Chem.* 97, 4128–4140.

Zisman, W. A. (1963), "Influence of Constitution on Adhesion," *Ind. Eng. Chem.* 55(10), 19–38.

Zisman, W. A. and Fox, J. (1952), "The Spreading of Liquids on Low-Energy Surfaces. III. Hydrocarbon Surfaces," *J. Colloid Sci.* 7, 428–442.

10

Nanomechanical Properties of Solid Surfaces and Thin Films

Bharat Bhushan

10.1 Introduction

Mechanical properties of the solid surfaces and surface thin films are of interest as the mechanical properties affect the tribological performance of surfaces. Among the mechanical properties of interest, one or more of which can be obtained using commercial and specialized hardness testers, are elastic–plastic deformation behavior, hardness, Young's modulus of elasticity, scratch resistance, film-substrate adhesion, residual stresses, time-dependent creep and relaxation properties, fracture toughness, and fatigue. Hardness measurements can assess structural heterogeneities on and underneath the surface such as diffusion gradients, precipitate, presence of buried layers, grain boundaries, and modification of surface composition.

Hardness implies the resistance to local deformation. For example, with materials that go through plastic deformation, a hard indenter is pressed into the surface and the size of the permanent (or plastic) indentation formed for a given load is a measure of hardness. With rubberlike materials (which do not go through plastic deformation), an indenter is pressed into the material and how far it sinks under load is measured. With brittle materials (which do not go through plastic deformation), hardness is measured by scratching it by a harder material. Hardness signifies different things to different people, for instance, resistance to penetration to a metallurgist, resistance to scratching to a mineralogist, and resistance to cutting to a machinist, but all are related to the plastic flow stress of material.

Hardness measurements usually fall into three main categories: scratch hardness, rebound or dynamic hardness, and static indentation hardness (Tabor, 1951). Scratch hardness is the oldest form of hardness measurement. It depends on the ability of one material to scratch another or to be scratched by another solid. The method was first put on a semiquantitative basis by Friedrich Mohs in 1822, who selected ten minerals as standards, beginning with talc and ending with diamond. The Mohs scale is widely used by mineralogists and lapidaries (Tabor, 1951). Today, solid and thin-film surfaces are scratched by a sharp stylus made of hard material typically diamond, and either the loads required to scratch or fracture the surface or delaminate the film or the normal/tangential load–scratch size relationships are used as a measure of scratch hardness and/or interfacial adhesion (Heavens, 1950; Tabor, 1951, 1970; Benjamin and Weaver, 1960; Campbell, 1970; Ahn et al., 1978; Mittal, 1978; Perry, 1981, 1983; Jacobson et al., 1983; Valli, 1986; Bhushan, 1987; Steinmann et al., 1987; Wu, 1991; Bhushan et al., 1995, 1996, 1997; Bhushan and Gupta, 1995; Gupta and Bhushan, 1995a,b; Patton and Bhushan, 1996; Bhushan and Li, 1997; Li and Bhushan, 1998b,c).

Another type of hardness measurement is rebound or dynamic hardness involving the dynamic deformation or indentation of the surface. In this method, a diamond-tipped hammer (known as tup) is dropped from a fixed height onto the test surface and the hardness is expressed in terms of the energy of impact and the size of the remaining indentation. For example, in the shore rebound scleroscope, the hardness is expressed in terms of the height of rebound of the indenter.

The methods most widely used in determining the hardness of materials are *(quasi) static indentation methods*. Indentation hardness is essentially a measure of their plastic deformation properties and only to a secondary extent with their elastic properties. There is a large hydrostatic component of stress around the indentation, and since this plays no part in plastic flow the indentation pressure is appreciably higher than the uniaxial flow stress of the materials. For many materials, it is about three times as large, but if the material shows appreciable elasticity, the yielding of the elastic hinterland imposes less constraint on plastic flow and the factor of proportionality may be considered less than 3. Indentation hardness depends on the time of loading and on the temperature and other operating environmental conditions. In the indentation methods, a spherical, conical, or pyramidal indenter is forced into the surface of the material which forms a permanent (plastic) indentation in the surface of the material to be examined. The hardness number (GPa or kg/mm^2), equivalent to the average pressure under the indenter, is calculated as the applied normal load divided by either the curved (surface) area (Brinell, Rockwell, and Vickers hardness numbers) or the projected area (Knoop and Berkovich hardness numbers) of the contact between the indenter and the material being tested, under load (Lysaght, 1949; Berkovich, 1951; Tabor,

1951, 1970; Mott, 1957; O'Neill, 1967; Westbrook and Conrad, 1973; Anonymous, 1979; Johnson, 1985; Blau and Lawn, 1986; Bhushan and Gupta, 1997).

Macrohardness tests are widely used because of availability of inexpensive testers, simplicity of measurement, portability, and direct correlation of the hardness with service performance. For applications with ultrasmall loads (few mN to nN) being applied at the interface, nanomechanical properties of the skin (as thin as a monolayer) of a solid surface or a surface film are of interest. Furthermore, ultrathin films as thin as a monolayer are used for micromechanical applications and their mechanical properties are of interest. Hardness tests can be performed on a small amount (few mg) of material and with the state-of-the-art equipment it is possible to measure hardness of the few surface layers on the sample surface.

In a conventional indentation hardness test, the contact area is determined by measuring the indentation size by a microscope after the sample is unloaded. At least, for metals, there is a little change in the size of the indentation on unloading so that the conventional hardness test is essentially a test of hardness under load, although it is subject to some error due to varying elastic contraction of the indentation (Stilwell and Tabor, 1961). More recently, in depth-sensing indentation hardness tests, the contact area is determined by measuring the indentation depth during the loading/unloading cycle (Pethica et al., 1983; Blau and Lawn, 1986; Wu et al., 1988; Bravman et al., 1989; Doerner et al., 1990; Nix et al., 1992; Pharr and Oliver, 1992; Oliver and Pharr, 1992; Nastasi et al., 1993; Townsend et al., 1993; Bhushan et al., 1995, 1996, 1997; Bhushan and Gupta, 1995; Gupta and Bhushan,1995a, b; Bhushan, 1996; Patton and Bhushan, 1996; Bhushan and Li, 1997; Li and Bhushan, 1998b,c). Depth measurements have, however, a major weakness arising from "piling-up" and "sinking-in" of material around the indentation. The measured indentation depth needs to be corrected for the depression (or the hump) of the sample around the indentation, before it can be used for calculation of the hardness (Doerner and Nix, 1986; Doerner et al., 1986; Wu et al., 1988; Nix, 1989; Oliver and Pharr, 1992; Fabes et al., 1992; Pharr and Oliver, 1992). Young's modulus of elasticity is the slope of the stress–strain curve in the elastic regime. It can obtained from the slope of the unloading curve (Nix, 1989; Oliver and Pharr, 1992; Pharr and Oliver, 1992). Hardness data can be obtained from depth-sensing instruments without imaging the indentations with high reproducibility. This is particularly useful for small indents required for hardness measurements of extremely thin films.

In addition to measurements of hardness and Young's modulus of elasticity, static indentation tests have been used for measurements of a wide variety of material properties such as elastic–plastic deformation behavior (Pethica et al., 1983; Doerner and Nix, 1986; Stone et al., 1988; Fabes et al., 1992; Oliver and Pharr, 1992), flow stress (Tabor, 1951), scratch resistance and film–substrate adhesion (Heavens, 1950; Tabor, 1951; Benjamin and Weaver, 1960; Campbell, 1970; Ahn et al., 1978; Mittal, 1978; Perry, 1981, 1983; Jacobson et al., 1983; Valli, 1986; Bhushan, 1987; Steinmann et al., 1987; Stone et al., 1988; Wu et al., 1989, 1990b; Wu, 1990, 1991; Bhushan et al., 1995, 1996, 1997; Bhushan and Gupta, 1995; Gupta and Bhushan, 1995a, b; Patton and Bhushan, 1996; Bhushan and Li, 1997; Li and Bhushan, 1998b, c), residual stresses (Swain et al., 1977; Marshall and Lawn, 1979; LaFontaine et al., 1991), creep (Westbrook, 1957; Mulhearn and Tabor, 1960/61; Atkins et al., 1966; Walker, 1973; Chu and Li, 1977; Hooper and Brookes, 1984; Li et al., 1991), stress relaxation (Hart and Solomon, 1973; Chu and Li, 1980; Hannula et al., 1985; Mayo et al., 1988a, 1990; LaFontaine et al., 1990a,b; Raman and Berriche, 1990, 1992; Wu, 1991; Nastasi et al., 1993), fracture toughness and brittleness (Palmquist, 1957; Lawn et al., 1980; Chantikul et al., 1981; Mecholsky et al., 1992; Lawn, 1993; Pharr et al., 1993; Bhushan et al., 1996; Li et al., 1997, 1998a), and fatigue (Li and Chu, 1979; Wu et al., 1991).

The extended load range of static indentation hardness testing is shown schematically in Figure 10.1. We note that only the lower micro- and ultramicrohardness or nanohardness load range can be employed successfully for measurements of extremely thin (submicron-thick) films. The intrinsic hardness of surface layers or thin films becomes meaningful only if the influence of the substrate material can be eliminated. It is therefore generally accepted that the depth of indentation should never exceed 30% of the film thickness (Anonymous, 1979). The minimum load for most commercial microindentation testers

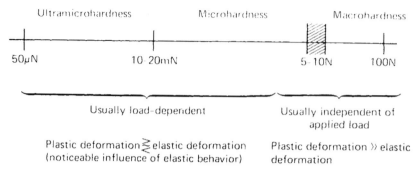

FIGURE 10.1   Extended load range of static indentation hardness testing.

available is about 10 mN. Loads on the order of 50 μN to 1 mN are desirable if the indentation depths are to remain few tens of a nanometer. In this case, the indentation size sometimes reaches the resolution limit of a light microscope, and it is almost impossible to find such a small imprint if the measurement is made with a microscope after the indentation load has been removed. Hence, either the indentation apparatuses are placed *in situ* and a scanning electron microscope (SEM) or *in situ* indentation depth measurements are made. The latter measurements, in addition, would offer the advantages to observe the penetration process itself. In viscoelastic/visoplastic materials, since indentation size changes with time, *in situ* measurements of the indentation size are particularly useful, which can, in addition, provide more complete creep and relaxation data of the materials.

In this chapter, we will review various prototype and commercial nanoindentation hardness test apparatuses and associated scratch capabilities for measurements of mechanical properties of surface layers of bulk materials and extremely thin films (submicron in thickness). A commercial depth-sensing nanohardness test apparatus will be described in detail followed by data analysis and use of nanohardness apparatuses for determination of various mechanical properties of interest.

10.2   Nanoindentation Hardness Measurement Apparatuses

In this section, we review nanoindentation hardness apparatuses in which the indent is imaged after the load has been removed as well as the depth-sensing indentation apparatuses in which the load-indentation depth is continuously monitored during the loading and unloading processes. Earlier work by Alekhin et al. (1972), Ternovskii et al. (1973), and Bulychev et al. (1975, 1979) led to the development of depth-sensing apparatuses. Both prototype and commercial apparatuses are reviewed. A commercial depth-sensing nanoindentation hardness test apparatus manufactured by Nano Instruments, Inc., is extensively used and is described in detail.

10.2.1   Commercial Nanoindentation Hardness Apparatuses with Imaging of Indents after Unloading

For completeness, we first describe a commercially available microindentation hardness apparatus (Model No. Micro-Duromet 4000) that uses a built-in light optical microscope for imaging of indents after the sample is unloaded. It is manufactured by C. Reichert Optische Werke AG, A-1171, Vienna, Box 95, Austria, Figure 10.2 (Pulker and Salzmann, 1986). The case of the indenter is of the size of a microscope objective mounted on the objective revolver. The load range for this design is from 0.5 mN to 2 N; therefore, it is used for thicker films.

A commercial nanoindentation hardness apparatus for use inside an SEM (Model No. UHMT-3) for imaging the indents after the sample is unloaded, is manufactured by Anton Paar K.G., A-8054, Graz, Austria. The apparatus is mounted on the goniometer stage of the SEM. In this setup, the indenter is

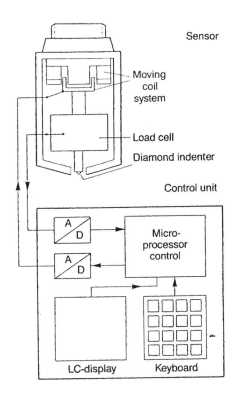

Sensor

Moving coil system

Load cell

Diamond indenter

Control unit

A/D

Micro-processor control

A/D

LC-display

Keyboard

FIGURE 10.2 Schematic of the microindentation hardness apparatus for use in a light optical microscope. (From Pulker, H.K. and Salzmann, K., 1986, *SPIE Thin Film Technol.* 652, 139–144. With permission.)

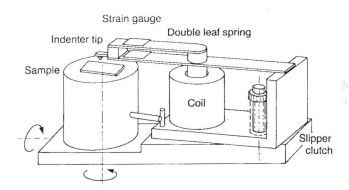

Strain gauge

Indenter tip

Double leaf spring

Sample

Coil

Slipper clutch

FIGURE 10.3 Schematic of the nanoindentation hardness apparatus for use in an SEM by Anton Parr K.G., Graz, Austria. (From Bangert, H. et al., 1981, *Colloid Polym. Sci.* 259, 238–242. With permission.)

mounted on a double-leaf spring cantilever and is moved against the sample by an electromagnetic system to attain the required indentation load, which is measured by strain gauges mounted on the leaf springs, Figure 10.3 (Bangert et al., 1981; Bangert and Wagendristel, 1986). Tilting the stage with respect to the electron beam allows observation of the tip during the indentation process. The indentation cycle is fully programmable and is controlled by the strain gauge signal. The motion of the indenter, perpendicular to the surface, is performed by increasing the coil current until a signal from the strain gauges is detected. Further, an increase of the current up to a certain gauge signal leads to the desired indentation force ranging from 50 μN to 20 mN. After the required load has been reached and the dwell time has elapsed, the sample is unloaded, and the indentation diagonal is measured by an SEM.

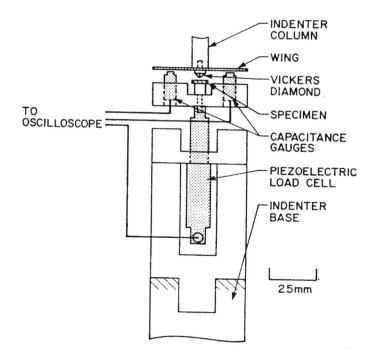

INDENTER COLUMN

WING

VICKERS DIAMOND

SPECIMEN

CAPACITANCE GAUGES

PIEZOELECTRIC LOAD CELL

INDENTER BASE

TO OSCILLOSCOPE

25mm

FIGURE 10.4 Schematic of a modified commercial microhardness test apparatus for load-penetration depth measurements. (From Pharr, G.M. and Cook, R.F., 1990, *J. Mater. Res.* 5, 847–851. With permission.)

10.2.2 Prototype Depth-Sensing Nanoindentation Hardness Apparatuses

Of all the nanohardness apparatuses described in this section, the apparatus designs by IBM Almaden Research Center and Nano Instruments, Inc., are the most modern apparatuses with the largest range of test capabilities. However, the IBM Almaden apparatus is not commercially available. The apparatus built by MTS Nano Instruments Innovation Center which is called the Nanoindenter, is commercially available and is comparable to the IBM Almaden design with complete software. Nanoindenter is most commonly used by the industrial and academic research laboratories. It will be described in some detail. The NEC design is commercially available; however, this has limited capabilities and is not popular.

10.2.2.1 IBM T.J. Watson Research Center Microhardness Tester Design

The apparatus to be described here is a "microhardness apparatus" and is only included here for completeness. Pharr and Cook (1990) instrumented a conventional microhardness tester to measure indentation load and penetration depth during the entire indentation process, Figure 10.4. This modified machine has the advantage of being relatively inexpensive since many of its components are standard equipment.

Pharr and Cook used a Buehler Micromet II machine with a load range of 0.1 to 10 N although other commercially available units can be used. In their modifications, load was measured with a piezoelectric load cell (Kistler model 9207) with a resolution of 0.5 mN and a maximum load of 50 N. The load cell was conditioned by a Kistler model 5004 charge amplifier with a frequency response of about 180 kHz. Displacement was measured with two capacitec model HPC-75 capacitance gauges with matching amplifier and conditioner (model 3201). These gauges have high resolution of about 0.05 μm and frequency response of 10 kHz. The sample stage was replaced with an assembly in which the load cell could be rigidly supported. A mount for the displacement gauges was then connected directly to the top of the

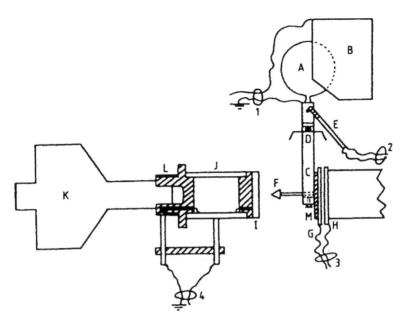

FIGURE 10.5   Schematic of a depth-sensing nanoindentation hardness apparatus by Newey et al. (From Newey, D. et al., 1982, *J. Phys. E: Sci. Instrum.* 15, 119–122. With permission.)

load cell. The specimen was attached to the center of this mount with the displacement gauges flanking it on either side. The gauges sensed the motion of a thin aluminum wing rigidly attached to the base of the moving diamond and its mount. The outputs from the displacement gauges were averaged so as to negate any displacements caused by bending in the system. The load and displacement outputs were measured using a storage oscilloscope.

In a typical experiment, the load and displacement signals were recorded as a function of time, with the load–displacement curve derived subsequently from these data. This modified apparatus can only be used for loads as low as about 100 mN, making it useful for only microhardness measurements.

10.2.2.2   AERE Harwell/Micro Materials Design

Newey et al. (1982) developed an apparatus capable of continuously monitoring the penetration depth as the load is applied, Figure 10.5. The test sample I is mounted on a piezoelectric barrel transducer J, their horizontal position being controlled by a micrometer movement K. A high-voltage supply is connected to the transducer J by means of a commutator arrangement. The indenter assembly C is made from folded tantalum foil to give a light structure, and is fitted with tungsten pivots seated in jeweled bearings D, from which it is suspended. Force is applied electrostatically by increasing the potential on the two plates B; force plate A is part of indenter assembly C and is kept at ground potential. The resulting force causes A to move into B and indenter F to move toward the specimen. The indentation depth is measured with a capacitor bridge arrangement. Plates G and H for measurement of indenter motion, are concentric with the axis of the indenter holder and form part of a capacitor bridge arrangement (plate G is insulated from C by mica sheet M). E is a piezoelectric bimorph transducer used to restrain the indenter assembly C when the specimen is being moved toward the indenter. In the modified design reported by Pollock et al. (1986), the specimen can be transferred between two locations (test and microscopic observation). A particular area of interest may therefore be identified in the microscope and then transferred to the test position.

This instrument is commercially available as Nano Test 550 from Micro Materials, Unit 3, The Byre, Wrexham Technology Park, Wrexham, Clywd, U.K. In this apparatus, an indentation load up to 500 mN with a resolution of 10 μN can be applied and the depth resolution measurement is better than 0.1 nm.

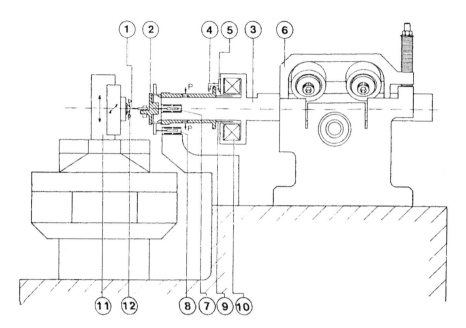

FIGURE 10.6 Schematic of a depth-sensing nanoindentation hardness apparatus by Philips Laboratory, Eindhoven, The Netherlands: (1) indenter, (2) indenter holder, (3) central shaft, (4) and (5) stops, (6) linear drive mechanism, (7) and (8) displacement transducers, (9) coil, (10) electromagnet, (11) sample holder, (12) accessory for holding the sample (From Wierenga, P.E. and van der Linden, J. H. M., 1986, in *Tribology and Mechanics of Magnetic Storage Systems*, Vol. 3 (B. Bhushan and N.S. Eiss, eds.), pp. 31–37, SP-21, ASLE, Park Ridge, IL. With permission.)

10.2.2.3 Philips Research Laboratory Design

Wierenga and Franken (1984) built a nanoindentation apparatus that measures *in situ* the indenter penetration as a function of time (relaxation testing) or load with a resolution of 5 nm, Figure 10.6. The indentation force can be varied from 10 μN to 5 mN. Indenter (1) is clamped to the holder (2) and can easily be exchanged. The indenter holder (2) supported on an air bearing can be moved virtually frictionlessly along a horizontal shaft (3). Two stops (4 and 5) attached to the shaft limit the movement of the holder. The shaft is supported by a linear drive mechanism (6) making use of friction wheels and guide rollers. The apparatus is equipped with two inductive displacement transducers (7 and 8) which measure the displacement of the indenter with respect to the shaft and of the shaft with respect to the surroundings, respectively. The signal from transducer 8 is automatically corrected for changes in the ambient temperature during an experiment. This is done by the application of a temperature-sensing element mounted on transducer 8. The indenter force is adjusted by means of an electromagnetic system. A coil (9) is attached to the indenter holder and can move in the annular gap of an electromagnet (10), which is mounted on the shaft. The sample holder (11) can be moved in two directions perpendicular to the axis of the shaft (3). Samples are held by using an accessory (12) which is held by suction to the sample holder. (Also see Wierenga and van der Linden, 1986.)

For an indentation experiment, the indenter is first brought into contact with the sample. For this purpose, transducer 7 is adjusted to give a zero signal when the stylus holder is somewhere between stops 4 and 5. By switching on the motor of the linear drive mechanism, the shaft and indenter are moved toward the sample. After the indenter has touched the sample surface, movement of the shaft is automatically halted when the signal from transducer 7 equals zero. The starting position for an indentation experiment is achieved by moving the sample a short distance sideways at a minimum indenter force. If the indenter force is increased, the signal from transducer 7 is kept to zero with the control system by moving the shaft. Thus, the penetration depth can be determined with transducer 8, which measures the displacement of the shaft with respect to the frame.

10.2.2.4   IBM Corporation/University of Arizona Tucson Design

Bhushan et al. (1985, 1988) and Williams et al. (1988) built a nanoindentation apparatus that can independently control and measure indentation depths with a resolution of 0.2 nm and loads with a resolution of 30 μN *in situ*, Figure 10.7. Samples and indenter positions are measured with a specially designed polarization interferometer. A minimum load of about 0.5 mN can be applied. In the apparatus, the test sample is clamped to the top of a mirror that is kinematically mounted to the moving stage of a damped parallel spring guide. The spring guide ensures smooth, low-friction, vertical motion. A linear actuator nested inside the parallel spring guide drives the moving stage vertically. The indenter is suspended above the sample and screws into the bottom of the moving stage of another damped parallel spring guide. Another mirror is kinematically mounted to the top of this stage. The indenter spring guide is independently calibrated and checked for linearity so that the indenter load can be correctly inferred. Both spring guides are damped to prevent oscillations and utilize auxiliary counterbalance springs to keep the spring guides close to their neutral, unstressed position, where their motion is linear.

The vertical positions of the sample and indenter mirrors and thus the positions of the sample and indenter are monitored independently by the polarization interferometer. Light from a helium–neon laser enters the polarization interferometer where the light beam is separated by a diffraction grating into seven separate beams, six equally spaced beams on a 12-mm-diameter circle and one in the center of the circle, which serves as the reference (Williams et al., 1988). Leaving the interferometer, the beams pass through a beam expander to enlarge the beam circle diameter. Three of the outer beams strike the indenter mirror and the other three pass through holes in the indenter mirror and stage and strike the sample mirror. Because of the large beam circle diameter, the beams avoid striking the central obstructions, the sample, and the indenter. The light reflected from both mirrors then returns to the interferometer. Thus, the positions of the sample and indenter mirrors are continuously monitored by comparing the relative phases of the light beams returning from the mirrors to the central reference beam. The computer subtracts the positions of the two mirrors to determine the resulting indentation depth and multiplies the indenter mirror position and the spring constant of the indenter parallel spring guide to determine the indentation load.

To initiate a test, the actuator slowly raises the sample toward the indenter until motion is registered by the interferometer, implying that contact has been made between the sample and the indenter. The control loop then takes over and performs the chosen test — it either keeps the load constant and measures the penetration depth as a function of time or it keeps the depth constant and measures the load as a function of time.

10.2.2.5   NEC, Kawasaki Design

Tsukamoto et al. (1987) and Yanagisawa and Motomura (1987, 1989) developed a nanoindentation hardness apparatus. NEC Corp., Kawasaki 216, Japan is attempting to commercialize it, although it is not popular, Figure 10.8. It consists of three parts: an indenter actuator, a load detector, and a displacement sensor. Indenter (1) with a diamond tip is attached to stylus (2) which is clamped on holder (3). The holder is attached to a piezoelectric actuator (4), which drives the holder up and down controlled by a personal computer (5) through an amplifier (6), a regulated power source (7), and an interface (8). Indentation load is detected by a digital electrobalance (9) with a 1-μN resolution at loads of up to 300 mN. The output signal is fed to the X-axis of an X–Y recorder (10). A sample (11) is placed on a sample disk (12). Penetration depth is detected by a fiber-optic displacement instrument (13) with a 4-nm displacement resolution. Light from a tungsten lamp in the displacement instrument is irradiated onto a mirror (15) through an optical fiber (14). The intensity of the reflected light from the mirror on the sample disk is measured by a photodetector in the displacement instrument and reduced to a displacement between the indenter and the sample. An output signal from the displacement instrument is connected to the Y-axis of the X–Y recorder. The apparatus is surrounded by a metal box (20) to minimize the influence of air currents and heat radiation. It is placed on a vibration-isolation air table (16).

For an indentation experiment, the indenter is first brought into contact with the sample by a micrometer (17). When contact is detected with the sample by the electrobalance, the distance between the

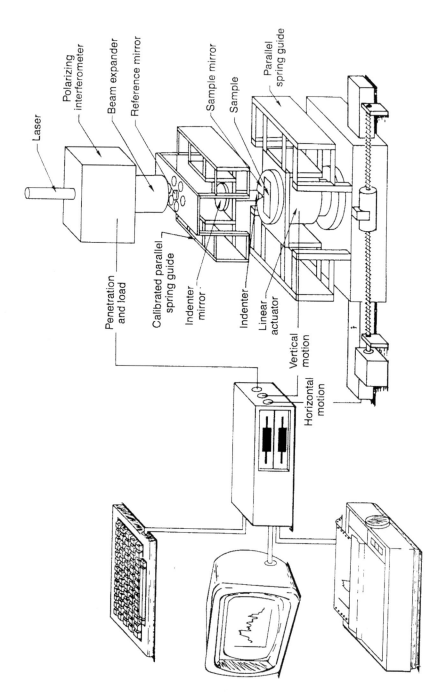

FIGURE 10.7 Schematic of a depth-sensing nanoindentation hardness apparatus by IBM Corporation, Tucson, and University of Arizona, Tucson, AZ. (From Bhushan, B. et al., 1988, *ASME J. Tribol.* 110, 563–571. With permission.)

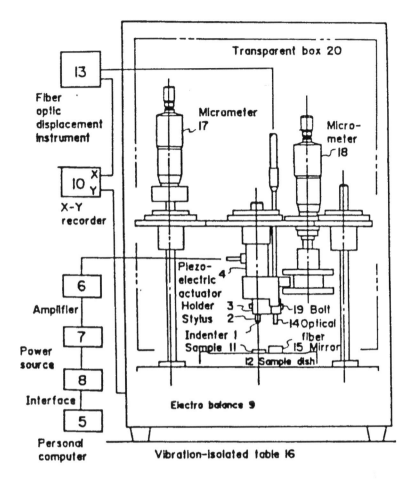

FIGURE 10.8 Schematic of a depth-sensing nanoindentation hardness apparatus by NEC Corp., Tokyo. (From Yanagisawa, M. and Motomura, Y., 1987, *Lubr. Eng.* 43, 52–56. With permission.)

optical fiber and the mirror is adjusted to a region with linearity by a micrometer (18). A bolt (19) is loosened in this adjustment and is fastened after the adjustment to make the optical fiber move together with the indenter. Measurement begins with increasing the voltage applied to the piezoelectric actuator. After the indenter touches and penetrates the sample (loading process), the voltage applied to the piezoelectric actuator is reduced (unloading process).

10.2.2.6   Ecole Central of Lyon Design

The surface force apparatus commonly used for molecular rheology of thin lubricant films, was modified by Loubet et al. (1993) to conduct nanoindentation studies. Figure 10.9 shows the schematic of their nanoindenter design which uses piezoelectric crystal for indenter motion and two capacitance probes for measurement of the load and the displacement. The indenter is fixed to the piezoelectric crystal and the specimen is supported by double-cantilever spring whose stiffness can be adjusted between 4×10^3 and 6×10^6 N/m. The double-cantilever spring prevents the surfaces from rolling and shearing during loading. Two capacitance displacement probes are used. One of the capacitive sensor C_1 measures the elastic deflection of the cantilever and thus the force transmitted to the sample. Another capacitive sensor C_2 measures the relative displacement between the indenter and the sample (or indentation depth).

In an indentation experiment, for the coarse approach, the translation motion is obtained by a differential micrometer. The desired displacement is controlled by a negative proportional integral (PI) feedback loop acting on the piezoelectric crystal via a high-voltage amplifier. The reference displacement

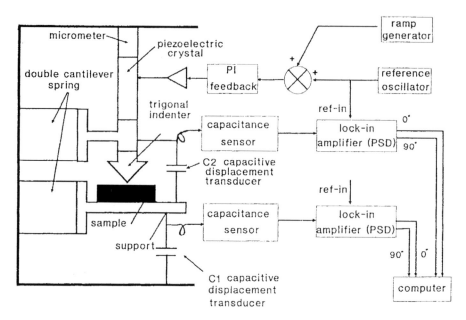

FIGURE 10.9 Schematic of a depth-sensing nanoindentation hardness apparatus by Ecole Central of Lyon. (From Loubet, J.L. et al., 1993, in *Mechanical Properties and Deformation of Materials Having Ultra-Fine Microstructures* (M. Nastasi et al., eds.), pp. 429–447. Kluwer Academic, Dordrecht. With permission.)

signal consists of two ramp reference signal and a sinusoidal signal. Ramp reference signal allows the use of a constant speed from 50 to 0.005 nm/s (typical speed of 0.5 nm/s). The sinusoidal motion designed to determine the dynamic behavior of the solids is obtained using a two-phase lock-in analyzer. It is generally set of about 0.26 nm rms in the frequency range of 0.01 to 500 Hz (typically 38 Hz). The displacement resolution is 0.015 nm.

10.2.2.7 Cornell University Design

Hannula et al. (1986) developed the nanoindentation apparatus shown in Figure 10.10. It is used to measure indenter penetration and load as a function of time during loading and unloading cycles. The apparatus is constructed with two load trains such that both large (up to 50 mm) and small (up to 12 μm) displacements can be applied accurately and independently to the same specimen. The large displacement is produced by a moving crosshead. The small displacements are made possible by using a piezoelectric translator. A load as small as 0.5 mN can be applied.

A specimen is attached to the (moving) crosshead and the diamond tip is attached to a part, which also serves as a counter plate for two capacitance probes. These probes are used either for controlling the position of the tip or for measuring the indentation depth. The load cell is placed between the part and the piezoelectric translator and is used to measure the normal load. The specimen can be aligned by using an x–y stage while observing the specimen directly with the microscope.

10.2.2.8 IBM Almaden Research Center Design

Wu et al. (1988, 1989) built a nanoindentation hardness apparatus based on the Cornell design. Their apparatus uses a piezoelectric transducer (PZT) for indenter motion, capacitance probe for the displacement, a servo-control circuitry for precise control of PZT motion, a multichannel data acquisition system, and a closed-loop TV camera for viewing the interface between the indenter tip and sample surface. (Also see Wu, 1991.) Figure 10.11a shows the block diagram of the apparatus which is composed of an indenter assembly, a load cell assembly and a fully automated precision X–Y–Z stage. Figure 10.11b shows schematically the indenter assembly and the load cell assembly. The indenter (5) is driven by a PZT transducer stack (9) which is monitored by a servo system. The servo mechanism allows great flexibility

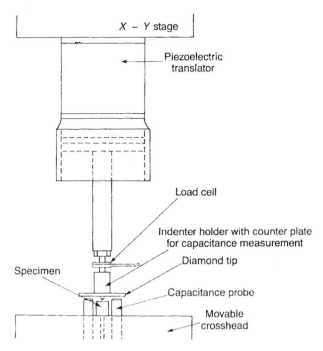

FIGURE 10.10 Schematic of a depth-sensing nanoindentation hardness apparatus by Cornell University, Ithaca, NY. (From Hannula, S.P. et al., 1986, in *The Use of Small-Scale Specimens for Testing Irradiated Materials* (W.R. Corwin and G.E. Lucas, eds.) pp. 233–251, STP 888, ASTM, Philadelphia. With permission.)

in controlling the motion of the indenter. The applied load can be calculated using the output voltage of the load cell capacitance probe (1) and the calibrated load cell spring constant. The total depth penetrated by the indenter with respect to the sample surface can be obtained either directly from the sample capacitance gauge (6) or from the difference of the displacement measurements between indenter gauge (7) and the load cell gauge (1). The load cell has loading ranges from a few tens of micronewtons to 2 N with a resolution of about 30 µN. Indentations with a depth of as low as 20 nm with a depth resolution of 1 nm can be made. For hardness measurements, the apparatus is operated in continuous loading and unloading modes with indenting speeds of 2 to 20 nm s^{-1}. A three-sided pyramidal diamond indenter, known as the Berkovich indenter, is used for measurements.

The PZT stack is driven by a voltage amplifier to follow a predetermined reference pattern and is monitored by closed-loop servo-circuitry. Either the indenter displacement (IND) output (7) or the normal load cell (LC) output (1) can be employed as a servo-input signal and in turn different testing modes can be generated using an IND servo, constant indenter rate testing (typically used in the constant loading and unloading tests), or constant indenter position (used in the load relaxation tests), or any programmed displacement pattern for the indenter can be performed. But using an LC servo, constant loading rate indentation (used in the continuous loading and unloading tests, or constant load indentation (used in the indentation creep test), or cyclic loading (using sawtooth or sinusoidal references, in the indentation fatigue tests) can be performed. Furthermore, under the LC servo mode, the microindenter can be used as an *in situ* profiler, which is used to measure the scratch track depth. Wu (1993) modified the nanoindenter for continuous stiffness measurements using dynamic loading based on the work by Oliver and Pethica (1989) and Pethica and Oliver (1989). The dynamic loading was accomplished by superimposing a sinusoidal waveform with small amplitude to the linear ramping DC voltage. This technique will be described in detail in a later section on nanoindenters.

The nanoindenter was modified to perform scratch tests, Figure 10.12a (Wu et al., 1989, 1990b; Wu, 1991). PZT-driven indenter assembly exhibits excellent rigidity and hence is mechanically stable in the X–Y plane. This extremely rigid design along the horizontal plane is crucial to performing scratch tests.

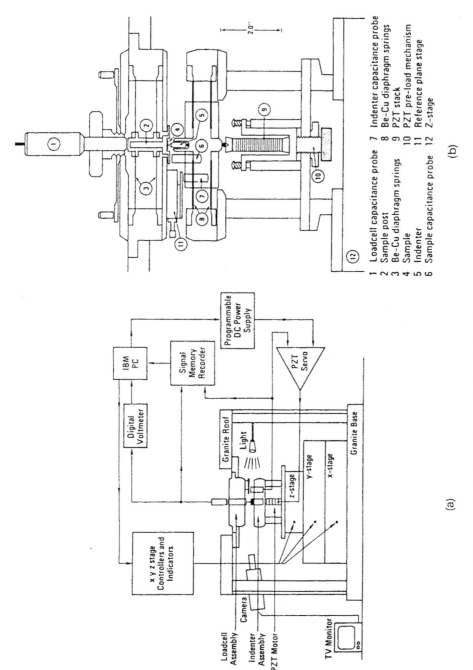

(b)

1 Loadcell capacitance probe   7   Indenter capacitance probe
2 Sample post   8   Be–Cu diaphragm springs
3 Be–Cu diaphragm springs   9   PZT stack
4 Sample   10 PZT pre-load mechanism
5 Indenter   11 Reference plane stage
6 Sample capacitance probe   12 Z-stage

(a)

FIGURE 10.11 (a) Block diagram of a depth-sensing nanoindentation hardness apparatus by IBM Almaden Research Center; (b) schematic diagram of the indenter assembly and normal load cell assembly: (1) load cell capacitance probe, (2) sample post, (3) Be–Cu diaphragm springs; (4) sample, (5) indenter, (6) sample capacitance probe, (7) indentation capacitance probe, (8) Be–Cu diaphragm, (9) PZT stack, (10) PZT preload mechanism, (11) reference plane stage, (12) Z-stage. (From Wu, T.W. et al., 1988, *Thin Solid Films* 166, 299–308. With permission.)

446

In their modified design, a tangential load cell (6, 9) as well as acoustic emission sensor (8) were added. An additional capacitance probe (TG, 9) was placed to monitor the displacement of the indenter holder, which is subsequently used to calculate the tangential force that the indenter applies on the sample surface. The tangential load cell has a loading range of 750 mN with a resolution of about 15 μN. Another capacitance probe (SD, 16) was added to measure scratch distance.

Figure 10.12b shows schematically the working principle of a nanoscratch test carried out by the upgraded apparatus (Figure 10.12a). To perform a scratch test, the indenter is first placed about 0.1 μm away from the sample surface. This step allows a scratch to begin with a zero applied load. Next the traveling range and speed of the X-translation stage are set usually at 150 μm and 1 μm/s, respectively; then the motion is started. Finally, the PZT motor is activated to drive the indenter toward the sample surface at the speed of about 15 nm/s. With this instrument, the following measurements can be made simultaneously during a scratch test: applied load and tangential load along the scratch length (coefficient of friction); critical load, i.e., applied normal load corresponding to an event of coating failure during a scratch process, total depth and plastic depth along the scratch length; the accumulated acoustic emission (AE) counts vs. the scratch length. In addition to the mechanical data, scratch morphology analysis is always available. Examples will be shown later in the chapter.

10.2.3 Commercial Depth-Sensing Nanoindentation Hardness Apparatus and Its Modifications

10.2.3.1 General Description and Principle of Operation

Although the NEC Corp. design and Micro Materials design presented in the previous section are commercially available, these are not popular. The most commonly used commercial depth-sensing nanoindentation hardness apparatus is manufactured by MTS Nano Instruments Innovation Center, 1001 Larson Drive, Oak Ridge, TN 37830. Ongoing development of this apparatus have been described by Pethica et al. (1983), Oliver et al. (1986), Oliver and Pethica (1989), Pharr and Oliver (1992), and Oliver and Pharr (1992). This instrument is called the Nanoindenter. The most recent model is Nanoindenter II (Anonymous, 1991). The apparatus continuously monitors the load and the position of the indenter relative to the surface of the specimen (depth of an indent) during the indentation process. The area of the indent is then calculated from a knowledge of the geometry of the tip of the diamond indenter. The load resolution is about ±75 nN and position of the indenter can be determined to ±0.1 nm. Mechanical properties measurements can be made at a minimum penetration depth of about 20 nm (or a plastic depth of about 15 nm) (Oliver et al., 1986). Specifications for the Nanoindenter are given in Table 10.1. The description of the instrument that follows is based on Anonymous (1991).

The nanoindenter consists of three major components: the indenter head, an optical microscope, and an X–Y–Z motorized precision table for positioning and transporting the sample between the optical microscope and indenter, Figure 10.13a. The loading system used to apply the load to the indenter consists of a magnet and coil in the indenter head and a high precision current source, Figure 10.13b. A coil is attached to the top of the indenter (loading) column and is held in a magnetic field. The passage of the current through the coil is used to raise or lower the column and to apply the required force to make an indent. The current from the source, after passing through the coil, passes through a precision resistor across which the voltage is measured and is displayed. During measurement, voltage is controlled by a computer. Two interchangeable indenter heads are available: the standard head, which features four load ranges 0 to 4 mN, 0 to 20 mN, 0 to 120 mN, and 0 to 350 mN, and a high-load head, which has a load range of 0 to 840 mN. The load resolution for the standard head in the most sensitive range is about ±75 nN, while the load resolution for the high load head is ±90 μN.

The displacement-sensing system consists of a special three-plate capacitive displacement sensor, used to measure the position of the indenter. All three plates are circular disks approximately 1.5 mm thick. The two outer plates have a diameter of 50 mm, and the inner, moving plate is half that size. The indenter column is attached to the moving plate. This plate-and-indenter assembly is supported by two leaf springs cut in such a fashion to have very low stiffness. The motion is damped by airflow around the central

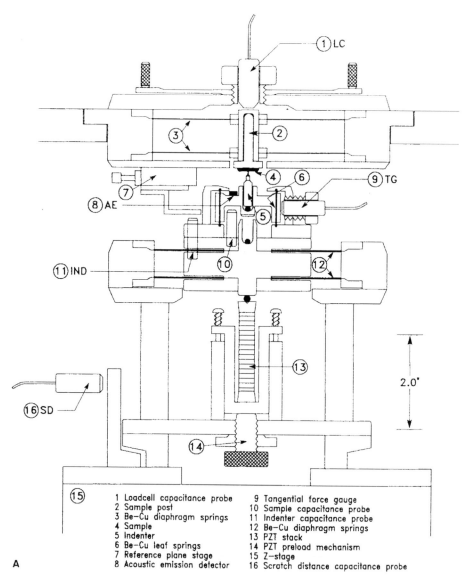

FIGURE 10.12

1 Loadcell capacitance probe	9 Tangential force gauge
2 Sample post	10 Sample capacitance probe
3 Be–Cu diaphragm springs	11 Indenter capacitance probe
4 Sample	12 Be–Cu diaphragm springs
5 Indenter	13 PZT stack
6 Be–Cu leaf springs	14 PZT preload mechanism
7 Reference plane stage	15 Z–stage
8 Acoustic emission detector	16 Scratch distance capacitance probe

plate of the capacitor, which is attached to the loading column. The load coil is used to raise or lower the plate and the indenter assembly through its 100-μm travel between the outer plates of the capacitor. Depth resolution of the systems is about ±0.04 nm. As seen in the plot at the right of Figure 10.13c, a load voltage of 1.7 V will just lift the indenter off its bottom stop, and 1.8 V suffice to bring it to the top of its travel. It should be emphasized that only the motion of the indenter column as controlled by the load coil is used in the actual making of an indent. The voltage output range of the displacement sensing (capacitance) system is –2.5 to +2.5 V.

At the bottom of the indenter rod, a three-sided pyramidal diamond tip (Berkovich indenter, to be discussed later) is generally attached.

The indenter head assembly is rigidly attached to the "U" beam below which the X–Y–Z table rides, Figure 10.13a. The optical microscope is also attached to the beam. The position of an indent on a specimen is selected using the microscope (maximum magnification of 1500¥). The remote-operation option provides a TV camera that is mounted atop the microscope, which permits the image of the

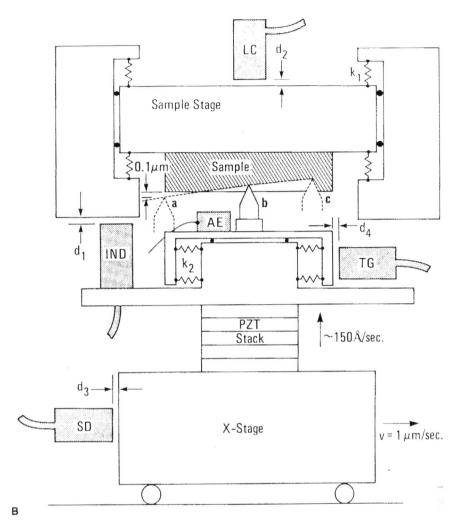

B

FIGURE 10.12 Schematic diagram of the upgraded nanoindentation hardness apparatus with the tangential load cell assembly: IND, indenter probe; LC, normal load cell probe; TG, tangential load cell probe; SD, scratch distance probe; and AE, acoustic emission detector, and (b) schematic illustration of the working principle of the nanoscratch test. From Wu, T.W., 1991, *J. Mater. Res.* 6, 407–426. With permission.)

specimen to be viewed remotely. The specimens are held on an *X–Y–Z* table whose position relative to the microscope or the indenter is controlled with a joystick. The spatial resolution of the position of the table in the *X–Y* plane is ±400 nm and its position is observed on the CRT. The specimen holder is a rectangular metal plate (150 ¥ 150 ¥ 28.5 mm) with ten 31.8-mm-diameter holes for mounting of standard metallographic samples. Samples can also be glued to special metal disks. The three components just described are enclosed in a heavy wooden cabinet to ensure the thermal stability of the samples. The apparatus should be housed in a laboratory in which the temperature is controlled to ±0.5°C. The entire apparatus is placed on a vibration-isolation table. The operation of the apparatus is completely (IBM-PC-compatible) computer controlled. Through an IEEE interface, the computer is connected to data acquisition and control system.

The nanoindenter also comes with a continuous stiffness measurement device (Oliver and Pethica, 1989; Pethica and Oliver, 1989). This device makes possible the continuous measurement of the stiffness of a sample, which allows the elastic modulus to be calculated as a continuous function of time (or indentation depth). Useful data can be obtained from indents with depths as small as 20 nm. Because of

TABLE 10.1 Specification of the Commercial Nanoindenter by Nano Instruments, Inc.

Load range	
Standard head	0–4 mN
	0–20 mN
	0–120 mN
	0–350 mN
High load head	0–840 mN
Load resolution	
Standard head	±75 nN
High load head	±90 nN
Vertical displacement range	0–100 μm
Vertical displacement resolution	±0.1 nm
Typical approach rate	10 nm/s
Typical indentation load rate	10% of peak load/s
Typical indentation displacement rate	10% peak diplacement/s
Optical microscope magnification	up to 1500¥
Spatial resolution of the X–Y–Z table	±400 nm in the X- and Y-directions
Area examined in a single series of indentations	150 ¥ 150 mm
Minimum penetration depth	~20 nm
Continuous stiffness option	
Frequency range	10–150 Hz
Time constant	0.33 s
Smallest measurable distance	0.1 nm
Scratch and tangential force option	
Scratch velocity	max. 100 μm/s with 20 points/mm
Tangential displacement range	2 mm
Tangential displacement resolution	400 nm
Tangential load resolution	50 μN
Minimum measurable tangential load	0.5 mN

the relatively small time constant of the measurements, the device is particularly useful in studies of time-dependent properties of materials.

10.2.3.2   Calibration Procedures

Calibration of the loading system involves the accurate measurement of the voltage through the force coil required to support a series of precalibrated hook weights so as to establish the change in force per volt. The typical value for this constant is 26,876.5 μN/V.

Calibration of the displacement system is carried out by correlating the voltage output of the displacement capacitor with the number of rings generated as the indenter tip is pressed against a lens-and-plate system designed to produce newton rings. A mirror mounted at the bottom on the indenter tip is pressed against a partially reflected lens and an He–Ne laser observed during the test on a video camera. The relationship between displacement voltage and displacement is linear in the range of interest.

Other important calibrations include microscope-to-indenter distance and spring constant of indenter support springs.

10.2.3.3   The Berkovich Indenter

The main requirements for the indenter are high elastic modulus, no plastic deformation, low friction, smooth surface, and a well-defined geometry that is capable of making a well-defined indentation impression. The first four requirements are satisfied by choosing the diamond material for the tip. A well-defined perfect tip shape is difficult to achieve. Berkovich is a three-sided pyramid and provides a *sharply pointed tip* compared with the Vickers or Knoop indenters, which are four-sided pyramids and have a slight offset (0.5 to 1 μm) (Tabor, 1970; Bhushan, 1996). Because any three nonparallel planes intersect at a single point, it is relatively easy to grind a sharp tip on an indenter if Berkovich geometry is used. However, an indenter with a sharp tip suffers from a finite but an exceptionally difficult-to-measure tip bluntness. In addition, pointed indenters produce a virtually constant plastic strain impression and there is the additional problem of assessing the elastic modulus from the continuously varying unloading slope. Spherical indentation overcomes many of the problems associated with pointed indenters. With a spherical indenter, one is able to follow the transition from elastic to plastic behavior and thereby define the yield stress (Bell et al., 1992). However, a sharper tip is desirable, especially for extremely thin films requiring shallow indentation. Therefore, Berkovich indenter is most commonly used for measurements of nanomechanical properties. Experimental procedures have been developed to correct for the tip shape, to be described later.

In the construction of the Berkovich indenter, an octahedron piece of diamond with large dimension of ½ ¥ ½ ¥ ½ mm, is directly brazed to a 304 stainless steel holder and the tip is ground to Berkovich shape. The Berkovich indenter is a three-sided (triangular-based) pyramidal diamond, with a nominal angle of 65.3° between the (side) face and the normal to the base at apex, an angle of 76.9° between edge and normal, and with a radius of the tip less than 0.1 μm (Figure 10.14a and b) (Berkovich, 1951). The typical indenter is shaped to be used for indentation (penetration) depths of 10 to 20 μm. The indents appear as equilateral triangles (Figure 10.14c) and the height of triangular indent ℓ is related to the depth h as

$$\frac{h}{\ell} = \left(\frac{1}{2}\right) \cot 76.9 = \frac{1}{8.59} \tag{10.1a}$$

The relationship $h(\ell)$ is dependent on the shape of the indenter. The height of the triangular indent ℓ is related to the length of one side of the triangle a as

$$\ell = 0.866 \, a \tag{10.1b}$$

and

$$\frac{h}{a} = \frac{1}{7.407} \tag{10.1c}$$

The projected contact area (A) for the assumed geometry is given as

$$A = 0.433 \, a^2 = 23.76 \, h^2 \tag{10.2}$$

The exact shape of the indenter tip needs to be measured for determination of hardness and Young's modulus of elasticity. Since the indenter is quite blunt, direct imaging of indentations of small size in the SEM is difficult. Determination of tip area function will be discussed later.

10.2.3.4   Indentation Procedure

The indenter procedure in this section is based on Anonymous (1991). An indentation test involves moving the indenter to the surface of the material and measuring the forces and displacements associated during indentation. The surface is located for each indentation by lowering the indenter at a constant loading rate against the suspending springs and detecting a change in velocity on contact with the surface. In the testing mode, the load is incremented in order to maintain a constant loading rate or constant displacement rate. The load and indentation depths are measured during indentation both in the loading and unloading cycles. The force contribution of the suspending springs and the displacements associated with the measured compliance of the instrument are removed.

Prior to indentation of test region on the sample surface, the scheme for the indentation pattern is selected. Typical indentation experiment consists of combination of several segments e.g., approach, load, hold, and unload, which can be programmed, Figure 10.15. Two typical examples are shown in Table 10.2 (Oliver and Pharr, 1992). A maximum of 12 segments can be programmed for an experiment. After

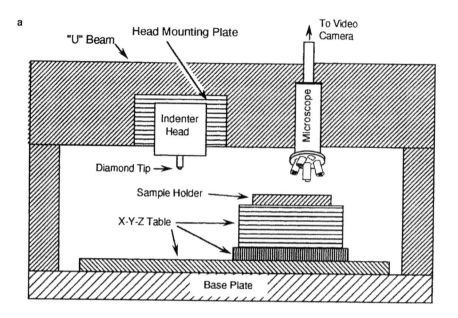

a

"U" Beam

Head Mounting Plate

To Video Camera

Indenter Head

Microscope

Diamond Tip →

Sample Holder →

X-Y-Z Table →

Base Plate

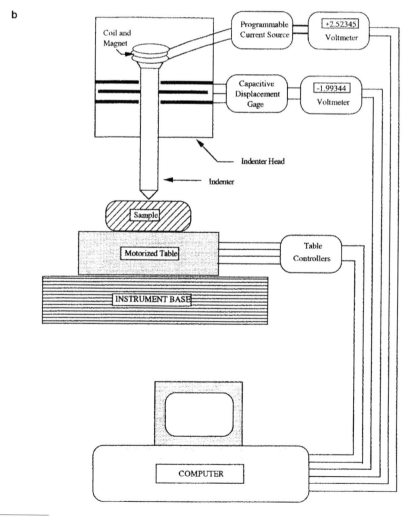

b

Coil and Magnet

Programmable Current Source

+2.52345 Voltmeter

Capacitive Displacement Gage

-1.99344 Voltmeter

Indenter Head

Indenter

Sample

Motorized Table

Table Controllers

INSTRUMENT BASE

COMPUTER

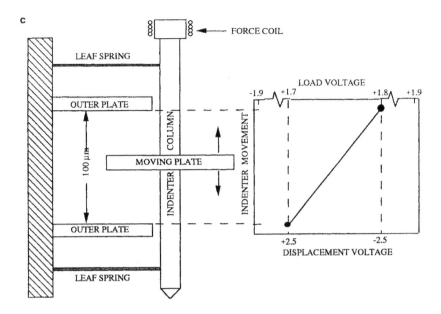

FIGURE 10.13 Schematics of the Nanoindenter II, (a) showing the major components — the indenter head, an optical microscope, and an *X–Y–Z* motorized precision table, (b) showing the details of indenter head and controls (microscope, which is directly behind the indenter, and the massive U bar are not shown for clarity), (c) showing the three-plate capacitor and indenter column that form the displacement-sensing system (not to scale). For the sake of clarity, only the left halves of the 50-mm-diameter, circular outer plates of the capacitor are shown. (Courtesy of Nano Instruments, Knoxville, TN.)

approach at a constant loading rate, the indenter is first loaded and unloaded typically three times in succession at a constant loading rate or displacement rate with each of the unloadings terminated at about 10 to 20% of the peak loading or displacement, respectively, to assure that contact is maintained between the specimen and the indenter. In a typical indentation experiment, it is usual to have two hold segments, the first one at the end of unloading to 10 to 20% after multiple loading/unloading cycles and the second one at the peak load just before final unloading. The reason for performing multiple loadings and unloadings is to examine the reversibility of the deformation (hysteresis) and thereby making sure that the unloading data used for calculation of the modulus of elasticity are mostly elastic. In some materials, there may be a significant amount of creep during the first unloading; thus, displacement recovered may not be entirely elastic, and because of this, the use of first unloading curves in the analysis of elastic properties can sometimes lead to inaccuracies. One way to minimize nonelastic effects is to include peak load hold periods in the loading sequence to allow time-dependent plastic effects to diminish. In addition, after multiple loadings, the load is held constant for a period of typically 100 s at 10 to 20% of the peak value while the displacement is carefully monitored to establish the rate of displacement produced by thermal expansion in the system. To account for thermal drift, the rate of displacement is measured during the last 80 s of the hold period, and the displacement data are corrected by assuming that this drift rate is constant throughout the test. Following this hold period, the specimen is loaded for a final time, with another 100 s hold period inserted at peak load to allow any final time-dependent plastic effects to diminish, and the specimen is fully unloaded. The final unloading curve is used for calculations of modulus of elasticity.

For an indentation experiment, the sample is placed on the mounting block. An appropriate region is selected by observing through the optical microscope. In a typical experiment, the tip of the indenter is moved toward the surface of the sample by gradually increasing the load on the indenter shaft. With a constant loading rate of typically 1 μN/s, the tip of the indenter travels downward at a velocity (approach rate) of about 10 nm/s. When the tip contacts the surface, its velocity drops below 1 nm/s, and the

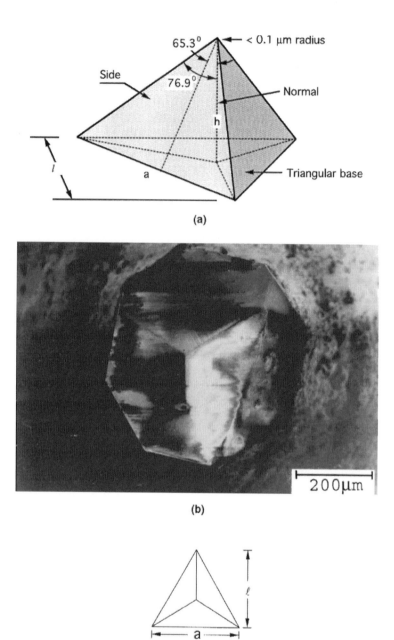

FIGURE 10.14   (a) Schematic, (b) photograph of the shape of a Berkovich indenter, and (c) indent impression.

computer records the displacement of the tip. This is the point at which the indentation experiment begins. The load is incremented in order to maintain a constant loading rate of about 10% of peak load/s or to maintain a constant displacement rate of 10% of peak displacement/s. The drift rate of the nanoindenter is about 0.01 nm/s; therefore, a loading duration of 10 s minimizes the measurement errors. Loading is followed by unloading and multiple loading/unloading and hold cycles.

Now we describe various steps in some detail based on Anonymous (1991). The first step is always the "Approach segment" in which the tip makes contact with the sample surface. The purpose of the approach segment is to determine accurately the "zero" of the indenter tip, that is, the values of the load

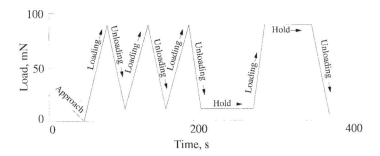

FIGURE 10.15 Different segments of a typical constant-loading indentation experiment.

TABLE 10.2 Examples of Two Typical Indentation Experiments

Segments[a]	Rate	Maximum Value
(a) Constant Loading Experiment		
Approach	10 nm/s	—
Loading at constant loading rate	12 mN/s	120 mN
Unloading at constant unloading rate	12 mN/s	10–20% of 120 mN
Multiple loading/unloading cycles (typically three times)		
•		
•		
Hold for 100 s	Data rate 1/s	100 points
Loading at constant loading rate	12 mN/s	120 mN
Hold for 100 s	Data rate 1/s	100 points
Unloading at constant unloading rate	12 mN/s	0 mN
(b) Constant Displacement Experiment		
Approach	10 nm/s	—
Loading at constant displacement rate	20 nm/s	200 nm
Unloading at constant displacement rate (typically three times)	20 nm/s	10–20% of 200 nm
•		
•		
Hold for 100 s	1 point/s	100 points
Loading at constant displacement rate	10 nm/s	200 nm
Hold for 100 s	1 point/s	100 points
Unloading at constant displacement rate	10 nm/s	0 nm

[a]In routine experiments the number of segments does not exceed more than 12, but if needed this number could be high.

and displacement at the point where the tip just touches the sample surface. Based on Anonymous (1991), these values are obtained in the following manner. The computer moves the sample from the microscope to a point below the indenter such that the position selected for the initial indent is offset from the indenter by a user-selected distance and angle (the default values are 50 μm and 180°). With the center plate of the capacitor (to which the indenter is attached) on its bottom stop, the table moves upward at a relatively high rate of speed until the indenter contacts the surface and makes the initial surface-finding indent. When contact occurs, the indenter is pushed upward, tripping the Z-motor interrupts and stopping the Z-motor of the table. The table is then moved downward at slow speed for 15 s before being moved in the X, Y plane until the point on the sample halfway between the initial surface-finding indent and the location of the first indent is under the indenter. The table is now raised once more, but at slow speed until contact with the specimen is made once more. This second contact gives the best estimate of the surface elevation that can be obtained by moving the table alone.

After this second surface-finding indent is made, the indenter is left in contact with the surface under a very small load with the displacement-sensing capacitor near the center of its travel. At this point the system is allowed to monitor changes in indenter displacement under constant load, and when the drift rate becomes smaller than the user-prescribed maximum (usually 0.05 nm/s), the displacement of the indenter is recorded, establishing an initial estimate of the elevation of the sample surface.

The indenter is then raised to near the top of its travel using the coil/magnet assembly (the elevation of the table remains fixed for the rest of the experiment), and the table is moved so that the chosen location for the first indent of the specified shape is under the indenter. The indenter is now lowered toward the surface at a rate of several hundred nanometers per second until the "surface search distance" is reached. The surface search distance is a user-specified distance (usual 1000 to 2000 nm) above the estimated elevation of the sample surface. At this point, the rate of approach to the surface is decreased to approximately 10 nm/s, and the load–displacement values that are constantly recorded are used to calculate the stiffness of the system as reflected initially in the stiffness of the very flexible leaf springs that support the indenter shaft. When the indenter finally reaches the surface, a large increase in stiffness is sensed, and when the stiffness increases by a factor of 4, the approach phase of the indentation process is complete.

The computer now discards all but the last 50 sets of load–displacement data taken during the approach. A plot of load vs. displacement for these data reflects the point of contact of the indenter with the sample surface in terms of a very sharp change in slope of the load–displacement plot (see Figure 10.16). For an approach rate of 10 nm/s and factor of four increase in stiffness, experience has shown that surface contact is made at the 13th or 14th data point from the end of the 50-data-point set. The zero points for both load and displacement are then taken as the averages of the loads and displacements of 12th and 13th data sets from the end of the approach data. For many materials this procedure identifies the sample surface to within 0.1 to 0.2 nm. However, for very soft materials such as many polymers or for other approach rates and stiffness-factor increases, the user may find it advisable to plot the approach segment data and, if necessary, change the algorithm used to define the precise point of contact with the sample surface.

Once surface contact is established, the other segments of the indentation process are carried out as prescribed in the programmed indentation experiment. The final segment always involves load removal. When the voltage on the indenter coil passes the displacement voltage at which the surface was detected in the approach portion of the cycle, the current through the coil is fixed while the raw data are recorded on the hard disk, and plotted on the computer monitor. The indenter is then raised well away from the surface in preparation for moving the sample to the position of the next indent. For subsequent indents in a given series of indents, the initial estimate of surface position used is that found in making the previous indent.

For each indentation step, load voltages, displacement (penetration depth or indentation depth) voltages, and real time are recorded in separate files. These raw voltage data are converted to load vs. displacement data by using load and displacement calibration constants. From the displacement data, the contact depth is calculated for calculations of the hardness. The slope of the unloading curve is used to calculate the modulus of elasticity.

10.2.3.5 Acoustic Emission Measurements during Indentation

AE measurement is a very sensitive technique to monitor cracking of the surfaces and subsurfaces. The nucleation and growth of cracks result in a sudden release of energy within a solid; then some of the energy is dissipated in the form of elastic waves. These waves are generated by sudden changes in stress and in displacement that accompany the deformation. If the release of energy is sufficiently large and rapid, then elastic waves in the ultrasonic frequency regime (AE) will be generated and these can be detected using PZTs via expansion and compression of the PZT crystals (Yeack-Scranton, 1986; Scruby, 1987; Bhushan, 1996).

Weihs et al. (1992) used an AE sensor to detect cracking during indentation tests using the nanoindenter. The energy dissipated during crack growth can be estimated by the rise time of the AE signal. They mounted a commercial transducer with W-impregnated epoxy backing for damping underneath

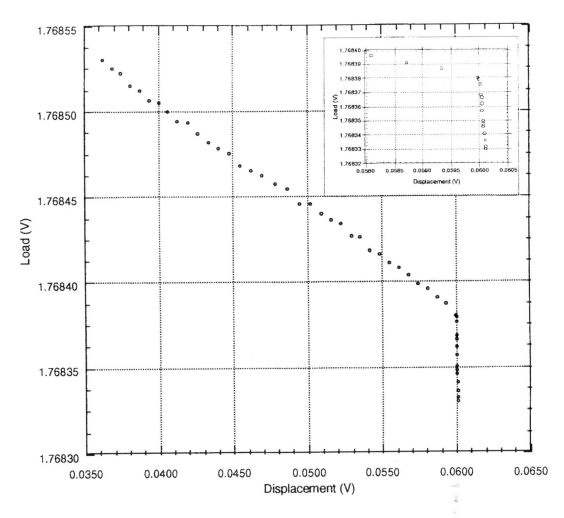

FIGURE 10.16 A plot of load vs. displacement (expressed in volts) for a typical approach segment of an indentation obtained using the lowest load range and with an approach rate of about 10 nm/s. A factor of four increase in stiffness was used as the criterion for terminating this segment. The very sharp "knee" at the right end of the plot indicates surface contact. Data near the knee of the curve are shown on an expanded scale in the inset. Note that the displacement voltage changes by only 0.1 mV (corresponding to a displacement of about 1.3 nm) between the surface contact and the end of the approach segment. Thus, it is obvious that the position of the surface (i.e., displacement voltage corresponding to the surface contact) can be specified to about ±0.1 nm. (Courtesy of Nano Instruments, Knoxville, TN.)

the sample as shown in Figure 10.17. The transducer converts the AE signal into voltage that is amplified by oscilloscopes and used for continuous display of the AE signal. Any correlation between the AE signal and the load–displacement curves can be observed. (Also see Wu et al., 1990b and Wu, 1991.)

10.2.3.6 Nanoscratch and Tangential Force Measurements

Several micro- and nanoscratch testers are commercially available, such as the Taber shear/scratch tester model 502 with a no. 139-58 diamond cutting tool (manufactured by Teledyne Taber, North Tonawanda, NY) for thick films; Revetest automatic scratch tester (manufactured by Centre Suisse d' Electronique et de Microtechnique S.A, CH-2007, Neuchatel, Switzerland) for thin films (Perry, 1981, 1983; Steinmann et al., 1987; Sekler et al., 1988), and nanoindenter for ultrathin films (Wu et al., 1989, 1990b; Wu, 1990, 1991; Anonymous, 1991; Bhushan et al., 1995, 1996, 1997; Bhushan and Gupta, 1995; Gupta and Bhushan, 1995a,b; Patton and Bhushan, 1996; Bhushan and Li, 1997; Li and Bhushan, 1998b,c).

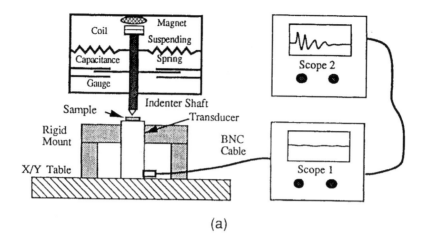

(a)

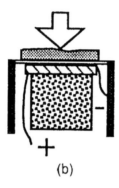

(b)

FIGURE 10.17 (a) Schematic of nanoindenter with an AE transducer, (b) schematic of commercial transducer with W-impregnated epoxy backing for damping. (From Weihs, T.P. et al., 1992, in *Thin Films: Stresses and Mechanical Properties III, Symp. Proc.*, W.E. Nix, et al., eds., Vol. 239, pp. 361–370, Materials Research Society, Pittsburgh. With permission.)

We describe the nanoscratch and tangential force option which allows making of the scratches of various lengths at programmable loads. Tangential (friction) forces can also be measured simultaneously (Anonymous, 1991). The additional hardware for the tangential force option includes a set of proximity (capacitance) probes for measurement of lateral displacement or force in the two lateral directions along x and y, and a special "scratch collar" which mounts around the indenter shaft with hardness indenter, Figure 10.18. The scratch collar consists of an aluminum block, mounted around the indenter shaft, with four prongs descending from its base. Two of these prongs hold the proximity probes and the set screws set them in place, while the other two prongs hold position screws (and corresponding set screws). The position screws serve a dual purpose; they are used to limit the physical deflection of the indenter shaft, and they are used to lock the indenter shaft in place during tip change operations. A scratch block is mounted on the end of the indenter shaft, in line with the proximity probes and the positioning screws. Finally, the scratch tip itself is mounted on the end of the indenter shaft, covering the scratch tip. The scratch tip is attached to the scratch block with two Allen head screws. The scratch tip can be a Berkovich indenter or a conventional conical diamond tip with a tip radius of about 1 to 5 μm and an included angle of 60 to 90° (typically 1 μm of tip radius with 60° of included angle, Wu et al., 1990a,b; Wu, 1991). A larger included angle of 90° may be desirable for a more durable tip. The tip radius should not be very small as it will get blunt readily.

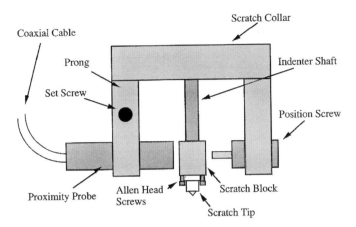

FIGURE 10.18 Schematic of the tangential force option hardware (not to scale and the front and rear prongs not shown). (Courtesy of Nano Instruments, Knoxville, TN.)

During scratching a load is applied up to a specified indentation load or up to a specified indentation depth, and the lateral motion of the sample is measured. In addition, of course, load and indentation depth are monitored. Scratches can be made either at the constant load or at ramp-up load. Measurement of lateral force allows the calculations of the coefficient of friction during scratching. The resolution of the capacitance proves to measure tangential load is about 50 μN; therefore, a minimum load of about 0.5 mN can be measured or a minimum normal load of about 5 mN should be used for a sample with coefficient of friction of about 0.1. Microscopy of the scratch produced at ramp-up load allows the measurement of critical load required to break up of the film (if any) and scratch width and general observations of scratch morphology.

Additional parameters that are used to control the scratch are scratch length (μm), draw acceleration (μm/s^2), and draw velocity (μm/s). The latter parameters control the speed with which the scratch is performed. The default values of 10 μm/s^2 and 10 μm/s provide safe rates for performing the scratch. Draw velocity is limited by the maximum rate of data acquisition (during a scratch the maximum rate is approximately 2/s) and the length of the desired scratch. Thus, a scratch with a desired 20 points over 1 mm must have a draw velocity no greater than 100 μm/s.

The lateral deflection calibration is performed with a calibrated cantilever beam (of known stiffness) mounted on the specimen tray. Lateral force applied to the indenter shaft with the beam can be used to determine response of the indenter shaft in micrometers per volt of probe or newton per volt (from known lateral stiffness of the indenter shaft). Calibration for the cross talk of the probe (resulting from the probe surface not being parallel to the axis of motion in the vertical direction) also needs to be performed. For this calibration, the scratch tip is moved up and down and any output of the proximity probe in volts of vertical motion per volt of proximity probe is measured.

10.3   Analysis of Indentation Data

An indentation curve is the relationship between load W and displacement (or indentation depth or penetration depth) h, which is continuously monitored and recorded during indentation. Stress–strain curves, typical indentation curves, the deformed surfaces after tip removal, and residual impressions of indentation for ideal elastic, rigid–perfectly plastic and elastic–perfectly plastic and real elastic–plastic solids are shown in Figure 10.19. For an elastic solid, the sample deforms elastically according to Young's modulus, and the deformation is recovered during unloading. As a result, there is no impression of the indentation after unloading. For a rigid–perfectly plastic solid, no deformation occurs until yield stress is reached, when plastic flow takes place. There is no recovery during unloading and the impression

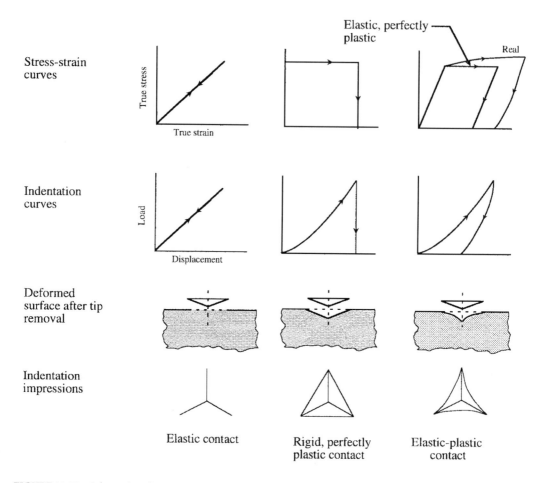

FIGURE 10.19 Schematics of stress–strain curves, typical indentation curves, deformed surfaces after tip removal, and residual impressions of indentation, for ideal elastic, rigid–perfectly plastic, elastic–perfectly plastic (ideal), and real elastic–plastic solids.

remains unchanged. In the case of elastic–plastic solid, it deforms elastically according to Young's modulus and then it deforms plastically. The elastic deformation is recovered during unloading. In the case of an elastic–perfectly plastic solid, there is no work hardening.

All engineering surfaces follow real elastic–plastic deformation behavior with work hardening (Johnson, 1985). The deformation pattern of a real elastic–plastic sample during and after indentation is shown schematically in Figure 10.20. In this figure we have defined the contact depth (h_c) as the depth of indenter in contact with the sample under load. The depth measured during the indentation (h) includes the depression of the sample around the indentation in addition to the contact depth. The depression of the sample around the indentation ($h_s = h - h_c$) is caused by elastic displacements and must be subtracted from the data to obtain the actual depth of indentation or actual hardness. At peak load, the load and displacement are W_{max} and h_{max}, respectively, and the radius of the contact circle is a. Upon unloading, the elastic displacements in the contact region are recovered and, when the indenter is fully withdrawn, the final depth of the residual hardness impression is h_f.

Schematic of a load–displacement curve is shown in Figure 10.21. Based on the work of Sneddon (1965) to predict the deflection of the surface at the contact perimeter for a conical indenter and a paraboloid of revolution, Oliver and Pharr (1992) developed an expression for h_c at the maximum load (required for hardness calculation) from h_{max},

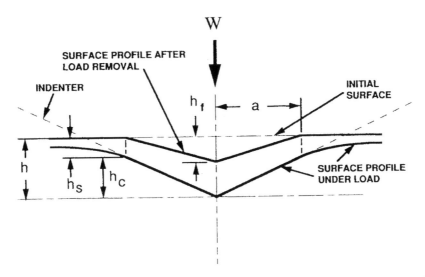

FIGURE 10.20 Schematic representation of the indenting process illustrating the depression of the sample around the indentation and the decrease in indentation depth upon unloading. (From Oliver, W.C. and Pharr, G.M., 1992, *J. Mater. Res.* 7, 1564–1583. With permission.)

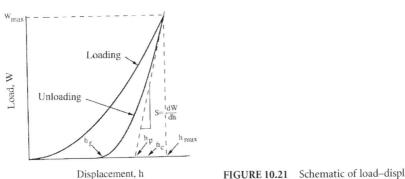

FIGURE 10.21 Schematic of load–displacement curve.

$$h_c = h_{max} - \varepsilon \, W_{max} / S_{max} \tag{10.3a}$$

where $\varepsilon = 0.72$ for the conical indenter, $\varepsilon = 0.75$ for the paraboloid of revolution, and $\varepsilon = 1$ for the flat punch; S_{max} is the stiffness (= 1/compliance) equal to the slope of unloading curve (dW/dh) at the maximum load. Oliver and Pharr assumed that behavior of Berkovich indenter is similar to that of conical indenter, since cross-sectional areas of both types of indenters varies as the square of the contact depth and their geometries are singular at the tip. Therefore, for Berkovich indenter, $\varepsilon \sim 0.72$. Thus, h_c is slightly larger than plastic indentation depth (h_p), which is given by

$$h_p = h_{max} - W_{max} / S_{max} \tag{10.3b}$$

We note that Doerner and Nix (1986) had underestimated h_c by assuming that $h_c = h_p$. Based on the finite element analysis of the indentation process, Laursen and Simo (1992) showed that h_c cannot be assumed equal to h_p for indenters which do not have flat punch geometry.

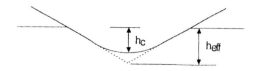

FIGURE 10.22 Schematic of an indenter tip with a nonideal shape. The contact depth and the effective depth are also shown.

For a Vickers indenter with ideal pyramidal geometry (*ideally sharp tip*), *projected* contact-area-to-depth relationship is given as (Doerner and Nix, 1986; Bhushan, 1996)

$$A = 24.5 \, h_c^2 \tag{10.4a}$$

Since the area-to-depth relationship is equivalent for both typical Berkovich and Vickers pyramids, Equation 10.4 holds for the Berkovich indenter as well. Although we have derived a slightly different expression for $A(h)$ presented in Equation 10.2 for the assumed Berkovich indenter geometry, we use Equation 10.4 for $A(h)$ in this chapter, as this relationship is most commonly used in the analysis of the indentation hardness data.

The indenter tip is generally rounded so that ideal geometry is not maintained near the tip (Figure 10.22). To study the effect of tip radius on the elastic–plastic deformation (load vs. displacement curve), Shih et al. (1991) modeled the blunt-tip geometry by a spherical tip of various radii. They derived a geometric relationship (assuming no elastic recovery) between the projected contact area of the indenter to the actual contact depth. Figure 10.23a shows the measured contact area vs. indentation depth data by Pethica et al. (1993) for nickel and by Doerner and Nix (1986) for annealed a-brass. From this figure, it seems that a tip radius of 1 μm fits the data best. If there is elastic recovery, the experimental data are smaller than what they should be, and then the tip radius would be even larger than 1 μm. Shih et al. (1991) used the finite-element method to simulate an indentation test. They showed that load–indentation depth data obtained using nanoindenter for nickel by Pethica et al. (1983) can be fitted with a simulated profile for a tip radius of about 1 μm, Figure 10.23b.

As shown in Figure 10.20, the actual indentation depth, h_c, produces a larger contact area than would be expected for an indenter with an ideal shape. For the real indenter used in the measurements, the nominal shape is characterized by an area function $F(h_c)$, which relates projected contact area of the indenter to the contact depth (Equation 10.4a),

$$A^{1/2} = F\!\left(h_c\right) \tag{10.4b}$$

The functional form must be established experimentally prior to the analysis (to be described later).

10.3.1 Hardness

Berkovich hardness HB (or H_B) is defined as the load divided by the projected area of the indentation. It is the mean pressure that a material will support under load. From the indentation curve, we can obtain hardness at the maximum load as

$$HB = W_{\max} / A \tag{10.5}$$

where $W_{\max}$ is the maximum indentation load and A is the projected contact area at the peak load. The contact area at the peak load is determined by the geometry of the indenter and the corresponding contact depth h_c using Equation 10.3a and 10.4b. A plot of hardness as a function of indentation depth for polished single-crystal silicon (111), with and without tip shape calibration, is shown in Figure 10.24.

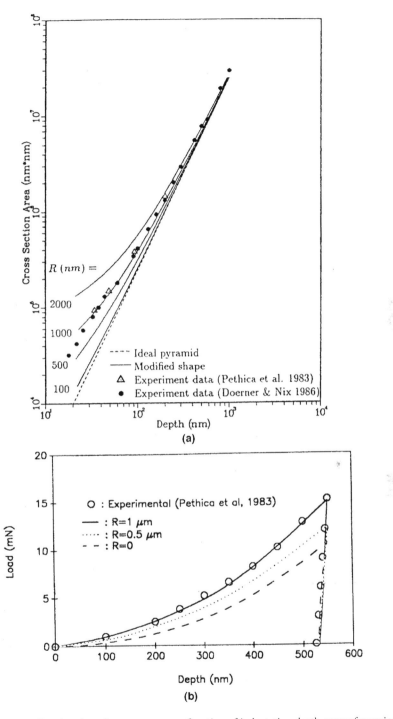

FIGURE 10.23 (a) Predicted projected contact area as a function of indentation depth curves for various tip radii and measured data; (b) predicted load as a function of indentation depth curves for various tip radii and measured data. (From Shih, C.W. et al., 1991, *J. Mater. Res.* 6, 2623–2628. With permission.)

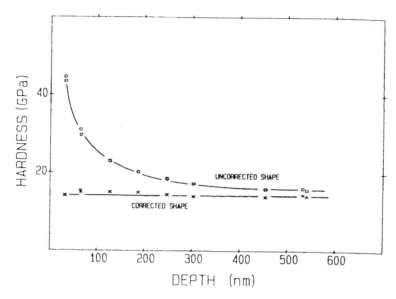

FIGURE 10.24 Hardness as a function of indentation depth for polished single-crystal silicon (111) calculated from the area function with and without tip shape calibration. (From Doerner, M.F. and Nix, W.D., 1986, *J. Mater. Res.* 1, 601–609. With permission.)

We note that, for this example, tip shape calibration is necessary and the hardness is independent of corrected depth. The hardness at any load used during indentation can be calculated, although generally hardness only at peak load is calculated. For measurement of hardness at various loads, indentation experiments at various peak loads corresponding to desired loads are generally carried out.

It should be pointed out that hardness measured using this definition may be different from that obtained from the more conventional definition in which the area is determined by direct measurement of the size of the residual hardness impression. The reason for the difference is that, in some materials, a small portion of the contact area under load is not plastically deformed, and, as a result, the contact area measured by observation of the residual hardness impression may be less than that at peak load. However, for most materials, measurements using two techniques give similar results.

10.3.2 Modulus of Elasticity

Even though during loading a sample undergoes elastic–plastic deformation, the initial unloading is an elastic event. Therefore, the Young's modulus of elasticity or, simply, the elastic modulus of the specimen can be inferred from the initial slope of the unloading curve (dW/dh) called stiffness (1/compliance) (at the maximum load) (Figure 10.21). It should be noted that the contact stiffness is measured only at the maximum load, and no restrictions are placed on the unloading data being linear during any portion of the unloading.

If the area in contact remains constant during initial unloading, an approximate elastic solution is obtained by analyzing a flat punch whose area in contact with the specimen is equal to the projected area of the actual punch. Based on the analysis of indentation of an elastic half space by a flat cylindrical punch by Sneddon (1965), Loubet et al. (1984) calculated the elastic deformation of an isotropic elastic material with a flat-ended cylindrical punch. They obtained an approximate relationship for compliance (dh/dW) for the Vickers (square) indenter. King (1987) solved the problem of flat-ended cylindrical, quadrilateral (Vickers and Knoop), and triangular (Berkovich) punches indenting an elastic half-space. He found that the compliance for the indenter is approximately *independent of the shape* (with a variation of at most 3%) if the projected area is fixed. Pharr et al. (1992) also verified that compliance of a paraboloid of revolution of a smooth function is the same as that of a spherical or a flat-ended cylindrical punch.

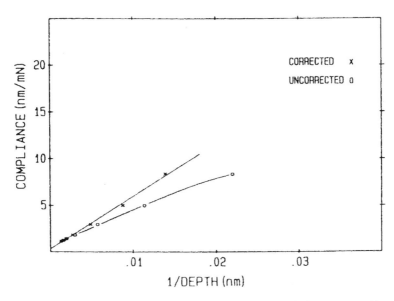

FIGURE 10.25 Compliance as a function of the inverse of indentation depth for tungsten with and without tip shape calibration. A constant modulus with 1/depth would be indicated by the straight line. The slope of the corrected curve is 480 GPa, which compares reasonably well to the known modulus of tungsten (420 GPa). The small *y*-intercept of about 0.3 nm/mN is attributed to load-frame compliance, not removed. (From Doerner, M.F. and Nix, W.D., 1986, *J. Mater. Res.* 1, 601–609. With permission.)

The relationship for the compliance *C* (inverse of stiffness *S*) for an (Vickers, Knoop, and Berkovich) indenter is given as

$$C = \frac{1}{S} = \frac{dh}{dW} \sim \frac{1}{2E_r}\left(\frac{\pi}{A}\right)^{1/2} \tag{10.6}$$

where

$$\frac{1}{E_r} = \frac{1-\nu_s^2}{E_s} + \frac{1-\nu_i^2}{E_i},$$

dW/dh is the slope of the unloading curve at the maximum load (Figure 10.21), E_r, E_s, and E_i are the reduced modulus and elastic moduli of the specimen and the indenter, and n_s and n_i are the Poisson's ratios of the specimen and indenter. *C* (or *S*) is the experimentally measured compliance (or stiffness) at the maximum load during unloading, and *A* is the projected contact area at the maximum load.

The contact depth h_c is related to the projected area of the indentation *A* for a real indenter by Equation 10.4b. A plot of the measured compliance (dh/dW) vs. the reciprocal of the corrected indentation depth obtained from various indentation curves (one data point at maximum load for each curve) should yield a straight line with slope proportional to $1/E_r$ (Figure 10.25) (Doerner and Nix, 1986). E_s can then be calculated, provided Poisson's ratio with great precision is known to obtain a good value of the modulus. For a diamond indenter, $E_i = 1140$ GPa and $n_i = 0.07$ are taken. In addition, the *y*-intercept of the compliance vs. the reciprocal indentation depth plot should give any additional compliance that is independent of the contact area. The compliance of the loading column is generally removed from the load–displacement curve, whose measurement techniques will be described later.

We now discuss a preferred method to measure initial unloading stiffness (*S*). Doerner and Nix (1986) measured *S* by fitting a straight line about one third upper portion of the unloading curve. The problem with this is that for nonlinear loading data, the measured stiffness depends on how much of the data is used in the fit. Oliver and Pharr (1992) proposed a new procedure. They found that the entire unloading data are well described by a simple power law relation

$$W = B\left(h - h_f\right)^m \tag{10.7}$$

where the constants B and m are determined by a least-square fit. The initial unloading slope is then found analytically, differentiating this expression and evaluating the derivative at the maximum load and maximum depth. As we have pointed out earlier, unloading data used for the calculations should be obtained after several loading/unloading cycles and with peak hold periods.

10.3.3 Determination of Load Frame Compliance and Indenter Area Function

As stated earlier, measured displacements are the sum of the indentation depths in the specimen and the displacements of suspending springs and the displacements associated with the measuring instruments, referred to as load frame compliance. Therefore, to determine accurately the specimen depth, load frame compliance must be known. This is especially important for large indentations made with high modulus for which the load frame displacement can be a significant fraction of the total displacement. The exact shape of the diamond indenter tip needs to be measured because hardness and Young's modulus of elasticity depend on the contact areas derived from measured depths. The tip gets blunt (Figure 10.22) and its shape significantly affects the prediction of mechanical properties (Figures 10.23 through 10.25).

The method used in the past for determination of the area function has been to make a series of indentations at various depths in materials in which the indenter displacement is predominantly plastic and to measure the size of the indentations by direct imaging. Optical imaging cannot be used to measure submicron-size impressions accurately. Because of the shallowness of the indent impressions, SEM results in poor contrast.

A method consists of making two-stage carbon replicas of indentations in a soft material and imaging them in the transmission electron microscope (TEM), was used initially by Pethica et al. (1983). Doerner and Nix (1986) produced a series of indentations of varying size in annealed a–brass. Cellulose acetate replicating tape was applied to the sample with a drop of acetone. Platinum with 20% palladium was used as a shadowing agent. The indentations were oriented such that one side of the triangular perimeter of each indentation was perpendicular to the shadowing direction. A shadowing angle of 19° was used. Following shadowing, a carbon film was evaporated onto the replica and the cellulose acetate removed by dissolving in acetone. Doerner and Nix then imaged the prepared replicas at zero tilt in the transmitted electron mode in the TEM. The areas of the indentations were measured and compared to the contact depths as measured using the nanoindenter. An example of the calibration curve is shown in Figure 10.23a. We can clearly see that use of the ideal geometry results in a large overestimate of the hardness and modulus at small depths since the indenter tip is considerably more blunt than the ideal pyramid. A calibration curve of the type shown in Figure 10.23a is used to determine the effective indentation depth, h_{eff}. The effective indentation depth is the depth needed for a pyramid of ideal geometry to obtain a projected contact area equivalent to that of a real pyramid. Since the projected area of ideal geometry of a Berkovich indenter is 24.5 h_c^2, the effective indentation depth can be obtained from the following equation:

$$h_{eff} = \left(\frac{\text{Area}}{24.5}\right)^{1/2} \tag{10.8}$$

where the area is obtained from the shape calibration and the true contact depth (Doerner and Nix, 1986). It is observed that ideal geometry underestimates the contact area which leads to overestimation

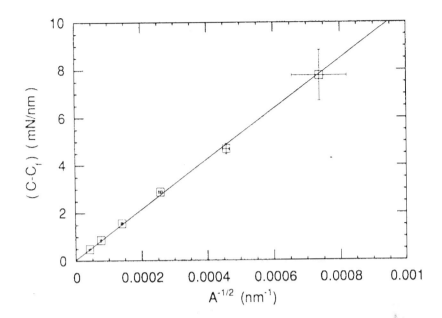

FIGURE 10.26 Plot of $(C - C_f)$ as a function $A^{-1/2}$ for aluminum. The error bars are two standard deviations in length. (From Oliver, W.C. and Pharr, G.M., 1992, *J. Mater. Res.* 7, 1564–1583. With permission.)

of hardness and modulus of elasticity especially at small indentation depths. For an example, see Figures 10.24 and 10.25.

Oliver and Pharr (1992) proposed an easier method for determining area functions that requires no imaging. Their method is based only on *one* assumption that Young's modulus is independent of indentation depth. They also proposed a method to determine load frame compliance. We first describe the methods for determining of load frame compliance followed by the method for area function. They modeled the load frame and the specimen as two springs in series; thus,

$$C = C_s + C_f \qquad (10.9)$$

where C, C_s, and C_f are the total measured compliance, specimen compliance, and load frame compliance, respectively. From Equations 10.6 and 10.9, we get

$$C = C_f + \frac{1}{2E_r}\left(\frac{\pi}{A}\right)^{1/2} \qquad (10.10)$$

From Equation 10.10, we note that if the *modulus of elasticity is constant*, a plot of C as a function of $A^{1/2}$ is linear and the vertical intercept gives C_f. It is obvious that most accurate values of C_f are obtained when the specimen compliance is small, i.e., for large indentations.

To determine the area function and the load frame compliance, Oliver and Pharr made relatively large indentations in aluminum because of its low hardness. In addition, for the larger aluminum indentations (typically 700 to 4000 nm deep), the area function for a perfect Berkovich indenter (Equation 10.4a) can be used to provide a first estimate of the contact. Values of C_f and E_r are thus obtained by plotting C as a function of $A^{-1/2}$ for the large indentations, Figure 10.26.

Using the measured C_f value, they calculated contact areas for indentations made at shallow depths on the aluminum with measured E_r and/or on a harder fused silica surface with published values of E_r, by rewriting Equation 10.10 as

TABLE 10.3 Peak Loads and Loading/Unloading Rates Used in the Load Frame Compliance and Area Function Calibration Procedures

Indentation Numbers	Peak Load (mN)	Loading/Unloading Rate (μN/s)	Nanoindenter Load Range
For Load-Frame Compliance			
1–10	120	12000	High
11–20	60	6000	High
21–30	30	3000	High
31–40	15	1500	High
41–50	7.5	750	High
51–60	3	300	High
For Area Function			
61–70	20	2000	Low
71–80	10	1000	Low
81–90	3	300	Low
91–100	1	100	Low
101–110	0.3	30	Low
111–120	0.1	10	Low

From Oliver, W.C. and Pharr, G.M., 1992, *J. Mater. Res.* 7, 1564–1583. With permission.

$$A = \frac{\pi}{4} \frac{1}{E_r^2} \frac{1}{\left(C - C_f\right)^2} \tag{10.11}$$

from which an initial guess at the area function was made by fitting A as a function h_c data to an eighth-order polynomial

$$A = 24.5\, h_c^2 + C_1\, h_c + C_2\, h_c^{1/2} + C_3\, h_c^{1/4} + \ldots + C_8\, h_c^{1/128} \tag{10.12}$$

where C_1 through C_8 are constants. The first term describes the perfect shape of the indenter; the others describe deviations from the Berkovich geometry due to blunting of the tip. A convenient fitting routine is that contained in the Kaleidagraph software for Apple Macintosh computers. A weighted procedure can be used to assure that data points with small and large magnitudes are of equal importance. An iterative approach can be used to refine the values of C_f and E_r further.

Now we describe the step-by-step procedure in detail, recommended by Oliver and Pharr (1992), for determination of load frame compliance and indenter area function. It involves making a series of indentations in two standard materials — aluminum and fused quartz — and relies on the facts that both these materials are elastically isotropic, their moduli are well known, and their moduli are independent of indentation depth. The first step is to determine precisely the load frame compliance. This is best accomplished by indenting a well-annealed, high-purity aluminum which is chosen because it is readily available, has a low hardness, and is nearly elastically isotropic. Some care must be exercised in preparing the aluminum to assure that its surface is smooth and unaffected by work hardening. A series of indentations are made in the aluminum using the first six peak loads, and loading rates shown in Table 10.3. Typically indentation depths range from 700 to 4000 nm. The load time history recommended by Oliver and Pharr is as follows: (1) approach and contact surface, (2) load to peak load, (3) unload to 90% of peak load and hold for 100 s, (4) reload to peak load and hold for 10 s, and (5) unload completely at half the rate shown in Table 10.3. The lower hold is used to establish thermal drift and the upper hold to minimize time-dependent plastic effects. The final unloading data are used to determine the unloading compliances using the power law fitting procedure described earlier. The load frame compliance is

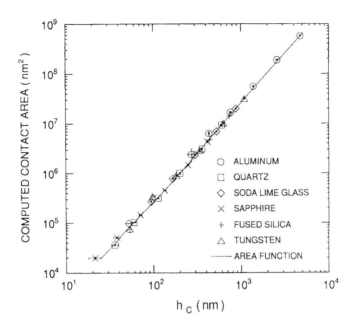

FIGURE 10.27 The computed contact areas as a function of contact depths for six materials. The error bars are two standard deviations in length. (From Oliver, W.C. and Pharr, G.M., 1992, *J. Mater. Res.* 7, 1564–1583. With permission.)

determined from the aluminum data by plotting the measured compliance as a function of area calculated, assuming the ideal Berkovich indenter. Calculated E_r is checked with known elastic constants for aluminum, $E = 70.4$ GPa and $n = 0.347$. The values of the elastic constants we use for the diamond indenter are $E_i = 1141$ GPa and $n_i = 0.07$.

The problem with using aluminum to extend the area function to small depths is that because of its low hardness, small indentations in aluminum require very small loads, and a limit is set by the force resolution of the indentation system. This problem can be avoided by making the small indentations in fused quartz, a much harder, isotropic material available in optically finished plate form. The standard procedure that Oliver and Pharr recommend for determining the area function involves making a series of indentations in fused quartz using the second set of six peak loads shown in Table 10.3. For the loads outlined in Table 10.3, the minimum contact depth is about 15 nm and the maximum about 4700 nm. (Typically measurements are made at depths ranging from about 15 to 700 nm. Above 700 nm of depth, indenter can be assumed to have a perfect shape.) The contact areas and contact depths are then determined using Equation 10.12 and h_c in conjunction with the reduced modulus computed from the elastic constant for fused quartz, $E = 72$ GPa and $n = 0.170$. The machine compliance is known from the aluminum analysis. The area function is only good for the depth range used in the calculations. Typical data of contact areas as a function of contact depths for six materials is shown in Figure 10.27.

10.3.4 Hardness/Modulus² Parameter

Calculations of hardness and modulus described so far require the calculations of the indent projected area from the indentation depth, which are based on the assumption that the test surface be smooth to dimensions much smaller than the projected area. Therefore, data obtained from rough samples show considerable scatter. Joslin and Oliver (1990) developed an alternative method for data analysis *without* requiring the calculations of the projected area of the indent. This method provides measurement of a parameter hardness/modulus², which provides a *measure of the resistance of the material to plastic penetration.*

They showed that for several types of rigid punches (cone, flat punch, parabola of revolution, and sphere) as long as there is a single contact between the indenter and the specimen,

$$H/E_r^2 = \left(4/\pi\right)\left(W/S^2\right) \tag{10.13}$$

where S is the stiffness obtained from the unloading curve. E_r is related to E_s by a factor of $1 - n_s^2$ for materials with moduli significantly less than diamond (Equation 10.6). H/E_s^2 parameter represents a materials resistance to plastic penetration. We clearly see that calculation of projected area and knowledge of area function are not required. However, this method does not give the hardness and modulus values separately.

10.3.5 Continuous Stiffness Measurement

Oliver and Pethica (1989) and Pethica and Oliver (1989) developed a dynamic technique for continuous measurement of sample stiffness during indentation without the need for discrete unloading cycles, and with a time constant that is at least three orders of magnitude smaller than the time constant of the more conventional method of determining stiffness from the slope of an unloading curve. Furthermore, the measurements can be made at exceedingly small penetration depths. (Also see Wu, 1993.) Thus, their method is ideal for determining the stiffness and, hence, the elastic modulus and hardness of films a few tens of a nanometers thick. Furthermore, its small time constant makes it especially useful in measuring the properties of some polymeric materials.

Measurement of continuous stiffness is accomplished by the superposition of a very small AC current of a known relatively high frequency (typically 69.3 Hz) on the loading coil of the indenter. This current, which is much smaller than the DC current that determines the nominal load on the indenter, causes the indenter to vibrate with a frequency related to the stiffness of the sample and to the indenter contact area. A comparison of the phase and amplitude of the indenter vibrations (determined with a lock-in amplifier) with the phase and amplitude of the imposed AC signal allows the stiffness to be calculated either in terms of amplitude or phase. Figure 10.28 is a schematic wiring diagram illustrating the operation of continuous-stiffness option. The displacement as small as 0.001 nm can be measured using frequency-specific amplification. The time constant of about 0.33 s provides a good combination of low noise and dynamic response.

To calculate the stiffness of the contact zone, the dynamic response of the indentation system has to be determined. The relevant components are the mass m of the indenter, the spring constant K_0 of the leaf springs that support the indenter, the stiffness of the indenter frame K_f, and the damping constant C due to the air in the gaps of the capacitor plate displacement-sensing system. These combine with the stiffness of the contact zone (sample stiffness) S, as shown schematically in Figure 10.29 to produce the overall response. If the imposed driving force is $F = F_0 \exp(iwt)$ and the displacement response of the indenter is $a = a_0 \exp(iwt + f)$, the ratio of amplitudes of the imposed force and the displacement response is given by (Pethica and Oliver, 1989)

$$\frac{F_0}{a_0} = \left[\omega^2 c^2 + \left(K - m\omega^2\right)^2\right]^{1/2} \tag{10.14}$$

and the phase angle, f, between the driving force and the response is

$$\tan\phi = \omega c/\left(K - m\omega^2\right) \tag{10.15}$$

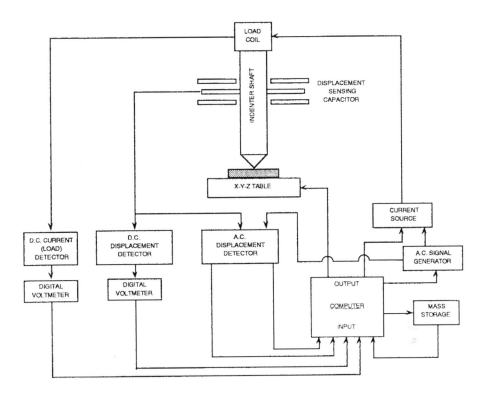

FIGURE 10.28 Schematic wiring diagram of the nanoindenter with the continuous stiffness measurement option. (Courtesy of Nano Instruments, Knoxville, TN.)

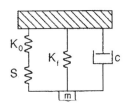

FIGURE 10.29 Components of the dynamic model of the indentation system. (From Pethica, J.B. and Oliver, W.C., 1989, in *Thin Films: Stresses and Mechanical Properties*, J.C. Bravman et al., eds., Vol. 130, pp. 13–23, Materials Research Society, Pittsburgh. With permission.)

where w is the frequency of the imposed force, c is the damping constant for the central plate of the displacement capacitor (damping due to the air in the capacitor gaps), and m is the mass of the indenter assembly. K, the combined spring constant for the system, is given by

$$\frac{1}{K} = \frac{1}{K_f} + \frac{1}{K_0 + S} \tag{10.16}$$

With the exception of S, all the terms in Equations 10.14 and 10.15 can be measured independently. Thus, the displacement signal resulting from the imposition of the DC current is measured with a phase-sensitive detector (lock-in amplifier), which yields both the amplitude and phase angle of the displacement signal. The AC input to the force coil is generated with a standard AC signal generator, and any frequency between about 10 and 150 Hz may be selected. The stiffness S can be determined either from phase angle f or from amplitude a_0 of response. Pethica and Oliver (1989) pointed out that f will be most sensitive to small values of S and a_0 will be best for larger values of S.

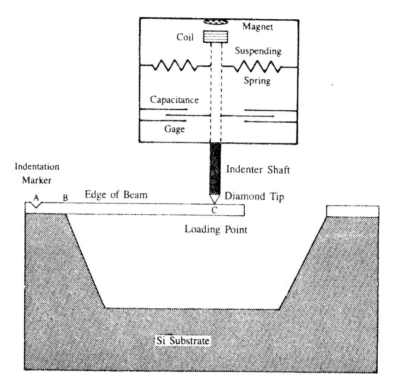

FIGURE 10.30 Schematic of a cantilever microbeam deflected by nanoindenter loading mechanism for measurement of modulus of elasticity and yield strength of microbeam material. (From Weihs, T.P. et al., 1988, *J. Mater. Res.* 3, 931–942. With permission.)

10.3.6 Modulus of Elasticity by Cantilever Deflection Measurement

The submicron indentation of thin films on substrates can lead to quick and accurate measurement of hardness, but the large pressures under the indenter may alter the structure of thin films being tested. For submicron (<200 nm) films, the influence of the substrate must be considered. With wafer curvature techniques, the average stress and strain in a film can be measured, but the range of stresses is limited by the thermal expansion and/or growth mismatch of the substrate and the film. Thus, for a given film and substrate, one cannot dictate the stress to be applied to the film. In an effort to avoid some of these difficulties, Weihs et al. (1988) developed a new technique based on the deflection of free-standing cantilever microbeams. Briefly, the technique involves the fabrication of the beams using silicon micromachining techniques and the deflection of the beams using a nanoindenter. The nanoindenter mechanically bends the beams while continuously monitoring the loading and the deflection of the beams. Using this technique, both elastic modulus and yield strength can be studied.

Weihs et al. measured properties of free-standing beams of SiO_2, LTO, and gold coatings used in VLSI fabrication cut into small sections (15 ¥ 15 mm). These beams were mounted on aluminum blocks with crystal bond wax, Figure 10.30. The microbeams were deflected at velocities ranging from 3 to 60 nm/s. Typical dimensions of the beam free to deflect were 40-μm length, 20-μm width, and 1-μm thickness. The modulus of elasticity is calculated from the following equation valid for small deflection of a thin film:

$$h = 4W\ell^3 \left(1 - v^2\right) \big/ \left(bt^3 E\right) \qquad (10.17)$$

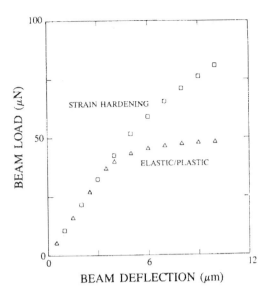

FIGURE 10.31 Theoretical load–deflection curves for cantilever beam deflections. The lower and upper curves represent an elastic–plastic material and a material that strain hardens, respectively. (From Weihs, T.P. et al., 1988, *J. Mater. Res.* 3, 931–942. With permission.)

where h is the vertical deflection, W is the force applied, ℓ is the effective length, n is Poisson's ratio, b is the width, t is the thickness, and E is the Young's modulus of the beam. As mentioned previously, the nanoindenter records the load required to displace its own spring and the microbeam. Consequently, the load required by the springs of the nanoindenter must be substrated from the total applied load to determine the load on the beam. For good accuracy, the spring constant of the microbeam must be approximately equal to or greater than 10% of the spring constant of the nanoindenter.

The application of simple beam theory to the load deflection data of beams also enable one to determine the yield strength of the microbeam material. For a homogeneous cantilever beam under load, the strain in the beam varies linearly through the thickness such that the maximum strain at a given length occurs at the top and at the bottom of the beam. For the loading shown in Figure 10.30, the top of the beam is in tension and the bottom of the beam is in compression. In addition, the maximum stress in the beam is located at the fixed end of the beam (B) where the applied moment is greatest. When this maximum stress reaches the yield strength of the material, the beam begins to deform plastically. The onset of such deformation can be recognized in the plot of load vs. deflection by a deviation from linearity, Figure 10.31. The load that marks this deviation is defined as the yield load W_y, and the yield strength is given by

$$\sigma_y = 6W_y\ell\big/\left(bt^2\right) \tag{10.18}$$

Following yielding, the beam continues to bend as more load is applied, and some of the strain in the beam is plastic and thus unrecoverable during unloading. The shape of the load vs. deflection curves after yielding and prior to unloading depends on the elastic and plastic properties of the material.

10.3.7 Determination of Hardness and Modulus of Elasticity of Thin Films from the Composite Response of Film and Substrate

As mentioned previously, for a thin film on a substrate, if the indentation exceeds about 30% of the film thickness, measured hardness is affected by the substrate properties. A number of researchers have attempted to derive expressions that relate thin-film hardness to substrate hardness, composite hardness (measured on the coated substrate) and film thickness, and so allow the calculations of these quantities given the remaining three (Buckle et al., 1973; Jonsson and Hogmark, 1984; Sargent, 1986; Burnett and Rickerby, 1987b; Bhattacharya and Nix, 1988a,b; Bull and Rickerby, 1990). Here we discuss two models based on the volume law of mixtures (volume fraction model) and finite-element simulation.

Sargent (1986) suggested that the hardness of a film/substrate composite is determined by a weighted average of the volume of plastically deformed material in the film (V_f) and that in the substrate (V_s),

$$H = H_f \frac{V_f}{V} + H_s \frac{V_s}{V} \tag{10.19}$$

where $V = V_f + V_s$. The deformed volumes of film and substrate can be calculated using the expanding spherical cavity model (Johnson, 1985). Burnett and Rickerby (1987b) found it necessary to incorporate a further weighting factor to deforming volume to obtain a reasonable fit to experimental data. This factor accounts for the differences in relative sizes of the plastic zones in the film and substrate. Equation 10.19 is modified as for a soft film on the hard substrate,

$$H = H_f X^3 \frac{V_f}{V} + H_s \frac{V_s}{V} \tag{10.20a}$$

and for the hard film on the soft substrate,

$$H = H_s \frac{V_f}{V} + H_s X^3 \frac{V_s}{V} \tag{10.20b}$$

where X is the ratio of plastic zone volumes given as

$$X = \left(\frac{E_f H_s}{H_f E_s} \right)^n \tag{10.20c}$$

They found that n, determined empirically, ranged from ½ to ⅓.

Bhattacharya and Nix (1988a) modeled the indentation process using the finite-element method to study the elastic–plastic response of materials. Bhattacharya and Nix (1988b) calculated elastic and plastic deformation associated with submicron indentation by a conical indenter of thin films on substrates, using the finite-element method. The effects of the elastic and plastic properties of both the film and substrate on the hardness of the film/substrate composite were studied by determining the average pressure under the indenter as a function of the indentation depth. They developed empirical equations for film/substrate combinations for which the substrate is either harder or softer than the film. For the case of a soft film on a harder substrate, the effect of substrate on film hardness can be described as

$$\frac{H}{H_s} = 1 + \left(\frac{H_f}{H_s} - 1 \right) \exp \left[- \frac{\left(\sigma_f / \sigma_s \right)}{\left(E_f / E_s \right)} \left(h_c / t_f \right)^2 \right] \tag{10.21a}$$

where E_f and E_s are the Young's moduli, s_f and s_s are the yield strengths, and H_f and H_s are the hardnesses of the film and substrate, respectively. H is the hardness of the composite, h_c is the contact indentation depth, and t_f is the film thickness. Similarly, for the case of a hard film on a softer substrate, the hardness can be expressed as

$$\frac{H}{H_s} = 1 + \left(\frac{H_f}{H_s} - 1\right)\exp\left[-\frac{\left(H_f/H_s\right)}{\left(\sigma_f/\sigma_s\right)\left(E_f/E_s\right)^{1/2}}\left(h_c/t_f\right)\right] \qquad (10.21b)$$

Composite hardness results were found to depend only very weakly on Poisson's ratio (n), and for this reason, this factor was not considered in the analysis. In Figure 10.32, they show the composite hardness results as a function of (h_c/t_f) for cases in which the film and substrate have different yield strengths. We note that hardness is independent of the substrate for indentation depths less than about 0.3 of the film thickness, after which the hardness slowly increases/decreases because of the presence of the substrate. In Figure 10.33, they show the composite hardness results for cases in which the film and substrate have different Young's moduli. It is observed that the variation of hardness with depth of indentation in these cases is qualitatively similar to cases in which the film and substrate have different yield strengths, although the hardness changes more gradually than in the previous cases. Burnett and Rickerby (1987c) and Fabes et al. (1992) have applied Equations 10.19 through 10.21 to calculate the film hardness from the measured data for various films and substrates.

Doerner and Nix (1986) empirically modeled the influence of the substrate on the elastic measurement of very thin film in an indentation test using the following expression for the compliance:

$$C = \frac{dh}{dW} = \frac{1}{2}\left(\frac{\pi}{A}\right)^{1/2}\left\{\frac{1-v_f^2}{E_f}\left[1-\exp\left(\frac{-\alpha t_f}{\sqrt{A}}\right)\right] + \frac{1-v_s^2}{E_s}\exp\left(\frac{-\alpha t_f}{\sqrt{A}}\right) + \frac{1-v_i^2}{E_i}\right\} + b \qquad (10.22)$$

where the subscripts f, s, and i refer to the film, substrate, and indenter, respectively. The term $\sqrt{A}$ is equal to $(24.5)^{1/2}\,h_c$ for the Vickers or Berkovich indenter. The film thickness is t_f, and b is the y-intercept for the compliance vs. 1/depth plot, obtained for the bulk substrate, which can be neglected in most cases. The weighing factors $[1 - \exp(\alpha t_f/\sqrt{A}]$ and $\exp(-\alpha t_f/\sqrt{A})$ have been added to account for the changing contributions of the substrate and film to the compliance. The factor α can be determined empirically.

King's analysis (1987) verified that Equation 10.22 is an excellent functional form for describing the influence of the substrate and theoretically determined the values of α for various indenter shapes. The value of α was found to depend on the indenter shape and size and film thickness and was found to be independent of E_i/E_s. The values of α as a function of $\sqrt{A}/t_f$ for Berkovich (triangular) indenters are shown in Figure 10.34. The values of α are found to be similar for square and triangular indenters. Bhattacharya and Nix (1988b) analyzed the deformations of a layered medium in contact with a conical indenter using the finite-element method. Their analysis also verified the relationship given in Equation 10.22.

10.4 Examples of Measured Mechanical Properties of Engineering Materials

To illustrate the usage of nanoindentation techniques, we present typical data obtained on various materials, coatings, and surface treatments.

10.4.1 Load — Displacement Curves

A variety of mechanical phenomena, such as transition from elastic to plastic deformation, creep deformation, formation of subsurface cracks, and crystallographic phase transition, can be studied by the load–displacement curves obtained at different loading conditions (Pethica et al., 1983; Doerner and Nix, 1986; Stone et al., 1988; LaFontaine et al., 1990c, 1991; Pharr et al., 1990; Page et al., 1992; Oliver and

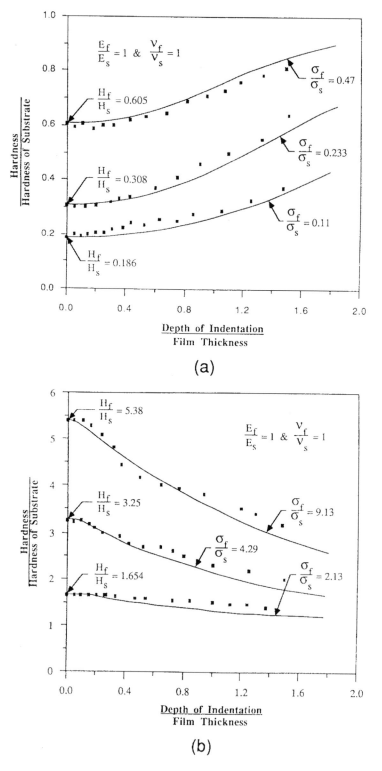

FIGURE 10.32 Effect of relative yield strengths of the film and the substrate on the composite hardness for (a) a soft film on a hard substrate and (b) a hard film on a soft substrate. (From Bhattacharya, A.K. and Nix, W.D., 1988, *Int. J. Solids Struct.* 24, 1287–1298. With permission.)

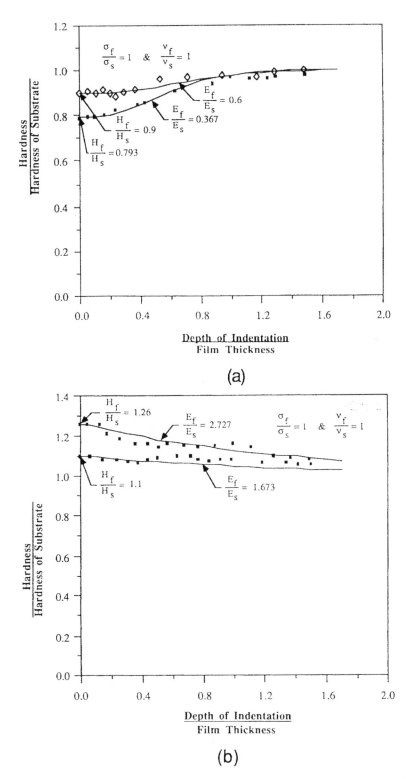

FIGURE 10.33 Effect of relative Young's moduli of the film and the substrate on the composite hardness for (a) a soft film on a hard substrate and (b) a hard film on a soft substrate. (From Bhattacharya, A.K. and Nix, W.D., 1988, *Int. J. Solids Struct.* 24, 1287–1298. With permission.)

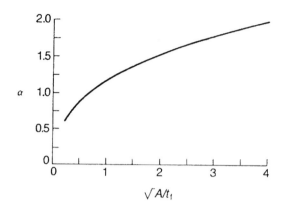

FIGURE 10.34 Parameter a as a function of normalized indenter size for a Berkovich indenter indenting a layered solid surface. (From King, R.B., 1987, *Int. J. Solids Struct.* 23, 1657–1664. With permission.)

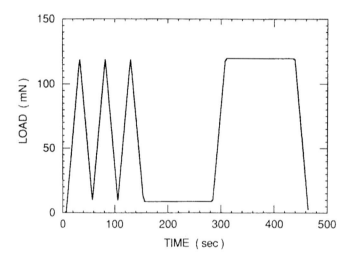

FIGURE 10.35 A typical load–time sequence; peak load = 120 mN. (From Oliver, W.C. and Pharr, G.M., 1992, *J. Mater. Res.* 7, 1564–1583. With permission.)

Pharr, 1992; Pharr, 1992; Whitehead and Page, 1992; Gupta et al., 1993, 1994; Gupta and Bhushan, 1994, 1995a,b; Bhushan et al., 1995, 1996, 1997; Bhushan and Gupta, 1995; Patton and Bhushan, 1996; Bhushan and Li, 1997; Li and Bhushan, 1998b, c). The load–time sequence used by Oliver and Pharr (1992) for study of various materials included three loading–unloading cycles, hold for 100 s at 10% of the peak load, reload, hold for 100 s, and unload, Figure 10.35. Load–displacement curves for mechanically polished single-crystal aluminum, electropolished single-crystal tungsten, soda lime glass, fused silica and (110) single-crystal silicon are shown in Figures 10.36 through 10.40. Silicon included only two loading–unloading cycles (Figure 10.40). The softest material is aluminum, with peak depth of almost 5000 nm at 120 mN load, while the hardest is silicon which was penetrated to a depth of only about 800 nm. Both the aluminum and tungsten data are typical of materials in which the hardness is relatively small compared to the modulus, as is observed in most metals; most of the indenter displacement in these metals is accommodated plastically and only a small portion is recovered on unloading. Soda lime glass, fused silica, and silicon are harder which show larger elastic recovery during unloading, the largest being that for fused silica.

The unloading/reloading behaviors of the various materials are different. For aluminum, the expanded view in Figure 10.36b shows that the peak load displacements shift to higher values in successive

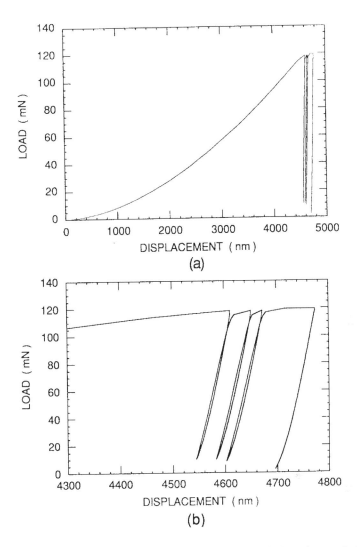

FIGURE 10.36 (a) Load vs. displacement plot for an mechanically polished single-crystal aluminum, (b) an expanded view of the unloading/reloading portion of the load vs. displacement data. (From Oliver, W.C. and Pharr, G.M., 1992, *J. Mater. Res.* 7, 1564–1583. With permission.)

loading/unloading cycles. In addition, the relatively large displacement just prior to final unloading is due to creep during the 100-s hold period at peak load. At 120 mN, the indentation depth in tungsten is about one fourth of that in aluminum, Figure 10.37a. However, the behavior of tungsten is similar to that of aluminum. Indentation at a very low load of 0.5 mN caused only elastic displacements, Figure 10.37b. At higher peak loads, the indentation is not just elastic, Figure 10.37c. When a threshold load of about 1 mN is reached, a sudden jump in displacement corresponding to the onset of plasticity is observed, and a permanent hardness impression is formed. Soda lime glass (Figure 10.38) and tungsten at low loads (Figure 10.37c) exhibit distinct hysteresis loops, as might be expected if there were a small amount of reverse plasticity upon loading. However, the looping degenerates with cycling after three or four cycles, the load–displacement behavior is largely elastic. The unloading/reloading curves for fused silica are nearly the same, Figure 10.39. The near-perfect reversibility suggests that at peak loads of 120 and 4.5 mN, deformation after initial unloading is almost entirely elastic.

The behavior of silicon shown in Figure 10.40, is in sharp contrast to other materials (e.g., fused silica data shown in Figure 10.39) (Pharr et al., 1989, 1990; Page et al., 1992; Pharr, 1992). The data presented

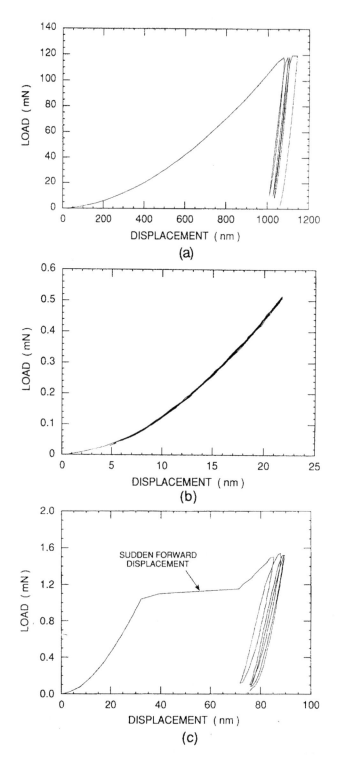

FIGURE 10.37 Load vs. displacement plot for an electropolished single-crystal tungsten (a) at a peak load of 120 mN, (b) at a peak load of 0.5 mN (elastic contact), and (c) at a peak load of 1.5 mN showing the yield point. (From Oliver, W.C. and Pharr, G.M., 1992, *J. Mater. Res.* 7, 1564–1583. With permission.)

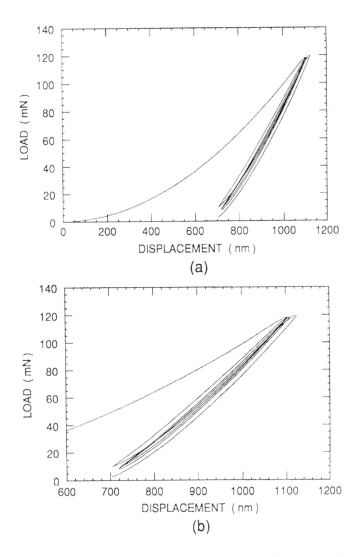

FIGURE 10.38 (a) Load vs. displacement plot for a soda lime glass and (b) an expanded view of the unloading/reloading portion of the load vs. displacement data. (From Oliver, W.C. and Pharr, G.M., 1992, *J. Mater. Res. 7*, 1564–1583. With permission.)

in Figure 10.40 were taken over two cycles of loading and unloading. At high peak loads, the initial unloading curve for silicon is not at all smooth, but exhibits a sharp discontinuity caused by the surface of the specimen being suddenly thrust outward, Figure 10.40a. At lower peak loads, the behavior changes, and below some critical value the discontinuity is no longer observed, Figure 10.40b. However, at this load the load–displacement behavior shows another anomalous feature — a large hysteresis which shows no sign of degeneration through several cycles of deformation. The fact that the curves are highly hysteretic implies that deformation is not entirely elastic. The discontinuity at high loads and the nondegenerative hysteresis at low loads are quite unique to silicon and are observed in each of the (100), (110), and (111) orientations. The load below which the discontinuity disappears and the hysteresis becomes apparent is generally in the range 5 to 20 mN. Pharr (1992) concluded that larger hysteresis observed in the unloading curve at low loads is due to a pressure-induced phase transformation from its normal diamond cubic form to a b–tin metal phase (Figure 10.41). At some point in the transformation, an amorphous phase is formed whose evidence is reported by Callahan and Morris (1992). The

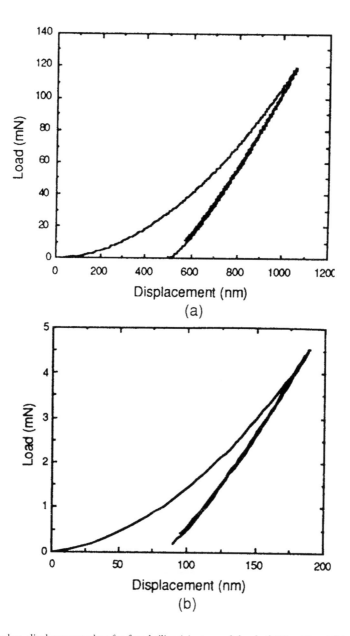

FIGURE 10.39 Load vs. displacement plots for fused silica (a) at a peak load of 120 mN and (b) at a peak load of 4.5 mN. (From Pharr, G.M., 1992, in *Thin Film: Stresses and Mechanical Properties III*, W.D. Nix et al., eds., Vol. 239, pp. 301–312, Materials Research Society, Pittsburgh. With permission.)

discontinuity in displacement observed during unloading at peak loads of greater than about 15 mN is due to formation of a lateral crack which forms at the base of the median crack (Figure 10.41). Lateral cracking is aided by the phase transformations.

Weihs et al. (1992) indented Si(100) with 120 mN of load. The load–displacement curve is shown in Figure 10.42a and the first AE signal that was recorded for this test is shown in Figure 10.42b. The AE signals were sharp and they correlated with small jumps in tip displacement. The rise time for the signal in Figure 10.42b is 1.5 μs. After testing, radial cracks were visible at the corners of indentation. AE events such as the one shown in Figure 10.42b were recorded at applied load as low as 48 mN on loading. In the final stages of unloading, small AE signals that had an inverted shape compared with Figure 10.42b

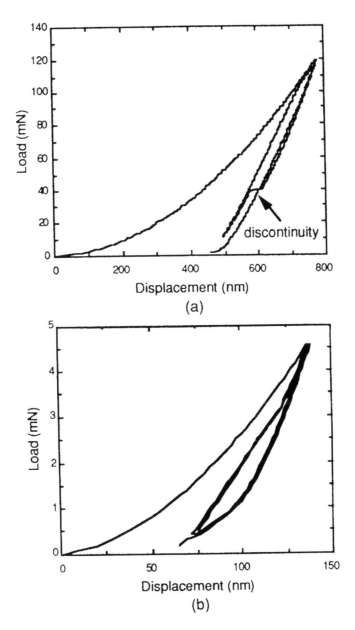

FIGURE 10.40 Load vs. displacement plots for (110) silicon (a) at a peak load of 120 mN and (b) at a peak load of 4.5 mN. (From Pharr, G.M., 1992, in *Thin Film: Stresses and Mechanical Properties III,* W.D. Nix et al., eds., Vol. 239, pp. 301–312, Materials Research Society, Pittsburgh. With permission.)

were detected occasionally. However, no AE signals were detected during the distinct load–displacement discontinuity on unloading which is seen in Figure 10.42a at 60 mN. Since no emission was detected, the source of the discontinuity is more likely to be a sluggish transformation (Hu et al., 1986; Clarke et al., 1988; Pharr, 1992) than a rapid crack growth.

Gupta et al. (1993, 1994) and Gupta and Bhushan (1994) reported that hysteresis in cyclic indentation and discontinuity kinks in the unloading curve are considerably reduced by ion implantation of compound forming species 0^+ and N^+ into single-crystal silicon. They further reported that amorphous silicon films did not exhibit either hysteresis in the cyclic indentation or a discontinuity kink in the indentation loads ranging from 1 to 90 mN. This suggests that for a perfect amorphous structure, the hysteresis does

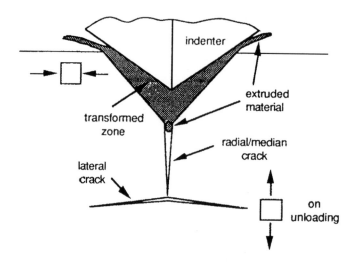

FIGURE 10.41 A schematic illustration of the coupling of the phase transformation and cracking during indentation of single-crystal silicon. (From Pharr, G.M., 1992, in *Thin Film: Stresses and Mechanical Properties III*, W.D. Nix et al., eds., Vol. 239, pp. 301–312, Materials Research Society, Pittsburgh. With permission.)

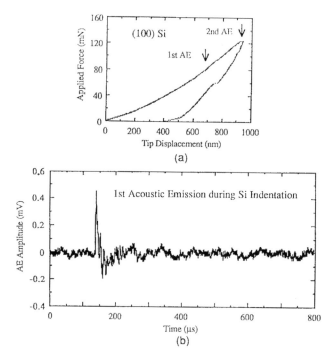

FIGURE 10.42 (a) Applied force as a function of tip displacement for Si(100) and (b) AE amplitude as a function of time during loading. (From Weihs, T.P. et al., 1992, in *Thin Films: Stresses and Mechanical Properties III, Symp. Proc.*, W.E. Nix, et al., eds., Vol. 239, pp. 361–370, Materials Research Society, Pittsburgh. With permission.)

not occur because of absence of crystallographic pressure-induced phase transition during cyclic loading and unloading. In addition, the disordered structure does not allow the nucleation and propagation of the lateral cracks beneath the indentation. They have also reported that ion implantation has an insignificant effect on hardness and elastic modulus. Improvement in the hardness is about 10% as a result

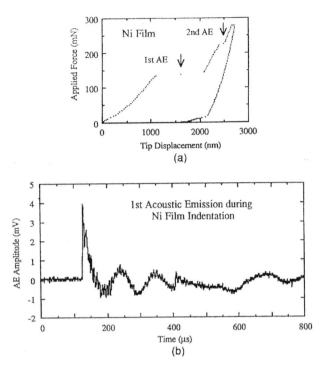

FIGURE 10.43 (a) Applied force as a function of tip displacement for Ni film on glass substrate and (b) AE amplitude as a function of time during loading. (From Weihs, T.P. et al., 1992, in *Thin Films: Stresses and Mechanical Properties III, Symp. Proc.,* W.E. Nix, et al., eds., Vol. 239, pp. 361–370, Materials Research Society, Pittsburgh. With permission.)

of implantation with carbon and boron ions. Gupta et al. (1993, 1994) reported that ion-bombarded silicon exhibits a very low coefficient of friction (~0.05) and a low wear factor (~10^{-6} mm^3 N^{-1}m^{-1}) when slid against alumina and 52100 steel balls.

Weihs et al. (1992) indented Ni films on a glass substrate. The Ni films debonded from their substrates at forces ranging between 130 and 250 mN. The debonding events were marked by the indenter tip jumping downward as "chunks" of the Ni film buckled away from the underneath. The indenter tip jumped a distance equal to the film thickness as it initially debonded. Figure 10.43 also shows the corresponding AE trace with a rise time of 1.8 μs. In this particular test, debonding continued at higher forces and a second AE event was recorded. After each test, optical microscopy confirmed the delamination of the film from underneath the indenter.

Gupta and Bhushan (1995) indented 20-nm-thick amorphous carbon (a–C, also known as diamond-like carbon or DLC) films deposited by four different deposition techniques, sputtered Aℓ_2O$_3$ and sputtered SiC films on Aℓ_2O$_3$–TiC substrate, Figure 10.44. The indentation depths are larger than the film thickness, which means that the tip had penetrated through the films into the substrate material during the indentation process. Most of the films exhibit a discontinuity or pop-in marks in the loading curve, shown by arrows in Figure 10.44, which indicate a sudden penetration of the tip into the sample. A nonuniform penetration of the tip into a thin film possibly results due to formation of cracks in the film, formation of cracks at the film–substrate interface, or debonding or delamination of the film from the substrate. The Aℓ_2O$_3$–TiC substrate exhibits a smooth loading curve free from any pop-in marks, which indicates that a 0.5-mN peak indentation load is insufficient to damage the surface. All of the films exhibited indentation depths at the peak load higher than that of the Aℓ_2O$_3$–TiC substrate.

Oliver and Pharr (1992) have reported that load–displacement curves during unloading are not linear, rather each unloading curve is slightly concave over its entire span. They showed that the unloading

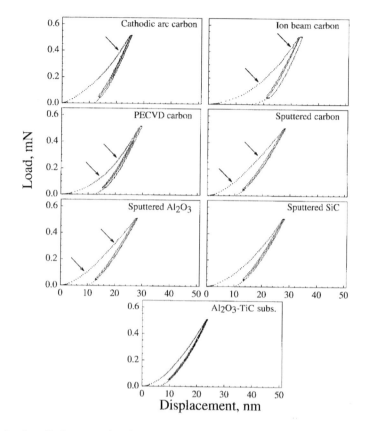

FIGURE 10.44 Load vs. displacement plots for various 20-nm-thick amorphous carbon films and ion beam–sputtered Al$_2$O$_3$ and RF-sputtered SiC coatings on Al$_2$O$_3$–TiC substrate and bare Al$_2$O$_3$–TiC substrate. (From Gupta, B.K. and Bhushan, B., 1995, *Wear* 190, 110–122. With permission.)

curves are well described by a power law relation like that of Equation 10.7 presented earlier. For the six materials tested, the exponents varied from 1.25 to 1.50 and were not distinctly greater than 1 which implies that none of the data is consistent with flat punch behavior.

10.4.2 Continuous Stiffness Measurements

Continuous stiffness measurements are used for continuous measurement of sample stiffness or compliance (related to elastic modulus) during indentation. Oliver and Pharr (1992) measured continuous stiffness of electropolished single-crystal tungsten at low peak loads of 0.5 mN or less, for purely elastic contact. Continuous stiffness data for the indentation data in Figure 10.37b are presented in Figure 10.45a. Comparisons of these data with the load–time history, shown in Figure 10.35, show that the measured contact stiffness and thus the contact area does increase and decrease in the way that would be expected based on the loading history. The continuous stiffness data for the indentation at 1.5-mN load (plastic contact, Figure 10.37c) is shown in Figure 10.45b. It is seen that for each of the four unloadings, the contact stiffness changes *immediately and continuously* as the specimen is unloaded. Thus, the contact area, which varies in that same way as the contact stiffness, is not constant during the unloading of the plastic hardness impression, even during the initial stages of unloading.

10.4.3 Hardness and Elastic Modulus Measurements

Nanoindenter is commonly used to measure surface mechanical properties of bulk materials (Pharr et al., 1990; Lucas et al., 1991; Oliver and Pharr, 1992; Bhushan and Gupta, 1995; Bhushan et al., 1996).

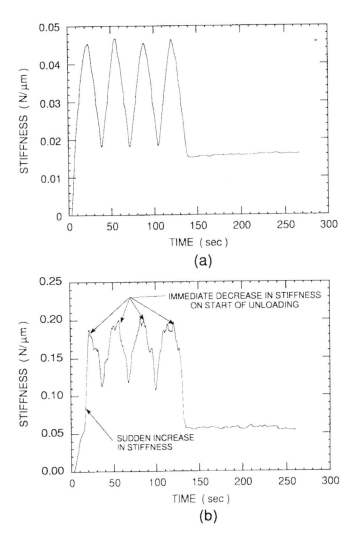

FIGURE 10.45 Contact stiffness vs. time. (a) for a fully elastic contact and (b) for a fully plastic contact on an electropolished single-crystal tungsten measured with the continuous stiffness technique. (From Oliver, W.C. and Pharr, G.M., 1992, *J. Mater. Res.* 7, 1564–1583. With permission.)

Hardness and elastic moduli for six bulk materials and three materials used as a substrate for the construction of magnetic rigid disks and single-crystal silicon are presented in Figure 10.46. The data show that there is a very little indentation size effect in several materials on the hardness values. In the case of aluminum and tungsten in Figure 10.46a, there is a modest increase in hardness at low loads. According to Oliver and Pharr (1992), this could be due to surface-localized cold-work resulting from polishing. The modulus data also show that there is very little evidence for an indentation size effect; i.e., the moduli remains more or less constant over the entire range of load. Oliver and Pharr (1992) reported that the hardness and modulus values at the two highest loads are comparable with the literature values. Comparison of published values and measured values of magnetic disk substrates in Figure 10.46b is shown in Table 10.4. (Fracture toughness data will be discussed later.)

Pharr et al. (1990) reported a nanoindentation hardness value of (110) single-crystal silicon to be about 11.4 GPa at a peak normal load of 120 mN and corresponding peak indentation depth of 750 nm. This value was slightly higher than the macrohardness value of silicon of 9 to 10 GPa (Anonymous, 1988). Using an atomic force microscope, Bhushan and Koinkar (1994) measured hardness of single-crystal

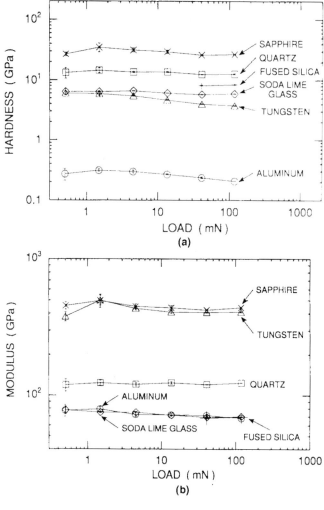

FIGURE 10.46

silicon at an ultrasmall normal load of 65 µN and corresponding indentation depth of 2.5 nm and reported a value of 16.6 GPa. It is clear that the hardness of silicon increases with a decrease in the load and corresponding indentation depth. An increase in hardness at lower indentation depths may result from contributions of the surface films. At smaller volumes, there is a high probability that indentation would be made into a region that was initially dislocation free. Furthermore, at small volumes, there is an increase in the stress necessary to operate dislocation sources (Gane and Cox, 1970; Sargent, 1986). These are some of the plausible explanations for the increase in hardness at smaller volumes.

A number of investigators have used the nanoindenter to study the effect of ion implantation on the hardness and elastic modulus of metals (Nastasi et al., 1988; Lee and Mansur, 1989; Bourcier et al., 1990, 1991; Was, 1990), silicon (Gupta et al., 1993, 1994; Gupta and Bhushan, 1994), ceramics (McHargue, 1989; McHargue et al., 1990; O'Hern et al., 1990; Was and DeKoven, 1991) and polymers (Lee et al., 1992, 1993; Rao et al., 1993). Generally, the ion-implanted zone is shallow and the nanoindenter is capable of measuring the changes in mechanical properties.

The nanoindenter is ideal for measurement of mechanical properties of thin films and composite structures. A number of investigators have reported hardness of composite structures with thin films on a substrate (Pethica et al., 1985; Burnett and Rickerby, 1987c; Stone et al., 1988; Wu et al., 1988, 1989, 1990b; O'Hern et al., 1989; Bhushan and Doerner, 1989; Jiang et al., 1989; Joslin et al., 1989; Baker et al.,

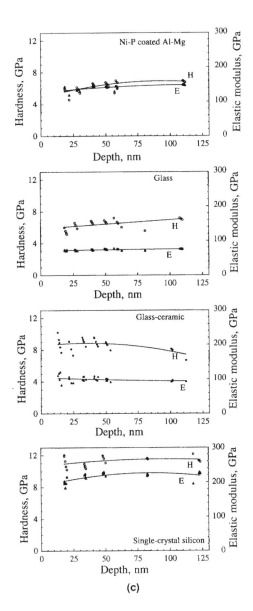

(c)

FIGURE 10.46 Hardness and elastic modulus (a) as a function of load for six bulk materials — mechanically polished single-crystal aluminum, electropolished single-crystal tungsten, soda lime glass, fused silica, (001) single-crystal sapphire, and (001) single-crystal quartz (Oliver and Pharr, 1992), and (b) as a function of indentation depth (load ranging from 0.1 to 2.5 mN) for mechanically polished 10-μm-thick electroless Ni–P film on Al–Mg alloy 5086, chemically strengthened alkali-aluminosilicate glass, chain silicate glass–ceramic (polycrystalline) (Canasite by Corning), and single-crystal silicon (111). (From Bhushan, B. and Gupta, B.K., 1995, *Adv. Info. Storage Syst.* 6, 193–208. With permission.)

TABLE 10.4 rms Roughness Values Measured by an Atomic Force Microscope, Hardness and Elastic Modulus Values Measured by Using a Nanoindenter (at an indentation depth of about 20 nm), and Fracture Toughness Measured by Using a Vickers Microindenter

Material	rms Roughness (nm)	Hardness[a] (GPa)	Elastic Modulus[b] (GPa)	Crack Length, c (μm)	Fracture Toughness[c] (MPa·m$^{1/2}$)
Ni–P/Al–Mg	3.6	6.0 (5.5)	130 (200)	No cracks	—
Chemically strengthened glass	1.1	6.0 (5.8)	85 (73)	Significant cracking	— (0.9)
Glass–ceramic	6.1	8.5 (5.5)	100 (83)	59.4	0.65 (4.0)
Single-crystal silicon (111)	0.95	11.0 (9-10)	200 (180)	—	—

[a]Values in parentheses are the reported values measured by conventional Vickers indentation method.
[b]Values in parentheses are the reported values measured by conventional tensile pull test method.
[c]Values in parentheses are the reported values measured by the chevron notched short bar method.

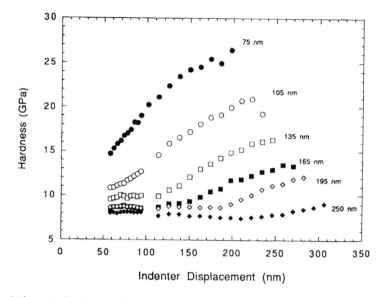

FIGURE 10.47 Indentation hardness as a function of indenter displacement for Ti coatings on sapphire substrates. Numbers next to each set of data correspond to the coating thickness. (From Fabes, B.D. et al., 1992, *J. Mater. Res.* 7, 3056–3064. With permission.)

1990; Bull and Rickerby, 1990; Cammarata et al., 1990; Cho et al., 1990; Cornett et al., 1990; Rubin et al., 1990; Chou et al., 1991; Cooper, 1991; Cooper and Beetz, 1991; Gissler et al., 1991; Schlesinger et al., 1991; Stone et al., 1991; Bhushan et al., 1992, 1995, 1997; Fabes et al., 1992; Knight et al., 1992; Lucas and Oliver, 1992; Savvides and Bell, 1992; Wang et al., 1992; Whitehead and Page, 1992; Li et al., 1993; Nastasi et al., 1993; Vancoille et al., 1993; Gupta and Bhushan, 1995a,b; Patton and Bhushan, 1996; Bhushan and Li, 1997; Li and Bhushan, 1998b, c).

True hardness of the films can be obtained if the indentation depth does not exceed about 30% of the film thickness. At higher indentation depths, the composite hardness changes with the indentation depth. Measured hardness values of soft Ti films on a hard sapphire substrate are presented in Figure 10.47. We note that hardness increases with a decrease in the film thickness or increase in the indentation depth, as expected. The film hardness is the steady-state hardness independent of the indentation depth.

Hardness and elastic modulus profiles as a function of indentation depth at the peak load for 400-nm-thick amorphous carbon films deposited by various deposition techniques, SiC films, and single-crystal silicon substrates are presented in Figure 10.48. We note that the hardness and elastic modulus of cathodic arc carbon and radio frequency (RF)-sputtered SiC films tend to decrease with the indentation depth, which is attributed to increased contributions of the silicon substrate with a lower hardness of 11 GPa, at larger depths. Cathodic arc carbon film exhibits the highest hardness of 38 GPa and elastic modulus of 300 GPa as compared with that of other coatings. The high hardness and elastic modulus of cathodic arc film are followed by RF-sputtered SiC, ion beam carbon, and plasma enhanced-chemical vapor deposition/direct current (PECVD/DC) sputtered carbon films. High hardness and elastic modulus of cathodic arc carbon film are attributed to the high kinetic energy of the carbon species involved in the cathodic arc deposition. The differences in the hardness and elastic modulus of carbon films is attributed to their varying sp^3- to sp^2-bonding ratio and the amount of hydrogen (Gupta and Bhushan, 1995a,b). The hardness and elastic modulus data are summarized in Table 10.5 (Gupta and Bhushan, 1995a,b).

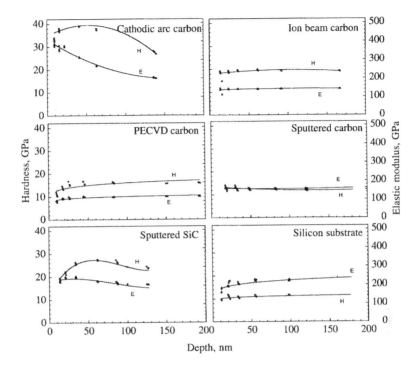

FIGURE 10.48 Hardness and elastic modulus as a function of indentation depth at the peak load of 400-nm-thick carbon films deposited by cathodic arc, ion beam, PECVD, and DC-sputtered and 400-nm-thick and RF-sputtered SiC film on single-crystal silicon substrates. Hardness and elastic modulus profiles for single-crystal (100) silicon are also included for comparison. (From Gupta, B.K. and Bhushan, B., 1995, *Thin Solid Films* 270, 391–398. With permission.)

TABLE 10.5 Comparison of Hardness and Elastic Modulus of Various 400-nm-Thick Coatings Measured by Nanoindentation and Bulk Samples Data Reported by Other Researchers

Coating	Hardness (GPa)	Elastic Modulus (GPa)
Cathodic arc carbon (a–C)	38	300
Ion beam carbon (a–C:H)	19	150
PECVD carbon (a–C:H)	17	140
DC sputtered carbon (a–C:H)	15	140
Bulk graphite (for comparison)	Very soft	9–15
Diamond (for comparison)	80–104	900–1050
Ion beam sputtered Al_2O_3	8.3	140
RF sputtered SiC	27	255
Single-crystal silicon (bulk)	11	220

10.5 Microscratch Resistance Measurement of Bulk Materials Using Micro/Nanoscratch Technique

Microscratch technique is commonly used for screening of bulk materials for wear resistance. As stated earlier, normal load applied to the scratch tip is gradually increased during scratching until the material is damaged. Friction force is sometimes measured during the scratch test (Bhushan and Gupta, 1995; Bhushan et al., 1996). After the scratch test, the morphology of the scratch region including debris is

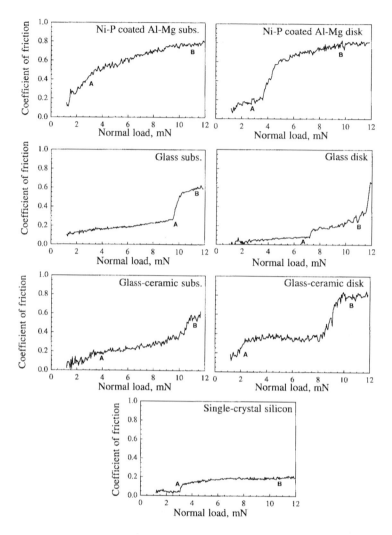

FIGURE 10.49 Coefficient of friction profiles as a function of normal load for 500-μm-long scratches made using a diamond tip with 1-μm tip radius, at a normal load ranging from 1 to 12 mN on various ceramic substrates, corresponding magnetic disks, and single-crystal silicon. (From Bhushan, B. and Gupta, B.K., 1995, *Adv. Info. Storage Syst.* 6, 193–208. With permission.)

observed in an SEM. Based on the combination of the changes in the friction force as a function of normal load and SEM observations, the critical load is determined and the deformation mode is identified. Any damage to the material surface as a result of scratching at a critical ramp-up load results in an abrupt or gradual increase in friction. The material may deform either by plastic deformation or fracture. Ductile materials (all metals) deform primarily by plastic deformation, resulting in significant plowing during scratching. Tracks are produced whose width and depth increase with an increase in the normal load. Plowing results in a continuous increase in the coefficient of friction with an increase in the normal load during scratching. Debris is generally ribbonlike or curly, whereas brittle materials deform primarily by brittle fracture with some plastic deformation. In the brittle fracture mode, the coefficient of friction increases very little until a critical load is reached at which the material fails catastrophically and produces fine debris, which is rounded, and the coefficient of friction increases rapidly above the critical load.

The coefficient of friction profiles as a function of normal load for scratches made on various ceramic substrates and corresponding magnetic disks (substrates coated with 75-nm-thick sputtered Co–Pt–Ni magnetic film and 20-nm-thick sputtered hydrogenated, DLC film) are compared in Figure 10.49. Figure 10.49 also includes the friction force profile for a single-crystal silicon substrate for comparison. SEM images of two regions of 500-μm-long scratches made at 1 to 12 mN normal load on various samples are compared in Figure 10.50. The upper images in the sets of two images for each sample correspond to a region where friction increased abruptly. These are the points indicated by "A" in Figure 10.50. The lower images in each set correspond to the region that is very close to the end of a scratch. These are the points indicated by 'B' in Figure 10.50. The extent of a damage in a scratch is estimated by the width and depth of the scratch and by the amount of debris generated toward the end of the scratch.

Single-crystal silicon exhibited the lowest friction with little plowing at a low load and cracking at higher loads. This observation suggests that scratching of silicon took place primarily by brittle deformation. In the case of the Ni–P-coated Al–Mg substrate, friction increase was continuous from the beginning of the scratching, Figure 10.49. SEM images of the Ni–P-coated Al–Mg substrate presented in Figure 10.50, show the material removal occurred by plowing with formation of curly ductile chips. It is evident that scratching took place primarily by plastic deformation typical of ductile materials. Plowing is responsible for the continuous increase in the friction for this substrate. Glass and glass–ceramic substrates and corresponding disks and the Ni–P-coated Al–Mg disk exhibited relatively low friction with a sudden increase at higher load. The glass substrate exhibited the lowest friction followed by the glass–ceramic substrate. In the case of the Ni–P-coated Al–Mg disk, the load at which friction increased was lower than that of the glass and glass–ceramic substrates. SEM images of these samples exhibit plowing in addition to the formation of fine debris. There is no evidence of cracking of ceramic substrates or the ceramic overcoats used in all disks at magnifications as high as 50,000¥. Glass is chemically strengthened in order to produce significant compressive stresses in the glass surface. Glass–ceramic consists of fine-grained polycrystalline material in a glass matrix. Chemical strengthening and the crystals add to the fracture toughness of the material. Thus, both ceramic substrate materials are expected to deform with ductile and brittle deformation modes. Ductile deformation results in plowing, whereas brittle deformation aids in debris generation. Lower values of the coefficient of friction before a sudden increase, as compared with Ni–P-coated Al–Mg substrate, suggest that brittle fracture contributes to overall deformation. Hard overcoats generally consist of significant compressive residual stresses. It is these compressive stresses that allow ductile deformation with little cracking. We further note that a sudden increase in the coefficient of friction for ceramic substrates and for all disks at some load results from significant damage to the bulk material or to the coating surface (Figure 10.50).

Based on the friction data, the width and depth of scratches, the amount of debris generated, and scratch morphology, glass substrates and corresponding disks exhibit a lower coefficient of friction against a diamond tip and a superior resistance to scratch, followed by glass–ceramic substrates and corresponding disks.

This example clearly suggests that deformation modes and critical load to failure can be identified using the scratch technique.

10.6 Nanoindentation and Microscratch Techniques for Adhesion Measurements, Residual Stresses, and Materials Characterization of Thin Films

Adhesion describes the sticking together of two materials. Adhesion strength, in a practical sense, is the stress required to remove a coating from a substrate. Indentation and scratch on the micro- and nanoscales are the two commonly used techniques to measure adhesion of thin hard films with good adhesion to the substrate (>70 MPa)(Campbell, 1970; Mittal, 1978; Blau and Lawn, 1986; Bhushan, 1987; Bhushan and Gupta, 1997). Nearly all coatings, by whatever means they are produced, and surface layers of treated parts are found to be in a state of residual (intrinsic or internal) stress. These are elastic stresses that exist in the absence of external forces and are produced through the differential action of plastic flow, thermal contraction, and/or changes in volume created by phase transformation. Microindentation and nanoin-

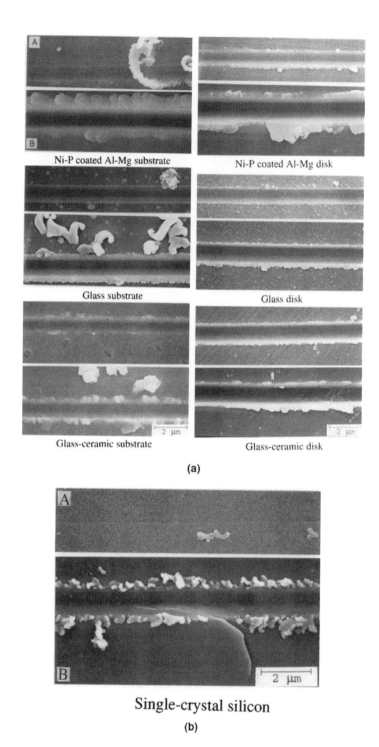

(a)

Ni-P coated Al-Mg substrate   Ni-P coated Al-Mg disk

Glass substrate   Glass disk

Glass-ceramic substrate   Glass-ceramic disk

Single-crystal silicon

(b)

FIGURE 10.50 SEM images of two regions on 500-µm-long scratches made at 1 to 12 mN load on (a) various ceramic substrates and corresponding magnetic disks and (b) single-crystal silicon. The scratching direction was from left to right. The upper images in the sets of two images for each sample correspond to a location or normal load where the friction increased abruptly and/or damage began to occur. These are the points indicated by "A" in Figure 10.49. The lower images correspond to a location close to the end of the scratch (~11 mN). These are the points indicated by "B" in Figure 10.49. (From Bhushan, B. and Gupta, B.K., 1995, *Adv. Info. Storage Syst.* 6, 193–208. With permission.)

dentation techniques are also used to measure residual stresses (Bhushan and Gupta, 1997). Microscratch and nanoscratch techniques (using nanoindenter) are also used to measure scratch resistance of surfaces of bulk materials (Bhushan and Gupta, 1995; Bhushan et al., 1996). The nanoindenter has also been modified to conduct microwear studies (Wu and Lee, 1994).

In this section, we describe adhesion measurements, residual stress measurements, and microwear measurements using nanoindentation and microscratch apparatuses. We also present typical examples.

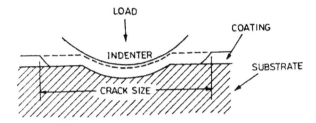

FIGURE 10.51 Schematic illustration of the indentation method for adhesion measurement.

10.6.1 Adhesion Strength and Durability Measurements Using Nanoindentation

In the indentation test method, the coating sample is indented at various loads. At low loads, the coating deforms with the substrate. However, if the load is sufficiently high, a lateral crack is initiated and propagated along the coating–substrate interface. The lateral crack length increases with the indention load. The minimum load at which the coating fracture is observed is called the *critical load* and is employed as the measure of coating adhesion (Figure 10.51). For relatively thick films, the indentation is generally made using a Brinell hardness tester with a diamond sphere of 20-μm radius (Tangena and Hurkx, 1986), Rockwell hardness tester with a Rockwell C 120° cone with a tip radius of 200 μm (Mehrotra and Quinto, 1985) or a Vickers pyramidal indenter (Chiang et al., 1981; Lin et al., 1990; Alba et al., 1993). However, for extremely thin films, a Berkovich indenter (Stone et al., 1988) or a conical diamond indenter with a tip radius of 5 mm and 30° of included angle (Tsukamoto et al., 1992) is used in a nanoindenter.

It should be noted that the measured critical load W_{cr} is a function of hardness and fracture toughness in addition to the adhesion of coatings. Chiang et al. (1981) have related the measured crack length during indentation, the applied load, and the critical load (at which coating fracture is observed) to the fracture toughness of the substrate–coating interface. A semianalytical relationship derived between the measured crack length c and the applied load W:

$$c = \alpha\left(1 - \frac{W_{cr}}{W}\right)^{1/2} W^{1/4} \qquad (10.23)$$

where

$$\alpha^2 = \frac{\alpha_1 t_c^{3/2} H^{1/2}}{\left(K_{Ic}\right)_{\text{interface}}},$$

a_1 is a numerical constant, t_c is the coating thickness, H is the mean hardness, and $(K_{Ic})_{\text{interface}}$ is the fracture toughness of the substrate–coating interface. Mehrotra and Quinto (1985) used this analysis to calculate fracture toughness of the interface.

Marshall and Oliver (1987) estimated adhesion of composites by measuring the magnitude of shear (friction) stresses at fiber/matrix interfaces in composites. They used a Berkovich indenter to push on the end of an individual fiber, and measured the resulting displacement of the surface of the fiber below the matrix surface (due to sliding). The shear stress was calculated from the force–displacement relation obtained by analysis of the frictional sliding. The force and displacement measurements were obtained only at the peak of the load cycle, and the sliding analysis was based on sliding at constant shear resistance at the interface. These experiments provided measurements of average shear stresses at individual fibers.

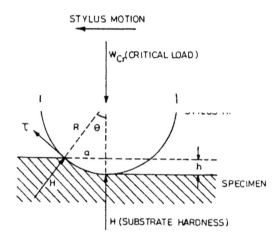

FIGURE 10.52 Geometry of the scratch.

10.6.2 Adhesion Strength and Durability Measurements Using Microscratch Technique

Scratching the surface with a fingernail or a knife is probably one of the oldest methods for determining the adhesion of paints and other coatings. In 1822, Friedrich Mohs used resistance to scratch as a measure of hardness. Scratch tests to measure adhesion of films was first introduced by Heavens in 1950 (Heavens, 1950). A smoothly round chrome-steel stylus with a tungsten carbide or Rockwell C diamond tip (in the form of 120° cone with a hemispherical tip of 200-mm radius) (Perry, 1981, 1983; Mehrotra and Quinto, 1985; Valli, 1986; Steinmann et al., 1987) or Vickers pyramidal indenter (Burnett and Rickerby, 1987a; Bull and Rickerby, 1990) for macro- and microscratching a conical diamond indenter (with 1 or 5 μm of tip radius and 60° of included angle) for nanoscratching (Wu et al., 1989, 1990b; Wu, 1991) is drawn across the coating surface. A normal load is applied to the scratch tip and is gradually increased during scratching until the coating is completely removed. The minimum or critical load at which the coating is detached or completely removed is used as a measure of adhesion (Benjamin and Weaver, 1960; Campbell, 1970; Greene et al., 1974; Ahn et al., 1978; Mittal, 1978; Perry, 1981, 1983; Laugier, 1981; Jacobson et al., 1983; Mehrotra and Quinto, 1985; Je et al., 1986; Valli, 1986; Burnett and Rickerby, 1987a; Steinmann et al., 1987; Sekler et al., 1988; Wu et al., 1989, 1990b; Wu, 1990, 1991; Bull and Rickerby, 1990; Cheng et al., 1990; Julia-Schmutz and Hintermann, 1991; White et al., 1993; Bhushan et al., 1995, 1998; Gupta and Bhushan, 1995a,b; Patton and Bhushan, 1996; Bhushan and Patton, 1996; Bhushan and Li, 1997; Li and Bhushan, 1998b,c). It is a most commonly used technique to measure adhesion of hard coatings with strong interfacial adhesion (>70 MPa).

For a scratch geometry shown in Figure 10.52, surface hardness H is given by

$$H = \frac{W_{cr}}{\pi a^2}$$

(10.24)

and adhesion strength t is given by (Benjamin and Weaver, 1960)

$$\tau = H \tan \theta$$

$$= \frac{W_{cr}}{\pi a^2} \left[\frac{a}{\left(R^2 - a^2 \right)^{1/2}} \right] \qquad (10.25a)$$

or

$$\tau = \frac{W_{cr}}{\pi a R} \quad \text{if} \quad R \gg a \qquad (10.25b)$$

where W_{cr} is the critical normal load, a is the contact radius, and R is the stylus radius.

Burnett and Rickerby (1987a) and Bull and Rickerby (1990) analyzed the scratch test of a coated sample in terms of three contributions: (1) a plowing contribution which will depend on the indentation stress field and the effective flow stress in the surface region, (2) an adhesive friction contribution due to interactions at the indenter–sample interface, and (3) an internal stress contribution since any internal stress will oppose the passage of the indenter through the surface, thereby effectively modifying the surface flow stress. They derived a relationship between the critical normal load W_{cr} and the work of adhesion W_{ad}

$$W_{cr} = \frac{\pi a^2}{2} \left(\frac{2 E W_{ad}}{t} \right)^{1/2} \qquad (10.26)$$

where E is the Young's modulus of elasticity and t is the coating thickness. Plotting of W_{cr} as a function of $a^2/t^{1/2}$ should give a straight line of the slope $\pi (2 E W_{ad}/t)^{1/2}/2$ from which W_{ad} can be calculated. Bull and Rickerby suggested that either the line slope (interface toughness) or W_{ad} could be used as a measure of adhesion.

An accurate determination of critical load W_{cr} sometimes is difficult. Several techniques, such as (1) microscopic observation (optical or SEM) during the test, (2) chemical analysis of the bottom of the scratch channel (with electron microprobes), and (3) acoustic emission, have been used to obtain the critical load (Perry, 1981, 1983; Je et al., 1986; Valli, 1986; Steinmann et al., 1987; Sekler et al., 1988; Wu et al., 1990b; Wu, 1991). In some instruments, tangential (or friction) force is measured during scratching to obtain the critical load (Jacobson et al., 1983; Valli, 1986; Wu et al., 1990b; Wu, 1990, 1991; Anonymous, 1991; Bhushan et al., 1995, 1997; Gupta and Bhushan, 1995a,b; Patton and Bhushan, 1996; Bhushan and Patton, 1996; Bhushan and Li, 1997; Li et al., 1998b,c). The AE and friction force techniques have been reported to be very sensitive in determining critical load. AE and friction force start to increase as soon as cracks begin to form perpendicular to the direction of the moving stylus.

Wu (1991) has used the scratch technique to study the adhesion and scratch resistance of diamondlike carbon and zirconia coatings deposited on Si(100) substrates. Figure 10.53 shows the scratch morphology at increasing normal loads and typical scratch data (normal load, tangential load and acoustic emission as well as calculated apparent coefficient of friction). We note that all three monitored outputs (LC, TG and AE) detected the first spallation event of the carbon coating by showing sudden changes in their output signals.

Bhushan et al. (1995, 1997), Bhushan and Li (1997), Gupta and Bhushan (1995a,b), Patton and Bhushan (1996), and Li and Bhushan (1998b,c) have used the scratch technique to study adhesion and scratch resistance (mechanical durability) of various ceramic films. Scratch tests conducted with a sharp diamond tip simulate a sharp asperity contact. Bhushan et al. have also conducted accelerated friction

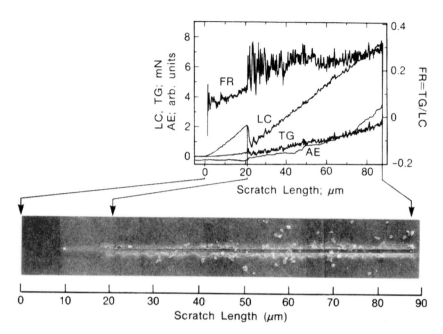

FIGURE 10.53 Scratch morphology and typical scratch data of 0.11-μm-thick DC-sputtered DLC film on an Si substrate. The monitored outputs are the normal load (LC), tangential load (TG), acoustic emission (AE). The apparent coefficient of friction (FR = TG/LC) is also shown. (From Wu, T.W. et al., 1990, in *Symp. Proc.*, Vol. 188, pp. 207–212, Materials Research Society, Pittsburgh. With permission.)

and wear (ball-on-coated disk) and functional tests and have found a good correlation between the scratch resistance and wear resistance measured using accelerated tests (Bhushan et al., 1995a, 1997; Gupta and Bhushan, 1995b; Patton and Bhushan, 1996; Bhushan and Patton, 1996; Bhushan and Li, 1997; Li and Bhushan, 1998b,c) and functional tests (Patton and Bhushan, 1996; Bhushan and Patton, 1996; Bhushan et al., 1997). Based on this work, scratch tests can be successfully used to screen materials and coatings for wear applications.

Gupta and Bhushan (1995b) conducted scratch tests on 20-nm-thick DLC films deposited by several deposition techniques and sputtered Al_2O_3 and sputtered SiC films on Al_2O_3–TiC (70–30 wt%) substrate. Films were scratched using a 1-μm-radius conical tip at ramping loads ranging from 2 to 25 mN with a scratch length of 500 μm. The friction profiles as a function of increasing normal load and SEM images of three regions over scratches — at the beginning of the scratch (indicated by A on friction profile), at the point of initiation of damage at which coefficient of friction increases abruptly to a very high value (indicated by B on friction profile), and toward the end of scratch (indicated by C on friction profile) — made on various carbon, ion beam-sputtered Al_2O_3, and DC-sputtered SiC films on Al_2O_3–TiC substrates and uncoated substrate are compared in Figure 10.54. The cathodic arc carbon film exhibits a low coefficient of friction of about 0.20 during scratch and a critical load of about 19 mN. All films except cathodic arc carbon on Al_2O_3–TiC substrates exhibit a continuous increase in the coefficient of friction during scratching and were damaged at much lower normal loads. Cathodic arc carbon film exhibited scratches with a width of about 0.9 μm (at about 22 to 25 mN normal load) comparable to that of ion beam carbon and sputtered carbon coating but lower than that PECVD carbon coating. A qualitative comparison of the amount of debris generated during scratching reveals that minimum debris was formed in the case of cathodic arc carbon coating, Table 10.6. These observations suggest that cathodic arc carbon coating on Al_2O_3–TiC substrate has a superior scratch resistance compared with other coatings. The RF-sputtered SiC coating on Al_2O_3–TiC exhibited lower scratch resistance than that of cathodic arc carbon coating on Al_2O_3–TiC and higher scratch resistance than that of all other carbon coatings, Table 10.6.

Accelerated friction tests were conducted by sliding a diamond tip (20-μm radius) and a single-crystal sapphire ball (3-mm diameter) against the films. The data are compared in Figure 10.55. We note that the coefficients of friction of all of the coatings are low (≈0.06 to 0.13) when these were slid against a diamond tip, whereas these coatings exhibited higher coefficients of friction from 0.20 to 0.50 when slid against a sapphire ball. SiC coating exhibited the lowest coefficients of friction among other coatings, when it was slid against a sapphire ball. Based on optical examination of worn samples, there were no wear tracks formed when cathodic arc carbon, ion beam carbon, PECVD carbon, and SiC coatings were slid for 2 h against a diamond tip at 10 mN load. Sputtered carbon and sputtered Al_2O_3 coatings exhibited a small amount of wear debris, because of their lower scratch resistance. On the other hand, for the coatings slid for 2 h against a sapphire ball, the cathodic arc carbon and SiC coatings on Al_2O_3–TiC substrate did not form any wear track, Figure 10.56. The higher wear performances of cathodic arc carbon, ion beam carbon, and SiC coatings are attributed to their high resistance to scratch and high hardnesses and elastic moduli.

Magnetic thin-film head sliders made with Al_2O_3–TiC substrate are used in magnetic storage applications (Bhushan, 1996). Multilayered thin-film pole-tip structure present on the head slider surface wears more rapidly than the Al_2O_3–TiC substrate, which is much harder. Pole-tip recession (PTR) is a serious concern in magnetic storage. Two of the DLC coatings superior in mechanical properties — ion beam and cathodic arc carbon — were deposited on the air bearing surfaces of Al_2O_3–TiC head sliders. The functional tests were conducted by running a metal-particle (MP) tape in a computer tape drive. Average PTRs as a function of sliding distance data are presented in Figure 10.57. We note that PTR increases for the uncoated head, whereas for the coated heads there is a slight increase in PTR in early sliding followed by little change. Thus, coatings provide protection.

This example clearly suggests that material characterizations (hardness, elastic modulus, and scratch resistance) are powerful ways of screening materials and the data correlates well with the functional friction and wear performance.

10.6.3   Residual Stress Measurements Using Nanoindentation

Indentation measurements similar to those used to determine the hardness and elastic modulus of a film can also be used to measure the residual stresses in it. When a compressive force on a biaxially stressed film during indentation is applied in a direction perpendicular to the film, yielding will occur at a smaller applied compressive force while a film is stressed in biaxial tension as compared with the unstressed film. Thus, the biaxial tension decreases hardness and the biaxial compression increases hardness (Swain et al., 1977; Vitovec, 1986). LaFontaine et al. (1990c, 1991) used the nanoindentation technique to measure the effect of residual stresses on the hardness of thin films. For samples that do not undergo large structural changes, changes in hardness with time reflect a change in residual stress in the film (LaFontaine et al., 1990c, 1991). LaFontaine et al. (1991) measured the stress relaxation in thin aluminum films and the residual stresses on identical films using an X-ray stress measurement technique. Results of the indentation and X-ray stress measurements compared closely, implying that decrease in hardness with time resulted from the relaxation of residual stresses. Thus, the indentation measurements can be used to investigate stresses in thin films. Due to the presence of a stress gradient in thicker (>1 μm) films, the technique is not applicable.

Tsukamoto et al. (1987) measured the deflection at the center of the bent beam (bent as a result of residual stresses in the film) by pressing the beam flat with a nanoindenter. The bent beam is placed on a flat glass surface supported by two fulcrums, and a load–deflection curve is generated (Figure 10.58). The distance h_a can be estimated from the inflection point in the curve. Because of limited flatness of most substrates, the film is removed from the substrate, and then the initial deflection is measured. The true deflection resulting from residual stresses in the film equal to $h_a - h_b$. The curvature $(1/R)$ of the substrate can then be calculated by the geometric relationship,

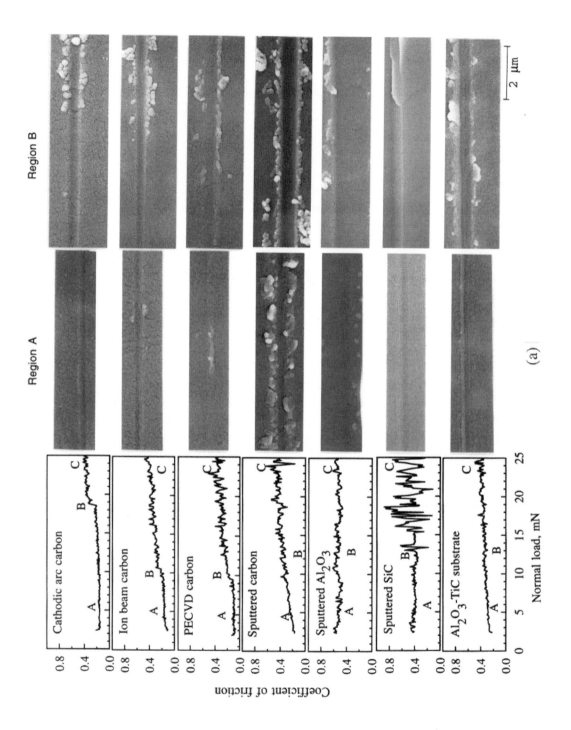

(a)

Region C

Cathodic arc carbon

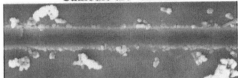

Ion beam carbon

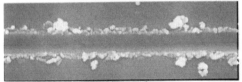

PECVD carbon

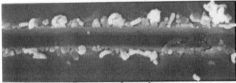

Sputtered carbon

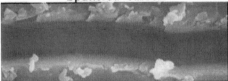

Sputtered Al₂O₃

Sputtered SiC

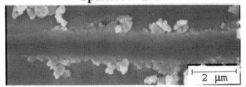

Al₂O₃-TiC substrate

(b)

FIGURE 10.54   (a) Coefficient of friction profiles as a function of normal load and SEM images of three regions over scratches; at the beginning of the scratch (indicated by A on friction profile), at the point of initiation of damage at which coefficient of friction increases abruptly to a very high value (indicated by B on friction profile) and (b) toward the end of scratch (indicated by C on friction profile), made on various DLC, Al_2O_3, and SiC films on Al_2O_3–TiC substrates and uncoated Al_2O_3–TiC substrate. (From Gupta, B.K. and Bhushan, B., 1995, *Wear* 190, 110–122. With permission.)

TABLE 10.6 Coefficient of Friction (at the beginning), Critical Load Estimated from an Abrupt Increase in Friction during Scratching, and Width of Scratches at about 23–25 mN Normal Load, Measured from SEM Images, and Qualitative Comparison of the Amount of Debris Generated during Scratching for Various 20-nm-Thick Coatings on Al_2O_3–TiC Substrates

| | | Coefficient of Friction during Scratching | SEM Observations of Scratches | |
| | | | Width of Scratch (μm) | Amount of Debris Generated[b] |
Coating	Critical Load (mN)			
Cathodic arc carbon	19.0	0.20	0.9	Low
Ion beam carbon	10.5	0.15	0.9	Medium
PECVD carbon	9.0	0.13	1.0	Medium
DC-sputtered carbon	6.0–9.0	0.20	0.9	Medium
Ion beam–sputtered Al_2O_3	Plowing[a]	0.55	1.6	Very large
RF-sputtered SiC	12.5	0.40	1.1	Medium
Al_2O_3–TiC substrate	7.0	0.30	1.5	Large

[a]Tip plowed into the coating right at the beginning.
[b]A qualitative comparison of debris generated toward the end of the scratch after catastrophic damage.

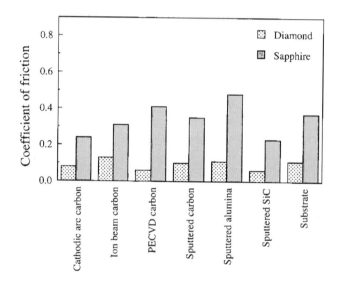

FIGURE 10.55 Initial coefficient of friction of various DLC, ion beam–sputtered Al_2O_3, and RF-sputtered SiC films deposited on Al_2O_3–TiC substrates and the substrate, against a diamond tip (tip radius = 20 μm) and a single-crystal sapphire ball (diameter = 3 mm) at a 10 mN normal load, 3.5 mm stroke length, 0.1 Hz frequency, 0.7 mm s^{-1} average linear speed, and under ambient temperature of 22 ± 1°C and humidity of 45 ± 5% RH. (From Gupta, B.K. and Bhushan, B., 1995, *Wear* 190, 110–122. With permission.)

$$R = \frac{L^2}{8\left(h_a - h_b\right)} \tag{10.27}$$

where L is the span.

Hong et al. (1990) used another deflection measurement technique. In this technique, a circular section of the substrate is removed from beneath the film to produce a drum-headlike membrane and the load is applied at its center. The stiffness of the membrane (film) is a sensitive function of the biaxial tension in it. The deflection h is related to load W as

Cathodic arc carbon

Ion beam carbon

PECVD carbon

Sputtered carbon

Sputtered Al₂O₃

Sputtered SiC

⊢ 100 μm ⊣

Al₂O₃-TiC substrate

FIGURE 10.56 Optical images of worn DLC, ion beam–sputtered Al$_2$O$_3$, and RF-sputtered SiC films deposited on Al$_2$O$_3$–TiC substrates and the bare substrate, slid against a single-crystal sapphire ball (diameter = 3 mm) for 2 h at a 10 mN normal load, 3.5 mm stroke length, 0.1 Hz frequency, 0.7 mm s^{-1} average linear speed, and under ambient temperature of 22 ± 1°C and humidity of 45 ± 5% RH. (From Gupta, B.K. and Bhushan, B., 1995, *Wear* 190, 110–122. With permission.)

$$h = \frac{Wa^2}{16\pi D} g(k) \qquad (10.28)$$

where

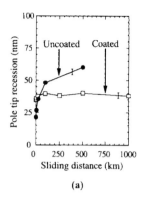

(a)

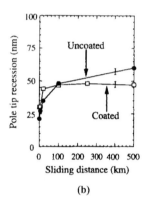

(b)

FIGURE 10.57 Pole tip recession as a function of sliding distance as measured with an atomic force microscope for (a) uncoated and ion beam carbon-coated, and (b) uncoated and cathodic arc carbon-coated Al_2O_3–TiC heads run against MP tapes.

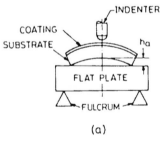

(a)

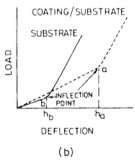

(b)

FIGURE 10.58 (a) Schematic diagram of the deflection measurement of a bent beam using a nanoindenter and (b) load–deflection curve for a warped composite-beam substrate.

$$D = \frac{Et^3}{12\left(1 - \nu^2\right)}$$

where a is the radius of the membrane, t is its thickness, and the function $g(k)$ depends on the membrane and its geometry. If the geometry and elastic constants of the membrane are known, the tension can be accurately evaluated. This technique can only be used to study tensile residual stresses since compressive stresses buckle the membrane when the substrate is removed.

10.6.4 Microwear Measurements Using Modified Nanoindentation

Wu and Lee (1994) modified the nanoindenter/nanoscratch technique for microwear studies. A piezo-electric pusher with a servo control was employed to implement a reciprocating horizontal motion at the indenter tip for implementing a microwear test. Through a lock-in detection scheme, they measured friction force.

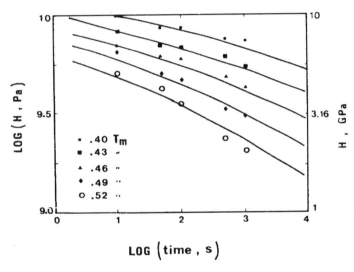

FIGURE 10.59 Indentation creep data for Si at different temperatures. (From Li, W.B. et al., 1991, *Acta Metall. Mater.* 39, 3099–3110. With permission.)

10.7 Other Applications of Nanoindentation Techniques

Nanoindentation techniques have been used for measurement of time-dependent viscoelastic/plastic properties (creep/relaxation), nanofracture toughness, and nanofatigue.

10.7.1 Time-Dependent Viscoelastic/Plastic Properties

Most materials including ceramics and even diamond are found to creep at temperatures well below half their melting points, even at room temperature. Indentation creep and indentation load relaxation (ILR) tests are used for measurement of the time-dependent flow of materials. These offer an advantage of being able to probe the deformation properties of a thin film as a function of indentation depth and location.

In the indentation creep test, the hardness indenter maintains its load over a period of time under well-controlled conditions, and changes in indentation size are monitored (Westbrook, 1957; Mulhearn and Tabor, 1960/61; Atkins et al., 1966; Walker, 1973; Hooper and Brookes, 1984; Li et al., 1991). The analysis of creep is more complex than that of creep data obtained using a conventional technique because of the shape of the tip, indentation stress acting on the sample decreases with time as the contact area increases. Chu and Li (1977, 1979, 1980) developed an impression creep test on a macroscale which used a circular tip with a flat end. Log (hardness) decreases (linearly for most metals) with log (time at load). Nanoindentors are also used for indentation creep studies (Li et al., 1991; Lucas and Oliver, 1992; Raman and Berriche, 1990, 1992). Figure 10.59 shows the plot of log (indentation hardness) as a function of log (time) for silicon at different temperatures. Indentation creep is influenced by a large number of variables such as the plastic deformation properties of the material, diffusion constants, normal load of indenter, duration of the indentation, and the test temperature. Li et al. (1991) reported that for temperatures between 27°C and melting, the mechanism of dislocation glide plasticity dominates the indentation creep process.

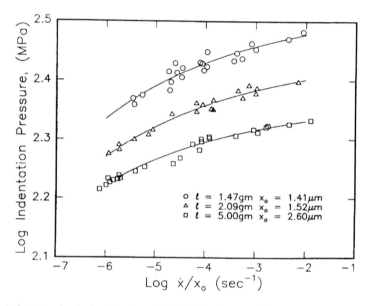

FIGURE 10.60 Indentation load relaxation data of a (111) single-crystal aluminum sample at 25°C; l is the load, x_0 is the plastic indentation depth at the beginning of the relaxation, and $\dot{x}$ is the plastic indentation rate. (From LaFontaine, W.R. et al., 1990, *J. Mater. Res.* 5, 2100–2106. With permission.)

In a typical ILR test, the indenter is first pushed into the sample at a fixed displacement rate until a predetermined load or displacement is achieved and the position of the indenter is then fixed. The material below the indenter is elastically supported and will continue to deform in a nonelastic manner, thereby tending to push the indenter farther into the sample. Load relaxation is achieved by conversion of elastic strain in the sample into inelastic strain in the sample. During the test, the load and position of the indenter and the specimen are continuously monitored. Normally the indenter motion is held constant and the changes in the load are monitored as a function of time. It is possible to obtain the plastic indentation rate from the indentation load and total depth information during the relaxation run (Hart and Solomon, 1973; Chu and Li, 1980; Nastasi et al., 1993). The resulting load relaxation data are reported in the form of log (indentation pressure) as a function of log (plastic indentation strain rate) (Hannula et al., 1985; LaFontaine et al., 1990a,b; Wu, 1991).

The indentation pressure is calculated by dividing the load by the projected area of the indenter. Once the plastic indentation depth is known as a function of time, the projected area is determined experimentally as described earlier. The plastic indentation strain rate — $[(1/h)(dh/dt)$, where h is the current indentation depth] — is calculated in the manner similar to that for bulk relaxation data. First, the plastic indentation depth is calculated which is equal to the total depth minus its elastic depth based on load vs. time data. The plastic indentation depth vs. time data is divided into segments. Over each segment, the plastic indentation depth is assumed to vary linearly with time. At the midpoint of each segment the slope of the plastic indentation depth curve is determined to obtain the plastic indentation rate. The average plastic indentation depth is used along with the corresponding average load over the same time interval to determine the indentation pressure (LaFontaine et al., 1990b). Figure 10.60 shows typical log (indentation pressure) vs. log (normalized plastic indentation rate normalized with the depth at the beginning of the relaxation experiment) flow curves for a (111) single-crystal aluminum.

Another technique to measure strain rate sensitivity of submicron films was developed by Mayo and Nix (1988a,b). They developed two procedures. In the first procedure, known as the constant rate of loading test, individual indentations are performed at a prescribed loading rate that is varied from one indentation to another by about a factor of 2. The values of the indentation pressure and strain rates from tests performed at different loading rates were compared at a common indentation depth. In a second related technique known as the loading rate change test, the loading rate is held constant at a specified depth, but is suddenly changed to a new value and the subsequent changes in pressure and strain rate are monitored.

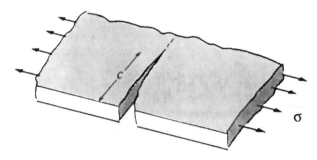

FIGURE 10.61 Schematic of a standard specimen used for measurement of fracture toughness of materials in tension.

Mayo et al. (1990, 1992) also developed yet another procedure for determining the strain rate sensitivity of nanophase materials in order to conduct tests at the high loading rates required for very hard materials. In this procedure, as soon as the indenter contacts the sample surface, the indenter loading rate is instantaneously increased to a high value and this value is maintained until the indenter reaches a prescribed displacement. The load is held constant at this point, and the load is monitored as a function of time. The initial fast descent rate of the indenter produces a substantial amount of creep during a constant-hold period. A range of descent rates are realized ranging from high values at the beginning of the hold period to smaller values as the material stops deforming. These correspond to a range of strain rates, and several stress–strain pairs can be obtained (Raman and Berriche, 1990, 1992).

The strain rate sensitivity of materials is measured in terms of stress exponent, n, which is defined by the equation,

$$\text{Plastic indentation rate} = A \, (\text{indentation pressure})^n \qquad (10.29)$$

where A and n are the constants. The stress exponent is found as a slope of a log–log plot of plastic indentation rate (or strain rate) and indentation pressure. In the ILR test, the continuous change in the contact area results in continuous changes in both plastic indentation rate and pressure. Thus, data from a single indentation test, which may span several orders of magnitude in both strain rate and pressure, are sufficient to determine the stress exponent. Stress exponent can be used to define the superplasticity of a material. The variations in stress exponent reflect the changes that may take place when the substructure generated at high strain rate approaches equilibrium condition (Mayo and Nix, 1988a).

10.7.2 Nanofracture Toughness

Fracture toughness, K_{1c} of a material is a measure of its resistance to the propagation of cracks and the ratio H/K_{1c} is an index of brittleness, where H is the hardness. Resistance to fracture is a strong function of crack pattern. It is typically measured in a test in which a specimen containing a sharp crack of known length, c, is subjected to an applied stress S, which is increased during test until the sample fractures (Lawn, 1993), Figure 10.61. The magnitude of the stresses near the crack tip are determined by the stress intensity factor K_I which, in turn, depends on S, c, and the specimen geometry (A)

$$K_I = A\sigma\sqrt{\pi c} \tag{10.30}$$

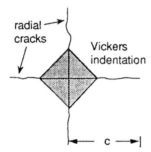

FIGURE 10.62   Schematic of Vickers indentation with radial cracks.

The term A provides correction for the thickness-to-width ratio of the material. Units of stress-intensity factor are MPa$\sqrt{m}$. With more intense stress or with deeper cracks, the stress intensity becomes sufficient for the fracture to progress spontaneously. This threshold stress intensity is a property of the material and is called the *critical stress intensity factor, K_{1c}, or the *fracture toughness* of the material. Ceramics generally have relatively low fracture toughness, consequently, it is an important property to be considered for the selection of ceramics for industrial applications.

Indentation fracture toughness is a simple technique for determination of fracture toughness (Palmquist, 1957; Lawn and Wilshaw, 1975; Evans and Charles, 1976; Lawn and Evans, 1977; Lawn and Marshall, 1979; Lawn et al., 1980; Antis et al., 1981; Chantikul et al., 1981; Chiang et al., 1981, 1982; Henshall and Brookes, 1985; Cheng et al., 1990; Cook and Pharr, 1990; Choi and Salem, 1993; de Boer et al., 1993; Lawn, 1993; Pharr et al., 1993; Bhushan et al., 1996). The indentation cracking method is especially useful for measurement of fracture toughness of thin films or small volumes. This method is quite different from conventional methods in that no special specimen geometry is required. Rather, the method relies on the fact that when indented with a sharp indenter, most brittle materials form radial cracks and the lengths of the surface traces of the radial cracks (for definition of crack length, see Figure 10.62) have been found to correlate reasonably well with fracture toughness. By using simple empirical equations, fracture toughness can then be determined from simple measurement of crack length.

In microindentation, cracks at relatively high indentation loads of several hundred grams are on the order of 100 μm in length and can be measured optically. However, to measure toughness of very thin films or small volumes, much smaller indentations are required. However, a problem exists in extending the method to nanoindentation regime in that there are well-defined loads, called cracking thresholds, below which indentation cracking does not occur in most brittle materials (Lankford, 1981). For a Vickers indenter, cracking thresholds in most ceramics are about 25 g. Pharr et al. (1993), Li et al. (1997), and Li and Bhushan (1998a) have found that the Berkovich indenter (a three-sided pyramid) with the same depth-to-area ratio as a Vickers indenter (a four-sided pyramid) has a cracking of the thresholds very similar to that of the Vickers indenter. They showed that cracking thresholds can be substantially reduced by using sharp indenters, i.e., indenters with smaller included tip angles, such as a three-sided indenter with the geometry of the corner of a cube. Studies using a three-sided indenter with the geometry of a corner of a cube have revealed that cracking thresholds can be reduced to loads as small as 0.5 g, for which indentations and crack lengths in most materials are submicron in dimension.

Based on fracture mechanics analysis, Lawn et al. (1980) developed a mathematical relationship between fracture toughness and indentation crack length, given as

$$K_{Ic} = B\left(\frac{E}{H}\right)^{1/2}\left(\frac{W}{c^{3/2}}\right) \tag{10.31}$$

where W is the applied load and B is an empirical constant depending upon the geometry of the indenter (also see Lawn, 1993; Pharr et al., 1993). Antis et al. (1981) conducted a study on a number of brittle materials chosen to span a wide range of toughnesses. They indented with a Vickers indenter at several loads and measured crack length optically. They found a value of $B = 0.016$ to give good correlation between the toughness values measured from the crack length and the ones obtained using more conventional methods. Mehrotra and Quinto (1985) used a Vickers indenter to measure fracture toughness of the coatings. Pharr et al. (1993) tested several bulk ceramics listed in Table 10.7 using Vickers, Berkovich,

TABLE 10.7 Typical Mechanical Properties of Materials Tested by Pharr et al. (1993)

Material	E(GPa)	H(GPa)	K_{Ic}(MPa√m)
Soda lime glass	70	5.5	0.70
Fused quartz	72	8.9	0.58
(111) Silicon	168	9.3	0.70
(111) Sapphire	403	21.6	2.2
Si_3N_4	300	16.3	4.0

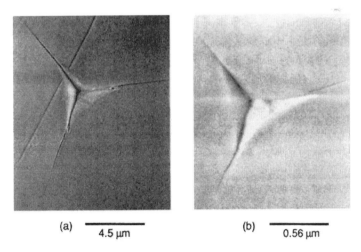

(a) 4.5 µm   (b) 0.56 µm

FIGURE 10.63   Indentations in fused quartz made with the cube corner indenter showing radial cracking at indentation loads of (a) 12 g and (b) 0.45 g. (From Pharr, G.M. et al., 1993, in *Mechanical Properties and Deformation Behavior of Materials Having Ultra-Fine Microstructures*, M. Nastasi et al., eds., pp. 449–461, Kluwer Academic, Dordrecht. With permission.)

and cube corner indenters. They found that the fracture toughness equation can be applied for the data obtained with all three indenters provided a different empirical constant was used for a cube corner indenter. The constant B for the Vickers and Berkovich indenter was found to be about 0.016 and for cube corner it was about 0.032. Pharr et al. (1993) further reported that predominant cracks formed with Vickers or Berkovich indenters are cone cracks and with cube corner indenter, predominant cracks were radial cracks, Figure 10.63. Bhushan et al. (1996) reported that cracks formed in microcrystalline ceramic material (glass–ceramic) with Vickers indenter are radial cracks, Figure 10.64. Note that cracks

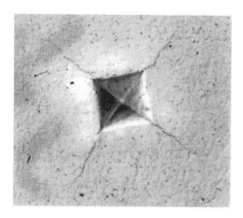

FIGURE 10.64   Optical images of Vickers indentation made on a glass–ceramic substrate at 500 g load.

propagate in a zigzag manner. The interlocked crystal morphology is responsible for propagation in a zigzag manner. By using Equation 10.31, the fracture toughness for this material is calculated and presented in Table 10.4.

Chantikul et al. (1981) developed a relationship between fracture toughness and the indentation fracture strength and the applied load given as,

$$K_{Ic} = c \left(\frac{E}{H} \right)^{1/8} \left(\sigma_f W^{1/3} \right)^{3/4}$$

(10.32)

where s_f is the fracture strength after indentation at a given load and c is an empirical constant (0.59). Advantage of this analysis is that the measurement of crack length is not required. Mecholsky et al. (1992) used this analysis to calculate fracture toughness of diamond films on silicon. They indented the films at various indentation loads of 3 to 9 kg and then fractured in four-point flexure to measure fracture strength. The data were then used to get fracture toughness. Equation 10.32 was found to hold for the measurements. They reported a fracture toughness of 6 MPa√m and 12-μm-thick diamond films on silicon on the order of 2 MPa√m.

For fracture toughness measurement of ultra-thin films ranging from 100 nm to few micrometers, indentation or four-point flexure techniques cannot be used. Because of shallow indentation depths required in the indentation technique, it is difficult to measure a radial crack length under even SEM. Li et al. (1997, 1998a) developed a novel technique based on nanoindentation in which through-thickness cracking in the coating is detected from a discontinuity observed in the load–displacement curve and energy released during the cracking is obtained from the curve. Based on the energy released, fracture mechanics analysis is then used to calculate fracture toughness. A cube corner is preferred because the through-thickness cracking of hard films can be accomplished at lower loads (Li and Bhushan, 1998a).

Load–displacement curves of indentations made at 30, 100, and 200 mN peak indentation loads together with the SEM micrographs of indentations on the cathodic arc carbon coating on silicon are shown in Figure 10.65. Steps are found in all loading curves as shown by arrows in Figure 10.65a. In the 30-mN SEM micrograph, in addition to several radial cracks, ringlike through-thickness cracking is observed with small lips of material overhanging the edge of indentation. The step at about 23 mN in

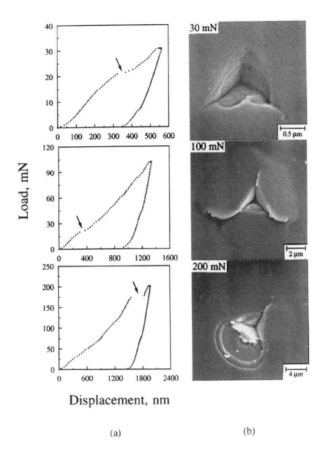

(a)   (b)

FIGURE 10.65 (a) Load-displacement curves of indentations made at 30, 100, and 200 mN peak indentation loads using the cube corner indenter and (b) the SEM micrographs of indentations on the cathodic arc carbon film on silicon. Arrows indicate steps during loading portion of the load–displacement curve. (From Li, X. et al., 1997, *Acta Mater.* 45, 4453–4461. With permission.)

the loading curves of indentations made at 30 and 100 mN peak indentation loads result from the ringlike through-thickness cracking. The step at 175 mN in the loading curve of indentation made at 200 mN peak indentation load is caused by spalling.

No steps were observed in the loading curve of indentation made at 20 mN peak indentation load which suggests that the coating under the indenter was not separated instantaneously from the bulk coating via the ringlike through-thickness cracking but occurred over a period of time. At 30 mN peak indentation load, partial ringlike spalling is observed around the indenter and the other parts of the film bulged upward. This partial ringlike spalling is believed to result in the step in the loading curve. Absence of long steps in the loading curve for uncoated silicon reported by Li et al. (1997, 1998a) suggest that the steps in the loading curve on the coating result from the film cracking. Based on their work, the fracture process progresses in three stages: (1) first ringlike through-thickness cracks form around the indenter by high stresses in the contact area, (2) delamination and buckling occur around the contact area at the film/substrate interface by high lateral pressure, (3) second ringlike through-thickness cracks and spalling are generated by high bending stresses at the edges of the buckled film, see Figure 10.66. In the first stage, if the film under the indenter is separated from the bulk film via the first ringlike through-thickness cracking, a corresponding step will be present in the loading curve. If discontinuous cracks form and the film under the indenter is not separated from the remaining film, no step appears in the loading curve because the film still supports the indenter and the indenter cannot suddenly advance into the material. In the second stage, for the films used in the present study, the advances of the indenter

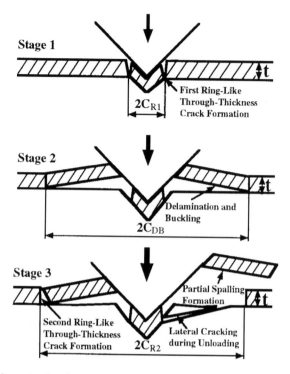

FIGURE 10.66 Schematic of various stages in nanoindentation fracture for the film/substrate system.

during the radial cracking, delamination, and buckling are not big enough to form steps in the loading curve because the film around the indenter still supports the indenter, but generate discontinuities which change the slope of the loading curve with increasing indentation loads. In the third stage, the stress concentration at the end of the interfacial crack cannot be relaxed by the propagation of the interfacial crack. With an increase in indentation depth, the height of the bulged film increases. When the height reaches a critical value, the bending stresses caused by the bulged film around the indenter will result in the second ringlike through-thickness crack formation and spalling at the edge of the buckled film as shown in Figure 10.66, which leads to a step in the loading curve. This is a single event and results in the separation of the part of the film around the indenter from the bulk film via cracking through films. The step in the loading curve is totally from the film cracking and not from the interfacial cracking or the substrate cracking.

The area under the load–displacement curve is the work performed by the indenter during elastic–plastic deformation of the film/substrate system. The strain energy release in the first/second ringlike cracking and spalling can be calculated from the corresponding steps in the loading curve. Figure 10.67 shows a

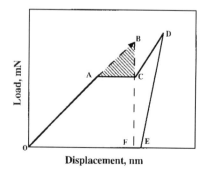

FIGURE 10.67 Schematic of a load–displacement curve, showing a step during the loading cycle and associated energy release.

modeled load–displacement curve. OACD is the loading curve. DE is the unloading. Since the first ringlike through-thickness cracking does not always lead to a step in the loading curve in some films, the second ringlike through-thickness crack should be considered. It should be emphasized that the edge of the buckled film is far from the indenter; therefore, it does not matter if the indentation depth exceeds the film thickness or if deformation of the substrate occurs around the indenter when we measure fracture toughness of the film from the released energy during the second ringlike through-thickness cracking (spalling). Suppose that the second ringlike through-thickness cracking occurs at AC. Now, let us consider the loading curve OAC. If the second ringlike through-thickness crack does not occur, it can be understood that OA will be extended to OB to reach the same displacement as OC. This means that the crack formation changes the loading curve OAB into OAC. For point B, the elastic–plastic energy stored in the film/substrate system should be OBF. For point C, the elastic–plastic energy stored in the film/substrate system should be OACF. Therefore, the energy difference before and after the crack generation is the area of ABC; i.e., this energy stored in ABC will be released as strain energy to create the ringlike through-thickness crack. According to the theoretical analysis by Li et al. (1997), the fracture toughness of thin films can be written as

$$K_{Ic} = \left[\left(\frac{E}{\left(1 - v^2\right) 2\pi C_R} \right) \left(\frac{U}{t} \right) \right]^{1/2} \tag{10.33}$$

where E is the elastic modulus, n is Poisson's ratio, $2\pi C_R$ is the crack length in the film plane, U the strain energy difference before and after cracking, and t is the film thickness.

By using Equation 10.33, the fracture toughness of the 0.4-μm-thick cathodic arc carbon coating is calculated. U of 7.1 nNm is assessed from the steps in Figure 10.65a at the peak indentation loads of 200 mN. The loading curve is extrapolated along the tangential direction of the loading curve from the starting point of the step up to reach the same displacement as the step. The area between the extrapolated line and the step is the estimated strain energy difference before and after cracking. C_R of 7.0 μm is measured from SEM micrographs in Figure 10.65b. Second ringlike crack is where the spalling occurs. For $E = 300$ GPa measured using nanoindenter (Table 10.5) and an assumed value of 0.25 for n, fracture toughness values is calculated as 10.9 MPa√m.

10.7.3 Nanofatigue

Delayed fracture resulting from extended service is called fatigue. Fatigue fracturing progresses through a material via changes within the material at the tip of a crack, where there is a high stress intensity. There are several situations: cyclic fatigue, stress corrosion, and static fatigue. Cyclic fatigue results from cyclic loading of machine components; e.g., the stresses cycle from tension and compression occurs in a loaded rotating shaft. Fatigue also can occur with fluctuating stresses of the same sign, as occurs in a leaf spring, in a dividing board. In a low flying slider in a head–disk interface, isolated asperity contacts occur during use and the fatigue failure occurs in the multilayered thin-film structure of the magnetic disk (Bhushan, 1996). Asperity contacts can be simulated using a sharp diamond tip in an oscillating contact with the thin-film disk.

Li and Chu (1979) developed an indentation fatigue test, called impression fatigue. In this test, a cylindrical indenter with a flat end was pressed onto the surface of the test material with a cyclic load and the rate of plastic zone propagation was measured.

Wu et al. (1991) developed a nanoindentation fatigue test by modifying their nanoindenter. The cyclic indentation was implemented by servo controlling the PZT stack to drive the indenter so that the load cell output followed a 0.1-Hz sinusoidal loading pattern, and the latter was specified by the cyclic frequency and the lower and upper limits for the desired load range. The lower limit for all the tests was set at about 0.2 mN to perform a full load cycle indentation fatigue. A nonzero limit was required in order to activate the load cell servo control mode. Several maximum loads, namely, 4, 16 and 24 mN, were used. The following results can be obtained: (1) endurance limit, i.e., the maximum load below which there is no coating failure for a preset number of cycles; (2) number of cycles at which the coating failure occurs; and (3) changes in contact stiffness measured by using the unloading slope of each cycle which can be used to monitor the propagation of the interfacial cracks during cyclic fatigue process. They used a conical diamond indenter with a nominal 1-μm tip. Applied load and penetration depth were simultaneously monitored during the entire test.

Typical nanoindentation fatigue results for a 0.11-μm DC planar magnetron-sputtered amorphous carbon films on (100) silicon substrate deposited at argon pressure of 30 mtorr is shown in Figure 10.68

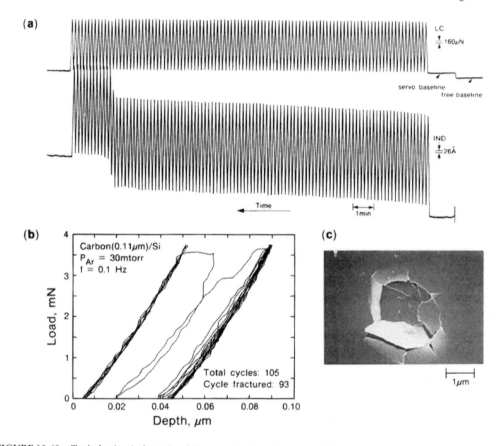

FIGURE 10.68 Typical microindentation fatigue results from 0.11-μm-thick DC-sputtered amorphous carbon on (100)Si, (a) the direct outputs from a strip chart recorder of applied load (LC) and indenter position (IND) (Maximum load = 4 mN, frequency = 0.1 Hz) (b) a plot of the indentation fatigue loading curve vs. total penetration depth (the plot includes only the load cycles from 91 to 100; note the abrupt depth increase started from the 93rd cycle), and (c) SEM micrograph of the fatigue indent. (From Wu, T.W. et al., 1991, *Surf. Coat. Technol.* 47, 696–709. With permission.)

(Wu et al., 1991). The test was run at a maximum cyclic load of 4 mN and 0.1 Hz for a total of 105 cycles with fracture in the 93rd cycle. An SEM micrograph of the damaged zone (Figure 10.68c) shows that the plastic deformation attributed primarily to the silicon substrate occurred in the central indent area, and moreover the carbon coating spalled around the indent and resulted in several isolated carbon flakes as well as cantilevered flakes. Wu et al. reported that in a single indentation test, the critical indentation load required to crack the carbon coating was about 6 mN. The critical indentation load was extracted by using the criterion of the applied load at which a load drop appeared along a loading curve. Evidently, the endurance limit can be significantly lower than the critical load of a single indentation. This scenario is analogous to the fatigue strength vs. tensile strength in macrotests. Wu et al. further reported that the carbon films deposited at an argon pressure of 6 mtorr exhibited an endurance limit of about 24 mN, about a factor of six higher than the film deposited at 30 mtorr. Scratch resistance of the film at 30 mtorr was also poorer as compared to that deposited at 6 mtorr (Wu et al., 1989, 1900b; Wu, 1991).

10.8   Closure

Depth-sensing nanoindentation hardness measurement techniques are commonly used to measure nano-mechanical properties of surface layers of bulk materials and of ultrathin coatings. The nanoindentation apparatus continously monitors the load and the position of the indenter relative to the surface of the specimen (depth of an indent) during the indentation process. The area of indent is then calculated from a knowledge of the geometry of the tip of the diamond indenter. Mechanical property measurements can be made at a minimum penetration depth of about 20 nm (or a plastic depth of about 15 nm). Typically, a three-sided pyramidal (Berkovich) diamond indenter is used for measurements. Mechanical properties that can be measured are elastic–plastic deformation behavior, hardness, Young's modulus of elasticity, scratch resistance, film–substrate adhesion, residual stresses, time-dependent creep and relaxation properties, fracture toughness, and fatigue.

References

Ahn, J., Mittal, K.L., and Macqueen, R.H. (1978), "Hardness and Adhesion of Filmed Structures as Determined by the Scratch Technique," *Adhesion Measurement of Thin Films, Thick Films, and Bulk Coatings* (K.L. Mittal, ed.), pp. 134–157, STP 640, ASTM, Philadelphia.

Alba, S., Loubet, J.L., and Vovelle, L. (1993), "Evaluation of Mechanical Properties and Adhesion of Polymer Coatings by Continous Hardness Measurements," *J. Adhesion Sci. Technol.* 7, 131–140.

Alekhin, V.P., Berlin, G.S., Isaev, A.V., Kalei, G.N., Merkulov, V.A., Skvortsov, V.N., Ternovskii, A.P., Krushchov, M.M., Shnyrev, G.D., and Shorshorov, M. Kh. (1972), "Micromechanical Testing by Micromechanical Testing of Materials by Microcompression," *Zavod. Lab.* 38, 619–621.

Anonymous (1979), "Standard Test Method for Microhardness of Materials," ASME Designation: E384–73, pp. 359–379.

Anonymous (1988), "Properties of Silicon," in *EMIS Data Reviews Series No. 4*, INSPEC, The Institution of Electrical Engineers, London.

Anonymous (1991) "NanoIndenter™II Operating Instructions," Nano Instruments, Inc., P.O. Box 14211, Knoxville, Tennessee 37914.

Antis, G.R., Chantikul, P., Lawn, B.R., and Marshall, D.B. (1981), "A Critical Evaluation of Indentation Techniques for Measuring Fracture Toughness: I Direct Crack Measurements," *J. Am. Ceram. Soc.* 64, 533–538.

Atkins, A.G., Silverio, A., and Tabor, D. (1966), *J. Inst. Met.* 94, 369.

Baker, S.P., Jankowski, A.F., Hong, S., and Nix, W.D. (1990), "Mechanical Properties of Compositionally Modulated Au-Ni Thin Films Using Indentation and Microbeam Deflection Techniques," in *Thin Films: Stresses and Mechanical Properties II* (M.F. Doerner, W.C. Oliver, G.M. Pharr, and F.R. Brotzen, eds.), pp. 289–294, Materials Research Society, Pittsburgh.

Bangert, H. and Wagendristel, A. (1986), "Ultralow Load Hardness Testing of Coatings in a Scanning Electron Microscope," *J. Vac. Sci. Technol. A* 4, 2956–2958.

Bangert, H., Wagendristel, A., and Aschinger, H. (1981), "Ultramicrohardness Tester for Use in a Scanning Electron Microscope," *Colloid Polym. Sci.* 259, 238–240.

Bell, T.J., Field, J.S., and Swain, M.V. (1992), "Elastic-Plastic Characterization of Thin Films with Spherical Indentation," *Thin Solid Films* 220, 289–294.

Benjamin, P. and Weaver, C. (1960), "Measurement of Adhesion of Thin Films," *Proc. R. Soc. London A* 254, 163–176.

Berkovich, E.S. (1951), "Three-Faceted Diamond Pyramid for Micro-Hardness Testing," *Ind. Diamond Rev.* 11, 129–132.

Bhattacharya, A.K. and Nix, W.E. (1988a), "Finite Element Simulation of Indentation Experiments," *Int. J. Solids Struct.* 24, 881–891.

Bhattacharya, A.K. and Nix, W.D. (1988b), "Analysis of Elastic and Plastic Deformation Associated with Indentation Testing of the Thin Films on Substrates," *Int. J. Solids Struct.* 24, 1287–1298.

Bhushan, B. (1987), "Overview of Coating Materials, Surface Treatments, and Screening Techniques for Tribological Applications — Part 2: Screening Techniques," in *Testing of Metallic and Inorganic Coatings* (W.B. Harding and G.A. DiBari, eds.), pp. 310–319, STP 947, ASTM, Philadelphia.

Bhushan, B. (1996), *Tribology and Mechanics of Magnetic Storage Devices,* 2nd ed., Springer-Verlag, New York.

Bhushan, B. and Doerner, M.F. (1989), "Role of Mechanical Properties and Surface Texture in the Real Area of Contact of Magnetic Rigid Disks," *ASME J. Tribol.* 111, 452–458.

Bhushan, B. and Gupta, B.K. (1995), "Micromechanical Characterization of Ni-P Coated Aluminum-Magnesium, Glass, and Glass Ceramic Substrates and Finished Magnetic Thin-Film Rigid Disks," *Adv. Info. Storage Syst.* 6, 193–208.

Bhushan, B. and Gupta, B.K. (1997), *Handbook of Tribology: Materials, Coatings and Surface Treatments,* McGraw Hill, New York (1991); reprint edition (1997), Krieger Publishing Co., Malabar, FL.

Bhushan, B. and Koinkar, V.N. (1994), "Nanoindentation Hardness Measurements Using Atomic Force Microscopy," *Appl. Phys. Lett.* 64, 1653–1655.

Bhushan, B. and Li, X. (1997), "Micromechanical and Tribological Characterization of Doped Single-Crystal Silicon and Polysilicon Films for Microelectromechanical Systems Devices," *J. Mater. Res.* 12, 54–63.

Bhushan, B. and Patton, S.T. (1996), "Pole Tip Recession Studies of Hard Carbon-Coated Thin-Film Tape Heads," *J. Appl. Phys.* 79, 5916–5918.

Bhushan, B., Landesman, A.L., Shack, R.V., Vukobratovich, D., and Walters, V.S. (1985), "Instrument for Testing Thin Films Such as Magnetic Tapes," *IBM Tech. Disclosure Bull.* 28, 2975–2976.

Bhushan, B., Williams, V.S., and Shack, R.V. (1988), "In-Situ Nanoindentation Hardness Apparatus for Mechanical Characterization of Extremely Thin Films," *ASME J. Tribol.* 110, 563–571.

Bhushan, B., Kellock, A.J., Cho, N.H., and Ager, J.W. (1992), "Characterization of Chemical Bonding and Physical Characteristics of Diamond-like Amorphous Carbon and Diamond Films," *J. Mater. Res.* 7, 404–410.

Bhushan, B., Gupta, B.K., and Azarian, M.H. (1995), "Nanoindentation, Microscratch, Friciton and Wear Studies of Coatings for Contact Recording Applications," *Wear* 181–183, 743–758.

Bhushan, B., Chyung, K., and Miller, R.A. (1996), "Micromechanical Property Measurements of Glass and Glass-Ceramic Substrates for Magnetic Thin-Film Rigid Disks for Gigabit Recording," *Adv. Info. Storage Syst.* 7, 3–16.

Bhushan, B., Theunissen, G.S.A.M., and Li, X. (1997) "Tribological Studies of Chromium Oxide Films for Magnetic Recording Applications," *Thin Solid Films* 311, 67–80.

Blau, P.J. and Lawn, B.R. (eds.) (1986), *Microindentation Techniques in Materials Science and Engineering,* STP 889, ASTM, Philadelphia.

Bourcier, R.J., Myers, S.M., and Polonis, D.H. (1990), "The Mechanical Reponse of Aluminum Implanted with Oxygen Ions," *Nucl. Inst. Meth. Phys. Res. B* 44, 278–288.

Bourcier, R.J., Follstaedt, D.M., Dugger, M.T., and Myers, S.M. (1991), "Mechanical Charaterization of Several Ion-Implanted Alloys: Nanoindentation Testing, Wear Testing and Finite Element Modelling," *Nucl. Inst. Meth. Phys. Res. B* 59/60, 905–908.

Bravman, J.C., Nix, W.D., Barnett, D.M., and Smith, D.A. (eds.) (1989), *Thin Films: Stresses and Mechanical Properties, Symp. Proc.,* Vol. 130, Materials Research Society, Pittsburgh.

Buckle, H. (1973), *The Science of Hardness Testing and its Research Applications* (J.W. Westbrook and H. Conrad, eds.), pp. 453–491, American Society of Metals, Metals Park, OH.

Bull, S.J. and Rickerby, D.S. (1990), "New Developments in the Modelling of the Hardness and Scratch Adhesion of Thin Films," *Surf. Coat. Technol.* 42, 149–164.

Bulychev, S.I., Alekhin, V.P., Shorshorov, M. Kh., Ternovskii, A.P., and Shnyrev, G.D. (1975), *Zavod Lab.* 41, 9.

Bulychev, S.I., Alekhin, V.P., and Shorshorov, M. Kh. (1979), "Studies of Physico-Mechanical Properties in Surface Layers and Microvolumes of Materials by the Method of Continuous Application of an Indenter," *Fiz. Khim. Obrab. Mater.* No. 5.

Burnett, P.J. and Rickerby, D.S. (1987a), "The Relationship between Hardness and Scratch Adhesion," *Thin Solid Films* 154, 403–416.

Burnett, P.J. and Rickerby, D.S. (1987b), "The Mechanical Properties of Wear Resistant Coatings I: Modelling of Hardness Behavior," *Thin Solid Films* 148, 41–50.

Burnett, P.J. and Rickerby, D.S. (1987c), "The Mechanical Properties of Wear-Resistant Coatings II: Experimental Studies and Interpretation of Hardness," *Thin Solid Films* 148, 51–65.

Callahan, D.L. and Morris, J.C. (1992), "The Extent of Phase Transformation in Silicon Hardness Indentations," *J. Mater. Res.* 7, 1614–1617.

Cammarata, R.C., Schlesinger, T.E., Kim, C., Qadri, S.B., and Edelstein, A.S. (1990), "Nanoindentation Study of the Mechanical Properties of Copper-Nickel Multilayered Thin Films," *Appl. Phys. Lett.* 56, 1862–1864.

Campbell, D.S. (1970), "Mechanical Properties of Thin Films," in *Handbook of Thin Film Technology* (L.I. Maissel and R. Glang, eds.), McGraw-Hill, New York, Chap. 12.

Chantikul, P., Anstis, G.R., Lawn, B.R., and Marshall, D.B. (1981), "A Critical Evaluation of Indentation Techniques for Measuring Fracture Toughness: II, Strength Method," *J. Am. Ceram. Soc.* 64 539–543.

Cheng, W., Ling, E., and Finnie, I. (1990), "Median Cracking of Brittle Solids Due to Scribing with Sharp Indenters," *J. Am. Ceram. Soc.* 73, 580–586.

Chiang, S.S., Marshall, D.B., and Evans, A.G. (1981), "Simple Method for Adhesion Measurement," *Surfaces and Interfaces in Ceramics and Ceramic-Metal Systems* (J. Pask and A.G. Evans, eds.), pp. 603–612, Plenum, New York.

Chiang, S.S., Marshall, D.B., and Evans, A.G. (1982), "The Response of Solids to Elastic/Plastic Indentation: I. Stresses and Residual Stresses," *J. Appl. Phys.* 53, 298–311.

Cho, N.H., Krishnan, K.M., Veirs, D.K., Rubin, M.B., Hopper, C.B., Bhushan, B., and Bogy, D.B. (1990), "Chemical Structure and Physical Properties of Diamond-like Amorphous Carbon Films Prepared by Magnetron Sputtering," *J. Mater. Res.* 5, 2543–2554.

Choi, S.R. and Salem, J.A. (1993), "Fracture Toughness of PMMA as Measured with Indentation Cracks," *J. Mater. Res.* 8, 3210–3217.

Chou, T.C., Adamson, D., Mardinly, J., and Nieh, T.G. (1991), "Microstructural Evolution and Properties of Nanocrystalline Alumina Made by Reactive Sputtering Deposition," *Thin Solid Films* 205, 131–139.

Chu, S.N.G. and Li, J.C.M. (1977), "Impression Creep: A New Creep Test," *J. Mater. Sci.* 12, 2200–2208.

Chu, S.N.G. and Li, J.C.M. (1979), "Impression Creep of b-Tin Single Crystals," *Mater. Sci. Eng.* 39, 1–10.

Chu, S.N.G. and Li, J.C.M. (1980), "Localized Stress Relaxation by Impression Testing," *Mater. Sci. Eng.* 45, 167–171.

Clarke, D.R., Kroll, M.C., Kirchner, P.D., and Cook, R.F. (1988), "Amorphization and Conductivity of Silicon and Germanium Induced by Indentation," *Phys. Rev. Lett.* 60, 2156–2159.

Cook, R.F. and Pharr, G.M. (1990), "Direct Observation and Analysis of Indentation Cracking in Glasses and Ceramics," *J. Am. Ceram. Soc.* 73, 787–817.

Cooper, C.V. (1991), "The Indentation and Wear Properties of a Thermal-Diffusion-Boronized Co-Cr-W-C Alloy," *Thin Solid Films* 195, 89–98.

Cooper, C.V. and Beetz, C.P. (1991), "The Effect of Methane Concentration in Hydrogen on the Microstructure and Properties of Diamond Films Grown by Hot-Filament Chemical Vapor Deposition," *Surface and Coat. Technol.* 47, 375–387.

Cornett, K.D., Fabes, B.D., and Oliver, W.C. (1990), "Mechanical Properties of Silica Coatings," in *Thin Films: Stresses and Mechanical Properties II* (M.F. Doerner, W.C. Oliver, G.M. Pharr and F.R. Brotzen, eds.), Vol. 188, pp. 133–138, Materials Research Society, Pittsburgh.

de Boer, M.P. Huang, H., Nelson, J.C., Jiang, Z.P., and Gerberich, W.W. (1993), "Fracture Toughness of Silicon and Thin Film Microstructures by Wedge Indentation," in *Thin Films: Stresses and Mechanical Properties IV* (P.H. Townsend, T.P. Weihs, J.E. Sanchez, and P. Borgesen, eds.), Vol. 308, pp. 647–652, Materials Research Society, Pittsuburgh.

Doerner, M.F. and Nix, W.D. (1986), "A Method for Interpreting the Data from Depth-Sensing Indentation Instruments," *J. Mater. Res.* 1, 601–609.

Doerner, M.F. Gardner, D.S., and Nix, W.D. (1986), "Plastic Properties of Thin Films on Substrates as Measured by Submicron Indentation Hardness and Substrate Curvature Techniques," *J. Mater. Res.* 1, 845–851.

Doerner, M.F., Oliver, W.C., Pharr, G.M., and Brotzen, F.R. (eds.) (1990), *Thin Films: Stresses and Mechanical Properties II* (M.F. Doerner, W.C. Oliver, G.M. Pharr, and F.R. Brotzen, eds.), Vol. 188, Materials Research Society, Pittsburgh.

Evans, A.G. and Charles, E.A. (1976), "Fracture Toughness Determination by Indentation," *J. Am. Ceram. Soc.* 59, 371–372.

Fabes, B.D., Oliver, W.C., McKee, R.A., and Walker, F.J. (1992), "The Determination of Film Hardness from the Composite Response of Film and Substrate to Nanometer Scale Indentation," *J. Mater. Res.* 7, 3056–3064.

Gane, N. and Cox, J.M. (1970), "The Micro-Hardness of Metals at Very Low Loads," *Philos. Mag.* 22, 881–891.

Gissler, W., Haupt, J., Crabb, T.A., Gibson, P.N., and Rickerby, D.G. (1991), "Evidence for Mixed-Phase Nanocrystalline Boron Nitride Films," *Mater. Sci. Eng. A* 139, 284–289.

Greene, J.E., Woodhouse, J., and Pestes, M. (1974), "A Technique for Detecting Critical Loads in the Scratch Test for Thin-Film Adhesion," *Rev. Sci. Instrum.* 45, 747–749.

Gupta, B.K. and Bhushan, B. (1994), "The Nanoindentation Studies of Ion Implanted Silicon," *Surface Coat. Technol.* 68/69, 564–570.

Gupta, B.K. and Bhushan, B. (1995a), "Micromechanical Properties of Amorphous Carbon Coatings Deposited by Different Deposition Techniques," *Thin Solid Films* 270, 391–398.

Gupta, B.K. and Bhushan, B. (1995b), "Mechanical and Tribological Properties of Hard Carbon Coatings for Magnetic Recording Heads," *Wear* 190, 110–122.

Gupta, B.K., Chevallier, J., and Bhushan, B. (1993), "Tribology of Ion Bombarded Silicon for Micromechanical Applications," *ASME J. Tribol.* 115, 392–399.

Gupta, B.K., Bhushan, B., and Chevallier, J. (1994), "Modification of Tribological Properties of Silicon by Boron Ion Implantation, *Tribol. Trans.* 37, 601–607.

Hannula, S.P., Stone, D., and Li, C.Y. (1985), "Determination of Time-Dependent Plastic Properties of Metals by Indentation Load Relaxation Techniques," *Electronic Packaging Materials Science* (E.A. Giess, K.N. Tu, and D.R. Uhlmann, eds.), Vol. 40, 217–224, Materials Research Society, Pittsburgh.

Hannula, S.P., Wanagel, J., and Li, C.Y. (1986), "A Miniaturized Mechanical Testing System for Small Scale Specimen Testing," in *The Use of Small-Scale Specimens for Testing Irradiated Material* (W.R. Corwin and G.E. Lucas, eds.), pp. 233–251, STP 888, ASTM, Philadelphia.

Hart, E.W. and Solomon, H.D. (1973), *Acta Metall.* 21, 195–200.

Heavens, O.S. (1950), *J. Phys. Rad.* 11, 355.

Henshall, J.L. and Brookes, C.A. (1985), "The Measurement of K_{Ic} in Single Crystal SiC Using the Indentation Method," *J. Mater. Sci. Lett.* 4, 783–786.

Hong, S., Weihs, T.P., Bravman, J.C., and Nix, W.D. (1990), *J. Electron. Mater.* 19, 903.

Hooper, R.M. and Brookes, C.A. (1984), *J. Mater. Sci.* 19, 4057.

Hu, J.H, Merkle, M.C., Menoni, C.S., and Spain, I.L. (1986), "Crystal Data for High-Pressure Phases of Silicon," *Phys. Rev. B* 34, 4679–4684.

Jacobson, S., Jonsson, B., and Sundquist, B. (1983), "The Use of Fast Heavy Ions to Improve Thin Film Adhesion," *Thin Solid Films* 107, 89–98.

Je, J.H., Gyarmati, E., and Naoumidis, A. (1986), "Scratch Adhesion Test of Reactively Sputtered TiN Coatings on a Soft Substrate," *Thin Solid Films* 136, 57–67.

Jiang, X., Reichelt, K., and Stritzker, B. (1989), "The Hardness and Young's Modulus of Amorphous Hydrogenated Carbon and Silicon Films Measured with an Ultralow Load Indenter," *J. Appl. Phys.* 66, 5805–5808.

Johnson, K.L. (1985), *Contact Mechanics*, Cambridge University Press, Cambridge, U.K.

Jonsson, B. and Hogmark, S. (1984), "Hardness Measurements of Thin Films," *Thin Solid Films* 114, 257–269.

Joslin, D.L. and Oliver, W.C. (1990), "A New Method for Analyzing Data from Continuous Depth-Sensing Microindentation Tests," *J. Mater. Res.* 5, 123–126.

Joslin, D.L., O'Hern, M.E., McHargue, C.J., Clausing, R.E., and Oliver, W.C. (1989), "Hardness of Amorphous Hard Carbon Films Determined by the Ultra-Low Load Microindentation Technique," *SPIE Vol.* 1050 *Infrared Syst. and Comp. III*, 218–225.

Julia-Schmutz, C. and Hintermann, H.E. (1991), "Microscratch Testing to Characterize the Adhesion of Thin Layers," *Surf. Coat. Technol.* 48, 1.

King, R.B. (1987), "Elastic Analysis of Some Punch Problems for Layered Medium," *Int. J. Solids Struct.* 23, 1657–1664.

Knight, J.C., Whitehead, A.J., and Page, T.F. (1992), "Nanoindentation Experiments on Some Amorphous Hydrogenated Carbon (a-C:H) Thin Films on Silicon," *J. Mater. Sci.* 27, 3939–3952.

LaFontaine, W.R., Yost, B., Black, R.D., and Li, C. (1990a), "Indentation Load Relaxation Experiments on Al/Si Metallizations," *Symp. Proc.*, Vol. 188, pp. 165–170, Materials Research Society, Pittsburgh.

LaFontaine, W.R., Yost, B., Black, R.D., and Li, C.Y. (1990b), "Indentation Load Relaxation Experiments with Indentation Depth in the Submicron Range," *J. Mater. Res.* 5, 2100–2106.

LaFontaine, W.R., Yost, B., and Li, C.Y. (1990c), "Effect of Residual Stress and Adhesion on the Hardness of Copper Films Deposited on Silicon," *J. Mater. Res.* 5, 776–783.

LaFontaine, W.R., Paszkiet, C.A., Korhonen, M.A., and Li, C.Y. (1991), "Residual Stress Measurements of Thin Aluminum Metallizations by Continuous Indentation and X-ray Stress Measurement Techniques," *J. Mater. Res.* 6, 2084–2090.

Lankford, J. (1981), "Threshold-Microfracture during Elastic/Plastic Indentation of Ceramics," *J. Mater. Sci.* 16, 1177–1182.

Laugier, M. (1981), "The Development of Scratch Test Technique for the Determination of the Adhesion of Coating," *Thin Solid Films* 76, 289–294.

Laursen, T.A. and Simo, J.C. (1992), "A Study of the Mechanics of Microindentation Using Finite Elements," *J. Mater. Res.* 7, 618–626.

Lawn, B. (1993), *Fracture of Brittle Solids*, 2nd ed., Cambridge University Press, Cambridge, U.K.

Lawn, B.R. and Evans, A.G. (1977), "A Model for Crack Initiation in Elastic/Plastic Indentation Fields," *J. Mater. Sci.* 12, 2195–2199.

Lawn, B.R. and Marshall, D.B. (1979), "Hardness, Toughness and Brittleness: An Indentation Analysis," *J. Am. Ceram. Soc.* 62, 347–350.

Lawn, B. and Wilshaw, R. (1975), "Review Indentation Fracture: Principles and Applications," *J. Mater. Sci.* 10, 1049–1081.

Lawn, B.R., Evans, A.G., and Marshall, D.B. (1980), "Elastic/Plastic Indentation Damage in Ceramics: The Median/Radial Crack System," *J. Am. Ceram. Soc.* 63, 574–581.

Lee, E.H. and Mansur, L.K. (1989), "Effect of Simultaneous B^+ and N_2^+ Implantation on Microhardness, Fatigue Life, and Microstructure in Fe-13Cr-15 Ni Base Alloys," *J. Mater. Res.* 4, 1371–1378.

Lee, E.H., Rao, G.R., and Mansur, L.K. (1992), "Improved Hardness and Wear Properties of B-Ion Implanted Polycarbonate," *J. Mater. Res.* 7, 1900–1911.

Lee, E.H., Lee, Y., Oliver, W.C., and Mansur, L.K. (1993), "Hardness Measurements of Ar^+-Beam Treated Polyimide by Depth-Sensing Ultra Low Load Indentation," *J. Mater. Res.* 8, 377–387.

Li, D., Chung, Y.W., Wong, M.S., and Sproul, W.D. (1993), "Nano-Indentation Studies of Ultrahigh Strength Carbon Nitride Thin Films," *J. Appl. Phys.* 74, 219–223.

Li, J.C.M. and Chu, S.N.G. (1979), "Impression Fatigue," *Scr. Metall.* 13, 1021–1026.

Li, W.B., Henshall, J.L., Hooper, R.M., and Easterling, K.E. (1991), "The Mechanism of Indentation Creep," *Acta Metall. Mater.* 39, 3099–3110.

Li, X. and Bhushan, B. (1998a), "Measurement of Fracture Toughness of Ultra-Thin Amorphous Carbon Films," *Thin Solid Films* 315, 214–221.

Li, X. and Bhushan, B. (1998b), "Micromechanical and Tribological Characterization of Hard Amorphous Carbon Coatings as Thin as 5 nm for Magnetic Recording Heads," *Wear* (in press).

Li, X. and Bhushan, B. (1998c), "Micro/nanomechanical Characterization of Ceramic Films for Microdevices," *Thin Solid Films* (in press).

Li, X., Diao, D., and Bhushan, B. (1997), "Fracture Mechanisms of Thin Amorphous Carbon Films in Nanoindentation," *Acta Mater.* 45, 4453–4461.

Lin, M.R., Ritter, J.E., Rosenfeld, L., and Lardner, T.J. (1990), "Measuring the Interfacial Shear Strength of Thin Polymer Coatings on Glass," *J. Mater. Res.* 5, 1110–1117.

Loubet, J.L., Georges, J.M., Marchesini, O., and Meille, G. (1984), "Vickers Indentation Curves of Magnesium Oxide (MgO)," *ASME J. Tribol.* 106, 43–48.

Loubet, J.L., Bauer, M., Tonck, A., Bec, S., and Gauthier-Manuel, B. (1993), "Nanoindentation with a Surface Force Apparatus," in *Mechanical Properties and Deformation Behavior of Materials Having Ultra-Fine Microstructures* (M. Nastasi, D.M. Parkin, and H. Gleiter, eds.), pp. 429–447, Kluwer Academic, Dordrecht, Netherlands.

Lucas, B.N. and Oliver, W.C. (1992), "The Elastic, Plastic and Time Dependent Properties of Thin Films as Determined by Ultra Low Load Indentation," in *Thin Films: Stresses and Mechanical Properties III* (W.E. Nix, J.C. Bravman, E. Arzt, and L.B. Freund, eds.), Vol. 239, pp. 337–341, Materials Research Society, Pittsburgh.

Lucas, B.N., Oliver, W.C., Williams, R.K., Brynestad, J., and O'Hern, M.E. (1991), "The Hardness and Young's Modulus of Bulk $YBa_2Cu_3O_{7-x}$ (1:2:3) and $YBa_2Cu_4O_8$ (1:2:4) as Determined by Ultra Low Load Indentation," *J. Mater. Res.* 6, 2519–2522.

Lysaght, V.E. (1949), *Indentation Hardness Testing*, Reinhold, New York.

Marshall, D.B. and Lawn, B.R. (1979), "Residual Stress Effects in Sharp Contact Cracking Part 1 Indentation Fracture Mechanics," *J. Mater. Sci.* 14, 2001–2012.

Marshall, D.B. and Oliver, W.C. (1987), "Measurement of Interfacial Mechanical Properties in Fiber-Reinforced Ceramic Composites," *J. Am. Ceram. Soc.* 70, 542–548.

Mayo, M.J. and Nix, W.D. (1988a), "A Micro-Indentation Study of Superplasticity in Pb, Sn, and Sn-38 wt% Pb," *Acta Metall.* 36, 2183–2192.

Mayo, M.J. and Nix, W.D. (1988b), in *Proc. 8th Int. Conf. on the Strength of Metals and Alloys* (P.O. Kettunen, T.K. Lepisto, and M.E. Lehtonen, eds.), p. 1415, Pergamon Press, New York.

Mayo, M.J., Siegel, R.W., Narayanasamy, A., and Nix, W.D. (1990), "Mechanical Properties of Nanophase TiO_2 as Determined by Nanoindentation," *J. Mater. Res.* 5, 1073–1082.

Mayo, M.J., Siegel, R.W., Liao, Y.X., and Nix, W.D. (1992), "Nanoindentation of Nanocrystalline ZnO," *J. Mater. Res.* 7, 973–979.

McHargue, C.J. (1989), "The Mechanical Properties of Ion Implanted Ceramics," in *Structure-Property Relationships in Surface-Modified Ceramics* (C.J. McHargue et al., eds.), pp. 253–273, Kluwer Academic, Dordrecht.

McHargue, C.J., O'Hern, M.E., and Joslin, D.L. (1990), "The Influence of Ion Implantation of the Near-Surface Mechanical Properties of Ceramics," *Symp. Proc.*, Vol. 188, pp. 111–120, Materials Research Society, Pittsburgh.

Mecholsky, J.J., Tsai, Y.L., and Drawl, W.R. (1992), "Fracture Studies of Diamond on Silicon," *J. Appl. Phys.* 71, 4875–4881.

Mehrotra, P.K. and Quinto, D.T. (1985), "Techniques for Evaluating Mechanical Properties of Hard Coatings," *J. Vac. Sci. Technol. A* 3, 2401–2405.

Mittal, K.L. (ed.) (1978), *Adhesion Measurements on Thin Coatings, Thick Coatings and Bulk Coatings*, STP 640, ASTM, Philadelphia.

Mott, B.W. (1957), *Microindentation Hardness Testing*, Butterworths, London.

Mulhearn, T.O. and Tabor, D. (1960/61), *J. Inst. of Metals* 87, 7.

Nastasi, M., Hirvonen, J.P., Jervis, T.R., Pharr, G.M., and Oliver, W.C. (1988), "Surface Mechanical Properties of C Implanted Ni," *J. Mater. Res.* 3, 226–232.

Nastasi, M., Parkin, D.M., and Gleiter, H. (eds.) (1993), *Mechanical Properties and Deformation Behavior of Materials Having Ultra-Fine Microstructures*, Kluwer Academic, Dordrecht.

Newey, D., Wilkins, M.A., and Pollock, H.M. (1982), "An Ultra-Low-Load Penetration Hardness Tester," *J. Phys. E: Sci. Instrum.* 15, 119–122.

Nix, W.D. (1989), "Mechanical Properties of Thin Films," *Metall. Trans. A* 20, 2217–2245.

Nix, W.D., Bravman, J.C., Arzt, E., and Freund, L.B. (eds.) (1992), *Thin Films: Stresses and Mechanical Properties III, Symp Proc.*, Vol. 239, Materials Research Society, Pittsburgh.

O'Hern, M.E., McHargue, C.J., White, C.W., and Farlow, G.C. (1990), "The Effect of Chromium Implantation on the Hardness, Elastic Modulus and Residual Stress of Al_2O_3," *Nucl. Inst. Meth. Phys. Res. B* 46, 171–175.

O'Hern, M., Parrish, R.H., and Oliver, W.C. (1989) "Evaluation of Mechanical Properties of TiN Films by Ultralow Load Indentation," *Thin Solid Films* 181, 357–363.

Oliver, W.C. and Pethica, J.B. (1989), "Methods for Continuous Determination of the Elastic Stiffness of Contact between Two Bodies," U.S. Patent No. 4,848,141, July 18.

Oliver, W.C. and Pharr, G.M. (1992), "An Improved Technique for Determining Hardness and Elastic Modulus Using Load and Displacement Sensing Indentation Experiments," *J. Mater. Res.* 7, 1564–1583.

Oliver, W.C., Hutchings, R., and Pethica, J.B. (1986), "Measurement of Hardness at Indentation Depths as Small as 20 nm," in *Microindentation Techniques in Materials Science and Engineering* (P.J. Blau and B.R. Lawn, eds.), STP 889, pp. 90–108, ASTM, Philadelphia.

O'Neill, H. (1967), *Hardness Measurement of Metals and Alloys*, Chapman and Hall, London.

Page, T.F., Oliver, W.C., and McHargue, C.J. (1992), "The Deformation Behavior of Ceramic Crystals Subjected to Very Low Load (Nano)indentations," *J. Mater. Res.* 7, 450–473.

Palmquist, S. (1957), *Jernkontorets Ann.* 141, 300.

Patton, S.T. and Bhushan, B. (1996), "Micromechanical and Tribological Characterization of Alternate Pole Tip Materials for Magnetic Recording Heads," *Wear* 202, 99–109.

Perry, A.J. (1981), "The Adhesion of Chemically Vapour-Deposited Hard Coatings on Steel—The Scratch Test," *Thin Solid Films* 78, 77–93.

Perry, A.J. (1983), "Scratch Adhesion Testing of Hard Coating," *Thin Solid Films* 197, 167–180.

Pethica, J.B. and Oliver, W.C. (1989), "Mechanical Properties of Nanometer Volumes of Material: Use of the Elastic Response of Small Area Indentations," in *Thin Films: Stresses and Mechanical Properties* (J.C. Bravman, W.D. Nix, D.M. Barnett, and D.A. Smith, eds.), Vol. 130, pp. 13–23, Materials Research Society, Pittsburgh.

Pethica, J.B. Hutchings, R., and Oliver, W.C. (1983), "Hardness Measurements at Penetration Depths as Small as 20 nm," *Philos. Mag A* 48, 593–606.

Pethica, J.B., Koidl, P., Gobrecht, J., and Schuller, C. (1985), "Micromechanical Investigations of Amorphous Hydrogenated Carbon Films on Silicon," *J. Vac. Sci. Technol A* 3, 2391–2393.

Pharr, G.M. (1992), "The Anomalous Behavior of Silicon during Nanoindentation," in *Thin Film: Stresses and Mechanical Properties III* (W.D. Nix, J.C. Bravman, E. Arzt, and L.B. Freund, eds.), Vol. 239, pp. 301–312, Materials Research Society, Pittsburgh.

Pharr, G.M. and Cook, R.F. (1990), "Instrumentation of a Conventional Hardness Tester for Load-Displacement Measurement during Indentation," *J. Mater. Res.* 5, 847–851.

Pharr, G.M. and Oliver, W.C. (1992), "Measurement of Thin Film Mechanical Properties Using Nanoindentation," *MRS Bull.*, July, 28–33.

Pharr, G.M., Oliver, W.C., and Clarke, D.R. (1989), "Hysteresis and Discontinuity in the Indentation Load-Displacement Behavior of Silicon," *Scr. Metall.* 23, 1949–1952.

Pharr, G.M., Oliver, W.C., and Clarke, D.R. (1990), "The Mechanical Behavior of Silicon during Small-Scale Indentation," *J. Electron. Mater.* 19, 881–887.

Pharr, G.M., Oliver, W.C., and Brotzen, F.R. (1992), "On the Generality of the Relationship among Contact Stiffness, Contact Area, and Elastic Modulus During Indentation," *J. Mater. Res.* 7, 613–617.

Pharr, G.M., Harding, D.S., and Oliver, W.C. (1993), "Measurement of Fracture Toughness in Thin Films and Small Volumes Using Nanoindentation Methods," in *Mechanical Properties and Deformation Behavior of Materials Having Ultra-Fine Microstructures* (M. Nastasi, D.M. Parkin, and H. Gleiter, eds.), pp. 449–461, Kluwer Academic, Dordrecht, Netherlands.

Pollock, H.M., Maugis, D., and Barquins, M. (1986), "Characterization of Submicrometre Surface Layers by Indentation," in *Microindentation Techniques in Materials Science and Engineering* (P.J. Blau and B.R. Lawn, eds.), STP 889, pp. 47–71, ASTM, Philadelphia.

Pulker, H.K. and Salzmann, K. (1986), "Micro-/Ultramicro Hardness Measurements with Insulating Films," *SPIE Thin Film Technol.* 652, 139–144.

Raman, V. and Berriche, R. (1990), "Creep Behavior of Sputtered TiN Films Using Indentation Testing," in *Thin Films: Stresses and Mechanical Properties II* (M.F. Doerner, W.C. Oliver, G.M. Pharr, and F.R. Brotzen, eds.), pp. 171–176, Materials Research Society, Pittsburgh.

Raman, V. and Berriche, R. (1992), "An Investigation of the Creep Processes in Tin and Aluminum Using a Depth-Sensing Indentation Technique," *J. Mater. Res.* 7, 627–638.

Rao, G.R., Lee, E.H., and Mansur, L.K. (1993), "Structure and Dose Effects on Improved Wear Properties of Ion-Implanted Polymers," *Wear* 162–164, 739–747.

Rubin, M.B., Hopper, C.B., Cho, N.H., and Bhushan, B. (1990), "Optical and Mechanical Properties of dc Sputtered Carbon Films," *J. Mater. Res.* 5, 2538–2542.

Sargent, P.M. (1986), "Use of the Indentation Size Effect on Microhardness of Materials Characterization," in *Microindentation Techniques in Materials Science and Engineering* (P.J. Blau and B.R. Lawn, eds.), STP 889, pp. 160–174, ASTM, Philadelphia.

Savvides, N. and Bell, T.J. (1992), "Microhardness and Young's Modulus of Diamond and Diamondlike Carbon Films," *J. Appl. Phys.* 72, 2791–2796.

Schlesinger, T.E., Cammarata, R.C., Gavrin, A., Xiao, J.Q., Chien, C.L., Ferber, M.K., and Hayzelden, C. (1991), "Enhanced Mechanical and Magnetic Properties of Granular Metal Thin Films," *J. Appl. Phys.* 70, 3275–3280.

Scruby, C.B. (1987), "An Introduction to Acoustic Emission," *J. Phys. E: Sci. Instrum.* 20, 946–953.

Sekler, J., Steinmann, P.A., and Hintermann, H.E. (1988), "The Scratch Test: Different Critical Load Determination Techniques," *Surf. Coat. Technol.* 36, 519–529.

Shih, C.W., Yang, M., and Li, J.C.M. (1991), "Effect of Tip Radius on Nanoindentation," *J. Mater. Res.* 6, 2623–2628.

Sneddon, I.N. (1965), "The Relation between Load and Penetration in the Axisymmetric Boussinesq Problem for a Punch of Arbitrary Profile," *Int. J. Eng. Sci.* 3, 47–57.

Steinmann, P.A., Tardy, Y., and Hintermann, H.E. (1987), "Adhesion Testing by the Scratch Test Method: The Influence of Intrinsic and Extrinsic Parameters on the Critical Load," *Thin Solid Films* 154, 333–349.

Stilwell, N.A. and Tabor, D. (1961), "Elastic Recovery of Conical Indentation," *Proc. Phys. Soc.* 78, 169–179.

Stone, D., LaFontaine, W.R., Alexopoulos, P.S., Wu, T.W., and Li, C.Y. (1988), "An Investigation of Hardness and Adhesion of Sputter-Deposited Aluminum on Silicon by Utilizing a Continuous Indentation Test," *J. Mater. Res.* 3, 141–147.

Stone, D.S., Yoder, K.B., and Sproul, W.D. (1991), "Hardness and Elastic Modulus of TiN Based on Continuous Indentation Technique and New Correlation," *J. Vac. Sci. Technol.* A 9, 2543–2547.

Swain, M.V., Hagan, J.T., and Field, J.E. (1977), "Determination of the Surface Residual Stresses in Tempered Glasses by Indentation Fracture Mechanics," *J. Mater. Sci.* 12, 1914–1917.

Tabor, D. (1951), *The Hardness of Metals,* Clarendon Press, Oxford, UK.

Tabor, D. (1970), "The Hardness of Solids," *Rev. Phys. Technol.* 1, 145–179.

Tangena, A.G. and Hurkx, G.A.M. (1986), "The Determination of Stress-Strain Curves of Thin Layers Using Indentation Tests," *ASME J. Eng. Mater. Technol.* 108, 230–232.

Ternovskii, A.P., Alekhin, V.P., Shorshorov, M. Kh., Khrushchov, M.M., and Skvortsov, V.N. (1973), "Micromechanical Testing of Materials by Depression," *Zavod. Lab.* 39, 1620–1624.

Townsend, P.H., Weihs, T.P., Sanchez, J.E., and Borgesen, P. (eds.) (1993), *Thin Films: Stresses and Mechanical Properties IV*, Vol. 308, Materials Research Society, Pittsburgh.

Tsukamoto, Y., Yamaguchi, H., and Yanagisawa, M. (eds.) (1987), "Mechanical Properties of Thin Films: Measurements of Ultramicroindentation Hardness, Young's Modulus and Internal Stresses," *Thin Solid Films* 154, 171–181.

Tsukamoto, Y., Kuroda, H., Sato, A., and Yamaguchi, H. (1992), "Microindentation Adhesion Tester and Its Applications to Thin Films," *Thin Solid Films* 213, 220–225.

Valli, J. (1986), "A Review of Adhesion Test Method for Thin Hard Coatings," *J. Vac. Sci. Technol. A* 4, 3007–3014.

Vancoille, E., Celis, J.P., and Roos, J.R. (1993), "Mechanical Properties of Heat Treated and Worn TiN, (Ti, Al)N, (Ti, Nb)N and Ti(C,N) Coatings as Measured by Nanoindentation," *Thin Solid Films* 224, 168–176.

Vitovec, F.H. (1986), "Stress and Load Dependence of Microindentation Hardness," in *Microindentation Techniques in Materials Science and Engineering* (P.J. Blau and B.R. Lawn, eds.), STP 889, pp. 175–185, ASTM, Philadelphia.

Walker, W.W. (1973), *The Science of Hardness Testing and its Research Applications* (J.H. Westbrook and H. Conrad, eds.), pp. 258–273, American Society of Metals, Metals Park, OH.

Wang, M., Schmidt, K., Reichelt, K., Dimigen, H., and Hubsch, H. (1992), "Characterization of Metal-Containing Amorphous Hydrogenated Carbon Films," *J. Mater. Res.* 7, 667–676.

Was, G.S. (1990), "Surface Mechanical Properties of Aluminum Implanted Nickel and Co-Evaporated Ni-Al on Nickel," *J. Mater. Res.* 5, 1668–1683.

Was, G.S. and DeKoven, B.M. (1991), "Hardness and Friction of N_2^+ and Al^+ Implanted B_4C," *Nucl. Inst. Meth. Phys. Res. B* 59/60, 802–805.

Weihs, T.P., Hong, S., Bravman, J.C., and Nix, W.D. (1988), "Mechanical Deflection of Cantilever Microbeams: A New Technique for Testing the Mechanical Properties of Thin Films," *J. Mater. Res.* 3, 931–942.

Weihs, T.P., Lawrence, C.W., Derby, C.B., and Pethica, J.B. (1992), "Acoustic Emissions during Indentation Tests," *Thin Films: Stresses and Mechanical Properties III, Symp. Proc.* (W.D. Nix, J.C. Bravman, E. Arzt, and L.B. Freund, eds.), Vol. 239, pp. 361–370, Materials Research Society, Pittsburgh.

Westbrook, J.H. (1957), *Proc. Am. Soc. Test. Mater.* 57, 873.

Westbrook, J.H. and Conrad, H. (eds.) (1973), *The Science of Hardness and Its Research Applications*, American Society of Metals, Metals Park, OH.

White, R.L., Nelson, J., and Gerberich, W.W. (1993), "Residual Stress Effects in the Scratch Adhesion Testing of Tantalum Thin Films," in *Thin Films: Stresses and Mechanical Properties IV* (P.H. Townsend, T.P. Weihs, J.E. Sanches, and P. Borgesen, eds.), Vol. 239, pp. 141–146, Materials Research Society, Pittsburgh.

Whitehead, A.J. and Page, T.F. (1992), "Nanoindentation Studies of Thin Film Coated Systems," *Thin Solid Films* 220, 277–283.

Wierenga, P.E. and Franken, A.J.J. (1984), "Ultramicroindentation Apparatus for the Mechanical Characterization of Thin Films," *J. Appl. Phys.* 55, 4244–4247.

Wierenga, P.E., and van der Linden, J.H.M. (1986), "Quasistatic and Dynamic Indentation Measurements on Magnetic Tapes," in *Tribology and Mechanics of Magnetic Storage Systems*, Vol. 3 (B. Bhushan and N.S. Eiss, eds.), pp. 31–37, SP-21, ASLE, Park Ridge, IL.

Williams, V.S., Landesman, A.L., Shack, R.V., Vukobratovich, D., and Bhushan, B. (1988), "In Situ Microviscoelastic Measurements by Polarization-Interferometric Monitoring of Indentation Depth," *Appl. Opt.* 27, 541–546.

Wu, T.W. (1990), "Microscratch Test for Ultra-Thin Films," in *Thin Films: Stresses and Mechanical Properties II* (M.F. Doerner, W.C. Oliver, G.M. Pharr, and F.R. Brotzen, eds.), *Symp. Proc.*, Vol. 188, pp. 191–205, Materials Research Society, Pittsburgh.

Wu, T.W. (1991), "Microscratch and Load Relaxation Tests for Ultra-Thin Films," *J. Mater. Res.* 6, 407–426.

Wu, T.W. (1993), "The a.c. Indentation Technique and Its Applications," *Mater. Chem. Phys.* 33, 15–30.

Wu., T.W. and Lee, C.K. (1994), "The Micro-Wear Technique and Its Application to Ultrathin Film Systems," *J. Mater. Res.* 9, 805–811.

Wu, T.W., Hwang, C., Lo. J., and Alexopoulos, P. (1988), "Microhardness and Microstructure of Ion-Beam-Sputtered, Nitrogen Doped NiFe Films," *Thin Solid Films* 166, 299–308.

Wu, T.W., Burn, R.A., Chen, M.M., and Alexopoulos, P.S. (1989), "Micro-Indentation and Micro-Scratch Tests on Sub-Micron Carbon Films," *Symp. Proc.*, Vol. 130, pp. 117–121, Materials Research Society, Pittsburgh.

Wu, T.W. Moshref, M., and Alexopoulos, P.S. (1990a), "The Effect of the Interfacial Strength on the Mechanical Properties of Aluminum Films," *Thin Solid Films* 187, 295–307.

Wu, T.W., Shull, A.L., and Lin, J. (1990b), "Microscratch Test on Carbon Films as Thin as 20 nm," *Symp. Proc.*, Vol. 188, pp. 207–212, Materials Research Society, Pittsburgh.

Wu, T.W., Shull, A.L., and Berriche, R. (1991), "Microindentation Fatigue Tests on Submicron Carbon Films," *Surf. Coat. Technol.* 47, 696–709.

Yanagisawa, M. and Motomura, Y. (1987), "An Ultramicro Indentation Hardness Tester and Its Application to Thin Films," *Lubr. Eng.* 43, 52–56.

Yanagisawa, M. and Motomura, Y. (1989), "Apparatus for Determining Microhardness," U.S. Patent 4, 820,051.

Yeack-Scranton, C.E. (1986), "Novel Piezoelectric Transducer to Monitor Head-Disk Interactions," *IEEE Trans. Magn.* 22, 1011–1016.

11

Atomic-Scale Simulation of Tribological and Related Phenomena

Judith A. Harrison, Steven J. Stuart, and Donald W. Brenner

11.1 Introduction

Understanding and ultimately controlling friction and wear have long been recognized as important to many areas of technology. Historical examples include the Egyptians, who had to invent new technologies to move the stones needed to build the pyramids (Dowson, 1979); Coulomb, whose fundamental studies of friction were motivated by the need to move ships easily and without wear from land into the water (Dowson, 1979); and Johnson et al. (1971), whose study of automobile windshield wipers led to a better understanding of contact mechanics, including surface energies. Today, the development of microscale (and soon nanoscale) machines continues to challenge our understanding of friction and wear at their most fundamental levels.

Our knowledge of friction and related phenomena at the atomic scale has rapidly advanced over the last decade with the development of new and powerful experimental methods. The surface force apparatus (SFA), for example, has provided new information related to friction and lubrication for many liquid and solid systems with unprecedented resolution (Israelachvili, 1992). The friction force and atomic force microscopes (FFM and AFM) allow the frictional and mechanical properties of solids to be characterized with atomic resolution under single-asperity contact conditions (Binnig et al., 1986; Mate et al., 1987; Germann et al., 1993; Carpick and Salmeron, 1997). Other techniques, such as the quartz crystal microbalance, are also providing exciting new insights into the origin of friction (Krim et al., 1991; Krim, 1996). Taken together, the results of these studies have revolutionized the study of friction, wear, and mechanical properties, and have reshaped many of our ideas about the fundamental origins of friction.

Concomitant with the development and use of these innovative experimental techniques has been the development of new theoretical methods and models. These include analytic models, large-scale molecular dynamics (MD) simulations, and even first-principles total-energy techniques (Zhong and Tomanek, 1990). Analytic models have had a long history in the study of friction. Beginning with the work of Tomlinson (1929) and Frenkel and Kontorova (1938), through to recent studies by McClelland and Glosli (1992), Sokoloff (1984, 1990, 1992, 1993, 1996), and others (Helman et al., 1994; Persson, 1991), these idealized models have been able to break down the complicated motions that create friction into basic components defined by quantities such as spring constants, the curvature and magnitude of potential wells, and bulk phonon frequencies. The main drawback of these approaches is that simplifying assumptions must be made as part of these models. This means, for example, that unanticipated defect structures may be overlooked, which may strongly influence friction and wear even at the atomic level.

Molecular dynamics computer simulations, which are the topic of this chapter, represents a compromise between analytic models and experiment. On the one hand, this method deals with approximate interatomic forces and classical dynamics (as opposed to quantum dynamics), so it has much in common with analytic models (for a comparison of analytic and simulation results, see Harrison et al., 1992c). On the other hand, simulations often reveal unanticipated events that require further analysis, so they also have much in common with experiment. Furthermore, a poor choice of simulation conditions, as in an experiment, can result in meaningless results. Because of this danger, a thorough understanding of the strengths and weaknesses of MD simulations is crucial to both successfully implementing this method and understanding the results of others.

On the surface, atomistic computer simulations appear rather straightforward to carry out: given a set of initial conditions and a way of describing interatomic forces, one simply integrates classical equations of motion using one of several standard methods (Gear, 1971). Results are then obtained from the simulations through mathematical analysis of relative positions, velocities, and forces; by visual inspection of the trajectories through animated movies; or through a combination of both (Figure 11.1). However, the effective use of this method requires an understanding of many details not apparent in this simple analysis. To provide a feeling for the details that have contributed to the success of this approach in the study of adhesion, friction, and wear as well as other related areas, the next section provides a brief review of MD techniques. For a more detailed overview of MD simulations, including computer algorithms, the reader is referred to a number of other more comprehensive sources (Hoover, 1986; Heermann, 1986; Allen and Tildesley, 1987; Haile, 1992).

The remainder of this chapter presents recent results from MD simulations dealing with various aspects of mechanical, frictional, and wear properties of solid surfaces and thin lubricating films. Section 11.2 summarizes several of the technical details needed to perform (or understand) an MD simulation. These range from choosing an interaction potential and thermodynamic ensemble to implementing temperature controls. Section 11.3 describes simulations of the indentation of metals and nonmetals, as well as the machining of metal surfaces. The simulations discussed reveal a number of interesting phenomena and trends related to the deformation and disordering of materials at the atomic scale, some (but not all) of which have been observed at the macroscopic scale. Section 11.4 summarizes the results of

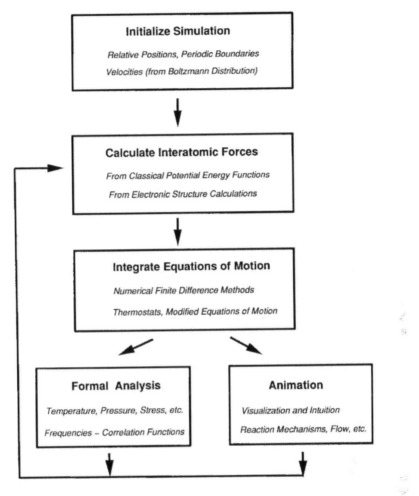

FIGURE 11.1 Flow chart of an MD simulation.

simulations that probe the properties of liquid films confined to thicknesses on the order of atomic dimensions. These systems are becoming more important as demands for lubricating moving parts approach the nanometer scale. In these systems, fluids have a range of new properties that bear little resemblance to liquid properties on macroscopic scales. In many cases, the information obtained from these studies could not have been obtained in any other way, and is providing unique new insights into recent observations made by instruments such as the SFA. Section 11.5 discusses simulations of the tribological properties of solid surfaces. Some of the systems discussed are sliding diamond interfaces, Langmuir–Blodgett films, self-assembled monolayers, and metals. The details of several unique mechanisms of energy dissipation are discussed, providing just a few examples of the many ways in which the conversion of work into heat leads to friction in weakly adhering systems. In addition, simulations of molecules trapped between, or chemisorbed onto, diamond surfaces will be discussed in terms of their effects on the friction, wear, and tribology of diamond. A summary of the MD results is given in the final section.

11.2   Molecular Dynamics Simulations

Atomistic computer simulations are having a major impact in many areas of the chemical, physical, material, and biological sciences. This is largely due to enormous recent increases in computer power, increasingly clever algorithms, and recent developments in modeling interatomic interactions. This last development, in particular, has made it possible to study a wide range of systems and processes using large-scale MD simulations. Consequently, this section begins with a review of the interatomic interactions that have played the largest role in friction, indentation, and related simulations. For a slightly broader discussion, the reader is referred to a review by Brenner and Garrison (1989). This is followed by a brief discussion of thermodynamic ensembles and their use in different types of simulations. The section then closes with a description of several of the thermostatting techniques used to regulate the temperature during an MD simulation. This topic is particularly relevant for tribological simulations because friction and indentation do work on the system, raising its kinetic energy.

11.2.1 Interatomic Potentials

Molecular dynamics simulations involve tracking the motion of atoms and molecules as a function of time. Typically, this motion is calculated by the numerical solution of a set of coupled differential equations (Gear, 1971; Heermann, 1986; Allen and Tildesley, 1987). For example, Newton's equation of motion,

$$\mathbf{F} = m\mathbf{a} = m\frac{d\mathbf{v}}{dt}, \tag{11.1}$$

where $\mathbf{F}$ is the force on a particle, m is its mass, $\mathbf{a}$ its acceleration, $\mathbf{v}$ its velocity, and t is time, yield a set of $3n$ (where n is the number of particles) second-order differential equations that govern the dynamics. These can be solved with finite-time-step integration methods, where time steps are on the order of $^1/_{25}$ of a vibrational frequency (typically tenths to a few femtoseconds) (Gear, 1971). Most current simulations then integrate for a total time of picoseconds to nanoseconds. The evaluation of these equations (or any of the other forms of classical equations of motion) requires a method for obtaining the force $\mathbf{F}$ between atoms.

Constraints on computer time generally require that the evaluation of interatomic forces not be computationally intensive. Currently, there are two approaches that are widely used. In the first, one assumes that the potential energy of the atoms can be represented as a function of their relative atomic positions. These functions are typically based on simplified interpretations of more general quantum mechanical principles, as discussed below, and usually contain some number of free parameters. The parameters are then chosen to closely reproduce some set of physical properties of the system of interest, and the forces are obtained by taking the gradient of the potential energy with respect to atomic positions. While this may sound straightforward, there are many intricacies involved in developing a useful potential energy function. For example, the parameters entering the potential energy function are usually determined by a limited set of known system properties. A consequence of that is that other properties, including those that might be key in determining the outcome of a given simulation, are determined solely by the assumed functional form. For a metal, the properties to which a potential energy function might be fit might include the lattice constant, cohesive energy, elastic constants, and vacancy formation energy. Predicted properties might then include surface reconstructions, energetics of interstitial defects, and response (both elastic and plastic) to an applied load. The form of the potential is therefore crucial if the simulation is to have sufficient predictive power to be useful.

The second approach, which has become more useful with the advent of powerful computers, is the calculation of interatomic forces directly from first-principles (Car and Parrinello, 1985) or semiempirical (Menon and Allen, 1986; Sankey and Allen, 1986) calculations that explicitly include electrons. The advantage of this approach is that the number of unknown parameters may be kept small, and, because the forces are based on quantum principles, they may have strong predictive properties. However, this does not guarantee that forces from a semiempirical electronic structure calculation are accurate; poorly chosen parameterizations and functional forms can still yield nonphysical results. The disadvantage of this approach is that the potentials involved are considerably more complicated, and require more computational effort, than those used in the classical approach. Longer simulation times require that both the system size and the timescale studied be smaller than when using more approximate methods. Thus, while this approach has been used to study the forces responsible for friction (Zhong and Tomanek, 1990), it has not yet found widespread application for the type of large-scale modeling discussed here.

The simplest approach for developing a continuous potential energy function is to assume that the binding energy E_b can be written as a sum over pairs of atoms,

$$E_b = \sum_i \sum_{j(>i)} V_{\text{pair}}\left(r_{ij}\right). \tag{11.2}$$

The indexes i and j are atom labels, r_{ij} is the scalar distance between atoms i and j, and $V_{\text{pair}}(r_{ij})$ is an assumed functional form for the energy. Some traditional forms for the pair term are given by

$$V_{\text{pair}}\left(r_{ij}\right) = D \cdot X\left(X-1\right), \tag{11.3}$$

where the parameter D determines the minimum energy for pairs of atoms. Two common forms of this expression are the Morse potential ($X = e^{-\beta r_{ij}}$), and the Lennard–Jones (LJ) "12-6" potential ($X = (\sigma/r_{ij})^6$). (β and σ are arbitrary parameters that are used to fit the potential to observed properties.)

The short-range exponential form for the Morse function provides a reasonable description of repulsive forces between atomic cores, while the $1/r^6$ term of the LJ potential describes the leading term in long-range dispersion forces. A compromise between these two is the "exponential-6," or Buckingham, potential. This uses an exponential function of atomic distances for the repulsion and a $1/r^6$ form for the attraction. The disadvantage of this form is that as the atomic separation approaches zero, the potential becomes infinitely attractive.

For systems with significant Coulomb interactions, the approach that is usually taken is to assign each atom a fractional point charge q_i. These point charges then interact with a pair potential

$$V_{\text{pair}}\left(r_{ij}\right) = \frac{q_i q_j}{r_{ij}}.$$

Because the $1/r$ Coulomb interactions act over distances that are long compared to atomic dimensions, simulations that include them typically must include large numbers of atoms, and often require special attention to boundary conditions (Ewald, 1921; Heyes, 1981).

Other forms of pair potentials have been explored, and each has its strengths and weaknesses. However, the approximation of a pairwise-additive binding energy is so severe that in most cases no form of pair potential will adequately describe every property of a given system (an exception might be rare gases). This does not mean that pair potentials are without use — just the opposite is true! A great many general principles of many-body dynamics have been gleaned from simulations that have used pair potentials, and they will continue to find a central role in computer simulations. As discussed below, this is especially true for simulations of the properties of confined fluids.

A logical extension of the pair potential is to assume that the binding energy can be written as a many-body expansion of the relative positions of the atoms

$$E_b = \frac{1}{2}\sum_i \sum_j V_{2\text{–body}} + \frac{1}{3!}\sum_i \sum_j \sum_k V_{3\text{–body}} + \frac{1}{4!}\sum_i \sum_j \sum_k \sum_l V_{4\text{–body}} + \dots. \tag{11.4}$$

Normally, it is assumed that this series converges rapidly so that four-body and higher terms can be ignored. Several functional forms of this type have had considerable success in simulations. Stillinger and Weber (1985), for example, introduced a potential of this type for silicon that has found widespread

529

use, an example of which is discussed in Section 11.3.3. Another example is the work of Murrell and co-workers (1984) who have developed a number of potentials of this type for different gas-phase and condensed-phase systems.

A common form of the many-body expansion is a valence force field. In this approach, interatomic interactions are modeled with a Taylor series expansion in bond lengths, bond angles, and torsional angles. These force fields typically include some sort of nonbonded interaction as well. Prime examples include the molecular mechanics potentials pioneered by Allinger and co-workers (Allinger et al., 1989; Burkert and Allinger, 1982). A common variation of the valence-force approach is to assume rigid bonds, and allow only changes in bond angles. Because the angle bends generally have smaller force constants, there tend to be larger variations in angles than in bond lengths, so this approximation often gives accurate predictions for the shapes of large molecules at thermal energies. The advantage of this approximation is that because the bending modes have lower frequencies than those involving bond stretching modes, time steps may be used that are an order of magnitude larger than those required for flexible bonds, with no larger numerical errors in the total energy.

One method of including many-body effects in Coulomb systems is to account for electrostatic induction interactions. Each point charge will give rise to an electric field, and will induce a dipole moment on neighboring atoms. This effect can be modeled by including terms for the atomic or molecular dipoles in the interaction potential, and solving for the values of the dipoles at each step in the dynamics simulation. An alternative method is to simulate the polarizability of a molecule by allowing the values of the point charges to change directly in response to their local environment (Streitz and Mintmire, 1994; Rick et al., 1994). The values of the charges in these simulations are determined by the method of electronegativity equalization and may either by evaluated iteratively (Streitz and Mintmire, 1994) or carried as dynamic variables in the simulation (Rick et al., 1994).

Several potential energy expressions beyond the many-body expansion have been successfully developed and are widely used in MD simulations. For metals, the embedded atom method (EAM) and related methods have been highly successful in reproducing a host of properties, and have opened up a range of phenomena to simulation (Finnis and Sinclair, 1984; Foiles et al., 1986; Ercollessi et al., 1986a,b). These have been especially useful in simulations of the indentation of metals, as discussed in Section 11.3. This approach is based on ideas originating from effective medium theory (Norskov and Lang, 1980; Stott and Zaremba, 1980). In this formalism, the energy of an atom interacting with surrounding atoms is approximated by the energy of the atom interacting with a homogeneous electron gas and a compensating positive background. The EAM assumes that the density of the electron gas can be approximated by a sum of electron densities from surrounding atoms, and adds a repulsive term to account for core–core interactions. Within this set of approximations, the total binding energy is given as a sum over atomic sites

$$E_b = \sum_i E_i,$$ (11.5)

where each site energy is given as a pair sum plus a contribution from a functional (called an embedding function) that, in turn, depends on the sum of electron densities at that site:

$$E_i = \frac{1}{2} \sum_j \Phi(r_{ij}) + F\left(\sum_j \rho(r_{ij})\right).$$ (11.6)

The function $\Phi(r_{ij})$ represents the core–core repulsion, F is the embedding function, and $\rho(r_{ij})$ is the contribution to the electron density at site i from atom j. For practical applications, functional forms are assumed for the core–core repulsion, the embedding function, and the contribution of the electron densities from surrounding atoms. For a more complete description of this approach, including a much more formal derivation and discussion, the reader is referred to Sutton (1993).

For close-packed metals, it has been found that the electron densities in the solid can be adequately approximated by a pairwise sum of atomic-like electron densities. With this approximation, the computing time required to evaluate an EAM potential scales with the number of atoms in the same way as a pair potential. The quantitative results of the EAM, however, are dramatically better than those obtained with pair potentials. Energies and structures of solid surfaces and defects in metals, for example, match experimental results (or more-sophisticated calculations) reasonably well despite the relatively simple analytic form. Because of this, EAM potentials have been used extensively in studies of indentation.

Schemes similar to EAM but based on other levels of approximation have also been developed. For example, three-body terms in the electron-density contributions have been studied for modeling a range of metallic and covalently bonded systems (Baskes, 1992). Another example is the work of DePristo and co-workers, who have studied various hierarchies of approximation within effective medium theory by including additional correction terms to Equation 11.6 (Raeker and DePristo, 1991), as have Norskov and co-workers (Jacobson et al., 1987).

A potential that is similar to the EAM but based on bond orders has also been developed (Abell, 1985). Originally adapted by Tersoff (1986) to model silicon, the approach has found use in computer simulations of a wide range of covalently bonded systems (Tersoff, 1989; Brenner, 1989a, 1990; Khor and Das Sarma, 1988). Like the embedded atom potentials, Tersoff potentials begin by approximating the binding energy of a system as a sum over sites:

$$E_b = \sum_i E_i. \tag{11.7}$$

In this case, however, each site energy is given by an expression that resembles a pair potential,

$$E_i = \sum_{j(>i)} \left[V_R\left(r_{ij}\right) + B_{ij} \cdot V_A\left(r_{ij}\right) \right]. \tag{11.8}$$

The functions $V_A(r_{ij})$ and $V_R(r_{ij})$ represent pairwise-additive attractive and repulsive terms, respectively, and B_{ij} represents an empirical bond-order expression that modulates the attractive pair potential. This bond-order term is where the many-body effects are introduced. As in the other potentials, functional forms for the bond order and the pair terms are fit to a range of properties for the systems of interest. Applications of this approach have assumed functional forms for the bond order that decrease its value as the number of nearest-neighbors of a given pair of bonded atoms increases. Physically, the attractive pair term can be envisioned as bonding due to valence electrons, with the bond order destabilizing the bond as the valence electrons are shared among more and more neighbors. Tersoff, Abell, and others have shown that this simple expression can capture a range of bond energies, bond lengths, and related properties for group IV solids. Also, if properly parameterized, the potential yields Pauling's bond-order relations (Abell, 1985; Khor and Das Sarma, 1988; Tersoff, 1989). These properties have been shown to make this expression very powerful for predicting covalently bonded structures and, therefore, useful for predicting new phenomena through MD simulations.

Although they are based on different principles, the EAM and Tersoff expressions are quite similar. In the EAM, binding energy is defined by electron densities through the embedding function. The electron densities are, in turn, defined by the arrangement of neighboring atoms. Similarly, in the Tersoff expression the binding energy is defined directly by the number and arrangement of neighbors through the bond-order expression. In fact, the EAM and Tersoff approaches have been shown to be identical for simplified expressions provided that angular interactions are not used (Brenner, 1989b).

Although not all potential energy expressions widely used in MD simulations have been covered, this section provides a brief introduction to those that have found the most use in the simulations of friction, wear, and related phenomena. As mentioned above, methods that explicitly incorporate semiempirical and first-principles electronic structure calculations have not been discussed. As computer speeds continue to increase and as new algorithms are developed, we expect that these more exact methods will play an increasingly important role in tribological simulations.

11.2.2 Thermodynamic Ensemble

When performing a MD simulation, a choice must be made as to which thermodynamic ensemble to study. These ensembles are distinguished by which thermodynamic variables are held constant over the course of the simulation. (For a broader and more rigorous treatment of ensemble averaging, the reader should consult any statistical mechanics text, (e.g., McQuarrie, 1976)).

Without specific reasons to do otherwise, it is quite natural to keep the number of atoms (N) and the volume of the simulation cell (V) constant over the course of a MD simulation. In addition, for a system without energy transfer, integrating the equations of motion (Equation 11.1) will generate a trajectory over which the energy of the system (E) will also be conserved. A simulation of this type is thus performed in the constant-NVE, or microcanonical, ensemble.

Systems undergoing sliding friction or indentation, however, require work to be performed on the system, which raises its energy and causes the temperature to increase. In a macroscopic system, the environment surrounding the region of tribological interest acts as an infinite heat sink, removing excess energy and helping to maintain a fairly constant temperature. Ideally, a sufficiently large simulation would be able to model this same behavior. But while the thousands of atoms at an atomic-scale interface are within reach of computer simulation, the $O(10^{23})$ atoms in the experimental apparatus are not. Thus, a thermodynamic ensemble that will more closely resemble reality will be one in which the temperature (T), rather than the energy, is held constant. These simulations are performed in the constant-NVT, or canonical, ensemble.

A constant temperature is maintained in the canonical ensemble by using any of a large number of thermostats, many of which are described in the following section. What is often done in simulations of indentation or friction is to apply the thermostat only in a region of the simulation cell that is well removed from the interface where friction is taking place. This allows for local heating of the interface as work is done on the system, while also providing a means for efficient dissipation of excess heat. These "hybrid" NVE/NVT simulations, although not rigorously a member of any true thermodynamic ensemble, are very useful and quite common in tribological simulations.

A particularly troublesome system for MD simulations is the nonequilibrium dynamics of confined thin films (see Section 11.4.2). In these systems, the constraint of constant atom number is not necessarily applicable. Under experimental conditions, a thin film under shear or tension is free to exchange molecules with a reservoir of bulk liquid molecules, and the total atom number is certainly conserved. But the number of atoms in the film itself is subject to rather dramatic changes. According to some studies, as many as half the molecules in an ultrathin film will exit the interfacial region upon a change in registry of the opposing surfaces (i.e., with no change in interfacial volume) (Schoen et al., 1989). Changes in the film particle number can be equally large under compression.

The proper conserved quantity in these simulations is not the particle number N, but the chemical potential μ. During a simulation performed in the constant-μVT, or grand canonical, ensemble, the number of atoms or molecules fluctuates to keep the chemical potential constant. A true grand canonical MD simulation is too difficult to perform for all but the simplest of liquid molecules, however, due to the difficulties associated with inserting or removing molecules at bulk densities. An alternative chosen by some authors is to mimic the experimental reservoir of bulk liquid molecules on a microscopic scale. This involves performing a constant-NVT (Wang et al., 1993a,b, 1994) or constant-NPT (Gao et al., 1997) simulation that explicitly includes a collection of molecules that are external to the interfaces (see Figure 11.14). As liquid molecules drain into or are drawn from the reservoir region, the number of particles directly between the interfaces is free to change. This method is then an approximation to the grand canonical ensemble when only a subset of the system is considered. Two drawbacks to this method are that the interface can extend infinitely (via periodic boundary conditions) in only one dimension instead of two, and also that a significant number of extraneous atoms must be carried in the simulation.

An interesting alternative to the grand canonical ensemble is that chosen by Cushman and co-workers (Schoen et al., 1989; Curry et al., 1994). They performed a series of grand canonical Monte Carlo simulations at various points along a hypothetical sliding trajectory. These simulations were used to calculate the correct particle numbers at a fixed chemical potential, which were in turn used as inputs to nonsliding, constant-NVE MD simulations at each of the chosen trajectory points. Because the system was fully equilibrated at each step along the sliding trajectory, the sliding speed can be assumed to be infinitely slow. This offers a useful alternative to continuous MD simulations, which are currently restricted to sliding speeds of roughly 1 m/s or greater — orders of magnitude larger than most experimental studies.

11.2.3 Temperature Regulation

As was discussed in the previous section, some method of controlling the system temperature is required in simulations involving friction or indentation. In this section, we discuss some of the many available thermostats that are used for this purpose. For a more formal discussion of heat baths and the trajectories that they produce, the reader is referred to Hoover (1986).

The most straightforward method for controlling heat production is simply to rescale intermittently the atomic velocities to yield a desired temperature (Woodcock, 1971). This approach was widely used in early MD simulations, and is often effective at maintaining a given temperature during the course of a simulation. However, it has several disadvantages that have spurred the development of more-sophisticated methods. First, there is little formal justification. For typical system sizes, averaged quantities, such as pressure, do not correspond to those obtained from any particular thermodynamic ensemble. Second, the dynamics produced are not time reversible, again making results difficult to analyze in terms of thermodynamic ensembles. Finally, the rate and mode of heat dissipation are not determined by system properties, but instead depend on how often velocities are rescaled. This may influence dynamics that are unique to a particular system.

A more-sophisticated approach to maintaining a given temperature is through Langevin dynamics. Originally used to describe Brownian motion, this method has found widespread use in MD simulations. In this approach, additional terms are added to the equations of motion, corresponding to a frictional term and a random force (Schneider and Stoll, 1978; Hoover, 1986; Kremer and Grest, 1990). The equations of motion (see Equation 11.1) are given by

$$m\mathbf{a} = \mathbf{F} = m\xi\mathbf{v} + R(t),\qquad(11.9)$$

where $\mathbf{F}$ are the forces due to the interatomic potential, the quantities m and $\mathbf{v}$ are the particle mass and velocity, respectively, ξ is a friction coefficient, and $R(t)$ represents a random "white noise" force. The friction kernel is defined in terms of a memory function in formal applications; kernels developed for harmonic solids have been used successfully in MD simulations (Adelman and Doll, 1976; Adelman, 1980; Tully, 1980).

As with any thermostat, the atom velocities are altered in the process of controlling the temperature. It is important to keep this in mind when using a thermostat, because it has the potential to perturb any dynamic properties of the system being studied. To help avoid this problem, one effective approach is to add Langevin forces only to those atoms in a region away from where the dynamics of interest occurs. In this way, coupling to a heat bath is established away from the important action, and simplified approximations for the friction term can be used without unduly influencing the dynamics produced by the interatomic forces. For heat flux via nuclear (as opposed to electronic) degrees of freedom in solids, it has been shown that a reasonable approximation for the friction coefficient ξ is $6/\pi$ times the Debye frequency β (Adelman and Doll, 1976). Lucchese and Tully (1983) have shown that with this approximation and a sufficiently large reaction zone, the vibrational modes of atoms away from the bath atoms are well described by the interatomic potential.

The random force can be given by a Gaussian distribution where the width, which is chosen to satisfy the fluctuation–dissipation theorem, is determined from the equation

$$\langle R(0) \cdot R(t) \rangle = 2mkT\xi\delta(t).$$

(11.10)

The function R is the random force in Equation 11.9, m is the particle mass, T is the desired temperature, k is Boltzmann's constant, t is time, and ξ is the friction coefficient. The random forces are uncoupled from those at the previous steps (as denoted by the delta function), and the width of the Gaussian distribution from which the random force is chosen depends on the temperature.

The simplified Langevin approach outlined above does not require any feedback from the current temperature of the system; instead, the random forces are determined solely from Equation 11.10. A slightly different expression has been developed that eliminates the random forces and replaces the constant-friction coefficient with one that depends on the ratio of the desired temperature to current kinetic energy of the system (measured as a temperature) (Berendsen et al., 1984). The resulting equations of motion are

$$m\mathbf{a} = \mathbf{F} + m\xi\left(\frac{T_0}{T} - 1\right)\mathbf{v},$$

(11.11)

where $\mathbf{F}$ is the force due to the interatomic potential, T_0 and T are the desired and actual temperatures, respectively, and $\mathbf{v}$ is again the particle velocity. The advantage of this approach is that it requires no evaluation of random forces, which can be expensive for a large number of bath atoms. One disadvantage in practice is that if the system is not pre-equilibrated to populate properly the vibrational modes, or if nonrandom external forces are applied to the system, it can be slow to fully equilibrate the system. For example, a simulated indentation of a solid surface requires that the bottom layers of the solid be held rigid (Section 11.3). Compression of the surface during indentation may cause sound waves to propagate into the bulk, reflect from the bottom layers, and continue to reflect between the surface and rigid layers. Because the Berendsen thermostat uses a frictional force that depends on the average kinetic energy, it would only reduce the total kinetic energy of the system, and not help dissipate a traveling wave. On the other hand, the Langevin approach using a random force on each atom does not require feedback from the system, and thermostats each atom individually. Thus, the random forces are much more efficient in eliminating these nonphysical reflecting waves.

Nonequilibrium equations of motion have also been developed to maintain a constant temperature (Hoover, 1986). Like the Berendsen thermostat, this approach adds a friction to the interatomic forces. However, it is derived from Gauss' principle of least constraint, which maintains that the sum of the squares of any constraining forces on a system should be as small as possible. Using a Lagrange multiplier, a frictional force on each atom i of the form

$$\mathbf{F}_i^{\text{friction}} = -\varsigma m_i \mathbf{v}_i,$$

(11.12)

where

$$\varsigma = \sum_i \left(\mathbf{F}_i \cdot \mathbf{v}_i\right) \bigg/ \sum_i \left(m\mathbf{v}_i^2\right),$$

(11.13)

can be derived that maintains a constant temperature. The quantity m_i is the mass of atom i, $\mathbf{v}_i$ is its velocity, and $\mathbf{F}_i$ is the total force on atom i due to the interatomic potential. Note that there is no target temperature in this expression; instead, the temperature of the system when the constraint is initiated is maintained for all time. This approach has several obvious advantages. First, it does not rely on an approximated input such as the Debye frequency as in the simplified Langevin or Berendsen thermostats. Heat loss and gain are therefore determined only by implicit system properties. Second, because a random force is not required, it does not significantly increase computational time. Third, the equations of motion are time reversible. Finally, by differentiating total energy with respect to time, the heat loss (or gain) due to the thermostat can be calculated directly (this is also true of the Langevin thermostat). However, as with the Berendsen thermostat, coupling of the frictional to global properties of the system may be slow to randomize nonphysical vibrational disturbances.

A thermostat that corresponds rigorously to a canonical ensemble has been developed by Nosé (1984a,b). This significant advance also adds a friction term, but one that maintains the correct distribution of vibrational modes. It achieves this by adding a new dimensionless variable to the standard classical equations of motion that can be thought of as a large heat bath, which couples to each of the physical degrees of freedom. The actual effect of the variable, however, is to scale the coordinates of either time or mass in the system. The dynamics of the expanded system correspond to the microcanonical ensemble, but when projected onto only the physical degrees of freedom they generate a trajectory in the canonical ensemble. Sampling problems associated with very small or very stiff systems can be overcome by attaching a series of these Nosé-Hoover thermostats to the system (Martyna et al., 1992). The resulting equations of motion are time reversible, and the trajectories can be analyzed exactly with well-established statistical mechanical principles (Martyna et al., 1992). For a complete description of the Nosé thermostat, its relation to other formalisms for generating classical equations of motion, and a comparison of the dynamics generated with this approach and the others that have been reviewed here, the reader is referred to Hoover (1986).

Each of the potential energy functions, thermodynamic ensembles, and thermostats outlined above has advantages and disadvantages. The optimum choice depends strongly on the particular system and process being simulated as well as on the type of information in which one is interested. For example, general principles related to liquid lubrication in confined areas may be most easily understood and generalized from simulations that use pair potentials and may not require a thermostat. On the other hand, if one wants to study the wear or indentation of a surface of a particular metal, then EAM or other semiempirical potentials, together with a thermostat, may yield more reliable results. Even more-detailed studies, including the evaluation of electronic degrees of freedom, may require interatomic forces derived from some level of electronic structure calculation. The best way to make this choice is to understand carefully the strong points of each of these approaches, decide what one wishes to learn from the simulation, and form conclusions based on this careful understanding.

11.3 Nanometer-Scale Material Properties: Indentation, Cutting, and Adhesion

Understanding material properties at the nanometer scale is crucial to developing the fundamental ideas needed to design new coatings with tailor-made friction and wear properties. One of the ways in which these properties is being characterized is through the use of the AFM. This technique is proving to be a very versatile tool that can provide a rich variety of atomic-scale information pertaining to a given tip–sample interaction (Burnham and Colton, 1993). For example, when an AFM tip (the radius of curvature of AFM tips typically ranges from 100 Å to 100 μm) is rastered across a sample substrate and the force on the tip perpendicular to the substrate is measured at each point, a force map of the surface is obtained that can be related to the actual topography of the surface (Meyer et al., 1992). Rastering the AFM tip across a substrate in the same way, but measuring the deflection of the tip in the lateral direction instead, produces a friction map of the surface (Germann et al., 1993). Finally, by moving the tip perpendicular to the surface of the substrate, the AFM can be used as a nanoindenter that probes the mechanical properties of various substrates and thin films (Burnham and Colton, 1989; Burnham et al., 1990).

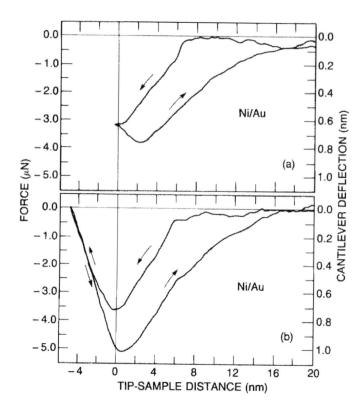

FIGURE 11.2 Experimentally measured force vs. tip-to-sample distance curves for an Ni tip interacting with an Au substrate for contact followed by separation in (a) and contact, indentation, then separation in (b). These curves were derived from AFM measurements taken in dry nitrogen. (From Landman, U. et al. (1990), *Science* 248, 454–461. With permission.)

Adhesive contact can be examined by gradually moving the tip closer to the substrate until the two come into contact. After contact, the tip is retracted from the surface, and any differences between the force curves for contact and retraction reflect characteristics of adhesion. A plot of normal force on the tip vs. tip–sample separation (i.e., a force curve) is typically made to record this sequence of events. A force curve for the adhesive contact of a Ni tip with an Au substrate (denoted Ni/Au) is shown in Figure 11.2a. This is a typical curve possessing the same qualitative features as most AFM force curves. For instance, as the tip and the sample begin to interact, a small attractive well, due to long-range forces (Burnham and Colton, 1993) is apparent in the force curve at large tip–sample separations (the well is centered at a tip–sample separation of approximately 17 nm). The distance between the tip and the sample is gradually decremented until the tip comes into contact with the sample. After contact the tip is retracted, and adhesion between the tip and the substrate manifests itself as a hysteresis in the force curve.

After contact, if the tip is moved farther toward the substrate, rather than away from the substrate, indentation of the sample by the tip results. This indentation is reflected as a dramatic increase in force as the tip is moved farther into the substrate (Figure 11.2b). This region of the force curve is known as the repulsive wall region (Burnham and Colton, 1993), or when considered without the rest of the force curve, an indentation curve. Retraction of the tip subsequent to indentation results in an enhanced adhesion, therefore, in a larger hysteresis in the force curve. The origin of this enhanced adhesion is discussed later.

Many types of adhesion at a tip–substrate interface are possible. Adhesion might result from the formation of covalent chemical bonds between the tip and the sample. Alternatively, real surfaces usually have a layer of liquid contamination on the surface that can lead to capillary formation and adhesion.

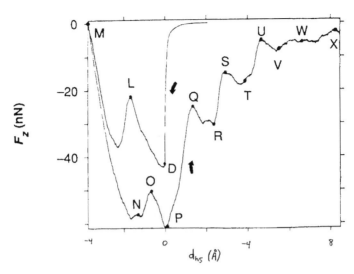

FIGURE 11.3 Computationally derived force F_z vs. tip-to-sample distance d_{hs} curves for approach, contact, indentation, then separation using the same tip–sample pair as in Figure 11.2. These data were calculated from an MD simulation. (From Landman U. et al. (1990), *Science* 248, 454–461. With permission.)

In metallic systems a different sort of wetting is possible; specifically, the sample can be wet by the tip (or vice versa). The result of this wetting is the formation of "connective neck" of metallic atoms between the tip and the sample and a consequent adhesion. Finally, entanglement of molecules that are anchored on the tip with molecules anchored on the sample could also be responsible for an observed hysteresis.

Molecular dynamics simulations of indentations were first employed in an effort to shed light on the physical phenomena that are responsible for the qualitative shape of AFM force curves. In addition to succeeding at this task, MD simulations have revealed an abundance of atomic-scale phenomena that occur during the indentation process. In the remainder of this section, MD simulations related to indentation and related processes are discussed.

11.3.1 Indentations of Metals

Landman et al. (1990) were among the first groups to use MD to simulate the indentation of a metallic substrate with a metal tip. In an early simulation, a Ni tip was used to indent a Au(001) substrate. The tip was originally arranged as a pyramid and contained 1400 dynamic atoms and 1176 rigid atoms used as a holder. The substrate was composed of 11 layers of Au atoms containing 450 atoms each. These constant-temperature simulations were carried out at 300 K. The forces governing the interatomic motion of the system were derived from EAM potentials for Ni and Au.

After equilibration of the tip and substrate to 300 K, the tip was brought into contact with surface by moving the tip holder 0.25 Å closer to the surface every 1525 fs. This rate (or a tip velocity of approximately 16 m/s), while fast compared with experiment, is much smaller than the speed of sound in Au, and allowed the system to evolve dynamically such that only natural fluctuations of system properties occurred. This process was continued throughout the indentation. By calculating the force on the rigid layers of the tip while moving the tip closer to the sample, a plot of force vs. tip sample separation was generated (Figure 11.3).

The shape of the computer-generated force curve for the indentation of the Au substrate by the Ni tip showed qualitative agreement with the experimentally derived force curve (Figure 11.2b) with two exceptions. First, there was no attractive well at large tip–sample separations in the computer-generated force curve. This was due to the lack of long-range attractive interactions, such as dispersion forces, in the simulations. Second, the computer-generated force curve contained a fine structure not present in the force curve generated from experimental data.

FIGURE 11.4 Illustration of atoms in the MD simulation of an Ni tip being pulled back from an Au substrate. This causes the formation of a connective neck of atoms between the tip and the surface. Red spheres represent Ni atoms. The first layer of Au substrate is colored yellow, the second blue, the third green, the fourth yellow, and so on. (From Landman U. et al. (1990), *Science* 248, 454–461. With permission.)*

One advantage of simulations is that the shape of the force curve, along with its fine structure, can be related to specific atomic-scale events. For instance, Landman et al. (1990) reported the observation of a jump to contact (JC) that corresponded to that region of the force curve where there was a precipitous drop in the force just prior to tip–substrate contact (Figure 11.3, point D). The JC phenomenon was previously observed by Pethica and Sutton (1988) and by Smith et al. (1989). In this region of the force curve, the gold atoms "bulged up" to meet the tip. Deformation occurred in the gold substrate because its modulus is much lower than that of the nickel tip. This deformation occurred in a short time span (approximately 1 ps) and was accompanied by wetting of the Ni tip by several Au atoms. Landman et al. (1990) concluded that the JC phenomenon in metallic systems is driven by the tendency of the interfacial atoms of the tip and the substrate to optimize their embedding energies while maintaining their individual material cohesive binding.

Advancing the tip past the JC point caused indentation of the gold substrate accompanied by the characteristic increase in force with decreasing tip–substrate separation (Figure 11.3, points D to M). This region of the computer generated force curve had a maximum not present in the force curve generated from experimental data (Figure 11.3, point L). The origin of this variation in force was tip-induced flow of the Au atoms. This flow caused "piling-up" of Au atoms around the edges of the Ni indenter.

The force curve was completed by reversal of the tip motion (Figure 11.3, points M to X). The hysteresis in this force curve was due to adhesion between the tip and the substrate. As the tip was retracted from the sample, a "connective neck" of atoms between the tip and the substrate formed (Figure 11.4). While this connective neck of atoms was largely composed of Au atoms, some Ni atoms did diffuse into the neck. Retraction of the tip caused the magnitude of the force to increase (i.e., become more negative) until, at some critical force, the atoms in adjacent layers of the connective neck rearranged so that an

---

* Color reproduction follows page 16.

additional row of atoms formed in the neck. These rearrangement events were the essence of the elongation process, and they were responsible for the fine structure (apparent as a series of maxima) present in the retraction portion of the force curve. These elongation and rearrangement steps were repeated until the connective neck of atoms was severed.

When the Ni tip was coated with an epitaxial gold monolayer (Landman et al., 1992) and the indentation of an Au(001) substrate repeated, the adhesive contact between the tip and the substrate was reduced. The JC instability, formation of an adhesive contact, and hysteresis during subsequent retraction were all observed. In contrast to the Ni/Au study, complete separation of the tip and substrate resulted in the transfer of a smaller number of substrate atoms to the tip when the connective neck of atoms, composed entirely of Au, was severed. Because the tip was covered with Au, the interaction between the tip and the substrate was composed mostly of Au–Au interactions, and Au possesses less of a tendency to wet itself than it does to wet Ni. This accounted for the insignificant number of substrate atoms transferred to the tip.

When a hard Ni tip indented a soft Au substrate, the substrate sustained most of the damage. Conversely, damage was predominantly done to the tip when a soft tip was used to indent a hard substrate. For example, Landman and Luedtke (1991) used a pyramidal Au(001) tip to indent a Ni(001) substrate. These constant-temperature indentations were carried out in the same manner described above for the Ni/Au system. Force curves generated from the indentation of the Ni substrate by the Au tip had the same qualitative shape as the Ni/Au force curves. However, there were differences in the fine structure of these force curves, therefore, in the atomic-scale events responsible for this structure. For instance, in this case the tip bulged toward the substrate during the JC, rather than the substrate bulging toward the tip. Thus, the softer material (i.e., the one with the lower modulus) was displaced during the JC. The adhesive contact between the tip and the substrate caused large structural rearrangements in the interfacial region of the Au tip. The closest three or four Au layers to the Ni substrate exhibited a marked tendency toward a (111) reconstruction, consistent with an increase in interlayer spacing. In fact, this reconstruction persisted throughout the separation process.

When the tip was pushed farther toward the substrate subsequent to the JC, it became flattened (or compressed) increasing its contact area. This flattening involved structural rearrangements of the outer layers of the tip that reduced the number of crystalline layers, leaving an interstitial-layer defect in the core of the tip. The interstitial defect was annealed away upon further compression of the tip. Upon separation, a connective neck of Au atoms was formed due to adhesion between the Au and the Ni. This connective neck of atoms underwent a series of elongation events, as in the Ni/Au study, until the tip–sample distance was such that the neck became thin and broke.

Other metallic tip–substrate systems were examined with interesting results. For instance, Tomagnini et al. (1993) studied the interaction of a pyramidal Au tip with a Pb(110) substrate using MD. These constant-energy simulations were carried out at approximately room temperature and again at temperatures high enough to initiate surface melting of the Pb substrate (600 K). The forces were calculated using a many-body potential, called the glue model, that is very similar to the EAM potentials (Ercolessi et al., 1988).

At room temperature (300 K), when the Au tip was brought into close proximity to the Pb substrate, a JC was initiated by a few Pb atoms wetting the Au tip. The connective neck of atoms between the tip and the surface was composed almost entirely of Pb. The tip became deformed because the inner-tip atoms were pulled more toward the sample surface than toward atoms on the tip surface. Because these were constant-energy simulations, the energy released due to the wetting of the tip caused an increase in temperature of the tip (of approximately 15 K). Extensive structural rearrangements in the tip occurred when the tip–sample distance was decremented further. Results for the retraction of the tip from the Pb substrate were not reported.

Increasing the substrate temperature to 600 K caused the formation of a liquid Pb layer (approximately four layers thick) on the surface of the substrate. During the indentation, the distance at which the JC occurred increased by approximately 1.5 Å. Due to the high diffusivity of Pb surface atoms at this temperature, the contact area also increased. Eventually, the Au tip dissolved in the liquid Pb atom "bath."

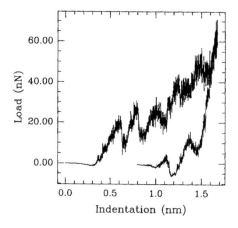

FIGURE 11.5 Computationally derived load vs. indentation depth curve for the indentation of Cu(111) with a rigid tip. These data were calculated from an MD simulation. (From Belak, J. and Stowers, I. F. (1992), in *Fundamentals of Friction: Macroscopic and Microscopic Processes* (I. L. Singer and H. M. Pollock, eds.), 511–520, Kluwer, Dordrecht. With permission.)

This liquid-like connective neck of atoms followed the tip upon retraction. As a result, the liquid–solid interface moved farther back into the bulk Pb substrate, increasing the length of the connective neck. Similar elongation events have been observed experimentally. For example, scanning tunneling microscopy (STM) experiments on the same surface demonstrated that the neck can elongate approximately 2500 Å without breaking.

Similar atomic-scale phenomena were observed for an Ir tip indenting a Pb substrate (Raffi-Tabar et al., 1992). These constant-temperature MD simulations made use of long-range, many-body Finnis-Sinclair potentials formulated to model the interatomic forces between atoms in face-centered-cubic (fcc) metallic alloys (Raffi-Tabar and Sutton, 1991). The method developed by Nosé (1984a) was used to regulate the temperature of the simulation. This simulation was unique in that periodic boundary conditions were also employed in the indentation direction. Therefore, images of the substrate–tip system were located in the cells above and below the computational cell. Indentation of the substrate by the tip was achieved by decrementing the computational cell length normal to the substrate surface.

During the indentation process, the Pb substrate wetted the Ir tip subsequent to the JC. Significant structural rearrangement of the Pb substrate was brought about when the tip was pushed closer to the substrate after the JC. This structural rearrangement led to a "piling up" of the Pb atoms around the edges of the tip and was brought about by the local diffusional flow of the Pb atoms in much the same way as it was when an Ni tip was used to indent a soft Au substrate (Landman et al., 1990). In both the Ni/Au and the Ir/Pb systems, the tip (Ni and Ir) retained most of its shape because it was harder than the substrate. Plastic flow in the Ir/Pb system resulted in a hysteresis in the force curve upon retraction of the tip. The nonmonotonic features in the force curves generated from the Ir/Pb simulation were associated with discrete, local, atomic movements; however, the precise nature of these movements was not elucidated.

The large-scale indentation of approximately 70,000-atom Cu and Ag(111) surfaces with a rigid, triangular-shaped tip has been simulated by Belak and co-workers (Belak and Stowers, 1992; Belak et al., 1993). The forces between metal atoms in these large-scale MD simulations were derived from EAM potentials and interaction between the tip and the metal substrate was modeled by an LJ potential. Indentations were performed by moving the tip closer to the substrate at constant velocities of 1, 10, and 100 m/s. A Nosé thermostat (1984a) was used to control the temperature of the simulation.

For the 1 m/s indentation of Cu(111), the load increased linearly with indentation depth until the tip indented approximately 0.6 nm into the substrate (Figure 11.5). At this point, the surface yielded plastically and the force dropped suddenly. This plastic yielding was concomitant with a single atom "popping" out onto the surface of the substrate from beneath the tip. Continued indentation caused several of these events to occur. At the maximum indentation depth (1.7 nm) atoms from the substrate were "piled up" around the edges of the tip (Figure 11.6) and plastic deformation was limited to a few lattice spacings surrounding the tip. The piling up of atoms is typical in cases where a hard tip is used to indent

FIGURE 11.6 Illustration of Cu(111) substrate atoms in an MD simulation after indentation by a rigid tip at 1 m/s. The piling-up of the substrate atoms (gray spheres) around the edge of the tool tip after indentation is evident in this picture. (From Belak, J. and Stowers, I. F. (1992), in *Fundamentals of Friction: Macroscopic and Microscopic Processes* (I. L. Singer and H. M. Pollock, eds.), 511–520, Kluwer, Dordrecht. With permission.)

a soft substrate. Retracting the tip from the substrate caused the load to drop quickly to zero with only small oscillations, presumably due to some plastic events at the surface. Both the pile up of substrate atoms and the plastic nature of the indentation were apparent from analysis of the Cu(111) substrate subsequent to indentation (Figure 11.6).

Plastic deformation in this system during the 1 m/s indentation occurred via the motion of point defects. The formation of defects or dislocations was driven by the need to release stored elastic energy. Plots of the energy required to create a dislocation and the elastic energy vs. length of the radius of the indenter cross at length values of a few nanometers. In contrast, at the faster indentation rates of 10 m/s, the system did not have time to relax completely and the stored elastic energy was much greater. Eventually, the surface yielded and a small dislocation loop was observed on the surface.

Even though the tip was much sharper for the indentation of the Ag(111), the force curves generated from that indentation were similar to those generated from the indentation of Cu(111). For the 1 m/s indentation, the initial plastic events corresponded to the "popping" of single atoms out of the substrate. The hardness value obtained from this simulation, estimated from the load divided by contact area, was approximately 4 GPa, or approximately four times larger than the experimentally determined hardness value of approximately 1 GPa (Pharr and Oliver, 1989).

11.3.2 Indentation of Metals Covered by Thin Films

Using simulation procedures similar to those in their earlier work, Raffi-Tabar and Kawazoe (1993) used an Ir tip to indent an Ir substrate that was covered with a monolayer Pb film. As the tip approached the Pb/Ir substrate, the Pb atoms directly below the tip strained upward to wet the Ir tip and a JC was observed. The disruption of the Pb monolayer caused by the JC also resulted in local deformation of the Ir substrate beneath the monolayer. Further indentation resulted in penetration of the Pb film at only one atomic site. As a result, no Ir–Ir adhesion was observed. Because Ir–Ir adhesion is stronger than Ir–Pb adhesion, the separation force was less in this case than it was in the absence of the Pb monolayer (Raffi-Tabar et al., 1992). In addition, the Ir substrate was not deformed as a result of the indentation because the Ir tip did not wet the substrate.

When a Pb tip was used to indent an Ir-covered Pb substrate, the Pb tip atoms wetted the Ir monolayer during the JC. As a result of the JC, the contact area between the tip and the Ir monolayer was larger and there was no discernible crystal structure present in the Pb tip. Instead, the tip appeared to have the structure and properties of a liquid drop wetting a surface. Because of the presence of the Ir monolayer, continued indentation of the tip did not result in the formation of any adhesive Pb–Pb bonds. During pull off, the Pb tip formed a connective neck, which decreased in width, as it separated from the monolayer–substrate system. This was largely due to the Pb–Pb interaction that is small compared with the Ir–Pb interaction. The radius of this connective neck of atoms was smaller than it was in the absence of the Ir layer. As a result, the pull-off force (i.e., the force of adhesion) was less when the Ir layer was present.

In summary, a reduction in the force of adhesion was observed when a monolayer film was placed between the tip and the substrate. In the Ir/Pb/Ir case, formation of strong Ir–Ir adhesion was prevented by the presence of the Pb film; therefore, the pull-off force was reduced. In the Pb/Ir/Pb case, the smaller radius of the connective neck between the tip and substrate was responsible for the reduction in the force of adhesion.

Molecular dynamics has also been used to simulate indentation of an *n*-hexadecane-covered Au(001) substrate with an Ni tip (Landman et al., 1992). The forces governing the metal–metal interactions were derived from EAM potentials. A so-called united-atom model (Ryckaert and Bellmans, 1978) was used to model the *n*-hexadecane film. In this model, the hydrogen and carbon atoms were treated as one united atom and the bonds between united atoms were held rigid. The interchain forces and the interaction of the chain molecules with the metallic tip and substrate were both modeled using an LJ potential energy function. The size of the metallic tip and substrate were the same as in a previous study (Landman et al., 1990). The hexadecane film consisted of 73 alkane molecules. The film was equilibrated on a 300 K Au surface and indentation was performed as described earlier (Landman et al., 1990).

Equilibration of the film with the Au surface resulted in a partially ordered film where molecules in the layer closest to the Au substrate were oriented parallel to the surface plane. When the Ni tip was lowered, the film "swelled up" to meet and partially wet the tip. Continued approach of the tip toward the film caused the film to flatten and some of the alkane molecules to wet the sides of the tip. Lowering the tip farther caused drainage of the top layer of alkane molecules from underneath the tip, increased wetting of the sides of the tip, "pinning" of hexadecane molecules under the tip, and deformation of the gold substrate beneath the tip. At this stage of the simulation, the force between the Ni tip and the film/Au substrate was repulsive. In contrast, the force between the tip and the substrate had been attractive when the alkane film was not present (Landman et al., 1990). Further lowering of the tip resulted in the drainage of the pinned alkane molecules, inward deformation of the substrate, and eventual formation of an intermetallic contact by surface Au atoms moving toward the Ni tip, which was concomitant with the force between the tip and the substrate becoming attractive.

The effect of indenter shape on the compression of self-assembled monolayers was examined using MD simulations by Tupper et al. (1994); 64 chains of hexadecanethiol were chemisorbed on an Au(111) substrate composed of 192 atoms. A flat compressing surface, also composed of 192 atoms, and an asperity, which was ¼ the size of the flat surface at the point of contact, were used to compress the hexadecanethiol film in separate studies. Equilibration of the hexadecanethiol films, prior to compression, resulted in highly ordered films in which the sulfur head group was bound to the threefold hollow sites in a hexagonal array with a nearest-neighbor distance of 4.99 Å. The temperature of the system was maintained at 300 K while the films were compressed by moving the flat surface (or the asperity) closer to the films at a constant velocity of 100 m/s. The potential energy, load, and average tilt angle of the film molecules were monitored during the simulation.

For the compression of the monolayer film by the flat surface, there was a reversible change in the tilt angle of the chain-tail group normal to the surface. Prior to compression, the tail groups were tilted uniformly at an angle of 28° with respect to the surface normal. As the surface and the film were brought into close proximity, the tilt angle decreased to approximately 20° during the JC. Compression of the film to a load of 1.0 nN/molecule caused the tilt angle to increase steadily to 48°. This uniform change in tilt angle of the tail groups suggested that a structural change had occurred. This was confirmed by examination of the structure factors for the sulfur head groups. The structure factors indicated that the head groups had converted from their original hexagonal packing to oblique packing. Because this change was also observed in the tail groups, this suggested that the film had undergone a uniform structural change to form an ordered structure (Tupper and Brenner, 1994). This structural rearrangement was largely due to the uniform compression of the bonds in the chains accompanied by the formation of few gauche defects in the hexadecanethiol film. When the compressing surface was retracted, the average tilt angle of the tail groups returned to its equilibrium value. Thus, this structural rearrangement was shown to be reversible. The signature of this transition appeared as a subtle slope change in the force vs. distance curve. A similar effect was observed in experimental data generated by Joyce et al. (1992).

Conversely, a large number of random structural changes of the film resulted when an asperity was used to compress the film. During compression, the average tilt angle of the tail groups varied nonuniformly between 20° and 30°. In addition, the distribution of bond lengths was much broader than for the flat surface compression. The average bond angle also decreased and a large number of gauche defects were formed in the film. Finally, the structure factors calculated for the sulfur head groups suggested that the sulfur head groups were disordered. By comparing the force profiles from these two indentations Tupper et al. (1994) concluded that an asperity can approach the substrate much closer than the flat surface before disrupting the film. This conclusion agrees with the hypothesis that surface asperities that penetrate these thick insulating films may play a crucial role in the STM imaging of these films (Liu and Salmeron, 1994).

11.3.3   Indentation of Nonmetals

Large-scale MD simulations have also been used to investigate nanometer-sized indentation processes in nonmetallic systems. For example, Kallman et al. (1993) examined the microstructure of amorphous and crystalline silicon before, during, and after indentation. This was done in an effort to examine the assertion that Si directly beneath the indenter undergoes two different pressure-induced, solid–solid phase transformations during indentation involving the high-pressure β-Sn structure (above 100 kbars) and a thermodynamically unstable amorphous phase. The constant-temperature MD simulations contained over 350,000 substrate silicon atoms. By using both a "smooth" continuum tip and rigid, but atomically "rough" tetrahedral tip, both amorphous and crystalline silicon were indented using two different indentation speeds (2.7 km/s, or ⅓ the longitudinal speed of sound in Si, and approximately 0.3 km/s). Interatomic forces governing the motion of the silicon atoms were derived from the many-body silicon potential of Stillinger and Weber (1985). Both substrates were also indented at a number of temperatures, the highest temperature being close to the melting point of silicon.

Phase transitions during the simulated indentation were monitored by calculating both a diffraction pattern and the angle-averaged pair distribution function $G(r)$. Because bar-code plots of $G(r)$ differ qualitatively for different phase of silicon, they reveal the structural state of the silicon during indentation, when examined in conjunction with the calculated diffraction patterns.

At the highest indentation rate and the lowest temperature, Kallman et al. (1993) determined that amorphous and crystalline Si have similar yield strengths (138 and 179 kbar). Near the melting temperature and at the slowest indentation rate, lower yield strengths (30 kbar) were observed for both amorphous and crystalline Si. Thus, the simulated nanoyield strength of Si depended on structure, rate of deformation, and temperature of the sample. Amorphous silicon did not show any sign of crystallization upon indentation. Conversely, indentation of the crystalline silicon close to its melting point did show a tendency to transform to the amorphous phase near the indenter surface. No evidence of the transformation to the β-Sn structure under warm indentation was found.

Ionic solids also show interesting behavior upon indentation. The interaction of a CaF_2 tip with a CaF_2 substrate was examined using constant-temperature MD simulations by Landman et al. (1992). The substrate was composed of 242 static and 2904 dynamic ions. The stacking sequence was ABAABA ..., where A and B correspond to F^- and Ca^{+2} layers, respectively. The tip was a (111)-faceted microcrystal that contained nine (111) layers. Indentation was performed by repeatedly moving the tip 0.5 Å closer to the substrate and allowing the system to equilibrate. The energy was described as a sum of pairwise interactions between ions, where the potential between ions was composed of both Coulomb and repulsive contributions, as well as a van der Waals dispersion interaction that was parameterized by fitting to experimental data. The long-range Coulomb interactions were treated via the Ewald summation method and the temperature was maintained at 300 K.

The attractive force between the tip and the substrate increased gradually as the tip approached the substrate. At the critical distance of 2.3 Å, the attractive force increased dramatically; this was accompanied by increased interlayer spacing (i.e., elongation) in the tip. This process was similar to the JC phenomenon observed in metals; however, the amount of elongation (0.35 Å) is much smaller in this case. Decrementing the distance between the tip holder and the substrate further caused an increase in the attractive energy until an extremum value was reached. Continued indentation resulted in a repulsive tip–substrate interaction, compression of the tip, and ionic bonding between the tip and substrate. These bonds were responsible for the observed hysteresis in the force curve and upon retraction from the substrate ultimately led to plastic deformation of the tip and its fracture.

As noted earlier, the AFM can be used to measure nanomechanical properties (such as elastic modulus and hardness) with depth and force resolution superior to other methods (Burnham et al., 1990; Burnham and Colton, 1993). One way to do this is to relate the shape and the slope of the repulsive wall region of the force curve for an elastic indentation to the Young's modulus of the material that is indented. With this in mind, MD simulations have been used to simulate indentation at the atomic scale to elucidate the strengths, limitations, and interpretation of nanoindentation for characterizing materials and thin-film properties.

The simulated indentation of various hydrocarbon substrates using an sp^3-hybridized indenter was first performed by Harrison et al. (1992a). In a series of 300 K simulations, a hydrogen-terminated, sp^3-hybridized tip was used to indent the (111) surface of a (1 × 1) hydrogen-terminated diamond substrate, the (100) surface of a (2 × 1) hydrogen-terminated diamond substrate, and the basal plane of a graphite substrate. The indentation rate was approximately 0.3 km/s. This rate is orders of magnitude faster than experimental indentation rates, but much slower than the propagation of sound in diamond (12 to 18 km/s). The particle forces were derived from a reactive empirical bond-order potential (REBO) that is unique among hydrocarbon potentials in its ability to model chemical reactions (Brenner, 1990; Brenner et al., 1991).

Elastic indentations were performed on all three substrates and force curves were generated. The loading portion of the force curves for the elastic indentation of each of these substrates were all approximately linear, with slopes $S_{111} > S_{100} \gg S_{graphite}$. Experimental indentation of diamond (100) and graphite using an AFM also resulted in linear loading curves and the curve for diamond (100) had a larger slope than the curve for graphite. Because the indentation of diamond (111) involved both the compression of, and the changing of angles between, carbon–carbon bonds, while indentation of the (100) surface only involved the latter (Harrison et al., 1991), the (100) surface was softer. While these simulations provided the correct qualitative information regarding the relative hardness of the diamond and graphite substrates, the quantitative values of Young's moduli were not in good agreement with experimental values.

Classical elasticity theory predicts that the slope of the indentation curve is proportional to the modulus of the material (Sneddon, 1965). For cubic systems, the Young's moduli $E(111)$ for a (111) surface and $E(100)$ for a (100) surface are related to the elastic constants c_{ij} of the substrate via the following relationships:

$$E(111) = 6c_{44}(c_{11} + 2c_{12})/(c_{11} + 2c_{12} + 4c_{44}) \qquad (11.14)$$

$$E(100) = c_{11} + c_{12} - 2c_{12}^2/c_{11}. \qquad (11.15)$$

Substitution of the bulk elastic constants for diamond into Equations 11.14 and 11.15 yields values of 1267 GPa for $E(111)$ and 1172 GPa for $E(100)$. When the elastic constants calculated from Brenner's potential (Brenner, 1990; Brenner et al., 1991) were substituted into Equations 11.14 and 11.15 values of 1102 and 488 GPa were obtained for $E(111)$ and $E(100)$, respectively. Because the hydrocarbon potential was developed principally to model chemical vapor deposition of diamond films, little attention was paid to fitting of the elastic constants. As a result, the values of $E(111)$ and $E(100)$ calculated from the hydrocarbon potential differ by 13% and 58%, respectively, from the values calculated using the experimentally determined elastic constants. With this in mind, efforts focused upon extracting *qualitative* information from the indentation of diamond (111) surfaces and upon refitting the hydrocarbon potential energy function.

The plastic indentation of diamond (111) substrates, with and without hydrogen termination, using a hydrogen-terminated sp^3-bonded tip, was investigated by Harrison et al. (1992b). The depth at which the diamond (111) substrate incurred a plastic deformation due to indentation was determined by examination of the total potential energy of the tip–substrate system as a function of tip–substrate separation (Figure 11.7a to c). No hysteresis in the potential energy vs. distance curve was observed (Figure 11.7a) when the maximum normal force on the tip holder was 200 nN or less. Thus, the indentation was nonadhesive and elastic; therefore, the tip–substrate system did not sustain any permanent damage due to the indentation. This was also apparent from a comparison of initial and final tip–substrate geometries.

In contrast, increasing the maximum normal force on the tip holder to 250 nN prior to retraction caused plastic deformation of the tip and the substrate. This was apparent from the marked hysteresis

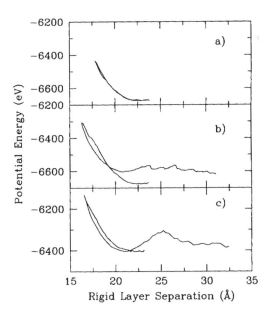

FIGURE 11.7 Potential energy as a function of rigid-layer separation generated from an MD simulation of an elastic (nonadhesive) indentation (a) and plastic (adhesive) indentation (b) of a hydrogen-terminated diamond (111) surface using a hydrogen-terminated, sp^3-hybridized tip. Panel (c) shows the results of the same tip indenting a diamond (111) surface after the hydrogen termination layer was removed. (Data from Harrison, J. A. et al. (1992), *Surf. Sci.* 271, 57–67.)

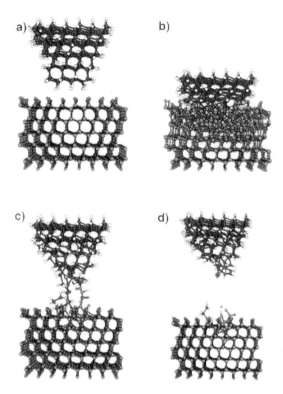

FIGURE 11.8 Illustrations of atoms in the MD simulation of the indentation of a hydrogen-terminated diamond (111) substrate with a hydrogen-terminated, sp^3-hybridized tip at selected time intervals. The tip–substrate system at the start of the simulation (a), at maximum indentation (b), as the tip was withdrawn from the sample (c), and at the end of the simulation (d). Large and small spheres represent carbon and hydrogen atoms, respectively. (Data from Harrison, J. A. et al. (1992b), *Surf. Sci.* 271, 57–67. With permission.)*

in the plot of potential energy vs. distance (Figure 11.7b) and from the initial and final tip–substrate geometries (Figures 11.8a to d). Examination of the atomic coordinates as a function of time allowed for specific, atomic-scale motions to be associated with certain features in the plot of potential energy as a function of distance. As the tip was withdrawn from the substrate, connective strings of atoms formed (Figure 11.8c). Increasing the distance between the tip and crystal caused these strings to break one by one. Each break was accompanied by a sudden drop in the potential energy at large positive values of tip–substrate separation (Figure 11.7b).

During indentation, the end of the tip twisted to minimize interatomic repulsions between hydrogen atoms chemisorbed to the tip and those chemisorbed to the substrate. This twisting allowed the tip atoms to form chemical bonds with the carbon atoms below the first layer of carbon atoms in the substrate. As a result, the indentation was disordered and ultimately led to the formation of connective strings of atoms between the substrate and tip as the tip was retracted.

When hydrogen was removed from the substrate surface and indentation to approximately the same value of maximum force was repeated, plastic deformation was also observed. However, the atomic-scale details, including the degree of damage, differed. The absence of hydrogen on the surface of the substrate minimized repulsive interactions during indentation and, therefore, allowed the tip to indent the substrate without twisting (Harrison, et al., 1992b). Because carbon–carbon bonds were formed between the tip and the first layer of substrate, the indentation was ordered (i.e., the surface was not disrupted as much by the tip) and the eventual fracture of the tip during retraction resulted in minimal damage to the

* Color reproduction follows page 16.

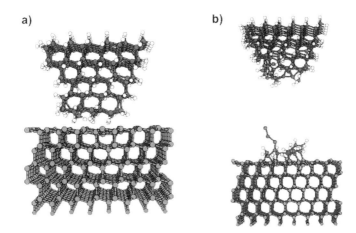

FIGURE 11.9 Illustrations of atoms in the MD simulation of the indentation of a non-hydrogen-terminated, diamond (111) substrate with a hydrogen-terminated, sp^3-hybridized tip. The tip–substrate system prior to indentation (a) and subsequent to indentation (b). The large, dark gray spheres represent carbon atoms and the small, white spheres represent hydrogen atoms. (Data from Harrison, J. A. et al. (1992b), *Surf. Sci.* 271, 57–67.)

substrate (Figures 11.9a and b). The concerted fracture of all bonds in the tip gave rise to the single maximum in the potential vs. distance curve at large distance (Figure 11.7c).

The hydrocarbon potential developed by Brenner (Brenner, 1990; Brenner et al., 1991) has also been used by Glosli et al. (1995) to examine the evolution of the microstructure of amorphous-carbon films during indentation. A blunt rigid tip was used to indent films, which were 4 nm thick, at 35 m/s. The tip interacted with the film via a truncated LJ potential. A thermostat was applied to the middle layers of the amorphous film to maintain the desired temperature.

The force curves in these simulations had steps that were indicative of plastic events, likely due to rapid rearrangement of the bonding network during the indentation. The reported hardness of the films calculated from these simulations was 75 ± 25 GPa.

In an effort to glean *quantitative* information from the indentation of hydrocarbons, the REBO potential was recently improved to reproduce the elastic constants of diamond and graphite accurately while maintaining all of its original properties (Brenner et al., unpublished). The elastic constants for diamond calculated using this improved potential are in good agreement with experimentally determined values (Sinnott et al., 1997). Substitution of the elastic constants calculated from the improved potential into Equations 11.14 and 11.15 yields values of 1367 and 1177 GPa for $E(111)$ and $E(100)$, respectively.

Recently, Sinnott et al. (1997) used an sp^3-hybridized carbon tip to indent a hydrogen-terminated (111) face of diamond and a thin film of amorphous carbon on (111) diamond. Comparison of pictures before and after indentation (Figure 11.10a to c) confirmed a lack of significant adhesion between the tip and substrate, so this indentation was classified as elastic. Thus, the simulated force curves could be used to calculate the reduced elastic modulus of the tip–sample system. For an elastic indentation, the loading portion of the force curve is related to the reduced modulus, although the exact relationship varies depending upon the geometry of the indenter. For example, classical elasticity theory predicts a linear relationship for flat-ended indenters and a nonlinear relationship for spherical or conical indenters (Sneddon, 1965).

For the force curve depicted in Figure 11.11, the force curve in the loading region is linear. The relationship of the slope to the reduced elastic modulus is given by

$$F = 2E_r ah, \qquad (11.16)$$

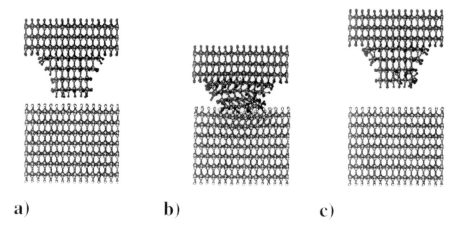

a) **b)** **c)**

FIGURE 11.10 Illustrations of atoms in the MD simulation of a hydrogen-terminated, diamond (111) surface indented by an sp^3-hybridized tip. The dark gray, light gray, white, and black spheres represent tip carbon atoms, surface carbon atoms, surface hydrogen atoms, and tip hydrogen atoms, respectively. Simulation times are 0.0 ps in (a), 6.0 ps in (b), and 15.0 ps in (c). (From Sinnott, S. B. et al. (1997), *J. Vac. Sci. Technol. A* 15, 936–940. With permission.)

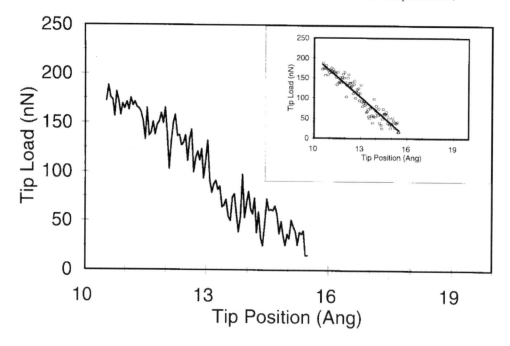

FIGURE 11.11 Force vs. position of the tip rigid layer for the indentation of a diamond (111) surface by the sp^3-hybridized tip using MD simulations. (Only the loading portion of the curve is shown.) The tip first contacts the surface when the tip position is approximately 14 Å. The inset shows the line of best fit to these data. (From Sinnott, S. B. et al. (1997), *J. Vac. Sci. Technol. A* 15, 936–940. With permission.)

where F is the force, E_r is the reduced elastic modulus, a is the radius of the contact area (approximately 6 Å), and h is the penetration depth. The reduced modulus is given by

$$1/E_r = \left(1 - v_1^2\right)/E_1 + \left(1 - v_2^2\right)/E_2, \qquad (11.17)$$

where v is Poisson's ratio and the subscripts 1 and 2 denote tip and substrate, respectively.

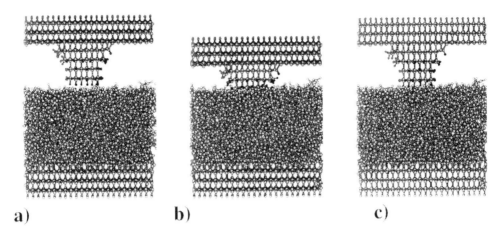

a)   b)   c)

FIGURE 11.12   Illustrations of atoms in the MD simulation of an amorphous carbon film indented by sp^3-hybridized tip. The dark gray, light gray, white, and black spheres represent tip carbon atoms, film carbon atoms, surface hydrogen atoms, and tip hydrogen atoms, respectively. Simulations times are 0.0 ps (a), 8.0 ps (b), and 14.9 ps (c). (From Sinnott, S. B. et al. (1997), *J. Vac. Sci. Technol. A* 15, 936–940. With permission.)

When the slope of 33.5 nN/Å, obtained from the data in Figure 11.11, was substituted into Equation 11.16 it yielded a reduced modulus of 279 GPa. Using the value of 1367 GPa for Young's modulus of the substrate (Equation 11.14) and Equation 11.17, the elastic modulus of the tip was determined to be 357 GPa. Thus, the sp^3-hybridized carbon tip, with no extensive stabilizing lattice, was found to be significantly softer than the diamond (111) substrate.

The indentation of diamond (111) covered by an amorphous-carbon film was performed with the same sp^3-hybridized carbon tip. Prior to indentation, the film was composed of approximately 21% sp^3-hybridized carbon and approximately 58% sp^2-hybridized carbon. Less than 2% of the carbon atoms had two nearest neighbors, and approximately 0.1% had five nearest neighbors. The remaining atoms were at the film edges (Figure 11.12a to c). For comparison purposes, the amorphous-carbon film was indented to approximately the same depth as the diamond (111) substrate discussed above. Analysis of the atomic positions as a function of time indicated that no significant depression was left in the film subsequent to indentation. In addition, the distribution of carbon-atom hybridizations was approximately the same as it was at the start of the indentation. Thus, this indentation was considered to be elastic.

The loading portion of the force curve was again linear and was related to the reduced modulus of the tip and the film (Figure 11.13). The slope of 18.4 nN/Å, when substituted into Equation 11.16, yielded a value of 153 GPa for E_r. By using the E_1 value of 357 GPa determined for the 696-atom tip, the elastic modulus of the amorphous carbon film was found to be 243 GPa, well within the range of experimentally determined modulus values (Davanloo et al., 1992; Robertson, 1992; Rossi et al., 1994). These simulations demonstrated that MD simulations, with an accurate potential, can be used to calculate quantitative information from indentations.

11.3.4 Cutting of Metals

The fabrication of high-tolerance metal parts involves the precision machining of metal surfaces. Single-point diamond tools are currently generating components with nanometer tolerances. However, the mechanisms by which tools wear, tools and substrates interact, and surfaces are cut are not currently known. With that in mind, Belak and co-workers (Belak and Stowers, 1990; Belak et al., 1993) have examined the orthogonal cutting of Cu(111) substrates using a rigid cutting tool.

In those simulations, a static diamond-like tip was placed into contact with the Cu surface. Cutting was performed by continuously moving the tool closer to the plane of the surface while the surface was moved in a direction perpendicular to the surface normal at 100 m/s. This process formed a Cu chip in

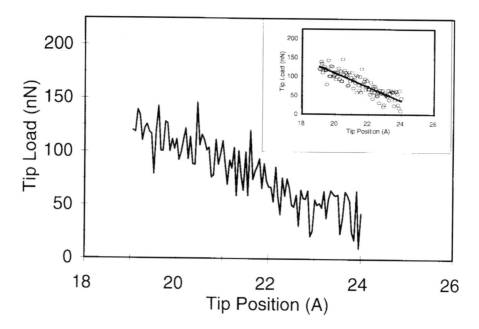

FIGURE 11.13 Force on the tip vs. position of the tip-rigid layer for the indentation of an amorphous-carbon film by the sp^3-hybridized tip using MD simulations. (Only the loading portion of the curve is shown.) The tip first contacts the film when the tip position is approximately 23 Å. The inset shows the line of best fit to these data. (From Sinnott, S. B. et al. (1997), *J. Vac. Sci. Technol. A* 15, 936–940. With permission.)

front of the tool (Figure 11.14). This chip was crystalline in nature and possessed a different orientation than the Cu substrate. Additionally, regions of disorder were formed in front of the tool tip and on the substrate surface in front of the chip. Finally, dislocations originating from the tool contact point were formed in the substrate.

The cutting force showed a strong dependence on tool edge radius, as it does experimentally. Tools with large radii require larger forces to achieve the same depth of cut as sharp tools. The specific energy (work per unit volume of material removed) determined from the simulation (and from micro-diamond-cutting experiments) has a power dependence on the depth of cut, with a power coefficient of −0.6. Macroscopic machining yields the same qualitative dependence of specific energy with depth of cut, but with a coefficient of −0.2. A change in slope in the simulations occurred at length scales of a few microns and larger. This transition has been interpreted as a change in the mechanism of plastic deformation from intergranular at macroscopic lengths (moving existing dislocations) to intragranular at the microscopic lengths (creating new dislocations). The simulations, among the largest yet performed, illustrate that the computer simulations are beginning to invite comparison with both nanoscale and microscale experiments.

11.3.5 Adhesion

Adhesion can be studied by bringing two materials into contact and then separating them or by separating two ends of a system already in contact. Pethica and Sutton (1988) performed one of the earliest theoretical, atomistic examinations of adhesion. Using both a continuum and an atomistic model (based on LJ pair potentials), they were able to demonstrate the JC phenomenon. The critical separation where the JC occurred was typically for separations less than 2 Å. When the surfaces were separated, they did not come apart as the critical jump separation was exceeded, thus, exhibiting hysteresis due to adhesion. The authors pointed out that one of the consequences of the existence of this region of inaccessible separation is that there will be significant differences between contact and noncontact AFM experiments.

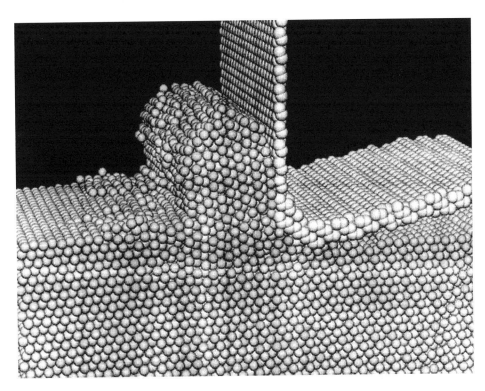

FIGURE 11.14 Illustrations of atoms from an MD simulation of the orthogonal cutting of a Cu(111) substrate with a rigid cutting tool. (From Belak, J. and Stowers, I. F. (1992), in *Fundamentals of Friction: Macroscopic and Microscopic Processes* (I. L. Singer and H. M. Pollock, eds.), 511–520, Kluwer, Dordrecht. With permission.)

Smith et al. (1989) also studied the JC phenomenon, using equivalent crystal theory (Smith and Banerjea, 1987) as the basis for their investigation. This method, based on a perturbation theory approach, had been demonstrated to give accurate surface energies and relaxed atomic positions for a number of transition metal surfaces. In a 1989 work of theirs, the JC or the "avalanche in adhesion" between two Ni(100) crystals was examined. The critical distance where the "avalanche" occurred was found to depend on the balance between the favorable energy of attraction between surface layers on opposing substrates and the unfavorable energy involved in pulling these surface layers away from their corresponding substrates. The avalanche process itself was rapid, on the order of 100 fs. The authors pointed out that an avalanche was not inevitable; its occurrence depends on film thickness, film stiffness, and the strength of the adhesive force. In some cases, the adhesive forces may be to weak relative to the restoring forces for these forces to ever be equal at some separation.

Subsequently, Lupkowski and Maguire, (1992) investigated the nature of the avalanche using MD simulations of two-dimensional, LJ solids. These simulations demonstrated that the structure of the surfaces after contact (after the avalanche) depends on temperature. At low temperatures, the avalanche was qualitatively similar to that predicted by energy minimization calculations (Pethica and Sutton, 1988). At higher temperatures, there were qualitative changes in the nature of the avalanche. As the approach was made in steps, individual atoms were stripped off, creating a bridge between the surfaces. The formation of the bridge took place prior to the contacting of the surfaces. Eventually, the two surfaces collapsed to form a strained solid with no defects. Approaches at even higher temperatures caused the bridging to become more pronounced and occurred at larger separations. Subsequent structures had a number of defects, including vacancies and dislocations in the interior and near what was previously the surface of the slab. Thus, the authors concluded that the structure of the surfaces at contact varied qualitatively with temperature.

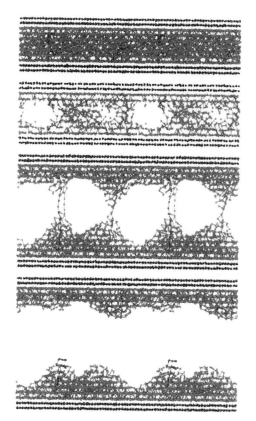

FIGURE 11.15 Snapshots of a glassy film during rupture at $v = 0.003\sigma/\tau$, where σ and τ are characteristic length and timescales of interaction. The 800 atoms in each wall are black, and monomers of the 128 chains are colored. Different colors were used to make the chains forming the bridges visible. The vertical direction corresponds to the z-direction and periodic boundary conditions were applied in the xy plane. Only a thin cross section of the film is shown in the second panel from the top. (From Baljon A. R. C. and Robbins, M. O. (1997), *Mater. Res. Soc. Bull.* 22, 22–26. With permission.)*

Other studies have concentrated not on the formation of adhesive bonds, but on the process by which these bonds are broken as two surfaces are separated. The failure of adhesive bonds is a complicated phenomenon that involves both interfacial attraction and kinetic effects. Energy dissipation mechanisms play a dominant role in the function of adhesives because the mechanical energy required to break a bond, G, can be 10^4 times greater than the reversible work of adhesion, W. Thus, the excess work, $G - W$, is dissipated during rupture of the adhesive.

Baljon and Robbins (1996, 1997) used MD simulations to follow the movement of individual atoms and energy dissipation during the rupture of a thin adhesive film. The model system contained two rigid solid walls joined by a thin adhesive film. In separate studies, films composed of linear-chain molecules of between 2 and 32 monomers (Figure 11.15) were investigated. The monomers interacted via a truncated LJ potential and adjacent monomers along each chain were also coupled through an attractive potential that prevented chain crossing and breaking (Kremer and Grest, 1990). The rigid solid walls consisted of two (111) planes of an fcc lattice. The wall atoms were coupled to lattice sites by stiff springs and their nearest-neighbor spacing was fixed at 80% of the equilibrium monomer spacing. The temperature was kept constant by coupling the wall atoms to a heat bath. For the results discussed here, the chains were composed of 16 monomers, the adhesive contained 2048 monomers, and each wall consisted of 800 atoms. Rupture was simulated by separating the walls with a uniform velocity. The particle motions, forces, potential and kinetic energies, and heat flow toward the bath were all monitored throughout the course of the simulation.'

The excess of work $G - 2\gamma$ (γ is the surface tension) as a function of rupture velocity and the shear response (friction) vs. shear velocity were examined at three different temperatures (Figure 11.16). The excess work and the shear response exhibited different behavior in each temperature regime. In the low-

---

* Color reproduction follows page 16.

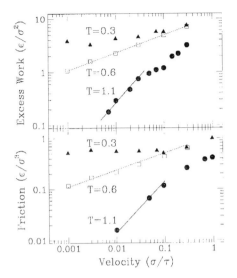

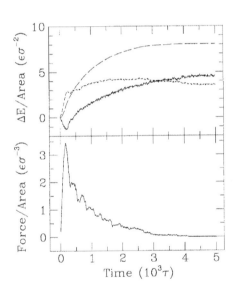

FIGURE 11.16 Excess work as a function of rupture velocity (upper panel) and mean frictional force on the tip wall as a function of shear velocity in the (100) direction at zero pressure (lower panel). Lines with slope ⅓ (dashed) and 1 (solid) indicate the power law scaling observed at T = 0.6 ε and 1.1 ε, respectively. (From Baljon, A. R. C. and Robbins, M. O. (1996), *Science* 271, 482–484. With permission.)

FIGURE 11.17 Time dependence of energy changes ΔE (upper panel) and the force on the walls (lower panel) during rupture of the glassy film shown in Figure 11.15. Lines in the upper panel indicate the external work (long dashes), the potential energy increase (short dashes), and the total heat flow (solid). Because the mean kinetic energy remains constant at fixed temperature, the work equals the sum of the potential energy change and heat flow. (From Baljon, A. R. C. and Robbins, M. O. (1996), *Science* 271, 482–484. With permission.)

temperature glassy state, the excess work and friction approached constant values at low velocities, v. These limits correspond to the amount of adhesion hysteresis and static friction, respectively. Just above the glass-transition temperature both excess work and friction increased as v^x, where $x = ⅓$. At lower velocities and higher temperatures, there was a newtonian regime in which both quantities increased linearly with velocity.

Above the glass-transition temperature, rupture produced smooth viscous flows in the film that resulted in a correspondence between the excess work and the shear stress. In addition, both flow velocities and dissipation approached zero at low velocities. In contrast, there was no correspondence between the excess work and the shear stress in glassy films. The shear was confined to the region between the film and the wall, and rupture occurred in the film through a sequence of rapid structural rearrangements.

As the glassy film was separated, the film deformed elastically, and the force needed to separate the walls and the distances between layers increased linearly with time (Figure 11.17). Eventually, a threshold layer separation was reached and the film became unstable against density fluctuations. Small cavities formed and grew, allowing the remainder of the film to relax to its original density. The interlayer separation where cavities formed was found to be independent of chain length and, thus, depended only on the force between individual monomers. Each time the internal stress in the film exceeded the local yield stress, further increases in the wall separation resulted in sudden structural rearrangements of the film (Figure 11.15, second panel from top). In the latter stages of rupture, the cavities coalesced (Figure 11.15, second panel from bottom). The lengths of the bridges connecting the walls grew to nearly that of a fully stretched chain before one end of the chain pulled free and collapsed onto the opposite surface (Figure 11.15, bottom panel). The final surfaces were very rough. Energy stored in the excess surface area was part of the unrecoverable work, which was gradually converted to heat as the surface annealed.

By examining hysteresis loops, the authors determined that the excess work was dissipated evenly among cavitation, plastic yield, and bridge rupture. All of these processes dissipated more energy with increased film thickness and chain length. The adhesive energy G for glassy films was approximately twice the reversible work.

In a similar type of study, Streitz and Mintmire (1996) studied the elastic and yield response of bulk and thin-film α-alumina. The technological importance of this material at metal–metal oxide interfaces was the driving force behind its selection. A thin film of α-alumina was constructed from a slab ten layers thick of α-alumina (about 25 Å thick) with two free (0001) surfaces. Periodic boundary conditions were applied in the plane of the free surfaces, but not normal to the surfaces. The forces between atoms were modeled using a variable-charge electrostatic model plus an EAM potential (ES+ method) (Streitz and Mintmire, 1994). The film was equilibrated; then a sequence of strains was applied by moving the outermost layers of the film a specified amount normal to the free surfaces. The remainder of the atoms were allowed to relax, and at each subsequent increment of total strain the internal atoms were allowed to reach equilibrium.

Streitz and Mintmire (1996) calculated a stress–strain curve for the application of compressive and tensile strains to both the bulk and thin-film systems. The slab responded elastically during initial loading and unloading, and the appropriate elastic constant could be calculated from these data. In addition, the variation of the elastic constant c_{33} was also examined as a function of strain. In general, c_{33} (the modulus) increased and the material became stiffer for compressive strains, while for tensile strains the modulus decreased and the material became softer. For the bulk system, c_{33} varied approximately linearly with strain, while marked deviations from linearity were apparent for the thin film. The value of c_{33} at zero strain was calculated to be 498 GPa for the thin film, in close agreement with the bulk calculation that yielded 509 GPa. Last, a theoretical yield stress of 44.5 GPa was calculated. This value is not far outside the theoretically determined range of 35 to 40 GPa. Thus, the authors concluded that the ES+ method is capable of describing the elastic response of α-alumina in situations that are far from thermodynamic equilibrium. In addition, these simulations showed that the elastic response of the α-alumina as a thin film might differ substantially from the elastic behavior of a bulk crystal. This conclusion is quite relevant in view of the prominent use of α-alumina as a coating.

11.4 Lubrication at the Nanometer Scale: Behavior of Thin Films

Experiments have shown that the properties of fluids confined between solid surfaces are drastically altered as the separation between the solid surfaces decreased (Horn and Israelachvili, 1981; Chan and Horn, 1985; Gee et al., 1990). For instance, at separations of a few molecular diameters, liquid viscosities increase by several orders of magnitude (Israelachvili et al., 1988; Van Alsten and Granick, 1988). Continuum hydrodynamic and elastohydrodynamic theories, which have been successful in describing lubrication by micron-thick films, begin to break down when the thickness of the liquid approaches the thickness of a few molecular diameters. Because an increasing number of applications involve lubricants in such confined geometries, the need to understand this sort of system through modeling has become increasingly important.

Molecular dynamics simulations have begun to fill the void created by the breakdown of continuum theories. These simulations have revealed a number of new phenomena, several of which have explained

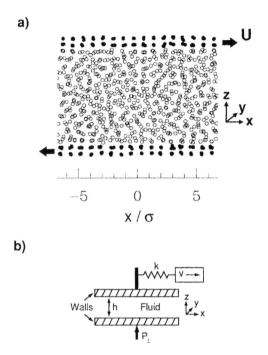

a)

-5   0   5

x / σ

b)

Walls h Fluid

$P_\perp$

FIGURE 11.18 Schematic representation of the simulation geometry used to model the confinement of liquids between parallel solid walls. Projections of liquid particle positions on the xz plane are represented as open circles and the wall molecules as filled circles in (a). (From Thompson P. A. and Robbins, M. O. (1990), *Phys. Rev. A* 41, 6830–6837. With permission.) There are 672 fluid atoms and 192 wall atoms. The walls are moved at a constant velocity U and in opposite directions along the x-axis. A sketch of a slightly different simulation geometry is shown in (b). The walls are held together by a constant load $P_\perp$. The upper wall is attached by a spring to a stage that is moved at constant velocity v. (From Thompson, P. A. and Robbins, M. O. (1990), *Science* 250, 792–794. With permission.) Periodic boundary conditions are imposed in the xy plane.

experimental observations pertaining to the behavior of confined films. The equilibrium properties of films of various types, such as spherical molecules, straight-chain alkanes, and branched alkanes confined between solid parallel walls have been examined. Spherical molecules, for example, have been shown to order both normal and parallel to the solid walls. Film properties, such as viscosity, have also been examined. Finally, while macroscopic experiments are consistent with one of the fundamental assumptions of newtonian flow, namely, the "no-slip" boundary condition (BC), recent microscopic experiments are not. (The "no-slip" BC requires that the tangential component of the fluid velocity be equal to that of the solid surface.) Therefore, flow BCs have been examined using MD simulations (Thompson and Robbins, 1990a,b; Thompson et al., 1992; Robbins et al., 1993). Some of the more recent work in these areas is discussed in the following sections.

11.4.1   Equilibrium Properties of Confined Thin Films

A number of groups have studied the equilibrium properties of spherical molecules (interacting through LJ potentials) confined between solid walls using both Monte Carlo methods (Schoen et al., 1987) and MD simulations (Bitsanis et al., 1987; Thompson and Robbins, 1990b; Sokol et al., 1992; Diestler et al., 1993). These studies have demonstrated that, irrespective of the atomic-scale roughness of the pore walls, when a fluid of spherical particles is placed inside a pore, the fluid layers are layered normal to the pore walls (Bitsanis et al., 1990).

    The typical signature used to identify the ordering of the liquid is the liquid density plotted as a function of distance from the pore walls (termed a density profile). For example, Thompson and Robbins (1990b) used MD simulations to examine the structure of LJ liquids confined between two solid walls that consisted of (001) planes of an fcc lattice. The simulation geometry (a slit pore consisting of two parallel solid surfaces) was chosen to closely resemble an SFA (Figure 11.18a). Oscillations in the calculated density profiles corresponded to well-defined, liquid layers (Figures 11.19a to c). (In Figure 11.19 the center of the pore corresponds to a z/σ value of zero and the distance z has been normalized by the characteristic LJ length parameter σ.) In the middle of the pore no oscillations in the density profile were present; thus, the liquid possessed the unstructured density appropriate for a bulk liquid in this region of the pore.

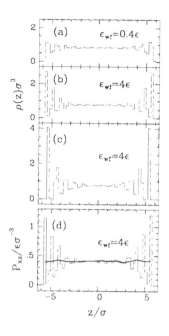

FIGURE 11.19 Liquid density $\rho(z)$ and the xz component of the microscopic pressure-stress tensor P_{xz} as a function of distance between the walls for a number of wall fluid interaction strengths $(\varepsilon_{wf}/\varepsilon)$. (The wall velocity $U = 0$.) The solid line in (d) represents P_{xz} averaged within the fluid layers. All quantities have been normalized using the appropriate variables so that they are dimensionless. (From Thompson, P. A. and Robbins, M. O. *Phys. Rev. A* 41, 6830–6837. With permission.)

Normal ordering of the fluid and even a phase transition to a solid structure where layers of the liquid become "locked" to the solid walls can be induced by increasing the strength of the interaction between the walls and the fluid (the ratio $\varepsilon_{wf}/\varepsilon$ in Figure 11.19) (Schoen et al., 1989; Thompson and Robbins, 1990b; Sokol et al., 1992). This phenomenon manifests itself as larger oscillations in the density profiles (Figure 11.19b and c). This effect was also observed when *n*-alkanes were trapped between structured walls (Ribarsky and Landman, 1992).

Schoen et al. (1987) were the first to observe that structure in the walls of the pore induces transverse order (parallel to the walls) in a confined atomic fluid. Using grand canonical MD studies of an atomic fluid confined between fcc (100) planes of like atoms, they demonstrated that for pore thicknesses of approximately 1 to 6 atomic diameters, the fluid alternatively freezes and thaws as a function of pore thickness. The solid formed epitaxially in distorted fcc (100) layers. This epitaxial effect decreased with increasing pore thickness but persisted indefinitely in the layer nearest to the pore wall. In a related work, a detailed analysis of the structure of the fluid within a layer, or epitaxial ordering, as a function of wall densities and wall–fluid interaction strengths was undertaken (Thompson and Robbins, 1990b). For small ratios of wall-to-liquid well depth $(\varepsilon_{wf}/\varepsilon = 0.4)$, fluid atoms were more likely to sit over gaps in the adjacent solid layer; however, self-diffusion within this layer was approximately the same as in the bulk liquid. In other words, although the solid induced order in the adjacent liquid layer, it was not sufficient to crystallize the liquid layer. Increasing the strength of the wall–fluid interactions by a factor of 4.5 resulted in epitaxial locking of the first liquid layer to the solid. This epitaxial ordering was confirmed from an analysis of the two-dimensional structure factors, the spatial probability distribution, mean-square displacement of the atoms within the layer, and the diffusion within the layer. While diffusion in the first layer was too small to measure, diffusion in the second layer was approximately half of its value in the bulk fluid. The second layer of liquid crystallized and became locked to the first "liquid" layer when the strength of the wall–liquid interaction was increased by approximately an order of magnitude over its original value. A third layer never crystallized.

The confinement of linear-chain molecules has also been examined by a number of groups (Ribarsky and Landman, 1992; Thompson et al., 1992; Wang et al., 1993a,b). For example, using a simulation geometry similar to that shown in Figure 11.18b, Thompson et al. (1992) examined the confinement of linear-chain molecules between two (111) fcc planes. The linear-chain molecules were modeled via the bead-spring model, which has been shown to yield realistic dynamics for polymer melts (Kremer and

Grest, 1990). Adjacent monomers were coupled via an attractive potential and non-nearest-neighbor monomers interacted via a repulsive, truncated LJ potential.

Confinement of the polymer between solid walls was shown to have a number of effects on the equilibrium properties of the static polymer films. The film thickness decreased as the normal pressure on the upper wall increased. At the same time, the degree of layering and in-plane ordering increased, and the diffusion constant parallel to the walls decreased. In contrast to films of spherical molecules, where there was a sudden drop in the diffusion constant associated with a phase transition to an fcc structure, films of chain molecules remained highly disordered and the diffusion constant dropped steadily as the pressure increased. This indicated the onset of a glassy phase at a pressure below the bulk transition pressure. This wall-induced glass phase has provided a natural explanation for the dramatic increases in measured relaxation times and viscosities of thin films (Gee et al., 1990; Van Alsten and Granick, 1988).

The confinement of n-octane between parallel, crystalline solid walls was examined by Wang et al. (1993a,b) using MD. A more realistic liquid potential energy function (Jorgensen et al., 1984) was used and rigid Langmuir–Blodgett (LB) monolayers were used to model the walls of the pore (Hautman and Klein, 1990). The pore was finite in one direction (typically 2.5 nm long) and made infinite in the other direction by the application of periodic boundary conditions. In this geometry, liquid exited the pore and collected as a droplet in the finite direction (Figure 11.20). Liquid vapor from these droplets interacted with vapors from the other side of the pore via the periodic boundaries (Figure 11.20a to f). The confined fluid was in equilibrium with the bulk-like droplet at 1 atm and pore widths ranged from 1.0 to 2.4 nm.

For the smallest pore size examined (1.0 nm) the film formed a layered structure with the molecules lying parallel to the pore walls (Figure 11.20a). At larger pore widths (Figure 11.20b to f), there was always a layered structure on each wall surface and more poorly defined layers in the center of the pore, with the exception of the 1.25-nm pore (Figure 11.20c). In that case, the film ordered so that the alkane molecules were oriented perpendicular to the walls. The oscillatory nature of the liquid density profiles (Figures 11.21a to h) confirmed the layered structure of the n-octane films; computed diffusion coefficients for the films, which were approximately equal to bulk values, confirmed the liquid nature of the films.

The layering of these films had profound effects on other equilibrium properties. For example, Wang et al. (1993a) showed that the solvation force of n-octane thin films increased dramatically as the pore size decreased. Surface force apparatus experiments have also shown that the nature of the film has an effect on the solvation force. It is well known that linear alkane molecules tend to layer close to a surface. This layering gives rise to oscillations in the density profile (Christenson et al., 1989). While early experiments indicated that the surface force oscillations vanish for branched alkanes such as 2-methyl-loctadecane (Israelachvili et al., 1989), more recent experiments (Granick et al., 1995) have shown oscillations in the force profiles of branched hydrocarbon molecules containing a single-pendant methyl group that are similar to those of linear hydrocarbons. Wang et al. (1993a,b, 1994) carried out MD studies on confined n-octane and 2-methylheptane and reached a similar conclusion.

In contrast, experimental studies that examined the confinement of highly branched hydrocarbons such as squalane showed that the surface force oscillations disappear (Granick et al., 1995). In an effort to shed light on this, Balasubramanian et al. (1996) used both Monte Carlo and MD to examine the adsorption of linear and branched alkanes on a flat Au(111) surface. In particular, they examined the adsorption of films of n-hexadecane, three hexadecane isomers (6-pentylundecane, 7,8-dimethyltetradecane, and 2,2,4,4,6,8,8-heptamethylnonane), and squalane. The alkane molecules were modeled using the united atom approach with an LJ potential used to model the interactions between united atoms. The alkane–surface interactions were modeled using an external 12-3 potential with the parameters appropriate for a flat Au(111) substrate (Hautman and Klein, 1989). The heptamethylnonane and squalane films were investigated using constant-NVT MD simulations. Other films were examined using configuration biased Monte Carlo (Siepmann and McDonald, 1993a,b; Siepmann and Frenkel, 1992).

The Monte Carlo calculations yielded density profiles for n-hexadecane and 6-pentylundecane that were nearly identical with experiment and previous simulations. In contrast, the density profiles of the

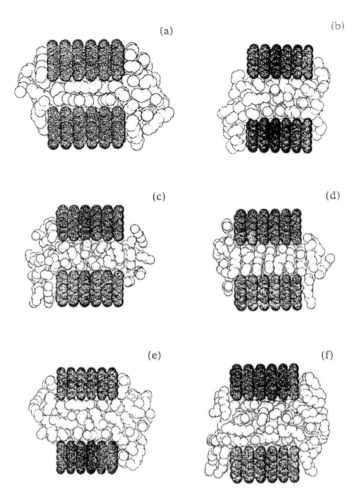

FIGURE 11.20 Atomic configurations adopted by *n*-octane when confined between parallel, rigid LB substrates of various pore widths. The pore widths and liquid temperatures are 1.0 nm and 297 K in (a), 1.6 nm and 297 K in (b), 1.25 nm and 297 K in (c), 1.25 nm and 250 K in (d), 1.8 nm and 297 K in (e), and 1.8 nm and 250 K in (f). The light spheres represent fluid (united) atoms and the dark spheres represent substrate atoms. (From Wang, Y. et al. (1993), *J. Phys. Chem.* 97, 9013–9021. With permission.)

more highly branched alkanes such as heptamethylnonane and 7,8-dimethyltetradecane exhibited an additional peak. These peaks arose from methyl branches that could not be accommodated in the first liquid layer next to the Au surface. That is, the branched hydrocarbons adsorbed in layered structures with interdigitation of the molecules. For thicker films, the oscillations in the density profiles for heptamethylnonane were out of phase with those for *n*-hexadecane, in agreement with the experimental observations (Granick et al., 1993).

Granick et al. (1993) did not observe force oscillations for squalane films confined between mica or organic monolayers of octadecyltriethoxysilane when the surfaces of the SFA were separated by more than 18 Å. In contrast, the MD simulations (Balasubramanian et al., 1996) yielded a density profile for the squalane film that was very similar to the density profile for 7,8-dimethyltetradecane. The most likely reason for this discrepancy between theory and experiment was the fact that the film was adsorbed, rather than confined, in the MD simulations.

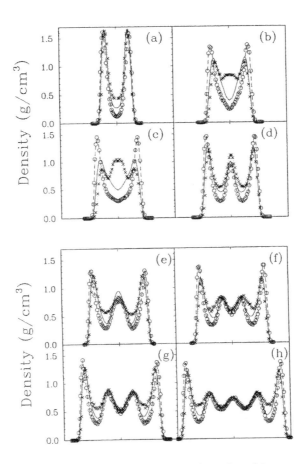

FIGURE 11.21 Liquid density profiles in the direction normal to the surface of the pore walls as a function of pore separation. The center of the pore is indicated by the single tick mark on the x-axis. Density profiles for all hydrocarbon groups within the liquid molecules are represented with a solid line, for methyl groups 1 and 8 with open circles, and for methylene groups 4 and 5 by crosses. The pore widths are 1.0 nm in (a), 1.2 nm in (b), 1.25 nm in (c), 1.4 nm in (d), 1.6 nm in (e), 1.8 nm in (f), 2.0 nm in (g), and 2.4 nm in (h). All liquid films are at 297 K. (From Wang, Y. et al. (1993), *J. Phys. Chem.* 97, 9013–9021. With permission.)

11.4.2 Behavior of Thin Films under Shear

The behavior of confined spherical and chain molecular films under shear has been examined using both Monte Carlo methods (Schoen et al., 1987; Curry et al., 1994; Curry and Cushman, 1995a,b) and MD simulations (see, for example, Schoen et al., 1989; Bitsanis et al., 1990; Bitsanis and Pan, 1993; Sokol et al., 1992; Diestler et al., 1993; Thompson and Robbins, 1990b; Thompson et al., 1992; Curry et al., 1994; Curry and Cushman, 1995a,b).

For instance, in an early work Bitsanis et al. (1990) examined the effect of shear on spherical, symmetric molecules, confined between planar, parallel walls that lacked atomic-scale roughness. Both Couette (simple shear) and Poiseuille (pressure-driven) flows were examined. The presence of flow had no effect on the density profile of the liquids as they were identical with the equilibrium density profiles for both types of flow and all pore widths. Velocity profiles, defined as the velocity of the liquid parallel to the wall as a function of distance from the center of the pore, should be linear and parabolic for Couette and Poiseulle flow, respectively, for a homogeneous liquid. In the simulations of Bitsanis et al. (1990) the velocity profiles for both types of flow deviate from the shape expected for homogeneous flow in the

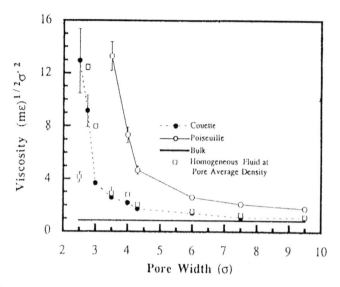

FIGURE 11.22 Effective viscosity for Couette and Poiseuille flows of an LJ liquid as a function of pore width. Pore width is given in units of the LJ parameter σ (parameters used in this simulation were appropriate for argon). The pore walls lack atomic-scale roughness. The viscosity of a homogeneous fluid at the pore average density and that of a bulk liquid are also shown. (From Bitsanis I. et al. (1990), *J. Chem. Phys.* 93, 3427–3431. With permission.)

regions occupied by the two outermost fluid monolayers. These outermost layers behaved like a fluid of very high viscosity.

The profoundly different nature of flow in molecularly thin films was further demonstrated by plotting the effective viscosity vs. pore width (Figure 11.22). For a bulk material the viscosity was independent of pore size. However, under both types of flow, the viscosity increased slightly as the pore size decreased. For ultrathin films, the viscosity increased dramatically.

The effect of transverse order on confined thin films under shear, induced by atomic structure in the walls, was also examined. Schoen et al. (1989) examined the behavior of argon LJ films under shear, confined between fcc (100) walls, using both grand canonical Monte Carlo and microcanonical MD simulations. At small wall separations of 2.0 σ, the pore accommodated two commensurate solid-like layers when the walls were fully out of registry with one another. Shearing caused the stress to increase linearly as the walls came closer into registry with one another. The number of atoms in the pore remained essentially constant until a critical stress was reached, at which point continued shearing caused almost an entire layer of fluid to exit from the pore, thereby decreasing the stress. These results are in agreement with experimental data, which indicates that solid surfaces slide past one another while separated by a discrete number of layers. Further, a critical shear stress was required to initiate sliding under appropriate conditions.

In contrast, with wall separations appropriate for an odd number of fluid layers (3.10 σ and 4.90 σ) the pore supported either three or five layers when the opposing surfaces were in registry. For the system with three solid-like layers, the shear stress increased linearly when the system was sheared. As in the previous case, this trend continued until a critical stress was achieved. Continued strain on the walls resulted in the drainage of one of the solid layers.

The flow of LJ liquids confined between two solid walls was also examined by Thompson and Robbins (1990b). In this case, the walls were composed of (001) planes of an fcc lattice. A number of wall and fluid properties, such as wall–fluid interaction strength, fluid density, and temperature, were examined. The geometry of the simulations closely resembled the configuration of an SFA (Figure 11.18b). Each wall atom was attached to a lattice site with a spring to maintain a well-defined solid structure with a minimum number of solid atoms. The spring constant controls the thermal roughness of the wall and its responsiveness to the fluid, and it was adjusted so that the mean-square displacement about the lattice

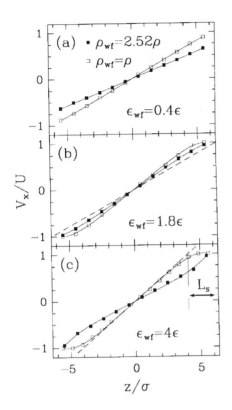

FIGURE 11.23 The component of the liquid velocity in the shearing direction V_x vs. the position between the pore walls z. The locations of the first layers of solid atoms on each side of the fluid correspond to the vertical borders. Data points indicate averages of V_x within the layers and the solid lines represent polynomial fits to these data. The dashed line in (b) represents the flow expected from hydrodynamics with a no-slip BC. The liquid velocity has been normalized by the velocity of the walls and the pore width by the LJ length parameter σ. (From Thompson, P. A. and Robbins, M. O. (1990), *Phys. Rev. A 41*, 6830–6837. With permission.)

sites was less than the Lindemann criterion for melting. The interactions between fluid atoms and between wall and fluid atoms were modeled by different LJ potentials. Couette flow was simulated by moving the walls at a constant velocity in opposite directions along the x-axis (Figure 11.18). The heat generated by the shearing of the liquid was dissipated using a Langevin thermostat. In most of the simulations performed by Thompson and Robbins (1990b), the fluid density and temperature were indicative of a compressed liquid about 30% above its melting temperature, with 672 fluid atoms, and either 192 or 352 wall atoms.

A number of interesting phenomena were observed in these simulations. First, the well-defined lattice structure of the solid walls induced both normal and parallel ordering in the adjacent liquid. The normal ordering of the films was apparent from the oscillations in the density profiles (Figure 11.19a to c). Density oscillations in the liquid also induced oscillations in other microscopic quantities normal to the walls such as the fluid velocity in the flow direction V_x and the microscopic pressure–stress tensor P_{xz} (Figure 11.19d). The position of the peaks in P_{xz} corresponded to the peak positions in the liquid density, while peak positions in dV_x/dz are between the layers. These observations were contrary to the predictions of the Navier–Stokes equations. However, it was determined that averaging dV_x/dz and P_{xz} over length scales that are larger than the molecular length σ produced smoothed quantities that satisfied the Navier–Stokes equations (smooth line in Figure 11.19d).

While the ordering of the fluid normal to the walls affected fluid flow, two-dimensional ordering of the liquid parallel to the walls affected the flow more significantly. The velocity of the fluid parallel to the wall was examined as a function of distance from the wall for a number of wall–fluid interaction strengths and wall densities (Figure 11.23). Analysis of velocity profiles showed that flow near solid boundaries is strongly dependent on the strength of the wall–fluid interaction and on wall density. For instance, when the wall and fluid densities were equal and wall–fluid interactions strengths were small ($\varepsilon_{wf}/\varepsilon = 0.4$), the velocity profile was linear with a no-slip BC (Figure 11.23a). As the wall–fluid interaction strength increased, the magnitude of the liquid velocity in the layers nearest the wall increased; thus, the

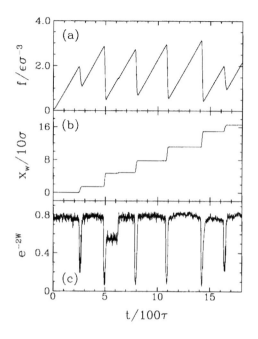

FIGURE 11.24 Force per unit area f, wall displacement X_w, and the Debye Waller factor e^{-2W} ($e^{-2W} = 0.6$ for a solid) as a function of time t during stick-slip motion for a wall velocity of 0.1. The simulation contained 288 wall atoms and 144 fluid atoms. The walls were fcc solids with (111) surfaces, and the shear direction was (100). All quantities have been normalized using the appropriate variables so that they are dimensionless. (From Thompson, P. A. and Robbins, M. O. (1990). *Science* 250, 792–794. With permission.)

velocity profiles became curved (Figure 11.23b). Increasing the wall–fluid interaction strength further (Figure 11.23c) caused the first two liquid layers to become locked to the solid wall.

For unequal wall and fluid densities, the flow BC changed dramatically. At the smallest wall–fluid interaction strengths examined, the velocity profile was linear; however, the no-slip BC was not present. The magnitude of the liquid velocity at the wall was less than the wall velocity; therefore, slip occurred (i.e., $|V_x|U < 1$) (Figure 11.23a). The magnitude of this slip decreased as the strength of the wall–fluid interaction increased. For an intermediate value of wall–fluid interaction strength ($\varepsilon_{wf}/\varepsilon = 1.8$), the first fluid layer was partially locked to the solid wall. Sufficiently large values of wall-fluid interaction strength ($\varepsilon_{wf}/\varepsilon = 15$) led to the locking of the second fluid layer to the wall.

Using a simulation geometry similar to Figure 11.18b, Thompson and Robbins (1990a) examined the origin of stick-slip motion when an LJ liquid is sheared between two solid walls using MD. Instead of two solids in contact moving smoothly, the solids may alternatively stick and then slip past one another, this is known as stick-slip motion. Simulation details are described above (Thompson and Robbins, 1990b), except that the walls in this case were composed of fcc solids with (111) surfaces and the density of the wall atoms was usually equal to the density of the fluid atoms. The wall separation usually corresponded to a few layers (i.e., 2σ). To mimic boundary-layer lubrication experiments, the simulations were performed at constant load by varying the distance between the walls.

The upper wall was coupled through a spring to a stage that advanced at constant velocity. Initially, the force on the spring was zero when the upper wall was at rest. As the stage moved forward, the spring stretched and the forced increased. When the film was in a liquid state, the upper wall accelerated until the force applied by the spring balanced the viscous dissipation. When the film was crystalline in nature, stick-slip behavior was observed (Figure 11.24). Initially, the film responded elastically, the wall remained stationary, and the forced increased linearly (Figure 11.24a). When the force exerted by the spring exceeded the yield stress of the film, the film melted and the tip wall began to slide. The wall accelerated to catch up with the stage, decreasing the force and creating the sawtooth pattern indicative of stick-slip motion. The stick-slip behavior was dependent on the velocity of the stage, with the degree of stick-slip increasing as the velocity decreased. Analysis of the two dimensional structure factor during the course of the simulation confirmed the existence of the solid–liquid phase transition as the film proceeded from static to sliding states (Figure 11.24c).

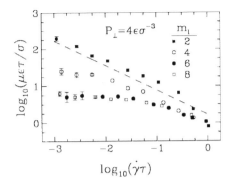

FIGURE 11.25 Dependence of time-averaged liquid viscosity μ ($\mu = fh/v$, where f is the mean frictional force per unit area, h is the wall separation, and v is the sliding velocity) on the shear rate γ ($\gamma = v/h$). The variable $P_\perp$ represents the pressure on the walls and m_l corresponds to the number of fluid layers. The dashed line has a slope of $-\frac{2}{3}$. (From Thompson, P. A. et al. (1992), *Phys. Rev. Lett.* 68, 3448–3451. With permission.)

While simulations with spherical molecules were successful in explaining some experimental phenomena, they were unable to reproduce other features of the experimental data. For instance, the calculated relaxation times and viscosities remained near bulk fluid values until the films crystallized. Experimentally, these quantities typically increase many orders of magnitude before a yield stress is observed (Israelachvili et al., 1988; Van Alsten and Granick, 1988). With this in mind, Thompson et al. (1992) repeated their earlier shearing simulations (Thompson and Robbins, 1990b) using freely jointed, linear-chain molecules instead of spherical molecules.

The behavior of the viscosity of the films as a function of shear rate was examined for films of different thicknesses. The relationship between film viscosity and shear rate was examined (Figure 11.25). The response of films that were six to eight molecular diameters m_l thick was approximately the same as it is for bulk systems. When the thickness of the film was reduced, the viscosity of the film increased dramatically, particularly at low shear rates, consistent with the aforementioned experimental observations.

The structure of liquids confined to smooth pores was compared with the structure of liquids in irregularly shaped pores by Curry et al. (1994) using grand canonical Monte Carlo and microcanonical MD methods. Both the confining pore walls were comprised of rigid atoms fixed in a (100) fcc arrangement. One wall was atomically flat and the other was scored with regularly spaced rectilinear grooves (i.e., corrugated) (Figure 11.26). The structure of the confined film as a function of groove dimension and frequency was examined. In each case, the behavior of the film was studied as the registry α was systematically varied at fixed temperature, chemical potential, and pore width.

When the groove was five lattice constants wide and the walls of the corrugated pore were in registry ($\alpha = 0.0$) the one-layer liquid in the narrow region of the pore was in equilibrium with the two-layer liquid in the wide region of the pore. Snapshots of the MD simulation taken together with the broad peaks in the density profile confirmed this conclusion (Figure 11.27). Changing the registry of the pore walls significantly affected the structure of the liquid. It was clear from examination of the density profiles that increasing the registry to $\alpha = 0.325$ caused the liquid in the narrow region of the pore to begin to separate into two layers, while the liquid in the wide region of the pore remained unaffected. At $\alpha = 0.4$, the liquid in the narrow portion of the pore had solidified into two distinct layers while there were still two liquid layers in the wide region of the pore. Thus, the confined film consisted of parallel, rectilinear liquid-filled nanocapillaries separated by solid strips. Increasing the registry to $\alpha = 0.5$ (exactly out of registry) caused the entire film to solidify and resulted in sharp peaks in the density profile.

The structural character within the film was ascertained from examination of the in-plane pair-correlation function $g^{(2)}$ (Figure 11.27). For registries corresponding to $\alpha = 0.0$ and $\alpha = 0.325$, the configurations were disordered or liquid-like as evidenced by the lack of structure in $g^{(2)}$. The in-plane pair correlation functions calculated when $\alpha = 0.4$ clearly showed a different degree of transverse order in the narrow portion of the pore than in the wide region of the pore. This supported the conclusion that nanocapillaries had formed. When the registry was increased to $\alpha = 0.5$ the in-plane pair correlation functions for both regions of the pore were identical and very structured, indicating total solidification of the film.

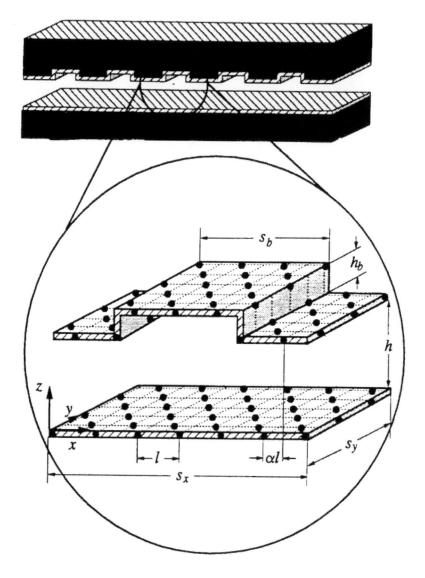

FIGURE 11.26 Schematic representation of a corrugated slit pore. The walls consist of atoms rigidly fixed in the (100) plane of the fcc lattice. The lattice constant corresponds to l ($l = 1.5985\sigma$), the cell dimensions are defined by s_y and s_x, the separation of the walls is h in the narrow regions and $h + h_b$ in the wide regions, and α determines the alignment of the walls. When $\alpha = 0$ the walls are in registry. The origin of the coordinate system lies at the intersection of the vectors labeled x, y, and z. (From Curry, J. B. et al. (1994). *J. Chem. Phys.* 101, 10824–10832. With permission.)

Plots of the stress in the film, T_{zx}^*, as a function of α also yielded insights into the sequence of phase transitions experienced by the film (Figure 11.28). When $\alpha = 0.5$ the corrugated-pore film consisted of two solid layers in the narrow region and three in the wide region. In comparison, the pore with no groove contained two solid layers. Because the surfaces are symmetrically arranged when $\alpha = 0.5$, the shear stress was equal to zero. Pushing the walls into registry (i.e., decreasing α) caused the shear stress to increase rapidly. Because the corrugated-pore film was, on average, thicker than the film in the atomically flat pore, the shear stress was less for the corrugated-pore film (Schoen et al., 1994). The shear stress in the ungrooved pore rose rapidly with increasing α and then dropped precipitously when α fell below 0.33. This sharp drop was due to abrupt melting and the concomitant loss of nearly an entire layer of the film. The result was a one-layer liquid film that continued to support substantial shear stresses

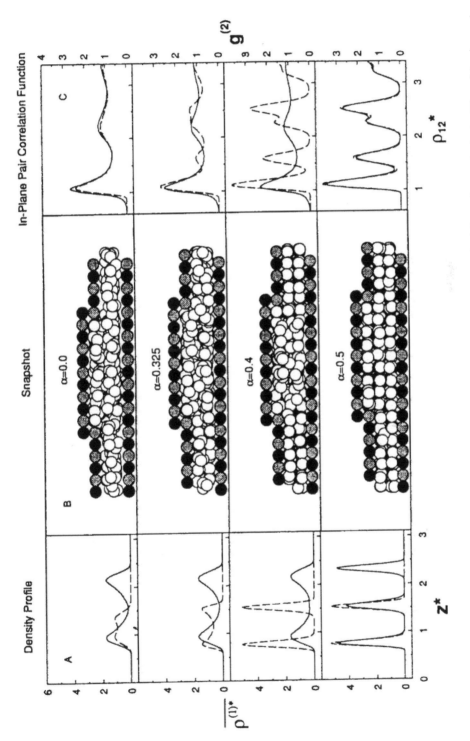

FIGURE 11.27 Local density profiles $\rho^{(1)*}(z)$ in (A), snapshots looking down the y-axis of the pore fluid for various registries in (B), and in-plane pair correlation functions $g^{(2)}$ in (C). The dotted and solid lines correspond to the functions calculated separately in the narrow and wide regions of the pore. In the snapshots black and gray shaded spheres correspond to alternating rows of the fcc lattice and unshaded spheres correspond to film atoms. Starred variables are reported in reduced units, that is, $z^* = z/\sigma$ and $\rho^* = \rho \times \sigma^3$. (From Curry, J. B. et al. (1994), *J. Chem. Phys.* 101, 10824–10832. With permission.)

565

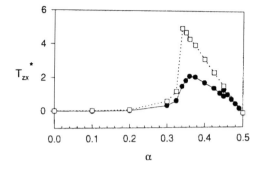

FIGURE 11.28 Shear stress T_{zx}^* in the x-direction vs. registry for the ungrooved slit pore (open squares) with $h^* = 2.3$ and for the corrugated pore shown in Figure 11.27 (filled circles) with $h^* = 2.3$ in the narrow region. Starred variables are reported in reduced units, that is, $T_{zx}^* = T_{zx} \times \sigma^3/\varepsilon$. (From Curry, J. B. et al. (1994), *J. Chem. Phys.* 101, 10824–10832. With permission.)

down to $\alpha = 0.30$. In contrast, the corrugated-pore film did not all melt at once as α was varied. First, when $\alpha = 0.45$ the solid in the wide region of the pore melted and almost an entire layer of liquid exited the wide region of the pore. This was accompanied by a drop in the shear stress of the film (Figure 11.28). The shear stress decreased approximately linearly with decreasing α until approximately equal to 0.35. Beyond this point the solid in the narrow region of the pore melted and one layer of liquid exited the pore; consequently, the shear stress decreased significantly. Because the film in the narrow region of the pore melted gradually, the shear stress in the region of $\alpha = 0.35$ was rounded (Figure 11.28). The melting of the film in the narrow region of the pore began at the interface between the narrow and wide regions and progressed toward the middle of the narrow region with increasing α.

Finally, the range of registry values where the solid (narrow region of pore) and liquid (wide region of pore) coexist was shown to increase with increasing groove width. In the limit of very large groove width, the fluid in the wide region of the pore acted like an ungrooved or smooth pore.

Curry and Cushman (1995a,b) also used the grand canonical Monte Carlo method to study binary mixtures of LJ atoms confined to a corrugated pore that were in thermodynamic equilibrium with their bulk-phase counterpart. The geometry of the corrugated pore was the same as it was in a previous work (Curry et al., 1994) (Figure 11.26). The fluid and wall atoms were spherical, nonpolar, LJ atoms characterized by diameters σ_i and interaction energies ε_i, where i represents the type of atom. The cutoff distance for the shifted force LJ potential used in this work depended on σ_{ij}, where $\sigma_{ij} = \frac{1}{2}(\sigma_i + \sigma_j)$. Thus, the cutoff distance varied depending on the identity of the atoms. The values of σ for the two different-sized atoms (large and small) were chosen so that the mixture resembled a mixture of octamethyltetracyclosiloxane and cyclohexane because this mixture has been used in SFA experiments. Two different pores were examined: one with the walls composed of large atoms and one with the walls composed of small atoms. For each pore, two binary mixtures were examined, with either 86% of 15% large atoms. Thus, four combinations of wall size and fluid atom mixture were studied.

When the wall atoms were large and the bulk fluid was composed of 86% large atoms, the small atoms were nearly eliminated from both regions of the pore (Figure 11.29) when the walls were in registry ($\alpha = 0.0$). The pore width was such that one solid-like fcc layer of large atoms fit into the narrow region of the corrugated pore while the wide region accommodated two solid-like fcc layers of large atoms. Altering the alignment of the pore walls gradually liquefied the film in both regions of the pore, as evidenced by the broadening of the local density profiles and the reduced structure in the in-plane pair-correlation functions. In addition, as the film melted a significant number of small atoms were able to enter both the narrow and the wide regions of the pore. This effect was most marked when the walls were totally out of alignment ($\alpha = 0.5$). The reason for this is that the alignment of the wall atoms effectively decreased the pore width. This had the effect of excluding large atoms from discrete regions of the pore.

Decreasing the percentage of large atoms (15%) in the bulk dramatically affected the behavior of the mixture in the pore. For example, when $\alpha = 0.0$, small atoms were no longer excluded from the pore. Indeed, in the wide region of the pore a two-layer, liquid-like film composed of small and large atoms was formed. The liquid-like nature of the film was confirmed using both the density profiles (Figure 11.30) and the in-plane pair-correlation function. This film was only slightly affected by changing the registry

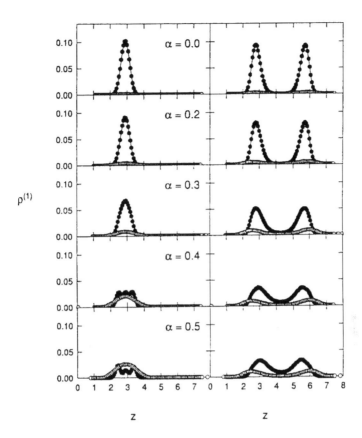

FIGURE 11.29 Local density profiles $\rho^{(1)}$ for large atoms (filled circles) and small atoms (open circles) calculated separately for the narrow (left) and wide (right) regions of the pore at various registries α. (The variable z is defined in Figure 11.27.) The bulk fluid contained 86% large atoms and the period of the corrugation s_x (Figure 11.26) was $10l$, where l is the lattice constant. The value of l used in these simulations was $1.5985\sigma_w$, where σ_w corresponds to the diameter of the wall atoms and was equal to the diameter of the large fluid species. The variables s_y and s_b (Figure 11.27) were both set to $5l$. (From Curry J. E. and Cushman, J. H. (1995a), *J. Chem. Phys.* 103, 2132–2189.)

of the walls. In the narrow region of the pore, the large atoms formed a one-layer solid and the number of large atoms participating in the structure was small compared with the number of small atoms. Decreasing the registry of the surfaces (increasing α) caused the large atoms to be driven from the narrow region of the pore and resulted in the formation of a two-layer small-atom solid. Examination of the in-plane correlation function confirmed the existence of the two-layer solid at $\alpha = 0.5$.

When the wall atoms were small and the mixture was composed of 86% large atoms, the narrow and the wide regions of the pore contained a one-layer and two-layer liquid, respectively (Figure 11.31). In the narrow region of the pore the one-layer liquid was composed mostly of large atoms, and changing the registry of the pore walls had very little effect. In contrast, the two-layer liquid in the wide region of the pore was composed of both large and small atoms. Altering the registry of the pore walls drove the small atoms from the wide region of the pore and caused the film to solidify. The two-layer solid in the wide region was in equilibrium with the one-layer liquid in the narrow region. This freezing in the wide region while the narrow region remained liquid was not observed in the other cases considered. Indeed, if the component that froze was the same size as the wall atoms, the atoms in the wide region of the pore froze only after the atoms in the narrow region were frozen (i.e., after nanocapillaries were formed).

Last, decreasing the percentage of large atoms in the mixture while keeping the size of the wall atoms small resulted in liquid-like structures, composed of large and small atoms, in both the wide and narrow regions of the pore. Altering the registry of the walls had little effect on the liquid in the wide region of

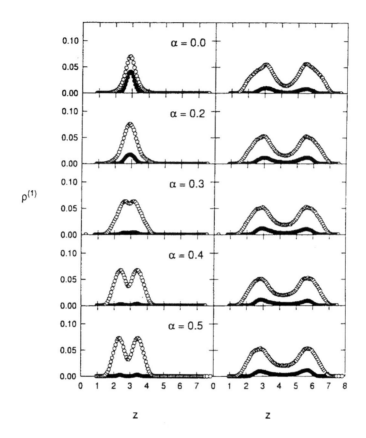

FIGURE 11.30 Local density profiles $\rho^{(1)}$ for large atoms (filled circles) and small atoms (open circles) calculated separately for the narrow (left) and wide (right) regions of the pore at various registries α. The bulk fluid contained 15% large atoms and all other pore parameters were the same as in Figure 11.29. (From Curry, J. E. and Cushman, J. H. (1995a), *J. Chem. Phys.* 103, 2132–2139.)

the pore. In contrast, the small atoms, which possessed a one-layer liquidlike structure in the narrow region of the pore, developed a two-layer structure as the registry was changed.

The corrugated-pore studies of Curry and co-workers (Curry et al., 1994; Curry and Cushman, 1995a,b) provided unique insight into the behavior of mixed LJ liquids confined between *nonuniform* surfaces. The behavior of liquids between nonuniform surfaces under shear was also investigated by Gao et al. (1995) using MD simulations. In that work, Gao et al. (1995) confined hexadecane (n–$C_{16}H_{34}$) between two Au substrates with exposed (111) faces. Topographical nonuniformities (or asperities) were modeled by flat-topped pyramidal Au structures of height h_a from the underlying Au surface. The asperities had a finite length in the sliding direction and an infinite length in the direction perpendicular to sliding. In this respect, they were very similar to the corrugated slit pore geometry of Curry et al. (1994). The interactions between Au atoms were modeled by EAM (Foiles et al., 1986), the alkane molecules were modeled by the united-atom model, and the interactions between the molecular segments and gold atoms were modeled by an LJ potential.

Sliding was performed by moving the opposing Au surfaces in opposite directions so that the relative sliding speed was 10 m/s. the simulations typically lasted 1.2 ns and were carried out at 350 K. Three simulation geometries were examined. The first consisted of 422 hexadecane molecules with the separation between asperity tops held at 17.5 Å. The second, "near-overlap," case consisted of 243 hexadecane molecules with the relative separation between asperity tops set to 4.6 Å. The third, "overlap," case consisted of 201 fluid molecules with a relative separation between asperity tops of –6.7 Å. (It should be noted that the height of the asperities was 9.3 Å in geometries one and two. In the third geometry the

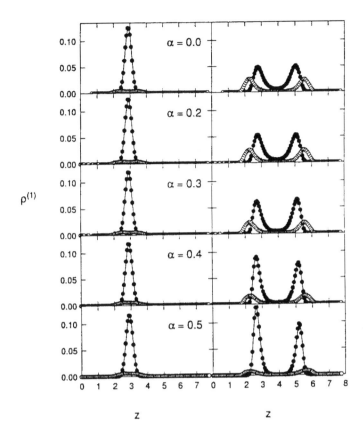

FIGURE 11.31 Local density profiles $\rho^{(1)}$ for large atoms (filled circles) and small atoms (open circles) calculated separately for the narrow (left) and wide (right) regions of the pore at various registries α. The bulk fluid contained 86% large atoms and the period of the corrugation s_x (Figure 11.26) was $14l$, where l is the lattice constant. The value of l equaled $1.5985\sigma_w$, where σ_w corresponds to the diameter of the wall atoms and was equal to the diameter of the small fluid species. The parameters s_y and s_b (Figure 11.26) were both set to $7l$. (From Curry, J. E. and Cushman, J. H. (1995a), *J. Chem. Phys.* 103, 2132–2139.)

asperities were larger at 14 Å.) Prior to the start of the sliding simulations, each system was equilibrated with the asperities well separated in the horizontal direction such that steady-state flow was established.

From an examination of the atomic and molecular configurations recorded as a function of time during the shearing process, Gao et al. (1995) were able to elucidate two main patterns for all three simulation geometries. First, there was an evolution of layered liquid structures in the region between the colliding asperities. In particular, for large transverse separations of the asperities, layering of the liquid in the interasperity region was initially absent and developed dynamically as the asperities approached each other. Furthermore, the number of layers in the interasperity region evolved in each case in a quantized manner as successive layers were squeezed out as the asperities drew near. Finally, severe deformation of the asperities resulted for the near-overlapping and overlapping asperity height cases.

This interasperity layering phenomenon was also apparent from examination of the time variation of the shear (f_x) and normal (f_z) forces. For example, for the near-overlap case, plots of both f_x and f_z vs. time had oscillatory patterns prior to the collision between asperities. This oscillatory behavior of the forces was reminiscent of solvation-force oscillations observed between smooth solid surfaces under equilibrium conditions.

Gao et al. (1995) also commented on the crucial importance of the dynamic mechanical response of the substrate (and the asperities) in determining the evolution and properties of the sheared systems.

When the asperities were separated vertically by large distances (the first simulation geometry), the shear flow was accompanied by a structuring of the liquid with no distortions of the metal surfaces. In contrast, for the overlapping and near-overlapping cases, shearing resulted in pressurization of the liquid in the interasperity region, accompanied by a significant increase in effective viscosity. This resulted in plastic deformations of the gold asperities.

It is clear, based on the discussion in this section, that fluids confined to areas of atomic-scale dimensions do not necessarily behave like liquids on the macroscopic scale. In fact, depending on the condition, they may often behave more like solids in terms of structure and flow. This presents a new set of concerns for lubricating moving parts at the nanometer scale. However, with the aid of simulations like the ones mentioned here, plus experimental studies using techniques such as the SFA, general properties of nanometer-scale fluids are being characterized effectively with profound precision. This, in turn, will allow scientists and engineers to design new materials and interfaces that will have specific interactions with confined lubricants, effectively controlling friction (and wear) at the atomic scale.

11.5   Friction

Friction at sliding solid interfaces results from the conversion of the work of sliding into some other, less-ordered, form. For strongly adhering systems, the work of sliding may be converted into damage within the bulk. As adhesive forces are decreased between the solid surfaces, the conversion of work changes to mechanical damage at or near the surface, leading to, for example, the formation of wear debris and transfer films (Singer, 1991; Singer et al., 1991). For still weaker adhesive forces, friction can occur through the conversion of work to heat at the interface with no permanent damage to the surface. The latter regime is the topic of this section.

While the thermodynamic principles of the conversion of work to heat are well known, many of the detailed mechanisms at sliding interfaces are just beginning to be understood. This understanding is the by-product of new scientific instrumentation, such as the SFA and the AFM, analytic theories, and simulations. A detailed understanding of these processes is important for producing interfaces with specific friction (and wear) properties.

Atomic-scale friction has been investigated using MD simulations for systems composed of a variety of materials, in a number of geometries. For instance, the atomic-scale friction and wear of diamond surfaces, both atomically flat and rough, was examined using MD by Harrison et al. (1992c, 1993a,b,c). In addition, atomic-scale friction between monolayers composed of alkane chains bound to rigid substrates (Glosli and McClelland, 1993), between perfluorocarboxylic acid and hydrocarboxylic LB monolayers (Koike and Yoneya, 1996), between contacting Cu surfaces (Hammerberg et al., 1995; Sorensen et al., 1996), between a silicon tip and a silicon substrate (Landman et al., 1989a,b), and between contacting diamond surfaces in the presence of third-body molecules (Perry and Harrison, 1997) have all been examined using MD simulations.

Molecular dynamics simulations have also provided the first insights into tribochemical reactions that might occur when two diamond surfaces are in sliding contact (Harrison and Brenner, 1994).

11.5.1   Solid Lubrication

Harrison et al. (1992c, 1993a,b,c) examined the atomic-scale phenomena that occurred when two (111) diamond surfaces were placed in sliding contact. The effects of crystallographic sliding direction, temperature, sliding speed, applied normal load, and nature of the surface on atomic-scale friction were examined. The frictional properties of hydrogen-terminated diamond surfaces were compared with the frictional properties of alkyl-terminated diamond surfaces to elucidate the effects of surface condition on atomic-scale friction. The alkyl-terminated systems were composed of two diamond lattices, configured so that their (111) surfaces were in contact (Figures 11.32a to d). One eighth of the hydrogen atoms on the upper surfaces were then randomly replaced with methyl (Figure 11.32b), ethyl (Figure 11.32c), or n-propyl (Figure 11.32d) groups.

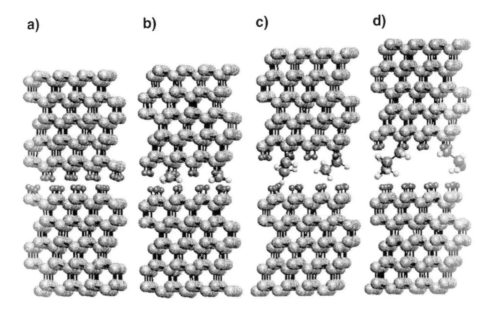

FIGURE 11.32 Illustration of atoms of a number of diamond-containing systems viewed along the [110] crystallographic direction at the start of MD simulations. The (111) faces of two hydrogen-terminated, diamond lattices are in contact in (a). Two of the upper-surface hydrogen atoms have been replaced by methyl groups in (b), ethyl groups in (c), and *n*-propyl groups in (d). Sliding was achieved by moving the top two (rigid) layers of the upper surfaces in the [112] crystallographic (right to left in the figure). Large spheres represent carbon atoms and small spheres represent hydrogen atoms. (From Harrison, J. A. et al. (1993), *Mater. Res. Soc. Bull.* 18, 50–53. With permission.)

In these simulations, the vertical separation of the atoms in the two outermost layers of the lattices were held constant. The molecular dynamics friction simulations were carried out by moving the atoms in the rigid layers of the upper surface at a constant velocity (usually 100 m/s) in the chosen crystallographic sliding direction. (Similar results were obtained with sliding speeds of 50 m/s.) Both the [110] and the [112] crystallographic directions were examined. (These correspond to moving the upper surface in Figure 11.32 in the page, or from left to right, respectively.) The remaining atoms were allowed to dynamically evolve in time according to classical equations of motion. The forces governing their motion were derived from the Brenner potential (Brenner, 1990). The temperature of the systems was maintained at 300 K by applying a thermostat (Berendsen et al., 1984) to the middle five layers of each lattice. The simulations were performed at average normal loads varying between 0 to 15 GPa. These pressures are well within the range of pressures that can be achieved experimentally using an AFM (Germann et al., 1993).

A constant value for the relative distance between the rigid layers in the top and bottom slabs was maintained during these simulations, and this distance was used to define the separation between the two slabs.

The average frictional forces for sliding in the [112] crystallographic direction were calculated as a function of load for the systems shown in Figure 11.32. For an individual sliding simulation, the frictional force was taken to be the sum of the forces in the sliding direction on the rigid-layer atoms, normalized by the number of rigid-layer atoms and averaged over the simulation. Because the relative lateral placement of the opposing diamond surfaces affects the frictional force, frictional force (and normal force) data were also averaged over simulations whose starting configurations differed in lateral registry of the upper surface. (For precise details, see Harrison et al., 1993c and Perry and Harrison, 1997.)

The frictional force was found to increase with normal load in all cases, usually in an approximately linear manner (Figure 11.33). This result is entirely reasonable, and was found to originate from the detailed atomic-scale sliding mechanisms and the associated pathways of energy dissipation. For instance,

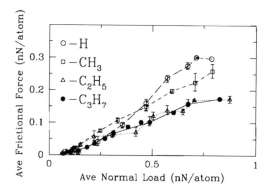

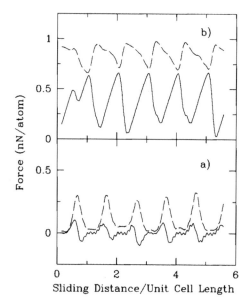

FIGURE 11.33 Average frictional force per rigid-layer atom as a function of average normal load per rigid-layer atom for sliding the upper surfaces shown in Figure 11.32 in the [112] crystallographic direction at 100 m/s and 300 K. The upper and lower lines drawn from these points extend to the maximum and minimum frictional force values obtained from independent starting configurations. Thus, the lines represent the range of frictional force values obtained at a given load. Data points that do not have lines extending above and below them were calculated from one sliding simulation. (Data from Harrison et al., 1993c; Perry and Harrison, 1997).

FIGURE 11.34 Frictional force (solid lines) and normal force (dashed lines) per rigid-layer atom vs. sliding distance when the system shown in Figure 11.32a is slid in the [112] crystallographic direction. The average normal and frictional forces per rigid-layer atom are 0.11 and 0.0041 nN/atom in (a) (corresponding to a pressure of 1.4 GPa) and 0.85 and 0.36 nN/atom (corresponding to a pressure of 10.3 GPa) in (b). (Data from Harrison, J. A. et al. (1995), *Thin Solid Films* 260, 205–211.)

for some starting configurations, if the hydrogen atoms were held rigid as the upper surface slid over the lower surface they would have passed directly above and below one another. Analysis of video sequences of the dynamics revealed that as the hydrogen atoms in each surface approached or collided, they would revolve around each other in the sliding plane. This motion resembled more of a "zigzag" pattern when the opposing hydrogen atoms were not directly in line in the sliding direction at the start of the simulation. This motion caused oscillations in the normal and frictional forces as a function of sliding distance (or, alternatively, sliding time). The oscillations possessed the periodicity of the unit cell, and their magnitude depended on the magnitude of the applied load (Figures 11.34a and b). Similar oscillations have been observed experimentally using an AFM (Mate et al., 1987).

For all loads, the approach of opposing hydrogen atoms caused in increase in both the normal and the frictional forces. As the hydrogen atoms moved past one another, both the normal and the frictional forces decreased. At low loads, the hydrogen atoms passed by one another with ease. Thus, the frictional force oscillated about zero, yielding values of average frictional force close to zero (Figure 11.34a). At higher loads, it was more difficult for the hydrogen atoms to pass by one another and they became "stuck." Once stuck, continued sliding of the upper lattice resulted in a linear increase in the frictional force as a function of sliding distance until some critical force was reached. At the critical force the hydrogen atoms "slipped" past one another and the frictional force dropped very sharply. This oscillatory, non-symmetric shape of the frictional force curve has been observed both experimentally (Mate et al., 1987;

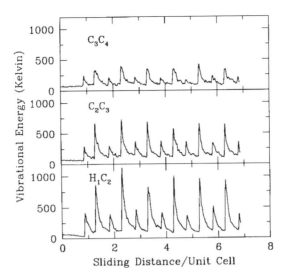

FIGURE 11.35 Average vibrational energy for each pair of oscillator atoms as a function of sliding distance. These energies are derived from an MD simulation of two hydrogen-terminated, diamond (111) surfaces in sliding contact (system shown in Figure 11.32a). The vibrational energy between the first and second layers of the lower diamond surface is shown in the lower panel, between the second and third layers in the middle panel, and between the third and fourth layers in the upper panel. (From Harrison, J. A. et al. (1995), *Thin Solid Films* 260, 205–211. With permission.)

German et al., 1993) and in other simulations (Glosli and McClelland, 1993; Landman et al., 1989a,b). In all of these cases, this behavior was attributed to atomic-scale stick-slip or ratcheting.

The collisions suffered by the interface hydrogen atoms resulted in mechanical excitation of the interface hydrogen atoms (and subsurface layer atoms) on opposing surfaces. Computation of the vibrational energy between layers as a function of sliding distance confirmed this behavior. The vibrational energy between layers for the lower diamond lattice as a function of sliding distance typically shows "ringing" of the surface bonds, followed by a propagation of this vibrational energy away from the surface, until it is absorbed by the heat bath (Figure 11.35). This conversion of the work of sliding into heat dissipated into the bulk was shown to be the mechanism of wearless atomic-scale friction.

When some of the hydrogen atoms on the upper surface were replaced with chemically bound methyl groups, the average frictional force behaved in much the same way as for the hydrogen-terminated case. In this case, the remaining hydrogen atoms, as well as the chemisorbed methyl groups, suffered collisions with the hydrogen atoms on the lower surface. In addition to excitation of carbon–hydrogen bonds, the methyl groups as a whole were able to rotate like turnstiles as they interacted with the hydrogen atoms on the lower surface. This motion introduced a novel mechanism of energy dissipation, although one that appeared to have little influence on the total amount of energy dissipated by the diamond lattices.

Extending the length of the chemisorbed hydrocarbon chains to ethyl and *n*-propyl caused a dramatic reduction in the average frictional force at higher loads. The size and flexibility of these larger hydrocarbon chains, in conjunction with the conformation they adopted when sliding, were responsible for this reduction. The carbon–carbon bond of a given ethyl group (or the chain backbone) was not typically parallel to the sliding direction at the start of the simulation. However, under small applied loads sliding usually induced alignment of the chain backbone with the sliding direction (Figure 11.36a). In this conformation, both the hydrogen atoms and the ethyl groups underwent "collisions" with hydrogen atoms on the lower surface. These collisions were similar to those that occurred in the hydrogen- and methyl-terminated systems. Because atomic-scale friction is a direct result of the mechanical excitation caused by these collisions, the friction was approximately the same as it was in the hydrogen-terminated and methyl-terminated systems.

a)

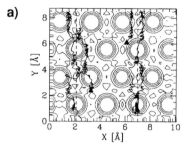

b)

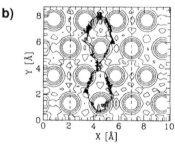

FIGURE 11.36 The solid lines represent the center of mass trajectories of the –CH₃ portion of ethyl groups (Figure 11.32c) attached to a hydrogen-terminated, diamond (111) surface as it slides over a hydrogen-terminated, diamond (111) surface. The trajectories are plotted on a potential energy contour map of a hydrogen-terminated, diamond (111) surface. Sliding corresponds to moving from the bottom of the figure to the top of the figure. Trajectories that disappear from the plots at large values of Y reappear at the bottom of the plot (near $Y = 0$) due to periodic boundary conditions. Filled triangles represent the starting points of individual trajectories and the dashed lines represent the path of the tethered portion of the ethyl group (–CH₂–). The closer the triangle is to the dashed line, the more aligned the backbone of the ethyl group is in the sliding direction. The average normal load per atom on the rigid-layer atoms of the upper surface was 0.065 nN/atom in (a) and 0.42 nN/atom in (b). The contour values are 0.271, 0.771, 1.271, 1.771, and 2.271 eV. (From Harrison, J. A. et al. (1993), *J. Phys. Chem.* 97, 6573–6576. With permission.)

In contrast, at higher applied loads the chains were not able to align when sliding, but rather were forced to occupy potential energy valleys between the hydrogen atoms on the lower surface (Figure 11.36b). This reduced the mechanical excitation of the system; thus, it reduced the friction force.

Recently, Sorensen et al. (1996) carried out simulation studies of atomic-scale friction for a number of tip–surface and surface–surface contacts consisting of Cu atoms. One of the systems studied was composed of a flat tip of Cu atoms placed on a Cu(111) substrate. The tip consisted of stacked (111) layers of Cu with 25 atoms (corresponding to a contact area of 140 Å²) touching the Cu substrate. The interatomic potentials used to model the interactions between atoms were derived from effective medium theory (EMT) (Jacobsen et al., 1987). Low-temperature (0 K) simulations were performed by moving rigid atoms in the top of the tip 0.05 Å in the sliding direction, and then allowing the dynamic atoms to relax by a local energy minimization procedure. MD simulations were used for 300 K investigations. In these simulations, the rigid atoms were displaced in the sliding direction before each time step and the temperature of the system was controlled by a Langevin thermostat.

When the Cu(111) tip was slid in the [101] direction on the Cu(111) substrate at 0 K, the lateral force on the tip exhibited the characteristic sawtooth pattern indicative of stick-slip motion. This behavior had the periodicity of the substrate (2.6 Å), which has been observed in previous atomic-scale friction simulations (Landman et al., 1989b; Harrison et al., 1992c, 1993a). There were two maxima in the friction force per 2.6 Å unit cell, because the Cu tip atoms at the interface jumped discontinuously in a zigzag pattern from fcc to hexagonal closest-packed positions on the Cu(111) substrate. At higher loads, the slips occurred at larger displacements. As a result, the frictional force increased with increasing load. Similar trends with increasing load have also been observed in other simulations (Harrison et al., 1992c).

Sorensen et al. (1996) also studied the effect of sliding velocity on the atomic-scale friction of contacting Cu(111) surfaces. The time-dependent frictional force in this system was examined at temperatures of both 12 and 300 K, and for sliding speeds of both 2 and 10 m/s. Superimposed on the sawtooth pattern indicative of stick-slip motion were the random oscillations due to thermal fluctuations of the system. For both the low- and high-temperature 2-m/s sliding simulations, the fluctuations in the frictional force curve increased substantially after each slip event and decayed during the subsequent slick event. The fluctuation patterns were almost identical for each stick-slip event. These fluctuations were too large to be due entirely to thermal motion. Indeed, the authors attributed these fluctuations to systematic motion,

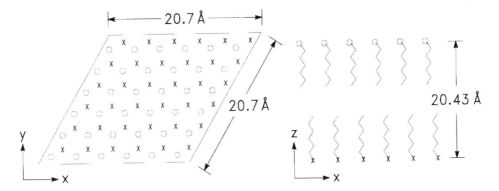

FIGURE 11.37 Initial configuration of the computational cell viewed from the top and from the side. The *x*-axis is the sliding direction. Periodic boundary conditions are applied in the *x*- and *y*-directions. (From Glosli, J. N. and McClelland, G. M. (1993), *Phys. Rev. Lett.* 70, 1960–1963. With permission.)

namely, the excitation of phonons. This observation is similar to the results found in the atomic-scale friction of diamond (Harrison et al., 1995; Perry and Harrison, 1996b).

For the low-temperature 2 m/s sliding simulations, the authors concluded that almost all the energy released after the slip event was dissipated by the thermostat during the long stick events. From the magnitude of the fluctuations in frictional force, it was clear that this was not the case when the sliding velocity was increased to 10 m/s. In this case, the fluctuations increased significantly after the first slip and remained large for the remainder of the simulation. In this case, the stick event was shorter; therefore, there was not sufficient time for the energy to be dissipated by the thermostat. The result was that the maximum values of frictional force were lower than they were at slower velocities. The implication was that phonons excited at previous slips promoted subsequent slip events.

Sorensen et al. (1996) also investigated the effects of contact area, contact geometry (tip–surface system vs. two contacting surfaces) and surface alignment on the atomic-scale friction.

Sliding friction at metallic interfaces has also been studied by Hammerberg et al. (1995) who investigated Cu interfaces in sliding contact for pressures in the kilobar range. These two-dimensional MD simulations contained 65,000 atoms, which are modeled with EAM potentials (Holian, 1991). The simulations were carried out at a range of different densities and sliding velocities. The calculated frictional force was found to be an increasing function of density, as expected from the adhesive model of asperity interactions. In this model, microscopic asperities were in contact under the local load required for mechanical stability at shear strains sufficient to initiate plastic flow. In addition, the simulations showed that the coefficient of friction was a weaker function of density in this pressure regime than the adhesive model would predict. The friction coefficient was also found to be a decreasing function of velocity, and above a critical velocity the temporal behavior of the tangential force became oscillatory. Above a critical velocity, the mechanical response of the underdamped stick-slip system scaled with time and the structure of the interface changed. In this region, a band of nanocrystalline material (approximately 16 nm wide) formed that had a significant dislocation density.

Boundary lubrication of close-packed hydrocarbon chains attached to rigid substrates has also been studied using MD simulations (Figure 11.37) (Glosli and McClelland, 1993). The films were modeled as a periodic 6 × 6 array of monomer alkane chains with rigid-bond lengths. The CH_3 and CH_2 entities were treated as united atoms (Ryckaert et al., 1978). The forces between pairs of united atoms on chains more distant than third nearest neighbors and the forces for all interactions between chains were derived from an LJ potential. Sliding was modeled by moving the substrate of the upper films at a 2.2 m/s velocity in the *x*-direction.

The frictional behavior of the films as a function of interaction strength between chains on opposing surfaces was investigated. The time average of the shear stresses τ/τ_0, (where $\tau_0 = 15.6$ MPa), given as the

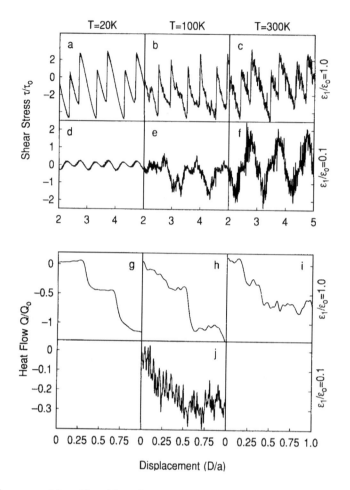

FIGURE 11.38 Shear stress (a) to (f) and heat flow (g) to (j) as a function of sliding displacement for normal ($\varepsilon_1/\varepsilon_0 = 1.0$) and reduced ($\varepsilon_1/\varepsilon_0 = 0.1$) interfacial interaction strengths at three temperatures. In (a) to (f) the horizontal scale extends over three lattice periods, in (g) to (j), over one period. In (h) and (i) the results for 10 and 20 cycles, respectively, have been averaged together to reduce thermal noise. The velocity v/v_0 equals 0.0112. Quantities have been normalized by $\tau_0 = 15.6$ MPa, $a = 4.44$ Å, and $Q_0 = \sigma/a^2 = 6$ mJ/m². (From Glosli, J. N. and McClelland, G. M. (1993), *Phys. Rev. Lett.* 70, 1960–1963. With permission.)

total frictional force divided by the area of the computational cell, was examined as a function of displacement at several temperatures (Figure 11.38). The shear stress possessed the characteristic saw-tooth pattern associated with forces alternating between aiding and opposing the sliding motion. When the force changed sign, the ends of the chains on each film abruptly jumped 0.66 Å (a relative motion of 1.11 Å). As a result, the two films vibrated in synchronization in opposite lateral directions. The vibrational excitation of the films is apparent from a thickening of the line in Figure 11.38a. This effect became more pronounced as the temperature of the system was increased (Figures 11.38b and c). The heat flow into the thermostat clearly showed that the pluck corresponded to the conversion of mechanical energy, stored as strain, to thermal energy. The interaction strength had a profound effect on the dynamics of these films (Figure 11.38a and d). For small interaction strengths (Figure 11.38d), the lateral force varied smoothly with position. Thus, a region of viscous dissipation existed. Indeed, at a simulation temperature of 20 K, an interaction strength threshold to plucking of approximately 28 K was identified.

For larger interaction strengths and for temperatures below 70 K, friction resulted from the plucking motion and showed little temperature dependence. Above 70 K the friction was strongly affected by changes in the mechanical characteristics of the film, examples of which include rotational melting and associated loss of the nearly herringbone structure near 118 K. The friction increased dramatically after 70 K, then declined steadily as the temperature was increased.

Molecular dynamics simulations have also been used to study the friction in monolayers of perfluorocarboxylic acid and hydrocarboxylic acid on SiO_2 (Koike and Yoneya, 1996). This system was chosen because AFM (Overney et al., 1994) and SFA (Briscoe and Evans, 1982) experiments have shown that the friction force on fluorocarbon-covered areas was three to five times larger than it was on hydrocarbon-covered areas. In the simulations, surfaces of $SiO_2(001)$ were composed of 510 to 714 atoms and were used both as the base and the slider. The monolayers were composed of 36 chains of $C_{14}F_{29}COOH$ and $C_{14}H_{29}COOH$ adsorbed on the SiO_2 surface. A valence force field was used to calculate the forces between atoms (Dauber-Osguthorpe et al., 1988).

Computational cell sizes corresponded to areas of 18.4 and 22.1 $Å^2$/molecule. These areas corresponded to experimental surface pressures between 20 and 40 mN/m. The LB films were close packed and contained almost no gauche defects. The films were equilibrated and then compressed at a constant velocity of 100 m/s. The slider was then moved at a constant velocity of 100 m/s in the [010] direction under a constant load of 0.60 nN.

Glosli and McClelland (1993) had previously observed two types of energy dissipation, resulting in friction of bilayers, continuous or viscous dissipation and discontinuous or plucking dissipation. These two types of dissipation were also observed in Koike and Yoneya's 1996 simulations. For the sliding of a gold surface on a long-chain thiol monolayer, Tupper and Brenner (1994b) found that the period of oscillation in the frictional force corresponded to the distance between the gold atoms. In contrast, because the SiO_2 slider was smooth compared with the in-plane interactions of the monolayer, Koike and Yoneya (1996) found the period of the oscillations in the frictional force were consistent with the periodicities of the tilt angles of the films and potential energy changes in the two systems.

The average frictional force calculated for the $C_{14}F_{29}COOH$ film was approximately three times larger than for the $C_{14}H_{29}COOH$. Thus, these simulations were in agreement with experimentally obtained trends (Overney et al., 1994). The authors concluded that it is impossible to explain the differences in frictional force using only the stiffness of the films as has been suggested previously. Rather, the difference in frictional force was largely due to differences in the intermolecular interactions. In addition, the frictional force was found to be approximately proportional to the difference in the potential energy fluctuations between shear and equilibrium conditions.

Potential energy fluctuations were also found to be important in earlier simulations of atomic-scale friction (Harrison et al., 1995). These simulations showed that two hydrogen-terminated diamond surfaces in sliding contact were deformed in the stick phase of motion, and released this energy as heat after the slip event. Thus, large amounts of energy were dissipated subsequent to sticking events, and the more the surfaces deformed the more energy was dissipated.

Simulations of adsorbed monolayers of rare gas atoms are also found in the literature. These studies typically make comparisons with quartz crystal microbalance (QCM) experiments. Early QCM experiments by Krim and co-workers (Watts et al., 1990; Krim et al., 1991) examined the friction between adsorbed gas atoms (Ar, Kr, Xe) and a (111)-oriented noble metal substrate (Au or Ag). Krypton adsorbed on Au first as islands of liquid. These islands grew in size until the gas atoms were adsorbed as an entire monolayer. (This monolayer existed as a disordered, two-dimensional liquid at low doses.) For Kr, as more atoms were adsorbed, the liquid became denser and eventually underwent a phase change into a crystalline state that was incommensurate with the Au substrate. Krim and co-workers observed that both the solid and liquid Kr monolayers exhibited a viscous force law, meaning that no threshold force was needed to initiate the sliding and the friction was proportional to the relative velocity. This behavior is more commonly associated with the macroscopic friction of liquids. In addition, Krim and co-workers determined that the solid monolayers slid more easily than the liquid monolayers. That is, the frictional force on the solid monolayer was less than the force on the liquid monolayer and the force on the liquid monolayer was approximately three orders of magnitude *weaker* than that between Kr layers in a bulk fluid.

In an effort to understand these counterintuitive results, Robbins and co-workers (Cieplak et al., 1994; Smith et al., 1996) performed MD simulations, in conjunction with perturbation theory, for a layer of

577

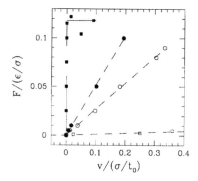

FIGURE 11.39 Variation of the steady-state velocity v with force per atom F. Results for the incommensurate case that models Kr on Au ($\varepsilon' = 1.19\varepsilon$, $\sigma' = 1.19\sigma$, $N_{ad} = 49$, and $N_{ad}/A_{surf} = 0.068$ Å^2) at $k_B T/\varepsilon = 0.385$ (solid phase) and 0.8 (fluid phase) are shown by open and filled circles, respectively. Results for $\varepsilon' = \varepsilon$, $\sigma' = \sigma = 2^{1/6} d_{nm}$, $k_B T/\varepsilon = 0.385$, and $N_{ad} = 64$ are shown by the squares. For $f = 0.3$ (filled squares) the adsorbate is a commensurate solid, and for $f = 0.1$ (open squares) the adsorbate is incommensurate. (From Smith, E. D. et al. (1996), *Phys. Rev. B* 54, 8252–8260. With permission.)

adsorbed molecules sliding over a substrate. The Au surface was modeled as a rigid substrate with atoms "fixed" in their ideal positions. This produced a fixed periodic potential that acted on the adsorbate. The interactions between the noble gas adsorbate atoms was modeled by an LJ potential with parameters, ε and σ, appropriate for Kr. The potential proposed by Steele et al. (1973) was modified by addition of a multiplicative factor f to model the adsorbate–substrate interactions. The multiplicative factor, f, allowed the corrugation of the potential (its variation with the plane of the substrate) to be varied without changing the energy of adsorption. Standard MD algorithms (Thompson and Robbins, 1990b) were used to follow the position and velocity of all the atoms interacting with the aforementioned potentials.

These simulations showed that the form of the frictional law was dependent on the equilibrium phase of the adsorbed layer (Figure 11.39). A static frictional force (i.e., nonzero frictional force when the velocity equals zero) was only found when the adsorbed layer locked into a commensurate solid phase and when f was sufficiently large. A viscous frictional law was found whenever the layer was in a fluid or incommensurate solid phase. (The crystalline monolayer was incommensurate due to the mismatch between the atomic sizes of Kr and Au.) In addition, the friction was lower for the crystalline monolayer in agreement with the findings of Krim and co-workers.

The slip time τ, which characterizes the rate of momentum transfer between the substrate and the adsorbate, was also calculated in the MD simulations. (Lower values of τ imply larger friction.) Perturbation theory was used to show that, for the symmetry of these simulations, τ was inversely proportional both to the structure factor and to the phonon lifetime. The calculated values of τ increased with coverage in the same way as the experimentally determined values. That is, τ had a small value in the fluid phase and a large value in the solid phase. In addition, the simulations showed that the order within the film, as measured by the structure factor, was responsible for the changes in τ with coverage. Thus, the monolayer solid, because it is a rigid network of atoms that cannot deform itself readily to an external potential, has a smaller degree of order and thus longer slip time (lower friction). Because a fluid layer could easily deform and lock itself to the substrate, inhibiting its motion, the degree of order is large, and the slip time is small (larger friction).

The various examples discussed here — sliding diamond interfaces and self-assembled monolayers — illustrate specific mechanisms where energy is dissipated at sliding interfaces resulting in a frictional force. Furthermore, they show how variables such as temperature and interface structure can have a profound effect on friction. As in the simulations of liquid lubrication discussed in the preceding section, the insights being generated by these simulations will help to guide scientists and engineers interested in designing materials with specific friction and wear properties. Clearly, though, much more can be done to characterize and understand friction at nanometer-scale solid surfaces, and we expect simulations (in conjugation with new experimental methods) to continue to play a leading role in developing this understanding.

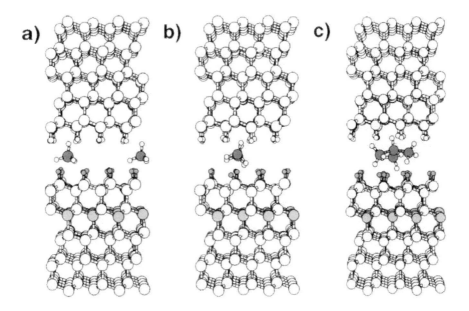

FIGURE 11.40 Initial configurations at low load for the diamond plus third-body molecule systems. These systems are composed of two diamond surfaces, viewed along the [110] direction, and two methane molecules in (a), one ethane molecule in (b), and one isobutane molecule in (c). Large white and dark gray spheres represent carbon atoms of the diamond surfaces and the third-body molecules, respectively. Small gray spheres represent hydrogen atoms of the lower diamond surface. Hydrogen atoms of the upper diamond surface and the third-body molecules are both represented by small white spheres. Large light gray spheres represent fourth-layer carbon atoms of the lower diamond surface. Sliding is achieved by moving the rigid layers of the upper surface from left to right in the figure. (From Perry, M. D. and Harrison, J. A. (1997), *J. Phys. Chem.* 101, 1364–1373. With permission.)

11.5.2 Friction in the Presence of a Third Body

Free particles at an interface can have effects on friction that are just as remarkable as the effects of surface roughness or of an adsorbed monolayer. This is exemplified by the results of Perry and Harrison (1996a, 1997) who used MD simulations to investigate the atomic-scale friction and wear when small hydrocarbon molecules are trapped between two (111) crystal faces of diamond. These trapped molecules, sometimes referred to as third-body molecules, might model hydrocarbon contamination trapped between contacting surfaces prior to a sliding experiment. Alternatively, these molecules might model hydrocarbon debris formed when two diamond surfaces are in sliding contact.

In separate studies, three simple hydrocarbon molecules were trapped between the hydrogen-terminated (111) faces of two diamond surfaces and atomic-scale friction was measured as a function of normal load. The third-body molecules examined were methane (CH_4), ethane (C_2H_6), and isobutane ((CH_3)$_3CH$) (Figure 11.40). The sliding simulations were carried out as described in Section 11.5.1.

Examination of the average frictional force as a function of load revealed that for all three systems the frictional force generally increased as a function of load (Figure 11.41). However, the presence of each of the third-body molecules markedly reduced the average frictional force compared with the hydrogen-terminated diamond (111) system in the absence of third-body molecules. The methane system exhibited the lowest frictional force, while the ethane and isobutane systems had somewhat larger frictional force values over the load range examined, although still lower than the bare surface.

The third-body molecules reduced the frictional force by acting as a boundary layer between the two diamond surfaces (Figure 11.42). The interfacial separation, hence, average distance between hydrogen atoms on opposing diamond surfaces, was always larger in the presence of the third-body molecules.

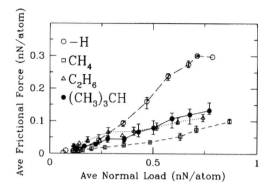

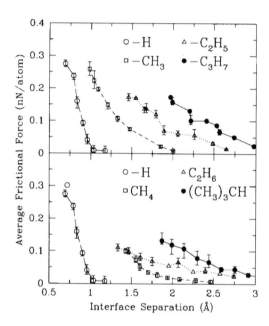

FIGURE 11.41 Average frictional force per rigid-layer atom as a function of average normal load per rigid-layer atom for the third-body systems shown in Figure 11.40 while sliding the upper diamond surface in the [112] crystallographic direction. Open squares, open triangles, filled circles, and open circles represent data for the methane (CH_4) system, the ethane (C_2H_6) system, the isobutane (($CH_3)_3CH$) system, and diamond surfaces in the absence of third-body molecules, respectively. Lines have been drawn to aid the eye. (Data from Perry, M. D. and Harrison, J. A. (1997), *J. Phys. Chem.* 101, 1364–1373.)

FIGURE 11.42 Average frictional force per rigid-layer atom as a function of interface separation for the third-body systems shown in Figure 11.40 while sliding the upper diamond surface in the [112] crystallographic direction. The symbols are the same as in Figure 11.41. Lines have been drawn to aid the eye. (Data from Perry, M. D. and Harrison, J. A. (1997), *J. Phys. Chem.* 101, 1364–1373.)

Thus, the third-body molecules caused a reduction in the interaction of hydrogen atoms on opposing diamond surfaces. The principal mechanism of energy dissipation in the system was the interaction of the third-body molecules directly with the diamond surfaces. Because the submonolayer coverage of third-body molecules meant that they did not interact with all of the surface hydrogens, the total vibrational excitation of the surfaces was less in the presence of these molecules. The larger surface area of ethane and isobutane molecules caused them to collide with a greater number of surface hydrogens, thereby increasing the friction.

A detailed description of the interaction of the third-body molecules with the diamond surfaces depends on their size, their shape, and the alignment of the opposing diamond surfaces. The spherical shape and small size of the methane molecules allowed them to tumble between the diamond surfaces traversing regions of low potential energy between hydrogen atoms on opposing diamond surfaces (Perry and Harrison, 1996a,b). This led to very little vibrational excitation of the hydrogen atoms terminating the diamond surfaces (Figure 11.43). The methane molecules underwent this sort of motion regardless of load or diamond–surface alignment. Thus, the frictional force was low for all loads examined and the range of frictional force values obtained at a given load was small (Perry and Harrison, 1997).

In contrast, the larger ethane and isobutane molecules suffered frequent collisions as they slid between hydrogen sites on the diamond surfaces. In addition, because these molecules had nonspherical shapes,

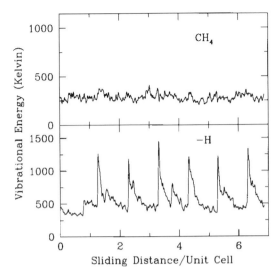

FIGURE 11.43 Average vibrational energy between the (C–H) bonds of the upper hydrogen-terminated, diamond (111) surface as a function of sliding distance for the hydrogen-terminated (H) (Figure 11.32a) and methane (CH_4) (Figure 11.40a) systems. For the hydrogen-terminated system, the average normal load and average frictional force were 0.79 nN/atom and 0.32 nN/atom, respectively. For the methane-debris system, the average normal load and the average frictional force were 0.85 nN/atom and 0.094 nN/atom, respectively. (From Perry, M. D. and Harrison, J. A. (1996), *Thin Solid Films* 290/291, 211–215. With permission.)

several initial configurations of the molecules relative to the sliding direction were possible. The effects of initial molecular orientation, size of molecule, and load on frictional force were ascertained by examining frictional force as a function of sliding distance (a so-called friction trace) for individual simulations (Figure 11.44).

Each maximum and minimum in the friction trace was assigned to specific atomic-scale motions. The shapes of these extrema were indicative of the type of atomic-scale motion and thus the amount of interaction with the diamond surfaces. At low loads, the ethane molecule typically began the simulations with its C–C bond at a 45° angle to the sliding direction (Figure 11.44, bottom panel). In this configuration, the ethane molecule "fits" well between hydrogen atoms on the lower diamond surface and therefore interacted weakly with the surface. During sliding, the ethane molecule was able to adopt configurations similar to this throughout the course of the simulations. As a result, the friction trace was periodic and regular and the average frictional force was low.

At high loads, however, it became more difficult for the ethane molecule to orient its C–C bond at 45° to the sliding direction during sliding. Indeed, the friction trace became irregular, indicating the presence of many different configurations during sliding. Analysis of animated sequences of the simulations also confirmed this conclusion. These configurations led to additional interactions with the diamond surfaces and increased frictional force. The number of configurations adopted during sliding depended on the relative alignment of the opposing diamond and the starting orientation of the molecule in addition to the load. For example, when the ethane molecule began the simulation with its C–C bond parallel to the sliding direction, its interactions with the diamond were much different than when the initial orientation was 45° to the sliding direction (Figure 11.44, top panel).

Because the isobutane molecule is also nonspherical, it had a number of possible starting configurations (Figure 11.45). These different starting configurations led to different interactions with the opposing diamond surfaces during sliding. The different types of interactions experienced by the isobutane molecule in the arrow vs. the sled orientation (Figure 11.45) are apparent from examination of the friction traces shown in Figure 11.46. In addition, for the same starting configuration, the interaction with the

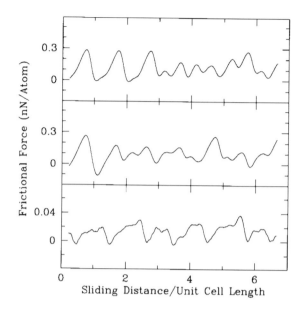

FIGURE 11.44 Frictional force as a function of sliding distance divided by unit cell length for three ethane system (Figure 11.40b) simulations. The average normal load and frictional force were 0.12 and 0.013 nN/atom (lower panel), 0.64 and 0.090 nN/atom (middle panel), and 0.63 and 0.11 nN/atom (upper panel). The sliding direction is the [112] crystallographic direction. (Data from Perry, M. D. and Harrison, J. A. (1997), *J. Phys. Chem.* 101, 1364–1373. With permission.)

diamond surfaces was also changed by altering the relative surface alignment (changing the space available to the third-body molecule) and the load (changing the distance between the diamond surfaces).

Despite the fact that the ethane and isobutane molecules forced the opposing diamond surfaces to be farther apart than the methane molecules (Figure 11.42), the friction was larger with ethane and isobutane molecules present because the molecules interacted more with the diamond surfaces while sliding.

It is interesting to note that the presence of the trapped methane molecules lowered the frictional force more than the presence of the chemisorbed methyl groups (Figure 11.42). Because the methyl groups were chemically bonded to the diamond surface, their interaction with the opposing diamond surface was greater because the chemisorbed molecules could not "avoid" the interaction with the hydrogen atoms on the diamond surfaces. However, the frictional force data were not dramatically different for the methane and methyl-terminated systems when analyzed as a function of interface separation. In fact, in the region between 1.5 and 2.0 Å, the curves overlap (Figure 11.42). The marked difference in frictional force when measured as a function of normal load arose from the higher load required to bring the methane system to the same interface separation as the methyl-terminated system.

11.5.3 Tribochemistry

A number of interesting phenomena result when two bodies are placed in sliding contact. One such phenomenon involves the formation of nascent particles via the wear of the contacting surfaces or via chemical reactions. Chemical reactions can occur between wear particles or between wear particles and contamination present on the contacting surfaces. (These reactions can lead to the formation of a distinct layer between the contacting surfaces, known as a transfer layer.) This interaction of chemistry and sliding is known as tribochemistry and can have profound effects on the friction between the contacting surfaces.

Experimental characterization of tribochemical reactions (Singer, 1991; Fischer, 1992) using traditional analysis methods such as optical microscopy, electron microscopy, cathodoluminescence, and infrared spectroscopy (Wilks and Wilks, 1979; Feng and Field, 1991; Tabor and Field, 1992) has proved difficult because, depending upon the geometry and materials, these techniques are usually restricted to examination of the contact region subsequent to sliding. This limitation has historically led to a lack of

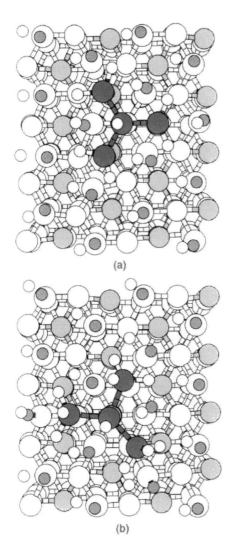

(a)

(b)

FIGURE 11.45 Sample starting configurations for two isobutane system sliding simulations. The "arrow" in (a) and the "sled" orientations in (b). For clarity, the carbon atoms of the upper diamond surface (Figure 11.40) are not shown and the color coding is as in Figure 11.40. The system is viewed along the [111] direction, the [112] direction is from left to right, and the [110] direction is from bottom to top. (From Perry, M. D. and Harrison, J. A. (1997), *J. Phys. Chem.* 101, 1364–1373. With permission.)

information regarding the mechanisms of tribochemical reactions. Indeed, even new atomic-scale proximal probe techniques, such as AFMs (Binnig et al., 1986) and FFMs, have yet to characterize specific, atomic-scale, tribochemical reaction mechanisms. The need for experiments that can probe the interface between sliding bodies as sliding progresses has been recognized (Singer, 1992). Recent experiments have attempted to probe the contact region as the sliding progresses. Hiller (1995) has used methods based on optical light scattering to detect particle sizes during sliding. More recent experiments have used ultraviolet Raman spectroscopy to monitor the formation of amorphous carbon when a steel ball is in sliding contact with a sapphire window in the presence of a polyperfluorinated ether lubricant (Cheong and Stair, 1997). Preliminary results have shown that the measured amount of amorphous carbon during sliding differs from the amount measured after sliding.

The recent experiments of Cheong and Stair (1997) underscore the need to monitor reactants and the formation of intermediates as the sliding progresses. Because the precise positions of all atoms are known as a function of time, MD simulations are uniquely suited to this task. Indeed, Harrison and Brenner (1994) reported tribochemical reaction sequences that occurred when two diamond surfaces were placed in sliding contact. These simulations provided the first glimpse into the rich, nonequilibrium tribochemistry that is possible in covalently bonded systems.

The MD simulations were carried out as described in Section 11.5.1 using the hydrogen-terminated, ethyl-terminated (Figure 11.47), and methyl-terminated diamond systems. The temperature of the system

583

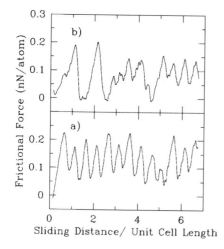

FIGURE 11.46 Frictional force as a function of sliding distance divided by unit cell length for two isobutane system simulations initially in the arrow orientation (Figure 11.45a). The average normal load and frictional force were 0.59 and 0.13 nN/atom (lower panel) and 0.59 and 0.080 nN/atom (upper panel). The sliding direction was the [112] crystallographic direction. (Data from Perry, M. D. and Harrison, J. A. (1997), *J. Phys. Chem.* 101, 1364–1373.)

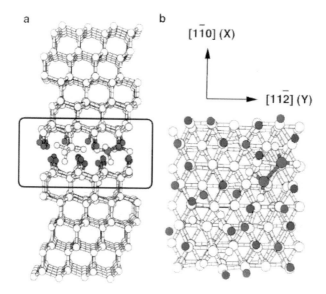

FIGURE 11.47 Typical starting configuration for the system with two chemisorbed ethyl (–CH₂CH₃) groups at the interface. Large white and dark spheres represent carbon atoms and small white and dark spheres represent hydrogen atoms. The system is viewed along the [110] direction in (a) and the [111] direction in (b). Box in (a) denotes region of interface illustrated in Figure 11.48. In (b) the carbon atoms of the upper lattice are not shown for clarity. (From Harrison, J. A. and Brenner, D. W. (1994), *J. Am. Chem. Soc.* 116, 10399–10402. With permission.)*

was maintained at 70 K using a thermostat (Berendsen et al., 1984) and the average normal force (load) on the systems was ≈ 1.3 nN/atom or ≈ 23 GPa. This pressure was well below the pressure that causes plastic deformation of diamond on the macroscopic scale (100 GPa). In addition, this pressure is comparable to the pressures (up to 20 GPa) achieved in AFM experiments and in macroscopic friction experiments on diamond in which wear was observed (Germann et al., 1993).

For both the hydrogen-terminated and the methyl-terminated surfaces examined, sliding did not result in any chemical reactions at the loads examined (Harrison et al., 1992b, 1993c). In contrast, when the ethyl groups were present, the simulations revealed a number of possible tribochemical reactions. For

---

* Color reproduction follows page 16

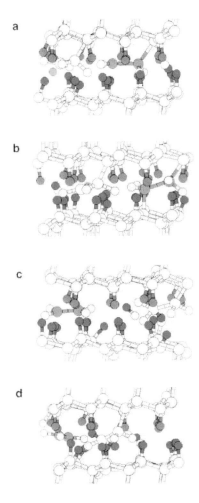

a

b

c

d

FIGURE 11.48 Simulation configurations at different times for sliding the upper diamond surface (in Figure 11.47) in the [112] direction with the same color coding as in Figure 11.47. The system is shown after hydrogen atoms have been sheared from the ends of both ethyl groups in (a). These free hydrogen atoms are colored dark to help distinguish them from bound atoms. Adhesion of the two diamond surfaces occurred in (b) via the formation of a carbon–carbon bond from the ethyl group at the right to the lower diamond surface. A hydrogen molecule is visible at the left of the figure. The system is shown after the adhesive bonds between the surfaces have broken in (c). The resulting ethylene and hydrogen molecules are visible at the left and center of the panel, respectively. At the end of the simulation the ethylene and hydrogen molecules were still present at the interface (d). Hydrogen atoms that originated from the ethyl groups became chemically bound to the upper (at the left) and lower (at the right) diamond surfaces. One of the ethyl groups abstracted two hydrogen atoms from the lower diamond surface. (From Harrison, J. A. and Brenner, D. W. (1994), *J. Am. Chem. Soc.* 116, 10399–10402. With permission.)*

instance, in one simulation, sliding in the [112] crystallographic direction resulted in the shearing of a hydrogen atom from the tail portion ($-CH_3$) of each ethyl group. Thus, radicals on each ethyl group along with free hydrogen atoms were formed. The free hydrogen atoms became trapped at the sliding interface. One of these free hydrogen atoms abstracted a hydrogen atom from the lower diamond surface to form a hydrogen molecule, H_2 (Figure 11.48c). This molecule did not undergo any further reaction during the 20-ps simulation. (The simulation geometry kept the free atoms trapped at the interface for the duration of the simulation. However, if the geometry was different, e.g., a tip and a surface, it would have been possible for an atom to escape the interface region during the sliding.)*

Further sliding of the upper surface caused one of the radical-containing ethyl groups to pass over the radical site on the lower surface created by the abstraction of hydrogen to form H_2. These two radicals combined to form a carbon–carbon bond. As a result, the upper and lower diamond surfaces became chemically joined by a D_u–CH_2CH_2–D_l linkage (green entity in Figure 11.48b). (Here, D_u and D_l represent the upper and lower diamond surfaces, respectively.) The upper and lower diamond surfaces remained bound until the shear stress was large enough to sever the carbon–carbon bonds connecting the two surfaces. The carbon–carbon bonds to the upper and lower surfaces broke almost simultaneously, creating an ethylene molecule, CH_2CH_2 (Figure 11.48c) and leaving radical sites on both the upper and lower surfaces. The planar ethylene molecule remained trapped at the interface for the remainder of the

---

* Color reproduction follows page 16.

---

simulation (Figure 11.48d). The nascent radical site on the lower surface was filled via the addition of a free hydrogen atom that originated from the tail of the second ethyl group (Figure 11.48d).

Note that the sequence of reactions differed for the two ethyl groups within the same simulation. Subsequent to shearing of the hydrogen atom from the second ethyl group, the ethyl radical abstracted a hydrogen atom from the lower surface (Figure 11.48c) forming a fully saturated ethyl group on the upper surface and a radical on the lower surface. Another hydrogen atom was "worn" from the tail of the ethyl group upon continued sliding. The nascent hydrogen atom became trapped at the interface and eventually added to the radical site on the upper surface created by the shearing of the D_u–CH_2CH_2–D_l connection (Figure 11.48d). Eventually, the radical-containing ethyl group abstracted a second hydrogen atom from the lower surface to become saturated (Figure 11.48d). This group did not undergo any additional reactions by the end of the 30-ps simulation. Each of these observations was consistent with conclusions inferred (but not directly measured) from experimental studies of the friction and wear of diamond surfaces.

The tribochemistry of the ethyl-terminated diamond surfaces was also shown to be anisotropic. For instance, sliding the upper surface of the same system described above in the [110] direction caused tail of one ethyl group to become lodged between several lower-surface hydrogen atoms. Because one end of the molecule was "fixed," continued sliding resulted in the breakage of the C–C bond of the ethyl group. Although the simulation evidence has too few of these rare events to allow a quantitative conclusion, these different reaction mechanisms imply different wear rates for each direction, consistent with the experimental observation that wear rates depend on the direction of abrasion (Wilks and Wilks, 1979; Feng and Field, 1991).

11.6 Summary

In this chapter, we have tried to present a relatively complete discussion of the contribution that MD and related simulations are making in the area of nanotribology. Based on the examples discussed above, it is clear that these approaches are providing new and exciting insights into friction, wear, and related processes at the atomic scale that could not have been obtained in any other way. Furthermore, the synergy between these simulations and new and exciting experimental techniques such as the SFA and AFM is producing a revolution in our understanding of the origin of friction at its most fundamental atomic level.

In the section on indentation, examples were discussed that illustrated many of the unexpected phenomena that are being characterized using MD simulations. These include atomic-scale jumps to contact in both metallic and ionically bonded systems that deform either tip or substrate depending on surface energies; the influence of very thin hydrocarbon films on inhibiting this jump to contact; and the indentation-induced disordering of covalently bonded silicon and diamond substrates. Furthermore, these results are closely related to experimental observations, often providing clear explanations for results such as hysteresis in loads during indentation and pull back.

The section on thin liquid films discussed the deviations of liquid film behavior from bulk fluids, again using specific examples from the literature. From this discussion it is clear that many of the ideas regarding liquid lubrication based on macroscopic experimental observations may not be valid for liquids as small as a few atomic layers across. Indeed, liquid films on this scale may not be good lubricants at all, forcing us to reevaluate how friction and wear can be controlled in soon-to-be-developed nanometer-scale machines. The answers to these and related questions may come from the types of studies discussed in the final section on solid bodies in contact. These simulations are beginning to provide the kind of knowledge needed to tailor new solid interfaces with desired atomic-scale friction and wear properties.

Finally, we note that simulations of the type discussed in this chapter have really just begun to make substantial contributions to the field of tribology. We fully expect that the range of processes and systems that can be characterized with this method to grow as more realistic and sophisticated models are developed and implemented on increasingly powerful computers. These, in turn, will continue to make important contributions to current and future technologies.

Acknowledgments

This work was supported by the U.S. Office of Naval Research under contracts #N00014-97-WR-20019 (USNA) and #N00014-95-1-270 (NCSU). David W. Brenner is supported by a National Science Foundation Early Career Development Grant. The authors would like to thank Richard J. Colton, Susan B. Sinnott, Irwin L. Singer, and Carter T. White for many helpful discussions and Mark O. Robbins, Jonathan G. Harris, Richard J. Colton, Joan E. Curry, James F. Belak, and Gary M. McClelland for providing figures from their work. Judith A. Harrison would also like to thank John Foerster and Jeanette Walsh for technical assistance.

References

Abell, G. C. (1985), "Empirical Chemical Pseudopotential Theory of Molecular and Metallic Bonding," *Phys. Rev. B* 31, 6184–6196.

Adelman, S. A. (1980), "Generalized Langevin Equations and Many-Body Problems in Chemical Dynamics," *Adv. Chem. Phys.* 44, 143–253.

Adelman, S. A. and Doll, J. D. (1976), "Generalized Langevin Equation Approach for Atom/Solid-Surface Scattering: General Formulation for Classical Scattering off Harmonic Solids," *J. Chem. Phys.* 64, 2375–2388.

Allen, M. P. and Tildesley, D. J. (1987), *Computer Simulation of Liquids*, Clarendon Press, Oxford.

Allinger, N. L., Yuh, Y. H., and Lii, J.-H. (1989), "Molecular Mechanics. The MM3 Force Field for Hydrocarbons," *J. Am. Chem. Soc.* 111, 8551–8566.

Balasubramanian, S., Klein, M., and Siepmann, J. I. (1996). "Simulation Studies of Ultrathin Films of Linear and Branched Alkanes on a Metal Substrate," *J. Phys. Chem.* 100, 11960–11963.

Baljon, A. R. C. and Robbins, M. O. (1996), "Energy Dissipation during Rupture of Adhesive Bonds," *Science* 271, 482–484.

Baljon, A. R. C. and Robbins, M. O. (1997), "Adhesion and Friction of Thin Films, *Mater. Res. Soc. Bull.* 22, 22–26.

Baskes, M. I. (1992), "Modified Embedded-Atom Potentials for Cubic Materials and Impurities," *Phys. Rev. B* 46, 2727–2742.

Belak, J. and Stowers, I. F. (1990), "A Molecular Dynamics Model of the Orthogonal Cutting Process," *Proc. of the American Society for Precision Engineering*, pp. 76–79.

Belak, J. and Stowers, I. F. (1992), "The Indentation and Scraping of a Metal Surface: A Molecular Dynamics Study," in *Fundamentals of Friction: Macroscopic and Microscopic Processes* (I. L. Singer and H. M. Pollock, eds.), pp. 511–520, Kluwer Academic Publishers, Dordrecht.

Belak, J., Boercker, D. B., and Stowers, I. F. (1993), "Simulation of Nanometer-Scale Deformation of Metallic and Ceramic Surfaces," *Mater. Res. Soc. Bull.* 18, 55–59.

Berendsen, H. J. C., Postman, J. P. M., Van Gunsteren, W. F., DiNola, A., and Haak, J. R. (1984), "Molecular Dynamics with Coupling to an External Bath," *J. Chem. Phys.* 81, 3684–3690.

Binnig, G., Quate, C. F., and Gerber, Ch. (1986), "Atomic Force Microscope," *Phys. Rev. Lett.* 56, 930–933.

Bitsanis, I. and Pan, C. (1993), "The Origin of Glassy Dynamics at Solid–Oligomer Interfaces," *J. Chem. Phys.* 99, 5520–5527.

Bitsanis, I., Magda, J., Tirrell, M., and Davis, H. (1987), "Molecular Dynamics of Flow in Micropores," *J. Chem. Phys.* 87, 1733–1750.

Bitsanis, I., Somers, S. A., Davis, T., and Tirrell, M. (1990), "Microscopic Dynamics of Flow in Molecularly Narrow Pores," *J. Chem. Phys.* 93, 3427–3431.

Brenner, D. W. (1989a), "Tersoff-Type Potentials for Carbon, Hydrogen, and Oxygen," *Mater. Res. Soc. Symp. Proc.* 141, 59–65.

Brenner, D. W. (1989b), "Relationship between the Embedded-Atom Method and Tersoff Potentials," *Phys. Rev. Lett.* 63, 1022.

Brenner, D. W. (1990), "Empirical Potential for Hydrocarbons for Use in Simulating the Chemical Vapor Deposition of Diamond Films," *Phys. Rev. B.* 42, 9458–9471.

Brenner, D. W. and Garrison, B. J. (1989), "Gas-Surface Reactions: Molecular Dynamics Simulations of Real Systems," in *Molecule Surface Interactions*, (K. P. Lawley, ed.), Vol. 76 of Advances in Chemical Physics, Wiley, Chichester, pp. 281–334.

Brenner, D. W., Harrison, J. A., White, C. T., and Colton, R. J. (1991), "Molecular Dynamics Simulations of the Nanometer-Scale Mechanical Properties of Compressed Buckminsterfullerene," *Thin Solid Films* 206, 220–223.

Brenner, D. W., Sinnott, S. B., Shenderova, O. A., and Harrison, J. A. (unpublished).

Briscoe, B. J. and Evans, D. C. (1982), "The Shear Properties of Langmuir–Blodgett Layers," *Proc. R. Soc. A* 380, 389–407.

Burkert, U. and Allinger, N. L. (1982), *Molecular Mechanics*, American Chemical Society, Washington, D.C.

Burnham, N. A. and Colton, R. J. (1989), "Measuring the Nanomechanical Properties and Surface Forces of Materials Using an Atomic Force Microscope," *J. Vac. Sci. Technol. A* 7, 2906–2913.

Burnham, N. A. and Colton, R. J. (1993), "Force Microscopy," in *Scanning Tunneling Microscopy and Spectroscopy: Theory, Techniques, and Applications* (D. A. Bonnell, ed.), pp. 191–249, VCH Publishers, New York.

Burnham, N. A., Dominguez, D. D., Mowery, R. L., and Colton, R. J. (1990), "Probing the Surface Forces of Monolayer Films with an Atomic-Force Microscope," *Phys. Rev. Lett.* 64, 1931–1934.

Car, R. and Parrinello, M. (1985), "Unified Approach for Molecular Dynamics and Density-Functional Theory," *Phys. Rev. Lett.* 55, 2471–2474.

Carpick, R. and Salmeron, M. (1997), "Scratching the Surface: Fundamental Investigations of Tribology with Atomic Force Microscopy," *Chem. Rev.* 97, 1163–1194.

Chan, D. Y. C. and Horn, R. G. (1985), "The Drainage of Thin Liquid Films between Solid Surfaces," *J. Chem. Phys.* 83, 5311–5324.

Cheong, C. U. and Stair, P. C. (1997), "The Drainage of Thin Liquid Films between Solid Surfaces," *Tribol. Lett.* in preparation.

Christenson, H. K., Gruen, D. W., Horn, R. G., and Israelachvili, J. (1989), "Structuring in Liquid Alkanes between Solid Surfaces: Force Measurements and Mean-Field Theory," *J. Chem. Phys.* 87, 1834–1841.

Cieplak, M., Smith, E. D. and Robbins, M. O. (1994), "Molecular Origins of Friction: The Force on Adsorbed Monolayers," 1209–1212.

Curry, J. E. and Cushman, J. (1995a). "Nanophase Coexistence and Sieving in Binary Mixtures Confined between Corrugated Walls," *J. Chem. Phys.* 103, 2132–2139.

Curry, J. E. and Cushman, J. H. (1995b), "Binary Mixtures of Simple Fluids in Structured Slit Micropores," *Mol. Phys.* 85, 173–192.

Curry, J. E., Zhang, F., Cushman, J., Schoen, M., and Diestler, D. J. (1994), "Transient Coexisting Nanophases in Ultrathin Films Confined between Corrugated Walls," *J. Chem. Phys.* 101, 10824–10832.

Dauber-Osguthorpe, P., Roberts, V. A., Osguthorpe, D. J., Wolff, J., Genest, M., and Hagler, A. T. (1988), "Structure and Energetics of Ligand-Binding to Proteins: *Escherichia coli* Dihydrofolate Reductase Trimethoprim, a Drug Receptor Systems," *Protein: Structure, Function, and Genetics*, 4, 31–47.

Davanloo, F., Lee, T. J., Jander, D. T., Park, H., You, J. H., and Collins, C. B. (1992), "Adhesion and Mechanical Properties of Amorphic Diamond Films Prepared by a Laser Plasma Discharge Source," *J. Appl. Phys.* 71, 1446–1453.

Diestler, D. J., Schoen, M., and Cushman, J. H. (1993), "On the Thermodynamic Stability of Confined Thin Films under Shear," *Science* 262, 545–547.

Dowson, D. (1979), *History of Tribology*, p. 215, Longman, London; Apparently Coulomb was also motivated by a prize of 2,000 louis d'or sponsored by the French navy for work on friction.

Ercolessi, F., Parrinello, M., and Tosatti, E. (1986a), "Au(100) Reconstruction in the Glue Model," *Surf. Sci.* 177, 314–328.

Ercolessi, F., Tosatti, E., and Parrinello, M. (1986b), "Au(100) Surface Reconstruction," *Phys. Rev. Lett.* 57, 719–722.

Ercolessi, F., Parrinello, M., and Tosatti, E. (1988), "Simulation of Gold in the Glue Model," *Philos. Mag. A* 58, 213–226.

Ewald, P. (1921), "Die Berechnung optischer und elektrostatischer Gitterpotentiale," *Ann. Phys.* 64, 253–287.

Feng, Z. and Field, J. E. (1991), "Friction of Diamond on Diamond and Chemical Vapor-Deposition Diamond Coatings," *Surf. Coat. Technol.* 47, 631–645.

Finnis, M. W. and Sinclair, J. F. (1984), "A Simple N-body Potential for Transition Metals," *Philos. Mag. A* 50, 45–56.

Fischer, T. E. (1992), "Chemical Effects in Friction," in *Fundamentals of Friction: Macroscopic and Microscopic Processes* (I. L. Singer and H. M. Pollock, eds.), pp. 299–321. Kluwer Academic Publishers, Dordrecht.

Foiles, S. M., Baskes, M. I., and Daw, M. S. (1986), "Embedded-Atom-Method Functions for the FCC Metals Cu, Ag, Au, Ni, Pd, Pt, and their Alloys," *Phys. Rev. B* 33, 7983–7991.

Frenkel, F. C. and Kontorova, T. (1938), "On the Theory of Plastic Demortation and Twinning," *Zh. Eksp. Teor. Fiz.* 8, 1340.

Gao, J., Luedtke, W. D. and Landman, U. (1995), "Nano-Elastohydrodynamics: Structure, Dynamics, and Flow in Nonuniform Lubricated Junctions," *Science* 270, 605–608.

Gao, J., Luedtke, W. D., and Landman, U. (1997), "Structure and Solvation Forces in Confined Films: Linear and Branched Alkanes," *J. Chem. Phys.* 106, 4309–4318.

Gear, C. W. (1971), *Numerical Initial Value Problems in Ordinary Differential Equations*, Prentice-Hall, Englewood Cliffs, NJ.

Gee, M. L., McGuiggan, P. M., Israelachvili, J. N., and Homola, A. M. (1990). "Liquid to Solidlike Transitions of Molecularly Thin Films under Shear," *J. Chem. Phys.* 93, 1895–1906.

Germann, G. J., Cohen, S. R., Neubauer, G., McClelland, G. M., Seki, H., and Coulman, D. (1993), "Atomic Scale Friction of a Diamond Tip on Diamond (100) and (111) Surfaces," *J. Appl. Phys.* 73, 163–167.

Glosli, J. N. and McClelland, G. M. (1993), "Molecular Dynamics Study of Sliding Friction of Ordered Organic Monolayers," *Phys. Rev. Lett.* 70, 1960–1963.

Glosli, J. N., Philpott, M. R., and McClelland, G. M. (1995), "Molecular Dynamics Simulation of Mechanical Deformation of Ultra-Thin Amorphous Carbon Films," *Mater. Res. Soc. Symp. Proc.* 383, 431–435.

Granick, S., Damirel, A. L., Cai, L. L., and Peanasky, J. (1995), "Soft Matter in a Tight Spot: Nanorheology of Confined Liquids and Block Copolymers," *Isr. J. Chem.* 35, 75–84.

Haile, J. M. (1992), *Molecular Dynamics Simulation: Elementary Methods*, John Wiley and Sons, New York.

Hammerberg, J. E., Holian, B. L., and Zhuo, S. J. (1995), "Studies of Sliding Friction in Compressed Copper," in *Shock Compression of Condensed Matter, Proceedings of the Conference of the American Physical Society Topical Group on Shock Compression of Condensed Matter*, Seattle, (Schmidt, S. C. and Tao, W. C., eds.), AIP Press, Woodbury, NY, API Conference Proceedings 370, Part I.

Harrison, J. A. and Brenner, D. W. (1994), "Simulated Tribochemistry: An Atomic-Scale View of the Wear of Diamond," *J. Am. Chem. Soc.* 116, 10399–10402.

Harrison, J. A., Brenner, D. W., White, C. T., and Colton, R. J. (1991), "Atomistic Mechanisms of Adhesion and Compression of Diamond Surfaces," *Thin Solid Films* 206, 213–219.

Harrison, J. A., Colton, R. J., White, C. T., and Brenner, D. W. (1992a), "Atomistic Simulation of the Nanoindentation of Diamond and Graphite Surfaces," *Mater. Res. Soc. Symp. Proc.* 239, 573–578.

Harrison, J. A., White, C. T., Colton, R. J., and Brenner, D. W. (1992b), "Nanoscale Investigation of Indentation, Adhesion, and Fracture of Diamond (111) Surfaces," *Surf. Sci.* 271, 57–67.

Harrison, J. A., White, C. T., Colton, R. J., and Brenner, D. W. (1992c), "Molecular-Dynamics Simulations of Atomic-Scale Friction of Diamond Surfaces," *Phys. Rev. B* 46, 9700–9708.

Harrison, J. A., Colton, R. J., White, C. T., and Brenner, D. W. (1993a), "Effect of Atomic-Scale Surface Roughness on Friction: A Molecular Dynamics Study of Diamond Surfaces," *Wear* 168, 127–133.

Harrison, J. A., White, C. T., Colton, R. J., and Brenner, D. W. (1993b), "Atomistic Simulations of Friction at Sliding Diamond Interfaces," *Mater. Res. Soc. Bull.* 18, 50–53.

Harrison, J. A., White, C. T., Colton, R. J., and Brenner, D. W. (1993c), "Effects of Chemically-Bound, Flexible Hydrocarbon Species on the Frictional Properties of Diamond Surfaces," *J. Phys. Chem.* 97, 6573–6576.

Harrison, J. A., White, C. T., Colton, R. J., and Brenner, D. W. (1995), "Investigation of the Atomic-Scale Friction and Energy Dissipation in Diamond Using Molecular Dynamics," *Thin Solid Films*, 260, 205–211.

Hautman, J. and Klein, M. L. (1989), "Simulation of a Monolayer of Alkyl Thiol Chains," *J. Chem. Phys.* 91, 4994–5001.

Hautman, J. and Klein, M. L. (1990), "Molecular Dynamics Simulation of the Effects of Temperature on a Dense Monolayer of Long-Chain Molecules," *J. Chem. Phys.* 93, 7483–7492.

Heermann, D. W. (1986), *Computer Simulation Methods in Theoretical Physics,* Springer-Verlag, Berlin.

Helman, J. S., Baltensperger, W., and Holyst, J. A. (1994), "Simple Model for Dry Friction," *Phys. Rev. B* 49, 3831–3838.

Heyes, D. M. (1981), "Electrostatic Potentials and Fields in Infinite Point Charge Lattices," *J. Chem. Phys.* 74, 1924–1929.

Hiller, B. (1995), "Contaminant Particles: The Microtribology Approach," in *Handbook of Micro/Nanotribology* (B. Bhushan, ed.), pp. 505–557, CRC Press, Boca Raton, FL.

Hirano, M. and Shinjo, K. (1984), "Atomistic Locking and Friction," *Phys. Rev. B* 41, 11837–11851.

Hirano, M. Shinjo, K., Kaneko, R., and Murata, R. (1991), "Anisotropy of Frictional Forces in Muscovite Mice," *Phys. Rev. Lett.* 67, 2642–2645.

Holian, B. L. (1991), "Effects of Pairwise versus Many-Body Forces on High-Stress Plastic Deformation," *Phys. Rev. A.* 43, 2655–2661.

Hoover, W. G. (1986), *Molecular Dynamics,* Springer-Verlag, Berlin.

Horn, R. G. and Israelachvili, J. N. (1981), "Direct Measurement of Structural Forces between Two Surfaces in a Nonpolar Liquid," *J. Chem. Phys.* 75, 1400–1411.

Israelachvili, J. N. (1992), "Intermolecular and Surface Forces," 2nd ed., pp. 169–172, Academic Press, London.

Israelachvili, J. N., McGuiggan, P. M., and Homola, A. M. (1988), "Dynamic Properties of Molecularly Thin Liquid Films," *Science* 240, 189–191.

Israelachvili, J. N., Kott, S. J., Gee, M., and Witten, T. A. (1989), "Forces between Mica Surfaces across Hydrocarbon Liquids: Effects of Branching and Polydispersity," *Macromolecules* 22, 4247–4253.

Jacobsen, K. W., Norskov, J. K., and Puska, M. J. (1987), "Interatomic Interactions in the Effective Medium Theory," *Phys. Rev. B* 35, 7423–7442.

Johnson, K. L., Kendell, K., and Roberts, A. D. (1971), "Surface Energy and the Contact of Elastic Solids," *Proc. R. Soc. London A* 324, 301–313.

Jorgensen, W. L., Madura, J. D., and Swenson, C. J. (1984), "Optimized Intermolecular Potential Functions for Liquid Hydrocarbons," *J. Am. Chem. Soc.* 106, 6638–6646.

Joyce, S. A., Thomas, R. C., Houston, J. E., Michalske, T. A., and Crooks, R. M. (1992), "Mechanical Relaxation of Organic Monolayer Films Measured by Force Microscopy," *Phys. Rev. Lett.* 68, 2790–2793.

Kallman, J. S., Hoover, W. G., Hoover, C. G., De Groot, A. J., Lee, S. M., and Wooten, F. (1993), "Molecular Dynamics of Silicon Indentation," *Phys. Rev. B* 47, 7705–7709.

Khor, K. E. and Das Sarma, S. (1988), "Proposed Universal Interatomic Potential for Elemental Tetrahedrally Bonded Semiconductors," *Phys. Rev. B* 38, 3318–3322.

Koike, A. and Yoneya, M. (1996), "Molecular Dynamics Simulations of Sliding Friction of Langmuir–Blodgett Monolayers," *J. Chem. Phys.* 105, 6060–6067.

Kremer, K. and Grest, G. (1990), "Dynamics of Entangled Linear Polymer Melts: A Molecular-Dynamics Simulation," *J. Chem. Phys.* 92, 5057–5086.

Krim, J. (1996), "Friction at the Atomic Scale," *Sci. Am.* 275, 74–84.

Krim, J., Solina, D. H., and Chiarello, R. (1991), "Nanotribology of a Kr Monolayer: A Quartz-Crystal Microbalance Study of Atomic-Scale Friction," *Phys. Rev. Lett.* 66, 181–184.

Landman, U. and Luedtke, W. D. (1991), "Nanomechanics and Dynamics of Tip–Substrate Interactions," *J. Vac. Sci. Technol. B* 9, 414–423.

Landman, U., Luedtke, W. D., and Ribarsky, M. W. (1989a), "Structural and Dynamical Consequences of Interactions in Interfacial Systems," *J. Vac. Sci. Technol. A* 7, 2829–2839.

Landman, U., Luedtke, W. D., and Ribarsky, M. W. (1989b), "Dynamics of Tip–Substrate Interactions in Atomic Force Microscopy," *Surf. Sci. Lett.* 210, L117–L184.

Landman, U., Luedtke, W. D., Burnham, N. A., and Colton, R. J. (1990), "Atomistic Mechanisms and Dynamics of Adhesion, Nanoindentation, and Fracture," *Science* 248, 454–461.

Landman, U., Luedtke, W. D., and Ringer, E. M. (1992), "Atomistic Mechanisms of Adhesive Contact Formation and Interfacial Processes," *Wear* 153, 3–30.

Landman, U., Luedtke, W. D., Ouyang, J., and Xia, T. K. (1993), "Nanotribology and the Stability of Nanostructures," *Jpn. J. App. Phys.* 32, 1444–1462.

Liu, G.-Y. and Salmeron, M. B. (1994), "Reversible Displacement of Chemisorbed *n*-Alkanethiol Molecules on Au(111) Surface: An Atomic Force Microscopy Study," *Langmuir* 10, 367.

Lucchese, R. R. and Tully, J. C. (1983), "Trajectory Studies of Rainbow Scattering from the Reconstructed Si(100) Surface," *Surf. Sci.* 137, 570–594.

Lupkowski, M. and Maguire, J. F. (1992), "Molecular-Dynamics Simulations of Avalanche in Adhesion in a Two-Dimensional Solid," *Phys. Rev. B* 45, 13733–13736.

Martyna, G. J., Klein, M. L., and Tuckerman, M. (1992), "Nosé-Hoover Chains: The Canonical Ensemble via Continuous Dynamics," *J. Chem. Phys.* 97, 2635–2643.

Mate, C. M., McClelland, G. M., Erlandsson, R., and Chiang, S. (1987), "Atomic-Scale Friction of a Tungsten Tip on a Graphite Surface," *Phys. Rev. Lett.* 59, 1942–1945.

McClelland, G. M. and Glosli, J. N. (1992), "Friction at the Atomic Scale," in *Fundamentals of Friction: Macroscopic and Microscopic Processes* (I. L. Singer and H. M. Pollock, eds.), pp. 405–422, Kluwer Academic Publishers, Dordrecht.

McQuarrie, D. A. (1976), *Statistical Mechanics,* Harper and Row, New York.

Menon, M. and Allen, R. E. (1986), "New Technique for Molecular-Dynamics Computer Simulations: Hellmann-Feyman Theorem and Subspace Hamiltonian Approach," *Phys. Rev. B* 33, 7099–7101.

Menzel, D. H. (1960), *Fundamental Formulas of Physics,* Dover Publications, New York.

Meyer, E., Overney, R., Brodbeck, D., Howald, L., Luthi, R., Frommer, J., and Guntherodt, H.-J. (1992), "Friction and Wear of Langmuir–Blodgett Films Observed by Friction Force Microscopy," *Phys. Rev. Lett.* 69, 1777–1780.

Minowa, K. and Sumino, K. (1992), "Stress-Induced Amorphization of a Silicon Crystal by Mechanical Scratching," *Phys. Rev. Lett.* 69, 320–323.

Murrell, J. N., Carter, S., Farantos, S. C., Huxley, P., and Varandas, A. J. C. (1984), *Molecular Potential Energy Functions,* Wiley, New York.

Norskov, J. K. and Lang, N. D. (1980), "Effective Medium Theory of Chemical Binding: Application to Chemisorption," *Phys. Rev. B* 21, 2131–2136.

Nosé, S. (1984a), "A Unified Formulation of the Constant-Temperature Molecular Dynamics Method," *J. Chem. Phys.* 81, 511–519.

Nosé, S. (1984b), "A Molecular Dynamics Method for Simulations in the Canonical Ensemble," *Mol. Phys.* 52, 255–268.

Overney, R. M., Meyer, E., Frommer, J., Guntherodt, H.-J., Fujihira, M., Takano, H., and Gotoh, Y. (1994), "Force Microscopy Study of Friction and Elastic Compliance of Phase-Separated Organic Thin Films," *Langmuir* 10, 1281–1286.

Perry, M. D. and Harrison, J. A. (1996a), "Molecular Dynamics Studies of the Frictional Properties of Hydrocarbon Materials," *Langmuir* 12, 4552–4556.

Perry, M. D. and Harrison, J. A. (1996b), "Molecular Dynamics Investigations of the Effects of Debris Molecules on the Friction and Wear of Diamond," *Thin Solid Films* 290/291, 211–215.

Perry, M. D. and Harrison, J. A. (1997), "Friction between Diamond Surfaces in the Presence of Small Third-Body Molecules," *J. Phys. Chem.* 101, 1364–1373.

Persson, B. N. J., Schumacher, D., and Otto, A. (1991), "Surface Resistivity and Vibrational Damping in Adsorbed Layers," *Chem. Phys. Lett.* 178, 204–212.

Pethica, J. B. and Sutton, A. P. (1988), "On the Stability of a Tip and Flat at very Small Separations," *J. Vac. Sci. Technol. A* 6, 2490–2494.

Pharr, G. M. and Oliver, W. C. (1989), "Nanoindentation of Silver — Relations between Hardness and Dislocation Structure," *J. Mater. Res.* 4, 94–104.

Raeker, T. J. and DePristo, A. E. (1991), "Theory of Chemical Bonding Based on the Atom-Homogeneous Electron Gas System," *Int. Rev. Phys. Chem.* 10, 1–54.

Raffi-Tabar, H. and Kawazoe, Y. (1993), "Dynamics of Atomically Thin Layers — Surface Interactions in Tip–Substrate Geometry," *Jpn. J. Appl. Phys.* 32, 1394–1400.

Raffi-Tabar, H. and Sutton, A. P. (1991), "Long-Range Finnis-Sinclair Potentials for FCC Metallic Alloys," *Philos. Mag. Lett.* 63, 217–224.

Raffi-Tabar, H., Pethica, J. B., and Sutton, A. P. (1992), "Influence of Adsorbate Monolayer on the Nano-Mechanics of Tip–Substrate Interactions," *Mater. Res. Soc. Symp. Proc.* 239, 313–318.

Ribarsky, M. W. and Landman, U. (1992), "Structure and Dynamics of *n*-Alkanes Confined by Solid Surfaces. I. Stationary Crystalline Boundaries," *J. Chem. Phys.* 97, 1937–1949.

Rick, S. W., Stuart, S. J., and Berne, B. J. (1994), "Dynamical Fluctuating Charge Force Fields: Applications to Liquid Water," *J. Chem. Phys.* 101, 6141–6156.

Robbins, M. O., Thompson, P. A., and Grest, G. A. (1993), "Simulations of Nanometer-Thick Lubricating Films," *Mater. Res. Soc. Bull.* 18, 45–49.

Robertson, J. (1992), "Properties of Diamond-Like Carbon," *Surf. Coat. Technol.* 50, 185–203.

Rossi, F., Andre, B., Van Veen, A., Mijnarends, P. E., Schut, H., Gissler, W., Haupt, J., Lucazeau, G., and Abello, J. (1994), "Effect of Ion Beam Assistance on the Microstructure of Nonhydrogenated Amorphous Carbon," *J. Appl. Phys.* 75, 3121–3129.

Ryckaert, J. P. and Bellmans, A. (1978), "Molecular Dynamics of Liquid Alkanes," *Discuss. Faraday Soc.* 66, 95–106.

Sankey, O. F. and Allen, R. E. (1986), "Atomic Forces from Electronic Energies via the Hellmann-Feynman Theorem, with Application to Semiconductor (110) Surface Relaxation," *Phys. Rev. B* 33, 7164–7171.

Schneider, T. and Stoll, E. (1978), "Molecular Dynamics Study of a Three-Dimensional One-Component Model for Distortive Phase Transitions," *Phys. Rev. B* 17, 1302–1322.

Schoen, M., Rhykerd, C. L., Diestler, D. J., and Cushman, J. H. (1987), "Fluids in Micropores. I. Structure of a Simple Classical Fluid in a Slit-Pore," *J. Chem. Phys.* 87, 5464–5476.

Schoen, M., Rhykerd, C. L., Diestler, D. J., and Cushman, J. H. (1989), "Shear Forces in Molecularly Thin Films," *Science* 245, 1223–1225.

Siepmann, J. I. and Frenkel, D. (1992), "Configurational Bias Monte Carlo: A New Sampling Scheme for Flexible Chains," *Mol. Phys.* 75, 59–70.

Siepmann, J. I. and McDonald, I. R. (1993a), "Monte Carlo Study of the Properties of Self-Assembled Monolayers Formed by Adsorption of $CH_3(CH_2)_{15}SH$ on the (111) Surface of Gold," *Mol. Phys.* 79, 457–473.

Siepmann, J.I. and McDonald, I.R. (1993b), "Monte Carlo Simulation of the Mechanical Relaxation of a Self-Assembled Monolayer, "*Phy. Rev. Lett.* 70, 453-456.

Singer, I. L. (1991), "A Thermochemical Model for Analyzing Low Wear Rate Materials," *Surf. Coat. Technol.* 49, 474–481.

Singer, I. L. (1992), "Epiloque to the Nato ASI on Fundamentals of Friction," in *Fundamentals of Friction: Macroscopic and Microscopic Processes* (I. L. Singer and H. M. Pollock, eds.), pp. 569–588, Kluwer Academic Publishers, Dordrecht.

Singer, I. K., Fayeulle, S., and Ehni, P. D. (1991), "Friction and Wear Behavior of TiN in Air: The Chemistry of Transfer Films and Debris Formation," *Wear* 149, 375–394.

Sinnott, S. B., Colton, R. J., White, C. T., Shenderova, O. A., Brenner, D. W., and Harrison, J. A. (1997), "Atomistic Simulations of the Nanometer-Scale Indentation of Amorphous-Carbon Thin Films," *J. Vac. Sci. Technol. A* 15, 936–940.

Smith, E. D., Robbins, M. O., and Cieplak, M. (1996), "Friction on Adsorbed Monolayers," *Phys. Rev. B* 54, 8252–8260.

Smith, J. R. and Banerjea, A. (1987), "A New Approach to Calculation of Total Energies of Solids with Defects: Surface-Energy Anisotropies," *Phys. Rev. Lett.* 59, 2451–2454.

Smith, J. R., Bozzolo, G., Banerjea, A., and Ferrante, J. (1989), "Avalanche in Adhesion," *Phys. Rev. Lett.* 63, 1269–1272.

Sneddon, I. N. (1965), "The Relation between Load and Penetration in the Axisymmetric Boussinesq Problem for a Punch of Arbitrary Profile," *Int. J. Eng. Sci.* 3, 47–58.

Sokol, P. E., Ma, W. J., Herwig, K. W., Snow, W. M., Wang, Y., Koplik, J., and Banavar, J. R. (1992), "Freezing in Confined Geometries," *Appl. Phys. Lett.* 61, 777–779.

Sokoloff, J. B. (1984), "Theory of Dynamical Friction between Idealized Sliding Surfaces," *Surf. Sci.* 144, 267–272.

Sokoloff, J. B. (1990), "Theory of Energy Dissipation in Sliding Crystal Surfaces," *Phys. Rev. B.* 42, 760–765.

Sokoloff, J. B. (1992), "Theory of Atomic Level Sliding Friction between Ideal Crystal Interfaces," *J. Appl. Phys.* 72, 1262–1270.

Sokoloff, J. B. (1993), "Possible Nearly Frictionless Sliding for Mesoscopic Solids," *Phys. Rev. Lett.* 71, 3450–3453.

Sokoloff, J. B. (1996), "Theory of Electron and Phonon Contributions to Sliding Friction," in *Physics of Sliding Friction* (B. N. J. Persson and E. Tosatti, eds.), pp. 217–229, Kluwer Academic Publishers, Dordrecht.

Sorensen, M. R., Jacobsen, K. W., and Stoltze, P. (1996), "Simulations of Atomic-Scale Friction Surfaces," *Phys. Rev. B* 53, 2101–2113.

Stillinger, F. H. and Weber, T. A. (1985), "Computer Simulation of Local Order in Condensed Phases of Silicon," *Phys. Rev. B* 31, 5262–5271.

Stott, M. J. and Zaremba, E. (1980), "Quasiatoms: An Approach to Atoms in Nonuniform Electronic Systems," *Phys. Rev. B* 22, 1564–1583.

Steele, W.A. (1973), "Physical Interaction of FASES with Crystalline Solids. I. Gas–Solid Energies and Properties of Isolated Adsorbed Atoms," *Surf. Sci.* 36, 317–352.

Streitz, F. H. and Mintmire, J. W. (1994), "Electrostatic Potentials for Metal-Oxide Surfaces and Interfaces," *Phys. Rev. B* 50, 11996–12003.

Streitz, F. H. and Mintmire, J. W. (1996), "Molecular Dynamics Simulations of Elastic Response and Tensile Failure of Alumina," *Langmuir* 12, 4605–4609.

Sutton, A. P. (1993), *Electronic Structure of Materials,* Clarendon, Oxford.

Tabor, D. and Field, J. E. (1992), in *The Properties of Natural and Synthetic Diamond* (Field, J. E., ed.), Academic Press, London, pp. 547–571 and references therein.

Tersoff, J. (1986), "New Empirical Potential for the Structural Properties of Silicon," *Phys. Rev. Lett.* 56, 632–635.

Tersoff, J. (1989), "Modeling Solid-State Chemistry: Interatomic Potentials for Multicomponent Systems," *Phys. Rev. B* 39, 5566–5568.

Thompson, P. A. and Robbins, M. O. (1990a), "Origin of Stick-Slip Motion in Boundary Lubrication," *Science* 250, 792–794.

Thompson, P. A. and Robbins, M. O. (1990b), "Shear Flow Near Solids: Epitaxial Order and Flow Boundary Conditions," *Phys. Rev. A* 41, 6830–6837.

Thompson, P. A., Grest, G. S., and Robbins, M. O. (1992), "Phase Transitions and Universal Dynamics in Confined Films," *Phys. Rev. Lett.* 68, 3448–3451.

Tomagnini, O., Ercolessi, F., and Tosatti, E. (1993), "Microscopic Interaction between a Gold Tip and a Pb(110) Surface," *Surf. Sci.* 287/288, 1041–1045.

Tomlinson, G. A. (1929), "A Molecular Theory of Friction," *Philos. Mag. Ser. 7*, 7, 905–939.

Tully, J. C. (1980), "Dynamics of Gas-Surface Interactions: 3D Generalized Langevin Model Applied to FCC and BCC Surfaces," *J. Chem. Phys.* 73, 1975–1985.

Tupper, K. J. and Brenner, D. W. (1994a), "Compression-Induced Structural Transition in a Self-Assembled Monolayer," *Langmuir* 10, 2335–2338.

Tupper, K. J. and Brenner, D. W. (1994b), "Molecular Dynamics Simulations of Friction in Self-Assembled Monolayers," *Thin Solid Films* 253, 185–188.

Tupper, K. J., Colton, R. J., and Brenner, D. W. (1994), "Simulations of Self-Assembled Monolayers under Compression: Effect of Surface Asperities," *Langmuir* 10, 2041–2043.

Van Alsten, J. and Granick, S. (1988), "Molecular Tribometry of Ultrathin Liquid Films," *Phys. Rev. Lett.* 61, 2570–2573.

Wang, Y., Hill, K., and Harris, J. G. (1993a), "Thin Films of *n*-Octane Confined between Parallel Solid Surfaces. Structure and Adhesive Forces vs. Film Thickness from Molecular Dynamics Simulations," *J. Phys. Chem.* 97, 9013–9021.

Wang, Y., Hill, K., and Harris, J. G. (1993b), "Comparison of Branched and Linear Octanes in the Surface Force Apparatus. A Molecular Dynamics Study," *Langmuir* 9, 1983–1985.

Wang, Y., Hill, K., and Harris, J. G. (1994), "Confined Thin Films of a Linear and Branched Octane. A Comparison of the Structure and Solvation Forces Using Molecular Dynamics," *J. Phys. Chem.* 100, 3276–3285.

Watts, E. T., Krim, J., and Widom, A. (1990), "Experimental Observation of Interfacial Slippage at the Boundary of Molecularly Thin Films with Gold Substrates," *Phys. Rev. B* 41, 3466–3472.

Wilks, J. and Wilks, E. M. (1979), in *The Properties of Diamond* (Field, J. E., ed.), pp. 351–381, Academic Press, London.

Woodcock, L. V. (1970), "Isothermal Molecular Dynamics Calculations for Liquid Salts," *Chem. Phys. Lett.* 10, 257–261.

Yan, W. and Komvopoulos, K. (1998), "Three-Dimensional Molecular Dynamics Analysis of Atomic-Scale Indentation," *J. Tribol.* 120, 385–392.

Zhong, W. and Tomanek, D. (1990), "First-Principles Theory of Atomic-Scale Friction," *Phys. Rev. Lett.* 64, 3054–3057.

Part II

Applications

12

Design and Construction of Magnetic Storage Devices

Hirofumi Kondo, Hiroshi Takino,
Hiroyuki Osaki, Norio Saito,
and Hiroshi Kano

12.1 Introduction

Magnetic recording is the most common technology used to store many different types of signals. Analog recording of sound was the first and is still a major application. Digital recording of encoded computer data on disk and tape recorders has evolved as another major use. Hard disk drives use high signal frequencies coupled with high medium speeds, and emphasize small access times together with high reliability. A third large application area is video recording for professional or consumer use. The high video frequencies are normally recorded using rotary-head drums. Despite the availability of other methods of storing data, such as optical recording and semiconductor devices, magnetic recording media has the following advantages: (1) inexpensive media, (2) stable storage, (3) relatively high data rate, (4) high volumetric density.

In principle, a magnetic recording medium consists of a permanent magnet and a pattern of remanent magnetization can be formed along the length of a single track, or a number of parallel tracks on its surface. Magnetic recording is accomplished by relative motion between a magnetic medium (tape or

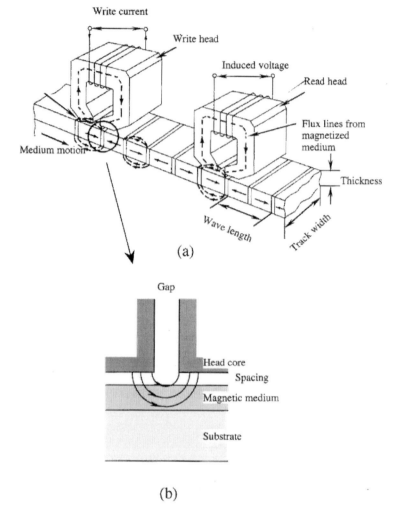

Write current

Write head

Induced voltage

Read head

Medium motion

Flux lines from
magnetized
medium

Thickness

Wave length

Track width

(a)

Gap

Head core

Spacing

Magnetic medium

Substrate

(b)

FIGURE 12.1 (a) Illustration of the recording and reproducing process. (b) Schematic of cross-sectional view showing the magnetic field at the gap.

disk) against a stationary or rotatory read/write head. The one track example is given in Figure 12.1a. The medium is in the form of a magnetic layer supported on a nonmagnetic substrate. The recording or the reproducing head is a ring-shaped electromagnet with a gap at the surface facing the medium. When the head is fed with a current representing the signal to be recorded, the fringing field from the gap magnetizes the medium as shown in Figure 12.1b. For a constant medium velocity, the spatial variations in remanent magnetization along the length of the medium reflect the temporal variations in the head current, and constitute a recording of the signal.

The recording magnetization creates a pattern of external and internal fields, in the simplest case, to a series of contiguous bar magnets. When the recorded medium is passed over the same head, or a reproducing head of similar construction, the flux emanating from the medium surface is intercepted by the head core, and a voltage is induced in the coil proportional to the rate of change of this flux. The voltage is not an exact replica of the recording signal, but it constitutes a reproduction of it in that information describing the recording signal can be obtained from this voltage by appropriate electrical processing. The combination of a ring head and a medium having longitudinal anisotropy tends to produce a recorded magnetization. This combination has been the one used traditionally, and it still

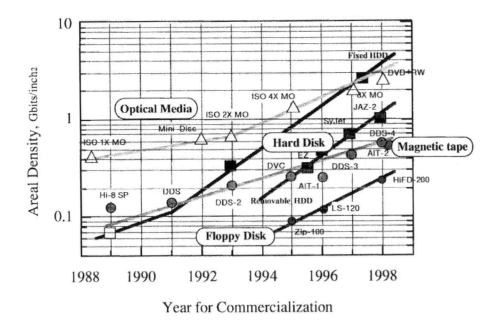

FIGURE 12.2 Areal density migration of magnetic recording media. Optical media shown for comparison.

dominates all major analog and digital applications. Ideally, the pattern of magnetization created by a square-wave recording signal would be like that shown in Figure 12.1a.

In between recording and reproduction, the recorded signal can be stored indefinitely even if the medium is not exposed to magnetic fields comparable in strength to those used in recording. Whenever recording is no longer required, it can be erased by means of a strong field applied by the same head as that used for recording or by a separate erase head. After erasure, the medium is ready for a new recording. Overwriting an old signal with a new one, without a separate erase step, is available for writing.

Figure 12.2 shows a road map of magnetic storage devices including hard disks (fixed and removable), magnetic tapes, and floppy disks and of optical storage device. The recording density has been increasing continuously over the years and a plot of logarithm of the areal density vs. year almost gives a straight line. The areal density of the hard disk is almost the same as the optical medium. For high areal recording density, the linear flux density and the track density should be as high as possible. Reproduced signal amplitude decreases rapidly with a decrease in the recording wavelength and track width. The signal loss is a function of magnetic properties and thickness of the magnetic coating, read gap length, and head–medium spacing. For high recording densities, high magnetic flux density and coercivity of a medium are needed. Regarding the materials, metal magnetic powder (MP) and a monolithic cobalt alloy thin film of higher magnetic saturation and coercivity have been launched in recent media. So as to a magnetic head, higher frequency response and sensitivity are required.

It is known that the signal loss as a result of spacing can be reduced exponentially by reducing the separation between the head and medium. A physical contact between the medium and the head occurs during starting and stopping operation and a load-carrying air film is developed at the interface in the relative motion. Closer flying heights lead to undesirable collision of asperities and increased wear so that this air film should be thick enough to mitigate any asperity contacts; on the contrary it must be thin enough to attain a large reproduced signal. Thus, the head–medium interface should be designed with optimum conditions.

The achievement of higher recording densities requires smoother surfaces. The ultimate objective is to use two smooth surfaces in contact for recording provided the tribological issues can be resolved. Smooth surfaces lead to an increase in adhesion, friction, and interface temperatures. Friction and wear issues are resolved by appropriate selection of interface materials and lubricants, by controlling the

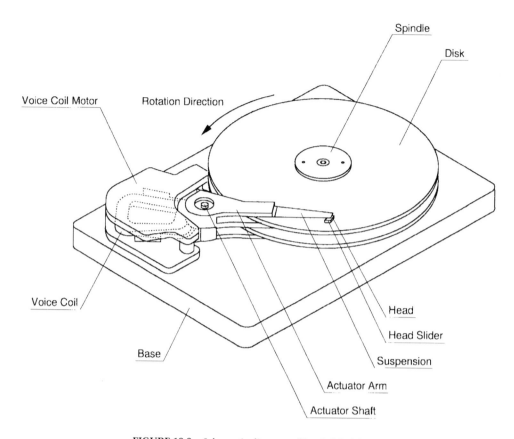

FIGURE 12.3 Schematic diagram of hard disk drive.

dynamics of the head and medium, and the environment. A fundamental understanding of the tribology of the magnetic head–medium interface becomes crucial for the continuous growth of the magnetic storage industry.

In this chapter materials and construction used in the modern media and heads are reviewed. Selected interesting fabrication processes of these devices are also described.

12.2 Hard Disk Files

Magnetic heads for rigid disk drives are discussed in this section. Figure 12.3 shows the schematic of the rigid disk drive. A 3.5-in.-diameter disk is widely used and two to three disks are typically stacked in one hard disk drive. For very high storage density drives, up to about ten disks are stacked. Writing and reading are done with magnetic heads attached to a spring suspension. The slider surface (air-bearing surface) is designed to develop a hydrodynamic force to maintain an adequate spacing (~50 nm) between a head slider and a disk surface. The magnetic head assembly is actuated by a stepper motor or voice coil motor to access the data on the disk. The magnetic head-suspension assembly is high, and the fast access speed can be achieved. From these characteristics, hard disk drives have an advantage of fast access speed and high storage density.

12.2.1 Heads

The areal density of the rigid disk drives have been increasing 60% per year; the magnetic recording head performance must be improved continuously to maintain this high growth rate of the areal recording density. The track width of the recording head must be narrower and narrower and the transfer rate

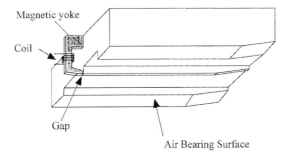

FIGURE 12.4   The schematic diagram of the ferrite monolithic head.

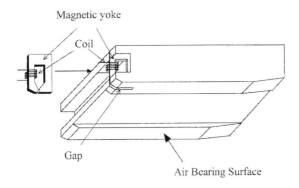

FIGURE 12.5   The schematic diagram of the composite head.

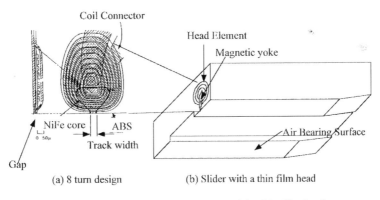

(a) 8 turn design   (b) Slider with a thin film head

FIGURE 12.6   The schematic diagram of the thin-film head.

becomes higher and higher. The ferrite bulk head (monolithic head, Figure 12.4) and the composite MIG head (metal-in-gap head Figure 12.5) were widely used for the rigid disk drives. Since these two types of bulk recording heads are fabricated mainly by conventional machining processes, it is difficult to control a narrow track width down to 10 μm. On the other hand, thin-film inductive heads are fabricated by using the same photolithography processes that are used for semiconductor devices, which allows control of a narrow track width. The coil inductance must be reduced for the high transfer rate application. The yoke size of the monolithic head is almost the same as that of the MIG head shown in Figures 12.4 and 12.5 (Jones, 1980). Figure 12.6a shows the eight-turn thin-film inductive head and Figure 12.6b shows the slider with a thin-film head. Minimizing the total magnetic ring yoke size of the film head, the coil inductance of the thin-film head can be reduced. Film heads have an advantage of the

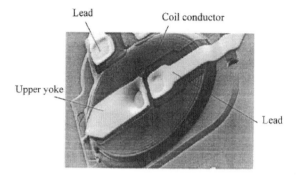

FIGURE 12.7   SEM image of the thin-film inductive head.

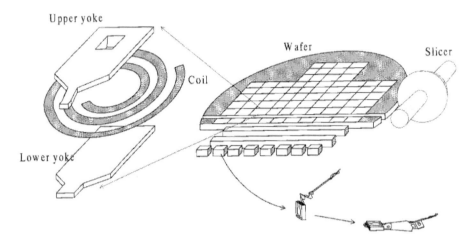

FIGURE 12.8   The schematics of the slider fabrication process.

high-frequency response and reduced inductance due to a small volume of magnetic yoke and allows higher transfer rate. Very high recording density drives require the use of a magnetoresistive (MR) head which will be described later.

12.2.1.1   Structure and Fabrication Process of Thin-Film Inductive Heads

Figure 12.6a shows the cross section and the planar view of the thin-film inductive head. The magnetic gap is located on the air-bearing surface (ABS). The track width is defined in the planar view. In order to achieve high magnetic yoke efficiency, the track width should be narrower compared with the width of the recessed yoke area. Figure 12.7 shows the SEM image of a thin-film head. Film heads must be deposited on a substrate for which a hard Al_2O_3–TiC ceramic is usually employed. With few exceptions, permalloy, which is the alloy of approximately 80 wt% Ni with 20 wt% Fe, is used for the magnetic layer of the film head, because an annealing process is not necessary to obtain the high permeability (1000 to 3000) and the low coercivity (3 to 5 Oe). As indicated in Figure 12.8, the heat-cured photoresist materials are used for insulation layers. After plating the coil layer, the surface of coil layer is not smooth; therefore, the photoresist is coated. The photoresist insulation layer also makes the surface of the upper coil layer smooth. Figure 12.8 shows the fabrication process of a thin-film head element. Several thousands of the head elements are fabricated on the same substrate at the same time. The thin-film heads are fabricated by stacking thin-film layers. First, the magnetic layer is deposited; then the coil layer and the upper magnetic layer are plated subsequently. The passivation layers are also deposited between the coil layer

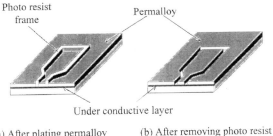

(a) After plating permalloy (b) After removing photo resist

FIGURE 12.9 Schematics of the framed permalloy plating: (a) after plating permalloy; (b) after removing photoresist.

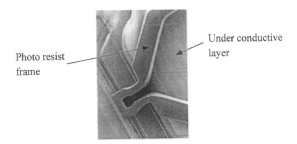

FIGURE 12.10 SEM image of the photoresist frame for plating permalloy upper yoke.

and both the upper and lower permalloy magnetic layers. Finally, the thick protective Al_2O_3 layer (30 to 50 μm) is sputtered for protecting the head element. Then the wafer is sliced into the head sliders.

In a thin-film head fabrication process, permalloy and copper can be deposited by evaporation, sputtering, or plating. In a photolithography process, a deposited film is etched physically or chemically through a patterned photoresist. In a electroplating process, a material is plated only on the conductive layer. Materials cannot be plated where a conductive layer is not exposed. Figure 12.9 shows this electroplating process. First, a conductive layer is deposited on all areas of a wafer and a photoresist is coated and patterned. The patterned photoresist covers a part of a conductive layer. An electroplating material (permalloy or copper) can be plated only on the exposed area. After removing this frame, patterned permalloy or copper can be obtained in Figure 12.9. Figure 12.10 shows the SEM images of the frame of an upper permalloy layer for an electroplating. The copper and the upper permalloy layer is also plated by using a photoresist frame. This frame is patterned on a conductive layer; permalloy is plated only on the exposed area of the under conductive layer. After removing the resist frame, the patterned upper permalloy layer is obtained as shown in Figure 12.7 The top pole width is controlled by the photoresist patterning width and the track width tolerance of the upper permalloy yoke can be reduced.

12.1.1.2 Head Slider Manufacturing Process

After finishing the wafer process, the wafer must be sliced into the head sliders. First, a wafer (Figure 12.11a) is sliced to a row of bars (Figure 12.11b). The sliced surface (surface A in Figure 12.11b) is lapped very carefully, because this surface will be an ABS and the head throat height is controlled through this lapping process. The throat height of thin-film head is about 1 μm, the tolerance of this row bar lapping process is required less than 1 μm. The row bar is attached to the toolings for lapping an ABS (Figure 12.11c) of row bars. This tooling can be bent for obtaining a precise throat height for all head tips in a row bar. After finishing the throat height lapping, many row bars are aligned and the lapped surfaces are etched to make the ABS at the same time (Figure 12.11d). Recently, in order to obtain a constant flying height for all disk radii, a negative pressure air bearing has been widely used. The shape of a negative-pressure air bearing is not simple; an ion-etching process must be used to make a air-bearing

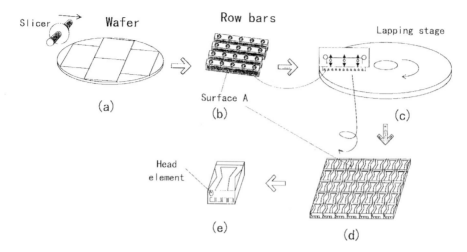

FIGURE 12.11 The schematics of the slider fabrication process.

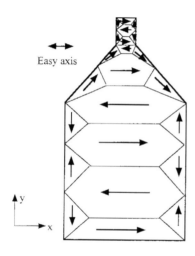

FIGURE 12.12 Typical domain configurations in an inductive film head yoke.

shape. After the ion-etching process, the head slider can be obtained by dividing a row bar into the head sliders (Figure 12.11e).

12.2.1.3 Domain Structure in a Thin-Film Head

Magnetic materials are composed of individual domains with local magnetization which is equal to the saturation magnetization of the materials. Magnetic domain structure is defined by minimizing the total magnetic energy of the domain wall, the magnetic anisotropy, and the magnetostriction energy. In general, when the size of the magnetic film is reduced to several hundred microns, the magnetic domain structure becomes clear. Since the size of a magnetic yoke of a film head is almost the same size, domain structure affects the read-and-write characteristics and the stability of the film head. A typical domain structure of the upper magnetic yoke is shown in Figure 12.12. The easy axis is indicated by the arrow direction, and the magnetization direction of most domain patterns is parallel to the easy axis. The domains are separated by a 180° Bloch wall. When the magnetic easy axis is in the x-direction, a large portion of a domain aligns the x-direction. To reduce total magnetic energy, a domain whose magnetization direction aligns in the y-direction appears in the edge region, because domains of the y-direction cancel surface magnetic charges. Also magnetostriction effects must be considered for designing a film

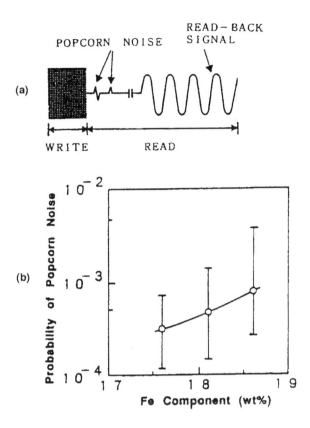

FIGURE 12.13 (a) Schematic of the noise just after a write mode and (b) probability of popcorn noise vs. Fe composition in Ni.

head. An anisotropy energy can be changed when length of a magnetic material changes. A magneto-striction coefficient (λ) is a ratio of anisotropy energy change to material length change. In general, λ is a very small value of 10^{-6}, but change in length of the film head magnetic material is rather large, because a thickness of the film material is very thin compared with the substrate thickness. The small distortion of the substrate affects the large distortion to the magnetic film materials. Consequently, the domain structure changes with the distortion of the substrate. Magnetostriction coefficient λ which is a function of a composition of Ni and Fe is an inherent characteristic of materials. The domain wall could not move smoothly due to an impurity and a void in the magnetic film. Magnetic energy rapidly changes when the magnetic domain wall moves through this impurity and the defect (Mallinson, 1994). The magnetic domain wall moves irregularly, if its energy change is large enough. This phenomenon results in two types of instabilities of an inductive head, so-called write instability and read instability. Write instability occurs after the termination of a write operation. A spike noise appears just after a write mode. Such noise is particularly detrimental in the drives, which employs sector head positioning servoing, since servoing must occur immediately after writing (Klassen and van Peppen, 1989). Figure 12.13a shows the schematics of the noise just after a write mode (Morikawa et al., 1991). Write and read modes in a rigid disk drive change very frequently, a film head must read the signals immediately after writing a signal. When a write current is large enough to saturate the magnetization of the film magnetic yoke, a magnetic domain wall disappears. After the write mode, the magnetic yoke forms the domain structure for reduction of the total magnetic energy. During this process, domain walls move to make a domain structure stable. If some domain wall moves irregularly, and the magnetic flux of the yoke changes irregularly, the coil undesirably detects this irregular flux change. This noise just after a write mode is

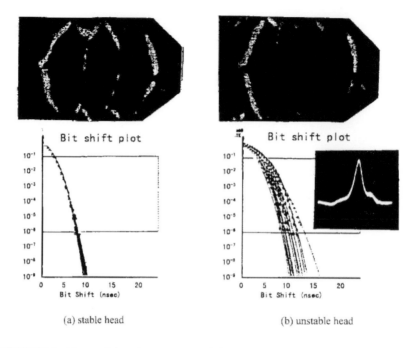

| (a) stable head | (b) unstable head |

FIGURE 12.14 Measured domain structures and the error rate of the inductive thin-film head.

called "popcorn noise," which is controlled by the magnetostriction energy. By controlling the composition of Ni and Fe, the magnetostriction of the magnetic yoke material can be optimized. Figure 12.13b shows the probability of popcorn noise vs. the Fe composition. The probability of popcorn noise is the probability of popcorn noise divided by the cycle of the read/write mode. Ni and Fe content must be controlled to reduce popcorn noise. "Read instability" associated with distortions of read-back pulse is called "wiggle." A distorted waveform is shown in Figure 12.14b, which shows a small pulse just after the main peak (Williams and Lambert, 1989). A window margin test (error rate as changing a detecting window) is also shown in Figure 12.13. The stable head in Figure 12.14a shows a very repeatable error rate characteristic, but a head (Figure 12.14b) shows excessive variability of the window margin test. Domain configurations of these heads are also shown in Figure 12.14 (Corb, 1990). By controlling the magnetostriction coefficient λ, the domain structure of the film head is designed to exhibit stable read and write characteristics.

12.2.1.4 Edge-Eliminated Head

Thin-film heads can take advantage of the improved wavelength response due to the use of finite pole lengths. But the finite pole length also shows undershoot signals on both sides of the main peak. The magnetic field distribution near the gap corner is shown in Figure 12.15 (van Herk, 1980). The magnetic field distribution itself exhibits undershoot on the both outer edges of the poles. Therefore, the reproduced pulse also has undershoot at both sides of the center pulse. These undershoot signals degrade the high-linear-density response because the undershoot signal can affect the adjacent signal (Singh and Bischoff, 1985). In order to eliminate these undershoot signals, a pole edge-eliminated head is proposed to remove the undershoot signal. Edges of top and under poles are trimmed as shown in Figure 12.16a (Yoshida, 1993). Figure 12.17a shows the ABS of the conventional thin-film inductive head. The isolated read-back pulse of the conventional head is shown in Figure 12.17b. For the isolated read-back pulse of an edge-eliminated head shown in Figure 12.16b, the read-back pulse has a little undershoot signal.

12.2.1.5 Thin-Film Silicon Head

Two major head design approaches had been developed: one uses the films perpendicular to the recording media and the other uses the films parallel to the recording media, as shown in Figure 12.18a and b. The

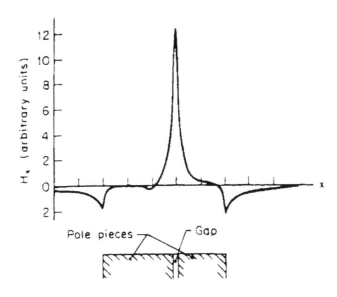

FIGURE 12.15   The magnetic field distribution near the recording gap.

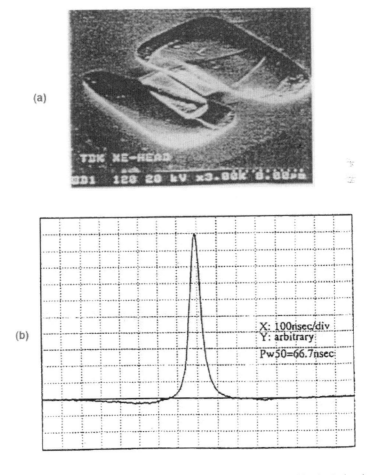

FIGURE 12.16   (a) SEM image of the pole-tip area of an edge-eliminated head. (b) The isolated read-back pulse of an edge-eliminated head.

(a)

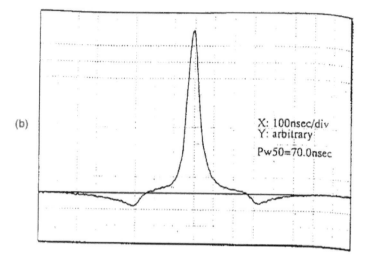

(b)

X: 100nsec/div
Y: arbitrary
Pw50=70.0nsec

FIGURE 12.17 (a) SEM image of the pole-tip area of the conventional thin-film head. (b) The isolated read-back pulse of the conventional head.

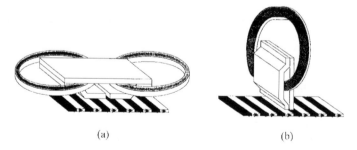

(a)   (b)

FIGURE 12.18 (a) Schematics of the silicon head and (b) the conventional vertical head.

construction in Figure 12.18b, which has become conventional, is known as the "vertical" configuration. The vertical configuration is widely used for rigid disk drive heads, which requires a precision lapping technique to provide the throat height on the order of 1 μm. The construction, shown in Figure 12.18a, is known as the "horizontal" configuration. This horizontal configuration was used for the earliest thin-film head design. Recently, these type heads have been introduced in some disk drives to eliminate the costly precision lapping process (Figure 12.19) (Lazzari, 1989).

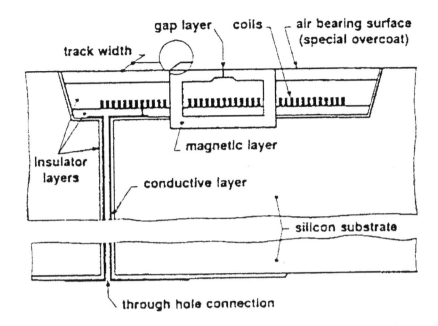

FIGURE 12.19   The cross section of the planer head.

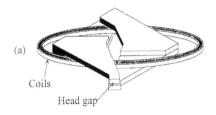

(a)

Coils

Head gap

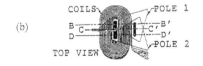

(b)

COILS   POLE 1
B
  C
D
        POLE 2
TOP VIEW

FIGURE 12.20   (a) Schematic diagram of the diamond head and (b) top view of the diamond head.

12.2.1.6   Diamond Head

A unique head design, which is called "diamond head" has been proposed for high-performance film heads. Figure 12.20 shows the schematic diagram and a planar view of a diamond head (Mallary and Ramaswamy, 1993). With diamond head, the magnetic yoke is twisted one more around the back part of the coil. The magnetic flux from the media goes through the magnetic yoke twice around the coil. The read efficiency of the diamond head is ideally twice as large as that of a conventional inductive head.

12.2.2   Construction of the Magnetoresistive Head

An MR head was proposed by R. Hunt in 1971 for the reproduce head of magnetic recording systems (Hunt, 1971). The MR head belongs to a group of reproduce heads that utilizes direct magnetic flux-sensing as a means of read back. The reproduce signal amplitude of the MR head is independent of the relative velocity between a head and a media, and the inductance of the MR head is very low compared

609

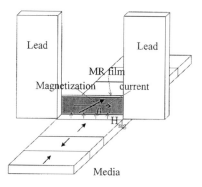

FIGURE 12.21 Schematic of the MR reproduce head.

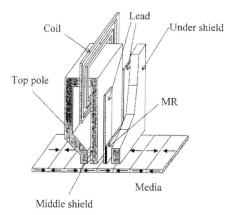

FIGURE 12.22 Combined MR head.

to that of the thin-film inductive head. An MR head is suitable for stationary tape head and rigid disk applications with high transfer rate. IBM first introduced the MR head (Tsang et al., 1990) for the rigid disk drives and the MR heads are now widely used.

12.2.2.1 MR Sensor Structure

An MR head belongs to a group of reproduce heads; therefore, a recording head and the MR reproduce head must be combined. Figure 12.21 shows the schematic of the MR reproduce head, and a combined MR head is shown in Figure 12.22. As shown in Figure 12.21 an MR sensor film is located perpendicular to the medium surface, and the leads are located on both sides of the MR sensor to supply the sense current for detecting the sensor resistance changes. The read sensor is made of an MR ferromagnetic film conductor such as permalloy ($Ni_{80}Fe_{20}wt\%$), whose resistance can be modulated by the angle between its magnetic moment and the current-flow direction. A resistivity of permalloy film changes is shown in Equations 12.1 and 12.2:

$$(12.1)$$

$$\Delta\rho = \rho_{\parallel} - \rho_{\perp} \qquad (12.2)$$

where $\rho_{\parallel}$ is a resistivity whose sensor magnetization is parallel to the current-flow direction, $\rho_{\perp}$ is a resistivity whose sensor magnetization is perpendicular to the current-flow direction, θ is the angle between the sensor magnetization and the current-flow direction. The resistivity of the MR element shows a quadratic change vs. $\cos\theta$. Permalloy thin film has been used for the MR sensor, because it has a high permeability ($\mu = 2000$) and a high MR ratio ($\Delta\rho/\rho = 2\%$). Equation 12.1 shows the relation

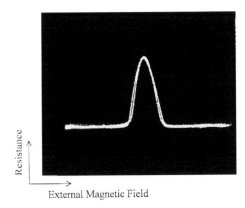

Resistance

External Magnetic Field

FIGURE 12.23   *R–H* curve of MR film
$(1 \times 1$ in. size film$)$.

between the resistivity and the angle θ, and also the relation between the resistivity and the external magnetic field is needed to investigate the reproduce characteristics. An MR element is a soft magnetic material with a uniaxial magnetic anisotropy. Total magnetic energy E_T is

$$E_\mathrm{T} = E_\mathrm{ex} + E_u \qquad (12.3)$$

where E_ex is a magnetic energy from an external magnetic field and E_u is a magnetic anisotropy energy. E_ex and E_u can be described as follows:

$$E_\mathrm{ex} = -MH_\mathrm{ex}\sin\theta \qquad (12.4)$$

$$E_u = K_u \sin^2\theta \qquad (12.5)$$

M is the magnetization of the MR element, H_ex is the external magnetic field whose direction is parallel to the signal magnetic field, and K_u is the uniaxial magnetic anisotropy constant. The quasi-stable state is obtained from $\partial E_T/\partial\theta = 0$. With using Equations 12.3 through 12.5, the resistivity of the relation between the resistivity and the external magnetic field is described in Equation 12.6,

$$\rho(\theta) = \rho(H_\mathrm{ex}) = \Delta\rho\left(1 - \left(\frac{H_\mathrm{ex}}{H_k}\right)^2\right) + \rho_\perp \qquad (12.6)$$

where $H_k = 2K_u/M$ is an anisotropy field. Therefore, the resistivity of the MR element also shows the quadratic change vs. the external magnetic field. Figure 12.23 is an R–H curve (relation between a resistance of an MR element and an external magnetic field) of a large 1-in.-square MR element. An MR sensor needs a bias magnetic field to obtain the linear response. The resistivity shows a quadratic change vs. the signal field (Figure 12.24). Without a bias magnetic field, the output waveform deforms as shown in Figure 12.24 (bottom). The linear output waveform can be obtained by applying the DC magnetic field to the MR sensor. With the optimum bias state, the positive amplitude and the negative amplitude are almost the same. Therefore, the optimum magnetization angle θ_0 satisfies the following equation:

$$\Delta\rho\left(1 - \left(\frac{H_\mathrm{ex}}{H_k}\right)^2\right) = \Delta\rho\cos^2\theta_0 = \frac{\Delta\rho}{2} \qquad (12.7)$$

611

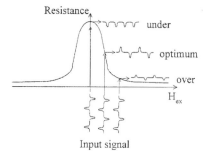

Input signal

FIGURE 12.24 *R–H* curve of MR element.

From this equation, $\cos \theta_0 = 1/\sqrt{2} = 0.7$, and θ_0 is found to be 45°. The optimum angle between the sensor magnetization and the current-flow direction should be 45°. The MR sensor needs a bias technique to obtain this optimum biased state.

12.2.2.2 Bias Technique

There are many bias techniques to linearize an MR signal response. Setting the initial magnetization of the MR element to 45°, the optimum bias magnetic field must be applied to the same direction of the signal magnetic field. This bias is called "transverse bias," because the bias field direction is transverse to the MR element. Three bias techniques are summarized in Figure 12.25a, b, and c. The bias techniques are summarized by Jeffers (1986).

12.2.2.2.1 Shunt Bias

The schematic diagram of the shunt bias technique is shown in Figure 12.25a (Shelledy and Brock, 1975). A nonmagnetic conductor film such as Ti is located adjacent to the MR element, which applies the bias magnetic field to the MR film. A sense current flows through an MR film and a shunt film and generates a magnetic field whose direction is transverse to the MR sensor. This field can be utilized to apply the bias field to the MR element. But the distribution of the shunt bias field is nonuniform across the height of the MR element, diminishing rapidly near the upper and lower sensor edges. Both edge regions are underbiased by a shunt bias technique. Since the MR sensor height varies through the lapping process, the sense current must be optimized for each element whose stripe height is different.

12.2.2.2.2 Self-Adjacent-Layer Bias

In order to improve the shunt bias technique, SAL (self-adjacent-layer) bias technique has been designed for the MR reproduce heads (Beaulieu and Nepala, 1975). Figure 12.25b shows the schematic diagram of the SAL bias technique whose structure is the same as the shunt bias technique. Instead of the shunt layer, the soft magnetic film is placed adjacent to the MR sensor film. The sense current flows both SAL and MR film; these sense currents generate the magnetic fields which are parallel to the external signal magnetic field. Moreover, a thickness of SAL layer is 70% of the MR sensor layer. If this SAL film magnetization is saturated with sufficient sense current, the magnetization of an MR sensor is not saturated. The value cos 45° is about 0.7, and the angle of the magnetization of the MR sensor may be 45° from the current-flow direction, which is roughly the optimum bias state of the MR sensor as mentioned before. For a SAL bias film, an MR magnetization is automatically magnetized 45° optimum bias state with any sense current. Also the magnetization distribution of an MR film is uniform across the height of the MR element, because the demagnetization magnetic field of the SAL film is high at the edges of the SAL film and diminishes rapidly at the center of the element. Therefore, the SAL bias is suitable for a bias method of an MR head.

12.2.2.2.3 Self-Bias

A simple bias technique has been proposed in Figure 12.25c. There is no extra layer for applying a bias magnetic field to the MR element, but the placement of the MR element is not symmetrical between the shields. If the MR element is placed in the center, the magnetic fields generated from two image currents

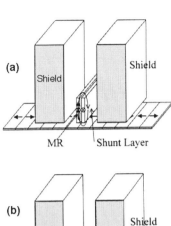

(a)

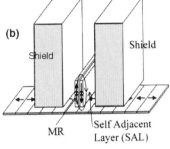

(b)

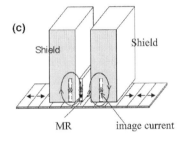

(c)

FIGURE 12.25 (a) Shunt bias MR head.
(b) Self-adjacent layer (SAL) MR head.
(c) Self-bias MR head.

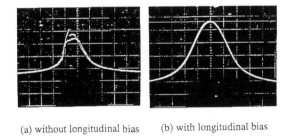

(a) without longitudinal bias   (b) with longitudinal bias

FIGURE 12.26   *R–H* curve without (a) and with (b) longitudinal bias magnetic field.

cancel each other. But if it is not in the center, the magnetic field generated from two image currents is not canceled and a transverse magnetic field is applied to the MR sensor. This magnetic field can utilize a bias magnetic field.

12.1.2.3   Barkhausen Noise

A magnetic material shows a domain structure to reduce the total magnetic energy. As indicated before, these domain walls do not move smoothly and magnetization changes irregularly. Figure 12.26a shows the *R–H* curve of the small permalloy element. There are many jumps and kinks on the *R–H* curves, and

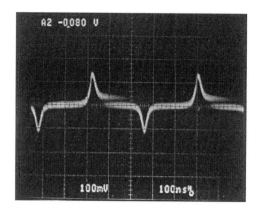

FIGURE 12.27 Read-back waveform with Barkhausen noise.

the output signal also shows irregularity (Tsang and Decker, 1982). Figure 12.27 shows the noisy output signals. The output signal deforms and has jumps. This irregular response is called "Barkhausen noise." To suppress Barkhausen noise, an MR sensor should be a single-domain state. The single-domain state can be obtained by applying a small magnetic field in the longitudinal direction (the same direction as the sensor current) of the MR sensor. Figure 12.26a and b shows curves with and without this longitudinal magnetic (or bias) field, respectively. Without the longitudinal bias field, $R–H$ curves show a large Barkhausen noise (Figure 12.26a). But with the longitudinal bias field, $H = 10$ Oe, the $R–H$ curve has the smooth and regular response to the external field (Figure 12.26b). Many techniques have been proposed to apply the longitudinal bias field to the MR sensor. Three techniques, namely, (1) hard magnet, (2) antiferromagnetic film, and (3) vertical and double layer, are shown in Figure 12.28a, b, and c. The hard magnets are located on both sides of the MR element and the hard magnet thin film is magnetized to the longitudinal direction (Figure 12.28a). Therefore the $R–H$ curve of this element shows a smooth response (Hannon et al., 1994). Figure 12.28b shows the technique for suppressing Barkhausen noise that uses antiferromagnetic thin film. When this element is annealed and cooled with a magnetic field from a blocking temperature, the contact part of the MR element to the antiferromagnetic layer is magnetized to the same direction of the annealing magnetic field. Figure 12.29 shows the $R–H$ curve after field annealing (Tsang, 1981, 1984). This figure indicates that when a longitudinal bias field is applied to an MR element, the response of the MR element can be stabilized. Therefore, the magnetization of the MR element rotates smoothly as the external magnetic field. Figure 12.28c shows the vertical and double-layer MR head. This structure also suppresses Barkhausen noise (van Ooyen et al., 1982; Jagielinsky et al., 1986; Saito et al., 1987). The sense current from the rear lead to the front lead generates the magnetic field whose direction is parallel to the track width direction (x-axis). This direction is the same as the longitudinal direction of the conventional MR head in Figure 12.28a and b. The magnetic field generated from the sense current acts as a longitudinal bias magnetic field of the conventional MR head. As shown in Figure 12.28c, this longitudinal bias field is antiparallel to the each MR element, and the direction of the MR magnetization is also antiparallel to each other. Therefore, the magnetization rotates in tandem by changing the signal magnetic field. The vertical and double-layer MR element shows a smooth $R–H$ curve (Figure 12.30).

12.1.2.4 MR Head Structure

As mentioned before, an MR head belongs to a group of reproduce heads that utilize direct magnetic flux sensing as a means of read back, a thin-film inductive recording head must be combined together. An MR head is more suitable for high-track-density and high-linear-density applications.

12.1.2.4.1 SAL Head

The MR head with the SAL bias film is widely used for the rigid disk drive application, and is called a SAL head. Figure 12.31 shows the schematic diagram of the SAL head. Similar to the thin-film inductive head, Al_2O_3–TiC ceramic is used for the substrate material. The inductive recording head is combined

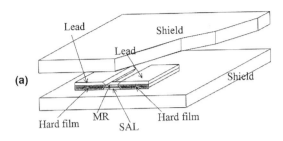

(a)

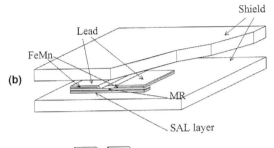

(b)

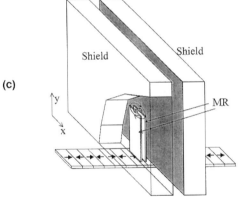

(c)

FIGURE 12.28 (a) MR head with hard film. (b) MR head with antiferromagnetic (FeMn) layer. (c) Vertical MR head.

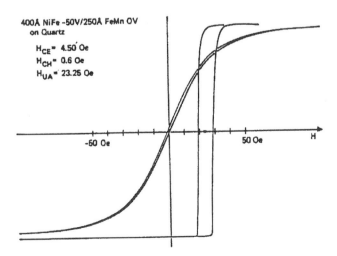

400Å NiFe –50V/250Å FeMn OV on Quartz

H_{CE} = 4.50 Oe
H_{CH} = 0.6 Oe
H_{UA} = 23.25 Oe

FIGURE 12.29 *R–H* curve with antiferromagnetic layer.

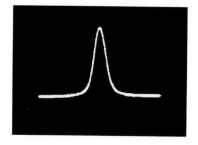

FIGURE 12.30 Smooth R–H curve of vertical MR element with a sense current (10 mA).

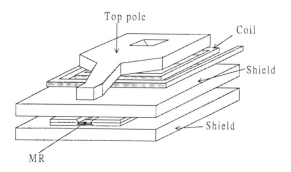

FIGURE 12.31 Structure of SAL head.

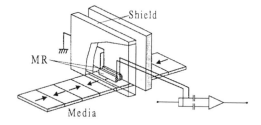

FIGURE 12.32 Structure of the dual-stripe MR head.

with the MR head. As explained before, the hard magnet layers are placed on both sides of the MR element to suppress Barkhausen noise. The MR sensor leads are located on the hard magnet layer. The upper shield layer is also used for the lower recording yoke. An electroplated permalloy film is used for a shield layer and Sendust (AlFeSi) or permalloy thin film is used for an under shield layer. The electroplated permalloy is used for a top pole material, but a high-B_s (saturation magnetization) material can also be used for recording high-coercivity media. A plated $Ni_{45}Fe_{55}$ (Robertson et al., 1997) and amorphous CoZr–X film and FeN film have been investigated for a high-B_s material for a top pole.

12.2.2.4.2   Dual-Stripe Head

A unique MR head was proposed in which two MR elements are connected differentially (Voegeli, 1975). This type of MR head is called a dual-stripe MR head. Figure 12.32 shows the schematic diagram of a dual-stripe MR head. There are the two same MR elements placed adjacent to each other between the shields. The sense current flows in the same direction, and the transverse bias field is applied antiparallel to the both MR elements. No extra bias layer is needed to optimize a linearity of the MR sensor response. The advantages of this sensor structure are that the output signal amplitude is practically double that of the SAL head because of two MR films and the differential voltage detection (Bhattacharya et al., 1987) and that thermal-induced noise can be canceled (Hempstead, 1975; Anthony et al., 1994). But this head needs three wires to connect two MR sensors differentially.

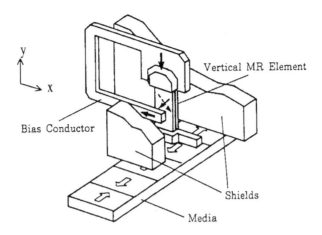

FIGURE 12.33 Structure of vertical MR head.

12.2.2.4.3   Vertical MR Head

The vertical MR head was designed in 1988 for high-track-density rigid disk drives (Suyama et al., 1988). The schematic diagram is shown in Figure 12.33. A conductor layer is needed to make the MR response linear, which generates the transverse bias field to the MR element. The major advantages of the vertical MR sensor configuration are that the output signal amplitude is independent of the track width (Takada et al., 1997), and the sensor might be safe from an electrical shorting problem at the ABS (Saito et al., 1993). But several problems are to be considered; read-efficiency reduction due to a longitudinal magnetic field from the sense current and a longer path of the sensor (Wang, 1993). In order to improve the read-efficiency, two vertical MR elements are insulated from each other (Shibata et al., 1996) so that the sensor current may flow only in one vertical MR sensor. Another improvement has been proposed for a vertical MR head. One of the vertical MR elements is to change the hard magnetic thin film to improve the stability of the sensor. The hard magnetic film whose magnetization direction is parallel to the ABS (x-axis in Figure 12.33) is placed adjacent to the MR element. The demagnetization magnetic field from this hard magnet acts as a longitudinal bias field, which stabilizes the MR element.

12.2.2.4.4   Horizontal Head

A horizontal MR head with a planar inductive write head has been designed as shown in Figure 12.34 (Chapman, 1989). This horizontal head has two MR elements that are connected differentially. A bias conductor layer is located above the MR element to apply a bias magnetic field in the same direction to the both elements. This horizontal MR head also cancels thermal-induced noise.

12.2.2.5   Thermal Asperity

The MR sensor needs a sensor current to detect the resistance change of its MR sensor, because the cross section of the MR element is very small and the sensor current density is very high, about 1×10^{11} A/m^2. The temperature of the MR sensor becomes several tens to 100°C. When the sensor hits an asperity on the disk, the MR sensor temperature and the resistance of the MR sensor also changes. Figure 12.35a and b show the base-level variation of the output signal (Sawatzky, 1997). This base-level variation is called "thermal asperity." If the MR sensor passes through a small emboss on the disk, the temperature of the sensor might rise. Therefore, the base level varies to the positive side immediately (Figure 12.35b), and the resistance of the MR sensor becomes high. On the other hand, if the MR sensor hits an asperity on the disk, the MR sensor can be cooled because the heat in the MR element is scattered by the asperity on the disk. The base level of the MR sensor varies to the negative side (Figure 12.35a). Dual-stripe head (Figure 12.32) can cancel this thermal asperity because the two MR sensors are connected differentially and also the horizontal head cancels a thermal asperity.

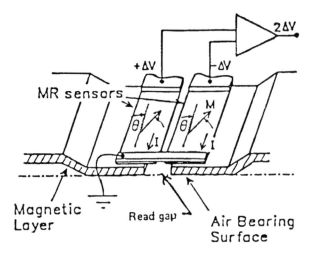

FIGURE 12.34   Structure of horizontal MR head.

12.1.2.6   Electrostatic Discharge Damage

The MR head needs to be protected against an electrostatic discharge (ESD). The cross section of the MR element is very small so that the MR element can be burned out from a very small ESD (Tian and Lee, 1995). ESD damage to the MR sensor results from excessive joule heating through electrical contact of the lead with a statically charged object. Figure 12.36 shows an SEM photograph from the ABS of a damaged MR element. The MR element is burned out by discharging electrostatic charges to both leads of the MR element.

12.1.2.7   GMR Head (Spin Valve Head)

Baibich et al. (1988) has demonstrated the new MR element, the giant magnetoresistive (GMR) effect which shows the very high MR ratio, using the synthetic superlattice. The new GMR head has been proposed for ever-increasing high density rigid disk drives. The GMR element is composed of over ten thin films. To apply this GMR elements for the reproduce head of rigid disk drives, film layers may be reduced to less than ten layers. This GMR head is called the spin valve head (Heim et al., 1994), because the magnetic free layer acts as a valve to the sense current. Figure 12.37 shows the structure of the spin valve head. First, a magnetic free layer is deposited; then a pinned layer is deposited. The magnetization direction of the pinned layer is the transverse direction, and the antiferromagnetic layer (FeMn, etc.) is used for magnetizing the pinned layer to the transverse direction. The adjacent hard magnet layer is needed to stabilize the behavior of a magnetic free layer. Any bias film is not necessary, because the resistance of the spin valve head varies linearly to the external magnetic field. The output signal is about double that of the SAL head; the spin valve head is expected to be used widely for high-density rigid disk drives.

12.2.2   The Disk

A hard disk drive occupies the major position in the external memory field for a computer at present because of the high recording capacity, the fast accessing speed, and the high data transfer rate. Especially, recording density is rapidly increasing owing to the appearance of thin-film disks and the MR heads. The disk(s) is mounted on a spindle which rotates inside a hard disk drive. The read/write head(s) is mounted on slider(s) which is attached to a spring suspension set on a swing-arm electromagnetic

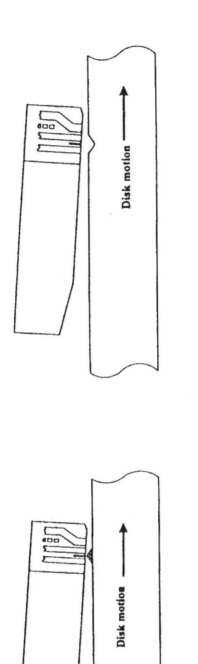

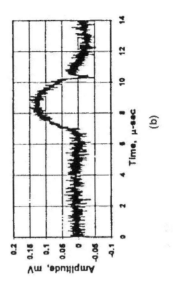

(a)

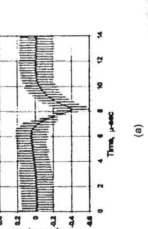

(b)

FIGURE 12.35 Thermal asperity of MR head. (a) MR head path through an asperity. (b) MR head hits an emboss on the disk.

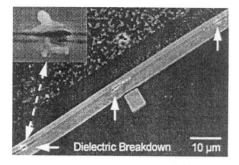

FIGURE 12.36   Micrograph of ESD-damaged MR head.

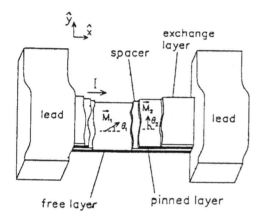

FIGURE 12.37   Schematic of spin valve element.

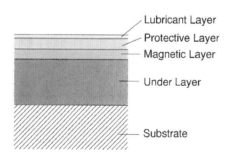

FIGURE 12.38   A construction of a hard disk.

actuator. The head is flown on the disk by a hydrodynamic bearing, which is generated at the interface between the slider and the disk. This head–disk interface is the primary factor of the fast accessing speed because the flying head mechanism performs a friction-free system in principle. Performance of a hard disk media is mainly determined from two points of view which are the recording density and the reliability for head–disk friction and clash. The former depends on both of the magnetic characteristics of a recording thin film and the flyability of a disk (substrate). The latter depends on the head–disk interface and the mechanical characteristics of a substrate/disk.

12.2.2.1   Construction of Thin-Film Disks

Figure 12.38 shows the construction of a typical hard disk with a recording layer of a thin magnetic film. A hard disk consists of a flat rigid substrate, a recording layer (thickness of about 20 nm) usually with an underlayer in thickness of about 100 nm, a protective layer of about 20 nm, and a lubricant layer of less than 4 nm.

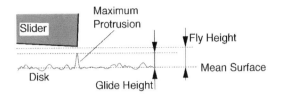

FIGURE 12.39 A head/disk interface (HDI).

There are two kinds of substrate for practical use. One is an aluminum-magnethium alloy disk on which a nickel–phosphorus film is plated. Another is a glass disk which is surface-strengthened chemically or crystallized to improve the mechanical strength. For example, diameters of disks are 5.25, 3.5, 2.5, and 1.8 in., and thicknesses are 1.2, 0.89, 0.635, and 0.389 mm, respectively. Figure 12.39 shows the head–disk interface (HDI) schematically to explain the flyability and the reliability for the head–disk friction and collision. A glide height means the minimum distance without collision between a head and a disk. Taking mechanical clearance of a drive into consideration, the glide height must be less than 40 nm to achieve the flying height of 50 nm. In addition, dents must be also removed from a disk surface because some defects of a disk causes errors of the electric signals.

Recording density of the disk is determined by reproducing voltage and media noise. The reproducing voltage depends on the magnetic characteristics, the thickness of the magnetic thin film, and the spacing loss which depends on the flying height of a head slider, the thickness of a protective layer, and a lubricant layer. A waveform of a reproduced isolated magnetic transition by an inductive head is expressed as follows in Equation 12.8 (Bertram, 1994):

$$V(x) = \frac{2}{\pi g} NWE\nu\mu_0 M_r \delta \left(\tan^{-1}\right)\left(\frac{g/2+x}{d+a} + \tan^{-1}\left(\frac{g/2-x}{d+a}\right)\right) \qquad (12.8)$$

where x is a position along a track, N is a turn number of the read-back head, W is a width of a head, E is an efficiency of a head, ν is a relative velocity between disk and head, μ_0 is the permeability of free space, g is a gap length of a head, d is a head/disk spacing, and a is a transition length of a magnetization. A transition length a is represented theoretically by Williams and Comstock in 1971 as Equation 12.9 (Williams and Comstock, 1971):

$$a = f + \sqrt{f^2 + \frac{M_r \delta d'}{H_c \, \pi r Q}} \qquad (12.9)$$

where

$$d' = \sqrt{d(d+\delta)}$$

$$f = \frac{d'(1-S^*)}{\pi Q}$$

$$Q = \frac{\sqrt{3}}{2}\left[1 - \frac{1}{4}\exp\left(\frac{-5d'}{3g}\right)\right]$$

$$r = \frac{3+S^*}{4}$$

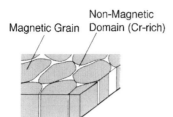

Non-Magnetic
Magnetic Grain Domain (Cr-rich)

FIGURE 12.40 A microstructure of a magnetic film.

The peak voltage of the signal at the transition center ($x = 0$) is expressed by Equation 12.10:

$$V_p^{\text{Ind}} = \frac{4}{\pi g}\, NWE\nu\mu_0 M_r \delta \tan^{-1}\left(\frac{g}{2(d+a)}\right) \tag{12.10}$$

For an MR head, the peak voltage of a reproduced signal is expressed as below:

$$V_p^{\text{MR}} = \frac{9\Delta\rho JWM_r\delta(g+t)}{8\sqrt{2}tgM_s^{\text{MR}}} \tan^{-1}\left(\frac{g}{2(d+a)}\right) \tag{12.11}$$

where $\Delta\rho$ is a coefficient of resistance change (Ω-m), J is a current density of MR element, W is a track width, t is a thickness of an MR element, and M_s^{MR} is a saturation magnetization of an MR element.

These equations show that the reproducing voltage depends on the three magnetic parameters of media, which are magnetic remanent magnetization (M_r), media thickness (δ), and coercive force (H_c).

There is another convenient expression for estimating a spacing loss (Bertram, 1994):

$$L_d = 20\log_{10} e^{2\pi d/\lambda} = 54.6\,\frac{d}{\lambda}\ \text{dB} \tag{12.12}$$

It means that an amplitude of reproduced signal shows logarithmic decay in proportion to a head/disk spacing (d) or a recorded wave number ($1/\lambda$). It can be easily understood that these parameters M_r, H_c, δ, and d must be optimized to obtain the highest reproducing voltage over the wide range of the recording wavelength, taking account of the recording ability of a head.

On the other hand, the media noise depends on the microstructure of the magnetic film. Figure 12.40 shows the structure of the magnetic film, which consists of many grains. Equation 12.13 shows the relation between a signal-to-noise ratio (S/N) and the number of magnetic particles which are included in a unit of volume (ν), where W is a track width, v is a relative velocity, and f is a frequency of recorded signal. An increase of ν has an equal effect to a decrease in the size of magnetic particles. The size of magnetic particles means the grain size of the magnetic film in the case of the thin-film media. Accordingly, it is necessary to reduce the media noise by decreasing the grain size.

$$S/N \propto v\sqrt{\frac{\nu W}{f}} \tag{12.13}$$

The material of the recording layer is cobalt alloy with which several kinds of metal elements are added to increase magnetic coercive force H_c and/or to reduce the grain size. Usually nickel, chromium, tantalum, and platinum are used for the additional metals. In addition, an underlayer consisting of a

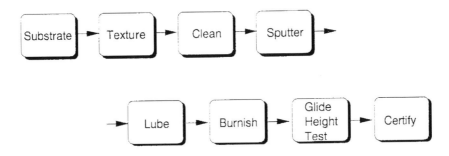

FIGURE 12.41   A manufacturing process of hard disks.

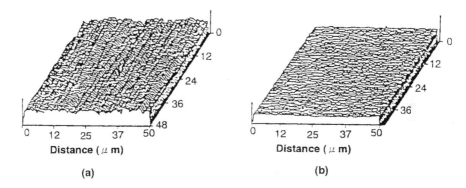

FIGURE 12.42   Textured disk surfaces by (a) mechanical lapping (in circumferential direction) and (b) chemical or physical etching (isotropically).

chromium–metal(s) alloy, molybdenum, or tungsten is also applied between a substrate and a recording layer to improve the magnetic characteristics.

It is important to obtain high reproducing voltage in the short recording wavelength area; the thickness of the lubricant and the protective film must be as small as possible, without deteriorating reliability against the head–disk collision. Amorphous carbon or SiO_x is usually applied for the protective layer and, recently, diamondlike carbon is also considered. Liquid fluorinated lubricant, perfluoropolyether, is utilized.

12.2.2.2   Manufacturing Process Step

Figure 12.41 shows the typical manufacturing process of a hard disk. Every step is performed automatically in an extremely clean environment to avoid generating dents and protrusions on the disk surface.

12.2.2.2.1   Texturing

In the hard disk drive, the head starts flying from the disk surface and stops flying to the disk surface; it is called contact-start-stop (CSS). As the disk surface becomes smooth, it may cause high friction force and stiction between a head slider and a disk. The head finally sticks to the disk surface when the friction force becomes larger than the rotating torque which can be supplied by a spindle motor. In order to reduce friction, the small random roughness is formed on the disk surface by the "texture" processing. Texture is performed in a circumferential direction by mechanical lapping, or is formed isotropically by chemical or physical etching as shown in Figure 12.42. It can be clearly understood that the CSS performance conflicts with the recording performance. Accordingly, the zone texture disk, which has the textured area only in the CSS zone and the nontextured area in the data zone, has been recently developed. The substrates are cleaned before the next sputtering process (Bhushan, 1996a,b).

12.2.2.2.2 Sputtering

The underlayer, the recording layer, and the protective layer are deposited on the substrate by the sputtering method. There are three important parameters that determine the sputtering condition. The argon gas pressure influences the angles of incidence and the kinetic energy of the sputtering atoms because of the collisions between atoms of argon and sputtering material. The substrate temperature relates to the diffusion energy of the sputtering atoms. The sputtering power is generally in proportion to the number of the sputtering atoms. For these reasons, the grain size and the magnetic characteristics of the recording film can be closely related with these sputtering parameters.

12.2.2.2.3 Lubrication

The lubricant is applied above the protective layer to improve lubrication performance between a head and a disk. The lubricant film is usually formed by the dipping method. The disks are dipped into the solution of perfluoropolyethers in fluorinated solvents. After the disks emerge from the solution, the solvent evaporates quickly resulting in a uniform lubricant film on the disk surface (Bhushan, 1996a,b).

12.2.2.2.4 Burnishing and Glide Height Testing

Burnishing and the glide height testing are usually performed on the same test stand. In a burnishing process, the protrusions on the disk surface are removed by a burnishing head which flies, maintaining a certain spacing to the disk surface. Almost all the protrusions larger than flying height of the burnishing head are removed. In succession, the glide height, which is defined as a minimum flying height so that a slider flies without collisions against a disk, is inspected by using the piezoelectric method (Ashar, 1997). The mechanical energy of a head–disk collision is transformed into an electric energy by the piezoelectric sensor which is embedded into a glide testing slider. When the electric output signal is observed from the piezoelectric sensor which is flying on a disk with a criteria of glide height, the disk is regarded as a glide failure. Recently, the thermal asperity method has been proposed for glide height testing, instead of the piezoelectric method. When a slider hits protrusions on a disk, a mechanical energy of collision generates heat. It changes a resistance between both ends of a conductive strip fabricated on a flying slider. When the constant current flows from one end of the strip to another, the collision signal is obtained as a change of the voltage between both ends of the strip.

12.2.2.2.5 Certification

In general, a track average amplitude, missing pulses, super pulses, extra pulses, positive modulation, negative modulation, and PW_{50} are used as a measure of certification. Testing methods and purposes follow.

To measure a track average amplitude (V_{ta}), the specified high-frequency (HF) signal is recorded on a disk for a revolution. A V_{ta} is defined as the average of all peak-to-peak amplitude for each reproduced pulse during one entire revolution. Accordingly, a V_{ta} is calculated referring to amplitudes of reproduced HF pulses. A V_{ta} gives not only a signal output level which is one of fundamental disk characteristics but also a reference value for the tests shown below. Next, the read-back HF signal is compared simultaneously with a specified missing pulse and/or superpulse threshold. A missing pulse is defined as any peak that does not reach the specified missing pulse threshold. A superpulse is defined as any peak that exceeds a specified superpulse threshold. These thresholds are usually expressed as a percentage of the V_{ta} (Figure 12.43).

The HF signal recorded for previous tests is erased with a DC current. The DC-erased track is reproduced comparing the residual signal with a specified extra pulse threshold. An extra pulse is defined as any peak that exceeds the specified extra pulse threshold. The extra pulse threshold is usually expressed as a percentage of the V_{ta} (Figure 12.44). A missing pulse, a superpulse, and an extra pulse are the results of local defects on the disk surface which originate in contamination, a pinhole of the magnetic thin film, texture failure, defects of the substrate, sputter spitting, and so on.

A positive modulation and a negative modulation are evaluated with an envelope of a reproduced signal (see Figure 12.45). A specified constant-frequency signal is recorded on a disk. The recorded track is reproduced comparing its envelope with a specified positive and/or negative modulation threshold. If

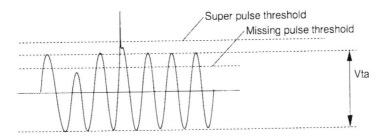

FIGURE 12.43 Missing pulse and superpulse.

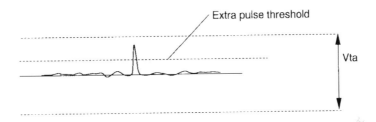

FIGURE 12.44 Extra pulse.

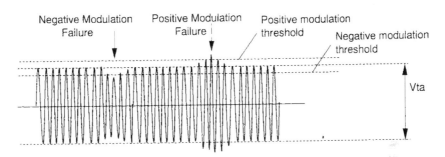

FIGURE 12.45 Positive and negative modulations.

an amplitude is larger or smaller than a specified positive modulation threshold, the disk is rejected for positive or negative modulation failure, respectively. These modulations are typically caused by macroscopic imperfection such as a nonuniformity of magnetic film or a waviness of a substrate.

A PW_{50} is defined as a pulse width at the 50% level of a peak amplitude of a signal reproduced from an isolated magnetic transition. A PW_{50} is a parameter that evaluates the linear recording density for the head/disk combination under test. A magnetic transition M is usually represented by the following expression, which is known as the arctangent model (Figure 12.46),

$$M = \left(2/\pi\right)M_{r} \cdot \tan^{-1}\left(x/a\right)$$

where the parameter x is a position along a track and the parameter a is an important parameter relating to a transition length.

In 1966, Middleton proposed an expression for a PW_{50} (Middleton, 1996):

$$PW_{50} = \left[g^2 + 4\left(d+a\right)\left(d+a+\delta\right)\right]^{1/2}$$

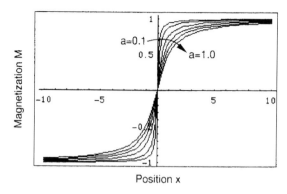

FIGURE 12.46 An arctangent transition model $M = (2/\pi)Mr \cdot \arctan(x/a)$.

where g is the gap length of the head, d is the head/disk spacing, and δ is the thickness of the magnetic film. This expression is convenient for estimating the transition parameter a for any head/disk combination knowing g, d, and δ. The transition parameter a is able to be calculated by using that expression and the PW_{50}, which is obtained experimentally.

12.2.3 The Head–Disk Interface

Earlier hard disk drives avoided the contact between a magnetic head and a disk medium even when the disk did not rotate to obtain high reliability, since rubbing often causes lots of tribological problems. This was accomplished by the use of hydrostatic bearing effect. These head designs were bulky and later flying heads were introduced. In the so-called dynamic loading system adopted to initial hard drives, the flying head was loaded on the disk medium after the disk rotating speed becomes constant and, consequently, uniform, and enough air bearing film is formed on the rotating disk surface to separate the head and the disk surfaces.

As a more precise assembling of drives was required to avoid the severe contact between a disk and a head during loading of the head slider with the decrease in flying height for higher recording densities, the CSS system was adopted to hard disk drives as one of the "Winchester" hard disk technologies. The flying head, which consists of a magnetic core and air sliders, contacts the disk surface while the disk is stopped, and is separated from the disk surface by air bearing film while the disk is rotating. Therefore, the head slider rubs against the disk surface during takeoff and landing on the disk surface. In most cases, the landing zone for CSS is separated from the data zone.

Figure 12.47 shows the relationship between the disk rotating velocity and the acoustic emission (AE) signal, which indicates the contact between the disk and the head sliders by an AE sensor attached to the base of the suspension mount (Tago et al., 1980). In range I, the head slider is dragged on the disk surface. The leading edge of the slider begins to fly in range II. The transition occurs from boundary lubrication to hydrodynamic lubrication at the end of range III. The slider flies without rubbing in range IV.

Therefore, the head slider rubs against the disk surface during takeoff and landing in CSS cycles. Because the disk and the head should not be damaged until about 100 thousand CSS cycles in most cases, the rubbing durability of the disk is required. The head–disk interface in hard disk files is explained widely and in detail by Bhushan (1996a,b).

As the particulate media are replaced by thin-film media and flying height is being reduced into near-contact condition to achieve increased recording density, microtribological analyses have been introduced to understand and improve their tribological properties instead of macrotribological ones.

12.2.3.1 Friction and Durability of Hard Disk Medium in CSS

Superior recording characteristics are realized by introducing thin-film magnetic disks. As the thin-film layer is damaged by rubbing against a head slider easily, the magnetic surface should be covered with a

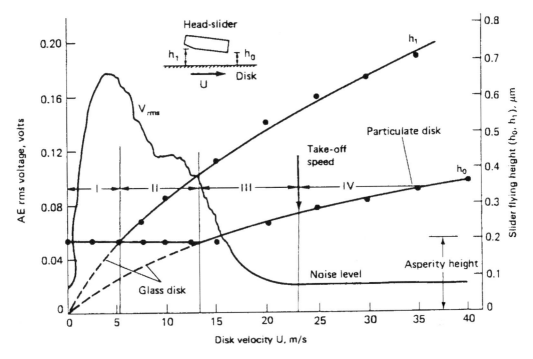

FIGURE 12.47 Ranges of contact between a head slider and a disk according to the disk-rotating velocity. (From Tago, A. et al., 1980, *Rev. Electron. Commun. Lab.* 28(5–6), 405–414. With permission.)

protective layer to obtain good durability. Such a protective layer also shows corrosion resistance. The typical materials of a protective layer for thin-film disks are sputtered carbon, carbon nitride, spin-coated or sputtered SiO_2, sputtered ZrO_2–Y_2O_3, or plasma polymerized films (Harada, 1981; Dimigen and Hubsch, 1983/1984; Yanagisawa, 1985a; Ishikawa et al., 1986; Yamashita et al., 1988). Of all of these films, sputtered carbon and carbon nitride are commonly used. The thickness of these protective layers is 10 to 30 nm. The protective layer is also covered with lubricants such as perfluoropolyether (Yanagisawa, 1985b; Bhushan, 1996a,b).

Wear gradually occurs in the mild adhesive wear process during takeoff and landing modes in CSS cycles. The wear debris of a typical combination between a protective layer and a lubricant usually shows low tackiness. It will not transfer onto the head slider strongly. However, wear debris piled up on the head sliders and dropped on the disk surface is agglomerated at the head–disk interface; head crash can be caused by dynamically unstable flying (Kawakubo, 1984).

The wear of the disk surface also makes the surface smoother, which results in an increase in friction coefficient. The protective layer or magnetic layer could be peeled off, and the segment may play the same role as the agglomerated wear debris into head crash, if the adhesion strength is small. A smoother surface often enlarges the difference between static friction and kinetic friction which causes stick-slip.

A hard surface of the disk is required to minimize damage, when it is hit by the head slider during unstable flying or stick-slip. The protective layer such as amorphous carbon by sputtering or chemical vapor deposition shows high hardness of 1000 to 3000 kg/mm^2. The hardness of diamondlike carbon (DLC) coating can be increased by introducing hydrogen during deposition of carbon coating (McKenzie et al., 1982; Miyasato et al., 1984; Nyaiesh and Holland, 1984; Pethica et al., 1985). The hardness of a substrate is also important, because the wear depth of a thin-film disk is affected by the hardness of the substrate which can be varied by the thickness of alumite in the case of a aluminum substrate (Ohta et al., 1985).

Lubricants are effective in reducing the shear stress and restraining the surface from smoothing. Polar lubricants such as Fomblin AM 2001 and Z-dol are more effective than nonpolar lubricants such as Z25.

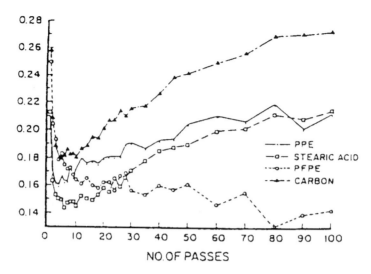

FIGURE 12.48 Variations of friction coefficients for carbon films lubricated with various lubricants. (From Timsit, R. S. et al., 1987, in *Tribology and Mechanics of Magnetic Recording Systems,* Vol. 4, SP-22 (B. Bhushan and N. S. Eiss, eds.), ASLE, Park Ridge, IL, pp. 98–104. With permission.)

It is found that perfluoroalkylpolyether (PFPE) adheres on the carbon film more strongly than polyphenylether (PPE) and stearic acid by IR analyses (Timsit et al., 1987). The variations of friction coefficients for several kinds of lubricants with increasing number of rubbing passes are shown in Figure 12.48. PFPE maintains a low friction coefficient for a long time compared with other lubricants.

Lubricant thickness is also an important factor for the friction coefficient and for durability of the disk. Optimum thickness of a lubricant should be designed to obtain low kinetic and static friction coefficients and good durability. The disk with less lubricant than the optimum value shows high kinetic friction coefficient and poor durability. The disk with more lubricant than the optimum value shows high static friction coefficient (Kondo and Kaneda, 1994). The maximum value of optimum range of lubricant thickness decreases with decrease in roughness of the slider and the disk surfaces. The atmosphere, such as temperature and relative humidity, has significant influences on the tribological properties of heads and media. At high humidity, adsorbed water causes the stiction between a head slider and a disk, as well as an excessive amount of lubricant. At low humidity, the wear debris of the head slider and the disk often transfers onto the slider surface more easily in general, compared with ambient or high humidity. The wear debris transferred onto the head slider could cause unstable flying and scratches in the disk surface during CSS as mentioned earlier. It is found that the wear life of a disk covered with DLC coating becomes poorer at lower humidity, whether it is lubricated or not (Enke et al., 1980). However, this phenomenon depends on the performance of the lubricant strongly and is not necessarily applicable to most cases.

12.2.3.2   Stiction

Stiction is the term commonly used when the force required to initiate sliding at the head–medium interface is excessive and is one of the most serious tribological problems in hard disk systems. The thickness of adsorbed water on a disk surface and a slider surface increases with increase in humidity.

There are some hypotheses for the mechanism of the adsorbed water behavior. "Meniscus theory" is the macroscopic hypothesis that the adsorbed water behaves like a liquid (McFarlane and Tabor, 1950), and the "interface layer of water theory" is a microscopic hypothesis that the adsorbed water is treated as uniformly arranged and layered molecules (Uedaira and Ousaka, 1989). In general, thick adsorbed water or an excessive amount of lubricant increases the static friction coefficient but keeps the kinetic friction coefficient small. The increased value of the difference between the static friction coefficient and

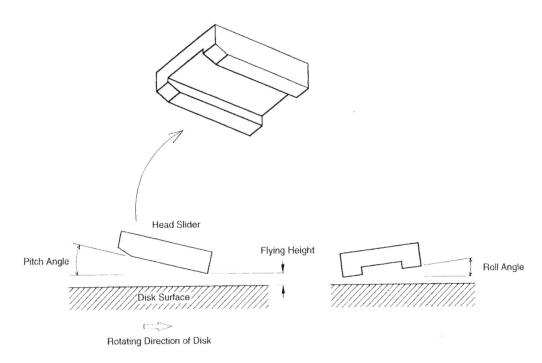

Head Slider

Pitch Angle

Flying Height

Roll Angle

Disk Surface

Rotating Direction of Disk

FIGURE 12.49   Schematic drawing of typical self-acting air-bearing slider.

the kinetic friction coefficient will cause stick-slip phenomenon. The disk could be damaged by the head clash that is caused by stick-slip during CSS mode.

The limiting value of the lubricant or adsorbed water thickness at which stiction occurs increases with increase in roughness of the slider and the disk surfaces. Texture is very effective to control roughness.

12.2.3.3   Flyability

A conventional self-acting air-bearing slider was introduced in IBM 3370 drives at first whose schematic diagram is shown in Figure 12.49. It consists of two rails with a taper on each rail and a magnetic core, which is located at the trailing edge of one rail to minimize the spacing with a disk surface. The viscous fluid, such as air, filled between a plate parallel to the direction of relative motion and inclined another plate produces a pressure between them. Pressure is produced between the taper of the rail and the disk surface, and the leading edge of the rail is lifted up. The rail, except for the taper area, is inclined by the lifted leading edge, and forms another taper by itself. This angle is called the pitch angle.

The typical parameters for the flying attitude of the slider are film thickness at the trailing edge (flying height), pitch angle, and roll angle. The flying height should be reduced to decrease spacing loss, keeping these parameters proper. As shown in Figure 12.50a and b, the reduction of slider rail width is effective to decrease the film thickness at the trailing edge, and the reduction of the rail width reduces the pitch angle (Nishihara et al., 1988). By the reduction of the film thickness and the pitch angle at the same time the average film thickness is reduced, which makes the contact between the slider and the disk easily. When the pitch angle increases as the trailing edge film thickness is decreased, the drop in the average film thickness could be reduced or eliminated, which reduces the contact between the slider and the disk surface during constant flying. An increased pitch angle design can be achieved by offsetting the pivot point toward the trailing edge of the slider (Bhushan, 1996a,b).

The disk rotating speed is also a very important parameter for flying height. The flying height at the outer area of a disk is higher than that at the inner, by the difference in relative speed. The skew angle also has a significant effect on flying height. The flying height decreases with increase in skew angle. By

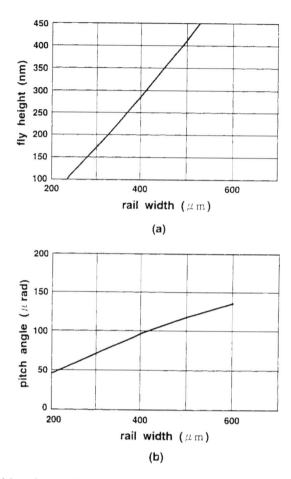

FIGURE 12.50 Effects of the reduction of rail width on (a) trailing-edge film thickness and (b) pitch angle. (From Nishihara, H. S. et al., 1988, in *Tribology and Mechanics of Magnetic Recording Systems,* Vol. 5, SP-25 (B. Bhushan and N. S. Eiss, eds.), ASLE, Park Ridge, IL, pp. 117–123. With permission.)

setting the skew angle larger at the outer area of the disk, the increase in flying height can be approximately canceled, as shown in Figure 12.51. This canceling can be controlled precisely now by processing the ABS using physical or chemical etching technology. The design of the ABS can be simulated by theoretical analyses (White, 1984a,b).

12.3 Tape Systems

Magnetic tape recording system was invented by V. Poulsen in 1898, and is widely used for various products, such as an audiocassette recorder, a videocassette recorder (VCR), and a computer data backup drive. The current magnetic tape systems are divided into two mechanisms. One is stationary head system and the other is helical scan system with rotating heads.

The stationary head system inherits the head–tape arrangement from the first Poulsen system. This system has advanced from open reel to cassette tape, and from single track to multitrack systems. This improvements provide easy tape handling, a faster data transfer rate, and higher recording capacity. Many computer backup drives uses this system because of its reliability (Figure 12.52).

The helical scan system has several rotating heads on a rotating drum, as shown in Figure 12.53. This system was branched off the stationary head to realize high-frequency video signal recording by using high-speed rotating heads and low-speed tape motion. The helical scan is commonly used in present

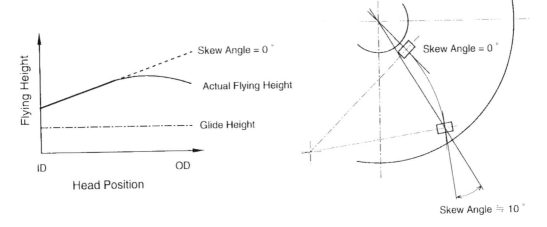

FIGURE 12.51   Schematic drawings of the adjustment of flying height at OD by skew angle.

VCRs. The areal recording density of this system is relatively high compared with the stationary head system because there is no guard band beside recorded tracks. This scheme is suitable for attaining high volumetric recording density, and some computer tape backup systems, for instance, digital data storage (DDS) drives, use this system (Bhushan, 1992).

12.3.1   The Recording Head

Why multitrack? In 1953 the IBM 727, a commercial multitrack tape system with seven tracks was introduced. The track pitch and the linear density of this system was only 14 TPI (tracks per inch) and 100 BPI (bits per inch), respectively. An IBM 3480 multitrack system which used a ½-in. cassette tape was introduced in 1984. Figure 12.54 shows a head for this system which has a read and a write head with 18 tracks for each. Today's multitrack heads are inherited from these systems. At the present time, many multitrack systems are available for data backup. The recent desktop ½-in. cartridge tape system (Quantum, DLT-7000) realizes more than 86 KBPI and 416 TPI. Also, the track density of the 16-GB ¼-in. cartridge (QIC) tape drive (Tandberg, MLR-1) exceeds 600 TPI.

The most important technical requirements for a tape system are high volumetric storage capacity and fast data transfer rate. Multitrack systems are able to obtain even faster data transfer rates. If data read or write operation is performed in more than two channels concurrently, the data transfer rate is multiplied by the number of data channels compared to that of a single-channel system. The above-mentioned DLT-7000 drive achieves 5 MBytes/s data transfer rate by a multitrack head with four tracks.

Another reason for using multitrack heads comes from the need for very high recording density. Most stationary tape systems use a burst or sample servo systems, in which a head is positioned by the servo signals written on a certain region of the tape blocks; then the head does not move to follow the track during read and write operations. A single-track head is usable in this system because reading both a servo block and a data block does not occur at the same time. When a track width becomes narrow in comparison with a tape running instability in the track direction, a continuous servo is necessary. This multitrack system can read the servo signals which are written in the certain tracks and follows a designated track continuously during read and write operations. The $^{13}\!/_{16}$ GB QIC drives realized 19 μm read track width by using a continuous servo/track following.

12.3.1.1   Design of Multitrack Head

12.3.1.1.1   Stationary Tape System Head
Many multitrack heads use thin-film heads rather than bulk heads because of their good performance and productivity. The MR reproduce head is most often used in recent multitrack systems, such as the

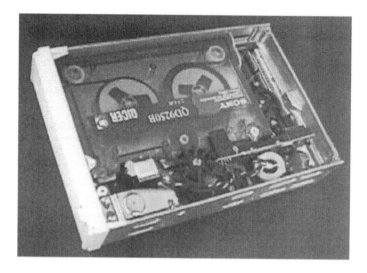

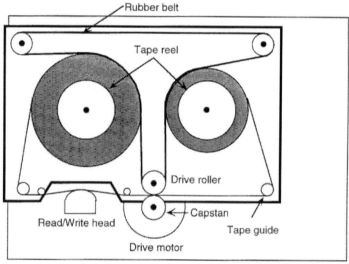

FIGURE 12.52 Stationary head system in a data cartridge tape drive.

IBM 3590 (follow-on 3480) and the 16-GB QIC. The MR head has many advantages compared with the inductive head. Larger signal output and a simple wafer manufacturing process are suitable for a multi-track head. However, the MR head is known for its difficulties in a machining process. Its depth (MR height: 2 to 3 μm) should be less than a few microns and the depth tolerance is only <10% of the depth; also an antiwear characteristic is strongly required for the heads because the distortion and the asymmetry of the output signal depend upon the depth.

Decreased wear is also required for a write head because a tape head always contacts a tape media at the write core. A thin-film write head uses polymer in the coil-leveling layer for which the deterioration temperature is about 300°C. For a bulk head core, many kinds of soft magnetic materials such as sendust (FeAlSi) alloy are used; however, most of them cannot be used for a thin-film head because they require annealing temperatures of more than 500°C. That is why the electrodeposited permalloy (NiFe alloy) which does not need to be annealed at high temperature has been used for a write core in spite of its

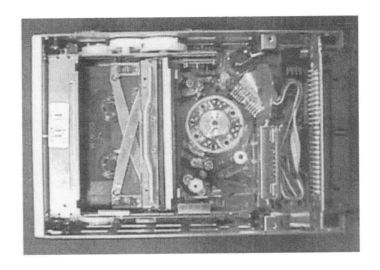

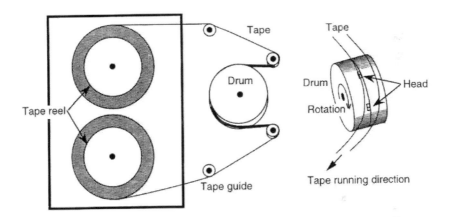

FIGURE 12.53 Helical scan head system in an 8-mm videotape drive.

FIGURE 12.54 Head schematic for an IBM 3480 data tape drive using a stationary head.

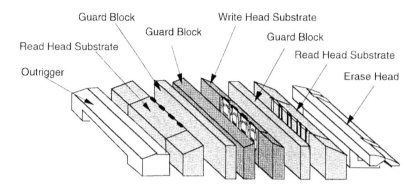

FIGURE 12.55 Multitrack head structure for a 16-GB QIC system.

poor wear characteristics. A thin protective layer, such as DLC, is commonly used in a hard disk head. However, it is not common in tape heads because the protective layer must be less than 100 nm to avoid a spacing loss of signal and is easily worn out by tape rubbing (Mallinson, 1994).

Recently, the requirement for replacement of the write core materials from permalloy to another material has been realized. Recording media material for a stationary head system is changing from $Co-\gamma Fe_2O_3$ to metal particle (MP) to obtain higher recording density. The coercive force (H_c) of $Co-\gamma Fe_2O_3$ and MP tape is about 900 and 1700 Oe, respectively. The saturation magnetization (M_s) of permalloy (400 emu/cc) is enough to obtain an appropriate overwrite margin for a $Co-\gamma Fe_2O_3$ tape, which is, however, insufficient for an MP tape. Recent high-performance heads use an amorphous or an NiFe/FeN multilayer for a write core which has more than 500 emu/cc. Since these materials are relatively corrosive, a good protective layer is also required from this point.

12.3.1.1.2   Head Contour Design

A schematic view of a multitrack head for a 16-GB QIC system is shown in Figure 12.55. This head consists of one write head and two MR read heads. Also, this head has an outrigger as a tape guide, and an erase head. As you can see from this figure, a multitrack head has complicated shapes. The most important issue of the head design is how to keep good head tape interface (HTI). For more than 40 years various types of head contour design have been proposed, and several types of contour design are shown in Figure 12.56. The recent demand for increases in linear recording density require reduction of head media spacing for minimizing spacing loss of reproduced signal.

Computer simulations are utilized for the design of head contours (e.g., Bhushan, 1996a; Wu and Talke, 1996). The effect of transverse slots on HTI is shown in Figure 12.57. This shows that the slots result in an increase in contact pressure and a decrease in head–media spacing.

12.3.1.1.3   Cross Feed and Cross Talk: Interferences from Another Channel or Head

As described above, when the data are read in more than two channels concurrently, interferences between each channel, so-called a cross talk, occurs. Also, many multitrack systems use a verify-while-write mode rather than a verify-after-write mode to shorten backup time, where an undesired signal from the leakage of a write head becomes a problem. This noise is called a cross feed. These interferences must be small enough as compared to the signal.

Reducing cross feed is relatively difficult because of the large magnetic and electric signal amplitude of a write head. Commonly, an electrical shield is inserted between a read and a write head to reduce cross feed, as shown in Figure 12.58. This shield is arranged to reduce undesirable emission from a write head itself. Recent hard disk heads use a merged-type MR/inductive combination head which has a read and a write element on the same wafer. However, many multitrack tape heads consist of a discrete block of read and write heads, because it is difficult for a multitrack head to insert a shield just like a hard disk–type two-in-one head.

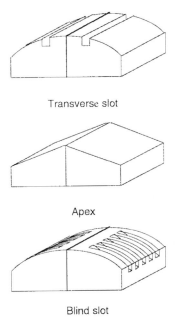

Transverse slot

Apex

Blind slot

FIGURE 12.56   Head contour design.

12.3.1.2   Head Fabrication

Fabricating a thin-film head is roughly divided into three processes; wafer process, machining, and assembly process. These technologies are quite different. However, the head characteristics depend on all the processes. From the production viewpoint, the head should be designed taking all these processes into consideration.

12.3.1.2.1   Wafer Process

A thin-film head is fabricated on a substrate by photolithography technology, which is almost the same as for the semiconductor process previously discussed. However, there are some differences. The first difference is in a wide variety of the film thickness of the magnetic head. For example, thickness of the write core is more than 3 μm to avoid magnetic saturation and the MR head uses a 20 to 40 nm thickness NiFe film for the MR element. A second difference is materials composition. The magnetic head uses Fe, Ni, and/or Co which is unable to be etched by reactive etching; therefore, a lift-off process and an ion-beam etching process are utilized for the purpose of patterning these materials. The material difference also requires change of the deposition methods. Vacuum evaporation and molecular beam epitaxy are popular deposition methods in a semiconductor process, whereas sputtering deposition and ion beam etching is commonly used in processing magnetic heads because good magnetic characteristics can be easily obtained.

A typical process for making an MR element and transverse bias films is shown in Figure 12.59. The key technology of this process is a tapered or a double-layered photoresist for the lifting-off process. Also, an oblique ion-beam etching and a sputtering deposition are used for making a transverse bias scheme for suppression of Barkhausen noise. This process decreases redeposition of etched materials and makes a desired tapered shape between the MR and the bias film.

One example for a wafer-processing procedure of the MR reproducing head and the write head is shown in Figure 12.60. The substrate for the head is Ni–Zn ferrite, Mn–Zn ferrite, or Al_2O_3–TiC. More than 100 chips are laid out in a single wafer.

12.3.1.2.2   Machining Process

After completion of the wafer process, the wafer is diced into individual chips, which are machined into the designated shape. First the wafer is set on a slicing machine and is cut to chips as shown in Figure 12.61.

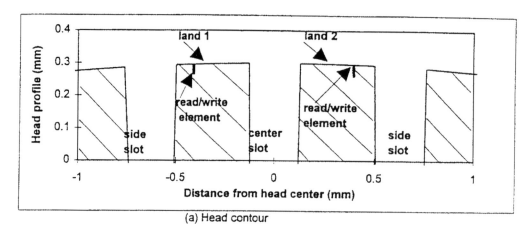

(a) Head contour

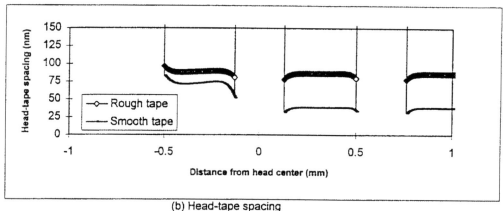

(b) Head-tape spacing

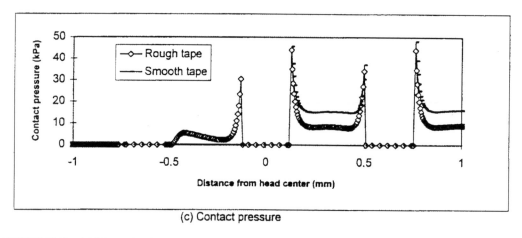

(c) Contact pressure

FIGURE 12.57 (a) Schematic of slotted head, (b) numerical results for head–tape spacing, and (c) contact pressure.

Next, a guard block which protects thin films from pealing off by tape rubbing is glued on the chip. Also side bars which support the tape running are glued beside the chip (Figure 12.62). After connecting a flexible cable to electrode terminals on the chip, head contour shaping and depth control lapping are performed.

The final tape contacting surface finish is done by a fine lapping tape to obtain smooth surfaces which are necessary for good tape contact. However, if the lapping speed of the final lapping is slow, then an

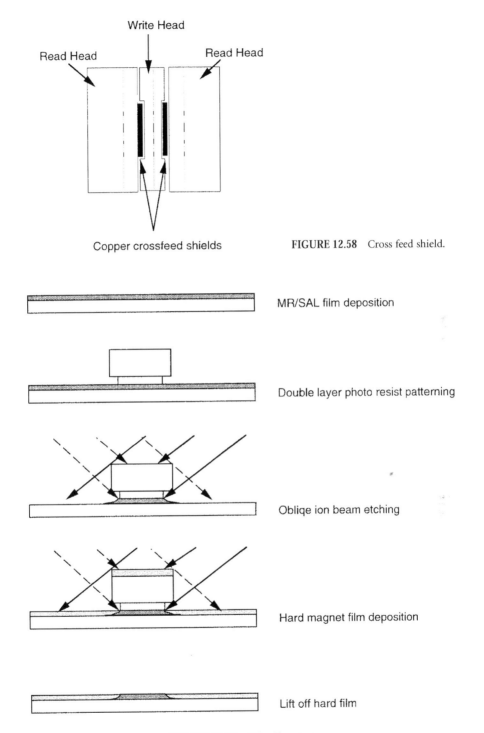

Write Head

Read Head   Read Head

Copper crossfeed shields

FIGURE 12.58   Cross feed shield.

MR/SAL film deposition

Double layer photo resist patterning

Obliqe ion beam etching

Hard magnet film deposition

Lift off hard film

FIGURE 12.59   Lift-off process.

external cylindrical grinder is used to improve production efficiency before the lapping. In the contour-shaping process, precise depth control within 0.2 μm tolerance is needed for an MR head. To achieve this requirement, the depth should be controlled within 0.3 μm after the grinding process. Also the difference between the depth of the right and the left side of the chip must be within 0.1 μm. In a usual

Read head processWafer process	Write head processWafer process

Read head processWafer process
1. Substrate polishing
2. Lower shield deposition and patterning
3. Lower gap deposition
4. MR film deposition
5. Lift off patterning
6. Hard magnetinc film deposition and lift off
7. Electrode deposition and patterning
8. Upper gap layer deposition and patterning
9. Protective layer deposition

Write head processWafer process
1. Substrate polishing
2. Lower core deposition and patterning
3. Gap deposition
4. Coil deposition and patterning
5. Coil leveling patterning
6. Upper core deposition and patterning
7. Electrode deposition and patterning
8. Protective layer deposition

Machining process
1. Chip dicing
2. Attach flexible cable
3. Guard plate bonding
4. Side guide bonding
5. Grinding by cylindrical grinder
6. Tape lapping

Machining process
1. Chip dicing
2. Attach flexible cable
3. Guard plate bonding
4. Side guide bonding
5. Grinding by cylindrical grinder
6. Tape lapping

Combining process
1. Positioning parts
2. Glue each other
3. Mount on base plate

FIGURE 12.60   Fabrication process.

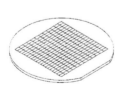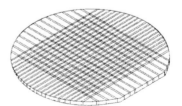

FIGURE 12.61   Chip cutting.

inductive head process, depth is checked by an optical microscope. However, it is not possible to attain the required accuracy by this depth control method. Usually, the depth of an MR head is controlled by an electrical depth sensor. Figure 12.63 shows an MR head pattern. The depth sensors are located on both sides of the head. The electrical resistance is inversely proportional to the cross-section of the sensor, which is proportional to the depth. Figure 12.64 shows a sample holder of an automatic grinding system. The holder has two piezoactuators for the depth control. The first is for along the in-feed direction and the other is for balancing control of the right and left depth. The control circuit of this system is shown in Figure 12.65. The actuators are controlled by the signals from the depth sensors. By using this grinding system, less than 0.25 μm depth accuracy is obtained.

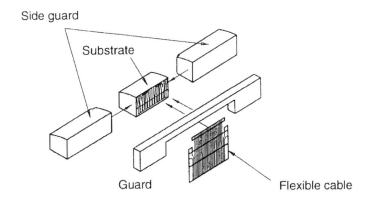

FIGURE 12.62 Head chip structure.

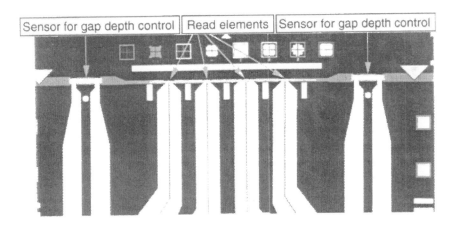

FIGURE 12.63 Water pattern of a multitrack head.

12.3.1.2.4 Assembly Process

Two MR heads, a write head, an erase head, and a tape guide are combined for one complete multitrack head as shown in Figure 12.66. The position of these pieces should be adjusted in three directions as shown in Figure 12.67. The alignment is necessary for positioning the track center of each head, and the azimuth adjustment is required for minimizing azimuth loss. The HTI is controlled by the tilt adjustment and the penetration of each head. Higher recording density and a more accurate positioning technique are required. The tolerance of the alignment, the azimuth, and the tilt is 2.5 μm, 3 min., and 3 min., respectively. These requirements are almost the limit of optical measurement systems.

12.3.2 Magnetic Tapes

The first commercially available magnetic recording tapes were produced in 1947. Since that time magnetic tapes developed rapidly for use in audio, video, and digital data recording systems. The linear analog technique is commonly used for most audio recorders. The magnetic tape is transported at speeds ranging from several centimeters per second to several meters per second over a fixed stationary head. On the other hand, helical-scanning rotary heads were developed for video recording, which afforded a high head-to-tape speed of more than several meters per second as high as 10 m/s and high recording density capabilities. However, high relative speed is liable to deteriorate the wear property of the tape. The success of a tape in actual use depends critically on its tribological properties.

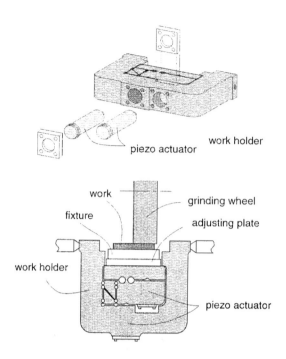

FIGURE 12.64   Head holder for automatic cylindrical grinder.

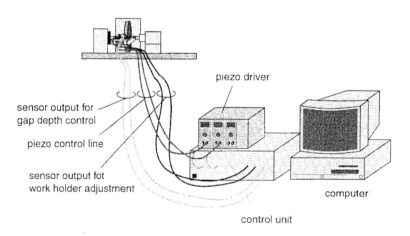

FIGURE 12.65   Automatic gap depth control system for grinding.

Magnetic tapes are divided into two groups: (1) particulate tapes where magnetic particles are dispersed in a polymer binder with some additives, and coated onto the substrate, and (2) thin film media of which monolithic magnetic thin films are deposited onto the substrate in vacuum. The overwhelming preponderance of media fabricated to date have been coating media. However, continuous demand for increasingly high recording density has led to the use of thin-film tapes. In this chapter, we review not only the construction of tape, but also the materials used in the general-purpose particulate tapes and the recent development of both monolithic thin-film tapes and double-coated particulate metal tapes (Bhushan, 1992).

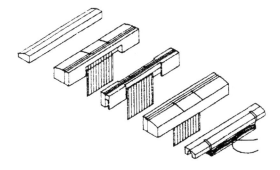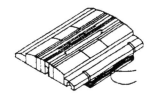

FIGURE 12.66   Combining heads.

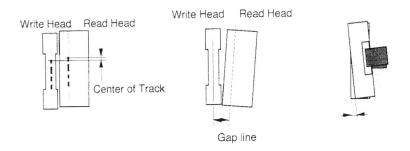

Write Head   Read Head

Write Head   Read Head

Center of Track

Gap line

FIGURE 12.67   Head adjustment.

12.3.2.1   Particulate Tapes

12.3.2.1.1   Structure of the Magnetic Tapes

Figure 12.68 shows a schematic and TEM photograph of the cross-sectional view of particulate media. The coating formulation (binder, magnetic particles, solvents, cross-linker, lubricant, and abrasives), which is mixed by a disperser, such as sandmills, is coated onto the polyethylene terephthalate (PET) substrate. A back coat is optionally applied to improve runnability. Thickness of the coating is commonly several micrometers.

12.3.2.1.2   Manufacturing Process

The magnetic tape is fabricated as follows, Figure 12.69. First, magnetic particles and a small amount of binder solution are kneaded, then dispersed with the remaining binder solution by means of sand mills (mixing disperser) for several hours. The coating slurry to which lubricant and cross-linker are added passes through the filter, and is coated onto a wide polymer web. After orientation, the coated web is dried in the oven and the tape surface is made smoother by the calendering process using hard and soft rollers under high pressure and high temperature. Then, the thermosetting resin binder in the coating is cured at elevated temperature to obtain optimal mechanical properties. After curing, the coating web is slit to the desired tape width.

12.3.2.1.3   Materials Used for Tapes

Magnetic coating consists of magnetic particles, polymer binder, abrasives, lubricant, and dispersant. The interaction of the particles and the organic compounds becomes important as particles are made smaller. Mechanical properties depend on the substrate as well as the magnetic coating. Lubricant is a key technology in determining the tribological performance.

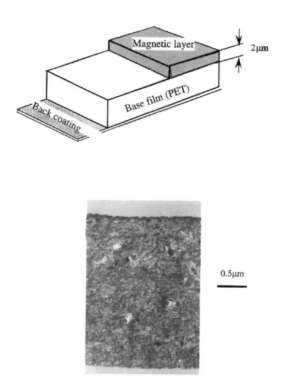

FIGURE 12.68 Transmission electron micrographs of cross-sectional view and schematic diagram of particulate media.

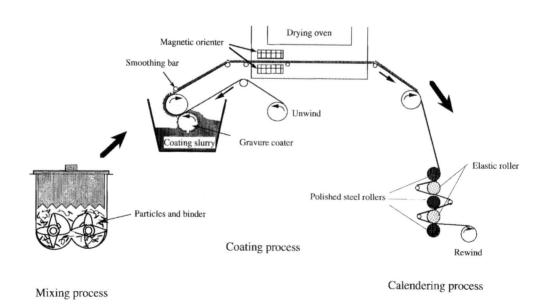

FIGURE 12.69 Schematic arrangement for the coating process.

TABLE 12.1 Properties of Magnetic Particles

	γ-Fe$_2$O$_3$	Co-γ-Fe$_2$O$_3$	Co-γ-Fe$_3$O$_4$	Fe
Saturation magnetization (σ_s), Am^2/kg	72–73	75–78	83	120–170
Coercivity, kA/M	24–32	28–68	48–76	95–200
Specific surface area, m^2/g	20–40	25–45	25–45	30–60
Length, μm	0.3–0.5	0.15–0.7	0.15–0.7	0.05–0.3
Magnetism	ferri	ferri	ferri	ferro

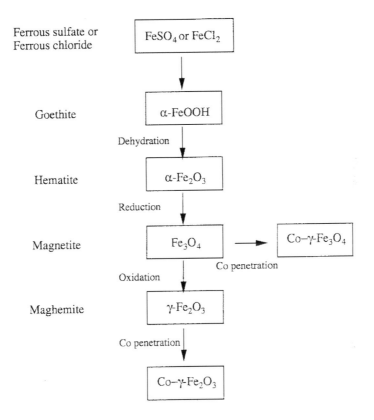

FIGURE 12.70 Synthetic scheme of oxide powder.

1. Magnetic Particles

Properties of magnetic particles currently used in magnetic media are listed in Table 12.1. The first commercially available magnetic recording tapes used γFe$_2$O$_3$ (iron oxide particles). γFe$_2$O$_3$ is still a widely used magnetic recording material because of its great chemical and physical stability. By doping cobalt to γFe$_2$O$_3$ particles, higher coercivity (H_c) and saturation magnetization (σ_s) can be obtained. γFe$_2$O$_3$ has been produced in the following scheme (Figure 12.70). First, needle-shaped goethite, FeOOH, is grown on precipitate from a solution of iron salts, and then dehydrated to the nonmagnetic hematite, αFe$_2$O$_3$, which is reduced to magnetite, Fe$_3$O$_4$ and then oxidized to γFe$_2$O$_3$. Magnetite seems to be a desirable recording material; however, it has some chemical and magnetic instabilities.

Further requirements in magnetic tapes call for higher values of coercivity than can be obtained from pure oxide. Particles could be made containing 2 to 3% cobalt having coercivity up to 60 kA/m (Kishimoto et al., 1981). Co is impregnated just near the surface of the γFe_2O_3 to avoid temperature dependence of coercivity with time, which results in uniaxial magnetic anisotropy, having a single preferred axis of magnetization like the oxides, and demagnetization is more difficult. Recently, by overcoming chemical instability of Fe_3O_4, Co-modified magnetite has been rapidly replacing Co–γFe_2O_3 because of its higher σ_s and its simple production process. It is the predominant material currently used in video tapes and some audio tape applications.

The principle attractive reason for using metal and alloys for recording applications rather than oxides is that pure iron has the highest magnetization of 222 Am^2/kg in the ferromagnetic elements, about three times that of γFe_2O_3. When the anisotropy of the particles is due to shape, the coercivity is directly proportional to the magnetization and should be about seven times higher than that of oxides (Table 12.1). Their big disadvantage is the need for protection against the tendency to oxidation. Stabilization can be attained by a controlled oxidation of their surfaces and decreasing pores from both oxide and metal cores (Okamoto et al., 1991). The resulting oxidized surface layer is typically 2 to 4 nm thick, which unfortunately reduces the average magnetization intensity of the particles from 160 to 120 emu/g. The resulting stable media can still have signal output amplitudes well above those of Co–γFe_2O_3 (Chubachi and Tamagawa, 1984).

The commercially significant process today is the reduction of acicular particles of oxides or oxyhydroxides. The technology of particles production involves a variety of additives and alloying elements. Silica and aluminum are common elements to prevent sintering of the particles during the required heating process (Sueyoshi et al., 1984) where oxide particles are reduced with hydrogen. Furthermore, the partial replacement of aluminum with yttrium at the surface has an effect on dispersibility of smaller particles as well as on antisintering (Okamoto et al., 1996).

Metal tapes were first introduced in 1979 in audio application and in 1985 in 8-mm videocassette. The particles used for the tapes have coercivities well above 100 kA/m and length less than 0.25 μm. With the use of MIG heads, metal particulate tapes have been widely applied for high-density recording media such as digital audio, Hi8 video, and data recording digital tapes. The magnetic properties of iron-based metallic particles continue to improve, with coercivity exceeding 200 kA/m, σ_s 160 Am^2/kg, and particle length not much more than 0.05 μm having been achieved (Figure 12.71). The frequency response of the tape using these particles exceeds 4 dB, a higher output level at the shorter wavelength than the Hi8-ME (Richter and Veitch, 1995; Okamoto et al., 1996).

Barium ferrite crystallizes in a complex hexagonal structure whose easy axis is the hexagonal axis (Figure 12.71). The interest of barium ferrite comes from the perpendicular magnetization (hexagonal axis). Potential benefits of perpendicular recording have been claimed to be the reduction of self-demagnetization and retention of sharp transitions, which might be an attractive candidate for advanced applications. However they have been slow to find commercial applications. Barium ferrite has not yet been used in products principally because of the poor dispersibility, the rather low M_s, and high price.

2. Base Film

Physical properties of magnetic tape mainly depend on the base film so that high Young's modulus and long-term dimensional stability (shrinkage) are required. PET is almost exclusively used for base film of the flexible media. At the outset of magnetic tape development, paper and cellophane were used; then acetate and polyvinyl chloride (PVC) were developed. Since the 1950s, PET has been used commonly for its mechanical properties, low shrinkage, surface roughness, and also its low price. PET films are generally oriented biaxially, which improves their mechanical properties as shown in Table 12.2.

Considering high recording density and long-time recording, the point in designing PET film is (1) surface smoothness, (2) high Young's modulus, and (3) dimensional stability (Bhushan, 1992; 1996a). Higher recording density can be achieved by the smoother magnetic surface because of its low spacing loss. Therefore, the design of the surface which affects the smoothness of magnetic coatings becomes important. Metal evaporated (ME) and MP tapes require the surface roughness of less than 5 nm and 10 nm, respectively. A biaxially oriented PET film exhibits significant in-plane anisotropy, Young's modulus, breaking strength, and coefficient of thermal expansion. Orientation involves stretching the film in

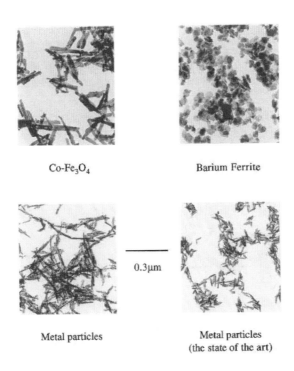

Co-Fe₃O₄ Barium Ferrite

0.3μm

Metal particles Metal particles (the state of the art)

FIGURE 12.71 Transmission electron micrographs of magnetic particles, Co-magnetite, barium ferrite, and two sizes of metal particles (Fe).

TABLE 12.2a Mechanical and Thermal Properties and Molecular Structure of the Base Film

	PET	PEN	Aramid
Mechanical Properties			
Tensile strength, MPa	245	373	392–490
Tensile elongation, %	130	48	50–70
Young's modulus, GPa	3.9–5.9	6.9	8.8–12.7
Thermal Properties			
Melting point, °C	263	274	None
T_g, °C	69	113	350
Thermal expansion coefficient, cm/°C	1.5×10^{-5}	1.3×10^{-5}	0.5–2.5×10^{-5}
Long-term heatproof temp., °C	120	155	230
Heat shrinkage (200°C × 5 min), %	5–10	1.5	0.1

the transverse (width) and machine (longitudinal) directions to produce balanced mechanical and thermal properties. By means of suitable chemical and structural modifications, PET films are fabricated with a broad range of properties for many applications and are commonly used as a substrate for magnetic media.

New base film materials are summarized in Table 12.2. PEN is one of the polyesters which is polycondensed with ethylene glycol and 2,6-naphthalene-dicarboxylic acid. PEN has higher mechanical properties, such as Young's modulus and breaking strength, and higher T_g and lower thermal expansion coefficient than PET and is used for several video and data processing tapes. In the fabrication process of thin-film tapes, heat is generated and high substrate temperature becomes higher during the magnetic films deposition process. Aramid, an aromatic polyamide, which can withstand high temperature of at least 200°C is considered to be suitable. Another advantage of these films are the isotropic properties.

645

TABLE 12.2b   Molecular Structure of the Base Film

Materials	Molecular Structure
PET	
PEN	
Aramid	

Aramid film is used for thinner data processing media (DDS II and III tape) and nontracking (NT) ME tape with film thickness of less than 5 μm because of its high mechanical strength. However, widespread use of these films is also restricted by its cost.

3. Binder

The magnetic particles are surrounded by the polymeric binder in the coatings. Tribological behavior of the head–media interface is also affected by the physical and chemical properties of the polymer binder. Selection of polymer binder is based on the following: (1) dispersibility of the magnetic particles, (2) adequate viscosity of the coating slurry, (3) adequate viscoelastisity of the coating, (4) stability for degradation (mechanical bond scission or hydrolysis), and (5) antiwear property.

The materials used for tape binders are usually thermoplastic elastomers that can be melted and cooled reversibly without major changes in their chemical and physical properties. PVC, polyester-polyurethane, and nitrocellulose are commonly blended and used. The most commonly used PVC for magnetic tapes has a molecular weight of about 30,000 and epoxy groups in end group to trap hydrochloric acid eliminated during manufacturing (Nakayama et al., 1985). T_g of 70°C is appropriate for magnetic media and the alkali sulfate polar group at the end of the molecule is adsorbed on the magnetic particles and disperses them well.

Another commonly used binder is polyester-polyurethane. Typically the polyester-polyurethanes contain a soft segment (ester moiety) connected to the hard segment (urethane moiety). The polyester segments are formed by the condensation of bifunctional carboxylic acid and alcohol such that the ester is terminated substantially with an alcohol end group followed by additional reaction with diisocyanate. For tape applications, a balance of these segments is sought to adapt to the requirements. Nowadays, polyester-polyurethanes used for magnetic tape application have a polar group such as an alkali sulfonate group (Mizunuma et al., 1978), an ammonium salt group (Endo et al., 1991), or an ammonium phosphate group (Farkes et al., 1994).

4. Abrasives

For low tape wear and to keep the head clean, particulate media use load-bearing abrasives, such as Al_2O_3 or Cr_2O_3, and most contacts occur at these particles, which significantly reduce the tape wear. These load-bearing abrasives are also known as head-cleaning agents. However, abrasives remove the mating head materials; then the amount of abrasives is considered to maintain low head wear. Thus, the media must have adequate durability and must not cause an unacceptable rate of head wear. The size and the shape of abrasives mainly determine abrasivity. The sizes are from 0.2 to 0.7 μm and angular-shaped abrasives are normally used.

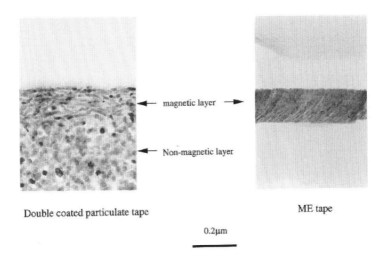

magnetic layer

Non-magnetic layer

Double coated particulate tape

ME tape

0.2μm

FIGURE 12.72 Transmission electron micrograph cross-sectional view of double-coating particulate media and thin-film media.

5. Lubricant

Properties of interest in selection of lubricant for low friction and wear can be summarized as follows (Klaus and Bhushan, 1985; Bhushan, 1992, 1996a): (1) good boundary lubricity, (2) affinity to the surface (ME tape), (3) low volatility, (4) thermal stability, (5) low surface tension, (6) resistance to both mechanical and chemical degradation, (7) chemical inertness to the magnetic coating and drive component.

There are two ways to incorporate the lubricant into the magnetic media: internal and topical lubrication. Internal lubrication where the lubricant is incorporated within the magnetic coating has an advantage, in which it replenishes the lubricant at the medium surface and can save cost by eliminating the topical coating process. Internal lubrication is normally used in particulate media; on the other hand, lubricant is topically coated onto the ME tape.

Lubricants for the particulate media have evolved material products, such as long-chain ester, fatty acid, and silicon oil. Both fatty acids and synthetic esters or their combination are widely used today as internal lubricants. The fatty acids and their esters include those with carbon numbers between 12 and 18 and with alcohols ranging from 2 to 18. There are several major chain properties which determine the lubrication performance: chain length, polarity, and chain substituents. Polar hydrocarbons, which usually have one carboxyl group at the end of a molecule, provide a significantly lower friction. Both the static and dynamic friction coefficient decrease with increasing chain length. The carboxylic acid reduces friction and the ester is effective on durability such as still-frame work.

12.3.2.1.4   Double-Coating Tape

One of the disadvantages of particulate tapes is that it is difficult to be coated thinner than 1 μm with high yields. Reducing the thickness of the magnetic coatings increases the performance for the high density from the point of self-demagnetization. Thin magnetic layer particulate tapes can be obtained only with a simultaneous double-coating technology with a nonmagnetic underlayer. These tapes have already been put to practical use in Hi 8 video applications, several data-processing tapes, and high-density floppy disks.

Recently, the thickness of upper magnetic layer of under 0.2 μm has been achieved (Inaba et al., 1993). The cross-sectional view of the double-coating layer is shown in Figure 12.72. The point of the double coating is for the surface smoothness and the tribological performance, which are attributed to the underlayer of nonmagnetic fine particles coatings such as TiO_2 and αFe_2O_3 (Saitoh et al., 1995). The smoothness depends on the dispersibility of the nonmagnetic particles in the underlayer as the upper

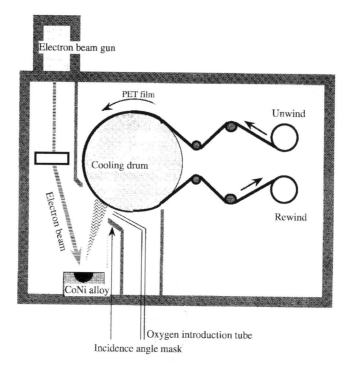

FIGURE 12.73 Schematic arrangement for deposition of evaporated metal films onto PET.

layer becomes thinner (Sasaki et al., 1997). The lubricant is mainly retained in the microvoids distributed throughout the underlayer, which replenish the lubricant at the upper layer surface.

12.3.2.2   Metal Evaporated (ME) Tape

12.3.2.2.1   Structure of the ME Tape

ME tapes fabricated using a vacuum evaporation process have high-density recording characteristics and low noise properties. These characteristics originated from obliquely aligned ferromagnetic fine particles or grains which are shown in Figure 12.72. The thickness of the magnetic layer is almost 200 nm and the most recent ME tapes such as DVC are covered with a 10-nm-thick carbon protective layer on top of the surface just the same as thin-film rigid disks, which leads to higher durability.

12.3.2.2.2   Manufacturing Process and Recording Characteristics

Manufacturing processes of ME tapes are shown in Figure 12.73. The magnetic properties, recording characteristics, and corrosion resistance are mainly controlled by changing the oxygen gas flow rate during evaporation. With a sufficient amount of oxygen, fine Co particles are formed and they are oriented with the axes parallel to the column, which leads to large magnetic anisotropy with oblique axis to the surface. The magnetic anisotropy and easy axis can be measured with a torque magnetometer. The apparent anisotropy constant, which contains the shape anisotropy constant, is 2.34×10^5 J/m^3, and it lies at an angle of 21° from the longitudinal direction of the thin film. By increasing the number of magnetic layers, the coercivity and squareness are increased. Also, an oxidized interlayer reduces the exchange coupling and tape noise (Kawana et al., 1995). In the case of a four-layer medium, in-plane coercivity is 155 kA/m and squareness of 0.84.

Fluorocarbons are by far the most widely used for thin-films media as a lubricant, both disks and tapes, because of low surface tension, chemical stability, and moisture repellency. Among the fluorocarbons, polar perfluoropolyethers (PFPEs) are superior to other lubricants in providing surface lubricity and low volatility (Lin and Wu, 1990). The performance of PFPEs depends upon the molecular structure

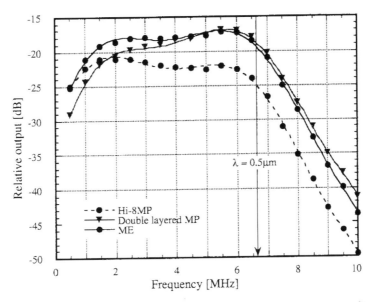

FIGURE 12.74 Frequency response of double-layered MP and ME tape. Hi8 MP tape is shown for comparison.

of the compound. Several specific PFPEs are commercially available, which differ in molecular weight, backbone structure, and polar end group. Fomblin Z is a random copolymer of CF_2O and CF_2CF_2O, Krytox a homopolymer of iso-hexafluoro-propylene oxide, Demnum a homopolymer of hexafluoro-propylene oxide. Demands on the lubricant for thin-film media are very stringent, and there are no completely satisfactory candidates. Recently, the synthesized lubricants that possess a polar end group have shown great promise as state-of-the-art thin-film media lubricants (Kondo et al., 1989). For the given PFPE chain structure, the extent to which a surface is covered depends on the polar group.

Figure 12.74 shows a frequency response of various ME and MP tapes. The thin-film tapes, both ME and MP, have higher reproduced output. The excellent performance of ME tape is attributed to the oblique anisotropy and the fine particle structure of the magnetic layer and that of MP is dependent on large magnetic energy, that is, large σ_s and H_c of magnetic flux.

12.3.2.3 Trends

The choice of available magnetic particles for recording is broad and is only limited by the magnetic heads. Advantage of the particulate coatings is that these are produced in wide webs at high coating speeds at high yields, resulting in low cost. The most pronounced trend in particulate tape materials is toward smaller particles, enabling the number of particles per recording wavelength to increase so that the signal-to-noise ratio remains high. Another important reason for the smaller particles is that the H_c for the particles increases. However, for the finer particles, the manufacturing process leads to agglomeration, which makes it difficult to achieve uniformity of the switching field. Furthermore, the orientation of particles with higher coercivity is inefficient. Double coating is capable of realizing thin-layer particulate tapes with increased performance for high-density recording. The aim is for dispersibility and adequate viscoelasticity of the coating slurry for double coating.

A disadvantage of particulate tapes is that magnetic particles represent only 40% of the volume of the coating. The monolithic magnetic thin films are recognized to have a significant improvement in noise. The durability of the thin-film tapes is significantly increased by overcoating with a carbon protective layer. On the other hand, the thickness of the protective layer is not negligible for short wavelength recording, that is, high-density recording. Furthermore, it requires an additional fabrication process, and the productivity of the ME tapes is therefore lower than that of the particulate tapes. The challenge remains to fabricate ME tapes economically.

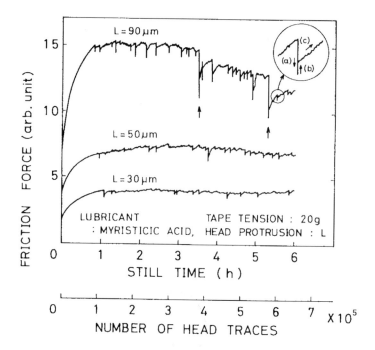

FIGURE 12.75 Variations of friction force between rotating head and metal particulate tape under various conditions of head protrusion. (From Osaki, H. et al., 1992, *IEEE Trans. Magn.* MAG-28(1), 76–83. With permission.)

12.3.3 The Head–Tape Interface

The typical structure of a helical scan videotape recorder (VTR) or a helical scan tape drive for data storage is shown in Figure 12.53, and the one of a linear tape drive for data storage is shown in Figure 12.52. As the loss in recording or reproducing a signal increases with a spacing between a magnetic tape and a magnetic head, what is called "spacing loss," the tape and the head are obliged to contact each other. Thus, the tribological problems between the tape and the head are very important issues for tape systems, such as VTRs and tape drives for data storage (see Figures 12.52 and 12.53).

A tape also rubs against tape-transport components such as stationary drum and guideposts, when it is transported in the helical scan tape system. Thus, the tribological problems between a tape and tape-transport components are also very important (see Figure 12.53). The head–tape interface in tape systems is explained widely and in detail by Bhushan (1996a,b).

12.3.3.1 Tribological Problems between Tapes and Rotary Heads

12.3.3.1.1 Tape Damages by Rotary Head

The pause picture mode in a VTR is called the "still-frame mode." The most severe use for a magnetic tape in a VTR is the still-frame mode, in which rotary heads trace the same track on a tape surface at the rate of 90 cycles/s for a three-head Hi8 VTR and VHS system or at the rate of 450 cycles/s for a three-head consumer digital VTR system. This mode is not common in a tape drive for data storage. This test in the drives, which is called the "dwell test," is sometimes performed just to evaluate tape durability.

1. Metal Particulate Tapes

Typical variations of the friction force between a metal particulate tape and a head with the number of head traces under various conditions of head protrusion are shown in Figure 12.75. The micrograph of the tape surface just before the first small drop of the friction force (approximately 1×10^5 traces) in Figure 12.75 is shown in Figure 12.76(a). White particles are magnetite (Fe_3O_4) particles as the marker,

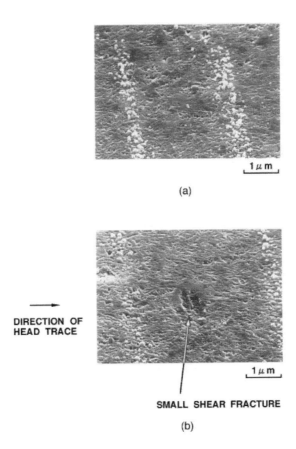

(a)

DIRECTION OF
HEAD TRACE

1 μ m

SMALL SHEAR FRACTURE

(b)

FIGURE 12.76   Scanning electron micrograph of metal particulate tape. (a) Before first small drop of friction force. (b) After first small drop of friction force. (From Osaki, H. et al., 1992, *IEEE Trans. Magn.* MAG-28(1), 76–83. With permission.)

which indicates that this area was definitely traced by the rotary heads. It seems that the magnetic surface became smoother by the plastic flow of the polymer binder around the magnetic powder. It is assumed that the increase in the friction force is caused by the increase in the real contact area, which is attributed to the plastic flow of the polymer binder (Osaki et al., 1992).

Small regions with shear fracture of the smooth surface can be seen in Figure 12.76b, after the first sudden drop of the friction force at approximately 1×10^5 traces in Figure 12.75. These shear fractures are supposed to be caused by the increase in the friction force. The wear debris from these fractured areas gathers together to form larger, retransferred wear debris. When the large fractured segment transfers onto the head surface, it scratches the tape surface catastrophically (Osaki et al., 1992).

Plastic flow caused by head trace deteriorated the durability in the still-frame mode of particulate tapes. Although a lubricant is effective for that purpose, some kinds of lubricants with polar functional groups like fatty acid, which makes polymers plastic in general, accelerate plastic flow of polymer binder. The variations of the friction force for the metal particulate tape with and without ester lubricant are shown in Figure 12.77. Plastic flow could not be seen with the ester tape, although the tape without a lubricant was soon damaged. Thus, the lubricants, which have poor compatibility to the binder resin, do not result in plastic flow of binder and improve the durability of still-frame mode of metal particulate tapes (Osaki, 1992).

Abrasives in a magnetic layer improve the durability of metal particulate tapes (Doshita and Mukaida, 1995). It is believed that the abrasives prevent the plastic flow of polymer binder by supporting the contact load between the tape and the head. In addition to the above reason, the resin binder transferred

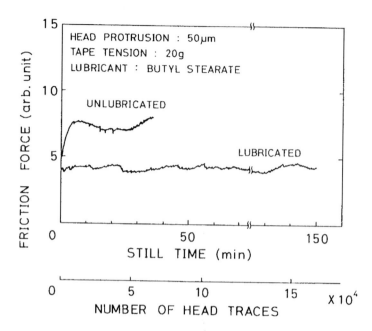

FIGURE 12.77 Variations of friction force between rotating head and particulate tape without lubricant and with lubricant. (From Osaki, H. et al., 1992, Doctorate thesis, Tohoku University, pp. 18–57.)

on the head surface, which may cause the increase in friction force, can be removed by abrasives mechanically.

2. Metal Evaporated Tapes

Tape damage in the still frame mode of a metal evaporated tape is caused by an increase in the friction force between the tape and the video heads, which peel the magnetic layer of the metal evaporated tape (Osaki, 1990). Therefore, the friction force should be maintained low and the adhesive strength between a magnetic layer and a base film should be high in order to improve the durability of the metal evaporated tapes (Wheeler and Osaki, 1990; Guo et al., 1990).

As ME tape has a thickness of the magnetic layer less than 0.2 μm, the wear and lubrication mechanism should be different from that of the MP tape. For example, lubricant is applied topically for ME tape, but internally for particulate media. The increase in friction force seems to be caused by the removal of lubricants and by the increase in real contact area which is attributed to smoothing of the tape surface by wear (Osaki et al., 1993).

The reduction of the real contact area can be performed by introducing surface asperities (see Figure 12.78a). The effect of the surface asperities of ME tapes on the reduction of the friction force between the tape and the head is shown in Figure 12.79. The friction force of the ME tape with surface asperities should be maintained at a low value for a long time without severe damage (Osaki, 1993). However, these surface asperities will be worn out and lose the effect on reduction of friction force after a practical lifetime, as shown in Figure 12.78b (Osaki, 1993; Berar et al., 1995). Thus, higher surface asperities show longer lifetime until they are worn out, and improve the durability. On the other hand, higher surface asperities decrease the reproduced output signal by spacing loss. The height of surface asperities was optimized for a commercial ME tape. To improve these conflictive refinements, a protective layer coating on the magnetic layer is very promising.

It is known that the durability in the still-frame mode of ME tapes can be improved by not only lubricants and surface asperities but also by a subsurface oxidized layer in the magnetic layer formed by introducing oxygen during evaporation (Odagiri and Tomago, 1985; Kunieda, 1985; Tomago et al., 1988). It is presumed that the surface hardening occurs at the subsurface of oxidized layer and the consequent

(a)

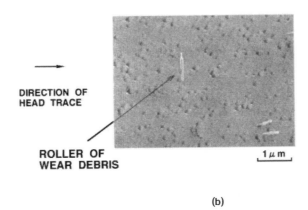

DIRECTION OF
HEAD TRACE

ROLLER OF
WEAR DEBRIS

(b)

FIGURE 12.78   Scanning electron micrograph of ME tape surface which artificial spherical asperities. (a) Before head trace. (b) After head trace. (From Osaki, H. 1993, *IEEE Trans. Magn.* MAG-29(1), 11–20. With permission.)

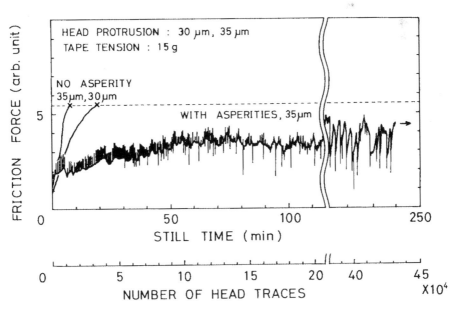

FIGURE 12.79   Effect of artificial asperities of tape surface on friction between rotating head and tape. (From Osaki, H., 1993, *IEEE Trans. Magn.* MAG-29(1), 11–20. With permission.)

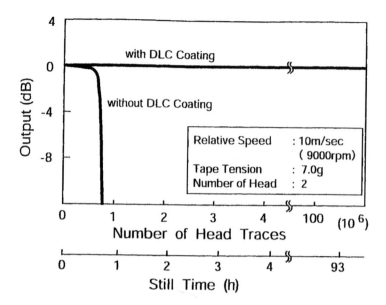

FIGURE 12.80 Improvement of durability in still-frame mode by DLC coating. Variations of the reproduced output for ME tape with and without DLC coating. (From Osaki, H., 1996, *Wear* 200, 244–251. With permission.)

improvement of the resistance against abrasion. And also high-wear-resistant coating had been introduced in addition to the current technologies. A total spacing between a tape and a head has to be reduced to be on the order of 10 nm to obtain higher recording density for this new system. DLC is one of the most promising materials for this requirement (Kurokawa et al., 1986; Shinohara et al., 1994; Kaneda, 1994; Osaki and Sato, 1995; Osaki, 1996). The variations of the reproduced output for the ME tape with and without DLC coating are shown in Figure 12.80. It is recognized that the durability in the still-frame mode has been tremendously improved by DLC coating and no damage can be observed even after such a severe application. From the point of recording density, the thickness of a protective coating should be decreased without losing durability. For this purpose, further research for superior materials of protective coatings is required.

12.3.3.1.2   Head Clogging

Head clogging is the problem that the reproduced output signal decreases suddenly due to the transferred wear debris of a tape onto the head (Osaki et al., 1994). Head clogging is caused mainly via two different mechanisms according to the tackiness of the wear debris from tapes. When the tackiness of wear debris is low, the wear debris piles up on the exit side of each head. When the wear debris becomes large enough to contact the tape surface, it contacts the tape surface and is deposited on the tape surface. In subsequent sliding, the head picks it up on the rubbing surface in the form of a large agglomerates. A spacing loss occurs with consequential decrease in reproduced output. When the tackiness of wear debris is high, the wear debris piles up on the exit side of each head. The wear debris transfers directly onto the rubbing surface and decreases the reproduced output.

12.3.3.1.3 Wear of Rotary Heads

1. Uniform Wear

The harder material of a magnetic head does not necessarily show better wear resistance, as compared to materials in thin-film structure. Atmosphere also has a significant effect on the wear of a magnetic head. In general, higher humidity and lower temperature are more severe conditions for the wear of magnetic heads (Mizoh, 1996), which is caused by the change in mechanical stiffness of the tape and the increase in friction coefficient between the tape and the transport components.

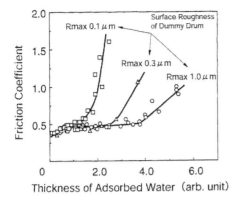

FIGURE 12.81 Relationship between friction coefficient and thickness of adsorbed water (ME tape/dummy drum). (From Kawakami, K. and Osaki, H., 1992, *Proc. Tribol. Conf.*, JAST, Tokyo, pp. 475–478. With permission.)

Crystallographic plane and direction of single-crystal Mn–Zn ferrite, which is the most common material for heads, are also important for wear of the heads (Miyoshi and Buckley, 1984). But the best plane and direction of Mn–Zn ferrite for a certain tape are not always the best choice for different tapes. Wear characteristics of magnetic heads depend on the combination of a head and a tape. The change of crystalline state of a single-crystal head and the deformed subsurface layer of the head caused by rubbing, which result in decrease in magnetic performances and consequently reproduced output signal, should be considered as well as wear of head itself (Miyoshi and Buckley, 1985).

2. Recess Wear

When the wear resistances of the each material in the head surface are different, various head materials wear nonuniformly, and form a recess. In the case that the magnetic gap is located in the recessed area, a spacing loss is increased. The combination between the materials of gap area and the tape should be selected to reduce the amount of recess. The lapping tape which has high abrasion effect is useful to clear the recess temporarily.

12.3.3.2 Tribological Problems between Tapes and Tape Transport Components in Helical Scan Systems

12.3.3.2.1 Friction and Stiction

As a tape rubs against a stationary drum and stationary guideposts when it is transported in VTRs or tape drives, the friction coefficient should be low to realize smooth transportation. If the friction coefficient is very high, "stiction" occurs and a tape is stopped being transported; sometimes it stops the rotation of rotary drum.

The friction coefficient between a tape and a stationary drum (or guideposts) is liable to be influenced by temperature and humidity, which seems to be caused by adsorbed water on the tape and drum surfaces. A relationship between the friction coefficient and the thickness of the adsorbed water was observed with an ME tape under various combinations of temperature and humidity (see Figure 12.81) (Kawakami and Osaki, 1992). This phenomenon is also observed when a tape has excessive lubricant.

The critical thicknesses where the friction force increases quickly become larger with rougher drum surfaces, as shown in Figure 12.81. In other words, the quick increase in friction force does not occur at the rough drum surface. The runnability can be greatly improved (friction coefficient can be decreased) by surface asperities of ME tapes (Tomago et al., 1988). This improvement is supposed to be achieved by the decrease in the real contact area, just like the case of durability in the still-frame mode.

12.3.3.2.2 Damage of Tape Edge

The edge of a tape is pressed onto the flange of a roller guide and is finally damaged after a certain amount of running, when the tape is transported along with the roller guide which is inclined slightly from the direction of tape transportation, as shown in Figure 12.82. When the static friction force is low, the tape will slip on the roller guide surface in the axis direction. In this case, the edge should not be

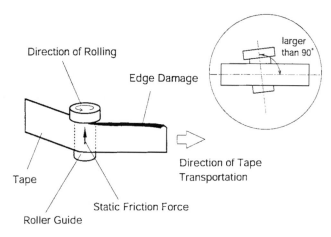

FIGURE 12.82 Schematic drawing of edge damage at roller guide. (From Osaki, H., 1996, *Wear* 200, 244–251. With permission.)

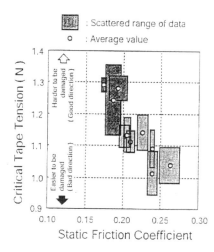

FIGURE 12.83 Relationship between critical tape tension and static friction coefficient. (From Itaya, T. et al., 1997, *IEICE General Conf.* 6-7-30, p. 74. With permission.)

damaged. The higher tape tension, which increases the static friction force between the tape and the roller guide, readily causes edge damage. The relationship between the critical tape tension, at which the tape edge is damaged at a roller guide, and the static friction coefficient is shown in Figure 12.83 (Itaya et al., 1997). It is observed that the static friction coefficient, not the kinetic friction coefficient, should be lower to avoid edge damage of a tape. Tape squeal caused by stick-slip is also controlled by the static friction coefficient.

12.3.3.3   Tribological Problems in Linear Recording System

The tribological problems in the linear recording systems between a tape and a tape transport elements or a linear heads are similar to that in the helical scan recording systems. A high friction coefficient prevents smooth tape transport; and wear debris transfers on a linear head causing clogging. The real contact area of magnetic media can be estimated by using a glass slide and are analyzed experimentally and theoretically by Bhushan (1984; 1985). Excessive friction can be found in some computer tapes after repeated use in tape drives, which results in physical (Bhushan et al., 1984) or chemical (Bradshaw and Bhushan, 1984) changes in the tape surface caused by repeated use (continuously or intermittently)

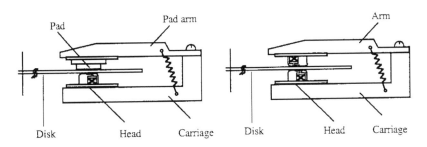

FIGURE 12.84 Schematics of single-sided head (left) and dual-sided head.

and/or by debris accumulated on the drive components. Debris can be either loose or adherent. Loose debris is also packed on the drive components after repeated use, but can be removed easily (Bhushan and Phelan, 1986). Adherent debris that transferred to the drive components lead to polymer-polymer contact, whose friction is higher than that of rigid material-polymer contact, and sometimes cause a serious problem (Bhushan and Phelan, 1986). Tapes having a high propensity for debris generation in addition to the physical changes of the surfaces may cause frictional problems depending on the design of the tape drive.

When the wear debris transfers on the stationary heads, the reproduced output signal decreases and the recorded data cannot be read. The resin binder transferred on the head surface causes the increase in friction force, which is removed by abrasive additives (Doshita and Mukaida, 1995). The friction force can be also reduced by acoustic excitation (Tam and Bhushan, 1997).

12.4 Floppy Disk Files

It was 1972 that the very first system of the floppy disk drive, also known as the diskette drive, was introduced to the market from IBM. The IBM 3740 system utilized the 8-in. diskette with its formatted capacity of 243 kB. The track density was 48 TPI and the linear density was 3268 BPI. Later on, both linear density and the track density were increased and the single-sided magnetic head became a dual-sided head (see Figure 12.84). However, as the drive with a small form factor began to appear in the market, the 8-in. drive shrank in volume.

In 1974, Shugart announced the 5.25-in. floppy drive, the SA400 with 48 TPI and 2581 BPI. The first model adopted the same technology inherited from the 8-in. drive, and the magnetic head was single sided at the time of introduction, but later changed its shape to dual sided as in the case of the 8-in. drive. Then in 1980, Sony announced the 3.5-in. drive in its word processor, which was first sold in the U.S. However, until 1982, detailed information about format was not revealed, so almost no compatible systems appeared until 1982. The other manufacturers also proposed and shipped out their original drive, but the movement of standardization accelerated the 3.5-in. drive to become the global standard.

12.4.1 Floppy Disk Heads

12.4.1.1 Head Design of 3.5-in. Floppy Disk Head

12.4.1.1.1 Structure of Magnetic Heads

One of the distinctive features that the floppy disk head has is its unique structure, the combination of read–write head and erase head. Though, there are several types of head, as shown in Figure 12.85, tunnel-erase head is the most commonly adopted structure; it was first patented by IBM in 1961. Even now most of the 3.5-in. floppy disk heads are based on the concept of that patent. Historically, the floppy disk head had adopted the tunnel-erase head because the flexible disk is a removable media and the chucking of the disk is not so accurate; therefore, in order to gain enough margin, a wide-write and

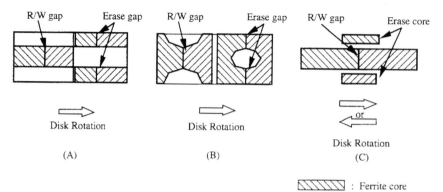

FIGURE 12.85 Gap and core geometry of ferrite head: (A) Tunnel-erase laminate type, (B) tunnel-erase bulk type, (C) straddle-erase type.

narrow-read head was desired. Generally, it requires two gaps, one for write only and the other is for read only. To delete the previous data, another erase gap is necessary. These structures are quite complex in the manufacturing process. The laminate-type structure was popular at the beginning and it inherited the structure from the magnetic heads used for audio recording. As the name indicates, read–write heads are sandwiched by two erase heads and glued by epoxy resin. However, as the track density increased, the mechanical tolerances for the alignment of read–write gap and erase gaps became tight and it reduced the process margin to control the adhesive layer. Nevertheless, the tunnel-erase head has a drawback caused by its structure — when the distance between the read–write gap and the erase gap is not negligible, it generates the delay to open or to close the gate of the read–write circuit and erase circuit. During operation, this delay forces the system to sacrifice the recording area. A straddle erase head was originally adopted in the 5.25-in. drive and it has a benefit that the distance of the read–write gap and erase core is negligibly small, although the manufacturing process was quite complicated. The bulk head uses the manufacturing process similar to video heads. Again, same as in the case of laminate head structure, a certain recording area is wasted due to the distance between the read–write gap and the erase gap along with the direction parallel to the head motion. Currently, the bulk head is the most popular structure due to the demand of higher recording density and lower production cost.

The most important and difficult part of the floppy disk head is a design of head core geometry. Equally to the other magnetic recording system with inductive magnetic head, the dominant parameters of the floppy disk head are gap length, throat height, and track width. Track width of both the tunnel-erase head and the read–write head are defined by its recording format. Therefore, the throat height and gap length are the only tunable parameters when designing head core geometry. Currently, the high-density floppy disk with formatted capacity of 1.44 MB (unformatted 2 MB) is the most familiar one, but in the past, several formats did exist and the latest format has to support some of those previous formats. This compatibility is quite important in the field of removable storage devices, but at the same time it makes the head design very complicated. The read–write gap and the throat height should be optimized considering a lot of characteristics, such as overwrite, window margin, resolution, etc., at different condition depending on the compatible format and even on the media with different magnetic parameters. Typically, the read–write gap length is chosen to be in the range from one third to one fourth of the minimum recording wavelength, and the throat height is chosen to be less than a few tens of a micrometer.

12.4.1.1.2   Slider Design

Besides the design of the magnetic transducer, the surface design or contour design of floppy disk head is quite important. In the 3.5-in. drive, the disk rotates at the speed of 300 rpm and the head should be

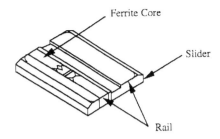

Ferrite Core

Slider

Rail

FIGURE 12.86 Structure of slider (bulk-type head).

in good contact with the disk during operation but should not give any damage to the media. As shown in Figure 12.86, two relatively wide rails are formed in parallel along with the direction of disk motion and the outer edges of the rails are rounded. Unlike the hard disk head, the magnetic transducer is located at the center of the one rail so as to obtain stable contact with media. The body is fixed to the gimbal with 1° or 2° of freedom in the plane parallel to the disk surface and is designed to trace the microroughness of the floppy disk.

12.4.1.2   Manufacturing Process of Floppy Disk Heads

12.4.1.2.1   Materials

A material used for a magnetic core should be capable of satisfying not only the electromagnetic characteristics but also good wear resistance. Also, from a manufacturing point of view it should be easy to process. The most adequate material for above requirements is a magnetic ferrite. Although, there are two candidates, Ni–Zn ferrite and Mn–Zn ferrite, the latter one is commonly used because of its high saturation magnetization and the ease of its machining process. Although floppy disk heads have inherited a lot of manufacturing processes from those of the video head, the choice of bonding glasses is more complicated, since, in its manufacturing process a floppy disk head has to use multiple types of glasses. Adding to the first glass-bonding process to glue the read–write gap and the erase gap, a floppy disk head has to experience other bonding processes with different types of glasses. For instance, bonding between the read–write core and the erase core, and sometimes bonding between the ferrite core and the slider, may require additional glass-bonding processes. One should be aware that, when handling second or third glasses in the bonding process, the maximum temperature should not exceed a certain point, since otherwise the first glass will deform and may change the gap length. Also, high-temperature treatment may degrade or corrode the ferrite core, but the glass with low working temperature generally shows poor durability against the environment. So a lot of attention should be paid when glass is used for the purpose of bonding. The slider is used to support the ferrite core to maintain good and stable head/disk contact. The property of this material should be close to or superior to the ferrite core in wear resistance, thermal expansion coefficient, and the ease in machining process, and needless to say, it should be nonmagnetic with low material cost. For those purposes, barium titanate or calcium titanate are commonly used.

12.4.1.2.2   Machining Process

The manufacturing process of the magnetic core of the floppy disk head is much like that of the video head. Two separate ferrite cores, with different shapes, called C core and I core, are bonded by glass with a gap spacer formed by deposition or sputtering. The track width is formed beforehand by a slicing machine. A read–write ferrite block and an erase ferrite block are again bonded with glass and then slices into tiny chips. The head chip is then sandwiched by the slider. Owing to the difficulty in winding a coil directly to the winding hole, sometimes the backside of the head core is cut off and then a bobbin coil is inserted. Finally, the backside of the core is bonded with the ferrite bar or C-shaped back core so as to close the magnetic path and gain core efficiency. The simplified flow of the machining process is illustrated in Figure 12.87.

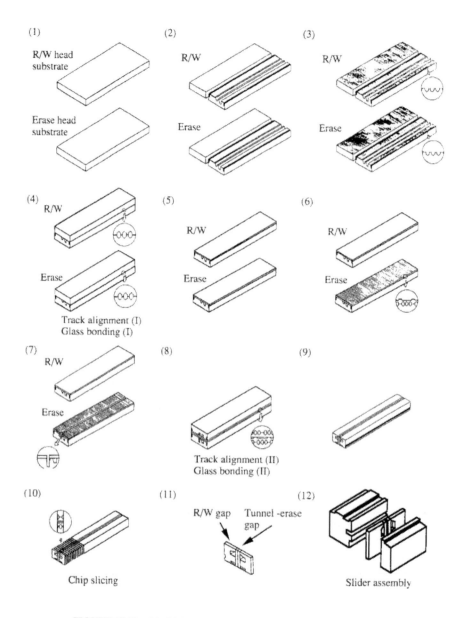

FIGURE 12.87 Machining process flow of bulk-type ferrite head.

12.4.1.2.3 Assembling Process

In the assembly of the head, the slider is combined with a bobbin coil and then the back core is glued with the slider. When the coil is directly wound on the winding hole, the above process is altered by the winding process. Then, the head is glued on the gimbal surrounded by the shield and a flexible printed circuit is attached (see Figure 12.88). The shape and the structure of the gimbal differs depending on the drive/device manufacturer, but the key thing is to maintain stable contact with disk and avoid unwanted vibration caused by the friction between head and disk. A flexible printed circuit should withstand the millions of bending motions in the lifetime.

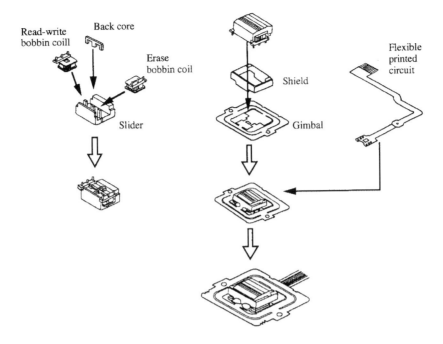

FIGURE 12.88 Floppy disk head assembly process.

12.4.2 Floppy Disks

The floppy disk typical 76-μm thickness consists of a 70-μm PET base film coated on both sides with a magnetic coating. The almost exclusively used disk diameter today for data-processing drive is an 86 mm (3.5-in.) microfloppy disk or MFD. Typical formatted capacity for MFD with Co–Fe$_2$O$_3$ is 1.4 MB. In the case of a floppy disk, since the head is in contact with the magnetic coating, higher linear recording density can be achieved. However, track densities are relatively low compared with rigid disks because of the poor dimensional stability of the 70-μm PET substrate and the lack of a track servo system.

The manufacturing process of floppy disks is almost similar to the tape with several exceptions: almost all floppy disk substrates are about 70-μm-thick PET; the coatings are oriented as randomly as possible; after the calendering process, the web is punched out into the disk form and is buffed by cleaning (abrasive) tape to remove higher asperities. Since the back coating is not deposited for the floppy disk, electrical conductivity of the magnetic layer is needed.

12.4.3 High-Storage-Capacity Floppy Disks

In the late 1980s, after the maturation of the 3.5-in. floppy drive market, a lot of drive manufacturers tried to break into the market with higher-storage-capacity floppy drive systems. The first attempt was to increase the linear density while maintaining compatibility with the existing 2-MB format. The unformatted storage capacity was doubled to 4 MB in the same 3.5-in. diskette. However, due to the small read–write gap length desired for the 4 MB format, it was quite challenging to satisfy the specification of overwrite. Then, there was another attempt to increase the linear density by coping with the metal-in-gap head (MIG head) that was again transferred from the video head technology. However, almost all of the those attempts did not lead to commercial products.

TABLE 12.3   Typical Characteristics of High-Density Floppy Disk

	MFD	ZIP	LS-120	UHC	HiFD
First customer shipment	1980	1995	1995	1998	1998
Disk diameter, mm	86	90	86	86	86
Magnetic particles	Co-γ-Fe$_2$O$_3$	Metal	Metal	Metal	Metal
Rotational speed, rpm	360	2945	720	3600	3600
Data capacity, MB	1.4	100	120	130	200
Linear density, kbpi	8.7	45.2	44.8	70	72–91
Track density, TPI	135	2188	2490	2700	2822
Data transfer rate, MB/s	0.5	0.79–1.4	0.4–0.66	2.45	3.6
Magnetic head	Single gap	Single gap	Dual gap	Dual gap	Dual gap
Backward compatibility	—	No	Yes	Yes	Yes
Drive manufacture	Sony	Iomega	MKE, Mitsubishi, ORT	Swan, Mitsumi	Sony

Realizing the limit of the existing format, a new attempt from a different aspect was investigated. Metal particle media or barium ferrite disks were used for higher recording densities. Target capacity was about 10 to 20 MB and the compatibility with the 3.5-in. format was sought. However, again most of those products did not end in success. At that time, there was not very much demand from the market, and also the impact of price competition against the existing floppy disk drive was too severe to break in with a new and costly drive.

Recently, thin-layered particulated disks, previously discussed in the particulate tape section, with a capacity of more than 100 MB are available for high-density applications. Rotation speed is so high that these drive systems use a flying hard disk head not a spherical surface contour which has been used for MFD. Over the years, the floppy disks used to be a major distribution tool, but nowadays CD-ROM has taken their role; therefore, the need for an MFD with more than 100 MB in capacity has become pressing. Table 12.3 shows typical characteristics of four new drives that are now available or will soon be available and the 3.5-in. floppy disk drive. Double-coating metal particulated disk previously discussed in the particulate tape section, are used for these applications. The technology adopted in Zip drive was mostly transferred from the hard disk drive, and that made the entire system achieve its storage capacity as much as 100 MB. Higher rotational speed and linear density increased the data transfer rate and the unique servo system and narrow track width gained the area density. If one considers of the hard disk drive with conventional MIG head, its capacity is nearly 1 GB on a single 3.5-in. platter, ten times that of the Zip drive in capacity. Even though the floppy disk is removable media and easy to deform by environmental changes in a microscopic sense, still there is room to get close to the hard disk drive system in its capacity.

While the Zip is an independent format, the other drives have supported compatibility with a 3.5-in. floppy disk drive. The magnetic head is designed separately for each format, which means there are two ferrite cores on two rails. Since continuous demand for higher recording capacity brings into competition of emerging removable storage system, the advance of the floppy disk drive will be accelerated more than ever. Magnetic disks as well as heads have to be developed in harmony with the head–disk interface.

12.4.4   Head–Floppy Disk Interface

The head–disk interface in some flexible disk systems are different depending on the disk rotational speed. In the system, which is compatible with the 3.5-in. floppy disk drive, the interface is the same as long as it is operated under compatible mode. Also, the LS120 drive has a similar head–disk interface with 3.5-in. floppy disk drive, since the rotational speed of the disk is not so different. While LS120 has its speed of 720 rpm, the ordinary 3.5-in. floppy disk drive has 300 rpm. In the high-capacity mode of UHC or Hi-FD drive, the rotational speed of the disk is set to 3600 rpm. The speed of the disk rotation is comparable to a bit old hard disk system, so it can be speculated that the head files over the floppy disk. However, in reality, unlike the hard disk, a floppy disk has a rather tough surface, and it makes the head–disk interface rather like a near-contact mode.

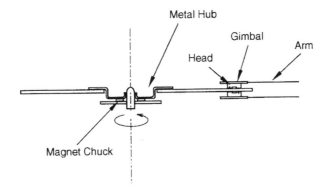

FIGURE 12.89   Schematic configuration of single-sided floppy disk drives.

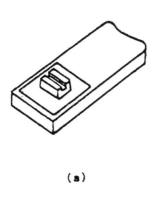

(a)

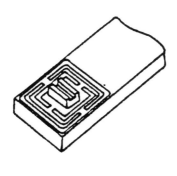

(b)

FIGURE 12.90   Typical heads for double-sided floppy drives. (a) Fixed type for side 0, (b) two-directional gimbal type for side 1.

A typical schematic drawing of the interfaces of the 3.5-in. floppy disk drive system is shown in Figure 12.89. A flexible disk is sandwiched between two spherical heads which have two rails on each head to obtain a good contact. The head slider are supported by a gimbal which is fixed to an arm. A combination of a fixed-type head slider and another head slider supported by a two-directional gimbal is used in a double-sided floppy disk system in general, as shown in Figure 12.90a and b, respectively.

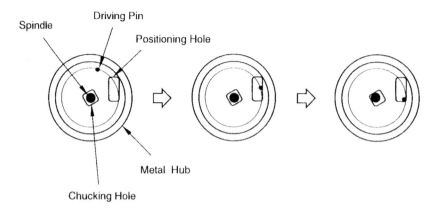

FIGURE 12.91 Typical chucking mechanism adopted in 3.5-in. floppy disk drive.

A chucking mechanism for flexible disk media is very important, since interchangeability is indispensable for floppy drives. A typical chucking mechanism adopted in a 3.5-in. floppy disk system is shown in Figure 12.91.

12.4.4.1 Durability of Media

As the head slider flies in boundary lubrication during operation, the contact pressure between a head and a medium in floppy disk systems is less than that in tape systems. However, the disk surface will be damaged after long use. The damage mechanisms are almost similar to the ones of particulate tapes. The interfacial shear stress between a magnetic layer and a substrate, which is caused by friction between a head and a disk surface, is increased with the decrease in the thickness of magnetic layer performed for the improvement of the recording characteristics. Therefore, if the friction coefficient exceeds a critical value, the magnetic layer will be peeled. The reduction of the amount of abrasives improves the recording characteristics, but causes an increase in the friction coefficient; because it is supposed that the additives prevent the plastic flow of polymer binder in magnetic layer by supporting the contact load between the tape and the head.

Figure 12.92 shows output signal level vs. pass number for 3.5-in. floppy disk drive (Teramura, 1985). These long-life wear tests were carried out in actual operating conditions. The output level after 2000×10^4 passes remains more than 80 percent of their initial signal level in both cases. Wear test tracks on which the head is sliding are from Track 5 to Track 60. However, it should be noted that the signal level in Track 0 decreases to the same amount as that of the head sliding tracks. It is considered that this decrease in the signal level on the track without head sliding is caused by contact with the liner material inside the jacket or case (Teramura, 1985).

12.4.4.2 Flyability

Very low flying height (less than 0.25 μm) has been obtained by introducing many techniques in floppy disk files, such as high head penetration, opposite pressure pads, opposite heads, pressure relief slots, and specially contoured nonspherical heads. The clearance of flying height should be nearly uniform to obtain enough reliability, because a head partially contacts a disk in floppy disk files. A schematic of the experimental apparatus used to measure the flying height of flexible disk is shown in Figure 12.93 (Talke and Tseng, 1984). The contact system in this apparatus is a little different from that in a real floppy disk, in which a flexible disk is sandwiched between two heads or between a head and a pressure pad. The contact load is produced by the protrusion against the flexible disk without a pressure pad. Results are shown in Figures 12.94 and 12.95. It is found that the main parameter in determining the flying height

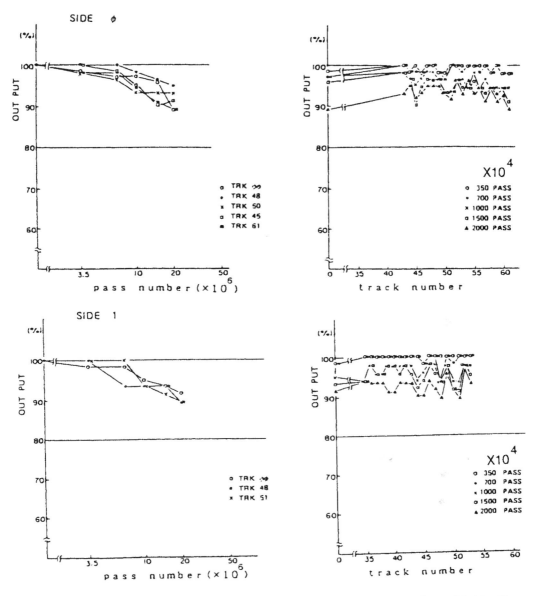

FIGURE 12.92 Variations of output signal level with increasing number of passes in 3.5-in. floppy disk drive. (From Teramura, N., 1985, in *Tribology and Mechanics of Magnetic Recording Systems*, Vol. 2, SP-19 (B. Bhushan and N. S. Eiss, eds.), ASLE, Park Ridge, IL. With permission.)

is the physical width of the head. In particular, as the head narrows, side flow increases and flying heights of the order on 0.125 μm can be achieved. A decrease in flying height is likewise observed by increasing the head protrusion or decreasing the radius of the head. Stability of the air bearing can be improved by incorporating a side rail adjacent to the head (Talke and Tseng, 1984).

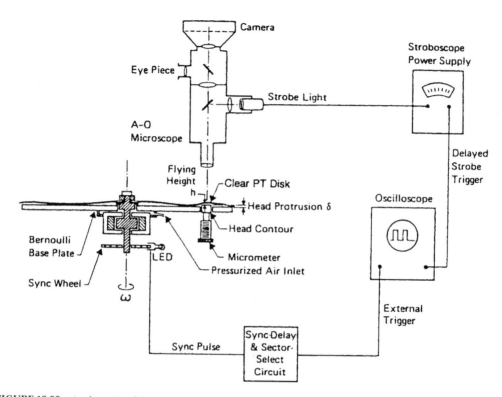

FIGURE 12.93 A schematic of the experimental apparatus used to measure the flying height of flexible disk. (From Talke, F. E. and Tseng, R. C., 1984, in *Tribology and Mechanics of Magnetic Recording Systems,* Vol. 1, SP-16 (B. Bhushan and N. S. Eiss, eds.), ASLE, Park Ridge, IL. With permission.)

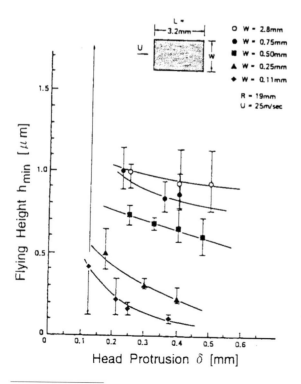

FIGURE 12.94 Flying height vs. protrusion for spherical head as a function of head width ($R = 19$ mm). (From Talke, F. E. and Tseng, R. C., 1984, in *Tribology and Mechanics of Magnetic Recording Systems,* Vol. 1, SP-16 (B. Bhushan and N. S. Eiss, eds.), ASLE, Park Ridge, IL. With permission.)

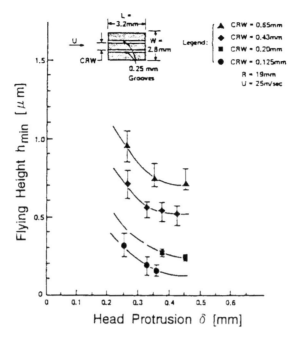

FIGURE 12.95 Flying height vs. protrusion for tri-rail spherical head as a function of center rail width ($R = 19$ mm). (From Talke, F. E. and Tseng, R. C., 1984, in *Tribology and Mechanics of Magnetic Recording Systems*, Vol. 1, SP-16 (B. Bhushan and N. S. Eiss, eds.), ASLE, Park Ridge, IL. With permission.)

References

Anthony, T. C., Naberhuis, S. L., Brug, J. A., Bhattacharyya, M. K., Tran, L. T., Hesterman, V. W., and Lopatin, G. G. (1994), "Dual Stripe Magnetoresistive Heads for High Density Recording," *IEEE Trans. Magn.*, 30(2), 303–308.

Ashar, K. G. (1997), "Magnetic Disk Drive Technology," IEEE Press, pp. 187.

Baibich, M., Broto, J., Fert, A., Nguyen Van Dau, F., Petroff, F. (1988), "Giant Magnetoresistance of (001) Fe/(001)Cr Magnetic Superlattices," *Phys. Rev. Lett.* 61, 2472.

Beaulieu, T. and Nepala, D. (1975), "Induced Bias Magnetoresistive Read Transducer," U.S. Patent, 864,751, February 4.

Berar, P., Kato, K., Osaki, H., and Kapsa, P. (1995), "Wear Mechanisms of Carbon-Coated Metal Evapo-rated Magnetic Tape in VCR," in *Proc. Int. Tribo. Cong.* pp. 1817–1822.

Bertram, H. N. (1994), *Theory of Magnetic Recording,* Cambridge University Press,

Bhattacharyya, M. K., Davidson, R. J., and Gill, H. S. (1987), "Bias Scheme Ability of Shielded MR Heads for Rigid Disk Recording," *J. Appl. Phys.* 61(8), 15, 4167–4169.

Bhushan, B. (1984), "The Real Area of Contact in Polymeric Magnetic Media: Critical Assessment of Experimental Techniques," ASLE Reprint No. 84-AM-1A-1, pp. 1–12.

Bhushan, B. (1985), "The Real Area of Contact in Polymeric Magnetic Media II: Experimental Data and Analysis," *ASLE Trans.* 28(2), 181–197.

Bhushan, B. (1992), *Mechanics and Reliability of Flexible Magnetic Media,* Springer-Verlag, New York.

Bhushan, B. (1995), *Handbook of Micro/Nanotribology,* CRC Press, Boca Raton, FL.

Bhushan, B. (1996a), *Tribology and Mechanics of Magnetic Storage Devices,* 2nd ed., Springer-Verlag, New York.

Bhushan, B. (1996b), "Tribology of the Head-Medium Interface," in *Magnetic Recording Technology* (Mee, C. D. and Daniel, E. D., eds.), 2nd ed., pp. 7.1–7.66, McGraw-Hill, New York.

Bhushan, B. and Phelan, R. M. (1986), "Frictional Properties as a Function of Physical and Chemical Changes in Magnetic Tapes During Wear," *ASLE Trans.* 29(3), 402–413.

Bhushan, B., Bradshaw, R. L., and Sharma, B. (1984), "Friction in Magnetic Tapes: Role of Physical Properties," *ASLE Trans.* 27(2), 89–100.

Bradshaw, R. L. and Bhushan, B. (1984), "Friction in Magnetic Tapes III: Role of Chemical Properties," *ASLE Trans.* 27(3), 207–219.

Chapman, D. W. (1989), "A New Approach to Making Thin Film Head-Slider Devices," *IEEE Trans. Magn.* MAG-25, 3686–3688.

Chubachi, R. and Tamagawa, N. (1984), "Characteristics and Applications of Metal Tape," *IEEE Trans. Magn.* 20, 45–47.

Corb, B. (1990), "Correlation between Recording Performance and Domain Structure for Thin Film Heads," *Dig. Intermagn. Conf.,* AA-06.

Cutiongco, E. C., Li, D., Chung, Y. W., and Bhatia, C. S. (1996), "Tribological Behavior of Amorphous Carbon Nitride Overcoats for Magnetic Thin-Film Rigid Disks," *J. Tribol.,* 118, 543–548.

Dimigen, H. and Hubsch, H. (1983/1984), "Applying Low-Friction Wear-Resistant Thin Solid Films by Physical Vapor Deposition," *Phillips Tech. Rev.* 41, 186–197.

Doshita, H. and Mukaida, Y. (1995), "Tribology of Particulate-Coated Video Tape," *J. Jpn. Soc. Tribol.,* 40, 987–991 (in Japanese).

Endo, M., Nakamura, K., and Konishi, S. (1991), JP03-188178.

Enke, K., Dimigen, H., and Hubsch, H. (1980), "Friction Properties of Diamond-Like Carbon Layers," *Appl. Phys. Lett.* 36, 291–292.

Farkas, J., Hall, D. R., Kim, K. J., and Vaduls, R. R. (1994), "Polyurethan Composition for Use as a Dispersing Binder," USP 5371166.

Guo, Q., Osaki, H., Keer, L. M., and Wheeler, D. R. (1990), "Measurement of the Intrinsic Bond Strength of Brittle Thin Films on Flexible Substrate," *J. Appl. Phys.,* 68(4), 15, 1649–1654.

Hannon, D., Krounbi, M., and Christner, J. (1994), "Allicat Magnetoresistive Head Design and Performance," *IEEE Trans. Magn.* 30(2), 298–300.

Harada, K. (1981), "Plasma Polymerized Protective Films for Plated Magnetic Disks," *J. Appl. Polym. Sci.* 26, 3707–3718.

Heim, D. E., Fontana, R. E., Jr., Tsang, C., Speriosu, V. S., Gurney, B. A., and Williams, M. L. (1994), "Design and Operations of Spin Valve Sensors," *IEEE Trans. Magn.* 30(2), 316–318.

Hempstead, R. D. (1975), "Analysis of Thermal Noise Spike Cancelation," *IEEE Trans. Magn.* MAG-11(5), 1224–1226.

Herk, A. van (1980), "Three-Dimensional Computation of the Field of Magnetic Recording Heads," *IEEE Trans. Magn.* MAG-16, 890–892.

Hunt, R. (1971), "Magnetoresistive Readout Transducer," *IEEE Trans. Magn.* MAG-7, 150–152.

Inaba, H., Ejiri, K., Abe, N., Masaki, K., and Araki, H. (1993), "The Advantages of the Thin Layer Particulate Recording Media," *IEEE Trans. Magn.* 29, 3607–3612.

Ishikawa, M., Tani, N., Yamada, T., Ota, Y., Nakamura, K., and Itoh, A. (1986), "Dual Carbon, a New Surface Protective Film for Thin Film Hard Disks," *IEEE Trans. Magn.* MAG-22, 999–1001.

Itaya, T., Kawakami, K., Kanou, T., and Osaki, H. (1997), "Effect of Static Friction on Tape Edge Damage in VTR," in *Proc. 1997 IEICE General Conf.,* C-7-30, p. 74 (in Japanese).

Jagielinski, T., Doyle, T., and Smith, N. (1986), *IEEE Trans. Magn.* MAG-22(5), 680–682.

Jeffers, F. (1986), "High-Density Magnetic Recording Heads," *Proc. IEEE* 74(11), 1540.

Jones, R. E., Jr., (1980), "IBM 3370 Film Head Design and Fabrication," IBM Disk Storage Technol., GA 26-1665-0, p. 6.

Kaneda, Y. (1994), "Tribology of Metal-Evaporated Tape for High-Density Magnetic Recording," *IEEE Trans. Magn.* MAG-33, 1058–1068.

Kawakami, K. and Osaki, H. (1992), "Relationship between Friction Coefficient of Magnetic Tape and Thickness of Adsorbed Water," in *Proc. Tribol. Conf.,* pp. 475–478, JAST, Tokyo, (in Japanese).

Kawakubo, Y., Ishihara, H., Seo, H., and Hirano, Y. (1984), "Head Crash Process of Magnetic Coated Disk Contact Start/Stop Operations," *IEEE Trans. Magn.* MAG-20, 933–935.

Kawana, T., Onodera, S., and Samoto, T. (1995), "Advanced Metal Evaporated Tape," *IEEE Trans. Magn.* 31(6), 2865–2870.

Kishimoto, M., Kitaoka, S., Andoh, H., Amemiya, M., and Hayama, F. (1981), "On the Coercivity of Cobalt-Ferrite Epitaxial Iron Oxide," *IEEE Trans.* MAG-17(6), 3029–3031.

Klassen, K. B. and van Peppen, J. C. L. (1989), "Delayed Relaxation in Thin-Film Heads," *IEEE Trans. Magn.* 25(5), 3212–3214.

Klaus, E. E. and Bhushan, B. (1985), "Lubrication in Magnetic Media — A Review," *ASLE* SP-19, 2, 7–15.

Kondo, H. and Kaneda, Y. (1994), "Development of Modified Perfluoropolyether," in *Proceedings of International Tribology Conference,* pp. 415–420,

Kondo, H., Seto, J., Ozawa, K., and Haga, S. (1989), "Novel Lubricant for Magnetic Thin Film Media," *J. Magn. Soc. Jpn.* 13(S1), 213–218.

Kunieda, T., Shinohara, K., and Tomago, A. (1985), "Metal Evaporated Video Tape," *IERE,* 55(6), 217–221.

Kurokawa, H., Mitsuya, T., and Yonezawa, T. (1986), "Application of Diamond Like Carbon Film to Magnetic Recording Media," *Tech. Mtg. Magn. Soc. Jpn.* 46–8, 67–75 (in Japanese).

Lazzari, J. P. and Deroux-Dauphin, P. (1989), "A New Thin Film Head Generation," *IEEE Trans. Magn.* MAG-25, 3190.

Lin, J. and Wu, A. W. (1990), "Lubricants for Magnetic Rigid Disks," *Proc. Jpn. Int. Tribol. Conf.* pp. 599–604.

Mallary, M. L. and Ramaswamy, S. (1993), "A New Thin Film Head Which Doubles the Flux through the Coil," *IEEE Trans. Magn.* 29(6), 3832–3834.

Mallinson, J. C. (1994), *The Foundation of Magnetic Recording,* 2nd ed., Academic Press, New York.

McFarlane, J. S. and Tabor, D. (1950), "Adhesion of Solids and the Effects of Surface Films," *Proc. R. Soc.* A202, 224–243.

McKenzie, D. R., McPhedran, R. C., Botten, L. C., Savvides, N., and Netterfield, R. P. (1982), "Hydrogenated Carbon Films Produced by Sputtering in Argon-Hydrogen Mixture," *Appl. Opt.* 21, 3615–3617.

Mee, C. D. and Daniel, E. D. (1996), *Magnetic Storage Handbook,* McGraw-Hill, New York.

Middleton, B. K. (1966), "The Dependence of Recording Characteristics of Thin Metal Tapes on Their Magnetic Properties and on the Replay Head," *IEEE Trans. Magn.* MAG-2, 225–227.

Miyasato, T., Kawakami, Y., Kawano, T., and Hiraki, A. (1984), "Preparation of Sp^3-Rich Amorphous Carbon Film by Hydrogen Gas Reactive rf-Sputtering of Graphite, and Its Properties," *Jpn. J. Appl. Phys.* 23, L234–L237.

Miyoshi, K. and Buckley, D. H. (1984), "Properties of Ferrites Important to Their Function and Wear Behavior," in *Tribology and Mechanics of Magnetic Recording Systems,* Vol. 1, SP-16 (B. Bhushan and N. S. Eiss, eds.), pp. 13–20, ASLE, Park Ridge, IL.

Miyoshi, K. and Buckley, D. H. (1985), "Effect of Wear on Structure-Sensitive Magnetic Properties of Ceramic Ferrite in Contact with Magnetic Tape," in *Tribology and Mechanics of Magnetic Recording Systems,* Vol. 2, SP-19 (B. Bhushan and N. S. Eiss, eds.), pp. 112–118, ASLE, Park Ridge, IL.

Mizoh, Y. (1996), "Wear of Tribo-Elements of Video Tape Recorders," *Wear* 200, 252–264.

Mizunuma, Y., Miyake, H., and Hiura, N. (1978), U.S. Patent 4152485.

Morikawa, K., Matsuura, T., Horibata, S., and Shibata, H. (1991), "A Study of Popcorn Noise for Thin Film Heads," *IEEE Trans. Magn.* 27(6), 4939–4941.

Nakayama, A., Nakamura, K., Hata, K., and Yamamoto, M. (1985), "Magnetic Paint for Magnetic Recording Media," U.S. Patent 4707411.

Narishige, S., Hanazono, M., Tokagi, M., and Kuwatsuka, S. (1984), "Measurements of the Magnetic Characteristics of Thin Film Heads Using Mageto-Optical Method," *IEEE Trans. Magn.* MAG-20, 848–850.

Nishihara, H. S., Dorius, L. K., Bolasna, S. A., and Best, G. L. (1988), "Performance Characteristics of the IBM 3380K Air Bearing Design," in *Tribology and Mechanics of Magnetic Recording Systems,* Vol. 5, SP-25 (B. Bhushan and N. S. Eiss, eds.), pp. 117–123, ASLE, Park Ridge, IL.

Nyaiesh, A. and Holland, H. (1984), "The Growth of Amorphous and Graphitic Carbon Layers under Ion Bombardment in an RF Plasma," *Vacuum* 34, 519–522.

Odagiri, M. and Tomago, A. (1985), "Magnetic Tape by Vacuum Deposition — Output Performance and Durability." *Tech. Mtg. Magn. Soc. Jpn.* 39, 21–30.

Ohta, S., Yoshimura, F. Kimachi, Y., and Terada, A. (1987), "Wear Properties of Sputtered γFe_2O_3 Thin Film Disks," in *Tribology and Mechanics of Magnetic Recording Systems,* Vol. 4, SP-22 (B. Bhushan and N. S. Eiss, eds.), ASLE, Park Ridge, IL, pp. 110–115.

Okamoto, K. (1991), "Stability of Metal Particle and Metal Particulate Media," in *NSSDC Conf. on Mass Storage Systems and Technologies for Space and Earth Science Applications,* Vol. 2, pp. 160–168.

Okamoto, K., Okazaki, Y., Nagai, N., and Uedaira, S. (1996), "Advanced Metal Particles Technologies for Magnetic Tapes," *J.M.M.M.* 155, 60–66.

Osaki, H. (1992), "Wear Mechanisms of Particulate Magnetic Tapes in Still Frame Mode," in Research on the Tribology of Magnetic Tapes and Heads in VTRs, Doctoral thesis, Tohoku University, Chap. 2, pp. 18–57 (in Japanese).

Osaki, H. (1993), "Role of Surface Asperities on Durability of Metal-Evaporated Magnetic Tapes," *IEEE Trans. Magn.* MAG-29(1), 11–20.

Osaki, H. (1996), "Tribology of Videotapes," *Wear* 200, 244–251.

Osaki, H. and Sato, K. (1995), "Improvement of Durability of Metal Evaporated Tape by DLC Coating," *New Diamond,* 39, 26–27 (in Japanese).

Osaki, H., Oyanagi, E., Kanou, T., and Kurihara, J. (1992), "Wear Mechanisms of Particulate Magnetic Tapes in Helical Scan Video Tape Recorders," *IEEE Trans. Magn.* MAG-28(1), 76–83.

Osaki, H., Uchiyama, H., and Honda, N. (1993), "Wear Mechanism of Co-Cr Tape for Perpendicular Magnetic Recording," *IEEE Trans. Magn.* MAG-29(1), 41–58.

Osaki, H., Kurihara, J., and Kanou, T. (1994), "Mechanisms of Head-Clogging by Particulate Magnetic Tapes in Helical Scan Video Tape Recorders," *IEEE Trans. Magn.* MAG-30(4), 1491–1498.

Pethica, J. B., Koidl, P., Gobrecht, J., and Schuler, C. (1985), "Micromechanical Investigations of Amorphous Hydrogenated Carbon Films on Silicon," *J. Vac. Sci. Technol.* 6, 2391–2393.

Richter, H. J. and Veitch, R. J. (1995), "Advances in Magnetic Tapes for High Density Information Storage," *IEEE Trans. Magn.* 31, 2883–2888.

Robertson, N., Hu, B., and Tsang, T. (1997), "High Performance Write Head Using NiFe 45/55," *IEEE Trans. Magn.* 33(5), 2818–2820.

Saito, N., Imakoshi, S., Takino, H., Suyama, H., and Wakabayashi, N. (1987), "The Characteristics of a Double Layer Magnetoresistive Element," in *Proc. Int. Symp. Phys. Magn. Mat.,* Sendai, Japan, 130–133.

Saito, N., Fukuyama, M., Suyama, H., Soda, Y., Wakabayashi, N., and Sekiya, T. (1993), "Development of a Magnetoresistive/Inductive Head and Low Noise Amplifier IC for High Density Rigid Disk Drives," *IEICE Trans. Fundam.* E76-A(7), 1167–1169.

Saitoh, S., Inaba, H., and Kashiwagi, A. (1995), "Developments and Advantages in Thin Layer Particulate Recording Media," *IEEE Trans. Magn.* 31, 2859–2864.

Sasaki, Y., Okamoto, K., Okazaki, Y., Maeda, N., and Kond, H. (1997), "Effect of Under Layer on Surface Smoothness of Double Layered Metal Tapes," *IEEE Trans. Magn.* 33(5), 3064–3066.

Sawatzky, E. (1997), "Thermal Asperity Fundamentals," in *Proc. Diskcon Conf.,* September.

Scarati, A. M. and Caporiccio, G. (1987), "Frictional Behavior and Wear Resistance of Rigid Disks Lubricated with Neutral and Functional Perfluoropolyethers," *IEEE Trans. Magn.* MAG-23, 106–108.

Shelledy, F. and Brock, G. (1975), "A Linear Self-Biased Magnetoresistive Head," *IEEE Trans. Magn.* MAG-11, 1206–1208.

Shibata, T., Takada, A., Narisawa, H., Fukuyama, M., and Soda, Y. (1996), "Advanced Vertical MR Head," *IEEE Trans. Magn.* 32(5), 3413–3415.

Shinohara, K., Yoshida, H., Kunieda, T., and Murai, M. (1994), *J. Mag. Soc. Jpn.* 18, 299–302.

Singh, A. and Bischoff, P. G. (1985), "Optimization of Thin Film Heads for Resolution, Peak Shift and Overwrite," *IEEE Trans. Magn.* MAG-21(5), 1572–1574.

Suyama, H., Tsunewaki, K., Fukuyama, M., Saito, N., Yamada, T., and Karamon, H. (1988), "Thin Film MR Head for High Density Rigid Disk Drive," *IEEE Trans. Magn.* 24, 2612–2614.

Sueyoshi, T., Naono, H., Kawanami, M., Amemiya, M., and Hayama, H. (1984), "Morphology of Iron Fine Particles," *IEEE Trans. Magn.* MAG-20(1), 42–44.

Tago, A., Satoh, I., Kogure, K., and Kita, T. (1980), "Methods of Estimating Mechanical Characteristics for Magnetic Recording Disks," *Rev. Elec. Commun. Lab.* 28(5–6), 405–414.

Takada, A., Honda, T., Abe, M., Kanno, Y., Shibata, T., and Soda, Y. (1997), "Vertical AMR Sensor with New Magnetic Stabilizing Design," *IEEE Trans. Magn.* 33(5), 2932–2934.

Talke, F. E. and Tseng, R. C. (1984), " A Study of Elastohydrodynamic Lubrication between a Magnetic Recording Head and a Rotating Flexible Disk," in *Tribology and Mechanics of Magnetic Recording Systems,* Vol. 1, SP-16 (B. Bhushan and N. S. Eiss, eds.), ASLE, Park Ridge, IL, pp. 107–114.

Tam, A. C. and Bhushan, B. (1987), "Reduction of Friction Between a Tape and a Smooth Surface by Acoustic Excitation," *J. Appl. Phys.* 61(4), 15, 1646–1648.

Teramura, N. (1985), "Recent Progress in Floppy Disk Recording Technology (Review Paper)," in *Tribology and Mechanics of Magnetic Recording Systems,* Vol. 2, SP-19 (B. Bhushan and N. S. Eiss, eds.), ASLE, Park Ridge, IL, 27–35.

Tian, H. and Lee, J. (1995), "Electrostatic Discharge Damage of MR Heads," *IEEE Trans. Magn.* 31, 2624–2626.

Timsit, R. S., Stratford, G., and Fairlee, M. (1987), "Characterization of Lubricant/Solid Interface by FTIR," in *Tribology and Mechanics of Magnetic Recording Systems,* Vol. 4, SP-22 (B. Bhushan and N. S. Eiss, eds.), ASLE, Park Ridge, IL, pp. 98–104.

Tokuyama, M. and Hirose, S. (1994), "Dynamic Flying Characteristics of Magnetic Head Slider with Dust," *Trans. ASME, J. Tribol.* 116, 95–100.

Tomago, A., Murai, M., and Enomoto, S. (1988), "Tribology of Vacuum Deposited Magnetic Thin Film Recording Tape," in *IECE Tech. Group Meet. Magn. Rec. Jpn.,* OME-27-59, pp. 29–34 (in Japanese).

Tsang, C. (1981), "Exchange Induced Unidirectional Anisotropy at FeMn-Ni80Fe20 Interfaces," *J. Appl. Phys.* 52(3), 2471–2473.

Tsang, C. (1984), "Magnetics of Small Magnetoresistive Sensors," *J. Appl. Phys.* 55(6), 2226–2231.

Tsang, C. and Decker, S. (1982), "Study of Domain Formation in Small Permalloy Magnetoresistive Elements," *J. Appl. Phys.* 53, 2602–2604.

Tsang, C., Chen, M., Yogi, T., and Ju, K. (1990), "Gigabit Density Recording Using Dual-Element MR/Inductive Heads on Thin-Film Disks," *IEEE Trans. Magn.* MAG-26, 1689–1693.

Tsukamoto, Y., Yamaguchi, H., and Yanagisawa, M. (1988), "Mechanical Properties and Wear Characteristics of Various Thin Films for Rigid Magnetic Disks," *IEEE Trans. Magn.* MAG-24, 2644–2646.

Uedaira, H. and Ousaka, A. (1989), *Water in Living Body System,* Kodansha, Tokyo, pp. 152–160 (in Japanese).

Van Ooyen, J. A. C., Druyvesteyn, W. F., and Postma, L. (1982), *J. Appl. Phys.* 53(3), 2596–2598.

Voegeli, O. (1975), "Magnetoresistive Read Head Assembly Having Matched Elements for Common Mode Rejection," U.S. Patent 3,860,965.

Wang, P. (1993), "Sensitivity of Orthogonal Magnetoresistive Heads," *IEEE Trans. Magn.* 29(6), 3820–3822.

Wheeler, D. R. and Osaki, H. (1990), "A New Method to Measure the Intrinsic Bond Strength of Brittle Thin Films on Flexible Substrates," in *American Chemical Society Books,* No. 440 (*Metallization in Polymers*), Chap. 36, pp. 500–512, American Chemical Society, Washington, D.C.

White, J. W. (1984a), "Flying Characteristics of the 3370-Type Slider on a 5¼-inch disk — Part I: Static Analysis," in *Tribology and Mechanics of Magnetic Recording Systems.* Vol. 1, SP-16 (B. Bhushan and N. S. Eiss, eds.), ASLE, Park Ridge, IL, pp. 72–76.

White, J. W. (1984b), "Flying Characteristics of the 3370-Type Slider on a 5¼-inch Disk — Part I: Dynamic Analysis," in *Tribology and Mechanics of Magnetic Recording Systems,* Vol. 1, SP-16 (B. Bhushan and N. S. Eiss, eds.), ASLE, Park Ridge, IL, pp. 77–84.

White, J. W. (1984c), "On the Design of Low Flying Heads for Floppy Disk Magnetic Recording," in *Tribology and Mechanics of Magnetic Recording Systems,* Vol. 1, SP-16 (B. Bhushan and N. S. Eiss, eds.), ASLE, Park Ridge, IL, pp. 126–131.

White, J. W., Raad, P. E., and Euler, J. A. (1986), "An Inverse Procedure for the Air Bearing Design of a Pair of Opposed Magnetic Heads in a Floppy Disk Drive," in *Tribology and Mechanics of Magnetic Recording Systems,* Vol. 3, SP-21 (B. Bhushan and N. S. Eiss, eds.), ASLE, Park Ridge, IL, pp. 138–143.

Williams, M. L. and Comstock, R. L. (1971), "An Analytical Model of the Write Process in Digital Magnetic Recording," in *17th Annu. AIP Conf. Proc.,* Part 1, pp. 738–742.

Williams, M. L. and Lambert, S. E. (1989), "Film-Head Pulse Distortion Due to Microvariation of Domain Wall Energy," *IEEE Trans. Magn.* 25(5), 3206–3208.

Wu, Y. and Talke, F. E. (1996), "Design of a Head-Tape Interface for Ultra Low Flying," *IEEE Trans. Magn.* 32(1), 160–162.

Yamashita, T., Chen, G. L., Shir, J., and Chen. T. (1988), "Sputtered ZrO_2 Overcoat with Superior Corrosion Protection and Mechanical Performance in Thin Film Rigid Disk Application," *IEEE Trans. Magn.* MAG-24, 2629–2634.

Yanagisawa, M. (1985a), "Tribological Properties of Spin-Coated SiO_2 Protective Film on Plated Magnetic Recording Disks," in *Tribology and Mechanics of Magnetic Recording Systems,* Vol. 2, SP-19 (B. Bhushan and N. S. Eizs, eds.), ASLE, Park Ridge, IL, pp. 7–15.

Yanagisawa, M. (1985b), "Lubricants on Plated Magnetic Disks," in *Tribology and Mechanics of Magnetic Recording Systems,* Vol. 2, SP-19 (B. Bhushan and N. S. Eiss, eds.), ASLE, Park Ridge, IL, pp. 7–15.

Yoshida, M., Sakai, M., Fukuda, K., Yamanaka, N., Koyanagi, T., and Matsuzaki, M. (1993), "Edge Eliminated Head," *IEEE Trans. Magn.* 29(6), 3837–3879.

13

Microdynamic Systems in the Silicon Age

Richard S. Muller

ABSTRACT

With the proven record of accomplishment in very large integrated circuits (VLSI) brought about by batch-fabrication technology for electronic devices of ever-decreasing size, there is widespread enthusiasm about emerging opportunities to design mixed micromechanical and microelectronic systems. New development, based heavily on integrated-circuit-related technologies, have led to rapid progress in the development of *microdynamic systems*. These systems are based upon the science, technology, and design of *moving* micromechanical devices, and are a subclass of what is known in the U.S. as microelectromechanical systems (MEMS). The dimensions of the micromechanical devices in MEMS are typically smaller than 100 μm. Recent rapid progress gives promise for new designs of integrated sensors, actuators, and other devices that can be combined with on-chip microcircuits to make possible high-performing, compact, portable, low-cost engineering systems. Mechanical materials for some of the microdynamic systems that have thus far been demonstrated consist of deposited thin films of polycrystalline silicon, silicon nitride, aluminum, polyimide, and tungsten among other materials. To make mechanical elements using thin-film processing, microstructures are freed from the substrate by etching a *sacrificial layer* of silicon dioxide. First demonstrated as a means to produce electrostatically driven, doubly supported beam bridges, this sacrificial-layer technology has proved very versatile and has been used to make, among other structures, laterally vibrating doubly folded bridges, gears, springs, and impacting microvibromotors. Recently, micromirrors that consist of multiple hinged plates which

Updated, expanded, and revised version of article first published in *Micro/Nanotribology and Its Applications* (B. Bhushan, ed.), NATO ASI Series, Kluwer Academic Publishers, Dordrecht, 1997.

fold out of the surface plane in which they are formed (reaching vertical heights of millimeter dimensions) have been demonstrated. The polycrystalline silicon cross section for these mirrors is thinner than 2 μm. The mirrors can be moved using electrostatic comb drives or microvibromotors. Continued research on the mechanical properties of the electrical materials forming microdynamic structures (which previously had exclusively electrical uses), on the scaling of mechanical design, on tribological effects, on coatings, and on the effective uses of computer aids is now under way. This research promises to provide the engineering base that will exploit this promising technology.

13.1   Introduction

Microdynamic systems, that is, systems in which micromechanical elements undergo controlled motion, are, from one very useful perspective, a logical consequence in the continued evolution of microelectronic systems. Microdynamics offer special opportunities for the production of extremely miniaturized, highly complex systems. The opportunities presented by microdynamics can be compared with those that enabled the field of electronics to be revolutionized by the achievement of large-scale electronic-device integration. The VLSI revolution has, of course, affected all of society and its multi-billion-dollar industrial component continues robust growth after three decades of development. The linkage between micromechanical methods and typical integrated-circuit processes has been developing strongly since it was first demonstrated in seminal work done in 1982 when R. T. Howe demonstrated means to make microbeams from polycrystalline silicon films (Howe and Muller, 1982). Following this demonstration, Howe built the prototype polysilicon MEMS to be used as a chemical vapor sensor (Howe and Muller, 1986); see Figure 13.1.

Through the 15 ensuing years, engineers have focused on employing the key steps of batch processing and self-assembly that underlie modern microelectronics for the development of systems composed of micromechanical as well as microelectronic components to make microelectromechanical systems, now typically designated by the acronym MEMS. The impact of MEMS is very broad, potentially affecting engineering design in areas as diverse as sensing, biological, environmental, and process instrumentation, robotics, engine control, and guidance. MEMS techniques also make possible the introduction of new test specimens and new testing protocols that can uncover fundamental information about material properties. As specimen dimensions shrink toward lower and lower multiples of single-crystal and molecular sizes, their controlled fabrication and manipulation in MEMS begin to offer exciting prospects for proving fundamental theory. The MEMS field promises to impact a wide swath of research and industry and affect not only material scientists, but also engineers, physicists, chemists, and biologists.

13.1.1   Origins

The modern integrated circuit (IC) is the direct descendant of the planar process which, when introduced in the 1950s, combined the arts of photolithography and silicon chemical technology to extraordinary advantage. The productivity of this combination was made evident very early by the rapidity with which circuits of ever-increasing complexity appeared. It was therefore easy to justify large and repeated capital expenditures for technology and equipment and to put new ideas into development shortly after their conception so that ICs advanced rapidly. Today, it is commonplace to find ICs built with millions of devices; in fact IC engineers speak of the immediate future as *the Gigabit Era*. There is no doubt that *integrated microelectronics* are keystone elements of information processing in the engineering systems of our modern silicon age.

For the complete design of an engineering system, however, it is not only necessary to process information. The information itself must be exchanged between a largely nonelectrical world and the electrical processing medium of the IC. This is the job of the *sensors* and *actuators* or, taken as one, of the *transducers*. In transducers we find the formidable partnership between micromechanics and microelectronics taking shape to make possible the new level of integration into MEMS.

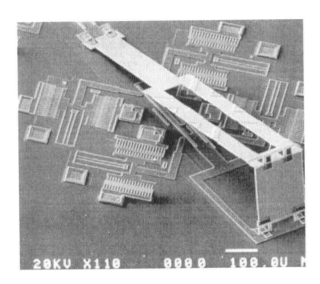

FIGURE 13.1 Detail from the first polysilicon MEMS. SEM of an apertured microbridge that is resonated by electrostatic forces under the control of on-chip circuits. (From Howe. R. T. and Muller, R. S., 1982, *Extended Abstracts of the 1982 Spring Meeting of the Electrochemical Society,* Montreal, Canada, 82-1, May 9–14. With permission.)

Some basic MEMS ideas had surfaced very early in the IC era. In fact, among the creative ideas tried out within the first decade of IC history was the employment of microfabrication technologies to produce a microdynamic element by a group of researchers at the Pittsburgh Westinghouse Research Laboratories. This group employed the silicon planar process to fabricate a resonant-gate, field-effect transistor (FET) (Nathanson et al., 1967) which consisted of a conventional silicon FET having a metal gate that was cantilevered over a surface channel covered only by silicon dioxide. An electrostatic field pumped the cantilever and its mechanical resonant frequency had applications as a timing source. The Westinghouse team also designed a large-screen optical projector based upon electrostatically driven cantilever reflectors. This work was very early in the development cycle of silicon microfabrication which was then (in the 1960s) revolutionizing all aspects of circuit design. The technology for the Westinghouse microdynamic elements was not refined sufficiently for widespread commercial adoption and the research can now be seen as an early harbinger rather than as a catalyst for new directions in microfabrication.

In the early 1980s, with two decades of IC development completed, the sophistication of electronics had advanced markedly. Circuit performance that would have been astonishing in the 1960s had become commonplace. The *interfaces* of these circuits with the largely nonelectrical world had become the logical point of focus for engineering design. The advantages of silicon micromechanics had begun to show themselves in its applications to ink-jet printing at companies like IBM, Hewlett-Packard, Canon, and Texas Instruments and in miniature pressure sensors such as those made at Honeywell Corporation. The impressive review of applications of silicon as a mechanical material published by Kurt Petersen (1982) focused the interest of many engineers on this area. Petersen's review paper established itself almost immediately as a prime reference for the practice of silicon micromechanics.

13.2   Micromachining

The earliest methods used to build structures from silicon making use of lithography and etch technology that was mastered as a part of solid-state device research was via *substrate micromachining*.

13.2.1   Substrate Micromachining

The technology for substrate micromachining is based upon orientation-dependent chemical etching of the silicon substrate. When used inventively together with etch-stopping techniques and masking films, this directed-etch technique can produce surprisingly complex structures. Some significant applications have been to a number of different sensing mechanics, such as silicon diaphragms (for pressure sensors) and cantilever beams (for accelerometers). Details about surface micromachining are not included in this chapter.

13.2.2   Surface Micromachining

In the early 1980s, research being carried out at Berkeley showed that polycrystalline silicon had use as a mechanical material with very good characteristics for compatible IC processing. The design and fabrication by Howe of polysilicon resonant beams together with on-chip MOS circuitry demonstrated the practicality of what has become known as *surface micromachining* (Howe, 1985). Howe designed his beams to be driven by electrostatic forces like those at Westinghouse, mentioned earlier. The polysilicon beams were fabricated from patterned thin films by etching an underlying *sacrificial* silicon dioxide layer. The sacrificial layer was heavily doped with phosphorus to enhance its etch rate in hydrofluoric acid. To carry out surface micromachining with polysilicon, it is necessary that the sacrificial silicon dioxide be etched considerably faster than is the polysilicon mechanical material, itself. Laboratory studies have shown that polysilicon is etched in hydrofluoric acid (HF) at a negligible rate while silicon dioxide is etched at rates of 100 nm/min to 1 μm/min, depending on composition. This high ratio of etch rates is the key to successful surface micromachining. In contrast to substrate micromachining in which mechanical parts are sculpted from the wafer itself, surface micromachining makes mechanical elements out of materials deposited on the wafer surface.

13.2.3   Polycrystalline Silicon Properties

For surface micromachining with polycrystalline silicon (polysilicon) deposited by low-pressure chemical vapor deposition (LPCVD) techniques, the polysilicon is often also used as a mechanoelectric transducer through its piezoresistivity. The use of CVD polysilicon strain gauges has advantages because the transducing layer is dielectrically isolated from the substrate by a silicon nitride layer. Gauges made of polysilicon layers typically exhibit higher coupling factors and superior temperature characteristics to those made of single-crystal silicon.

Electrical Properties. The electrical properties of polysilicon have been thoroughly studied and can be reviewed in several reference works (Kamins, 1988). Piezoresistance in polysilicon is a consequence of strain effects on the passage of carriers across the barriers between the crystalline grain boundaries, as well as the contributions of the bulk piezoresistivity of the silicon crystallites (Seto, 1976; French and Evans, 1985).

Mechanical Properties. To use polysilicon effectively as a mechanical material, it is necessary to know well its mechanical properties. For most applications, the most important of these are the Young's modulus, Poisson's ratio, residual strain, and ultimate strength. In work over the past 10 years, considerable dependence has been found in these properties dependent on the details of the polysilicon fabrication process.

At typical IC-LPCVD conditions, polysilicon films are in a residual compressive state after cool down to room temperature. This is generally undesirable and can cause multiply constrained structures, such as diaphragms, plates, or doubly fixed bridges to buckle. The compressive strain can have several possible sources including crystallites impinging during growth (Guckel and Burns, 1986) or entrapped gases such as oxygen (Muraka and Retajczyk, 1983). Studies of residual strain have been carried out using measurements of substrate curvature. The strain can be reduced significantly by annealing (typically, at 850°C) (Choi and Hearn, 1984). Residual strain can be reduced significantly and even brought to a tensile state if LPCVD polysilicon is grown at temperatures very near the amorphous boundary (580°C). By annealing for 3 h at 1150°C, very low residual strains can be achieved (a typical value is -1.4×10^{-4}) (Guckel et al., 1987).

Young's modulus E and Poisson's ratio ν for polysilicon have been measured using the curvature of different substrates coated with a given thin film over a range of temperatures. From these measurements, the ratio $E/(1 - \nu)$ and the thermal expansion coefficient α are found. Some reported results from polysilicon heavily doped with phosphorus are $E/(1 - \nu) \approx 140$ Gpa, about 70% as large as is found in single-crystal silicon.

13.2.4 Tribology in MEMS

Tribology takes center stage when one considers the motions of very small bodies in which micrograms are proper units of mass. At this scale, as we know from observing the insect world, inertial effects play far smaller (in many cases negligible) roles in the relationships between driving force and motion. The nature of friction on this tiny scale is in need of fundamental study as systems for a multitude of applications are explored.

A major challenge for MEMS designers employing surface micromachining is to understand and overcome the effects of stiction. The term *stiction* is used to describe two circumstances: (1) sticking of a "freed" member together with the newly exposed surface underneath the sacrificial layer after the final "freeing" etch step of the micromachining process, and (2) adherence of two mechanical microparts that approach or touch one another when the microsystem is in operation.

MEMS engineers employing surface micromachining frequently encounter fatal stiction effects when they attempt to release structures in the final etch step. Research into this problem has identified the major role of surface tension during the drying stage after a final liquid-etch release step (Mulhern et al., 1993). As MEMS designs have incorporated larger-area, very thin, low-stiffness members (lateral surfaces tens to hundreds of a micrometer on a side and 1 or 2 μm thick), the release-stiction problem becomes of first order. When sacrificial layers are dissolved by the etch, adjacent surfaces within micrometer dimensions result in menisci that lead to attractive forces. These forces can, in turn, buckle the "freed" member, and possibly cement it to the underlying surface. Mechanical analysis of release-stiction effects has been performed (Mastrangelo, 1997). By using mechanisms to sublimate the final etch, release stiction can be avoided, and large surface-micromachined structures processed successfully (Guckel et al., 1990; Lebouitz et al., 1995).

Already with the first microdynamic devices in which surfaces slide past one another (Fan et al., 1998), stiction was observed to account for the major retarding forces, and techniques, such as the production of "dimples" in moving polycrystalline silicon elements, were demonstrated. Recently, methods that enable quantitative study of stiction effects have been described. Mastrangelo and co-workers have shown a method for quantifying the post-release stiction of structures using an array of released cantilever beams having graduated lengths (Mastrangelo, 1997). By applying electrostatic forces, the beams are attracted to a surface. When the force is removed, beams longer than a critical "detachment" length remain adhered to the surface while shorter beams are freed because of their internal stiffness (Mastrangelo and Hsu, 1992). The detachment length can be related to an adhesion energy. By observing the influences of different surface coatings on adhesion-energy values, one can form comparative evaluations about the coatings (Houston et al., 1995, 1996).

The MEMS field has developed sufficiently to produce micromechanisms as complex as gear boxes capable of two-and-three levels of speed reduction (Sniegowski, 1997). For such structures to enter commercial applications, surface-lubricating films are a necessity. At Texas Instruments, Inc., surface micromachining has been used to produce a micromirror array composed of movable aluminum structures. Results with lubricants used on this digital micromirror array are reported by Henck (1997). Research results on films for polysilicon MEMS mechanisms are discussed by Maboudian and Howe (1997).

13.3 MEMS Structures and Systems

Both surface and bulk micromachining have been employed to make devices and systems having very surprising complexity. Some of these are now incorporated into commercial products, while others are still the subject of research. Because of the particular interests of the author and the need to limit the length of this discussion, the following examples will be drawn exclusively from surface-micromachined MEMS.

An impressive example of a commercial MEMS that makes use of surface micromachining to build a fully integrated accelerometer is the AD-XL50 produced by Analog Devices, Inc., for use in automobile airbag-deployment systems. This accelerometer integrates a tiny seismic mass in the midst of formidable bipolar-CMOS circuitry. The system employs force feedback using coulombic force to hold the seismic mass fixed and it senses position making use of a differential capacitance measurement. The AD-XL50 has been widely described in the popular literature (Goodenough, 1991) and it now accounts for a large and growing share in a very competitive marketplace. Figure 13.2 lists some of the surprising properties of this monolithic MEMS.

Interesting Facts

- ▷ 10mm² Die (3mm x 3mm)
- ▷ 1mm² Sensor Area
- ▷ 0.1μgrams Proof Mass
- ▷ 0.1pF per Side for the Differential Capacitor
- ▷ 20aF (10⁻¹⁸f) for Minimum Detectable Capacitance Change
- ▷ Total Capacitance Change for 50g is 10fF
- ▷ 1.3μm Gaps Between Capacitor Plates
- ▷ 0.2Å Minimum Detectable Beam Deflection
- ▷ 1.6 μm Between the Suspended Beam and Substrate
- ▷ 4000g for Beam to Touch Substrate
- ▷ 22kHz Resonant Frequency of Beam

FIGURE 13.2 Facts about the Analog Devices AD-XL50 accelerometer. (Courtesy of Richard Payne, Analog Devices, Inc.)

A coulombic actuating force (as employed in the AD-XL50) is used frequently in the MEMS field, especially for surface-micromachined embodiments because coulombic force (alternatively, electrostatic force) is easily incorporated into the system and employs only an imposed voltage between two elements. The force increases linearly with the capacitance between the electrodes and quadratically with the applied voltage. For typical surface-micromachined parts and spacings, the force is quite small (of the order of micronewtons) but still adequate for many applications. To increase the range of achievable forces in MEMS, a number of alternative actuation mechanisms have been used or are under study at this time. Most of these alternatives to coulombic force introduce complication into the fabrication technology for the MEMS. Nonetheless, to obtain millinewton forces and higher, it is likely that these other forcing

Linear Resonant Structure Layout

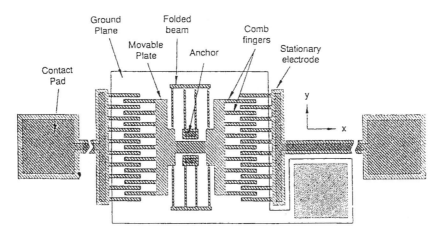

FIGURE 13.3 Comb-drive resonator. (Courtesy of William Tang, Jet Propulsion Laboratory.)

techniques will be necessary. A listing of other actuation means that have already been applied is given below.

13.3.1 MEMS Actuation Forces

13.3.1.1 Electrostatic Comb Drive

As already discussed, actuation in MEMS using coulombic force has distinct advantages for compatibly processed MEMS. Howe, for example, made use of coulombic force to power his resonating beam in the first polysilicon micromechanical structure (Howe and Muller, 1982). In further work with electrostatic forcing, Howe and student W. Tang first constructed a polysilicon comb-drive resonant actuator which employs coulombic force in a flexure structure to cause it to move parallel to the substrate surface (Tang et al., 1989). The layout of this comb drive, which is supported by only two pedestals attached to the substrate, is shown in Figure 13.3.

The comb drive has become a very frequently used microactuator both for resonant and nonresonant systems. Because the actuated part is a flexing structure, there is no surface friction in its motion, and it can be driven in resonance with a very high mechanical Q value (100,000 in vacuum). Much work is continuing on comb-drive resonators and their applications and they have been made by substrate as well as by surface micromachining. Figure 13.4 compares aspects of the behavior of the comb-drive to that of a parallel-plate actuator. A photograph of one of the original moving comb structures undergoing oscillation is shown in Figure 13.5.

13.3.1.2 Vertical vs. Lateral Oscillation

Vibrating micromechanical structures are useful for a variety of sensors and actuators. For sensing, one can make use of the dependence of the frequency of a mechanical resonator to physical or chemical parameters which affect the vibrational energy. Microfabricated resonant structures for sensing pressure, acceleration, and vapor concentration have been demonstrated. An elegant example of resonant drive for actuators is provided by the successful Bulova accutron wristwatch movement in which an electronic tuning fork is coupled mechanically to a rotating mechanism. Pisano has pointed out the possible applications of this design to microdynamics (Pisano, 1989). Recently, some very spectacular mechanisms have been made using polysilicon surface micromachining at the Sandia Laboratories in New Mexico. The Sandia process makes use of chemical-mechanical polishing in order to planarize wafers after the

Vertical vs. Lateral Oscillation

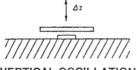 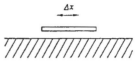

VERTICAL OSCILLATION

Squeeze film damping
[Low Q]

Restricted design freedom

Δz vs. applied voltage nonlinear

Driven by parallel plate capacitor
[Millivolt drives in vacuum]

LATERAL OSCILLATION

Reduced Drag
[High Q]

Sophisticated 2D geometry
*[Stress-relief design, comb drive,
torsional resonant structures, etc]*

Δx vs. applied voltage can be linear

Driven by fringing fields
[Larger drive voltage in vacuum]

FIGURE 13.4 Comparative behavior of a vertical resonator with a comb-driven substrate-parallel resonator. (Courtesy of William Tang, Jet Propulsion Laboratory.)

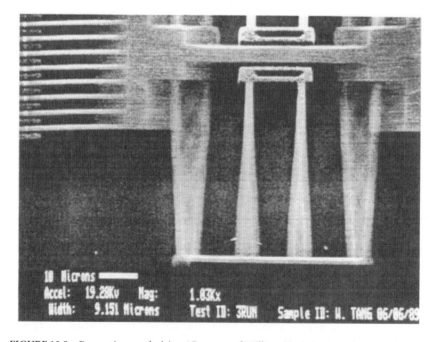

FIGURE 13.5 Resonating comb drive. (Courtesy of William Tang, Jet Propulsion Laboratory.)

micromechanisms have been made in a recessed area of the chip. Then, in a subsequent CMOS process, very high quality electronic circuits can be fabricated in a continuous-batch process.

Figure 13.6 shows a gear assemblage driven by resonant comb drives that was made at Sandia. The tiny rotor is 55 μm in diameter and it has achieved rotation rates as high as 300,000 rpm. It can be rotated either counterclockwise or clockwise. To gain a measure of the size of these structures, Figure 13.7

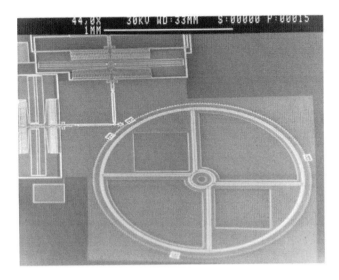

FIGURE 13.6 Batch-fabricated polysilicon assemblage made at Sandia Laboratories. (Courtesy of P. McWhorter.)

FIGURE 13.7 Rotor driver on the wheel assemblage of Figure 13.6 made at Sandia Laboratories. (Courtesy of P. McWhorter.)

shows a photograph of the tiny driven rotor of Figure 13.6. Next to the rotor in Figure 13.7 are clumps of red blood cells and, adjacent to the side arms, a grain of sand.

The MEMS field has begun to expand in several directions where full miniature "systems on a chip" are being realized. Important commercial MEMS applications to accelerometry, to environmental sensing, and to display have received fairly wide coverage in the technical and industrial product literature. An area in which the author has worked that is just now developing into fully engineered systems is the area of miniature integrated optical systems for communications, instrumentation, and sensing applications. The remainder of this chapter will focus on developments in this area called *microphotonics*.

13.3.2 MEMS for Microphotonics

The incorporation of micromechanical structures into fiber-optic systems holds promise of reducing costs and providing new opportunities for systems applications. Research groups around the world are

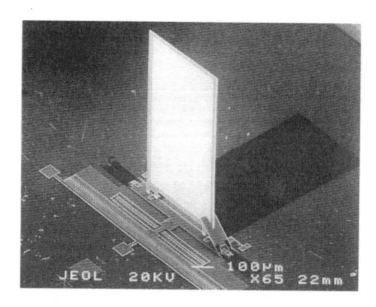

FIGURE 13.8 Micromirror made from polysilicon folded into an upright position. An attached comb-driven actuator makes the structure useful in a Fabrey–Perot interferometer. (Courtesy of M.-H. Kiang, UC Berkeley.)

instituting programs to meet these opportunities using a range of technologies. Silicon-based micromachining, both surface and bulk, has been exploited. Work at the Berkeley Sensor & Actuator Center (BSAC) at UC Berkeley and at UCLA has concentrated on surface micromachining with polysilicon and has developed actuation techniques appropriate both for beam switching and for scanning using micromirrors made by folding micromachined polysilicon structures out of the surface plane and covering them with a gold reflecting surface. Figure 13.8 shows such a folded mirror suitable for use in an interferometer.

The electrostatic comb drive has been used to move the micromirror at frequencies ranging from DC to more than 1 kHz. Impact-actuated linear vibromotors allow mirrors to travel over large (>100 μm) ranges with smaller than 1 μm positioning precision. These actuated mirrors have been built and operated as components on a laser-to-fiber coupling chip and in a bar-code reader. As an alternative to electrostatic actuation, a magnetically actuated mirror responsive to an off-chip field has also been demonstrated at Berkeley. The mirror can be conveniently deactivated even in the presence of the field by clamping it with an on-chip electrostatic clamp. By this means one can address individual mirrors among an array of them (Judy and Muller, 1996).

The use of MEMS technologies has the potential of adding very significant impetus to the already strong market for fiber-optic systems. With their proven high performance for communications and sensing, fiber-optic systems are in high demand despite typical high costs and relatively large size. The positive impact of MEMS on both cost and size reduction will open new opportunities that are now unreachable because of economical or portability limitations. Researchers around the world have sensed the opportunities in this area and have launched a research field that can broadly be labeled *microphotonics*. By using silicon substrates and well-known IC technologies, such as anisotropic etching, "silicon-optical bench" assembly is already an important commercial activity. Until now, however, the term *silicon-optical bench* has referred to a passive assemblage of optical components that remains fixed as configured in the final step of manufacture. Recent advances in MEMS technologies make possible a major advance toward the design of dynamic microphotonic systems.

To build a dynamic "optical bench on chip," high-aspect-ratio structures are required to interact with laser beams that typically travel in free space on paths that are parallel to the surface of the silicon substrate. To make high-aspect-ratio structures from the typically planar elements produced in surface-

Fabrication Process

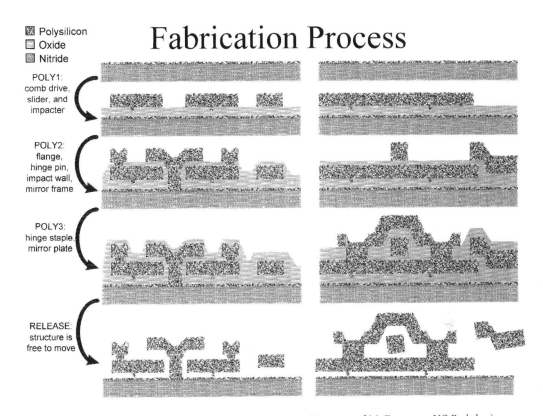

Polysilicon
Oxide
Nitride

POLY1:
comb drive,
slider, and
impacter

POLY2:
flange,
hinge pin,
impact wall,
mirror frame

POLY3:
hinge staple,
mirror plate

RELEASE:
structure is
free to move

FIGURE 13.9 Processing steps to form micromirrors. (Courtesy of M. Daneman, UC Berkeley.)

micromachining technology, we use hinges (Pister et al., 1992) that permit the rotation of optical elements out of the plane of the substrate (Tien et al., 1995; Kiang et al., 1996). Figure 13.9 shows the important steps in the fabrication process for the hinged mirrors.

To fabricate the structures as shown in Figure 13.9, three layers of structural polysilicon above a polysilicon ground plane are used. A layer of sacrificial LPCVD phosphosilicate glass (PSG) separates the polysilicon layers. After deposition of sequential passivation layers of silicon oxide and silicon nitride, a 0.5 μm n^+ polysilicon ground plane is deposited and patterned. The first sacrificial layer of PSG is then deposited and patterned. Depressions are wet etched into the PSG using 5:1 hydrofluoric acid to create dimples under the mirror (for reduced surface-to-surface contact). The dimples lessen stiction between these structures and the substrate. Openings are also etched (usually in a fluorine-based, CF^4 plasma) for anchors between the next structural polysilicon and the substrate. Next, the first structural polysilicon layer is deposited by LPCVD and patterned using a chlorine-based plasma etch. This layer forms the comb-drive actuator. A second sacrificial layer is then deposited and patterned. After deposition, the second structural polysilicon layer is etched to form the pins of the hinges. The third sacrificial layer and the third structural layer are then deposited and patterned to complete the hinge structures and form the mirror. The polysilicon structural layers are 2 to 5 μm thick, and the sacrificial silicon dioxide layers range from 1 to 2 μm in thickness. The sacrificial layers are removed through wet etching in hydrofluoric acid to release the movable structures. The last step is the evaporation of 50 nm of gold to increase mirror reflectivity—resulting in measured reflectivities of 85% at 1.35 μm. The reduction from the theoretical 96% reflectivity of a perfect gold mirror is due to scattering caused by polysilicon surface roughness and by etch holes and dimples in the mirror.

Movable Micromirror. Applications of these folded mirrors are for beam steering, optical alignment, scanning, and switching functions. Figure 13.10 illustrates the concept of the optical bench on a chip.

Actuated Silicon Optical Bench on a Chip

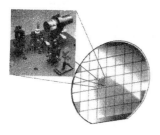

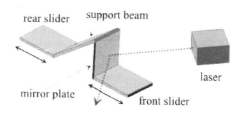

rear slider   support beam

laser

mirror plate   front slider

- ## Movable microreflector
  - ### Functions
    - Beam steering
    - Optical alignment
    - Scanning
    - Switching
  - ### Systems
    - Laser-to-fiber couplers
    - External cavity lasers (ECL)
    - Fiber switches
    - Interferometers

FIGURE 13.10   Concept and uses for a silicon optical bench on a chip using folded micromirrors. (Courtesy of M. Daneman, N. Tien, O. Solgaard, and K. Lau, UC Berkeley.)

Typical micromirror sizes are 0.5×0.5 mm². If they are driven directly using comb drives as shown in Figure 13.8, maximum movement is in the order of 10 μm. This drive is useful for harmonic cavities as, for example, in an external-cavity laser or to scan a light source in a bar-code reader.

Precision. The positioning precision of the micromirrors by the integrated actuators has been measured using visual observation of the micromirror displacement as the voltage applied to the comb drive is varied. The measurements were made for two identical mirrors having differing lengths of restoring springs in their actuators. The standard deviation of the mirror displacement with respect to applied DC bias was lower than 0.2 μm (limited by the resolution of the measuring equipment) for the longer (and, therefore, more flexible) spring. In either case, this mirror system is capable of submicron positioning precision. Figure 13.11 shows results obtained with the mirrors used as scanners. Tests run at Berkeley have exhibited 13 cycles of constructive and destructive interference in a Fabry–Perot interferometer at a 1.3-μm wavelength. The corresponding displacement of the mirror is 8.45 μm.

The frequency response of these comb-driven mirrors in air extends to about 1 kHz. Although studies are yet to be completed, it is expected that this upper frequency will increase markedly for mirrors encapsulated in a low-pressure ambient. By using several comb drives to move mirrors at more than one support point, it is possible to scan beams at angles situated in planes parallel to the substrate as well as perpendicular to it. An example of a mirror with these capabilities is shown in Figure 13.12.

Larger movements are required for some of the other applications shown in Figure 13.10. For a laser-to-fiber coupler or for switching applications, typical requirements are total displacements in the order of 100 μm or greater with submicron precision. The program at Berkeley has addressed this need by building microvibromotors — motors in which resonant energy is delivered to the load in the form of collisions. Earlier research at BSAC had shown that polysilicon could withstand multiple collisions without fracture and that elements could be made to slide on a silicon surface when actuated by a microvibromotor (Houston et al., 1996). The driving energy for the vibromotors is derived from comb drives which are built in a similar process to that described earlier. The motor and mirror carriage are built using only two layers of polysilicon.

Beam Steering and Control

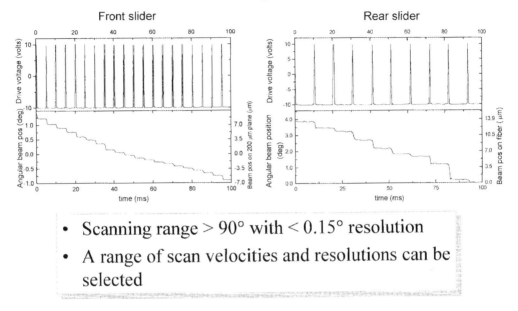

FIGURE 13.11 Measured performance of folded micromirrors used as scanners. (Courtesy of M. Daneman, O. Solgaard, and M. H. Kiang, UC Berkeley.)

FIGURE 13.12 Scanning micromirror with comb drives for scanning around three axes. (Courtesy of M. H. Kiang, and O. Solgaard, UC Berkeley.)

The impact force in a vibromotor turns out to be a useful way to overcome the sticking forces between two polysilicon layers. These forces dominate over any inertial effects and lead to results that can be surprising to the MEMS neophyte. Tests conducted at Hewlett-Packard Company facilities by Berkeley researchers showed, for example, that the tiny carriages holding the folded micromirrors would not move even under 500g loading in shock tests. Direct hammering on the carriage, however, is able to move it

Vibromotor

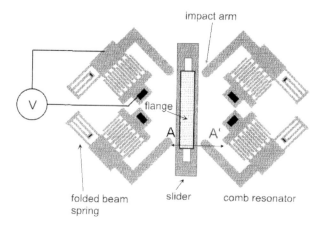

FIGURE 13.13 Microvibromotors design using four drivers to move a slider 100 m. (Courtesy of M. Daneman, O. Solgaard, and N. Tien, UC Berkeley.)

FIGURE 13.14 Folded micromirror driven by four microvibromotor drivers. (Courtesy of M. Daneman, O. Solgaard, and N. Tien, UC Berkeley.)

in steps that are only a fraction of a micrometer per blow. In the BSAC design, pairs of vibromotors strike the carriage at 45° relative to its guiding keyway slot.

Figure 13.13, shows the concept for a microvibromotor system to drive a slider (ultimately the mount for a micromirror) over about 100 μm displacement. Pairs of these vibromotors are used to position the support vane of a folded mirror in the structure shown in Figure 13.14. Some details of the microvibro-motor-driven micromirror system are indicated in Figure 13.15.

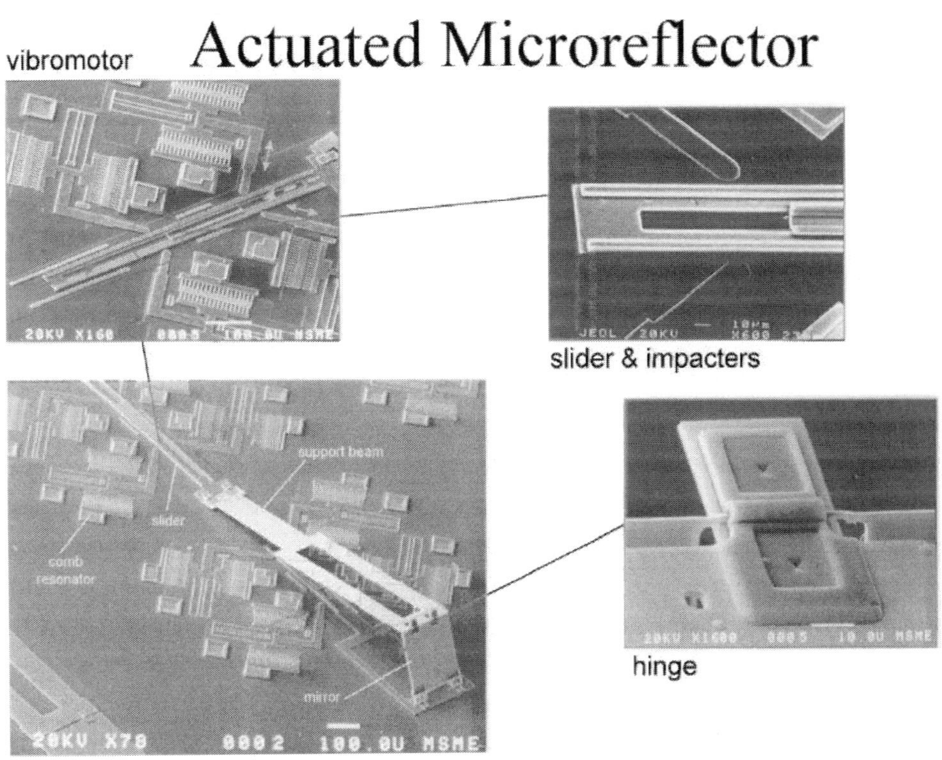

Actuated Microreflector

vibromotor

slider & impacters

support beam

slider

comb resonator

mirror

hinge

FIGURE 13.15 Features of an actuated microreflector system. (Courtesy of M. Daneman, UC Berkeley.)

13.3.3 Coupling Efficiency

Coupling to single-mode optical fiber using the actuated for microreflector fine alignment should provide coupling efficiencies as high as 70% according to theory.

The research on microphotonics at BSAC has been reported in a series of papers that are listed separately in the following bibliography.

13.4 Berkeley Microphotonics Research

1. O. Solgaard, M. Daneman, N.C. Tien, A. Friedberger, R.S. Muller, and K.Y. Lau, "Optoelectronic Packaging Using Silicon Surface-Micromachined Alignment Motors," *IEEE Photonics Technol. Lett.*, 7, 41–43, January 1995.
2. M.-H. Kiang, O. Solgaard, R.S. Muller, and K.Y. Lau, "Surface Micromachined Micromirrors with Integrated High-Precision Actuators for External-Cavity Semiconductor Lasers," *IEEE Photonics Technol. Lett.*, 8, 95–100, January 1996.
3. M. Daneman, O. Solgaard, N.C. Tien, K.Y. Lau, and R.S. Muller, "Laser-to-Fiber Coupling Module Using A Micromachined Alignment Mirror," *IEEE Photonics Technol. Lett.*, 8, 396–398, March 1996.
4. O. Solgaard, M. Daneman, N.C. Tien, A. Friedberger, R.S. Muller, and K.Y. Lau, "Micromachined Alignment Mirrors for Active Optoelectronic Packaging," in *Late News Paper, 1994 Conference on Lasers and Electro-Optics (CLEO)*, 9–13 May, Anaheim, CA, 1994.

5. N.C. Tien, O. Solgaard, M. Daneman, A. Friedberger, K. Lau, and R.S. Muller, "Movable, High-Aspect Ratio Micromirror for Optoelectronic Applications," *1994 Sensor and Actuator Workshop, Late News Poster Paper*, Transducer Research Foundation, Hilton Head Island, SC, 13–16 June, 1994.

6. N.C. Tien, M. Daneman, O. Solgaard, K.Y. Lau, and R.S. Muller, "Impact-Actuated Linear Microvibromotor for Micro-Optical Systems on Silicon," *1994 IEEE Int. Electr. Dev. Mtg.*, San Francisco, CA, Tech Digest 924–926, 12–14 December, 1994.

7. M.J. Daneman, N.C. Tien, O. Solgaard, A.P. Pisano, K.Y. Lau, and R.S. Muller, "Linear Microvibromotor for Positioning Optical Components," MEMS '95, 1995 Int. Conference on Microelectromechanical Systems, Amsterdam, The Netherlands, January, 1995.

8. O. Solgaard, M. Daneman, N.C. Tien, A. Friedberger, R.S. Muller, and K.Y. Lau, "Precision and Performance of Polysilicon Micromirrors for Hybrid Integrated Optics," 1995 International Symposium on Optoelectronic, Microphotonics, and Laser Technologies, Conference 2383A, SPIE, The International Society for Optical Engineering, San Jose, CA, 4–10 February, 1995.

9. M. Daneman, O. Solgaard, N.C. Tien, R.S. Muller, and K.Y. Lau, "Integrated Laser to Fiber Coupling Module using a Micromachined Alignment Mirror," *Conf. on Lasers and Electro-Optics (CLEO)*, Baltimore, MD, May, 1995.

10. M.-H. Kiang, O. Solgaard, M. Daneman, N.C. Tien, R.S. Muller, and K.Y. Lau, "High-Precision Silicon-Micromachined Micromirrors with On-Chip Actuation for External-Cavity Semiconductor Lasers," *Conf. on Lasers and Electro-Optics (CLEO)*, pp. 248–249, Vol. 15, May 1995, Baltimore, MD.

11. N.C. Tien, O. Solgaard, M.-H. Kiang, M. Daneman, K.Y. Lau, and R.S. Muller, "Surface-Micromachined Mirrors for Laser-Beam Positioning," *1995 IEEE Int. Conf. on Solid-State Sensors and Actuators, TRANSDUCERS '95, Tech. Digest.* 352–355, 25–29 June, 1995, Stockholm, Sweden.

12. N.C. Tien, M.-H. Kiang, M.J. Daneman, O. Solgaard, K.Y. Lau, and R.S. Muller, "Actuation of Polysilicon Surface-Micromachined Mirrors," *SPIE Proc.*, 2687, SPIE Conf., 27 Jan.–2 Feb. 1996, San Jose, CA.

13. M.-H. Kiang, O. Solgaard, R.S. Muller, and K.Y. Lau, "Surface Micromachined Electrostatic-Comb-Driven Scanning Micromirrors for Barcode Scanners," IEEE MEMS '96 Workshop, 11–15 February, 1996, San Diego, CA.

14. M.-H. Kiang, O. Solgaard, R.S. Muller, and K.Y. Lau, "Design and Fabrication of High-Performance Silicon Micromachined Resonant Microscanners for Optical Scanning Applications," *Integrated Photonics Research, 1996 Technical Digest Series*, Vol. 6, pp. 545–548, April 29–May 2, 1996, Boston, MA.

15. M.J. Daneman, N.C. Tien, O. Solgaard, K.Y. Lau, and R.S. Muller, "Actuated Micromachined Microreflector with Two Degrees of Freedom for Integrated Optical Systems," Integrated Photonics Research Conference, Boston, MA, May, 1996.

16. M.J. Daneman, N.C. Tien, O. Solgaard, K.Y. Lau, and R.S. Muller, "Linear Vibromotor-Actuated Micromachined Microreflector for Integrated Optical Systems," 1996 Sensor and Actuator Workshop, 109–112, Transducer Research Foundation, Hilton Head Island, SC, 2–6 June 1996.

17. M.-H. Kiang, O. Solgaard, R.S. Muller, and K.Y. Lau, "High-Precision Silicon Micromachined Micromirrors for Laser-Beam Scanning and Positioning," *1996 Sensor and Actuator Workshop*, (late news) Transducer Research Foundation, Hilton Head Island, SC, 2–6 June 1996.

Acknowledgments

Many colleagues have contributed their thoughts, suggestions, and work to this brief review of the MEMS field. The author especially thanks R.T. Howe, W. Tang, N. Tien, K. Lau, K. Pister, P. McWhorter, R. Payne, O. Solgaard, M. Daneman, M.-H. Kiang, J. Judy, A. Friedberger, C. Keller, and the staff of the Berkeley Micro Laboratory. The author's work and time has been partially supported by the National Science Foundation through A. Schwartzkopf, D. Crawford and M. White, by the Defense Advanced Research Projects Agency through K. Gabriel, and by the Hewlett-Packard Company through W. Ishak, K. Carey, and G. Trott.

References

Choi, M.S. and Hearn, E.W. (1984), *J. Electrochem. Soc.,* 131, 2443.

Fan, L.-S., Tai, Y.-C., and Muller, R.S. (1988), "Integrated Movable Micromechanical Structures for *Sensors Actuators,*" *IEEE Trans. Electron Devices,* ED-35, 724–730.

French, P.J. and Evans, A.G.R. (1985), *Sensors Actuators,* 8.

Goodenough, F. (1991), "Airbags Boom When IC Accelerometer Sees 50G," *Electron. Design,* August 8.

Guckel, H. and Burns, D.W. (1986), *Technical Digest, IEEE Int. Electron Devices Mtg.,* Los Angeles, CA, 176.

Guckel, H., Burns, D.W., Rutigliano, C.R., Showers, D.K., and Uglow, J. (1987), *Technical Digest,* Int. Conf. on Solid-State Sensors Actuators, Tokyo, Japan, 277.

Guckel, H., Sniegowski, J.J., Christenson, T.R., and Raissi, F. (1990), *Sensors Actuators,* A21–23, 346.

Henck, S.A. (1997), "Lubrication of Digital Micromirror Devices," *Tribol. Lett.,* 3, 239–247.

Houston, M.R., Howe, R.T., and Maboudian, R. (1995), in *Proc. 8th Int. Conf. on Solid-State Sensors and Actuators, TRANSDUCERS'95,* pp. 210–213, Stockholm, Sweden, June.

Houston, M.R., Maboudian, R., and Howe, R.T. (1996), *Technical Digest,* Solid-State Sensor and Actuator Workshop, Hilton Head Island, SC, pp. 42–47, June.

Howe, R.T. (1985), "Surface Micromachining," in *Micromachining and Micropackaging of Transducers,* (C.D. Fung, ed.), Elsevier Science Publishers, New York, 169.

Howe, R.T. and Muller, R.S. (1982), "Polycrystalline Silicon Micromechanical Beams," *Extended Abstracts of the 1982 Spring Meeting of the Electrochemical Society,* Montreal, Canada, 82-1, May 9–14.

Howe, R.T. and Muller, R.S. (1986), "Resonant Microbridge Vapor Sensor," *IEEE Trans. Electron Devices,* ED-33, 499–507.

Judy, J.W. and Muller, R.S. (1996), "Batch-Fabricated, Addressable, Magnetically Actuated Microstructures," in *Solid-State Sensor and Actuator Workshop,* pp. 187–190, Transducer Research Foundation, Hilton Head Island, SC, 2–6 June.

Kamins, T.I. (1988), *Polycrystalline Silicon for Integrated Circuit Applications,* Kluwer Academic Publishers, Dordrecht.

Kiang, M-H., Solgaard, O., Muller, R.S., Lau, K.Y. (1996), "Silicon-Micromachined Micromirrors with Integrated High-Precision Actuators for External-Cavity Lasers," *IEEE Photonics Technol. Lett.,* 8.

Lebouitz, K.S., Howe, R.T., and Pisano, A.P. (1995), in *Proc. 8th Int. Conf. on Solid-State Sensors and Actuators (TRANSDUCERS'95),* pp. 224–227, 25–29 June, Stockholm, Sweden.

Maboudian, R. and Howe, R.T. (1997), "Stiction reduction processes for surface micromachines," *Tribol. Lett.,* 3, pp. 215–221.

Mastrangelo, C.H., (1997), "Adhesion-Related Failure Mechanisms in Micromechanical Devices," *Tribol. Lett.,* 3, 223–238.

Mastrangelo, C.H. and Hsu, C.H. (1993), IEEE Solid-State Sensor and Actuator Workshop, Hilton Head Island, SC, 22–25, June 1992; JMEMS, pp. 208–213.

Mulhern, G.T., Soane, D.S., and Howe, R.T. (1983), in *Proc. 6th Int. Conf. on Solid-State Sensors and Actuators (TRANSDUCERS'93),* pp. 296–299, Yokohama, Japan, 7–10 June.

Muraka, S.P. and Retajczyk, T.F., Jr. (1983), "Effect of Phosphorus Doping on Stress in Silicon and Polycrystalline Silicon," *J. Appl. Phys.,* 54, 2069.

Nathanson, H.C., Newell, W.E., Wickstrom, R.A., and Davis, J.R., Jr. (1967), "The resonant-gate transistor," *IEEE Trans. Electron Devices,* ED-14, 117–133.

Petersen, K.E. (1982), "Silicon as a Mechanical Material," *Proc. IEEE,* 70, 420–457.

Pisano, A.P. (1989), "Resonant-Structure Micromotors," *1989 IEEE Micro-Electro-Mechanical Systems Conference,*" pp. 44–48, Salt Lake City, UT, Feb. 20–22.

Pister, K.S.J., Judy, M.W., Burgett, S.R., and Fearing R.S. (1992), "Microfabricated Hinges," *Sensors Actuators (A)* 33.3, 249–256.

Seto, J.Y.W. (1976), *J. Appl. Phys.,* 47, 4780.

Sniegowski, J.J. (1997), "MEMS: A New Approach to Micro-Optics," IEEE/LEOS Int. Conf. on Optical MEMS and Their Applications, pp. 209–214, Nara, Japan, Nov. 18–21.

Tang, W.C., Nguyen, T-C.H., and Howe, R.T. (1989), "Laterally Driven Polysilicon Resonant Microstructures," *1989 IEEE Micro-Electro-Mechanical Systems Conference*, pp. 53–59, Salt Lake City, UT, Feb. 20–22.

Tien, N.C., Solgaard, O., Kiang, M.-H., Daneman, M., Lau, K.Y., and Muller, R.S. (1995), "Surface Micro-Machined Mirrors for Laser-Beam Positioning," in *Proc. 8th Int. Conf. Solid-State Sensors and Actuators* (*Transducers'95*) *and Eurosensors IX*, Stockholm, Sweden, June 25–29, pp. 352–355.

14

Micro/Nanotribology and Micro/Nanomechanics of Magnetic Storage Devices

Bharat Bhushan

14.1 Introduction

Micro/nanotribological studies are needed to develop fundamental understanding of interfacial phenomena on a small scale and to study interfacial phenomena in micro- and nanostructures used in magnetic storage systems, microelectromechanical systems (MEMS), and other industrial applications (Bhushan, 1992, 1993, 1994, 1995a,b, 1996a, 1997, 1998b). The components used in micro- and nanostructures are very light (on the order of few micrograms) and operate under very light loads (on the order of few micrograms to a few milligrams). As a result, friction and wear (on a nanoscale) of lightly loaded micro/nanocomponents are highly dependent on the surface interactions (few atomic layers). These structures and magnetic storage devices are generally lubricated with molecularly thin films. Micro- and nanotribological techniques are ideal to study the friction and wear processes of micro- and nanostructures and molecularly thick lubricant films (Bhushan et al., 1994a–e, 1995a–g, 1997a–c; Koinkar and Bhushan, 1996a,b, 1997a,b, 1998; Sundararajan and Bhushan, 1998). Although micro/nanotribological studies are critical to study micro- and nanostructures, these studies are also valuable in fundamental understanding of interfacial phenomena in macrostructures to provide a bridge between science and engineering. At interfaces of technological applications, contact occurs at multiple asperity contacts. A sharp tip of tip-based microscopes (atomic force/friction force microscopes or AFM/FFM) sliding on a surface simulates a single asperity contact, thus allowing high-resolution measurements of surface interactions at a single asperity contacts. AFMs/FFMs are now commonly used for tribological studies (Bhushan, 1998a).

In this chapter, we present the state of the art of micro/nanotribology of magnetic storage devices including surface roughness, friction, adhesion, scratching, wear, indentation, transfer of material detection, and lubrication.

14.2 Experimental

14.2.1 Experimental Apparatus and Measurement Techniques

AFM/FFM used in the studies conducted in our laboratory has been described in detail in Chapter 1 of this book. (Also see Ruan and Bhushan, 1993, 1994a–c; Bhushan, 1995a,b, 1998a; Bhushan et al., 1994a–e, 1995a–g, 1997a,c, 1998; Koinkar and Bhushan, 1996a,b, 1997a,b; Sundararajan and Bhushan, 1998.) Briefly, the sample is mounted on a piezoelectric transducer (PZT) tube scanner to scan the sample in the X–Y plane and to move the sample in the vertical (Z) direction. A sharp tip at the end of a flexible cantilever is brought in contact with the sample. Normal and frictional forces being applied at the tip–sample interface are measured using a laser beam deflection technique. Simultaneous measurements of surface roughness and friction force can be made with this instrument. For surface roughness and friction force measurements, a microfabricated square pyramidal Si_3N_4 tip with a tip radius of about 30 nm on a cantilever beam (with a normal beam stiffness of about 0.4 N/m) (Chapter 1) is generally used at normal loads ranging from 10 to 150 nN. A preferred method of measuring friction and calibration procedures for conversion of voltages corresponding to normal and friction forces to force units is described by Ruan and Bhushan (1994a). For roughness measurements, the AFM is generally used in a tapping mode as compared to conventional contact mode, to yield better lateral resolution (Chapter 1; Bhushan et al., 1997c). During the tapping mode, the tip is oscillated vertically on the sample with small oscillations on the order of 100 nm near the resonant frequency of the cantilever on the order of 300 kHz. The tapping tip is only in intermittent contact with the sample with a reduced average load. This minimizes the effects of friction and other lateral forces in roughness measurements for improved lateral resolution and to measure roughness of soft surfaces without small-scale plowing. For roughness and friction measurements, the samples are typically scanned over scan areas ranging from 200×200 nm to 10×10 µm, in a direction orthogonal to the long axis of the cantilever beam (Bhushan et al., 1994a, c–e, 1995a–g, 1997a,c, 1998; Ruan and Bhushan, 1994a–c; Koinkar and Bhushan, 1996a,b, 1997a,b, 1998; Sundararajan and Bhushan, 1998). The samples are generally scanned with a scan rate of 1 Hz and the sample scanning speed of 1 µm/s, for example, for a 500×500 nm scan area.

For adhesion force measurements, the sample is moved in the Z-direction until it contacts the tip. After contact at a given load, the sample is slowly moved away. When the spring force exceeds the adhesive force, the tip suddenly detaches from the sample surface and the spring returns to its original position. The tip displacement from the initial position to the point where it detaches from the sample multiplied by the spring stiffness gives the adhesive force.

In nanoscale wear studies, the sample is initially scanned twice, typically at 10 nN to obtain the surface profile, then scanned twice at a higher load of typically 100 nN to wear and to image the surface simultaneously, and then rescanned twice at 10 nN to obtain the profile of the worn surface. No noticeable change in the roughness profiles was observed between the initial two scans at 10 nN, two profiles scanned at 100 nN, and the final two scans at 10 nN. Therefore, changes in the topography between the initial scans at 10 nN and the scans at 100 nN (or the final scans at 10 nN) are believed to occur as a result of local deformation of the sample surface (Bhushan and Ruan, 1994e).

In picoscale indentation studies, the sample is loaded in contact with the tip in the force calibration mode. During loading, tip deflection (normal force) is measured as a function of vertical position of the sample. For a rigid sample, the tip deflection and the sample traveling distance (when the tip and sample come into contact) equal each other. Any decrease in the tip deflection as compared to vertical position of the sample represents indentation. To ensure that the curvature in the tip deflection–sample traveling distance curve does not arise from PZT hysteresis, measurements on several rigid samples including single-crystal natural diamond (IIa) were made. No curvature was noticed for the case of rigid samples. This suggests that any curvature for other samples should arise from the indentation of the sample (Bhushan and Ruan, 1994e).

For microscale scratching, microscale wear, and nanoscale indentation hardness measurements, a three-sided pyramidal single-crystal natural diamond tip with an apex angle of 80° and a tip radius of about 100 nm (determined by scanning electron microscopy imaging) is used at relatively higher loads (1 – 150 μN). The diamond tip is mounted on a stainless steel cantilever beam with normal stiffness of about 30 N/m (Chapter 1). For scratching and wear studies, the sample is generally scanned in a direction orthogonal to the long axis of the cantilever beam (typically at a rate of 0.5 Hz) so that friction can be measured during scratching and wear. The tip is mounted on the beam such that one of its edge is orthogonal to the beam axis; therefore, wear during scratching along the beam axis is higher (about two to three times) than that during scanning orthogonal to the beam axis. For wear studies, typically an area of 2×2 μm is scanned at various normal loads (ranging from 1 to 100 μN) for a selected number of cycles (Bhushan et al., 1994a,c,d, 1995a–e, 1997a, 1998; Koinkar and Bhushan, 1996a, 1997b). For nanoindentation hardness measurements the scan size is set to zero and then the normal load is applied to make the indents (Bhushan et al., 1994b). During this procedure the diamond tip is continuously pressed against the sample surface for about 2 s at various indentation loads. Sample surface is scanned before and after the scratching, wear, or indentation to obtain the initial and the final surface topography, at a low normal load of about 0.3 μN using the same diamond tip. An area larger than the scratched worn or indentation region is scanned to observe the scratch or wear scars or indentation marks.

Nanohardness is calculated by dividing the indentation load by the projected residual area of the indents (Bhushan et al., 1994a–d, 1995a–e, 1997a,b, 1997a; Koinkar and Bhushan, 1996a, 1997b). From the image of the indent, it is difficult to identify the boundary of the indentation mark with great accuracy. This makes the direct measurement of contact area somewhat inaccurate. A nano/picoindentation technique with the dual capability of depth sensing as well as *in situ* imaging is most appropriate (Bhushan et al., 1996). This indentation system provides load–displacement data and can be subsequently used for *in situ* imaging of the indent. Hardness value is obtained from the load–displacement data. Young's modulus of elasticity is obtained from the slope of the unloading curve. This system is described in detail in Chapter 7 in this book.

The force modulation technique is used to obtain surface elasticity maps (Maivald et al., 1991; DeVecchio and Bhushan, 1997; Scherer et al., 1997). An oscillating tip is scanned over the sample surface in contact under steady and oscillating loads. The oscillations are applied to the cantilever substrate with a bimorph, consisting of two piezoelectric transducers bonded to either side of a brass strip, which is located on the substrate holder, Figure 14.1. For measurements, the tip is first bright in contact with a sample under a static load of 50 to 300 nN. In addition to the static load applied by the sample piezo, a small oscillating (modulating) load is applied by a bimorph generally at a frequency (about 8 kHz) far below that of the natural resonance of the cantilever (70 to 400 kHz). When the tip is brought in contact with the sample, the surface resists the oscillations of the tip, and the cantilever deflects. Under the same applied load, a stiff area on the sample would deform less than a soft one; i.e., stiffer surfaces cause greater deflection amplitudes of the cantilever, Figure 14.2. The variations in the deflection

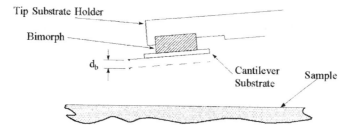

FIGURE 14.1 Schematic of the bimorph assembly used in AFM for operation in tapping and force modulation modes.

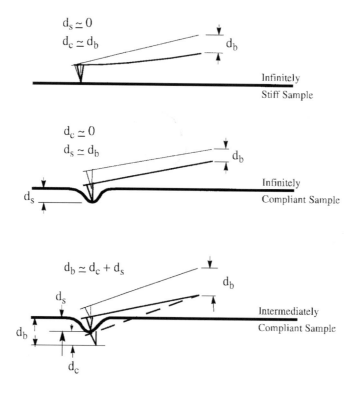

FIGURE 14.2 Schematics of the motion of the cantilever and tip as a result of the oscillations of the bimorph for an infinitely stiff sample, an infinitely compliant sample, and an intermediately compliant sample. The thin line represents the cantilever at the top of the cycle; and the thick line corresponds to the bottom of the cycle. The dashed line represents the position of the tip if the sample was not present or was infinitely compliant. d_c, d_s, and d_b are the oscillating (AC) deflection amplitude of the cantilever, penetration depth, and oscillating (AC) amplitude of the bimorph, respectively. (From DeVecchio, D. and Bhushan, B., 1997, *Rev. Sci. Instrum.* 68, 4498–4505. With permission.)

amplitudes provide a measure of the relative stiffness of the surface. Contact analyses (Bhushan, 1996b) can be used to obtain quantitative measure of localized elasticity of soft and compliant samples (DeVecchio and Bhushan, 1997). The elasticity data are collected simultaneously with the surface height data using a so-called negative lift mode technique. In this mode, each scan line of each topography image (obtained in tapping mode) is retraced with the tapping action disabled and with the tip lowered into steady contact with the surface.

A variant of this technique, which enables one to measure stiffer surfaces, has been used to measure the elastic modulus of hard and rigid surfaces quantitatively (Scherer et al., 1997). This latter technique engages the tip on the top of the sample which is then subjected to oscillations at the frequencies near the cantilever resonances, up to several megahertz, by a PZT beneath the sample. These sample oscillations create oscillations in the tip. The resonance frequencies of these tip oscillations depend on the surface elasticity. The high-frequency technique is useful for stiffer materials (like metals and ceramics) without the need for special tips, but requires the extra piezo and driving equipment and it is more complicated in its theory and application.

All measurements are carried out in the ambient atmosphere ($22 \pm 1°C$, $45 \pm 5\%$ RH, and Class 10,000).

14.2.2   Test Specimens

In this chapter, data on various head slider materials, magnetic media and silicon materials with and without various treatments are presented. Al_2O_3–TiC (70/30 wt%) and polycrystalline and single-crystal (110) Mn–Zn ferrite are commonly used for construction of disk and tape heads. Al_2O_3–TiC, a single-phase material, is also selected for comparisons with the performance of Al_2O_3–TiC, a two-phase material. A α-type SiC is also selected which is a candidate slider material because of its high thermal conductivity and attractive machining and friction and wear properties.

Two thin-film rigid disks with polished and textured substrates, with and without a bonded perfluoropolyether, are selected. These disks are 95 mm in diameter made of Al–Mg alloy substrate (1.3 mm thick) with a 10-μm-thick electroless plated Ni–P coating, 75-nm-thick ($Co_{79}Pt_{14}Ni_7$) magnetic coating, 20-nm-thick amorphous carbon or diamondlike carbon (DLC) coating (microhardness ~ 1500 kg/mm^2 as measured using a Berkovich indenter), and with or without a top layer of perfluoropolyether lubricant with polar end groups (Z-Dol) coating. The thickness of the lubricant film is about 2 nm. The metal particle (MP) tape is a 12.7 mm wide and 13.2 μm thick — poly(ethylene terephthalate (PET) base thickness of 9.8 μm, magnetic coating of 2.9 μm with Al_2O_3 and Cr_2O_3 particles, and back coating of 0.5 μm. The barium ferrite (BaFe) tape is a 12.7-mm-wide and 11-μm-thick (PET base thickness of 7.3 μm, magnetic coating of 2.5 μm with Al_2O_3 particles, and back coating of 1.2 μm). Metal-evaporated (ME) tape is a 12.7-mm-wide tape with 10-μm-thick base, 0.2-μm-thick evaporated Co–Ni magnetic film, and about 10-nm-thick perfluoropolyether lubricant and a backcoat. PET film is a biaxially oriented, semicrystalline polymer with particulates. Two sizes of nearly spherical particulates are generally used: submicron (~0.5 μm) particles of typically carbon and larger particles (2 to 3 μm) of silica.

Virgin single-crystal and polycrystalline silicon samples and thermally oxidized (under both wet and dry conditions) plasma-enhanced chemical vapor deposition (PECVD) oxide-coated and ion-implanted single-crystal pins of orientation (111) are measured. Thermal oxidation of silicon pins was carried out in a quartz furnace at temperatures of 900 to 1000°C in dry oxygen and moisture-containing oxygen ambients. The latter condition was achieved by passing dry oxygen through boiling water before entering the furnace. The thicknesses of the dry oxide and wet oxides are 0.5 and 1 μm, respectively. PECVD oxide was formed by the thermal oxidation of silane at temperatures of 250 to 350°C and was polished using a lapping tape to a thickness of about 5 μm. Single-crystal silicon (111) was ion implanted with C$^+$ ions at 2 to 4 mA cm^{-2} current densities, 100 keV accelerating voltage, and at a fluence of 1×10^{17} ion cm^{-2}.

14.3   Surface Roughness

Solid surfaces, irrespective of the method of formation, contain surface irregularities or deviations from the prescribed geometric form. When two nominally flat surfaces are placed in contact, surface roughness causes contact to occur at discrete contact points. Deformation occurs in these points, and may be either elastic or plastic, depending on the nominal stress, surface roughness, and material properties. The sum of the areas of all the contact points constitutes the real area that would be in contact, and for most materials at normal loads, this will be only a small fraction of the area of contact if the surfaces were perfectly smooth. In general, real area of contact must be minimized to minimize adhesion, friction, and wear (Bhushan, 1996a,b, 1998c).

Characterizing surface roughness is therefore important for predicting and understanding the tribological properties of solids in contact. The AFM has been used to measure surface roughness on length

(a)

$\sigma = 4.16$nm $R_p = 18.3$nm
P-V $= 39.9$nm $\beta^* = 0.20\mu$m

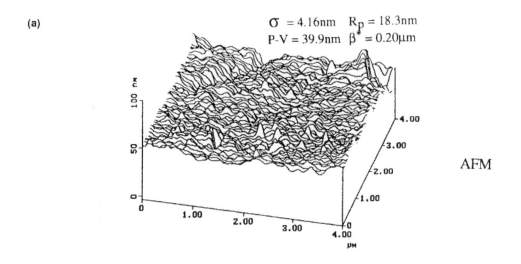

AFM

$\sigma = 2.51$nm $R_p = 5.34$nm
P-V $= 10.9$nm $\beta^* = 12.6\mu$m

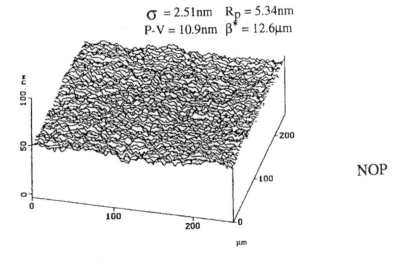

NOP

$\sigma = 3.50$nm $R_p = 14.0$nm P-V $= 25.0$nm $\beta^* = 4.52\mu$m   SP

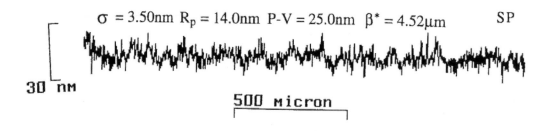

30 nm

500 micron

FIGURE 14.3

scales from nanometers to micrometers. Roughness plots of a glass–ceramic disk measured using an AFM (lateral resolution of ~15 nm), noncontact optical profiler (lateral resolution ~1 μm), and stylus profiler (lateral resolution of ~0.2 μm) are shown in Figure 14.3a. Figure 14.3b compares the profiles of the disk obtained with different instruments at a common scale. The figures show that roughness is found at scales ranging from millimeter to nanometer scales. The measured roughness profile is dependent on the lateral and normal resolutions of the measuring instrument (Bhushan and Blackman, 1991; Oden

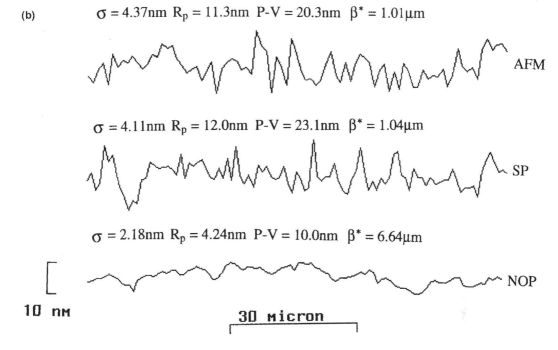

FIGURE 14.3 Surface roughness plots of a glass–ceramic disk (a) measured using an AFM (lateral resolution ~ 15 nm), NOP (lateral resolution ~ 1 μm), and stylus profiler (SP) with a stylus tip of 0.2-μm radius (lateral resolution ~ 0.2 μm), and (b) measured using an AFM (~150 nm), SP (~0.2 μm), and NOP (~1 μm) and plotted on a common scale. (From Poon, C.Y. and Bhushan, B., 1995, *Wear* 190, 89–109. With permission.)

et al., 1992; Ganti and Bhushan, 1995; Poon and Bhushan, 1995a,b). Instruments with different lateral resolutions measure features with different scale lengths. It can be concluded that a surface is composed of a large number of length of scales of roughness that are superimposed on each other.

Surface roughness is most commonly characterized by the standard deviation of surface heights, which is the square roots of the arithmetic average of squares of the vertical deviation of a surface profile from its mean plane. Due to the multiscale nature of surfaces, it is found that the variances of surface height and its derivatives and other roughness parameters depend strongly on the resolution of the roughness-measuring instrument or any other form of filter, hence not unique for a surface (Ganti and Bhushan, 1995; Poon and Bhushan, 1995a,b; Koinkar and Bhushan, 1997a); see, for example, Figure 14.4. Therefore, a rough surface should be characterized in a way such that the structural information of roughness at all scales is retained. It is necessary to quantify the multiscale nature of surface roughness.

A unique property of rough surfaces is that if a surface is repeatedly magnified, increasing details of roughness are observed right down to nanoscale. In addition, the roughness at all magnifications appear quite similar in structure, as qualitatively shown in Figure 14.5. That statistical self-affinity is due to similarity in appearance of a profile under different magnifications. Such a behavior can be characterized by fractal analysis (Majumdar and Bhushan, 1990; Ganti and Bhushan, 1995; Poon and Bhushan, 1995a,b; Koinkar and Bhushan, 1997a). The main conclusions from these studies are that a fractal characterization of surface roughness is *scale independent* and provides information of the roughness structure at all length scales that exhibit the fractal behavior.

Structure function and power spectrum of a self-affine fractal surface follow a power law and can be written as (Ganti and Bhushan model)

$$S(\tau) = C\eta^{(2D-3)}\tau^{(4-2D)},\tag{14.1}$$

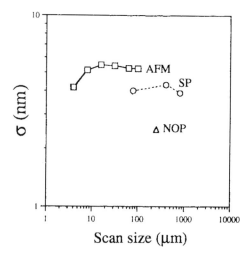

FIGURE 14.4 Scale dependence of standard deviation of surface heights for a glass–ceramic disk, measured using AFM, SP, and NOP.

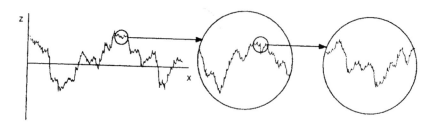

FIGURE 14.5 Qualitative description of statistical self-affinity for a surface profile.

$$P(\omega) = \frac{c_1 \eta^{(2D-3)}}{\omega^{(5-2D)}}, \tag{14.2a}$$

and

$$c_1 = \frac{\Gamma(5-2D)\sin\left[\pi(2-D)\right]}{2\pi}C. \tag{14.2b}$$

The fractal analysis allows the characterization of surface roughness by two parameters D and C, which are instrument independent and unique for each surface. D (ranging from 1 to 2 for surface profile) primarily relates to relative power of the frequency contents and C to the amplitude of all frequencies. η is the lateral resolution of the measuring instrument, τ is the size of the increment (distance), and ω is the frequency of the roughness. Note that if $S(\tau)$ or $P(\omega)$ are plotted as a function of τ or ω, respectively, on a log–log plot, then the power law behavior would result in a straight line. The slope of line is related to D and the location of the spectrum along the power axis is related to C.

Figure 14.6 presents the structure function of a thin-film rigid disk measured using AFM, noncontact optical profiler (NOP), and stylus profiler (SP). A horizontal shift in the structure functions from one scan to another arises from the change in the lateral resolution. D and C values for various scan lengths are listed in Table 14.1. We note that fractal dimension of the various scans is fairly constant (1.26 to 1.33); however, C increases/decreases monotonically with σ for the AFM data. The error in estimation

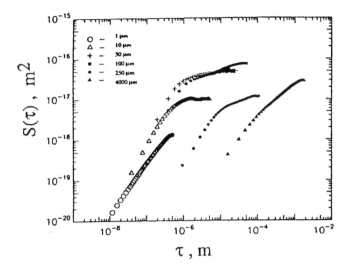

FIGURE 14.6 Structure functions for the roughness data measured at various scan sizes using AFM (scan sizes: 1 × 1 μm, 10 × 10 μm, 50 × 50 μm, and 100 × 100 μm), NOP (scan size: 250 × 250 μm), and SP (scan length: 4000 μm), for a magnetic thin-film rigid disk. (From Ganti, S. and Bhushan, B., 1995, *Wear* 180, 17–34. With permission.)

TABLE 14.1 Surface Roughness Parameters for a Polished Thin-Film Rigid Disk

Scan size (μm x μm)	σ (nm)	D	C (nm)
1 (AFM)	0.7	1.33	9.8×10^{-4}
10 (AFM)	2.1	1.31	7.6×10^{-3}
50 (AFM)	4.8	1.26	1.7×10^{-2}
100 (AFM)	5.6	1.30	1.4×10^{-2}
250 (NOP)	2.4	1.32	2.7×10^{-4}
4000 (NOP)	3.7	1.29	7.9×10^{-5}

AFM = atomic force microscope; NOP = noncontact optical profiler.

of η is believed to be responsible for variation in C. These data show that the disk surface follows a fractal structure for three decades of length scales.

Majumdar and Bhushan (1991) and Bhushan and Majumdar (1992) developed a fractal theory of contact between two rough surfaces. This model has been used to predict whether contacts experience elastic or plastic deformation and to predict the statistical distribution of contact points. For a review of contact models, see Bhushan (1996b, 1998c).

Based on the fractal model of elastic–plastic contact, whether contacts go through elastic or plastic deformation is determined by a critical area which is a function of D, C, hardness, and modulus of elasticity of the mating surfaces. If the contact spot is smaller than the critical area, it goes through the plastic deformations and large spots go through elastic deformations. The critical contact area for inception of plastic deformation for a thin-film disk was reported by Majumdar and Bhushan (1991) to be about 10^{-27} m², so small that all contact spots can be assumed to be elastic at moderate loads.

The question remains as to how large spots become elastic when they must have initially been plastic spots. The possible explanation is shown in Figure 14.7. As two surfaces touch, the nanoasperities (detected by AFM-type of instruments) first coming into contact have smaller radii of curvature and are therefore plastically deformed instantly, and the contact area increases. When load is increased, nanoasperities in the contact merge, and the load is supported by elastic deformation of the large-scale asperities or microasperities (detected by optical profiler type of instruments) (Bhushan and Blackman, 1991).

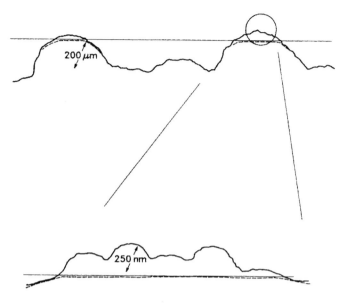

FIGURE 14.7 Schematic of local asperity deformation during contact of a rough surface, upper profile measured by an optical profiler and lower profile measured by AFM; typical dimensions are shown for a polished thin-film rigid disk against a flat slider surface. (From Bhushan, B. and Blackman, G.S., 1991, *ASME J. Tribol.* 113, 452–458. With permission.)

Majumdar and Bhushan (1991) and Bhushan and Majumdar (1992) have reported relationships for cumulative size distribution of the contact spots, portions of the real area of contact in elastic and plastic deformation modes, and the load–area relationships.

14.4   Friction and Adhesion

14.4.1   Nanoscale Friction

Ruan and Bhushan (1994b) measured friction on the nanoscale using FFM. They reported that atomic-scale friction of a freshly cleaved, highly oriented pyrolytic graphite (HOPG) exhibited the same periodicity as that of corresponding topography (also see Mate et al., 1987), Figure 14.8. However, the peaks in friction and those in corresponding topography profiles were displaced relative to each other, Figure 14.9. Using Fourier expansion of the interaction potential, they calculated interatomic forces between the FFM tip and the graphite surface. They have shown that variations in atomic-scale lateral force and the observed displacement can be explained by the variations in intrinsic interatomic forces in the normal and lateral directions.

14.4.2   Microscale Friction and Adhesion

Friction and adhesion of magnetic head sliders, magnetic media, virgin, treated and coated Si(111) wafers, and graphite on a microscale have been measured by Kaneko et al. (1988, 1991a), Miyamoto et al. (1990, 1991a,c), Mate (1993a,b), Bhushan et al. (1994a–c,e, 1995a–g, 1997c, 1998), Ruan and Bhushan (1994a–c), Koinkar and Bhushan (1996a,b, 1997a,b), and Sundararajan and Bhushan (1998).

Koinkar and Bhushan (1996a,b) and Poon and Bhushan (1995a,b) reported that rms roughness and friction force increase with an increase in scan size at a given scanning velocity and normal force. Therefore, it is important that while reporting friction force values, scan sizes and scanning velocity should be mentioned. Bhushan and Sundararajan (1998) reported that friction and adhesion forces are a function of tip radius and relative humidity (also see Koinkar and Bhushan, 1996b). Therefore, relative

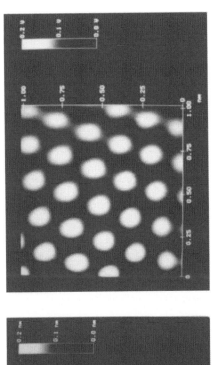

(a)

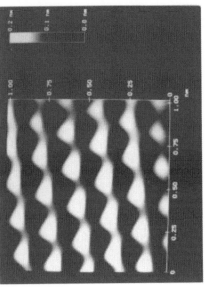

(b)

FIGURE 14.8 Gray-scale plots of (a) surface topography and (b) friction force maps of a 1 70× 1 nm area of a freshly cleaved HOPG showing the atomic-scale variation of topography and friction. Higher points are shown by lighter color. (From Ruan, J. and Bhushan, B., 1994, *J. Appl. Phys.* 76, 5022–5035. With permission.)

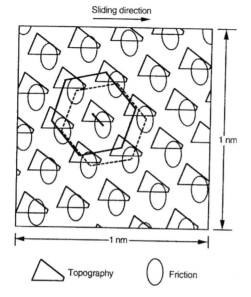

Sliding direction

1 nm

1 nm

△ Topography ○ Friction

FIGURE 14.9 Schematic of surface topography and friction force maps shown in Figure 14.8. The oblate triangles and circles correspond to maxima of topography and friction force, respectively. There is a spatial shift between the two. (From Ruan, J. and Bhushan, B., 1994, *J. Appl. Phys.* 76, 5022–5035. With permission.)

humidity should be controlled during the experiments. Care also should be taken to ensure that tip radius does not change during the experiments.

14.4.2.1   Head Slider Materials

Al_2O_3–TiC is a commonly used slider material. In order to study the friction characteristics of this two phase material, friction of Al_2O_3–TiC (70/30 wt%) surface was measured. Figure 14.10 shows the surface roughness and friction force profiles (Koinkar and Bhushan, 1996a). TiC grains have a Knoop hardness of about 2800 kg/mm², which is higher than that of Al_2O_3 grains of about 2100 kg/mm². Therefore, TiC grains do not polish as much and result in a slightly higher elevation (about 2 to 3 nm higher than that of Al_2O_3 grains). Based on friction force measurements, TiC grains exhibit higher friction force than Al_2O_3 grains. The coefficients of friction of TiC and Al_2O_3 grains are 0.034 and 0.026, respectively, and the coefficient of friction of Al_2O_3–TiC composite is 0.03. Local variation in friction force also arises from the scratches present on the Al_2O_3–TiC surface. A good correspondence between surface slope (also shown in Figure 14.10) and friction force at scratch locations is observed. (Reasons for this correlation will be discussed later.) Thus, local friction values of two-phase materials can be measured. Ruan and Bhushan (1994c) reported that local variation in the coefficient of friction of cleaved HOPG was significant, which arises from structural changes occurring during the cleaving process. The cleaved HOPG surface is largely atomically smooth but exhibits line-shaped regions in which the coefficient of friction is more than an order of magnitude larger. Meyer et al. (1992) and Overney et al. (1992) also used FFM to measure structural variations of a composite surface. They measured friction distribution of mixed monolayer films produced by dipping into a solution of hydrocarbon and fluorocarbon molecules. The resulting film consists of discrete islands of hydrocarbon in a sea of fluorocarbon. They reported that FFM can be used to image and identify compositional domains with a resolution of ~0.5 nm. These measurements suggest that friction measurements can be used for *structural mapping* of the surfaces. FFM measurements can also be used to map chemical variations, as indicated by the use of the FFM with a modified FFM tip to map the spatial arrangement of chemical functional groups in mixed monolayer films (Frisbie et al., 1994). Here, sample regions that had stronger interactions with the functionalized FFM tip exhibited larger friction.

Surface roughness and coefficient of friction of various head slider materials were measured by Koinkar and Bhushan (1996a). For typical values, see Table 14.2. Macroscale friction values for all samples are higher than microscale friction values; the reasons are presented in the following subsection.

Miyamoto et al. (1990) measured adhesive force of four tips made of tungsten, Al_2O_3–TiC, Si_3N_4, and SiC tips in contact with unlubricated, polished SiO_2-coated thin-film rigid disk and a disk lubricated with 2-nm functional lubricant (with hydroxyl end groups, Z-Dol). Nominal radii for all tips were about 5 μm. Adhesive force data are presented in Table 14.3. Mean adhesive forces of the tungsten, Al_2O_3–TiC, Si_3N_4, and SiC tips on a disk medium coated with the functional lubricant are about 10% of those for an unlubricated disk surface. The adhesive force of the ceramic tips is lower than that for the tungsten tip. The adhesive forces of the SiC tip show very low values, even for an unlubricated disk. A good correlation was found between adhesive forces measured by the AFM and the coefficient of macroscale static friction. They also reported that adhesive force increased almost linearly with an increase in the tip radius. (Also see Sugawara et al., 1993; Bhushan et al., 1998).

14.4.2.2   Magnetic Media

Bhushan and co-workers measured friction properties of magnetic media including polished and textured thin-film rigid disks, MP, BaFe and ME tapes, and PET tape substrate. For typical values of coefficients of friction of polished and textured, thin-film rigid disks and MP, BaFe and ME tapes, PET tape substrate, see Table 14.4. In the case of magnetic disks, similar coefficients of friction are observed for both lubricated and unlubricated disks, indicating that most of the lubricant (although partially thermally bonded) is squeezed out from between the rubbing surfaces at high interface pressures, consistent with liquids being poor boundary lubricant (Bowden and Tabor, 1950). Coefficient of friction values on a microscale are much lower than those on the macroscale. When measured for the small contact areas and very low loads used in microscale studies, indentation hardness and modulus of elasticity are higher than at the macroscale (data to be presented later). This reduces the real area of contact and the degree of wear. In addition, the small apparent areas of contact reduces the number of particles trapped at the interface, and thus minimizes the "plowing" contribution to the friction force (Bhushan et al., 1995d,f).

Miyamoto et al. (1991b) reported the coefficient of friction of an unlubricated disk with amorphous carbon and SiO_2 overcoats against the diamond tip to be 0.24 and 0.36, respectively. The coefficients of friction of disks lubricated with 2-nm-thick perfluoropolyether lubricant films were 0.08 for functional lubricant (with hydroxyl end groups, Z-Dol) on SiO_2 overcoat, 0.10 for functional lubricant on carbon overcoat, and 0.19 for nonpolar lubricant (Krytox 157FS L) on carbon overcoat. They found that the coefficient of friction of a 4-nm-thick lubricant film was about twice that of a 2-nm-thick film. Mate (1993a) measured the coefficient of friction of unlubricated polished and textured disks and with a lubricant film with ester end groups (Demnum SP) against a tungsten tip with a tip radius of 100 nm. The coefficients of friction of unlubricated polished disks and with 1.5-nm-thick lubricant film were 0.5 and 0.4, respectively, and of unlubricated textured disks and with 2.5-nm-thick lubricant film were 0.5 and 0.2, respectively.

Coefficient of microscale friction values reported by Miyamoto et al. (1991b), by Mate (1993a) and by Bhushan et al. (1995g) (to be reported later in this section) are larger than those reported by Bhushan et al. (1994a–c,e, 1995a–f, 1997c) in Table 14.4. Miyamoto et al. made measurements with a three-sided pyramidal diamond tip at large loads of 500 nN to tens of micronewtons and Mate et al. made measurements with a soft tungsten tip from 30 to 300 nN, as compared to Bhushan et al.'s measurements made using the Si_3N_4 tip at lower loads ranging from 10 to 150 nN. High values reported by Miyamoto et al. and Mate et al. may arise from plowing contribution at higher normal loads and differences in the friction properties of different tip materials. Bhushan et al. (1995f) have reported that the coefficient of friction on microscale is a strong function of normal load. The critical load at which an increase in friction occurs corresponds to surface hardness. At high loads, the coefficient of friction on microscale increases toward values comparable with those obtained from macroscale measurements. The increase in the value of microscale friction at higher loads is associated with interface damage and associated plowing.

In order to elegantly show any correlation between local values of friction and surface roughness, Bhushan (1995b) measured the surface roughness and friction force of a gold-coated ruling with rectangular girds. Figure 14.11 shows the surface roughness profile, the slopes of roughness profile taken along the sliding direction, and the friction force profile for the ruling. We note that friction force changes

Surface Topography

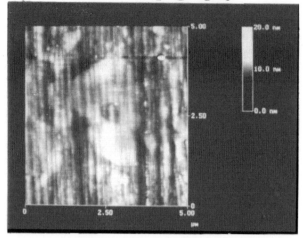

Surface Slope

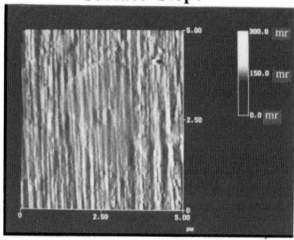

Friction Force

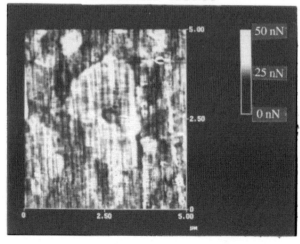

704

TABLE 14.2 Surface Roughness (σ and P-V distance), Micro- and Macroscale Friction, Microscratching/Wear, and Nano- and Microhardness Data for Various Samples

| Sample | Surface Roughness nm (1 × 1 µm) | | Coefficient of friction | | | Scratch Depth at 60 µN (nm) | Wear Depth at 60 µN (nm) | Hardness (GPa) | |
| | σ | P-V[a] | Microscale | Macroscale[b] | | | | Nano at 2 mN | Micro |
				Initial	Final				
Al_2O_3	0.97	9.9	0.03	0.18	0.2–0.6	3.2	3.7	24.8	15.0
Al_2O_3–TiC	0.80	9.1	0.05	0.24	0.2–0.6	2.8	22.0	23.6	20.2
Polycrystalline Mn–Zn ferrite	2.4	20.0	0.04	0.27	0.24–0.4	9.6	83.6	9.6	5.6
Single–crystal (110) Mn–Zn ferrite	1.9	13.7	0.02	0.16	0.18–0.24	9.0	56.0	9.8	5.6
SiC (α-type)	0.91	7.2	0.02	0.29	0.18–0.24	0.4	7.7	26.7	21.8

[a] Peak-to-valley distance.

[b] Obtained using silicon nitride ball with 3 mm diameter in a reciprocating mode at a normal load of 10 mN, reciprocating amplitude of 7 mm, and average sliding speed of 1 mm/s. Initial coefficient of friction values were obtained at first cycle (0.007 m sliding distance) and final values at a sliding distance of 5 m.

TABLE 14.3 Mean Values and Ranges of Adhesive Forces between Unlubricated and Lubricated SiO_2-Coated Disks and Tips Made of Various Materials

| Tip Material | Adhesive Force (µN) | |
	Without Lubricant	Functional Liquid Lubricant (2.0 nm)
Tungsten	12.1 (2.32–13.9)	1.09 (0.67–1.60)
Al_2O_3–TiC	5.17 (3.64–6.92)	0.25 (0.078–0.47)
Si_3N_4	1.88 (1.00–2.82)	0.07 (0–0.16)
SiC	0.21 (0.13–0.32)	0.030 (0–0.09)

From Miyamoto, T. et al., 1990, *ASME J. Tribol.* 112, 567–572. With permission.

significantly at the edges of the grid. Friction force is high locally at the edge of grid with a positive slope and is low at the edge of the grid with a negative slope. Thus, there is a strong correlation between the slope of the roughness profiles and the corresponding friction force profiles.

Bhushan et al. (1994c,e, 1995a–f, 1997c) examined the relationship between local variations in microscale friction force and surface roughness profiles for magnetic media. Figures 14.12 and 14.14 show the surface roughness map, the slopes of roughness profile taken along the sliding direction, and the friction force map for textured and lubricated disks, and an MP tape, respectively. Bhushan and Ruan (1994a) noted that there is no resemblance between the friction force maps and the corresponding roughness maps; e.g., high or low points on the friction force map do not correspond to high or low points on the roughness map. By comparing the slope of roughness profiles taken in the tip sliding direction and friction force map, we observe a strong correlation between the two. (For a clearer correlation, see grayscale plots of surface roughness slope and friction force profiles for FFM tip sliding in either directions in Figures 14.13 and 14.15).

FIGURE 14.10 Gray-scale plots of surface topography ($\sigma = 1.12$ nm), slope of the roughness profiles taken along the sliding direction (the horizontal axis) (mean = -0.003, $\sigma = 0.015$), and friction force map (mean = 28.5 nN, $\sigma = 4.0$ nN; Al_2O_3 grains: mean = 24.8 mN, $\sigma = 1.85$ nN and TiC grains: mean = 32.7 nN, $\sigma = 2.6$ nN) for a Al_2O_3–TiC slider for a normal load of 950 nN.

TABLE 14.4 Surface Roughness (σ), Microscale and Macroscale Friction, and Nanohardness Data of Thin-Film Magnetic Rigid Disk, Magnetic Tape, and Magnetic Tape Substrate (PET) Samples

Sample	σ (nm) NOP 250 × 250 μm^a	AFM 1 × 1 μm^a	AFM 10 × 10 μm^a	Coefficient of Microscale Friction 1 × 1 μm^a	Coefficient of Microscale Friction 10 × 10 μm^a	Coefficient of Macroscale Friction Mn–Zn Ferrite	Coefficient of Macroscale Friction Al_2O_3–TiC	Nanohardness (GPa)/ Normal Load (μN)
Polished, unlubricated disk	2.2	3.3	4.5	0.05	0.06	—	0.26	21/100
Polished, lubricated disk	2.3	2.3	4.1	0.04	0.05	—	0.19	—
Textured, lubricated disk	4.6	5.4	8.7	0.04	0.05	—	0.16	—
MP tape	6.0	5.1	12.5	0.08	0.06	0.19	—	0.30/50
Barium-ferrite tape	12.3	7.0	7.9	0.07	0.03	0.18	—	0.25/25
ME tape	9.3	4.7	5.1	0.05	0.03	0.18	—	0.7 to 4.3/75
PET tape substrate	33	5.8	7.0	0.05	0.04	0.55	—	0.3/20 and 1.4/20[b]

[a] Scan area; NOP = noncontact optical profiler; AFM = atomic force microscope.
[b] Numbers are for polymer and particulate regions, respectively.

To further verify the relationship between surface roughness slope and friction force values, to eliminate any effect resulting from nonuniform composition of disk and tape surfaces, Bhushan and Ruan (1994a) measured a polished natural (IIa) diamond. Repeated measurements were made along one line on the surface. Highly reproducible data were obtained, Figure 14.16. Again, the variation of friction force correlates to the variation of the slope of the roughness profiles taken along the sliding direction of the tip. This correlation has been shown to hold for various magnetic disks, magnetic tapes, polyester tape substrates, silicon, graphite and other materials (Bhushan et al., 1994a,c–e, 1995a–d, 1997a, 1998; Ruan and Bhushan, 1994b,c).

We now examine the mechanism of microscale friction, which may explain the resemblance between the slope of surface roughness profiles and the corresponding friction force profiles (Bhushan and Ruan, 1994a; Ruan and Bhushan, 1994b,c). There are three dominant mechanisms of friction: adhesive, adhesive and roughness (ratchet), and plowing. As a first order, we may assume these to be additive. The adhesive mechanism alone cannot explain the local variation in friction. Let us consider the ratchet mechanism. According to Makinson (1948), we consider a small tip sliding over an asperity making an angle θ with the horizontal plane, Figure 14.17. The normal force (normal to the general surface) applied by the tip to the sample surface W is constant. Friction force F on the sample varies as a function of the surface roughness. It would be a constant $\mu_0 W$ for a smooth surface in the presence of "adhesive" friction mechanism. In the presence of a surface asperity, the local coefficient of friction μ_1 in the ascending part is

$$\mu_1 = F/W = \left(\mu_0 + \tan\theta\right) / \left(1 - \mu_0 \tan\theta\right). \tag{14.3}$$

Since $\mu_0 \tan\theta$ is small on a microscale, Equation 14.3 can be rewritten as

$$\mu_1 \sim \mu_0 + \tan\theta, \tag{14.4}$$

indicating that in the ascending part of the asperity one may simply add the friction force and the asperity slope to one another. Similarly, on the right-hand side (descending part) of the asperity,

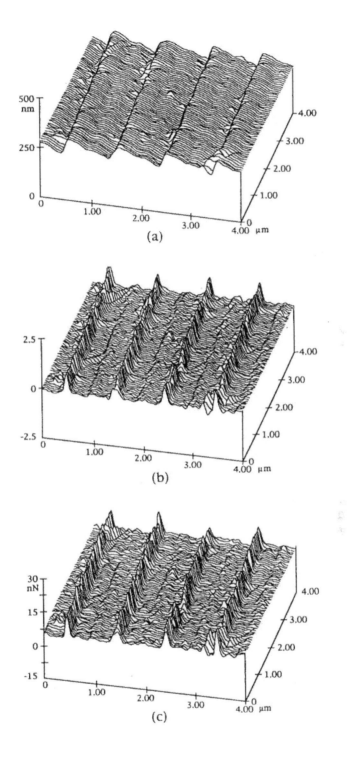

FIGURE 14.11 (a) Surface roughness map, (b) slope of the roughness profiles taken in the sample sliding direction (the horizontal axis), and (c) friction force map for a gold-coated ruling at a normal load of 155 nN. (From Bhushan, B., 1995, *Tribol. Int.* 28, 85–95. With permission.)

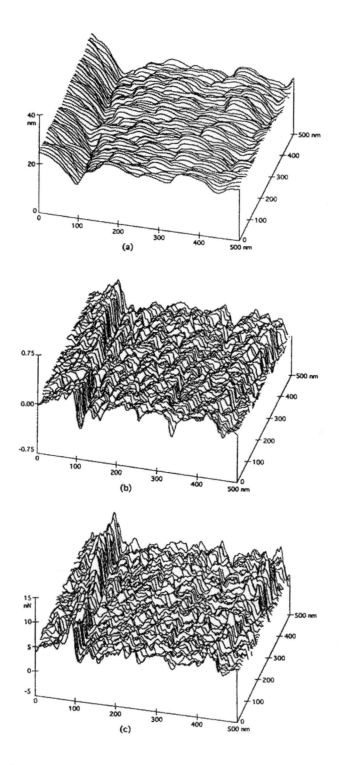

FIGURE 14.12 (a) Surface roughness map (σ = 4.4 nm), (b) slope of the roughness profiles taken in the sample sliding direction (the horizontal axis) (mean = 0.023, σ = 0.197), and (c) friction force map (mean = 6.2 nN, σ = 2.1 nN) for a textured and lubricated thin-film rigid disk for a normal load of 160 nN. (From Bhushan, B. and Ruan, J., 1994, *ASME J. Tribol.* 116, 389–396. With permission.)

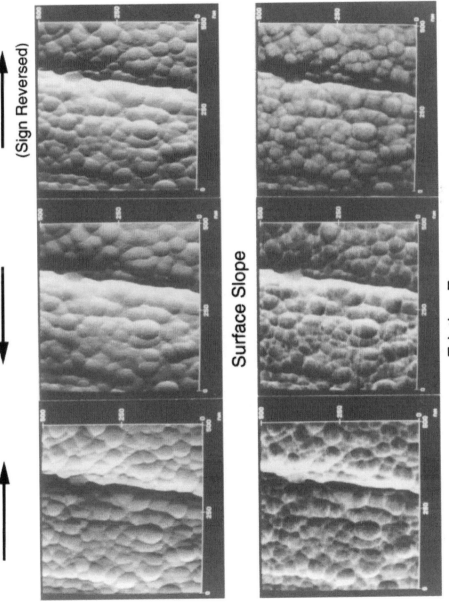

FIGURE 14.13 Gray-scale plots of the slope of the surface roughness and the friction force maps for a textured and lubricated thin-film rigid disk. Arrows indicate the tip sliding direction. Higher points are shown by lighter color.

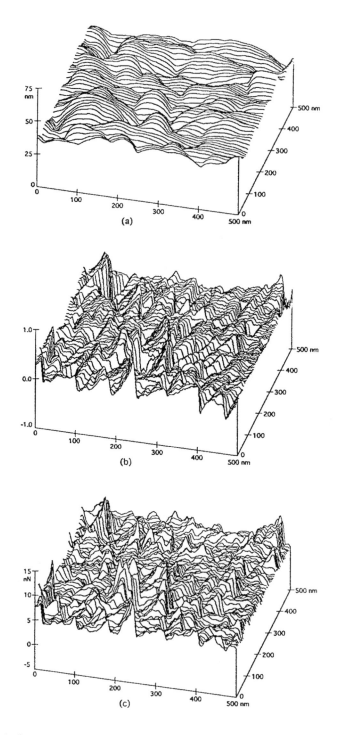

FIGURE 14.14 (a) Surface roughness map ($\sigma = 7.9$ nm), (b) slope of the roughness profiles taken along the sample sliding direction (mean = –0.006, $\sigma = 0.300$), and friction force map (mean = 5.5 nN, $\sigma = 2.2$ nN) of an MP tape at a normal load of 70 nN. (From Bhushan, B. and Ruan, J., 1994, *ASME J. Tribol.* 116, 389–396. With permission.)

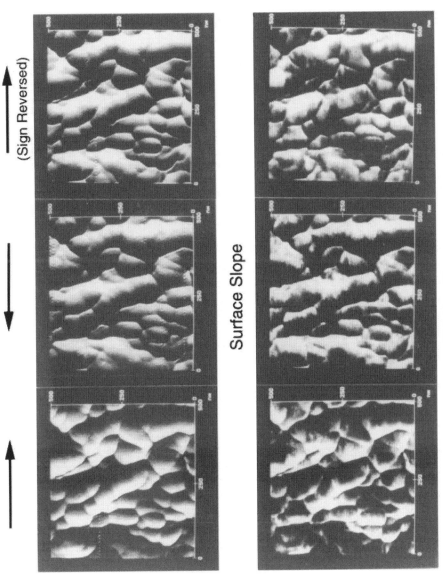

FIGURE 14.15 Gray-scale plots of the slope of the roughness and the friction force maps for an MP tape. Arrows indicate the tip sliding direction. Higher points are shown by lighter color.

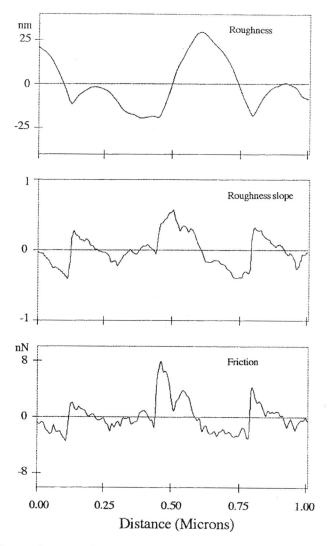

FIGURE 14.16 Surface roughness map (σ = 15.4 nm), slope of the roughness map (mean = −0.052, σ = 0.224), and the friction force map (σ = 2.1 nN) of a polished natural (IIa) diamond crystal. (From Bhushan, B. and Ruan, J., 1994, *ASME J. Tribol.* 116, 389–396. With permission.)

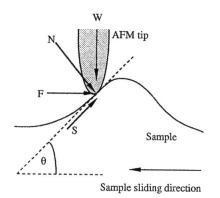

FIGURE 14.17 Schematic illustration showing the effect of an asperity (making an angle θ with the horizontal plane) on the surface in contact with the tip on local friction in the presence of "adhesive" friction mechanism. *W* and *F* are the normal and friction forces, respectively. *S* and *N* are the force components along and perpendicular to the local surface of the sample at the contact point, respectively.

$$\mu_2 = \left(\mu_0 - \tan\theta\right) \big/ \left(1 + \mu_0 \tan\theta\right)$$

$$\sim \mu_0 - \tan\theta,$$

(14.5)

if $\mu_0 \tan\theta$ is small. For a symmetrical asperity, the average coefficient of friction experienced by the FFM tip traveling across the whole asperity is

$$\mu_{ave} = \left(\mu_1 + \mu_2\right) \big/ 2$$

$$= \mu_0 \left(1 + \tan^2\theta\right) \big/ \left(1 - \mu_0^2 \tan^2\theta\right)$$

$$\sim \mu_0 \left(1 + \tan^2\theta\right),$$

(14.6)

if $\mu_0 \tan\theta$ is small.

The plowing component of friction (Bowden and Tabor, 1950) with tip sliding in either direction is

$$\mu_p \sim \tan\theta.$$

(14.7)

Since in the FFM measurements, we notice little damage of the sample surface, the contribution by plowing is expected to be small and the ratchet mechanism is believed to be the dominant mechanism for the local variations in the friction force profile. With the tip sliding over the leading (ascending) edge of an asperity, the slope is positive; it is negative during sliding over the trailing (descending) edge of the asperity. Thus, friction is high at the leading edge of asperities and low at the trailing edge. The ratchet mechanism thus explains the correlation between the slopes of the roughness profiles and friction profiles observed in Figures 14.11 to 14.16. We note that in the ratchet mechanism, the FFM tip is assumed to be small compared to the size of asperities. This is valid since the typical radius of curvature of the tips is about 30 nm. The radius of curvature of the asperities of the samples measured here (the asperities that produce most of the friction variation) is found to be typically about 100 to 200 nm, which is larger than that of the FFM tip (Bhushan and Blackman, 1991).

We also note that the variation in attractive adhesive force (W_{att}) with topography can also contribute to observed variation in friction (Mate, 1993a,b). The total force in the normal direction is the intrinsic force (W_{att}) in addition to the applied normal load (W). Thus, friction force

$$F = \mu\left(W + W_{att}\right),$$

where μ is the coefficient of friction. Based on Mate (1993a), major components of the attractive adhesive force are the van der Waals force (between the tip and the summits and valleys of the mating sample surface) and meniscus or capillary forces. Approximating the tip–sample surface geometry as a sphere on a flat, the magnitude of the attractive van der Waals force can be expressed as (Derjaguin et al., 1987)

$$W_{vdW} = \frac{AR}{6D^2}$$

(14.8a)

and

$$A = 24\pi D_0^2 \left(\gamma_t \gamma_d\right)^{1/2},$$

(14.8b)

713

where A is the Hamaker constant, R is the tip radius, D is the separation distance by the surface roughness of the means of the tip and the sample surfaces, γ_t and γ_d are the surface energies of the tip and sample surfaces, and $D_0 \sim 0.2$ nm (Israelachvili, 1992). As a consequence of the strong $1/D^2$ dependence, the tip should experience a much weaker van der Waals force on the top of a summit as compared with that of a valley. Mate (1993a) reported that a separation change ΔD of 5 nm would give a variation in the van der Waals force by a factor of 5 if the distance of closest approach, approximately the amount of roughness separation between the two surfaces, is 4 nm. Another component of the attractive adhesive force in the presence of liquid film is the meniscus force. The meniscus force for a sphere on a flat in the presence of liquid is

$$W_M = 4\pi R \gamma_l, \qquad (14.9)$$

where γ_l is the surface tension of the liquid. Meniscus force is generally much stronger than the van der Waals force. Thus, the contribution of adhesion mechanism to the friction force variation is relatively small for samples used in this study. Furthermore, the correlation between the surface and friction force profiles is poor; therefore, an adhesion mechanism cannot explain the topography effects. The ratchet mechanism already quantitatively explains the variation of friction.

Since the local friction force is a function of the local slope of sample surface, the local friction force should be different as the scanning direction of the sample is reversed. Figures 14.13 and 14.15 show the gray-scale plots of slope of roughness profiles and friction force profiles for a lubricated textured disk and an MP tape, respectively. The left side of the figures corresponds to the tip sliding from the left toward the right (or the sample sliding from the right to the left). We again note a general correspondence between the surface roughness slope and the friction profiles. The middle figures in Figures 14.13 and 14.15 correspond to the tip sliding from the right toward left. We note that generally the points that have high friction force and high slope in the left-to-right scan have low friction and low slope as the sliding direction is reversed (Meyer and Amer, 1990; Grafstrom et al., 1993; Overney and Meyer, 1993; Bhushan and Ruan, 1994e; Ruan and Bhushan, 1994b). This results from the slope being of opposite sign as the direction is reversed, which reverses the sign of friction force contribution by the ratchet mechanism. This relationship is not true at all locations. The right-side figures in Figures 14.13 and 14.15 correspond to the left-hand set with sign reversed. On the right, although the sign of friction force profile is the reverse of the left-hand profile, some differences in the right two friction force profiles are observed which may result from the asymmetrical asperities and/or asymmetrical transfer of wipe material during manufacturing of the disk. This *directionality in microscale friction force* was first reported by Bhushan et al. (1994a,c,e, 1995a–d, 1997a, 1998).

If asperities in a sample surface have a preferential orientation, this directionality effect will be manifested in macroscopic friction data; that is, the coefficient of friction may be different in one sliding direction from that in the other direction. Such a phenomenon has been observed in rubbing wool fiber against horn. It was found that the coefficient of friction is greatest when the wool fiber is rubbed toward its tip (Mercer, 1945; Lipson and Mercer, 1946; Thomson and Speakman, 1946). Makinson (1948) explained the directionality in the friction by the "ratchet" effect. Here, the ratchet effect is the result of large angle θ, where instead of true sliding, rupture or deformation of the fine scales of wool fibers occurs in one sliding direction. We note that the frictional directionality can also exist in materials with particles having a preferred orientation.

The directionality effect in friction on a macroscale is also observed in some magnetic tapes. In a macroscale test, a 12.7-mm-wide MP tape was wrapped over an aluminum drum and slid in a reciprocating motion with a normal load of 0.5 N and a sliding speed of about 60 mm/s. The coefficient of friction as a function of sliding distance in either direction is shown in Figure 14.18. We note that the coefficient of friction on a macroscale for this tape is different in different directions.

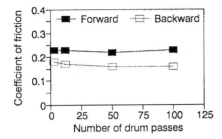

FIGURE 14.18 Coefficient of macroscale friction as a function of sliding cycles for an MP tape sliding over an aluminum drum in a reciprocating mode in both directions. Normal load = 0.5 N over 12.7-mm-wide tape, sliding speed = 60 mm/s.

TABLE 14.5 Roughness (σ), Microfriction, Microscratching/Microwear, and Nanoindentation Hardness Data for Various Virgin, Coated, and Treated Silicon Samples

Material	σ (nm) 500×500 nm[a]	Coefficient of Friction	Scratch Depth at $40\ \mu N$ (nm)	Wear Depth at $40\ \mu N$ (nm)	Hardness at $100\ \mu N$ (GPa)
Si(111)	0.11	0.03	20	27	11.7
Si(110)	0.09	0.04	20	—	—
Si(100)	0.12	0.03	25	—	—
Polysilicon	1.07	0.04	18	—	—
Polysilicon (lapped)	0.16	0.05	18	25	12.5
PECVD oxide-coated Si(111)	1.50	0.01	8	5	18.0
Dry-oxidized Si(111)	0.11	0.04	16	14	17.0
Wet-oxidized Si(111)	0.25	0.04	17	18	14.4
C[+]-implanted Si(111)	0.33	0.02	20	23	18.6

[a] Scan area.

From Bhushan, B. and Koinkar, V.N., 1994, *J. Appl. Phys.* 75, 5741–5746. With permission.

14.4.2.3   Silicon

Coefficient of microscale friction data for virgin, treated, and coated Si(111) samples are presented in Table 14.5 (Bhushan et al., 1994a). (Also see Bhushan et al., 1993c, 1997a,b; Sundararajan and Bhushan, 1998.) We note that crystalline orientation of silicon has little effect on the coefficient of friction. PECVD oxide-coated Si(111) exhibits a coefficient of friction value lower than that of any other silicon sample.

14.5   Scratching and Wear

14.5.1   Nanoscale Wear

Bhushan and Ruan (1994e) conducted nanoscale wear tests on MP tapes at a normal load of 100 nN. Figure 14.19 shows the topography of the MP tape obtained at two different loads. For a given normal load, measurements were made twice. There was no discernible difference between consecutive measurements for a given normal load. However, as the load increased from 10 to 100 nN, topographical changes were observed; material (indicated by an arrow) was pushed toward the right side in the sliding direction of the AFM tip relative to the sample. The material movement is believed to occur as a result of plastic deformation of the tape surface. Similar behavior was observed on all tapes. Magnetic tape coating is made of magnetic particles and polymeric binder. Any movement of the coating material can eventually lead to loose debris. Debris formation is an undesirable situation as it may contaminate the head, which may increase friction and/or wear between the head and tape, in addition to the deterioration of the tape itself. With disks, they did not notice any deformation under a 100 nN normal load.

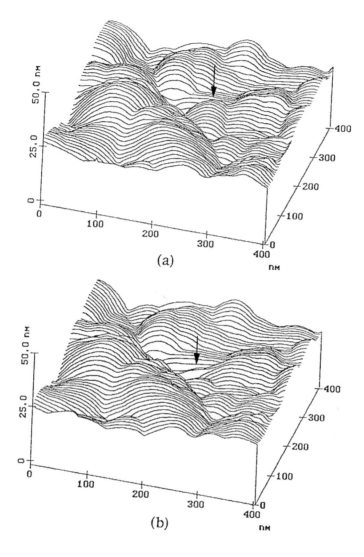

FIGURE 14.19 Surface roughness maps of an MP tape at applied normal load of (a) 10 nN and (b) 100 nN. Location of the change in surface topography as a result of nanowear is indicated by arrows. (From Bhushan, B. and Ruan, J., 1994, *ASME J. Tribol.* 116, 389–396. With permission.)

14.5.2 Microscale Scratching

Microscratches have been made on various potential head slider materials (Al_2O_3, Al_2O_3–TiC, Mn–Zn ferrite, and SiC), various magnetic media (unlubricated polished thin-film disk, MP, BaFe, ME tapes, PET substrates) and virgin, treated, and coated Si(111) wafers at various loads (Miyamoto et al., 1991c, 1993; Bhushan et al., 1994a,c,d, 1995a–e, 1997a, 1998; Koinkar and Bhushan, 1996a, 1997b; Sundararajan and Bhushan, 1998). As mentioned earlier, the scratches are made using a diamond tip.

14.5.2.1 Head Slider Materials

Scratch depths as a function of load and representative scratch profiles with corresponding two-dimensional gray scale plots at various loads after a single pass (unidirectional scratching) for Al_2O_3, Al_2O_3–TiC, polycrystalline and single-crystal Mn–Zn ferrite and SiC are shown in Figures 14.20 and 14.21, respectively. Variation in the scratch depth along the scratch is about ±15%. The Al_2O_3 surface could be scratched

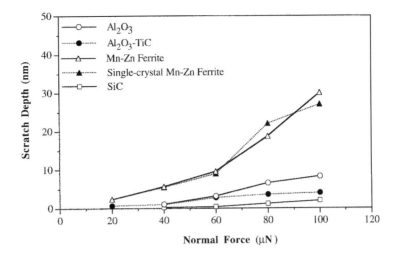

FIGURE 14.20 Scratch depth as a function of normal load after one unidirectional cycle for Al_2O_3, Al_2O_3–TiC, polycrystalline Mn–Zn ferrite, single-crystal Mn–Zn ferrite, and SiC. (From Koinkar, V.N. and Bhushan, B., 1996, *Wear* 202, 110–122. With permission.)

at a normal load of 40 µN. The surface topography of polycrystalline Al_2O_3 shows the presence of porous holes on the surface. The two-dimensional gray scale plot of the scratched Al_2O_3 surface shows one porous hole between scratches made at normal loads of 40 and 60 µN. Regions with defects or porous holes present, exhibit lower scratch resistance (see region marked by the arrow on two-dimensional gray-scale plot of Al_2O_3). The Al_2O_3–TiC surface could be scratched at a normal load of 20 µN. The scratch resistance for TiC grains is higher than that of Al_2O_3 grains. The scratches generated at normal loads of 80 and 100 µN show that the scratch depth of Al_2O_3 grains is higher than that of TiC grains (see corresponding gray-scale plot for Al_2O_3–TiC). Polycrystalline and single-crystal Mn–Zn ferrite could be scratched at a normal load of 20 µN. The scratch width is much larger for the ferrite specimens as compared with other specimens. For SiC there is no measurable scratch observed at a normal load of 20 µN. At higher normal loads very shallow scratches are produced. Table 14.2 presents average scratch depth at 60 µN normal load for all specimens. SiC has the highest scratch resistance followed by Al_2O_3–TiC, Al_2O_3, and poly-crystalline and single-crystal Mn–Zn ferrite. Polycrystalline and single-crystal Mn–Zn ferrite specimens exhibit comparable scratch resistance.

14.5.2.2   Magnetic Media

Scratch depths as a function of load and scratch profiles at various loads after ten scratch cycles for unlubricated, polished disk, and MP tape are shown in Figures 14.22 and 14.23, respectively. We note that scratch depth increases with an increase in the normal load. Tape could be scratched at about 100 nN. With disk, gentle scratch marks under 10 µN load were barely visible. It is possible that material removal did occur at lower load on an atomic scale which was not observable with a scan size of 5 µm square. For disk, scratch depth at 40 µN is less than 10 nm deep. The scratch depth increased slightly at the load of 50 µN. Once the load is increased in excess of 60 µN, the scratch depth increased rapidly. Bhushan et al. (1994c) believed that the DLC coating cracked at about 60 µN. These data suggest that the carbon coating on the disk surface is much harder to scratch than the underlying thin-film magnetic film. This is expected since the carbon coating is harder than the magnetic material used in the construction of the disks.

Since tapes scratch readily, for comparisons in scratch resistance of various tapes, Bhushan et al. (1995c) made scratches on three tapes with one cycle. Figure 14.24 presents the scratch depths as a function of normal load after one cycle for three tapes — MP, BaFe, and ME tapes. For the MP and BaFe particulate tapes, Bhushan et al. (1995c) noted that the scratch depth along (parallel) and across (perpendicular)

717

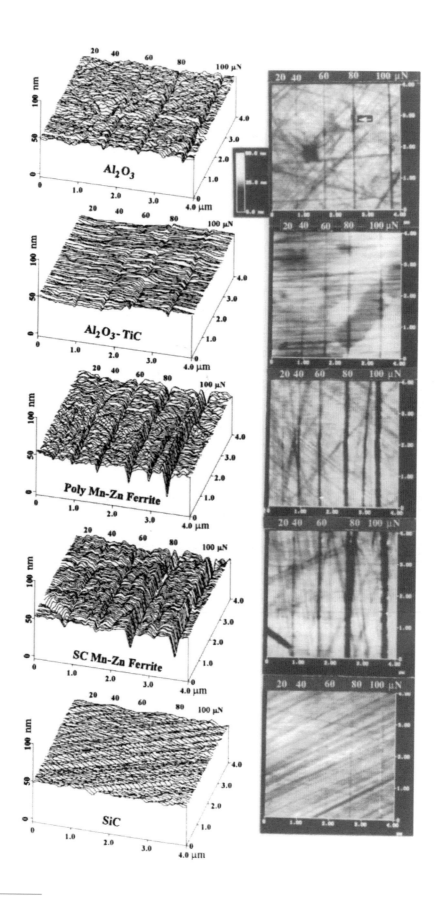

the longitudinal direction of the tapes is similar. Between the two tapes, MP tape appears to be more scratch resistant than BaFe tape, which depends on the binder, pigment volume concentration (PVC), and the head-cleaning agent (HCA) contents. ME tapes appear to be much more scratch resistant than the particulate tapes. However, the ME tape breaks up catastrophically in a brittle mode at a normal load higher than the 50 µN (Figure 14.25), as compared to particulate tapes in which the scratch rate is constant. They reported that the hardness of ME tapes is higher than that of particulate tapes; however, a significant difference in the nanoindentation hardness values of the ME film from region to region (Table 14.4) was observed. They systematically measured scratch resistance in the high- and low-hardness regions along and across the longitudinal directions. Along the parallel direction, load required to crack the coating was lower (implying lower scratch resistance) for a harder region, than that for a softer region. The scratch resistance of the high-hardness region along the parallel direction is slightly poorer than that for along perpendicular direction. Scratch widths in both low- and high-hardness regions is about half (~2 µm) than that in perpendicular direction (~1 µm). In the parallel direction, the material is removed in the form of chips and lateral cracking also emanates from the wear zone. ME films have columnar structure with the columns lined up with an oblique angle on the order of about 35° with respect to the normal to the coating surface (Bhushan, 1992; Hibst, 1993). The column orientation may be responsible for the directionality effect on the scratch resistance. Hibst (1993) have reported the directionality effect in the ME tape–head wear studies. They have found that the wear rate is lower when the head moves in the direction corresponding to the column orientation than in the opposite direction.

PET films could be scratched at loads of as low as about 2 µN, Figure 14.22. Figure 14.26a shows scratch marks made at various loads. Scratch depth along the scratch does not appear to be uniform. This may occur because of variations in the mechanical properties of the film. Bhushan et al. (1995a) also conducted scratch studies in the selected particulate regions. Scratch profiles at increasing loads in the particulate region are shown in Figure 14.26b. We note that the bump (particle) is barely scratched at 5 µN, and it can be scratched readily at higher loads. At 20 µN, it essentially disappears.

14.5.2.3   Silicon

A summary of microscratching data for various silicon samples is presented in Table 14.5. Virgin and modified silicon surfaces could be scratched at 10 µN load, see Figures 14.27 and 14.28 and Table 14.5 (Bhushan et al., 1994a). (Also see Bhushan et al., 1993c; 1997a–b; Sundararajan and Bhushan, 1998.) Scratch depth increased with an increase in load. We note that crystalline orientation of silicon has little influence on the scratch depth. Virgin silicon is poor in scratch resistance as compared with treated samples; PECVD oxide samples had the largest scratch resistance followed by dry-oxidized, wet-oxidized, and ion-implanted samples. Ion implantation showed no improvements on the scratch resistance.

14.5.3   Microscale Wear

By scanning the sample (in two dimensions) while scratching, wear scars are generated on the sample surface (Bhushan et al., 1994a,c,d, 1995a–e, 1997a, 1998; Koinkar and Bhushan,1996a, 1997b; Sundararajan and Bhushan, 1998). The major benefit of a single-cycle wear test over a scratch test is that scratch/wear data can be obtained over a large area.

14.5.3.1   Head Slider Materials

Figure 14.29 shows the wear depth as a function of load for one cycle for different slider materials. Variation in the wear depth in the wear mark is dependent upon the material. It is generally within ±5%. The mean wear depth increases with the increase in normal load. The representative surface profiles showing the wear marks (central 2×2 µm region) at a normal load of 60 µN for all specimens are shown

---

FIGURE 14.21   Surface profiles (left column) and two-dimensional gray-scale plots (right column) of scratched Al$_2$O$_3$, Al$_2$O$_3$–TiC, polycrystalline Mn–Zn ferrite, single-crystal Mn–Zn ferrite, and SiC surfaces. Normal loads used for scratching for one unidirectional cycle are listed in the figure. (From Koinkar, V.N. and Bhushan, B., 1996, *Wear* 202, 110–122. With permission.)

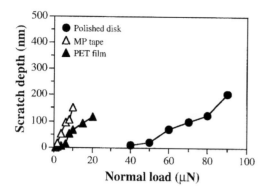

FIGURE 14.22 Scratch depth as a function of normal load after ten scratch cycles for an unlubricated polished thin-film rigid disk, MP tape, and PET film.

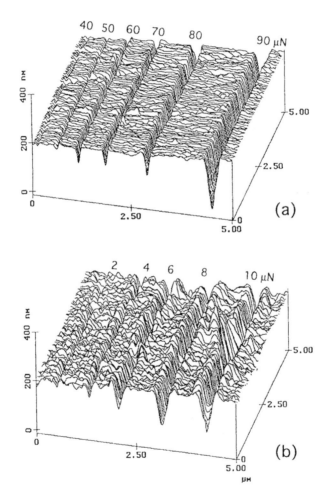

FIGURE 14.23 Surface profiles for scratched (a) unlubricated polished thin-film rigid disk and (b) MP tape. Normal loads used for scratching for ten cycles are listed in the figure. (From Bhushan, B. et al., 1994, *Proc. Inst. Mech. Eng. Part J: J. Eng. Tribol.* 208, 17–29. With permission.)

in Figure 14.30. The material is removed uniformly in the wear region for all specimens. Table 14.2 presents average wear depth at 60 µN normal load for all specimens. Microwear resistance of SiC and Al_2O_3 is the highest followed by Al_2O_3–TiC, single-crystal, and polycrystalline Mn–Zn ferrite.

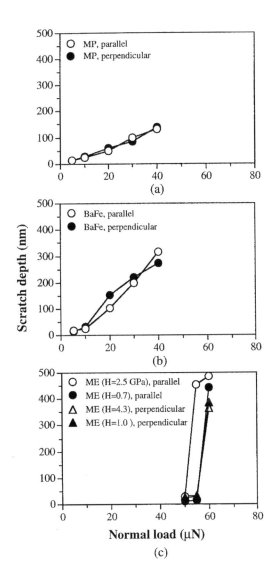

FIGURE 14.24 Scratch depth as a function of normal load after one scratch cycle for (a) MP, (b) BaFe, and (c) ME tapes along parallel and perpendicular directions with respect to the longitudinal axis of the tape. (From Bhushan, B. and Koinkar, V.N., 1995, *Wear* 180, 9–16. With permission.)

Next, wear experiments were conducted for multiple cycles. Figure 14.31 shows the two-dimensional gray-scale plots and corresponding section plot (on top of each gray-scale plot), taken at a location shown by an arrow for Al_2O_3 (left column) and Al_2O_3–TiC (right column) specimen obtained at a normal load of 20 μN and at a different number of scan cycles. The central regions (2 × 2 μm) show the wear mark generated after a different number of cycles. Note the difference in the vertical scale of the gray scale and section plots. The Al_2O_3 specimen shows that wear initiates at the porous holes or defects present on the surface. Wear progresses at these locations as a function of number of cycles. In the porous hole free region, microwear resistance is higher. In the case of the Al_2O_3–TiC specimen for about five scan cycles, the microwear resistance is higher at the TiC grains and is lower at the Al_2O_3 grains. The TiC grains are removed from the wear mark after five scan cycles. This indicates that microwear resistance of multiphase materials depends upon the individual grain properties. Evolution of wear is uniform within the wear mark for ferrite specimens. Figure 14.32 shows a plot of wear depth as a function of number of cycles at a normal load of 20 μN for all specimens. The Al_2O_3 specimen then reveals highest microwear resistance followed by SiC, Al_2O_3–TiC, polycrystalline and single crystal Mn–Zn ferrite. Wear resistance of Al_2O_3–TiC is inferior to that of Al_2O_3. Chu et al. (1992) studied friction and wear behavior of the single-phase and

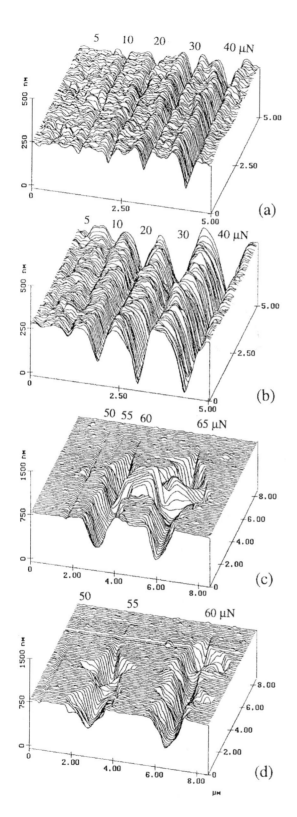

FIGURE 14.25 Surface maps for scratched (a) MP, (b) BaFe, (c) ME ($H = 0.7$ GPa), and (d) ME ($H = 2.5$ GPa) tapes along parallel direction. Normal loads used for scratching for one cycle are listed in the figure. (From Bhushan, B. and Koinkar, V.N., 1995, *Wear* 180, 9–16. With permission.)

multiphase ceramic materials and found that wear resistance of multi-phase materials was poorer than single-phase materials. Multiphase materials have more material flaws than the single-phase material. The differences in thermal and mechanical properties between the two phases may lead to cracking during processing, machining, or use.

14.5.3.2 Magnetic Media

Figure 14.33 shows the wear depth as a function of load for one cycle for the polished, unlubricated, and lubricated disks (Bhushan et al., 1994c). Figure 14.34 shows profiles of the wear scars generated on unlubricated disk. The normal force for the imaging was about $0.5 \, \mu N$ and the loads used for the wear were 20, 50, 80, and $100 \, \mu N$ as indicated in the figure. We note that wear takes place relatively uniformly across the disk surface and essentially independent of the lubrication for the disks studied. For both lubricated and unlubricated disks, the wear depth increases slowly with load at low loads with almost the same wear rate. As the load is increased to about $60 \, \mu N$, wear increases rapidly with load. The wear depth at $50 \, \mu N$ is about 14 nm, slightly less than the thickness of the carbon film. The rapid increase of wear with load at loads larger than $60 \, \mu N$ is an indication of the breakdown of the carbon coating on the disk surface.

Figure 14.35 shows the wear depth as a function of number of cycles for the polished disks (lubricated and unlubricated). Again, for both unlubricated and lubricated disks, wear initially takes place slowly with a sudden increase between 40 and 50 cycles at $10 \, \mu N$. The sudden increase occurred after 10 cycles at $20 \, \mu N$. This rapid increase is associated with the breakdown of the carbon coating. The wear profiles at various cycles are shown in Figure 14.36 for a polished, unlubricated disk at a normal load of $20 \, \mu N$. Wear is not uniform and the wear is largely initiated at the texture grooves present on the disk surface. This indicates that surface defects strongly affect the wear rate.

Hard amorphous carbon coating controls the wear performance of magnetic disks. A thick coating is desirable for long durability; however, to achieve ever-increasingly high recording densities, it is necessary to use as thin a coating as possible. Bhushan and Koinkar (1995e) studied the effect of coating thickness of sputtered carbon on the microwear performance. The critical number of cycles (wear life) above which wear increases rapidly increases with an increase in the carbon film thickness, Figure 14.37. Film as thin as 5 nm does provide some wear protection. As expected, a thicker film is superior in wear protection. The concern with films of thicknesses 5 and 10 nm is whether these ultrathin films are continuous or deposited as islands, which is undesirable from corrosion point of view. Based on surface mapping of coatings using Auger electron spectroscopy, they concluded that even the thinnest 5-nm-thick film is essentially continuous with $0.2 \, \mu m$ spatial resolution. Koinkar and Bhushan (1997b) compared the microtribological properties of 20-nm-thick hard amorphous carbon coatings deposited by sputtering, ion beam, and filtered cathodic arc processes. Wear depths as a function of number of cycles for various coatings are plotted in Figure 14.38. The data for silicon are plotted for comparison. Cathodic arc coating exhibits highest wear resistance followed by ion beam, sputtered, and silicon. Differences in kinetic energy of deposition species in different deposition processes affect the coating hardness and adhesion between coating and substrate, which in turn affect tribological and mechanical properties. Hardness data of various coatings are presented in a later section.

Wear depths as a function of normal load for MP, BaFe, and ME tapes along the parallel direction are plotted in Figure 14.39 (Bhushan et al., 1995d). For the ME tape, there is negligible wear until the normal load of about $50 \, \mu N$; above this load the magnetic coating fails rapidly. This observation is consistent with the scratch data. Wear depths as a function of number of cycles for MP, BaFe, and ME tapes are shown in Figure 14.40. For the MP and BaFe particulate tapes, wear rates appear to be independent of the particulate density. Again, as observed in the scratch testing, wear rate of BaFe tapes is higher than that for MP tapes. ME tapes are much more wear resistant than the particulate tapes. However, the failure of ME tapes is catastrophic as observed in scratch testing. Wear studies were performed along and across the longitudinal tape direction in high- and low-hardness regions. At the high-hardness regions of the ME tapes, failure occurs at lower loads. A directionality effect, again, may arise from the columnar structure of the ME films (Bhushan, 1992; Hibst, 1993). Wear profiles at various cycles at a normal load

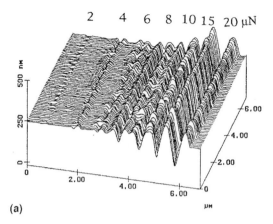

2   4   6   8   10   15   20 μN

500 nm

250

0

(a)

0   2.00   4.00   6.00   μM

2.00

4.00

6.00

FIGURE 14.26 Surface profiles for scratched PET film (a) polymer region, (b) ceramic particulate region. The loads used for various scratches at ten cycles are indicated in the plots. (From Bhushan, B. and Koinkar, V.N., 1995, *Tribol. Trans.* 38, 119–127. With permission.)

of 2 μN for MP and at 20 μN for ME tapes are shown in Figure 14.41. For the particulate tapes, we note that polymer gets removed before the particulates do (Figure 14.41a). Based on the wear profiles of the ME tape shown in Figure 14.41a, we note that most wear occurs between 50 to 60 cycles which shows the catastrophic removal of the coating. It was also observed that wear debris generated during wear test in all cases is loose and can easily be removed from the scan area at light loads (~0.3 μN).

The average wear depth as a function of load for a PET film is shown in Figure 14.42. Again, the wear depth increases linearly with load. Figure 14.43 shows the average wear depth as a function of number of cycles. The observed wear rate is approximately constant. PET tape substrate consists of particles sticking out on its surface to facilitate winding. Figure 14.44 shows the wear profiles as a function of number of cycles at 1 μN load on the PET film in the nonparticulate and particulate regions (Bhushan et al., 1995a). We note that polymeric materials tear in microwear tests. The particles do not wear readily at 1 μN. Polymer around the particles is removed but the particles remain intact. Wear in the particulate region is much smaller than that in the polymer region. We will see later that nanohardness of the particulate region is about 1.4 GPa compared with 0.3 GPa in the nonparticulate region (Table 14.4).

14.5.3.3   Silicon

Wear data on selected Si samples are presented in Table 14.5 and the wear profiles at 40 μN of load are shown in Figure 14.45 (Bhushan et al., 1994c). (Also see Bhushan et al., 1993c, 1997a,b; Sundararajan and Bhushan, 1998). Virgin silicon is poor in wear resistance. It clearly needs to be treated for wear applications. PECVD oxide samples had the largest wear resistance followed by dry-oxidized, wet-oxidized and ion-implanted samples. Bhushan et al. (1994c) observed wear debris in the wear zone just after the wear test which could be easily removed by scanning the worn region. It suggests that wear debris is loose. They further studied the wear resistance of ion-implanted samples, Figure 14.46. For tests conducted at various loads on Si(111) and C$^+$-implanted Si(111), they found that wear resistance of implanted sample is slightly poorer than that of virgin Si up to about 80 μN. Above 80 μN, the wear resistance of implanted Si improves. As they continued to run tests at 40 μN for a larger number of cycles, an implanted sample exhibits higher wear resistance than an unimplanted sample. Miyamoto et al. (1993) have also reported that damage from the implantation in the top layer results in poorer wear resistance; however, an implanted zone at the subsurface is more wear resistant than the virgin Si.

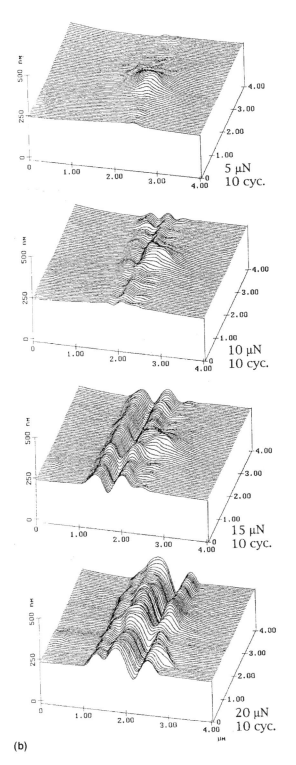

(b)

FIGURE 14.26

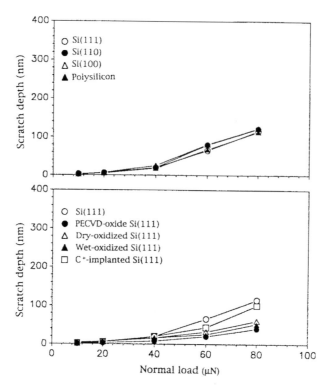

FIGURE 14.27 Scratch depth as a function of normal load after ten cycles for virgin, treated, and coated Si surfaces. (From Bhushan, B. and Koinkar, V.N., 1994, *J. Appl. Phys.* 75, 5741–5746. With permission.)

14.6 Indentation

14.6.1 Picoscale Indentation

Bhushan and Ruan (1994a) measured indentability of magnetic tapes at increasing loads on a picoscale, Figure 14.47. In this figure, the vertical axis represents the cantilever tip deflection and the horizontal axis represents the vertical position (Z) of the sample. The "extending" and "retracting" curves correspond to the sample being moved toward or away from the cantilever tip, respectively. In this experiment, as the sample surface approaches the AFM tip a fraction of a nm away from the sample (point A), the cantilever bends toward the sample (part B) because of attractive forces between the tip and sample. As we continue the forward position of the sample, it pushes the cantilever back through its original rest position (point of zero applied load) entering the repulsive region (or loading portion) of the force curve. As the sample is retracted, the cantilever deflection decreases. At point D in the retracting curve, the sample is disengaged from the tip. Before the disengagement, the tip is pulled toward the sample after the zero deflection point of the force curve (point C) because of attractive forces (van der Waals forces and longer-range meniscus forces). A thin layer of liquid, such as liquid lubricant and condensations of water vapor from ambient, will give rise to capillary forces that act to draw the tip toward the sample at small separations. The horizontal shift between the loading and unloading curves results from the hysteresis in the PZT tube.

The left portion of the curve shows the tip deflection as a function of the sample traveling distance during sample–tip contact, which would be equal to each other for a rigid sample. However, if the tip indents into the sample, the tip deflection would be less than the sample traveling distance, or, in other words, the slope of the line would be less than 1. In Figure 14.47, we note that line in the left portion of

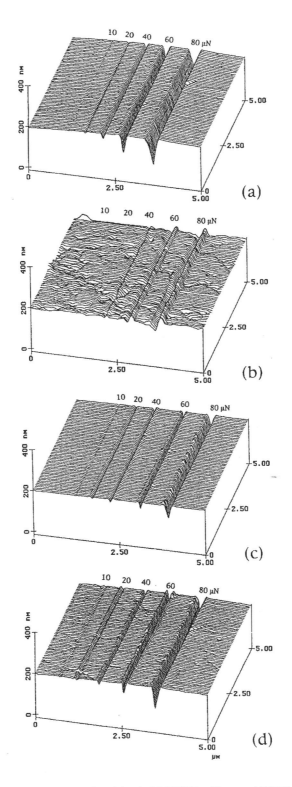

FIGURE 14.28 Surface profiles for scratched (a) Si(111), (b) PECVD oxide-coated Si(111), (c) dry-oxidized Si(111), and (d) C^+-implanted Si(111). Normal loads used for various scratches at ten cycles are indicated in the plot. (From Bhushan, B. and Koinkar, V.N., 1994, *J. Appl. Phys.* 75, 5741–5746. With permission.)

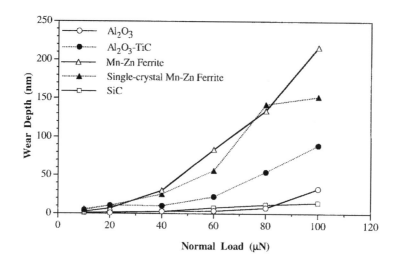

FIGURE 14.29 Wear depth as a function of normal load after one scan cycle for Al_2O_3, Al_2O_3–TiC, polycrystalline Mn–Zn ferrite, single-crystal Mn–Zn ferrite, and SiC. (From Koinkar, V.N. and Bhushan, B., 1996, *Wear* 202, 110–122. With permission.)

the figure is curved with a slope of less than 1 shortly after the sample touches the tip, which suggests that the tip has indented the sample. Later, the slope is equal to 1 suggesting that the tip no longer indents the sample. This observation indicates that the tape surface is soft locally (polymer-rich) but it is hard (as a result of magnetic particles) underneath. Since the curves in the extending and retracting modes are identical, the indentation is elastic up to at a maximum load of about 22 nN used in the measurements.

A shown in Figure 14.48 for unlubricated and lubricated textured disks, the slope of the deflection curves is 1 and remains constant as the disks touch and continue to push the AFM tip. The disks are not indented.

14.6.2 Nanoscale Indentation

Nanoscale mechanical properties of magnetic media and single-crystal silicon have been measured by Bhushan et al. (1994a–d, 1995a–e, 1997a, 1998) and Koinkar and Bhushan (1996a, 1997b). Bhushan et al. (1994b) have reported that indentation hardness with a penetration depth as low as 1 nm can be measured using AFM. Figures 14.49a and show the gray-scale plots and line plots of inverted images of indents made on the as-received Si(111) at normal loads of 60, 65, 70, and 100 μN. Triangular indents can be clearly observed with very shallow depths. It is found that below a normal load of 60 μN indents are unobservable. At a normal load of 60 μN indents are observed and the depth of penetration is about 1 nm. As we increase the normal load, the indents become clearer and indentation depth increases. Figure 14.50 shows a plot of hardness and normal load as a function of indentation depth. The indentation depth increases with an increase in normal load. We note that the hardness at a small indentation depth of 2.5 nm is 16.6 GPa and it drops to a value of 11.7 GPa at a depth of 7 nm and normal load of 100 μN. The observed drop in the hardness values at higher loads (Figure 14.50) is comparable to the trends in the nanohardness data measured using a commercial nanoindenter by Pharr et al. (1990) and Page et al. (1992). Pharr et al. (1990) have reported a value of 11.4 GPa at large indentation depths (>50 nm) and Page et al. (1992) have reported a value of about 14.2 GPa at an indentation depth of about 20 nm and monotonic decrease in hardness to about 7.5 GPa at indentation depths of about 500 nm. Reported values of microhardness values of silicon is about 9 to 10 GPa (Anonymous, 1988). If the silicon material is used at very light loads such as in microsystems, the high hardness of surface films would protect the surface until it is worn.

Higher hardness values obtained in low-load indentation may arise from the observed pressure-induced phase transformation during the nanoindentation (Pharr, 1991; Callahan and Morris, 1992). Additional increase in the hardness at an even lower indentation depth of 2.5 nm reported here may arise from the contribution by complex chemical films (not from native oxide films) present on the silicon surface. At small volumes, there is a high probability that indentation would be made into a region that was initially dislocation free. Furthermore, at small volumes, it is believed that there is an increase in the stress necessary to operate dislocation sources (Gane and Cox, 1970; Sargent, 1986). These are some of the plausible explanations for the increase in hardness at smaller volumes.

Nanohardness values of virgin, coated, and treated silicon samples are presented in Table 14.5. Coatings and treatments improved nanohardness of silicon. We note that dry-oxidized and PECVD oxide films are harder than wet-oxidized films, as these films may be porous (Bhushan and Venkatesan, 1993c). High hardness of oxidized films may be responsible for lower wear and scratch resistance (Table 14.5). Hardness values of virgin and C^+-implanted Si(111) at various indentation depths (normal loads) are presented in Figure 14.51. We note that the surface layer of the implanted zone is much harder than that of the subsurface, and may be brittle leading to higher wear on the surface. Subsurface of the implanted zone is harder than the virgin silicon, resulting in high wear resistance (data presented earlier, Figure 14.46).

Bhushan et al. (1994c) measured nanohardness of polished thin-film disks at loads of 80, 100, and 140 µN loads, Figure 14.52. Hardness values were 21 GPa (10 nm), 21 GPa (15 nm), and 9 GPa (40 nm); the depths of indentation are shown in the parenthesis. The hardness value at 80 and 100 µN is much higher than at 140 µN. This is expected since the indentation depth is only about 15 nm at 100 µN, which is smaller than the thickness of DLC coating (~20 to 30 nm). The hardness value at lower loads is primarily the value of the carbon coating. The hardness value at higher loads is primarily the value of the magnetic film, which is softer than the carbon coating. This result is consistent with the scratch and wear data discussed previously.

For the case of hardness measurements made on thin-film rigid disk at low loads, the indentation depth is on the same order as the variation in the surface roughness. For accurate measurements of indentation size and depth, it is desirable to subtract the original (unindented) profile from the indented profile. Bhushan et al. (1994b,c, 1995d), and Lu and Bogy (1995) developed an algorithm for this purpose. Because of hysteresis, a translational shift in the sample plane occurs during the scanning period, resulting in a shift between images captured before and after indentation. Therefore, the image needs to be shifted for perfect overlap before subtraction can be performed. For this purpose, a small region on the original image was selected and the corresponding region in the indented image was found by maximizing the correlation between the two regions. (Profiles were plane-fitted before subtraction.) Once two regions were identified, overlapped areas between the two images were determined and the original image was shifted with the required translational shift and then subtracted from the indented image. An example of profiles before and after subtraction is shown in Figure 14.53. The indent on the subtracted image can be measured easily. At a normal load of 140 µN the hardness value of polished, unlubricated magnetic thin-film rigid disk (σ roughness = 3.3 nm) is 9.0 GPa and the indentation depth is 40 nm.

Figure 14.54a shows the hardness as a function of residual depth for three types of 100-nm-thick amorphous carbon coatings deposited on silicon by sputtering, ion beam, and cathodic arc processes (Kulkarni and Bhushan, 1997). Data on uncoated silicon are also included for comparisons. The cathodic arc carbon coating exhibits highest hardness of about 24.9 GPa, whereas the sputtered and ion beam carbon coatings exhibit hardness values of 17.2 and 15.2 GPa, respectively. The hardness of Si(100) is 13.2 GPa. High hardness of the cathodic arc carbon coating explains its high wear resistance, reported earlier. Figure 14.54b shows the elastic modulus as a function of residual depth for various samples. The cathodic arc coating exhibits the highest elastic modulus. Its elastic modulus decreases with an increasing residual depth, while the elastic moduli for the other carbon coatings remain almost constant. In general, hardness and elastic modulus of coatings are strongly influenced by their crystalline structure, stoichiometry, and growth characteristics which depend on the deposition parameters. Mechanical properties of carbon coatings have been known to change over a wide range with sp^3–sp^2 bonding ratio and amount

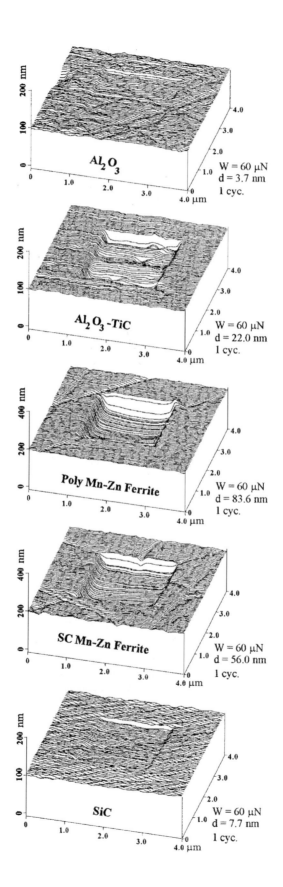

of hydrogen. Hydrogen is believed to play an important role in the bonding configuration of carbon atoms by helping to stabilize tetrahedral coordination of carbon atoms.

14.6.3 Localized Surface Elasticity

AFM can be used in the force modulation mode to measure surface elasticities in a nondestructive manner (Maivald et al., 1991; DeVecchio and Bhushan, 1997; Scherer et al., 1997). By using this technique, it is possible to measure quantitatively the elasticity of soft and compliant materials with penetration depths of less than 100 nm. This technique has been successfully used to get localized elasticity maps of particulate magnetic tapes. An elasticity map of a tape can be used to identify relative distribution of hard magnetic/nonmagnetic ceramic particles and the polymeric binder on the tape surface, which has an effect on friction and stiction at the head-tape interface.

Figure 14.55 shows surface height and elasticity maps on a sample of metal particle magnetic tape. The elasticity image reveals sharp variations in surface elasticity due to the composite nature of the film. As can be clearly seen, regions of high elasticity do not always correspond to high or low topography. Figure 14.56 shows measurements on three different formulations of magnetic tapes with a larger scan size. It is seen that the number of stiff regions on the second two tapes is higher than that on the first tape. The trend in increasing number of stiff regions has been correlated to reduced stiction problems in these three tapes (Bhushan et al., 1997c). Based on a Hertzian elastic-contact analysis, the static indentation depth of these samples during the force modulation scan is estimated to be about 1 nm. We conclude that the contrast seen is influenced most strongly by material properties in the top few nanometers, independent of the composite structure beneath the surface layer.

14.7 Detection of Material Transfer

Ruan and Bhushan (1993) conducted indentation on C_{60}-rich fullerene films (Bhushan et al., 1993a,b) using AFM in the force calibration mode. They observed transfer of fullerene molecules to the AFM tip during indentation. The fullerene molecules transferred to the AFM tip were subsequently transported to a diamond surface when the diamond sample was scanned with the contaminated tip, Figure 14.57. The discontinuity in the tip deflection shown in Figure 14.57a is due to the transferred fullerene molecules to the AFM tip. After repeated sliding of the tip against the diamond surface, the discontinuity in the tip deflection disappeared as shown in Figure 14.57b. This demonstrates the capability of detection of material transfer on a molecular scale using AFM.

14.8 Lubrication

The boundary films are formed by physical adsorption, chemical adsorption, and chemical reaction. The physisorbed film can be either monomolecular or polymolecular thick. The chemisorbed films are monomolecular, but stoichiometric films formed by chemical reaction can have a large film thickness. In general, the stability and durability of surface films decrease in the following order: chemical reaction films, chemisorbed films, and physisorbed films. A good boundary lubricant should have a high degree of interaction between its molecules and the sliding surface. As a general rule, liquids are good lubricants when they are polar and thus able to grip solid surfaces (or be adsorbed). Polar lubricants contain reactive functional groups with low ionization potential or groups having high polarizability (Bhushan, 1993). Boundary lubrication properties of lubricants are also dependent upon the molecular conformation and lubricant spreading (Novotny et al., 1989; Novotny, 1990; Mate and Novotny, 1991; Mate, 1992a).

FIGURE 14.30 Surface profiles showing the worn region (center 2×2 µm) after one scan cycle at a normal load of 60 µN for Al_2O_3, Al_2O_3–TiC, polycrystalline Mn–Zn ferrite, single-crystal Mn–Zn ferrite, and SiC. (From Koinkar, V.N. and Bhushan, B., 1996, *Wear* 202, 110–122. With permission.)

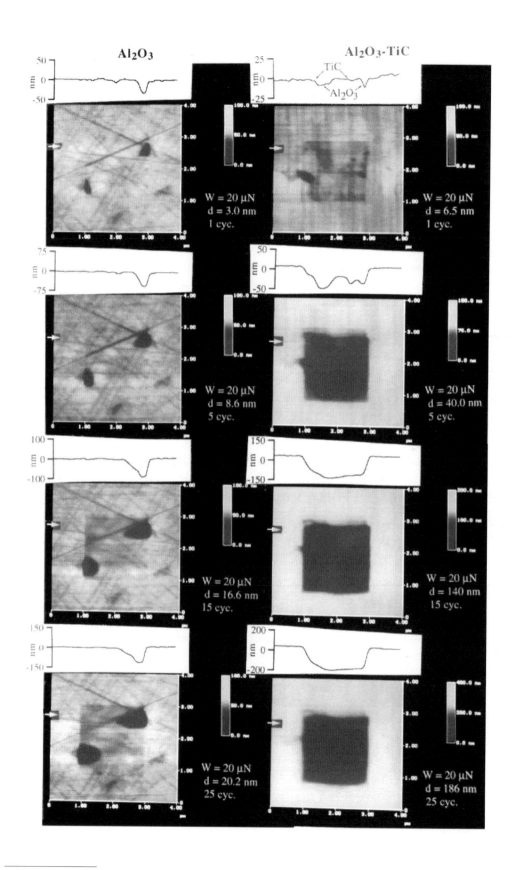

Al₂O₃ Al₂O₃-TiC

Mechanical interactions between the magnetic head and the medium in magnetic storage devices are minimized by the lubrication of the magnetic medium. The primary function of the lubricant is to reduce the wear of the medium and to ensure that friction remains low throughout the operation of the drive. The main challenge, however, in selecting the best candidate for a specific surface is to find a material that provides an acceptable wear protection for the entire life of the product, which can be several years in duration. There are many requirements that a lubricant must satisfy in order to guarantee an acceptable life performance. An optimum lubricant thickness is one of these requirements. If the lubricant film is too thick, excessive stiction and mechanical failure of the head–disk is observed. On the other hand, if the film is too thin, protection of the interface is compromised, and high friction and excessive wear will result in catastrophic failure. An acceptable lubricant must exhibit properties such as chemical inertness, low volatility, high thermal, oxidative and hydrolytic stability, shear stability, and good affinity to the magnetic medium surface.

Fatty acid esters are excellent boundary lubricants, and esters such as tridecyl stearate, butyl stearate, butyl palmitate, butyl myristate, stearic acid, and myrstic acid are commonly used as internal lubricants, roughly 1 to 7% by weight of the magnetic coating in particulate flexible media (tapes and particulate flexible disks). The fatty acids involved include those with acid groups with an even number of carbon atoms between C_{12} and C_{22}, with alcohols ranging from C_3 to C_{13}. These acids are all solids with melting points above the normal surface operating temperature of the magnetic media. This suggests that the decomposition products of the ester via lubrication chemistry during a head–flexible medium contact may be the key to lubrication.

Topical lubrication is used to reduce the wear of rigid disks and thin-film tapes. Perfluoropolyethers (PFPEs) are chemically the most stable lubricants with some boundary lubrication capability, and are most commonly used for topical lubrication of rigid disks. PFPEs commonly used include Fomblin Z and Fomblin Y lubricants, made by Ausimont, Milan, Italy; Krytox 143 AD, made by Dupont, U.S.; and Demnum S, made by Diakin, Japan; and their difunctional derivatives containing various reactive end groups e.g., hydroxyl or alcohol (Fomblin Z-Dol), piperonyl (Fomblin AM 2001), isocyanate (Fomblin Z-Disoc), and ester (Demnum SP). The difunctional derivatives are referred to as reactive (polar) PFPE lubricants. The chemical structures, molecular weights, and viscosities of various types of PFPE lubricants are given in Table 14.6. We note that rheological properties of thin films of lubricants are expected to be different from their bulk properties. Fomblin Z and Demnum S are linear PFPE, and Fomblin Y and Krytox 143 AD are branched PFPE, where the regularity of the chain is perturbed by –CF$_3$ side groups. The bulk viscosity of Fomblin Y and Krytox 143 AD is almost an order of magnitude higher than the Z-type. Fomblin Z is thermally decomposed more rapidly than Y (Bhushan, 1993). The molecular diameter is about 0.8 nm for these lubricant molecules. The monolayer thickness of these molecules depends on the molecular conformations of the polymer chain on the surface (Novotny et al., 1989; Mate and Novotny, 1991).

The adsorption of the lubricant molecules on a magnetic disk surface is due to van der Waals forces, which are too weak to offset the spin-off losses, or to arrest displacement of the lubricant by water or other ambient contaminants. Considering that these lubricating films are on the order of a monolayer thick and are required to function satisfactorily for the duration of several years, the task of developing a workable interface is quite formidable. An approach aiming at alleviating these shortcomings is to enhance the attachment of the molecules to the overcoat, which, for most cases, is sputtered carbon. There are basically two approaches which have been shown to be successful in bonding the monolayer

---

FIGURE 14.31 Gray-scale two-dimensional plots showing the worn region (center 2×2 μm) at a normal load of 20 μN and different number of scan cycles for Al_2O_3 and Al_2O_3–TiC. The two-dimensional section plots taken at a location shown by an arrow are shown on the top of corresponding gray-scale plot. Note the change in vertical scale for gray-scale and two-dimensional section plots. (From Koinkar, V.N. and Bhushan, B., 1996, *Wear* 202, 110–122. With permission.)

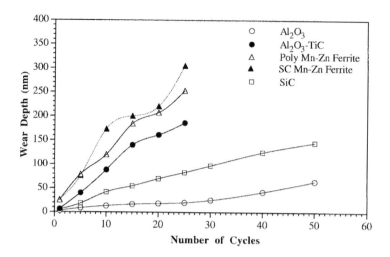

FIGURE 14.32 Wear depth as a function of number of cycles at a normal load of 20 μN for Al_2O_3, Al_2O_3–TiC, polycrystalline Mn–Zn ferrite, single-crystal (SC) Mn–Zn ferrite, and SiC. (From Koinkar, V.N. and Bhushan, B., 1996, *Wear* 202, 110–122. With permission.)

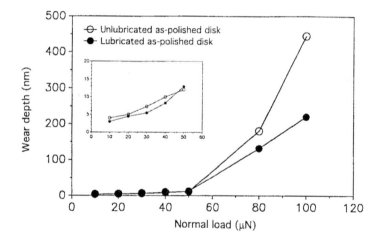

FIGURE 14.33 Wear depth as a function of normal load for polished, lubricated and unlubricated thin-film rigid disks after one cycle. (From Bhushan, B. et al., 1994, *Proc. Inst. Mech. Eng. Part J: J. Eng. Tribol.* 208, 17–29. With permission.)

to the carbon. The first relies on exposure of the disk lubricated with neutral PFPE to various forms of radiation, such as low-energy X ray (Heidemann and Wirth, 1984), nitrogen plasma (Homola et al., 1990), or far ultraviolet (e.g., 185 nm) (Saperstein and Lin, 1990). Another approach is to use chemically active PFPE molecules, where the various functional (reactive) end groups offer the opportunity of strong attachments to specific interface. These functional groups can react with surfaces and bond the lubricant to the disk surface, which reduces its loss due to spin off and evaporation. Bonding of lubricant to the disk surface depends upon the surface cleanliness. After lubrication, the disk is generally heated at 150°C for 30 min to 1 h to improve the bonding. If only a bonded lubrication is desired, the unbonded fraction can be removed by washing it off for 60 s with a nonfreon solvent (FC-72). Their main advantage is their ability to enhance durability without the problem of stiction usually associated with weakly bonded lubricants (Bhushan, 1996a).

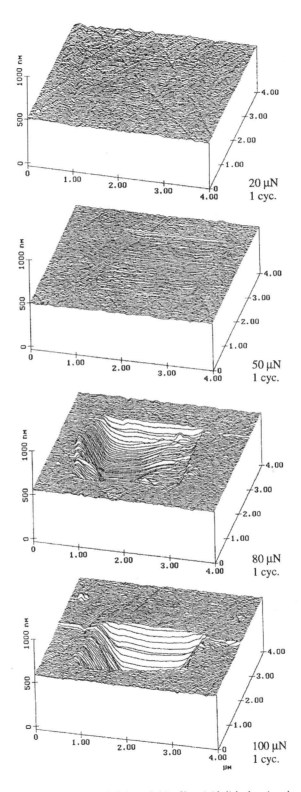

FIGURE 14.34 Surface maps of a polished, unlubricated thin-film rigid disk showing the worn region (center 2 × 2 μm) after one cycle. The normal loads are indicated in the figure. (From Bhushan, B. et al., 1994, *Proc. Inst. Mech. Eng. Part J: J. Eng. Tribol.* 208, 17–29. With permission.)

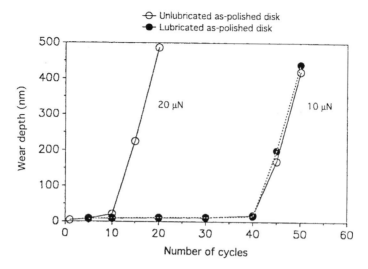

FIGURE 14.35 Wear depth as a function of number of cycles for polished, lubricated, and unlubricated thin-film rigid disks at 10 μN and for polished, unlubricated disk at 20 μN. (From Bhushan, B. et al., 1994, *Proc. Inst. Mech. Eng. Part J: J. Eng. Tribol.* 208, 17–29. With permission.)

Fomblin Y and Z are most commonly used for lubrication of particulate and thin-film rigid disks, respectively. Lubricants with lower viscosity (such as Fomblin Z-types) with polar end groups (Z-Dol and AM 2001) are most commonly used in thin-film disks in order to minimize stiction.

14.8.1 Imaging of Lubricant Molecules

Kaneko et al. (1991b) and Andoh et al. (1992) used scanning tunneling microscope (STM) to observe the configuration, adsorption, and mobility of perfluoropolyether lubricants on HOPG (defect-free and inactive surface), MoS_2, and sputtered carbon (active surface). (Also see Foster and Frommer, 1988.) Their study revealed that the lubricant molecules can easily change position and are mobile. They found that adsorption of the lubricant molecules on a sputtered carbon surface is more than that on the HOPG.

14.8.2 Measurement of Localized Lubricant Film Thickness

The local lubricant thickness is measured by Fourier transform infrared spectroscopy (FTIR), ellipsometry, angle-resolved X-ray photon spectroscopy (XPS), STM, and AFM. Ellipsometry and angle-resolved XPS have excellent vertical resolution, on the order of 0.1 nm, but their lateral resolutions are on the order of 1 and 0.2 mm, respectively. AFM can measure the thickness of the liquid film with a lateral resolution on the order of the tip radius, about 100 nm, which it is not possible to achieve by other techniques (Bhushan, 1993).

The schematic of AFM used for measurement for localized lubricant-film thickness by Mate et al. (Mate et al., 1989, 1990; Bhushan and Blackman, 1991) is shown in Figure 14.58. The lubricant thickness is obtained by measuring the forces on the tip as it approaches, contacts, and pushes through the liquid film. In the left part of Figure 14.58 is a diagram of an AFM tip interacting with a lubricant-covered particulate rigid disk. A typical force-vs.-distance curve for a tungsten tip of radius ~100 nm dipped into a disk surface coated with a perfluoropolyether lubricant is shown in Figure 14.59. As the surface approaches the tip, the liquid wicks up, causing a sharp onset of attractive force. The so-called meniscus force experienced by the tip is $\sim 4\pi R\gamma$, where R is the radius of the tip and γ is the surface tension of the liquid. In Figure 14.59, the attractive force measured is about 5×10^{-8} N. While in the liquid film, the forces on the lever remain constant until repulsive contact with the disk surface occurs. The distance between the sharp snap-in at the liquid surface and the hard wall of the substrate is proportional to the

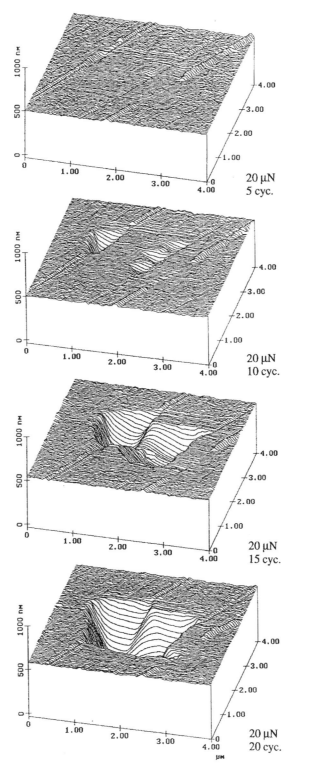

FIGURE 14.36 Surface maps of a polished, unlubricated thin-film rigid disk showing the worn region (center 2 × 2 μm) at 20 μN. The number of cycles are indicated in the figure. (From Bhushan, B. et al., 1994, *Proc. Inst. Mech. Eng. Part J: J. Eng. Tribol.* 208, 17–29. With permission.)

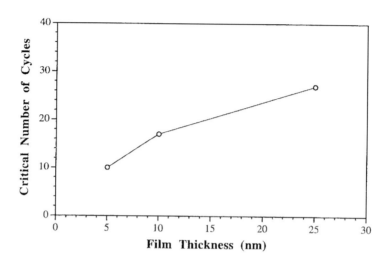

FIGURE 14.37 Critical number of cycles (above which wear increases rapidly) as a function of thickness of sputtered carbon coatings on magnetic disks at 20 μN load.(From Bhushan, B. and Koinkar, V.N., 1995, *Surf. Coat. Technol.* 76–77, 655–669. With permission.)

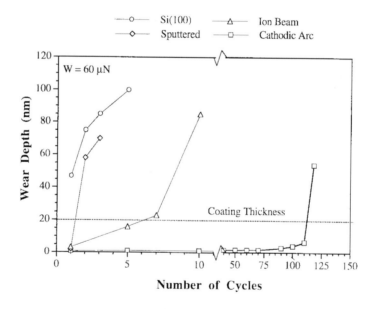

FIGURE 14.38 Wear depth as a function of number of cycles at 60 μN load for uncoated single-crystal silicon and 20-nm-thick coatings deposited by sputtering, ion beam, and cathodic arc carbon processes. (From Koinkar, V.N. and Bhushan, B., 1997, *Proc. Inst. Mech. Eng. Part J: J. Eng. Tribol.* 211, 365–372. With permission.)

lubricant thickness at that point. (The measured thickness is about 2 nm larger than the actual thickness due to a thin layer of lubricant wetting the tip, Mate et al., 1990). When the sample is withdrawn, the forces on the tip slowly decrease to zero as a long meniscus of liquid is drawn out from the surface.

Particulate disks were mapped by Mate et al. (1990) and Bhushan and Blackman (1991). The distribution of lubricant across an asperity was mapped by collecting force-vs.-distance curves with the AFM in a line across the surface. A particulate disk was coated with nominally 20 to 30 nm of lubricant, but by AFM the average thickness was 2.6 ± 1.2 nm. (The large standard deviation reflects the huge variation of lubricant thickness across the disk.) A large percentage of the lubricant is expected to reside below the

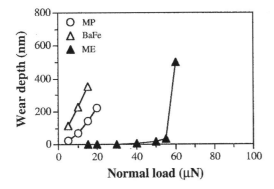

FIGURE 14.39 Wear depth as a function of normal load for three tapes in the parallel direction after one cycle. (From Bhushan, B. and Koinkar, V.N., 1995, *Wear* 181–183, 360–370. With permission.)

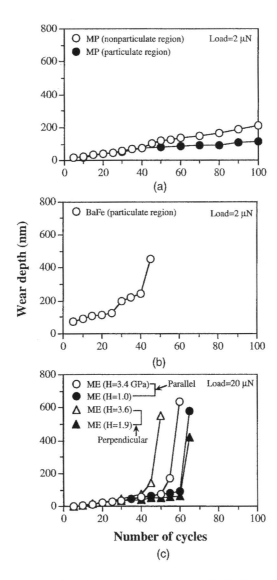

FIGURE 14.40 Wear depth as a function of number of cycles for (a) MP, (b) BaFe, and (c) ME tapes in different regions at normal loads indicated in the figure. Note a higher load used for the ME tape in (c). (From Bhushan, B. and Koinkar, V.N., 1995, *Wear* 181–183, 360–370. With permission.)

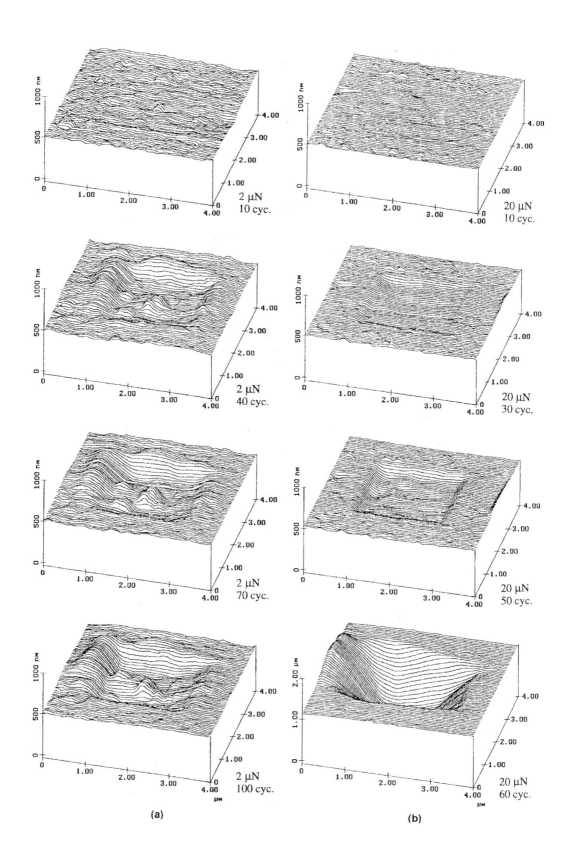

(a)　　　　　　　　　　　　　　　　(b)

surface in the pores. In Figure 14.60, we show histograms of lubricant thickness across three regions on the disk. The light part of the bar represents the hard wall of the substrate and the dark part of the top is the thickness of the lubricant. Each point on the histogram is from a single force-vs.-distance measurement, with points separated by 25-nm steps. The lubricant is not evenly distributed across the surface. In regions 1 and 2 there is more than twice as much lubricant than there is on the asperity (region 3). There are some points on the top and the side of the asperity which have no lubricant coating at all (Bhushan and Blackman, 1991).

14.8.3 Boundary Lubrication Studies

To study lubricant depletion during microscale measurements, Koinkar and Bhushan (1996b) conducted nanowear studies using Si_3N_4 tips. They measured friction as a function of number of cycles for virgin silicon and silicon surface lubricated with Z-15 and Z-Dol lubricants, Figure 14.61. An area of 1×1 μm was scanned at a normal force of 300 nN. Note that the friction force in a virgin silicon surface decreases in a few cycles after the natural oxide film present on silicon surface gets removed. In the case of Z-15-coated silicon sample, the friction force starts out to be low and then approaches that of an unlubricated silicon surface after a few cycles. The increase in friction of the lubricated sample suggests that the lubricant film gets worn and the silicon underneath is exposed. In the case of the Z-Dol-coated silicon sample, the friction force starts out to be low and remains low during the 100-cycle test. It suggest that Z-Dol does not get displaced/depleted as readily as Z-15. (Also see Bhushan et al., 1995g.) Microwear studies were also conducted using the diamond tip at various loads. Figure 14.62 shows the plots of wear depth as a function of normal load and Figure 14.63 shows the wear profiles of worn samples at 40 μN normal load. The Z-Dol lubricated sample exhibits better wear resistance than unlubricated and Z-15-lubricated silicon samples. Wear resistance of the Z-15-lubricated sample is little better than that of the unlubricated sample. The Z-15-lubricated sample shows the debris inside wear track. Since the Z-15 is a liquid lubricant, debris generated is held by the lubricant and it becomes sticky which moves inside the wear track and does damage, Figure 14.63.

14.9 Closure

AFM/FFM have been successfully used for measurements of surface roughness, friction, adhesion, scratching, wear, indentation, detection of material transfer, and lubrication on micro- to nanoscales. Commonly measured roughness parameters are scale dependent, requiring the need of scale-independent fractal parameters to characterize surface roughness. A generalized fractal analysis is presented which allows the characterization of surface roughness by two scale-independent parameters. Measurements of nanoscale friction of a freshly cleaved, highly oriented pyrolytic graphite exhibited the same periodicity as that of corresponding topography. However, the peaks in friction and those in corresponding topography were displaced relative to each other. Variations in atomic-scale friction and the observed displacement can be explained by the variations in interatomic forces in the normal and lateral directions. Local variation in microscale friction force is found to correspond to the local surface slope, suggesting that a ratchet mechanism is responsible for this variation. Directionality in the friction is observed on both micro- and macroscales which results from the surface preparation and asymmetrical asperities on the surface. Microscale friction is generally found to be smaller than the macroscale friction as there is less plowing contribution in microscale measurements.

FIGURE 14.41 Surface maps showing the worn region (center 2×2 μm) after various cycles of wear at (a) 2 μN for MP (particulate region) and at (b) 20 μN for ME ($H = 3.4$ GPa, parallel direction) tapes. Note a different vertical scale for the bottom profile of (b). (From Bhushan, B. and Koinkar, V.N., 1995, *Wear* 181–183, 360–370. With permission.)

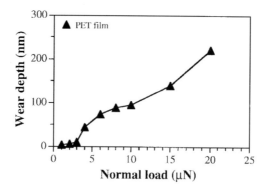

FIGURE 14.42 Wear depth as a function of normal load (after one cycle) for a PET film. (From Bhushan, B. and Koinkar, V.N., 1995, *Tribol. Trans.* 38, 119–127. With permission.)

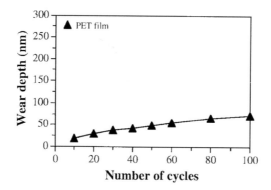

FIGURE 14.43 Wear depth as a function of number of cycles at 1 μN for a PET film. (From Bhushan, B. and Koinkar, V.N., 1995, *Tribol. Trans.* 38, 119–127. With permission.)

Wear rates for particulate magnetic tapes, polyester tape substrates, and single-crystal silicon are approximately constant for various loads and test durations. However, for magnetic disks and magnetic tapes with a multilayered thin-film structure, the wear of the DLC overcoat in the case of disks and magnetic layer in the case of tapes is catastrophic. Breakdown of thin films can be detected with AFM. Evolution of the wear has also been studied using AFM. We find that the wear is initiated at nanoscratches. Amorphous carbon films as thin as 5 nm are deposited as continuous films and exhibit some wear life. Wear life increases with an increase in film thickness. Carbon coatings deposited by cathodic arc process are superior in wear and mechanical properties followed by ion beam and sputtering processes. AFM has been modified for nanoindentation hardness measurements with depth of indentation as low as 1 nm. Scratching and indentation on nanoscales are powerful ways of evaluation of the mechanical integrity of ultrathin films. Detection of material transfer on a nanoscale is possible with AFM.

Measurement of lubricant film thickness with a lateral resolution on a nanoscale and boundary lubrication studies have been conducted using AFM. AFM/FFM friction experiments show that lubricants with polar (reactive) end groups dramatically increase the load or contact pressure that a liquid film can support before solid–solid contact and thus exhibit long durability.

Friction and wear on micro- and nanoscales have been found to be generally smaller compared with that at macroscales. Therefore, micro/nantribological studies helps define regimes for ultralow friction and zero wear.

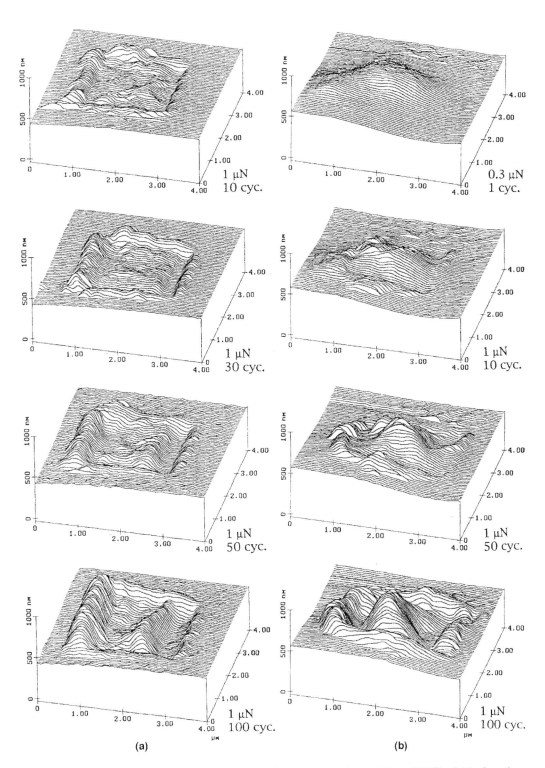

FIGURE 14.44 Surface maps showing the worn region (center 2×2 μm) at 1 μN for a PET film (a) in the polymer region, and (b) in the particulate region. The number of cycles are indicated in the figure. (From Bhushan, B. and Koinkar, V.N., 1995, *Tribol. Trans.* 38, 119–127. With permission.)

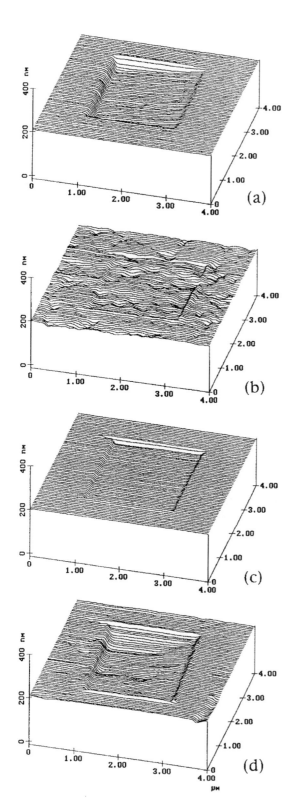

FIGURE 14.45 Surface maps showing the worn region (center 2×2 μm) after one cycle of wear at 40 μN load (a) Si(111), (b) PECVD-oxide-coated Si(111), (c) dry-oxidized Si(111), and (d) C^+-implanted Si(111). (From Bhushan, B. and Koinkar, V.N., 1994, *J. Appl. Phys.* 75, 5741–5746. With permission.)

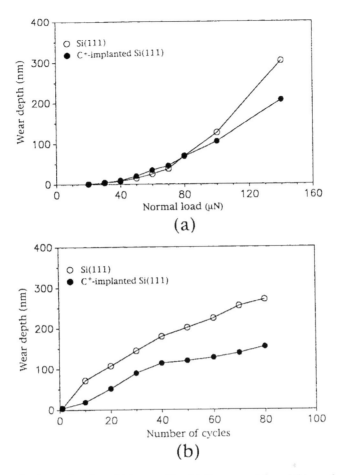

FIGURE 14.46 Wear depth as a function of (a) normal load (after one cycle), and (b) number of cycles (normal load = 40 μN) for Si(111) and C⁺-implanted Si(111). (From Bhushan, B. and Koinkar, V.N., 1994, *J. Appl. Phys.* 75, 5741–5746. With permission.)

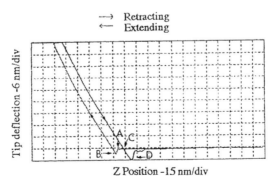

FIGURE 14.47 Tip deflection (normal force) as a function of Z (separation distance) curve for an MP tape. The spring constant of the cantilever used was 0.4 N/m. (From Bhushan, B. and Ruan, J., 1994, *ASME J. Tribol.* 116, 389–396. With permission.)

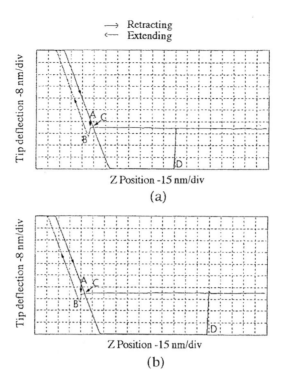

$\longrightarrow$ Retracting
$\longleftarrow$ Extending

Tip deflection -8 nm/div

Z Position -15 nm/div

(a)

Tip deflection -8 nm/div

Z Position -15 nm/div

(b)

FIGURE 14.48 Tip deflection (normal force) as a function of Z (separation distance) curve for (a) unlubricated and (b) lubricated textured thin-film rigid disks. The pull-off force is 42 nN for the unlubricated disk and 64 nN for the lubricated disk calculated from the horizontal distance between points C and D and the cantilever spring constant of 0.4 N/m. (From Bhushan, B. and Ruan, J., 1994, *ASME J. Tribol.* 116, 389–396. With permission.)

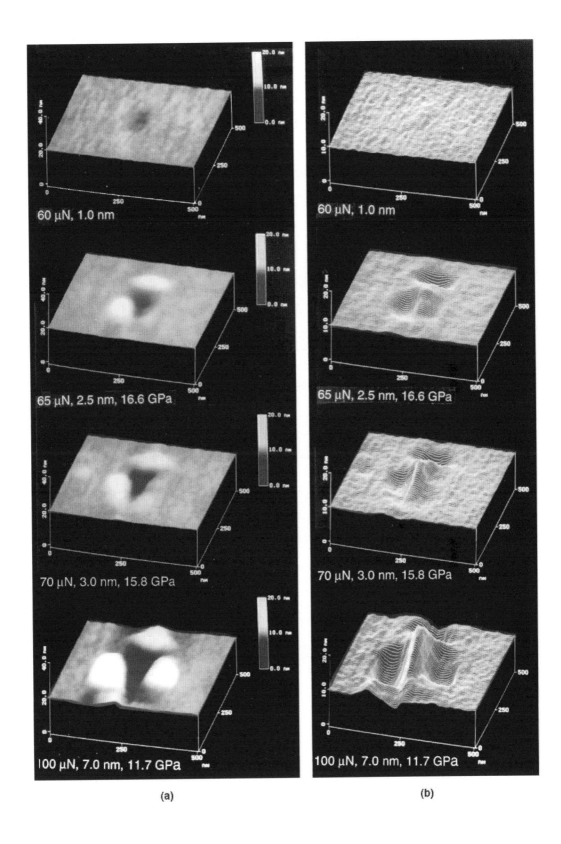

FIGURE 14.49 (a) Gray-scale plots and (b) line plots of inverted images of indentation marks on the as-received Si(111) sample at various normal loads. Loads, indentation depths, and hardness values are listed in the figure. (From Bhushan, B. and Koinkar, V.N., 1994, *Appl. Phys. Lett.* 64, 1653–1655. With permission.)

747

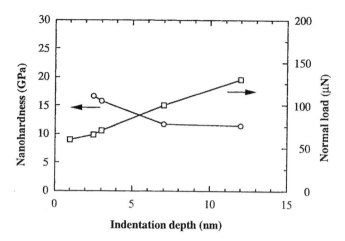

FIGURE 14.50 Nanohardness and normal load as a function of residual indentation depth for the as-received Si(111) sample. (From Bhushan, B. and Koinkar, V.N., 1994, *Appl. Phys. Lett.* 64, 1653–1655. With permission.)

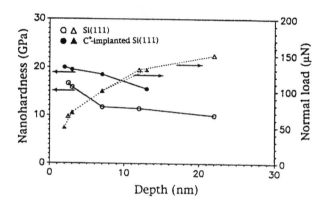

FIGURE 14.51 Nanohardness and normal load as function of residual indentation depth for virgin and C⁺-implanted Si(111) sample. (From Bhushan, B. and Koinkar, V.N., 1994, *J. Appl. Phys.* 75, 5741–5746. With permission.)

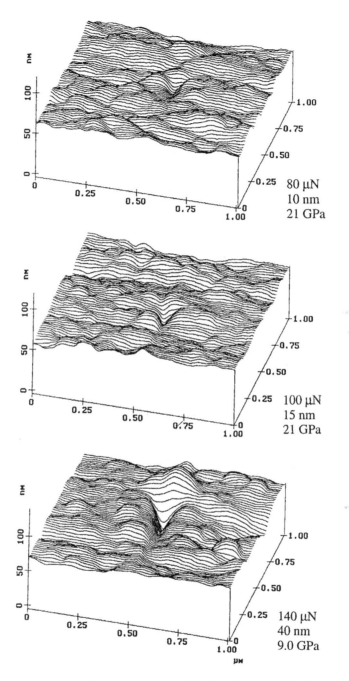

FIGURE 14.52 Nanoindentation marks generated on a polished, unlubricated thin-film rigid disk. The normal load used in the indentation, the indentation depths, and the hardness values are indicated in the figure. (From Bhushan, B. et al., 1994, *Proc. Inst. Mech. Eng. Part J: J. Eng. Tribol.* 208, 17–29. With permission.)

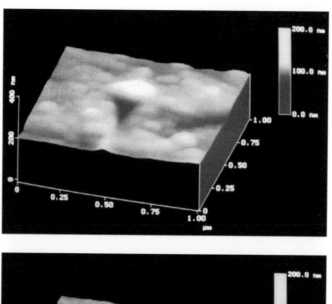

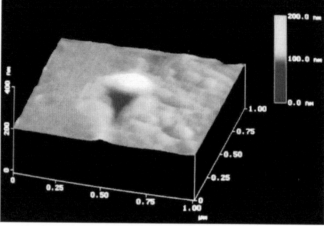

FIGURE 14.53 Images with nanoindentation marks generated on a polished, unlubricated thin-film rigid disk at 140 μN (a) before subtraction and (b) after subtraction. (From Bhushan, B. et al., 1994, *Proc. Inst. Mech. Eng. Part J: J. Eng. Tribol.* 208, 17–29. With permission.)

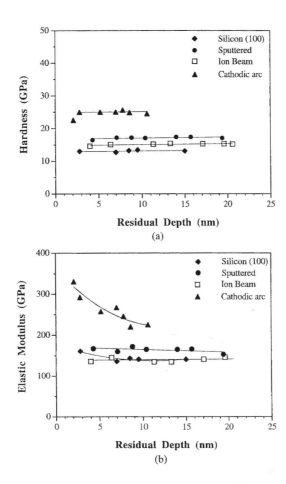

FIGURE 14.54   Nanohardness and elastic modulus as a function of residual indentation depth for Si(100) and 100-nm-thick coatings deposited by sputtering, ion beam, and cathodic arc processes. (From Kulkarni, A.V. and Bhushan, B., 1997, *J. Mater. Res.* 12, 2707–2714. With permission.)

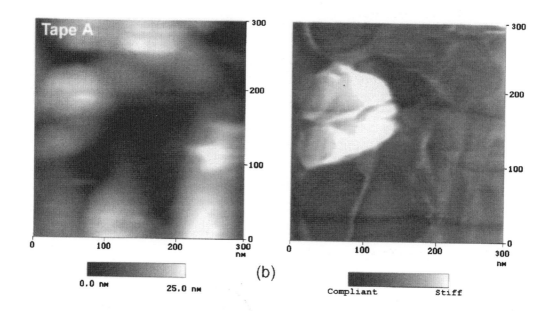

(b)

FIGURE 14.55 (a) Surface height and elasticity maps for an MP tape A (σ = 6.72 nm and P-V = 31.7 nm). σ and P-V refer to standard deviation of surface heights and peak-to-valley distance, respectively. The gray scale on the elasticity map is arbitrary. (From DeVecchio, D. and Bhushan, B., 1997, *Rev. Sci. Instrum.* 68, 4498–4505. With permission.)

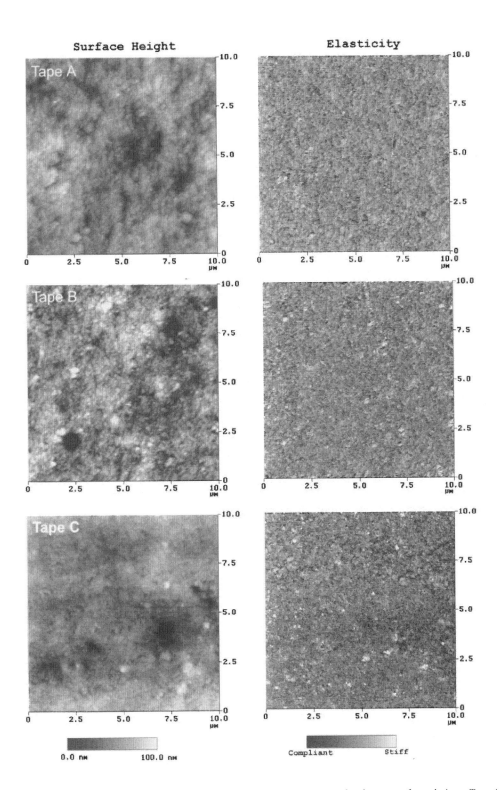

FIGURE 14.56 Surface height and elasticity maps for a scan size 10×10 μm for three tape formulations: Tape A ($\sigma = 15.1$ nm and P-V = 171 nm), tape B ($\sigma = 10.4$ nm and P-V = 136 nm), tape C ($\sigma = 14.4$ nm and P-V = 117 nm). (From DeVecchio, D. and Bhushan, B., 1997, *Rev. Sci. Instrum.* 68, 4498–4505. With permission.)

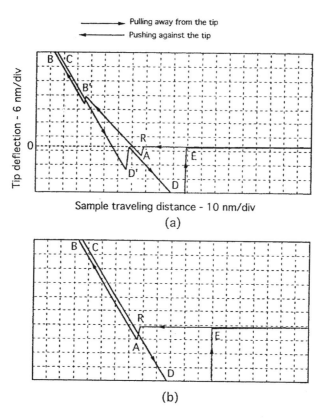

FIGURE 14.57 Tip deflection as a function of sample traveling distance curves for a natural diamond (a) right after the tip has been indented into as-deposited fullerene film and (b) after the tip has been scanned over the diamond surface for a short time. (From Ruan, J. and Bhushan, B., 1993, *J. Mater. Res.* 8, 3019–3022. With permission.)

References

Andoh, Y., Oguchi, S., Kaneko, R., and Miyamoto, T. (1992), "Evaluation of Very Thin Lubricant Films," *J. Phys. D: Appl. Phys.* 25, A71–A75.

Anonymous (1988), "Properties of Silicon," *EMIS Data Reviews Series*, No. 4, INSPEC, The Institution of Electrical Engineers, London.

Bhushan, B. (1992), *Mechanics and Reliability of Flexible Magnetic Media*, Springer-Verlag, New York.

Bhushan, B. (1993), "Magnetic Recording Surfaces," in *Characterization of Tribological Materials* (W.A. Glaeser, ed.), pp. 116–133, Butterworth-Heinemann, Boston.

Bhushan B. (1994), "Tribology of Magnetic Storage Systems," in *Handbook of Lubrication and Tribology*, Vol. 3, pp. 325–374, CRC Press, Boca Raton, FL.

Bhushan, B. (1995a), "Nanotribology and Its Applications to Magnetic Storage Devices and MEMS," in *Forces in Scanning Probe Methods* (H.J. Guntherodt, D. Anselmetti, and E. Meyer, eds.), Vol. E 286, pp. 367–395, Kluwer Academic, Dordrecht, The Netherlands..

Bhushan, B. (1995b), "Micro/Nanotribology and Its Application to Magnetic Storage Devices and MEMS," *Tribol. Int.* 28, 85–95.

Bhushan, B. (1996a), *Tribology and Mechanics of Magnetic Storage Devices*, 2nd ed., Springer-Verlag, New York.

Bhushan, B. (1996b), "Contact Mechanics of Rough Surfaces in Tribology: Single Asperity Contact," *Appl. Mech. Rev.* 49, 275–298.

Bhushan, B. (1997), *Micro/Nanotribology and Its Applications*, E330, Kluwer Academic Publishers, Dordrecht, Netherlands.

TABLE 14.6 Chemical Structure, Molecular Weight, and Viscosity of Perfluoropolyether Lubricants

Lubricant	Formula	Molecular Weight (Daltons)	Kinematic Viscosity cSt(mm²/s)
Fomblin Z-25	$CF_3-O-(CF_2-CF_2-O)_m-(CF_2-O)_n-CF_3$	12800	250
Fomblin Z-15	$CF_3-O-(CF_2-CF_2-O)_m-(CF_2-O)_n-CF_3$ (m/n ~2/3)	9100	150
Fomblin Z-03	$CF_3-O-(CF_2-CF_2-O)_m-(CF_2-O)_n-CF_3$	3600	30
Fomblin Z-DOL	$HO-CH_2-CF_2-O-(CF_2-CF_2-O)_m-(CF_2-O)_n-CF_2-CH_2-OH$	2000	80
Fomblin AM2001	Piperonyl$-O-CH_2-CF_2-O-(CF_2-CF_2-O)_m-(CF_2-O)_n-CF_2-O-$piperonyl[a]	2300	80
Fomblin Z-DISOC	$O-CN-C_6H_3-(CH_3)-NH-CO-CF_2-O-(CF_2-CF_2-O)n-(CF_2-O)m-CF_2-CO-NH-C_6H_3-(CH_3)-N-CO$	1500	160
Fomblin YR	CF_3 | $CF_3-O-(C-CF_2-O)_m(CF_2-O)_n-CF_3$ (m/n~40/1) | F	6800	1600
Demnum S-100	$CF_3-CF_2-CF_2-O-(CF_2-CF_2-CF_2-O)_m-CF_2-CF_3$	5600	250
Krytox 143AD	CF_3 CF | $CF_3-CF_2-CF_2-O-(C-CF_2-O)_m-CF_2-CF_3$ | F	2600	—

[a] 3,4-Methylenedioxybenzyl.

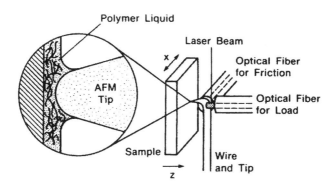

FIGURE 14.58 AFM (right) and the enlarged region and cross-sectional view of the particulate rigid disk surface and the AFM tip (left) during lubricant film thickness measurement. (From Mate, C.M., 1992, *Phys. Rev. Lett.* 68, 3323–3326. With permission.)

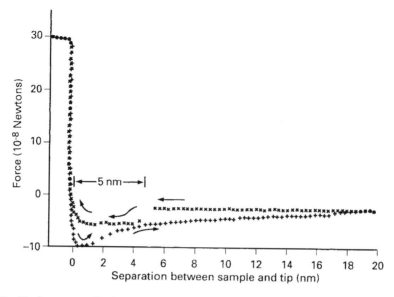

FIGURE 14.59 The force perpendicular to the surface acting on the tip as a function of the particulate disk sample position, as the sample is first brought into contact with the tip (x) and then pulled away (+). The sample is moved with a velocity of 100 nm/s, and the zero sample position is defined to be the position where the force on the tip is zero when in contact with the sample. A negative force indicates an attractive force. (From Bhushan, B. and Blackman, G.S., 1991, *ASME J. Tribol.* 113, 452–458. With permission.)

Bhushan, B. (1998a), "Micro/Nanotribology Using Atomic Force/Friction Force Microscopy: State of the Art," *Proc. Inst. Mech. Eng. Part J: J. Eng. Tribol.* 212, 1–18.

Bhushan, B. (1998b), *Tribology Issues and Opportunities in MEMS*, Kluwer Academic, Dordrecht, Netherlands.

Bhushan, B. (1998c), "Contact Mechanics of Rough Surfaces in Tribology: Multiple Asperity Contact," *Tribol. Lett.* 4, 1–35.

Bhushan, B. and Blackman, G.S. (1991), "Atomic Force Microscopy of Magnetic Rigid Disks and Sliders and Its Applications to Tribology," *ASME J. Tribol.* 113, 452–458.

Bhushan, B. and Majumdar, A. (1992), "Elastic-Plastic Contact Model for Bifractal Surfaces," *Wear* 153, 53–64.

Bhushan, B., Gupta, B.K., Van Cleef, G.W., Capp, C., and Coe, J.V. (1993a), "Sublimed C_{60} Films for Tribology," *Appl. Phys. Lett.* 62, 3253–3255.

Bhushan, B., Ruan, J., and Gupta, B.K. (1993b), "A Scanning Tunneling Microscopy Study of Fullerene Films," *J. Phys. D: Appl. Phys.* 26, 1319–1322.

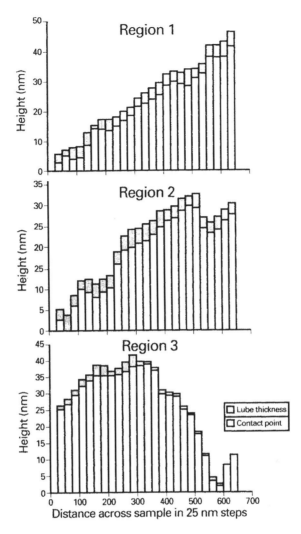

FIGURE 14.60 Histograms of lubricant thickness distribution at three different regions of a particulate disk. (From Bhushan, B. and Blackman, G.S., 1991, *ASME J. Tribol.* 113, 452–458. With permission.)

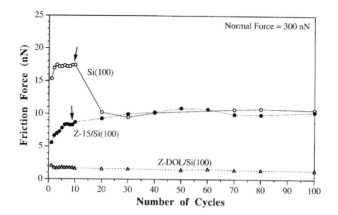

FIGURE 14.61   Friction force as a function of number of cycles using Si_3N_4 tip at a normal force of 300 nN for unlubricated and lubricated samples in ambient environment. Arrows in the figure indicate significant changes in the friction force because of removal of surface or lubricant film. (From Koinkar, V.N. and Bhushan, B., 1996, *J. Vac. Sci. Technol. A* 14, 2378–2391. With permission.)

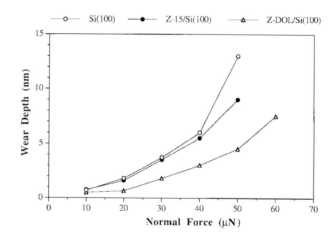

FIGURE 14.62   Wear depth as a function of normal force using diamond tip for unlubricated and lubricated samples after one cycle. (From Koinkar, V.N. and Bhushan, B., 1996, *J. Vac. Sci. Technol. A* 14, 2378–2391. With permission.)

Bhushan B. and Venkatesan, S. (1993c), "Friction and Wear Studies of Silicon in Sliding Contact with Thin-Film Magnetic Rigid Disks," *J. Mater. Res.* 8, 1611–1628.

Bhushan, B., and Koinkar, V.N. (1994a), "Tribological Studies of Silicon for Magnetic Recording Applications," *J. Appl. Phys.* 75, 5741–5746.

Bhushan, B. and Koinkar, V.N. (1994b), "Nanoindentation Hardness Measurements Using Atomic Force Microscopy," *Appl. Phys. Lett.* 64, 1653–1655.

Bhushan B., Koinkar, V.N., and Ruan, J. (1994c), "Microtribology of Magnetic Media," *Proc. Inst. Mech. Eng. Part J: J. Eng. Tribol.* 208, 17–29.

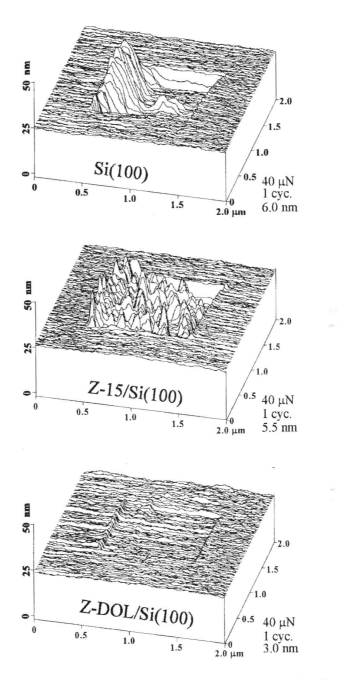

FIGURE 14.63 Wear maps for unlubricated and lubricated samples after microwear using diamond tip. Normal force used and wear depths are listed in the figure. (From Koinkar, V.N. and Bhushan, B., 1996, *J. Vac. Sci. Technol. A* 14, 2378–2391. With permission.)

Bhushan, B., Koinkar, V.N., and Ruan, J. (1994d), "Microtribological Studies by Using Atomic Force and Friction Force Microscopy and Its Applications," in *Determining Nanoscale Physical Properties of Materials by Microscopy and Spectroscopy* (M. Sarikaya, H.K. Wickramasinghe, and M. Isaacson, eds.), Vol. 332, pp. 93–98, Materials Research Society, Pittsburgh.

Bhushan, B. and Ruan, J. (1994e), "Atomic-Scale Friction Measurements Using Friction Force Microscopy: Part II — Application to Magnetic Media," *ASME J. Tribol.* 116, 389–396.

Bhushan, B. and Koinkar, V.N. (1995a), "Microtribology of PET Polymeric Films," *Tribol. Trans.* 38, 119–127.

Bhushan, B. and Koinkar, V.N. (1995b), "Macro and Microtribological Studies of CrO_2 Video Tapes," *Wear* 180, 9–16.

Bhushan, B. and Koinkar, V.N. (1995c), "Microtribology of Metal Particle, Barium Ferrite and Metal Evaporated Magnetic Tapes," *Wear* 181–183, 360–370.

Bhushan, B., Israelachvili, J.N., and Landman, U. (1995d), "Nanotribology: Friction, Wear and Lubrication at the Atomic Scale," *Nature* 374, 607–616.

Bhushan, B. and Koinkar, V.N. (1995e), "Microscale Mechanical and Tribological Characterization of Hard Amorphous Carbon Coatings as Thin as 5 nm for Magnetic Disks," *Surf. Coat. Technol.* 76–77, 655–669.

Bhushan, B. and Kulkarni, A.V. (1995f), "Effect of Normal Load on Microscale Friction Measurements," *Thin Solid Films* 278, 49–56.

Bhushan, B., Miyamoto, T., and Koinkar, V.N. (1995g), "Microscopic Friction between a Sharp Diamond Tip and Thin-Film Magnetic Rigid Disks by Friction Force Microscopy," *Adv. Info. Storage Syst.* 6, 151–161.

Bhushan, B., Kulkarni, A.V., Bonin, W., and Wyrobek, J.T. (1996), "Nanoindentation and Picoindentation Measurements Using a Capacitance Transducer System in Atomic Force Microscopy," *Philos. Mag.* 74, 1117–1128.

Bhushan, B. and Koinkar, V.N. (1997a), "Microtribological Studies of Doped Single-Crystal Silicon and Polysilicon Films for MEMS Devices," *Sensors Actuators A* 57, 91–102.

Bhushan, B. and Li, X. (1997b), "Micromechanical and Tribological Characterization of Doped Silicon and Polysilicon Films for MEMS Devices," *J. Mater. Res.* 12, 54–63.

Bhushan, B., Sundararajan, S., Scott, W.W., and Chilamakuri, S. (1997c), "Stiction Analysis of Magnetic Tapes," *IEEE Trans. Magn.* 33, 3211–3213.

Bhushan, B. and Sundararajan, S. (1998), "Micro/Nanoscale Friction and Wear Mechanisms of Thin Films Using Atomic Force and Friction Force Microscopy," *Acta Mater.* 46, 3793–3804.

Bowden, F.P. and Tabor, D. (1950 and 1964), *The Friction and Lubrication of Solids*, Part 1 (1950) Part 2 (1964), Clarendon Press, Oxford, U.K.

Callahan, D.L. and Morris, J.C. (1992), "The Extent of Phase Transformation in Silicon Hardness Indentations," *J. Mater. Res.* 7, 1612–1617.

Chu, M.Y., Bhushan, B., and DeJonghe, L. (1992), "Wear Behavior of Ceramic Sliders in Sliding Contact with Rigid Magnetic Thin-Film Disks," *Tribol. Trans.* 35, 603–610.

Derjaguin, B.V., Chuarev, N.V., and Muller, V.M. (1987), *Surface Forces*, Consultants Bureau, New York.

DeVecchio, D. and Bhushan, B. (1997), "Localized Surface Elasticity Measurements Using an Atomic Force Microscope," *Rev. Sci. Instrum.* 68, 4498–4505.

Foster, J. and Frommer, J. (1988), "Imaging of Liquid Crystal Using a Tunneling Microscope," *Nature* 333, 542–547.

Frisbie, C.D., Rozsnyai, L.F., Noy, A., Wrighton, M.S., and Lieber, C.M. (1994), "Functional Group Imaging by Chemical Force Microscopy," *Science* 265, 2071–2074.

Gane, N. and Cox, J.M. (1970), "The Micro-Hardness of Metals at Very Light Loads," *Philos. Mag.* 22, 881–891.

Ganti, S. and Bhushan, B. (1995), "Generalized Fractal Analysis and Its Applications to Engineering Surfaces," *Wear* 180, 17–34.

Grafstrom, S., Neitzert, M., Hagen, T., Ackermann, J., Neumann, R., Probst, O., and Wortge, M. (1993), "The Role of Topography and Friction for the Image Contrast in Lateral Force Microscopy," *Nanotechnology* 4, 143–151.

Heideman, R. and Wirth, M. (1984), "Transforming the Lubricant on a Magnetic Disk into a Solid Fluorine Compound," *IBM Tech. Disclosure Bull.* 27, 3199–3205.

Hibst, H. (1993), "Metal Evaporated Tapes and Co-Cr Media for High Definition Video Recording," in *High Density Digital Recording* (K.H.J. Buschow, G.J. Long, and F. Grandjean, eds.), Vol. E229, pp. 137–159, Kluwer Academic, Dordrecht.

Homola, A.M., Lin, L.J., and Saperstein, D.D. (1990), "Process for Bonding Lubricant to a Thin Film Magnetic Recording Disk," U.S. Patent 4,960,609, October 2.

Israelachvili, J.N. (1992), *Intermolecular and Surface Forces*, 2nd ed., Academic Press, New York.

Kaneko, R. and Miyamoto, T. (1988), "Friction and Adhesion on Magnetic Disk Surface," *IEEE Trans. Magn.* Mag. 24, 2641–2643.

Kaneko, R., Miyamoto, T., and Hamada, E. (1991a), "Development of a Controlled Frictional Force Microscope," *Adv. Info. Storage Syst.* 1, 267–277.

Kaneko, R., Oguchi, S., Andoh, Y., Sugimoto, I., and Dekura, T. (1991b), "Direct Observation of the Configuration, Adsorption, and Mobility of Lubricants by Scanning Tunneling Microscopy," *Adv. Info. Storage Syst.* 2, 23–34.

Koinkar, V.N. and Bhushan, B. (1996a), "Microtribological Studies of Al_2O_3, Al_2O_3–TiC, Polycrystalline and Single-Crystal Mn–Zn Ferrite and SiC Head Slider Materials," *Wear* 202, 110–122.

Koinkar, V.N. and Bhushan, B. (1996b), "Microtribological Studies of Unlubricated and Lubricated Surfaces Using Atomic Force/Friction Force Microscopy," *J. Vac. Sci. Technol.* A 14, 2378–2391.

Koinkar, V.N. and Bhushan, B. (1997a), "Effect of Scan Size and Surface Roughness on Microscale Friction Measurements," *J. Appl. Phys.* 81, 2472–2479.

Koinkar, V.N. and Bhushan, B. (1997b), "Microtribological Properties of Hard Amorphous Carbon Protective Coatings for Thin-Film Magnetic Disks and Heads," *Proc. Inst. Mech. Eng. Part J: J. Eng. Tribol.* 211, 365–372.

Kulkarni, A.V. and Bhushan, B. (1997), "Nanoindentation Measurements of Amorphous Carbon Coatings," *J. Mater. Res.* 12, 2707–2714.

Lipson, M., and Mercer, E.H. (1946), "Frictional Properties of Wool Treated with Mercuric Acetate," *Nature* 157, 134–135.

Lu, C.J. and Bogy, D.B. (1995), "Sub-Microindentation Hardness Tests on Thin Film Magnetic Disks," *Adv. Info. Storage Syst.* 6, 163–175.

Maivald, P., Butt, H.J., Gould, S.A.C., Prater, C.B., Drake, B., Gurley, J.A., Elings, V.B., and Hansma, P.K. (1991), "Using Force Modulation to Image Surface Elasticities with the Atomic Force Microscope," *Nanotechnology* 2, 103–106.

Majumdar, A. and Bhushan, B. (1990), "Role of Fractal Geometry in Roughness Characterization and Contact Mechanics of Surfaces," *ASME J. Tribol.* 112, 205–216.

Majumdar, A. and Bhushan, B. (1991), "Fractal Model of Elastic-Plastic Contact Between Rough Surfaces," *ASME J. Tribol.* 113, 1–11.

Makinson, K.R. (1948), "On the Cause of the Frictional Difference of the Wool Fiber," *Trans. Faraday Soc.* 44, 279–282.

Mate, C.M. (1992a), "Application of Disjoining and Capillary Pressure to Liquid Lubricant Films in Magnetic Recording," *J. Appl. Phys.* 72, 3084–3090.

Mate, C.M. (1992b), "Atomic-Force-Microscope Study of Polymer Lubricants on Silicon Surface," *Phys. Rev. Lett.* 68, 3323–3326.

Mate, C.M. (1993a), "Nanotribology of Lubricated and Unlubricated Carbon Overcoats on Magnetic Disks Studied by Friction Force Microscopy," *Surf. Coat. Technol.* 62, 373–379.

Mate, C.M. (1993b), "Nanotribology Studies of Carbon Surfaces by Force Microscopy," *Wear* 168, 17–20.

Mate, C.M., McClelland, G.M., Erlandsson, R., and Chiang, S. (1987), "Atomic-Scale Friction of a Tungsten Tip on a Graphite Surface," *Phys. Rev. Lett.* 59, 1942–1945.

Mate, C.M., Lorenz, M.R., and Novotny, V.J. (1989), "Atomic Force Microscopy of Polymeric Liquid Films," *J. Chem. Phys.* 90, 7550–7555.

Mate, C.M., Lorenz, M.R., and Novotny, V.J. (1990), "Determination of Lubricant Film Thickness on a Particulate Disk Surface by Atomic Force Microscopy," *IEEE Trans. Magn.* Mag. 26, 1225–1228.

Mate, C.M. and Novotny, V.J. (1991), "Molecular Conformation and Disjoining Pressures of Polymeric Liquid Films," *J. Chem. Phys.* 94, 8420–8427.

Mercer, E.H. (1945), "Frictional Properties of Wool Fibers," *Nature* 155, 573–574.

Meyer, G., and Amer, N.M. (1990), "Simultaneous Measurement of Lateral and Normal Forces with an Optical-Beam-Deflection Force Microscope," *Appl. Phys. Lett.* 57, 2089–2091.

Meyer, E., Overney, R., Luthi, R., Brodbeck, D., Howald, L., Frommer, J., Guntherodt, H.-J, Wolter, O., Fujihira, M., Takano, T., and Gotoh, Y. (1992), "Friction Force Microscopy of Mixed Langmuir-Blodgett Films," *Thin Solid Films* 220, 132–137.

Miyamoto, T., Kaneko, R., and Ando, Y. (1990), "Interaction Force between Thin Film Disk Media and Elastic Solids Investigated by Atomic Force Microscope," *ASME J. Tribol.* 112, 567–572.

Miyamoto, T., Kaneko, R., and Andoh, Y. (1991a), "Microscopic Adhesion and Friction between a Sharp Diamond Tip and Al_2O_3–TiC," *Adv. Info. Storage Syst.* 2, 11–22.

Miyamoto, T., Kaneko, R., and Andoh, T. (1991b), "Microscopic Friction on Silicon Oxide and Amorphous Carbon Films Investigated by Point Contact Microscopy and Frictional Force Microscopy," *Adv. Info. Storage Syst.* 3, 137–146.

Miyamoto, T., Kaneko, R., and Miyake, S. (1991c), "Tribological Characteristics of Amorphous Carbon Films Investigated by Point Contact Microscopy," *J. Vac. Sci. Technol. B* 9, 1336–1339.

Miyamoto, T., Miyake, S., and Kaneko, R. (1993), "Wear Resistance of C^+-Implanted Silicon Investigated by Scanning Probe Microscopy," *Wear* 162–164, 733–738.

Novotny, V.J. (1990), "Migration of Liquid Polymers on Solid Surfaces," *J. Chem. Phys.* 92, 3189–3196.

Novotny, V.J., Hussla, I., Turlet, J.M., and Philpott, M.R. (1989), "Liquid Polymer Conformation on Solid Surfaces," *J. Chem. Phys.* 90, 5861–5868.

Oden, P.I., Majumdar, A., Bhushan, B., Padmanabhan, A., and Graham, J.J. (1992), "AFM Imaging, Roughness Analysis and Contact Mechanics of Magnetic Tape and Head Surfaces," *ASME J. Tribol.* 114, 666–674.

Overney, R.M., Meyer, E., Frommer, J., Brodbeck, D., Luthi, R., Howard, L., Guntherodt, H.-J., Fujihira, M., Takano, H., and Gotoh, Y. (1992), "Friction Measurements on Phase-Separated Thin Films with a Modified Atomic Force Microscope," *Nature* 359, 133–135.

Overney, R. and Meyer, E. (1993), "Tribological Investigations Using Friction Force Microscopy," *MRS Bull.* 18, 26–34.

Page, T.F., Oliver, W.C., and McHargue, C.J. (1992), "The Deformation Behavior of Ceramic Crystals Subjected to Very Low Load Indentation," *J. Mater. Res.* 7, 450–473.

Pharr, G.M., Oliver, W.C., and Clarke, D.R. (1990), "The Mechanical Behavior of Silicon during Small-Scale Indentation," *J. Electron. Mater.* 19, 881–887.

Pharr, G.M. (1991), "The Anomalous Behavior of Silicon during Nanoindentation," in *Thin Films: Stresses and Mechanical Properties III* (W.D. Nix, J.C. Bravman, E. Arzt, and L.B. Freund, eds.), Vol. 239, pp. 301–312, Materials Research Society, Pittsburgh.

Poon, C.Y. and Bhushan, B. (1995a), "Comparison of Surface Roughness Measurements by Stylus Profiler, AFM and Non-Contact Optical Profiler," *Wear* 190, 76–88.

Poon, C.Y. and Bhushan, B. (1995b), " Surface Roughness Analysis of Glass-Ceramic Substrates and Finished Magnetic Disks, and Ni–P Coated Al–Mg and Glass Substrates," *Wear* 190, 89–109.

Ruan, J. and Bhushan, B. (1993), "Nanoindentation Studies of Fullerene Films Using Atomic Force Microscopy," *J. Mater. Res.* 8, 3019–3022.

Ruan, J. and Bhushan, B. (1994a), "Atomic-Scale Friction Measurements Using Friction Force Microscopy: Part I — General Principles and New Measurement Techniques," *ASME J. Tribol.* 116, 378–388.

Ruan, J. and Bhushan, B. (1994b), "Atomic-Scale and Microscale Friction of Graphite and Diamond Using Friction Force Microscopy," *J. Appl. Phys.* 76, 5022–5035.

Ruan, J. and Bhushan, B. (1994c), "Frictional Behavior of Highly Oriented Pyrolytic Graphite, *J. Appl. Phys.* 76, 8117–8120.

Saperstein, D.D. and Lin, L.J. (1990), "Improved Surface Adhesion and Coverage of Perfluoropolyether Lubricant Following Far-UV Irradiation," *Langmuir* 6, 1522–1524.

Sargent, P.M. (1986), "Use of the Indentation Size Effect on Microhardness for Materials Characterization," in *Microindentation Techniques in Materials Science and Engineering* (P.J. Blau and B.R. Lawn, eds.), STP 889, pp. 160–174, ASTM, Philadelphia.

Scherer, V., Bhushan, B., Rabe, U., and Arnold, W. (1997), "Local Elasticity and Lubrication Measurements Using Atomic Force and Friction Force Microscopy at Ultrasonic Frequencies," *IEEE Trans. Magn.* 33, 4077–4079.

Sugawara, Y., Ohta, M., Konishi, T., Morita, S., Suzuki, M., and Enomoto, Y. (1993), "Effect of Humidity and Tip Radius on the Adhesive Force Measured with Atomic Force Microscopy," *Wear* 168, 13–16.

Sundararajan, S. and Bhushan, B. (1998), "Micro/Nanotribological Studies of Polysilicon and SiC Films For MEMS Applications," *Wear* (in press).

Thomson, H.M.S. and Speakman, J.B. (1946), "Frictional Properties of Wool," *Nature* 157, 804.

15

Mechanical Properties of Materials in Microstructure Technology

Fredric Ericson and Jan-Åke Schweitz

15.1   Introduction

Mechanical properties are of critical importance to any material that is used for transmission of forces or moments, or just for sustaining loads. The gradual introduction of microcomponents in practical applications within microstructure technology (MST) has instigated an increasing demand for insight into the fundamental factors that determine the mechanical integrity of such elements, for instance, their long-term reliability or how to choose proper safety limits in design and use. The mechanical integrity of a microsystem is not only of importance in mechanical applications. It is not unusual that mechanical (or thermomechanical) integrity is a prerequisite for reliable performance of microsystems with primarily nonmechanical functions, e.g., electric, optic, or thermal.

Do we have enough knowledge about mechanical properties to determine the long-time reliability of a micromechanical component or to choose proper safety limits? In general, the answer to this question is negative. Are the properties of bulk materials applicable to microsystems? Again, we do not know for certain in every case; we cannot even be absolutely sure that bulk data on the fundamental elastic constants are valid for a micromachined element. Furthermore, the materials used in semiconductor technology are usually well characterized from an electronic viewpoint, but from a mechanical viewpoint they are in many cases more or less uncharted. In some cases we do not even have access to bulk data. This leads to the conclusion that much more work is required on the systematic exploration of the mechanical properties of microsized elements, as well as on the influence of the manufacturing processes on these properties.

This chapter aims to define some basic concepts concerning mechanical properties, and to relate these concepts to experimental procedures and to some practial design aspects. For many properties we give numerical examples, if they exist, mostly concerning silicon and related materials. Sometimes comparisons of these materials with other types of materials are made. Silicon-based micromechanics is predominant today. In the future, mechanically high-performing materials like SiC may be frequently used in micromachined structures in high-temperature applications, for instance. For this reason, a number of thermomechanical phenomena are briefly defined and discussed in this chapter.

15.2 Cohesion and Crystal Structures

15.2.1 System Energy and Interatomic Binding

Mechanical properties such as elasticity, plasticity, fracture strength, adhesion, internal stresses, etc., depend on the fundamental mechanisms of cohesion between atoms. Basically, the atoms in a solid material are held together by electrostatic attraction between charges of opposite signs. Magnetic forces are of minor importance to the cohesion. The potential energy of interatomic binding consists of terms of classical electrostatic interaction as well as terms of electrostatic quantum interaction (exchange effects). At equilibrium, the attractive potential energy U_o is balanced by the repulsive kinetic energy T_o in a state of minimum system energy E_o:

$$\mathsf{E}_o = U_o + T_o, \tag{15.1}$$

see Figure 15.1. Neglecting boundary effects, the balance between potential and kinetic energy at equilibrium is given by the virial theorem:

$$2T_o + U_o = 0, \tag{15.2}$$

where T_o is positive (repulsive) and U_o is negative (attractive). In Figure 15.1 the parameter a represents some measure that is proportional to the average interatomic distance, for instance, the lattice parameter. If the state of equilibrium is shifted by external forces, compressive or tensile, the value of a will decrease or increase, and the system will move along the curve of Figure 15.1 away from the state of minimum system energy. The virial theorem is then modified into

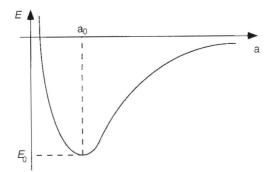

E

a_0

a

E_0

FIGURE 15.1 Crystal system energy E vs. lattice spacing parameter a.

$$2T + U = -a \frac{\partial E}{\partial a},$$ (15.3)

where $F = -\partial E / \partial a$ is the force striving to restore equilibrium at minimum energy. For small deviations from equilibrium, E is a harmonic function of a in most materials, and the restoring force F is a linear function of the change in a. This is one way of expressing *Hooke's law* of elasticity:

$$\sigma = E\varepsilon,$$ (15.4)

where σ is the applied stress (force per area unit), ε is the resulting strain $\Delta a / a_0$, and E is *Young's modulus* of the material. Hence, Young's modulus (= linear modulus of elasticity) is proportional to the second derivative of the system energy E with respect to the lattice parameter a at equilibrium ($a = a_0$).

Depending on the electron structure of the constituent atoms, the mechanism of cohesion can vary from weak dipole interaction (van der Waals interaction) to strong covalent binding. In silicon the latter type is predominant, but some materials of interest for micromachining also exhibit ionic binding or metallic binding. In III-V semiconductors, for instance, the cohesion is of a mixed covalent and ionic nature, and in tungsten metallic binding is predominant.

Ionic binding is found in chemical compounds such as common salt, NaCl. One or more electrons are transferred from one type of atom to the other, whereby electrostatic attraction between ions of opposite electric charge occurs.

Metallic binding occurs in metallic elements or compounds. In this case the loosely bound valence electrons are disconnected from the atoms, and form a quasi-free electron gas, the conduction electron gas. These electrons are at liberty to move between the ion cores and to a large extent also through them. The high mobility of these electrons gives rise to the good electric conductivity found in most metals. Hence, the positive ions are immersed in a sea of negative conduction electrons, which act as a fluid cement holding the ions together.

Covalent binding is found in some elemental solids (e.g., silicon) as well as in very complicated molecular structures (e.g., polymers). The cohesive mechanism is complex, but in a very simplified picture it can be described in terms of negative-valence electrons preferring to locate themselves in the regions between a positive ion and its nearest neighbors, by which an electrostatic coupling occurs. This type of binding is usually strong and "directional"; i.e., the interatomic bonds are formed at specific angles, resulting in well-defined molecular or crystalline structures.

The strength of the interatomic bonds is decisive for the stiffness and the brittleness of the crystal. Strong and directional covalent bonds give silicon high stiffness and strength. In GaAs the bonds are of a mixed covalent and ionic type, making this material less stiff and more fragile than Si. Also the melting points (Table 15.1) are affected by the bond strength.

TABLE 15.1   Melting Points (°C) of a Number of Semiconductors and Other Materials

Si	1412	Nylon	137–150
Ge	937	Teflon	290
SiC	2537	Stainless steel	1400–1500
BN	4487	Al_2O_3	2050
AlAs	1737	TiC	3100
GaAs	1238	HfC	3890
GaP	1467	SiO_2	1610
InP	1070	Glass	~700
InSb	536		

15.2.2   Lattice Structures and Structural Defects

Crystalline as well as amorphous (disordered) materials are regularly used in MST. One important material in micromechanics of the latter type is glass. Another important material is silicon dioxide, SiO_2, which is used in crystalline form (quartz) as well as in amorphous form (for instance, low-temperature oxide, LTO). Also silicon and most other relevant materials can be grown in both forms. From a mechanical-strength viewpoint, an amorphous structure is sometimes preferred, due to the lack of active slip planes for dislocation movement in such structures. In general, however, the strength performance is more related to the distribution and geometry of microscopic flaws in the material, especially surface flaws.

It would lead too far in the present context to define all crystalline lattice structures of interest in MST. For this reason we will confine ourselves to very brief descriptions of two important lattice types: the diamond lattice type found in crystalline silicon and the zinc blend (ZnS) lattice type found in III-V semiconductors.

The *diamond structure* is one of the simplest and most symmetric lattice types, and is found in Si and Ge, for example. It consists of two face-centered cubic (fcc) lattices which are inserted into each other in such a manner that they are shifted relative to each other by one quarter of a cube edge along all three principal axes. Each atom is surrounded by four other atoms in a tetragonal configuration.

The *zinc blende structure* is found in III-V compounds such as GaAs, InP, and InSb. It is identical with the diamond structure apart from the fact that one of the two overlapping fcc lattices consists entirely of the type III element (e.g., Ga) and the other entirely of the type V element (e.g., As). Every atom of one kind is tetragonally surrounded by four atoms of the other kind, and crystallographic planes of any chosen orientation are periodically arranged in parallel pairs consisting of one III-type and one V-type atomic plane (in some orientations the parallel planes of a pair coincide).

Common crystal defects are *point defects* such as vacancies (one atom is missing), substitutionals (one atom is replaced by an impurity atom), or interstitials (one atom is "squeezed in" between the ordinary atoms). Other frequent crystal imperfections are *line defects*, such as dislocations, and more *complex defects*, such as stacking faults or twins. All types of lattice defects affect the mechanical properties of a crystal to a greater or lesser extent, but dislocations are the most detrimental of the lattice defects from a mechanical-strength viewpoint due to their extremely high mobility (when a critical load limit, *the yield limit*, has been exceeded).

Beyond the basic crystalline lattice structure (and the various types of lattice defects that may be present in it), a number of *superstructures* can be of major importance to the mechanical behavior. The grain structure of a polycrystalline material is one superstructure influencing the hardness and the yield limit of the material, and precipitates of impurities, alloying substances, or intermediary phases are other examples. The size and shape distribution of geometric flaws, for instance, voids or cracks in the micron or submicron range, is of crucial importance to the fracture strength of a brittle material. These super-structures will be discussed in further detail in following sections.

Foreign atoms of dopants, or contaminants such as oxygen, nitrogen, and carbon, commonly occur in semiconductor materials, and are of great importance to their electronic properties (Hirsch, 1983;

Sumino, 1983a). At "normal" levels of doping or contamination in electronic components, the influence of such impurities on the mechanical behavior is fairly limited, however. For extreme doping levels, some influence on the plasticity behavior can be observed, especially at elevated temperatures, as will be exemplified later on.

15.3   Elasticity Properties

15.3.1   Isotropic Elasticity

For small deformations at room temperature most metals and ceramics (including conventional semi-conductors) display a linear elastic behavior, i.e., they obey Hooke's law, Equation 15.4, for the relation between applied normal stress (σ) and resulting normal strain (ε). The corresponding relationship between shear stress (τ) and shear strain (γ) is given by

$$\tau = G\gamma, \tag{15.5}$$

where G is the *shear modulus* of the material. The Young's modulus and the shear modulus are anisotropic in crystalline materials. For fine-grained polycrystalline materials, however, isotropic (averaged) E and G values are sometimes sufficient.

When a linear-elastic material is subjected to a uniaxial strain (relative elongation) $\varepsilon = (L - L_0)/L_0$, its cross-sectional dimension will diminish by a relative contraction $\varepsilon_c = (d_0 - d)/d_0$. The ratio of these two strains is a materials constant called the *Poisson's ratio*:

$$\nu = \varepsilon_c / \varepsilon. \tag{15.6}$$

In isotropic media the elastic parameters are related by

$$G = E / \left[2(1 + \nu) \right]. \tag{15.7}$$

The relative *volume change* caused by a uniaxial stress σ is given by

$$\Delta V / V_o = (1 - 2\nu) \sigma / E = K\sigma / 3, \tag{15.8}$$

where K is the *compressibility*:

$$K = 3(1 - 2\nu) / E. \tag{15.9}$$

The *bulk modulus* is defined as the inverse value of the compressibility:

$$B = 1/K = E / \left[3(1 - 2\nu) \right]. \tag{15.10}$$

Multilayer structures consisting of different materials are frequent within micromechanics. For such layered composites the Young's moduli in the lateral and the transverse directions can be calculated from

$$E_{\parallel} = \sum_{n=1}^{N} f_n E_{\parallel n}, \tag{15.11}$$

769

TABLE 15.2 Values of Elastic Stiffness
Constants of a Number of Semiconductors
at 300 K (in units of GPa)

	C_{11}	C_{12}	C_{44}
Si	165.78	63.94	79.62
Ge	129.11	48.58	67.04
GaAs	118.80	53.80	59.40
InP	102.20	57.60	46.00
InAs	83.29	45.26	39.59
Diamond	1076.4	125.2	577.4

The values are results published by different
workers, as compiled by Simmons et al.
(1971). The values for diamond were published by van Enckevort (1994).

$$1/E_\perp = \sum_{n=1}^{N} f_n / E_{\perp n}, \tag{15.12}$$

where $E_{\|n}$ and $E_{\perp n}$ are the Young's moduli of the constituent materials in the two directions, and f_n are the relative thickness fractions (= relative volume fractions) of the layers. Equations 15.11 and 15.12 are applicable to stress-free multilayer structures built up of layers of individual thicknesses of ~100 nm or more. In some superlattice structures the existence of a "supermodulus effect" has been suggested, i.e., the composite E values are supposed to radically deviate from the values predicted by conventional elastic theories for multilayer structures or for homogeneous alloys. The existence of this effect is at present under debate, and no physical model for it has been generally accepted as yet.

15.3.2 Anisotropic Elasticity

In single-crystalline materials the anisotropic elasticity is described by the elastic stiffness constants C_{ij} (i,j = 1, 2, ... 6) or, alternatively, by the elastic compliance constants S_{ij}. These matrices are symmetric, and in cubic crystals their number of elements is reduced by symmetry considerations to three independent constants: C_{11}, C_{12}, and C_{44} (or S_{11}, S_{12}, and S_{44}). Table 15.2 gives typical room-temperature values in gigapascals of these constants for a number of materials (Simmons and Wang, 1971).

The stiffness and compliance constants of cubic crystals are related by

$$C_{11} - C_{12} = \left(S_{11} - S_{12} \right)^{-1}, \tag{15.13}$$

$$C_{11} + 2C_{12} = \left(S_{11} + 2S_{12} \right)^{-1}, \tag{15.14}$$

$$C_{44} = S_{44}^{-1}. \tag{15.15}$$

Anisotropic values of E and ν can be calculated from these elastic constants. The Young's modulus in the crystallographic direction $\langle lmn \rangle$ is given by

$$1/E = S_{11} - 2\left(S_{11} - S_{12} - S_{44}/2 \right) k_1, \tag{15.16}$$

TABLE 15.4 Poisson Ratios ν at 300 K for Tension along [*lmn*] and Contraction in the Perpendicular <*ijk*> Direction

System	Si	Ge	GaAs	InP	InAs	Diamond
[100]<010>	0.278	0.273	0.312	0.360	0.352	0.104
[100]<011>	0.278	0.273	0.312	0.360	0.352	0.104
[110]<001>	0.362	0.367	0.443	0.555	0.543	0.115
[110]<1$\bar{1}$0>	0.062	0.026	0.021	0.015	0.001	0.008
[110]<1$\bar{1}$1>	0.162	0.139	0.162	0.195	0.182	0.044
[111]<1$\bar{1}$0>	0.180	0.157	0.188	0.238	0.222	0.045
[110]<1$\bar{1}$2>	0.262	0.253	0.303	0.375	0.362	0.079
[111]<11$\bar{2}$>	0.180	0.157	0.188	0.238	0.222	0.045
Poly	0.222	0.208	0.243	0.294	0.283	0.070

The poly values have been calculated by the method indicated in Table 15.3.

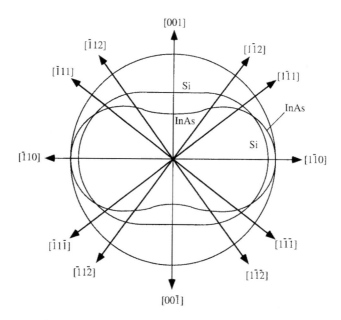

FIGURE 15.2 Illustration of the anisotropy of the Poisson ratio ν for a normalized (hypothetical) case of 100% elastic straining along the [110] axis in Si and InAs. The outer contour illustrates a circular cross section of an unstrained rod (a {110} plane), and the two inner contours illustrate cross sections in hypothetical states of 100% elastic strain.

often strongly textured, and theories or expressions for elastic averaging over such morphologies also exist (Brantley, 1973; Guckel et al., 1988; Maier-Schneider, 1995a). The resulting Young's modulus of a textured polycrystalline film can deviate up to 10% from a nontextured polycrystalline film.

15.4 Internal Stresses

The presence or nonpresence of internal stresses in a layered structure can be of great importance to the mechanical behavior of the component, so for this reason some aspects of internal stresses will be summarily discussed also in the present context. For instance, internal stresses can cause loss of adhesion between the film and the substrate and, consequently, lead to delamination failure of the composite. They can have a beneficial or detrimental effect on the fracture properties of the structure, by inhibiting or promoting crack propagation in film or substrate (Johansson et al., 1989). Furthermore, various

TABLE 15.5 Linear Coefficients of Thermal Expansion α (in units of 10^{-6} K^{-1})

	Isometric		Anisometric ($\perp$ Axis)	Anisometric ($\parallel$ Axis)
Si	2.4	SiO$_2$ (quartz)	13.7	7.5
GaAs	6.0	Graphite	1.0	27.0
AlAs	5.2	Al$_2$O$_3$	8.3	9.0
SiC	4.5–5.0	Al$_2$TiO$_5$	−2.6	11.5
Diamond	1.3	LiAlSi$_2$O$_6$	6.5	−2.0
Glass	8.0	LiAlSiO$_4$	8.2	−17.6
Cu	16.2			

mechanisms of relaxation of internal stresses can have rather a drastic influence on the morphology of ductile films (Smith et al., 1991). Internal stresses in layered structures are of two fundamentally different origins: *thermal* stresses caused by thermal mismatch between two adhering layers and *intrinsic*, or microstructural, stresses generated during the deposition process.

15.4.1 Thermal Film Stress

One of the most important parameters in the generation of thermal stresses is the linear coefficient of thermal expansion (α), or, to be more precise, the difference in α for two adhering layers. The materials parameter α is defined as the relative elongation of a body per degree temperature rise:

$$\alpha = \frac{1}{L_o} \frac{dL}{dT},$$

(15.23)

which can be expressed as

$$\varepsilon^{\text{therm}} = \frac{\Delta L}{L_o} = \alpha \Delta T,$$

(15.24)

where $\varepsilon^{\text{therm}}$ is the thermal strain and ΔT is the difference between the initial and the final temperatures. In cubic (isometric) single crystals, as well as in amorphous or polycrystalline materials, α is nearly isotropic. In noncubic (anisometric) single crystals, α can be strongly anisotropic. In certain extreme cases, e.g., Al$_2$TiO$_5$, LiAlSi$_2$O$_6$, and LiAlSiO$_4$, α is positive in one direction and negative in a perpendicular direction. This anisotropy can be utilized in micromechanical structures to control the spatial dimensions by temperature variation. Some selected α values are found in Table 15.5.

A thermal stress is generated when the thermal expansion (or contraction) of one layer is prevented by external forces of constraint, for instance, by adjacent layers with differing α values or differing temperatures. In the case of a uniaxially clamped structure, the thermal stress caused by a temperature difference ΔT can easily be calculated from Hooke's law, Equation 15.4, and Equation 15.24:

$$\sigma^{\text{therm}} = E\varepsilon^{\text{therm}} = E\alpha \, \Delta T.$$

(15.25)

For a thin film on a thick substrate, we have biaxial stress conditions, and Hooke's law is

$$\sigma^{\text{therm}} = \frac{E_f}{1 - \nu_f} \Delta\alpha \, \Delta T,$$

(15.26)

where E_f and v_f are the Young's modulus and Poisson ratio of the film material, and $\Delta\alpha$ is the difference in α between film and substrate. It is apparent from Equation 15.26 that thermal stresses are minimized by low E_f values as well as by small differences in the expansion coefficient and the temperature. The latter difference is minimized by high thermal conductivities (k values) of the constituent materials.

In interfaces between differently oriented crystals, for instance, in grain boundaries, thermal stresses can also be caused by anisotropy effects in E, α, and k. If the thermal stresses are not completely relaxed upon cooling, which commonly occurs in thin-film deposition, residual thermal stresses will be present in the structure.

15.4.2 Intrinsic Film Stress

Internal stresses of intrinsic origin, on the other hand, are of a more complex physical nature and cannot be expressed in terms of fundamental materials parameters. These nonthermal stresses are generated during the film growth process and strongly depend on which deposition technique is used and on various process parameters. The magnitude of the intrinsic stresses can be very high, sometimes exceeding the yield or fracture strengths of the corresponding bulk materials. Many theories to explain these stresses have been suggested, and a summary is given in a review by Windischmann (1992). Intrinsic *tensional* stresses have been attributed to grain boundary formation, to constrained shrinkage of disordered material buried behind the advancing film surface, and to attractive interatomic forces acting between detached grains separated by a few atomic distances. Intrinsic *compressive* stresses, on the other hand, have been attributed to impurity or working gas incorporation, increased defect density, and to recoil implantation of film atoms.

The *total residual stress* in a film after deposition hence can be expressed as a sum of the thermal residual and the intrinsic stresses:

$$\sigma^{res} = \sigma^{therm} + \sigma^{intr},\tag{15.27}$$

where nonthermal stresses induced by lattice mismatch at the interface have been included among the intrinsic stresses.

15.4.3 Substrate and Interface Stresses

Residual stresses in a film will induce balancing stresses of opposite sign in the substrate. Using equilibrium relationships for forces and moments, the stress response induced in the substrate surface can be expressed as

$$\sigma_s^{res} = -4 \frac{t_f}{t_s} \sigma_f^{res},\tag{15.28}$$

where t_f and t_s are the thicknesses of film and substrate, respectively. Hence, in very thick substrates ($t_f \ll t_s$), negligible stress response is induced. Equation 15.28 is derived for low t_f/t_s ratios. If this ratio is larger than 0.01, the error in σ_s^{res} becomes noticeable (>5%).

All stresses in film and substrate discussed so far are *normal* tensile or compressive stresses oriented parallel to the interface. In a well-adhered film–substrate composite no *shear* stresses are present in the inner parts of the interface. Along the edges of a coated region, on the other hand, shear stresses may be present for moment balance reasons. In thin, layered structures this effect is sometimes manifested by a visible buckling of the edges. If nonbonded areas exist in the interior part of an interface, shear stresses may be present along their boundaries and contribute in lowering the mechanical strength of the interface.

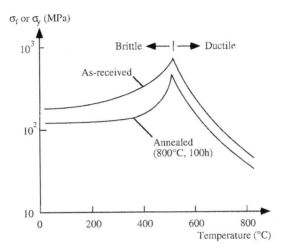

FIGURE 15.3 Variation of fracture limit σ_f and yield limit σ_y of silicon as a function of temperature. The upper curve is as-received CZ or FZ silicon (difference negligible), and the lower curve is CZ silicon annealed at 800°C for 100 h. For temperatures below 525°C (curves maxima) the curves illustrate the fracture limit σ_f, above this transition temperature they illustrate the plastic yield limit σ_y. (From Yasutake, K. et al., 1982a.)

15.5 Plasticity and Thermomechanical Properties

At room temperature, silicon, quartz, the III-V compounds, and many other materials used in MST display a linear-elastic response to tensile stresses, all the way to brittle fracture. Hence, the plastic yield limit is never reached, and plastic deformation or other effects based on dislocation slip will not occur under tension at room temperature. At elevated temperature, however, or for high compressive loads at room temperature, many of these materials may reach their yield limits before they reach the fracture limit, in which case dislocation slip is activated and eventually they will deform plastically. Figure 15.3 illustrates the variation of the fracture limit σ_f or the yield limit σ_y of silicon as a function of temperature (Yasutake et al., 1982a). In most applications plastic yield is undesirable and is, therefore, avoided by ample dimensioning or by a "safe" materials selection. In some cases, however, the room temperature yield limit is locally exceeded in high-compressive-stress fields which may be generated internally during processing of multilayer structures or by, e.g., unintentional microscratches or microindentations. In yet other cases dislocation slip may be activated by high operating temperatures and induce time-dependent processes such as creep, aging, or fatigue.

15.5.1 Elastic–Ductile Response

The difference between elastic–brittle response and elastic–ductile response of a material is illustrated by Figures 15.4a and b. The first diagram illustrates the case when the fracture limit is lower than the critical load limit for dislocation slip. The second diagram illustrates the opposite case; i.e., the dislocations (if they exist from the start) are immobile until the critical resolved shear stress is reached in some slip systems, where dislocation slip and eventually dislocation multiplication are initiated. The resulting plastic deformation — contrary to elastic deformation — is irreversible upon unloading. The plastic curve segment in Figure 15.4b is not as steep as the elastic curve segment, but still displays a positive slope corresponding to a *strain hardening* effect. This effect is primarily due to the gradually increasing density of dislocations, which tend to get entangled and obstruct further dislocation slip, hence demanding a gradual increase of the applied stress in order to maintain the straining process.

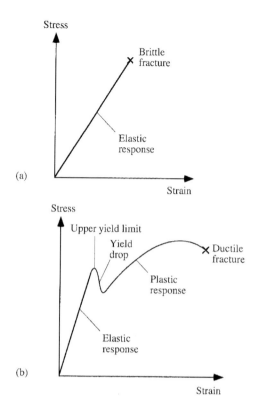

(a)

(b)

FIGURE 15.4 (a) Linear elastic–brittle response. (b) Elastic–ductile response.

Much work has been devoted to the study of dislocations and plastic flow in silicon and other semiconductors. Essential parts of this work are surveyed in well-known articles by Alexander and Haasen (1968) and by Hirsch (1985). In the plastic interval the *dopant* and *impurity levels* play a certain role (Alexander and Haasen, 1968; Yonenaga and Sumino, 1978, 1984; Sumino et al., 1980, 1983, 1985; Sumino and Imai, 1983; Imai and Sumino, 1983; Sumino, 1983b; Hirsch, 1985). At high levels of *n*-doping in Si, the mobility of the dislocations is increased; i.e., the strain hardening effect is diminished. Impurities such as oxygen or nitrogen atoms, on the other hand, tend to gather around the dislocations and hamper their motion; so-called *Cottrell atmospheres* are formed around the dislocations. This means that higher loads are required to "tear loose" the dislocations from these atmospheres, implying increased yield limit. When the dislocations have been torn loose, they regain their high mobility, resulting in a marked *yield drop*, see Figures 15.4b and 15.5.

The *strain rate* $\dot{\varepsilon}$ during plastic deformation of semiconductors has been the subject of many studies (Alexander and Haasen, 1968; Sumino, 1983a). Its dependency on temperature, effective resolved shear stress, and dislocation velocity has been investigated in detail. Also the influence of doping levels, impurities, and growth process (float-zone growth, FZ, or Czochralski growth, CZ) have been studied, and found to be considerable (Imai and Sumino, 1983; Sumino and Imai, 1983). Micromechanical elements normally are designed to function below the yield limit, so the strain rate will only be discussed here in connection with creep.

Microhardness is a complex materials parameter involving several properties of a more fundamental nature, in particular plasticity properties. From a practical viewpoint the microhardness is a simple and convenient measure of the susceptibility of a material to contact damage in the micron range (microindentations, microscratches, etc.), and it plays a major role in most models describing tribological processes. The microhardness can be measured by several methods, among which Vickers indentation and Knoop indentation are most commonly used. In both methods a diamond stylus is pressed into the surface of the body by a given load and at a given loading rate. Upon unloading, the size of the residual

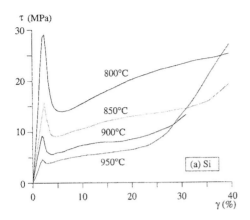

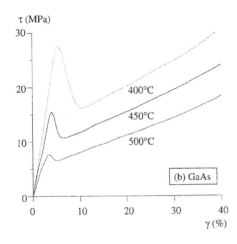

FIGURE 15.5 Resolved shear stress τ vs. shear strain γ for tensile testing of single crystalline specimens at various temperatures for (a) Si (Adapted from Yonenaga, I. and Sumino, K., 1978, *Phys. Status Solidi* 50, 685.) and (b) GaAs (Adapted from Sumino, K. et al., 1985, in Proc. 27th Meeting, 145 Committee of JSPS, p. 91.)

indentation mark in the surface is measured and related to a characteristic hardness parameter (dimension: force/area). In the Vickers method the diamond stylus has the shape of a low profile, square pyramid, whereas the Knoop stylus has a rhombic pyramid shape.

In single-crystalline surfaces, the *Vickers hardness* (H_v) is weakly anisotropic (Ericson et al., 1988). Table 15.6 gives a few characteristic H_v values for semiconductor surfaces of different crystallographic orientations.

15.5.2 Time-Dependent Effects

Creep is a thermally activated deformation process which occurs under constant load (below the yield limit) and during an extended period of time (Alexander and Haasen, 1968). Creep testing of brittle materials is usually performed under compression. In Figure 15.6 two sets of typical creep curves (strain vs. time) for Si under various loads and temperatures are displayed. In these curve sets, the points of maximum creep rate $\dot{\varepsilon}_{max}$ (i.e., the points of inflection) obey an exponential type of relationship (Reppich et al., 1964):

$$\dot{\varepsilon}_{max} = C\tau^n \exp\left(-U/kT\right),$$ (15.29)

where U is an activation energy and T is the temperature. This general creep behavior of Si is typical also for Ge and many III-V compounds. Doping has a major influence on $\dot{\varepsilon}_{max}$. Doping of Si with As to

777

Table 15.6 Microhardness Values Obtained by Vickers Indentation at 50g Load (polarity effect neglected)

	H_v (GPa)	Indentation orientation 1	2
Si(111)	10.8 ± 1.3	$[\bar{1}10]$	$[11\bar{2}]$
Si(100)	11.2 ± 1.0	$[011]$	$[0\bar{1}1]$
Si(110)	11.3 ± 1.1	$[001]$	$[1\bar{1}0]$
Ge(111)	9.2 ± 0.5	Undefined	
GaAs(100)	6.9 ± 0.3	$[011]$	$[0\bar{1}1]$
GaAs(100)(doped)	6.9 ± 0.1	Undefined	
GaAs(111)	7.0 ± 0.2	$[\bar{1}10]$	$[11\bar{2}]$
InP(100)	4.3 ± 0.2	$[011]$	$[0\bar{1}1]$
InAs(polycrystalline)	3.5 ± 0.1	Undefined	

Orientations 1 and 2 are the two diagonals of the indentation mark.
Data from Ericson, F. et al., 1988, *Mater. Sci. Eng.* A 105/106, 131.

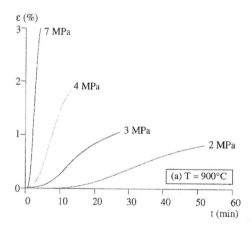

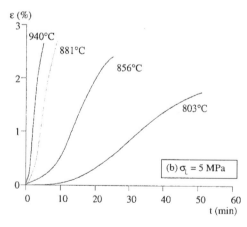

FIGURE 15.6 Creep curves for silicon (Reppich et al., 1964). (a) Varying applied stress at 900°C, and (b) varying temperature at 5 MPa applied stress. (Adapted from Reppich, B. et al., *Acta Met.* 12, 1283.)

a level of $2 \cdot 10^{18}$ cm^{-3} will cause an increase of $\dot{\varepsilon}_{max}$ by a factor of 10, whereas doping with B to the same level will lower the strain rate by a factor of 0.5 (Milvidskij et al., 1966).

Aging means that the mechanical properties of a material are changed by thermal, thermochemical, or thermomechanical processes during a period of time. If a semiconductor is *thermally aged* at elevated temperature in some process step, the Cottrell atmospheres around the dislocations may dissolve and the yield limit will be correspondingly lowered. On return to lower temperature, the Cottrell atmospheres

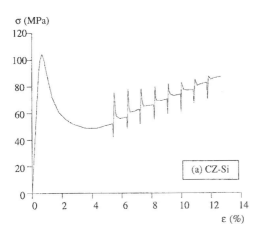

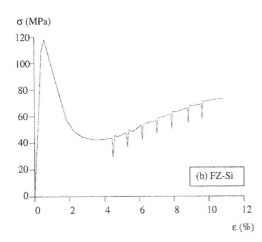

FIGURE 15.7 Strain aging behavior at 800°C of (a) Czochralski-grown Si and (b) float-zone-grown Si. (From Yasutake et al., 1982b.)

will not always reestablish themselves, but the impurities instead prefer to form small, particle-like precipitates, resulting in a maintained low yield limit, see Figure 15.5. Also, during plastic deformation these precipitates may act as sources for generation of new dislocations, hence affecting the strain hardening behavior of the material. For these reasons, the *thermal history* of a semiconductor material is of great importance to its plasticity behavior.

Strain aging is a thermomechanical effect causing recovery of a plasticized material after unloading, and a raised yield limit with yield drop upon reloading. Figure 15.7 shows examples of the strain aging behavior of CZ-Si and FZ-Si at 800°C (Yasutake et al., 1982b). Other thermomechanical phenomena are *thermal chock* and *thermal fatigue*. These effects are discussed in the Section 15.6.

As was previously mentioned, plastic deformation of a micromachined construction element usually is detrimental to the structure, and is therefore avoided by a proper choice of materials, ample dimensioning, or simply by avoiding overloads. On the other hand, micromachined structures offer one of the few existing means of investigating the plasticity and the thermomechanical behavior of thin films (Smith et al., 1991; Kristensen et al., 1991a,b; Ericson et al., 1991).

15.6 Fracture Properties

15.6.1 Fracture Limit and Fracture Toughness

First of all, it is important to clarify the fact that the *fracture limit* is not a materials property, but essentially a design property. This implies that the "intrinsic" strength properties of a material, for instance, expressed

779

in terms of the interatomic bond strength, are of less importance to the overall strength of a component than pure design factors such as geometric shape, surface roughness, how the load is applied, etc. Obviously, the intrinsic strength is not completely without importance, and is sometimes used to define an upper, theoretical strength limit, the so-called *theoretical fracture limit,* commonly given by

$$\sigma_{th} = \left(E\gamma / a_o \right)^{1/2}. \tag{15.30}$$

E is the Young's modulus, γ is the surface energy, and a_o is the distance between atomic planes parallel to the crack plane. In a generalized representation, the γ value in Equation 15.30 should be the *fracture surface energy,* i.e., one half of a cleavage energy including the broken bond energies as well as the energy of the dynamic elastic stress field around the propagating crack (this energy is eventually dissipated as heat) and energy dissipated or stored by plastic deformation or other irreversible processes during cracking. Usually σ_{th} is of an order of $E/10$ or $E/5$, but the practical fracture limit often is several orders of magnitude lower due to design factors.

Hence, the practical fracture limit, although important in design, is not a useful measure of the general fracture strength of a material. A more useful concept is the *fracture toughness* (= the critical stress intensity factor) K_{Ic}, which is considered to be a true, or nearly true, materials constant. For instance, for a body containing a stress-concentrating sharp crack of length c perpendicular to the applied load, K_{Ic} is related to the effective fracture limit σ_c of the body by:

$$K_{Ic} = Y\sqrt{\pi c}\ \sigma_c. \tag{15.31}$$

Y is a dimensionless factor which equals 1.12 for a surface crack and $1/\sqrt{2}$ for an interior crack. Both Y and σ_c are geometry dependent, but according to practical experience they are related in such a manner that K_{Ic} of Equation 15.31 becomes independent of crack geometry. Expressions similar to Equation 15.31 exist also for stress-concentrating geometries other than sharp cracks. The subscript "I" refers to fracture of mode I type, i.e., tensile cracking perpendicularly to the applied load. Modes II and III are cracking during longitudinal or transverse shearing, i.e., shearing in the crack propagation direction or transversely to it, respectively. In brittle materials, modes II and III rarely occur.

Similarly to σ_{th} of Equation 15.30, K_{Ic} can be related to the surface energy and the Young's modulus. For plane strain, we have

$$K_{Ic} = \left[2\gamma E / \left(1 - v^2 \right) \right]^{1/2}, \tag{15.32}$$

and, for plane stress,

$$K_{Ic} = \left[2\gamma E \right]^{1/2}. \tag{15.33}$$

Combining Equations 15.30 and 15.33, we obtain a formal and simple relationship between the theoretical fracture limit and the fracture toughness:

$$K_{Ic} = \sigma_{th} \sqrt{2a_o}. \tag{15.34}$$

TABLE 15.7 Fracture Data for Different Cleavage Planes in Si (Johansson et al.,1989)

Fracture Plane	E (GPa)	a_o (Å)	K_{Ic} (MPa m$^{1/2}$)	σ_{th} (GPa)	γ (Jm^{-2})
{100}	130	1.36	0.95	58	3.52
{110}	171	1.92	0.90	46	2.38
{111}	190	3.14	0.82	33	1.80

The K_{Ic} values are taken from Chen et al. (1980).

TABLE 15.8 K_{Ic} Values and γ Values Deduced from Erosion Results

Specimen	E (GPa)	ν	K_{Ic} (MPa m$^{1/2}$)	γ (Jm^{-2})
Si	163	0.22	0.94	2.58
Ge	132	0.21	0.60	1.30
GaAs	116	0.24	0.44	0.79
InP	89	0.29	0.36	0.67
InAs	76	0.28	0.32	0.62

The E and ν values are for polycrystalline specimens.
Data from Ericson, F. et al., 1988, *Mater. Sci. Eng.* A105/106, 131.

15.6.2 Some Fracture Data

In single-crystalline materials fracture data are anisotropic. In Table 15.7 fracture data for differently oriented cleavage planes in Si are listed (Johansson et al., 1989). These data are interrelated by Equations 15.33 and 15.34. It is seen that (111) is the low-energy cleavage plane, in agreement with common experience. In practice, cleavage is sometimes observed also in (110) planes in Si, whereas cleavage along (100) is difficult to achieve. In GaAs the preferential cleavage planes are of (110) type (Blakemore, 1982). Averaged (isotropic) fracture toughness data for a number of semiconductors have been determined by indentation and solid particle erosion techniques (Ericson et al., 1988). Fracture data from erosion experiments are listed in Table 15.8. In this case, the γ values have been calculated by means of Equation 15.32.

Measured fracture stresses for single-crystalline Si microbeams can be very high: 6.1 GPa on the average was found in one investigation (Ericson and Schweitz, 1990). For comparison, an extremely high-strength steel has a fracture limit of ~1 GPa. Introduction of a thermal oxide film 0.53 μm thick (which consumed the top 0.23 μm of the Si surface) was found to raise the fracture strength to 7.2 GPa; subsequent removal of the oxide by HF treatment lowered the strength marginally to 6.6 GPa, i.e., still 10% higher than the original strength. This behavior is explained by the fact that all submicron damage present in the original silicon surface was incorporated in the oxide film. Possibly some of the damage was "healed" in the oxide by the formation of oxygen bridges, which could explain the increased fracture limit of the oxidized beam. But more important is that the elastic modulus of SiO$_2$ is much lower than for Si, and the stress concentrations at the surface damage were correspondingly reduced. Subsequent removal of the oxide by HF treatment also removed the original surface defects. The high fracture limit after removal indicates that the resulting Si surface was less imperfect than the original surface. Tensile tests performed on surface micromachined polysilicon structures show a widespread in the presented data. Greek et al. (1997) presented measured fracture strength values of 566 and 768 MPa for two different polysilicon films, while Tsuchiya et al. (1997) presented values in the range of 2.0 to 2.7 GPa. The large difference between the two investigations is mainly due to differences in techniques used to etch out the test specimens. The reactive ion-etching process in the case of Greek et al. (1997) introduced more severe surface defects on their test specimens. Walker et al. (1990) exposed thin polysilicon membranes to HF solutions of various concentrations, and found opposite behavior; the fracture limits were drastically reduced. This result might be because no surface defects were removed (since no oxidization was performed); on the contrary, new defects might have been introduced by HF attacks on the grain boundaries.

The fracture strength of carefully bulk-micromachined GaAs has been found by Hjort et al. (1994) to be 2.7 GPa, which is about one half of the Si value, but certainly high enough to satisfy most demands on mechanical strength (several times higher than steel). The general notion that GaAs is extremely fragile as compared to Si stems from the fact that wafers or components of GaAs often contain more serious surface damage. This is a problem that is possible to overcome by a careful processing and gentle handling of the specimens.

15.6.3 Fracture Initiation

Knowing the fracture toughness K_{Ic}, and having experimentally determined the effective fracture limit σ_c of the body in question, Equation 15.31 opens the possibility of a postfailure evaluation of the size c of the defect which initiated the fracture. In brittle materials like micromachined silicon, fracture is almost always initiated at some surface defect, and sharp, vertical cracks in the surface are the most detrimental type of defects in this respect. Although we usually do not know the exact geometry of the initiating surface flaw, Equation 15.31 (with Y equal to 1.12) yields the c value of a defect which is equivalent to the real defect from a fracture viewpoint, but with an assumed, known geometry.

For microbeams, bulk micromachined by conventional KOH etching in a (001) Si wafer, the initiating defect size typically is of the order $c \sim 10$ nm (Johansson et al., 1988a, 1989; Ericson and Schweitz, 1990). In beams oriented along an $\langle 110 \rangle$ direction, the initiating flaws have been found to be twice as large as in beams oriented along $\langle 100 \rangle$ type directions. The effective fracture limit was found to be correspondingly lower in $\langle 110 \rangle$ beams than in $\langle 100 \rangle$ beams (~ 4 and ~ 6 GPa, respectively) (Johansson et al., 1989). These results are consistent with the observation that (110)-cleavage is energetically more favorable than (100)-cleavage (see Table 15.7), implying that the unintentional submicron cracks which initiate fracture might be more extended in (110) planes. In the final fracture process, on the other hand, "clean" fracture along crystallographic planes only occurs for low-strength beams, i.e., beams containing a serious crack-initiating defect. High-strength beams usually exhibit strongly irregular and "hackled" fracture surfaces. This is in agreement with the theory of crack propagation, stating that low-velocity cracking (caused by a low fracture limit) results in a smooth fracture surface — a so-called fracture mirror — whereas high-velocity cracking results in hackled fractures. Hence, simple fractographic observations may yield some qualitative information on the magnitude of the initiating defect.

15.6.4 Weibull Statistics

It is obvious from the discussion above that the size and shape distributions of the surface defects are of crucial importance to the strength properties of a brittle component, and that a probabilistic approach is called for. The most popular statistical theory of brittle fracture is due to Weibull (1939, 1951), and is based on a weakest link argument.

The *Weibull probability distribution function*,

$$P_f = 1 - \exp\left[-\int_v \left(\frac{\sigma_a(x,y,z) - \sigma_u}{\sigma_0} \right)^m dV \right],$$

(15.35)

gives the probability of failure of a body which is exposed to a stress distribution $\sigma_a(x,y,z)$. The stressed volume of the body is V, the stress σ_u is a lower limit which is usually equal to zero in brittle materials,

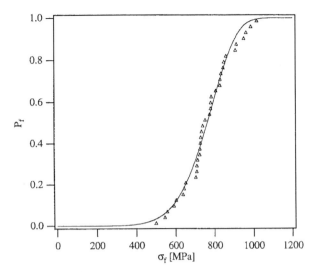

FIGURE 15.8 A plot showing the fracture probability as a function of the fracture stress from tensile tests of micromachined polysilicon beams. The Weibull modulus m, σ_0, as well as the expected mean fracture strength can be determined from the fitted curve. (Data from Greek, S. et al., 1997, *Thin Solid Films*, 292, 247–254.

σ_0 is a parameter related to the average fracture stress, and m is the *Weibull modulus*, which is a measure of the statistical scatter displayed by the fracture events. A low m value indicates a large scatter, whereas a high m value means low scatter. The Weibull modulus is an important measure of the engineering reliability of a material used in design. For simple analysis, the stress distribution within a structure can be expressed as

$$\sigma_a\big(x, y, z\big) = \sigma_a g \tag{15.36}$$

where σ_a is the maximum applied stress and g is a function that depends on the geometry of the structure and the load distribution. The fracture probabilities are calculated as

$$P_f = \frac{i - \frac{1}{2}}{n}, \tag{15.37}$$

where n is the total number of fracture tests and i is the index of the fracture stress result in an array where the values are written in ascending order.

A plot presenting the fracture probability as a function of applied stress for tensile tests on polysilicon film structures (Greek et al., 1997) is shown in Figure 15.8. The solid curve in Figure 15.8 is fitted by the chi-square method to the measured data, and the Weibull modulus, m, as well as σ_0 are derived from this fit. The expected mean fracture strength of the tests can be calculated from

$$\overline{\sigma}_a = \frac{1}{h^{1/m}}\,\Gamma\!\left(1 + \frac{1}{m}\right), \tag{15.38}$$

where $\Gamma(z)$ is the gamma function and h is given by

$$h = \frac{1}{\sigma_0^m} \int_V g^m \, dV. \tag{15.39}$$

783

TABLE 15.9 Typical Weibull Moduli m
of Different Ceramic Specimens

Material	m
Glass	2–3
China	8
Silicon wafer	10
Si_3N_4 (reaction sintered)	10
SiC (pressure sintered)	18
Si_3N_4 (isotropic pressure sintered)	25–40

The expected mean fracture strength can also be deduced from the fitted curve, see Figure 15.8, for $P_f =$ 0.5, and it is not equal to the arithmetical mean value. A Weibull modulus of 7 and an expected mean fracture strength of 768 MPa was deduced from the tensile test experiments shown in Figure 15.8.

Use of probabilistic design allows a trade-off in material selection between high strength and low scatter. An application requiring 0.99 reliability could be satisfied by, for example, a material with an effective strength of 120 MPa and an m of 8, or a material with a strength of 94 MPa and an m of 10, or a material with a strength of only 61 MPa and an m of 16. The advantage of probabilistic design is that these trade-offs are possible and can be integrated into the design analysis. This flexibility is not available in empirical and deterministic approaches. One drawback of the Weibull method is that a large number of tests (usually 100 or more) is required in order to determine the Weibull parameters with a reasonable degree of statistical confidence. Examples of typical Weibull moduli for different ceramic specimens are given in Table 15.9.

15.6.5 Fatigue

Static fatigue is a slow fracture process that is given many names in the literature: *delayed fracture, slow crack growth, subcritical crack growth, secondary crack growth,* or *stress rupture*. These names fairly well describe the phenomenon: at some constant load (below the nominal fracture limit of the body) stress concentrations at local defects in the material may cause an insidious crack growth. When the crack reaches a certain critical size, the effective cross-sectional area of the specimen has been sufficiently reduced as to make the specimen reach its nominal fracture limit, and a disastrous residual failure occurs. This phenomenon is well known in many ceramic materials, e.g., glass and SiO_2, and is partly considered to be due to *stress corrosion* at the crack tip. Static fatigue of Si has been reported, but this result has been disputed by others.

Dynamic fatigue is a fracture process caused by progressive crack growth during cyclic or intermittent loading. Similarly to static fatigue, the crack growth is initiated at some local point of weakness (usually in the surface) at a stress level which clearly falls below the nominal static fracture limit of the component. The crack will grow a small distance with every load cycle, and sometimes cause a characteristic pattern (so-called striations) in the fracture surface. When the effective cross section has been sufficiently reduced, a sudden, residual rupture will occur. The dynamic fatigue behavior of a material can be determined in a series of tests on identical specimens, where the amplitude of an applied, periodic load is systematically varied, and the corresponding lifetimes of the specimens are measured. Such load-vs.-lifetime plots are named Wöhler diagrams. Dynamic fatigue is a statistic phenomenon, and the theory is based on probabilistic arguments. In extensive fatigue tests, failure probability distributions can be determined and included in the diagram. Such Wöhler diagrams of PSN type (Probability of failure, Stress level, Number of cycles) are sometimes used to adjust the load amplitude to a desired probability of failure during a given lifetime. Dynamic fatigue is a fairly common phenomenon in metallic materials, where it is attributed to dislocation slip in intersecting slip systems and to cross-slip. In brittle ceramic materials, these mechanisms seldom operate at normal working temperatures, and consequently dynamic fatigue is rare in this type of material. It has never been reported for silicon, and is unlikely to occur in this material except perhaps under extreme thermal or environmental conditions.

Thermal fatigue is a special case of dynamic fatigue where the load cycling is caused by temperature changes and an associated variation in the thermal stress. The fatigue process is accelerated by the periodic temperature peaks, which causes thermal destabilization of the material, activates corrosion processes, speeds up diffusion processes, etc. Contrary to "normal" dynamic fatigue, which usually is of a high-cycle character, thermal fatigue often is of low-cycle character. *Thermal chock* is a well-known problem in brittle materials used for high-temperature applications. A rapid temperature change gives rise to a steep thermal gradient in the body, which fractures under the influence of the high local thermostress generated. Silicon and other conventional semiconductors are normally used in low-temperature applications and in chemically protected environments, and there is little reason to worry about thermal fatigue or thermal chock under these conditions. In the near future, however, materials like Si_3N_4 and SiC will be frequently used in micromachined structures for high-temperature applications, and phenomena such as thermal fatigue and thermal chock will have to be taken into consideration.

15.7 Adhesive Properties and Influence of Coatings

15.7.1 Adhesion

Adhesion is the single most important mechanical property of a film–substrate composite because without it the composite could not exist. It is of interest not only in the context of thin-film adhesion (Mittal, 1978; Valli, 1986), but equally so in all types of bonded or sealed structures (Johansson et al., 1988b,c). In MST, flaking is a frequently observed problem such as in silicon nitride films on silicon substrates. Similarly, the limited strength of bonded structures produced by, for example, field-assisted bonding (anodic bonding) or Si direct bonding (fusion bonding) has been a problem to many workers. Sometimes a deposited film will flake or spall off spontaneously as a result of strong internal stresses in the system. In other cases flaking is the result of external bending moments. In yet other cases neither internal nor external causes seem to be able to detach the film; it just sticks to the substrate for good. Sometimes a careful pretreatment of the surface combined with an equally careful optimization of the deposition parameters is sufficient to produce a well-adhered film.

Bad "mating" properties often are due to contaminants or oxides in the surfaces to be bonded, or to factors like crystalline misfit and thermal or elastic mismatch. Also the fundamental chemical affinity between the two materials, i.e., their mutual attraction or repulsion when it comes to forming interatomic bonds, is of importance. The latter factor, however, is not as important as one might assume at first sight, as will be further discussed below.

Reports of micromechanical adhesion tests are scarce in the literature, due partly to experimental difficulties and evaluation problems. However, since adhesion testing frequently turns up as an issue in discussions concerning layered structures, it is useful to highlight a few aspects of theory and practice. Note that unintentional sticking or stiction, due to electrostatic, van der Waals, or other weak interaction, is a different problem which is not included in this context.

Adhesive strength is commonly associated with the work required to break up the interface into two free surfaces, and therefore it is sometimes defined as the sum of the two free surface energies. This is an improper definition, however, since it excludes the chemical affinity. Disregarding all types of oxides and interfacial contaminants, a better definition is

$$\gamma_{AB}^{adh} = -\kappa\left(\gamma_A^s + \gamma_B^s\right) + \gamma_{AB}^{chem}, \tag{15.40}$$

which expresses the adhesive strength γ_{AB}^{adh} between materials A and B as a negative energy per unit area. This definition comprises one large, negative (attractive) term involving the two free surface energies γ_A^s and γ_B^s, and another smaller term accounting for the chemical affinity γ_{AB}^{chem} between A and B. The latter term can be either negative (attractive) or positive (repulsive) and, hence, tends to strengthen or weaken the interfacial bond. In the theory of alloy formation, chemical affinity is commonly expressed in terms of the heat of solution of A in B. The factor κ included in the first term accounts for the atomic misfit at the interface. For no atomic misfit κ equals unity; otherwise it is less than unity. For example, in a large-angle grain boundary κ is about 0.85 (Öberg et al., 1985). Since this way of defining adhesive strength is based on the work required to separate the two materials, the surface energies γ_A^s and γ_B^s are fracture surface energies which were previously defined in Equation 15.30.

The observation that the last term of Equation 15.40 is nearly always much smaller than the first term illustrates the interesting fact that a well-adjusted deposition or bonding process can produce strong adhesion even when the chemical affinity between the two materials is positive (repulsive).

Commonly in adhesion or bond-strength tests, a high-quality bond does not fracture in the interface but instead fractures nearby. The cracking usually occurs in the weaker of the two constituent materials, close to the interface, and runs parallel to it (Johansson et al., 1988b). A customary interpretation of this behavior is that the interface is "stronger than the weaker of the two materials." However, this is usually a misconception. Closer investigation of the fracture surface normally reveals that the cracking has indeed initiated at a flaw somewhere in the interface, and then has rather abruptly veered into one of the two bulk materials and, at some small distance from the interface, turned again to run parallel to it. This behavior does not reflect the high strength of the interface, but rather the asymmetric stress that develops during interfacial cracking between dissimilar materials.

15.7.2   Influence of Coatings

When discussing adhesive strength, interfacial cracking or delamination comes naturally into the argument. However, even in composites that do not delaminate under an applied load the fracture properties of the coatings are of interest from two different aspects: the fracture strength of the films themselves and their influence on the fracture strength of the substrate or composite. In this context we concentrate on transverse fracture, i.e., crack propagation in planes more or less orthogonal to the interface between film and substrate.

Surface coatings may have drastic effects on the strength of a component (Johansson et al., 1989). For instance, a hard brittle surface layer can reduce the fracture strength of an Si microbeam by as much as a factor of 10, whereas a ductile surface layer instead may increase the fracture strength. These effects depend on where in the layer–substrate composite fracture is initiated. An early cracking of a brittle surface layer may generate significant stress concentrations in the silicon substrate at the interface, and hence cause premature failure of the component. A ductile surface layer, on the other hand, may reduce the stress concentrations at microdefects in the silicon surface, and hence tend to postpone failure to higher loads. In Table 15.10 the effect of a number of surface layers on the fracture strength of Si beams

TABLE 15.10   Reinforcement Factors f for Various Coatings on Si Cantilever Beams Oriented along $\langle 110 \rangle$ and $\langle 100 \rangle$, Respectively Johansson et al. (1989)

Coating	Thickness (μm)	Si<110> f	Si<100> f
SiO$_2$	0.53	1.1	0.8
Al	0.41	—	1.0
Al	0.60	1.3	1.0
Ti	0.40	0.4	0.2
Ti	0.84	0.9	0.5
TiN	0.40	0.2	0.1
TiN	0.82	0.2	0.1

$f > 1$ implies strengthening, $f < 1$ implies weakening.
Data from Johansson, S. et al., 1989, *J. Appl. Phys.* 65, 122.

is illustrated in terms of a reinforcement factor f, which is larger than unity for strengthening and lower than unity for weakening (Johansson et al., 1989). From this table it is seen that magnetron-sputtered TiN can have a disastrous effect on the strength of the beams, whereas Al deposited by the same method has shown a somewhat strengthening effect in one case. It should be pointed out, however, that the effect of an Al coating is temperature dependent. High deposition or annealing temperatures may cause diffusion of Al into the Si substrate — so-called spiking — which is detrimental to the strength. This effect can be avoided by special measures.

Various types of crack initiation in layer composites are discussed by Johansson et al. (1989) in terms of the fracture toughness concept. Fracture toughnesses are not easily determined for thin coatings. There are, however, two interesting reports on a fracture toughness measurement on a released polysilicon film (Fan et al.,1990a; Kahn et al., 1996). In the experiment of Fan et al. (1990a) the test load was provided by the internal stress of the system itself and in the other case by an external probe.

15.8 Testing

15.8.1 General Test Structures and Testing Methods

In micromechanical property characterization the most common specimen structures are free-standing *cantilever beams*, *bridges*, or *membranes*. They are either a single-layer type, bulk or surface micromachined, or a two-layer type. In the latter case a thin substrate structure, e.g., a cantilever beam of silicon, may have been produced by bulk or surface micromachining prior to the deposition of the film. Alternatively, the film can first be deposited on the whole wafer, and the two-layer test structure fabricated by surface micromachining or combined surface and bulk micromachining.

Evaluation of elasticity properties, internal stresses, fracture properties, and, to some extent, plasticity properties by measurements on micromachined structures is performed by three main methods: (1) *static deflection*, i.e., a micromachined layer is deflected out-of-plane by external or internal forces, (2) *static tension or compression*, i.e., the structure is longitudinally strained in its plane, and (3) *dynamic testing*, usually of resonant structures, i.e., different modes of vibration are somehow excited in the structure.

Since the mechanical properties are of major concern within the field of MST, most measurements using micromachined structures have been made on materials used for MST, primarily silicon. However, investigations on polymers, metals, and ceramic materials have also been performed, and the present subsection aims to review these. We start with static or quasi-static measuring techniques, then proceed to dynamic testing.

15.8.2 Elasticity Testing by Static Techniques

Several different techniques for measuring elastic properties of micromachined structures have been devised, for example, bending experiments on cantilever beams. Such experiments require high-resolution micromechanical test equipment capable of simultaneous measurement of applied force and beam deflection on microsized specimens. Equipment capable of this is the Nanoindenter (Doerner et al., 1987; Weihs et al., 1988, 1989; Hong et al., 1989; Vinci and Braveman, 1991;) with a load resolution of 0.25 μN and a displacement resolution of 0.2 nm.

Using the Nanoindenter on single-layer beams, E values in reasonable agreement with literature data have been measured for Si, thermal and low-temperature SiO_2, and for Au (Weihs et al., 1988, 1989). The same technique has also been used to determine E values of nitride films in SiN_x/SiO_2 two-layer beams (Hong et al., 1989).

As will be further discussed below, from an error analysis viewpoint the best way to measure elastic moduli is by direct tensile testing. Several investigations have been carried out in this field during the last years. The most difficult part during a tensile test on a microscale is to measure the true strain in the test specimen. A solution to that problem was presented by Sharpe et al. (1997) where thin gold lines were deposited in a square shape onto surface-micromachined polysilicon test specimens. By using an

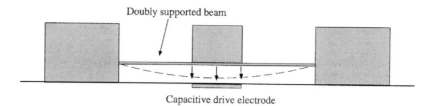

Doubly supported beam

Capacitive drive electrode

FIGURE 15.9 A principle sketch showing a doubly supported bridge structure for Young's modulus measurements. (Najafi, K. and Suzuki, K., 1989, in *Proc. IEEE Micro Electro Mechanical Systems*, Salt Lake City, UT, February 20–22, p. 96. With permission.)

optical interferometric technique the relative displacement between the gold lines could be measured yielding both the axial and the transverse strain. By combining the measured strain with the applied stress, both the Young's modulus and the Poisson ratio for polysilicon were determined. The measured data agree well with the calculated bulk values given in Table 15.3 and 15.4 in Section 4.3.2.

Young's modulus was also determined by Ogawa et al. (1996) from tensile tests of thin Al and Ti films, where the elongation was determined by measuring the relative displacement of two gauge marks on the specimens by a double-field-of-view light microscope combined with image analysis.

Other static techniques are concerned with deflection of membranes, bridges (doubly supported beams), or released structures of more complex geometries. In the membrane technique a thin diaphragm, circular or square, of the material to be tested is supported by a surrounding rigid frame. A uniform pressure is applied to one side of the membrane, and the bulging of the membrane is detected. This technique has been applied to membranes of polyimide (Allen et al., 1987), polysilicon (Maier-Schneider et al., 1995), Au and Al (Paviot et al., 1995) and SiC (Tong and Mehregany, 1992; Yamaguchi et al., 1995). Polysilicon membranes environmentally exposed to HF solutions during a period of time have been investigated in a similar way (Walker et al., 1990), and composite SiN_x/poly-Si membranes have been used to determine the Young's moduli of nitride films on polysilicon substrate (Tabata et al., 1989). Measurements on thin SiN_x single-layer membranes by applying a point load with a Nanoindenter have also been performed (Hong et al., 1990).

Young's modulus measurement by static deflection of a single-crystalline Si bridge has been carried out (Najafi and Suzuki, 1989). In this case the doubly supported beam was deflected electrostatically by a capacitive drive electrode at the middle of the beam (Figure 15.9). In another experimental setup, a low-stress SiN_x bridge was deflected by a stylus-type surface profiler (Tai and Muller, 1990). A related test configuration is the so-called bridge-slider, a long beam fixed at one end and released at the other. The released end is enclosed in a housing which allows movement only in the length direction of the beam. When the movable end is slid toward the fixed end, the beam bulges out-of-plane due to the compression, and Young's modulus can be evaluated from this deformation. Such experiments have been reported for polysilicon (Tai and Muller, 1990).

Other test geometries, such as ring-and-beam structures (Guckel et al., 1988) or T-shape structures (Allen et al., 1987) have also been suggested for static measurement of the elastic properties of released micromachined structures.

15.8.3   Elasticity Testing by Dynamic Techniques

Elastic data have been extracted from dynamic experiments with cantilever beam or bridge structures, and lately also with comb-drive structures suspended in elastic springs. Petersen and Guarnieri, 1979, pioneers in the field of dynamic microtesting, measured the transverse mechanical resonant frequencies of cantilever beams, micromachined in a number of thin insulating films deposited by various methods. The beam vibration was electrostatically excited, and the resonant frequency was measured by detection of the movement of a reflected laser beam. By using a similar technique, elastic modulus measurements have been carried out on annealed polysilicon cantilever beams (Putty et al., 1989) and on boron-doped, single-crystalline, $\langle 110 \rangle$-oriented silicon beams (Ding et al., 1990). Dynamic modulus measurement using microbridges (doubly supported beams) have been made on polysilicon (Guckel et al., 1988) and on boron-doped $\langle 100 \rangle$-oriented single-crystalline silicon beams (Zhang et al., 1991) Also, laterally moving comb-drive structures have been used to extract E values for polysilicon (Hirano et al., 1991; Pratt et al., 1991; Biebl et al., 1995) and electroless plated Ni films (Roy et al., 1995).

Determinations of the linear-elastic temperature coefficients for single-crystalline silicon were made by Bourgeois et al. (1995) with the aid of three different resonating structures.

15.8.4 Testing of Other Properties

Micromechanical testing of the *fracture strength* is performed by static loading of microbeams (cantilevers or bridges), but occasional strength tests on membranes have also been performed (Walker et al., 1990). Resonant experiments for micromechanical strength testing have not been reported so far, but a resonant micromechanical device for fatigue testing has been described (Connally and Brown, 1991). Strength tests have been reported for single-layer, single-crystalline beams of silicon (Johansson et al., 1988a; Ericson and Schweitz, 1990; Ding et al., 1990) and GaAs (Hjort et al., 1994), as well as on released single-layer beams of polysilicon (Tai and Muller, 1988), silicon oxide (Weihs et al., 1988; Nix, 1989), polyimide (Allen et al., 1987; Mehregany et al., 1987), and gold (Weihs et al., 1988). The fracture behavior of two-layer beams has been studied for SiO_2/Si (Johansson et al., 1989; Ericson and Schweitz, 1990) TiN/Si (Johansson et al., 1989) Ti/Si (Johansson et al., 1989; Ljungcrantz et al., 1993) and for Al/Si (Johansson et al., 1989).

Besides the beam-bending method, uniaxial tensile testing methods have recently been published. Greek et al. (1997) performed tensile tests on surface-micromachined polysilicon beams using a micro-manipulator in an SEM. The free end of the beam has the form of a ring that is gripped by the testing probe and moved by piezoelectric actuators. Other methods for gripping of the beam during a tensile test have been used by Tsuchiya et al. (1997) who utilized electrostatic forces built up by an applied voltage, and Read and Marshall (1996) who simply glued the test probe to the specimen. Gripping the test beam with a probe in a ring seems to be the best method since the rounded surfaces interact to align the test structure in the direction of the motion.

In principle, it is possible to study the *plasticity* behavior of thin elements micromachined into suitable test geometries by the same means as the elasticity or fracture behavior is studied. Elasticity and fracture, however, are the two extremes of deformation, whereas plasticity is a deviation from linear stress–strain behavior between these extremes. Hence, high-resolution measuring equipment is required to detect, for example, the onset of plastic flow (the yield limit) and the strain-hardening behavior of the material beyond that limit. Such measurements have been reported for free-standing gold cantilevers (Weihs et al., 1988, 1989), and the technique has been described in review articles by Nix (1989) and Vinci and Bravman (1991). The plastic yield limit of a gold film deposited by electron beam evaporation was determined to be 340 MPa, and a clear strain hardening effect (increased yield limit) was detected by repeated load–unload cycling.

Numerous methods for *internal stress* measurement exist, including X-ray diffraction techniques, techniques for measuring curvature, techniques for load deflection of membranes using buckling effects, and techniques utilizing surface-micromachined indicator structures giving an output depending on the internal stress state.

The most commonly used method is the wafer curvature technique, where the curvature is obtained by optical or mechanical surface-profiling measurements. The micromechanical beam curvature technique has been used by several authors, e.g., Johansson et al. (1989) and Ljungcrantz et al. (1993), who studied sputter-deposited coatings of Al, Ti, and TiN on Si cantilever beams. The beam deflections in these cases were measured optically and gave the biaxial stress state in the film materials. One general comment on a fundamental difference between the commonly used method of X-ray diffraction (XRD) for stress analysis, on the one hand, and the micromechanical beam curvature technique, on the other, should be emphasized in the present context. In the XRD method the lattice dilation (elastic strain) is

789

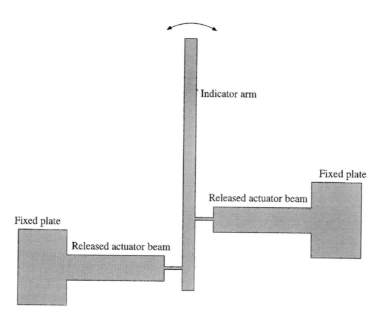

FIGURE 15.10 A principle sketch of a rotating structure for measurement of the internal stress (van Drieënhuizen et al., 1993; Lin et al., 1993; Ericson et al., 1997; Zhang et al., 1997). Internal strain in the two actuator beams is converted to a rotation of the indicator arm when the whole structure (except for the fixed plates) is released from the substrate.

measured, and knowledge of the elastic constants is required to evaluate the elastic stress. Often such information is missing for thin films, in which cases the two-layer beam curvature method is an attractive alternative. In this method the strain is evaluated from the observed curvature, and the stress from considerations of the equilibrium of forces and moments, without the use of any materials parameters for the film. Hence, the internal stress and strain can be obtained for a completely unknown thin-film material on a known substrate material.

To determine the internal stress state in a free-standing thin film other methods have to be applied because the stress state in the film changes when it is freed from the substrate. Bulge testing is one of these methods where a uniform pressure is applied to a membrane and the measured deflection not only yields the Young's modulus but also the internal film stress. Tabata et al. (1989) made measurements on polysilicon and silicon nitride, Maier-Schneider et al. (1995) on polysilicon, and Paviot et al. (1995) investigated Au and Al thin films.

Different kinds of free-standing micromachined structures have also been utilized to derive the stress state more locally than the other techniques. van Drieënhuizen et al. (1993) compared different structures, for instance, a so-called diamond structure capable of measuring both compressive and tensile stress. Also a rotating structure, where an internal strain is converted to a rotation of an indicator arm, was presented and realized in polysilicon (Figure 15.10). Rotating structures of different designs have also been employed by other authors (Lin et al., 1993; Ericson et al., 1997; Zhang et al., 1997) to study thin films and thick films of polysilicon and thin films of silicon nitride.

If the stress in the film material not only is of a biaxial nature but also varies through the thickness of the film, it could be disastrous for micromechanical components. To measure these internal stress gradients, Fan et al. (1990b) presented spiral microstructures freed from the substrate which bent into different shapes depending on the sign of the gradient. Also cantilever beams etched out of the film material bending upward or downward have been used in the same sense as the spirals (Ericson et al., 1997).

Several types of *adhesion* tests have been devised, for example, the "Scotch tape" test, the scratch test, the linear peel test, and the inflated membrane test (Valli, 1986; Mittal, 1978). The latter test has been performed on micromachined structures (Senturia et al., 1987), but in general little work on adhesion testing has been done within micromechanics. This is probably because in most practical cases a simple scratch test is sufficient to determine if a coating is prone to flake off. If the coating flakes off, it is a good idea to try to modify the surface pretreatment and the deposition process until it does not. Some type of proof testing might be necessary. But usually it is not necessary, or even possible, to evaluate such a test in terms of a well-defined adhesive strength parameter. Normally, a simple ranking in terms of some externally applied load, for instance, a critical pressure in an inflated membrane test or a critical load in a scratch test, is quite sufficient.

15.9   Modeling and Error Analysis

The adequacy of an evaluation model is just as important as the accuracy and reproducibility of the experimental procedure. A theoretical model describing a real physical process will always be idealized to some extent. For instance, simplifying assumptions are usually made regarding boundary conditions and deformation states. Care must be taken to define the geometric structure properly, to use the correct constitutive relationships between stress and strain, and to introduce a realistic overall deformation behavior into the model, etc. Such care has not always been taken in the past, resulting in contradictory results from, e.g., different elastic property measurements on micromachined structures.

Naturally, the choice of evaluation model is governed by the choice of experimental method, the geometric test configuration, the testing conditions, etc. The existing models are too numerous to allow a detailed analysis of their merits and limitations. As one example, however, bending of a micromachined cantilever beam will be discussed, and some basic relationships given. These relationships will be used for comparisons with other types of tests, and taken as starting points for some fundamental considerations concerning the error analysis.

15.9.1   Single-Layer Beam

The maximum stress σ_m, suffered by a cantilever beam of length L which is deflected by a perpendicular force F applied to its free end is

$$\sigma_m = FL/\left(I/c\right), \tag{15.41}$$

where I is the moment of inertia of the beam cross section and c is the distance between the neutral layer (the zero-stress plane during bending) and the upper beam surface. For a rectangular beam cross section, we have

$$I = wt^3/12, \tag{15.42}$$

$$c = t/2. \tag{15.43}$$

It is important to note that no material properties (e.g., Young's modulus) appear in the stress evaluation model of Equation 15.41. The single experimental input variable that enters this expression is the applied load F; all other parameters are geometric constants. Not even the deflection of the beam, caused by F, need be known. From an error analysis viewpoint, one observes that the evaluated stress depends linearly (or inverse linearly) on all input parameters, except on the beam thickness which enters as t^{-2}. Consequently, a relative thickness error will be doubled in the evaluated stress, whereas all other relative input

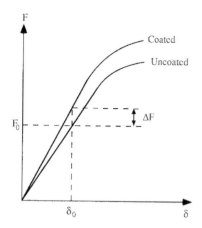

FIGURE 15.11 Schematic illustration of force vs. deflection curves for uncoated and coated beams.

errors add up in the stress error without amplification. This is an important observation, since it is not trivial to make an accurate measurement of the thickness of a micromachined element.

From a beam-bending experiment, the Young's modulus can be determined from the following expression:

$$E = FL^3\left(1-v^2\right)\Big/\left(3\delta I\right), \quad \left(\delta \ll L\right), \tag{15.44}$$

where δ is the deflection of the beam at the loading point. F and δ must be independently measured in the experiment, and both values enter Equation 15.44 to the first power. Two geometric parameters, L and t (via I), enter the expression to the third power. For this reason this evaluation model puts high demands on the numerical accuracy of the geometric parameters. In micromachined structures these demands are not always easy to meet, especially not concerning the thickness. For this reason, bending experiments on cantilever beams are not well suited for determining E. Pure tensile tests are better since in this configuration the thickness enters the E expression to the first power.

The above discussion is centered on the problem of the thickness error, which in practice is one of the most important sources of errors. There are, however, several other such sources which should be taken into consideration in the evaluation of a beam-bending experiment; for instance, erroneous input data or errors due to oversimplifying assumptions in the deformation model. These errors are systematically reviewed by Schweitz (1992), and suitable corrections are suggested. In general, the evaluation of an experiment benefits greatly from numerical simulations (FEA), as complements to the analytical modeling.

15.9.2 Two-Layer Beam

Analytical models for evaluating bending experiments with two-layer beams also exist (Johansson et al., 1989). In these models the elastic properties of the substrate material usually must be known, in order to evaluate the properties of the deposited film material. There exists one comparative test method, however, which does not require explicit knowledge of the substrate properties (Schweitz, 1991). According to this method, one beam of the substrate material (or one set of beams) is first tested in an uncoated condition, and force-vs.-deflection curves are registered. Next the beam is coated by, e.g., sputtering or evaporation techniques, and then tested again. The *difference* between the $F(\delta)$ curves of the coated and the uncoated beams, respectively, is caused by the film (Figure 15.11). For any given deflection δ_0 the film stress can be expressed in terms of the *increase* in force (ΔF) required to deflect the coated beam to the same δ_0 value as the uncoated beam:

$$\sigma_f = \frac{2L\Delta F}{wt_f\left(t_s+t_f\right)}, \quad \left(t_f \ll t_s\right). \tag{15.45}$$

σ_f is the maximum film stress, and w is the beam width. In this method the stress can be determined without using constitutive relationships for the two materials, and hence without previous knowledge of the elasticity or plasticity parameters of any of the materials. From Equations 15.41 and 15.45 and Hooke's law, the Young's modulus of the film can be evaluated. Assuming that $t_f \ll t_s$, and E_f *not* $\gg E_s$, a first approximation is (Schweitz, 1992)

$$E_f \approx \frac{1}{3}\left(\frac{t_s}{t_f}\right)\left(\frac{\Delta F}{F_0}\right)E_s, \tag{15.46}$$

where F_0 is the force required to give the uncoated beam the designated deflection δ_0.

15.9.3 Resonant Beam

In resonant experiments with single-layer cantilever beams, the resonant frequency ω is related to the beam geometry and its material properties by

$$\omega = 0.162\,\frac{t}{L^2}\left(\frac{E}{\rho}\right)^{1/2}, \tag{15.47}$$

where ρ is the mass density of the beam material. E can be evaluated using this relationship. In this evaluation model, E is proportional to t^{-2} (static beam bending: t^{-3}) and to L^4 (static beam bending: L^3). Hence, the dynamic model is less sensitive to thickness errors than the model of static deflection, but still more sensitive to the same error than the model of static in-plane tension.

15.9.4 Micro vs. Bulk Results

It has been pointed out that different measurements of the Young's modulus on micromachined structures yield contradictory results, and that "micro" results sometimes disagree with generally accepted bulk values (Schweitz, 1992). In some cases, when the E values measured on micromachined silicon structures have been found to differ drastically (up to 30%) from the bulk value, the influence of the dopants has been suggested to be the cause. However, according to alloying theory, not even the most extreme doping levels in silicon would affect the E value to more than a few percent (Hall, 1967; Kim, 1981). Dislocation networks generated by the doping process have also been suggested as possible causes. But dislocations do not in general affect the E value much, not even in high density, except under very special circumstances which are unlikely to prevail in doped silicon at room temperature. The high surface-to-volume ratio of thin microelements has also been proposed as an explanation. But the influence of surface effects on the elastic properties becomes negligible for thicknesses exceeding a few nanometers. Grain size or texture may have some effect (see Section 15.4.3.2) for narrow structures, but this effect still is too small and too scarce to explain the alleged difference between micro and bulk results.

In fact, to date no convincing physical explanation to the observed discrepancies between the accepted bulk E value, on the one hand, and recently measured "micro" values on the other, has been put forward. But many of the reported micromechanical experiments appear to have been evaluated by means of oversimplified theoretical models. This, in combination with uncertain input values, at present seems to be the most probable cause of the observed scatter in the "micro" values, as well as of the discrepancy between some "micro" results and bulk values.

15.10   Summary and Conclusions

This chapter is not primarily intended for readers who are specialists in materials science, but rather for workers who are electronics engineers or other types of specialists within the multidisciplinary field of

MST. For this reason, some basics of crystal cohesion, elasticity, plasticity, thermomechanics, and fracture have been reviewed, along with a few mechanical systems properties of particular interest within micromechanics: internal stresses and adhesion.

Mechanical testing of micromachined structures and systems is of paramount significance to the development of microdevices with high demands on mechanical integrity and long-term reliability. In this context the accuracy and reproducibility of the testing procedure, as well as the adequacy of the evaluation method, is of central importance. The most serious problem today is perhaps the poor reproducibility of test results. The experimental techniques used are not consistent, and other workers find it difficult to verify results. The best way to bring this problem under control probably is to introduce a limited number of well-defined, standardized test procedures that can be performed and evaluated by independent researchers. This could be an important task for, e.g., an alert, international research organization in the field of microstructure technology.

References

Alexander, H. and Haasen, P. (1968), "Dislocations and Plastic Flow in the Diamond Structure," *Solid State Phys.* 22, 27.

Allen, M.G., Mehregany, M., Howe, R.T., and Senturia, S.D. (1987), "Microfabricated Structures for the in Situ Measurement of Residual Stress, Young's Modulus, and Ultimate Strain of Thin Films," *Appl. Phys. Lett.* 51, 241.

Biebl, M., Brandl, G., and Howe, R.T. (1995), "Young's Modulus of in Situ Phosphorus-Doped Polysilicon," in *Digest of Technical Papers from the 8th Int. Conf. on Solid-State Sensors Actuators, and Eurosensors IX*, Stockholm, Sweden, June 25–29, pp. 80–83.

Blakemore, J.S. (1982), "Semiconducting and Other Major Properties of Gallium Arsenide," *J. Appl. Phys.* 53, R123.

Bourgeois, C., Hermann, J., Blanc, N., deRooij, N.F., and Rudolf, F. (1995), "Determination of the elastic temperature coefficients of monocrystalline silicon," in *Digest of Technical Papers from the 8th Int. Conf. on Solid-State Sensors Actuators, and Eurosensors IX*, Stockholm, Sweden, June 25–29, pp. 92–95.

Brantley, W.A. (1973), "Calculated Elastic Constants for Stress Problems Associated with Semiconductor Devices," *J. Appl. Phys.* 44, 534.

Chen, C.P. and Leipold, M.H. (1980), "Fracture Toughness of Silicon," *Am. Ceram. Soc. Bull.* 59, 469.

Connally, J.A. and Brown, S.B. (1991), "Micromechanical Fatigue Testing," in *Digest of Technical Papers, Int. Conf. on Solid-State Sensors and Actuators (Transducers '91)*, p. 953.

Ding, X., Ko, W.H., and Mansour, J.M. (1990), "Residual Stress and Mechanical Properties of Boron-Doped p$^+$-Silicon Films," *Sensors Actuators* A21–A23, pp. 866.

Doerner, M.F., Gardner, D.S., and Nix, W.D. (1987), "Plastic Properties of Thin Films on Substrates as Measured by Submicron Indentation Hardness and Substrate Curvature Techniques," *J. Mater. Res.*, 1, 845.

Ericson, F. and Schweitz, J.-Å. (1990), "Micromechanical Fracture Strength of Silicon," *J. Appl. Phys.* 68, 5840.

Ericson, F., Johansson, S., and Schweitz, J.-Å. (1988), "Hardness and Fracture Toughness of Semiconducting Materials Studied by Indentation and Erosion Techniques," *Mater. Sci. Eng.* A105/106, 131.

Ericson, F., Kristensen, N., Schweitz, J.-Å., and Smith, U. (1991), "A Transmission Electron Microscopy Study of Hillocks in Thin Aluninum Films," *J. Vac. Sci. Technol.* B9, 58.

Ericson, F., Greek, S, Söderkvist, J., and Schweitz, J.-Å. (1997), "High-Sensitivity Surface Micromachined Structures for Internal Stress and Stress Gradient Evaluation," *J. Micromech. Microeng.* 7, 30–36.

Fan, L.S., Howe, R.T., and Muller, R.S. (1990a), "Fracture Toughness Characterization of Brittle Thin Films," *Sensors Actuators* A21–A23, 872.

Fan, L.S., Muller, R.S., Yun, W., Howe, R.T., and Huang, J. (1990b), "Spiral Microstructures for the Measurement of Average Strain Gradients in Thin Films," in *Proc. IEEE Micro Electro Mechanical Systems*, Napa Valley, CA, February 11–14, pp. 177–181.

Greek, S., Ericson, F., Johansson, S., and Schweitz, J.-Å. (1997), "In Situ Tensile Strength Measurements and Weibull Analysis of Thick Film and Thin Film Micromachined Polysilicon Structures," *Thin Solid Films* 292, 247–254.

Guckel, H., Burns, D.W., Tilmans, H.A.C., deRoo, D.W., and Rutigliano, C.R. (1988), "Mechanical Properties of Fine Grained Polysilicon — The Repeatability Issue," *Technical Digest of IEEE Solid-State Sensor and Actuator Workshop*, Hilton Head Island, June 6–9, pp. 96.

Hall, J.J. (1967), "Electronic Effects in the Elastic Constants of *n*-Type Silicon," *Phys. Rev.* 161, 756.

Hirano, T., Furuhata, T., Gabriel, K.J., and Fujita, H. (1991), "Operation of Sub-Micron Gap Electrostatic Comb-Drive Actuators," in *Digest of Technical Papers, Int. Conf. on Solid-State Sensors and Actuators (Transducers '91)*, pp. 873.

Hirsch, P.B. (1983), "Effect of Doping on Mechanical Properties, Recrystallisation and Diffusion in Semiconductors," *Inst. Phys. Conf. Ser.* No. 67, 1.

Hirsch, P.B. (1985), "Dislocations in Semiconductors," *Mater. Sci. Technol.* 1, 666.

Hjort, K., Ericson, F., and Schweitz, J.-Å. (1994), "Micromechanical Fracture Strength of Semi-Insulating GaAs," *Sensors Mater.* 6, 359–367.

Hong, S., Weihs, T.P., Bravman, J.C., and Nix, W.D. (1989), "The determination of mechanical parameters and residual stresses for thin films using micro-cantilever beams," in *Thin Films: Stresses and Mechanical Properties, Symp. Proc.* (J.C. Bravman, W.D. Nix, D.M. Barnett, and D.A. Smith, eds.), Vol. 130, pp. 93, Materials Research Society, Pittsburgh.

Hong, S., Weihs, T.P., Bravman, J.C., and Nix, W.D. (1990), "Measuring Stiffnesses and Residual Stresses of Silicon Nitride Thin Films," *J. Electron. Mater.* 19, 903.

Imai, M. and Sumino, K. (1983), "In Situ X-Ray Topographic Study of the Dislocation Mobility in High-Purity and Impurity-Doped Silicon Crystals," *Philos. Mag.* A47, 599.

Johansson, S., Schweitz, J.-Å., Tenerz, L., and Tirén, J. (1988a), "Fracture Testing of Silicon Microelements in Situ in a Scanning Electron Microscope," *J. Appl. Phys.* 63, 4799.

Johansson, S., Gustafsson, K., and Schweitz, J.-Å. (1988b), "Strength Evaluation of Field-Assisted Bond Seals between Silicon and Pyrex Glass," *Sensors Mater.* 3, 143.

Johansson, S., Gustafsson, K., and Schweitz, J.-Å. (1988c), "Influence of Bonded Area Ratio on the Strength of FAB Seals between Silicon Microstructures and Glass," *Sensors Mater.* 4, 209.

Johansson, S., Ericson, F., and Schweitz, J.-Å. (1989), "Influence of Surface Coatings on Elasticity, Residual Stresses, and Fracture Properties of Silicon Microelements," *J. Appl. Phys.* 65, 122.

Kahn, H., Stemmer, S., Nandakumar, K., Heuer, A.H., Mullen, R.L., Ballarini, R., and Huff, M.A. (1996), "Mechanical Properties of Thick Surface Micromachined Polysilicon Films," in *Proc. IEEE Micro Electro Mechanical Systems*, San Diego, February 11–15, pp. 343–348.

Kim, C.K. (1981), "Electronic Effect on the Elastic Constant C_{44} of n-Type Silicon," *J. Appl. Phys.* 52, 3693.

Kristensen, N., Ericson, F., Schweitz, J.-Å., and Smith, U. (1991a), "Grain Collapses in Strained Aluminum Thin Films," *J. Appl. Phys.* 69, 2097.

Kristensen, N., Ericson, F., Schweitz, J.-Å., and Smith, U. (1991b), "Hole Formation in Thin Aluminium Films under Controlled Variation of Strain and Temperature," *Thin Solid Films* 197, 67.

Lin, L., Howe, R.T., and Pisano, A.P. (1993), "A Passive, in Situ Micro Strain Gauge," in *Proc. IEEE Micro Electro Mechanical Systems*, Fort Lauderdale, February 7–10, pp. 201–206.

Ljungcrantz, H., Hultman, L., Sundgren, J.-E., Johansson, S., Kristensen, N., Schweitz, J.-Å., and Shute, C.J. (1993), "Residual Stresses and Fracture Properties of Magnetron Sputtered Ti Films on Si Microelements," *J. Vac. Sci. Technol.* A11, 543–553.

Maier-Schneider, D. (1995), *LPCVD-polysilizium in der mikromechanik: Bestimmung der elstischen Eigenschaften*, Reihe 18, VDI-Verlag GmbH, Düsseldorf.

Maier-Schneider, D., Maibach, J., Obermeier, E., and Schneider, D. (1995), "Variations in Young's Modulus and Intrinsic Stress of LPCVD-Polysilicon Due to High Temperature Annealing," *J. Micromech. Microeng.* 5, 121–124.

Mehregany, M., Howe, R.T., and Senturia, S.D. (1987), "Novel Microstructures for the in Situ Measurement of Mechanical Properties of Thin Films," *J. Appl. Phys.* 62, 3579.

Milvidskij, M.G., Osvenskij, V.B., and Stolyarov, O.G. (1966), "Effect of Alloying on the Creep of Single-Crystal Silicon," *Neorg. Mater.* 2, 585.

Mittal, K.L., ed. (1978), *Adhesion Measurement of Thin Films, Thick Films, and Bulk Coatings*, STP 640, ASTM, Philadelphia.

Najafi, K. and Suzuki, K. (1989), "A Novel Technique and Structure for the Measurement of Intrinsic Stress and Young's Modulus of Thin Films," in *Proc. IEEE Micro Electro Mechanical Systems*, Salt Lake City, UT, February 20–22, p. 96.

Nix, W.D. (1989), "Mechanical Properties of Thin Films," *Met. Trans.* 20A, 2217.

Öberg, Å., Mårtensson, N., and Schweitz, J.-Å. (1985), "Fundamental Aspects of Formation and Stability of Explosive Welds", *Met. Trans.* 16A, 841.

Ogawa, H., Ishikawa, Y., and Kitahara, T. (1996), "Measurements of Stress–Strain Diagrams of Thin Films by a Developed Tensile Machine," in *Proceedings of SPIE, Microlithography and Metrology in Micromachining II*, Austin, TX, October 14–15, 2880, 272–277.

Paviot, V.M., Vlassak, J.J., and Nix, W.D. (1995), "Measuring the Mechanical Properties of Thin Metal Films by Means of Bulge Testing of Micromachined Windows," *Mater. Res. Soc. Symp. Proc.* 356, 579–584.

Petersen, K.E. and Guarnieri, C.R. (1979), "Young's Modulus Measurements of Thin Films Using Micromechanics," *J. Appl. Phys.* 50, 6761.

Pratt, R.I., Johnson, G.C., Howe, R.T., and Chang, J.C. (1991), "Micromechanical Structures for Thin Film Characterization," in *Digest of Technical Papers, Int. Conf. on Solid-State Sensors and Actuators (Transducers '91)*, pp. 205.

Putty, M.W., Chang, S.C., Howe, R.T., Robinson, A.L., and Wise, K.D. (1989), "One-Port Active Polysilicon Resonant Microstructures," in *Proc. IEEE Micro Electro Mechanical Systems*, Salt Lake City, Utah, February 20–22, pp. 60.

Read, D.T. and Marshall, J.C. (1996), "Measurements of Fracture Strength and Young's Modulus of Surface Micromachined Polysilicon," in *Proceedings of SPIE, Microlithography and Metrology in Micromachining II*, Austin, TX, October 14–15, Vol. 2880, pp. 56–63.

Reppich, B., Haasen, P., and Ilschner, B. (1964), "Creep of Silicon Single Crystals," *Acta Met.* 12, 1283.

Roy, S., Furukawa, S., Miyaima, H., and Mehregany M. (1995), "In Situ Measurement of Young's Modulus and Residual Stress of Thin Electroless Nickel Films for MEMS Applications," *Mater. Res. Soc. Symp. Proc.* 356, 573–578.

Schweitz, J.-Å. (1991), "A New and Simple Micromechanical Approach to the Stress–Strain Characterization of Thin Coating," *J. Micromech. Microeng.* 1, 10–15.

Schweitz, J.-Å. (1992), "Mechanical Characterization of Thin Films by Micromechanical Techniques," *MRS Bull.* XVII, 34–45.

Senturia, S.D. (1987), "Microfabricated Structures for the Measurement of Mechanical Properties and Adhesion of Thin Films," in *Digest of Technical Papers, 4th Int. Conf. on Solid-State Sensors and Actuators* (Inst. of Electrical Engineers of Japan), p. 11.

Sharpe, W.N., Jr., Yuan, B., and Vaidyanathan, R. (1997), "Measurements of Young's Modulus, Poisson's Ratio, and Tensile Strength of Polysilicon," in *Proc. IEEE Micro Electro Mechanical Systems*, Nagoya, Japan, January 26–30, pp. 424–429.

Simmons, G. and Wang, H. (1971), *Single Crystal Elastic Constants and Calculated Aggregate Properties — A Handbook*, 2nd ed., The M.I.T. Press, Cambridge, MA.

Smith, U., Kristensen, N., Ericson, F., and Schweitz, J.-Å. (1991), "Local Stress Relaxation Phenomena in Thin Aluminum Films," *J. Vac. Sci. Technol.* A9, 2527.

Sumino, K. (1983a), "Interactions between Dislocations and Impurities in Silicon," *J. Phys.* C4, 44.

Sumino, K. (1983b), "Interaction of Dislocations with Impurities and Its Influence on the Mechanical Properties of Silicon Crystals," *Mater. Res. Soc. Symp. Proc.* 14, 307.

Sumino, K. and Imai, M. (1983), "Interaction of Dislocations with Impurities in Silicon Crystals Studied by in Situ X-Ray Topography," *Philos. Mag.* A47, 753.

Sumino, K., Harada, H., and Yonenaga, I. (1980), "The Origin of the Difference in the Mechanical Strengths of Czochralski-Grown Silicon and Float-Zone Grown Silicon," *Jpn. J. Appl. Phys.* 19, L49.

Sumino, K., Yonenaga, I., Imai, M., and Abe, T. (1983), "Effects of Nitrogen on Dislocation Behavior and Mechanical Strength in Silicon Crystals," *J. Appl. Phys.* 54, 5016.

Sumino, K., Yonenaga, I., Shibata, H., and Onose, U. (1985), in *Proc. 27th Meeting,* 145 Committee of JSPS, p. 91.

Tabata, O., Kawahata, K., Sugiyama, S., and Igarashi, I. (1989), "Mechanical Property Measurements of Thin Films Using Load-Deflection of Composite Rectangular Membranes," *Sensors Actuators,* 20, 135.

Tai, Y.C. and Muller, R.S. (1988), "Fracture Strain of LPCVD Polysilicon," in *Technical Digest, IEEE Solid-State Sensor and Actuator Workshop,* Hilton Head Island, SC, June 6–9, pp. 88.

Tai, Y.-C. and Muller, R.S. (1990), "Measurement of Young's Modulus on Microfabricated Structures Using a Surface Profiler," in *Proc. IEEE Micro Electro Mechanical Systems,* Napa Valley, CA, February 11–14, pp. 147.

Thokala, R. and Chaudhuri, J. (1995), "Calculated Elastic Constants of Wide Band Gap Semiconductor Thin Films with a Hexagonal Crystal Structure for Stress Problems," *Thin Solid Films* 266, 189–191.

Tong, L. and Mehregany, M. (1992), "Mechanical Properties of 3C Silicon Carbide," *Appl. Phys. Lett.* 60, 2992–2994.

Tsuchiya, T., Tabata, O., Sakata, J., and Taga, Y. (1997), "Specimen Size Effect on Tensile Strength of Surface Micromachined Polycrystalline Silicon Thin Films," in *Proc. IEEE Micro Electro Mechanical Systems,* Nagoya, Japan, January 26–30, pp. 529–534.

Valli, J. (1986), "A Review of Adhesion Test Methods for Thin Hard Coatings," *J. Vac. Sci. Technol.* A4, 3007.

van Drieënhuizen, B.P., Goosen, J.F.L., French, P.J., and Wolffenbuttel, R.F. (1993), "Comparison of Techniques for Measuring Both Compressive and Tensile Stress in Thin Films," *Sensors Actuators,* A37–38, 756–765.

van Enckevort, W.J.P. (1994), "Physical, Chemical and Microstructural Characterization and Properties of Diamond," in *Synthetic Diamond: Emerging CVD Science and Technology* (K.E. Spear and J.P. Dismukes, eds), John Wiley and Sons, New York, pp. 307–353.

Vinci, R.P. and Bravman, J.C. (1991), "Mechanical Testing of Thin Films," in *Digest of Technical Papers, Int. Conf. on Solid-State Sensors and Actuators (Transducers '91),* p. 943.

Walker, J.A., Gabriel, K.J., and Mehregany, M. (1990), "Mechanical Integrity of Polysilicon Films Exposed to Hydrofluoric Acid Solutions," in *Proc. IEEE Micro Electro Mechanical Systems,* Napa Valley, CA, Feb. 11–14, pp. 56.

Weibull, W. (1939), *Ingenjörsvetenskapsakad. Handl.* 151, 44.

Weibull, W.A. (1951), "A Statistical Distribution Function of Wide Applicability," *J. Appl. Mech.* 18, 293.

Weihs, T.P., Hong, S., Bravman, J.C., and Nix, W.D. (1988), "Mechanical Deflection of Cantilever Microbeams: A New Technique of Testing the Mechanical Properties of Thin Films," *J. Mater. Res.* 3, 931.

Weihs, T.P., Hong, S., Bravman, J.C., and Nix, W.D. (1989), "Measuring the Strength and Stiffness of Thin Film Materials by Mechanically Deflecting Cantilever Microbeams," in *Thin Films: Stresses and Mechanical Properties, Symp. Proc.* (J.C. Bravman, W.D. Nix, D.M. Barnett, and D.A. Smith, eds.), Vol. 130, p. 87, Materials Research Society, Pittsburgh.

Windischmann, H. (1992), "Intrinsic Stress in Sputter-Deposited Thin Films," *Crit. Rev. Solid State Mater. Sci.* 17, 547–596.

Yamaguchi, Y., Nagasawa, H., Shoki, T., and Annaka, N. (1995), "Properties of Heteroepitaxial 3C-SiC Films Grown by LPCVD," in *Digest of Technical Papers from the 8th Int. Conf. on Solid-State Sensors Actuators, and Eurosensors IX,* Stockholm, Sweden, June 25–29, pp. 190–193.

Yasutake, K., Murakami, J., Umeno, M., and Kawabe, H. (1982a), "Mechanical Properties of Heat-Treated CZ-Si Wafers from Brittle to Ductile Temperature Range," *Jpn. J. Appl. Phys.* 21, L288.

Yasutake, K., Umeno, M., and Kawabe, H. (1982b), "Compression Tests of Heat-Treated Czochralski-Grown Silicon Crystals," *Phys. Status Solidi (a)* 69, 333.

Yonenaga, I. and Sumino, K. (1978), "Dislocation Dynamics in the Plastic Deformation of Silicon Crystals," *Phys. Status Solidi (a)* 50, 685.

Yonenaga, I. and Sumino, K. (1984), "Role of Carbon in the Strengthening of Silicon Crystals," *Jpn. J. Appl. Phys.* 23, L590.

Zhang, L.M., Uttamchandari, D., and Culshaw, B. (1991), "Measurement of the Mechanical Properties of Silicon Microresonators," *Sensors Actuators,* A29, 79.

Zhang, X., Zohar, Y., and Zhang, T.Y. (1997), "Measurements of Residual Stresses in Low-Stress Silicon Nitride Thin Films Using Micro-Rotating Structures," *Mater. Res. Soc. Symp. Proc.* 444, 111–116.

16

Micro/Nanotribology and Micro/Nanomechanics of MEMS Devices

Bharat Bhushan

16.1 Introduction

16.1.1 Background

The advances in silicon photolithographic process technology since 1960s have led to the development of microcomponents or microdevices, known as microelectromechanical systems (MEMS). More recently, lithographic processes have been developed to process nonsilicon materials. These lithographic processes are being complemented with nonlithographic micromachining processes for fabrication of milliscale components or devices. Using these fabrication processes, researchers have fabricated a wide variety of miniaturized devices, such as acceleration, pressure and chemical sensors, linear and rotary actuators, electric motors, gear trains, gas turbines, nozzles, pumps, fluid valves, switches, grippers, tweezers, and optoelectronic devices with dimensions in the range of a couple to a few thousand microns (for an early review, see Peterson, 1982; for recent reviews, see Muller et al., 1990; Madou, 1997; Trimmer, 1997; and Bhushan, 1998a). MEMS technology is still in its infancy and the emphasis to date has been on the fabrication and laboratory demonstration of individual components. MEMS devices have begun to be commercially used, particularly in the automotive industry. Silicon-based high-*g* acceleration sensors are used in airbag deployment (Bryzek et al., 1994). Acceleration sensor technology is slightly less than a billion-dollar-a-year industry dominated by Lucas NovaSensor and Analog Devices. Texas Instruments uses deformable mirror arrays on microflexures as part of airline-ticket laser printers and high-resolution projection devices.

Potential applications of MEMS devices include silicon-based acceleration sensors for anti-skid braking systems and four-wheel drives, silicon-based pressure sensors for monitoring pressure of cylinders in automotive engines and of automotive tires, and various sensors, actuators, motors, pumps, and switches in medical instrumentation, cockpit instrumentation, and many hydraulic, pneumatic, and other consumer products (Fujimasa, 1996). MEMS devices are also being pursued in magnetic storage systems (Bhushan, 1996a), where they are being developed for supercompact and ultrahigh-recording-density magnetic disk drives. Horizontal thin-film heads with a single-crystal silicon substrate, referred to as silicon planar head (SPH) sliders are mass-produced using integrated-circuit technology (Lazarri and Deroux-Dauphin, 1989; Bhushan et al., 1992). Several integrated head/suspension microdevices have been fabricated for contact recording applications (Hamilton, 1991; Ohwe et al., 1993). High-bandwidth servo-controlled microactuators have been fabricated for ultrahigh-track-density applications which serve as the fine-position control element of a two-stage, coarse/fine servo system, coupled with a conventional actuator (Miu and Tai, 1995; Fan et al., 1995b). Millimeter-sized wobble motors and actuators for tip-based recording schemes have also been fabricated (Fan and Woodman, 1995a). In some cases, MEMS devices are used primarily for their miniature size, while in others, as in the case of the air bags, because of their high reliability and low-cost manufacturing techniques. This latter fact has been possible since semiconductor-processing costs have reduced drastically over the last decade, allowing the use of MEMS in many previously impractical fields.

The fabrication techniques for MEMS devices employ photolithography and fall into three basic categories: bulk micromachining, surface micromachining, and LIGA a German acronym (Lithographie Galvanoformung Abformung) for lithography, electroforming, and plastic molding. The first two approaches, bulk and surface micromachining, use planar photolithographic fabrication processes developed for semiconductor devices in producing two-dimensional (2D) structures (Jaeger, 1988; Madou, 1997; Bhushan, 1998a). Bulk micromachining employs anisotropic etching to remove sections through the thickness of a single-crystal silicon wafer, typically 250 to 500 μm thick. Bulk micromachining is a proven high-volume production process and is routinely used to fabricate microstructures such as acceleration and pressure sensors and magnetic head sliders. Surface micromachining is based on depositing and etching structural and sacrificial films to produce a free-standing structure. These films are typically made of low-pressure chemical vapor deposition (LPCVD) polysilicon film with 2 to 20 μm thickness. Surface micromachining is used to produce surprisingly complex micromechanical devices such as motors, gears, and grippers. LIGA is used to produce high-aspect ratio (HAR) MEMS devices that are up to 1 mm in height and only a few microns in width or length (Becker et al., 1986). The LIGA process yields very sturdy 3D structures due to their increased thickness. The LIGA process is based on the combined use of X-ray photolithography, electroforming, and molding processes. One of the limitations of silicon microfabrication processes originally used for fabrication of MEMS devices is lack of suitable materials which can be processed. With LIGA, a variety of nonsilicon materials such as metals, ceramics and polymers can be processed. Nonlithographic micromachining processes, primarily in Europe and Japan, are also being used for fabrication of millimeter-scale devices using direct material microcutting or micromechanical machining (such as micromilling, microdrilling, microturning) or removal by energy beams (such as microspark erosion, focused ion beam, laser ablation, and machining, and laser polymerization) (Friedrich and Warrington, 1998; Madou, 1998). Hybrid technologies including LIGA and high-precision micromachining techniques have been used to produce miniaturized motors, gears, actuators, and connectors (Lehr et al., 1996, 1997; Michel and Ehrfeld, 1998). These millimeter-scale devices may find more immediate applications.

TABLE 16.1 Selected Bulk Properties[a] of 3C (β- or cubic) SiC and Si(100)

Sample	Density (kg/m³)	Hardness (GPa)	Elastic Modulus (GPa)	Fracture Toughness (MPa m^{1/2})	Thermal Conductivity[b] (W/m K)	Coeff. of Thermal Expansion[b] (× 10⁻⁶/°C)	Melting Point (°C)	Band-Gap (eV)
β–SiC	3210	23.5–26.5	440	4.6	85–260	4.5–6	2830	2.3
Si(100)	2330	9–10	130	0.95	155	2–4.5	1410	1.1

[a]Unless stated otherwise, data shown were obtained from Bhushan and Gupta (1997).
[b]Obtained from Shackelford et al. (1994).

Silicon-based MEMS devices lack high-temperature capabilities with respect to both mechanical and electrical properties. Recently, researchers have been pursuing SiC as a material for high-temperature microsensor and microactuator applications (Tong et al., 1992; Shor et al., 1993). SiC is a likely candidate for such applications since it has long been used in high-temperature electronics, high-frequency and high-power devices, such as SiC metal–semiconductor field effect transistors (MESFETS) (Spencer et al., 1994) and inversion-mode metal-oxide-semiconductor field effect transistors (MOSFETS). Many other SiC devices have also been fabricated including ultraviolet detectors, SiC memories, and SiC/Si solar cells. SiC has also been used in microstructures such as speaker diaphragms and X-ray masks. For a summary of SiC devices and applications, see Harris (1995). Table 16.1 compares selected bulk properties of SiC and Si(100). Because of the large band gap of SiC, almost all devices fabricated from SiC have good high-temperature properties. This high-temperature capability of SiC combined with its excellent mechanical properties, thermal dissipative characteristics, chemical inertness, and optical transparency makes SiC an ideal choice for complementing polysilicon (polysilicon melts at 1400°C) in MEMS devices. Since MEMS devices need to be of low cost to be viable in most applications, researchers have found low-cost techniques of producing single-crystal 3C-SiC (cubic or β-SiC) films via epitaxial growth on large area silicon substrates (Zorman et al., 1995). This technique allows high-volume batch processing and has the advantage of having silicon as the substrate, an inexpensive material for which microfabrication and micromachining technologies are well established. It is believed that these films will be well suited for MEMS devices.

16.1.2 Tribological Issues

In MEMS devices, various forces associated with the device scale down with the size. When the length of the machine decreases from 1 mm to 1 μm, the area decreases by a factor of a million and the volume decreases by a factor of a billion. The resistive forces such as friction, viscous drag, and surface tension that are proportional to the area, increase a thousand times more than the forces proportional to the volume, such as inertial and electromagnetic forces. The increase in resistive forces leads to tribological concerns, which become critical because friction/stiction (static friction), wear and surface contamination affect device performance and in some cases, can even prevent devices from working.

Examples of two micromotors using polysilicon as the structural material in surface micro-machining — a variable capacitance side drive and a wobble (harmonic) side drive — are shown in Figures 16.1 and 16.2, which can rotate up to 100,000 rpm. Microfabricated variable-capacitance side-drive micromotor with 12 stators and a 4-pole rotor shown in Figure 16.1 is produced using a three-layer polysilicon process and the rotor diameter is 120 μm and the air gap between the rotor and stator is 2 μm (Tai et al., 1989). It is driven electrostatically to continuous rotation (by electrostatic attraction between positively and negatively charged surfaces). The intermittent contact at the rotor–stator interface and physical contact at the rotor–hub flange interface result in wear issues, and high stiction between the contacting surfaces limits the repeatability of operation or may even prevent the operation altogether. Figure 16.2 shows the SEM micrograph of a microfabricated harmonic side-drive (wobble) micromotor

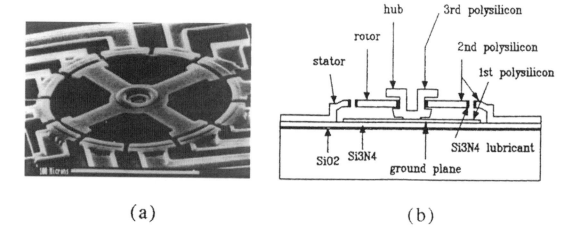

FIGURE 16.1 (a) SEM micrograph, and (b) schematic cross-section of a variable capacitance side-drive micromotor fabricated of polysilicon film. (From Tai et al., 1989, *Sensors Actuators* A21–23, 180–83. With permission.)

FIGURE 16.2 SEM micrograph of a harmonic side-drive (wobble) micromotor. (From Mehregany, M. et al., 1990, in *Proc. IEEE Micro Electromechanical Systems*, pp. 1–8, IEEE, New York. With permission.)

(Mehregany et al., 1988). In this motor, the rotor wobbles around the center bearing post rather than the outer stator. Again friction/stiction and wear of rotor-center bearing interface are of concern. There is a need for development of bearing/bushing materials that are both compatible with MEMS fabrication processes and which provide superior friction and wear performance. Monolayer lubricant films are also of interest. Figure 16.3 shows the SEM micrograph of an air turbine with gear or blade rotors, 125 to 240 μm in diameter, fabricated using polysilicon as the structural material in surface micromachining. The two flow channels on the top are connected to the two independent input ports and the two flow channels at the bottom are connected to the output port. Wear at the contact of gear teeth is a concern. In microvalves used for flow control, the mating valve surfaces should be smooth enough to seal while

FIGURE 16.3 SEM micrograph of a gear train with three meshed gears, in an air turbine. (From Mehregany, M. et al., 1988, *IEEE Trans. Electron Devices* 35, 719–723. With permission.)

maintaining a minimum roughness to ensure low friction/stiction (Bhushan, 1996a, 1998b). Studies have been conducted to measure the friction/stiction in micromotors (Tai and Muller, 1990), gear systems (Gabriel et al., 1990) and polysilicon microstructures (Lim et al., 1990) to understand friction mechanisms. Several studies have been conducted to develop solid and liquid lubricant and hard films to minimize friction and wear (Bhushan et al., 1995b; Deng et al., 1995; Beerschwinger et al., 1995; Koinkar and Bhushan, 1996a,b; Bhushan, 1996b; Henck, 1997).

In a silicon planar head slider for magnetic disk drives shown in Figure 16.4, wear and friction/stiction are an issue because of the close proximity between the slider and disk surfaces during steady operation and continuous contacts during start and stops (Lazzari and Deroux-Dauphin, 1989; Bhushan et al., 1992). Hard diamondlike carbon (DLC) coatings are used as an overcoat for protection against corrosion and wear. Two electrostatically driven rotary and linear microactuaters (surface-micromachined, poly-silicon microstructure) for a magnetic disk drive shown in Figure 16.5, consist of a movable plate connected only by springs to a substrate, on which there are two sets of mating interdigitated electrodes which activate motion of the plate in opposing directions. Any unintended contacts may result in wear and stiction.

Figure 16.6 shows an SEM micrograph of a micromechanical switch (Peterson, 1979). As the voltage is applied between the deflection electrode and the p^+ ground plane, the cantilever beam is deflected and the switch closes, connecting the contact electrode and the fixed electrode; wear during contact is of concern. Figure 16.7 shows an SEM micrograph of a pair of tongs (Mehregany et al., 1988). The jaws open when the linearly sliding handle is pushed forward, demonstrating the linear slide and the linear-to-rotary motion conversion; for this pair of tongs, the jaws open up to 400 μm in width. Wear at the teeth is of concern.

As an example of nonsilicon components, Figure 16.8a shows a DC brushless permanent magnet millimotor (diameter = 1.9 mm, length = 5.5 mm) with an integrated milligear box which is produced with parts obtained by hybrid fabrication processes including the LIGA process, micromechanical machining, and microspark erosion techniques (Lehr et al., 1996, 1997; Michel and Ehrfeld, 1998). The motor can rotate up to 100,000 rpm and deliver a maximum torque of 7.5 μNm. The rotor, supported on two ruby bearings, consists of a tiny steel shaft and a diametrically magnetized rare earth magnet. The rotational speed of the motor can be converted by the use of a milligear box to increase the torque for a specific application. Gears are made of metal (e.g., electroplated Ni–Fe) or injected polymer materials (e.g., POM) using the LIGA process, Figures 16.8b and c. Optimum materials and liquid and solid lubrication approaches for bearings and gears are needed.

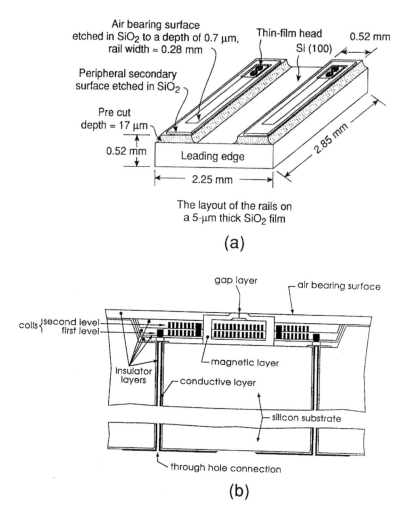

Air bearing surface
etched in SiO₂ to a depth of 0.7 μm,
rail width ≈ 0.28 mm

Thin-film head
Si (100)

0.52 mm

Peripheral secondary
surface etched in SiO₂

Pre cut
depth = 17 μm

0.52 mm

Leading edge

2.85 mm

2.25 mm

The layout of the rails on
a 5-μm thick SiO₂ film

(a)

gap layer

air bearing surface

coils { second level
first level

Insulator
layers

magnetic layer

conductive layer

silicon substrate

through hole connection

(b)

FIGURE 16.4 Schematic (a) of a silicon planar head slider and (b) of cross section of the slider for magnetic disk drive applications. (From Bhushan, B. et al., 1992, *IEEE Trans Magn.* 28, 2874–2876. With permission.)

There are tribological issues in the fabrication processes as well. For example, in surface microma-chining, the suspended structures can sometimes collapse and permanently adhere to the underlying substrate, Figure 16.9 (Guckel and Burns, 1989). The mechanism of such adhesion phenomena needs to be understood (Mastrangelo, 1997).

Friction/stiction and wear clearly limit the lifetimes and compromise the performance and reliability of microdevices. Since microdevices are designed to small tolerances, environmental factors, surface contamination, and environmental debris affect their reliability. There is a need for development of a fundamental understanding of friction/stiction, wear, and the role of surface contamination and envi-ronment in microdevices (Bhushan, 1998a). A few studies have been conducted on the tribology of bulk silicon and polysilicon films used in microdevices (Bhushan and Venkatesan, 1993a,b; Gupta et al., 1993; Venkatesan and Bhushan, 1993, 1994; Gupta and Bhushan, 1994; Bhushan and Koinkar, 1994; Bhushan, 1996b). Mechanical properties of polysilicon films are not well characterized (Mehregany et al., 1987; Ericson and Schweitz, 1990; Schweitz, 1991; Guckel et al., 1992; Bhushan, 1995; Fang and Wickert, 1995). The advent of atomic force/friction force microscopy (AFM/FFM) (Bhushan, 1995, 1997; Bhushan et al., 1995a) has allowed the study of surface topography, adhesion, friction, wear, lubrication, and measure-ment of mechanical properties, all on a micro- to nanometer scale. Recently, microtribological studies

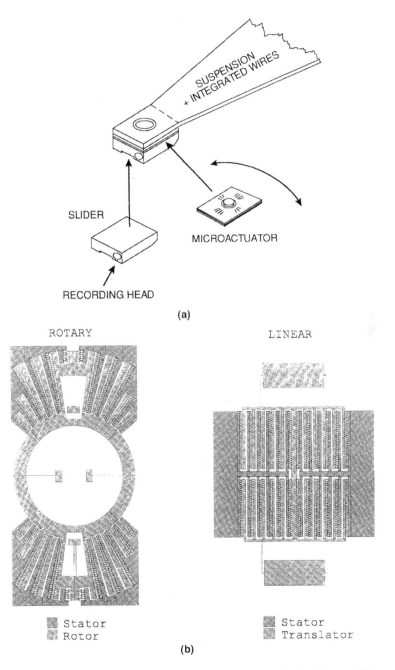

FIGURE 16.5 Schematics of (a) a microactuator in place with magnetic head slider, and (b) top view of two electrostatic, rotary and linear microactuators (electrode tree structure). (From Fan, L.S. et al., 1995; *IEEE Trans. Ind. Electron.* 42, 222–233. With permission.)

have been conducted using the AFM/FFM on undoped and doped silicon and polysilicon films and SiC films that are used in MEMS devices (Bhushan, 1996b, 1997, 1998; Bhushan et al., 1994, 1997a,b, 1998; Li and Bhushan, 1998; Sundararajan and Bhushan, 1998).

This chapter presents a review of macro- and micro/nanotribological studies of single-crystal silicon and polysilicon, oxidized and implanted silicon, doped and undoped polysilicon films and SiC films. A summary of limited component-level tests is also presented.

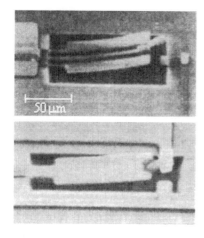

FIGURE 16.6 SEM micrograph of single-contact and double-contact (with two orientations of the fixed electrodes) designs of micromechanical switches (Peterson, 1979). (From Peterson, K.E., 1979, *IBM J. Res. Dev.* 23, 376. With permission.)

FIGURE 16.7 SEM micrograph of a partially released pair of tongs. (From Mehregany, M. et al., 1988, *IEEE Trans. Electron. Devices* 35, 719–723. With permission.)

16.2 Experimental Techniques

16.2.1 Description of Apparatus and Test Procedures

16.2.1.1 Micro/Nanoscale Tests

A modified AFM/FFM (Nanoscope III, Digital Instruments, Santa Barbara, CA), was used for the micro/nanotribological studies. Surface roughness and microscale friction measurements were simultaneously made over a scan size of $10 \times 10 \ \mu m$ with an Si_3N_4 tip (tip radius ~ 50 nm, cantilever stiffness ~ 0.6 N/m) sliding over the sample surface orthogonal to the long axis of the cantilever at 25 μm/s. A coefficient of friction and conversion factors for converting the friction signal voltage to force units (nN) were obtained through the methods developed previously by Bhushan and co-workers (Bhushan, 1995). The normal loads used in the friction measurements varied between 50 to 300 nN. The reported values are each an average of six separate measurements.

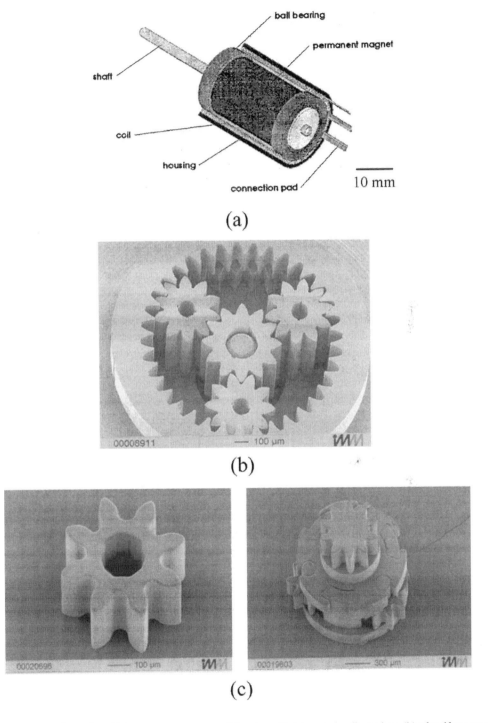

(a)

(b)

(c)

FIGURE 16.8 Schematics of (a) permanent magnet millimotor with integrated milligear box, (b) of wolfrom-type system made of Ni–Fe metal (Lehr et al., 1996), and (c) of multistage planetary gear system made with microinjected POM plastic showing a single gear and the gear system. (From Thurigen, C. et al., 1998, in *Tribology Issues and Opportunities in MEMS*, B. Bhushan, ed., Kluwer Academic, Dordrecht. With permission.)

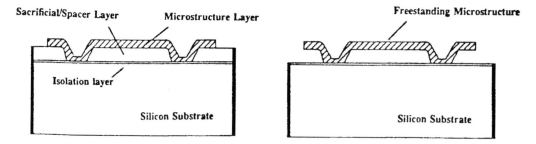

FIGURE 16.9 Schematics of microstructures during fabrication using surface micromachining before and after removal of sacrificial/spacer layer.

For the scratch and wear tests, specially fabricated diamond microtips were used (Bhushan et al., 1997a; Sundararajan and Bhushan, 1998). These microtips consisted of single-crystal natural diamond, ground to the shape of a three-sided pyramid, with an apex angle of 60° and tip radius of about 70 nm, mounted on a platinum-coated stainless steel cantilever beam whose stiffness was 50 N/m. Samples were scanned orthogonal to the long axis of the cantilever with loads ranging from 20 to 100 μN to generate scratch/wear marks. Scratch tests consisted of generating scratches in a reciprocating mode at a given load for 10 cycles over a scan length (stroke length) of 5 μm at 10 μm/s. Wear marks were generated over a scan area of 2×2 μm at 4 μm/s and the wear marks were observed by scanning a larger 4×4 μm area with the wear mark at the center. Imaging scans of both scratch and wear tests were done at a low normal load of 0.5. The reported scratch/wear depths are an average of three runs at separate instances. All measurements were performed in an ambient environment (21 ± 1°C, 45 ± 5% RH).

Hardness and elastic modulus were calculated from load–displacement data obtained by nanoindentation using a commercially available nanoindenter (Bhushan, 1995; Bhushan et al., 1997b; Li and Bhushan, 1998). The instrument monitored and recorded dynamic load and displacement of a three-sided pyramidal diamond (Berkovich) indenter with a force resolution of about 75 nN and displacement resolution of about 0.1 nm. Multiple loading and unloading were performed to examine reversibility of the deformation and thereby ensuring that the regime was elastic.

The fracture toughness measurements were made using a microindentation technique. A Vickers indenter (four-sided diamond pyramid) was used to indent samples in a microhardness tester at a normal load of 0.5 N. The indentation impressions were examined in an optical microscope to measure the length of median-radial cracks to calculate the fracture toughness (Li and Bhushan, 1998).

16.2.1.2 Macroscale Tests

Macroscale studies were conducted using either a ball-on-flat tribometer under reciprocating motion or a magnetic rigid disk drive. In the ball-on-flat tribometer tests, a 5-mm diameter alumina ball (hardness ~ 21 GPa) was slid in a reciprocating mode (2 mm amplitude and 1 Hz frequency) under a normal load of 1 N in the ambient environment (Gupta et al., 1993). The coefficient of friction was measured during the tests using a strain gauge ring. Wear volume was measured by measuring the wear depth using a stylus profiler.

In the magnetic disk drive tests, a modified disk drive was used. The silicon pins or magnetic head slider specimens to be tested were slid in a unidirectional sliding mode against a magnetic thin-film disk under a normal load of 0.15 N and the rotational speed of 200 rpm. The sliding speeds at track radii ranging from 45 to 55 mm varied from 0.9 to 1.2 m/s (Bhushan and Venkatesan, 1993). At these speeds, the pin or slider specimen remained in contact throughout the period of testing. The coefficient of friction was measured during the tests using a strain gauge beam. Samples were examined using scanning electron microscopy to detect any wear. Chemical analyses of the samples were also carried out to study failure mechanisms.

16.2.2 Test Samples

Materials of most interest for planar fabrication processes using silicon as the structural material are undoped and boron-doped (p^+-type) single-crystal silicon and phosphorus-doped (n^+-type) LPCVD polysilicon films. For tribological reasons, silicon needs to be coated with a solid and/or liquid overcoat or be surface treated, which exhibits low friction and wear.

Studies have been conducted on various types of virgin silicon samples: undoped (lightly doped) single-crystal Si(100), Si(111), and Si(110) and the following types of treated/coated silicon samples: PECVD-oxide-coated Si(111), dry-oxidized, wet-oxidized, and C^+-implanted Si(111) (Bhushan and Venkatesan, 1993; Bhushan and Koinkar, 1994). Studies have also been conducted on heavily doped (p^+-type) single-crystal Si(100), undoped polysilicon film, heavily doped (n^+-type) polysilicon film and 3C-SiC (cubic or β-SiC) film (Bhushan et al., 1997a,b, 1998; Li and Bhushan, 1998; Sundararajan and Bhushan, 1998). A 10×10 mm coupon of each sample was ultrasonically cleaned in methanol for 20 min and dried with a blast of dry air prior to measurements. The undoped Si(100) was a p-type material grown by the CZ process. It had a boron concentration of 1.7×10^{15} ions/cm^3 from intrinsic doping during the manufacturing process. The doped wafer (p^+-type single-crystal silicon) was heavily doped with boron ions (from a solid source of oxide of boron) with concentration of 7×10^{19} ions/cm^3 down to a depth of 5.5 μm using thermal diffusion. The grain size of polysilicon wafer was about 5 mm. The polysilicon film was produced as follows: (1) The substrate used was thermally oxidized Si(100) wafers with the oxide layer grown using a standard wet oxidation recipe to a nominal thickness of about 100 nm; (2) the polysilicon film was grown on the substrate using an LPCVD process (deposition temperature, 610°C; silane flow rate, 285 sccm; deposition pressure, 230 mtorr), using the thermal decomposition of silane vapor. The films were about 3 μm thick, with columnar grains and a grain size of about 750 nm. X-ray diffraction and transmission electron microscope characterization showed the film to be highly oriented (110). The n^+-doped polysilicon film was obtained by doping the polysilicon film with phosphorus ions from a solid source of P_2O_5 by thermal diffusion at 875°C for 90 min. The 3C-SiC films were grown through an atmospheric pressure chemical vapor deposition (APCVD) process on an Si(100) substrate. To grow the SiC film on the wafer by carbonizing its surface, the wafer is placed on an SiC-coated graphite susceptor, which is induction-heated by an RF-generator to the growth temperature of 1360°C in the presence of propane and silane at 1 atm. Prior to film growth, the wafer is heated to 1000°C in the presence of hydrogen, which etches the native oxide from the wafer surface (Zorman et al., 1995). The films obtained were about 2 μm thick. Both as-deposited and polished versions of the undoped polysilicon and SiC films were studied. The polysilicon film was chemomechanically polished in a Struers Planopol-3 polishing machine using 100 ml colloidal silica dispersion (Rippey Corporation, particle size of 30 to 100 nm) mixed in 2000 ml deionized water at a force of 210 N for 15 min, with the pads running at 150 rpm in the same direction. The SiC and doped polysilicon films were polished in a Buehler Ecomet-3 polishing machine with diamond slurry (General Electric Company, particle size of 100 to 500 nm) for 30 min for SiC and 12 min for doped polysilicon film at a load of 50 N, with the pads running at 10 rpm in the same direction. Doped polysilicon film was polished using Fuji film lapping tape, LT-2, the main lapping agent being 37-μm-sized Cr_2O_3 particles.

Boundary lubrication studies have been conducted on silicon samples coated with perfluoropolyether lubricants (Koinkar and Bhushan, 1996a,b) and Langmuir–Blodgett and chemically grafted self-assembled monolayer films (Bhushan et al., 1995b).

16.3 Results and Discussion

Reviews of five studies are presented in this section. The first study compares micro/nanotribological properties of various forms of virgin, coated, and treated silicon samples. The second study is composed of similar studies conducted on SiC film and compares this material to other materials currently used in MEMS devices. The third study compares the macroscale friction and wear data of virgin, coated, and

TABLE 16.2 rms, Microfriction, Microscratching/Microwear and Nanoindentation Hardness Data for Various Virgin, Coated, and Treated Silicon Samples

Material	rms Roughness[a] (nm)	Microscale Coefficient of Friction[b]	Scratch Depth[c] at 40 µN (nm)	Wear Depth[c] at 40 µN (nm)	Nanohardness[c] at 100 µN (GPa)
Si(111)	0.11	0.03	20	27	11.7
Si(110)	0.09	0.04	20	—	—
Si(100)	0.12	0.03	25	—	—
Polysilicon	1.07	0.04	18	—	—
Polysilicon (lapped)	0.16	0.05	18	25	12.5
PECVD-oxide coated Si(111)	1.50	0.01	8	5	18.0
Dry-oxidized Si(111)	0.11	0.04	16	14	17.0
Wet-oxidized Si(111)	0.25	0.04	17	18	14.4
C+-implanted Si(111)	0.33	0.02	20	23	18.6

[a] Scan size of 500×500 nm.

[b] Versus Si_3N_4 ball, ball radius of 3 mm at a normal load of 0.1 N (0.3 GPa) at an average sliding speed of 0.8 mm/s.

[c] Measured using an AFM with a diamond tip of radius of 100 nm.

TABLE 16.3 Surface Roughness and Coefficients of Micro- and Macroscale Friction of Selected Samples

Material	rms Roughness (nm)	Coefficient of Microscale Friction[a]	Coefficient of Macroscale Friction[b]
Si(111)	0.11	0.03	0.18
C+-implanted Si(111)	0.33	0.02	0.18

[a] Versus Si_3N_4 tip, tip radius of 50 nm in the load range of 10–150 nN (2.5–6.1 GPa) at a scanning speed of 5 µm/s over a scan area of 1×1 µm.

[b] Versus Si_3N_4 ball, ball radius of 3 mm at a normal load of 0.1 N (0.3 GPa) at an average sliding speed of 0.8 mm/s.

treated silicon samples. The fourth study discusses various forms of boundary lubrication that may be suitable for MEMS devices. Finally, the fifth study presents a review of component level studies.

16.3.1 Micro/nanotribological Studies of Virgin, Coated, and Treated Silicon Samples

Table 16.2 summarizes the results of the studies conducted on various silicon samples (Bhushan and Koinkar, 1994). Coefficient of microscale friction values of all the samples are about the same. Table 16.3 compares macroscale and microscale friction values for two of the samples. When measured for small contact areas and very low loads used in microscale studies, indentation hardness and elastic modulus are higher than that at the macroscale. This reduces wear. This, added to the effect of the small apparent area of contact reducing the number of trapped particles on the interface, results in less plowing contribution in the case of microscale friction measurements. Figure 16.10 and Table 16.2 show microscale scratch data for the various silicon samples (Bhushan and Koinkar, 1994). These samples could be scratched at 10 µN load. Scratch depth increased with normal load. Crystalline orientation of silicon has little influence on scratch resistance. PECVD-oxide samples showed the best scratch resistance, followed by dry-oxidized, wet-oxidized, and ion-implanted samples. Ion implantation does not appear to improve scratch resistance. Microscratching experiments just described can be used to study failure mechanisms on the microscale and to evaluate mechanical integrity (scratch resistance) of ultrathin films at low loads.

Wear data on the silicon samples are presented in Table 16.2 (Bhushan and Koinkar, 1994). PECVD-oxide samples showed superior wear resistance followed by dry-oxidized, wet-oxidized, and ion-implanted samples. This agrees with the trends seen in scratch resistance. In PECVD, ion bombardment

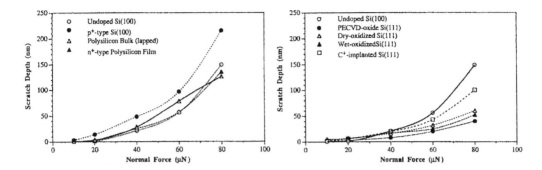

FIGURE 16.10 Scratch depth as a function of normal force after 10 cycles for various silicon samples, virgin, treated, and coated. (From Bhushan, B. and Koinkar, V.N., 1994, *J. Appl. Phys.* 75, 5741–5746. With permission.)

during the deposition improves the coating properties such as suppression of columnar growth, freedom from pinhole, decrease in crystalline size, and increase in density, hardness and substrate–coating adhesion. These effects may help in improving mechanical integrity of the sample surface.

The wear resistance of ion-implanted silicon samples was further studied, Figure 16.11 (Bhushan and Koinkar, 1994). For tests conducted at various loads on Si(111) and C$^+$-implanted Si(111), it is noted that wear resistance of implanted sample is slightly poorer than that of virgin silicon up to about 80 μN. Above 80 μN, the wear resistance of implanted Si improves. As one continues to run tests at 40 μN for a larger number of cycles, the implanted sample exhibits higher wear resistance than the unimplanted sample. Damage from the implantation in the top layer results in poorer wear resistance; however, the implanted zone at the subsurface is more wear resistant than the virgin silicon.

Nanoindentation hardness values of all samples are presented in Table 16.2. Coatings and treatments improved nanohardness of silicon. Note that dry-oxidized and PECVD films are harder than wet-oxidized films, as these films may be porous. High hardness of oxidized films may be responsible for measured low wear on the microscale and macroscale (data to be presented later). Figure 16.12 shows the indentation marks generated on virgin and C$^+$-implanted Si(111) at a normal load of 70 μN with a depth of indentation about 3 nm and hardness values of 15.8 and 19.5 GPa, respectively (Bhushan and Koinkar, 1994). Hardness values of virgin and C$^+$-implanted Si(111) at various indentation depths (normal loads) are presented in Figure 16.13 (Bhushan and Koinkar, 1994). Note that the hardness at a small indentation depth of 2.5 nm is 16.6 GPa and it drops to a value of 11.7 GPa at a depth of 7 nm and a normal load of 100 μN. Higher hardness values obtained in low-load indentation may arise from the observed pressure-induced phase transformation during the nanoindentation (Pharr, 1991; Callahan and Morris, 1992). Additional increase in the hardness at an even lower indentation depth of 2.5 nm reported here may arise from the contribution by complex chemical films (not from native oxide films) present on the silicon surface. At small volumes there is a high probability that indentation would be made into a region that was initially dislocation free. Furthermore, at small volumes, it is believed that there is an increase in the stress necessary to operate dislocation sources (Gane and Cox, 1970; Sargent, 1986). These are some of the plausible explanations for an increase in hardness at smaller volumes. If the silicon material is to be used at very light loads such as in microsystems, the high hardness of surface films would protect the surface until it is worn.

From Figure 16.13, hardness values of C$^+$-implanted Si(111) at a normal load of 50 μN is 20.0 GPa with an indentation depth of about 2 nm which is comparable to the hardness value of 19.5 GPa at 70 μN, whereas measured hardness value for virgin silicon at an indentation depth of about 7 nm (normal load of 100 μN) is only about 11.7 GPa. Thus, ion implantation results in an increase in hardness. Note that the surface layer of the implanted zone is much harder compared with the subsurface, and may be brittle leading to higher wear on the surface. The subsurface of the implanted zone is harder than the virgin silicon, resulting in higher wear resistance, which will be shown later in macroscale tests conducted at high loads.

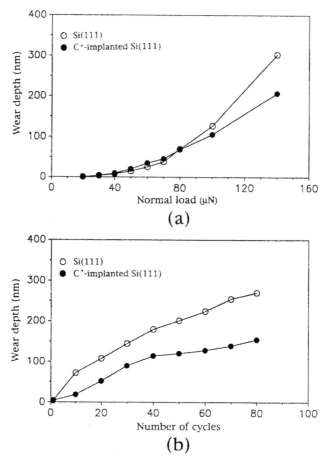

FIGURE 16.11 Wear depth as a function of (a) load (after one cycle), and (b) cycles (normal load = 40 μN) for Si(111) and C⁺-implanted Si(111). (From Bhushan, B. and Koinkar, V.N., 1994, *J. Appl. Phys.* 75, 5741–5746. With permission.)

16.3.2 Micro/Nanotribological Studies of Doped and Undoped Polysilicon Films, SiC Films, and Their Comparison to Single-Crystal Silicon

16.3.2.1 Surface Roughness and Friction

The surface roughness of various samples obtained with the AFM is compared in Figure 16.14a (Sundararajan and Bhushan, 1998). Polishing of the as-deposited polysilicon and SiC films drastically affect the roughness as the values reduce by two orders of magnitude. Si(100) appears to be the smoothest followed by polished undoped polysilicon and SiC films, which have comparable roughness. The doped polysilicon film shows higher roughness than the undoped sample, which is attributed to the doping process. The coefficients of microscale friction of the various samples are shown in Figure 16.14b (Bhushan et al., 1998; Sundararajan and Bhushan, 1998). Despite being the smoothest sample, Si(100) shows higher friction than the other samples. Doped polysilicon also shows high friction, which is due to the presence of grain boundaries and high surface roughness. In the case of the undoped polysilicon and SiC films, polishing did not affect the friction values by much. From the data, the polished SiC film shows the lowest friction followed by polished, undoped polysilicon film, which strongly supports the candidacy of SiC films for use in MEMS devices. Gray-scale top-view images of surface height and corresponding friction force of selected samples obtained with the AFM/FFM are shown in Figure 16.15. Brighter regions indicate higher height than darker regions in the case of the surface height images, while brighter regions indicate higher friction force experienced by the tip than the darker regions. The doped polysilicon film shows a large number of grain boundaries. It appears that regions of high and low surface height do not necessarily correspond to regions of high and low friction.

The low microscale friction exhibited by SiC compared to the other materials agrees with the fact that many ceramic–ceramic interfaces generally show low friction on the macroscale. In this study, the Si_3N_4 tip–SiC film interface also shows the same trend. In the case of ceramic materials, formation of tribochemical films on the surface due to sliding results in low values of friction. SiC can react with water vapor to form silicon hydroxide and oxide, which improve the frictional behavior (Xu and Bhushan, 1997). From this discussion, it can be seen that an obvious drawback with SiC is its readiness to form tribochemical films, which is environment dependent. Another factor that affects friction in ceramic materials is fracture, which leads to high friction because of the plowing contribution. But in the case of SiC film, fracture toughness is believed to follow the same trend as that of the bulk material, which is higher than that of silicon (see Table 16.1), thereby resulting in lower plowing contribution and hence lower friction. Macroscale friction measurements indicate that SiC film exhibits one of the lowest friction values as compared to the other samples. Doped polysilicon sample shows low friction on the macroscale as compared to the undoped polysilicon sample possibly due to the doping effect.

16.3.2.2 Scratch/Wear Tests

As explained earlier, the scratch tests consisted of making scratches for ten cycles with varying loads. Figure 16.16a shows a plot of scratch depth vs. normal load for various samples (Bhushan et al., 1998; Sundararajan and Bhushan, 1998). Error bars are given for the Si(100) data. The variation was typically about 12%. Scratch depth increases with increasing normal load. Si(100) and the doped and undoped polysilicon film show similar scratch resistance. From the data, it is clear that the SiC film is much more scratch resistant than the other samples. The increase in scratch depth with normal load is very small and all depths are less than 30 nm, while the Si(100) and polysilicon films reach depths in excess of 150 nm. Figure 16.17 shows 3D images of the scratch marks.

Wear tests were conducted on the samples for one cycle at normal loads ranging from 20 to 80 μN. The resulting wear depths are plotted against the normal loads in Figure 16.16b (Bhushan et al., 1998; Sundararajan and Bhushan, 1998). Again, the wear depth increases with increasing normal load. Similar to the scratch resistance data, Si(100) and the polysilicon samples also behave similarly in terms of wear resistance. The SiC film starts out showing comparable wear depth at 20 μN to all the other samples, but at higher loads SiC film shows superior wear resistance. Also there is hardly an increase in wear depth with increasing normal load in the case of SiC film. The evolution of wear on the various samples was also studied. This test was performed by wearing the same region for 20 cycles at a normal load of 20 μN, while observing wear depths at different intervals (1, 2, 10, 15, and 20 cycles), Figure 16.18. This would give information as to the progression of wear of the material. The wear depths observed are plotted against the number of cycles in Figure 16.16b. For all the materials, the wear depth increases almost linearly with increasing number of cycles. This suggests that the material is removed layer by layer in all the materials. Here also, SiC film exhibits lower wear depths than the other samples. Doped polysilicon film wears less than the undoped film. Not many debris particles are seen in the figures. This is due to the fact that the debris particles created are loose and are pushed outside the imaging area by the sliding tip during scanning. It can also be seen from the figure that the wear marks are uniform at the bottom, indicating that uniform wear has occurred with particle pile up on the edges of the wear mark.

The wear tests clearly show that SiC film possesses an extremely wear-resistant surface compared to the other samples and, together with the scratch tests results, indicate that SiC film has better surface mechanical properties than Si(100) and the polysilicon films. Table 16.4 shows a summary of the various properties of the samples measured in this study (Bhushan et al., 1998; Sundararajan and Bhushan 1998). The superior scratch/wear resistance of the SiC film is consistent with the hardness values near the surface measured with the nanoindenter. Higher hardness of SiC film is one of the factors responsible for its

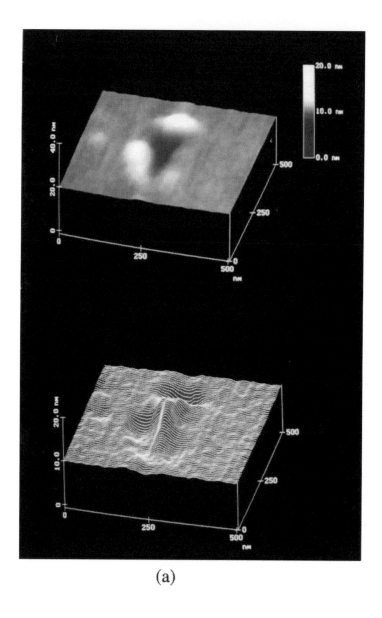

(a)

FIGURE 16.12

better scratch/wear resistance. Wear in ceramic materials in the case of asperity contacts occurs due to brittle fracture. As the AFM tip slides over the material surface, Hertzian cone cracks can occur when the normal stress exceeds a critical value (Hutchings, 1992). Friction (tangential) forces during sliding reduce this critical value. High fracture toughness and low coefficient of friction of SiC film help in reducing the chances of brittle fracture, which results in low wear. Therefore, abrasive wear due to plastic deformation and fracture govern the wear process. In the case of all samples, surface reactions result in the formation of tribochemical films that are different in nature than the underlying material. Therefore, at low loads and at the onset of wear, all samples would show similar wear depths as they all have comparable interfacial shear strengths and attack angles (attack angle is the included angle between the leading face of the asperity and the contact plane at the point of contact). This is seen in the case of the reported wear depths at 20 μN for 1 cycle (Figure 16.16b). In the case of the scratch test data (Figure 16.16a), SiC shows slightly lower scratch depth at 20 μN since the data is for ten cycles, during

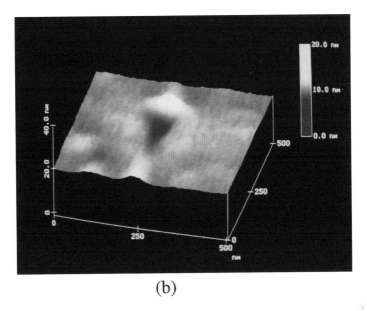

(b)

FIGURE 16.12 Gray-scale plot and line plot of the inverted nanoindentation mark on (a) Si(111) at 70 μN (hardness ~ 15.8 GPa), and (b) gray-scale plot of indentation mark on C⁺-implanted Si(111) at 70 μN (hardness ~ 19.5 GPa). The indentation depth of indent was about 3 nm.

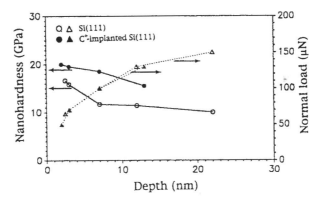

FIGURE 16.13 Nanohardness and normal load as function of indentation depth for virgin and C⁺-implanted Si(111). (From Bhushan, B. and Koinkar, V.N., 1994, *J. Appl. Phys.* 75, 5741–5746. With permission.)

which time it is possible that the bulk properties of the material start coming into play in addition to that of the tribochemical films. This is consistent with the wear depths at 20 μN for ten cycles and more (Figure 16.16c). As the normal load increases, Si(100) and polysilicon films show a sharp increase in degree of penetration (ratio of groove depth to radius of contact) and attack angle, which leads to higher wear (Hokkirigawa and Kato, 1988; Koinkar and Bhushan, 1997). Higher fracture toughness and higher hardness of SiC as compared to Si(100) is responsible for its lower wear. Also the higher thermal conductivity of SiC (see Table 16.1) as compared to the other materials leads to lower interface temperatures which generally results in less degradation of the surface (Bhushan, 1996a). Doping of the polysilicon does not affect the scratch/wear resistance and hardness much. The measurements made on the doped sample are affected by the presence of grain boundaries.

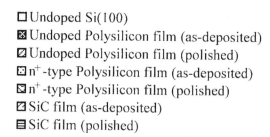

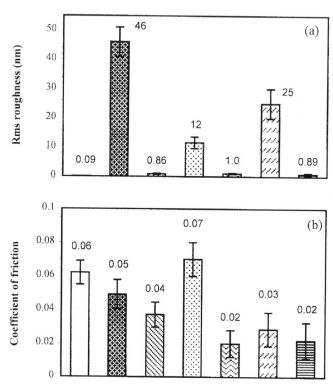

FIGURE 16.14 Comparisons of (a) rms surface roughness and (b) coefficient of microscale friction values for various samples. (From Bhushan, B. et al., 1998, in *Tribology Issues and Opportunities in MEMS*, B. Bhushan, ed., Kluwer Academic, Dordrecht. With permission.)

16.3.2.3   Hardness, Elastic Modulus, and Fracture Toughness Measurements

Representative load–displacement curves of indentations made at 0.2 and 15 mN peak indentation loads on various samples are shown in Figure 16.19. Measured hardness, elastic modulus, and fracture toughness values are given in Table 16.4 (Bhushan et al., 1998; Li and Bhushan, 1998; Sundararajan and Bhushan, 1998). Properties of undoped polysilicon films are comparable to that of bulk single-crystal silicon. Doping of the polysilicon film degrades its mechanical properties. SiC film shows hardness and elastic modulus higher than the other samples. Higher hardness of SiC is believed to be responsible for its superior scratch/wear resistance compared to the other samples.

16.3.3   Macroscale Tribological Studies of Virgin, Coated, and Treated Samples

To study the effect of ion implantation of silicon on friction and wear, single-crystalline and polycrystalline silicon samples were ion implanted by Gupta et al. (1993, 1994) and Gupta and Bhushan (1994) with various doses of C[+], B[+], N$_2^+$, and Ar[+] ion species at 200 keV energy. The coefficient of friction and wear factor of C[+]-implanted silicon samples as a function of ion dose is presented in Figure 16.20 (Gupta et al., 1993). The friction and wear tests were conducted using a ball-on-flat tribometer (Gupta et al., 1993). Each data bar represents the average value of four to six measurements. The coefficient of friction and wear factor decrease drastically with ion dose. Silicon samples bombarded above 10^{17} C[+] cm^{-2} exhibit extremely low values of coefficients of friction (typically 0.03 to 0.06 in air) and the wear factor (reduced by as much as four orders of magnitude). Gupta et al. (1993) reported that a decrease in coefficient of friction and wear factor of silicon as a result of C[+] ion bombardment occurred because of formation of silicon carbide rather than amorphization of silicon. Gupta et al. (1994) also reported an improvement in friction and wear with B[+] ion implantation.

For magnetic disk drive applications, macroscale friction and wear experiments have been performed using a magnetic disk drive with unoxidized, oxidized, and implanted pins sliding against amorphous carbon–coated magnetic disks lubricated with perfluoropolyether lubricant (Bhushan and Venkatesan, 1993; Venkatesan and Bhushan, 1993, 1994). The data of silicon samples were compared with the commonly used slider materials in disk drives: Al$_2$O$_3$–TiC and Mn–Zn ferrite. Representative profiles of the variation of the coefficient of friction with number of sliding cycles for Al$_2$O$_3$–TiC slider and uncoated and coated silicon pins are shown in Figure 16.21. Friction data obtained from the various tests conducted in ambient air in terms of the initial coefficient of friction, the contact life, i.e., the number of revolutions before the coefficient of friction increased by a factor of two, the maximum value of the coefficient of friction for cases when the increase was more than by a factor of two are presented in Table 16.5. For the case of oxidized samples, a significant increase (by a factor of two or more) was not observed and so the range of variation of the coefficient of friction for the duration of 50,000 cycles is indicated. It was found that crystalline orientation of silicon has no effect of friction and wear. For bare silicon, after initial increase in the coefficient of friction, it drops to a steady state value of 0.1 following the increase, as seen in Figure 16.21a. The rise in the coefficient of friction and damage on the pin surface for Si(111) is associated with the transfer of amorphous carbon from the disk to the pin, oxidation-enhanced fracture of pin material followed by tribochemical oxidation of the transfer film. The drop is associated with the formation of a transfer coating on the pin, Figure 16.22a. The mechanism of transfer and tribochemical oxidation was seen to be also operative for the friction increase for Al$_2$O$_3$–TiC and Mn–Zn ferrite pin/slider materials tested in ambient air. As seen in Table 16.5 and Figure 16.21b, dry-oxidized Si(111) exhibits excellent characteristics and this behavior has been attributed to the chemical passivity of the oxide and lack of transfer of DLC from the disk to the pin. The behavior of PECVD was comparable to that of dry oxide, but for the wet oxide there was some variation in the coefficient of friction (0.26 to 0.4). The difference between dry and wet oxide has been attributed to increased porosity of the wet oxide (Bhushan and Venkatesan, 1993; Pilskin, 1977).

Since tribochemical oxidation was determined to be a significant factor, experiments were conducted in dry nitrogen (Venkatesan and Bhushan, 1993, 1994). The variation of the coefficient of friction for a silicon pin sliding against a thin-film disk is shown in Figure 16.21a. For comparison the behavior in ambient is also shown. It is seen that in a dry nitrogen environment, the coefficient of friction of Si(111) sliding against a disk decreased from an initial value of about 0.2 to 0.05 with continued sliding. Similar behavior was also observed while testing with Al$_2$O$_3$–TiC and Mn–Zn ferrite pins and sliders. Based on SEM and chemical analysis this behavior has been attributed to the formation of a smooth amorphous-carbon/lubricant transfer patch and suppression of oxidation in a dry nitrogen environment, Figure 16.22b.

The experiments in dry nitrogen indicated that low friction conditions can be achieved in dry nitrogen although transfer of carbon from the disk to the pin occurs. An experiment was performed with a hydrogenated amorphous carbon–coated (~18 nm thick) silicon pin sliding against a lubricated thin-film disk in dry nitrogen. The friction variation with sliding for this experiment is shown in Figure 16.23. No damage for the pin and disk surfaces could be detected after this experiment.

Based on macrotests using disk drives, we find that the friction and wear performance of bare silicon is not adequate. With single and polycrystalline silicon sliding against an amorphous carbon–coated

Surface Height

Friction Force

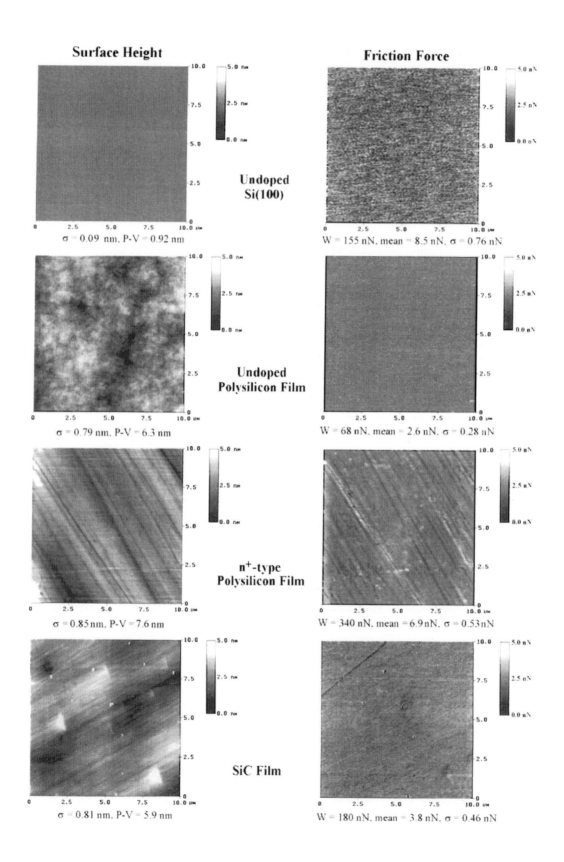

Undoped
Si(100)

$\sigma = 0.09$ nm, P-V = 0.92 nm

W = 155 nN, mean = 8.5 nN, $\sigma = 0.76$ nN

Undoped
Polysilicon Film

$\sigma = 0.79$ nm, P-V = 6.3 nm

W = 68 nN, mean = 2.6 nN, $\sigma = 0.28$ nN

n^+-type
Polysilicon Film

$\sigma = 0.85$ nm, P-V = 7.6 nm

W = 340 nN, mean = 6.9 nN, $\sigma = 0.53$ nN

SiC Film

$\sigma = 0.81$ nm, P-V = 5.9 nm

W = 180 nN, mean = 3.8 nN, $\sigma = 0.46$ nN

magnetic disk, transfer of amorphous carbon from the disk to the pin/slider and oxidation-enhanced fracture of pin/slider material followed by oxidation of the transfer coating (tribochemical oxidation) is responsible for degradation of the sliding interface and consequent friction increase in ambient air. With dry-oxidized or plasma-enhanced CVD SiO_2-coated silicon, no significant friction increase or interfacial degradation was observed in ambient air. In the absence of an oxidizing environment (in dry nitrogen), the coefficient of friction decreased from 0.2 to 0.05 following amorphous carbon transfer for the materials tested.

16.3.4 Boundary Lubrication Studies

The classical approach to lubrication uses freely supported multimolecular layers of liquid lubricants (Bhushan, 1995). The liquid lubricants are chemically bonded to improve their wear resistance. To study depletion of boundary layers, the microscale friction measurements were made as a function of number of cycles on virgin Si(100) surface and silicon surface lubricated with about 2-nm-thick Z-15 and Z-DOL perfluoropolyether (PFPE) lubricants, the data of which are shown in Figure 16.24 (Koinkar and Bhushan, 1996a). Z-DOL is a PFPE lubricant with hydroxyl end groups. Its lubricant film was thermally bonded at 150°C for 30 min and washed off with a solvent to provide a chemically bonded layer for the lubricant film. In Figure 16.24, the unlubricated silicon sample showed a slight increase in friction force followed by a drop to a lower steady-state value after 20 cycles. Depletion of native oxide and possible roughening of the silicon sample are believed to be responsible for the decrease in friction force after 20 cycles. The initial friction force for Z-15 lubricated sample is lower than that of unlubricated silicon and increases gradually to a friction force value comparable to that of unlubricated silicon after 20 cycles. This suggests depletion of the Z-15 lubricant in the wear track. In the case of the Z-DOL-coated sample, the friction force starts out to be low and remains low during the cycle of 100 tests. It suggests that Z-DOL does not get displaced or depleted as readily as Z-15. Additional studies of freely supported liquid lubricants showed that either increasing the film thickness or chemically bonding the molecules to the substrate with a mobile fraction improves the lubrication performance (Koinkar and Bhushan, 1996a,b).

For lubrication of microdevices, a more effective approach involves the deposition of organized, dense molecular layers of long-chain molecules on the surface contact (Bhushan et al., 1995b). Such monolayers and thin films are commonly produced by Langmuir–Blodgett (LB) deposition and by chemical grafting of molecules into self-assembled monolayers (SAMs). Based on measurements, SAMs of octodecyl (C_{18}) compounds based on aminosilanes on oxidized silicon exhibited a lower coefficient of friction (0.018) and greater durability (Figure 16.25) than LB films of zinc arachidate adsorbed on a gold surface coated with octadecylthiol (ODT) (coefficient of friction 0.03) (Bhushan et al., 1995b). LB films are bonded to the substrate by weak van der Waals attraction, whereas SAMs are chemically bound via covalent bonds. Because of the choice of the chain length and terminal linking groups that SAMs offer, they hold great promise for boundary lubrication of microdevices.

16.3.5 Component Level Studies

Friction and wear studies have been conducted on actual or simulated MEMS components. One of the MEMS devices in which friction and wear issues are critical, is a micromotor. To measure the *in situ* static friction of a rotor-bearing interface in a micromotor, Tai and Muller (1990) measured the starting torque (voltage) and pausing position for different starting positions under a constant-bias voltage. A friction-torque model was used to obtain the coefficient of static friction. To *in situ* measure the kinetic friction of the turbine and gear structures, Gabriel et al. (1990) used a laser-based measurement system

---

FIGURE 16.15   Gray-scale maps of surface height corresponding friction force maps of Si(100), undoped polysilicon film (polished), n^+-type polysilicon film (polished), and SiC film (polished) samples. (From Bhushan, B. et al., 1998, in *Tribology Issues and Opportunities in MEMS*, B. Bhushan, ed., Kluwer Academic, Dordrecht. With permission.)

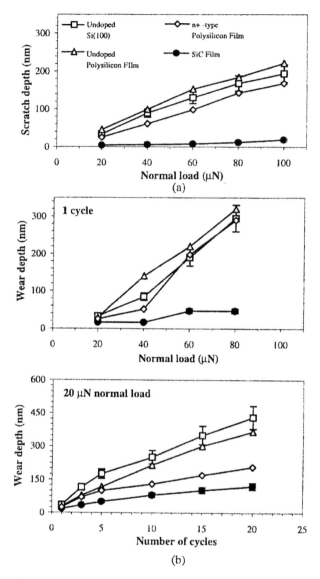

FIGURE 16.16 (a) Scratch depths for 10 cycles as a function of normal load and (b) wear depths as a function of normal load and as a function of number of cycles for various samples. (From Bhushan, B. et al., 1998, in *Tribology Issues and Opportunities in MEMS,* B. Bhushan, ed., Kluwer Academic, Dordrecht. With permission.)

to monitor the steady-state spins and decelerations. Lim et al. (1990) designed and fabricated a polysilicon microstructure to *in situ* measure the static friction of various films. The microstructure consisted of shuttle suspended above the underlying electrode by a folded beam suspension. A known normal force was applied and lateral force was measured to obtain the coefficient of static friction. Mehregany et al. (1992) developed a quantitative method for *in situ* wear measurements in micromotors. They used a wobble micromotor under electric excitation for quantitative wear measurement. Since the gear ratio of the wobble micromotor depends on the bearing clearance, changes in the gear ratio can be a direct measure of wear in the motor bearing.

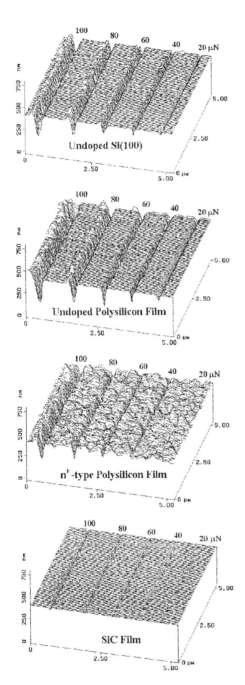

FIGURE 16.17 Scratch profiles for various samples. (From Bhushan, B. et al., 1998, in *Tribology Issues and Opportunities in MEMS*, B. Bhushan, ed., Kluwer Academic, Dordrecht. With permission.)

16.4 Closure

Silicon-based MEMS devices are made from single-crystal silicon, LPCVD polysilicon films and other ceramic films. For high-temperature applications, SiC films are being developed to replace polysilicon films. Tribology in MEMS devices requiring relative motion is of importance. AFM/FFM and nanoindentation techniques have been used for tribological studies on the micro- to nanoscale on materials of

Undoped Polysilicon Film   **SiC Film**

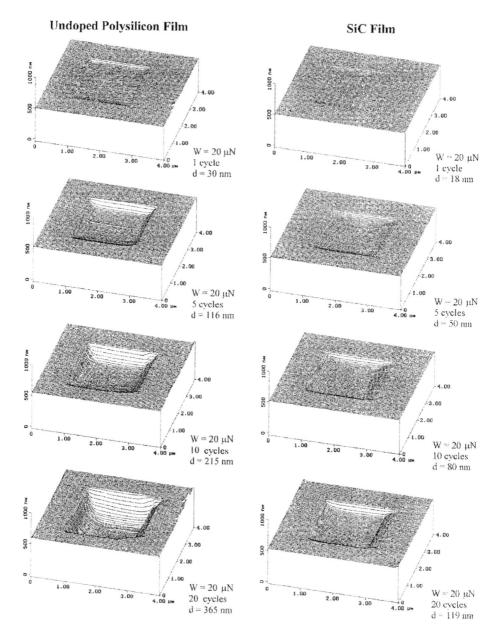

FIGURE 16.18 3D wear profiles of undoped polysilicon and SiC films (polished) after 1, 5, 10, and 20 wear cycles at 20 μN normal load. (From Bhushan, B. et al., 1998, in *Tribology Issues and Opportunities in MEMS*, B. Bhushan, ed., Kluwer Academic, Dordrecht. With permission.)

interest. These techniques have been used to study surface roughness, friction, scratching/wear, indentation, and boundary lubrication of bulk and treated silicon, polysilicon films, and SiC films. Macroscale friction and wear tests have also been conducted using the ball-on-flat tribometer.

Measurements of microscale and macroscale friction forces show that friction values on both scales of all the silicon samples are about the same among different silicon materials and higher than that of SiC. The microscale values are lower than the macroscale values as there is less plowing contribution in the microscale measurements. Surface roughness has an effect on friction. In microscale and macroscale

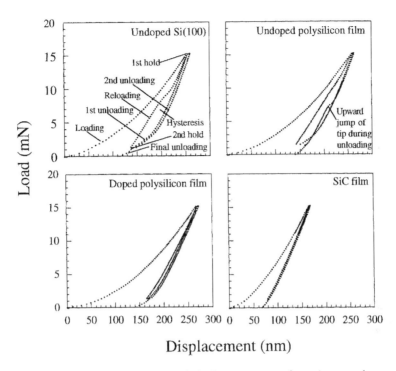

FIGURE 16.19   Representative load–displacement curves for various samples.

tests, C$^+$-implanted, oxidized, and PECVD oxide-coated single-crystal silicon samples exhibit much larger scratching and wear resistance as compared with untreated samples. Polysilicon films and undoped single-crystal silicon show similar friction and wear characteristics. Doping of polysilicon film does not affect its tribological properties. Microscratching, microwear and nanoindentation, and macroscale friction and wear studies indicate that SiC films are superior when compared to the other materials currently used in MEMS devices. Higher hardness and fracture toughness of the SiC film is believed to be responsible for its superior mechanical integrity and lower friction. Chemically grafted self-assembled monolayers and chemically bonded liquid lubricants show promising performance for boundary lubrication in MEMS devices.

TABLE 16.4 Summary of Micro/Nanotribological Properties of the Sample Materials

Sample	rms Roughness[a] (nm)	P-V Distance[a] (nm)	Coefficient of Friction Micro[b]	Coefficient of Friction Macro[c]	Scratch Depth[d] (nm)	Wear Depth[e] (nm)	Nanohardness[f] (GPa)	Young's Modulus[f] (GPa)	Fracture Toughness,[g] K_{Ic} (MPa$\sqrt{m}$)
Undoped Si(100)	0.09	0.9	0.06	0.33	89	84	12	168	0.75
Undoped polysilicon film (as deposited)	46	340	0.05	—	—	—	—	—	—
Undoped polysilicon film (polished)	0.86	6	0.04	0.46	99	140	12	175	1.11
n^+-type polysilicon film (as deposited)	12	91	0.07	—	—	—	—	—	—
n^+-type polysilicon film (polished)	1.0	7	0.02	0.23	61	51	9	95	0.89
SiC film (as deposited)	25	150	0.03	—	—	—	—	—	—
SiC film (polished)	0.89	6	0.02	0.20	6	16	25	395	0.78

[a] Measured using AFM over a scan size of 10×10 μm.

[b] Measured using AFM/FFM over a scan size of 10×10 μm.

[c] Obtained using a 3-mm-diameter sapphire ball in a reciprocating mode at a normal load of 10 mN and average sliding speed of 1 mm/s after 4 m sliding distance.

[d] Measured using AFM at a normal load of 40 μN for 10 cycles, scan length of 5 μm.

[e] Measured using AFM at normal load of 40 μN for 1 cycle, wear area of 2×2 μm.

[f] Measured using Nanoindenter at a peak indentation depth of 20 nm.

[g] Measured using microindenter with Vickers indenter at a normal load of 0.5 N.

TABLE 16.5 Macrofriction Data for Various Hemispherical Pins with Radius of 100 mm Sliding against Lubricated Thin-Film Rigid Disks in Ambient Air

Pin Material	Initial Coefficient of Friction	Cycles to Friction Increase by a Factor of 2	Max. or Ending Value of Coefficient of Friction
A_2O_3–TiC	0.20	~2,200	0.78
Mn–Zn ferrite	0.22	~5,500	0.45
Single-crystal silicon	0.20	~1,200	0.40
Polysilicon	0.20	~3,000	0.40
PECVD-oxide-coated Si(111)	0.28	>50,000	0.23–0.28
Dry-oxidized Si(111)	0.22	>50,000	0.20
Wet-oxidized Si(111)	0.26	>50,000	0.26–0.40
C^+-implanted Si(111)	0.16	~1,000	0.60

Normal load = 0.5 N; sliding speed = 0.9–1.2 m/s; ambient air (RH 45 ± 5%).

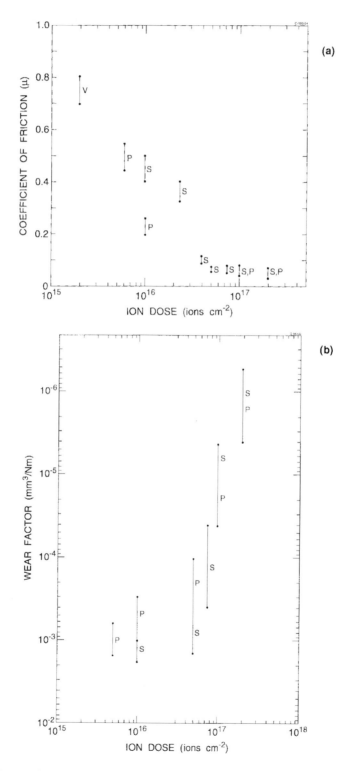

FIGURE 16.20   Influence of ion doses on the (a) coefficient of friction and (b) wear factor on C⁺ ion bombarded single-crystal and polycrystalline silicon slid against alumina ball. S and P denote tests that correspond to single- and polycrystalline silicon, respectively, and V corresponds to virgin single-crystal silicon. (From Gupta, B.K. et al., 1993, *ASME J. Tribol.* 115, 392–399. With permission.)

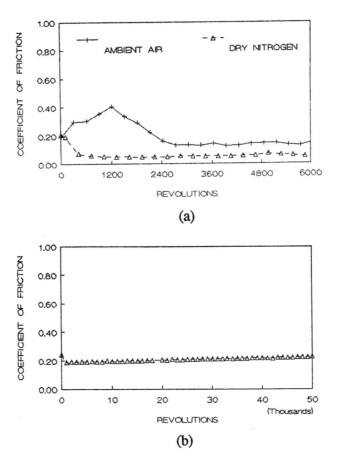

FIGURE 16.21 Coefficient of friction as a function of number of sliding revolutions in ambient air for (a) Si(111) pin in ambient air and dry nitrogen and (b) dry-oxidized silicon pin in ambient air. (From Bhushan, B. et al., 1993b, *J. Mater. Res.* 8,1611–1628.)

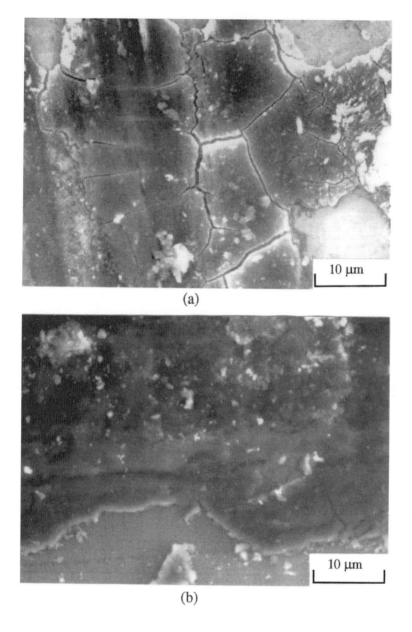

FIGURE 16.22 SEM of Si(111) after sliding against a magnetic disk in (a) ambient air for 6000 cycles and in (b) dry nitrogen after 15,000 revolutions.

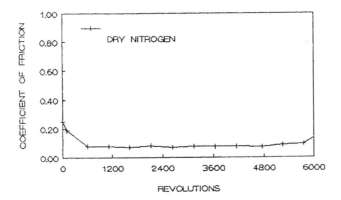

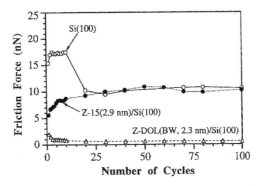

FIGURE 16.23 Variation of the coefficient of friction with number of revolutions in dry nitrogen for a DLC-coated silicon pin sliding against a magnetic disk. (Venkatesan and Bhushan, 1993, *Adv. Info. Storage Syst.* 5, 241–257).

FIGURE 16.24 Friction force as a function of number of cycles using an Si_3N_4 tip at 300 nN normal load for unlubricated and lubricated silicon samples. (From Koinkar, V.N. and Bhushan, B., 1996, *J. Appl. Phys.* 79, 8071–8075. With permission.)

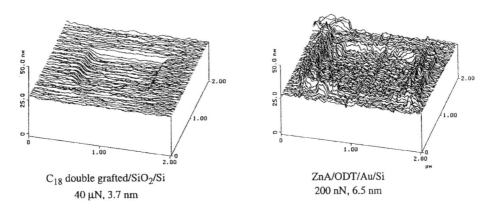

C_{18} double grafted/SiO$_2$/Si
40 μN, 3.7 nm

ZnA/ODT/Au/Si
200 nN, 6.5 nm

FIGURE 16.25 Surface profiles showing worn region after one scan cycle for self-assembled monolayers of octodecyl silanol (C_{18}) (left) and zinc arachidate (ZnA) (right). Normal load and wear depths are indicated.

References

Becker, E.W., Ehrfeld, W., Hagmann, P., Maner, A., and Munchmeyer, D. (1986), "Fabrication of Micro-structures with High Aspect Ratios and Great Structural Heights by Synchrotron Radiation Lithography, Galvanoforming, and Plastic Moulding (LIGA Process)," *Microelectr. Eng.* 4, 35–56.

Beerschwinger, U., Albrecht, T., Mathieson, D., Rueben, R.L., Yang, S.J., and Taghizadeh, M. (1995), "Wear at Microscopic Scales and Light Loads for MEMS Applications," *Wear* 181–183, 426–435.

Bhushan, B. (1992), *Mechanics and Reliability of Flexible Magnetic Media*, Springer-Verlag, New York.

Bhushan, B. (1995), *Handbook of Micro/Nanotribology*, CRC, Boca Raton, FL.

Bhushan, B. (1996a), *Tribology and Mechanics of Magnetic Storage Devices*, 2nd ed., Springer-Verlag, New York.

Bhushan, B. (1996b), "Nanotribology and Nanomechanics of MEMS Devices," in *Proc. Ninth Annual Workshop on Micro Electro Mechanical Systems*, pp. 91–98, IEEE, New York.

Bhushan, B. (1997) *Micro/Nanotribology and Its Applications*, E330, Kluwer Academic, Dordrecht, Netherlands.

Bhushan, B. (1998a), *Tribology Issues and Opportunities in MEMS*, Kluwer Academic, Dordrecht, Netherlands.

Bhushan, B. (1998b), "Contact Mechanics of Rough Surfaces in Tribology: Multiple Asperity Contact," *Tribol. Lett.* 4, 1–35.

Bhushan, B., Dominiak, M., and Lazarri, J.P. (1992), "Contact Start-Stop Studies with Silicon Planar Head Sliders Against Thin-film Disks," *IEEE Trans. Magn.* 28, 2874–2876.

Bhushan, B. and Venkatesan, S. (1993a), "Mechanical and Tribological Properties of Silicon for Micro-mechanical Applications: A Review," *Adv. Info. Storage Syst.* 5, 211–239.

Bhushan, B. and Venkatesan, S. (1993b), "Friction and Wear Studies of Silicon in Sliding Contact with Thin-Film Magnetic Rigid Disks," *J. Mater. Res.* 8, 1611–1628.

Bhushan B. and Koinkar, V.N. (1994), "Tribological Studies of Silicon for Magnetic Recording Applications," *J. Appl. Phys.* 75, 5741–5746.

Bhushan, B., Israelachvili, J.N., and Landman, U. (1995a), "Nanotribology: Friction, Wear and Lubrication at the Atomic Scale," *Nature* 374, 607–616.

Bhushan, B., Kulkarni, A.V., Koinkar, V.N., Boehm, M., Odoni, L., Martelet, C., and Belin, M. (1995b), "Microtribological Characterization of Self-Assembled and Langmuir–Blodgett Monolayers by Atomic Force and Friction Force Microscopy," *Langmuir* 11, 3189–3198.

Bhushan, B. and Gupta, B.K. (1997), *Handbook of Tribology: Materials, Coatings and Surface Treatments*, Reprint edition, Krieger, Malabar, FL.

Bhushan, B. and Koinkar, V.N. (1997a), "Microtribological Studies of Doped Single-Crystal Silicon and Polysilicon Films for MEMS Devices," *Sensors Actuators* A 57, 91–102.

Bhushan, B. and Li, X. (1997b), "Micromechanical and Tribological Characterization of Doped Single-Crystal Silicon and Polysilicon Films for Microelectromechanical Systems," *J. Mater. Res.* 12, 1–10.

Bhushan, B., Sundararajan, S., Li, X., Zorman, C.A., and Mehregany, M. (1998), "Micro/Nanotribological Studies of Single-Crystal Silicon and Polysilicon and SiC Films for Use in MEMS Devices," in *Tribology Issues and Opportunities in MEMS* (B. Bhushan, ed.), pp. 407–430, Kluwer Academic, Dordrecht, Netherlands.

Bryzek, J., Peterson, K., and McCulley, W. (1994), "Micromachines on the March," *IEEE Spectrum* May, 20–31.

Callahan, D.L. and Morris, J.C. (1992), "The Extent of Phase Transformation in Silicon Hardness Indentation," *J. Mater. Res.* 7, 1612–1617.

Deng, K., Collins, R.J., Mehregany, M., and Sukenik, C.N. (1995), "Performance Impact of Monolayer Coatings of Polysilicon Micromotors," in *Proc. MEMS* 95, Amsterdam, Netherlands, Jan–Feb.

Ericson, F. and Schweitz, J.A (1990), "Micromechanical Fracture Strength of Silicon," *J. Appl. Phys.* 68, 5840–5844.

Fan, L.S. and Woodman, S. (1995a), "Batch Fabrication of Mechanical Platforms for High-Density Data Storage," *8th Int. Conf. Solid State Sensors and Actuators (Transducers '95)/Eurosensors IX*, Stockholm, Sweden, 25–29 June, pp. 434–437.

Fan, L.S., Ottesen, H.H., Reiley, T.C., and Wood, R.W. (1995b), "Magnetic Recording Head Positioning at Very High Track Densities Using a Microactuator-Based, Two-Stage Servo System," *IEEE Trans. Ind. Electron.* 42, 222–233.

Fang, W. and Wickert, J.A. (1995), "Comments on Measuring Thin-Film Stresses Using Bi-Layer Micromachined Beams," *J. Micromech. Microeng.* 5, 276–281.

Friedrich, C.R. and Warrington, R.O. (1998), "Surface Characterization of Non-Lithographic Micromachining," *Tribology Issues and Opportunities in MEMS* (B. Bhushan, ed.), pp. 73–84, Kluwer Academic, Dordrecht, Netherlands.

Fujimasa, I. (1996), *Micromachines: A New Era in Mechanical Engineering*, Oxford University Press, Oxford, U.K.

Gabriel, K.J., Behi, F., Mahadevan, R., and Mehregany, M. (1990), "In Situ Friction and Wear Measurement in Integrated Polysilicon Mechanisms," *Sensors Actuators* A21–23, 184–188.

Gane, N. and Cox, J.M. (1970), "The Micro-Hardness of Metals at Very Light Loads," *Philos. Mag.* 22, 881–891.

Guckel, H. and Burns, D.W. (1989), "Fabrication of Micromechanical Devices from Polysilicon Films with Smooth Surfaces," *Sensors Actuators* 20, 117–122.

Guckel, H., Burns, D., Rutigliano, C., Lovell, E., and Choi, B. (1992), "Diagnostic Microstructures for the Measurement of Intrinsic Strain in Thin Films," *J. Micromech. Microeng.* 2, 86–95.

Gupta, B.K. and Bhushan, B. (1994), "Nanoindentation Studies of Ion Implanted Silicon," *Surf. Coat. Technol.* 68/69, 564–570.

Gupta, B.K., Chevallier, J., and Bhushan, B. (1993), "Tribology of Ion Bombarded Silicon for Micromechanical Applications," *ASME J. Tribol.* 115, 392–399.

Gupta, B.K., Bhushan, B., and Chevallier, J. (1994), "Modification of Tribological Properties of Silicon by Boron Ion Implantation," *Tribol. Trans.* 37, 601–607.

Hamilton, H. (1991), "Contact Recording on Perpendicular Rigid Media," *J. Magn. Soc. Jpn.* 15 (Suppl. S2), 483–481.

Harris, G.L. (ed.) (1995), *Properties of Silicon Carbide*, Inst. of Elect. Eng., London.

Henck, S.A. (1997), "Lubrication of Digital Micromirror Devices," *Tribol. Lett.* 3, 239–247.

Hokkirigawa, K. and Kato, K. (1988), "An Experimental and Theoretical Investigation of Ploughing, Cutting and Wedge Forming during Abrasive Wear," *Tribol. Int.* 21, 51–57.

Hutchings, I.M. (1992), *Tribology: Friction and Wear of Engineering Materials*, CRC Press, Boca Raton, FL.

Jaeger, R.C. (1988), *Introduction to Microelectronic Fabrication*, Vol. 5, Addison-Wesley, Reading, MA.

Koinkar, V.N. and Bhushan, B. (1996a), "Micro/Nanoscale Studies of Boundary Layers of Liquid Lubricants for Magnetic Disks," *J. Appl. Phys.* 79, 8071–8075.

Koinkar, V.N. and Bhushan, B. (1996b), "Microtribological Studies of Unlubricated and Lubricated Surfaces Using Atomic Force/Friction Force Microscopy," *J. Vac. Sci. Technol.* A14, 2378–2391.

Koinkar, V.N. and Bhushan, B. (1997), "Scanning and Transmission Electron Microscopies of Single-Crystal Silicon Microworn/Micromachined Using Atomic Force Microscopy," *J. Mater. Res,* 12, 3219–3224.

Lazarri, J.P. and Deroux-Dauphin, P. (1989), "A New Thin Film Head Generation IC Head," *IEEE Trans. Magn.* 25, 3190–3193.

Lehr, H. Abel, S., Doppler, J., Ehrfeld, W., Hagemann, B., Kamper, K.P., Michel, F., Schulz, Ch., and Thurigen, Ch. (1996), "Microactuators as Driving Units for Microrobotic Systems," *Proc. Microrobotics: Components and Applications* (A. Sulzmann, ed.), Vol. 2906, pp. 202–210, SPIE.

Lehr, H., Ehrfeld, W., Hagemann, B., Kamper, K.P., Michel, F., Schulz, Ch., and Thurigen, Ch. (1997), "Development of Micro-Millimotors," *Min. Invas. Ther. Allied Technol.* 6, 191–194.

Li, X. and Bhushan, B. (1998), "Micro/Nanomechanical Characterization of Ceramic Films for Microdevices," *Thin Solid Films* (in press).

Lim, M.G., Chang, J.C., Schultz, D.P., Howe, R.T., and White, R.M. (1990), "Polysilicon Microstructures to Characterize Static Friction," *Proc. IEEE Micro Electro Mechanical Systems,* 82–88.

Madou, M. (1997), *Fundamentals of Microfabrication,* CRC Press, Boca Raton, FL.

Madou, M. (1998), "Facilitating Choices of Machining Tools and Materials for Miniaturization Science: A Review," in *Tribology Issues and Opportunities in MEMS* (B. Bhushan, ed.), pp. 31–51, Kluwer Academic, Dordrecht, Netherlands.

Mastrangelo, C.H. (1997), "Adhesion-Related Failure Mechanisms in Micromechanical Devices," *Tribol. Lett.* 3, 233–238.

Mehregany, M., Howe, R.T., and Senturia, S.D. (1987), "Novel Microstructures for the in Situ Measurement of Mechanical Properties of Thin Films," *J. Appl. Phys.* 62, 3579–3584.

Mehregany, M., Gabriel, K.J., and Trimmer, W.S.N. (1988), "Integrated Fabrication of Polysilicon Mechanisms," *IEEE Trans. Electron Devices* 35, 719–723.

Mehregany, M., Nagarkar, P., Senturia, S.D., and Lang, J.H. (1990), "Operation of Microfabricated Harmonic and Ordinary Side-Drive Motors," *Proc. IEEE Micro Electromechanical Systems,* pp. 1–8, IEEE, New York.

Mehregany, M., Senturia, S.D., and Lang, G.H. (1992), "Measurement of Wear in Polysilicon Micromotors," *IEEE Trans. Electron Devices* 39, 1136–1143.

Michel, F. and Ehrfeld, W. (1998), "Microfabrication Technologies for High Performance Microactuators" in *Tribology Issues and Opportunities in MEMS* (B. Bhushan, ed.), pp. 53–72, Kluwer Academic, Dordrecht, Netherlands.

Miu, D.K. and Tai, Y.C. (1995), "Silicon Micromachined Scaled Technology," *IEEE Trans. Ind. Electron.* 42, 234–239.

Muller, R.S., Howe, R.T., Senturia, S.D., Smith, R.L., and White, R.M. (1990), *Microsensors,* IEEE Press, New York.

Ohwe, T., Mizoshita, Y., and Yonoeka, S. (1993), "Development of Integrated Suspension System for a Nanoslider with an MR Head Transducer," *IEEE Trans. Magn.* 29, 3924–3926.

Peterson, K.E. (1979), "Micromechanical Membrane Switches on Silicon," *IBM J. Res. Dev.* 23, 376.

Peterson, K.E. (1982), "Silicon as a Mechanical Material," *Proc. IEEE* 70, 420–457.

Pharr, G.M. (1991), "The Anomalous Behavior of Silicon during Nanoindentation," in *Thin Films: Stresses and Mechanical Properties III* (W.D. Nix, J.C. Bravman, E. Arzt and L.B. Freund, eds.), Vol. 239, pp. 301–312, Materials Research Society, Pittsburgh.

Pilskin, W.A. (1977), *J. Vac. Sci. Technol.* 14, 1064.

Sargent, P.M. (1986), "Use of the Indentation Size Effect on Microhardness for Materials Characterization," *Microindentation Techniques in Materials Science and Engineering* (P.J. Blau and B.R. Lawn, eds.), STP 889, pp. 160–174, ASTM, Philadelphia.

Schweitz, J.A. (1991), "A New and Simple Approach to the Stress-Strain Characteristics of Thin Coatings," *J. Micromech. Microeng.* 1, 10–15.

Shackelford, J.F., Alexander, W., and Park, J.S. (eds.) (1994), *CRC Material Science and Engineering Handbook,* 2nd ed., CRC Press, Boca Raton, FL.

Shor, J.S., Goldstein, D., and Kurtz, A.D. (1993), "Characterization of n-Type β-SiC as a Piezoresistor," *IEEE Trans. Electron Devices* 40, 1093–1099.

Spencer, M.G., Devaty, R.P., Edmond, J.A., Khan, M.A., Kaplan, R., and Rahman, M. (eds.) (1994), *SiC and Related Materials, Proc. Fifth Conf.,* Inst. of Physics Publishing Ltd., Bristol, U.K.

Sundararajan, S. and Bhushan, B. (1998), "Micro/Nanotribological Studies of Polysilicon and SiC Films for MEMS Applications," *Wear* (in press).

Tai, Y.C., Fan, L.S., and Muller, R.S. (1989), "IC-Processed Micro-Motors: Design, Technology and Testing," *Proc. IEEE Micro Electro Mechanical Systems,* 1–6.

Tai, Y.C. and Muller, R.S. (1990), "Frictional Study of IC Processed Micromotors," *Sensors Actuators* A21–23, 180–183.

Thurigen, C., Ehrfeld, W., Hagemann, B., Lehr, H., and Michel, F. (1998), "Development, Fabrication and Testing of a Multi-Stage Micro Gear System," *Tribology Issues and Opportunities in MEMS* (B. Bhushan, ed.), pp. 397–402, Kluwer Academic, Dordrecht, Netherlands.

Tong, L., Mehregany, M., and Matus, L.G. (1992), "Mechanical Properties of 3C Silicon Carbide," *Appl. Phys. Lett.* 60, 2992–2994.

Trimmer, W.S. (ed.) (1997), *Micromachines and MEMS, Classic and Seminal Papers to 1990,* IEEE Press, New York.

Venkatesan, S. and Bhushan, B. (1993), "The Role of Environment in the Friction and Wear of Single-Crystal Silicon in Sliding Contact with Thin-Film Magnetic Rigid Disks," *Adv. Info. Storage Syst.* 5, 241–257.

Venkatesan, S. and Bhushan, B. (1994), "The Sliding Friction and Wear Behavior of Single-Crystal, Polycrystalline and Oxidized Silicon, *Wear* 171, 25–32.

Xu, J. and Bhushan, B. (1997), "Friction and Durability of Ceramic Slider Materials in Contact with Lubricated Thin-Film Rigid Disks," *Proc. Inst. Mech. Eng., Part J: J. Eng. Tribol.* 211, 303–316.

Zorman, C.A., Fleischmann, A.J., Dewa, A.S., Mehregany, M., Jacob, C., Nishino, S., and Pirouz, P. (1995), "Epitaxial Growth of 3C-SiC Films on 4 in. Diam Si(100) Silicon Wafers by Atmospheric Pressure Chemical Vapor Deposition," *J. Appl. Phys.* 78, 5136–5138.